GERALD D. SCHMIDT & LARRY S. ROBERTS'

FOUNDATIONS OF PARASITOLOGY

Enlargement of liver and spleen (hepatospleenomegaly) in a laboratory mouse infected with the rodent malaria parasite, *Plasmodium berghei*. The liver and spleen enlarge and darken from the accumulation of parasite hemozoin pigment granules in reticuloendothelial cells of these organs (p. 153). Hepatospleenomegaly can also occur in humans infected with their malaria parasites. The mouse-*P. berghei* system, like other non-human models of parasitic diseases, has been important in improving understanding of the complex interactions between hosts and parasites.

NINTH EDITION

Gerald D. Schmidt & Larry S. Roberts'
FOUNDATIONS OF PARASITOLOGY

LARRY S. ROBERTS
Texas Tech University Emeritus

JOHN JANOVY, JR.
University of
Nebraska–Lincoln

STEVE NADLER
University of
California, Davis

McGraw Hill

Connect
Learn
Succeed™

GERALD D. SCHMIDT & LARRY S. ROBERTS' FOUNDATIONS OF PARASITOLOGY
NINTH EDITION
International Edition 2013

Exclusive rights by McGraw-Hill Education (Asia), for manufacture and export. This book cannot be re-exported from the country to which it is sold by McGraw-Hill. This International Edition is not to be sold or purchased in North America and contains content that is different from its North American version.

10 09 08 07 06 05 04 03 02 01
20 15 14 13
CTP   SLP

When ordering this title, use ISBN 978-007-132641-4 or MHID 007-132641-3

Printed in Singapore

www.mhhe.com

about the authors

LARRY S. ROBERTS

Larry S. Roberts, professor emeritus of biology at Texas Tech University, was professor of zoology at University of Massachusetts, Amherst, and was adjunct professor of biology at Florida International University and the University of Miami, where he had extensive experience teaching parasitology, invertebrate zoology, marine biology, and developmental biology. He received his Sc.D. in parasitology at the Johns Hopkins University and has coauthored Foundations of Parasitology from the first edition through this, the ninth edition. He is also coauthor of Integrated Principles of Zoology, Biology of Animals, and Animal Diversity, and is author of The Underwater World of Sport Diving.

Dr. Roberts has published many research articles and reviews. He has served as president of the American Society of Parasitologists, the Southwestern Society of Parasitologists, the Southeastern Society of Parasitologists, and the Helminthological Society of Washington. He received the Henry Baldwin Ward Medal from the American Society of Parasitologists. His hobbies include scuba diving, underwater photography, and tropical horticulture.

Dr. Roberts can be contacted at Lroberts1@compuserve.com

JOHN JANOVY, JR.

John Janovy, Jr. (PhD University of Oklahoma, 1965) is Professor Emeritus at the University of Nebraska where he was the Paula and D. B. Varner Distinguished Professor of Biological Sciences for much of his career. His research interest is parasitology, with particular focus on parasite ecology and life cycles. He has been director of the Cedar Point Biological Station, interim director of the University of Nebraska State Museum, an assistant dean of Arts and Sciences, and secretary-treasurer of the American Society of Parasitologists. He is currently (2012) President-Elect of the American Society of Parasitologists. His scholarly and creative accomplishments consist of approximately 100 scientific papers and book chapters; 14 books, including *Keith County Journal, On Becoming a Biologist, Teaching in Eden, Outwitting College Professors,* and *Foundations of Parasitology* (with Larry Roberts and Steve Nadler); the screenplay for the televised version of *Keith County Journal* (Nebraska Public Television); and numerous popular articles. His teaching experiences include almost continuous service in the large-enrollment freshman biology course; Field Parasitology (BIOS 487/887) at the Cedar Point Biological Station; Invertebrate Zoology (BIOS 381); Parasitology (BIOS 385); a decade in BIOS 103/204 (Organismic Biology/ Biodiversity); and numerous honors seminars. He has supervised 18 MS students, 14 PhD students, and approximately 50 undergraduate researchers, including 10 Howard Hughes scholars. His honors include the University of Nebraska Distinguished Teaching Award (1970), University Honors Program Master Lecturer (1986), American Health magazine book award (1987, for *Fields of Friendly Strife*), University of Nebraska Outstanding Research and Creativity Award (1998), The Nature Conservancy Hero recognition (2000), and the American Society of Parasitologists Clark P. Read Mentorship Award (2003).

GERALD D. SCHMIDT

Gerald D. Schmidt was professor of biology at the University of Northern Colorado (UNC) when he passed away. He received his PhD from Colorado State University. He was active in research and promoting research activities at UNC, and he published more than 160 research articles in scientific journals, as well as six books. He received awards from UNC for outstanding teaching and for distinguished scholarship. He was a board member of the World Federation of Parasitologists; a Fellow of the Royal Society of Tropical Medicine and Hygiene, London; and a Fellow of the Royal Society of South Australia.

Dr. Schmidt served the American Society of Parasitologists as secretary-treasurer for seven years. He was co-author of Foundations of Parasitology through the first four editions. His hobbies were hunting and fishing, especially fishing, and he wrote a book on fishing. Dr. Schmidt died on 16 October 1990; many more details of his life can be found in the Journal of Parasitology, 78:757–773.

STEVE NADLER

Steve Nadler (PhD in Medical Parasitology, Louisiana State University Medical Center, New Orleans) is Professor of Nematology in the Department of Entomology and Nematology at the University of California, Davis. His research interests concentrate on the systematics and evolutionary biology of nematodes, including both free-living and parasitic species. He served as chair of the Department of Nematology at UC Davis for  six years. Dr. Nadler was an associate editor of the Journal of Parasitology, and president of the American Society of Parasitologists (2007–08). His scholarly accomplishments include approximately 90 scientific papers, and his research has been supported by grants from the U.S. National Science Foundation and the National Institutes of Health. He currently serves on the editorial boards of the journals Parasitology, Systematic Parasitology, Zookeys, and Animal Cells and Systems. His research laboratory is supported by efforts of undergraduate and graduate students, along with visiting scientists and postdoctoral scholars. At UC Davis his undergraduate and graduate teaching includes courses in parasitology, nematology, and molecular phylogenetics.

brief contents

contents

20 Cestoidea: Form, Function, and Classification of Tapeworms   299

21 Tapeworms   325

22 Phylum Nematoda: Form, Function, and Classification   349

23 Nematodes: Trichinellida and Dioctophymatida, Enoplean Parasites   377

24 Nematodes: Tylenchina, a Functionally Diverse Clade   391

preface

We enthusiastically present the ninth edition of this book with numerous updates on topics of vigorous contemporary research. We continue to preserve essential qualities of the text that students and professors liked in the first eight editions. The reception accorded *Foundations of Parasitology* has been most gratifying. Your comments and suggestions are always welcome. Keep them coming.

SCOPE OF THIS BOOK

This textbook is designed specifically for upper division courses in general parasitology. It emphasizes principles, illustrating them with material on the biology, physiology, genetics, morphology, phylogeny and ecology of the major parasites of humans and domestic animals. We have found that they are of most interest to the majority of students. Other parasites are included as well, when they are of unusual biological interest.

The first three chapters delineate important definitions and principles in evolution, ecology, immunology, and pathology of parasites and parasitic infections. Chapters on specific groups follow, beginning with protozoa and ending with arthropods. Presentation of each group is not predicated on students having first studied groups presented in prior chapters; therefore, the order can vary as an instructor desires. As always, we have strived for readability, enhancing words with photographs, drawings, electron micrographs, and tables.

NEW TO THIS EDITION

This edition integrates a wealth of new discoveries and literature. Many areas of parasitology are theaters of intense research effort and fruitful results. As always, addition of material compelled us to prune out an equal amount of text and illustrations so as not to increase book length, but we hope that we have been judicious in our reshaping. We have continued to include trenchant quotations at the beginning of each chapter. Well, maybe some of them are not so trenchant. Nevertheless, we hope these observations of pioneering researchers, as well as references to literature and even pop culture, will broaden your view of parasitology. Their curiosity piqued, some readers have asked us for sources of quotations, so we have included these where possible.

The numerous changes in chapter 1 included updating the table on global prevalence of various human parasites. We have retained our section with the light-hearted title of "Parasitology for Fun and Profit" to emphasize how students can earn an income while studying the fascinating world of parasites. We are including some web links because many students enjoy taking advantage of those resources. Concepts in Chapters 2 and 3 are briefly covered, but understanding them is *essential* to understanding the rest of the book.

Chapter 2 has been further reorganized to include fascinating material on the role played by parasites in food webs and ecosystems. Our increased emphasis on molecular systematics and phylogenetics has been retained, and we provide some examples here and in chapters to follow. Propelled in large measure by modern molecular methods, immunologists continue their torrent of discoveries. The 1980s through 2000s saw enormous increases in our understanding of the role and mechanisms of cytokine function and witnessed our realization of the importance of immunopathology in parasitic diseases. Thus, chapter 3 has again undergone major surgery. It has been rewritten, reorganized, and expanded, including a section introducing antimicrobial peptides (defensins) and Toll-like receptors and tables listing the many ways that protozoan and helminth parasites evade host defenses. We added a figure in the 8th edition illustrating a JAK-STAT cell signaling pathway. In this edition we expanded the discussion of T_{reg} and dendritic cells, and added a section on the microbial deprivation hypothesis relating parasitism to immune system development.

"Form and Function" chapters on protozoan parasites, trematodes, cestodes, nematodes, and arthropods have again been updated and rewritten significantly to provide a stronger base of knowledge with which to investigate each group further. When available, we include phylogenies to show evolutionary relationships of some of the major groups.

We again modified the classification section of chapter 4, making it consistent with all the major taxonomic literature published since the seventh and eighth editions. We continue use of the words "protozoa" and "protozoans" as common names with no taxonomic status and that refer to a number of phyla. Chapter 5 on Kinetoplasta includes the latest information on antigenic variation in trypanosomes. *Leishmania*-host cell relationships, and the important new anti-leishmanial drug miltefosine. In chapter 6 we continue usage of *Giardia duodenalis* to be consistent with the latest nomenclatural decisions about this important parasite. Several examples in this chapter cite the importance of molecular techniques to diagnosis and contributions to the overall biology of the organisms. Other protistan chapters address the exploding body of knowledge about opportunistic parasitic infections in immunocompromised persons and the amazing diversity of coccidians as revealed by the active systematic research on these parasites.

Chapter 7 on amebas was reworked considerably in the 8th edition, and several new figures were added, including an *Acanthamoeba*-infected eye. Both chapters 8 and 9 have information on the important membranous organelle known as an apicomplast. Intense scrutiny of malaria continues, reflecting its widespread importance as a human disease, and chapter 9 has been revised accordingly. *Plasmodium knowlesi* has been included as one of the species that often causes human malaria. We retained the expanded table comparing *Plasmodium* spp. and updated methods of diagnosis, role of cytokines in pathogenesis and immunity, progress toward vaccines, and drug action and resistance. A figure illustrates fluctuations in body temperature (fever phases) in *falciparum* compared with *vivax* malaria and relationship of the temperature fluctuations to phases of schizogony.

In chapter 12, we recognize two phyla of mesozoans, Dicyemida and Orthonectida, in accord with recent literature, and the classification has been revised. In chapter 13 of the seventh edition, we introduced significant revision of flatworm systematics, which has been retained in this edition. We point out, in chapter 16, the potential for widespread increase in prevalence of *Schistosoma japonicum* resulting from the huge Three Gorges Dam on the Yangtze River in China. Other sections of this chapter have been rewritten, including pathology, control, and other *Schistosoma* spp.

In chapter 20 on cestode form and function we retain the revisions made for the seventh edition, and on the basis of its extremely unusual scolex, we recognize order Cathetocephalidea and include a figure of its scolex. We have followed the sensible suggestions of Kuchta et al. (2008) in suppressing the name Pseudophyllidea and recognizing two new orders of cestodes, Bothriocephalidea and Diphyllobothriidea. We have retained the numerous revisions in chapter 21 and rearranged several sections.

The most profound change in chapter 22 introducing nematodes (compared with the seventh and earlier editions) is the adoption of the phylogeny and nomenclature of De Ley and Blaxter (2002, in D. L. Lee (Ed.), *The biology of nematodes,* Taylor and Francis). This nomenclature has been updated to reflect molecular phylogenetic hypotheses published since the 7th edition. Revised and expanded phylogenies for nematodes have required some reorganization of the taxonomic groups covered in subsequent chapters. To ease the transition required by De Ley and Blaxter's classification, within certain chapters we retain common usage of nematode groups (e.g., families) used in earlier editions of this book.

We incorporated numerous other changes and updates in the nematode chapters. Throughout these chapters we have incorporated examples of how new approaches and tools, including gleaning information from nematode models such as *Caenorhabditis elegans,* are helping to advance understanding of parasites. Updated information on infection prevalence was added, when available. The eighth species of *Trichinella, T. zimbabwensis,* is covered and added to Table 23.1. We added the probable environmental cue that determines whether *Strongyloides* females will initiate homogonic or heterogonic cycle. In Chapter 24 we emphasized the diversity of nematodes in the suborder Tylenchina, which includes free-living species and important plant and animal parasites. In chapter 25 we remarked on the difficulties in distinguishing hookworm eggs from those of *Oesophagostomum bifurcum* and *Ternidens deminutus* in areas of Africa where they parasitize humans, and we recognize *Angiostrongylus vasorum* as an emerging infection of canids. In accord with updated molecular phylogenetic results, Camallanoidea was transferred to chapter 30 with Dracunculomorpha.

Chapter 31 of the seventh edition was an entirely new chapter on those amazing worms, Nematomorpha. This chapter brings together all findings of the most recent research on this group, especially the life cycle work. *Foundations of Parasitology* is the *only* text to date including invertebrate and zoology texts that has this information. Chapter 32 on Acanthocephala has an expanded discussion of recent molecular work linking this phylum to Rotifera.

Form and function of arthropods has now become chapter 33. We have added a discussion of Arthropoda phylogeny, including its position as a member of superphylum Ecdysozoa. Readers of the classification coverage in this chapter will find that we have included Pentastomida within Arthropoda as a subclass of crustacean class Maxillopoda. Chapter 34 adopts the currently most authoritative classification of Crustacea. In this chapter we include a photo of a shark embryo parasitized by trebiid copepods; these amazing organisms enter the uterus of pregnant sharks, attacking the uterine wall as well as the surface of the embryos, thus becoming endosymbiotic ectoparasites!

Chapter 35 covers Pentastomida and includes an explanation of its demotion from phylum status to a subclass of Crustacea. Much information was been added in the eighth edition to the remaining chapters on insects, such as use of endectocides for control of lice, potential for bed bugs to transmit hepatitis, and a dramatic picture of a strepsipteran emerging from a fire ant. The section on plague has been extensively reworked.

Chapter 41 on ticks and mites had new material on tick behavior in the eighth edition, especially their attraction to human breath, on dogs as carriers of various tick-borne infections, and on chorioptic mange as a veterinary problem.

INSTRUCTIVE DESIGN

Students using the ninth edition of *Foundations of Parasitology* are guided to a clear understanding of the topic through our careful use of study aids. Essential terms, many of which are defined in a complete glossary, are boldfaced in the text to provide emphasis and ease in reviewing. In response to student requests, we again provide pronunciation guides for glossary entries. Numbered references at the end of each chapter make supporting data and further study easily accessible. Clear labeling makes all illustrations approachable and self-explanatory to the student. Student learning outcomes are provided for each chapter, which can be used by instructors for assessment.

We have again been fortunate indeed to have William C. Ober and Claire W. Garrison draw new illustrations for this and the last several editions. Their artistic skills and knowledge of biology have enhanced other zoology texts coauthored by Larry Roberts. Bill and Claire bring to their work a unique perspective resulting from their earlier careers as physician and nurse, respectively.

ACKNOWLEDGMENTS

We are indebted to the numerous students and colleagues who have commented on previous editions. We especially wish to thank the following individuals who reviewed certain chapters or the entire text. The comments were enormously helpful.

Osman Bannaga, *Miles College*
Dale Clayton, *University of Utah*
William Dees, *McNeese State University*

Todd Huspeni, *University of Wisconsin,—Stevens Point*
Barry OConnor, *University of Michigan*
Martin Olivier, *McGill University*
Dennis Richardson, *Quinnipiac University*
Samuel Zeakes, *Radford University*
Dr. Janine Caira, *University of Connecticut*
Tatiana Rossolimo, *Dalhousie University*
Peter Kima, *University of Florida*
Ravinder Sehgal, *SF State University*
Kristin Michel, *Kansas State University*

We are indebted to students who aided in literature retrieval and review, correspondence, filing, and other office work associated with the ninth edition. These individuals include Stephanie Bitzes, Brittany Bunker, Shaye Sisneros, and Brittany Stork. We deeply appreciate the large number of photographs contributed by our many colleagues from around the world. We would like to especially recognize Dr. Gerald Esch, editor of the *Journal of Parasitology,* for permissions to use to many figures from that journal.

We thank the dedicated and conscientious staff of McGraw-Hill Higher Education, especially Lori Bradshaw, Developmental Editor, Rebecca Olson, Brand Manager. Finally, we (LSR and JJJ) extend a warm welcome to a third collaborator on *Foundations of Parasitology,* Dr. Steven Nadler. We are confident that his contributions to our efforts will make a fine textbook even better.

Larry S. Roberts
John Janovy, Jr.
Steve Nadler

Chapter 1

Introduction to Parasitology

So, Nat'ralist observe, a Flea
Hath smaller Fleas that on him prey,
And these have smaller Fleas to bite 'em;
And so proceed ad infinitum.

—J. Swift, *On Poetry*

Few people realize that there are far more kinds of parasitic than nonparasitic organisms in the world. Even if we exclude viruses and rickettsias, which are all parasitic, and the many kinds of parasitic bacteria and fungi, parasites are still in the majority. The bodies of free-living plants and animals represent rich environments, which have been colonized innumerable times throughout evolutionary history.

In general the parasitic way of life is so successful that it has evolved independently in nearly every phylum of animals, from protistan phyla to arthropods and chordates, as well as in many plant groups. Organisms that are not parasites are usually hosts. Humans, for example, can be infected with more than a hundred kinds of flagellates, amebas, ciliates, worms, lice, fleas, ticks, and mites. It is unusual to examine a domestic or wild animal without finding at least one species of parasite on or within it. Even animals reared under strict laboratory conditions are commonly infected with protozoa and other parasites. Often the parasites themselves are hosts of other parasites.

The relationships between parasites and hosts are typically quite intimate, biochemically speaking, and it is a fascinating, often compelling task to explain just why a species of parasite is restricted to one or a few host species. It is no wonder that the science of **parasitology** has developed out of efforts to understand parasites and their relationships with their hosts.

RELATIONSHIP OF PARASITOLOGY TO OTHER SCIENCES

The first and most obvious stage in the development of parasitology was the discovery of parasites themselves. **Descriptive parasitology** probably began in prehistory. **Taxonomy** as a formal science, however, started with Linnaeus's publication of the 10th edition of *Systema Naturae* in 1758. Linnaeus

himself is credited with the description of the sheep liver fluke, *Fasciola hepatica,* and over the next 100 years many common parasites, as well as their developmental stages, were described. The discovery and description of new parasite species continues today, just as does the description of new species in almost every group of organisms. Although biologists have a massive "catalog" of Earth's biota, this list is far from complete. Indeed, based on the rate of new published descriptions, scientists estimate that humans are destroying species faster than they are discovering them, especially in the tropics. There is every reason to believe this generality applies to parasites as well as butterflies.

Today **systematists** rely on published species descriptions, as well as on studies of DNA, proteins, ecology, and geographical distribution, to develop **phylogenies** (singular, *phylogeny*), or evolutionary histories, of parasites. On the practical side, an **epidemiologist** may need to understand sociological and political factors, climate, local traditions, and global economics, as well as pharmacology, pathology, biochemistry, and clinical medicine, to devise a scheme for controlling parasitic infections.

When people became aware that parasites were troublesome and even serious agents of disease, they began an ongoing effort to heal the infected and eliminate the parasites. Curiosity about routes of infection led to studies of parasite **life cycles;** thus it became generally understood in the last part of the 19th century that certain animals—for example, ticks and mosquitoes—could serve as **vectors** that transmitted parasites to humans and their domestic animals. As more and more life cycles became known, parasitologists quickly realized the importance of understanding these seemingly complex series of ecological and embryological events. It is naive to try to control an infection without knowledge of how an infectious agent, in this case a parasite, reproduces and gets from one host to another.

Parasite biology does not differ fundamentally from biology of free-living organisms, and parasite systems have provided outstanding models in studies of basic biological phenomena. In the 19th century van Beneden described meiosis and Boveri demonstrated the continuity of chromosomes, both in parasitic nematodes. In the 20th century refined techniques in physics and chemistry applied to parasites have added to our understanding of basic biological principles and mechanisms. For example, Keilin discovered cytochrome and the electron transport system during his investigations of parasitic worms and insects.[29] Today biochemical techniques are widely used in studies of parasite metabolism, immunology, and chemotherapy. Use of the electron microscope resulted in many new discoveries at the subcellular level. The techniques of modern molecular biology have contributed new diagnostic methods and new knowledge of relationships between parasites,[19, 33] and they offer much hope in the development of new vaccines. Certain parasitic protozoa (for example, trypanosomes) today serve as models for some of the most exciting research in molecular genetics and gene expression.[6, 15, 32]

Historically centered on animal parasites of humans and domestic animals, the discipline of parasitology usually does not include a host of other parasitic organisms, such as viruses, bacteria, fungi, and nematode parasites of plants. Thus, parasitology has evolved separately from virology, bacteriology, mycology, and plant nematology. Medical entomology, too, has branched off as a separate discipline, but it remains a subject of paramount importance to parasitologists, who must understand the relationships between arthropods and the parasites they harbor and disperse.

SOME BASIC DEFINITIONS

Parasitology is largely a study of **symbiosis,** or, literally, "living together." Although some authors restrict the term *symbiosis* to relationships wherein both partners benefit, we prefer to use the term in a wider sense, as originally proposed by the German scholar A. de Bary in 1879: *Any two organisms living in close association, commonly one living in or on the body of the other, are symbiotic, as contrasted with free living.* Usually the **symbionts** are of different species but not necessarily.

Symbiotic relationships can be characterized further by specifying the nature of the interactions between the participants. It is always a somewhat arbitrary act, of course, for people to assign definitions to relationships between organisms. But animal species participate in a wide variety of symbiotic relationships, so parasitologists have a need to communicate about these interactions and thus have coined a number of terms to describe them.

Interactions of Symbionts

Phoresis

Phoresis exists when two symbionts are merely "traveling together," and there is no physiological or biochemical dependence on the part of either participant. Usually one **phoront** is smaller than the other and is mechanically carried about by its larger companion (Fig. 1.1). Examples are bacteria on the

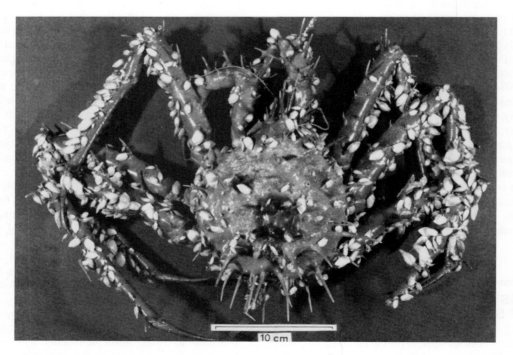

Figure 1.1 Gooseneck barnacles (*Poecilasma kaempferi*) growing on the legs and carapace of a crab (*Neolithodes grimaldi*). This is an example of phoresis since the two species are merely "traveling together." However, the relationship could grade into commensalism; some advantages probably accrue to the barnacles.

From R. Williams and J. Moyse, "Occurrence, distribution, and orientation of *Poecilasma kaempferi* Darwin (Cirripedia: Pedunculata) epizoic on *Neolithodes grimaldi* Milne-Edwards and Bouvier (Decapoda: Anomura) in the northeast Atlantic," in *J. Crust. Biol.* 8:177–186. Copyright © 1988. Reprinted with permission of publisher.

legs of a fly or fungal spores on the feet of a beetle. *Dermatobia hominis* is a fly whose larva lives beneath the skin of warmblooded animals (p. 596). The female does not attach her eggs directly to the host of the larva but rather to another insect, such as a mosquito. When the mosquito finds an animal upon which to feed, the eggs hatch rapidly, and the larvae drop onto the new host and burrow into its skin.

Mutualism

Mutualism describes a relationship in which both partners benefit from the association. Mutualism is usually obligatory, since in most cases physiological dependence has evolved to such a degree that one mutual cannot survive without the other. Termites and their intestinal protozoan fauna are an excellent example of mutualism. Termites cannot digest cellulose because they cannot synthesize and secrete the enzyme cellulase. The myriad flagellates in a termite's intestine, however, synthesize cellulase and consequently digest wood eaten by their host. The termite uses molecules excreted as a by-product of the flagellates' metabolism. If we kill the flagellates by exposing termites to high temperature or high oxygen concentration, then the termites starve to death, even though they continue to eat wood.

An astonishing variety of mutualistic associations can be found among animals, bacteria, fungi, algae, and plants. Blood-sucking leeches cannot digest blood, for example, but their intestinal bacteria, species that are restricted to leech guts, do the digestion for their hosts. At least 20% (perhaps as many as 70%) of insect species, as well as many mites, spiders, crustaceans, and nematodes, are infected with bacteria of genus *Wolbachia*.[46] Filarial nematodes such as *Wuchereria bancrofti* and *Onchocerca volvulus* (chapter 29), which cause serious human diseases, are infected with *Wolbachia*, and they can be "cured" of their bacterial infections by treating patients with antibiotics.[36] But then the worms die too! Although the nature of this relationship is not known, we presume it is metabolic; that is, the partners exchange needed molecules. As is the case with many such relationships, exploration of the basis for the mutualism would make an interesting doctoral dissertation project!

Mutualistic interactions are not restricted to physiological ones. For example, **cleaning symbiosis** is a behavioral phenomenon that occurs between certain crustaceans and small fish—the cleaners—and larger marine fish (Fig. 1.2) on coral reefs. Cleaners often establish stations, which the large fish visit periodically, and the cleaners remove ectoparasites, injured tissues, fungi, and other organisms. Some evidence exists that such associations may be in fact obligatory; when all cleaners are carefully removed from a particular area of reef, for example, all the other fish leave too. You can find other examples of mutualistic and related associations in the texts edited by Cheng[11] and Henry.[22]

Commensalism

In **commensalism** one partner benefits from the association, but the host is neither helped nor harmed. The term means "eating at the same table," and many commensal relationships involve feeding on food "wasted" or otherwise not consumed by the host. Pilot fish (*Naucrates* spp.) and remoras (Echeneidae) are often cited as examples of commensals.

Figure 1.2   Cleaning symbiosis.
Giant moray (*Gymnothorax javanicus*) and a cleaner wrasse (*Labroides dimidiatus; arrow*) on a coral reef in the Red Sea.
Photograph by Larry S. Roberts.

A remora is a slender fish whose dorsal fin is modified into an adhesive organ, which it attaches to large fish, turtles, and even submarines! The remora gets free rides and scraps, but it does not harm the host or rob it of food. Some remoras, however, are mutuals, because they clean the host of parasitic copepods (chapter 34).[12]

Commensalism may be **facultative,** in the sense that the commensal may not be required to participate in an association to survive. Stalked ciliates of the genus *Vorticella* are frequently found on small crustaceans, but they survive equally well on sticks in the same pond. Related forms, however, such as *Epistylis* spp., are evidently **obligate** commensals, because they are not found except on other organisms, especially crustaceans.

Humans harbor several species of commensal protozoans, such as *Entamoeba gingivalis* (chapter 7). This ameba lives in the mouth, where it feeds on bacteria, food particles, and dead epithelial cells but never harms healthy tissues. It has no cyst or other resistant stage in its life cycle. Adult tapeworms are universally regarded as parasites, yet some have no known ill effects on their host.[28]

Parasitism

Parasitism is a relationship in which one of the participants, the parasite, either harms its host or in some sense lives at the expense of the host. Parasites may cause mechanical injury, such as boring a hole into the host or digging into its skin or other tissues, stimulate a damaging inflammatory or immune response, or simply rob the host of nutrition. Most parasites inflict a combination of these conditions on their hosts.

If a parasite lives on the surface of its host, it is called an **ectoparasite;** if internal, it is an **endoparasite.** Most parasites are **obligate parasites;** that is, they cannot complete their life cycle without spending at least part of the time in a parasitic relationship. However, many obligate parasites have free-living stages outside any host, including some periods of time in the external environment within a protective eggshell or cyst. **Facultative parasites** are not normally parasitic but can become so when they are accidentally eaten or enter a wound or other body orifice. Two examples are certain free-living amebas, such as *Naegleria fowleri* (p. 114), and free-living nematodes belonging to genus *Halicephalobus.*[2] Infection of humans with either of these is extremely serious and usually fatal.

When a parasite enters or attaches to the body of a species of host different from its normal one, it is called an **accidental,** or **incidental, parasite.** For instance, it is common for nematodes, normally parasitic in insects, to live for a short time in the intestines of birds or for a rodent flea to bite a dog or human. Accidental parasites usually do not survive in the wrong host, but in some cases they can be extremely pathogenic (see sections on *Baylisascaris, Toxocara,* chapter 26). Parasitism is usually the result of a long, shared evolutionary history between parasite and host species. Accidental parasitism puts both host and parasite into environmental conditions to which neither is well adapted; it is not surprising that the result may be serious harm to either or both participants.

Some parasites live their entire adult lives within or on their hosts and may be called **permanent parasites,** whereas a **temporary,** or **intermittent, parasite,** such as a mosquito or bed bug, only feeds on the host and then leaves (chapters 37, 39). Temporary parasites are often referred to as **micropredators,** in recognition of the fact that they usually "prey" on several different hosts (or the same host at several discrete times).

Predation and parasitism are conceptually similar in that both the parasite and the predator live at the expense of the host or prey. A parasite, however, normally does not kill its host, is small relative to the size of the host, has only one host (or one host at each stage in its life cycle), and is symbiotic. The predator kills its prey, is large relative to the prey, has numerous prey, and is not symbiotic. **Parasitoids,** however, are insects, typically wasps or flies (orders Hymenoptera and Diptera, respectively, chapters 40, 39), whose immature stages feed on their host's body, usually another insect, but finally kill the host. Parasitoids resemble predators in this regard, but they only require a single host individual. **Protelean parasites** are insects in which only the immature stages are parasitic. Mermithid nematodes (p. 362) and hairworms (Phylum Nematomorpha, chapter 31) may also be considered protelean parasites.

Hosts

Parasitologists differentiate among various types of hosts according to the role the host plays in the life cycle of the parasite. A **definitive host** is one in which the parasite reaches sexual maturity. Sexual reproduction has not been clearly shown in some parasites—such as amebas and most trypanosomes—and in these cases we arbitrarily consider the definitive host the one most important to humans. An **intermediate host** is one that is required for parasite development but one in which the parasite does not reach sexual maturity. Definitive hosts are often but not necessarily vertebrates; malarial parasites, *Plasmodium* spp., reach sexual maturity and undergo fertilization in mosquitoes, which are therefore by definition their definitive hosts, whereas vertebrates are the intermediate hosts (see chapter 9).

A **paratenic** or **transport host** is one in which the parasite does not undergo any development but in which it remains alive and infective to another host. Paratenic hosts may bridge an ecological gap between the intermediate and definitive hosts. For example, owls may be parasitized by thorny-headed worms (chapter 32), which undergo development to infective stages in insects that pick up the worm eggs from owl feces. Large owls rarely, if ever, eat insects, but shrews eat them regularly, sometimes accumulating large numbers of juvenile worms that encyst in their mesenteries. Owls do catch shrews, however, sometimes getting heavily infected with the worms. In this case the shrew is a transport host between the insect intermediate and the owl definitive host. In an already cited example of phoresis, the mosquito would be a paratenic host of *Dermatobia hominis.*

Most parasites develop only in a restricted range of host species. That is, parasites exhibit varying degrees of **host specificity,** some infecting only a single host species, others infecting a number of related species, and a few being capable of infecting many host species. The pork tapeworm, *Taenia solium,* apparently can mature only in humans, so adult *T. solium* have absolute host specificity. The nematodes *Trichinella* spp. seem to be able to mature in almost any mammal.

Any animal that harbors an infection that can be transmitted to humans is called a **reservoir host,** even if the animal is a normal host of the parasite. Examples are rats and wild carnivores with *Trichinella spiralis* (p. 386), dogs with *Leishmania* spp., and armadillos with *Trypanosoma cruzi,* the causative agent of Chagas' disease (see chapter 5).

Finally, many parasites host other parasites, a condition known as **hyperparasitism.** Examples are *Plasmodium* spp. in mosquitoes, a tapeworm juvenile in a flea, a monogene (*Udonella caligorum*) on a copepod parasite of fish, and the many insects whose larvae parasitize other parasitic insect larvae.

PARASITOLOGY AND HUMAN WELFARE

Humans have suffered greatly through the centuries because of parasites. Fleas and their obligate symbiont bacteria together destroyed a third of the European population in the 17th century, and malaria, schistosomiasis, and African sleeping sickness have sent untold millions to their graves. Even today, after successful campaigns against yellow fever, malaria, and hookworm infections in many parts of the world, parasitic diseases in association with nutritional deficiencies are the primary killers of humans. Recent summaries of worldwide prevalence of selected parasitic diseases (Table 1.1) show that

Table 1.1	**Some Human Infections with Parasites**

Disease Category	Human Infections	Deaths Per Year
All helminths	4.46 billion	
Ascaris lumbricoides	1221 million	60 thousand
Hookworms	740 million	65 thousand
Trichuris trichiura	795 million	10 thousand
Filarial worms	657 million	20–50+ thousand
Schistosomes	200 million	20 million
Malaria	298–659 million	1–2 million
Entamoeba histolytica	50 million	40 thousand

there are more than enough existing infections for every living person to have one or more.[9, 13, 14, 16, 25, 38, 40, 45]

The parasites in Table 1.1 are, of course, only a few of the many kinds of parasites that infect humans, and in addition to causing many deaths, they complicate and contribute to other illnesses. The majority of the more serious infections occur in tropical regions, particularly in less-developed countries, so most dwellers within temperate, industrialized regions are unaware of the magnitude of the problem. The global prevalence (proportion of a population infected) of *Ascaris lumbricoides* was estimated in 2003 at 26%, that of *Trichuris trichiura* at 17%, and of hookworm at 15%.[14] These figures remained virtually unchanged for *50 years*, despite the fact that the earth's population had more than doubled in that period!

Money for research on tropical infections is very scarce because pharmaceutical companies are reluctant to spend money to develop drugs for treating people who cannot pay for them, and the less-developed countries have many other urgent financial problems. In 2003, $543 was spent for cancer research in the United States per person with a history of cancer by the National Cancer Institute alone, in addition to money spent by private philanthropies.[42, 43] For every case of cardiovascular disease in the United States, the National Heart, Lung, and Blood Institute spent over $32 per case on research.[42, 44] By way of contrast, the World Health Organization spent $0.004 per case for research on schistosomiasis (chapter 16) for each of the five years ending in 2002, although "it is at present difficult to determine the [total from all sources] invested in schistosomiasis research."[48]

The notion held by the average person that humans in the United States are free of worms is largely an illusion—an illusion created by the fact that the topic is rarely discussed because of our attitudes that worms are not the sort of thing that refined people talk about, the apparent reluctance of the media to disseminate such information, and the fact that poor people are the ones most seriously affected. Some estimates place the number of children in the United States infected with worms at about 55 million. This is a gross underestimate if one includes pinworms *(Enterobius vermicularis),* which infect people of all socioeconomic groups. Some authorities believe that infection with juveniles of dog roundworm *(Toxocara canis)* may be more common than pinworm infection in the United States and Canada.[26]

However, the public is becoming more conscious of some parasites. Some one-celled parasites, such as *Pneumocystis, Toxoplasma,* and *Cryptosporidium,* are among the most common opportunistic infections in patients with acquired immunodeficiency syndrome (AIDS). Although common, these parasites rarely cause serious disease in people with uncompromised immune responses, but *Cryptosporidium* was responsible for a widely publicized diarrhea epidemic affecting 403,000 people in Milwaukee, Wisconsin, in 1993.[20] Sales campaigns for heartworm medication have increased public awareness of this dangerous pathogen of dogs (p. 453). Stories on frog deformities reported all across the United States appeared in the media, including that the most important cause of the deformities was a trematode.[5]

We are witnessing emergence of "new" disease agents, some of which are parasitic or transmitted by arthropods, as well as development of drug resistance in long-known pathogens. The first infection of *Cyclospora cayetanensis* (p. 132) in humans was diagnosed in 1977, and it was reported only sporadically between 1977 and 1996. In 1996 and 1997 there were many outbreaks involving hundreds of people in the United States.[8] The most dangerous species of malarial organism, *Plasmodium falciparum,* has become drug resistant in many parts of the world, and there are numerous reports of drug resistance in *P. vivax* (p. 157).

Along with immigration from tropical countries into the United States, travel of U.S. residents to tropical countries is increasing. Many thousands of immigrants who are infected with schistosomes, malaria organisms, hookworms, and other parasites—some of which are communicable—currently live in the United States. Service personnel returning from abroad often bring parasite infections with them. In 1992, 302 of 917 U.S. Peace Corps volunteers in Malawi tested positive for *Schistosoma* infection.[7] There are documented cases of viable filariasis and *Strongyloides* 40 or more years after the initial infection![4, 34] A traveler may become infected during a short layover in an airport, and many pathogens find their way into the United States as stowaways on or in imported products. Travel agents and tourist bureaus are reluctant to volunteer information on how to avoid the tropical diseases that a tourist is likely to encounter—they might lose the customer.[21] Small wonder, then, that "exotic" diseases confront the general practitioner with frequency.

There are other much less obvious ways in which parasites affect humanity. For example, 500 million people in the world have protein-energy malnutrition, and 350 million have iron-deficiency anemia.[39] Malnutrition is exacerbated both by population increase and by environmental degradation. From 2 billion in 1927, the population of the earth doubled to 4 billion in 1974, passed 6 billion in 1999, and is expected to exceed 8.9 billion in 2030.[23] Meanwhile, environmental degradation such as erosion continues to decrease the available supply of cropland. The increasing scarcity of resources contributes to violent conflict in the world.[24] The contributions of parasites to malnutrition are important but are underestimated because of underreporting.[39] Hospitals usually list what appears to be the most obvious cause of death, but most patients have multiple infections that have contributed to their disease state.

Even where food is being produced, it is not always used efficiently. Considerable caloric energy is wasted by fevers caused by parasitic infections. Heat production of the human body increases about 7.2% for each degree rise in Fahrenheit. A single acute day of fever caused by malaria requires approximately 5000 calories, or an energy demand equivalent to two days of hard manual labor. To extrapolate, in a population with an average diet of 2200 calories per day, if 33% had malaria, 90% had a worm burden, and 8% had active tuberculosis (conditions that are repeatedly observed), there would be an energy demand equivalent to 7500 tons of rice per month per million people in addition to normal requirements. That is a waste of 25% to 30% of the total energy yield from grain production in many societies.[35]

Humans create many of their own disease conditions because of high population density and subsequent environmental pollution. Population shifts from rural to urban areas commonly overload water and sewage capabilities of even major cities. Industrialization has first priority in developing countries, with reduction in pollution being neglected.[30] Nightsoil (human feces and urine) is often used as fertilizer for food crops (Fig. 1.3). Millions of people, especially children, die each year from diseases that could be prevented with proper sanitation facilities.[30]

Parasites are also responsible for staggering financial loss. Malaria, for example, is usually a chronic, debilitating, periodically disabling disease. In situations where it is prevalent, the number of hours of productive labor lost multiplied by the number of malaria sufferers yields a figure that can be charged as loss in the manufacture of goods, in the production of crops, or in the earning of a gross national product. Nations that import goods from countries infected with malaria, schistosomiasis, hookworm, and many other parasitic diseases pay more for these products than they would had the products been produced without the burden of disease. Plant parasites further diminish the productive capacities of all countries.

National and international efforts to increase productivity and standard of living in less-developed countries sometimes inadvertently increase parasitic disease. Schistosomiasis in Egypt increased after construction of the Aswan High Dam on the Nile River (p. 246). Smaller dams for drainage and agriculture have promoted transmission of schistosomiasis, onchocerciasis, dracunculiasis, and malaria.[37] The World Bank loaned Brazil funds to pave highways into the Amazon region to settle poor urban workers for farming, despite contrary advice from their own agricultural experts.[30] This action of the World Bank and government of Brazil produced an increase in malaria and spread of malaria to new foci when the migrants returned to the cities after their farms failed.[31]

An important role of parasitologists, together with that of other medical disciplines, is to help achieve a lower death rate. However, it is imperative that this reduction be matched with a concurrent lower birth rate and higher quality of life. If not, we are faced with the "parasitologist's dilemma," that of sharply increasing a population that cannot be supported by the resources of the country. George Harrar, president of the Rockefeller Foundation, observed, "It would be a melancholy paradox if all the extraordinary social and technical advances that have been made were to bring us to the point where society's sole preoccupation would of necessity become survival rather than fulfillment." Harrar's paradox is already a fact for half the world. Parasitologists have a unique opportunity to break the deadly cycle by contributing to the global eradication of communicable diseases while making possible more efficient use of the earth's resources.

PARASITES OF DOMESTIC AND WILD ANIMALS

Both domestic and wild animals are subject to a wide variety of parasites. Although wild animals are usually infected with several species of parasites, they seldom suffer massive deaths, or **epizootics,** because of the normal dispersal and territorialism of most species. However, domesticated animals are usually confined to pastures or pens year after year, often in great numbers, so that parasite eggs, larvae, and cysts become extremely dense in the soil, and the burden of adult parasites within each host becomes devastating. For example, the protozoa known as *coccidia* thrive under crowded conditions; they may cause up to 100% mortality in poultry flocks, 28% reduction in wool in sheep, and 15% reduction in weight of lambs.[35] Infections in poultry are controlled by the costly method of prophylactic drug administration in feed. Unfortunately, coccidia have become resistant to one drug after another.[10] Many other examples could be given, some of which are discussed later in this book. Thanks to the continuing efforts of parasitologists around the world, identifications and life cycles of most parasites of domestic animals are well known. This knowledge, in turn, exposes weaknesses in the biology of these pests and

Figure 1.3   **"Nightsoil" is a logical use of human feces and urine.**

Here it is applied to a vegetable garden, a technique practiced in much of the world. Although sometimes controlled by government regulations, it still serves as a significant means for distribution of eggs of some helminths and certain protozoan cysts.

Photograph by Robert E. Kuntz.

suggests possible methods of control. Similarly, studies of the biochemistry of organisms continue to suggest modes of action for chemotherapeutic agents. We should bear in mind, however, that control of parasites in domestic animals may bear considerable ecological hazard.[37] Antiparasitic drugs may have important impact on numbers and diversity of dung fauna where treated animals are pastured.[27]

Less can be done to control parasites of wild animals. Most wild animals can tolerate their parasite burdens fairly well, but they will succumb when crowded and suffering from malnutrition, just as will domestic animals and humans. For example, the range of the bighorn sheep in Colorado has been reduced to a few small areas in the high mountains. The sheep are unable to stray from these areas because of human pressure. Consequently, lungworms have so increased in numbers that in some herds no lambs survive the first year of life. These herds seem destined for quick extinction unless a means for control of their parasites can be found in the near future.

Still another important aspect of animal parasitology is transmission to humans of parasites normally found in wild and domestic animals. The resultant disease is called a **zoonosis.** Many zoonoses are rare and cause little harm, but some are more common and important to public health. An example is trichinosis, a serious disease caused by a minute nematode, *Trichinella* spp. (chapter 23). These worms exist in several **sylvatic** cycles that involve wild animals and in an **urban** or **domestic** cycle chiefly among rats and swine. People become infected when they enter the cycles, such as by eating undercooked bear or pork. Another zoonosis is echinococcosis, or hydatid disease, in which humans accidentally become infected with juvenile tapeworms when they ingest eggs from dog feces (chapter 21). *Toxoplasma gondii,* which is normally a parasite of felines and rodents, is now known to cause many human birth defects (chapter 8).

We recognize new zoonoses from time to time. Lyme disease, a bacterial infection transmitted by ticks, was long present in deer and white-footed mice, but frequent transmission to humans began only in the 1970s.[3] It is the obligation of parasitologists to identify, understand, and suggest means of control of such diseases. The first step is always proper identification and description of existing parasites so that other workers can recognize and refer to them correctly by name in their work. Thousands of species of parasites of wild animals are still unknown and will occupy the energies of taxonomists for many years to come. Unfortunately, the numbers of parasites described each year has been declining, probably because of the decline in young taxonomists being trained.

Aside from their roles as causative agents of disease, parasites provide us with an almost unlimited supply of fascinating and challenging problems in ecology and evolution (chapter 2). Presence of a parasite species with a complex life cycle demonstrates unequivocally that intermediate hosts occupy an area and that an ecological relationship exists between hosts and parasites. Parasites also may be one of the factors, along with predation and abiotic events, that function to regulate host populations. Finally, virtually every species of animal is parasitized by at least one other species. Thus, much of the overall biodiversity found in any ecosystem can be attributed to parasitism. Biodiversity is essential

to human survival and quality of life, and the parasites they bear can tell us much about the biology and evolution of the hosts.[41, 47]

PARASITOLOGY FOR FUN AND PROFIT

In addition to having medical and economic importance, the study of parasites is fascinating (fun), and one can pursue such study as a career (profit). Most parasites are products of a long evolution as symbionts and are thus exquisitely adapted for life within the body of another organism. That there are more parasites than free-living organisms in the world is an indication of the success of parasitism. From a biological perspective they are interesting, beautifully adapted, and intricate organisms. Despite our effort to alleviate human affliction with the most serious pathogens, we should appreciate that parasites are a huge part of nature. Whether or not you become a career parasitologist or health-care professional, your study of parasites will be an adventure.

Careers in Parasitology

Parasitology offers an area of interest for every biologist. The field is large and encompasses so many approaches and subdivisions that anyone who is interested in biological research can find a lifetime career in parasitology.[1, 17] It is a satisfying career because each bit of progress made, however small, contributes to our knowledge of life and to the eventual conquering of disease. As in all scientific endeavor, every major breakthrough depends on many small contributions made, usually independently, by individuals around the world. Previously little-known parasites suddenly became life-threatening infections in AIDS patients. Had their identifications and life cycles been better understood, we would have saved much expense and time in recognizing the complex facets of AIDS.

The training required to prepare a parasitologist is rigorous. Modern researchers in parasitology are well grounded in physics, chemistry, and mathematics, as well as biology from the subcellular through the organismal and populational levels. They must be grounded firmly in medical entomology, histology, and basic pathology. Depending on their interests, they may require advanced work in physical chemistry, immunology, molecular biology, genetics, and systematics. Such intense training is understandable, because parasitologists must be familiar with the principles and practices that apply to over a million species of animals; in addition they need thorough knowledge of their fields of specialty. Most parasitologists hold a Ph.D. or other doctoral degree, but people with master's or bachelor's degrees have made many contributions, and undergraduates working on independent study projects or honors theses have also contributed. Once they have received their basic training, parasitologists continue to learn during the rest of their lives. Even after retirement, many remain active in research or writing for the sheer joy of it. Parasitology and parasitologists indeed have something for everyone, including a sense of humor and even fun.[17]

References

References for superscripts in the text can be found at the following Internet site: www.mhhe.com/robertsjanovynadler9e

Additional Readings

Ahmadjian, V., and S. Paracer. 1986. *Symbiosis. An introduction to biological associations.* Hanover, NH, and London: University Press of New England. Short text but includes consideration of symbioses not usually covered in parasitology courses, such as bacterial, viral, fungal, and algal.

Combes, C. (Transl. by D. Simberloff.) 2005. *The art of being a parasite.* Chicago and London: University of Chicago Press. Another well-written book intended for professionals and general audience. Recommended for all students.

Cox, F. E. G. (Ed.). 1993. *Modern parasitology* (2d ed.). Oxford: Blackwell Scientific Publications. Excellent for further reading in epidemiology, biochemistry, molecular biology, physiology, immunology, chemotherapy, and control.

Gallagher, R. B., J. Marx, and P. J. Hines. 1994. Progress in parasitology. *Science* 264:1827. This is the lead editorial in an issue of the journal *Science* featuring parasitology news and research.

Hyde, J. E. 1990. *Molecular parasitology.* New York: Van Nostrand Reinhold.

Marr, J. J., and M. Müller (Eds.). 1995. *Biochemistry and molecular biology of parasites.* London: Academic Press.

Wyler, D. J. (Ed.). 1990. *Modern parasite biology. Cellular, immunological, and molecular aspects.* New York: W. H. Freeman and Co.

Zimmer, C. 2000. *Parasite rex.* New York: The Free Press. A well-written book about parasites and parasitologists, highly recommended for professionals and general audience.

Parasitology on the World Wide Web

Recent years have witnessed the burgeoning of information easily available to anyone with a computer, modem, and connection to the Internet. Web surfers should use caution, however, because there is a great deal of misinformation on the Internet. Some sites that are authoritative, accurate, and helpful to students in parasitology follow. They all have links to sites with further information.

http://www.biosci.ohio-state.edu/~parasite/home.html has links to more than 550 images of parasites.

http://asp.unl.edu is the Web page of the American Society of Parasitologists, which has a section on Careers in Parasitology.

http://www.astmh.org is the Web page of the American Society of Tropical Medicine and Hygiene.

http://www.who.int/en/ is the home page of the World Health Organization.

http://www.histology.wisc.edu/histo/uw/histo.htm is the University of Wisconsin Medical School Histology Home Page. It provides an excellent review of the normal appearance of tissues in microscopical thin section.

http://www.cdc.gov/travel/index.htm is a site for travelers seeking health advice.

Chapter 2

Basic Principles and Concepts I: Parasite Systematics, Ecology, and Evolution

The host is an island invaded by strangers with different needs, different food requirements, different locations within which to raise their progeny.

—W. Taliaferro

Systematics is the study of biological diversity and classification, all within an evolutionary context.[36] Systematists seek to understand the origin of diversity at all levels of classification, from species to kingdom. Ecological and evolutionary research, including that done by parasitologists, ultimately depends on the accurate identification, complete inventories, and descriptions provided by systematists. Thus, the three subject areas—systematics, ecology, and evolution—are inextricably linked and interdependent. This linkage is especially important for parasitologists trying to control disease transmission because epidemiology requires knowledge of the causative organisms (systematics), understanding of environmental and life cycle factors contributing to infection (ecology), and the history of host-parasite relationships (evolution).

SYSTEMATICS AND TAXONOMY OF PARASITES

In general the world's invertebrates, of which parasites make up a sizeable fraction, are not nearly as well-known as vertebrates. Many new species of protozoa, helminths, and arthropods are described every year. Indeed, with a not unreasonable amount of work, almost anyone, undergraduate biology majors included, can find and describe a new species. Then the finder can pick the species name and be immortalized, through the name, in the parasitological literature. The monogenetic trematode *Salsuginus thalkeni*, for example, is named for a landowner who gave students permission to use his property for projects. *Actinocephalus carrilynnae*, a protozoan parasite of damselflies, was named in "honor" of a little sister by an older sister who threatened to "name a parasite after" her. Occasionally one hears parasitologists talk about naming parasites after politicians. But, in a more sober and dignified vein, a number of trematodes and tapeworms were named by Edward Adrian Wilson and Robert Leiper in honor of members of the ill-fated Robert F.

Scott expedition to the South Pole who died on the trip.[9] Campbell's story of that expedition is a compelling one, worth reading by any person who feels that scientific names are just biologists' way of separating Latin scholars from the rest of humanity.

Taxa (pl.; s. taxon) are groups, ranging from subspecies and species, to the increasingly inclusive genera, families, orders, classes, phyla, and kingdoms. Members of a taxon are considered evolutionarily related. **Taxonomy**, or the science of classification, is as vibrant an area of biology today as it was a hundred years ago. A good part of the activity is due to molecular techniques that have been adopted by taxonomists.

Parasitologists are constantly evaluating the criteria used to make taxonomic decisions and reexamining the genus, family, and order groupings of animals they study. Molecular techniques have proven to be exceedingly powerful tools for resolving taxonomic problems. An excellent example of such resolution is use of 18S ribosomal gene sequences to provide evidence that myxozoans (chapter 11), often considered protozoans, were in fact cnidarians.[47] Cladistic analysis of nonmolecular myxozoan characters, including ultrastructure and spore development, seemed to establish a link with multicellular phyla, but molecular data confirmed this relationship.[47]

Systematists today are expected to employ such techniques, use them in phylogenetic studies, and deposit DNA sequences in the globally available database GenBank. These sequences are assigned accession numbers and then become readily available to anyone with a computer and Internet access. In the case of new species descriptions, however, type specimens are also deposited in museums and assigned accession numbers but usually are not available for study except to qualified researchers.

Why is this work important? Taxonomy is a basic subdiscipline of biology. Scientific names carry with them massive amounts of information, some implied, some explicit, and all of value to ecologists, immunologists, epidemiologists, and evolutionary biologists. For example, a doctor or veterinarian cannot make a decision about treatment without knowing what kind of parasite is infecting a patient. And an epidemiologist looking for ways to control malaria or filariasis is stumped if unable to differentiate among species of mosquitoes.

Taxonomic criteria vary from parasite group to group. In arthropods, skeletal morphology is still of primary importance. Classification of Platyhelminthes is based to a large extent on reproductive organs—primarily their numbers, sizes, and relative positions in the body, although more inclusive taxonomic groupings are now based mostly on ultrastructural and molecular characters. Nematode taxonomists also must focus on reproductive structures, including those at the posterior end of males, but arrangements of sensory papillae and other cuticular features, especially around the mouth, are also considered. Protozoan taxonomic characters include cyst morphology (amebas, coccidia), number and arrangement of flagella, and biochemical properties. Members of genus *Leishmania,* for example, are "typed" using a variety of molecular methods.[34]

In the recent past, several types of macromolecules ranging from enzymes to genes for ribosomal subunits have been used in systematics research. Currently, however, there is an emphasis on the cytochrome-*c* oxidase subunit 1 (CO1) gene largely as a result of international efforts by various Barcode of Life organizations.[38] The intent is to develop a molecular library of CO1 sequences for all the world's biota, parasites included. In at least some invertebrates, taxonomic distinctions based on CO1 sequences generally correspond to those established on morphology.[4] Parasitologists continue to develop phylogenetic hypotheses, however, using not only morphology but also genes such as those for internal transcribed spacers (ITS) and both large and small RNA subunits, addressing questions of host-parasite co-evolution and geographic distribution.[38, 42]

Molecular parasitologists often can obtain identified material from the American Type Culture Collection in Manassas, Virginia, a living museum of microorganisms, including many species and strains of parasites. It is not only easier but also more advisable for experimental biologists to obtain described and documented organisms from such a collection than it is for them to do the taxonomy themselves. A parasitological ecologist, however, must be prepared to identify animals and to describe them if necessary. Thus, a researcher quickly becomes familiar with the massive body of literature, some of it published in obscure and foreign journals, that has accumulated since Linnaeus first described the sheep liver fluke *Fasciola hepatica.*

PARASITE ECOLOGY

The Host as an Environment

Ecology is the study of relationships between organisms and their environments (including other organisms), with a focus on those factors that regulate numbers and distributions of organisms. The host is, of course, a parasite's environment in both ecological and evolutionary senses. Thus, parasitologists often find themselves studying infective organisms from many different perspectives, including taxonomy, transmission, population dynamics, and evolutionary history.

Although a parasite's environment is primarily the host, transmission stages such as spores, eggs, and often juveniles must also survive abiotic conditions. A host represents a rich and highly regulated supply of nutrients. Most animals' body fluids have a wide array of dissolved proteins, amino acids, carbohydrates, and nucleic acid precursors, and virtually all animals have mechanisms for maintaining the chemical makeup and osmotic balance of their body fluids. Vertebrates and many invertebrates control body temperature as well either by metabolic or behavioral means. Parasites exhibit traits that allow them to exploit such living environments, and we should expect evolutionary changes in hosts to be accompanied by parallel, perhaps adaptive, changes in their parasites. But we should also expect adaptations—e.g., resistant cysts—that aid survival in the abiotic environment between hosts.

Hosts are relatively small patches within the vast matrix that is their own habitat. That is, they are islands in the sense of Taliaferro's quote, although these islands can move and defend themselves, such as through immune reactions. Thus, suitable parasite environments are dispersed in addition to being rich and regulated. For example, there is an enormous volume of water in a lake compared to the volume of fish in that same lake. This seemingly trivial observation points to a major problem for monogenean flatworms (p. 283) that must live on these fish: Unless the worms' reproductive stages are able to keep finding fish to infect, the parasites are likely to become locally extinct. Indeed, parasite control strategies often are based on reducing the probability of host and parasite encounter. Conversely, many parasites possess traits that evidently function to increase the probability of finding a host.

Throughout the following chapters, interactions between hosts and parasites are described for the parasite species discussed. These interactions can be thought of as ecological associations that sometimes result in changes to the environment, such as pathology, or an immune reaction that may affect host or parasite survival.

A Parasite's Ecological Niche

A parasite's ecological niche includes resources provided by the living body of another species as well as abiotic conditions encountered by transmission stages such as eggs, cysts, spores, and juveniles. Thus, most parasites encounter a wide variety of environmental conditions during their life cycles.

The human digestive tract is a good illustration of a resource that varies according to region, thus providing numerous microenvironments.[45] Food processing occurs in distinct phases, from chewing and salivary amylase action of the mouth, to acid pH and proteolytic enzyme reactions of the stomach, to more neutral pH and numerous amylases, proteases, lipases, and nucleases working in the small intestine, to reclamation of water in the large intestine and subsequent elimination of solid wastes. A trip through the gut could be described also in terms of different symbionts encountered along the way, from *Entamoeba gingivalis* in the mouth, to fourth-stage juvenile *Ascaris lumbricoides* in the stomach, to *Taenia saginata* (or many other helminths) in the small intestine, to *Dientamoeba fragilis, Entamoeba coli, Endolimax nana,* and *Trichuris trichiura* in the large intestine, and finally to pinworms *(Enterobius vermicularis)* crawling

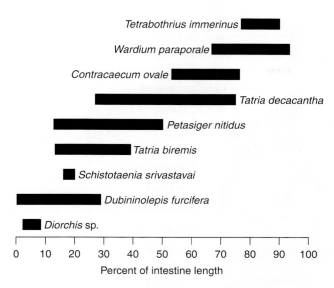

Figure 2.1   Distribution of intestinal nematodes, trematodes, and tapeworms in an aquatic bird (eared grebe).
The horizontal bars are the average position +/− one standard deviation, in terms of the relative distance from the stomach.

Redrawn by John Janovy Jr. from T. M. Stock and J. C. Holmes, "Functional relationships and microhabitat distributions of enteric helminths and grebes (Podicipedidae): The evidence for interactive communities," in *J. Parasitol.* 74:214–227, 1988.

around the anal orifice. Detours into the lungs, up the bile ducts, and through the mucosa into the portal system would also bring us into contact with site-specific parasites.

Host intestinal length is one example of an easily measured resource, and much research has been done on the distribution of cestodes, nematodes, and acanthocephalans within the gut. Intestinal worms usually occur within a particular region, although that distribution is sometimes influenced by host diet, physiological condition, and the presence of other helminths (Fig. 2.1). In addition, subtle differences occur in oxygen and carbon dioxide tension, pH, and other chemical and physical factors between the mucosa and the center of the lumen. Such differences occur even between the top of a villus and its base, making at least two different habitats available for colonization by parasites of suitable sizes. In a study of parasites in the turtle *Testudo graeca,* for example, Schad found eight species of nematode genus *Tachygonetria* living in the large intestine.[46] The species were differentially restricted along intestine length, as well as radially from center to mucosa. Schad concluded that even when two parasite species are found in the same area of the intestine, they may use different resources and thus occupy distinct niches.

Digestive systems also vary greatly between species (compare the stomachs of humans and cows) and even between life-cycle stages of a host, such as between tadpoles and adult toads and frogs. Tadpoles are typically herbivorous: Some are filter feeders; others graze on algae and detritus. Adult anurans, however, are carnivorous. Metamorphosis from tadpole into adult involves loss of intestinal epithelium and significant reduction in intestinal length. Metamorphosis in beetles also involves loss of larval gut tissue. Some protozoan parasites of beetles, such as four species of gregarine parasites (p. 122) in the common mealworm *Tenebrio*

molitor, are specific not only to their host species, but also to the life-cycle stage.[11] These parasites experience the larva as an environment distinct from that of an adult beetle. But even within a single life-cycle stage of a single host species, an intestine may offer many more places, or ways, for parasites to exist than might be suspected. Parasitologists have discovered over 40 species of tapeworms and tens of thousands of individual parasites in scaup ducks alone.[6]

Infection Sites

When viewed from a parasite's perspective, all organisms are complex environments with many separate habitats. Even the smallest insects and crustaceans offer many places, both internally and externally, that can be colonized by parasites. And larger animals, such as rodents, birds, and human beings, provide dozens of microenvironments capable of supporting parasites. Site specificity is actually evidence of parasite adaptation to a particular habitat within a host, and in the following chapters you will find again and again that parasites occur only in their characteristic infection sites. Beginning students are often surprised to discover how many different kinds of parasites can infect a single host species; parasitologists considering the rich opportunities provided by vertebrate bodies, however, might wonder why there are so few.

Although most endoparasites of vertebrates live in the digestive system, adult parasites are found in and on virtually all parts of the body, and juvenile stages often undergo elaborate migrations through the body before arriving at their definitive sites. Parasites that inhabit the lumen of the intestine or other hollow organs are said to be **coelozoic,** while those living within tissues are called **histozoic.** Parasites are generally adapted to and restricted to particular sites within or upon a host (Fig. 2.2). Examples of this phenomenon are malarial parasites living inside red blood cells (p. 147), filarial nematodes that congregate in the heart or beneath the skin (pp. 447, 453), bird mites that occur only on flight feathers, and monogeneans found in the urinary bladders of frogs (p. 294).

Such observations lead to hypotheses about the evolutionary forces that contributed to this circumstance. It is still a matter of some controversy whether parasites compete with one another for resources provided by the host. Some hosts are very heavily infected in nature, with up to several thousand individual worms of a dozen species. It is difficult to imagine that, under these circumstances, some competition is not occurring.

On the other hand, hosts may provide many more microenvironments than we realize. Consider the human eye, an organ not as obviously suited to infection by parasites as the intestine. The retina may be infected by the apicomplexan *Toxoplasma gondii* and juveniles of the filarial nematode *Onchocerca volvulus;* the chamber may harbor bladder worm metacestodes of the tapeworms *Taenia solium, T. crassiceps, T. multiceps,* or *Echinococcus granulosus;* the conjuctiva may host another wandering filarial nematode, *Loa loa;* and the orbit may be the home of nematodes in genus *Thelazia.* Parasitologically, the vertebrate body can be considered a mass of habitats that have been colonized by a great diversity of species. It has been said that if a host were infected with all the parasites capable of infecting it, and host tissues were then removed to leave only parasites, the host could still be recognized!

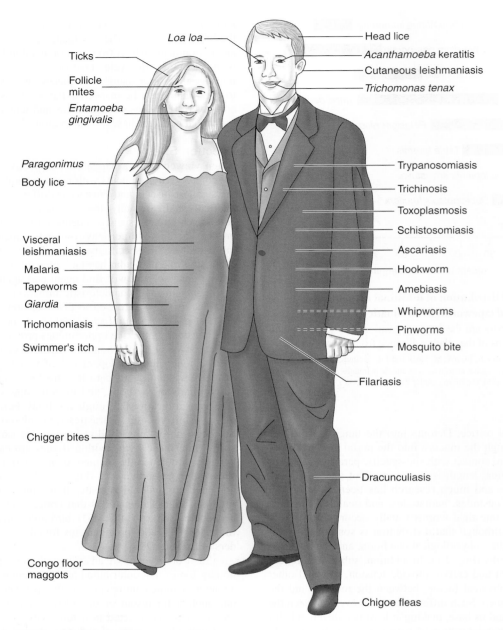

Figure 2.2   **Some parasites of humans.**

Two humans with some parasites they could easily acquire under appropriate circumstances, along with infection sites of those parasites.

Drawing by William Ober and Claire Garrison.

Parasite Populations

Quantitative Descriptors

Numbers of parasites are of major interest to epidemiologists, public health workers, ecologists, and evolutionary biologists. A scientist assessing the impact of parasitic diseases on a human population must know who is infected, whether infections are distributed equally among all age groups and both sexes, and whether certain individuals have unusually high numbers of parasites. Evolutionary biologists also are interested in parasite numbers because relative reproductive success (fitness) is usually described quantitatively.

Parasitologists have adopted a number of terms for describing parasite populations and communities of different parasite species. These terms are defined in Bush et al.[7] and summarized in Table 2.1. You are likely to encounter these words throughout this book, especially in epidemiological sections, because parasitologists use them frequently to communicate information about existing parasite burdens, factors that either enhance transmission or help sustain parasite populations, and problems of disease control. As a minimum, a student of parasitology should have a firm grasp of **prevalence, incidence, abundance,** and **aggregated populations** (Table 2.1) to understand factors influencing the maintenance and spread of disease.

Table 2.1	**Ecological terms as applied to parasite populations and communities**

Ecological term	**Definition**
Population structure	A frequency distribution graph in which numbers of hosts (dependent variable) are plotted against parasite/host classes (independent variable), plus the calculated quantitative descriptors of the frequency distribution. See Fig. 2.3.
Quantitative descriptors	Numbers such as mean, prevalence, etc., that can be calculated from the observed data on the number of parasites in individual hosts.
Sampling unit	One individual host animal in a collection of such hosts.
Infrapopulation	Number of parasites in an individual host (can take the value of zero).
Density	Average number of parasites per host in a sample of hosts, equal to the arithmetic mean.
Intensity	Number of parasites in an infected host (cannot be zero).
Mean intensity	Average number of parasites in infected hosts of a sample of hosts.
Metapopulation	All the infrapopulations in a single host species in an ecosystem.
Suprapopulation	All the parasites of a species regardless of developmental stage, in an ecosystem.
Infracommunity	All the parasites of all species in an individual host.
Compound community	All the parasites of all species in a sample of hosts of a single species in an ecosystem.
Prevalence	Fraction or percentage of a single host species infected at a given time.
Incidence	Number of new infections per unit time divided by the number of uninfected hosts at the beginning of the measured time.
Abundance	Another term sometimes used as synonymous with density or mean.
Aggregated	A situation in which most of the parasites occur in a relative minority of hosts and most host individuals are either uninfected or lightly infected.
Overdispersed	A term sometimes used as a synonym for aggregated.
Variance/mean ratio	Quotient of the variable (square of standard deviation of a frequency distribution) divided by the mean; sometimes used as a measure of aggregation.
k	The value of a parameter of the negative binomial distribution; usually k must be calculated to describe an aggregated parasite population by use of mathematical models.

As an illustration of how to use the terms in Table 2.1, consider a sample of 10 mice with a total of 75 pinworms. This sample would have a density (mean, abundance) of 7.5 worms per host. However, these 75 worms could all be in one mouse (in which case the prevalence would be 0.10) or distributed among all the mice (the prevalence would equal 1.00). Imagine that you are a veterinarian seeking to rid these mice of their worms with only a limited supply of antihelminthic drugs, and you can see immediately why parasite population structure is of major interest to scientists and clinicians. You do not want to waste your medicine by giving it to noninfected rodents!

Macro- and Microparasites

Large parasites that do not multiply (in the life-cycle stage of interest) in or on a host are called **macroparasites.** Examples of macroparasites are adult tapeworms, adult trematodes, most nematodes, acanthocephalans, and arthropods such as ticks and fleas. Macroparasites often, if not typically, occur in aggregated populations. That is, most of the parasites are in relatively few hosts of a species, while the majority of host species individuals are either uninfected or lightly infected (Fig. 2.3). This generality was recognized by H. D. Crofton in the early 1970s[13]; Crofton claimed that such population structure was so characteristic of parasites that it should be included in the definition of parasitism. Crofton also offered several explanations for the origin of this aggregation; as a result, he inspired a massive amount of both theoretical and empirical work on parasite population biology.

Small parasites that multiply within a host are called **microparasites,** and these include bacteria, rickettsia, and protozoan infections such as the malarial parasites (genus *Plasmodium,* p. 143), trypanosomes (p. 64), and amebas (p. 105). Whereas in the case of macroparasites one can generally assume that one parasite reflects a single encounter between host and infective stage, that assumption is not necessarily valid for microparasites. Thus, a population ecologist must use different methods for microparasites than for macroparasites when attempting to discover mechanisms that allow a parasite to maintain itself in a host species' population. The most fundamental questions, however—who is infected, who is resistant, and who is at risk—remain the same regardless of the parasite species involved.

Population Structure

Parasite **population structure** is a critical piece of information for those seeking to control infections. Population structure is often described by the density (mean, abundance), variance (a statistical parameter whose value is related to the shape of a frequency distribution), and curve of best fit. The last is really an equation that generates a theoretical frequency distribution of the parasites among hosts (see Fig. 2.3). A graph can be constructed by plotting parasite per host classes along the X-axis and numbers of hosts that fall into these classes on the Y-axis. The result is a frequency distribution that describes the parasite's population structure.

Figure 2.3 is an example of such a graph; it illustrates Crofton's general principle that most of the host individuals

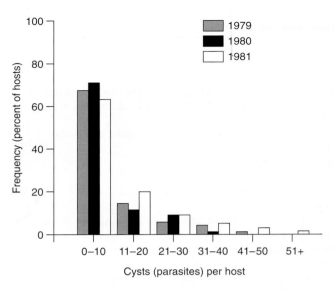

Figure 2.3 **Population "structure" of the trematode** *Uvulifer ambloplitis* **(larvae) in bluegill sunfish in North Carolina over a three-year period.**

Most fish are uninfected, while most parasites are in the relatively few heavily infected fish, those with more than 25 larval cysts. These frequency distributions match those predicted by the mathematical model (equation) known as the *negative binomial*.

Redrawn by John Janovy Jr. from D. A. Lemly and G. W. Esch, "Population biology of the trematode *Uvulifer ambloplitis* (Hughes, 1927) in juvenile bluegill sunfish, *Lepomis macrochirus,* and large mouth bass, *Micropterus salmoides,*" in *J. Parasitol.* 70:466–474, 1984.

are uninfected or only lightly infected, while most parasites are in a few host individuals. In addition to parameter values that dictate the shape of this graph, parasite population structure also includes fractions of juvenile, mature, and gravid parasites and sex ratios (in the case of dioecious parasites). A complete quantitative description of any population—parasites included—obviously involves a great deal of counting, measuring, and determination of sexes and ages of maturity. Parasitologists can quickly form mental pictures of a parasite population from such quantitative information.

There is some evidence that certain individuals within host populations are either genetically or behaviorally predisposed to heavy infections.[14] In studies conducted in both Kenya and Burma, individuals who were heavily infected with *Ascaris* before treatment of an entire village were most likely to be heavily infected again one or two years later. In populations of wild animals, unless specific reasons for a particular parasite population structure have been discovered, one should consider individual host differences, including genetic ones, ecological circumstances, and just plain bad luck all as factors producing heavy infections.

Multiple Species Infections

A single host individual can be infected with several parasite species; that is, it can contain a **parasite community.** These communities can be extraordinarily rich, as illustrated by the intestinal parasites of some endothermic (warm-blooded) vertebrates. In a series of studies by Holmes and his colleagues, 26 species of intestinal helminths were reported from a sample of 31 eared grebes, and 52 species, with "slightly less than 1 million individuals," were found in

45 scaup ducks.[49] Mammals such as coyotes and black bears may also be heavily and frequently infected.[41]

Parasites can interfere with one another in various ways, especially in heavy intestinal infections. This observation has led workers to postulate a variety of types of parasite communities, from interactive ones, in which competition may occur, to noninteractive ones, usually with few species, in which there appears to be little if any competition.[24]

Trophic Relationships

All parasites are heterotrophic, requiring their energy and carbon in the form of existing complex organic molecules and their nitrogen as a mixture of amino acids. In this respect parasites are no different from other kinds of animals. A parasite's feeding devices, however, may differ considerably from those seen in most free-living animals. For example, tapeworms (phylum Platyhelminthes, chapters 20, 21) and spinyheaded worms (phylum Acanthocephala, chapter 32) have no digestive tracts and absorb sugars and amino acids directly across their outer surface. These worms feed through uptake sites on the plasma membrane. The anticoagulant of tick saliva (chapter 41) is an integral part of the parasite's feeding mechanism, another illustration of an adaptation to a characteristic of the host—in this case the clotting property of blood.

Parasites always live at a higher trophic level than their hosts. Thus, all parasites are at least secondary consumers, and those infecting top predators such as hawks, owls, and carnivores live quite high on a typical food pyramid. Trophic relationships are direct and obvious for parasites that eat host tissue and fluids, such as hookworms (p. 397), frog lung flukes (p. 267), and ticks. But parasite use of the host can also be somewhat indirect. For example, all free-living animals spend significant amounts of energy regulating their internal milieus and producing offspring. Thus, parasites can be thought of as using homeostatic mechanisms and reproductive efforts of organisms at lower trophic levels.

Parasites are often said to "exploit" trophic relationships between the various host species. Figure 2.4 illustrates this idea with a few of the parasites found in and on a common sunfish, the bluegill. The fish typically feeds on a variety of invertebrates such as aquatic insects and small crustaceans, which may in turn be infected with larval flukes, larval tapeworms, and juvenile roundworms (*(g)* through *(k)*). These immature parasites can either mature in the fish's digestive tract (the "circles" *(m)*), or develop further in the fish's tissues; in the former case, the bluegill is the definitive host, in the latter, it is a second intermediate host. By eating the fish (*(f)*), birds such as the heron can acquire worms encysted as larvae or juveniles in the fish's tissues. None of these parasite life cycles can be completed unless the food web is intact.

Over the past decade, studies have shown that parasites can make up a significant fraction of biomass and add to the number of trophic levels in ecosystems.[2, 30] When parasites are taken into account, food webs can become extremely complex (Fig. 2.5), with most of the energy, and parasites, flowing through a group of "core" species. In estuaries, trematode biomass can be particularly high, being "comparable to that of . . . birds, fishes, burrowing shrimps and polychaetes."[30] Parasites with complex life cycles also can be used as indicators of overall biodiversity because the presence of certain species in a vertebrate community implies the presence of all

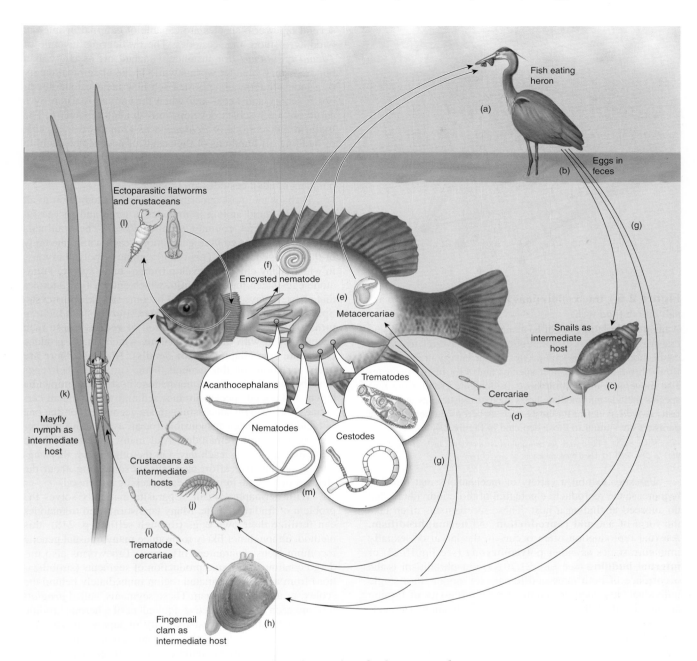

Figure 2.4   **Ecology of parasitism in a typical North American freshwater pond.**
Fish-eating birds such as herons are the habitat of several adult helminth parasites that use blugill as second intermediate hosts, although the fish also is a definitive host for other parasites such as crustaceans and monogenes on the gills and acanthocephalans in the cecea. All the parasite life cycles depend on trophic relationships in an intact food web.

Drawing by Bill Ober and Claire Garrison

intermediate hosts, and the presence of larval stages reveals use of a habitat by vertebrate definitive hosts.[21, 22]

Adaptations for Transmission

Parasite Reproduction

Among animals, parental care is one factor that tends to increase the chance of an offspring surviving. Parasites, on the other hand, exhibit little parental care, although **viviparity,** or live birth, such as occurs in some nematodes and

monogeneans, can be considered a more "caring" approach than indiscriminate scattering of eggs. But no amount of parental care can counter the fact that hosts are indeed islands separated by an often extensive abiotic environment. An inverse relationship between numbers of offspring and the probability of individual success is known for a wide variety of both plants and animals. Low individual reproductive success is considered an evolutionary force leading to high reproductive output in parasites. Thus, the high reproductive potential of parasites represents a heavy energy investment to counteract the low probability of individual offspring success.

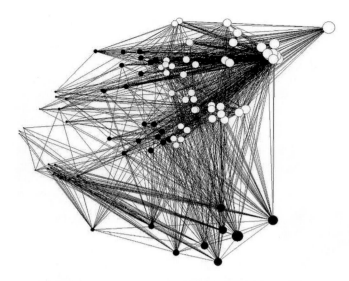

Figure 2.5   Interconnections between participants in a salt marsh food web.
A highly interacting core of host species (open circles) and several peripheral species (closed circles) that have relatively few connections to the core group. Most of the energy in this system flows through the core host species and 80% of the parasites also move only through the hosts in this core group. Core host species include intermediate hosts such as crustaceans and small fish, as well as definitive hosts such as herons, larger fish. The parasites are similar to those depicted in Figure 2.4.

From Sukhdeo, M., "Food webs for parasitologists: A review." In *J. Parasitol.* 96:273–284. © 2010. Used by permission.

Parasites exhibit a variety of mechanisms that function to increase the reproductive potential of those individuals that do succeed at finding a host. These mechanisms often take the form of **asexual reproduction** and **hermaphroditism.** Asexual reproduction often occurs in the larval or sexually immature stages as either **polyembryony** (see Fig.15.22) or **internal budding** (see Fig. 21.20). Hermaphroditism is the occurrence of both male and female sex organs in a single individual. It sometimes eliminates the necessity of finding an individual of the opposite sex for fertilization if gonads of both sexes function simultaneously and selffertilization is mechanically possible. Reproductive encounters result in two fertilized female systems. The specific manifestations of asexual reproduction and hermaphroditism, however, differ depending on the group of parasites.

Schizogony, or **multiple fission,** is asexual reproduction characteristic of some parasitic protozoa (see Fig. 8.6, chapter 9, plates 1 and 3). In schizogony the nucleus divides numerous times before cytokinesis (cytoplasmic division) occurs, resulting in simultaneous production of many daughter cells. A more detailed discussion of the role of schizogony in parasites' life cycles is found in chapter 8. Simple **binary fission** is also asexual reproduction. It is common among familiar free-living protozoa such as *Paramecium* species as well as some amebas, including parasitic ones (chapter 7). As with any process in which numbers double regularly, rapid fission can result easily in millions of offspring after only a few days.

Trematodes and some tapeworms reproduce asexually during immature stages. The juveniles (**metacestodes**) of

several tapeworm species are capable of external or internal budding of more metacestodes. The **cysticercus** juvenile of *Taenia crassiceps,* for instance, can bud off as many as a hundred small bladder worms while in the abdominal cavity of a mouse intermediate host. Each new metacestode develops a scolex and neck, and when the mouse is eaten by a carnivore, each scolex develops into an adult tapeworm. The **hydatid** metacestode of *Echinococcus granulosus* is capable of budding off hundreds of thousands of new scolices within a fluid-filled bladder (see Figs. 21.20 through 21.26). When such a packet of immature worms is eaten by a dog, vast numbers of adult cestodes are produced.

Perhaps the most remarkable asexual reproduction in all zoology is found among trematodes, a large and successful group of parasites commonly called *flukes*. These animals produce a series of embryo generations, each within the body of the prior generation. This is an example of polyembryony, in which many embryos develop from a single zygote. Trematode eggs hatch into **miracidia,** which enter a first intermediate host, always a mollusc, and become saclike **sporocysts.** Sporocysts may give rise to **daughter sporocysts,** which, in turn, may each produce a generation of **rediae.** These then become filled with **daughter rediae,** which finally produce **cercariae** (see chapter 15 for details). Although there are many variations on this general theme (in some, their eggs must be eaten by the first intermediate host before miracidia hatch, and not all species produce redia), by the time all cercariae from a single successful egg are accounted for, the one miracidium has been responsible for an astonishing number of potential adult trematodes. And many flukes give birth to thousands of eggs each day. On the other hand, this staggering reproductive effort also tells us something about the chances of a single trematode egg reaching adulthood.

With hermaphroditism, a parasite evidently solves the problem of finding a mate. Many tapeworms and trematodes can fertilize their own eggs (through selfing, p. 218); this method, although not likely to produce many unusual genetic recombinations, guarantees offspring. Tapeworms also undergo continuous asexual production of segments (strobilization) from an undifferentiated region immediately behind the scolex, or attachment organ. These segments, called *proglottids,* are each the reproductive equivalent of a hermaphroditic worm, at least in the vast majority of tapeworm species, because each contains both male and female reproductive organs. Each fertilized female system in each proglottid eventually becomes filled with eggs containing larvae. The result of this combination of asexual reproduction, hermaphroditism, and self-fertilization is a veritable tapeworm egg factory. Whale tapeworms of the genus *Hexagonoporus,* for example, are 100-foot reproductive monsters consisting of about 45,000 proglottids, each with 5 to 14 sets of male and female systems. There are not many whales and the ocean is truly a vast space, so perhaps this massive investment of energy in reproduction is the minimum necessary to ensure survival of a parasite whose ancestors colonized whales.

Many parasites increase reproductive potential through production of vast numbers of eggs. A common rat tapeworm, *Hymenolepis diminuta,* for example, produces up to 250,000 eggs a day for the life of its host. During a period of slightly over a year, a single tapeworm can thus generate a hundred million eggs. If all these eggs reached maturity in new hosts, they would represent more than 20 tons of tapeworm tissue. Female nematodes are also sometimes prodigious egg layers; a single *Ascaris lumbricoides* can produce more than 200,000 eggs a day for several months, and over the course of their lifetimes, members of the filarial genus *Wuchereria bancrofti* may release several million young into their host's blood. Such high reproductive potential, of course, ensures that such parasites will become medical and veterinary problems when host populations are crowded and transmission conditions are favorable.

Behavioral Adaptations

There are numerous examples of parasite attributes that presumably increase a species' chances of encountering new hosts. These attributes often influence an intermediate host in some way, making it more susceptible to predation by a definitive host. Trematodes of genus *Leucochloridium,* for example, infect land snails as first intermediate hosts and insectivorous birds as definitive hosts. Sporocysts of *Leucochloridium* species are elongated and have pigmented bands (brown or green). These sporocysts move into the snail's tentacles and pulsate, looking for all the world like caterpillars. Although we assume that predatory birds are more likely to select these snails than noninfected ones, field studies necessary to demonstrate this differential predation convincingly are not always conclusive.[37]

The immature stages of some thorny-headed worms (phylum Acanthocephala, chapter 32) infect freshwater crustaceans of order Amphipoda (side-swimmers). Some acanthocephalan juveniles appear as conspicuous white or orange spots in the hemocoel of the translucent amphipods, making infected ones stand out from the uninfected. Within genus *Acanthocephalus,* infection results in loss of body pigment in the isopod intermediate host, while in *Polymorphus paradoxus,* a parasite of ducks, not only is the juvenile orange, but infection alters an amphipod's behavior so that it becomes positively phototactic and swims closer to the water surface than it would otherwise. Ducks prey selectively on these infected, behaviorally altered amphipods.[3]

The most notorious behavior-changing infection involves trematodes of the genus *Dicrocoelium,* which infect large herbivores such as sheep (p. 265). The second intermediate host of *Dicrocoelium dendriticum* is an ant. A metacercaria lodges in the ant's brain, making the insect move to the top of a grass blade, where its likelihood of being accidentally ingested by a definitive host is greatly increased. It has been shown that the "brain worm" is not infective, but related metacercariae in other parts of the ant's body are infective. This difference may well reveal a case of kin selection, such as described in social insects, in which the "brain worm's" kin benefit from the altered ant behavior.[37] Among other arthropods, parasite behavior itself often promotes infection, as in the case of nymphal ticks, which climb up on vegetation, thus increasing chances of encountering a passing host. For students interested in further information on parasite-induced behavioral changes, Moore[37] provides an extensive and detailed analysis of this subject.

Epidemiology and Transmission Ecology

Epidemiology is the study of all ecological aspects of a disease to explain its transmission, distribution, prevalence, and incidence in a population. **Macroepidemiology** concerns large-scale problems of disease distribution, demographic and cultural factors that affect transmission, illness and death rates, and economic impacts. Collection of macroepidemiological data requires substantial funding, institutions such as hospitals or universities, trained personnel, and government policies that allow or even promote such data collection.[33] **Microepidemiology** concerns small-scale problems, for example, the effect of individual host-parasite interactions, parasite strains, host genetic variation, and immunity on disease distribution.[35] A complete understanding of disease transmission, especially when human behavioral factors are involved (as they typically are), requires study at both levels.

Any health-related events that influence the probability an individual will need health care, including, of course, parasitism, can be studied from an epidemiological perspective. In the United States, the Centers for Disease Control and Prevention (CDC) in Atlanta, Georgia, monitors national health statistics, issues a weekly *Morbidity and Mortality Report,* and responds to a variety of situations by seeking to discover the origin and transmission dynamics of infectious diseases. CDC also provides selected health statistics electronically, including some on parasitic infections (www.cdc.gov). The World Health Organization (WHO, www.who.int) provides information about a wide variety of health issuues, including parasitism, on a global scale.

The distribution of parasitism in a population may be influenced by a number of factors, including host age, sex, social and economic status, diet, and ecological conditions that favor completion of parasite life cycles. Pinworms (p. 425) are a good example of parasites whose distribution tends to be influenced by age, at least in developed countries, where children may serve as a source of parasites for the entire family. *Leishmania mexicana* infections often occur in agricultural workers, thus illustrating the influence of occupation on health; the name "chiclero's ulcer" is derived from this distribution (p. 82).

Among the most important epidemiological factors in parasitic infections are **vectors,** which are often snails or blood-sucking arthropods. Vectors are vehicles by which infections are transmitted from one host to another, although the term tends to be used most often to describe vehicles for which the hosts are of economic or personal interest to humans, such as ourselves and our domestic animals. Some of the most medically important vectors are anopheline mosquitoes, which transmit malarial parasites (chapter 9), and snails of certain genera, which carry infective larval blood flukes, or schistosomes (chapter 16). Malaria and schistosomiasis are still among the most serious human diseases, infecting nearly 700 million people, mostly in developing countries (p. 5).

Vector biology usually must be well understood before disease control measures can become effective. It is standard practice in malaria control efforts, for example, to eliminate mosquito breeding habitat—namely, standing water. Unfortunately, in some areas of the world, the only drinking and bathing water available is also a breeding ground of mosquitoes. Agricultural practices have sometimes added to disease-control problems, as in Egypt, where irrigation ditches are ideal environments for snails that serve as intermediate hosts for schistosomes. Vectors are actually hosts required for completion of parasites' life cycles. Thus epidemiology also involves the study of parasite life cycles, especially mechanisms by which parasites move from one host to the next.

Between World Wars I and II it was noted by the Russian school of Pavlovsky that certain parasitic infections occur in some ecosystems but not in others.[40] Components of these ecosystems can be categorized so that they can be recognized wherever they occur. Thus, each disease has a natural focus, or **nidus,** which is the set of ecological conditions under which it can be predicted to occur. Discovery of this natural nidality of infection was a landmark in the history of parasitology because it enabled epidemiologists to recognize "landscapes" where certain diseases could be expected to exist or, equally, where they might be effectively controlled.

Such **landscape epidemiology** requires thorough knowledge of all factors that influence transmission, such as climate, plant and animal population densities, geological conditions, and human activities within the nidus. This holistic approach is best applied to **zoonoses,** which are diseases of animals that are also transmissible to humans. Zoonoses can become of particular importance in areas experiencing environmental disturbance. However, the principles of landscape epidemiology can be applied equally well to pinworm or head louse transmission in day-care centers, whipworm infections in mental institutions, and *Giardia duodenalis* outbreaks at posh resorts.

Landscape epidemiology is now done with satellite-supported Geographic Information Systems (GIS) that can reveal vegetation and land use patterns. One study in Ethiopia, for example, showed that various commonly used analyses of vegetation cover, crop production, and a climate-based forecast that used growing degree days and water budgets could predict the occurrence of onchocerciasis (river blindness, chapter 29).[18] Crop production data most clearly revealed endemic zones, and climate-based forecast results were most closely matched to zones of high disease risk. But all the GIS analyses predicted suitable transmission conditions outside areas where onchocerciasis was known to occur. The authors thus recommended "ground-based validation" of the predictions, with possible community treatment programs.[18]

Molecular techniques and innovative diagnostic tools tools are now often used to address epidemiological problems. For example, the same finger-prick samples can be used not only for blood smears to diagnose malarial infections, but also for DNA analysis to reveal the species of *Plasmodium* involved and the presence of mixed infections.[52] Diagnostic aids such as IsoCode STIX® allow such samples to be collected in the field and transported to urban facilities, sometimes internationally, for processing. And the human genome project, as you might suspect, has opened up many new opportunites for epidemiologists to address public health problems.[27]

Theoretical Parasitology

Theoretical studies of parasitism include mathematical models of transmission, attempts to develop general principles from large data sets, efforts to determine the relative contributions of ecology versus phylogeny to parasite host specificity, and endeavors to explain the origin of complex life cycles.[16, 28, 29, 44] Models can generate predictions that in turn stimulate further research, a good example being the seminal paper by Crofton, mentioned above, that has kept at least a generation of parasitologists busy trying to explain aggregated distributions of parasites among host populations.[13]

Modern technology has helped, if not actually enabled, much theoretical work by making vast amounts of information accessible, as illustrated by Poulin and Leung's analysis of 419 published data sets in their study of relationship between fish body size and role in food webs,[43] and Poulin et al.'s efforts to find "hotspots of parasite diversity" through use of published surveys.[44] Such analyses serve as guides to the design of future studies because they reveal unanswered questions and suggest principles that may not have been recognized.

Theoretical work can also be of great practical value, especially when predictions are counterintuitive. For example, in the Philippines, *Schistosoma japonicum* parasitizes not only humans, but also dogs, pigs, and field rats as definitive hosts (see chapter 16). In a particularly illustrative study, Hairston used quantitative models to suggest that rats alone could support the suprapopulation of *S. japonicum* because of the high rate of contact between rats and snail intermediate hosts.[20] Thus, even if all humans were cured at once, only a few years would be required for human infections to reach their previous level. Hairston's prediction illustrates beautifully the conflict that can occur when science runs into deeply entrenched feelings about our fellow humans. If you are a physician in an endemic area of the Philippines and you are trying to relieve the human population of a parasite burden but are in possession of only limited resources, do you spend your time treating humans or killing rats?

PARASITE EVOLUTION

Evolutionary Associations Between Parasites and Hosts

An overriding concern among parasitologists studying evolution is the pattern of association among parasites, hosts, and the ecological and geographical distributions of each.[5] In general there are two factors that influence these patterns: descent and colonization. That is, a parasite may be associated with a host because the two share a long evolutionary history (descent), having undergone evolutionary change together, or host and parasite may be associated because the parasite has colonized the host in a manner analogous to colonization of an island. This type of colonization also is called **host switching** or **host capture.** To explain patterns of host/parasite association one therefore must discover whether these patterns are a product of descent, colonization, physical separation of populations, extinction, or some combination of the four.

Processes such as continental drift, orogeny (mountain building), and island formation have also influenced geographical distributions of both hosts and parasites. The interplay between evolution and long-term geological changes is termed **phylogeography.** A good illustration of this interaction can be found in Perkins's study of lizard malaria on Caribbean islands.[42] Using a combination of molecular techniques, life cycle characteristics, and morphology, she showed that two strains of *Plasmodium azurophilum* in *Anolis* lizards had quite different patterns of dispersal among the Lesser Antilles. The fact that this host-parasite system occurred on an archipelago explained much of the pattern of colonization by the lizards, parasites, and mosquito vectors involved.[42]

Parasitologists use cladistic methods, also called **phylogenetic systematics** or **phyletics**, to infer evolutionary histories of hosts and parasites. **Phylogenies** are actually evolutionary hypotheses, typically presented as treelike diagrams, with relationships between taxa shown in the branching patterns. Characters used to produce these phylogenies may be molecular, structural, ecological, or geographical. These characters are determined to be either **plesiomorphic** (primitive, shared among both ingroup and outgroup members), or **apomorphic** (derived, evolutionary novelties, present only in the ingroup) by comparison between an **ingroup** (a taxon of interest) and an **outgroup** (a related taxon chosen for the express purpose of comparison). Apomorphic characters shared by ingroup members are called **synapomorphies.** Characters are then analyzed by computer programs that generate phylogenies, typically doing so on the basis of synapomorphies.

A group defined by synapomorphies and containing a hypothetical ancestor and all descendants of that ancestor is termed **monophyletic.** A group that contains a hypothetical ancestor but only some of its descendants, however, is termed **paraphyletic,** and a group made up of taxa that do not share a closest common ancestor is called **polyphyletic.** Given that taxa ("groups") may well have been established on dubious criteria in the past, evolutionary biologists seek to discover monophyletic groups and resolve problems with the others.

The software used in these studies generates branching diagrams that only superficially resemble evolutionary trees often seen in older texts. A **cladogram** such as shown in Figure 2.6, for example, indicates only closest relatives based on numbers of shared derived traits, the latter taken as evidence of common ancestry. Phylogenetic hypotheses may be falsified by additional research.

The work of Janine Caira and her colleagues on tapeworms of elasmobranchs provides an excellent illustration of both the excitement and challenges of studying parasite evolution.[8] These parasitologists focused on a single tapeworm family, the Onchobothriidae. First, they established a set of criteria by which such studies should be judged: Both hosts and parasites must be in monophyletic groups, all taxa should be correctly identified to species, the host taxa must have been adequately sampled, phylogenies for both hosts and parasites should be available, and the parasites should be specific to their hosts. Their intent was to avoid paraphyletic groups or polyphyletic groups. Next, they assembled a data matrix on the tapeworms, using a large number of structural features, including those revealed by electron microscopy, and performed the cladistic analysis. Finally, they resolved identification issues of both tapeworms and sharks as best they could; the shark phylogeny was taken from the elasmobranch literature.

The tapeworm family Onchobothriidae contains about 200 species that are highly host specific, each occurring in only a single species of shark. Caira and her coworkers wanted to determine whether the distribution of parasite species among related hosts was due to common descent and speciation or to colonization of hosts by parasites without regard to host ancestry. The answer is seen in Figure 2.6. Overall, there was very little congruence between the parasites' evolutionary history and that of their hosts. The parasites' high degree of host specificity suggests association by descent, but lack of congruence between the phylogenies suggests extensive colonization. In addition, species assigned to a single genus did not necessarily turn out to be their own closest relatives, as in the case of the two *Acanthobothrium* species in Figure 2.6. Finally, the researchers determined that a single elasmobranch species can play host to worms of several genera as well as several species in the same genus.[8] Efforts to explain the origin of this wonderful menagerie could well occupy Janine Caira and her students for the remainders of their careers!

Molecular techniques can help resolve some of the questions raised by studies using only morphology, but molecular phylogenetic research often uncovers as many puzzles as it solves. A good illustration of this can be found in the literature on a common intestinal parasite, *Giardia duodenalis* (= *G. intestinalis* = *G. lamblia;* chapter 6). Several researchers have tried to use allozymes and nucleotide sequences to determine evolutionary relationships of *G. duodenalis* isolates from humans as well as dogs, cats, livestock, mice, and birds.[36] Parasites isolated from these various sources cannot be distinguished by morphology. The molecular work, however, revealed two main "assemblages" of flagellates, with several "groups" from humans. In some cases (groups I and II) the human parasites were most closely related to those of livestock; in other cases (group IV) the flagellates seemed most similar to those from dogs. The parasite we know as *G. duodenalis* may in fact be a number of cryptic species, differing not only in their molecular makeup, but also in their virulence and growth requirements.

These two papers illustrate the general techniques used, the types of questions that parasitologists ask, and some problems that arise in the study of parasite evolution. A rich literature has developed involving many groups of hosts and parasites, and Brooks and McLennan's book *Parascript*[5] provides a summary of the history, major questions, a bibliography, and the existing database on parasite evolution. The authors analyze case studies involving diverse groups of hosts and parasites in detail, and the results reveal some fascinating and ancient relationships among parasites, hosts, and global geological events. For example, some turtle blood flukes apparently enjoyed a long coevolutionary relationship with their hosts (since the Mesozoic), while others seem to have diversified following the breakup of Pangaea. The literature of parasite evolutionary biology suggests that many young scientists struggling with large problems and massive data sets eventually come to see the interactions among hosts, parasites, and global changes in climate and geography over geological time scales as part of a single grand picture of life on earth.

Parasitism and Sexual Selection

Some biologists believe that parasitism is a factor contributing to evolution of host reproductive biology. Indeed, sex itself has been explained as a mechanism for reducing the evolutionary impact of parasitism.[23] Negative effects of parasitism on reproductive behavior and success have been observed in a wide range of animals, from insects to mammals. The impact can be on both males and females, affecting both egg production and mate choice.[39] It also has been postulated that females select males according to immunocompetence of the males (see chapter 3). Mathematical models, however, show that when pathogen prevalence, or the kinds of pathogens present, fluctuate, then females no longer choose males based on disease resistance.[1] Some female birds, such as

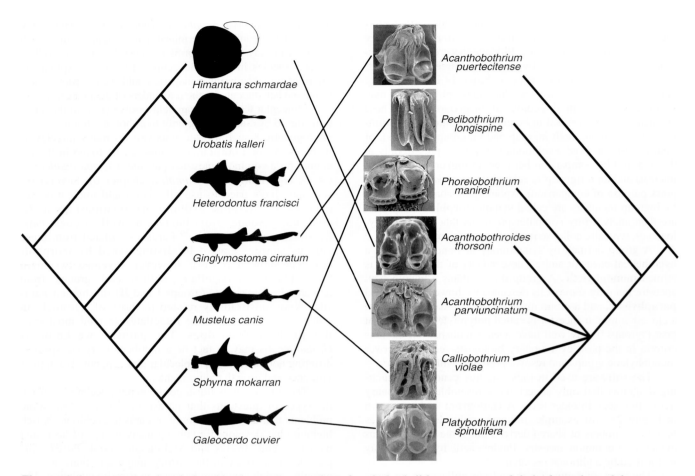

Figure 2.6   Phylogenetic relationships between a number of onchobothriid tapeworms and their elasmobranch hosts.
The cestodes are represented by their holdfast organs (scolices), which differ in structure. Analysis suggests that members of a cestode genus as currently defined are not necessarily their own closest relatives (cf. *Acanthobothrium puertecitense* and *A. parviuncinatum*) and that sister taxa among sharks (e.g., *Sphyrna mokarran* and *Galeocerdo cuveri*) do not necessarily have the most closely related worms. This figure illustrates some of the kinds of interesting problems encountered by those who study parasite evolution.

Graphic design by Janine Caira, Kirsten Jensen, and Claire Healy. *Phoreiobothrium manirei* from J. N. Caira, C. J. Healy, and J. Swanson, "A new species of *Phoreiobothrium* (Cestoidea: Tetraphyllidea) from the great hammerhead shark *Sphyrna mokarran* and its implications for the evolution of the onchbothriid scolex." in *J. Parasitol.* 82:458–462. Copyright © 1996. *Acanthobothrium puertecitense* from J. N. Caira and S. D. Zahner, "Two new species of *Acanthobothrium* Beneden, 1849 (Tetraphyllidea: Onchobothriidae) from horn sharks in the Gulf of California, Mexico," in *Systematic Parasitol.* 50:219–229. Copyright © 2001. *Acanthobothroides thorsoni, Acanthobothrium parviuncinatum, Platybothrium (=Dicranobothrium spinulifera,* and *Pedibothrium longispine* from J. N. Caira, K. Jensen, and C. J. Healy. "Interrelationships among tetraphyllidean and lecanicephalidean cestodes" in D. T. J. Littlewood and R. A. Bray (Eds.), *Interrelationships of the Platyhelminthes.* London, Taylor, and Francis. Copyright © 2001. *Calliobothrium violae* and *Phoreiobothrium manirei* photographs courtesy of Janine Caira. All figures reprinted by permission.

swallows, are evidently able to distinguish parasitized from nonparasitized ("nonhealthy" vs. "healthy"?) males and select mates accordingly.[10] However, the parasites in these cases are often those having a direct effect on plumage quality—namely, lice and acarines (chapters 36, 41).

Infection with haematozoans (chapter 9), which may have an indirect effect on plumage, is not so strongly associated with either vector attraction or plumage quality.[54] In one study using extensive published data sets, Yezerinac and Weatherhead[54] concluded that variations in local ecological conditions that affected parasite transmission could easily override any relationship between parasitism and mate selection. And in a study focused on a single species, the yellowhammer *(Emberiza citrinella),* Sundberg found no relationship between male color and number of fledglings produced by a pair of birds, and pairing itself was not related to haematozoan infections.[51]

It has been shown, however, that haematozoan infections acquired early in a bird's life can have a negative effect on its ability to later learn the songs so critical to mating success.[48] This situation is somewhat analogous to that demonstrated by Guerrant and his coworkers on human development, in which it was shown that frequent bouts of childhood diarrhea were correlated, years later, with lower scores on cognitive development tests.[19]

In contrast to birds, however, mammals depend strongly on odors, and some studies have shown that female mice avoid males infected with both coccidians and nematodes.[25] Fecal avoidance in large herbivores such as reindeer has also been postulated to be a behavior that reduces risks of nematode parasitism, but one study showed that soil moisture had more of an effect on parasite transmission than did fecal concentration.[53] In general, behavioral responses to parasitism are intriguing observations, but their evolutionary significance has yet to be established in a large number of cases.

Parasitism is also considered a factor that can maintain genetic diversity in host populations. In one study of *Capillaria hepatica* (Nematoda, p. 380) in deer mice, for example, parasite prevalence was lowest in host populations exhibiting the most heterozygosity, a result consistent with the prediction that inbred host populations would be most susceptible to infection.[35] Similar results were found in systems as disparate as Hirta Island (Scotland) sheep and New Zealand snails.[12, 32]

Evolution of Virulence

Life history traits (such as fecundity), life cycles themselves (loss or addition of stages), and virulence are all subject to evolutionary change. The question of why some parasites seem to be especially virulent while others are relatively benign has captured the attention of numerous investigators, although much of their research remains theoretical.[17, 31]

A long-established paradigm states that parasites should evolve into less virulent forms, mainly because death of a host should have a negative effect on parasite survival. However, according to some theories, parasites should evolve an optimal virulence that maximizes parasite numbers, with "optimal" depending on numerous factors such as pathogenicity and transmission dynamics.[31] Most, if not all, parasites are transmitted both vertically (between generations) and horizontally (among members of the same generation). Theoretical work suggests that vertical transmission tends to select for less virulent parasite strains, whereas horizontal transmission, especially when coupled with high transmission rates, selects for more virulent strains.[15] Not all studies support this idea, because factors such as genetic diversity of host and parasite, individual host-parasite interactions, and different time scales for transmission can also affect the evolution of virulence.[15]

Learning Outcomes

By the time a student has finished studying this chapter, he or she should be able to:

1. Define *incidence, prevalence,* and *intensity.* Explain how the three concepts are related and tell why each concept is important, but in a different way, for an understanding of parasite population dynamics.

2. Draw and label a graph that demonstrates an aggregated distribution of macroparasites.

3. Draw a simplified food web that includes several different kinds of parasitic relationships.

4. Draw hypothetical host and parasite phylogenies that demonstrate co-speciation and host switching.

5. Write a short paragraph that describes, in general, a complex parasite life cycle.

6. Tell the difference between macroparasites and microparasites.

7. Explain the role that vectors and intermediate hosts play in the maintenance of some representative life cycles.

8. Describe two adaptations for transmission.

9. Explain what is meant by behavioral adaptation (for transmission) and give at least one example of such an adaptation (or presumed adaptation).

10. Define landscape epidemiology and describe how molecular techniques can be used in epidemiological studies.

References

References for superscripts in the text can be found at the following Internet site: www.mhhe.com/robertsjanovynadler9e

Additional Readings

Clayton, D. H., and J. Moore. 1997. *Host-parasite evolution: General principles and avian models.* Oxford: Oxford University Press.

Croll, N. A., and E. Chadirian. 1981. Wormy persons: Contributions to the nature and patterns of overdispersion with *Ascaris lumbricoides, Ancylostoma duodenale, Necator americanus,* and *Trichiuris trichiura. Trop. Geogr. Med.* 33:241–248.

Esch, G. W., A. O. Bush, and J. M. Aho (Eds.). 1990. *Parasite communities: Patterns and process.* New York: Chapman and Hall.

Esch, G. W., and J. C. Fernandez. 1993. *A functional biology of parasitism.* New York: Chapman and Hall.

Gillett, J. D. 1985. The behavior of *Homo sapiens*, the forgotten factor in the transmission of tropical disease. *Trans. Roy. Soc. Trop. Med. Hyg.* 79:12–20.

Price, P. W. 1980. Evolutionary biology of parasites. *Monographs in population biology* 15. Princeton, NJ: Princeton University Press.

Schmidt, G. D. (Ed.). 1969. *Problems in systematics of parasites.* Baltimore, MD: University Park Press.

Smith, T. 1963. *Parasitism and disease.* New York: Hafner Publishing Co. A classic, originally published in 1934.

Chapter 3

Basic Principles and Concepts II: Immunology and Pathology

. . . in parasitic conditions, there often is limited pathology directly attributable to the organism. Most morbidity is related to the immunoinflammatory response of the host to the parasite.

—S. Michael Phillips[45]

A traditional view of host-parasite interaction asserts that as a symbiont becomes progressively more specialized through evolution, it increasingly limits the potential number of host species it can infect; that is, its host specificity increases. A vital component in this process is the habitat (host), which is a dynamic, living, and evolving partner in the relationship. The host reacts to the presence of a symbiont, mounting a defense against the foreign invader, and a successful symbiont must evolve strategies to evade host defenses. Parasitologists have come to recognize not only that host specificity is determined in great degree by which host individuals can mount an effective defense and which parasites can evade that defense, but also that much of the disease caused by parasites is directly related to host defense mechanisms. This chapter will explore host-parasite relationships by introducing concepts related to host defenses, the evasion of host defenses by parasites, and how parasites cause disease in their hosts.

Immunologists commonly utilize acronyms for many molecules of interest. For convenience of students, we are gathering those in this chapter in Table 3.1.

Table 3.1	Some Immunological Abbreviations and Acronyms Used in Chapter 3 and Other Places Throughout This Text			
ADCC	Antibody-dependent, cell-mediated cytotoxicity		AIDS	Acquired immune deficiency syndrome
APC	Antigen presenting cell		B cell	Bone marrow-derived lymphocyte
CD	Cluster of differentiation		CF	Complement fixation test
CTL	Cytotoxic T lymphocyte		DC	Dendritic cell
ELISA	Enzyme-linked immunosorbant assay		DTH	Delayed type hypersensitivity
Fc	Crystallizable fragment of antibody		Fab	Antigen-binding fragment of antibody
GPIs	glycophosphatidylinositols		GAS	Gamma activated sequences
IFA	Indirect fluorescent antibody test		HIV	Human immunodeficiency virus
Ig	Immunoglobulin		IFN	Interferon
IHA	Indirect hemagglutination test		IH	Immediate hypersensitivity
JAK	Janus kinase family of tyrosine kinases		IL	Interleukin
MAPK	Mitogen activated protein kinase		LAK	Lymphokine-activated killer cell
MyD88	Myeloid differentiation protein 88		MHC	Major histocompatibility complex
NFAT	Nuclear factor of activated T-cells		NF-κB	Nuclear factor kappa B
PAMP	Pathogen-associated molecular pattern		NK	Natural killer cell
PRR	Pattern recognition receptor		PMN	Polymorphonuclear leukocyte
RE	Reticuloendothelial		RAG	Recombination-activating gene
ROI	Reactive oxygen intermediate		RNI	Reactive nitrogen intermediate
T cell	Thymus-derived lymphocyte		STAT	Signal transducers and activators of transcription
T_H1	Cellular immune response, on cell surfaces only		TGF	Transforming growth factor
TLR	Toll-like receptor		T_H2	Humoral immune response, on cells and dissolved
T_{reg}	Regulatory T cells		TNF	Tumor necrosis factor

SUSCEPTIBILITY AND RESISTANCE

A host is **susceptible** to a parasite if the host cannot eliminate the parasite before the parasite can become established. The host is **resistant** if its physiological status prevents the establishment and survival of the parasite. Corresponding terms from the viewpoint of the parasite would be **infective** and **noninfective.**

These terms deal only with the success or failure of infection, not with the mechanisms producing the result. Mechanisms that increase resistance (and correspondingly reduce susceptibility and infectivity) may involve either attributes of the host not related to active defense mechanisms or specific defense mechanisms mounted by the host in response to a foreign invader. Furthermore, the terms are relative, not absolute; for example, one individual organism may be more or less resistant than another.

The term **immunity** has been, on the one hand, often used as synonymous with *resistance* and, on the other hand, associated with the sensitive and specific immune response exhibited by vertebrates. However, because invertebrates can be immune to infection with various agents, a more general yet concise statement is that *an animal demonstrates immunity if it possesses cells or tissues capable of recognizing and protecting the animal against nonself invaders.*[26]

All animals show some degree of **innate** immunity; that is, a mechanism of defense that does not depend on prior exposure to the invader (Fig. 3.1).[4] In addition to having innate immunity, jawed vertebrates (gnathostomes) develop **adaptive (acquired) immunity,** which is specific to the particular nonself material, requires time for development, and occurs more quickly and vigorously on secondary exposure. Many of the innate mechanisms discussed in the next section *are dramatically influenced and strengthened in vertebrates as a consequence of adaptive immune responses.*

Resistance provided by immune mechanisms may not be complete. In some instances a host may recover clinically and be resistant to specific challenge, but some parasites may remain and reproduce slowly, as in toxoplasmosis (p. 133), Chagas' disease (p. 71), and malaria (p. 155). The parasites are held in check by the host's immune system, and the host is asymptomatic. This condition is called **premunition.** In some infections a parasite may elicit a protection against reinfection, but the parasite itself may remain in the host, unaffected by the immune response **(concomitant immunity),** as in schistosomiasis (p. 246).[55] In this case the host may suffer significant morbidity (illness).

INNATE DEFENSE MECHANISMS

The unbroken surface of most animals provides a barrier to invading organisms. This surface may be tough and cornified, as in many terrestrial vertebrates, or sclerotized, as in arthropods. Soft outer surfaces are usually protected by a layer of mucus, which lubricates the surface and helps dislodge particles from it. Mucin in gastrointestinal tracts provides attachment sites for normal gut flora preventing potential pathogens in the gut lumen from attaching.[35] To reach the gut mucosa, pathogens must be able to breach the mucin lining.

Useful functioning of any system of defense requires distinction between cells in an animal's own body (self) and those of another individual or invader (nonself). A principal test of the ability of invertebrate tissues to recognize nonself is grafting of a piece of tissue from another individual of the same species **(allograft)** or a different species **(xenograft)** onto a host. If a graft grows in place with no host response, the host tissue is treating it as self, but if cell response and rejection of the graft occur, the host exhibits immune recognition. Most invertebrates tested in this way reject xenografts; and almost all can reject allografts to some degree.[21, 26]

Cell Signaling

In both innate and adaptive responses, cells detect many molecules in their environment that bind to receptors on their surface, resulting in initiation of intracellular signal cascades (series of linked events within the cell).[17] Depending on the receptor and the binding molecule **(ligand),** such cascades may trigger activation of transcription factors or other proteins that control gene induction, phagocytosis, apoptosis, or secretion. Ligands may be located on the surface of neighboring cells, dissolved in the blood **(cytokines)** or on the surface of or secreted by pathogens.

Cytokines and Cytokine Receptors
Cytokines (Table 3.2) are protein hormones that play important roles in both the innate and adaptive immune systems and are a major means by which immune cells communicate.

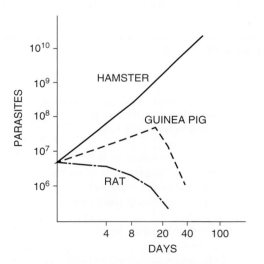

Figure 3.1   **Innate immunity to *Leishmania donovani* exhibited by three different rodent species.**
"Parasites" are total numbers of amastigotes (see p. 78) in the liver as determined by organ weight and amastigotes/host cell nucleus counts in tissue smears. "Days" are days post-infection with amastigotes by intracardial injection. Hamsters show no innate immunity (resistance) and parasite numbers increase until the host dies; guinea pigs and rats eventually control or eliminate the parasite.

Adapted from Host resistance to the Khartoum strain of *Leishmania donovani*. by L. A. Stauber, *Rice Institut. Pamph.* 45:80–96, 1958.

Table 3.2 Some Important Cytokines		
Cytokine	**Principal Source**	**Functions**
Type I interferon (IFN)	Activated macrophages, fibroblasts	Antiviral; antiproliferative; increases MHC I expression; activates NK cells
Interferon-γ (IFN-γ)	Some CD4$^+$ and almost all CD8$^+$ cells	Strong macrophage-activating factor; causes a variety of cells to express class II MHC molecules; promotes T and B cell differentiation; activates neutrophils and NK cells; activates endothelial cells to allow lymphocytes to pass through walls of vessels
Tumor necrosis factor (TNF)	Activated macrophages	Major mediator of inflammation; low concentrations activate endothelial cells, activate PMNs, stimulate macrophages and cytokine production (including IL-1, IL-6, and TNF itself); higher concentrations cause increased synthesis of prostaglandins, resulting in fever
Interleukin-1 (IL-1)	Activated macrophages	Mediates inflammation; activates T and B cells
Interleukin-2 (IL-2)	CD4$^+$ cells, some from CD8$^+$ cells	Major growth factor for T and B cells; enhances cytolytic activity of natural killer cells, causing them to become lymphokine-activated (LAK) cells
Interleukin-3 (IL-3)	CD4$^+$ cells	Multilineage colony-stimulating factor; promotes growth and differentiation of all cell types in bone marrow
Interleukin-4 (IL-4)	Mostly by T$_H$2 CD4$^+$ cells	Growth factor for B cells, some CD4$^+$ T cells, and mast cells; suppresses T$_H$1 differentiation
Interleukin-5 (IL-5)	Certain CD4$^+$ cells	Activates eosinophils; acts with IL-2 and IL-4 to stimulate growth and differentiation of B cells
Interleukin-6 (IL-6)	Macrophages, endothelial cells, fibroblasts, and T$_H$2 cells	Important growth factor for B cells late in their differentiation
Interleukin-8 (IL-8)	Antigen-activated T cells, macrophages, endothelial cells, fibroblasts, and platelets	Activating and chemotactic factor for neutrophils and, to a lesser extent, other PMNs
Interleukin-10 (IL-10)	T$_H$2 CD4$^+$ cells	Inhibits T$_H$1, CD8$^+$, NK, and macrophage cytokine synthesis
Interleukin-12 (IL-12)	Monocytes, macrophages, neutrophils, dendritic cells, B cells	Activates NK cells and T cells; potently induces production of IFN-γ; shifts immune response to T$_H$1
Interleukin-17 (IL-17)	T cells, mast cells, granulocytes, NKT cells, epithelial cells	Triggers inflammatory responses, including autoimmune ones.
Chemokines	Macrophages, endothelial cells, fibroblasts, T cells, platelets	Leucocyte activation and chemotaxis
Transforming growth factor-β (TGF-β)	Macrophages, lymphocytes, and other cells	Inhibits lymphocyte proliferation, CTL and LAK cell generation, and macrophage cytokine production
Migration inhibition factor	T cells	Converts macrophages from motile to immotile state

Modified from Abbas, A. K., A. H. Lichtman, and J. S. Pober. 1994. *Cellular and molecular immunology.* Philadelphia: W. B. Saunders Company.[1]

Cytokines can produce their effects on the same cells that produce them, on cells nearby, or on cells distant in the body from those that produced the cytokines. They exert their effects on target cells by ligation with a specific receptor, one end of which protrudes from the cell surface where it binds with the ligand. After ligation, the cytosolic end of the receptor attracts molecules that trigger an intracellular cascade (pathway) of activations. An essential result of many signaling cascades is activation of transcription factors (molecules that promote expression of particular genes), such as NF-kB (**n**uclear **f**actor kappa B) and NFAT (**n**uclear **f**actor of **a**ctivated **T**-cells).

One of the JAK-STAT pathways will serve for illustration of the cascade proccess (Fig. 3.2). Ligation of the cytokine to its receptor causes attraction of tyrosine kinases of the **Ja**nus **k**inase (JAK) family to the intracellular portions of the receptor. Then JAKs recruit other proteins, called STATs (for **s**ignal

transducers and **a**ctivators of **t**ranscription), and activate them by phosphorylation of their tyrosine residues. Activated STATs translocate into the nucleus, where they become transcription factors when they associate with GAS (for **g**amma **a**ctivated **s**equences) elements in inducible genes for interferon-γ (IFN-γ).[17] Differing JAKs and STATs result in other activations, for example, control of differentiation of T helper cells toward T$_H$1 or T$_H$2 arms of the adaptive immune response (p. 28).

Antimicrobial Molecules and Pattern Recognition Receptors (PRRs)

In the 1980s it was discovered that inoculation of moth larvae with bacteria caused release of a barrage of antimicrobial agents that killed the bacteria, even without prior exposure to the invader. Since that time, hundreds of antimicrobial peptides have been described from a broad spectrum

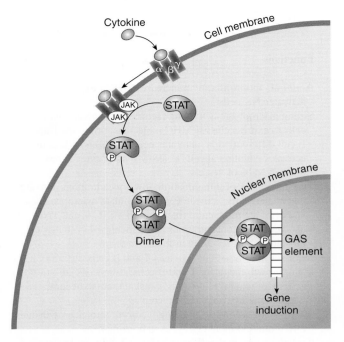

Figure 3.2 **Example of cell signaling by a JAK-STAT pathway.**

Following ligation of a cytokine to a cytokine receptor, JAKs are recruited and phosphorylate tyrosine residues in the β and γ chains of the receptor, which enables STAT recruitment and phosphorylation. Phosphorylated STATs form dimers, which translocate to the nucleus, where they become transcription factors by binding to GAS elements.

Redrawn by William Ober and Claire Garrison from H. S. Goodridge and M. M. Harnett, "Introduction to immune cell signalling," in *Parasitology* 130: S3-S9, 2005.

of invertebrates, vertebrates, and even plants.[15] They are especially important at surfaces where an organism meets the environment, such as skin or mucous membranes. They do not have such high specificity as does the adaptive immune response of vertebrates, but rather each peptide is effective against a certain category of microbe, for example Gram-positive bacteria (bacteria that stain with "Gram stain"), Gram-negative bacteria, and fungi.

Release of the peptides is immediate in the presence of a foreign organism and is not subject to prior immunizing experience with the microbe. **P**attern **r**ecognition **r**eceptors (**PRR**s) on host cells are stimulated by various **p**athogen-**a**ssociated **m**olecular **p**atterns (**PAMP**s) and mediate secretion of the various peptides. Some pattern recognition peptides are secreted, where they can bind to molecules on the surface of bacteria, fungi, protozoan parasites, or helminths. Such ligation may facilitate phagocytosis (see below) or activate **complement** by the **alternative pathway** or in some instances, by the **classical pathway.**[34]

Complement is an important innate defense against invasion by bacteria, some fungi, and some helminths. Complement is a series of enzymes that are activated in sequence. Activation by the so-called classical pathway usually depends on fixed antibody (antibody bound to antigen; see below) and so is an effector mechanism in the adaptive immune response. The classical and alternative pathways share some but not all

components. In the alternative pathway, the first component is activated spontaneously in the blood and binds to cell surfaces. This event initiates a cascade of activations, ultimately resulting in cell lysis. The host's own cells are not lysed because regulatory proteins rapidly inactivate the first component when it binds to host cells.

Release of peptides begins when PRRs on a cell's surface recognize a microbial molecule. Examples of PRRs are **scavenger receptors, complement receptors,** and **Toll-like receptors (TLRs).** Complement receptors recognize fragments of complement components released during the cascade of complement activation. They are found on surface membranes of a variety of cells and mediate various defense functions, including phagocytosis (in both innate and adaptive immunity).[35] Scavenger receptors bind many ligands, including lipoproteins and lipopolysaccharides from bacterial cells.

TLRs are an evolutionarily conserved family of receptors found in animals and plants.[32] Interaction between a TLR and a microbial component activates innate immunity, as well as initiation of adaptive immunity.[2] TLRs are vital for recognition of carbohydrates, polynucleotides, and proteins derived from viruses, bacteria, fungi, protozoa, and helminth parasites. At least ten TLRs (TLR1 through TLR10) have been described in humans,[63] each of which recognizes a specific pattern of molecules from a class of microbes.

Ligation of a particular TLR often requires an adaptor protein, such as MyD88 (**m**yeloid **d**ifferentiation factor **88**), which then initiates a cascade that leads to activation of one or more transcription factors, such as NF-kB or one of the MAPK families.[17] Activation of TLRs induces expression of a variety of antimicrobial peptides.[59] Activation of TLR4 by lipopolysaccharide from Gram-negative bacteria, for example, induces genes for several inflammatory cytokines and costimulatory molecules.

GPIs

GPIs (glycophosphatidylinositols) are glycolipids that are a ubiquitous feature of eukaryote cell membranes. Their principal function is to serve as anchors for proteins on membranes, although many are present that are not conjugated with a protein. Mammalian cells typically have about 10^5 copies of GPI anchors per cell, but parasitic protozoans usually display many more, up to 10^7 or more copies in kinetoplastids, for example.[37] GPI-anchored proteins are associated with a variety of pathogenic effects and interaction with host immune systems (p. 74; p. 152).

Other Chemical Defenses

Many vertebrates have a low pH in the stomach and vagina and hydrolytic enzymes in secretions of the alimentary tract that are antimicrobial in action. Mucus is produced by mucous membranes lining the digestive and respiratory tract of vertebrates and contains parasiticidal substances such as **IgA** (immunoglobulin A) and **lysozyme.** We now know that IgA is a class of antibody (p. 28) and so is actually part of the adaptive immune response. IgA can cross cellular barriers easily and is an important protective agent in mucus of the intestinal epithelium. It is present also in saliva and sweat and is also found in granules of polymorphonuclear leucocytes (see below). Lysozyme is an enzyme that attacks the cell wall of many bacteria.

Various cells, including those involved in the adaptive immune response, liberate protective compounds.

A family of low-molecular-weight glycoproteins, called **interferons** (see Table 3.2), are cytokines released by a variety of eukaryotic cells in response to invasion by intracellular parasites (including viruses) and other stimuli. Another cytokine, **tumor necrosis factor (TNF),** is produced mainly by macrophages but also by activated T cells and natural killer cells. It is a major mediator of inflammation (p. 32), and in sufficient concentration it causes **fever.** Fever in mammals is one of the most common symptoms of infection and is a fundamental defense mechanism. High body temperature may destabilize certain viruses and bacteria, and in vitro results indicate that it may have a beneficial effect in malaria.[30]

The normal intestine of vertebrates harbors a population of bacteria that do not seem to be harmed by the body's defenses, nor do they elicit any protective immune response. In fact, the normal intestinal microflora tends to inhibit establishment of pathogenic microbes.

Substances in normal human milk can kill intestinal protozoa such as *Giardia lamblia* (chapter 6) and *Entamoeba histolytica* (chapter 7), and these substances may be important in the protection of infants against these and other infections.[16] Antimicrobial elements in human breast milk include lysozyme, IgA, interferons, and leukocytes.

Cellular Defenses: Phagocytosis

Phagocytosis occurs in all metazoa and is a feeding mechanism in many single-celled organisms. A cell that has this ability is a **phagocyte.** Phagocytosis is a process of engulfment of an invading particle within an invagination of the phagocyte's cell membrane. The invagination becomes pinched off, and the particle becomes enclosed within an intracellular vacuole. **Lysosomes** pour digestive enzymes into the vacuole to destroy the particle. Lysosomes of many phagocytes also contain enzymes that catalyze production of cytotoxic **reactive oxygen intermediates (ROIs)** and **reactive nitrogen intermediates (RNIs).** Examples of ROIs are

superoxide radical (O_2^-), hydrogen peroxide (H_2O_2), singlet oxygen ($^1O_2$), and hydroxyl radical (OH•). RNIs include nitric oxide (NO) and its oxidized forms, nitrite (NO_2^-) and nitrate (NO_3^-). All such intermediates are potentially toxic to invasive microorganisms or parasites.

Phagocytes

Many invertebrates have specialized cells that function as itinerant troubleshooters within the body, acting to engulf or wall off foreign material and repair wounds. The cells are variously known as *archaeocytes, amebocytes, hemocytes, coelomocytes,* and so on, depending on the animals in which they occur. If a foreign particle is small, it is engulfed by phagocytosis; but if it is larger than about 10 μm, it is usually encapsulated. Arthropods can wall off the foreign object by deposition of melanin around it, either from the cells of the capsule or by precipitation from their **hemolymph** (blood).

In vertebrates several categories of cells are capable of phagocytosis. **Monocytes** arise from stem cells in the bone marrow (Fig. 3.3) and give rise to the **mononuclear phagocyte system** or **reticuloendothelial (RE) system.** As monocytes leave the blood and spread through a variety of tissues, they differentiate into active phagocytes. They become **macrophages** in lymph nodes, spleen, and lung; **Kupffer cells** in sinusoids of liver; and **microglial cells** in the central nervous system. Macrophages also have important roles in the specific immune response of vertebrates. Phagocytes show abundant expression of all TLRs.[59] **Dendritic cells** (DC; see below; p. 30) arise in bone marrow; immature dendritic cells then circulate in the blood as active phagocytes.[52] Phagocytic activity and TLR signals cause dendritic cells to mature and assume their critical role in stimulation of the adaptive response.[41]

Other phagocytes that circulate in blood are **polymorphonuclear leukocytes (PMNs),** a name that refers to the highly variable shape of their nucleus. Another name for these phagocytes is **granulocytes,** which alludes to the many small granules that can be seen in their cytoplasm after

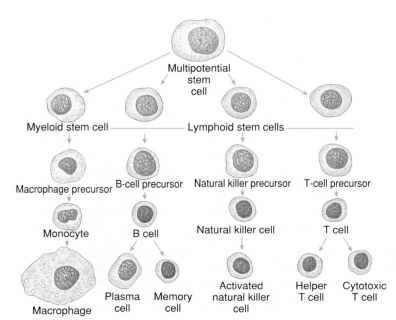

Figure 3.3   Lineages of some cells active in immune response.

These cells, as well as red blood cells and other white blood cells, are derived from multipotential stem cells in the bone marrow. B cells mature in bone marrow and are released into blood or lymph. Precursors of T cells go through a period in the thymus gland. Precursors of macrophages circulate in blood as monocytes.

treatment with appropriate stains. According to the granules' staining properties, granulocytes are further subdivided into **neutrophils, eosinophils,** and **basophils.** Neutrophils are the most abundant, and they provide the first line of phagocytic defense in an infection. Eosinophils in normal blood account for about 2% to 5% of the total leukocytes, and basophils are the least numerous at about 0.5%. **Eosinophilia** (a high eosinophil count in the blood) is often associated with allergic diseases and parasitic infections, especially with nematodes.

Several other kinds of cells, such as basophils, are not important as phagocytes but are important cellular components of the defense system. **Mast cells** are basophillike cells found in the dermis and other tissues. When they are stimulated to do so (in inflammation, p. 25), basophils and mast cells release a number of pharmacologically active substances that affect surrounding cells. **Lymphocytes** are crucial in the adaptive immune response. **Natural killer cells (NKs)** are lymphocyte-like cells that can kill virus-infected and tumor cells in the absence of antibody. They release substances onto the target-cell surface that lyse it.

ADAPTIVE IMMUNE RESPONSE OF VERTEBRATES

The specialized system of nonself recognition possessed by vertebrates results in increased resistance to *specific* foreign substances or invaders on repeated exposures. Research on the mechanisms involved is currently intense, and our knowledge of them is increasing rapidly. For concise accounts of general immunology, consult Murphy.[36] Reviews of various aspects of parasite immunology are in Warren.[68]

An adaptive immune response is stimulated by a specific foreign substance called an **antigen,** and, circularly, an antigen is any substance that stimulates an immune response. Antigens may be any of a variety of substances with a molecular weight of over 3000. They are most commonly proteins and are usually (but not always) foreign to the host; therefore, the number of potential antigens is huge. The types of antigen recognition molecules are **antibodies** and **T-cell receptors.** Antibodies are proteins called **immunoglobulins.** They are borne in the surface of **B lymphocytes (B cells)** or secreted by cells (**plasma cells**) derived from B cells.

T-cell receptors are, of course, borne on the surface of **T lymphocytes (T cells)** and also belong to the immunoglobulin superfamily of proteins. During development B cells and T cells rearrange their genes for immunoglobulin and T-cell receptors to encode 10^{11} different kinds of antigen receptors.[59] Each mature lymphocyte thus carries one kind of receptor, and following encounter with a matching antigen that lymphocyte is activated, leading to proliferation of cells with the same receptor, that is, formation of a clone of such cells within the body, a process known as **clonal selection.**[36] This astonishing diversity of receptors is made possible by two **r**ecombination **a**ctivating **g**enes (RAG 1 and 2) that catalyze rearrangement of preexisting immunoglobulin gene segments. Scientists believe that RAG 1 and 2 are evolutionary descendants of a prokaryote transposase gene inserted by horizontal transfer (not through descent) into the common ancestor of jawed vertebrates.[5]

Basis of Self and Nonself Recognition in Responses

Nonself recognition is very specific in vertebrates, much more so than in invertebrates. If tissue from one individual is transplanted into another individual of the *same species* (allograft), the graft will grow for a time and then die as immunity against it rises. In the absence of drugs that modify the immune system, tissue grafts will only grow successfully if they are between identical twins or between individuals of highly inbred strains of animals. The molecular basis for this specificity in nonself recognition involves certain proteins imbedded in the cell surface. These proteins are coded by genes known as the **major histocompatibility complex (MHC).**

MHC proteins are among the most variable known, and unrelated individuals almost always have different alleles. There are two types of MHC proteins: class I and class II. Class I proteins are found on the surface of virtually all cells in a vertebrate, whereas class II MHC proteins are found only on certain cells, such as lymphocytes and macrophages, participating in the immune responses. We discuss the role of MHC proteins in nonself recognition in the following text, but they are not themselves the molecules that recognize the foreign substance. This task falls to antibodies and T-cell receptors.

The two arms of adaptive responses are referred to as **cellular (T_H1)** and **humoral (T_H2).** Humoral immunity is based on **antibodies,** which are both on cell surfaces (bound) and dissolved in blood and lymph (circulating), whereas cellular immunity is entirely associated with cell surfaces. There is extensive communication and interaction among the cells of the two arms.

Antibodies

The basic antibody molecule consists of four polypeptide strands: two identical light chains and two identical heavy chains held together in a *Y*-shape by disulfide bonds and hydrogen bonds (Fig. 3.4). The amino acid sequence toward the ends of the *Y* varies in both the heavy and light chains, according to the specific antibody molecule (the **variable region**), and this variation determines with which antigen the antibody can bind. Each of the ends of the *Y* forms a cleft that acts as the antigen-binding site (see Fig. 3.4), and the specificity of the molecule depends on the shape of the cleft and the properties of the chemical groups that line its walls.

The remainder of the antibody is known as the **constant region,** although it also varies to some extent. The variable end of the antibody molecule is referred to as **Fab,** for **a**ntigen**b**inding **f**ragment, and the constant end is known as the **Fc,** for **c**rystallizable **f**ragment (see Fig. 3.4). The constant region of the light chains can be either of two types: kappa or lambda. The heavy chains may be any of five types: mu, gamma, alpha, delta, or epsilon. Each of these five is a **class** of antibody, referred to as **IgM, IgG** (now familiar to many people as *gamma globulin*), **IgA, IgD,** and **IgE,** respectively. The class of an antibody determines its role in the immune response (for example, the antibody may be secreted or held on a cell surface) but not the antigen it recognizes.

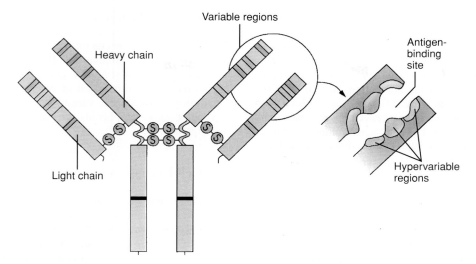

Figure 3.4   **Antibody molecule is composed of two shorter polypeptide chains (light chains) and two longer chains (heavy chains) held together by covalent disulfide bonds.**
The light chains may be either of two types: kappa or lambda. The class of antibody is determined by the type of heavy chain: mu (IgM), gamma (IgG), alpha (IgA), delta (IgD), or epsilon (IgE). The constant portion of each chain does not vary for a given type or class, and the variable portion varies with the specificity of the antibody. Antigen-binding sites are in clefts formed in the variable portions of the heavy and light chains. IgM normally occurs as a pentamer, five of the structures illustrated being bound together by another chain. IgA may occur as a monomer, dimer, or trimer.

Functions of Antibody in Host Defense

Antibodies can mediate destruction of an invader (antigen) in several ways.

1. **Opsonization.** Foreign particles, for example, bacteria or viruses, become coated with IgG molecules as their Fab regions become bound to the particle. Receptors for Fc on the surface of macrophages bind to the projecting Fc regions, thus stimulating the macrophage to engulf the particle.
2. **Neutralization.** IgG and IgM antibodies can neutralize toxins that are secreted by bacteria and prevent toxin molecules from binding to their target cells. IgA in secretions of the digestive and respiratory tracts neutralizes toxins produced by bacteria in these organs. Antibodies can bind to the envelope of viruses and prevent the viruses from attaching and penetrating host cells.
3. **Activation of complement.** An important process, particularly in destruction of bacterial cells, is interaction with complement activated by the classical pathway. As noted previously (p. 26), the first component in the classical pathway is activated by bound antibody. The end result in both classical and alternative pathways can be the same; that is, perforation of the foreign cell. Both pathways may also lead to opsonization or enhancement of inflammation. Binding of complement to antigenantibody complexes can facilitate clearance of these potentially harmful masses by phagocytic cells.
4. **Antibody-dependent, cell-mediated cytotoxicity (ADCC).** Antibody bound to the surface of an invader may trigger contact killing of the invader by host cells in what is known as antibody-dependent, cell-mediated cytotoxicity (ADCC), a particularly important mechanism against parasites. Eosinophils activated by IL-5 (p. 25) can be effector cells in ADCC. They,

as well as lymphoid cells and neutrophils, can destroy bloodstream forms of *Trypanosoma cruzi* in the presence of antibody against this organism.[23] The parasites are phagocytized, and granules in eosinophils and neutrophils fuse with the phagosome and kill the *T. cruzi* with H_2O_2.[65] In the presence of antibody, neutrophils and eosinophils kill newborn juveniles of *Trichinella spiralis* by release of reactive oxygen intermediates, but adults and muscle-stage juveniles are much more resistant to ADCC, evidently because they secrete antioxidant enzymes.[8] *Schistosoma mansoni* schistosomula (juveniles) are also killed by reactive oxygen intermediates released by neutrophils and eosinophils in the presence of antibody and complement.[8]

Lymphocytes

As already noted, B lymphocytes have antibody molecules in their surface and give rise to plasma cells that actively secrete antibodies into the blood, and T lymphocytes (T cells) have surface receptors that bind antigens. Lymphocytes are **activated** when they are stimulated to move from their recognition phase, in which they simply bind with particular antigens, to a phase in which they proliferate and differentiate into cells that function to eliminate the antigens. We also speak of activation of effector cells, such as macrophages, when they are stimulated to carry out their protective function.

Subsets of T Cells

Communication between cells in an immune response, regulation of the response, and certain effector functions are carried out by different kinds of T cells. Although morphologically similar, subsets of T cells can be distinguished by characteristic proteins in their surface membranes. Most T cells also bear other transmembrane proteins closely linked to the T-cell receptors, which serve as **accessory** or **coreceptor** molecules. These are of one of two types: **CD4** or **CD8**. Cells with the coreceptor protein CD4 (for **c**luster of **d**ifferentiation) are CD4$^+$ and those with CD8 are described as CD8$^+$. Immunologists once believed that certain CD4$^+$ cells (T helper or T$_H$) activated immune responses, and certain CD8$^+$ cells (T suppressors) downregulated such responses. Present evidence suggests a more complicated web of interactions (Fig. 3.5). Some T$_H$ cells (designated T$_H$1) activate cell-mediated immunity while suppressing the humoral response, and others (called T$_H$2) activate humoral and suppress cell-mediated immunity.

Cytotoxic T lymphocytes (CTLs) are CD8$^+$ cells that kill target cells expressing certain antigens. A CTL binds tightly to its target cell and secretes a protein that causes pores to form in the cell membrane. The target cell then lyses.[67]

A class of T cells known as regulatory T cells (T$_{reg}$) constitute 5% to 10% of CD4$^+$ T cells normally present. T$_{reg}$s are characterized by expression of the FoxP3 gene and they function to partially suppress the immune response, resulting in tolerance to self-antigens.[6, 13, 48] Autoimmune diseases are evidently due, at least in part, to these cells' failure to carry out their typical function. Th17 cells are a subset of T$_{reg}$s that produce the cytokine IL-17, which is actually a family of proteins, with various members playing diverse roles in both adaptive and auto immunity.[40] IL-17A and IL-17F promote inflammation and stimulate granulocyte production. Th17 cells have been implicated in both protection against, and pathology induced by, various parasites including amebas (chapter 7), kinetoplastids (chapter 5), and apicomplexans (chapters 8, 9).[7, 19, 42]

T-Cell Receptors

T-cell receptors are transmembrane proteins on the surfaces of T cells. Like antibodies, T-cell receptors have a constant region and a variable region. The constant region extends slightly into the cytoplasm, and the variable region, which ligates specific antigens, extends outward (Figs. 3.4, 3.6).

Generation of a Humoral Response

When an antigen is introduced into the body, some antigen is taken up by **antigen-presenting cells (APCs),** such as macrophages and dendritic cells (DC), that partially digest

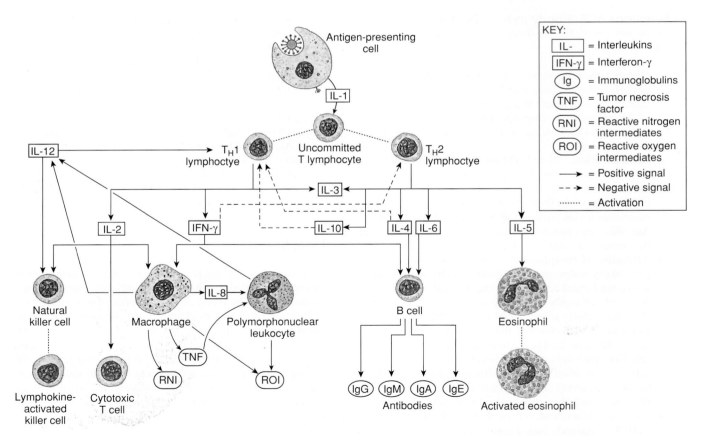

Figure 3.5   **Major pathways involved in the immune response to parasitic infections as mediated by cytokines.**

Solid arrows indicate positive signals and *broken arrows* indicate inhibitory signals. *Broken lines without arrows* indicate the path of cellular activation. *IFN-*γ, interferon-γ; *Ig,* immunoglobulin; *IL,* interleukin; *TNF,* tumor necrosis factor; *T$_H$1,* helper CD4$^+$ and CD8$^+$ cells that stimulate cell-mediated response; *T$_H$2,* helper CD4$^+$ and CD8$^+$ cells that stimulate humoral response; *RNI,* reactive nitrogen intermediates; *ROI,* reactive oxygen intermediates.

Redrawn by William Ober and Claire Garrison from F. E. G. Cox and E. Y. Liew, "T-cell subsets and cytokines in parasitic infections," in *Parasitol. Today* 8:371–374, 1992.

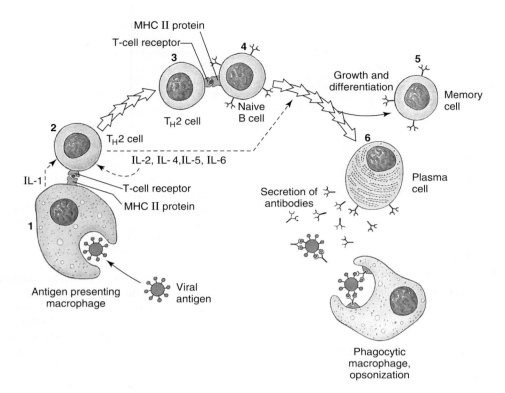

Figure 3.6   Humoral immune response.

(*1*) Macrophage consumes antigen, partially digests it, displays epitope on its surface, along with class II MHC protein, and secretes interleukin-1 (IL-1). (*2*) T helper cell, stimulated by IL-1, recognizes epitope and class II protein on macrophage, is activated, and secretes IL-4, IL-5, IL-6. (*3*) T helper then activates B cell, which carries antigen and class II protein on its surface. (*4*) Activated B cells finally produce many plasma cells that secrete antibody. (*5*) Some of B-cell progeny become memory cells. (*6*) Antibody produced by plasma cells binds to antigen and stimulates macrophages to consume antigen (opsonization).

the antigen. Dendritic cells, so named for their shape, were originally discovered in the spleen, but they occur in other tissues as well, especially in bone marrow. Their primary role is presentation of antigen acquired either through direct invasion or uptake of molecules released by parasites. DCs function in both innate and acquired immunity, although their actual cytokine products and participation in an immune response may vary according to parasite species, with impaired DC function being at least partly responsible for survival of some parasites.[52, 54]

APCs incorporate portions of the antigen into their own cell surface, bound in the cleft of MHC II protein (Fig. 3.6; see also Fig. 3.7). That portion of the antigen presented on the surface of the APC is called the **epitope** (or **determinant**). Macrophages also secrete IL-1, which stimulates T_H2 cells. The specific T-cell receptor for that particular epitope recognizes the epitope bound to the MHC II protein. Ligation of the T-cell receptor with the epitope-MHC II complex is enhanced by the coreceptor CD4, which itself binds to the constant portion of the MHC II protein (Fig. 3.6). The bound CD4 molecule also transmits a stimulation signal to the interior of the T cell. Activation of the T cell requires interaction of additional co-stimulatory and adhesion signals from other proteins on the surface of the APC and T cell. The CD8 coreceptor functions in a similar way on CD8$^+$ cells; that is, it enhances binding of the T-cell receptor and transmits a stimulatory signal into the T cell. The activated T_H2 cell secretes the cytokine IL-2, which stimulates that cell to proliferate.

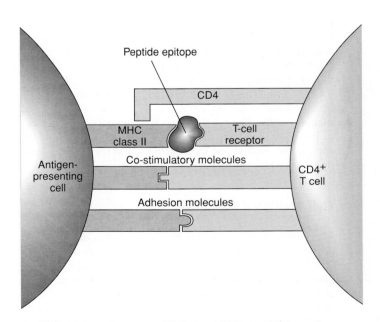

Figure 3.7   Interacting molecules during activation of a T helper cell.

Concurrently with processing and presentation of antigen by the APC, the B cell with the same antigen ligated to specific antibody on its surface is activated by TLR signaling.[41] It internalizes the antigen-antibody complex by

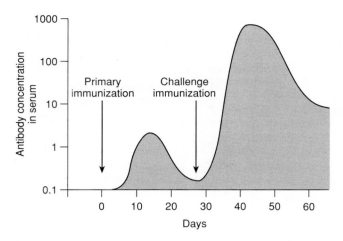

Figure 3.8 **Typical immunoglobulin response after primary and challenge immunizations.**

Secondary response is result of large numbers of memory cells produced after primary B-cell activation.

receptor-mediated endocytosis and itself partially digests the antigen. Epitopes of the antigen become associated with MHC II proteins, which are then moved to the surface and displayed. These epitopes are recognized by the antigen-specific T_H2 cells, which secrete IL-4, IL-5, and IL-6, stimulating that specific B cell to proliferate and differentiate. It multiplies rapidly and produces many plasma cells, which secrete large quantities of antibody for a period of time and then die.

Thus, if we measure the concentration of the antibody (**titer**) soon after the antigen is injected, we can detect little or none. The titer then rises rapidly as the plasma cells secrete antibody, and it may decrease somewhat as they die and the antibody is degraded (Fig. 3.8). However, if we give another dose of antigen (the **challenge**), there is no lag, and the antibody titer rises quickly to a higher level than after the first dose. This is the **secondary response,** and it occurs because some of the activated B cells gave rise to long-lived **memory cells.** There are many more memory cells present in the body than there are original B lymphocytes with the appropriate antibody on their surfaces, and they rapidly multiply to produce additional plasma cells.

Cell-Mediated Response

Many immune responses involve little, if any, antibody and depend on the action of cells only. In cell-mediated immunity (CMI), the epitope of the antigen is also presented by macrophages, but the T_H1 arm of the immune response is activated as the T_H2 arm is suppressed. Like humoral immunity, CMI shows a secondary response due to large numbers of memory T cells produced from the original activation. For example, a second tissue allograft (challenge) between the same donor and host will be rejected much more quickly than the first.

Four types of CMI are distinguished:

1. **Delayed type hypersensitivity (DTH).** T_H1 cells, activated by a specific antigen, secrete several cytokines

that lead to **inflammation,** considered in more detail in the following text. In DTH the principal effector cells are macrophages, but many cell types participate. Eggs of schistosomes serve as sources of antigen that cause DTH reactions (p. 246).

2. **Cytolytic T lymphocyte (CTL) responses.** CTL responses are important in organ transplant rejection and in viral infections. Activated T_H1 cells secrete IL-2 that causes CD8$^+$ T cells to become functional CTLs. CTLs lyse cells that display the target antigen on their surfaces. This response is important in protozoan infections in which parasites such as malarial organisms reproduce within host cells.

3. **Natural killer (NK) cell responses.** NK cells are large, granular lymphocytes that express neither T or B markers on their surface. The response is also important in organ transplantation and viral infection, but it tends to occur earlier than the CTL response. IL-2 and IL-12 stimulate differentiation of NK cells into **lymphokine-activated killer (LAK) cells** that lyse target cells.

4. **Immediate hypersensitivity (IH).** IH responses are in fact mediated by antibody (IgE) and the T_H2 arm. However, in the **late phase reaction** of IH, eosinophils are recruited into an area of inflammation to participate in an ADCC reaction (p. 29) that can kill parasites.

Inflammation

Inflammation is a vital process in the mobilization of body defenses against an invading organism or other tissue damage and in the repair of damage thereafter. Although inflammation is basically a sign of innate immunity, the course of events in the process is greatly influenced by prior immunizing experience and by duration of an invader's presence or its persistence in the body. The mechanisms by which an invader is actually destroyed, however, are themselves nonspecific. Manifestations of inflammation are delayed type hypersensitivity and immediate hypersensitivity, depending on whether the response is cell mediated or antibody mediated.

The term *delayed type hypersensitivity (DTH)* is derived from the fact that a period of 24 hours or more elapses between the time of antigen introduction and the response to it in an immunized subject. This delay occurs because the T_H1 cells with receptors in their surface for that particular antigen require some time to arrive at the antigen site, recognize the epitopes displayed by the APCs, and become activated and secrete IL-2, TNF, and IFN-γ. TNF causes endothelial cells of the blood vessels to express on their surface certain molecules to which leukocytes adhere: first neutrophils and then lymphocytes and monocytes.

TNF also causes the endothelium to secrete inflammatory cytokines such as IL-8, which increase the mobility of leukocytes and facilitate their passage through the endothelium. Finally, TNF and IFN-γ stimulate endothelial cells to change shape, favoring both leakage of macromolecules from blood into the tissues and passage of cells through the vascular system lining. Escape of fibrinogen from the blood vessels leads to conversion of fibrinogen to fibrin, and the area becomes swollen and firm.

As monocytes pass out of blood vessels, they become activated macrophages, which are the main effector cells

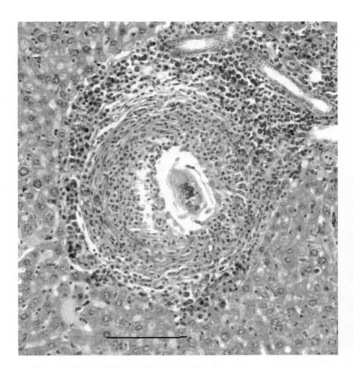

Figure 3.9   Circumoval granuloma in liver caused by hepatic schistosomiasis.

Centrally located remnants of a Schistosoma mansoni egg are surrounded by epithelioid cells, a middle region of fibrous connective tissue and an outer zone of lymphocytes. The inflammation resulting in the granuloma is induced by soluble egg antigens of the schistosome (see chapter 16). Bar = 100 μm.

Photograph courtesy Steven Nadler.

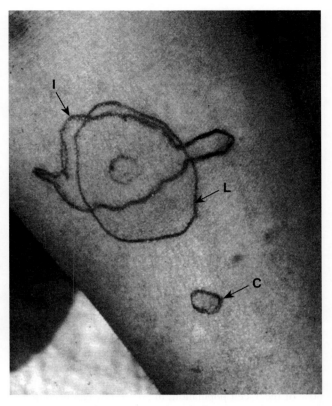

Figure 3.10   Immediate hypersensitivity reaction in an intradermal test for schistosomiasis.

Limits of swelling are outlined, following injection of an antigen or a non-antigenic control. The two small circles are controls (C). The immediate (15-minute) response is the irregular outline (I), and the larger late-stage reaction (L) has a smooth edge.

From I. G. Kagan, and S. E. Maddison, in G. T. Strickland (Ed.), *Hunter's tropical medicine* (7th ed.). © 1991 W. B. Saunders Co.

of the DTH. They phagocytize particulate antigen, secrete mediators that promote local inflammation, and secrete cytokines and growth factors that promote healing. If the antigen is not destroyed and removed, its chronic presence leads to deposition of fibrous connective tissue, or **fibrosis.** Nodules of inflammatory tissue called **granulomas** may accumulate around persistent antigen and are found in numerous parasitic infections (Fig. 3.9).

Immediate hypersensitivity is quite important in some parasitic infections.[29] This reaction involves degranulation of mast cells in the area. Their surfaces bear receptors for the Fc portions of antibody, especially IgE. Occupation of these membrane sites by antigen-specific antibodies enhances degranulation of the mast cells when the Fab portions bind the particular antigen. There is a rapid release of several mediators, such as histamine, that cause dilation of local blood vessels and increased vascular permeability. Escape of blood plasma into the surrounding tissue causes swelling **(wheal),** and engorgement of vessels with blood produces redness, the characteristic **flare** (Fig. 3.10). Widespread systemic, immediate hypersensitivity is **anaphylaxis,** which may be fatal if not treated rapidly.

Although the wheal and flare of many immediate hypersensitivity reactions resolve in about an hour, some elicit a late phase reaction in two to four hours. The swelling and change in permeability of the capillaries allow antibodies and leukocytes to move from the capillaries and easily reach an

invader. The first phagocytic line of defense is neutrophils, and abundance of these PMNs may last a few days. Macrophages (either fixed or differentiated from monocytes) then become predominant and secrete MIF, which modulates TLR4 and upregulates production of proinflammatory cytokines.[61] Eosinophils may kill parasites by an ADCC reaction.

Some degree of cell death (**necrosis**) always occurs in inflammation, but necrosis may not be prominent if the inflammation is minor. When necrotic debris is confined within a localized area, pus (spent leukocytes and tissue fluid) may increase in hydrostatic pressure, forming an **abscess.** An area of inflammation that opens out to a skin or mucous surface is an **ulcer.** Abscesses and ulcers can be a result of invasive amebiasis (see chapter 7).

Immediate hypersensitivity in humans is the basis for allergies and asthma, which are quite undesirable conditions, leading one to wonder why they evolved. Some workers believe that the allergic response originally evolved to help the body ward off parasites because only allergens and parasite antigens stimulate production of large quantities of IgE.[29] Avoidance of or reduction in effects of parasites would have conferred a selective advantage in human evolution. The hypothesis is that in the absence of heavy parasitic challenge, the immune system is free to react against other substances, such as ragweed pollen.[29] People now living where parasites remain abundant are less troubled with allergies than are those living in relatively parasite-free areas.

Acquired Immune Deficiency Syndrome (AIDS)

AIDS is an extremely serious disease in which the ability to mount an immune response is disabled completely. It is caused by the **human immunodeficiency virus (HIV).** The first case of AIDS was recognized in 1981, and by the end of 2000, over 920,000 people had contracted the disease in the North America alone.[46] It is estimated that 36 million people in the world, of which 25.3 million were in sub-Saharan Africa, were infected with HIV in 2000. HIV infection virtually always progresses to AIDS after a latent period of some years.

To the best of our current knowledge, AIDS is a terminal disease. AIDS patients are continuously plagued by infections with microbes and parasites that cause insignificant problems in persons with normal immune responses. HIV preferentially invades and destroys $CD4^+$ lymphocytes. CD4 protein is the major surface receptor for the virus.[69] To penetrate the T-cell, however, the virus requires one of numerous chemokine co-receptors, the most important of which are CCR5 and CXCR4 ("chemokine" is a contraction of chemotaxis and cytokine).[11] Normally, $CD4^+$ cells make up 60% to 80% of the T-cell population; in AIDS they can become too rare to be detected.[27] T_H1 cells are relatively more depleted than T_H2 cells, which upsets the balance of immunoregulation and results in persistent, nonspecific B-cell activation.

IMMUNODIAGNOSIS

Although we diagnose many parasitic infections most easily by finding the parasites themselves or their products, such as eggs in host feces, the organisms in many infections may be difficult to demonstrate. Thus, numerous tests have been developed that take advantage of a patient's immune response. Space permits only a few examples here, but every parasitology student should be aware of these extremely valuable diagnostic tools. You should also be aware of some difficulties. For example, false positives may arise when two related agents have antigens in common or that are similar enough to crossreact with antibodies raised against the other. This is often the case with **skin tests,** in which a small amount of antigen is injected into the skin of the patient. Many parasitic infections produce immediate or delayed type hypersensitivity reactions, which are easily observed (see Fig. 3.10).[70]

Some additional techniques are the **indirect hemagglutination (IHA)** test, the **indirect fluorescent antibody (IFA)** test, and the **complement fixation (CF)** test. In IHA, red blood cells are coated with parasite antigen and incubated with the patient's serum (test serum). Agglutination of the blood cells indicates the presence of antibody in the test serum. For an IFA test, parasites themselves are fixed to a microscope slide, incubated with test serum, washed, treated with antibody to human immunoglobulin (anti-Ig) that has been chemically bound to fluorescein, and washed again. If anti-Ig-fluorescein binds to Ig that was in the test serum,

in the first step it can be visualized under a fluorescence microscope.

The CF test is a bit more complicated; it is an ingenious method to determine whether complement has been bound to an antigen-antibody complex ("fixed"). Of course, fixation cannot be visualized directly unless the antibody is bound to surface antigens of cells that lyse. The test serum is incubated with parasite antigen in the presence of guinea pig complement. If antibody to the parasite antigen is present, its components bind to the antigen-antibody complex. If antibody against the parasite antigen is not present in the test serum, the components of complement remain free and inactivated. Sheep red blood cells are then added, along with antibody to sheep red blood cells. Lysis of the cells indicates that complement did not fix earlier and, therefore, that antibody to the antigen was not present in the test serum.

The **enzyme-linked immunosorbent assay (ELISA)** and its variants have become quite popular. They are good diagnostic tests and serve as powerful research tools. They are simple to perform and usually do not require sophisticated equipment. In the assay a small quantity of antigen is adsorbed to the bottom of a small cup in a plastic microplate (Fig. 3.11). Next, test serum is added to the cup (Fig. 3.12). The serum is removed, and the cup is rinsed several times. If the serum contained antibodies to the antigen, they will have bound to the antigen and will not be removed by rinsing. A solution containing antibodies to human Ig (anti-Ig) is added. The anti-Ig must be prepared beforehand and linked covalently to an enzyme. The enzyme can be any one of several whose reaction product is colored. This solution is then removed from the cup, the cup is rinsed again, and the substrate for the enzymatic reaction is added. If the tested serum contained antibodies against the antigen, anti-Ig will have been bound to them, and the enzymatic reaction will occur, producing a color.

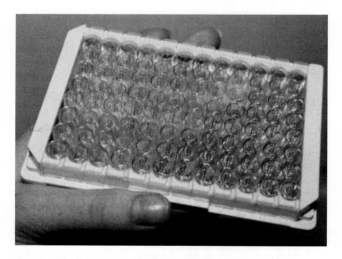

Figure 3.11 A microplate for an ELISA test.
In addition to serving as positive and negative controls, wells are available on the plate for testing several individuals. Positive controls are wells in which antibody is known to be present, and negative controls omit the enzyme-linked anti-Ig.
Photograph by Larry S. Roberts.

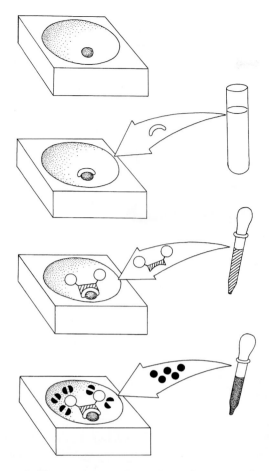

Figure 3.12   Sequence of steps in performance of an ELISA test.

(*a*) Known antigen is adsorbed to bottom of microplate well. (*b*) Serum from patient is added and then well is rinsed. (*c*) Enzyme-linked antibody against human immunoglobulin is added and then well is rinsed again. (*d*) Enzyme substrate is added. If colored products of enzyme reaction are observed, this indicates presence of bound anti-Ig, which, in turn, indicates presence of antibody against antigen. Thus, the test is positive.

Drawing by William Ober and Claire Garrison.

A variation is the "sandwich" ELISA, which can detect parasite antigen, rather than host antibody. In this case antibody to the antigen in question is adsorbed to the plastic cup, the test serum is added, the cup is rinsed, and additional antibody linked to an enzyme is added. Formation of a color indicates a positive result. Adaptations of the sandwich ELISA use a "dipstick" of acetate plastic, with the antibody adsorbed to a film of nitrocellulose.[3] This method can detect antigens of intestinal parasites in the patient's feces and is very convenient to use in the field.

The *Para*Sight®-F test for diagnosis of malaria caused by *Plasmodium falciparum* (p. 152) is a dipstick ELISA developed in the 1990s and subsequently made commercially available as a kit.[56] This test detects an antigen present in the blood of infected individuals. The nitrocellulose strip is prepared with a monoclonal antibody to the specific antigen applied in a line about 1 cm from the end of the strip and in a line of the antigen itself about 1 cm farther from the end (Fig. 3.13). The line of antigen serves as a reagent control. A drop of blood from a finger prick is hemolyzed (cells lysed) with detergent, and the end of the dipstick is immersed in the hemolyzed blood. The blood is absorbed quickly, and a solution containing antibody coupled to a colored reagent is applied. After a clearing reagent removes the hemolyzed blood, either one or two lines can be discerned, depending on whether the *P. falciparum* antigen was present in the test serum (see Fig. 3.13). The basic dipstick technology has now been engineered so that the entire system is contained in a plastic cassette that facilitates handling and diagnosis. Such cassettes are available commercially from several companies.

Although not an "IMMUNODIAGNOSTIC" method, detection of parasite DNA after amplification by the polymerase chain reaction is proving valuable.[47, 53] Such techniques are not currently adaptable to field use; they can be helpful when a laboratory is available and or eggs are scarce and difficult to find by microscopy.

PATHOGENESIS OF PARASITIC INFECTIONS

The pathogenic effects of a parasitic infection may be so subtle as to be unrecognizable, or they may be strikingly obvious. An apparently healthy animal may be host to hundreds of parasitic worms and yet show no obvious signs of distress. Another host may be so anemic, unthrifty, and stunted that parasites are undoubtedly the reason for its sad state. The pathogenic effects of parasites are many and varied, but for the sake of convenience they can be discussed under the headings of *trauma, nutrition robbing, toxin production,* and *interactions of the host immune/inflammatory responses.*

Physical trauma, or destruction of cells, tissues, or organs by mechanical or chemical means, is common in parasite infections. When an *Ascaris* or hookworm juvenile (pp. 411, 397) penetrates a lung capillary to enter an air space, it damages the blood vessel and causes hemorrhage and possible infection by bacteria that may have been inhaled. Hookworms, after completing migration to the small intestine, feed by biting deeply into the mucosa and sucking blood thus causing anemia in heavy infections. The dysentery ameba *Entamoeba histolytica* (p. 106) digests away mucosa of the large intestine, forming ulcers and abscessed pockets that can cause severe disease. These are but a few examples of known physical trauma caused by parasites. Many are discussed in later chapters on the particular parasites involved.

A less obvious but often pernicious pathogenic situation is diversion of the host's nutritive substances. Although most tapeworms absorb so little food in proportion to the amount eaten by the host that the host still manages very well, the broad fish tapeworm *Diphyllobothrium latum* has such strong affinity for vitamin B_{12} that it absorbs large amounts from the intestinal wall and contents of its host (p. 328). Since B_{12} is necessary for erythrocyte production, a severe anemia may result. The large nematode *Ascaris lumbricoides* (p. 411) inhabits the small intestine—often in large numbers—and consumes a good deal of food the host intends for itself. There is strong evidence that infection with *Ascaris* contributes to childhood malnutrition and retards growth.[20] Other studies showed that removal of the nematode *Trichuris trichiura* (p. 377) resulted in

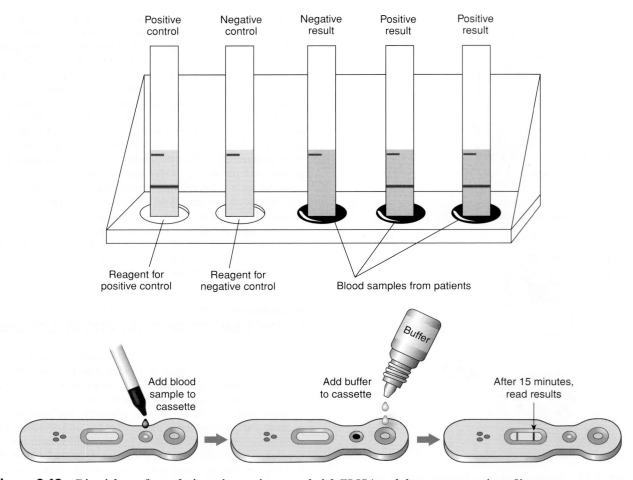

Figure 3.13 **Dipstick test for malaria antigen using a sandwich ELISA and the cassette version of it.**
(a) Monoclonal antibody to antigen is adsorbed to nitrocellulose strip in a line about 1 cm from end. Then the specific antigen is adsorbed in a line about 1 cm above the antibody. The strips are dipped in hemolyzed blood to be tested and antibody bound to a color reagent is applied. A colored band indicates the presence of antigen. (b) A cassette test; the patient's name is written on the back, a drop of blood from a finger prick is added to the small hole, buffer is added to the large hole, and the results read in 15 minutes. Lines in the window show whether the patient is infected or not, and whether the test itself is a valid one (test must be repeated if the results indicate an invalid one).

(a) Drawn by Bill Ober and Claire Garrison from C. J. Shiff et al., *Parasitol. Today* 10:494–495. (b) Drawn by Bill Ober and Claire Garrison from FIND (Foundation for Innovative New Diagnostics) instructional literature.

significant improvements in long- and short-term memory and much higher rate of growth in children.[10, 39] The most important forms of malnutrition are aggravated by infection with these and other helminths.[57] Note that helminths could contribute to malnutrition by decreasing host nutrient intake, increasing nutrient excretion, and/or decreasing nutrient utilization.

The tiny protozoan *Giardia* robs its host in a different way. It is concave on its ventral surface and applies this suction cup to the surface of an intestinal epithelial cell. When many of these parasites are present, they cover so much intestinal absorptive surface that they interfere with the host's absorption of nutrients. The unused nutrients then pass uselessly through the intestine and are wasted.[9] We mentioned the caloric cost of a day of fever caused by malaria in chapter 1. Chronic malaria is also associated with failure of children to gain weight and with iron deficiency anemia.[31]

A well-known example of effects traditionally attributed to toxins is found in malaria. The disease in humans is due to

species of *Plasmodium* (chapter 9). The parasites invade red blood cells and reproduce by multiple fission, bursting forth, with each new offspring parasite invading a new red cell and repeating the process. The reproductive cycles of many individual parasites are more or less synchronized, so that many erythrocytes burst at once. Lysis of the parasitized red blood cells unleashes large amounts of waste products and cell debris into the blood, to which the host responds with a sharp rise in TNF and other pro-inflammatory cytokines.[24] Synchrony of red cell lysis and consequent eruption of TNF accounts for the periodicity of the typical paroxysms of chills and fever in malaria. The effects are not caused by "toxins" in the strict sense, but rather by the many cell-membrane fragments bearing GPIs that interact with TLRs on macrophage surfaces, initiating signaling cascades and triggering release of a flood of pro-inflammatory cytokines (p. 152).[18]

Malaria parasites produce another toxin, hemozoin, which is the insoluble waste product of their digestion of hemoglobin. Macrophages and other phagocytes engulf particles of hemozoin produced when parasitized erythrocytes lyse, but, because it is insoluble, they cannot digest it, and it remains unchanged in their cytoplasm. The presence of hemozoin in macrophages reduces their capacity to perform further phagocytosis.[62]

In recent years we have come to realize that a great many—perhaps the most serious and most pervasive—pathogeneses are actually caused by the host's own defense system: the immune response and inflammation.[45] A number of cases previously thought due to toxins released by the parasite are now understood as caused by the host's reaction to parasite products. For example, the protozoan *Trypanosoma cruzi,* an intracellular parasite, develops clusters of infected cells in smooth and cardiac muscle cells of its host, and when the parasites degenerate—sometimes years later—the inflammatory response damages the supporting cells of the nerve ganglia that control peristalsis and heart contraction (p. 74). Parasite antigens on the host's own cells, particularly in the endocardium, cause autoimmune reactions, and the host's cells may be attacked as foreign by the immune system.[60] Some of the large amount of antigen-antibody complex formed in infections with the African trypanosomes *(T. brucei rhodesiense* and *T. b. gambiense)* adsorbs to the host's red blood cells, activating complement and causing lysis with resulting anemia (p. 68).[60]

The flow of blood carries many of the eggs laid by schistosomes to the liver where they lodge, leaking antigen and causing a chronic DTH reaction (p. 246). The formation of granulomas around the eggs eventually impedes blood flow through the liver, resulting in cirrhosis and portal hypertension.[43] Adults of the filarial nematode *Onchocerca volvulus* live in the dermis of humans. They release live juveniles, many of which wander into the eyes, including the cornea. Each degenerating juvenile in the cornea becomes a focus of inflammation, and over time sclerosing keratitis (hardening inflammation of the cornea) and other complications cause permanent blindness (p. 447).[22] More recent evidence suggests that the inflammation may actually be caused by bacteria *(Wolbachia)* living symbiotically within the nematodes.[50] Today there are villages in Africa where the majority of adults are blind because of this parasite.

These and many other diseases to be discussed in context are examples of the immune response gone wrong. We could scarcely do without the defenses of our immune system, but some manifestations of the immune response are responsible for much of the pathogeneses hosts suffer.

ACCOMMODATION AND TOLERANCE IN THE HOST-PARASITE RELATIONSHIP

In overall presentation to the immune system, there is substantial difference between viral and bacterial parasites compared with protistan and helminth parasites. Protists and helminths are much larger in size than viruses and bacteria and thus have many more antigenic molecules per parasite. These molecules can be borne on the surface or released as excretory/secretory (ES) antigens. Helminths usually do not reproduce within a vertebrate host, but they are often quite long-lived, and thus infections are chronic. They may go through developmental stages that are antigenically distinct from each other. These factors constitute effective challenges to the immune system.

Successful parasites have had to evolve one or more tactics to avoid the defenses of a given host. Otherwise, the host simply would not be susceptible. Parasites display an astonishing array of such tactics (Tables 3.3 and 3.4). We will examine only a few examples; for many others and more information see Warren,[68] and the volume introduced by Mitchell.[34]

The location of the parasite may provide some protection against host defenses. The intestinal lumen is one such site. Although IgA is secreted into the intestine, IgA is not a very potent effector molecule against worms, and complement and phagocytic cells are normally not found in the intestine. However, the rat nematode *Nippostrongylus braziliensis* can be expelled because inflammation and an immediate hypersensitivity reaction change the permeability of the mucosa and evidently allow IgG to leak into the lumen.[66] Many other intestinal parasites, not provoking such inflammation, are relatively long-lived. Numerous parasites, such as juvenile tapeworms (cestodes) in various tissues, achieve protection from the host response by envelopment with cystic membranes (p. 333). Others may be shielded by their location within a host cell. Recognition of the infected cell by the host's cell-mediated effector systems is precluded if no parasite antigens are present in the host cell's outer membrane, as seems to be the case in liver cells infected with malaria parasites.

Parasites that are constantly or frequently bathed in blood would seem particularly vulnerable to the range of host defenses, but they have evolved fascinating mechanisms for evasion. African trypanosomes display a "moving target"—that is, a continuing succession of variant antigenic types—so that just as the host mounts an antibody response to one, another type proliferates (p. 68).

Other important mechanisms of evasion are present in these infections as well. Antibody and cell-mediated responses are suppressed, apparently by some substance secreted by the trypanosomes. Suppression may be achieved by polyclonal B-cell activation early in the infection; many subtypes of B cells are stimulated to divide, leading to the production of nonspecific IgG and autoantibodies.[64] Polyclonal B-cell activation effectively exhausts the immune system without producing anything useful against the invader. Also in trypanosomiasis there is a suppression of IL-2 secretion and expression of IL-2 receptors, and T cells become refractory to normal signals.

Visceral leishmaniasis, caused by other protozoa, shows a kind of immunosuppression by misdirection of the immune response (p. 83). The organisms initially infect macrophages near the site of the infection and then invade cells of the reticuloendothelial system throughout the body. The CMI arm of the immune response is necessary to control proliferation of the protozoa; patients with positive DTH reaction to leishmanial antigens successfully resolve the infection. In other patients, however, there is a strong humoral response, and the CMI is suppressed.[44] In these patients continued reproduction of the parasites eventually leads to death (if untreated).

In addition to using immunosuppression, polyclonal lymphocyte activation, and other mechanisms, the blood fluke *Schistosoma* actually adsorbs many host antigens so that the host immune system "sees" only self, not recognizing

Table 3.3	Mechanisms Favoring Immune Evasion in Some Helminths

Parasite	Product	Result
Nematodes		
Dirofilaria immitis (dog heartworm)	Ig-cleaving protease on surface	Cleaves adherent antibodies
Heligmosomoides polygyrus (intestinal worm in mice)	Immunosuppressant	Macrophages incompetent as APCs
Nippostrongylus brasiliensis (intestinal worm in rats)	Acetylhydrolase secreted	Blocks neutrophil attraction by hydrolyzing platelet-activating factor
Onchocerca volvulus (dermal filiarial worm)	Cystatin on surface	Protease inhibitor, blocks antigen processing
Brugia pahangi (rodent filarial worm)	Secretes superoxide dismutase	Protects against ROI
	Glutathione peroxidase on surface	Protects against ROI from neutrophils and macrophages
Brugia malayi (lymphatic filarial worm)	Microfiliariae secrete prostaglandin E2	Anti-inflammatory
Platyhelminths		
Schistosoma mansoni (blood fluke)	Ig-cleaving protease on surface	Cleaves adherent antibodies
	Glutathione-*S*-transferase	Antioxidant, protects against ROI
	Immunosuppressant	Secreted
Schistosoma japonicum (blood fluke)	Glutathione-*S*-transferase	Antioxidant, protects against ROI
Fasciola hepatica (liver fluke)	Ig-cleaving protease on surface	Cleaves adherent antibodies
Echinococcus granulosus (hydatid tapeworm)	Elastase inhibitor in cyst fluid	Blocks neutrophil attraction by complement component
Taenia solium (human tapeworm)	Paramyosin secreted	Binds complement
Taenia taeniaeformis (cat tapeworm)	Secretes taeniaestatin	IL-2 and neutrophil chemotaxis inhibitor
	Secretes sulfated proteoglycan	Blocks complement
	Juveniles secrete prostaglandin E2	Anti-inflammatory

Data from Maizels et al. 1993. Immunological modulation and evasion by helminth parasites in human populations. *Nature* 365:797–805.

the parasite as foreign (p. 243).[38] For example, if adult worms are removed from mice and transferred surgically to monkeys, the worms stop producing eggs for a time but then recover and resume normal egg production. However, if the worms from mice are transferred to a monkey that has been previously immunized against mouse red blood cells, the worms are destroyed promptly. Interestingly, several antischistosomal drugs compromise the effectiveness of the worms' immune evasion. Praziquantel, for example, at concentrations too low to be directly lethal to the schistosomes, allows immunological destruction. The drug apparently alters the architecture of the tegumental surface, exposing epitopes to the immune system that are normally sequestered beneath host antigens.[38] Molecular characterization of the schistosome genome has shown that hundreds of genes have a remarkable identity of nucleotide sequences between host and parasite genes.[51] Whether this spectacular molecular mimicry evolved by an amazing evolutionary convergence or an appropriation of sequences is a question yet to be resolved.

THE MICROBIAL DEPRIVATION HYPOTHESIS

Humans, and presumably pre-humans, have been living with their parasites and prokaryotic symbionts for a very long time. According to the microbial deprivation hypothesis, this long association is of evolutionary significance. Thus proper development of the immune system depends on continuous exposure to a variety of antigens, among which are helminth parasites.[14] Studies have found an inverse relationship between some autoimmune diseases and parasitic infections, evidently resulting from a variety of mechanisms that affect T cell activation, cytokine levels, TLR signaling, dendritic cell function, and other aspects of the immune response.[71] Diseases involved in these studies include inflammatory bowel disease, multiple sclerosis, systemic lupus erythematosus, and type 1 diabetes in mice.[14, 71]

OVERVIEW

Living organisms have mechanisms to recognize and protect against invasion by foreign cells or organisms (nonself), that is, have some degree of immunity. Many such mechanisms do not depend on prior exposure to the invader and are, therefore, *innate*. Jawed vertebrates evolved abilities to recognize and repel *specific* molecular patterns on an invader which become stronger on repeated exposure: *adaptive* immunity. Innate and adaptive mechanisms strongly interact in vertebrates.

Immune cells comunicate by means of *cytokines,* the binding of which to cytokine receptors on a cell surface result in a cascade of reactions that stimulate many defense

Table 3.4	**Comparison of Main Evasion Mechanisms for Selected Protozoan Parasites**[a]

Parasite (Disease)	**Main Strategies of Evasion**	**Result**
Plasmodium falciparum (malaria)	Antigenic variation and/or polymorphisms	Evades the IR
	Induction of blocking antibodies	Blocks binding of real inhibitory antibodies
	Molecular mimicry	Alters immune recognition
	Anergy of T cells	Immunosuppression
	Altered peptide ligand	Alters functions of memory T cells
Trypanosoma brucei (African trypanoso-miasis or sleeping sickness)	Antigenic variation by VSG	Evades previously established IR
	Alteration of T- and B-cell populations	Immunosuppression
	Abnormal activation of macrophages	Impairs macrophage functions
	Induction of changes in pattern of cytokines released by CD8$^+$ T cells	Increases IFN-γ and decreases IL-2 and IL-2R; renders T cells unresponsive
	Production of a gp63-like protein	Resists complement
Trypanosoma cruzi (Chagas' disease)	Increased phagocytic activity	More CD8$^+$ T cells and reduced TDR and TIR
	Parasite mucin that binds to macrophages	Impairs macrophage functions
	Anergy of T cells	Immunosuppression
	Production of blocking IgM antibodies	Blocks binding of real inhibitory antibodies
	Turnover of surface molecules, phospholipases, and complement-regulating factors	Resists complement
Entamoeba histolytica (intestinal and liver amebiasis)	Cytolytic capacity	Damages host cells and tissues, interfering with IR
	Degradation of antibodies by proteases	Evades humoral immunity
	Acquisition of complement-regulating factors; shedding of immune complexes by capping and inactivation of complement components	Resists complement and protects against inflammatory response
	Anergy of T cells	Immunosuppression
	Release of products (MLIF and others) that act on macrophages; produces PGE2	Impairs macrophage function
	Induction of IL-4 and IL-10	Modulates the T_H1 response
Leishmania parasites (leishmaniasis: cutaneous, mucocutaneous, and visceral)	Inhibition of phagolysosome formation and proteolytic enzymes from lysosome	Evades macrophage proteolytic processes
	Abnormal activation of protein kinase C and scavenging of ROIs	Inhibits respiratory burst
	Shedding of MAC and some MAC components	Resists lysis by complement
	Represses MHC II gene expression	Prevents antigen presentation
	Inhibits production of IL-1, TNF, IL-12, IL-6, and various chemokines	Suppresses inflammation; blocks T_H1 response
	Induces production of TGF-β and IL-10	Immunosuppression
	Interferes with intracellular signaling, including a JAK/STAT pathway downstream from the IFN-γ receptor	Repression of IFN-inducible genes

[a]Abbreviations: IFN-γ, interferon-γ; IL, interleukin; IL-2R, interleukin-2 receptor; IR, immune response; MAC, membrane attack complex; MHC, major histocompatibility complex; MLIF, monocyte locomotion inhibition factor; PGE2, prostaglandin E2; TIR, thymus independent response; TNF-αR, tumor necrosis factor a receptor; VSG, variant surface glycoprotein

From S. Zambrano-Villa et al., How protozoan parasites evade the immune response, in *Trends in Parasitol.* 18:272–278, 2002; and M. Olivier et al., Subversion mechanisms by which *Leishmania* parasites can escape the host immune response: a signaling point of view, in *Clinical Microbiol. Rev.* 18:293–305, 2005.

responses by that cell. In many cases innate immunity is mediated by any of various *pattern recognition receptors* on host cells that recognize and bind with *pathogen-associated molecular patterns* on invading cells. Often this cascade results in production of antimicrobial peptides. *Phagocytosis,* engulfment and killing or digestion of invading particles, is an important component of both innate and adaptive immunity.

Adaptive immune responses result from exposure to *antigens,* which are most commonly proteins foreign to the host. The two types of antigen recognition molecules are *T-cell receptors* (found only on the surface of T cells) and *antibodies* (found on the surface of B cells and dissolved in the blood). Foreign proteins and cells are distinguished from a host's own cells by surface proteins encoded by genes of the *major histocompatibility complex (MHC)*. MHC I proteins differ in unrelated individuals and are found on all cells in the body, but MHC II proteins play a role in the immune response and are found only on certain cells with an immune role.

Arms of adaptive responses are T_H1 (based on T-cell receptors) and T_H2 (based on antibody). When one arm is active in a given response, the other arm tends to be downregulated. Cells activated in the two arms of adaptive responses defend a host in a variety of ways, including enhanced inflammatory reactions mediated by inflammatory cytokines. Inflammation is basically a manifestation of innate immunity, but the process is greatly affected by the past exposure of a host to antigens involved. Despite our need for the processes of inflammation in defense against disease agents, pathogenesis of some important parasitic diseases is a result of excessive inflammation. These diseases include malaria, schistosomiasis, filariasis, and onchocerciasis.

Parasites could not survive in their hosts if they could not evade the defenses mounted by immune responses. Mechanisms vary widely, including secretion of antiinflammatory agents, immunosuppresants, and enzymes to cleave ROIs and antibodies.

Learning Outcomes

By the time a student has finished studying this chapter, he/she should be able to

1. Draw and label a sketch of an antibody molecule and explain the role played by the various parts of this molecule.

2. Define the terms *cytokine* and *cytokine receptor* and explain the role that these molecules play in the immune response.

3. Explain the difference between humoral and cellular responses, using vocabulary that refers to cell types and cellular functions associated with these two types of responses.

4. Answer the question: How does an ELISA test work and what does it reveal?

5. Explain the mechanisms by which some parasites are able to evade host immune response.

References

References for superscripts in the text can be found at the following Internet site: www.mhhe.com/robertsjanovynadler9e

Additional Readings

Bourke, C. D., R. M. Maizels, and F. Mutapi. 2011. Acquired immune heterogeneity and its sources in human helminth infection. *Parasitol.* 138:139–159.

Carlier, Y., C. Truyens, P. Deloron, and F. Peyron. 2012. Congenital parasitic infections: a review. *Acta Tropica* 121:55–70.

Cox, F. E. G., and E. Y. Liew. 1992. T-cell subsets and cytokines in parasitic infections. *Parasitol. Today* 8:371–374. Very good diagrammatic summary of cytokine action.

Desowitz, R. S. 1987. *The thorn in the starfish. The immune system and how it works.* New York: W. W. Norton & Co. Principles of immunity told in Desowitz's inimitable style. Worthwhile reading, despite being out-of-date now.

Djuardi, Y., L. J. Wammes, T. Supali, E. Sartono, and M. Yazdanbakhsh. 2011. Immunological footprint: the development of a child's immune system in environments rich in microorganisms and parasites. *Parasitol.* 138 (sp.issu.SI):1508–1518.

Evering, T., and L. M. Weiss. 2006. The immunology of parasite infections in immunocompromised hosts. *Parasite Immunol.* 7:1379–1386.

Stephen, L. S. 1987. *The impact of helminth infections on human nutrition. Schistosomes and soil-transmitted helminthes.* London: Taylor and Francis.

Velavan, T. P., and O. Ojurongbe. 2011. Regulatory T Cells and Parasites. *J. Biomed. Biotechnol.* Article no. 520940.

Chapter 4

Parasitic Protozoa: Form, Function, and Classification

My excrement being so thin, I was at divers times persuaded to examine it; and each time I kept in mind what food I had eaten, and what drink I had drunk, and what I found afterwards. I have sometimes seen animalcules a-moving very prettily . . . .

—A. van Leeuwenhoek (November 4, 1681)

Because of their small size, heterotrophic, eukaryotic microorganisms were not detected until Antony van Leeuwenhoek developed his microscopes in the 17th century. He recounted his discoveries to the Royal Society of London in a series of letters covering a period between 1674 and 1716. Among his observations were oocysts in the livers of rabbits, a species known today as *Eimeria stiedai* (see p. 131). Another 154 years passed before a second apicomplexan was found, when in 1828 Delfour described gregarines from the intestine of beetles. Leeuwenhoek also observed *Giardia duodenalis* in his own diarrheic stools, and he discovered *Opalina* and *Nyctotherus* species in frog intestines. By mid-8th century other species were being reported at a rapid rate, and such discoveries have continued unabated to the present. Parasitic protozoa still kill, mutilate, and debilitate more people in the world than do any other group of disease organisms. For this reason, studies on these parasites occupy a prominent place in the history of parasitology.

The word Protozoa was once a phylum name, but today the term is used as a common noun referring to a number of phyla. Several other nouns, such as Archaezoa, Protoctista, and Protista, have been used to refer to this group of microscopic creatures. However, none of these terms, even when used as a taxon name, implies monophyly.[16, 20] Ultrastructural research and the accompanying life cycle and molecular work have shown that organisms once thought to be basically similar are in fact highly diverse and are organized structurally along a number of distinct lines. Thus, most current texts list at least seven phyla of protozoa, and some list over 30 phyla.[24] Our choice of the word *protozoa* as a common noun follows the practice of two recent sources, namely Hausmann and Hülsmann[16] and Lee et al.[20] Both of these references provide critical examinations of classification schemes, their basis, and their utility. All such schemes are plagued with uncertainty, and terms such as Protista and Protoctista are no more indicative of common ancestry than the familiar word protozoa. The classification section at the end of this chapter has a more detailed discussion of current taxonomic issues involving eukaryotic microorganisms.

FORM AND FUNCTION

Protozoa consist of a single cell, although many species contain more than one nucleus during all or portions of their life cycles. By mid-19th century, many protozoan genera had been described, and their enormous structural diversity, complexity, and even beauty were widely recognized. Early electron microscopists found unicellular eukaryotes fascinating subjects, and soon after World War II, researchers recognized that the group was a structurally complex and heterogenous assemblage whose members did not all conform to a single body plan (Figs. 4.1, 4.2). In 1980 a committee of the Society of Protozoologists revised the classification, recognizing seven phyla, and further revisions were recommended in 1985.[21] More recent classifications, incorporating molecular data, propose groupings that seem quite contrary to those of older systems. An example of the latter is superphylum Alveolata, which includes dinoflagellates, phylum Apicomplexa (coccidia and malarial parasites, chapters 8 and 9), phylum Ciliophora (ciliates, chapter 10), and Haplosporidia.[16, 29]

Ultrastructural studies have shown that regardless of how elaborate or elaborately arranged protozoa are, most components of their organelles do not differ in any basic way from those of metazoan cells.[16] Indeed, Pitelka[27] concluded "that the fine structure of protozoa is directly and inescapably comparable with that of cells of multicellular organisms." Much of the apparent upheaval in eukaryotic systematics is a result of such admission, with the distinctions between unicellular and multicellular organisms becoming quite blurred at the ultrastructural and molecular levels.[5, 16] On the other hand, to a student who first encounters them, protozoan

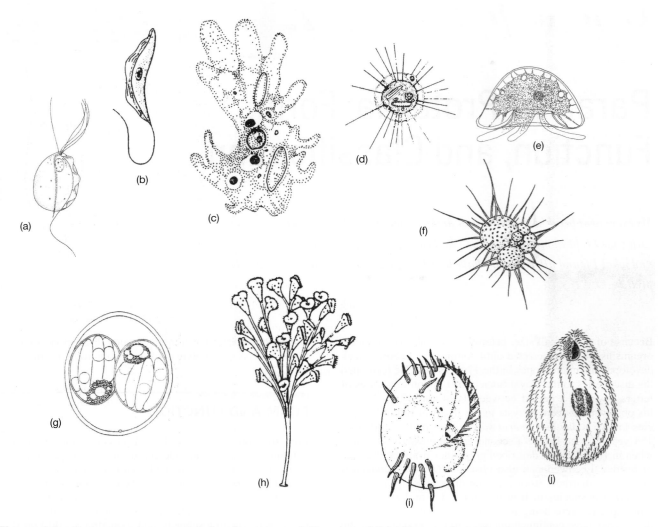

Figure 4.1 **Representative protozoa showing structural diversity exhibited by members of various groups.**
The organisms are not drawn to scale, nor are they all the same life-cycle stages. (*a*) *Pentatrichomonas hominis*, 8–20 μm long, a harmless commensal of the human digestive tract. (*b*) A species of *Trypanosoma*, 15–30 μm, from the bloodstream of vertebrates (both *a* and *b* have undulating membranes). (*c*) Free-living *Amoeba* sp., 100–150 μm, showing lobopodia. (*d*) *Actinosphaerium* sp., 200 μm (many species are much smaller), with actinopodia. (*e*) *Arcella vulgaris,* a freshwater shelled ameba, about 100 μm with lobopodia. (*f*) *Globigerina* sp., a marine foraminiferan up to 800 μm, with filopodia. (*g*) Oocyst of *Levineia canis,* (35–42) × (27–33) μm, a coccidian parasite of dogs. (*h*) *Zoothamnium* sp. colony, individuals 50–60 μm, colony up to 2 mm tall, an obligate ectocommensal ciliate of aquatic invertebrates. (*i*) *Euplotes* sp., 100–170 μm, a free-living ciliate with ventral cirri and prominent oral membranes. (*j*) *Tetrahymena* sp., ~60 μm, a free-living ciliate showing ciliary rows (kineties).

(*a*) drawn by William Ober. (*c* and *d*) Adapted from R. Kudo, *Protozoology* (5th ed.) 1966. Charles C. Thomas Publishers, Springfield, IL. (*e*) drawn by John Janovy, Jr. (*g*) From N. D. Levine and V. Ivens, "*Isospora* species in the dog" in *J. Parasitol.*, 51:859–864. Copyright © 1965. Reprinted by permission of the publisher.

structures can seem bizarre, often multiple versions and arrangements of the familiar organelles studied in introductory biology. If it seems like the words *may, usually, typically,* and *often* occur more frequently in this chapter than in others, then such use is a reflection of the great structural diversity found in single-celled eukaryotes (Figs. 4.1, 4.2).

Nucleus and Cytoplasm

Like all cells, the bodies of protozoa are covered by a **plasma membrane,** which is the lipid bilayer, fluid mosaic, described in introductory texts. Many protozoa have more

than one such membrane as part of their **pellicle.** Additional membranes may be present as **alveoli,** or sacs, which in some ciliates are enlarged, producing ridges and craters on the cell surface (see Fig. 4.6). Protozoa may also possess a thick **glycocalyx,** or *glycoprotein* surface coat, which, in the case of parasitic forms, has immunological importance (see chapter 5). Other membrane proteins may serve as binding sites that function during uptake of intracellular parasites by host cells.

Pellicular microtubules may course just beneath the plasma membrane, the number and arrangement of such tubules being typical of a group. The pellicle may be thrown into more or less permanent folds, supported by

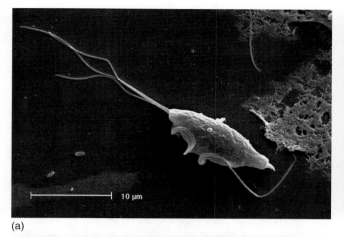

(a)

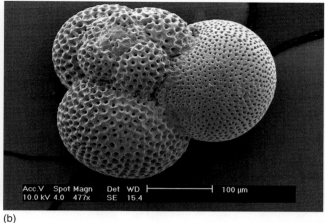

(b)

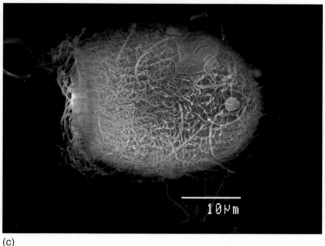

(c)

Figure 4.2   **Structural diversity of surface features in some representative protozoa.**

(*a*) *Tritrichomonas augusta,* a parasite of amphibians, from culture, showing three anterior flagella, undulating membrane, trailing flagellum, and axostyle. (*b*) A foraminferan skeleton (spiral side), genus *Subbotina,* from Eocene deposits in Tanzania. (*c*) *Balanion masanensis,* a marine ciliate from Korea, showing kineties and structural complexity of the cell surface. All figures are scanning electron micrographs. Bar in (*b*) = 1 mm.

(*a*) Courtesy of Geraldo A. De Carli; (*b*) courtesy of Paul Pearson; (*c*) courtesy of H. J. Jeong.

microtubules, as in gregarine parasites of insects (Fig. 4.3). Or such microtubules may underlie a flexible membrane, as in kinetoplastid flagellates (see Fig. 5.2, p. 62). The structural elaboration of membranes, through folding and addition of electron-dense materials, also occurs in tissue-dwelling cysts (see Fig. 8.4). Adjoining membranes may have an electrondense or fibrous connection between them, such as that between the body and **undulating membrane** of trypanosomes and trichomonads (see Figs. 4.1, 5.2, and 6.12).

Mitochondria, organelles that bear enzymes of oxidative phosphorylation and the tricarboxylic acid cycle, often have tubular rather than lamellar cristae. In addition, some amebas have branched tubular cristae, but in other protozoan groups cristae may be absent altogether. Mitochondria may be present as a single, large body, as in some flagellates, or arranged as elongated, sausage-shaped structures, as occur in pellicular ridges of some ciliates.

The **Golgi apparatus (dictyosome)** is quite elaborate in some flagellates, occurring as large and/or multiple **parabasal bodies** in association with **kinetosomes,** the "basal bodies" of flagella (Figs. 4.4 and 4.5), and is present, although not always as prominent, in amebas and ciliates. Dictyosomes can play diverse roles in the lives of protozoa—for example, they can be the source of skeletal plates in some amebas and polar filaments in microsporidian parasites.

Microbodies are usually, but not always, spherical membrane-bound structures with a dense, granular matrix.[10]

In most animal and many plant cells, microbodies contain oxidases and catalase. The oxidases reduce oxygen to hydrogen peroxide, and catalase decomposes hydrogen peroxide to water and oxygen. Microbodies in these cells are called **peroxisomes** because of this biochemical activity. Peroxisomes are found in many aerobic protozoa in which oxygen is a terminal electron acceptor in metabolism.[26] In at least some anaerobes, such as parasitic *Trichomonas* spp., microbodies produce molecular hydrogen and are called **hydrogenosomes** (see Fig. 4.5). Microbodies may also contain enzymes of the glyoxylate cycle, a series of reactions that function in the synthesis of carbohydrate from fat. Microbodies of Kinetoplastida are called **glycosomes** and contain most of the glycolytic enzymes (which in other eukaryotic cells are found in the cytosol).[25]

Other more unusual membrane-bound organelles include about a dozen kinds of **extrusomes,** which generally originate in the dictyosome and come to lie beneath the cell membrane. Upon proper stimulus extrusomes fuse with the cell membrane, releasing their contents to the exterior. Extrusomes as **toxosomes** may release toxic substances, evidently as a defensive mechanism,[16] or function as **kinetocysts** in food capture, as **haptocysts** to paralyze prey, or as **trichocysts** in mechanical resistance to predators. The dark (electron-dense), elongated bodies perpendicular to the cell membrane in Figure 4.3*b* are **mucocysts** of a parasitic ciliate, *Ichthyophthirius multifiliis.* Mucocysts are thought to provide

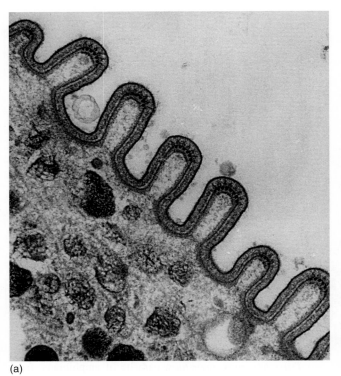

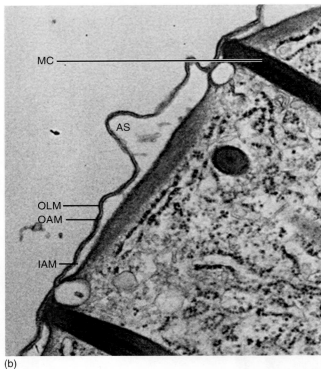

(a)

(b)

Figure 4.3   **Plasma membranes and their modifications in protozoa.**

(*a*) Epicytic folds of a gregarine parasite of damselflies. These folds extend along the body as ridges. (*b*) Membranes of *Ichthyophthirius multifiliis,* a parasite of fishes; the dark elongate bodies perpendicular to the membranes are mucocysts (*AS,* alveolar sac; *OLM,* outer limiting membrane; *OAM,* outer alveolar membrane; *IAM,* inner alveolar membrane; *MC,* mucocyst).

a coating that protects the cell against osmotic shock.[7] Not all extrusomes, however, have obvious functions.[16]

The cytoplasmic matrix consists of very small granules and filaments suspended in a low-density medium with the physical properties of a colloid; that is, with the capability of existing in a relatively fluid (sol state) or relatively solid (gel state) condition. Central and peripheral zones of cytoplasm can often be distinguished as **endoplasm** and **ectoplasm.** Endoplasm is in the sol state, and it bears the nucleus, mitochondria, Golgi bodies, and so on. Ectoplasm is often in the gel state; under the light microscope it appears more transparent than sol, and in this physical state cytoplasm functions to maintain cell shape.

Protozoa, like fungi, plants, and animals, are **eukaryotes;** that is, their genetic material—**deoxyribonucleic acid (DNA)**—is carried on well-defined **chromosomes** combined with basic proteins called **histones,** and the chromosomes are contained within a membrane-bound **nucleus.** At the light microscope level, protozoan nuclei are typically oval, discoid, or round, and they are usually vesicular, with an irregular distribution of chromatin material and "clear" areas in the nuclear sap. But in ciliates, which contain at least one micronucleus and one macronucleus, the latter may be dense, elongated, chainlike, or branched. Micronuclei are reproductive nuclei, undergoing meiosis prior to sexual reproduction (conjugation). Macronuclei are considered "somatic"; they function in cell metabolism and growth but do not undergo meiosis.

In electron micrographs nucleoplasm appears finely granular, with aggregations of dense chromatin. Chromosomes may remain as recognizable bodies throughout the cell cycle. Nucleoli are usually present, but they typically disappear during nuclear division. **Endosomes,** conspicuous internal bodies, are nucleoli, although they do not disappear during mitosis. Parasitic amebas and trypanosomes have endosomes. The term *endosome* may also be used in reference to vesicles arising by endocytosis.[16]

The **nuclear envelope** is similar to that of most eukaryotic cells, consisting of two membranes that fuse in the region of pores, but the envelope may be thickened by a fibrous layer or have strange honeycomblike tubes on the outer or inner face. The nuclear envelope may or may not persist during mitosis, again depending on the species, and mitotic spindles can be intra- or extranuclear.

Locomotor Organelles

Protozoa move by three basic types of organelles: pseudopodia, flagella, and cilia; flagella and cilia are also called **undulipodia.** Some amebas possess both flagella and pseudopodia, although transformation from flagellated to ameboid cell occurs in response to environmental conditions and is a recognized lifecycle event. Flagella may also occur in large numbers and in rows, thus superficially resembling

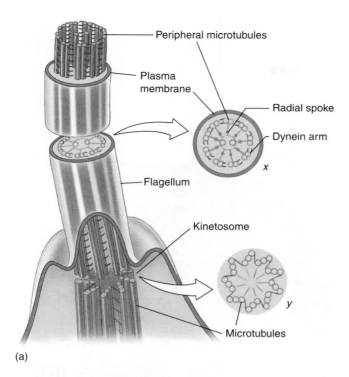

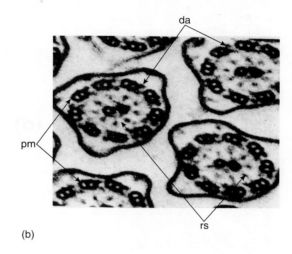

(a)

(b)

Figure 4.4   **Flagella (undulipodia).**

(*a*) General structure of a cilium or flagellum, showing a section through the axoneme within the cell membrane and a section through the kinetosome. The nine pairs of microtubules plus the central pair make up the axoneme. The central pair ends at about the level of the cell surface in a basal plate. Peripheral microtubules continue beneath the cell surface to compose two of each of the triplets in the kinetosome (or basal body, level y). (*b*) Electron micrograph of a section through several flagella, corresponding to level *x* in (*a*); *da,* dynein arm; *pm,* peripheral microtubules; *rs,* radial spoke.

cilia. In Ciliophora the cilia bases are connected by a complex fibrous network, or **infraciliature** (see below).

Flagella are slender, whiplike undulipodia, each composed of a central **axoneme** and an outer sheath that is a continuation of the cell membrane (Fig. 4.4). An axoneme consists of nine peripheral and one central pair of microtubules (the nine-plus-two arrangement found in cilia and flagella throughout the animal kingdom with a few exceptions). Central microtubules are singlets, but peripheral ones are often doublets or even doublets "with arms." The central two microtubules are bilateral, and peripheral ones can thus be numbered with reference to a plane perpendicular to the line between the central pair. The axoneme arises from a **kinetosome (basal body),** which is ultrastructurally indistinguishable from centrioles of other eukaryotic cells, being made up of the nine peripheral elements, typically microtubule triplets arranged in a cartwheel manner. Kinetosomes may lie at the bottom of **flagellar pockets** or **reservoirs** of differing depths, depending on the species. When a flagellate has at least two flagella with differing structures, the condition is termed **heterokont.**

The entire unit—flagellum, kinetosome, and associated organelles—is called a **mastigont** or a **mastigont system** (see Figs. 4.4, and 4.5). Kinetosomes are more or less fixed in position relative to other organelles; thus, flagella may be directed anteriorly, laterally, or posteriorly, independent of their movements. Most flagellates have more than one flagellum, and these may be inserted into the cell at different angles. The flagellum may also be bent back along and loosely attached to the lateral cell surface, forming a finlike **undulating membrane,** which may be an adaptation to life in relatively viscous environments.[16] Flagellar movements are generally helical waves that begin at either the base or tip, pushing fluids along the flagellar axis. The resulting body movement may be fast or slow, forward, backward, lateral, or spiral. In some cases, such as with trichomonad parasites, movement is highly characteristic and recognized instantly by most parasitologists who have previously studied these flagellates in fresh intestinal contents.

A mastigont system may also include a prominent, striated rod, or **costa,** that courses from one of the kinetosomes, under the pellicle and just beneath the recurrent flagellum and undulating membrane. A tubelike **axostyle,** formed by a sheet of microtubules, may run from the area of the kinetosomes to the posterior end, where it may protrude. In phylum Parabasalia, kinetosomes of the three anteriorly directed flagella are numbered 1, 2, and 3, and have **lamina** (sheets) of microtubules that in cross sections appear either as hooks (kinetosomes 1 and 3) or as sigmoid profiles (kinetosome 2). A Golgi body (dictyosome) may be present; if a periodic fibril, or **parabasal filament,** runs from the Golgi body to contact a kinetosome, the Golgi body is referred to as a **parabasal body.** A fibril running from a kinetosome to a point near the surface of the nuclear membrane is called a **rhizoplast,** and the entire complex of organelles and an associated nucleus is thus referred to as a **karyomastigont.**

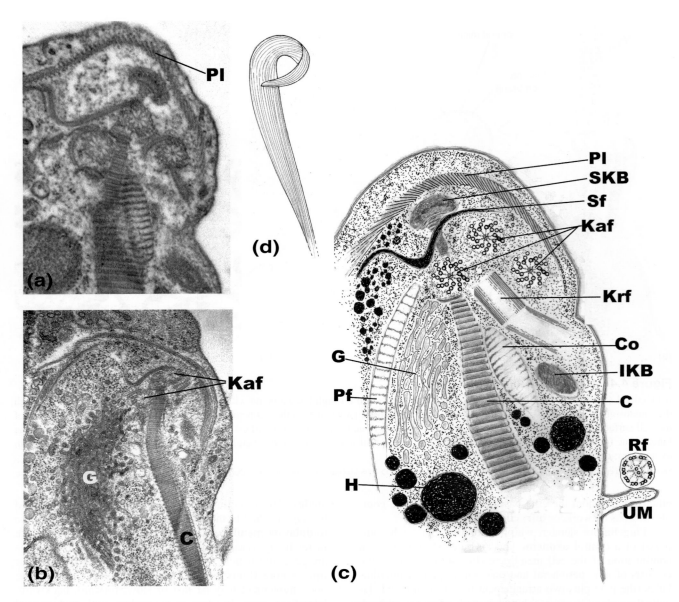

Figure 4.5   **Complex mastigont system as seen in trichomonad flagellates (see also chapter 6).**

(*a*) Anterior end of *Tritrichomonas foetus* from cattle. (*b*) Anterior end of *Tritrichomonas mobilensis,* a flagellate from squirrel monkeys. (*c*) Interpretive drawing, showing typical mastigont structures seen in trichomonads. (*d*) Three-dimensional view of the pelta, a curved sheet of microtubules that extends posteriorly to become the axostyle (see also Fig. 6.12). C, costa; Co, comb; G, Golgi body; H, hydrogenosomes; IKB, infrakinetosomal body; Kaf, kinetosomes of anterior flagella; Krf, kinetosome of recurrent flagellum; Pf, parabasal filament; Pl, pelta; sf, sigmoidal filament; Rf, recurrent flagellum; SKB, suprakinetosomal body; UM, undulating membrane.

(*a*), (*b*) photographs courtesy of Marlene Benchimol; (*c*), (*d*) drawings by John Janovy, Jr.

In class Kinetoplastida, which includes trypanosomes (chapter 5), a dark-staining body or **kinetoplast** is found near the kinetosome (see Fig. 5.1). The kinetoplast is actually a disc made of DNA circles, called kDNA, located within a single large mitochondrion. kDNA has different genetic properties from nuclear DNA. Kinetoplastids also have a **paraxial (crystalline rod)** that lies alongside the axoneme, within the flagellum. And, finally, many free-living flagellates possess fine fringes or hairlike **mastigonemes** on their flagella, making them look like motile test-tube brushes in electron micrographs. Tubular flagellar hairs with three fine filaments at the tip are a structural character that unites the so-called "stramenopiles" (see following taxonomic section). Students interested in evolutionary biology might find their life's work in trying to explain the origin of all this subcellular complexity; those intrigued by cellular function will discover an equally challenging task.

Cilia (also undulipodia) are structurally similar to flagella, with a kinetosome and an axoneme composed of two central and nine peripheral microtubules. Cilia typically appear to beat regularly, with a back-and-forth stroke in a twodimensional plane, whereas flagella often appear to beat irregularly, turning and coiling in a three-dimensional space. However, cilia may beat in a helical movement, some flagella beat in a plane, and both types of undulipodia beat in metachronal waves, reminiscent of a field of waving grain, when they occur in large numbers.[16]

In ciliates, body cilia (**somatic ciliature**) are arranged in rows, known as **kineties,** which in turn are composed of **kinetids,** the basic units of ciliate pellicular organization. *Monokinetids* contain a single kinetosome and associated fibers; *dikinetids* contain a pair of kinetosomes; and so on. The pellicle of *Dexiotricha media,* a ciliate found in an Illinois pig wallow, is simple enough to serve as an introduction to ciliate organization (Fig. 4.6). A kinetid consists of the kinetosome; a small membranous pocket, the **parasomal sac;** and a number of fibers or sheets, made from microtubules, that extend in various directions from the kinetosome. A tapering banded fiber, the **kinetodesma** (plural *kinetodesmata*), arises from the clockwise side of each kinetosome (when viewed from the anterior end of the cell), courses anteriorly, and joins a similar fiber from the adjoining cilium in the same row. The resulting compound fiber of kinetodesmata is called a **kinetodesmose.** Flat sheets of microtubules, the **postciliary microtubules,** run posteriorly from each kinetosome, and similarly constructed bands, the **transverse microtubules,** lie perpendicular to kineties. Kinetosomes and associated fibrils constitute the **infraciliature.** Ciliates differ significantly in the structure of their infraciliature, and such differences are of major taxonomic importance. Obviously, this great diversity in structure is assumed to reflect an equal diversity in function, but we still do not have much knowledge about the details.

Oral ciliature can be amazingly complex and is an outstanding example of the elaboration of familiar organelles (see *Euplotes* sp. in Fig. 4.1). **Oral membranes** are actually **polykinetids;** that is, fields or rows of cilia and their kinetosomes linked by electron dense fibrous networks. The **adoral zone of membranelles** is a series of such oral membranes located to the "left" of or counterclockwise from the side of the oral area of the more complex ciliates. Polykinetids may also be found on the body as **cirri** (singular **cirrus**), tufts of cilia that function together, usually in locomotion along a substrate. Much of the wonder that ciliates seem to produce in students comes from the action of polykinetids; for example, the "walking" motion of cirri tends to make the organism look as if it is behaving in a rather purposeful way. A group of kinetosomes forming a tuft of ciliary organelles in the aboral region of peritrich ciliates is called the **scopula.** It is involved in stalk formation.

Cilia beat with a powerful backstroke, pushing the surrounding fluid posteriorly, in metachronal waves. Membranelles have their own beat cycles that are usually independent of the somatic ciliature. When ciliates divide, the ciliature is reorganized according to a precise sequence of events. Reorganization of oral polykinetids is a complex process. This "embryological development" has been the basis for much of the class level taxonomy in the Ciliophora, but current classifications also rely heavily on ultrastructural details of body ciliature.

The mechanism by which flagella and cilia move requires ATP and involves the interaction of the arms of each microtubule pair (see Fig. 4.4) with the neighboring pair of microtubules. A motor protein, **ciliary dynein,** which is also an ATPase, functions to make the peripheral microtubule pairs slide back and forth relative to one another, producing

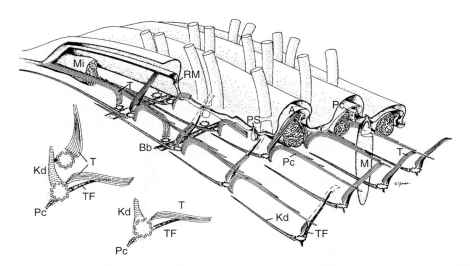

Figure 4.6   A diagram of the structure of a ciliate cortex (*Dexiotricha media*), reconstructed from electron micrographs, illustrating the relationships between the various elements of the ciliate cortex.

A, alveolar sac; *Bb,* basal filamentous bundle of fibers; *Kd,* kinetodesmata; *M,* mucocyst; *Mi,* sausage-shaped mitochondrion; *Pc,* postciliary microtubular ribbons; *PS,* parasomal sac; *RM,* single microtubule running through a pellicular ridge; *T,* transverse microtubule ribbon; *TF,* transverse fiber. The anterior end of the cell is to the upper left.

From R. K. Peck, "Cortical ultrastructure of the scuticocilates *Dexiotricha media* and *Dexiotricha colpidiopsis* (Hymenostomata)," in *J. Protozool.* 24:122–134, 1977. Copyright © 1977. The Society of Protozoologists. Reprinted by permission.

ciliary or flagellar movement. For reviews of this movement see Lindemann,[22] Sloboda,[33] and Vincensini et al.[34]

Pseudopodia are temporary extensions of the cell membrane and are found in amebas as well as in a variety of cell types in other organisms. Pseudopodia function in locomotion and feeding. In some amebas, movement is by flow of the entire body, with no definite extensions. Such amebas are called **limax forms** (see Fig. 7.9), after the slug genus *Limax.* Four general types of pseudopodia occur in amebas; three of these types are illustrated in Figure 4.1. **Lobopodia** are finger-shaped, round-tipped pseudopodia that usually contain both ectoplasm and endoplasm (Fig. 4.1(c)). Most free-living soil and freshwater amebas and all parasitic and commensal amebas of humans have this kind of pseudopodium. **Filopodia** (Fig. 4.1(f)) are slender, sharp-pointed organelles, composed only of ectoplasm. They are not branched like **rhizopodia,** which branch extensively and may fuse together to form netlike meshes. **Axopodia** (Fig. 4.1(d)) are like filopodia, but each contains a slender axial filament composed of microtubules that extends into the interior of the cell. Both pseudopod shape and the shapes of **uroids** (membranous extensions at the posterior end of the cell) are taxonomic characters in amebas. Uroids may be bulbous, spiny, morulate (like a grape cluster), or papillate.

Movement by means of pseudopodia is a complex form of protoplasmic streaming involving protrusion of the cell, adhesion to substrate, and subsequent contraction. Bereiter-Hahn[4] and Condeelis[8] give excellent reviews of the signaling systems and protein interactions involved in cell crawling. Evidence suggests that the mechanism requires coordinated structural modification, polymerization, and crosslinking of actin filaments, myosin-mediated filament sliding, adhesion, and deadhesion.[3, 8]

Although protoplasmic streaming is well studied, the mechanisms that determine pseudopod shape are not known. Amebas obviously have some characteristics that function to produce extensions of plasma membrane that are indeed temporary but are also consistent enough in structure so that they may be used in identification and classification. Pseudopod formation is certainly no less wondrous than polykinetid function.

In many apicomplexans (gregarines, coccidia, and malaria parasites, chapters 8 and 9), the merozoites, ookinetes, and sporozoites appear to glide through fluids with no subcellular motion whatever.[23] Gregarines (p. 121), for example, exhibit a variety of slow, sometimes almost snake-like movements, depending on the species and the kind of fresh tissue preparation that is examined. Electron microscope studies reveal longitudinal pellicular ridges (**epicytic folds**) on these cells, which often appear to have been fixed in the process of forming an undulatory wave. Subpellicular microtubules are found in the folds, and it has been proposed that these fibers function in the gliding locomotion (see Fig. 4.3). Experimental work, however, reveals that contact with a substrate is essential to gregarine movement and suggests that mucous secretion may also play a role in locomotion.[23]

Reproduction and Life Cycles

Protozoan reproduction may be either asexual or sexual, although many species alternate the two types in their life cycles or perform one or the other reproductive functions in response to environmental conditions. Most often asexual reproduction is by **binary fission,** in which one individual divides into two. The plane of fission is random in amebas, longitudinal in flagellates (between kinetosomes or flagellar rows; that is, **symmetrogenic**), and transverse in ciliates (across kineties, or **homothetogenic**). The sequence of division is (1) kinetosome(s), (2) kinetoplast (if present), (3) nucleus, and (4) cytokinesis.

Nuclear division during asexual reproduction is by mitosis, except in macronuclei of ciliates, which are highly polyploid and divide amitotically. However, patterns of mitosis are much more diverse among unicellular eukaryotes than among metazoa. An inventory of these patterns is beyond the scope of this book, but examples include nuclear membranes that persist through mitosis, spindle fibers that form within the nuclear membrane, missing centrioles, and chromosomes that may not go through a well-defined cycle of condensation and decondensation. Nevertheless, the essential features of mitosis—replication of chromosomes and regular distribution of daughter chromosomes to daughter nuclei—are always present.

Multiple fission (merogony, schizogony) occurs in some amebas and in Apicomplexa. In this type of division the nucleus and other essential organelles divide repeatedly before cytokinesis. Thus, a large number of daughter cells are produced almost simultaneously and are presumably in the same or similar physiological condition. Cells undergoing schizogony are called **schizonts, meronts,** or **segmenters.** Depending on the species, schizont daughter nuclei may arrange themselves peripherally, with membranes of daughter cells forming beneath the cell surface of the mother cell (Fig. 4.7). Daughter cells are **merozoites,** and they eventually break away from a small residual mass of protoplasm remaining from the mother cell to initiate another phase of merogony (schizogony producing more asexually reproducing merozoites) or to begin **gametogony** (gametocyte formation).

Another type of multiple fission often recognized is **sporogony,** which is meiosis immediately after the union of gametes, typically followed by mitosis. The products of merogony are additional parasites of the same life-cycle stage, such as those that invade red blood cells during a malarial infection. The products of sporogony, however, are usually of a completely different life-cycle stage, such as sporozoites in resistant oocysts ("spores") of gregarines.

Several forms of **budding** can be distinguished. In **plasmotomy,** sometimes regarded as budding, a multinucleate individual divides into two or more smaller but still multinucleate daughter cells. Plasmotomy itself is not accompanied by mitosis. *External budding* is found among some ciliates, such as suctorians. Here nuclear division is followed by unequal cytokinesis, resulting in a smaller daughter cell, which then swims away from the sessile parent and subsequently settles, metamorphoses, and grows to its adult size. *Internal budding,* or **endopolyogeny,** differs from schizogony only in the location where daughter cells are formed. In this process daughter cells begin forming within their own cell membranes, distributed throughout the mother cell's cytoplasm rather than at the periphery. The process occurs in schizonts of some coccidians. **Endodyogeny** is endopolyogeny in which only two daughter cells are formed (Fig. 4.8). Protozoans, it seems, are as varied and elaborate in their asexual reproduction as they are in their structure.

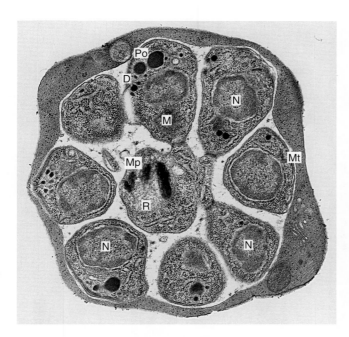

Figure 4.7 **Late stage in the development of**
***Plasmodium cathemerium* within the host erythrocyte.**
The segmentation has been almost completed, and paired organelles (*Po*), dense bodies (*D*), nucleus (*N*), mitochondrion (*M*),
pellicular complex with microtubules (*Mt*), and ribosomes are
observed in the new merozoites. A residual body (*R*) surrounded
by a rim of cytoplasm of the mother schizont contains a cluster
of malarial pigment (*Mp*) granules. (×30,000)

From M. Aikawa, "The fine structure of the erythrocytic stages of three avian
malarial parasites, *Plasmodium fallax, P. lophurae,* and *P. cathemerium*," in *Am. J.
Trop. Med. Hyg.* 15:449–471. Copyright © 1966.

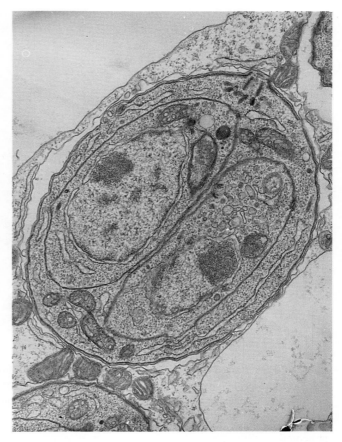

Figure 4.8 *Toxoplasma gondii* **exhibiting two daughter
cells in a mother cell, formed by endodyogeny.**

From E. Vivier and A. Petitprez, "Le complexe membranaire superficiel et son
evolution lors de l'elaboration des individus-fils chez *Toxoplasma gondii*," in
J. Cell Biol. 43:329–342, 1969.

Sexual reproduction involves reductional division in
meiosis, resulting in a change from diploidy to haploidy, with
a subsequent union of two cells to restore diploidy. Cells
that join to restore diploidy are **gametes,** and the process
of producing gametes is **gametogony.** Cells responsible for
gamete production are **gamonts** (Fig. 4.9). Reproduction may
be **amphimictic,** involving the union of gametes from two
parents, or **automictic,** in which one parent gives rise to both
gametes. Uniting gametes may be entire cells or only nuclei.
When gametes are whole cells, the union is called **syngamy.**

In syngamy, gametes may be outwardly similar (**isogametes**) or dissimilar (**anisogametes**). Although isogametes
look similar, they will fuse only with isogametes of another
"mating type." Anisogametes often differ in cytoplasmic contents, in size (sometimes markedly), and in surface proteins
that determine mating type. The larger, more quiescent of the
pair is a **macrogamete;** the smaller, more active partner is a
microgamete. It is tempting to call these forms *female* and
male, respectively, but it is debatable whether gender, in the
commonly used sense, can or even should be distinguished
in protozoa. Fusion of a microgamete and macrogamete produces a **zygote,** which may be a resting stage that overwinters
or forms spores that enable survival between hosts.

Conjugation, in which only nuclei unite, is found only
among ciliates, whereas syngamy occurs in all other groups

in which sexual reproduction is found. Two individuals
ready for conjugation unite, and their pellicles fuse at the
point of contact. The macronucleus in each disintegrates, and
their micronuclei undergo meiotic divisions into four haploid
pronuclei (of which two degenerate). A **migratory pronucleus** from each conjugant passes into the other to fuse with
a **stationary pronucleus,** restoring the diploid condition.
The cells separate, and subsequent nuclear divisions produce
one or more macronuclei. The **exconjugants,** which are now
genetic recombinants, then actively reproduce by fission.

The details of conjugation, including exconjugants' relationships, extent of cytoplasmic sharing, and fate of exconjugants, vary widely among ciliates. Under natural conditions
conjugating pairs are seen occasionally, especially when
environmental conditions deteriorate. Clone cultures, descended from single individuals, can be prepared in the lab
and stressed to produce cells that are ready to conjugate and
will do so en masse when mixed with other clones (the mating type reaction). This technique has been useful in the study
of mating specificity, genetics, and surface protein function
in ciliates. Variations of conjugation are **cytogamy,** in which
two individuals fuse but do not exchange pronuclei, with
two pronuclei in each cell rejoining to restore diploidy, and
autogamy, in which haploid pronuclei from the same cell fuse
but there is no cytoplasmic fusion with another individual.

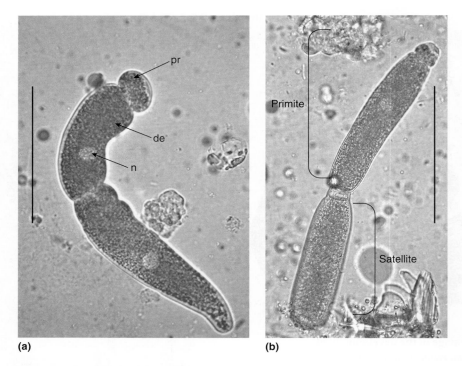

(a)   **(b)**

Figure 4.9   Paired gamonts of protozoan parasites from flour beetles.
(*a*) *Awrygregarina billmani* from *Tribolium brevicornis* larvae; (*b*) *Gregarina cloptoni from Tribolium freemani* larvae. Primite and satellite are the mated pair of gamonts; protomerite (pr) and deutomerite (de) are the cell compartments; n, nucleus. See chapter 8 for life cycle details. Bar = 100 μm for both figures.
Photographs by John Janovy, Jr.

In Apicomplexa, meiosis occurs in the first division of the zygote **(zygotic meiosis),**[15] and all other stages are haploid. **Intermediary meiosis,** which occurs only in the Foraminifera among protozoa but which is widespread in plants, results in a regular alternation of haploid and diploid generations.

Encystment

Many protozoa can secrete a resistant covering and enter a resting stage, or **cyst.** Cyst formation is particularly common among parasitic protozoa as well as among free-living protozoa found in temporary bodies of water that are subject to drying or other harsh conditions.[35] In addition to providing protection against unfavorable conditions, cysts may serve as sites for reorganization and nuclear division, followed by multiplication after excystation. In a few forms, such as *Ichthyophthirius multifiliis,* a ciliate parasite of fish, cysts fall from the host to the substrate and stick there until excystation occurs (chapter 10). Cellulose has been found in the cyst walls of some amebas, and others contain chitin.[2] Cysts can be highly complex and layered structures, as seen with an electron microscope in the filamentous cysts of *Giardia* species.[11] The outer layers may also react with immunodiagnostic reagents, although not always in a highly specific manner.[17]

Conditions favoring encystment are not fully understood, but they are thought in most cases to involve some adverse environmental events such as food deficiency, desiccation, increased tonicity, decreased oxygen concentration, or pH or temperature change. It is vitally important for parasitologists to understand the elusive factors that induce cyst formation within the host, the role that cysts play in completion of a parasite's life cycle, and factors that work to disseminate cysts. For example, human amebiasis, caused by *Entamoeba histolytica,* is spread by persons who often have no clinical symptoms but who pass cysts in their feces (chapter 7).

During encystment a cyst wall is secreted, and some food reserves, such as starch or glycogen, are stored. Projecting portions of locomotor organelles are partially or wholly resorbed, and certain other structures, such as contractile vacuoles, may be dedifferentiated. During the process or following soon thereafter, one or more nuclear divisions can give a cyst more nuclei than a trophozoite. In coccidians the cystic form is an **oocyst,** which is formed after gamete union and in which multiple fission **(sporogony)** occurs to produce **sporozoites.** In eimerian coccidians, oocysts containing sporozoites serve as resistant stages for transmission to new hosts, whereas in haemosporidians (including the causative agents of malaria, *Plasmodium* spp.) oocysts serve as developmental capsules for sporozoites within their insect host (see chapters 8 and 9).

In species in which the cyst is a resistant stage, a return of favorable conditions stimulates excystation. In parasitic forms some degree of specificity in the requisite stimuli provides that excystation will not take place except in the presence of conditions found in a host's gut. Mechanisms for excystation may include absorption of water with consequent swelling of the cyst, secretion of lytic enzymes by the protozoan, and action of host digestive enzymes on the cyst wall. Excystation must include reactivation of enzyme pathways that were "turned off" during the resting stage, internal reorganization, and redifferentiation of cytoplasmic and locomotor organelles.

Feeding and Metabolism

Some protozoa are photosynthetic and synthesize carbohydrates in chloroplasts, the organelles of "typical" plants. Such organisms are often considered algae, but some participate in symbiotic relationships of interest to parasitologists. Zooxanthellae (dinoflagellates) are very important mutuals living in cells of reef-forming corals and other invertebrates (including some other protozoa), contributing significant amounts of carbohydrates to their hosts. Students interested in the biochemistry or evolution of symbiosis can find a fertile field in the obligate relationships between animals and their algal symbionts.

Protozoa lacking chloroplasts are all **heterotrophic,** requiring their energy in the form of complex carbon molecules and their nitrogen in the form of a mixture of preformed amino acids. Protozoa are typically particle feeders—that is, grazers and predators—and many symbiotic species feed on host cells. Their mouth openings may be temporary, as in amebas, or permanent **cytostomes,** as in ciliates. A submicroscopic micropore is present in *Eimeria* and *Plasmodium* species and, in certain stages, is involved in taking in nutrients (Fig. 4.10).

Particulate food passes into a food vacuole, which is a digestive organelle that forms around any food thus ingested. Indigestible material is voided either through a temporary opening or through a permanent **cytopyge,** which is found in many ciliates. **Pinocytosis** is an important activity in many protozoa, as is **phagocytosis.** Both pinocytosis and phagocytosis are examples of **endocytosis,** differing only in that pinocytosis deals with droplets of fluid, whereas in phagocytosis, particulate matter is internalized.

Like most other eukaryotic cells, protozoa generally carry out the many reactions of glycolysis, Krebs (citric acid) cycle, pentose-phosphate shunt, electron transport, transaminations, lipid oxidations and syntheses, nucleic acid metabolism, and the multitude of other metabolic events that make biochemical pathways look like printed circuits of high-tech electronic equipment. ATP is the most common form of immediately usable energy, although a few parasites use inorganic pyrophosphate in a similar role. Polysaccharides, especially glycogen or related molecules, function as deep energy storage. Genes are transcribed in the nucleus, and polypeptides are synthesized on ribosomes, as in other cells. General biology and biochemistry texts include material on major catabolic and anabolic pathways; consult such references if you feel a need to refresh your memory of cellular biochemistry.

Comparative biochemical studies reveal that details of protozoan metabolism are as varied as details of protozoan sex. Some important biological factors to consider are that many parasites occupy environments in which the oxygen supply is quite limited. Others live in tissues, such as blood, where neither oxygen nor glucose is limited. In the latter case, there is no energy advantage in completely oxidizing glucose. Organisms that are adapted to such environments, including many protozoan parasites, often derive all their energy from glycolysis and excrete the partially oxidized products as waste. A complete Krebs cycle and cytochrome system then become excess metabolic machinery, at least in terms of energy production. However, the problem of reoxidation of reduced NAD remains, because oxidized compounds must be available for continuous functioning of glycolysis. In some parasites electrons are transferred to pyruvate, and the resulting ethanol or lactate is excreted, although many organisms excrete such compounds as succinate, acetate, and short-chain fatty acids as end products of glycolysis. Some metabolic solutions to the problem of NAD oxidation will be mentioned in subsequent chapters.

Metabolic flexibility is a feature of obligate heterotroph protozoa. For example, the Krebs cycle requires a continuous supply of the 4-carbon molecule oxaloacetate, one of the cycle's own end products, as an acceptor of 2-carbon units during formation of citric acid. Krebs cycle intermediates are routinely taken out of circulation and used in synthetic reactions such as transaminations. Thus, an alternate source of oxaloacetate is required, which in many protozoans is the **glyoxylate cycle,** a metabolic pathway especially important in those species that rely heavily on ethanol, fatty acids, and acetate for their energy and carbon skeletons. The glyoxylate cycle uses two acetyl-CoA molecules to make a single oxaloacetate molecule; the enzymes for this cycle are found in the **glyoxysomes** (peroxisomes).

Protozoa also may utilize a variety of hydrogen acceptors in the final oxidations coupled with ATP production. In aerobic metabolism of most animals, this final acceptor is molecular oxygen. Under anaerobic conditions protozoa may produce lactic acid or ethanol by using pyruvate as a hydrogen acceptor. Ciliates of genus *Loxodes* evidently use NO_3^- as

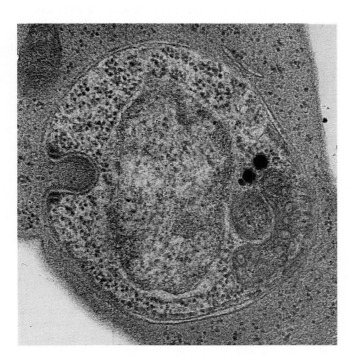

Figure 4.10 Uninucleate trophozoite of *Plasmodium cathemerium* ingesting host cell cytoplasm through a cytostome (micropore). (×52,000)

From M. Aikawa et al., "Feeding mechanisms of avian malarial parasites," in *J. Cell Biol.* 28:355–373. Copyright © 1966. The Rockefeller University Press.

a terminal hydrogen acceptor in the mitochondria and contain enzymes more typical of bacteria than eukaryotes to carry out this feat.[12] In parasitic protozoa without mitochondria—*Trichomonas vaginalis, T. foetus, Giardia duodenalis,* and *Entamoeba histolytica*—the final acceptor can be pyruvate, a key molecule in carbohydrate metabolism, in which case the end product is lactate or ethanol. These protozoa take up molecular oxygen, but availability of oxygen makes little or no difference in their energy metabolism. Absence of mitochondria has been variously interpreted as either a primitive character, reflecting an ancient evolutionary origin, or a derived character resulting from secondary loss.[32]

Odd and parasite-specific metabolic pathways are, of course, inviting targets for chemotherapy. Some of the more effective antimalarial drugs interfere with the parasites' ability to metabolize 1-carbon units during nucleic acid synthesis. Intracellular stages of the flagellate genus *Leishmania* do not build their nucleic acid precursors but instead salvage them from their host cells. Allopuranol, a purine analog, cannot be metabolized by the parasites but can be taken up from the host cell and used to build nucleic acids that do not function properly in the parasite. Needless to say, parasitologists leave few metabolic pathways unexplored in their efforts to find ways of treating diseases.

Many parasitic protozoa are intracellular. In some, entry into a host cell is by host phagocytosis of the parasite. An example is *Leishmania donovani* (see chapter 5), which is eaten by freeroaming macrophages and reticuloendothelial cells. The host cell forms a membrane-bound **parasitophorous vacuole** around the parasite, but instead of killing the parasite with digestive enzymes, as might be expected from a macrophage, the host cell provides it with nutrients. Members of the important apicomplexan genera *Babesia, Eimeria, Plasmodium,* and *Toxoplasma* are all intracellular at least at some stages in their lives, Mobile infective stages of these genera invade host cells, probably aided by digestive secretions.[28] Microsporidians (chapter 11) employ a different mode of entry into host cells. These parasites' cyst stages contain a coiled, hollow filament that evidently is under great pressure. When eaten by a host, usually an arthropod, this tubule is forcibly extruded from the cyst and penetrates host cell. The organism within the spore (**sporoplasm**) then crawls through the tube and enters its host. In this case the parasite's membrane is in direct contact with host cytoplasm, with no parasitophorous vacuole being formed.

Excretion and Osmoregulation

Most protozoa appear to be **ammonotelic;** that is, they excrete most of their nitrogen as ammonia, most of which readily diffuses directly through the cell membrane into the surrounding medium. Other sometimes unidentified waste products are also produced, at least by intracellular parasites. After these substances are secreted they accumulate within their host cell and, on the death of the infected cell, have toxic effects on the host. Carbon dioxide, lactate, pyruvate, and short-chain fatty acids are also common waste products.

Contractile vacuoles are probably more involved with osmoregulation than with excretion per se. Because freeliving, freshwater protozoa are hypertonic to their environment, they imbibe water continuously by osmosis. Contractile

vacuoles effectively pump out the water. Marine species and most parasites do not form these vacuoles, probably because they are more isotonic to their environment. However, *Balantidium* species (chapter 10) have contractile vacuoles.

Endosymbionts

Just as many protozoa live symbiotically in the bodies of larger animals, many organisms live within the bodies of protozoa. Zooxanthellae were mentioned earlier. It is now commonly accepted that chloroplasts and mitochondria and perhaps flagella arose from prokaryotes that came to live inside other cells (**endosymbiosis**). Corliss[9] proposed that any such structure or organism be referred to as a **xenosome,** which is a body or constituent organelle that contains DNA, is bounded by at least one membrane, lives within a cell, and is capable of reproducing itself. The term *xenosome* implies that "the symbiont once functioned as a free-living organism outside its present residence." In addition to the previous examples, zoochlorellae (green algal cells in protozoa and some multicellular animals), a variety of prokaryotes in protozoa, many intracellular protozoan parasites of multicellular animals, many hyperparasites of parasitic protozoa, and even nuclei of eukaryotic cells would be considered xenosomes. Endosymbionts living within protozoa are numerous and have been discussed by several authors.[6, 28] Their contribution to and interaction with the metabolism of their hosts undoubtedly varies, but in many cases is poorly understood.

CLASSIFICATION OF PROTOZOAN PHYLA

Classification of eukaryotic microorganisms is a monumental task that has occupied scientists for at least two centuries. Biologists who work with these organisms generally applaud each other's efforts to achieve monophyletic groupings while admitting that such efforts are often in vain. Unicellular eukaryotes are exceedingly diverse, but advances in molecular biology and comparative ultrastructure have allowed us to resolve some questions about homology and evolutionary significance of certain organelles. For example, it is doubtful that the various kinds of pseudopodia are homologous structures.[16, 20] Progress has been made, however, in terms of our altered perceptions of primitive (plesiomorphic) and derived (apomorphic) conditions. Thus, the presence of flagella (undulipodia) is now considered a plesiomorphic character of virtually all eukaryotes; in the past possession of a flagellum during much of the life cycle was used as a defining character for subphylum Mastigophora.[16]

The parasitologically important Apicomplexa present a particular problem for taxonomists (and textbook authors!) because of current work on alveolates in general and especially on some dinoflagellates. Thus, the apical complex is present in a variety of forms not only in parasites such as *Plasmodium* and *Eimeria* species, but also in free-living predators such as *Colpodella* species.[19] Molecular data, however, seem to show that *Perkinsus marinus* and *Parvilucifera infestans,* both parasites of molluscs and sometimes classified within Apicomplexa, comprise the sister group to dinoflagellates, whereas *Colpodella* species are the sister group to apicomplexans as typically defined.[19] We have chosen to use the Kuvardina et al.[19] research as justification for retaining *Perkinsus* and *Parvilucifera* as dinoflagellates, with the understanding that relationships among the major alveolate lineages are far from established conclusively.

Like all other recent protozoan classification schemes, the following one is a compromise between current evolutionary thinking and the practical need for a system of nomenclature that allows scientists to communicate with one another and retrieve information from older literature. This classification emphasizes groups with parasitic members and is based primarily on that of Lee et al.,[20] Hausmann and Hülsmann,[16] and Adl et al.[1] Neither Lee et al.[20] nor Adl et al.[1] provides a complete Linnaean taxonomy (phylum, class, order, etc.) for every group for two reasons: first, uncertainty about higher-level relationships had led to "taxonomic redundancy," or the establishment of taxa with only a single subtaxon (e.g., a class with only one order), and

second, ranks (e.g., class, order, family) are not necessarily equivalent in terms of diversity and inclusiveness.

Adl et al.[1] dispense with taxon designations (e.g. class, order) altogether and indicate subordinate groups by rank only (Figure 4.11). Although we acknowledge the problems associated with classification of eukaryotic microorganisms, we also believe that taxonomic names that are proper nouns are most easily remembered when embedded in an organized framework. These names are also very useful to students seeking additional information, especially from older literature or by using electronic search software. Thus we retain phylum and subordinate taxon names and arrangements, virtually all of which are consistent with the rankings of Adl et al.[1] These authors established six very inclusive "super-groups" defined by common structural characters. We have included these "supergroups" in the taxonomic listing below.

Main differences between the following classification scheme and ones found in other sources are: (1) This scheme

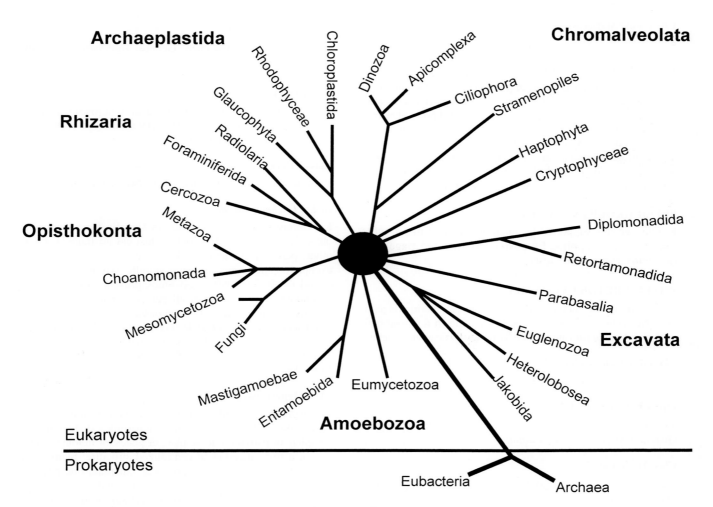

Figure 4.11   Phylogeny of the eukaryotes according to Adl at al.[1]
The tree is based largely on ultrastructural features and shows proposed relationships between various groups. Archaeplastida includes algae and green plants; other groups (e.g. Jakobida) may be free living and thus not mentioned in the text. Note that according to this phylogeny, amebas with lobose pseudopods (e.g. *Entamoeba* sp.) are not necessarily the closest relatives of those amebas with complex skeletons and often branching pseudopods (e.g. the foraminiferans).

Redrawn from the *J. Eukaryotic Microbiology*, volume 52, issue 5 cover illustrating the classification of Adl et al. 2006. The new higher level classification of eukaryotes with emphasis on the taxonomy of protists. *J. Euk. Microbiol.* 52:399–451.

includes only groups with parasitic members. (2) Groups are not listed in the same order as in Adl et al.[1] (3) Phylum Retortamonada as used by Hausmann and Hülsmann[16] is retained, as is order-level rank for Enteromonadida and Diplomonadida (Lee et al.[20] list these two groups as suborders of Diplomonadida). (4) The phylum name Axostylata is retained for those protozoa with mobile axostyles, and the remaining members of Hausmann and Hülsmann's[16] Axostylata are moved to Parabasalia according to Lee et al.[20] (5) We retain the phylum Euglenozoa and its subordinate groups, although the ranks of these groups differ from those in both Hausmann and Hülsmann[16] and Lee et al.[20] (6) We add information on "stramenopiles" of Lee et al.[20] but retain Hausmann and Hülsmann[16] phylum and classes for this group. Myxozoa are no longer considered protists.[18]

Most of the terminology in the following section has been covered already in this chapter; and other terms will be defined in upcoming chapters.

SUPER-GROUP OPISTHOKONTA
Unicellular stages with single posterior flagellum; no mastigonemes; flat cristae.

PHYLUM MICROSPORIDIA
Unicellular, spores, each with imperforate wall, containing one uninucleate or dinucleate sporoplasm and a polar filament; sporoplasm injected into host cells through extruded polar filament; without mitochondria, peroxisomes, or hydrogenosomes; with 70S ribosomes; now considered Fungi;[1, 13] intracellular parasites in nearly all major animal groups. Genera: *Amphiacantha, Metchnikovella, Encephalitozoon, Glugea, Pleistophora, Thelohania, Amblyospora, Nosema, Antonospora.*

SUPER-GROUP EXCAVATA
Feeding groove supported by microtubules and fibers, with intake current supplied by posteriorly directed flagellum (sometimes secondarily lost).

PHYLUM RETORTAMONADA
Mitochondria and dictyosomes absent; three anterior flagella and one recurrent flagellum, the latter lying in a cytostomal groove; intestinal parasites or free living in anoxic environments.

Class Retortamonadea
Intranuclear division spindle.

Order Retortamonadida
Two pairs of kinetosomes, large cytostome; cysts present. Genera: *Chilomastix, Retortamonas.*

Class Diplomonadea
One or two karyomastigonts; individual mastigonts with one to four flagella, typically one of them recurrent and associated with cytostome or with organelles forming cell axis; mitochondria and Golgi apparatus absent; semiopen mitosis; cysts present; free living or parasitic.

Order Enteromonadida
Single karyomastigont containing one to four flagella; one recurrent flagellum in genera with more than single flagellum; frequent transitory forms with two karyomastigonts; all parasitic. Genera: *Enteromonas, Trimitus.*

Order Diplomonadida
Two karyomastigonts; body with twofold rotational symmetry; each mastigont with four flagella, one recurrent; with variety of microtubular bands; free living or parasitic. Genera: *Giardia, Hexamita.*

PHYLUM AXOSTYLATA
With a mobile axostyle made of microtubules.

Class Oxymonadea
One or more karyomastigonts, each containing four flagella typically arranged in two pairs in motile stages; one to many axostyles per organism; mitochondria and Golgi apparatus absent; division spindle intranuclear; cysts in some; sexuality in some; all parasitic in termites and wood-eating cockroaches.

Order Oxymonadida
With characters of the class. Genera: *Monocercomonoides, Oxymonas, Pyrsonympha.*

PHYLUM PARABASALIA
With parabasal fibers originating at kinetosomes; large dictyosomes associated with karyomastigont; axostyle nonmotile; up to thousands of flagella.

Class Trichomonada
With two parabasal fibers and one or two dictyosomes.

Order Trichomonadida
Karyomastigonts with four to six flagella (one recurrent) but only one flagellum in one genus and no flagella in another; pelta and noncontractile axostyle in each mastigont, except for one genus; division spindle extranuclear; mitochondria absent; hydrogenosomes present; no sexual reproduction; true cysts rare; all parasitic. Families (genera): Monocercomonidae (*Dientamoeba, Monocercomonas, Histomonas*); Trichomonadidae (*Trichomonas, Tritrichomonas, Pentatrichomonas*); Devescovinidae (*Bullanympha, Mixotricha, Gigantimonas*); Calonymphidae (*Calonympha, Coronympha*).

Class Hypermastigia
Mastigont system with numerous flagella and multiple parabasal bodies; flagella-bearing kinetosomes distributed in complete or partial circle, in plate or plates, or in longitudinal or spiral rows meeting in centralized structure; many with microtubule sheets or peltoaxostylar lamellae; one nucleus per cell; mitochondria absent; division spindle extranuclear; cysts in some; sexuality in some; all symbiotic in wood-eating insects.

Order Lophomonadida
Extranuclear organelles arranged in one system; typically all old structures resorbed in division and new organelles formed in daughter cells. Genera: *Lophomonas, Microjoenia.*

Order Trichonymphida

Body divided into anterior rostral and posterior postrostral regions; two or, occasionally, four mastigont systems; typically equal separation of mastigont systems in division, with total or partial retention of old structures when new systems are formed. Genera: *Barbulanympha, Trichonympha.*

Order Spirotrichonymphida

Flagellar bands begin at anterior end and spiral in helical coil around body. Genus: *Spirotrichonympha.*

PHYLUM EUGLENOZOA

With cortical microtubules; flagella often with paraxial rod; mitochondria with discoid cristae; nucleoli persist during mitosis.

Class Euglenoidea

Two heterokont flagella arising from apical reservoir; with pellicular microtubules that stiffen pellicle; some species with light-sensitive stigma and chloroplasts; some ectocommensal; one species in tadpole gut. Genera: *Colacium, Euglenamorpha.*

Class Diplonemea

Two equal flagella without paraxial rods; cytostome supported by microtubular rods; one species in blood of lobster. Genera: *Diplonema, Rhynchopus.*

Class Kinetoplasta

With a unique mitochondrion containing a large disc of DNA, made from both mini- and maxicircles; paraxial rod; some with undulating membranes. Phylogenies based on molecular data (18S-RNA) suggest Kinetoplasta diverged from Euglenoidea about 1 billion years ago.

Order Bodonida

Typically two unequal flagella, one directed anteriorly and one posteriorly; no undulating membrane; kinetoplastic DNA in several discrete bodies in some, dispersed throughout mitochondrion in some; free living and parasitic. Genera: *Bodo, Cryptobia, Rhynchomonas, Ichthyobodo, Trypanoplasma.*

Order Trypanosomatida

Single flagellum either free or attached to body by undulating membrane; flagellum typically with paraxial rod that parallels axoneme; single mitochondrion (nonfunctional in some forms) extending length of body as tube, hoop, or network of branching tubes, usually with single conspicuous DNA-containing kinetoplast located near flagellar kinetosomes; Golgi apparatus typically in region of flagellar pocket, not connected to kinetosomes and flagella; all parasitic. Genera: *Blastocrithidia, Leptomonas, Herpetomonas, Crithidia, Leishmania, Trypanosoma.*

SUPER-GROUP AMEBOZOA

Locomotion by pseudopodia; mitochondria with tubular, often branched, cristae; flagellated stages, if present, typically with single flagellum.

Molecular and ultrastructural studies have shown that the ameboid body form is not primitive at all but evidently has arisen many times, probably from flagellated ancestors.[20] Consequently, former phylum Sarcodina (or Sarcomastigophora) is no longer recognized as valid, and many subordinate taxa also have been eliminated. Regardless of arguments over classification schemes, the organisms themselves often have been known for a long time and have been the subject of intensive research efforts. In addition, amebas fall into some fairly familiar groups based on structure, and, in many cases, identification is based on light-level morphology.[20] While we recognize the utility of these groups, we also have retained a few of the former taxon names as a means of connecting familiar species with the older literature.

Characters Generally Shared by Amebas

Pseudopodia or locomotive protoplasmic flow without discrete pseudopodia; flagella, when present, usually restricted to developmental or other temporary stages; body naked or with external or internal test or skeleton; asexual reproduction by fission; sexuality, if present, associated with flagellated or, more rarely, ameboid gametes; most free living.

Amebas of uncertain affinities, including some members of former subphylum Rhizopoda and class Lobosea

Locomotion by lobopodia, filopodia, or reticulopodia or by protoplasmic flow without production of discrete pseudopodia ("Rhizopoda"). Pseudopodia lobose and more or less filiform but produced from broader hyaline lobe; usually uninucleate; no sporangia or similar fruiting bodies ("Lobosea"). Parasitic forms typically uninucleate; when present, mitochondria with unbranched cristae; no tests; no flagellate stage; usually asexual; body branched or unbranched cylinder. Genera: *Entamoeba, Balamuthia, Iodamoeba, Endolimax.*

Ramicristate amebas (with branched mitochondrial cristae; including former Gymnamoebae, in part)

Pseudopods more or less finely tipped, sometimes filiform, often branched and hyaline, produced from a broad hyaline lobe; cysts common. Genus: *Acanthamoeba.*

Ameboflagellates (former Heterolobosea, in part)

Body with shape of monopodial cylinder, usually moving with more or less eruptive, hyaline, hemispherical bulges; typically uninucleate; temporary flagellate stages in most species. Genus: *Naegleria.*

PHYLUM PLASMODIOPHORA (CLASS MYCETOZOEA)

Uninuclear, multinucleate, or plasmodial; sporulation either by differentiation of single ameboid cells or by aggregates of amebas that form pseudomultinucleate plasmodia into fruiting body or by differentiation of spores from a truly multinucleate plasmodium.

Order Plasmodiophorida

Obligate intracellular parasites of plants, with minute plasmodia; zoospores produced in sporangia and bearing anterior pair of unequal flagella. Genus: *Plasmodiophora.*

Stramenopiles

The stramenopiles are a large, heterogeneous group of protists that share a single synapomorphy; namely, tubular mastigonemes that branch into three fine filaments at their tips ("tripartite hairs").[20] Some members of this group (e.g., diatoms, brown algae, and chrysophytes) possess chloroplasts and are autotrophic, others are almost funguslike heterotrophs. Parasitic forms are commonly found in the intestines of ectothermic vertebrates. Hausmann and Hülsmann[16] placed stramenopiles in a phylum Chromista, characterized by heterokont flagellar apparatus and mastigonemes. Lee et al.[20] placed the parasites of reptiles and amphibians in a stand-alone "order," Slopalinida, but did not provide either a class or phylum name. The group known as "labyrinthulids" have been classifed as either fungi or stramenopile protozoa; Lee et al.[20] do not mention them. We retain Hausmann and Hülsmann's[16] classification including proteromonads, opalinids, and labyrinthulids.

PHYLUM CHROMISTA

With heterokont flagella having mastigonemes derived from dictyosomes; plastids enveloped in endoplasmic cisternae.

Class Proteromonadea

One or two pairs of heterokont, heterodynamic flagella; with rhizoplast and dictyosome.

Order Proteromonadida

One or two pairs of unequal flagella without paraxial rods; single mitochondrion, distant from kinetosomes, curling around nucleus, not extending length of body, without kinetoplast; Golgi apparatus encircling band-shaped rhizoplast passing from kinetosomes near surface of nucleus to mitochondrion; cysts present; all species parasitic in amphibia, reptiles, and mammals. Genera: *Karotomorpha, Proteromonas.*

Class Opalinea

Numerous flagella in oblique rows over entire body and originating in anterior field of kinetosomes (falx); some fibrils associated with kinetosomes; cytostome absent; binary fission generally symmetrogenic; known life cycles involve syngamy with anisogamous flagellated gametes; all parasitic.

Order Opalinida (Slopalinida)

With characters of the class. Genera: *Opalina, Protoopalina, Cepedea.*

Class Labyrinthulea

Trophic stage as ectoplasmic network with spindle-shaped or spherical nonameboid cells; in some genera ameboid cells move within network by gliding; with sagenogenetosome (unique cell-surface organelle, associated with ectoplasmic network); inclusion in Chromista is based on heterokont structure of zoospores; saprozoic and parasitic on algae; mostly marine and estuarine.

Order Labyrinthulida

With characters of the class. Genus: *Labyrinthula.*

SUPER-GROUP CHROMALVEOLATA

With secondarily endosymbiotic plastids, sometimes lost or reduced; flat cristae.

SUPERPHYLUM ALVEOLATA

With micropores and membranous pellicular vesicles or alveoli.

PHYLUM DINOFLAGELLATA

Two flagella, typically one transverse and one trailing; body usually grooved transversely and longitudinally, forming a girdle and sulcus, each containing a flagellum; chromatophores usually yellow or dark brown, occasionally green or blue-green; nucleus unique among eukaryotes in having chromosomes that lack or have low levels of histones; mitosis intranuclear; flagellates, coccoid unicells, colonies, and simple filaments; sexual reproduction present; few parasites of invertebrates; one or more species (*Zooxanthella microadriatica*) very important mutuals in tissues of various marine invertebrates, especially cnidarians; some, such as *Perkinsus marinus,* of major economic importance to oyster industry.[30]

Class Perkinsasidea

Flagellated zoospores with curved anterior (apical) ribbon of microtubules surrounded by alveolar sheath; no sexual reproduction; homoxenous.

Order Perkinsorida

With characters of the class. Genera: *Parvilucifera, Perkinsus.*

Other classes

Noctiluciphyceae, Blastodiniphyceae, Syndiniophyceae.

PHYLUM APICOMPLEXA

Apical complex (generally consisting of polar ring, micronemes, rhoptries, subpellicular tubules, and conoid) present at some stage; micropore(s) usually present; sexuality by syngamy; all parasitic.

Class Conoidasida

Subclass Gregarinasina

Mature gamonts large, extracellular; mucron formed from conoid; generally syzygy of gamonts; gametes usually isogamous or nearly so; zygotes forming oocysts within gametocysts; locomotion of mature organisms by body flexion, gliding, or undulation of longitudinal ridges; in digestive tract or body cavity of invertebrates; generally homoxenous.

Order Archigregarinorida

Life cycle usually with merogony, gametogony, sporogony; in annelids, sipunculids, hemichordates, or ascidians. Genera: *Exoschizon, Selenidioides.*

Order Eugregarinorida

Merogony absent; gametogony and sporogony present; typically parasites of arthropods and annelids.

Suborder Blastogregarinorina

Gametogony by gamonts while still attached to intestine; no syzygy; gametocysts absent; gamont of single compartment with mucron, without definite protomerite and deutomerite; in polychaete annelids. Genus: *Siedleckia*.

Suborder Aseptatorina

Gametocysts present; gamont of single compartment, without definite protomerite and deutomerite but with mucron in some species; syzygy present. Genera: *Lecudina, Lankesteria, Monocystis, Selenidium, Diplocystis*.

Suborder Septatorina

Gametocysts present; gamont divided into protomerite and deutomerite by septum; with epimerite; in alimentary canal of invertebrates, especially arthropods. Genera: *Gregarina, Didymophyes, Leidyana, Actinocephalus, Stylocephalus, Acanthospora, Menospora*.

Order Neogregarinorida

Merogony (possibly acquired secondarily); in Malpighian tubules, intestine, hemocoel, or fat tissues of insects. Genera: *Gigaductus, Farinocystis (= Triboliocystis), Mattesia*.

Subclass Coccidiasina

Gamonts ordinarily present; mature gamonts small, typically intracellular, without mucron or epimerite; syzygy generally absent; life cycle characteristically consisting of merogony, gametogony, and sporogony; most species in vertebrates.

Order Agamococcidiorida

Merogony and gametogony absent. Genera: *Rhytidocystis, Gemmocystis*.

Order Ixorheorida

Sporogony present; merogony possibly present; gamogony absent; in holothuroideans. Genus: *Ixotheis*.

Order Protococcidiorida

Merogony absent; in invertebrates. Genera: *Eleutheroschizon, Grellia*.

Order Eucoccidiorida

Suborder Adeleorina

Syzygy between microand macrogamonts; sporozoites with envelope; in both invertebrates and vertebrates. Genera: *Adelina, Dactylosoma, Haemogregarina, Hepatozoon, Klossiella*.

Suborder Eimeriorina

Macrogamete and microgamont developing independently; no syzygy; microgamont typically producing many microgametes; zygote not motile; sporozoites typically enclosed in sporocyst within oocyst; homoxenous or heteroxenous. Genera: *Aggregata, Cryptosporidium, Cyclospora, Eimeria, Isospora, Lancasterella, Neospora, Sarcocystis, Toxoplasma*.

Class Aconoidasida

Conoid generally absent, although present in some species' ookinetes.

Order Haemosporida

Macrogamete and microgamont developing independently; no syzygy; conoid ordinarily absent; microgamont producing eight flagellated microgametes; zygote motile (ookinete); sporozoites naked, with three-membraned wall; heteroxenous, with merogony in vertebrates and sporogony in invertebrates; transmitted by bloodsucking insects. Genera: *Haemoproteus, Hepatocystis, Leucocytozoon, Plasmodium, Saurocytozoon*.

Order Piroplasmorida

Piriform, round, rod-shaped, or ameboid; conoid absent; no oocysts, spores, or pseudocysts; flagella absent; usually without subpellicular microtubules, with polar ring and rhoptries; asexual and probably sexual reproduction; parasitic in erythrocytes and sometimes also in other circulating and fixed cells; heteroxenous, with merogony in vertebrates and sporogony in invertebrates; sporozoites with single-membraned wall; known vectors are ticks. Genera: *Babesia, Theileria*.

PHYLUM CILIOPHORA

Simple cilia or compound ciliary organelles typical in at least one stage of life cycle; with subpellicular infraciliature present even when cilia absent; with pellicular alveoli; two types of nuclei, with rare exception; binary fission transverse, basically homothetogenic, but budding and multiple fission also occur; sexuality involving conjugation, autogamy, and cytogamy; contractile vacuole typically present; most species free living but many commensal and some parasitic. (Numerous taxa with commensal and some with parasitic species are characterized here because they are not covered in later chapters.)

SUBPHYLUM INTRAMACRONUCLEATA

Division of macronucleus involves intramacronuclear microtubules.

Class Spirotrichea

Somatic ciliature dikinetids with anterior or both kinetosomes ciliated or with polykinetids, with well-developed overlapping postciliar ribbons; generally with conspicuous oral and/or preoral ciliature with serial polykinetids.

Order Clevelandellida

Somatic ciliature well developed, sometimes separated into distinct areas by well-defined suture lines; several specialized unique fibers associated with kinetosomes; macronuclear karyophore (region of cytoplasm apparently supporting nucleus) and/or conspicuous dorsoanterior sucker characteristic of many species; endoparasitic in digestive tract of insects, other arthropods, and amphibians, occasionally in oligochaetes or molluscs. Genera: *Clevelandella, Nyctotherus*.

Class Litostomatea

Body monokinetids with tangential transverse ribbon and nonoverlapping laterally directed kinetodesmal fibrils; simple oral cilia usually not as polykinetids.

Order Vestibuliferida

Apical or near apical densely ciliated vestibulum commonly present; no polykinetids; free living or parasitic, especially in digestive tract of vertebrates and invertebrates. Genera: *Balantidium, Isotricha, Sonderia*.

Order Entodiniomorphida

Somatic ciliature in form of unique ciliary tufts or bands—otherwise body naked; pellicle generally firm, sometimes drawn out into processes; oral area often with retractile cilia and serial polykinetids; skeletal plates in many species; commensals in mammalian herbivores, including anthropoid apes.

Suborder Blepharocorythina

Somatic ciliature markedly reduced; oral ciliature inconspicuous; in herbivorous mammals, especially equids. Genera: *Blepharocorys, Ochoterenaia*.

Suborder Entodiniomorphina

Somatic ciliature as tufts, bands, or girdles; oral cilia usually as distinct polykinetids; pellicle rigid, firm, often spiny; in vertebrates, especially artiodactyls and perissodactyls. Genera: *Entodinium, Ophryoscolex*.

Class Phyllopharyngea

Somatic monokinetids; rudimentary transverse microtubule ribbons; laterally projecting kinetodesmal fibrils; leaflike microtubule ribbons in oral region.

Subclass Chonotrichia

Somatic ciliature absent; helical, funnel-like collar with ciliary rows inside; ectocommensal on Crustacea. Genera: *Helichona, Spirochona*.

Subclass Suctoria

With sucking tentacles; somatic ciliature absent except in free-swimming immature forms; some with endogenous budding. Some ectocommensal on aquatic invertebrates.

Class Oligohymenophorea

Somatic monokinetids with forwardly directed, distinctly overlapping fibrils, divergent postciliary ribbons, and radial transverse ribbons; oral apparatus generally well defined, in buccal cavity, with distinct paroral dikinetid and one to many polykinetids.

Subclass Hymenostomatia

Body ciliation often uniform and heavy; buccal cavity, when present, ventral; sessile forms, stalks, and colony formation relatively rare; freshwater forms predominant.

Order Hymenostomatida

Buccal cavity well defined; oral area on ventral surface, usually in anterior half of body; several species causing white spot disease in marine and freshwater fishes. Genera: *Ichthyophthirius, Ophryoglena*.

Subclass Peritrichia

Oral ciliary field prominent, covering apical end of body, bordered by a dikinetid file and polykinetid that originate in an infundibulum; paroral membrane and adoral membranelles present; somatic ciliature reduced to temporary posterior circlet of locomotor cilia; many stalked and sedentary, others mobile, all with aboral scopula; conjugation total, involving fusion of microconjugants and macroconjugants.

Order Sessilida

Mature trophonts usually sessile, attached, with stalk; some obligate ectosymbionts on aquatic invertebrates. Genera: *Epistylis, Lagenophrys, Rhabdostyla*.

Order Mobilida

Mobile forms, usually conical or cylindrical (or discoidal and orally aborally flattened), with permanently ciliated trochal band (ciliary girdle); complex thigmotactic apparatus at aboral end, often with highly distinctive denticulate ring; all ectoparasites or endoparasites of freshwater or marine vertebrates and invertebrates. Genera: *Trichodina, Urceolaria*.

Subclass Astomatia

Mouthless endocommensals, especially in gut of annelids but also in gastropods and amphibians. Genera: *Anoplophrya, Haptophrya, Radiophrya*.

Subclass Apostomatia

Ectocommensals on crustaceans, annelids, and cnidarians; somatic ciliature in helical rows; with "rosette" organelle. Genera: *Foettingeria, Spirophrya*.

SUPER-GROUP RHIZARIA

With axopodia or simple, branching, or anastomosing filopodia.

PHYLUM HAPLOSPORIDIA

Spore uninucleate, without polar capsule or filaments but with anterior opening (sometimes covered with operculum); all parasitic in invertebrates.[14, 20] Genera: *Haplosporidium, Minchinia, Urosporidium*.

▌Learning Outcomes

By the time a student has finished studying this chapter, he or she should be able to:

1. Draw a typical mastigont system and label the parts.
2. Explain the terms *merogony, schizogony, zygotic meiosis, binary fission,* and *budding.*
3. Explain how different locomotor systems in protozoa operate.
4. Write extended paragraphs, each outlining the criteria involved in classification of flagellates, ciliates, and apicomplexans respectively.
5. Write an extended paragraph outlining the problems and issues associated with classification of amebas.
6. Describe the various feeding methods in protozoa.
7. Describe the various kinds of cysts that are formed by parasitic protozoa and explain how these stages function in transmission.

References

References for superscripts in the text can be found at the following Internet site: www.mhhe.com/robertsjanovynadler9e

Additional Readings

Hyman, L. H. 1940. *The invertebrates, vol. I. Protozoa through Ctenophora.* New York: McGraw-Hill Book Co. An excellent reference to general aspects of the Protozoa.

Jahn, T. L., E. C. Bovee, and F. F. Jahn. 1979. *How to know the Protozoa* (2d ed.). Dubuque, IA: Wm. C. Brown Publishers. Identification keys to the common Protozoa.

Kreier, J. P., and J. R. Baker. 1987. *Parasitic protozoa.* Boston: Allen and Unwin.

Kudo, R. R. 1966. *Protozoology.* New York: C. Thomas.

Levine, N. D. 1973. *Protozoan parasites of domestic animals and man* (2d ed.). Minneapolis: Burgess Publishing Co.

Margulis, L. 1981. *Symbiosis in cell evolution.* San Francisco: W. H. Freeman and Co., Publishers. Presents the case for the symbiotic origin of the eukaryotes.

Scholtyseck, E. 1979. *Fine structure of parasitic protozoa.* Berlin: Springer-Verlag. Atlas of electron micrographs accompanied by labeled diagrams. Heavy on Apicomplexa.

Sleigh, M. A. 1989. *Protozoa and other protists.* New York: Edward Arnold.

Chapter 5

Kinetoplasta: Trypanosomes and Their Kin

Total, sheer or ruthless clearing means the destruction of all trees and shrubs in the area treated. It is a completely effective method of eliminating Glossina *and the oldest.*

—J. Ford, T. A. M. Nash, and T. R. Welch describing tsetse fly control methods.[32]

However, it is necessary to affirm that wholesale slaughter of the larger mammal populations is a completely effective method of eliminating Glossina morsitans *and* G. pallidipes.

—J. Ford, with additional assessment of tsetse control methods.[30]

The class Kinetoplasta contains species that parasitize everything from humans to plants. Members of this group are characterized by a single large mitochondrion containing a body—the **kinetoplast**—that stains darkly in histological preparations. The kinetoplast lies beside the kinetosome at the base of the flagellum, and, along with nearby parts of the mitochondrion, it remains in a more or less established relationship with the kinetosome throughout the parasite's life cycle (Fig. 5.1). The kinetoplast is actually a disc-shaped, DNA-containing organelle within the mitochondrion. Kinetoplast DNA (kDNA) is organized into a network of linked rings, quite unlike the organization of chromosomal DNA.[5] There are up to 20,000 tiny rings (minicircles) and 20 to 50 larger rings (maxicircles) in the kinetoplast network. Most electron micrographs show no physical connection between the kinetosome and kinetoplast, and the nature of their established association is unknown.

In addition to their distinctive mitochondrial structure, kinetoplastans have a cytoskeleton consisting of microtubules arranged at regular intervals beneath the plasma membrane (Figs. 5.1, 5.2). Other characteristics include a sizable flagellar pocket, sometimes elongated, and a latticelike crystalline **paraxial rod** alongside the axoneme that has short projections connecting it to the axonemal microtubules, an **undulating membrane** (depending on the species); and occasionally a prominent **glycocalyx,** or surface coat, visible in electron micrographs. Finally, kinetoplastans have two other unique features: first, **glycosomes,** organelles in which glycolytic reactions occur, and second, splicing of a short, characteristic RNA piece onto every molecule of mRNA.[106]

Kinetoplastan genera differ considerably in their host distribution, life cycles, and medical and veterinary importance. Two families are recognized: Bodonidae (order Bodonida, coprozoic and free living or parasites of fish and invertebrates) and Trypanosomatidae, some members of which are important human and veterinary pathogens. These organisms provide fascinating challenges for parasitologists, ranging from extraordinarily difficult control problems to dramatic pathological effects such as erosion of facial features (see Fig. 5.20). Some have been popular research organisms because of their ease of culture; others defied taxonomists until molecular biology began to reveal their relationships; and still others present us with such a diverse clinical picture that we have yet to dissect out the effects of parasite traits, human genetic makeup, and environmental factors from the parasites' overall public health impact.

FORMS OF TRYPANOSOMATIDAE

All species of Trypanosomatidae have a single nucleus and are either elongated with a single flagellum or rounded with a very short, nonprotruding flagellum. Many members of the family are **heteroxenous:** During one stage of their lives they live in the blood and/or fixed tissues of vertebrates and during other stages they live in the intestine of bloodsucking invertebrates. In addition, laboratory culture media for these parasites usually must contain blood. Thus, we call them **hemoflagellates.**

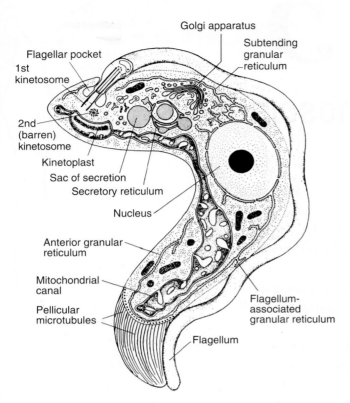

Figure 5.1 **Diagram to show principal structures revealed by electron microscopy in the bloodstream trypomastigote of a salivarian trypanosome, *Trypanosoma congolense*.**

It is shown cut in sagittal sections, except for most of the shaft of the flagellum and the anterior extremity of the body.

From K. Vickerman, "The fine structure of *Trypanosoma congolense* in its bloodstream phase," in *Journal of Eukaryotic Microbiology,* 16:54–69. Copyright © 1969 John Wiley & Sons. Reprinted with permission of the publisher.

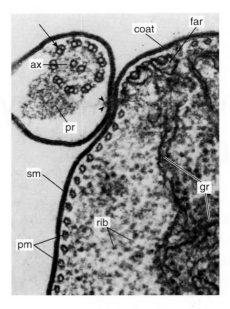

Figure 5.2 *Trypanosoma congolense.*

Transverse section of shaft of flagellum and adjacent pellicle in region of attachment. Both flagellum and body surface have a limiting unit membrane (*sm*) covered by a thick coating (*coat*) of dense material. The axoneme (*ax*) of the flagellum shows the partition (*arrow*) dividing one of the tubules of each doublet; alongside the axoneme lies the paraxial rod (*pr*). Pellicular microtubules (*pm*) underlie the surface membrane of the body, and a diverticulum (*far*) of the granular reticulum (*gr*) is always found embracing three or four of these microtubules close to the flagellum. Note the fibrous condensations (*arrowheads*) on either side of the opposed surface membranes, apparently "riveting" the flagellum to the body. A row of these "rivets" replaces a microtubule along the line of adherence. *rib,* ribosomes. (× 66,000)

From K. Vickerman, "The fine structure of *Trypanosoma congolense* in its bloodstream phase," in *J. Protozool.* 16:54–69. Copyright © 1969 The Society of Protozoologists. Reprinted with permission of the publisher.

Sexual phenomena have not been routinely observed in these organisms and most populations are probably collections of clones.[115] Nevertheless, there is a considerable amount of indirect evidence for sexuality.[35, 119] In experimental work parental stocks and hybrids can be identified using isoenzyme markers or fragment length distributions following enzymatic digestion of DNA. Within the tsetse fly vector "mating"—that is, a recombination of karyotype or isoenzyme phenotypes—has been demonstrated, although genetic recombination evidently is not obligatory in the life cycle. Segregation of allelic marker genes suggests meiosis, but, largely because of the parasites' small size and nonpredictable "mating," meiosis such as seen in larger eukaryote gametocytes has not been observed. Furthermore, at least experimentally, trypanosomatids may take up foreign DNA.[6] Thus, there exists a variety of mechanisms by which strains of these parasites may come to vary genetically, adding, no doubt, to the often confusing clinical picture seen in infections.

Trypanosomatids may have originally parasitized the digestive tract of insects and leeches, but researchers have proposed alternate and plausible scenarios in which vertebrates are the original hosts.[106] Although some species are still **monoxenous**—that is, parasitic only within a single

arthropod host[123]—most trypanosomatids are heteroxenous and pass through different morphological stages, depending on their life cycle phase and host they are parasitizing. In the past these stages were named after the genera they most resembled—for example, **leptomonad** for a stage resembling species of genus *Leptomonas*—but now we use a nomenclature referring to kinetoplast and nucleus positions (Fig. 5.3).

The **trypomastigote** stage is characteristic of *Trypanosoma* species' bloodstream forms as well as infective **metacyclic** stages in tsetse fly vectors. In trypomastigotes, both kinetoplast and kinetosome are near the posterior end of the body, and the flagellum runs along the surface, usually continuing as a free whip anterior to the body. The flagellar membrane is closely applied to the body surface, and, when the flagellum beats, this area of the pellicle is pulled up into a fold; the fold and flagellum constitute the undulating membrane. A second, "barren" kinetosome without a flagellum is usually found near the flagellar kinetosome.

In a typical bloodstream trypomastigote, a simple mitochondrion with or without tubular cristae runs anteriorly from the kinetoplast. In stages developing in insects, the mitochondrion is much larger and more complex, with lamellar

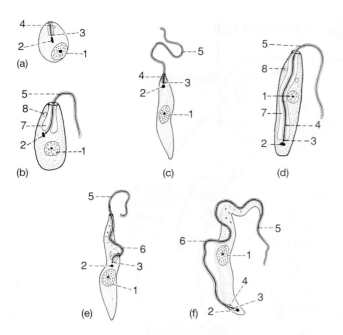

Figure 5.3 **Genera of Trypanosomatidae.**

(*a*) *Leishmania* (amastigote form); (*b*) *Crithidia* (choanomastigote); (*c*) *Leptomonas* (promastigote); (*d*) *Herpetomonas* (opisthomastigote); (*e*) *Blastocrithidia* (epimastigote); (*f*) *Trypanosoma* (trypomastigote). *1,* nucleus; *2,* kinetoplast; *3,* kinetosome; *4 and 5,* axoneme and flagellum; *6,* undulating membrane; *7,* flagellar pocket; *8,* contractile vacuole.

cristae. At the flagellar base, surrounding the kinetosome, is a **flagellar pocket** or reservoir. A system of **pellicular microtubules** spirals around the body just beneath the cell membrane (see Figs. 5.1 and 5.2). A rough endoplasmic reticulum is well developed, and a Golgi body lies between the nucleus and kinetosome.

Other trypanosomatid body forms differ in shape, position of kinetosome and kinetoplast, development of flagellum, or shape of flagellar pocket (see Fig. 5.3). A spheroid **amastigote** occurs in some species' life cycles and is definitive in genus *Leishmania.* The tiny (2–3μ) *Leishmania* amastigotes may be the smallest eukaryotic cells.[65] The flagellum is very short, projecting only slightly beyond the flagellar pocket. In the **promastigote** stage the elongated body has the flagellum extending forward as a functional organelle. The kinetosome and kinetoplast are located in front of the nucleus, near the anterior end of the body. Promastigote forms are found in life cycles of several species while they are in their insect hosts. It is the mature form in genus *Leptomonas.* If the flagellum emerges through a wide, collarlike pocket, the type is termed a **choanomastigote,** which is found in some species of *Crithidia.*

An **epimastigote** form occurs in some life cycles. Here the kinetoplast and kinetosome are still located between the nucleus and the anterior end, but a short undulating membrane lies along the proximal part of the flagellum. The genera *Crithidia* and *Blastocrithidia,* both parasites of insects, exhibit this form during their life cycles. Finally, the **paramastigote** and **opisthomastigote** forms are found in *Herpetomonas,* a widespread group of insect parasites that occur mainly in flies (order

Diptera). In paramastigotes the kinetosome and kinetoplast are beside the nucleus; in opisthomastigotes, these organelles are located between the nucleus and posterior end, but there is no undulating membrane, and the flagellum pierces a long reservoir that passes through the entire length of the body and opens at the anterior end. In genus *Herpetomonas* reproduction occurs only in the promastigote form, with other body forms appearing after populations have reached their peak, such as in culture. Despite their apparent structural simplicity, trypanosomatids are actually quite diverse, with much of their variability manifested in ultrastructural features and in internal distribution of organelles.

Trypanosomatid life cycles also vary with respect to host species, vectors, behavior of parasites in vectors and in vertebrate hosts, and life-cycle stages in which reproduction occurs (Fig. 5.4). *Leptomonas* species exhibit the simplest cycle in which an insect is the sole host, multiplication is by promastigotes in the gut, and transmission occurs by way of an ingested amastigotelike cyst. *Leishmania* species undergo multiplication as promastigotes in blood-sucking insects such as sand flies (chapter 39), but they are injected into a vertebrate host when the sand fly feeds, and they undergo additional multiplication, as amastigotes, in a variety of tissues.

Members of genus *Trypanosoma* exhibit the greatest diversity of forms during their life cycles, changing into multiplying epimastigotes in an insect vector's midgut and then into infective trypomastigotes (metacyclic forms) in either the hindgut or foregut, depending on the species. Metacyclic trypomastigotes are either passed in feces to contaminate a wound (e.g., *T. cruzi*) or injected with saliva during feeding (e.g., *T. brucei*). Tsetse flies of genus *Glossina* (Fig. 5.5) serve as vectors for the medically important *Trypanosoma brucei,* but fleas, horse flies, true bugs (order Hemiptera), and bats also function as vectors, depending on the species of *Trypanosoma.*

Members of genus *Leishmania* also occupy two strikingly different environments: the gut of their insect vector and the interior of a vertebrate host cell, typically a macrophage. In the vector (flies of family Psychodidae, subfamily Phlebotominae, chapter 39) or in culture at 25°C, the parasites are promastigotes and divide rapidly. But in a vertebrate host, promastigotes are phagocytized by macrophages. Within phagocytic (**parasitophorous**) vacuoles, promastigotes transform into amastigotes. Although they continue to multiply, they do so at a much slower rate than in culture. Whether inside a phlebotomine gut or in parasitophorous vacuoles, the parasites are living inside organs, or organelles, that usually function to digest foreign objects.

As in the case of trypanosomes, *Leishmania* species exhibit ultrastructural, metabolic, and antigenic changes as they pass from one life-cycle stage to another. Loss of external flagellum, change from elongated to round body form, rearrangement of subpellicular microtubules, reduction in oxygen consumption, and activation of metabolic pathways that function to use host cell nucleic acid precursors all accompany transformation from extracellular promastigote to intracellular amastigote.[91] Heat shocking of promastigotes has proven an effective technique for producing amastigotes, allowing continuous culture of amastigotes at elevated temperatures.[27] In some species (*L. mexicana* and *L. amazonensis*) stationary phase promastigotes are required in order to obtain amastigotes, but in all cases certain biochemical and infectivity criteria—including downregulation of β-tubulin genes and synthesis of amastigote-specific proteins—are employed to judge success of the culture techniques.[42]

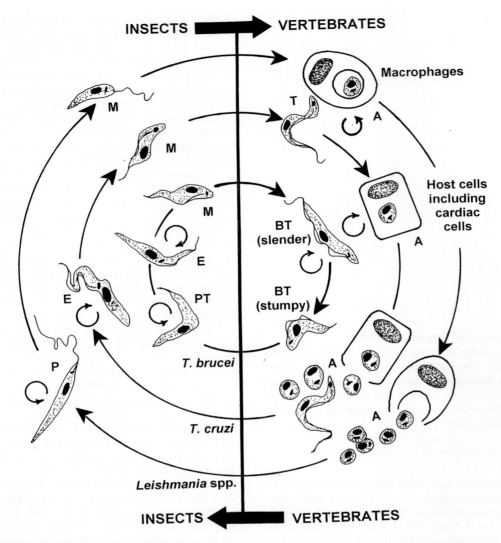

Figure 5.4 **Life cycles of trypanosomatids infective for humans.**
Three basic life cycle types are illustrated by *Trypanosoma brucei*, *T. cruzi*, and *Leishmania* species. Large arrows show the sequence of transformations; small circular arrows indicate the dividing stages. Abbreviations: **A**, amastigote; **BT**, bloodstream trypomastigote; **E**, epimastigote; **M**, metacyclic stages; **P**, promastigote; **PT**, procyclic trypomastigote; **T**, trypomastigote.

Redrawn from F. Bringaud et al., "Energy metabolism of trypansomatids: Adaptation to available carbon sources," in *Mol. Biochem. Parasitol.* 149:1–9. Copyright © 2006 used with permission from Elsevier.

Physiological, biochemical, and molecular studies on trypanosomatids have focused primarily on the disease-causing species, often with the intent of discovering unique metabolic pathways susceptible to antiparasite drugs. For example, trypanosomatids lack enzymes needed to build purines but nevertheless require these compounds, taking them up by means of "salvage" enzymes.[16] These enzymes are inviting targets for chemotherapy and especially for use of purine analogs.[20]

GENUS *TRYPANOSOMA*

All trypanosomes (except *T. equiperdum*) are heteroxenous or at least are transmitted by vectors. Various species pass through amastigote, promastigote, epimastigote, and/or trypomastigote stages, with other forms developing in the invertebrate hosts. Members of genus *Trypanosoma* are parasites of all vertebrate classes. Most live in blood and tissue fluids, but some important ones, such as *T. cruzi,* occupy intracellular habitats as well. The majority are transmitted by blood-feeding invertebrates, although other transmission mechanisms exist.

Much research has been conducted on *Trypanosoma* species because of their extreme importance to the health of humans and domestic animals. Reviews are available dealing with various aspects of the group, including host susceptibility,[81] epidemiology and control,[90] physiology and morphology,[121] chemotherapy,[126] taxonomy,[51] immunology,[1, 4, 110] evolution,[106] and vector relationships.[75]

A few species of trypanosomes are responsible for misery and privation of enormous proportions, and evidently this has been the case for centuries.[41, 74] In the Bible, Isaiah 7:18–19 is considered a reference to tsetse flies[31] ("And it

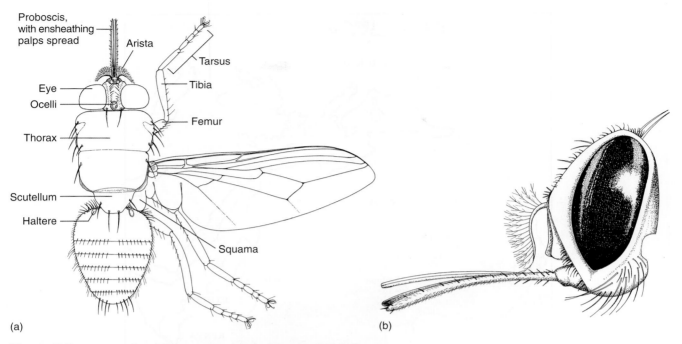

Figure 5.5   **A tsetse fly with general anatomical features labeled.**

(*a*) Dorsal view of *Glossina* showing general anatomical features. (*b*) Lateral view of head showing proboscis and palps.

Drawn by J. Janovy Jr. from a University of Nebraska State Museum specimen, provided by B.C. Ratcliffe, Curator of Entomology.

will come about in that day, that the Lord will whistle for the fly that is in the remotest part of the rivers of Egypt, and for the bee that is in the land of Assyria. And they will all come and settle on the steep ravines, on the ledges of the cliffs, on all the thorn bushes, and on all the watering places."), and Arabian historians described what were probably cases of sleeping sickness in the 14th century.[74]

In Africa alone, an area of 4.5 million square miles, larger than the United States, is incapable of supporting a cattle industry—not because the land is poor, since it is much like the grasslands of the American West, but because domestic livestock, up to 10,000 a day by some estimates, are killed by trypanosomes.[54] Native grazing animals are adapted to the parasites and do not experience severe pathology as a result of infection. Thus, semiarid lands that otherwise could support agronomy are denied to millions of persons who most need the protein afforded by the rich soil. More directly affected are millions of people in South America who have never known a day of good health because of *T.cruzi* infections.

Trypanosomes are divided into two broad groups, or sections—Salivaria and Stercoraria—based on characteristics of their development in their insect hosts. If a species develops in the anterior portions of the digestive tract, it is said to undergo **anterior station** development and is placed in section Salivaria, which contains several subgenera (taxonomic division of a genus). Species such as *Trypanosoma evansi,* which are transmitted mechanically without development in flies, are believed to have evolved from *T. brucei,* a member of Salivaria. When a species develops in a vector's hindgut, it is said to undergo **posterior station** development and is placed in section Stercoraria. Other developmental and morphological criteria separate the two sections, aiding in placement in the proper section of species that do not require development in an intermediate host (*T. equiperdum,*

T. equinum).[51] Classification of the various species into subgenera is based on their physiology, morphology, and biology.

Section Salivaria

Trypanosoma (Trypanozoon) brucei

The three subspecies of *Trypanosoma brucei*—*T. b. brucei, T. b. gambiense,* and *T. b. rhodesiense*—are morphologically indistinguishable but traditionally have been treated as separate species. They vary in infectivity for different species of hosts and produce somewhat different pathological syndromes. Biochemical studies of kDNA nucleotide sequences and isoenzyme variations suggest that *T. b. gambiense* is more likely a strain of *T. brucei* than a distinct species or subspecies. Regardless of their taxonomic status, *T. brucei*–type trypanosomes are widely distributed in tropical Africa between latitudes 15°N and 25°S, an area commonly known as the "fly belt" (Fig. 5.6), roughly corresponding in distribution with the trypanosomes' vectors, tsetse flies (*Glossina* spp.). Health statistics are difficult to assemble, but there probably are at least 60 million people at risk and probably about 100,000 new cases per year, with most of these in central Africa.[90]

Trypanosoma brucei brucei is a bloodstream parasite of native antelopes and other African ruminants, causing a disease called **nagana.** The parasite also infects introduced livestock, including sheep, goats, oxen, horses, camels, pigs, dogs, donkeys, and mules. It is pathogenic to these animals as well as to several native animal species. Humans, however, are not susceptible.

Trypanosoma brucei gambiense and *T. b. rhodesiense* are the etiological agents of African sleeping sickness, the human disease. There are physiological differences between the subspecies, and they differ in pathogenesis, growth rate,

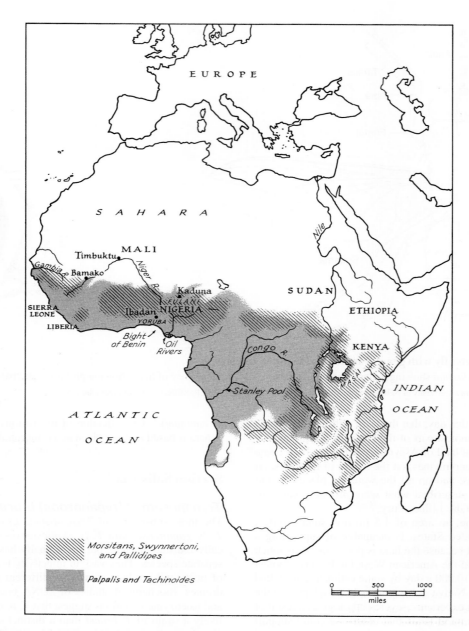

Figure 5.6   **Distribution of five *Glossina* species across Africa.**

From John J. McKelvey Jr., *Man against tsetse: Struggle for Africa.* Maps drawn by John Morris. Copyright ©1973 by Cornell University. Used by permission of the publisher, Cornell University Press.

and biology. As early as 1917 a German researcher named Taute showed that the nagana trypanosome would not infect humans. He repeatedly inoculated himself and native "volunteers" with nagana-ridden blood; none of them acquired sleeping sickness. His work was largely discounted by British experts for reasons that can only be guessed (maybe they could not believe a scientist would be curious enough about a parasite's behavior to try to infect himself with it, but this has happened more than once in the history of parasitology).

Trypanosoma brucei gambiense causes a chronic form of sleeping sickness. It is found in west central and central Africa, whereas *T. b. rhodesiense* occurs in central and east central Africa and causes a more acute type of infection. Native game animals serve as reservoirs for Rhodesian trypanosomiasis but not for Gambian trypanosomiasis.[90]

- **Morphology and Life History.**   *Trypanosoma brucei* in natural infections tends to be pleomorphic (polymorphic) in its vertebrate host, ranging from long, relatively slender trypomastigotes with a long free flagellum through intermediate forms to short, stumpy individuals with no free flagellum (Fig. 5.7). The small kinetoplast is usually very near the posterior end, and the undulating membrane is conspicuous.

    Insect vectors of *T. b. brucei* and *T. b. rhodesiense* are *Glossina morsitans*, *G. pallidipes*, and *G. swynnertoni*, whereas those of *T. b. gambiense* are *G. palpalis* and *G. tachinoides* (see Fig. 5.6). At least 90% of the flies are refractive to infection. When sucked up by a susceptible fly along with a blood meal, *T. brucei* locates in the posterior section of the midgut of the insect, where it multiplies in the trypomastigote form for about 10 days. At the end of this time the slender individuals produced migrate forward into the foregut, where they are found between the 12th and 20th days. They then migrate farther forward into the esophagus, pharynx, and hypopharynx and enter the salivary glands.

Once in the salivary glands trypomastigotes transform into epimastigotes and attach to host cells or lie free in the lumen. After several asexual generations the epimastigotes transform into **metacyclic trypomastigotes,** which are small and stumpy and lack a free flagellum. In the fly an entire cycle is completed in 15 to 35 days. Only metacyclic trypomastigotes are infective to a vertebrate host. When feeding, a tsetse fly may inoculate a host with up to several thousand flagellates in a single bite. Within a vertebrate, the parasites multiply as trypomastigotes in blood and lymph. In chronic trypanosomiasis, many parasites invade the central nervous system, multiply, and enter intercellular spaces within the brain.

Biochemical, ultrastructural, and immunological studies have added greatly to our understanding of trypanosomatids.[10, 121] Of considerable value is the fact that essentially pure preparations of certain morphological stages can be obtained. For example, when *Trypanosoma brucei* is passed by syringe from one vertebrate host to another, the strain tends to become monomorphic after a period of time, its population consisting only of slender trypomastigotes that are no longer infective to tsetse flies and cannot be cultivated in vitro. Their morphology and metabolism correspond to the slender trypomastigotes in natural infections. In contrast, when *T. brucei* is placed in certain in vitro culture systems, its morphology and metabolism revert to those found in a fly's midgut, with its kinetoplast farther from the posterior end and close to the nucleus.

The monomorphic, syringe-passed strain depends entirely on glycolysis for its energy production, degrading glucose only as far as pyruvate and having no tricarboxylic acid cycle or oxidative phosphorylation by way of the classical cytochrome system. The reduced NAD produced in glycolysis is reoxidized by a nonphosphorylating glycerophosphate oxidase system, which, although it requires oxygen, is not sensitive to cyanide. This respiratory system is inhibited by **suramin,** an antitrypanosomal drug, and is evidently localized in membrane-bound microbodies called *alpha-glycerophosphate oxidase bodies,* or **glycosomes.**[78] The long, slender trypomastigotes are very active, and consume substantial quantities of both glucose and oxygen in their inefficient energy production. Blood and lymph have such a plentiful supply of both glucose and oxygen that there is no selective value in conservation of either.

The situation is quite different when trypanosomes find themselves in a blood clot in their vector's midgut. In this case the parasites completely degrade glucose via glycolysis, the tricarboxylic acid cycle, and the cyanidesensitive cytochrome system. Oxygen and glucose consumption of the midgut (or culture) forms is only 1/10 that of bloodstream forms. The glycerophosphate oxidase system is also present in culture forms, but its activity now is sensitive to mitochondrial inhibitors.[78]

Ultrastructural observations on mitochondria in the respective forms (bloodstream vs. culture or vector)

correlate beautifully with the biochemical findings. Long, slender trypomastigotes have a single, simple mitochondrion extending anteriorly from their kinetoplast; cristae are few, short, and tubular. Midgut stages have elaborate mitochondria extending both posteriorly and anteriorly from the kinetoplast, and cristae are numerous and platelike. The curious movement of the kinetoplast away from the posterior end in the midgut trypomastigote and anterior to the nucleus in the epimastigote can now be understood as reflecting the elaboration of the posterior section of mitochondrion, which "pushes" the kinetoplast forward. Furthermore, the short, stumpy forms are the only ones infective to tsetse flies, and the intermediate forms are transitional from the long, slender noninfective forms (see Fig. 5.7). Electron microscopy has shown that this transition is marked by increasing elaboration of the mitochondrion; synthesis of mitochondrial enzymes has been shown by cytochemical means. Similarly, metacyclic

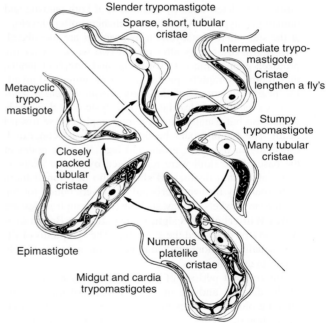

Figure 5.7   Diagram to show changes in form and structure of the mitochondrion of *Trypanosoma brucei* throughout its life cycle.

The slender bloodstream form lacks a functional Krebs cycle and cytochrome chain. Stumpy forms have a partially functional Krebs cycle but still lack cytochromes. The glycerophosphate oxidase system functions in terminal respiration of bloodstream forms. The fly gut forms have a fully functional mitochondrion with active Krebs cycle and cytochrome chain. Cytochrome oxidase may be associated with the distinctive platelike cristae of these forms. Reversion to tubular cristae in the salivary gland stages may, therefore, indicate loss of this electron transfer system.

From K. Vickerman, "Morphological and physiological considerations of extracellular blood protozoa," in A. M. Fallis (Ed.), *Ecology and physiology of parasites.* Copyright © 1971 University of Toronto Press. Reprinted by permission.

trypomastigotes have mitochondria much like those of bloodstream forms.

- **Pathogenesis.** In their vertebrate hosts these trypanosomes live in the blood, lymph nodes and spleen, and cerebrospinal fluid. They do not invade or live within cells but inhabit connective tissue spaces within various organs and the reticular tissue spaces of the spleen and lymph nodes. They are particularly abundant in lymph vessels and intercellular spaces of the brain.

  The clinical course in *T. b. brucei* infections depends on susceptibility of the host species. Horses, mules, donkeys, some ruminants, and dogs suffer acutely, and they may die in 15 days, although they may survive for up to four months. Symptoms include anemia, edema, watery eyes and nose, and fever. Within a few days the animals become emaciated, uncoordinated, and paralyzed, and they die shortly afterward. Blindness resulting from infection is common in dogs. Cattle are somewhat more refractory to the disease, often surviving for several months after the onset of symptoms. Swine usually recover.

  In human infections with *T. b. rhodesiense* or *T. b. gambiense,* virulence is determined by interaction of both parasite and human genotype.[105] A small sore (chancre) often develops at the site where metacyclic trypanosomes are inoculated. This lesion disappears after one to two weeks. After the protozoa gain entrance to the blood and lymph channels, they reproduce rapidly, producing a parasitemia and invading nearly all organs of the body. *Trypanosoma b. rhodesiense* rarely invades the nervous system as does *T. b. gambiense* but usually causes a more rapid course toward death. The lymph nodes become swollen and congested, especially in the neck, groin, and legs. Swollen nodes at the base of the skull were recognized by slave traders as signs of certain death, and slaves who developed them were routinely thrown overboard by slavers bound for the Caribbean markets. Today such swollen lymph nodes are called **Winterbottom's sign,** named after the British officer who first described the symptom. The symptoms of illness usually are more marked in newcomers than in people native to endemic areas.

  Intermittent periods of fever accompany early stages of the disease, and the number of trypanosomes in circulating blood increases greatly at these times. As previously noted, successive parasite populations represent different antigenic types. With fever there is an increase in swelling of lymph nodes, generalized pain, headache, weakness, and cramps. Infection by *T. b. rhodesiense* causes rapid weight loss and heart involvement. Death may occur within a few months of infection, but *T. b. rhodesiense* causes no somnambulism or other protracted nervous disorders found with *T. b. gambiense* because the host usually dies before these conditions can develop.

  When *T. b. gambiense* trypanosomes invade the central nervous system, they initiate the chronic, sleeping-sickness stage of infection. Increasing apathy, a disinclination to work, and mental dullness accompany disturbances of coordination. Cardiac involvement is common, but cardiopathy rarely causes severe congestive heart failure and subsides after treatment.[8] Tremor of the tongue, hands, and trunk is common, and paralysis or convulsions usually follow. Sleepiness increases, with the patient falling asleep even while eating or standing. Finally, coma and death ensue. Death may result from any one of a number of related causes, including malnutrition, pneumonia, heart failure, other parasitic infections, or a severe fall.

  The mechanism of pathogenesis is unclear, although recent research shows that *T. b. gamiense* can enter microvascular endothelial cells of the blood-brain barrier.[83] In acute infections of small mammals, in which death occurs rapidly with a high level of parasitemia, mortality probably results from overall disruption of normal physiological processes. In humans, neurological involvement results in demyelinating encephalitis, accompanied by dementia, occasional hallucinations, and decreased consciousness.[9] Melarsoprol, a trivalent arsenic compound, is typically used in treatment of such cases, but the drug also causes a potentially fatal encephalopathy in some patients. Magnetic resonance imaging (MRI) studies have shown brain lesions attributable to trypanosomiasis in the white and gray matter as well as cortex.[9] Subcurative treatment may increase the severity of central nervous system pathology.[28] *Trypanosoma brucei* significantly influences the production of host cytokines.

- **Immunology.** Trypanosomatids present parasitologists with a number of especially challenging immunological problems. In trypanosomes, for example, clinical course of infection varies according to the host infected, but in certain hosts (guinea pig, dog, cow, and rabbit) repeated remissions alternate with very high levels of parasitemia. That is, periods with few trypanosomes (and disease symptoms) evident are followed by a large increase in parasite population. This cycle tends to repeat itself until the host dies or becomes asymptomatic.[90, 98]

  The parasites have evolved an amazing mechanism for escaping obliteration by the host's defenses—namely, **antigenic variation,** resulting from the successive dominance of each of a series of **variable antigen types (VATs)** over time. Remissions result from generation of protective antibodies that destroy the homologous trypanosomes. But each time a host's antibodies are almost successful in eliminating infection, the trypanosomes elude destruction by expressing a new **variant-specific surface glycoprotein (VSG),** thus becoming a new VAT, and then rapidly multiplying.

  The means by which trypanosomes achieve this succession of antigenic types is a fascinating story of gene expression, and modern molecular biology has done much to explain this phenomenon.[4] The VSG recognized by a host's immune system is released through the trypanosome's flagellar reservoir and completely covers the organism as a surface coat. VSGs thus serve as a barrier to antibodies and complement that might act against nonvariant membrane proteins. Bloodstream trypanosomes actively internalize host antibody bound to them; antiVSG IgG (chapter 3) and transferrin, a host protein used to carry iron to organs like the liver, are both taken up by endocytosis and degraded in endosomes.[39] New VSGs are then built and exported onto the cell surface.[86] Each *T. brucei* individual possesses approximately 1000 genes coding for VSGs, making up at least 20% of the genome; but only one VSG gene is expressed at any time. The others are transcriptionally silent.[4]

VSG genes are expressed only when at the ends of chromosomes, in special telomeric positions called **VSG expression sites.**[4] There are evidently three mechanisms that produce the necessary rearrangements. In one of these the gene is duplicated and transposed to another position near the telomere (chromosome end), replacing the resident gene. The transposed DNA segment becomes the expression-linked copy; it is located downstream from a promoter region and is thus transcribed. A second mechanism involves duplicative transposition of an unexpressed telomeric VSG gene to a second telomeric site where it will be expressed. In a third mechanism, inactive telomeric genes become activated and expressed but not duplicated. Although these mechanisms involve telomeric positions, some evidence suggests that there may be two or more expression sites for VSG genes.[80]

Expression of the genes occurs in an imprecisely predictable order; that is, expression of a given VSG more commonly occurs after the expression of another, particular VSG but not invariably. Thus, the VSG genes in a population of trypanosomes are heterogeneous at any time in a chronic infection, but there is a single VAT that is predominant in the blood and against which a host mounts its antibody defense. Other dominant VATs can be found in such places as the brain and liver.[103] Adding to the complexity of the system, trypanosomes can lose VSG genes and add new genes to their repertoire by a mechanism of segmental gene conversion (nonreciprocal crossing over).[87, 88]

When trypanosomes are ingested by a tsetse fly, VSG is replaced by another protein, procyclin, that is proteaseresistant, and thus may provide protection against hydrolytic enzymes in the fly's gut. The VSG is removed by a combination of endocytosis and hydrolysis of the linkage between VSG and an anchoring protein in the parasite's surface membrane.[39] Expression resumes when the trypanosomes reach the metacyclic state and are then able to infect a mammalian host. A much smaller number of VATs characterizes the metacyclic trypanosomes; only eight VATs made up 60% to 80% of the population in one *T. b. rhodesiense* clone.[118] Although the first VSG gene expressed after infection of a mammalian host is one of the metacyclic VATs, within a few days the VSG found on the trypanosome ingested by the fly is expressed. This reexpression of the ingested VAT is referred to as **anamnestic** expression.

Although the VAT story is best known for *T. brucei,* antigen switching is also found in at least some other trypanosomes, such as *T. vivax.*[3] Through studies of immunology and VSG gene expression, parasitologists have kept alive the line of investigation that began in the early 1900s in an attempt to explain the course of trypanosome infections. Now we know that surface proteins are parasite counter-defenses against host defenses in several groups of parasites (pp. 91, 154, 230).

- **Diagnosis and Treatment.** Demonstration of parasites in blood, bone marrow, or cerebrospinal fluid unequivocally establishes diagnosis, but an inexpensive card agglutination (CATT) test to detect antibodies in whole blood or serum is available for use in control programs.[90] Sensitivity is improved by examination of the white cell fraction (buffy coat).[13]

Arsenical drugs historically have been used in treatment of African trypanosomiases, but these drugs have severe drawbacks.[36] They cause eye damage and are best administered intravenously; furthermore, trypanosomes rapidly become tolerant to them. Other drugs (suramin, pentamidine, and Berenil) have been used in recent years and have proved to be satisfactory in most early infections.[58] Prognosis, however, is poor if the nervous system has become involved. Melarsoprol, an arsenical, has been used in late-stage infections, but up to 10% of the patients die from its side effects.[90] Difluoromethylornithine (DFMO) is efficacious in the treatment of African trypanosomiasis, especially in brain infections, but certain parasite strains, especially of *T. b. rhodesiense,* may be innately resistant.[56] Other drugs, such as eflornithine and fexinidazole, are apparently effective and less toxic than melarsoprol.[12]

Nutrition of the vertebrate host can also affect the course of the disease. Adequate dietary lipid limits infectivity of *T. brucei* in rats and possibly also protects people against African sleeping sickness.[85]

- **Epidemiology and Control.** The most important factors influencing transmission are (1) reservoir hosts and (2) the presence of suitable vectors and the environments necessary for these vectors to reproduce. Tsetse flies occupy 4.5 million square miles of Africa (see Fig. 5.6). *Glossina* species that transmit *T. b. brucei* and *T. b. rhodesiense* occur in open country, pupating in dry, friable earth. Vectors of *T. b. gambiense* are riverine flies, breeding in shady, moist areas along rivers. Trypanosomes of the brucei group do not occur throughout the entire range of tsetse flies, and not all species of *Glossina* are vectors for them. Therefore, transmission varies locally, depending on coincidence of the trypanosome and proper fly species. Furthermore, there is an inheritance of susceptibility to trypanosomes in tsetse flies.[71]

Use of molecular techniques has shed some light on the dynamics of *T. b. rhodesiense* epidemics.[48] In Uganda, for example, sleeping sickness epidemics have been short, with long periods of low incidence in between. Isozyme analysis of the parasites suggested that new, perhaps relatively virulent strains were involved in these epidemics. Subsequent work, using DNA markers, however, showed that the parasite strains were ones present in the area for at least 30 years, perhaps even since the early 1900s, and were not likely a result of mutations or genetic exchanges with local *T. b. brucei* strains and that domestic cattle were probably the most important reservoirs.[48]

Control of trypanosomiasis brucei is conducted along several lines, most of which involve vectors. Tsetse flies are larviparous, and they deposit their young on the soil under brush. Because of this behavioral characteristic and because adults rest in bushes at certain heights above the ground and no higher, brush removal and trimming are very successful means of control. When wide belts of land are thus cleared, the flies seldom cross them and can be contained. However, this method is expensive and must be followed up every year to remove new growth. Elimination of wild game reservoirs has been proposed and practiced in some regions, stimulating an outcry among conservation-minded people all over the world. Programs have been established in which people simply sit and catch flies that try to bite them. Because the flies feed only during daytime, some farmers graze their livestock at night, moving them into enclosures during the day and protecting them from flies with switches. See the book edited by Mulligan and Potts[79] for an extensive discussion of fly catching, brush removal, and wild ungulate slaughter as control methods. The ability of the trypanosomes to call forth a seemingly inexhaustible variety of variant surface glycoproteins has led to the conclusion that prospects for control of African trypanosomiasis by vaccination are not good.[69]

Insecticide spraying from aircraft has also been used. DDT and benzene hexachloride are inexpensive and highly effective for this purpose. *Glossina pallidipes* was eradicated from Zululand in this manner at a cost of about $0.40 per acre. However, the possibilities of harmful side effects of DDT must be carefully weighed against the benefits gained by its use. Traps or screens impregnated with insecticides are also effective as control devices under certain environmental conditions, especially in the West African savannah. These devices have replaced spraying as the tsetse control method of choice in many locations, primarily because of reduced environmental impact and economy.[90]

One interesting approach to *T. brucei* control is the development and use of trypanotolerant cattle stocks, which may offer a practical alternative to control of the vectors or trypanosomes.[81] Breeds of cattle that have survived in trypanosome-infested regions of Africa show great promise as trypanotolerant animals. Breeding as few as three generations of trypanotolerant bulls to a trypanosensitive stock will produce a relatively trypanotolerant genotype.[26] However, breeding for tolerance to trypanosomes does not necessarily improve meat or milk production, nor does it confer resistance to other diseases.

Both the political situation in some African countries, often involving decreased cooperation between adjacent nations, and increased mobility of the human population contribute to wider and faster spread of the disease. Trypanosomiasis is an excellent illustration of the way complex interactions among parasite and vector biology, economics, and social factors can make disease control extraordinarily difficult.

Trypanosoma (Nannomonas) congolense and *Trypanosoma (Duttonella) vivax*

Nagana also is caused by *Trypanosoma congolense,* which is similar to *T. brucei* but lacks a free flagellum. It occurs in South Africa, where it is the most common trypanosome of large mammals.[59] The life cycle, pathogenesis, and treatment are as for *T. brucei.* The vascular damage reported in chronic *T. congolense* infections may be a result of the propensity of these trypanosomes to attach to the walls of small blood vessels by their anterior ends.[2]

Trypanosoma vivax is also found in the tsetse fly belt of Africa and has spread to the western hemisphere and Mauritius. Very similar to *T. brucei,* it causes a like disease in the same hosts. In the New World, transmission is mechanical and involves tabanid flies (p. 587). Pathogenesis and control are as for *T. brucei.* Also, changes in the mitochondrion

through the life cycle in *T. vivax* and *T. congolense* are similar to those in *T. brucei.* However, these two species appear to retain some mitochondrial respiratory function in the bloodstream form. A review of *T. vivax* is given by Jones and Dávila.[59]

Trypanosoma (Trypanozoon) evansi and *Trypanosoma (Trypanozoon) equinum*

Trypanosoma evansi causes a widespread disease of camels, horses, elephants, deer, and many other mammals. The disease goes by many different names in different languages and countries but is most often called **surra.** *Trypanosoma evansi* probably was originally a parasite of camels.[49] Today it is distributed throughout the northern half of Africa, Asia Minor, southern Russia, India, southwestern Asia, Indonesia, Philippines, and Central and South America. The Spaniards introduced the disease to the western hemisphere by way of infected horses in the 16th century.

This trypanosome is morphologically indistinguishable from *T. brucei.* Typically it is 15 µm to 34 µm long. Most cells are slender in shape, but stumpy forms occasionally appear. However, the biology of *T. evansi* is quite different from that of *T. brucei.* The life cycle does not involve *Glossina* spp. or development within an arthropod vector. In most areas where it occurs, mouthparts of horse flies (*Tabanus* spp.) transmit the disease mechanically, but flies of genera *Stomoxys, Lyparosia,* and *Haematopota* can also transmit it. In South America, vampire bats are common vectors of the disease, which is known there as **murrina.**[50] Surra is most severe in horses, elephants, and dogs, with nearly 100% fatalities in untreated cases. It is less pathogenic to cattle and buffalo, which may be asymptomatic for months. In camels, the disease is serious but tends to remain chronic. Pathogenesis, symptoms, and treatment are the same as for *T. brucei.*

Trypanosoma (Trypanozoon) equinum occurs in South America, where it causes **mal de caderas,** a disease in horses similar to surra. *Trypanosoma equinum* is similar to *T. evansi* except that it appears to lack a kinetoplast. Actually, a vestigial kinetoplast can be seen in electron micrographs, but it does not function in activation of the mitochondrion; this condition is known as **dyskinetoplasty.** *Trypanosoma brucei* and *T. evansi* can be rendered dyskinetoplastic with certain drugs, and the character is inherited as a mutation. Such altered organisms can survive as bloodstream parasites but no longer can infect flies. *Trypanosoma equinum* also is transmitted mechanically by tabanid flies. Pathogenesis, symptoms, and treatment are as for *T. evansi.*

Trypanosoma (Trypanozoon) equiperdum

Another trypanosome, *T. equiperdum,* also morphologically indistinguishable from *T. brucei,* causes a venereal disease called **dourine** in horses and donkeys. The organisms are transmitted during coitus, and no arthropod vector is known.

Dourine is found in Africa, Asia, southern and eastern Europe, Russia, and Mexico. It was once common in western Europe and North America but has been eradicated from these areas. The disease exhibits three stages. In the first the genitalia become edematous, with a discharge from the urethra and vagina. Areas of the penis or vulva may become depigmented. In the second stage a prominent rash appears on the sides of the body, remaining for three or four days. The third stage produces paralysis, first of the neck and nostrils and then of the hind body; the paralysis finally becomes general. Dourine is usually fatal unless treated.

Diagnosis depends on finding trypanosomes in the blood, genital secretions, or fluids from the large urticarious patches of the skin during the second stage. A complement fixation test is very reliable and was used by U.S. Department of Agriculture (USDA) personnel to ferret out infective horses during their successful campaign to eradicate the disease in the United States. All horses now entering the United States must be tested for dourine before being admitted.

Section Stercoraria

Trypanosoma (Schizotrypanum) cruzi

Trypanosoma cruzi (see Fig. 5.9) carries the unusual distinction of having been discovered and studied several years before it was known to cause a disease. In 1910 a 40-year-old Brazilian, Carlos Chagas, dissected a number of conenosed bugs (Hemiptera, family Reduviidae, subfamily Triatominae) and found their hindguts swarming with trypanosomes of the epimastigote type. The biology and habits of triatomines are discussed in chapter 37.

Chagas sent a number of the bugs to the Oswaldo Cruz Institute, where they were allowed to feed on marmosets and guinea pigs. Trypanosomes appeared in the blood of the animals within a month. Chagas thought the parasites went through a type of schizogony in the lungs, so he named them *Schizotrypanum cruzi.* The name *Schizotrypanum* still is employed by some workers, although most prefer to use it as a subgenus of *Trypanosoma.* By 1916 Chagas demonstrated that an acute, febrile disease, common in children throughout the range of conenosed bugs, was always accompanied by the trypanosome. Unfortunately, he thought that goiter and cretinism also were caused by this parasite, and, when these hypotheses were disproved, suspicion was cast on the rest of his work. Also, Chagas maintained to near the end of his life that transmission of the disease, which now bears his name, occurred through the bite of the insect. Not until the early 1930s was it shown that **Chagas' disease** was transmitted by way of the feces of conenosed bugs.

Trypanosoma cruzi is distributed throughout most of South and Central America, where an estimated 12 to 19 million persons were infected in the early 1990s, with an annual incidence of 561,000.[25] Subsequent intergovernmental eradication efforts evidently reduced that number to around 11 million, but nearly 25% of the people in Latin America (~120 million) remain at risk. Globally, Chagas' disease "represents the third-largest parasitic disease burden after malaria and schistosomiasis."[44] Molecular evidence indicates humans may have been suffering from Chagas' disease for at least 4000 years.[41] Many kinds of wild and domestic mammals serve as reservoirs (Fig. 5.9). Animals that live in proximity to humans, such as dogs, cats, opossums, armadillos, and wood rats, are particularly important in the epidemiology of Chagas' disease.

In the United States, *T. cruzi* has been found in Maryland, Georgia, Florida, Texas, Arizona, New Mexico, California, Alabama, and Louisiana. Fourteen species of infected mammals have been found in the United States.[60] The first indigenous infection in a human in the United States was reported in 1955.[128] Since then a number of cases have been reported in Arizona, mainly on Indian reservations. Several North American strains have been isolated. They are morphologically indistinguishable from any other *T. cruzi,* but they seem to be much less pathogenic. It is possible that this infection in humans is more widespread in the United States than is now known; surveys using immunological tests showed 0.8% positive cases among a random sample of 500 people from the lower Rio Grande Valley of Texas.[11]

- **Morphology.**   Trypomastigotes are found in circulating blood. They are slender, 16 μm to 20 μm long, and their posterior end is pointed. Their free flagellum is moderately long, and the undulating membrane is narrow, with only two or three undulations at a time along its length. The kinetoplast is subterminal and is the largest of any trypanosome; it sometimes causes the body to bulge around it. The protozoan commonly dies in a question mark shape, the appearance it retains in stained smears (Fig. 5.8).

  Amastigotes develop in muscles and other tissues. They are spheroid, 1.5 μm to 4.0 μm wide, and occur in clusters composed of many organisms. Intermediate forms are easily found in smears of infected tissues.

- **Biology.**   When reduviid bugs feed (Fig. 5.9), they often defecate on their host's skin. The feces may contain metacyclic trypanosomes, which gain entry into the body of a vertebrate host through the bite, through scratched skin, or, most often, through mucous membranes that are rubbed with fingers contaminated with the insects' feces. Also, reservoir mammals can become infected by eating infected insects.[129] Although trypomastigotes are abundant in the blood in early infections, they do not reproduce

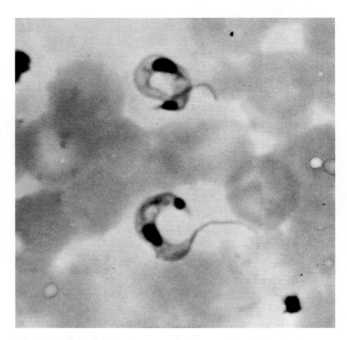

Figure 5.8   *Trypanosoma cruzi.*
Trypomastigote form in a blood film.
Courtesy of Ann Arbor Biological Center.

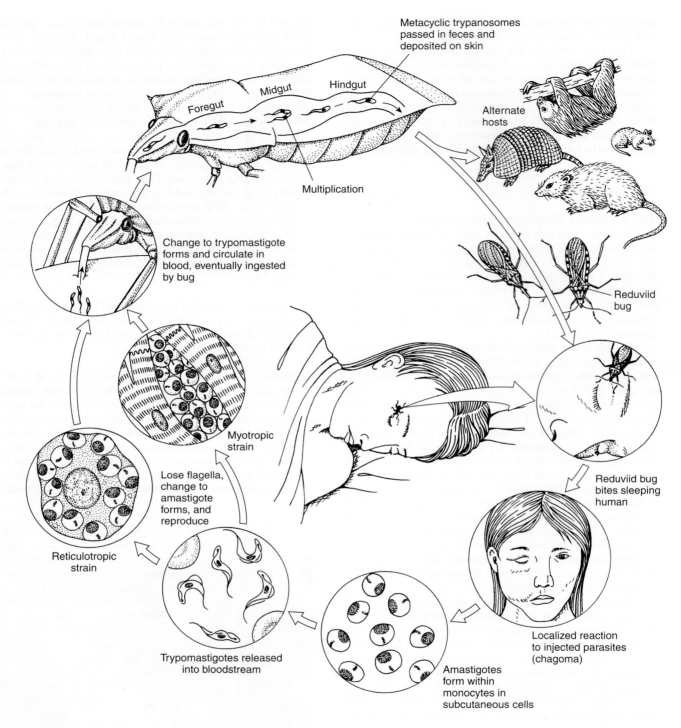

Metacyclic trypanosomes passed in feces and deposited on skin

Foregut

Midgut

Hindgut

Multiplication

Alternate hosts

Reduviid bug

Change to trypomastigote forms and circulate in blood, eventually ingested by bug

Myotropic strain

Lose flagella, change to amastigote forms, and reproduce

Reticulotropic strain

Reduviid bug bites sleeping human

Trypomastigotes released into bloodstream

Amastigotes form within monocytes in subcutaneous cells

Localized reaction to injected parasites (chagoma)

Figure 5.9   **Life cycle of *Trypanosoma cruzi*.**

Drawing by William Ober and Claire Garrison.

until they have entered a cell and have transformed into amastigotes. Most frequently invaded are cells of the spleen, liver, and lymphatics and cells in cardiac, smooth, and skeletal muscles. Nervous system, skin, gonads, intestinal mucosa, bone marrow, and placenta also are infected in some cases. There is some evidence that trypanosomes can actively penetrate host cells, but they may also enter through phagocytosis by host macrophages.

The undulating membrane and flagellum disappear soon after the parasite enters a host cell. Repeated binary fission produces so many amastigotes that the host cell soon is killed and lyses. When released, the protozoa attack other cells. Cystlike pockets of parasites, called **pseudocysts,** form in muscle cells (Fig. 5.10). Intermediate forms (promastigotes and epimastigotes) can be seen in the interstitial spaces. Some of these complete

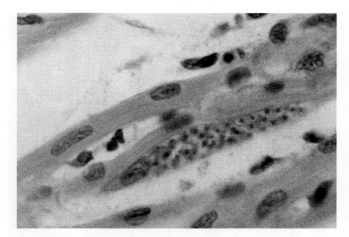

Figure 5.10   *Trypanosoma cruzi* **pseudocyst in cardiac muscle.** (× 780)

Courtesy of S. S. Desser.

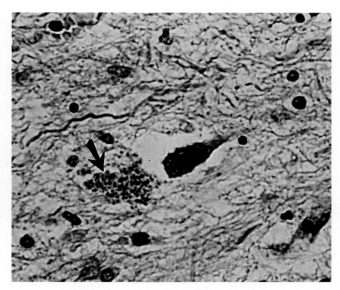

Figure 5.11   **Pseudocyst of** *Trypanosoma cruzi* **in brain tissue.**

AFIP neg. no. 67-5313.

metamorphosis into trypomastigotes and find their way into the blood.

Trypanosoma cruzi is a "partial aerobic fermenter"; some of the glucose carbon it consumes is degraded completely to carbon dioxide, but it excretes a substantial portion as succinate and acetate.[14] The oxygen consumption of blood and intracellular forms is the same as that of culture forms, and bloodstream forms apparently have a Krebs cycle and classical cytochrome system.[43] Thus, *T. cruzi* differs from African trypanosomes in the spectrum of metabolic properties displayed at various life-cycle stages.

Trypomastigotes that are ingested by a triatomine bug pass through to the posterior portion of the insect's midgut, where they become short epimastigotes, which in turn multiply by longitudinal fission to become long, slender epimastigotes. Short metacyclic trypomastigotes appear in the insect's rectum 8 to 10 days after infection. Metacyclic forms pass with feces and can infect a mammal if rubbed into a mucous membrane or wound. First-generation amastigotes in an insect's stomach group together to form aggregated masses. These masses fuse and may represent a primitive form of sexual reproduction, although some researchers dispute this interpretation.[114]

• **Pathogenesis.**   Entrance of metacyclic trypanosomes into cells of subcutaneous tissue produces an acute local inflammatory reaction. Within one to two weeks of infection, the parasites spread to regional lymph nodes and begin to multiply in the cells that phagocytose them. Intracellular amastigotes undergo repeated divisions to form large numbers of parasites, producing the so-called *pseudocyst.* After a few days some of the organisms retransform into trypomastigotes and burst out of the pseudocyst. A generalized parasitemia occurs then, and parasites invade almost every type of tissue in the body, although they show a particular preference for muscle and nerve cells (Fig. 5.11).

Reversion to amastigote, pseudocyst formation, retransformation to trypomastigote, and pseudocyst rupture are repeated in newly invaded cells; then the process

begins again. Rupture of a pseudocyst is accompanied by an acute, local inflammatory response, with degeneration and necrosis (cell or tissue death) of nerve cells in the vicinity, especially of ganglion cells. This degeneration is an important pathological change in Chagas' disease, and it appears to be the indirect result of parasitism of supporting cells, such as glial cells and macrophages, rather than of invasion of neurons themselves.[109]

Chagas' disease manifests *acute* and *chronic* phases. The acute phase is initiated by inoculation of trypanosomes from the bug's feces into the wound. The local inflammation produces a small red nodule, known as a **chagoma,** with swelling of the regional lymph nodes. In about 50% of cases, trypanosomes enter through the conjunctiva of the eye, causing edema of the eyelid and conjunctiva and swelling of the preauricular lymph node. This symptom is known as **Romaña's sign.** As the acute phase progresses, pseudocysts may be found in almost any organ of the body, although the intensity of attack varies from one patient to another. Heart muscle usually is invaded, with up to 80% of cardiac ganglion cells being lost. Symptoms of the acute phase include anemia, loss of strength, nervous disorders, chills, muscle and bone pain, and varying degrees of heart failure. Death may ensue three to four weeks after infection.

The acute stage is most common and severe among children less than five years old. The chronic stage, however, is most often seen in adults. Its spectrum of symptoms is primarily the result of central and peripheral nervous dysfunction, which may last for many years. Some patients may be virtually asymptomatic and then suddenly succumb to heart failure. Chagas' disease accounts for about 70% of cardiac deaths in young adults in endemic areas. Part of the inefficiency in heart function is caused by loss of muscle tone resulting from destroyed nerve ganglia (Fig. 5.12). The heart itself becomes greatly enlarged and flabby.

Host and parasite genetic makeup, sex, age, prior infection, and a variety of other factors influence disease

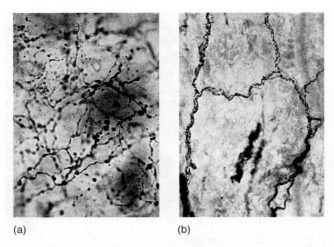

(a)   (b)

Figure 5.12   Diaphanised tricuspid valves with zincosmium impregnation of nerve fibers (*dark lines*).
(*a*) Normal heart; (*b*) Chagas' cardiopathy with marked reduction of nerve fibers.

From M. S. R. Hutt, F. Koberle, and K. Salfelder, in H. Spencer (Ed.), *Tropical pathology.* Springer Verlag, 1973.

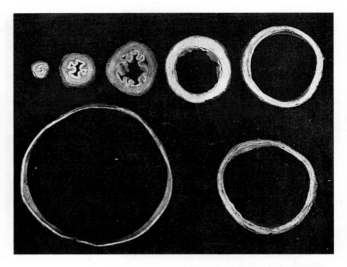

Figure 5.13   Different stages of Chagasic esophagopathy beginning with a normal organ and passing through hypertrophy and dilatation to the final megaesophagus.

From M. S. R. Hutt, F. Koberle, and K. Salfelder, in H. Spencer, editor, *Tropical Pathology,* 1973 Springer Verlag.

development, and relationships among these factors are still unresolved. Autoimmunity is still a controversial explanation for pathology of *T. cruzi* infections. Autoantibodies against host myosin appear in *T. cruzi* infections, but such antibodies also occur in a variety of patients, including some who have undergone bypass surgery and others who are healthy. Furthermore, such antibodies tend to appear within weeks after infection, but pathology associated with chronic infection tends to develop over much longer periods.[62]

Tarleton[111] reviewed an ingenious set of experiments involving heart transplants in mice. Syngeneic hearts (with the same genetic makeup) were accepted, but allogeneic hearts (with different genetic makeup) were rejected. Mice with chronic *T. cruzi* infections, however, rejected syngeneic hearts, a reaction that could be countered by depleting recipients' CD4+ cells. Later studies, however, produced different results, demonstrating that rejection of syngeneic hearts was a result of parasite-induced inflammation.[112] Thus the respective roles of autoantibodies and autoreactive T cells in Chagas' disease pathology remains unclear and somewhat controversial. For a recent review of this controversy, see Kierszenbaum.[62]

In some regions of South America it is common for autonomic ganglia of the esophagus or colon to be destroyed. This ruins the tone of the muscle layer, resulting in deranged peristalsis and gradual flabbiness of the organ, which may become huge in diameter and unable to pass materials within it. This advanced condition is called **megaesophagus** or **megacolon,** depending on the organ involved (Fig. 5.13). Advanced megaesophagus may be fatal when the patient can no longer swallow. It has been demonstrated experimentally that testis tubules and epididymis also atrophy in chronic cases.[29]

• **Immunology.**   *Trypanosoma cruzi,* spending most of its vertebrate host phase as an intracellular parasite, presents us with some immunological phenomena quite different

from those of the bloodstream trypanosomes. As in the case of leishmanial parasites, host reactions to *T. cruzi* infections are largely cellular, especially during the acute phase. It is still not clear exactly how these cellular reactions are involved in control of the disease, although studies show that parasite membrane glycoproteins stimulate host cytokine production, which, in turn, enhances macrophage killing capacity linked to NO production.[113]

As is also the case with *Leishmania* research, our most detailed information on immunology comes from a study of infections in mice. The overriding early event of an experimental *T. cruzi* infection is immunosuppression, which may be responsible also for some of the parasites' pathological effects.[110] But mice also evidently kill vast numbers of injected trypomastigotes, which means that some mechanism(s) is at work to help protect the host regardless of the level of prior exposure. Production of the cytokine interleukin-2 (IL-2) is suppressed during the acute phase, an event that, in turn, affects T-cell growth. Low levels of IL-2 are matched by those of the corresponding mRNA, so regulation of the cytokine is probably at the level of transcription.[82] Production of other cytokines is not generally suppressed, however, and IFN-γ levels are elevated.

Chronic infections (sometimes called *postacute* to distinguish them from long-lasting infections resulting in cardiac and gut pathology) are controlled mainly by humoral responses, and, in some mouse/parasite strain combinations, circulating IgG is protective. Such antibody fixes complement and lyses trypanosomatids. Nonprotective antibody is also produced, however, and remains in the blood even after drug cure. Such antibody can be used in serological tests for chronic or past infections.

There is disagreement among parasitologists on the sugnificance of autoimmunity in Chagas' disease. For example, infection results in both a strong blast transformation response (mitogenesis) in lymphocytes in general (polyclonal activation) and in elevated levels of circulating immuno-globulins.[93] Much of this immunoglobulin does not "recognize" parasite antigens, and the lymphocyte populations stimulated by infection may contain T- and B-cell clones that are autoreactive. Infected cardiac muscle cells eventually rupture, releasing amastigotes and provoking an inflammatory response with infiltrations of lymphocytes and macrophages. This process can lead eventually to fibrosis and loss of cardiac muscle's ability to conduct impulses. It still is not clear whether autoantibodies are involved in this pathology. The exchange of views on this subject by Kierszenbaum and Hudson[53, 61] is an excellent illustration of a gentlemanly debate, still unresolved, over a very complex subject.

• **Diagnosis and Treatment.**   Diagnosis usually is by demonstration of trypanosomes in blood, cerebrospinal fluid, fixed tissues, or lymph. Trypomastigotes are most abundant in peripheral blood during periods of fever; they may be difficult to find at other times or in cases of chronic infection. In these other cases blood can be inoculated into guinea pigs, mice, or other suitable hosts, and the animals in turn can be examined by heart smear or spleen impression. Another widely employed method is **xenodiagnosis.** Laboratory-reared triatomines are allowed to feed on a patient; after a suitable period of time (10 to 30 days) the bugs are examined for intestinal flagellates. This technique can detect cases in which trypanosomes in the blood are too few to be found by ordinary examination of blood films.

Complement fixation or other immunodiagnostic tests are effective in demonstrating chronic cases, although they may give false positive reactions if the patient is infected with a *Leishmania* species or another species of trypanosome. In experiments with infected opossums, *Didelphis marsupialis,* an indirect fluorescent antibody test (IFAT) was the most sensitive test for *T. cruzi,* followed by xenodiagnosis.[57] Antigens excreted or secreted by the parasites have been used in immunoblot assays.[120] Both flagellar and cytoplasmic *T. cruzi* proteins have been cloned in *Escherichia coli* and used in ELISAs; these assays are much more specific to *T. cruzi* as well as more sensitive than those using crude antigen preparations.[63] Furthermore, the use of recombinant technology reduces overall costs of diagnostic reagent production. Dot-immunobinding assays using antigen bound to nitrocellulose paper offer the advantage of requiring very small amounts of fluid and, because they need no expensive equipment, show promise for use under field conditions.[17]

Diagnostic methods based on detection of parasite DNA using polymerase chain reaction (PCR) techniques have also been developed, but so far they have not come into general use because of the problem of false negatives.[63]

Unlike other trypanosomes of humans, *T. cruzi* does not respond well to chemotherapy. The most effective drugs kill only extracellular protozoa, but intracellular forms defy the best efforts at eradication. Reproductive stages, inside living host cells, seem to be shielded from drugs. The lives and strength of millions of Latin American people depend on discovery of a drug or vaccine that is effective against *T. cruzi.* One hope is the drug ketoconazole, which completely cured 78.5% of otherwise fatally infected mice.[73] Nifurtimox and benznidazole have

been shown to be somewhat effective in curing acute infections, but they required long treatment duration and had significant side effects, and patients remained seropositive even after the disappearance of parasites from the blood.[25] Plasmids containing genes for parasite proteins, especially transialidase, have been used experimentally as DNA-based vaccines in mice. These vaccines produced the best results when used concurrently with plasmids containing various cytokine genes.[34] Pinazo and coworkers reported successful treatment of chronic Chagas disease with posaconazole.[92]

• **Epidemiology.**   The principal vectors of *Trypanosoma cruzi* in Brazil are *Panstrongylus megistus, Triatoma sordida,* and *Triatoma brasiliensis.* In Uruguay, Chile, and Argentina, *Triatoma infestans* is the primary culprit (Fig. 5.14). Argentina, Bolivia, Brazil, Chile, Paraguay, and Uruguay have joined forces in an attempt to eradicate *T. infestans.*[101] Two hundred million dollars have been spent in this effort, nearly 2 million houses have been sprayed, and obligate screening of blood donors has been initiated. *Rhodnius prolixus* is the main vector in northern South America and *Triatoma dimidiata* is the main vector in Central America. *Triatoma barberi* is an important vector in Mexico, and the world's largest triatomine— *Dipetalogaster maximus*—sucking up large quantities of blood, is a vector in southern Baja California.[44] Several other species of triatomines have been found infected throughout this range. Natural infections in *T. sanguisuga* have been found in the United States. The insects can become infected as nymphs or adults. Triatomines can infect themselves when they feed on each other, presumably by sucking the contents of the intestine. Ticks,

Figure 5.14   **Distribution of Chagas' disease in humans and of its four principal vectors.**

AFIP neg. no. 65-5015.

sheep keds, and bedbugs have been experimentally infected, but there is no evidence that they serve as natural vectors. Mammalian reservoirs of infection have been mentioned, but domestic dogs and cats probably are the most important to human health. Because the bugs hide by day, primitive or poor-quality housing favors their presence. Thatched roofs, cracked walls, and trash-filled rooms are ideal for the breeding and survival of the insects. Misery compounds itself.

Transmission from human to human during coitus or through breast milk may be possible, although this has yet to be documented. *Trypanosoma cruzi* can and does cross the placental barrier from mother to fetus. Newborn infants with advanced cases of Chagas' disease, including megaesophagus, have been described in Chile. In one study in Bolivia the prevalence among pregnant women was 42%, but the incidence among their newborns was 2.6%.[99] In some Mexican villages people believe that triatomines are aphrodisiacs; therefore, they are eaten, and the trypanosomes gain access through the oral mucosa.[100] The victim's age is important in Chagas' disease epidemiology, and most new infections are in children less than two years old. The acute phase is most often fatal in this age group. Finally, the hazard of transmission by blood transfusion from donors with cryptic infection should not be underestimated.[97] In the United States, blood donors who have traveled to endemic areas are routinely asked whether they have Chagas' disease, but unless they know about the parasite, this question may not be answered correctly.

Trypanosoma (Herpetosoma) rangeli

Trypanosoma rangeli first was found, as was *T. cruzi,* in a triatomine bug in South America. *Rhodnius prolixus* is the most common vector, but *Triatoma dimidiata* and other species will also serve. Development is in the hindgut, and the epimastigote stages that result are from 32 μm to more than 100 μm long. The kinetoplast is minute, and the species can thereby be differentiated from *T. cruzi,* with which it often coexists.

Trypanosoma rangeli is common in dogs, cats, and humans in Venezuela, Guatemala, Chile, El Salvador, and Colombia. It has been found in monkeys, anteaters, opossums, and humans in Colombia and Panama. The trypomastigotes, 26 μm to 36 μm long, are larger than those of *T. cruzi.* The undulating membrane is large and has many curves. The nucleus is preequatorial, and the kinetoplast is subterminal.

The method of transmission is unclear. Although development is by posterior station, transmissions both by fecal contamination and by feeding inoculation have been reported.[116] *Trypanosoma rangeli* multiplies by binary fission in a mammalian host's blood. No intracellular stage is known, and the organism is apparently not pathogenic either in humans or in experimentally infected dogs, monkeys, opossums, or raccoons.[51] However, infections with *T. rangeli* or mixed infections with *T. rangeli* and *T. cruzi* are potential problems for diagnosis.[40] Conventional immunofluorescence and ELISA assays, reinforced by immunoprecipitation and Western blot analysis, diagnose either or both infections.

Trypanosoma (Herpetosoma) lewisi

Trypanosoma lewisi (Fig. 5.15) is a cosmopolitan parasite of *Rattus* spp. Other rodents, including white-footed mice, deer mice, and kangaroo rats in the United States, are infected

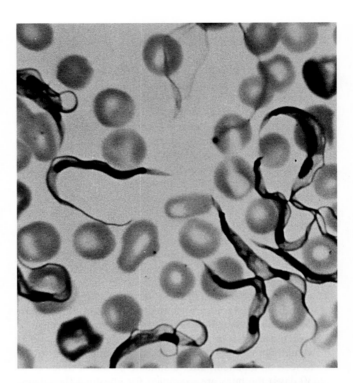

Figure 5.15 *Trypanosoma lewisi* **trypomastigotes in the blood of a rat.**

Courtesy of Turtox/Cambosco.

with lewisilike trypanosomes, but it is not completely clear whether these are the same species as that found in *Rattus* or a form more closely related to *T. musculi,* a species restricted to mice. The vector of *T. lewisi* is the northern rat flea, *Nosopsyllus fasciatus,* in which parasites develop inside posterior midgut cells. Metacyclic trypomastigotes appear in large numbers in the insect's rectum, infecting rats that eat fleas or their feces. The parasite seems to be nonpathogenic in most cases, but infection may contribute to abortion and arthritis.

Much research has been conducted on this species because of the ease of maintaining it in laboratory rats. One fascinating subject of this research is the "ablastin" phenomenon.[121] **Ablastin** is an antibody that arises during the course of an infection. After a rat is infected by metacyclic trypomastigotes the parasites begin reproducing as epimastigotes in the visceral blood capillaries. After about five days trypanosomes appear in peripheral blood as rather "fat" forms, and shortly thereafter a crisis occurs in which most of these trypanosomes are killed by a trypanocidal antibody. A small population of slender trypomastigotes remains; they are infective for fleas but do not reproduce further while in the rat. After a few weeks the host produces another trypanocidal antibody, which clears the remaining trypanosomes, and the infection is cured. The slender trypomastigotes are sometimes known as *adults.* Their reproduction is inhibited by the ablastin, a globulin with many characteristics of a typical antibody but that which inhibits reproduction. Nucleic acid and protein synthesis by the trypanosome is inhibited, as is uptake of nucleic acid precursors. However, it is still not clear how this antibody functions.[1]

Trypanosoma (Megatrypanum) theileri

Trypanosoma theileri is a cosmopolitan parasite of cattle. The vectors are horse flies of the genera *Tabanus* and *Haematopota*. Trypanosomes reproduce in the fly gut as epimastigotes.

The size of *T. theileri* varies with strain—from 12 µm to 46 µm, 60 µm to 70 µm, and even up to 120 µm in length. The posterior end is pointed, and the kinetoplast is considerably anterior to it. Both trypomastigote and epimastigote forms can be found in the blood. Reproduction in vertebrate hosts is in the epimastigote form and apparently occurs extracellularly in the lymphatics.

Trypanosoma theileri is usually nonpathogenic, but under conditions of stress it may become quite virulent. When cattle are stressed by immunization against another disease, undergo physical trauma, or become pregnant, the parasite may cause serious disease.

This parasite is rarely found in routine blood films. Detection usually depends on in vitro cultivation from blood samples. In fact, during tissue culture of bovine blood or cells, *T. theileri* is the most commonly found contaminant. Strong evidence points to transplacental transmission. In the United States a similar trypanosome is also common in deer and elk.

Other *Trypanosoma* Species

Other species of *Trypanosoma* are common in other classes of vertebrates—for example, *T. percae* in perch, *T. granulosum* in eels, *T. rotatorium* in frogs, *T. avium* in birds, and incompletely known species in turtles and crocodiles. Genetic analysis suggests that a species found in Brazilian caimans also is most closely related to a species found in African crocodiles.[122] Trypanosomes are commonly found in a variety of marine fishes.

GENUS *LEISHMANIA*

Like most trypanosomes, leishmanias are heteroxenous. Part of their life cycle is spent in a sand fly gut, where they become promastigotes; the remainder of their life cycle is completed in vertebrate tissues, where only amastigotes are found. Traditionally, amastigotes are also known as **Leishman-Donovan (L-D) bodies.** Vertebrate hosts of *Leishmania* spp. are primarily mammals. Nearly a dozen species have been reported from lizards, but those species now are placed in subgenus *Sauroleishmania,* based on their biochemical and immunological characteristics.[96] Mammals most commonly infected with *Leishmania* spp. are humans, dogs, and several species of rodents. The parasites cause a complex of diseases called **leishmaniasis.** In some cases, especially with many Old World cutaneous infections, leishmaniasis is a zoonosis, with a wild mammal reservoir, for example, gerbils.

Species in humans are widely distributed (Fig. 5.16). It is likely that transport of slaves to the Western world from Africa through the Middle East and Asia spread *Leishmania* species into previously uncontaminated areas, where they now evidently are evolving rapidly into new strains. As is the case with virtually all infectious diseases, air travel generates the potential for quick spread of leishmanial parasites.

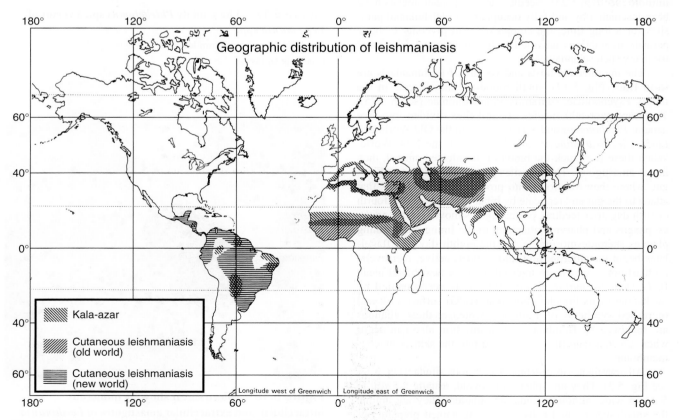

Figure 5.16   **Geographical distribution of leishmaniasis.**

AFIP neg. no. 68-1805-2.

It is not always easy to estimate the numbers of people infected or at risk of acquiring a parasitic disease, especially on a global basis. One relatively recent estimate suggests about 12 million people are infected, with at least a million new cases annually, in 88 countries.[24] Accurate public health records are not always easy to compile in developed nations, much less in those with less than ideal health-care delivery systems. Leishmaniasisinfected areas also broadly overlap areas in which human immunodeficiency virus (HIV) infections are increasing, and about a third of these patients die during their first visceral leishmaniasis episode.[24, 94] Co-infections with HIV and *Leishmania* spp. have been reported from 35 countries.

Leishmania species present us with some of the most baffling problems in immunology, many of which must be solved to diagnose, treat, or prevent infections. First, within a vertebrate's body, the parasites live inside macrophages, the very cells that in most cases function to kill invading organisms. Second, within macrophages, amastigotes reside in phagolysosomes, compartments that normally function directly to digest foreign particles. Third, *Leishmania* species differ markedly among themselves in terms of clinical manifestations, producing infections that range from self-healing cutaneous ones to fatal visceral involvements or to extremely disfiguring afflictions that erode facial features. Fourth, the contributions of human host genetic makeup and nutritional state to the course of infection have yet to be completely described. And finally, drug treatment may precipitate a subsequent clinical manifestation quite different from that of the original infection, such as the **post–kalaazar dermal leishmanoid** (see Fig. 5.23). Needless to say, parasitologists have been fascinated by and fully occupied with leishmanial parasites for a long time; the complexity and diversity of host/ parasite interactions have led to the nickname *leishmaniac* for many such scientists.

The intermediate hosts and vectors of leishmaniasis are **sand flies** (Fig. 5.17), small blood-sucking insects in the family Psychodidae, subfamily Phlebotominae (see p. 575). There are over 600 species of sand flies divided into five genera: *Phlebotomus* and *Sergentomyia* in the Old World and *Lutzomyia, Brumptomyia,* and *Warileya* in the New World. When these flies suck the blood of an infected animal, they ingest amastigotes. The parasites pass to the midgut or hindgut, where they transform into procyclic promastigotes that attach to the gut and replicate by binary fission. By the fourth or fifth day after feeding, promastigotes move forward to the esophagus and pharynx, attaching to the lining and forming plaques (hemidesmosomes). By the eighth day, the flagellates begin metamorphosing into slender, active, metacyclic promastigotes, which are injected with the next blood meal. In *Leishmania major,* metamorphosis is accompanied by thickening of its lipophosphoglycan (LPG) surface coat and increased synthesis of a surface protease; these changes are reviewed by Handman.[46] Transmission also can occur when infected sand flies are crushed into the skin or mucous membrane.

All amastigotes in vertebrate tissues look similar (Fig. 5.18; see Fig. 5.3). They are spheroid to ovoid, usually 2.5 μm to 5.0 μm wide, although some are smaller. They are among the smallest nucleated cells known. In stained preparations only the nucleus and a very large kinetoplast can be seen, and the cytoplasm appears vacuolated. Exceptionally a short

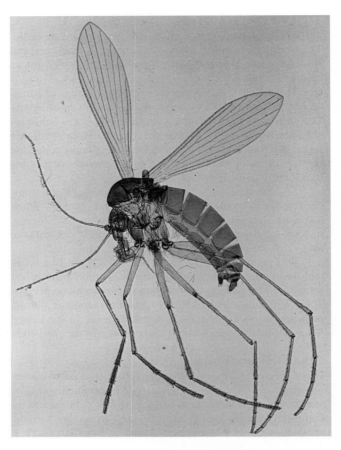

Figure 5.17   **The sand fly *Phlebotomus* sp., a vector of *Leishmania* spp.**

Sand flies are about 3 mm long.

Courtesy of Jay Georgi.

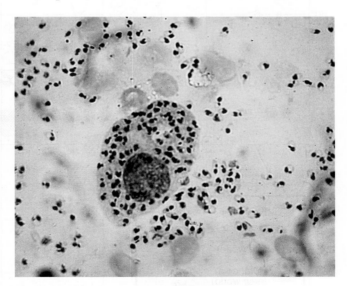

Figure 5.18   **Spleen smear showing numerous intracellular and extracellular amastigotes of *Leishmania donovani*.**

AFIP neg. no. 55-17580.

axoneme is visible within the cytoplasm under the light microscope.

Leishmania species' amastigotes differ in their biochemical properties, especially their membrane components, and the mechanisms by which amastigotes survive and proliferate may not be identical in every species.[46] There is evidence that membrane-bound lipophosphoglycans contribute to virulence, but species differ in this regard. For example, *Leishmania major* strains lost virulence when genes responsible for synthesis of these molecules were knocked out, but that was not the case with *L. mexicana*.[117]

Although all *Leishmania* spp. exhibit similar morphology, they differ clinically, biologically, and serologically. Even so, these characteristics often overlap, so distinctions between species are not always clear-cut. Leishmaniases that normally are visceral may become dermal; dermal forms can become mucocutaneous; and an immunodiagnostic test derived from the antigens of one species may give positive reactions in the presence of other species of *Leishmania* or even *Trypanosoma*.

The older literature is sometimes confusing because some researchers referred to several species while other researchers considered the same organisms as a single, widespread species with slightly different clinical manifestations but similar or identical immunological properties. As a result of the difficulty in species definition within *Leishmania*, strains and species have been characterized biochemically, through use of isoenzymes, RFLP and RAPD analysis, and various gene sequences.[33, 106]

The difficulty of identifying *Leishmania* species extends to the forms in the sand fly. Identification of parasites in their vectors is often critical if vector control is part of an overall disease control strategy. It does not help to focus an attack on one species of vector if it is not carrying the parasite or to be fooled into trying to eliminate a species that carries a parasite morphologically identical to those found in but not infective for humans. Attempts to solve this problem through biochemical methods such as tests based on hybridization of known kDNA with that of parasites have been partially successful.[102] Promising approaches involve use of PCR to amplify kinetoplast DNA from samples and parasite species-specific probes for use in dot hybridization tests.[23, 70] For example, there is evidence that such tools can be used to distinguish *L. donovani* strains producing the disfiguring post–kala-azar dermal leishmanoid from those that do not.[22]

Treatment of leishmanial infections varies according to the clinical manifestations. In earlier years trivalent antimonials were the only drugs available, but they were so toxic as to be downright dangerous. Pentavalent antimonials have been used extensively, but they also are toxic and usually must be administered under the care of a physician. Two pentavalent preparations are available: Pentostam and Glucantime; only Pentostam is available in the United States, through the Centers for Disease Control and Prevention parasite drug service. Drug resistance has been reported in some strains of some species.[38] Furthermore, relapses and post–kala-azar dermal leishmanoid may follow insufficient treatment.

Some of the most creative, although still largely experimental, approaches to treatment involve turning liability into an asset (so to speak)—the liability being the fact that in visceral infections the parasites are located within macrophages. Macrophages will eat foreign particles, so injected drugs such as amphotericin B are bound to artificial particles— liposomes or colloidal particles—to enhance their efficacy by delivering them to the cells where the amastigotes reside.[20]

It has long been known that tropical forests are a rich source of plant molecules with potential medicinal uses. The rapid disappearance of these forests has led to renewed interest in natural plant products, including those that may be effective against leishmanial infections. So far, antileishmanial activity has been found in a number of plant species, including those from the families Apocynaceae (dogbanes), Gentianaceae (gentians), and Euphorbiaceae.[104]

In late 2002, miltefosine (hexadecylphosphocholine), an orally administered drug originally developed for cancer patients, was reported to cure 98% of visceral leishmaniasis cases.[107] Miltefosine is now licensed for use in India. An oral, relatively nontoxic drug effective against a virtually fatal parasitic infection is a health professional's dream; miltefosine seems to fulfill that dream, although its mode of action is still not known.[107] Several studies indicate miltefosine also is effective against at least some cutaneous infections.[26a]

Species of flagellate that develop in the sand fly's midgut before moving anteriorly are placed in subgenus *Leishmania*. Those that develop in the hindgut first are placed in subgenus *Viannia*. Subgenus *Leishmania* includes causative organisms of both Old World and New World visceral and cutaneous leishmaniasis. Members of *Viannia* are New World cutaneous parasites including some of the most disfiguring ones.[65] Species and subspecies of *Leishmania* infecting mammals are listed in Table 5.1; most authors now refer to biochemically related groups as *species complexes*. The taxa are in general agreement with species separation according to isozyme patterns, mainly of glycolytic and Krebs cycle enzymes as well as transaminases.[65, 96] Of those species complexes we will consider the six most important to human welfare: *L. tropica, L. major, L. mexicana, L. braziliensis, L. donovani,* and *L. infantum*.

Cutaneous Leishmaniasis

Leishmania tropica and *L. major.*

Leishmania tropica and *L. major* produce cutaneous ulcers variously known as **oriental sore, cutaneous leishmaniasis, Jericho boil, Aleppo boil,** and **Delhi boil.** They are found in west central Africa, the Middle East, and Asia Minor into India. These two species have similar life cycles; however, *L. tropica* and *L. major* are found in different localities and have different reservoir and intermediate hosts. The lesions they cause also are somewhat different, although in humans the lesions may vary in severity according to age and other factors. The two species can be differentiated biochemically.

- **Morphology and Life Cycle.**   Amastigotes of *L. tropica* and *L. major* are similar to those of the other leishmanias (see Figs. 5.3 and 5.18). Sand flies of genus *Phlebotomus* are the intermediate hosts and vectors. When a fly takes a blood meal containing amastigotes, parasites multiply in the midgut and then move to the pharynx; they are then inoculated into the next mammalian victim. There they multiply in the reticuloendothelial system and lymphoid cells of the skin. Few amastigotes are found except in the immediate vicinity of the site of infection, so the sand flies

Table 5.1	Species and Subspecies of *Leishmania* Infecting Mammals

Parasite	Locality
SUBGENUS *LEISHMANIA* ROSS, 1903	
L. donovani phenetic complex	
L. donovani (Laveran and Mesnil, 1903)	India, China, Bangladesh
L. archibaldi Castellani and Chalmers, 1919	Sudan, Ethiopia
L. infantum phenetic complex	
L. infantum Nicolle, 1908	North central Asia, northwest China, Middle East, southern Europe, northwest Africa
L. chagasi Cunha and Chagas, 1937	South and Central America
L. tropica phenetic complex	
L. tropica (Wright, 1903)	Urban areas of Middle East and India
L. killicki Rioux, Lanotte, and Pratlong, 1986	Tunisia
L. major phenetic complex	
L. major Yakimoff and Schokhor, 1914	Africa, Middle East, Soviet Asia
L. gerbilli phenetic complex	
L. gerbilli Wang, Qu, and Guan, 1973	China, Mongolia
L. arabica phenetic complex	
L. arabica Peters, Elbihari, and Evans, 1986	Saudi Arabia
L. aethiopica phenetic complex	
L. aethiopica Bray, Ashford, and Bray, 1973	Ethiopia, Kenya
L. mexicana phenetic complex	
L. mexicana Biagi, 1953	Mexico, Belize, Guatemala, south central United States
L. amazonensis Lainson and Shaw, 1972	Amazon Basin, Brazil
L. venezuelensis Bonfante-Garrido, 1980	Venezuela
L. enrietti phenetic complex	
L. enrietti Muniz and Medina, 1948	Brazil
L. hertigi phenetic complex	
L. hertigi Herrer, 1971	Panama, Costa Rica
L. deanei Lainson and Shaw, 1977	Brazil
SUBGENUS *VIANNIA* LAINSON AND SHAW, 1987	
L. braziliensis phenetic complex	Brazil
L. braziliensis Viannia, 1911	Western Andes
L. peruviana Velez, 1913	
L. guyanensis phenetic complex	French Guiana, Guyana, Surinam
L. guyanensis Floch, 1954	Panama, Costa Rica
L. panamensis Lainson and Shaw, 1972	Brazil
L. shawi Lainson et al., 1986	
L. naiffi phenetic complex	Brazil, Ecuador, Peru
L. naiffi Lainson and Shaw, 1989	
L. lainsoni phenetic complex	
L. lainsoni Silveira et al., 1987	Brazil, Bolivia, Peru
L. colombiensis Kreutzer et al., 1991	Colombia
L. equatorensis Grimaldi et al., 1992	Ecuador

In other classifications, subspecies of *L. mexicana* have been recognized, and these names—e.g., *L. mexicana aristedesi, L. m. garnhami,* and *L. m. pifanoi*— appear in the literature, with the subspecific name sometimes used as a specific epithet (for example, *L. pifanoi*). The groupings in the table are based on molecular (isozyme and kDNA) data and cladistic analysis of Rioux et al.,[96] Cupolillo et al.,[21] and Corréa et al.[18]

must feed there to become infected. Sand fly saliva contains low molecular weight compounds, as well as a peptide, that serve as vasodilators and facilitate infection.[46]

- **Pathogenesis.** The incubation period lasts from a few days to several months. The first symptom of infection is a small, red papule at the site of the bite. This may disappear in a few weeks, but usually it develops a thin crust that hides a spreading ulcer underneath. Two or more ulcers may coalesce to form a large sore (Fig. 5.19). In uncomplicated cases the ulcer will heal within two months to a year, leaving a depressed, unpigmented scar. It is common, however, for secondary infection to occur, including, for example, yaws (a disfiguring disease caused by a spirochete) and myiasis (infection with fly maggots, p. 592).

Leishmania tropica is found in more densely populated areas. Its lesion is dry, persists for months before ulcerating, and has numerous amastigotes within it. By

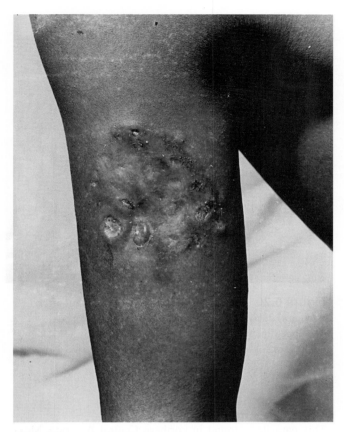

Figure 5.19   Oriental sore.

A complicated case with several lesions.

AFIP neg. no. A-43418-1.

contrast, *L. major* is found in sparsely inhabited regions. Its papule ulcerates quickly, is of short duration, and contains few amastigotes.

Most species and subspecies of *Leishmania* can produce cutaneous lesions. There is an astonishing variety of forms of such lesions, ranging from tiny sores to massive, diffused ulcers. Some even have been misdiagnosed as leprosy or tuberculosis. Diagnosis, then, is difficult at times, especially when two species occur in the same locality.

Leishmania tropica can also become viscerotropic, resulting in enlarged spleen and inflammation of lymph glands. Twelve out of a half-million military personnel surveyed following the 1990–1991 Persian Gulf War developed such viscerotropic *L. tropica* infections.[55] Leishmaniasis remained a problem for American military personnel deployed to Iraq, Afghanistan, and Kuwait, with over 600 cases of cutaneous infections and four visceral infections, due to *Leishmania infantum* (see below) diagnosed in 2003–04 following the invasion of Iraq in 2003.[125]

- **Immunology.**   In the case of Old World cutaneous leishmaniasis, protective immunity following medical treatment seems to be absolute, and immunity as a result of the natural course of the disease is 97% to 98% effective. Recognizing this, some native peoples deliberately inoculated their children on a part of their bodies normally hidden by clothing. This practice prevented a child from later developing a disfiguring scar on an exposed part of the body. Attempts at mass vaccination with controlled infections showed promising results in Israel, Iran, and the former Soviet Union, but these programs ended when it was discovered that parasites persisted in immune hosts.[45]

The gene that controls susceptibility to visceral *L. donovani* infection in mice (the *LSH* gene, now named *SLC11A1*) has no effect on resistance or susceptibility to the cutaneous species *L. major* and *L. mexicana*.[7] Instead, the severity of cutaneous infections is influenced by another gene, *Scl-1,* which is nonallelic to *Lsh* and controls healer and nonhealer phenotypes, and a third gene, *Scl-2,* in DBA/2 mice, which exerts a "no growth" lesion phenotype that mimics certain clinical pictures in humans.[7] In mouse strains resistant to infection with *L. major,* the T_H1 arm of the immune response (p. 31) is activated, with production of IFN-g and a delayed type hypersensitivity reaction.[67] However, in susceptible mouse strains activation of the T_H2 arm stimulates production of IL-4, hyperglobulinemia, and elevated IgE levels.[68] The T cells that respond to infection in both cases are those of the lymph node draining the infection site.

Without extensive use of inbred mouse strains of known genetic makeup, progress toward our understanding of leishmanial infections would be greatly slowed. Mice can be obtained with a variety of genotypes that affect their immune reactions to parasites. Furthermore, in such animals, antibodies that neutralize various cytokines can be used as "probes" to neutralize these molecules to follow the resultant course of infection.[68] For example, anti–INF-g antibody given to protectively immunized (against *L. major*) C3H mice can reduce levels of immunity, resulting in a disseminated infection. Conversely, nonhealing BALB/c mice can be converted into healers by administration of anti–IL-4 antibody. In both cases, however, treatment must be given within a week or 10 days of infection. But the interactions of T cells and cytokines within these mice is not a simple matter, for administration of the respective cytokine molecules themselves does not affect the outcome of experimental infections.

The house mouse *Mus musculus* runs through our folklore, poetry, nursery rhymes, and popular cultures, bringing us much delight. *Mus musculus* also has played a crucial role in development of our understanding of disease processes. Parasitologists especially owe a great deal to this lowly rodent.

- **Diagnosis.**   Diagnosis of infection is greatly facilitated by finding amastigotes. Scrapings from the side or edge of an ulcer smeared on a slide and stained with Wright's or Giemsa's stain will show the parasites in endothelial cells and monocytes, even though they cannot be found in the circulating blood. Cultures should be made in case amastigotes go undetected.

Leishmania braziliensis

Leishmania braziliensis produces a disease in humans variously known as **espundia, uta,** or **mucocutaneous leishmaniasis.** It is found throughout the vast area between central Mexico and northern Argentina, although its range does not extend into the high mountains, except for the south slope of the Andes. Clinically similar cases have been reported in northwest Africa, due to *L. donovani*. The clinical manifestations of the disease vary along its range, which has led to confusion regarding identity of the organisms responsible. Several species names have been proposed for different clinical and serological types (see Table 5.1). Once again it appears that the parasite is rapidly evolving into groups that are adapting to local populations of humans and flies. Morphologically, *L. braziliensis* cannot be differentiated from *L. tropica, L. mexicana,* or *L. donovani.* An interesting historical account of this disease, with evidence of its pre-Columbian existence in South America, is given by Hoeppli.[52]

- **Life Cycle and Pathogenesis.** The life cycle and methods of reproduction of *L. braziliensis* are identical to those of *L. donovani* and *L. tropica* except that the promastigotes reproduce in the hindgut of the sand fly, with several species of *Lutzomyia* serving as vectors. Inoculation of promastigotes by a sand fly's bite causes a small, red papule on the skin. This becomes an itchy, ulcerated vesicle in one to four weeks and is similar at this stage to oriental sore. This primary lesion heals within 6 to 15 months. The parasite never causes a visceral disease but often develops a secondary lesion on some region of the body.

  In Venezuela and Paraguay the lesions more often appear as flat, ulcerated plaques that remain open and oozing. The disease is called **pian bois** in that area. Sloths and anteaters are the primary reservoirs of pian bois in northern Brazil.[66] In the more southerly range of *L. braziliensis,* the parasites have a tendency to metastasize, or spread directly from the primary lesion to mucocutaneous zones. The secondary lesion may appear before the primary has healed, or it may be many years (up to 30) before secondary symptoms appear.[19]

  The secondary lesion often involves the nasal system and buccal mucosa, causing degeneration of the cartilaginous and soft tissues (Fig. 5.20). Necrosis and secondary bacterial infection are common. **Espundia** and **uta** are the names applied to these conditions. The ulceration may involve the lips, palate, and pharynx, leading to great deformity. Invasion of the larynx and trachea destroys the voice. Rarely genitalia may become infected. The condition may last for many years, and death may result from secondary infection or respiratory complications. A similar condition is known to occur in the Old World due to *L. major* or *L. infantum.*[37]

- **Diagnosis and Treatment.** Diagnosis is established by finding L-D bodies in affected tissues. Espundialike conditions are also caused by tuberculosis, leprosy, syphilis, and various fungal and viral diseases, and these must be differentiated in diagnosis. Skin tests are available for diagnosis of occult infections. Culturing the parasite in vitro is also a valuable technique when L-D bodies cannot be demonstrated in routine microscope preparation.

  Treatment is similar to that for kala-azar and tropical sore: antimonial compounds applied on lesions or injected intravenously or intramuscularly. Secondary bacterial infections should be treated with antibiotics. Mucocutaneous lesions are particularly refractory to treatment and require extensive chemotherapy. Relapse is common, but, once cured, a person usually has lifelong immunity. However, if the infection is not cured but merely becomes occult, there may be a relapse with onset of espundia many years later. Because this is primarily a sylvatic disease, there is little opportunity for its control.

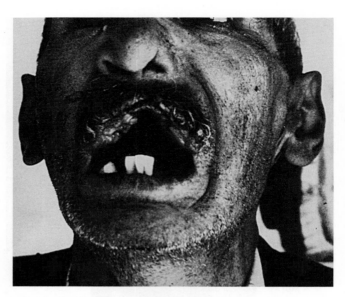

Figure 5.20   **Espundia of 2 years' development after 24 years' delay in onset.**

The upper lip, gum, and palate are destroyed.

From B. C. Walton et al., "Onset of espundia after many years of occult infection with *Leishmania braziliensis,*" in *Am. J. Trop. Med. Hyg.* 22:696–698. Copyright © 1973.

Leishmania mexicana

This parasite is found in northern Central America, Mexico, Texas, and possibly the Dominican Republic and Trinidad. Primarily a cutaneous form, it infects several thousand persons a year, especially agricultural or forest laborers. Three clinical manifestations are found—cutaneous, nasopharyngeal mucosal, and visceral—although some records probably are due to *L. braziliensis.* Traditionally, the cutaneous form of disease has been called *chiclero ulcer* because it is so common in "chicleros," forest-dwelling people who glean a living by harvesting the gum of chicle trees. In Belize, an English-speaking country, it is called *bay sore.*

- **Life Cycle and Pathogenesis.** As in other *Leishmania* species, sand flies are vectors of *L. mexicana.* Several species of *Lutzomyia* are involved. The disease is a zoonosis, and the main reservoirs are rodents. The most important reservoirs are those that live or travel at ground level. Obviously, arboreal reservoirs are less efficient sources of infection to humans. No domestic reservoir is known for chiclero ulcer.

  Cutaneous leishmaniasis due to the *L. mexicana* complex usually heals spontaneously in a few months except when the lesions are in the ear. Ear cartilage is poorly vascularized so immune responses are weak. Chronic lesions with a duration of 40 years are known. Considerable mutilation may result. Mucocutaneous and visceral manifestations are rare. At least eight cases of autochthonous infections of *L. mexicana* in humans and one on the ear of a cat are known in Texas.

- **Diagnosis and Treatment.** The diagnosis and treatment of *L. mexicana* are the same as for *L. tropica.*

Visceral Leishmaniasis

Leishmania donovani

In 1900 Sir William Leishman discovered *L. donovani* in spleen smears of a soldier who died of a fever at Dum-Dum, India. The disease was known locally as **Dum-Dum fever** or **kala-azar.** Leishman published his observations in 1903, the year that Charles Donovan found the same parasite in a spleen biopsy. The scientific name honors these men, as does the common name of the amastigote forms, Leishman-Donovan (L-D) bodies. The Indian Kala-Azar Commission (1931 to 1934) demonstrated the transmission of *L. donovani* by *Phlebotomus* spp.

- **Morphology and Life Cycle.**   *Leishmania donovani* amastigotes cannot be differentiated from other *Leishmania* species on the basis of morphology as seen in a light microscope; the rounded or ovoid bodies measure 2 μm to 3 μm, with a large nucleus and kinetoplast. They live within cells of the reticuloendothelial (RE) system, including spleen, liver, mesenteric lymph nodes, intestine, and bone marrow. Amastigotes have been found in nearly every tissue and fluid of the body.

  The life cycle is similar to that of *L. tropica* except that *L. donovani* is primarily a visceral infection. When a sand fly of genus *Phlebotomus* ingests amastigotes along with a blood meal, the parasites lodge in the midgut and begin to multiply. They transform into slender promastigotes and quickly block the insect's gut. Soon they can be seen in the esophagus, pharynx, and buccal cavity, from where they are injected into a new host with the fly's bite. Not all strains of *L. donovani* are adapted to all species and strains of *Phlebotomus.*

  Once in a mammalian host, parasites are immediately engulfed by macrophages, in which they divide by binary fission, eventually killing the host cell. Escaping the dead macrophage, parasites are engulfed by other macrophages, which they also kill; by this means they eventually severely damage the RE system, a system that plays a critical role in host defense. Interestingly, amastigotes engulfed by neutrophils and eosinophils are killed, but in untreated cases these polymorphonuclear leucocytes have little or no effect on the eventual outcome of the disease.

- **Pathogenesis.**   Clinically, *L. donovani* infections may range from asymptomatic to progressive, fully developed kala-azar. The incubation period in humans may be as short as 10 days or as long as a year but usually is two to four months. The disease typically begins slowly with low-grade fever and malaise and is followed by progressive wasting and anemia, protrusion of the abdomen from enlarged liver and spleen (Fig. 5.21), and finally death (in untreated cases) in two to three years. In some cases symptoms may be more acute in onset, with chills, fever up to 40°C (104°F), and vomiting; death may occur within 6 to 12 months. Accompanying symptoms are edema, especially of the face, bleeding of mucous membranes, breathing difficulty, and diarrhea. The immediate cause of death often is invasion of secondary pathogens that the body is unable to combat. A certain proportion of cases, especially in India, recover spontaneously.

Visceral leishmaniasis may be viewed essentially as a disease of the reticuloendothelial system. The phagocytic cells, which are so important in defending the host against invasion, are themselves the habitat of the parasites. Blood-forming organs, such as spleen and bone marrow, undergo compensatory production of macrophages and other phagocytes (hyperplasia) to the detriment of red cell production. The spleen and liver become greatly enlarged (hepatosplenomegaly, Fig. 5.22), while the patient becomes severely anemic and emaciated.

A skin condition known as *post–kala-azar dermal leishmanoid* develops in some cases (Fig. 5.23).[77] It is rare in the Mediterranean and Latin American areas but develops in 5% to 10% of cases in India. The condition usually becomes apparent about one to two years after inadequate treatment for kala-azar. It is marked by reddish, depigmented nodules that sometimes become quite disfiguring.

- **Immunology.**   In both experimental animals and in humans, the response to visceral leishmanial infections is different than it is to cutaneous species. Human patients also differ among themselves, depending on whether the disease is subclinical or progressive and symptomatic.[127] Clinical visceral leishmaniasis may not develop for some time, even years, after infection, and asymptomatic infections, of which there are many, may result from early activation of the T_H1 arm. Monocytes (p. 27) from people with subclinical infections respond to leishmanial antigens by proliferating and producing IL-2, IFN-γ, and IL-12,[127] whereas patients with symptomatic kala-azar do not develop T_H1 responses against *L. donovani,* and their macrophages do not secrete IFN-γ or IL-2 in the presence of leishmanial antigens (see above discussion of immune reactions to *L. major,* p. 27). However, these latter patients regularly have high titres of antileishmanial antibodies;[89] that is, their T_H2 arm is activated and the T_H1 arm is downregulated. *Leishmania donovani* also possess membrane lipophosphoglycans that may inhibit gene expression in macrophages. The inhibition is of protein kinase-dependent expression, such as that involved in macrophage activation by TNF and IFN-γ.[67]

  There is an intricate interplay between host immune response and progression of visceral leishmaniasis, and the outcome of this potentially deadly contest is likely influenced by host genotype. Mice strains certainly differ in susceptibility to leishmanial infections depending on genetic makeup. The *SLC11A1* gene (formerly known as *LSH*), controlling susceptibility to *L. donovani* infections in mice, is on chromosome 1, whereas the gene(s) influencing resistance to cutaneous leishmanias is on chromosome 11.[127] In human populations, the ratio of asymptomatic to symptomatic infections may differ significantly according to ethnicity, and familial aggregations of either symptomatic or asymptomatic infections have also been reported. Wilson et al.[127] provide an excellent review of the relationship between disease, immune response, and genetic makeup, as well as an extensive list of references on this subject.

- **Diagnosis and Treatment.**   As in *L. tropica,* diagnosis of *L. donovani* depends on finding L-D bodies in tissues or secretions. Spleen punctures, blood or nasal smears,

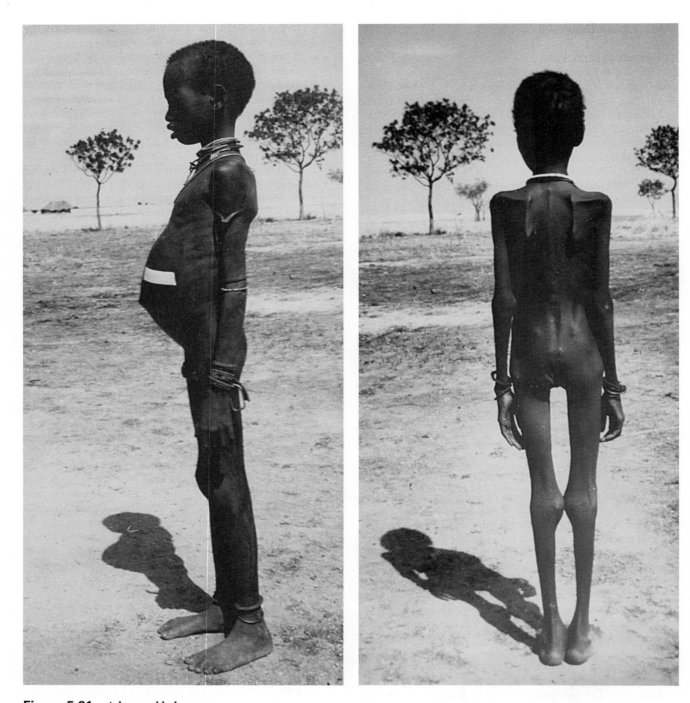

Figure 5.21 Advanced kala-azar.

Boy, about six years old, from Sudan, showing extreme hepatosplenomegaly and emaciation typical of advanced kala-azar.

From H. Hoogstraal and D. Heyneman, "Leishmaniasis in the Sudan Republic 30. Final epidemiologic report," in *Am. J. Trop. Med. Hyg.* 18:1091–1210. Copyright © 1969.

bone marrow, and other tissues should be examined for parasites, and cultures from these and other organs should be attempted. Immunodiagnostic tests are sensitive but cannot differentiate between species of *Leishmania* or between current and cured cases. The tests most frequently used are the enzyme-linked immunosorbent assay (ELISA) and the indirect fluorescent antibody test (IFA). Other diseases that might have symptoms similar to kala-azar are typhoid and paratyphoid fevers, malaria, syphilis,

tuberculosis, dysentery, and relapsing fevers. Each must be eliminated in the diagnosis of kala-azar.

Treatment consists of injections of various antimony compounds, as previously described for *L. tropica,* and good nursing care. The promising oral drug miltefosine is discussed on p. 79.

- **Epidemiology and Control.** Transmission of visceral leishmaniasis is related to both human activities and sand

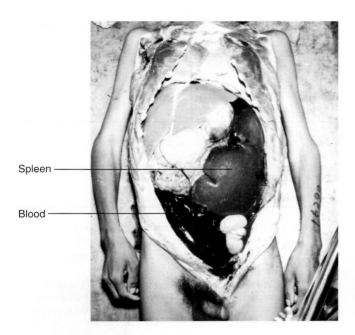

Figure 5.22 **A patient with kala-azar who died of hemorrhage after a spleen biopsy.**

Note the greatly enlarged spleen. (The dark matter in the lower abdominal cavity is blood.)

AFIP neg. no. A-45364.

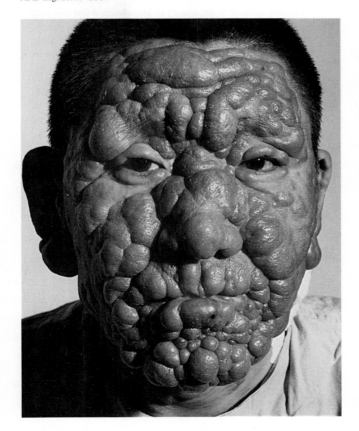

Figure 5.23 **Post–kala-azar dermal leishmanoid.**

This patient responded very well to treatment, regaining a nearly normal appearance.

Courtesy of Robert E. Kuntz.

fly biology. Control of sand flies and reservoir hosts is required in endemic areas. *Phlebotomus* spp. exist mainly at altitudes under 2000 feet, most commonly in flat plains areas. Even in desert areas such as in Sudan, the flies rest in cracks in the parched earth and under rocks, which offer protection. In such conditions the flies are active only during certain hours of the day. For humans to become infected, they must be in sand fly areas at these times.

A wide variety of animals can be infected experimentally, although dogs are the main important reservoir in most areas. Canine infection is less common in India, where it is believed that a fly-to-human relationship is maintained. Visceral cases in dogs in Oklahoma have been discovered; histories of these dogs suggest that canine leishmaniasis (due to what has been called the *OKD* or *Oklahoma dog strain*) may have become endemic in the United States.

Age of the victim is a factor in the course of the disease, and fatal outcome is most frequent in infants and small children. Males are more often infected than are females, most likely as the result of more exposure to sand flies. Poor nutrition, concomitant infection with other pathogens, and other stress factors predispose the patient to lethal consequences. *Leishmania infantum* (= *L. chagasi*) is a visceral form—part of the *L. donovani* species complex—around the Mediterranean basin, parts of China, and in South America. Vectors are *Phlebotomus* and *Lutzomyia* spp. in the Old and New Worlds respectively, especially *P. perniciosus* and *L. longipalpis*. *Leishmania infantum* is not as virulent as *L. donovani,* and dogs, especially pets, are the main reservoir. A "conservative estimate" suggests as many as 2.5 million dogs may be infected in countries surrounding the Mediterranean.[76] Symptoms, diagnosis, and treatment are similar to those of *L. donovani;* the two species can be distinguished using molecular techniques.[72]

Asymptomatic infections can occur with *L. infantum* as well as *L. donovani*. Failure to diagnose such cases results in underestimation of prevalence, but it is not always clear how important asymptomatic humans (as opposed to dogs) are to maintenance of leishmaniasis in a host population.[19]

• **Other *Leishmania* Species.** In recent years, molecular and immunological techniques have revealed a number of different lineages within genus *Leishmania*, especially in Latin America, and depending on the author, lineages are designated as species, subspecies, or strains.[21] *Leishmania naiffi, L. colombiensis, L. equatorensis,* and *L. shawi* are all species being isolated from humans, various other mammals, and vectors. Although they are all of subgenus *Viannia,* the molecular data suggest extensive and probably rapid evolutionary diversification.

OTHER TRYPANOSOMATID PARASITES

Because of their ease of culture and biochemical characteristics, several species of trypanosomatid flagellates from the following genera have been used as models to study a variety of cellular processes. Although these species are mostly found in insects, some can occur as transient infections in vertebrates, and several can be opportunistic parasites of immunosuppressed HIV/AIDS patients.[15]

Leptomonas species are parasitic in invertebrates and are of no medical importance. *Leptomonas* species are variously promastigotes and intracellular amastigotes throughout their monoxenous life cycle. Species are found in molluscs, nematodes, insects, and other protozoa. Transmission may be by way of amastigotelike cysts or even the flagellates which can survive for three days in water.[108]

Members of *Herpetomonas* also are characteristically monoxenous in insects. They pass through amastigote, promastigote, opisthomastigote, and possibly epimastigote stages in their life cycles. In the opisthomastigote the flagellum arises from a reservoir that runs the entire length of the body.

Crithidia species are choanoflagellates of insects. They are often clustered together against the inner lining of their host's intestine. They can assume the amastigote form and are monoxenous.

Blastocrithidia species are monoxenous insect parasites, usually found as epimastigotes and amastigotes in the intestines of their hosts. Species are common in water striders (family Gerridae).

Phytomonas species are parasites of milkweeds, euphorbias, and related plants. They pass through promastigote and amastigote phases in the intestines of certain insects and appear as promastigotes in the sap (latex) of their plant hosts.

▌ Learning Outcomes

By the time a student has finished studying this chapter, he or she should be able to:

1. Understand the morphological forms of trypanosomatids and the ultrastructure of trypomastigotes.

2. Understand the changes in form and structure of *Trypanoma brucei* throughout its life cycle.

3. Understand the significance of variable antigen types (VATs) in *Trypanosoma brucei* infection.

4. Name the most important vectors of *Trypanosoma brucei* and their means of control.

5. Identify the differences in morphology, hosts, life cycle, and geographic distribution between *T. brucei* and the causative agents of Chagas' disease.

6. Understand the differences between mucocutaneous and visceral leishmaniasis with regard to the following: causative agents, vectors, diseases caused, and geographic distribution.

7. Explain the differences between *T. brucei* and the causative agents of Chagas' disease in the following respects: vectors, diseases caused, and geographic distribution.

8. Explain how *T. brucei* and *T. cruzi* can be distinguished in blood smears.

9. Explain the difference in symptoms between those caused by *Leishmania tropica* and *L. major.*

▌ References

References for superscripts in the text can be found at the following Internet site: www.mhhe.com/robertsjanovynadler9e

▌ Additional Readings

Adler, S. 1964. Leishmania. In B. Dawes (Ed.), *Advances in parasitology* 2. New York: Academic Press, Inc., pp. 1–34.

Berriman, M. et al. 2005. The genome of the African trypanosome *Trypanosoma brucei.* Science 309:416–422.

Courtin, F., V. Jamonneau, G. Duvaliet, A. Garcia, B Coulibaly, J. P. Doumenge, G. Cuny, and P. Solano. 2008. Sleeping sickness in West Africa (1906–2006): Changes in spatial repartition and lessons from the past. *Trop. Med. Intl. Health* 13:334–344.

Desowitz, R. S. 1991. *The malaria capers.* New York: W. W. Norton & Co.

Ford, J. 1971. *The role of the trypanosomiases in African ecology.* Oxford: Clarendon Press.

Foster, W. D. 1965. *A history of parasitology.* Edinburgh: E. & S. Livingstone. Chapter 10, "The Trypanosomes," is a very interesting account of the history of knowledge about this group.

Hoogstraal, H., and D. Heyneman. 1969. Leishmaniasis in the Sudan Republic. 30. Final epidemiological report. *Am. J. Trop. Med. Hyg.* 18:1089–1210. An extensive account of the aspects of leishmaniasis by two men who have an unashamed love for humanity. It should be required reading for all students of parasitology, and it stands by itself as an example of what scientific writing should be.

Marsden, P. D. 1985. Clinical presentations of *Leishmania braziliensis braziliensis.* Parasitol. *Today* 1:129–133. An outstanding review of the subject with excellent illustrations.

Mauel, J., and R. Behin. 1982. Leishmaniasis: Immunity, immuno-pathology and immunodiagnosis. In S. Cohen and K. S. Warren (Eds.), *Immunology of parasitic infections.* Oxford: Blackwell Scientific Publications Ltd., pp. 299–355.

Mulligan, H. W., and W. H. Potts (Eds.). 1970. *The African trypanosomiases.* London: George Allen and Unwin, Ltd. The quotes at the beginning of the chapter are from this source.

Pays, E. 2005. Regulation of antigen gene expression in *Trypanosoma brucei. Trends in Parasitol.* 21:517–520.

The *Trypanosoma cruzi* Genome Consortium. 1997. The *Trypanosoma cruzi* genome initiative. *Parasitol. Today* 13:16–22.

Vickerman, K. 1985. Leishmaniasis–the first centenary. *Parasitol. Today* 1:149, 172.

Chapter 6

Other Flagellated Protozoa

Perhaps, science will have replaced the art when the addition of totally defined nutrients, the removal of metabolic wastes, monitoring of physical and chemical conditions of growth, and harvesting of the crop have become automated.

—Louis Diamond, on the challenge of "separating a protozoan from its habitat in the wild and inducing it to take up a new existence in the culture tube"[22]

Although kinetoplastans include some exceedingly important parasites whose economic impact is quite severe and whose pathology is dramatic, several other groups of flagellated protozoa also have members that are parasitic. These flagellates are likely to be found in every kind of animal, from cockroaches to humans. A few of them are structurally complex, and some, such as *Giardia duodenalis,* have become favorites of evolutionary biologists because of their biochemical characteristics. Space limitations prevent us from covering all of these parasites in detail. Consequently, representative species are drawn from four orders.

The following two orders, Retortamonadida and Diplomonadida, are members of phylum Retortamonada, classes Retortamonadea and Diplomonadea respectively (see chapter 4). Members of these orders lack mitochondria and dictyosomes (Golgi), possess a recurrent flagellum in a cytostomal groove, and occupy anoxic environments.

ORDER RETORTAMONADIDA

Family Retortamonadidae

Two species in family Retortamonadidae are commonly found in humans. Although they are evidently harmless commensals, they are worthy of note because they easily can be mistaken for pathogenic species.

Chilomastix mesnili

Chilomastix mesnili (Fig. 6.1) infects about 3.5% of the United States population and 6% of the world population.[5] It lives in the cecum and colon of humans, chimpanzees, orangutans, monkeys, and pigs. *Chilomastix* species also are known in other mammals, birds, reptiles, amphibians, fish, leeches, and insects.

A living trophozoite is pyriform, with the posterior end drawn out into a blunt point, and it is 6 μm to 24 μm by 3 μm to 10 μm. A longitudinal spiral groove occurs in the surface of the middle of the body, but this is usually visible only on living specimens. The sunken cytostomal groove is prominent near the anterior end. Along each side of the cytostome runs a cytostomal fibril, presumably strengthening cytostome lips. The cytostome leads into a cytopharynx, where endocytosis takes place. Four flagella, one longer than the others, emerge from kinetosomes at the anterior end, and the kinetosomes are interconnected by microfibrillar material.[10] One flagellum is very short and delicate, curving back into the cytostome, where it undulates. The large nucleus is located anteriorly.

A cyst stage occurs, especially in formed stools (Fig. 6.2). A typical cyst is thick-walled, 6.5 μm to 10.0 μm long, and pear or lemon shaped. It has a single nucleus and

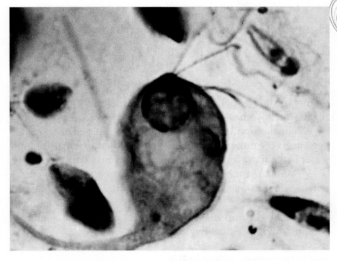

Figure 6.1 **Trophozoite of *Chilomastix caulleryi,* which is similar morphologically to *C. mesnili.***

Note the four flagella and the cytostomal fibrils. It is 6 μm to 24 μm long.

Photograph by Larry S. Roberts.

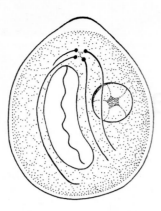

Figure 6.2   **Cyst of *Chilomastix mesnili* from a human stool.**

Note the characteristic lemon or pear shape. Also visible are the large, irregular karyosome and the cytostomal fibrils.

Drawing by William Ober.

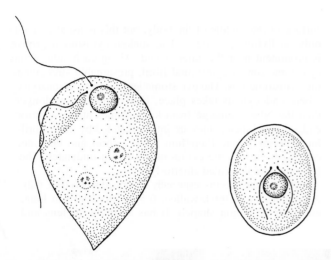

Figure 6.3   *Retortamonas intestinalis* **trophozoite and cyst.**

Drawing by William Ober.

retains all the cytoplasmic organelles, including cytostomal fibrils, kinetosomes, and axonemes.

Transmission is by ingestion of cysts; trophozoites cannot survive stomach acid. Fecal contamination of drinking water is the most important means of transmission.

Chilomastix mesnili usually is considered nonpathogenic, but it often co-occurs with other parasites that are pathogenic.[18] In such cases the flagellates are confirming what the presence of *Giardia duodenalis* reveals about local sanitary conditions and personal hygiene.

Retortamonas intestinalis

Retortamonas intestinalis (Fig. 6.3) is a tiny protozoan that is similar to *C. mesnili*, but the trophozoite is only 4 μm to 9 μm long. Furthermore, it has only two flagella, one of which extends anteriorly and the other of which emerges from the cytostomal groove and trails posteriorly. Living trophozoites

usually extend into a point at their posterior ends, but bend to round up in fixed specimens. Cysts are ovoid to pear shaped and contain a single nucleus.

Like *C. mesnili,* this species is probably a harmless commensal. It lives in the cecum and large intestine of monkeys, chimpanzees, and humans, and evidently is not a common symbiont anywhere in the world. *Retortamonas* species lack mitochondria, and molecular work suggests these flagellates are much more closely related to diplomonads than indicated in our classfication (see chapter 4).[76] Other members of genus *Retortamonas* have been reported from crickets, cockroaches, termites, guinea pigs, and toads, including the cane toad, *Bufo marinus,* imported into Australia where it evidently acquired *R. dobelli* from local anurans.[21]

ORDER DIPLOMONADIDA

Family Hexamitidae

Members of Hexamitidae are easily recognized because they have two identical nuclei lying side by side. There are several species in five genera; most of them are parasitic in vertebrates or invertebrates. One species, *Giardia duodenalis,* is a parasite of humans and will serve to illustrate genus *Giardia.* *Spironucleus meleagridis* is an example of a related species in domestic animals.

Genus *Giardia*

Members of genus *Giardia* have come to occupy a prominent place in both the parasitological and evolutionary biology literature. Their lack of mitochondria has been interpreted as a primitive trait, and phylogenetic analysis of ribosomal RNA has been used to place *Giardia* species near the point of divergence between pro- and eukaryotes.[39] However, both molecular and cladistic analysis of *Giardia* indicate the parasites are actually derived from more recent parasitic ancestors.[35, 75] Regardless of their origins, *Giardia* species will remain of interest to parasitologists and nonparasitologists alike because of their widespread occurrence and fairly frequent infections in people from all nations and socioeconomic levels.

More than 40 species of *Giardia* have been described, but only five are now considered valid:[85] *G. duodenalis* (= *intestinalis;* = *lamblia*) and *G. muris* from mammals, *G. ardeae* and *G. psittaci* from birds, and *G. agilis* from amphibians (Fig. 6.5).

Giardia duodenalis

Giardia duodenalis was first discovered in 1681 by Antony van Leeuwenhoek, who found it in his own stools. The species' taxonomy was confused in the 19th century, and that confusion remained unresolved through most of the 20th century. Most current literature refers to parasites from humans as *Giardia duodenalis,* although *G. intestinalis* and *G. lamblia* have been used as synonyms.[82, 85] The species is cosmopolitan but occurs most commonly in warm climates; children are especially susceptible. *Giardia duodenalis* is the most common flagellate of the human digestive tract.

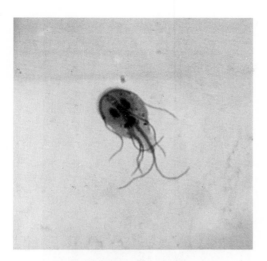

Figure 6.4 *Giardia duodenalis* **trophozoite in a human stool.**

It is 12 µm to 15 µm long.

Courtesy of Sherwin Desser.

• **Morphology.** Trophozoites (Figs. 6.4 and 6.5) are 12 µm to 15 µm long, rounded at their anterior ends and pointed at the posterior. The organisms are dorsoventrally flattened and convex on the dorsal surface. The flattened ventral surface bears a concave, bilobed **adhesive disc,** which actually is a rigid structure, reinforced by microtubules and fibrous ribbons, surrounded by a flexible, apparently contractile, striated rim of cytoplasm (Figs. 6.6 and 6.7). Application of this flexible rim to a host intestinal cell, working in conjunction with **ventral flagella,** found in a **ventral groove,** is responsible for the organism's remarkable ability to adhere to host cells (Fig. 6.8). The pair of ventral flagella as well as three more pairs of flagella arise from kinetosomes located between the anterior portions of the two nuclei (see Fig. 6.5). Axonemes of all flagella course through cytoplasm for some distance before emerging from the cell body; those of the anterior flagella actually cross and emerge laterally from the adhesive disc area on the side opposite their respective kinetosomes.

A pair of large, curved, transverse, dark-staining **median bodies** lies behind the adhesive disc. These bodies

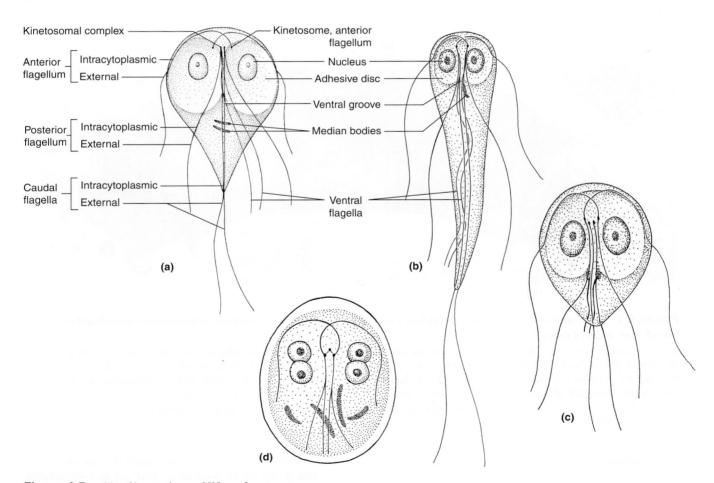

Figure 6.5 *Giardia* **species and life cycle stages.**

Giardia species differ in overall body shape and relative sizes of their adhesive discs. *(a) Giardia duodenalis* trophozoite, 10–15 µm long. *(b) Giardia agilis* from amphibians, ∼20 µm long. *(c) Giardia muris* from mice, approximately the same size as *G. duodenalis* but with a relatively broad body. *(d)* Cyst of *G. duodenalis.* These cysts are 8–12 µm long; karyosomes of all four cyst nuclei, as well as several intracytoplasmic axonemes and median bodies are visible.

Drawing by William Ober and Claire Garrison.

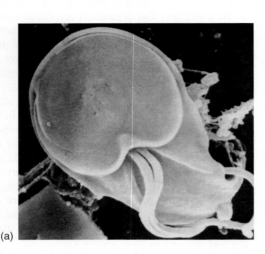
(a)

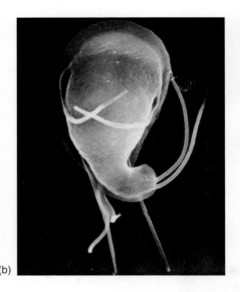
(b)

Figure 6.6   **Scanning electron micrograph of a *Giardia* species.**

(*a*) The ventral view shows the flat adhesive disc and the relationship of the ventral and posterior flagella and ventral groove, but the caudal flagella curve around to the other side in this photograph. (*b*) The dorsal view shows these flagella, as well as the anterior flagella. The organism is 12 μm to 15 μm long.

Courtesy of Dennis Feely.

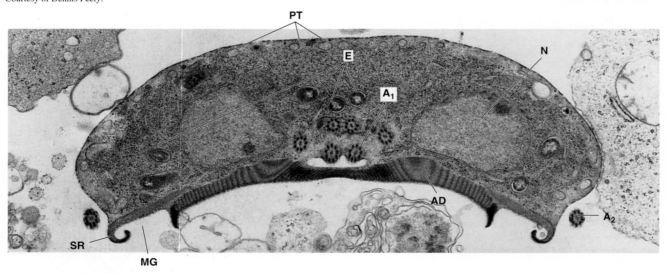

Figure 6.7   **Transmission electron micrograph of a transverse section of a *Giardia muris* trophozoite found in the small bowel of an infected mouse.**

The marginal groove is the space between the striated rim of cytoplasm and the lateral ridge of the adhesive disc. The beginning of the ventral groove can be seen dorsal to the central area of the adhesive disc. This specimen bears endosymbionts, which are evidently bacteria. *PT*, peripheral tubules; *E*, endosymbionts; *N*, nucleus; A_1, axonemes of posterior, ventral, and caudal flagella; A_2, axoneme of anterior flagellum; *AD*, adhesive disc; *MG*, marginal groove; *SR*, striated rim of cytoplasm. (×15,350)

From P. C. Nemanic et al., "Ultrastructural observations on giardiasis in a mouse model. II. Endosymbiosis and organelle distribution in *Giardia muris* and *Giardia lamblia*," in *J. Infect. Dis.* 140:222–228. Copyright © 1979 University of Chicago.

are unique to *Giardia*. Various authors have regarded them as parabasal bodies, kinetoplasts, or chromatoid bodies, but ultrastructural studies have shown they are none of these.[30] Their function is obscure, although it has been suggested that they may help support the posterior end of the organism, or they may be involved in its energy metabolism. There is no axostyle; the structure so described by previous authors is formed by the

intracytoplasmic axonemes of ventral flagella and associated groups of microtubules. There are no mitochondria, Golgi bodies, or lysosomes, and there is no smooth endoplasmic reticulum.[30]

The overall effect of the two nuclei behind the lobes of the adhesive disc and the median bodies is that of a wry little face that seems to be peering back at the observer.

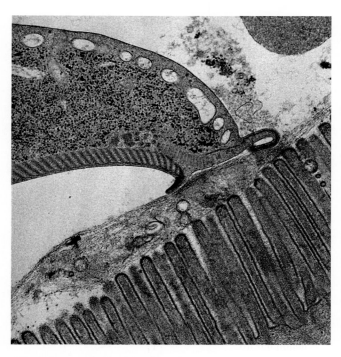

Figure 6.8   **Periphery of *Giardia muris* in contact with the mucous stream covering the microvilli of a duodenal epithelial cell.**

It appears that the peripheral flange of striated cytoplasm is the grasping organelle of the ventral surface. ($\times$ 33,000)

From D. S. Friend, "The fine structure of *Giardia muris,*" in *J. Cell Biol.* 29:317–332. Copyright © 1966 The Rockefeller University Press.

- **Life Cycle.**   *Giardia duodenalis* lives in the duodenum, jejunum, and upper ileum of humans, with the adhesive disc fitting over the surface of an epithelial cell. In severe infections the free surface of nearly every cell is covered by a parasite. The protozoa can swim rapidly using their flagella.

  Trophozoites divide by binary fission, although the replication of mastigont parts involves a complex set of events in which flagella "migrate, assume different position, and transform into different flagellar types in progeny."[60] Nuclei divide first, followed by the locomotor apparatus, sucking disc, and cytoplasm in that order. Three cell divisions are required before the protists become mature.[60] Enormous numbers of flagellates can build up rapidly. It has been calculated that a single diarrheic stool can contain 14 billion parasites and a stool from a moderately infected individual can contain 300 million cysts.[16] Obviously one infected individual can spread around a lot of misery.

  In the small intestine and in watery stools, only trophic stages can be found. However, as feces enter the colon and begin to dehydrate, the parasites become encysted. Experimental evidence suggests cholesterol deprivation is a trigger for encystment and that a "Golgilike complex" develops, producing vesicles that contain cyst wall material.[37] These secretory vesicles contain proteins that are specific to the cyst wall and become polymerized following exocytosis.[32] The cyst wall consists of a membranous inner and filamentous outer layer. Cysts (Fig. 6.5) are 8 μm to 12 μm by 7 μm to 10 μm in size. Newly formed cysts have two nuclei, but older ones have four. Soon the disc and locomotor apparatus are doubled, and the twinned flagellates are ready to emerge. When swallowed by a host, cysts pass safely through the stomach and the flagellates excyst in the duodenum, immediately completing cytoplasmic division. Flagella grow out, and the parasites are once again at home.

- **Metabolism.**   *Giardia duodenalis* is an aerotolerant anaerobe.[47] As mentioned earlier, these flagellates have no mitochondria. The tricarboxylic acid cycle and cytochrome system are absent, but the organisms avidly consume oxygen when it is present. Glucose is evidently the primary substrate for respiration, and the parasites store glycogen. But *G. duodenalis* also multiplies and generally produces the same metabolites when glucose is absent or present in low concentrations.[72] Principal end products are ethanol, acetate, and CO_2, both aerobically and anaerobically. In the absence of oxygen, reducing equivalents are transferred to acetaldehyde to produce ethanol. When oxygen is present the flagellates produce more acetate and less ethanol. All their energy is produced by substrate-level phosphorylation via a flavin, iron-sulfur, protein-mediated fermentative pathway.[47]

- **Pathogenesis.**   *Giardia duodenalis* strains differ in their pathogenicity and response to treatment.[80, 84] Many cases of infection show no evidence of disease. Some people are more sensitive than others to the presence of *G. duodenalis,* and considerable evidence suggests that some protective immunity can be acquired. In some individuals there is a marked increase of mucus production, diarrhea (sometimes incapacitating), dehydration, intestinal pain, flatulence, and weight loss. The stool is fatty but never contains blood. The parasite does not lyse host cells but appears to feed on mucous secretions. A dense coating of flagellates on the intestinal epithelium damages microvilli and interferes with absorption of fats and other nutrients, which probably triggers the onset of disease.[80] The gallbladder may become infected, which can cause jaundice and colic. The disease is not fatal but can be intensely discomforting.

  As in the case of trypanosomes, *Giardia duodenalis* exhibits antigenic variation, with up to about 180 different antigens being expressed over 6 to 12 generations, depending on the strain.[59] Infections are controlled mainly by humoral responses, the major antigens being cysteine-rich surface proteins, which are the same ones that vary antigenically during the course of infection.[1] The mechanism of variant specific protein expression differs from that of trypanosomes (p. 68); evidently no movement of genes is involved and epigenetic mechanisms may play a role in production of diverse antigens.[42] Not surprisingly, there also is evidence that these variable surface proteins are related to both infectivity and virulence.[84] See Prucca and Lujan[68] for a review of antigenic variation in *G. duodenalis.*

- **Diagnosis and Treatment.**   Recognition of trophozoites or cysts in stained fecal smears is adequate for diagnosis.[31] However, an otherwise benign infection may coexist with a peptic ulcer, enteritis, tumor, or strongyloidiasis, any of which could actually be causing the symptoms. In a small percentage of cases, cysts are not passed or are passed sporadically. Duodenal aspiration may be necessary for diagnosis by demonstrating trophozoites.

A variety of immunodiagnostic methods, relying on detection of serum antibodies or antigens in feces, are in use, although not all of them distinguish between current and past infections.[36] Efforts are being made to develop diagnostic methods based on molecular techniques; such methods may help in cases in which cysts are passed in very low numbers. PCR-based techniques can detect a single cyst and also distinguish between species and strains of differing pathogenicity.[48] Experimental vaccines have been tested in dogs and cats against infection by *G. duodenalis* strains originally isolated from humans. These vaccines reduced both the severity of disease and the number of cysts shed.[61]

Treatment with quinacrine or metronidazole (Flagyl, a 5-nitroimidazole compound) usually effects complete cure within a few days. Metronidazole is typically issued with strong warnings against concurrent consumption of alcoholic beverages. Several newer nitroimidazole derivatives have shown good antigiardial activity in single doses and against strains resistant to metronidazole.[70] All household occupants should be treated simultaneously to avoid reinfection of treated by untreated family members.

- **Epidemiology and Transmission Ecology.** Giardias is is highly contagious. If one member of a family becomes infected, others usually will also. Transmission depends on the swallowing of mature cysts. Prevention, therefore, depends on a high level of sanitation.

A summary of surveys of 134,966 people throughout the world showed that the prevalence of the infection ranged from 2.4% to 67.5%.[5] In 1984, 26,560 cases of giardiasis were reported in the United States.[14] The Centers for Disease Control in Atlanta, in its 1989–1990 summary of waterborne disease outbreaks, indicated that "*Giardia lamblia* was the most frequently identified etiologic agent . . . for the 11th and 12th consecutive years."[13] Estimates from the mid-1990s suggest 200 million people may be infected throughout Asia, Africa, and Latin America, with an incidence of half a million new cases a year.[80] Outbreaks continue to flare up in the United States, often without regard for the affluence of the people involved.[40, 58]

Although *G. duodenalis* is easily transmitted from human to human, giardiasis can also be a zoonosis.[80] Faunal surveys in watersheds that were known sources of infections to people have shown that numerous animals, including beavers, dogs, cats, and sheep, serve as reservoirs.[57] Around the world, farm livestock, especially calves, are infected, with prevalences ranging up to 100%.[88]

Among wild animals, beavers in particular are epidemiologically significant in human giardiasis. After hiking for miles in the wild on a hot day, a person is easily tempted to fill a canteen and drink from a crystal-clear beaver pond. Many infections have been acquired in just that way, including some in parasitologists' relatives. In 1980 numerous cases of giardiasis were diagnosed in the resort village of Estes Park, Colorado. Surprisingly, all were in one half of the town, with the other half remaining parasite free. Each half was served with water from a different river. Both rivers had beavers in abundance, but the municipal water filtration system had broken down for one source but not the other.[58]

Resorts are certainly not the only places where people can pick up giardiasis. In 1990 an outbreak among Wisconsin insurance company office workers was traced to an employee cafeteria where raw sliced vegetables had been prepared by an infected food handler.[55] Daycare centers also can become foci of transmission. There have been several reports of a late summer peak in transmission, although the exact reasons for this increased seasonal risk remain somewhat of a mystery.[28]

Research using molecular techniques reveals two main genotype assemblages among *Giardia* species.[82] These assemblages are referred to as A and B, with two "clusters" in A: A-I, including closely related isolates from both humans and other animal species, and A-II, isolated only from humans. Assemblage B is much more genetically diverse than A and includes isolates from both human and nonhuman sources. Organisms from cluster A-I likely have the most potential for being zoonoses.[82] Some strains may be restricted to nonhuman animals, however, and some wild animal infections may not be a public health hazard.[81]

Spironucleus meleagridis

Spironucleus meleagridis (*Hexamita meleagridis* in older literature) is a parasite of young galliform birds, including turkey, quail, pheasant, partridge, and peafowl. It occurs in the United States, Great Britain, and South America, although it is probably common elsewhere. Prior to 1950 in the United States, *S. meleagridis* caused about $1 million dollars in loss annually to the turkey industry, but drugs such as oxytetracycline, combined with proper flock management, have reduced this problem significantly.[52, 53]

Morphologically, *S. meleagridis* is elongated, with four pairs of flagella and nuclei that are tapered and wrapped around one another (thus the name: *Spironucleus*).[66, 67] Unlike *Giardia* spp., *S. meleagridis* has no sucking disc and contains no median bodies. (Fig. 6.9) The kinetosomes are grouped anterior to and between the nuclei, but three pairs of axonemes emerge anteriorly, and one pair courses within the cytoplasm, running posteriorly along granular lines and emerging to become posterior flagella.

The *S. meleagridis* life cycle is essentially the same as for *Giardia* spp., except that birds rather than mammals are normal hosts. Spironucleosis is mainly a disease of young animals. Symptomless adults are reservoirs of infection. Mortality in a flock may range from 7% to 80% in very young birds (Fig. 6.10). Survivors are somewhat immune but commonly are stunted in size. They become a ready source of infection for new broods. No completely satisfactory treatment is available, but prevention in domestic flocks is possible by proper management and sanitation. Separation of chicks from adult birds is mandatory.

Chickens and turkeys are not the only commercially important animals vulnerable to infection. *Spironucleus salmonis* is a pathogen of salmon, producing ascites and inflammation of liver and kidneys.[41] *Spironucleus* species also have been reported from frogs and implicated in health problems of cultured oysters.[51, 54]

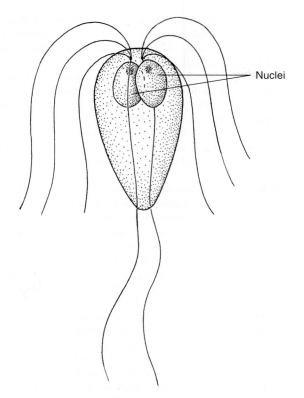

Nuclei

Figure 6.9   **Diagram of a trophozoite of *Spironucleus meleagridis.***

It is 6 μm to 12 μm long.

Drawing by William Ober.

TRICHOMONADS (CLASS TRICHOMONADA, ORDER TRICHOMONADIDA)

Trichomonads are now considered members of phylum Parabasalea, based on their complex mastigont system that includes parabasal fibers and a nonmotile axostyle (see chapter 4).

Family Trichomonadidae

Members of this family are rather similar to one another in structure (see Figs. 6.11 through 6.13). They are easily recognized because they have an anterior tuft of flagella, a stout median rod (the **axostyle**), and an **undulating membrane** along the recurrent flagellum. These structural features produce a characteristic jerky, twisting, locomotion that makes trichomonads easy to recognize in fresh preparations.

Trichomonads are found in intestinal or reproductive tracts of vertebrates and invertebrates, with one group occurring exclusively in the gut of termites. Phylogenetic studies show a number of distinct groups within the family, although relationships among some genera remain unclear.[19] Unlike other protozoa covered in this chapter, most members of this order do not form cysts. Three species are common in humans, and one is of extreme importance in domestic ruminants.

The three trichomonads of humans, *Trichomonas tenax, T. vaginalis,* and *Pentatrichomonas hominis,* are similar enough morphologically to have been considered conspecific by many taxonomists but differences between *P. hominis* and the other two are now recognized. As currently defined, *Trichomonas* contains only three species: *T. tenax, T. vaginalis,* and a species found in birds, *T. gallinae,* which is more like *T. tenax* than is *T. vaginalis.*[46]

Figure 6.10   **Young chukar partridges infected with *Spironuleus meleagridis.***

These five-week-old birds from a commercial game-rearing farm are afflicted with enteritis and dermatitis. Mortality was high (80%); treatment with neomycin and oxytetracycline was ineffective.

From G. L. Cooper, B. R. Charlton, A. A. Bickford, and R. Nordhausen. "*Hexamita meleagridis (Spironucleus meleagridis)* infection in chukar partridges associated with high mortality and intracellular trophozoites," in *Avian Dis.* 48:706–710. Reprinted with permission.

Trichomonas tenax

Trichomonas tenax (Fig. 6.11) was first discovered by O. F. Müller in 1773 when he examined an aqueous culture of tartar from teeth. *Trichomonas tenax* is now known to have worldwide distribution.

- **Morphology.** Like all species of *Trichomonas*, *T. tenax* has only a trophic stage. It is an oblong cell 5 μm to 16 μm long by 2 μm to 15 μm wide, with size varying according to strain. There are four anterior free flagella, with a fifth flagellum curving back along the margin of an undulating membrane and ending posterior to the middle of the body.[49, 65] The recurrent flagellum is not enclosed by an undulating membrane but is closely associated with it in a shallow groove. A densely staining lamellar structure (**accessory filament**) courses within the undulating membrane along its length. A **costa** arises in the kinetosome complex and runs superficially beneath and generally parallel to the undulating membrane's serpentine path. The costa, a rodlike structure with complex cross-striations, distinguishes Trichomonadidae from other families in its order. The costa probably serves as a strong, flexible support in the region of the undulating membrane.

  A parabasal body (Golgi body, dictyosome) lies near the nucleus, with a parabasal filament running from the kinetosome complex through or very near the parabasal body and ending in the posterior portion of the body (Fig. 4.5, Pf; page 46). A small, "minor" parabasal filament, which is inconspicuous in light microscope preparations, has been shown in other trichomonads, and it probably is present in *T. tenax* as well. A tubelike axostyle extends from near the kinetosomes posteriorly to protrude from the end of the body (covered by a cell membrane; Fig. 4.5, (d)).

  The axostylar tube is formed by a sheet of microtubules, and its anterior, middle, and posterior parts are known as **capitulum, trunk,** and **caudal tip,** respectively. Toward the capitulum, the tubular trunk opens out to curve around the nucleus, and microtubules of the capitulum slightly overlap the curving, collarlike **pelta** (Fig. 4.5, Pl and (d)). The pelta also comprises a sheet of microtubules and appears to function in supporting the "periflagellar canal," a shallow depression in the anterior end from which all flagella emerge. A cytostome is not present. *Trichomonas tenax* has concentrations of microbodies traditionally called **paracostal granules** along its costa, and other species of *Trichomonas* have **paraxostylar granules** along their axostyles (Fig. 4.5, H). These bodies are now called **hydrogenosomes** on the basis of their biochemical characteristics. We discuss the metabolic functions of hydrogenosomes below.

- **Biology.** *Trichomonas tenax* can live only in the mouth and, apparently, cannot survive passage through the digestive tract. Transmission, then, is direct, usually through kissing or common use of eating or drinking utensils; *T. tenax* can live for several hours in drinking water. Trophozoites divide by binary fission. They are considered harmless commensals, feeding on microorganisms and cellular debris, although there is one report of a submaxillary gland infection that defied diagnosis until flagellates were found in fluid removed by subcutaneous needle aspiration.[24] They are most abundant between the teeth and gums and in pus pockets, tooth cavities, and crypts of the tonsils, but they also have been found in the lungs and trachea. Although good oral hygiene is said to decrease or eliminate the infection, in one survey 15.7% of patients in a clinical practice in New York were positive, and none had oral hygiene rated as poor.[9]

Trichomonas vaginalis

This species (Figs. 6.12 and 6.13) was first found by Donné in 1836 in purulent vaginal secretions and in secretions from a male's urogenital tract. In 1837 he named it *Trichomonas vaginalis,* thereby creating the genus. It is a cosmopolitan species, found in reproductive tracts of both men and women the world over. Donné thought the organism was covered with hairs, which is what prompted the generic name (Greek *thrix* = hair).

- **Morphology.** *Trichomonas vaginalis* is very similar to *T. tenax* but differs in the following ways: It is somewhat larger, 7 μm to 32 μm long by 5 μm to 12 μm wide; its undulating membrane is shorter; and there are more granules along its axostyle and costa. In living and appropriately fixed and stained specimens, the constancy in presence and arrangement of hydrogenosomes is the best criterion for distinguishing *T. vaginalis* from other *Trichomonas* spp.[33] *Trichomonas vaginalis* frequently produces pseudopodia.

- **Biology.** *Trichomonas vaginalis* lives in the vagina and urethra of women and in the prostate, seminal vesicles, and urethra of men. It is transmitted primarily by sexual intercourse,[38] although it has been found in newborn infants. Its presence occasionally in very young children, including virginal females, suggests that the infection can be contracted from soiled washcloths, towels, and clothing. Viable cultures of *T. vaginalis* have been obtained from damp cloth as long as 24 hours after inoculation. Acidity of the normal vagina (pH 4.0 to 4.5) ordinarily discourages infection, but, once established, the organism itself causes a shift toward alkalinity (pH 5 to 6), which further encourages its growth.

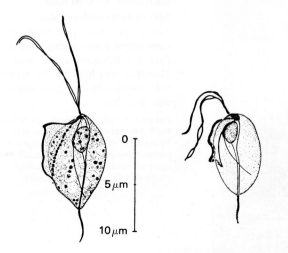

Figure 6.11 **Typical trophozoites of *Trichomonas tenax*.**

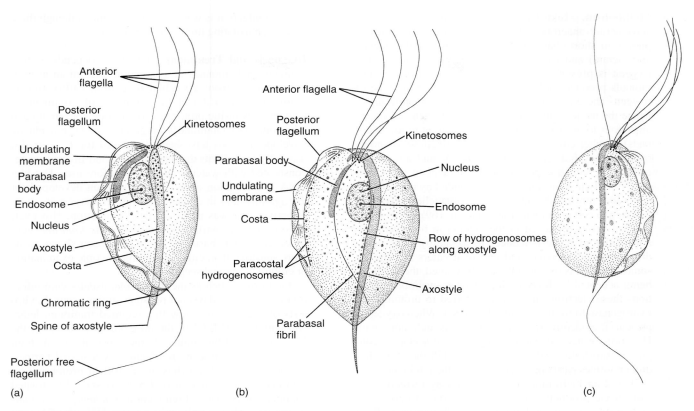

Figure 6.12   Morphology of trichomonads.

(*a*) *Tritrichomonas foetus*; 10–25 μm long. (*b*) *Trichomonas vaginalis*; 7–32 μm long. The hydrogenosomes are not always in a definite row; (*c*) *Pentatrichomonas hominis* trophozoite. This species ranges from 8–20 μm long.

(*a*) and (*b*) drawn by Bill Ober from D. H. Wenrich and M. A. Emerson, "Studies on the morphology of *Tritrichomonas foetus* (Riedmüller) from American cows," in *J. Morphol.* 55:195, 1933. (*c*) drawn by Bill Ober.

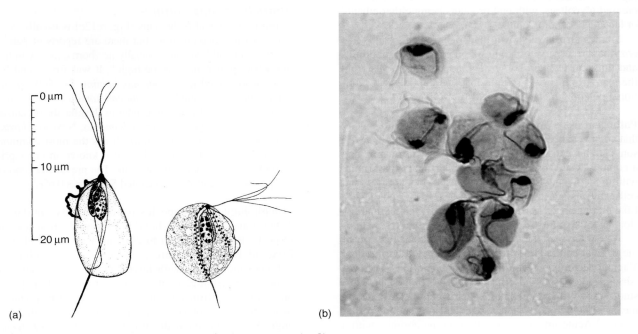

Figure 6.13   Typical trophozoites of *Trichomonas vaginalis.*

(*a*) Drawing, showing size and flagellar arrangements; (*b*) cultured flagellates as they appear on a stained slide.

(*a*) from B. M. Honigberg and V. M. King, "Structure of *Trichomonas vaginalis* Donné," in *J. Parasitol.* 50:345–364. Copyright © 1964. *Journal of Parasitology.* Reprinted by permission. (*b*) Photograph courtesy of John Janovy, Jr.

- **Metabolism.** Like *Giardia* species, trichomonads are aerotolerant anaerobes, degrading carbohydrates incompletely to short-chain organic acids (principally acetate and lactate) and carbon dioxide, regardless of whether oxygen is present.[27] Unlike *Giardia,* however, trichomonads produce molecular hydrogen in the absence of oxygen. These reactions take place in hydrogenosomes—hence the name of these organelles. Hydrogenosomes are analogous to mitochondria (which are absent in trichomonads) in other eukaryotes; but their distinctness is shown by their morphology, absence of DNA, and absence of cardiolipin, which is present in membranes of mitochondria.[62, 83] Hydrogenosomes are surrounded by two, closely apposed 6 nm membranes.[6] Similar organelles have been reported in certain rumen ciliates (see p. 169).

  Pyruvate is produced in the cytoplasm by glycolysis. Part of this pyruvate is reduced to lactate by lactic dehydrogenase and excreted, and part of it enters hydrogenosomes where it is oxidatively decarboxylated, the electrons being accepted by **ferredoxin.**[43] Under anaerobic conditions these electrons are then transferred to protons by a hydrogenase to form molecular hydrogen. When oxygen is present, it evidently accepts the electrons and, along with H^+, forms water. Oxidation of pyruvate to acetate is coupled to substrate-level generation of ATP; therefore, hydrogenosomes participate in energy production in the cell.

  The drug metronidazole is reduced by ferredoxin to form toxic products, thus explaining the effectiveness of this drug in chemotherapy for trichomoniasis. Both metronidazole-sensitive and -resistant strains occur, however, and resistant strains show higher glucose uptake rates, lower hydrogenase activity, and lower H_2 formation than do sensitive strains.[27] Drug resistance is attributed to loss of two oxidoreductases, enzymes necessary for reducing metronidazole, and subsequent metabolic switch to increased glycolysis, with end products being either ethanol or lactate, depending on the *Trichomonas* species.[43]

  Studies also have shown that trichomonads lack some enzymes necessary to synthesize complex phospholipids and thus must obtain some membrane components from their environment.[3, 64] Such observations suggest additional potential metabolic targets for drug action.

- **Pathogenesis.** Most strains are of such low pathogenicity that an infected person is virtually asymptomatic. However, some strains cause intense inflammation, with itching and a copious white discharge (**leukorrhea**) that is swarming with trichomonads. They feed on bacteria, leukocytes, and cell exudates and are themselves ingested by monocytes. Like all flagellates, *T. vaginalis* divides by longitudinal fission, and, like other trichomonads, it does not form cysts.

  A few days after infection there is a degeneration of the vaginal epithelium followed by leukocytic infiltration. Vaginal secretions become abundant and white or greenish, and the tissues become intensely inflamed. In vitro studies show that flagellates attach to epithelial cells by means of numerous cytoplasmic extensions and microfilaments.[56] Acute infections usually become chronic, with a lessening of symptoms, but occasionally flare up again. It should be noted, however, that leukorrhea is not diagnostic for trichomoniasis; indeed, at least half of patients even with severe leukorrhea are negative for *T. vaginalis.*[29] In men the infection is usually asymptomatic, although there may be an irritating urethritis or prostatitis.

- **Diagnosis and Treatment.** Diagnosis depends on recognizing trichomonads in a secretion or from an in vitro culture made from a vaginal irrigation. Cultivation is recommended to detect low numbers of organisms.[29] Culture of parasitic protozoa is often time-consuming and laborious (see Diamond's epigraph, p. 87), but plastic envelope methods have been developed for *T. vaginalis,* using dry ingredients that have a long shelf life and are reconstituted with water immediately before use.[4] Dot-blot DNA hybridization assays have also been developed for *T. vaginalis,* and in clinical trials these assays were more effective than was microscopic examination. However, cross-reactions with *Pentatrichomonas hominis* were observed.[71] PCR-based methods have been shown to be more sensitive than either direct microscopic examination or culture of vaginal secretions.[86]

  Oral drugs, such as metronidazole, usually cure infection in about five days, but resistant strains occur. In vitro tests of such strains show that required minimum lethal concentrations (MLC) of the drug are up to 11 times the MLC of susceptible strains.[8] Some apparently recalcitrant cases may be caused by reinfection by a sexual partner. Suppositories and douches are useful in promoting an acid pH of the vagina. Sexual partners should be treated simultaneously to avoid reinfection. *Trichomonas vaginalis* has been shown to survive cryopreservation of human semen, suggesting that infections could be contracted through artificial insemination.[74] PCR-based diagnostic research suggests that standard diagnostic methods do not detect all the cases and therefore result in significant undertreatment for vaginal trichomoniasis.[86]

Pentatrichomonas hominis

The third trichomonad of humans (Fig. 6.12*c*) is usually considered a harmless commensal, but there are reports of pathogenic effects in children and especially newborns, mostly in the tropics where prevalence can be high.[17] It was first found by Davaine, who named it *Cercomonas hominis* in 1860. Traditionally, it has been called *Trichomonas hominis,* but because most specimens actually bear five anterior flagella, the organism has been assigned to genus *Pentatrichomonas.* Next to *Giardia duodenalis* and *Chilomastix mesnili,* this is the most common intestinal flagellate of humans. It is also known in other primates and in various domestic animals. The prevalence among 13,517 persons examined in the United States was 0.6%.[5]

- **Morphology.** This species is superficially similar to *T. tenax* and *T. vaginalis* but differs from them in several respects. Its size is 8 μm to 20 μm by 3 μm to 14 μm. Five anterior flagella are present in most specimens, although individuals with fewer flagella are sometimes found. This arrangement is referred to as "four-plus-one," since the fifth flagellum originates and beats independently of the others. A recurrent (sixth) flagellum is aligned alongside the undulating membrane, as in *T. tenax* and *T. vaginalis,* but, in contrast to these two species, the recurrent flagellum in *P. hominis* continues as a long, free flagellum past the posterior end of the body. An axostyle, a pelta, a parabasal body, "major" and "minor" parabasal filaments, a costa, and paracostal hydrogenosomes are present. Paraxostylar hydrogeno somes are absent.

- **Biology.**   *Pentatrichomonas hominis* lives in the large intestine and cecum, where it divides by binary fission, often building up incredible numbers. It feeds on bacteria and debris, probably taking them in with active pseudopodia. The organism often is present in routine examinations of diarrheic stools, but it co-occurs with *Giardia duodenalis* and pathogenic bacteria, so may not be the primary cause of illness. There also are reports of *P. hominis* from liver abscesses.[17] In formed stools the flagellates are rounded and dormant but not encysted. They are difficult to identify at this stage because they do not move, and structures normally characteristic for the species cannot be distinguished.

    The organism apparently can survive acidic conditions of the stomach, and transmission occurs by contamination. Filth flies can serve as mechanical vectors. High prevalence is correlated with unsanitary conditions. Diagnosis depends on identification of the organism in fecal preparations, and prevention depends on personal and community sanitation. *Pentatrichomonas hominis* cannot establish in the mouth or urogenital tract.

Tritrichomonas foetus

Tritrichomonas foetus (see Fig. 6.12*a*) is responsible for a serious genital infection in cattle, zebu, and possibly other large mammals and is especially common in Europe and the United States. Molecular research provides strong evidence that *T. foetus* and *T. suis* from pigs are the same species.[78] It is one of the leading causes of abortion in cattle (along with brucellosis, leptospirosis, and *Neospora caninum* infection— see chapter 8). The USDA estimated losses from *T. foetus* in the United States between 1951 and 1960 at $8.04 million. More recent research suggests that in a herd with 40% of bulls infected, producers could expect up to 35% reduction in economic benefits per cow confined with an infected bull.[69]

- **Morphology.**   The cells are spindle to pear shaped, 10 μm to 25 μm long by 3 μm to 15 μm wide. There are three anterior flagella, and a fourth flagellum, which is recurrent, extends free from the posterior end of the body about the length of the anterior flagella. The mastigont system is generally similar in organization to those of trichomonads described previously, but it is even more complex and will not be described here.[34] The costa is prominent and, although similar in position and function to those of other trichomonads, differs in ultrastructural detail, resembling a parabasal filament in this respect. Its undulating membrane structure is curious, consisting of two parts. The proximal part is a foldlike differentiation of the dorsal body surface, and the distal part, which contains the axoneme of the recurrent flagellum, courses along the rim of the proximal part with no obvious physical connection to it. A thick axostyle protrudes from the posterior end of the body. Numerous paraxostylar hydrogenosomes are present in the posterior part of the organism, just anterior to the point of the axostyle, and these are apparent in the light microscope preparations as a "chromatic ring."

- **Biology.**   These trichomonads live in the preputial cavity of bulls, although testes, epididymis, and seminal vesicles also may be infected. In cows the flagellates first infect the vagina, causing a vaginitis, and then move into the uterus. After establishing in the uterus they may disappear from the vagina or remain there as a low-grade infection. Bovine genital trichomoniasis is a venereal disease transmitted by coitus, although transmission by artificial insemination is possible. The flagellates multiply by longitudinal fission and form no cyst.

- **Pathogenesis.**   The most characteristic sign of bovine trichomoniasis is early abortion, which usually happens 1 to 16 weeks after insemination. Because the fetus is quite small at that stage, it may not be evident that the cow has aborted and, therefore, that she had conceived. If all fetal membranes are passed after abortion, a cow may recover spontaneously. However, if they remain, she usually develops chronic endometritis, which may cause permanent sterility. The parasites release extracellular proteases that have the capacity to digest proteins, including immunoglobulins, that might otherwise function in host defense.[79] Normal gestation and delivery occasionally occur in an infected animal.

    Pathogenesis is not observable in bulls, but an infected bull is worthless as a breeding animal; unless treated, it usually remains infected permanently. Treatment is expensive, difficult, and not always effective. Because of the immense prices paid for top-quality bulls, the loss of a single animal may bankrupt a breeder.

- **Epizootiology.**   Experimental infections have been established in rabbits, guinea pigs, hamsters, dogs, goats, sheep, and pigs, but the epizootiological significance of such infec-tions has not been determined. Trichomonads can survive freezing in semen ampules, although some media are more detrimental than are others. This precludes use of semen from infected bulls for artificial insemination.

- **Diagnosis, Treatment, and Control.**   Direct identification of protozoa from smears or culture remains the only sure means of diagnosis, although molecular studies have shown at least three other trichomonad genera may be present and easily confused with *T. foetus*.[25] Smears can be obtained from amniotic or allantoic fluid, vaginal or uterine exudates, placenta, fetal tissues or fluids, or preputial washings from bulls. Flagellates fluctuate in numbers in bulls; in cows they are most numerous in the vagina two or three weeks after infection.

    No satisfactory treatment is known for cows, but the infection is usually self-limiting in them, with subsequent, partial immunity. Bulls can be treated if the condition has not spread to the inner genital tubes and testes. Treatment is usually attempted only on exceptionally valuable animals because it is a tedious, expensive task. Preputial infection is treated by massaging antitrichomonal salves or ointments into the penis, after it has been let down by nerve block or by injection of a tranquilizer into the penis retractor muscles. Repeated treatment is usually necessary. Systemic drugs show promise of becoming the standard method of treatment.

    Control of bovine genital trichomoniasis depends on proper herd management. Cows that have been infected should be bred only by artificial insemination to avoid infecting new bulls. Bulls should be examined before purchase, with a wary eye for infection in the resident herd. Unless they are extremely valuable, infected bulls should be killed. Like any venereal disease, trichomoniasis can be controlled and eventually eliminated with proper treatment and reporting, but the disease is likely to remain a problem for some time. Vaccines have been developed, some of which employ parasite surface proteins involved in attachment of the flagellates to vaginal epithelial cells.[7] Field trials of a polyvalent vaccine showed that 62.5% of vaccinated heifers bred to infected bulls produced calves as compared to 31.5% of controls.[44] These vaccines are most effective when used in conjunction with other control measures, including replacement of older bulls with younger ones.

Family Monocercomonadidae

Monocercomonadidae exhibit well-developed pseudopodia; an undulating membrane is absent, and flagella tend to be reduced. Most species are parasites of insects, but three genera infect domestic animals. One of these is economically important and has evolved a unique mode of transmission: in the egg of a nematode.

Histomonas meleagridis

Histomonas meleagridis (Fig. 6.14), a cosmopolitan parasite of gallinaceous fowl, including chickens, turkeys, peafowl, and pheasant, causes a severe disease known variously as **blackhead, infectious enterohepatitis,** and **histomoniasis.** The disease is more virulent in some host species than in others; chickens show the disease less often than do turkeys, for example. Economic loss in the United States resulting from histomoniasis in chickens and turkeys is not easy to determine but is estimated at about $2 million per year.[52]

The taxonomic history of *H. meleagridis* has been very confused because of its polymorphism in different situations. At various times it has been confused with amebas, coccidia, fungi, and *Trichomonas* spp. Even the disease that it causes has been attributed to different organisms, from amebas to viruses. Today much is known about the organism, and its biology and pathogenesis are less mysterious than they once were.

- **Morphology.**   *Histomonas meleagridis* is pleomorphic; its stages change size and shape in response to environmental factors. There is no cyst, only various trophic stages, in the life cycle, although "cystlike" forms have been produced experimentally using cultures subjected to stressful conditions.[90] When they are found in the lumen

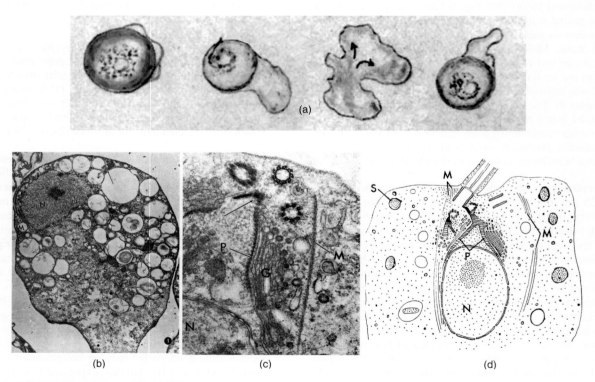

Figure 6.14   *Histomonas meleagridis.*
(*a*) Early drawings by E. E. Tyzzer showing ameboid movements of *H. meleagridis* in a hanging drop suspension. (*b*) Transmission electron micrograph of *H. meleagridis* trophozoite; endoplasm is vacuolated (top portion of the figure) and ectoplasm is more granulated (lower portion of figure). (*c*) TEM of mastigont system, showing cross sections of three kinetosomes, Golgi apparatus (G), parabasal fiber (P); and a tract of microtubules (M) extending toward the nucleus (N). (*d*) Interpretation of ultrastructural features; M, tracts of microtubules that correspond to the pelta and axostyle found in trichomonads; S, electron-dense structures, possibly hydrogenosomes.

(*a*) From E. E. Tyzzer, "The flagellate character and reclassification of the parasite producing "blackhead" in turkeys—*Histomonas* (gen. nov.) *meleagridis* (Smith)," in *J. Parasitol.*, 6:124–131; copyright © 1920, American Society of Parasitologists. (*b*)–(*d*) From F. L. Schuster, "Ultrastructure of *Histomonas meleagridis* (Smith) Tyzzer, a parasitic amebo-flagellate," in *J. Parasitol.*, 54:725-737, copyright © 1968, American Society of Parasitologists. All figures reprinted by permission.

of the cecum (which is rare) or in culture, the stages are ameboid, 5 μm to 30 μm in diameter, and almost always with only one flagellum. There are usually four kinetosomes, the basic number for trichomonads, although this condition has been attributed to duplication of the kinetic apparatus in preparation for mitosis.[73] The nucleus is vesicular and often has a distinct endosome. One can usually discern a clear ectoplasm and a granular endoplasm. Food vacuoles may contain host blood cells, bacteria, or starch granules.

Electron microscope studies have revealed a pelta, a V-shaped parabasal body, a parabasal filament, and a structure resembling an axostyle (Fig. 6.14). These structures cannot be seen with light microscopy, but their presence supports placement of *Histomonas* spp. in order Trichomonadida. No mitochondria have been observed, but hydrogenosomes are evidently present.[50] Forms within the tissues have no flagella, although kinetosomes are present near the nucleus.

- **Biology and Transmission Ecology.**   Like other flagellates, *H. meleagridis* divides by binary fission. No cysts or sexual stages occur in the life cycle. Trophozoites are fragile and cannot long survive in the external environment or a host's stomach acids. Certain factors can, and sometimes do, conspire to allow infection by trophozoites. If trophozoites are eaten with certain foods that raise the stomach pH, they may survive to initiate a new infection. This can be the means of an epizootic in a dense flock of birds. Turkeys can transmit infections among themselves evidently by way of "cloacal drinking," although trasmission to chickens usually involves a nematode vector (see below).[53]

The most important and by far the most interesting mode of transmission is within eggs of the cecal nematode *Heterakis gallinarum*. Because the protozoan undergoes development and multiplication in the nematode, the worm can be considered a true intermediate host.[45] After being ingested by a worm, flagellates enter the nematode's intestinal cells, multiply, and then break out into the pseudocoel and invade the germinative area of the nematode's ovary. There they feed and multiply extracellularly, move down the ovary with developing oogonia, and then penetrate oocytes (Fig. 6.15).

Feeding and multiplication continue in oocytes and newly formed eggs. Passing out of the mother worm and out of the bird with its feces, the parasite divides rapidly, invading tissues of the juvenile nematode, especially those of the digestive and reproductive systems. Interestingly, *H. meleagridis* also parasitizes the reproductive system of the male nematodes.[45] Presumably, it could be transmitted to a female during copulation, thus constituting a venereal infection of nematodes!

Infected nematode eggs can survive for at least two years in soil. If worm eggs are eaten by an appropriate bird, they hatch in the intestine, and juvenile *Heterakis gallinarum* pass down into the cecum, where *Histomonas meleagridis* is free to leave its temporary host to begin residence in a more permanent one.

Earthworms are important paratenic hosts of both *Heterakis gallinarum* and its contained *Histomonas meleagridis*. When eaten by an earthworm, nematode eggs

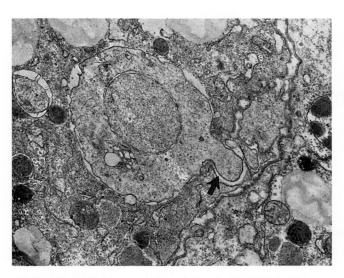

Figure 6.15   **Electron micrograph of a section through the growth zone of the ovary of *Heterakis gallinarum* to show *Histomonas meleagridis* in the process of entering an oocyte (*arrow*).** (×13,800)

From D. L. Lee, in A. M. Fallis (Ed.), *Ecology and physiology of parasites.* Toronto: University of Toronto Press, 1971.

will hatch, releasing second-stage juveniles that become dormant in the earthworm's tissues. When the earthworm is eaten by a gallinaceous fowl, *Heterakis gallinarum* juveniles are released, and the bird becomes infected by two kinds of parasites at once. Chickens are the most important reservoirs of infection because they are less often affected by *Histomonas meleagridis* than are turkeys. Because both *Heterakis gallinarum* eggs and infected earthworms can survive for so long in soil, it is almost impossible to raise uninfected turkeys in the same yards in which chickens have lived.

- **Pathogenesis.**   Turkeys are most susceptible between the ages of 3 and 12 weeks, although they can become infected as adults. In very young poults, losses may approach 100% of a flock. Chickens are less prone to the disease, but outbreaks among young birds have been reported. Quail and partridge show varying degrees of susceptibility.

The principal lesions of histomoniasis are found in the cecum and liver. At first pinpoint ulcers are formed in the cecum. These may enlarge until nearly the entire mucosa is involved. Ceca often become filled with cheesy, foul-smelling plugs that adhere to the cecal walls. Complete perforation of the cecum, with peritonitis and adhesions, can occur. Ceca are usually enlarged and inflamed. Liver lesions are rounded, with whitish or greenish areas of necrosis. Their size varies, and they penetrate deep into the parenchyma.

Infected birds show signs of droopiness, ruffled feathers, and hanging wings and tail. Yellowish diarrhea usually occurs. Skin of the head turns black in some cases, giving the disease its name *blackhead;* however, other diseases also can cause this symptom.

Histomonas meleagridis by itself is incapable of causing blackhead but does so only in the presence of intestinal bacteria of several species, particularly *Escherichia coli* and *Clostridium perfringens*. Birds that survive are immune for life. A related histomonad, *Parahistomonas wenrichi*, also is transmitted by *Heterakis gallinarum* but is not pathogenic.

- **Diagnosis, Treatment, and Control.** Cecal and liver lesions are diagnostic. Scrapings of these organs will reveal histomonads, thereby distinguishing the disease from coccidiosis. Several types of drugs have been used in prevention and treatment, including nitrofurans, nitroimidazoles, and phenylarsonic acid derivatives. These successfully inhibit, suppress, or cure the disease, but they may have undesirable side effects, such as delaying sexual maturity of the bird. Some of these compounds have been banned for veterinary use in the United States and Canada because of their persistence in meat.[53] Treatment of birds with nematocides, such as mebendazole, cambendazole, and levamisole, to eliminate *H. gallinarum* is effective in preventing future outbreaks, because *Histomonas meleagridis* cannot survive in soil by itself.

    Control depends on effective management techniques, such as rearing young birds on hardware cloth above the ground, keeping young birds on dry ground, and controlling *Heterakis gallinarum*. Pasture rotation of *Heterakis*-free flocks is also successful.

Dientamoeba fragilis

Dientamoeba fragilis (Fig. 6.16) has traditionally been considered a member of ameba family Endamoebidae, but its differences from other members of this family have long been recognized. For example, a large proportion of individuals have two nuclei, their nuclear structure is rather unlike that of other Endamoebidae, an extranuclear spindle is present during division, and cysts are not formed. More than 60 years ago Dobell believed that *D. fragilis* was closely related to ameboflagellates of genus *Histomonas*.[23] On the basis of ultrastructural and immunological evidence, Camp and coworkers placed *Dientamoeba* in a subfamily of Monocercomonadidae.[11] This change reflects the organism's phylogenetic relationship rather than the fact that it moves by pseudopodia instead of flagella. *Dientamoeba fragilis*, infecting about 4% of humans, is the only species known in the genus.

- **Morphology.** Only trophozoites are known in this species; cysts are not formed. Trophozoites (see Fig. 6.16) are very delicate and disintegrate rapidly in feces or water. They are 6 μm to 12 μm in diameter, and ectoplasm is somewhat differentiated from endoplasm. A single, broad pseudopodium usually is present. Food vacuoles contain bacteria, yeasts, starch granules, and cellular debris. About 60% of individuals contain two nuclei, which are connected to each other by a filament and are observable by light microscopy; the rest have only one nucleus. By electron microscopy one can discern that the filament connecting the nuclei is a division spindle composed of microtubules; binucleate individuals are, in reality, in an arrested telophase. The endosome is eccentric, sometimes fragmented or peripheral in the nucleus, and concentrations of chromatin are usually apparent. A filament and Golgi apparatus are present and are reminiscent of parabasal fibers and parabasal bodies found in *Histomonas meleagridis* and trichomonads. There are no kinetosomes or centrioles.

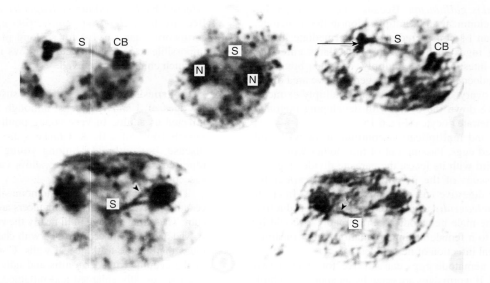

Figure 6.16 *Dientamoeba fragilis.*

Photomicrographs of binucleate organisms. Four chromatin bodies (*CB*) can be resolved within the telophase nucleus of the organism, shown in the first and third figures. The extranuclear spindle (*S*) extends between the nuclei (*N*) in all figures. Note the branching of the spindle (*arrowheads*) near the nucleus in the fourth and fifth figures (Bouin's fixative: first, second, and fourth figures—bright field [×4950]; Nomarski differential interference: third and fifth figures [×3650]).

- **Biology.** *Dientamoeba fragilis* lives in the large intestine, especially in the cecal area. It feeds mainly on debris and traditionally has been considered a harmless commensal. However, a study of 43,029 people in Ontario showed a high percentage of intestinal problems in those infected with *D. fragilis*.[89] Symptoms included diarrhea, abdominal pain, anal pruritus, abnormal stools, and other indications of abdominal distress. It is possible that *D. fragilis* is responsible for many such cases of unknown etiology, especially in small children. Iodoquinol, tetracycline, and metronidazole have all been used in treatment.[87] Because *D. fragilis* infections often co-occur with other species, it is not always easy to determine which, or which combination, of parasites is responsible for the most damage. In one study of 414 excised appendices, for example, Cerva et al.[15] found pinworms, ascaris eggs, *Endolimax nana*, *Entamoeba coli,* and *Giardia* cysts in addition to *D. fragilis*.

The mode of transmission is unknown; *D. fragilis* does not form cysts and it cannot survive in the upper digestive tract. The organism survives transmission in eggs of a parasitic nematode, as does its relative, *Histomonas meleagridis*. Small, ameboid organisms resembling *D. fragilis* have been found in eggs of the common human pinworm, *Enterobius vermicularis*, and there is strong epidemiological evidence that the nematode is the vector of the protistan.[89]

ORDER HYPERMASTIGIDA

These flagellates are all intestinal parasites of wood-eating insects such as termites and some cockroaches, as are members of order Oxymonadida (phylum Axostylata); the latter

is not discussed here. Hypermastigida are highly complex structurally (Fig. 6.17), with parabasal bodies, very many flagella (estimated over 52,000 in one case),[12, 13] often in zones and commonly of diverse lengths, cytoplasmic ridges between flagellar rows, microtubular complexes, and hydrogenosomes. Many also have ectosymbiotic bacteria that mimic flagella.[12] The flagellates actually digest wood for their hosts; thus the relationship is a true mutualistic one. The hypermastigid condition evidently has evolved more than once, with radiation into diverse forms (Fig. 6.17*a*),[12] so there must be something about the gut of wood-feeding insects that drives such evolutionary diversification.

ORDER OPALINIDA

The opalinids, commensals in the lower digestive tract of amphibians, are considered members of phylum Chromista, although that phylum is a very heterogeneous group of stramenopiles and is likely to undergo further taxonomic revision in the future (see chapter 4).

Family Opalinidae

There are about 150 species of opalinids, most of which live in the lower intestines of amphibians. They are of no economic or medical importance but are of zoological interest because of their peculiar morphology and the fact that their reproductive cycles evidently are controlled by host hormones.[26] Also, study of opalinids may contribute to our understanding of amphibian zoogeography and evolution,

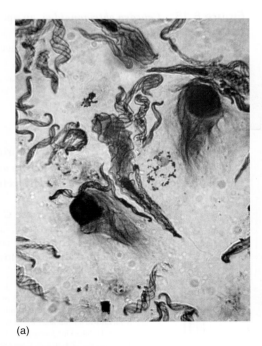

(a)

(b)

Figure 6.17 **Flagellates from termites.**

(*a*) Several genera as seen in a stained smear. (*b*) Scanning electron micrograph of *Pseudotrichonympha paulistana* from the termite *Heterotermes tenuis,* demonstrating the extreme number of flagella typical of Hypermastigida.

(*a*) Courtesy John Janovy, Jr. (*b*) Courtesy Juan Saldarriaga.

although proper identification of species is a problem.[2, 20] For example, geographic distribution of *Opalina* species evidently reveals an invasion of North America by way of Beringia (the prehistoric land bridge), carried by frogs of genus *Rana*, whereas members of genus *Zelleriella* probably invaded North America from the Neotropics.[20] Opalinids are commonly encountered in routine dissections of frogs in teaching laboratories; these protozoans' large size, and graceful movements make them exciting finds for students who never gave much thought to organisms that might live in a frog rectum.

Numerous oblique rows of short flagella occur over the body surface of opalinids, giving them a strong resemblance to ciliates (Fig. 6.18). At the anterior end is a sickle-shaped field of kinetosomes, called a **falx;** dorsal, ventral, and lateral parts of the body are defined with respect to the falx.[2] At the posterior end, the flagellar rows simply converge, although, in *Opalina* species, the convergence is in the form of a seam-like suture.[2] Opalinids possess only one type of nucleus, reproduce sexually by anisogamous syngamy, and asexually by binary fission between flagellar rows (instead of across them) thus differing from the ciliates they superficially resemble.

Ultrastructurally, all opalinid genera exhibit cortical folds, corticular ribbons of microtubules, mitochondria with long tubular cristae, and pinocytotic vesicles budding from the bases of the cortical folds (Fig. 6.19). Genera differ, however, in other ultrastructural features such as presence or absence of fibrous tracts alongside the kinetosomes.[63]

Adult opalinids reproduce asexually by binary fission in the rectum of frogs and toads during the summer, fall, and winter. In spring, which is their host's breeding season, they accelerate divisions and produce small, precystic forms, which then form cysts and pass out with feces. When the cysts are eaten by tadpoles, male and female gametes excyst and fuse to form zygotes, which resume asexual reproduction. The exact chemical identity of compound(s) that stimulate encystment is not known, but present evidence indicates that it is one or more breakdown products of steroid hormones excreted in the frog's urine. This is an interesting example of a physiological adaptation to ensure the production of infective stages at the time and place of new host availability. Effectiveness of this adaptation is attested to by the high prevalence of opalinids in frogs and toads.

In addition to members of genus *Opalina*, amphibians may also be infected with opalinid species of the genera *Protoopalina, Cepedea,* and *Zelleriella,* although *infected* is a rather strange word for a group of nonpathogenic symbionts so closely, commonly, and inextricably tied to their hosts. A curious symbiosis is found in *Zelleriella opisthocarya,* a parasite of toads, and *Entamoeba* sp., in which more than 200 cysts of the ameba may occur in one opalinid.[77]

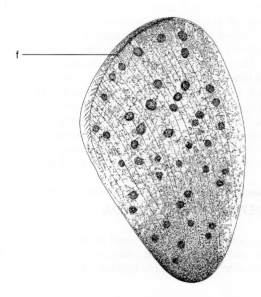

Figure 6.18 *Cepedea obtrigonoidea,* an opalinid from the toad *Bufo fowleri.*

Note falx (*f*) extending along the ventral surface and note the numerous nuclei.

Drawn after F. M. Affa'a and D. H. Lynn, "A review of the classification and distribution of five opalinids from Africa and North America," in *Can. J. Zool.* 72:665–674, 1994.

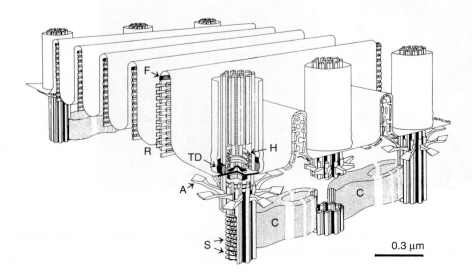

0.3 µm

Figure 6.19 The cortex of an opalinid, *Protoopalina australis,* as reconstructed from electron micrographs.

A, kinetosomal arms; *C,* interkinetosomal connectives; *F,* apical fibers; *H,* transitional helix; *R,* cortical ribbons of microtubules; *S,* kinetosomal shelves; *TD,* transitional disc.

From D. J. Patterson and Ben L. J. Delvinquier, "The fine structure of the cortex of the protist *Protoopalina australis* (Slopalinida, Opalindae) from *Litoria nasuta* and *Litoria inermis* (Amphibia: Anura: Hylidae) in Queensland, Australia," in *J. Protozool.* 37:449–455. Copyright © 1990. The Society of Protozoologists. Reprinted by permission.

Learning Outcomes

By the time a student has finished studying this chapter, he or she should be able to:

1. Describe the host and geographic distribution of opalinid flagellates.

2. Draw the mastigont systems of *Giardia duodenalis* and a typical trichomonad.

3. Explain the disease management strategy appropriate for control of *Tritrichomonas foetus*.

4. Write an extended paragraph describing the transmission mechanisms of *Histomonas meleagridis*.

5. Tell the drugs used to control flagellates discussed in this chapter, being certain to mention potential side effects.

6. Explain the transmission mechanisms and epidemiology of giardiasis.

References

References for superscripts in the text can be found at the following Internet site: www.mhhe.com/robertsjanovynadler9e

Additional Readings

Adam, R. D. 2000. The *Giardia lamblia* genome. *Int. J. Parasitol.* 30:475–484.

Honigberg, B. M. 1963. Evolutionary and systematic relationships in the flagellate order Trichomonadida Kirby. *J. Protozool.* 10:20–63.

Honigberg, B. M. 1978. Trichomonads of importance in human medicine. In J. P. Kreier (Ed.), *Parasitic protozoa* 3. New York: Academic Press, Inc.

Kolisko, M., I. Cepicka, V. Hampi, J. Kulda, and J. Flegr. 2005. The phylogenetic position of enteromonads: A challenge for the present models of diplomonad evolution. *Int. J. Syst. Evol. Microbiol.* 55:1729–1733.

Kulda, J., and E. Nohynkova. 1978. Flagellates of the human intestine and intestines of other species. In J. P. Kreier (Ed.), *Parasitic protozoa* 3. New York: Academic Press, Inc.

McDougald, I. R., and W. M. Reid. 1978. *Histomonas meleagridis* and its relatives. In J. P. Kreier (Ed.), *Parasitic protozoa* 3. New York: Academic Press, Inc.

Meyer, E. A., and S. Radulescu. 1979. Giardia and giardiasis. In W. H. R. Lumsden (Ed.), *Advances in parasitology* 17. New York: Academic Press, Inc.

Nadler, S. A., and B. M. Honigberg. 1988. Genetic differentiation and biochemical polymorphism among trichomonads. *J. Parasitol.* 74:797–804.

Thompson, R.C. A., J. A. Reynoldson, and A. H. W. Mendis. 1993. *Giardia* and giardiasis. In J. R. Baker and R. Muller (Eds.), *Advances in parasitology* 32. New York: Academic Press, Inc.

Wolfe, M. S. 1992. Giardiasis. *Clinical Microbiol.* Reviews 5:93–100.

The Amebas

*. . . he is shaking his head slowly in wonderment, looking at something brown
and gelatinous held in his hand, saying, "That is very interesting water."*

—Lewis Thomas, *Lives of a Cell*

Biology students are introduced to amebas early in their careers. Most are left with the impression that amebas are harmless, microscopic creatures that spend their lives aimlessly wandering about in mud, water, and soil, occasionally catching a luckless ciliate for lunch and unemotionally reproducing by binary fission. Actually, this is a pretty fair account of many amebas, although foraminiferans may have much more dramatic lives out in the ocean. A few amebas are parasites of other organisms, however, and one or two are responsible for much misery and death of humans. Still others are commensals, a characteristic that must be recognized to differentiate them from pathogenic species.

Amebas probably appeared early in eukaryote evolutionary history, and the ameboid body form may have arisen numerous times, most likely from various flagellates.[37] Structural characters used to suggest ancient evolutionary relationships include permanent cytostomes and both flagellate and ameboid stages, such as found in flagellate genus *Tetramitus*. The life cycle of *Naegleria* species also includes flagellate and ameboid stages, but no permanent cytostome is found in members of this genus. *Vahlkampfia* species have no flagellate stage, but their ameboid stages are like those of *Naegleria*.

At least one important parasite, *Entamoeba histolytica*, lacks mitochondria and therefore was thought by some to have diverged early from the eukaryotic line.[24] Later molecular studies, however, showed that *E. histolytica* was descended from ancestors that possessed mitochondria.[13] Of the many families of amebas, only Entamoebidae has species of great medical or economic importance. Three other families, Vahlkampfiidae, Hartmannelidae, and Acanthamoebidae, have species that can become facultatively parasitic in humans. Ameba taxonomy is extremely unsettled, especially at the "higher" levels (see chapter 4). Although there are no phylum names in this chapter, in places we use order and family names that are found in the current protozoological literature.[37]

AMEBAS INFECTING MOUTH AND INTESTINE

Family Entamoebidae

Species in Entamoebidae are parasites or commensals of the digestive systems of arthropods and vertebrates. Genera and species are differentiated microscopically on the basis of size and nuclear structure. Three genera contain known parasites or commensals of humans and domestic animals: *Entamoeba, Endolimax,* and *Iodamoeba.*

Genus *Entamoeba*

Entamoeba species possess a vesicular nucleus that has a small endosome at or near the center (Fig. 7.1). Chromatin granules are arranged around the periphery of the nucleus and, in some species, also around the endosome. The cytoplasm contains a variety of food vacuoles, often with particles of food being digested, usually bacteria or starch grains. On the ultrastructural level, the outer membrane possesses a "fuzzy coat," and the cytoplasm contains numerous vesicles, sometimes considered exocytotic because of their accumulation at the uroid (temporary posterior end).[40] Golgi bodies and mitochondria evidently are absent. Curious, small **helical bodies** can be seen widely distributed in the cytoplasm of some trophozoites. These bodies are 0.3 μm to 1.0 μm in length, contain up to 40 distinct ribonucleoproteins, and following encystment become crystallized into **chromatoid bodies** or **bars**[40, 42] (Figs. 7.1 and 7.2), that stain darkly with basic dyes. Chromatoidal bars may be blunt rods or splinter shaped, according to species, and in some species they are noticeable only in young cysts. As a cyst ages, the bars evidently are disassembled and disappear. *Entamoeba histolytica* is also sometimes infected with viruses.[40]

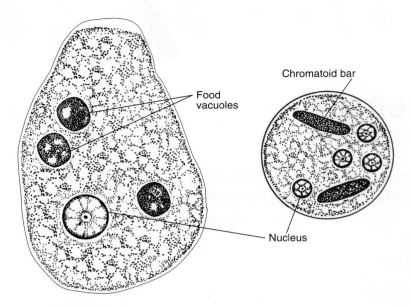

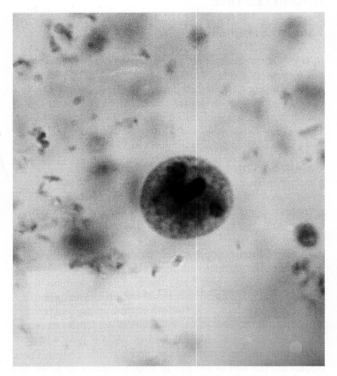

Figure 7.1 *Entamoeba histolytica* **trophozoite and cyst.**
The vesicular nucleus has light areas with strands of chromatin, in this case arranged as spokes. The endosome is the dark body at the nuclear center.

Drawing by Jeanne Robertson.

Figure 7.2 **Young cyst of *Entamoeba histolytica* containing two nuclei and a prominent chromatoidal bar.**
Usually, such a cyst is 10 μm to 20 μm wide.

Photograph by Larry S. Roberts.

Entamoeba species occur in both vertebrate and invertebrate hosts. Five species *(E. histolytica, E. dispar, E. hartmanni, E. coli,* and *E. gingivalis)* occur in humans and will be considered here; *E. polecki* is mentioned in passing.

Entamoeba histolytica. Dysentery, both bacterial and amebic, has long been known as a handmaiden of war, often inflicting more casualties than bullets and bombs. Accounts of dysentery epidemics accompany nearly every thorough account of war, from antiquity to the prison camp horrors of World War II and Vietnam. Captain James Cook's first voyage met with amebic disaster in Batavia, Java, and modern tourists, too, often find themselves similarly afflicted when visiting foreign ports.

Entamoeba histolytica (Fig. 7.3) is the ameba responsible for such misery. Close to 500 million people are believed infected at any one time, and up to 100,000 deaths occur per year (although see the *Diagnosis and Treatment* section). These numbers may increase as urban migration and deteriorating economies of some developing countries result in unhygienic conditions. In addition, high rates of infection exist in certain high-risk groups, such as people who practice anilingus, where infections can reach epidemic levels.

The history of our knowledge about *E. histolytica* is rampant with confusion and false conclusions. Foster[20] provides this interesting account: The ameba was first discovered in 1873 by a clinical assistant, D.F. Lösch, in St. Petersburg, Russia. The patient, a young peasant with bloody dysentery, was passing large numbers of amebas in his stools. Many of these, Lösch observed, contained erythrocytes in their food vacuoles. He successfully infected a dog by injecting amebas from his patient into the dog's rectum. On dissection Lösch found the dog's colonic mucosa riddled with ulcers that contained amebas. His human patient soon died, and at autopsy Lösch found identical ulcers in the intestinal mucosa. Despite these clear-cut observations, Lösch concluded that the ulcers were caused by some other agent and that the amebas merely interfered with their healing. Nearly 40 years passed before it was generally accepted that an intestinal ameba can cause disease.

A major cause of the 40-year delay was human ignorance about the fact that several species of amebas are found in the

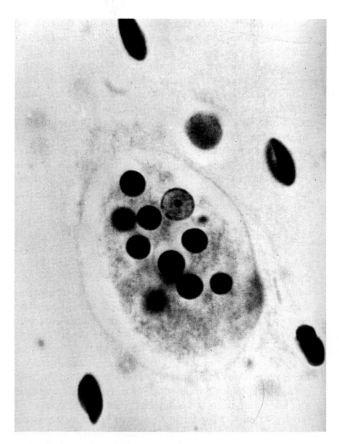

Figure 7.3   Trophozoite of *Entamoeba histolytica* with several erythrocytes in food vacuoles.

From M. Kenney and L. K. Eveland, "Transformation *in vivo* of a large race of *Entamoeba histolytica* into a small race," in *Bull. N. Y. Acad. Med.* 57:234–239. Copyright © 1981.

human intestine. Once this situation was recognized and non-pathogenic species were delineated, only one species complex remained that appeared—and only occasionally at that—to cause disease. Schaudinn named this group *Entamoeba histolytica* in 1903,[50] although the epithet *coli* was already applied to it by Lösch (as *Amoeba coli*). Schaudinn applied the epithet to a nonpathogenic species that he named *Entamoeba coli.*

Through the years it became obvious that *E. histolytica* occurs in two sizes. The smaller-sized amebas have trophozoites 12 μm to 15 μm in diameter and cysts 5 μm to 9 μm wide. This form is encountered in about a third of those who harbor amebas, and it is not associated with disease. The larger form has trophozoites 20 μm to 30 μm in diameter and cysts 10 μm to 20 μm wide. The small, nonpathogenic type is considered here as a separate species called *E. hartmanni*. Its life cycle, general morphology, and overall appearance, with the exception of size, are identical to those of *E. histolytica*. The task of proper identification is placed on the diagnostician, whose diagnosis may save the life of the patient or add the burden of unnecessary medication. A third species, *E. moshkovskii*, is identical in morphology to *E. histolytica*; it has not been considered a symbiont, but recent studies using molecular diagnostic techniques indicate that *E. moshkovskii* can establish in humans.[3]

In the past, strains of *E. histolytica,* differing in pathogenicity, were distinguished from nonpathogenic ones by

isozyme analysis. *Entamoeba histolytica* has now been divided into two species, the other one being the noninvasive *E. dispar,* based on molecular data. Previous claims of conversion of nonpathogenic *E. histolytica* into pathogenic and invasive forms are thus strongly disputed.[4, 16] Although *E. dispar* is considered nonpathogenic, it is evidently capable of producing intestinal lesions in experimental animals, and it is often found in captive primates.[16, 56]

Morphology and Life Cycle. Several successive stages occur in the life cycle of *E. histolytica:* **trophozoite, precyst, cyst, metacyst,** and **metacystic trophozoite.** Although the diameter of most trophozoites (see Fig.7.1 and Fig. 7.3) falls into a range of 20 μm to 30 μm, occasional specimens are as small as 10 μm or as large as 60 μm. In the intestine and in freshly passed, unformed stools, the parasites actively crawl about, their short, blunt pseudopodia rapidly extending and withdrawing. They also have filopodia, which are usually not discernible by light microscopy.[37] The clear ectoplasm is a rather thin layer but is differentiated from the granular endoplasm. The nucleus is difficult to discern in living specimens, but nuclear morphology may be distinguished after fixing and staining with iron-hematoxylin. The nucleus is spherical and is about one-sixth to one-fifth the cell's diameter. A prominent endosome is located in the center of the nucleus, and delicate, achromatic fibrils radiate from it to the inner surface of the nuclear membrane. Chromatin is absent from a wide area surrounding the endosome but is concentrated in granules or plaques on the inner surface of the nuclear membrane. This gives the appearance of a dark circle with a bull's-eye in the center. The nuclear membrane itself is quite thin.

Food vacuoles are common in the cytoplasm of active trophozoites and may contain host erythrocytes in samples from diarrheic stools (see Fig. 7.3). Granules typical for all amebas are numerous in the endoplasm. Chromatoidal bars are not found in this stage.

In a normal, asymptomatic infection, amebas are carried out in formed stools. As fecal matter passes posteriorly and becomes dehydrated, the parasites are stimulated to encyst. Cysts are neither found in stools of patients with dysentery nor formed by the amebas when they have invaded host tissues. Trophozoites passed in stools are unable to encyst. At the onset of encystment trophozoites disgorge any undigested food they may contain and condense into spheres called *precysts*. Precysts are so rich in glycogen that in young cysts large glycogen vacuoles may occupy most of the cytoplasm. Chromatoidal bars that form typically are rounded at their ends. These bars may be short and thick, thin and curved, spherical, or very irregular in shape, but they do not have the splinterlike appearance of those found in *E. coli.*

Precysts rapidly secrete a thin, tough hyaline **cyst wall** to form *cysts* that may be somewhat ovoid or elongate but usually are spheroid and 10 μm to 20 μm wide. Young cysts have only a single nucleus, but this rapidly divides twice to form two- and four-nuclei stages (Fig. 7.4). As nuclear division proceeds and cysts mature, the glycogen vacuole and chromatoidal bodies disappear. In semiformed stools one can find precysts and cysts with one to four nuclei, but quadrinucleate cysts *(metacysts)* are most common in formed stools (see Figs. 7.1 and 7.4). This stage can survive outside the

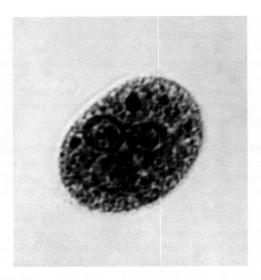

Figure 7.4 **Metacyst of** ***Entamoeba histolytica.***
Three of the four nuclei are in focus, and two small chromatoid bodies can be seen.

Photograph by Larry S. Roberts.

host and can infect a new one. After excysting in the small intestine, both the cytoplasm and nuclei divide to form eight small amebulas, or *metacystic trophozoites.* These are basically similar to mature trophozoites except in size.

- **Biology.** Trophozoites may live and multiply indefinitely within the crypts of the large intestine mucosa, apparently feeding on starches and mucous secretions and interacting metabolically with enteric bacteria. However, such trophozoites commonly initiate tissue invasion when they hydrolyze mucosal cells and absorb the predigested product. At this stage they no longer require the presence of bacteria to meet their nutritional requirements.

  The complex of environmental factors within a host's intestine is difficult to untangle because conditions mutually interact. The oxidation-reduction potential and pH of gut contents influence invasiveness, but these conditions are determined largely by the bacterial flora, which is, in turn, influenced by host diet and perhaps even overall nutritional state. Newcomers to endemic areas may suffer more from amebic infection than does the local population because of differences in their bacterial floras.

  Invasive amebas erode ulcers into the intestinal wall, eventually reaching the submucosa and underlying blood vessels. From there they may travel with the blood to other sites such as liver, lungs, or skin. Although these endogenous forms are active, healthy amebas that multiply rapidly, they are on a dead-end course. They cannot leave the host and infect others and so perish with their luckless benefactor.

  Mature cysts in the large intestine, on the other hand, leave the host in great numbers. An individual that produces such cysts is usually asymptomatic or only mildly afflicted. Cysts of *E. histolytica* can remain viable and infective in a moist, cool environment for at least 12 days, and in water they can live up to 30 days; however, they are rapidly killed by putrefaction, desiccation, and temperatures below 5°C

and above 40°C. They can withstand passage through the intestines of flies and cockroaches. The cysts are resistant to levels of chlorine normally used for water purification.

When swallowed, cysts pass through the stomach unharmed and show no activity while in an acidic environment. When they reach the alkaline medium of the small intestine, metacystic forms begin to move within their cyst walls, which rapidly weaken and tear. Quadrinucleate amebas emerge and divide into amebulas that are swept downward into the cecum. This is the organisms' first opportunity to colonize, and their success depends on one or more metacystic trophozoites making contact with the mucosa. Obviously, chances for establishment are improved when large numbers of cysts are swallowed.

Biochemistry and metabolism of *E. histolytica* have been reviewed by McLaughlin and Aley.[40] The amebas possess several hydrolytic enzymes, including phosphatases, glycosidases, proteinases, and an RNAse. Major metabolic end products are CO_2, ethanol, and acetate, whose proportions vary with the extent to which the parasites are deprived of oxygen. Although once thought to be a strict anaerobe, we now know that *E. histolytica* is more of a metabolic opportunist and able to utilize oxygen when it is present in the environment. Glucose, from external sources or stored glycogen, is metabolized via the Embden-Meyerhof pathway exclusively, and fructose phosphate is phosphorylated, prior to lysis, by enzymatic reactions unique to these amebas. Pyruvate is converted mostly to ethanol, even in the presence of oxygen, via coenzyme-A, and a pyruvate oxidase similar to the one found in the trichomonads (see p. 96). Terminal electron transfers are accomplished with ferredoxinlike iron-sulfur proteins, a trait that may contribute to the efficacy of metronidazole in treatment.[40] Similar metabolic traits in *Trichomonas vaginalis* and *Giardia duodenalis* also are metronidazole targets.

- **Pathogenesis.** To quote Elsdon-Dew, "Were one tenth, nay, one hundredth, of the alleged carriers of this parasite to suffer even in minor degree, then the ameba would rank as the major scourge of mankind."[18] Obviously, not every infected person shows symptoms of disease; Elsdon-Dew rendered that opinion, however, prior to our discoveries of the differences between *E. histolytica, E. hartmanni,* and *E. dispar.*[18, 27]

  Entamoeba histolytica is almost unique among amebas in its ability to hydrolyze and invade host tissues. Tiny cytoplasmic extensions from the surface, as seen in electron micrographs, are filopodia.[38] These structures could have functions related to pathogenesis, for example attachment to host cells, release of cytotoxic substances, or contact cytolysis of host cells. Both *E. histolytica* and *E. dispar* have galactose-specific membrane lectins that function in binding to host cells, but only the *E. histolytica* lectin produces a host inflammatory response through stimulation of host cytokine production.[54] Such inflammation can easily contribute to subsequent pathology (see chapter 3).

  Trophozoites have active cysteine proteases (CP) on their surfaces and these enzymes have been implicated as factors contributing to the parasites' invasive abilities.[2] At least three CP versions are found in *E. histolytica,* accounting for

90% of the enzymatic activity, but one, known as EhCP5 (*Entamoeba histolytica* cysteine protease #5) evidently is the most important. Enhanced expression of the EhCP5 gene, introduced into amebas by transfection (nonviral transfer of DNA into the cells), increased the parasites' ability to invade the liver in mice.[57] Some studies, however, have failed to make a clear association between phagocytic ability, proteinase activity, and pathogenicity.[41]

An **intestinal lesion** (Figs. 7.5 and 7.6) usually develops initially in the cecum, appendix, or upper colon and then spreads the length of the colon. Parasite numbers build up in the ulcer, increasing the speed of mucosal destruction. The muscularis mucosa is somewhat of a barrier to further progress, and pockets of amebas form, communicating with the intestinal lumen through a slender, ductlike opening. The lesion may stop at the basement membrane or at the muscularis mucosae and then begin eroding laterally, causing broad, shallow areas of necrosis. Tissues may heal nearly as fast as they are destroyed, or the entire mucosa may become pocked.

These early lesions usually are not complicated by bacterial invasion, and there is little cellular response by the host. In older lesions the amebas, assisted by bacteria, may break through the muscularis mucosae, infiltrate the submucosa, and even penetrate the muscle layers and serosa. This enables trophozoites to be carried by blood and lymph to ectopic sites throughout the body where secondary lesions then form. A high percentage of deaths results from perforated colons with concomitant peritonitis. Surgical repair of perforation is difficult because a heavily ulcerated colon becomes very delicate.

Sometimes a granulomatous mass, called an **ameboma,** forms in the intestinal wall and may obstruct the bowel. It is the result of cellular responses to a chronic ulcer and often still contains active trophozoites. The condition is rare except in Central and South America.

Secondary lesions have been found in nearly every organ of the body (see Fig. 7.6), but the liver is most commonly affected (about 5% of all cases). Regardless of the secondary site, the initial infection is an intestinal abscess, even though it may go undetected. **Hepatic amebiasis** results when trophozoites enter mesenteric venules and travel to the liver through the hepatoportal system. They digest their way through portal capillaries and enter the sinusoids, where they begin to form abscesses. Lesions thus produced may remain at a pinpoint size, or they may continue to grow, sometimes reaching the size of a grapefruit. The center of the abscess is filled with necrotic fluid, a median zone consists of liver stroma, and the outer zone consists of liver tissue being attacked by amebas, although it is bacteriologically sterile. The abscess may rupture, pouring debris and amebas into the body cavity, where they attack other organs.

Pulmonary amebiasis is the next most common secondary lesion. It usually develops by metastasis from a hepatic lesion but may originate independently. Most cases originate when a liver abscess ruptures through the diaphragm. Other ectopic sites occasionally encountered are the brain, skin, and penis (with the amebiasis possibly acquired venereally). Rare ectopic sites include kidneys, adrenals, spleen, male and female genitalia, and pericardium. As a rule all ectopic abscesses are bacteriologically sterile.

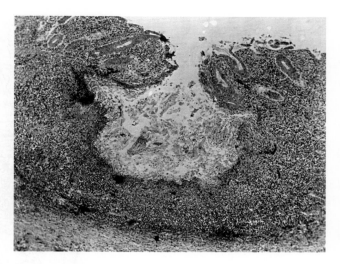

Figure 7.5   **Typical flask-shaped amebic ulcer of the colon.**

Extensive tissue destruction has resulted from invasion by *Entamoeba histolytica.*

AFIP neg. no. N–44718.

- **Symptoms.**   Symptoms of infection vary greatly among cases. The strain of *E. histolytica* present, the host's natural or acquired resistance to that strain, and the host's physical and emotional condition when challenged all affect the disease course in any individual. When conditions are appropriate, a highly pathogenic strain can cause a sudden onset of severe disease. This usually is the case with waterborne epidemics. More commonly disease develops slowly, with intermittent diarrhea, cramps, vomiting, and general malaise.

  Infection in the cecal area may mimic symptoms of appendicitis. Some patients tolerate intestinal amebiasis for years with no sign of colitis (although they are passing cysts) and then suddenly succumb to ectopic lesions. Depending on the number and distribution of intestinal lesions, a patient might experience pain in the entire abdomen, fulminating diarrhea, dehydration, and loss of blood. Amebic diarrhea is marked by bouts of abdominal discomfort with four to six loose stools per day but little fever.

  Acute amebic dysentery is a less common condition, but the sufferer from this affliction can best be described as miserable. The onset may be sudden after an incubation period of 8 to 10 days or after a long period in which the sufferer has been an asymptomatic cyst passer. In acute onset there may be headache, fever, severe abdominal cramps, and sometimes prolonged, ineffective straining at stool. An average of 15 to 20 stools, consisting of liquid feces flecked with bloody mucus, are passed per day. Death may occur from peritonitis, resulting from gut perforation, or from cardiac failure and exhaustion. Bacterial involvement may lead to extensive scarring of the intestinal wall, with subsequent loss of peristalsis. Symptoms arising from ectopic lesions are typical for any lesion of the affected organ.

- **Diagnosis and Treatment.**   Demonstration of trophozoites or cysts is usually necessary for diagnosis of *E. histolytica.* Examination of stool samples is the most effective

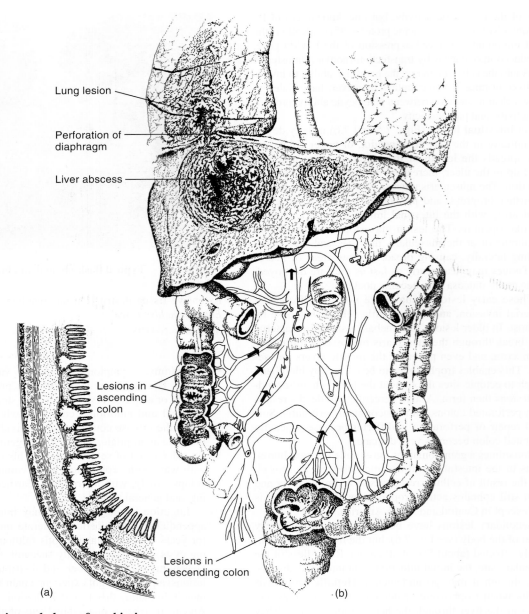

Lung lesion

Perforation of diaphragm

Liver abscess

Lesions in ascending colon

Lesions in descending colon

(a)

(b)

Figure 7.6 **Major pathology of amebiasis.**
Invasion of the intestinal mucosa occurs most commonly in the cecum and next most commonly in the rectosigmoid area. Small lesions develop into large, flask-shaped ulcers with ragged edges (*a*). Passage of trophozoites via the hepatic portal circulation (*arrows, b)* may result in liver abscess formation. Metastasis through the diaphragm may produce secondary abscess formation in the lungs. Trophozoites carried in the bloodstream may cause foci of infection anywhere in the body.

From J. Walter Beck and J. E. Davies, *Medical parasitology.* Copyright © 1976 Mosby Yearbook, St. Louis, MO. Reprinted by permission.

means of diagnosis of gut infection. A direct smear examined either as a wet mount or fixed and stained will usually reveal heavy infections. Even so, repeated examinations may be necessary.[26] One of us found abundant trophozoites in the stool of a hospital patient after negative findings on three previous days. Lighter infections of cyst passers may be detected with concentration techniques, such as zinc sulfate flotation.

Because of methods now available to distinguish between *E. dispar* and *E. histolytica*,[44, 63] some authors question the commonly accepted figure of 500 million

infections worldwide and suggest that the figure is closer to 50 million. Molecular techniques are now available for distinguishing between these two species in fresh and preserved stool samples, including those with mixed infections.[19, 22, 44, 63]

A large proportion of patients with extraintestinal amebiasis have no concurrent intestinal infection; diagnosis in such cases must occur, therefore, primarily by molecular and immunological means.[55] X-ray examination and other means of scanning the liver may be useful in detecting abscesses and ELISA assays for amebic lectin antigens, including those in saliva, have been developed for use in diagnoses.[1]

Many other diseases can easily be confused with amebiasis; on the hospital chart of the patient who tested negative on three days, a dozen possible explanations other than amebiasis for his persistent diarrhea had been listed. Hence, demonstration of the organisms and distinction between *E. histolytica* and *E. dispar* are mandatory for accurate diagnosis.

Several drugs have a high level of efficacy against colonic amebiasis. Most fall into the categories of arsanilic acid derivatives, iodochlorhydroxyquinolines, and other synthetic and natural chemicals. Antibiotics, particularly tetracycline, are useful as bactericidal adjuvants. These drugs are not as effective in ectopic infections, for which chloroquine phosphate and niridazole show promise of efficacy.

Metronidazole (a 5-nitroimidazole derivative) has become the preferred drug in treatment of amebiasis. It is low in toxicity and is effective against both extraintestinal and colonic infections, as well as cysts. However, metronidazole has been reported as being mutagenic in bacteria and carcinogenic in mice at doses not much higher than those given for the treatment of amebiasis. Furthermore, patients must be warned that the drug cannot be taken with alcohol because of its side effects (intense vasodilation, vomiting, and headache). Finally, its efficacy may not be as high as originally reported. Tetracycline in combination with diiodohydroxyquin results in a high rate of cures. Two other 5-nitroimidazole derivatives, ornidazole and tinidazole, have been reported to cure amebic liver abscess with a single dose.[35]

- **Epidemiology and Transmission Ecology.**   *Entamoeba histolytica* is found throughout the world. Although clinical amebiasis is most prevalent in tropical and subtropical areas, the parasite is well established from Alaska to the southern tip of Argentina. Prevalence of infection varies widely, depending on local conditions, from less than 1% in Canada and Alaska to 40% in many tropical areas. A survey of 216,275 stool specimens examined by U.S. state diagnostic laboratories in a single year (1987) revealed that 0.9% were infected with *E. histolytica.*[32]

Prevalence in the United States may be much higher among particular groups, such as persons in mental hospitals or orphanages. Age influences prevalence: Children younger than five have a lower infection rate than other age groups. In the United States the greatest prevalence occurs in the age group 26 to 30. Higher prevalence in the tropics results from lower standards of sanitation and greater longevity of cysts in a favorable environment. Onset of disease in persons who travel from temperate regions to endemic tropical areas may be partly the result of lessened resistance from the stress of travel and unaccustomed heat in addition to a change in bacterial flora in the gut, as mentioned previously. All races are equally susceptible.

In the late 1970s amebiasis was recognized as a sexually transmitted disease of increasing prevalence in New York City and a major health problem, particularly among gay men. In one study of 126 volunteers who participated in a gay men's health project, 39.7% were infected with *E. histolytica* and 18.3% with *Giardia duodenalis,* both fecal-borne organisms.[33] Authors of this study believed that, if multiple stools were examined, the figure of 39.7% might have increased at least to 50%. The primary mode of infection in these cases was oral to anal contact, and certainly the situation is not restricted to New York City. Thus a "new' health problem was discovered that probably has been fairly common since ancient times.

The manner of human waste disposal in a given area is the most important factor in *E. histolytica* epidemiology. Transmission depends heavily on contaminated food and water. Filth flies, particularly *Musca domestica,* and cockroaches also are important mechanical vectors of cysts. These insects' sticky, bristly appendages easily can carry cysts from a fresh stool to the dinner table, and the house fly habit of vomiting and defecating while feeding is an important means of transmission.

Polluted water supplies, such as wells, ditches, and springs, are common sources of infection. There have been instances of careless plumbing in which sanitary drains were connected to freshwater pipes with resultant epidemics. Carriers (cyst passers) handling food can infect the rest of their family groups or even hundreds of people if they work in restaurants. The use of human feces as fertilizer in Asia, Europe, and South America contributes heavily to transmission. Although humans are the most important reservoir of this disease, dogs, pigs, and monkeys are also implicated. A bizarre event occurred in Colorado in 1980 when an epidemic of amebiasis was caused by colonic irrigation with a contaminated enema machine in a chiropractic clinic. Ten patients had to have colonectomies; seven of them died.[8]

Entamoeba coli. *Entamoeba coli* often coexists with *E. histolytica* and, in the living trophozoite stage, is difficult to differentiate from it. Unlike *E. histolytica,* however, *E. coli* is a commensal that never lyses its host's tissues. It feeds on bacteria, other protozoa, yeasts, and occasionally blood cells. The diagnostician must identify this species correctly; if it is incorrectly diagnosed as *E. histolytica,* the patient may be submitted to unnecessary drug therapy.

Entamoeba coli is more common than *E. histolytica,* partly because of its superior ability to survive in putrefaction.

- **Morphology.**   *Entamoeba coli* trophozoites (Fig. 7.7) are 15 μm to 50 μm (usually 20 μm to 30 μm) in diameter and superficially identical to those of *E. histolytica,* but their nuclei differ. The *E. coli* endosome is usually eccentrically placed (but may appear central more often than expected because the nucleus may be turned a particular way at fixation), whereas that of *E. histolytica* is central. Also, chromatin lining the nuclear membrane is ordinarily coarser, with larger granules, than that of *E. histolytica.* Food vacuoles of *E. coli* are more likely to contain bacteria and other intestinal symbionts than are those of *E. histolytica,* although both may ingest available blood cells.

Encystment follows the same pattern as for *E. histolytica* Precysts are formed and a cyst wall is then rapidly secreted. Young cysts usually have a dense mass of chromatoidal bars that are splinter shaped, rather than blunt as in *E. histolytica.* As a cyst matures, its nucleus divides repeatedly to form eight nuclei (Fig. 7.8). Rarely as many as 16 nuclei may be produced. Cysts vary in diameter from 10 μm to 33 μm.

- **Biology.**   Infection and migration to the large intestine in the case of *E. coli* are identical to those of *E. histolytica.*

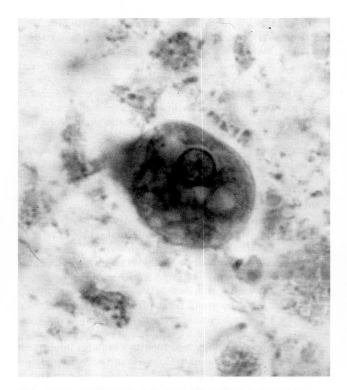

Figure 7.7 Trophozoite of *Entamoeba coli*, a commensal in the human digestive tract.
Note the characteristic eccentrically located endosome. The size is usually 20 μm to 30 μm.
Courtesy of Sherwin Desser.

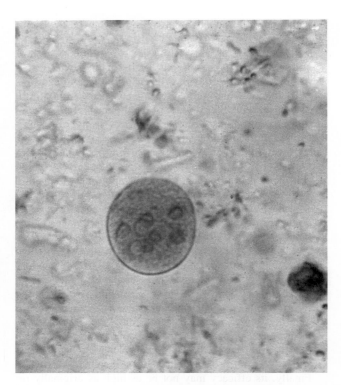

Figure 7.8 Metacyst of *Entamoeba coli*, showing eight nuclei.
The size is 10 μm to 33 μm.
Courtesy of David Oetinger.

The octanucleate metacyst produces 8 to 16 metacystic trophozoites, which first colonize the cecum and then the general colon. Infection is by contamination; in some areas of the world it nearly reaches 100%. Obviously, this widespread infection is a reflection of the level of sanitation and water treatment. Because *E. coli* is a commensal, no treatment is required. However, infection with this ameba indicates that opportunities exist for ingestion of *E. histolytica* or other parasites transmitted in a manner similar to *E. coli.*

Entamoeba gingivalis. *Entamoeba gingivalis* was the first ameba of humans to be described. It is present in all populations, dwelling only in the mouth. Like *E. coli,* it is a commensal and is of interest to parasitologists as another example of niche location and speciation.

• **Morphology.** Only trophozoites have been found, and encystment probably does not occur, although molecular studies indicate the ability to encyst has been lost over time.[12] Trophozoites (Fig. 7.9) are 10 μm to 20 μm (exceptionally 5 μm to 35 μm) in diameter and are quite transparent in life. They move rather quickly by means of numerous blunt pseudopodia. The spheroid nucleus is 2 μm to 4 μm in diameter and has a small, nearly central endosome. As in all members of this genus, chromatin is concentrated on the nuclear membrane's inner surface. Food vacuoles are numerous and contain cellular debris, bacteria, and occasionally blood cells.

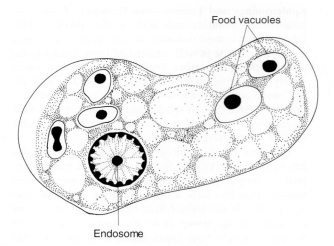

Food vacuoles

Endosome

Figure 7.9 *Entamoeba gingivalis* trophozoite.
The size usually is 10 μm to 20 μm.
Drawing by Ian Grant.

• **Biology.** *Entamoeba gingivalis* lives on the surface of teeth and gums, in gingival pockets near the base of teeth, and sometimes in the crypts of tonsils. The organisms often are abundant in cases of gum or tonsil disease, but no evidence shows that they cause these conditions. More likely, the protozoa multiply rapidly with an increased abundance of food. They even seem to fare well on dentures if the devices are not kept clean. *Entamoeba gingivalis* also infects other primates, dogs, and cats.

Because no cyst is formed, transmission must be direct from one person to another by kissing, by droplet spray, or by sharing eating utensils. Up to 95% of persons with unhygienic mouths may be infected, and up to 50% of persons with healthy mouths may harbor this ameba.[28]

Entamoeba polecki. *Entamoeba polecki* is usually a parasite of pigs and monkeys, although on rare occasions it occurs in humans. It is generally nonpathogenic in humans, but symptomatic cases may be difficult to treat.[49] It can be distinguished from *E. histolytica* by several morphological criteria, including the facts that *E. polecki* cysts have just one nucleus, with only about 1% of cysts ever reaching a binucleate stage, and that uninucleate cysts of *E. histolytica* are infrequent.

Genus *Endolimax*

Members of genus *Endolimax* live in both vertebrates and invertebrates. These amebas are small, each with a vesicular nucleus. The endosome is comparatively large and irregular and is attached to the nuclear membrane by achromatic threads. Encystment occurs in the life cycle.

Endolimax nana. *Endolimax nana* lives in the human large intestine, mainly near the cecum, and feeds on bacteria. Like *E. coli,* it is a commensal.

- **Morphology.**   The trophozoite of this tiny ameba (Fig. 7.10) measures 6 μm to 15 μm in diameter, but it is usually less than 10 μm. The ectoplasm is a thin layer surrounding the granular endoplasm. Pseudopodia are short and blunt, and the amebas move very slowly, two characteristics from which their name, "dwarf internal slug," is derived. The nucleus is small and contains a large centrally or eccentrically located endosome. Marginal chromatin is in a thin layer. Large glycogen vacuoles are often present, and food vacuoles contain bacteria, plant cells, and debris.

  Encystment follows the same pattern as in *E. coli* and *E. histolytica.* The precyst secretes a cyst wall, and the

young cyst thus formed includes glycogen granules and, occasionally, small curved chromatoidal bars. The mature cyst (see Fig. 7.10) is 5 μm to 14 μm in diameter and contains four nuclei.

- **Biology.**   As with other cyst-forming amebas that infect humans, mature cysts must be swallowed for infection to occur. Metacysts excyst in the small intestine, and colonization begins in the upper large intestine. Incidence of infection parallels that of *E. coli* and reflects the degree of sanitation practiced within a community. The cysts are more susceptible to putrefaction and desiccation than are those of *E. coli.* Although the protozoan is not a pathogen, its presence indicates that opportunities exist for infection by a variety of disease-causing organisms.

Genus *Iodamoeba*

Iodamoeba buetschlii. The genus *Iodamoeba* has only one species, and it infects humans, other primates, and pigs. Its distribution is worldwide. *Iodamoeba buetschlii* is the most common ameba of swine, which probably are its original host. The prevalence of *I. buetschlii* in humans is typically 4% to 8%, considerably lower than that of *E. coli* or *E. nana.*

- **Morphology.**   Trophozoites (Fig. 7.11) are usually 9 μm to 14 μm long but may range from 4 μm to 20 μm. They move slowly by means of short, blunt pseudopodia. The ectoplasm is not clearly demarcated from the granular endoplasm. The nucleus is relatively large and vesicular, containing a large endosome that is surrounded by lightly staining granules about midway between it and the nuclear membrane. Achromatic strands extend between the

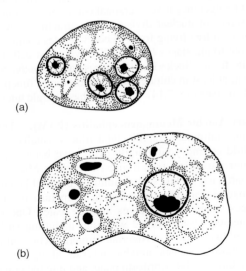

Figure 7.10   *Endolimax nana.*

(*a*) Cyst; (*b*) trophozoite. Note the large karyosome and thin layer of chromatin granules on the nuclear membrane.

Drawing by Ian Grant.

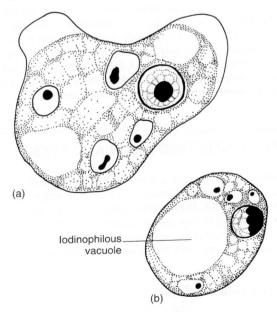

Iodinophilous vacuole

Figure 7.11   *Iodamoeba buetschlii.*

(*a*) Trophozoite; (*b*) cysts. Note the persistence of glycogen mass in the cyst and the large eccentric karyosome.

Drawing by Ian Grant.

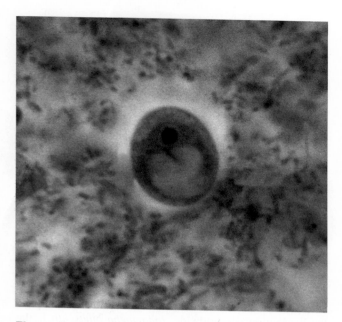

Figure 7.12 *Iodamoeba buetschlii* **cyst in human feces.**
Note the large iodinophilous vacuole. The size is 6 μm to 15 μm.
Courtesy of James Jensen.

endosome and nuclear membrane, which has no peripheral granules. Food vacuoles usually contain bacteria and yeasts.

The precyst is usually oblong and contains no undigested food. It secretes the cyst wall that also is usually oblong, measuring 6 μm to 15 μm long. The mature cyst (Fig. 7.12; see Fig. 7.11) nearly always has only one nucleus. A large conspicuous glycogen vacuole stains deeply with iodine—hence the generic name.

- **Biology.** *Iodamoeba buetschlii* lives in the large intestine, mainly in the cecal areas, where it feeds on intestinal flora. Infection spreads by contamination, since mature cysts must be swallowed to induce infection. It is possible that humans become infected through pig feces as well as human feces. A few reports of *I. buetschlii* causing ectopic abscesses like those of *E. histolytica* probably were actually misidentifications of *Naegleria fowleri*

AMEBAS INFECTING BRAIN AND EYES

A number of ameba species from three families are now recognized as opportunistic parasites that can cause serious illness and death in humans. These protists typically are free-living but, if provided access to host tissues, for example, through eyes or nasal membranes, can become invasive. Excellent reviews of the major culprits are given by Schuster and Visvesvara.[52, 53]

Family Vahlkampfiidae

Vahlkampfiids have both flagellate and ameboid stages in their life cycles, have eruptive pseudopod formation, and can produce cysts. These organisms also have flagella without mastigonemes and thus have been placed in a group called

Heterolobosea, which is either a phylum or class, depending on the authors.

Amebas of family Vahlkampfiidae are aerobic inhabitants of soil and water, are mainly bacteriophagous, and possess both a flagellated stage and an ameboid form. Binary fission seems to take place only in the ameboid form; thus, these are diphasic amebas, with ameboid stages predominating over flagellates. Although several genera and species in this family live in stagnant water, soil, sewage, and the like, a few are able to become facultative parasites in vertebrates. There has been confusion in the taxonomy of *Naegleria, Hartmannella, and Acanthamoeba,* with reports of all three genera as facultative parasites of humans. We take the view that *Naegleria* species are members of this family, whereas the other two genera belong in the Hartmannellidae and Acanthamoebidae respectively.[37]

Soil amebas have evidently been on earth for a long time. These amebas are among the exceedingly few shell-and skeletonless sarcodines evidently represented in the fossil record. Cysts virtually identical to those of *Naegleria gruberi* have been found in Cretaceous amber from Kansas.[60] Research on RNA has shown that some *Naegleria* species are as distantly related to one another as are mammals and frogs, but structurally they are very similar. Thus, evolutionary divergence has occurred at the molecular level without being matched by structural diversity.

Naegleria fowleri. *Naegleria fowleri* (Fig. 7.13) is also known in some literature as *N. aerobia.* It is the major cause of a disease called **primary amebic meningoencephalitis (PAM).** Other known species, *N. gruberi, N. lovaniensis,* and *N. australiensis,* apparently are harmless.

Flagellated stages of *N. fowleri* bear two long flagella at one end, are rather elongated, and do not form pseudopodia; ameboid stages usually have one blunt pseudopodium although pointed tips are visible by scanning electron microscope (Fig. 7.14).

Transformation from ameboid to flagellated form is quite rapid; once flagella develop, the organisms can swim rapidly. Their nucleus is vesicular and has a large endosome and peripheral granules. Dark polar masses are formed at mitosis, and Feulgen-negative interzonal bodies are present during late stages of nuclear division. A contractile vacuole is conspicuous in free-living forms. Food vacuoles contain bacteria in free-living stages but are filled with host cell debris in parasitic forms. Suckerlike structures called *amebastomes* are present; at least in culture forms amebastomes function in phagocytosis (see Fig. 7.14).[29] The cyst has a single nucleus.

- **Primary Amebic Meningoencephalitis (PAM).** This is an acute, fulminant, rapidly fatal illness usually affecting children and young adults who have been exposed to water harboring free-living *N. fowleri.* Most cases are contracted in lakes or swimming pools. Flagellated trophozoites probably are forced deep into the nasal passages when a victim dives into the water. One well-documented case involved washing, including sniffing water up his nose, by a Nigerian farmer.[36] After entrance to the nasal passages, amebas migrate along olfactory nerves, through the cribriform plate, and into the cranium. Death from brain destruction is rapid, and few cures have been reported. The mechanism of pathogenesis is not known, but the amebas produce cytolytic polypeptides similar to those of *E. histolytica.*[25]

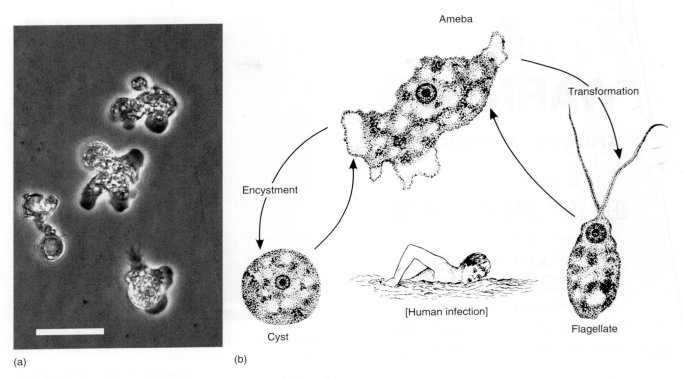

(a)

(b)

Figure 7.13 *Naegleria fowleri,* **ameba stages and life cycle.**

(*a*) Phase contrast photograph of *N. fowleri* in culture. Bulging ectoplasmic pseudopods (arrow) are a distinctive feature of the ameba forms. Bar is 20 μm. (*b*) Life cycle of *N. fowleri*. In the laboratory, and probably in nature, transformation from ameba to flagellate is stimulated by depletion of nutrients (David John, personal communication).

(*a*) From F. L. Schuster and G. S. Visvesvara, "Free-living amoebae as opportunistic and non-opportunistic pathogens of humans and animals," in *Int. J. Parasitol.* 34:1001–1027. Copyright © 2004. Reprinted by permission.

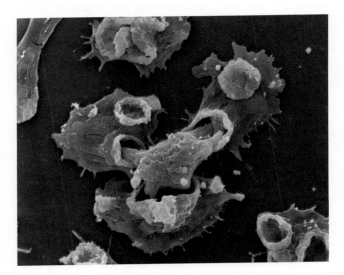

Figure 7.14 **Three** *Naegleria fowleri* **from axenic culture.**

They are attacking and beginning to devour or engulf a fourth, presumably dead ameba with their amebastomes. (× 2160)

From D. T. John et al., "Sucker-like structures on the pathogenic amoeba *Naegleria fowleri,*" in *Appl. Envir. Microbiol.* 47:12–14. Copyright © 1984.

These amebas kill a variety of laboratory animals when injected intranasally, intravenously, or intracerebrally.[11] They do not form cysts in the host. They have even been isolated from bottled mineral water in Mexico.[47] Up through the mid-1990s, 179 cases of PAM were recorded in widely separated parts of the world, including the United States, Czechoslovakia, Mexico, Africa, New Zealand, and Australia. From 1937–2007, 121 cases were reported in the United States, with a median age of 12 years; 78% of them were male.[10] Undoubtedly, many cases remain undiagnosed.

Naegleria amebas proliferate rapidly as water temperature rises, so thermal pools that are contaminated by rainwater runoff are particularly at risk (Fig. 7.15). However, one temperate-zone survey showed that natural populations are most widely distributed in spring and autumn.[30] Although these amebas are ubiquitous and common (the aforementioned survey found an average of one pathogenic ameba per 3.4 liters of water), the risk of acquiring this infection fortunately is small. In one modeling study, the risk was calculated at 8.5×10^{-8} when swimming in water with 10 amebas per liter.[7]

- **Treatment.** Unfortunately, most cases of PAM are diagnosed at autopsy. The disease is so rare and its course of brain destruction so rapid that only seldom has it been diagnosed in time for treatment to be attempted. Amphotericin B kills *N. fowleri* in vitro and has been used successfully in at least two human cases.[21] Recently *N. fowleri* was shown to be sensitive to qinghaosu (p. 156) in vitro. The lack of toxicity of this drug makes it potentially useful for therapy of PAM.[14] In nature, *N. fowleri* interacts with various organisms in the soil, including, evidently, bacteria that produce substances that, in turn, kill the amebas.[15] Such substances may hold promise as treatment in the future.

WARNING

AMOEBIC MENINGITIS

In all thermal pools

KEEP YOUR HEAD ABOVE WATER

to avoid the possibility of developing the serious illness called AMOEBIC MENINGITIS.

This disease can be caught in thermal pools if water enters the nose, while swimming or diving.

ISSUED BY THE DEPARTMENT OF HEALTH

Figure 7.15 Warning encountered in Rotorua, New Zealand, where several cases of primary amebic meningoencephalitis have been contracted in hot pools.

Courtesy of New Zealand Department of Health.

Family Acanthamoebidae

Members of a single genus, *Acanthamoeba,* are facultative parasites of humans in much the same manner as *Naegleria* species. *Acanthamoeba culbertsoni, A. polyphaga, A. hatchetti, A. castellanii,* and *A. rhysodes* have been identified in human tissues. Some of these have been reported as species of *Hartmannella,* but that genus is not pathogenic.[62] Biology of the free-living forms is similar to that of *Naegleria* species except that flagella are not known to be produced, and *Acanthamoeba* spp. cannot tolerate water as hot as can those of *Naegleria.*

Acanthamoeba spp. usually cause chronic infection of the skin or central nervous system in immunocompromised persons, although immunocompetent victims may also suffer corneal ulcers and keratitis. Through the late-1990s, 103 cases of meningoencephalitis due to *Acanthamoeba* species were reported although that number is now estimated to be closer to 200 worldwide.[53] *Acanthamoeba* keratitis (see below) is much more common, with more than 3000 cases distributed globally.[53]

Live trophozoites of *Acanthamoeba* species are easily differentiated from those of *Naegleria,* by their small spiky acanthopodia and very slow movement (Fig. 7.16). By contrast, *Naegleria* spp. have one blunt lobopodium and move rapidly (more than two body lengths per minute). *Acanthamoeba* species and strains differ in their invasive potential, with the more pathogenic ones exhibiting an enhanced ability

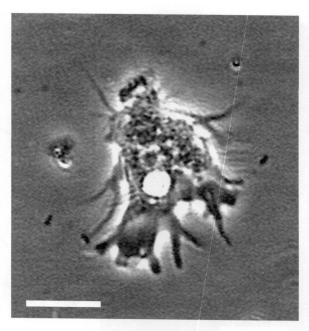

Figure 7.16 Phase contrast photograph of an *Acanthamoeba* sp. trophozoite.

The multiple fingerlike acanthopodia are a distinctive feature of this genus. The clear circular vesicle is a contractile vacuole. Bar is 10 μm.

From F. L. Schuster and G. S. Visvesvara, "Free-living amoebae as opportunistic and non-opportunistic pathogens of humans and animals," in *Int. J. Parasitol.* 34:1001–1027. Copyright © 2004. Reprinted by permission.

to attach to host cells. *Acanthamoeba* spp. strains are differentiated based on 18S ribosomal DNA and one, the T4 genotype, appears responsible for most cases of keratitis. This genotype evidently occurs commonly on beaches and grows in water with salinity ranging from 0 to 3%.[5]

Acanthamoeba species are causal agents of keratitis (corneal inflammation and opacity), with contact lenses, homemade saline washes containing amebas, and corneal abrasion considered to be contributing factors (Fig. 7.17).[9, 34] Among 24 *Acanthamoeba* keratitis cases reported in the mid-1980s, 22 patients were initially diagnosed as having corneal herpes simplex infections, 2 had an enucleated infected eye, and 12 underwent corneal transplantation.[9] Other recorded cases have usually involved some trauma to the cornea before exposure to parasites. Indeed, experimental work with animal systems has shown that corneal abrasion is a necessary condition for keratitis resulting from use of contaminated contact lenses.[58] Public swimming pools have been implicated as sources of infection,[48] but when one notes that *Acanthamoeba* is the most common ameba in fresh water and soil, it is surprising that more infections do not occur.

Treatment is difficult, but some cases have been treated successfully with ketoconazole, miconazole, and propamidine isethionate.[9] Some studies have shown that bacterial contamination in contact lens cleaning solution enhances ameba multiplication, possibly increasing the chances of eye infection.[6] Other studies have suggested that certain species of bacteria, such as *Pseudomonas aeruginosa,* produce toxins that are lethal to *Acanthamoeba species,*[45] and still other research has shown that the amebas are chemically attracted

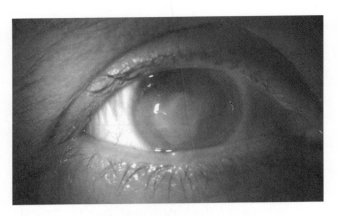

Figure 7.17   A case of *Acanthamoeba* keratitis.
The infected cornea has become fibrotic and dense. The patient was a contact lens wearer and amebas were cultured from corneal scrapings.
Courtesy of James P. McCulley, University of Texas Southwestern Medical Center, Dallas, TX.

to bacteria, even those species that produce toxins.[51] We now know that bacteria, including *Legionella pneumophila,* the causative organism of Legionnaire's disease, can infect various *Acanthamoeba* species. Thus, the amebas can play a role in epidemiology of human bacterial infections.[23]

As might be expected, *Acanthamoeba* species also can be opportunistic parasites in immunocompromised individuals. In one report, for example, five cases of skin infections, including ulcers, were observed in AIDS patients.[43]

Amebas of Uncertain Affinities

Balamuthia mandrillaris. *Balamuthia mandrillaris* (Fig. 7.18) was first isolated by culturing it from the brain of a baboon that had died from meningoencephalitis at the San Diego Zoo.[59] Although *B. mandrillaris* was originally placed in family Leptomyxiidae, that classification is now considered invalid.[37] Subsequent to its isolation from a baboon,

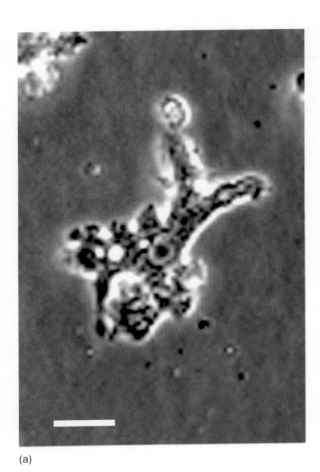

(a)

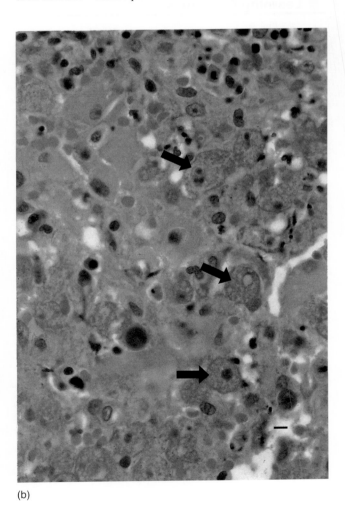

(b)

Figure 7.18   *Balamuthia mandrillaris* trophozoites in culture and *in situ*.
(*a*) Phase contrast of trophozoite in culture. The extended pseudopods are typical of this species in culture. *(Bar = 10 μm)* (*b*) *Balamuthia mandrillaris* in the central nervous system of a baboon. Numerous trophozoites (*arrows*) can be seen throughout the tissues, along with a darkly staining cyst. *(Bar = 10 μm)*

the species was shown to cause PAM in humans, including AIDS patients.[46] *Balamuthia mandrillaris* is a relatively large ameba (12 μm to 60 μm; average about 30 μm) that moves by broad pseudopodia but is also capable of forming fingerlike pseudopods and "walking" across a culture dish. In mammalian cell cultures, amebas actually enter cells and consume cytoplasm.[17]

Infections are probably acquired through the respiratory tract (mice can be infected by intranasal injection) or skin lesions. Both trophozoites and cysts can occur in central nervous system tissues. Because the patient exhibits a chronic granulomatous inflammatory response, the disease is called *granulomatous amebic encephalitis* (GAE). As of 1996 there were 63 reported cases of GAE due to *B. mandrillaris*,[39] but additional cases are reported sporadically, including one in a Great Dane that swam regularly in a stagnant water pond.[53]

Learning Outcomes

By the time a student has finished studying this chapter, he or she should be able to:

1. Describe the life cycle of *Entamoeba histolytica.*

2. Draw the nuclear structures of ameba species infecting humans and label the drawings with information needed to diagnose an infection microscopically.

3. Explain why diagnosis of *Entamoeba* species infections is a problem and how this problem is generally solved.

4. Write an extended paragraph describing the potential consequences and all the potential pathological effects of an infection with *Entamoeba histolytica.*

5. Describe some ecological situations in which fatal infections with *Naegleria fowleri* typically occur.

References

References for superscripts in the text can be found at the following Internet site: www.mhhe.com/robertsjanovynadler9e

Additional Readings

Band, R. N., et al. 1983. Symposium—the biology of small amoebae. *J. Protozool.* 30:192–214.

Chang, S. L. 1971. Small, free-living amebas: Cultivation, quantitation, identification, classification, pathogenesis, and resistance. In T. C. Cheng (Ed.), *Current topics in comparative pathobiology 1.* New York: Academic Press, Inc., 202–254 A review of the facultatively parasitic amebas.

Connor, D. H., R. C. Neafie, and W. M. Meyers. 1976. Amebiasis. In C. H. Binford, and D. H. Connor, (Eds.), *Pathology of tropical and extraordinary diseases.* Washington, DC: Armed Forces Institute of Pathology.

Culbertson, C. G. 1976. Amebic meningoencephalitides. In C. H. Binford, and D. H. Connor, (Eds.), *Pathology of tropical and extraordinary diseases.* Washington, DC: Armed Forces Institute of Pathology.

Hoare, C. A. 1958. The enigma of host-parasite relations in amebiasis. *Rice Inst. Pamphlet.* 45:23–35. Very interesting reading.

Lösch, F. A. 1875. Massive development of amebas in the large intestine (B. H. Kean, and K. E. Mott, Trans., 1975) *Am. J. Trop. Med. Hyg.* 24:383–392.

Schuster, F. L., and G. S. Visvesvara. 2004. Free-living amoebae as opportunistic and non-opportunistic pathogens of humans and animals. *Int. J. Parasitol.* 34:1001–1027.

Schuster, F. L., and G. S. Visvesvara. 2004. Amebae and ciliated protozoa as causal agents of waterborne zoonotic disease. *Vet. Parasitol.* 126:91–120.

Chapter 8

Phylum Apicomplexa: Gregarines, Coccidia, and Related Organisms

I began the study of the gregarines of insects in 1942, but I lost many data and manuscripts by the fire caused by the atomic bomb dropped on Hiroshima. After the Second World War, I came back to my work, and ressumed [sic] some parts of my previous study.

—Kinichiro Obata[71]

Phylum Apicomplexa contains organisms that possess a certain combination of structures, called an **apical complex,** distinguishable only by electron microscopy. All apicomplexans are parasitic, and all have a single type of nucleus and no cilia or flagella, except for the flagellated microgametes in some groups. The phylum contains two classes: Conoidasida, gregarines and coccidians, whose sporozoites have conoids (see below), and Aconoidasida, malarial parasites and piroplasms, generally lacking conoids.

Included in Apicomplexa are an astonishing array of organisms, some of which are of major veterinary and medical importance. For example, members of coccidian genus *Eimeria* cause a variety of intestinal diseases in poultry and cattle, and members of genus *Plasmodium* (see chapter 9) cause malaria, one of humankind's most persistent and prevalent public health problems. From an evolutionary perspective, order Eugregarinorida (gregarines) is one of the most speciose of the animal kingdom; well over a thousand species of gregarines have been described, mostly from annelids and arthropods, but only a tiny fraction of all invertebrate species have been studied parasitologically. Apicomplexans have cysts ("spores") that function in transmission; in some, however, the cyst wall has been eliminated, and development of infective stages (sporozoites) is completed within an invertebrate vector.

APICOMPLEXAN STRUCTURE

Ultrastructure of sporozoites and merozoites in class Conoidasida is typical of Apicomplexa.[52] These banana-shaped organisms are somewhat more attenuated at their anterior, apical complex end (Fig. 8.1) than at their posterior end, which often contains crystalline bodies or granules. An apical complex always includes one or two **polar rings,** electron-dense structures immediately beneath the cell membrane, which encircle the anterior tip. The **conoid** is a truncated cone of spirally arranged fibrillar structures just within these rings. **Subpellicular microtubules** radiate from the polar rings and run posteriorly, parallel to the body axis. These organelles probably serve as structural elements and may be involved with locomotion. Two to several elongated electron-dense bodies known as **rhoptries** extend to the cell membrane within the polar rings (and conoid, if present). Rhoptries participate in adhesion to and penetration of host cell, and formation of the subsequent parasitophorous vacuole[21] (see Fig. 8.1*b*). **Micronemes** are smaller, more convoluted elongated bodies that also extend posteriorly from the apical complex. Ducts of the micronemes run anteriorly into the rhoptries or join a common duct system with the rhoptries to lead to the cell surface at the apex. Contents of rhoptries and micronemes seem similar in electron micrographs, and this material is released during entry into a host cell.

Most if not all apicomplexans (but evidently not *Cryptosporidium* spp.) contain an organelle called the **apicoplast,** which is bound by four membranes, has a 35 kb genome, and is considered a vestigal plastid derived from a cyanobacterium by secondary endosymbiosis.[94] Genomic studies indicate this organelle is involved in fatty acid synthesis.[94] The apicoplast is essential for parasite survival, thus is a potential target for chemotherapy (see also chapter 9, p. 159).[69]

Along a sporozoite's side are one or more **micropores,** which function in ingestion of food material during the parasite's intracellular life. Micropore edges are marked by two concentric, electron-dense rings located immediately beneath the cell membrane. As host cytoplasm or other food matter within the parasitophorous vacuole is pulled through the rings, the parasite's cell membrane invaginates accordingly and finally pinches off to form a membrane-bound food vacuole.

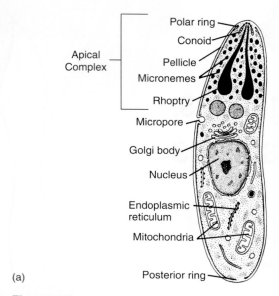

(a)

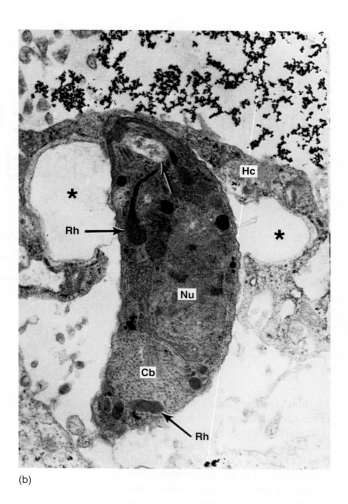

(b)

Figure 8.1   Apicomplexan structure.

(*a*) An apicomplexan sporozoite or merozoite illustrating the apical complex and other structures typical of this life-cycle stage. (*b*) Electron micrograph of a sporozoite of *Hammondia heydorni* penetrating a cultured cell. *Arrow* indicates empty rhoptry; *asterisk* shows host cell vacuole formed at point of sporozoite entry. *Hc,* host cell; *Nu,* parasite nucleus; *Rh,* rhoptry; *Cb,* crystalline body.

(*b*) From C. A. Speer and J. P. Dubey, Ultrastructure of sporozoites and zoites of *Hammondia heydorni,* in *J. Protozool.* 36:788–493, 1989. The Society of Protozoologists.

With the exception of micropores, structures described previously dedifferentiate and disappear after a sporozoite or merozoite penetrates a host cell to become a trophozoite.

Locomotor organelles are not as obvious as they are in other protozoan phyla. Pseudopodia are found only in some tiny, intracellular forms; flagella occur only on gametes of a few species, and a very few have cilia-like appendages. Various species have suckerlike depressions, knobs, hooks, myonemes, and/or internal fibrils that aid in limited locomotion. Myonemes and fibrils form tiny waves of contraction across the body surfaces; these can propel the parasite slowly through a liquid medium.

Both asexual and sexual reproduction is known in many apicomplexans. Asexual reproduction is either by binary or multiple fission or by endopolyogeny. Sexual reproduction is by isogamous or anisogamous fusion; in many cases this stage marks the onset of oocyst (spore) formation. Insofar as is known, meiosis is postzygotic, so that all life-cycle stages other than the zygote are haploid.

CLASS CONOIDASIDA, SUBCLASS GREGARINASINA

Members of Gregarinasina (gregarines) parasitize invertebrates, primarily annelids and arthropods, although species have been reported from many other phyla. Gregarine life cycles include a **gametocyst** stage, within which develop resistant **oocysts** (containing sporozoites), which, in turn, function to transmit infections between hosts (see Figs. 8.2 and 8.3). Because gregarines are widespread, common, and may be large in size, they are often used as instructional materials in zoology laboratories. Among the most frequently encountered gregarines are members of genus *Monocystis,* found in earthworm seminal vesicles, and species of *Gregarina,* which occur in mealworms. Both of these representative gregarines are discussed in some detail next.

In **acephaline** gregarines the body consists of a single unit that may have an anterior anchoring device, the **mucron.** In **cephaline** species the body is divided by a septum into an anterior **protomerite** and a posterior **deutomerite** that contains the nucleus. Sometimes the protomerite bears an anterior anchoring device, or **epimerite.** Mucrons and epimerites are considered modified conoids.

Multiple fission, or **schizogony** (or **merogony,** p. 48), occurs in a few families of gregarines (in the orders Archigregarinorida and Neogregarinorida). Most gregarines (order Eugregarinorida) have no schizogony but undergo multiple fission, within cysts, during gametogenesis. Oocyst production follows zygote formation. In both cephalines and acephalines, hosts become infected by swallowing oocysts. Most species parasitize the body cavity, intestine, or reproductive system of their hosts. Gregarines range in size from only a few micrometers to at least 1 mm long. Some are so large that 19th-century zoologists placed them among the worms!

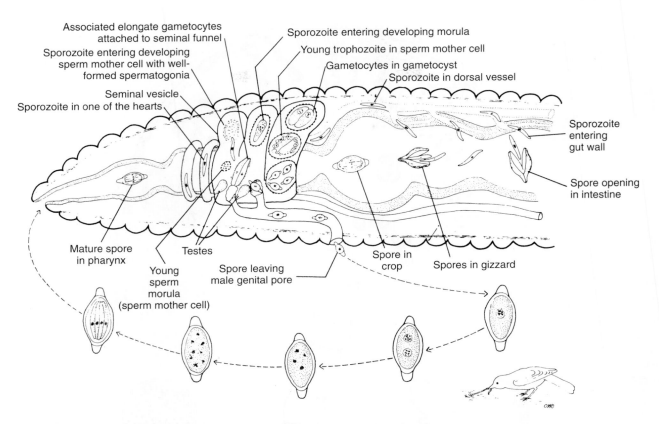

Figure 8.2   Life cycle of *Monocystis lumbrici,* an aseptate gregarine of earthworms.
In this figure, "spores" are actually oocysts (see Fig. 8.3). These oocysts pass through the digestive tract of predators that eat earthworms and are thus distributed widely.

From O. W. Olsen, *Animal parasites: Their life cycles and ecology.* Copyright © 1974 Dover Publications, Inc., New York, NY. Reprinted by permission.

Order Eugregarinorida

Suborder Aseptatorina

Monocystis lumbrici. *Monocystis lumbrici* (Fig. 8.2) lives in seminal vesicles of *Lumbricus terrestris* and related earthworms. Worms become infected when they ingest oocysts, each containing several sporozoites, which then emerge in the gizzard, penetrate the intestinal wall, enter a dorsal vessel, and move forward to the hearts. They then leave the circulatory system and penetrate seminal vesicles, where they enter sperm-forming cells (blastophores) in the vesicle wall.

After a short period of growth during which they destroy developing spermatocytes, sporozoites enter the vesicle lumen where they grow and mature into **gamonts (sporadins),** measuring about 200 μm long by 65 μm wide. Gamonts attach to cells near the sperm tunnel and undergo **syzygy,** in which two or more gamonts connect with one another.

After syzygy, gamonts surround themselves with a common cyst envelope, forming a **gametocyst.** Each gamont then undergoes numerous nuclear divisions. The many small nuclei move to the cytoplasm periphery and, taking a small portion of the cytoplasm with them, bud off to become gametes. Some cytoplasm of each gamont remains as a **residual body.** The gametes from each of the two gamonts are morphologically distinguishable and are thus **anisogametes.** Gametes fuse to form a zygote and then secrete an **oocyst membrane**

around that zygote. Three cell divisions (sporogony) follow to form eight sporozoites. Thus, each gametocyst now contains many oocysts, and the new host may become infected by eating a gametocyst or, if that body ruptures, an oocyst.

Only zygotes are diploid, and reductional division in sporogony (**zygotic meiosis**) returns sporozoites to the haploid condition. Gametocysts or oocysts pass out through the sperm duct to be ingested by other worms, although oocysts may also be passed by shrews, raccoons, and other predators that eat worms. The frequency with which one encounters infected earthworms reveals that this convoluted life history is no barrier to transmission.

Suborder Septatorina

Gregarina cuneata. *Gregarina cuneata* (Fig. 8.3) is a common parasite of the mealworm *Tenebrio molitor* and usually infects colonies of beetles maintained in a laboratory. Gamonts are cylindrical, up to 380 μm long by 105 μm wide, each with a small, conical epimerite that is inserted into a host cell. Gamonts associate in tandem; the anterior partner is the **primite,** and the posterior one is the **satellite.** In syzygy the satellite differs structurally from the primite (see Figs. 4.9a and 8.3), suggesting that the two gamont mating types are established early, perhaps during zygotic meiosis. A gametocyst wall is secreted; gametogenesis, fertilization, and oocyst production (sporulation) occur much as

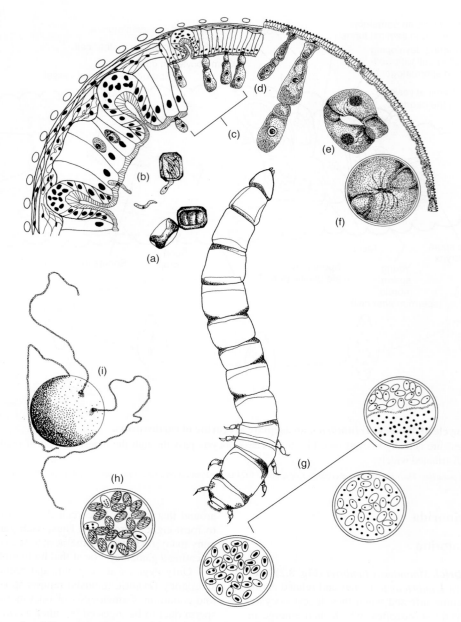

Figure 8.3   **Life cycle of *Gregarina cuneata* in yellow mealworms.**
(*a*) "Spores" (oocysts); (*b*) exsporulation in the insect's midgut and penetration of epithelial cell by sporozoite; (*c*) growth of the trophont; (*d*) pairing of gamonts; (*e*) syzygy; (*f*) secretion of the gametocyst wall; (*g*) gametogenesis and fertilization; (*h*) division of zygote into sporozoites; (*i*) dehiscence of the gametocyst, with spore chain formation. In the center is a *Tenebrio molitor* larva.
Drawing by Richard Clopton.

in *Monocystis* sp. Gametocysts pass out with the host's feces; oocysts are extruded through tubes and in long chains in a process called **dehiscence.**

Tenebrio molitor plays host to at least four species of genus *Gregarina,* beetle larvae are parasitized by *G. polymorpha* and *G. steini* in addition to *G. cuneata,* whereas adults are parasitized by *G. niphandrodes.* These species differ in size, in body proportions, and in details of their oocyst structures.

Gregarines of suborder Septatorina have been described from many insects, including roaches, dragonflies, and numerous beetle species, as well as from polychaetes, crustaceans, and myriapods. But only a small fraction of invertebrates has

been examined for apicomplexan parasites. Thus, septate gregarines are potentially one of the most diverse groups of organisms because their invertebrate hosts are so numerous.

GREGARINE-LIKE APICOMPLEXANS: *CRYPTOSPORIDIUM* SPECIES

Family Cryptosporidiidae
This family contains the single genus *Cryptosporidium,* parasites occupying brush borders of intestinal epithelia in fish, reptiles, birds, and mammals. Both molecular data and

developmental studies suggest that *Cryptosporidium* is more closely related to gregarines (p. 121) than to coccidians, a relationship that helps explain its resistance to anti-coccidial drugs.[90] Ten species of the genus are currently recognized,[35] although within the widespread and common *Cryptosporidium parvum,* there are eight distinct genotypes that could be host-adapted species.[90] The *C. parvum* genotype from cattle is of zoonotic importance, and it is genetically different from a recently named *C. hominis,* although the two species cannot be separated on the basis of structure alone.[70] Parasites morphologically identical to *Cryptosporidium parvum* have been reported from at least 150 mammal species, including a wide variety of pet and zoo animals, as well as poultry,[20, 44, 58] but many of these parasites have unique genotypes.[90] Obviously the species-level taxonomy of genus *Cryptosporidium* is still a challenging and unresolved problem, although one that has some very interesting evolutionary and epidemiological aspects.

Cryptosporidium parvum is an opportunistic parasite of humans, both immunodeficient and immunocompetent, and especially of young children. Cryptosporidiosis commonly occurs in patients with AIDS and can be an important contributory factor in their deaths.[97] For excellent reviews of this organism see Dubey et al.,[30] Fayer et al.,[35] Okhuysen and Chappell,[72] and Tentor et al.[90]

These coccidians are very small (2 μm to 6 μm) and live in the brush border or just under the free-surface membrane of host gastrointestinal or respiratory epithelial cells (Fig. 8.4). Oocysts are seen only in feces, and diagnosis is made using formalin-ethyl acetate and hypertonic sodium chloride flotation followed by Ziehl-Nielsen staining methods (fuchsin followed by methylene blue) or by use of Giemsa, nigrosin, or light-green. Methods for large-scale purification of oocysts and sporozoites, using differential centrifugation and sucrose or percoll gradients, have been described, especially for use in research.[3] Examination by differential interference or phasecontrast is often preferred over typical light microscopy.

- **Biology.** The tiny spherical oocysts (Fig. 8.5*a*) are 4 μm to 5 μm wide, are highly refractile, and contain one to eight prominent granules, usually in a small cluster near the cell's margin. Sporocysts are absent. Each oocyst contains four slender, fusiform sporozoites (Fig. 8.5*b*). Oocysts generally live a long time in water, including seawater, but they do not survive drying.

  When oocysts are swallowed, sporozoites excyst in the intestine and invade epithelial cells of either the respiratory system or intestine (from the ileum to the colon). Meronts are about 7 μm wide and produce eight banana-shaped merozoites and a small residuum. Microgamonts produce 16 rod-shaped, nonflagellated microgametes that are 1.5 μm to 2.0 μm long. Oocysts are passed as early as five days after infection. Virulence may be strain specific; calves experimentally infected with various human isolates developed infections that were significantly different in their severity.[76]

- **Pathogenesis and Treatment.** In patients with AIDS, the parasites cause profuse, watery diarrhea lasting for several months. Bowel-movement frequency ranges from 6 to 25 per day, and the maximal stool volume ranges from 1 to 17 liters per day. Evidently nitazoxanide is effective against cryptosporidial diarrhea, including that in AIDS patients, as well as against a number of other intestinal parasites, including amebas, tapeworms, and nematodes (p. 415).[22] Experiments using animal models have suggested that oral treatments with monoclonal

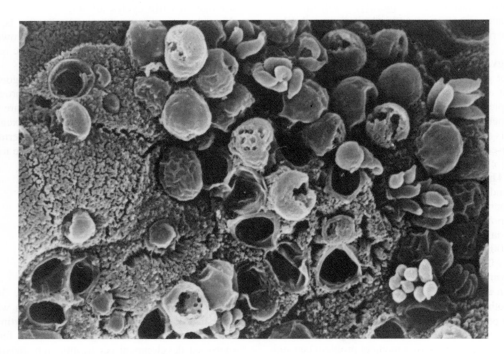

Figure 8.4   **Oocysts of *Cryptosporidium* in various stages of development in intestinal epithelium.**

The slender, elongated bodies are emerging sporozoites.

Courtesy of S. Tzipori, Royal Children's Hospital, Australia.

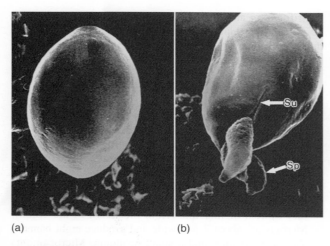

(a)   (b)

Figure 8.5 *Cryptosporidium parvum.*

(a) Oocyst. (b) Three sporozoites (*Sp*) emerging from a suture (*Su*) in an oocyst obtained from a calf and excysted experimentally in vitro.

From D. W. Reduker et al., "Ultrastructure of *Cryptosporidium parvum* oocysts and excysting sporozoites as revealed by high resolution scanning electron microscopy," in *J. Protozool.* 32:708–711. Copyright © 1985 by the Society of Protozoologists.

antibodies and hyperimmune colostrum were effective, but one study failed to demonstrate that antigens in human breast milk reduced the severity of infection.[36, 88] The infection is much less severe in immunocompetent patients, with no symptoms in some and with a self-limiting diarrhea and abdominal cramps lasting from 1 to 10 days in others.

Cryptosporidium parvum does not have typical mitochondria and there is evidence that its energy metabolism is mainly fermentative. Consequently, 5-nitrothiazole compounds active against anaerobic bacteria are also being used experimentally against *C. parvum*.[15]

- **Epidemiology.** Infection is by fecal-oral contamination. A number of animals can serve as reservoirs of infection. Current and his coworkers experimentally infected kittens, puppies, and goats with oocysts from an immunodeficient person.[18] They also infected calves and mice using oocysts from infected calves and humans. Finally, they diagnosed 12 infected immunocompetent persons who worked closely with calves that were infected with *C. parvum*. Thus, cryptosporidiosis should be considered a zoonosis and, in fact, may be a fairly common cause of short-term diarrhea in the population at large.

The zoonotic potential of *C. parvum* is illustrated by surveys on cattle. In one study Anderson[2] examined nearly 100,000 cattle and discovered that 65% of the dairies and 80% of the feedlots had infected animals. Although overall prevalence was low (less than 5% for any state), in some pens 31% of the cattle were passing oocysts. In other studies swimming pools have been implicated as a potential source of infection.[87]

Cryptosporidium infections dramatically illustrate the manner in which discovery of a medical problem can suddenly focus attention on organisms previously thought obscure and rare. Recognition of opportunistic parasites as a cause of disease in persons with AIDS led to interest in the distribution of such parasites in the immunocompetent and nonsymptomatic population. We now know from a large number of studies that cryptosporidiosis is a serious problem especially in the warmer parts of the world, and it may be one of the three most common causative agents of chronic diarrhea in humans.[17]

SUBCLASS COCCIDIASINA

In contrast to gregarines, members of Coccidiasina (coccidians, or coccidia) are small, with intracellular asexual reproduction and no epimerite or mucron. Some species are monoxenous, whereas others require two hosts to complete their life cycles. Coccidia live in digestive tract epithelium, liver, kidneys, blood cells, and other tissues of vertebrates and invertebrates.

A typical coccidian life cycle has three major phases: merogony, gametogony, and sporogony. The infective stage is a rodor banana-shaped sporozoite that enters a host cell. The parasite then becomes an ameboid trophozoite that multiplies by merogony to form more rodor banana-shaped **merozoites,** which escape from the host cell. These enter other cells to initiate further merogony or transform into gamonts (**gametogony**). Gamonts produce "male" **microgametocytes** or "female" **macrogametocytes.** Most species are thus anisogamous. Macrogametocytes develop directly into comparatively large, rounded macrogametes, which are ovoid bodies with a central nucleus and are filled with globules of a refractile material. Microgametocytes undergo multiple fission to form tiny, biflagellated microgametes.

Fertilization produces zygotes. Multiple fission of zygotes (sporogony) produces sporozoite-filled oocysts. In homoxenous life cycles all stages occur in a single host, although oocysts mature ("sporulate"—that is, complete sporozoite development) in the oxygen-rich, lower-temperature environment outside a host. Sporozoites are released when a sporulated oocyst is eaten by another host.

In some heteroxenous life cycles merogony and a part of gametogony occur in a vertebrate host. Sporogony, however, occurs in an invertebrate, and sporozoites are transmitted by the bite of the invertebrate. In other heteroxenous life cycles sporozoites are infective to a vertebrate intermediate host, in which are produced zoites that are infective to a carnivorous vertebrate host.

Order Eucoccidiorida

Suborder Adeleorina

Family Hepatozoidae

***Hepatozoon* species.** Approximately 300 species of *Hepatozoon* have been described from a wide variety of terrestrial vertebrates.[48] Blood-feeding arthropods are the definitive hosts, in which gametogenesis, fertilization, and sporocyst development occur, although transmission is typically by consumption of these hosts rather than by their bite (Fig. 8.6). Two species, *H. americanum* and *H. canis,* occur in dogs, especially in the southern United States, but *H. canis* also has been reported from South America, Europe, and Asia. Rodents have been implicated as paratenic hosts.[48]

In *H. americanum* infections, meronts occur in muscle, gamonts are in leukocytes, and ticks are definitive hosts. Nymphal ticks become infected by feeding on dogs. Gamonts penetrate the gut, develop into gametes, and fuse to form oocysts (Fig. 8.6), which are retained when the tick molts to become an adult. Sporocysts, containing sporozoites, develop within oocysts over the next 50 days.[49] Dogs become infected by eating an adult tick.

Clinical signs vary, depending on parasite species and strain, but can include elevated temperature, weight loss, anemia, lethargy, and restricted mobility, especially in hind limbs. Concurrent infections with *Babesia canis*, also transmitted by ticks (see p. 161), can compound a dog's health problems.[1]

Hepatozoon species differ in tissue involvement; for example, *H. canis* merogony occurs in lymphoidal tissue and other visceral organs; with *H. catesbianae* in bullfrogs, meronts are in the liver, gamonts are in erythrocytes, and mosquitoes are definitive hosts.[19]

Suborder Eimeriorina

In Eimeriorina, microand macrogametes develop independently without syzygy. Microgametocytes produce many active microgametes, which then encounter macrogametes, typically located within cells of a host's intestinal epithelium. This suborder is very large, with several families and thousands of species parasitizing most wild animals, although species in domestic animals are of major economic significance in agriculture. Some species also infect humans and are important zoonotic and opportunistic parasites.

Family Eimeriidae

In this family, following syngamy, oocysts develop resistant walls and contain one, two, four, or sometimes more sporocysts, each with one or more sporozoites. The organisms develop in the host cell proper and neither gamonts nor meronts have attachment organelles. Merogony and gemetogony occur within a host; sporogony typically, although not necessarily, occurs outside. Microgametes have two or three flagella.

Taxonomy of Eimeriidae is an area of active research interest, with new species being described annually from all classes of vertebrates. A complete review of the problems, issues, and recommended practices for such research can be found in Tenter et al.[90] Oocyst size, shape, and contents, presence or absence of several of the internal structures described next, and texture of the outer wall are important taxonomic characters. Parasites with similar oocysts can have distinct life cycles, however, and thus belong to different taxa[38] (see Table 8.1). Nevertheless, coccidian oocysts are remarkably constant in their morphology within a given species, so that identification can usually be made, at least tentatively, by examining oocysts, assuming the host has been accurately identified. Students interested in coccidian systematics should consult the Tenter et al.[90] review.

A typical oocyst (*Eimeria* sp.) is shown in Figure 8.7a. The oocyst wall has two layers, an outer one that is electron dense and varies in thickness among coccidian genera,

and an inner one that is 20–40 nm thick and not so dense. A membrane known as the **veil** surrounds the outer wall layer and can be seen in electron micrographs.[8] The wall is comprised mostly of lipids and proteins, and is resistant to proteolytic enzymes as well as a variety of chemicals; oocysts used in research, for example, are typically stored in 2% potassium dichromate. Belli et al.[8] provide an excellent review of the developmental biology of coccidian oocysts.

In many species there is a tiny opening at one end of the oocyst, the **micropyle,** and this may be covered by the **micropylar cap.** A refractile **polar granule** may lie somewhere within the oocyst. The oocyst wall (and probably the sporocyst wall, too) is of a resistant material that helps the organism survive harsh conditions in the external environment. Figure 8.7b shows an oocyst of an *Isospora* species. Comparison of the two oocysts of Figure 8.7 reveals a major difference between these two genera—namely, number of sporocysts contained within a sporulated oocyst.

Most species form sporocysts, which contain sporozoites, within the oocyst. During sporogony, cytoplasmic material not incorporated into sporozoites forms an **oocyst residuum.** In like manner some material may be left over within sporocysts to become a **sporocyst residuum.** However, the sporocyst residuum contains a large amount of lipid that seems to be an important source of energy for sporozoites during their sojourn outside a host.[98] The sporocyst wall consists of a thin outer granular layer surrounded by two membranes and a thick, fibrous inner layer. At one end of the sporocyst, a small gap in the inner layer is plugged with a homogeneous **Stieda body.** In some species additional plug material underlies the Stieda body and is designated the **substiedal body.** When sporocysts reach the intestine of a new host or are treated in vitro with trypsin and bile salt, the Stieda body is digested, the substiedal body pops out, and sporozoites wriggle through the small opening thus created.[80] In addition to having an apical complex, sporozoites may contain one or more prominent refractile bodies of unknown function.

A given species of *Eimeria* may be limited to certain organ systems, narrow zones in that system, specific kinds of cells in a zone, and even specific locations within the cells.[65] One species may be found only at the tips of intestinal villi, another in crypts at the bases of villi, and a third in the interior of the villi, all in the same host. Some species develop below the host-cell nucleus, others above it, and a few within it. Most coccidia inhabit the digestive tract but a few are found in other organs such as liver and kidneys.

Eimeria species vary in their pathogenicity. Individual hosts may not exhibit illness even when infected with multiple species, but in some cases the parasites are highly pathogenic, with almost every intestinal epithelial cell being infected (Figure 8.9).[31]

Infections with at least some *Eimeria* species are self-limiting and hosts may develop at least partial immunity to reinfection. Efforts to develop vaccines against *Eimeria* spp. or manage infections to stimulate immunity, however, have not been uniformly successful. For example, strains of mice differ naturally in their susceptibility to *E. vermiformis.* Oral vaccination with crude oocyst antigens increased resistance in susceptible mouse strains but reduced resistance in nonsusceptible strains.[81] Trickle doses of *E. alabamensis,* given to calves prior to release in contaminated pastures, did not protect them completely but prevented diarrhea.[89]

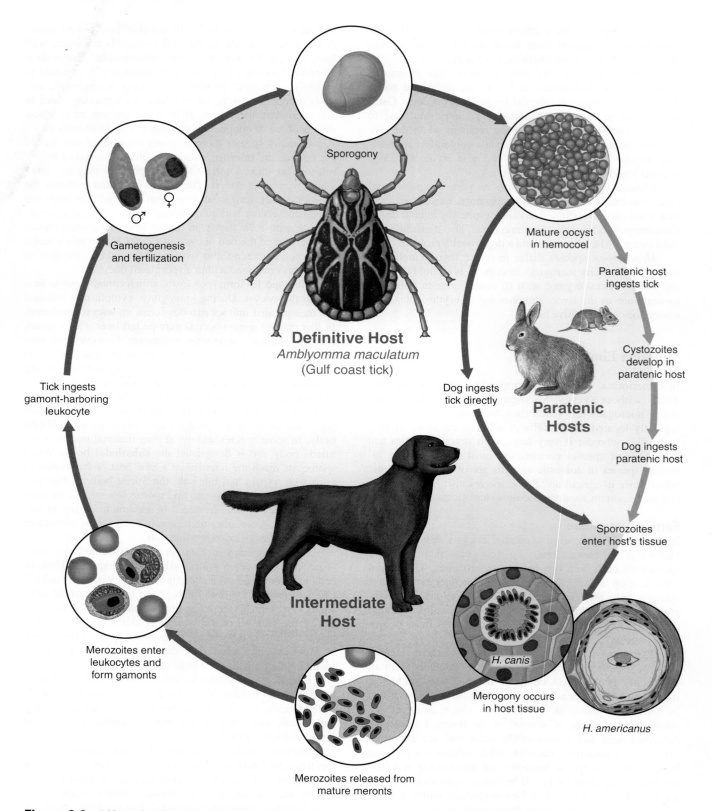

Definitive Host
Amblyomma maculatum
(Gulf coast tick)

Sporogony

Gametogenesis
and fertilization

Mature oocyst
in hemocoel

Paratenic host
ingests tick

Cystozoites
develop in
paratenic host

Dog ingests
paratenic host

**Paratenic
Hosts**

Tick ingests
gamont-harboring
leukocyte

Dog ingests
tick directly

Sporozoites
enter host's tissue

Merozoites enter
leukocytes and
form gamonts

**Intermediate
Host**

H. canis

Merogony occurs
in host tissue

H. americanus

Merozoites released from
mature meronts

Figure 8.6 **Life cycle of *Hepatozoon* species in dogs.**

The two common species in dogs, *H. americanum* and *H. canis*, have different geographical distributions, with *H. canis* more widely distributed globally, and different nuclear arrangements in the tissue meront stage. Different tick species may also be involved, depending on the geographical location.

Drawing by Bill Ober and Claire Garrison.

Table 8.1 Comparison of Homoxenous and Heteroxenous Coccidia with Four Sporozoites in Each of Two Sporocysts per Oocyst*

	Eimeriidae	Cystoisosporinae	Toxoplasmatinae			Sarcocystinae	
			Sarcocystidae				
Genus	*Isospora*	*Cystoisospora*	*Toxoplasma*	*Hammondia*	*Besnoitia*	*Sarcocystis*	*Frenkelia*
Definitive host	Specific	Specific	Variable	Specific	Carnivores	Specific (except in birds)	Variable
Proliferative stages in gut	Present	Present	Present	Present	Present	Absent	Absent
Where zygote formed	Epithelium (exceptions)	Epithelium	Epithelium	Epithelium	Epithelium	Lamina propria cell	Lamina propria cell
Oocysts shed	Unsporulated	Unsporulated	Unsporulated	Unsporulated	Unsporulated	Sporulated, usually as sporocysts	Sporulated, usually as sporocysts
Prepatent or patent period	Days	Days	Days	Days	Days	Weeks or months	Weeks or months
Intermediate host	None	Many hosts	Many hosts	Many hosts	Many hosts	Specific (except in birds)	Variable
Tissue cysts	None	Monozoic	Polyzoic	Polyzoic	Polyzoic	Polyzoic	Polyzoic
Cystozoites	None	Bradyzoite	Bradyzoite	Bradyzoite	Bradyzoite	Metrocytes, bradyzoite	Metrocytes, bradyzoite
Location of cyst	—	Lymphoid muscle	Many cells	Striated muscle	Many tissues	Striated muscle	Neurons
Cyst wall	—	Thin, within cell	Thin, within cell	Thin, within cell	Thick, surrounds cell	Thin or with spines of taxonomic value	Thin, within cell
Host cell nucleus	—	±	±	±	Undergoes hypertrophy and hyperplasia	±	Enlarged
Infectivity							
Oocyst to							
• definitive host	Yes	Yes	Yes	No	No	No	No
• intermediate host	None	Yes	Yes	Yes	Yes	Yes	Yes
Tissue cysts to							
• definitive host	—	Yes	Yes	Yes	Yes	Yes	Yes
• intermediate host	—	No	No	No	Variable	No	No
Of tissue cysts	—	Immediate	Immediate	Immediate	Immediate	After weeks, months	After weeks, months

*Genus *Arthrocystis*, with spherical zoites and cysts articulated like bamboo, has not been studied experimentally. Some believe it to be a stage of *Leucocytozoon.*

From J. Frenkel et al., "Beyond the oocyst: Over the molehills and mountains of coccidialand," in *Parasitol. Today* 3:250–251. Copyright © 1987 Elsevier Science Publishing, Cambridge, UK. Reprinted by permission.

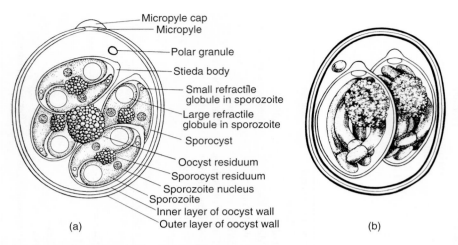

- Micropyle cap
- Micropyle
- Polar granule
- Stieda body
- Small refractile globule in sporozoite
- Large refractile globule in sporozoite
- Sporocyst
- Oocyst residuum
- Sporocyst residuum
- Sporozoite nucleus
- Sporozoite
- Inner layer of oocyst wall
- Outer layer of oocyst wall

(a)   (b)

Figure 8.7   Oocysts of two common genera of coccidians.

(a) Structure of sporulated *Eimeria* oocyst; (b) sporulated *Isospora* oocyst.

(a) From N. D. Levine, *Protozoan parasites of domestic animals and of man* (2d ed.). Minneapolis, MN: Burgess Publishing Co., 1961. (b) From McQuistion, T. E., "*Isospora daphnemsis* n. sp. (Apicomplexa: Eimeriidae) from the medium ground finch (*Geospiza fortis*) from the Galapagos Islands," in *J. Parasitol.* 76:30–32. Copyright © 1990 *Journal of Parasitology.* Reprinted by permission.

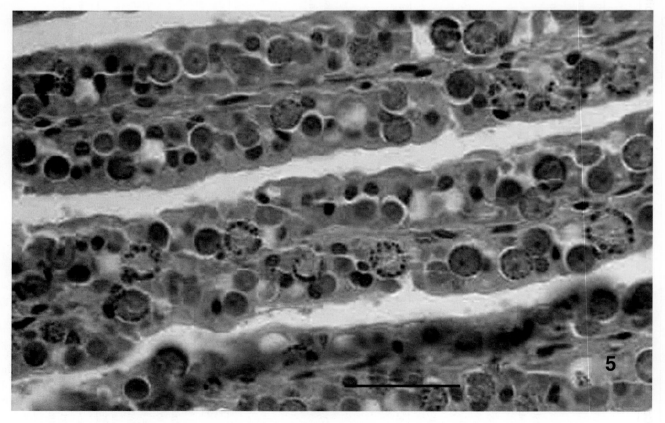

Figure 8.8   Intestinal epithelium of a pygmy rabbit infected with *Eimeria brachylagia*.

Sections of three villi with virtually every cell infected.

From Duszynski, D. W., L. Harrenstein, L. Couch, and M. M. Garner, "A pathogenic new species of *Eimeria* from the pygmy rabbit, *Brachylagus idahoensis*, in Washington and Oregon, with description of the sporulated oocysts and intestinal endogenous stages," in *J. Parasitol.* 91:618–623. Copyright © 2005. American Society of Parasitologists. Reprinted by permission.

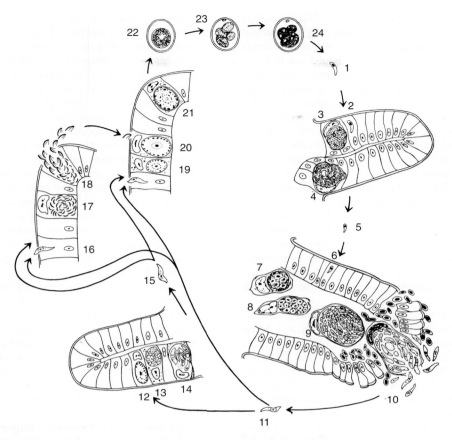

Figure 8.9   **Life cycle of the chicken coccidian *Eimeria tenella*.**

A sporozoite (*1*) enters an intestinal epithelial cell (*2*), rounds up, grows, and becomes a first-generation schizont (*3*). This produces a large number of first-generation merozoites (*4*), which break out of the host cell (*5*), enter new intestinal epithelial cells (*6*), round up, grow, and become second-generation schizonts (*7, 8*). These produce a large number of second-generation merozoites (*9, 10*), which break out of the host cell (*11*). Some enter new host intestinal epithelial cells and round up to become third-generation schizonts (*12, 13*), which produce third-generation merozoites (*14*). The third-generation merozoites (*15*) and the great majority of secondgeneration merozoites (*11*) enter new host intestinal epithelial cells. Some become microgametocytes (*16, 17*), which produce a large number of microgametes (*18*). Others turn into macrogametes (*19, 20*). The macrogametes are fertilized by the microgametes and become zygotes (*21*), which lay down a heavy wall around themselves and turn into young oocysts. These break out of the host cell and pass out in the feces (*22*). The oocysts then sporulate. The sporont throws off a polar body and forms four sporoblasts (*23*), each of which forms a sporocyst containing two sporozoites (*24*). When the sporulated oocyst (*24*) is ingested by a chicken, the sporozoites are released (*1*).

From N. D. Levine, *Protozoan parasites of domestic animals and of man* (2d ed.). Minneapolis, MN: Burgess Publishing Company, 1961.

Vaccination has been most successful with poultry, now being the coccidiosis control strategy of choice (see following discussion on *E. tenella*).[13]

The number of coccidian species is staggering. Levine and Ivens[57] recognized 204 species of *Eimeria* in rodents, but they estimated that there must be at least 2700 species of *Eimeria* in rodents alone. Only a small fraction of vertebrates have been studied parasitologically, however, so thousands of coccidian species probably remain to be discovered and described. One never knows where scientists are likely to discover new *Eimeria* species; recent descriptions have listed hosts as disparate as marine fish, tropical lizards, rodents, and even domestic animals whose parasites one would expect had been studied extensively for decades.

Eimeria tenella. *Eimeria tenella* (Fig. 8.9) lives in epithelium of intestinal ceca of chickens, where it destroys tissues, causing a high mortality rate in young birds. This

and related species are of such consequence that commercial feeds for young chickens now contain anticoccidial agents ("coccidiostats"). Drug resistance has been reported, however, and concerns over residual coccidiostats in food products makes vaccination the long-term goal in poultry production.[82]

• **Biology and Course of Infection.**   Chickens become infected when they swallow food or water that is contaminated with sporulated oocysts. The micropyle ruptures in the bird's gizzard, and activated sporozoites escape their sporocyst in the small intestine. Once in a cecum, sporozoites enter surface epithelium cells and pass through the basement membrane into the lamina propria. There they are engulfed by macrophages that carry them to the glands of Lieberkühn. They then escape the macrophages and enter into a glandular epithelial cell of the crypt, where they locate between the nucleus and basement membrane.

Sporozoites become trophozoites within epithelial cells, feeding on host cells and enlarging to become meronts. During merogony meronts separate into about 900 first-generation merozoites, each about 2 μm to 4 μm long. These break out into the cecum lumen about two and one half to three days after infection, destroying their host cells. Surviving first-generation merozoites enter other cecal epithelial cells to initiate a second endogenous generation. Merozoites develop into meronts that live between the nuclei and free borders of host cells. A great many merozoites will form meronts in the lamina propria under the basement membrane.

About 200 to 350 second-generation merozoites, each about 16 μm long, are then formed by merogony. These rupture the host cell and enter the cecal lumen about five days after infection. Some of these merozoites enter new cells to initiate a third generation of merogony below the nucleus, producing 4 to 30 third-generation merozoites, each about 7 μm long. Many merozoites are engulfed and digested by macrophages during these cycles of merogony.

Some second-generation merozoites enter new epithelial cells in the cecum to begin gametogony. Most develop into macrogametocytes. Both male and female gamonts lie between the host cell nucleus and the basement membrane. Microgametocytes bud to form many slender, biflagellated microgametes that leave their host cell and enter cells containing macrogametes, where fertilization takes place.

Macrogametes have many granules of two types. Immediately after fertilization these granules pass peripherally toward the zygote's surface, flatten out, and coalesce to form first the outer and then the inner layer of the oocyst wall. This coalescence takes place within the zygote's cell membrane, and the membrane thus becomes the outer-wall covering. Oocysts are then released from the host cells, move with cecal contents into the large intestine, and pass out of the body with feces. Oocysts appear in feces within six days of infection and are passed for several days because not all second-generation merozoites reenter host cells at the same time. Furthermore, oocysts often remain in the cecal lumen for some time before moving to the large intestine.

Freshly passed oocysts each contain a single cell, the *sporont*. Sporogony (often called *sporulation*), or development of the sporont into sporocysts and sporozoites, is exogenous (occurs outside the host). Sporonts are diploid, and the first division is reductional, a polar body being expelled. The haploid chromosome number is two. The sporont divides into four sporoblasts, each of which forms a sporocyst containing two sporozoites. Sporulation takes two days at summertime temperatures, whereupon the oocysts are infective.

Although unsporulated oocysts can survive anaerobic conditions, as might be found in freshly passed feces, metabolism of sporulation is an aerobic process and will not proceed in the absence of oxygen.[98] Oxygen consumption is high at first but falls steadily as sporulation is completed. The organisms have large amounts of glycogen, which is rapidly consumed, and measurements of their respiratory quotient indicate that they depend primarily on carbohydrate oxidation for energy during sporoblast formation and then change over to lipid for energy as sporulation is completed. Thus, their biochemistry suggests an interesting developmental control in metabolism: A rapid burst of energy fuels sporulation and then a shift to a low level of maintenance metabolism conserves resources until a new host is reached.

The number of oocysts produced in any infection can be astounding. Theoretically, one oocyst of *E. tenella*, containing eight sporozoites, can produce 2.52 million second-generation merozoites, most of which will become macrogametes and thereby oocysts. However, many merozoites and sporozoites are discharged with feces before they can penetrate host cells, and many are destroyed by host defenses. A complete replacement of cecal epithelium normally occurs about every two days, so any merozoite or sporozoite that invades a cell that is about to be sloughed is out of luck. Young chickens are more susceptible to infection and discharge more oocysts than do older birds.

Eimeria spp. infections are self-limiting; that is, asexual reproduction does not continue indefinitely. If the chicken survives through oocyst release, it recovers. It may become reinfected, but a primary infection usually imparts some degree of protective immunity to a host. *Eimeria tenella* is not the only coccidian infecting chickens, however, and one study showed that infection with *E. acervulina* or *E. adenoeides* actually enhanced invasion of epithelial cells by *E. tenella*.[4]

- **Pathogenesis and Economic Importance.** Cecal coccidiosis is a serious disease that causes a bloody diarrhea, sloughing of epithelium, and commonly death of the host.[61] Emergence of merozoites destroys tissues and cells. Large schizonts, especially when packed together, disrupt delicate capillaries that service the epithelium, further altering normal tissue physiology and also causing hemorrhage (Fig. 8.10). A hard core of clotted blood and cell debris often plugs up the cecum, causing necrosis of that organ (Fig. 8.11). Birds that are not killed outright by the infection become listless and are susceptible to predation and other diseases. The USDA estimated that loss to poultry farmers in the United States alone in

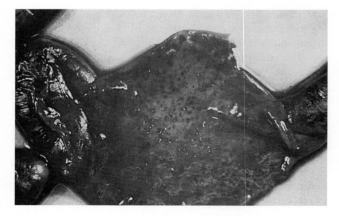

Figure 8.10 Cecum of chicken, opened to show patches of hemorrhage caused by *Eimeria tenella*.
Courtesy of James Jensen.

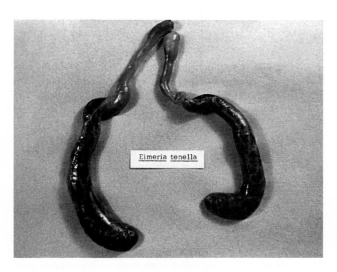

Figure 8.11 **Ceca of chicken infected by *Eimeria tenella*.**
Note distention caused by clotted blood and debris and dark color from hemorrhage.

Courtesy of James.

the mid-1980s was $80 million, counting the extra cost of medicated feeds (coccidiostats) and added labor. Annual broiler production in the United States is about 4.2 trillion birds.[62] Globally, the overall cost of poultry coccidiosis is estimated at $800 million annually.[82]

Many useful drugs are available as prophylaxes against coccidiosis. However, once infection is established, there is no effective chemotherapy. Therefore, if a coccidiostat is administered, it must be used continuously in food or water to prevent an outbreak of disease. These compounds affect the schizont primarily, so the host can still build up an immunity in response to invading sporozoites.

Vaccines used for poultry contain oocysts that are administered to young birds in drinking water, as gel tablets, or as an eye spray.[13, 68] Chicks get a measured and relatively low dose that allows them to build up resistance. At least seven *Eimeria* species parasitize chickens. Vaccines are effective but differ in terms of the species they protect against. Thus, commercial poultry operations may select a vaccine depending on the kind of stock being maintained (breeders, broilers, or layers).[13, 68]

Other *Eimeria* Species. Some of the most common *Eimeria* species in domestic animals are *E. auburnensis* and *E. bovis* in cattle, *E. ovina* in sheep, *E. debliecki* and *E. porci* in pigs, *E. stiedai* in rabbits, *E. necatrix* and *E. acervulina* in chickens, *E. meleagridis* in turkeys, and *E. anatis* in ducks. All have life cycles similar to that of *E. tenella* but differ in details of their courses of infection and effects on the host. In recent years additional species have been described from stingrays, marine bony fish, freshwater fish, frogs, lizards, snakes, turtles, doves, llamas, gazelles, and manatees. Clearly, members of genus *Eimeria* know few limits, evolutionarily speaking, on the types of hosts they colonize.

Isospora Group. Oocysts of *Isospora* species contain two sporocysts, each with four sporozoites. Oocysts of genera *Toxoplasma*, *Sarcocystis*, *Besnoitia*, and *Frenkelia*, are

similarly constructed, but these parasites are heteroxenous, with vertebrate intermediate hosts.[23] For this reason they are placed in family Sarcocystidae, discussed next.

For an enlightening discussion of the taxonomic problems surrounding *Isospora* and the sarcocystid genera, see the lively exchanges between Baker,[5] Frenkel et al.,[38] and Levine.[56] It is only within the past two decades that we have begun to understand the taxonomic position of many of these parasites. Taxonomic difficulties resulted from ignorance of life cycles, and the confusion was compounded through description of different life-cycle stages as different species (an event not unknown with other groups of parasites).

Isopsora contains far fewer species than *Eimeria* and most of them parasitize birds. The genus *Atoxoplasma* has had a long and convoluted taxonomic history but is now considered a synonym of *Isospora*.[7] Thus the latter genus now includes species that have merogony in a variety of host cells, including a variety of blood cells as well as those of the intestinal epithelium, gametogony in the intestinal epithelium, and sporogony outside the host. Infection is through ingestion of oocysts.

Coccidians formerly classified as *Isospora* species infecting mammals are now considered members of genus *Cystoisospora,* the change being based on molecular evidence.[7] Thus the parasite previously reported as *Isospora belli*, infecting humans, is now named *Cystoisospora belli* (Fig. 8.12). Most cases of *C. belli* have been reported from the tropics. The parasite can cause severe disease with fever, malaise, persistent diarrhea, and even death, especially in AIDS patients.[22]

Table 8.1 summarizes the present state of our knowledge about the major genera of Eimeriidae and Sarcocystidae that possess two sporocysts, each with four sporozoites per oocyst.

Cyclospora cayetanensis. *Cyclospora cayetanensis* (Fig. 8.13) is one of a growing list of parasites recognized as being of medical importance. Although genus *Cyclospora* was established in 1881 for a parasite of moles, the role of *C. cayetanensis* as a cause of human diarrhea was not established

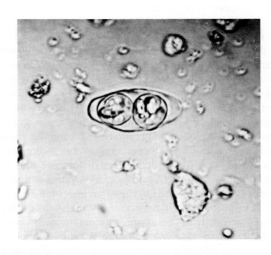

Figure 8.12 *Cystoisospora belli* **oocyst.**
It averages 35 μm by 9 μm.

From J. W. Beck and J. E. Davies, *Medical parasitology* (3d ed.). The C. V. Mosby Co.

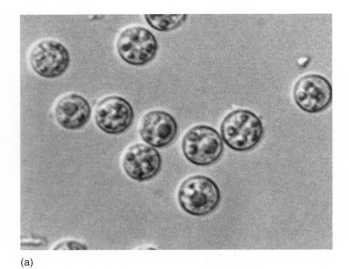

(a)

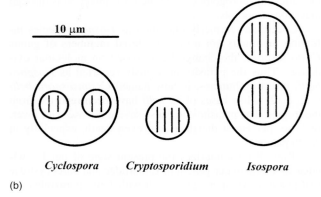

(b)

Figure 8.13 *Cyclospora* **and other oocysts.**
(*a*) Unsporulated *Cyclospora* oocysts from a human fecal specimen (×400). (*b*) Diagrammatic comparison of oocysts from *Cyclospora cayetanensis, Cryptosporidium parvum,* and *Isospora belli.* Outer circle or ellipse is the oocyst wall; inner circles, if present, are the sporocyst walls. Line is 10 μm.

(*a*) Courtesy of Ynes Ortega. (*b*) From R. Soave, *"Cyclospora*: An overview," in *Clin. Inf. Dis.* 23:429–437. © 1996. Reprinted with permission.

until the early 1990s.[86] Interest in *Cryptosporidium parvum* probably contributed to the ultimate discovery of *Cyclospora cayetanensis,* mainly because the acid-fast stains (e.g., the Ziehl-Nielsen technique) used to detect *Cryptosporidium* oocysts in fecal samples also stained larger oocysts in some patients. In fresh fecal samples, oocysts are 8 μm to 10 μm in diameter and contain membrane-bound refractile globules. Sporulation requires 5 to 11 days; mature oocysts contain two sporocysts about 4 μm in diameter and fluoresce blue-green under ultraviolet light (*Cryptosporidium parvum* and *Cystoisospora belli* oocysts do not fluoresce under UV light).[86]

Cyclosporosis is characterized by diarrhea, especially relapsing or cyclical, sometimes alternating with constipation. Patients may also exhibit fatigue, cramps, weight loss, and vomiting. Infection is typically concentrated in the jejunum, although in people with AIDS the bile duct may also be involved. The diarrhea is usually self-limiting in immunocompetent hosts but prolonged in AIDS patients.[99]

The first reported outbreak of cyclosporosis in the United States was evidently among staff physicians at a Chicago hospital in July 1990.[45] Symptoms included low-grade fever, explosive diarrhea, anorexia, and severe abdominal cramping. The source of infection could not be identified conclusively, but tap water at a physicians' dormitory was implicated. Because oocysts must sporulate before they are infective, direct human-to-human transmission is unlikely, and recent outbreaks have been blamed on contaminated fresh fruit, such as raspberries, typically served at social events.[12]

Quintero-Betancourt et al. provide an excellent review of the current methods of detecting both *C. cayetanensis* and *Cryptosporidium parvum* oocysts in water supplies.[78] Trimethoprim-sulfamethoxazole is the drug of choice against *C. cayetanensis*.[86]

Family Sarcocystidae

Members of this family differ from those of Eimeriidae principally in having heteroxenous life cycles (Table 8.1). Asexual development occurs in vertebrate intermediate hosts, whereas other vertebrates, mainly carnivorous mammals and birds, are definitive hosts. Oocysts contain two sporocysts, each with four sporozoites. There are well over a hundred described species of *Sarcocystis,* and new ones are being discovered regularly. One genus, *Toxoplasma,* is of importance to humans. Others are of veterinary importance.

Members of this family have life cycles with both intestinal and tissue stages (Fig. 8.14; see Fig. 8.18). Oocysts from a definitive host sporulate and are swallowed by an intermediate host. Sporozoites released from oocysts infect various tissues and rapidly undergo endodyogeny to form merozoites, also known as **tachyzoites.** These can infect other tissues such as muscles, fibroblasts, liver, and nerves. Asexual reproduction in these tissues is much slower than in the original site, and the parasites develop large, cystlike accumulations of merozoites that are called **bradyzoites.** The cyst itself is called a **zoitocyst,** or simply a **tissue cyst.** A definitive host is infected when it eats meat containing bradyzoites or, rarely, tachyzoites or, in some cases, when it swallows a sporulated oocyst. Tachyzoites and bradyzoites have antigenic differences.[63] When tissue cysts are ingested by a definitive host, bradyzoites invade enteroepithelial cells and undergo schizogony, then gametogenesis, and finally fertilization to produce oocysts (Fig. 8.14).

Toxoplasma gondii. *Toxoplasma gondii* (see Figs. 8.14 to 8.17) was first discovered in 1908 in a desert rodent, the gondi, in a colony maintained at the Pasteur Institute in Tunis. Since then the parasite has been found in almost every country of the world in many species of carnivores, insectivores, rodents, pigs, herbivores, primates, and other mammals as well as in birds. We now realize that it is cosmopolitan in the human population. The importance of *T. gondii* as a human pathogen has stimulated a huge amount of research. Since the mid-1980s *T. gondii* also has joined a number of other parasites recognized as complicating factors for immunosuppressed patients. This once obscure protozoan parasite of an obscure African rodent has become one of many exciting subjects whose importance has been revealed by research.

- **Biology.** *Toxoplasma gondii* is an intracellular parasite of many kinds of tissues, including muscle and intestinal

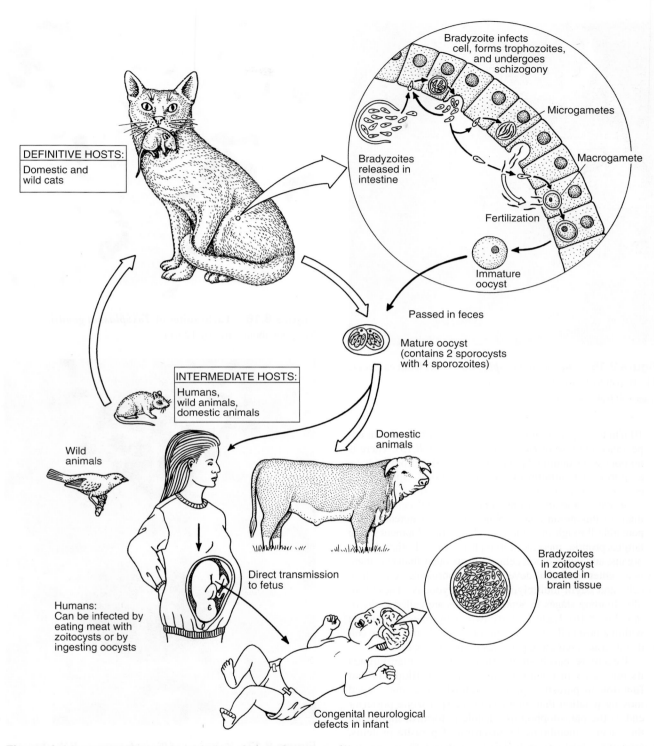

Figure 8.14   Life cycle and transmission of *Toxoplasma gondii*.

Drawing by William Ober and Claire Garrison.

epithelium. In heavy acute infections the organism can be found free in the blood and peritoneal exudate. It may inhabit the host cell nucleus but usually lives in the cytoplasm. The life cycle includes intestinal-epithelial **(enteroepithelial)** and **extraintestinal** stages in domestic cats and other felines but only extraintestinal stages in

other hosts. Sexual reproduction occurs in cats, and only asexual reproduction is known in other hosts.

Extraintestinal stages begin when a cat or other host ingests bradyzoites. Ingested tachyzoites or sporocysts also sometimes are infective. Intrauterine infection is possible (see the discussion of pathogenesis). Oocysts are

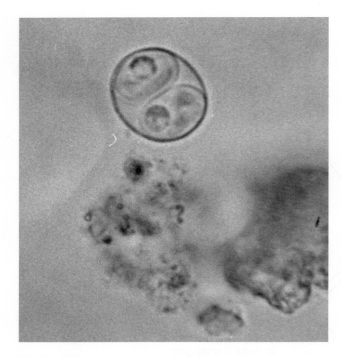

Figure 8.15 Oocyst of *Toxoplasma gondii* from cat feces.
It is 10 μm to 13 μm by 9 μm to 11 μm.
Courtesy of Harley Sheffield.

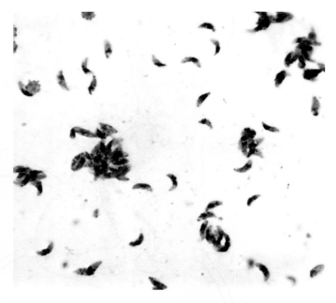

Figure 8.16 Tachyzoites of *Toxoplasma gondii.*
They are about 7 μm by 12 μm.

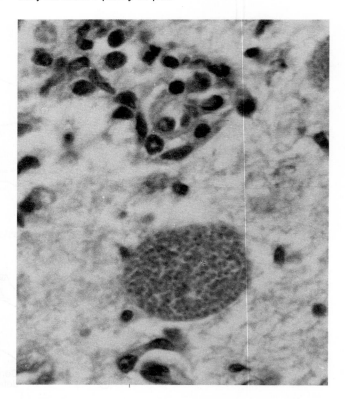

Figure 8.17 Zoitocyst of *Toxoplasma gondii* in the
brain of a mouse.
Courtesy of Sherwin Desser.

10 μm to 13 μm by 9 μm to 11 μm and are similar in appearance to those of *Isospora* species (Fig. 8.15). There is no oocyst residuum or polar granule, and sporocysts have a sporocyst residuum but no Stieda body. Sporozoites escape from sporocysts and oocysts in the small intestine.

In cats some sporozoites enter epithelial cells and remain to initiate an enteroepithelial cycle, whereas others penetrate through the mucosa to begin development in the lamina propria, mesenteric lymph nodes and other distant organs, and white blood cells. In hosts other than cats there is no enteroepithelial development; sporozoites enter host cells and begin multiplying by endodyogeny. These rapidly dividing stages in acute infections are called *tachyzoites* (Fig. 8.16). Eight to thirty-two tachyzoites accumulate within a host cell's parasitophorous vacuole before the cell disintegrates, releasing parasites to infect new cells.

Recent research shows that *T. gondii* manipulates its host cells in complex ways, inducing filopodia that function in parasite uptake, establishing a ring-shaped moving junction that migrates to the sporozoite posterior end as the parasitophorous vacuole is formed, modifying the vacuole membrane by insertion of parasite proteins, arresting a host cell in S (DNA-synthesis) phase, and reorganizing host cell cytoskeleton.[75] For a review of these, and other intriguing ways *T. gondii* interacts with host cells, see Peng et al.[75]

As infection becomes chronic, zoites infecting brain, heart, and skeletal muscles multiply much more slowly than they do during the acute phase. They are now called *bradyzoites,* and they accumulate in large numbers within a host cell. They become surrounded by a tough wall, resulting in *zoitocysts* or *tissue cysts* (Fig. 8.17). Cysts may persist for months or even years after infection,

particularly in nervous tissue. Cyst formation coincides with the time of development of immunity to new infection, which is usually permanent. If immunity wanes, released bradyzoites can boost the immunity to its prior level. This protection against superinfection by the presence of the infectious agent in the body is called **premunition** (p. 24).

Immunity to *Toxoplasma* involves both antibody (T_H2) and cell-mediated (T_H1) types; the latter is more important. Except when a cyst breaks down, the tough, thin cyst wall effectively separates the parasites from their host, and a cyst does not elicit an inflammatory reaction. The cyst wall and its bradyzoites develop intracellularly, but they may eventually become extracellular because of distention and rupture of the host cell. Bradyzoites are resistant to digestion by pepsin and trypsin, and when eaten they can infect a new host.

Enteroepithelial stages are initiated when a cat ingests zoitocysts containing bradyzoites, oocysts containing sporozoites, or occasionally tachyzoites. Another possible means of epithelial infection is by migration of extraintestinal zoites into the intestinal lining within the cat. Once inside an epithelial cell of the small intestine or colon, the parasites become trophozoites that grow and prepare for merogony. Strains differ in duration of stages, number of merozoites produced, shape, and other details.[37, 38] Anywhere from 2 to 40 merozoites are produced by merogony, endopolyogeny, or endodyogeny, and these initiate subsequent asexual stages.

The number of merogonous cycles is variable, but gametocytes are produced within 3 to 15 days of cyst-induced infection. Gametocytes develop throughout the small intestine but are more common in the ileum. From 2% to 4% of gametocytes are male; each produces about 12 microgametes. Oocysts appear in a cat's feces from three to five days of infection by cysts, with peak production occurring between days five and eight. Oocysts require oxygen for sporulation; they sporulate in one to five days. Extraintestinal development can proceed simultaneously with enteroepithelial development in cats. Ingested bradyzoites penetrate the intestinal wall and multiply as tachyzoites in the lamina propria. They may disseminate widely in a cat's extraintestinal tissues within a few hours of infection.[23]

One final note of interest is that at the Pasteur Institute in Tunis in 1908, when gondis were brought in from the field and died, the source of their infection was never established. However, it is known that at the time a cat had been roaming the laboratory.[47]

- **Pathogenesis.**   Antibody to *Toxoplasma* is widely prevalent in humans throughout the world yet clinical toxoplasmosis is less common, so it is clear that most infections are asymptomatic or mild. Several factors influence this phenomenon: virulence of the *Toxoplasma* strain, susceptibility of an individual host and host species, and a host's age and degree of acquired immunity. Pigs are more susceptible than cattle; white mice are more susceptible than white rats; chickens are more susceptible than most carnivores. The reasons for natural resistance or susceptibility to infection are not known. Occasionally, circumstances conspire to make a mild case important, as when Martina Navratilova lost the U.S. Open tennis championship and $500,000 in 1982 when she had toxoplasmosis.

In 127 surveys of women of childbearing age from 53 countries, conducted between 1986 and 1999, seroprevalence was 42%, suggesting that 2.5 billion people may have been infected at some time.[91] More recent studies show seroprevalence is >60% in some areas of South America, especially much of Brazil, 18% to 50% throughout the rest of the world, depending on the country, and around 11% in China, Vietnam, and the United States.[73] In the United States, about 3500 infants are born each year with severe infections.[79]

Tachyzoites proliferate in many tissues, and this rapid reproduction tends to kill host cells at a faster rate than does the normal turnover of such cells. Enteroepithelial cells, on the other hand, normally live only a few days, especially at the tips of the villi. Therefore, extraintestinal stages, particularly in sites such as the retina or brain, tend to cause more serious lesions than do those in intestinal epithelium.

Because there seems to be an age resistance, infections of adults or weaned juveniles are asymptomatic, although exceptions occur. Asymptomatic infections can suddenly become fulminating if immunosuppressive drugs such as corticosteroids are employed for other conditions. Symptomatic infections can be classified as acute, subacute, and chronic.

In most **acute infections** the intestine is the first site of infection. Cats infected by oocysts usually show little disease beyond loss of individual epithelial cells, and these are rapidly replaced. Actually, oocysts probably are of little importance in infecting cats compared to the feeding of infected prey to kittens by their mother. In massive infections, however, intestinal lesions can kill kittens in two to three weeks.

The first extraintestinal sites to be infected in both cats and other hosts, including humans, are mesenteric lymph nodes and liver parenchyma. These sites, too, experience rapid regeneration of cells and perform an effective preliminary screening of parasites. The most common symptom of acute toxoplasmosis is painful, swollen lymph glands in the cervical, supraclavicular, and inguinal regions. This symptom may be associated with fever, headache, muscle pain, anemia, and sometimes lung complications, a syndrome that can be mistaken easily for flu. Acute infection can, although rarely does, cause death. If immunity develops slowly, the condition can be prolonged and is then called *subacute*.

In **subacute infections** pathogenic conditions are extended. Tachyzoites continue to destroy cells, causing extensive lesions in the lung, liver, heart, brain, and eyes. Damage may be more extensive in the central nervous system than in unrelated organs because of lower immunocompetence in these tissues.

Chronic infection results when immunity builds up sufficiently to depress tachyzoite proliferation. This condition coincides with the formation of zoitocysts. These cysts can remain intact for years and produce no obvious clinical effect. Occasionally a cyst wall will break down, releasing bradyzoites; most of these are killed by host reactions, although some may form new cysts. Death of bradyzoites elicits an intense hypersensitive inflammatory reaction, the area of which, in the brain, is gradually replaced by nodules of glial cells. If many such nodules are formed, a host may develop symptoms of chronic encephalitis, with spastic paralysis in some cases. Chronic active or relapsing infections of retinal cells by tachyzoites causes blind spots and extensive infection of the central macular area, which may lead to blindness. Cysts and cyst rupture in the retina can also lead to blindness.

Other kinds of extensive pathological conditions such as myocarditis, with permanent heart damage and with pneumonia, can occur in chronic toxoplasmosis.

In an immunocompetent person, *T. gondii* ordinarily is kept at bay by cell-mediated immunity. When an infected person becomes immunosuppressed, the organism will disseminate rapidly, which may lead to ocular toxoplasmosis and to fatal disorders of the central nervous system such as encephalitis. Any long-term steroid therapy, such as is given to some cancer patients, can result in disseminated toxoplasmosis. *Toxoplasma gondii* is a serious opportunistic infection in AIDS. Death usually results from cyst rupture with continued multiplication of tachyzoites.

Another tragic form of this disease is **congenital toxoplasmosis.** If a mother contracts acute toxoplasmosis at the time of her child's conception or during pregnancy, the organisms often will infect her developing fetus. Fortunately, most neonatal infections are asymptomatic, but a significant number cause death or disability to newborns. It is generally assumed that *T. gondii* crosses the placental barrier from the mother's blood.

The transplacental transmission rate from a maternal infection is about 45%. Of those infected, about 60% are subclinical, 9% may die, and 30% may suffer severe damage such as hydrocephalus, intracerebral calcification, retinochoroiditis, and mental retardation. However, even subclinical cases may develop into ocular toxoplasmosis later in life.

Stillbirths and spontaneous abortions may result from fetal infection with *T. gondii* in humans and other animals. Sheep seem to be particularly susceptible, and abortions caused by *T. gondii* in this host often reach epidemic proportions. Congenital toxoplasmosis probably accounts for half of all ovine abortions in England and New Zealand.[9]

In a study of more than 25,000 pregnant women in France, no case of congenital toxoplasmosis was found whenever maternal infection occurred before pregnancy.[16] However, of 118 cases of maternal infection near the time of or during pregnancy, there were nine abortions or neonatal deaths without confirmation by examination of the fetus, 39 cases of acute congenital toxoplasmosis with two deaths, and 28 cases of subclinical infection. Maternal infection in the first three months of pregnancy results in more extensive pathogenesis, but transmission to a fetus is more frequent if maternal infection occurs in the third trimester.

In cases of twins one may have severe symptoms and the other no overt evidence of infection. In children who survive infection there is often congenital damage to the brain, manifested as mental retardation and retinochoroiditis. Thus, toxoplasmosis is a major cause of human birth defects, probably causing more congenital abnormalities in the United States than rubella, herpes, and syphilis combined.

- **Diagnosis and Treatment.** Specific diagnosis in humans is based on one or more laboratory tests. Demonstration of the organism at necropsy or biopsy is definitive. Intraperitoneal inoculation of a biopsy of lymph node, liver, or spleen into mice is useful and accurate as is culture of parasites in fibroblast cells in vitro. Demonstration

of specific antibody, using an enzyme-linked immunosorbent assay (ELISA), is also employed, and molecular methods are currently used in the preparation of antigen reagents. Attempts have been made to develop diagnostic techniques that rely on nucleic acid probes to detect small amounts of parasite DNA. These probes are made using PCR (polymerase chain reaction) amplification of parasite ribosomal DNA.[41] Such techniques allow for the detection of single organisms in tissue samples (0.1 pg *T. gondii* DNA).

Pyrimethamine and sulfonamides given together are widely used against *T. gondii.* They act synergistically by blocking a pathway involving *p*-aminobenzoic acid and the folic-folinic acid cycle respectively. Possible side effects of this treatment are thrombocytopenia and/or leukopenia, but these can be avoided by administration of folinic acid and yeast to a patient. Vertebrates can employ presynthesized folinic acid, whereas *T. gondii* cannot. Experimental chemotherapy may involve additional drugs in combination with the aforementioned compounds.[3] The apicoplast also has some distinct pathways, such as that of Type II fatty acid synthesis, that are potential targets for chemotherapy. Genes for the enzymes involved are homologous to those of bacteria, and *T. gondii* is susceptible to the antibacterial compound triclosan.[15]

- **Epidemiology.** In the United States the prevalence of chronic, asymptomatic toxoplasmosis is age related, increasing 0.5% to 1.0% per year of age.[51] Although clinical toxoplasmosis usually affects only scattered individuals, small epidemics occur from time to time, with raw meat evidently being a prime source of infection. For example, in the spring of 1968, several Cornell University Medical College students were infected simultaneously by wolfing down undercooked hamburgers between classes.[50] Considering the custom of backyard cooking and Americans' fondness for rare beef, many cases of toxoplasmosis may be acquired every day.

Although beef is certainly a potential source of infection, pork and lamb are much more likely to be contaminated. Freezing at −14°C for even a few hours apparently kills most cysts. To avoid a multitude of parasites, persons who insist on eating undercooked meat would do well to see that it has been hard frozen.

Feral and domestic cats will continue to be a source of infection of humans. Stray cats lead to problems of several kinds and are reservoirs of several diseases; efforts should be made to keep their numbers down. A more difficult problem to resolve is the household pet, the tabby that spends most of its time in a close relationship with its owners. Any cat, no matter how well fed and protected, may be passing *T. gondii* oocysts, although for only a few days after infection.

The possibilities are particularly alarming if someone in the house becomes pregnant. Certainly, a woman who knows she is pregnant should never empty a litterbox or clean up after a cat's occasional indiscretion. (Emptying the box every two days should help, but because cysts require one to three days to sporulate, it is better to have someone else do the job.) Having a cat tested for antibodies is impractical, for their presence does not correlate with shedding of oocysts. Also, because children's sand boxes become litter boxes for neighborhood cats, they should have tightly fitting covers. Covers also will protect children from larva migrans from hookworms and ascaridoid juveniles.

Filth flies and cockroaches are capable of carrying *T. gondii* oocysts from cat feces to the dinner table.[96] Earthworms may serve to move oocysts from where cats have buried them to the ground surface. Any soil reservoir of oocysts is a most important source of infection of humans. Tenter et al. provide an excellent review of *T. gondii* transmission, with particular focus on its zoonotic potential.[91]

Toxoplasma gondii tachyzoites have been isolated from human nasal, vaginal, and eye secretions; milk; saliva; urine; seminal fluid; and feces. The role of any of these in spreading infection is unknown, but it seems reasonable that any or all may be involved. Whole blood or leukocyte transfusions and organ transplants are also potential sources of serious infection, given that recipients may be immunodeficient because of disease or treatment.

***Sarcocystis* Species and Related Parasites.** *Sarcocystis* spp. have been known from their zoitocysts in muscle of reptiles, birds, and mammals since the late 19th century; but their life cycles remained obscure until 1972 when it was discovered that the bradyzoites would lead to development of coccidian gametes in cell culture and of oocysts after being fed to cats.[34] Since then it has been found that some species of what was called *Isospora* were in fact stages of *Sarcocystis* in their definitive hosts (for example, *S. bigemina* and *S. hominis*),[53] and what had been considered single species of *Sarcocystis* from particular hosts comprised several species in each. For example, oocysts of *Sarcocystis cruzi* (syn. *S. bovicanis*), *S. tenella* (syn. *S. ovicanis*), and *S. meischeriana* (syn. *S. suicanis*) cannot be distinguished morphologically. For a review of *Sarcocystis* taxonomy, see Levine.[55]

Sarcocystis spp. are obligately heteroxenous, with a herbivorous intermediate host—such as various species of reptiles, birds, small rodents, and hoofed animals—and a carnivorous definitive host (Fig. 8.18). When sporozoites are released from sporocysts consumed by an intermediate host, they penetrate the intestinal epithelium, are distributed through the body, and invade endothelial cells of blood vessels in many tissues. There they undergo merogony, and additional merogonous generations may ensue. Zoitocysts (tissue cysts, Fig. 8.19) then form in skeletal and cardiac muscle and occasionally the brain. The cysts are also known as **sarcocysts** or **Miescher's tubules.**

Some species' cysts are large enough to be seen by the unaided eye. They usually have internal septa and compartments and are elongated, cylindroid, or spindle shaped; but they also may be irregularly shaped. They lie within a muscle fiber, in the same plane as the muscle bundle. Their overall size varies, reaching 1 cm in diameter in some cases, but they usually are 1 mm to 2 mm in diameter and 1 cm or less long.

Cyst wall structure varies among "species" and among different stages of development. In some cases the outer wall is smooth; in others it has an outer layer of fibers, the **cytophaneres,** which radiate out into the muscle (Fig. 8.20). The cyst wall's origin is controversial: Some authors conclude that it is of host origin; others maintain that it is made by the parasite. It may well be derived from both sources. Two distinct regions can be distinguished in the cyst. The peripheral region is occupied by globular **metrocytes.** After several divisions the metrocytes give rise to more elongated bradyzoites, which resemble typical coccidian merozoites except that they have a larger number of micronemes. Metrocytes also lack rhoptries and micronemes. Only bradyzoites are infective to definitive hosts.

When a zoitocyst is consumed by a definitive host, its wall is digested away, and the bradyzoites penetrate the lamina propria of the small intestine. There they undergo gamogony without an intervening merogonic generation. Male gametes penetrate female gametes, and the resulting oocysts sporulate in the lamina propria. Oocyst walls are thin and are usually broken during passage through the intestine; thus, sporocysts rather than oocysts are normally passed in feces. Sporocysts can infect intermediate hosts but not definitive hosts.

Humans have been named definitive hosts for some species *(S. hominis, S. suihominis),* but zoitocysts of several unidentified species occasionally are found in human muscle.[74] Discovery of a generalized infection in dogs indicates that carnivores may develop infections typical of those found in intermediate hosts.[29] *Sarcocystis* species are globally distributed, being found in a wide variety of vertebrate animals. It is likely that many *Sarcocystis* species await discovery, especially in sylvatic prey-predator systems. Illustrations of the latter are parasites that utilize small owls and deer mice, king snakes and voles, and opossums and birds.[14, 33, 60] More than 50% of adult swine, cattle, and sheep probably are infected with *Sarcocystis* spp.[23] Some of these parasites are nonpathogenic (see Table 8.1), but some may cause serious symptoms, which may include loss of appetite, fever, lameness, anemia, weight loss, and abortion in pregnant animals. Heavily infected animals may die. Flies may act as sporocyst transport hosts.[64]

Sarcocystis neurona is an important parasite of horses, causing a neurological disease known as equine protozoal myoencephalitis (EPM) (Fig. 8.18).[26] Horses evidently are aberrant intermediate hosts because sarcocysts do not develop in them; EPM-like disease occurs in a variety of mammals, including mink, raccoons, and sea otters. Opossums are considered the definitive hosts.[26]

***Besnoitia* Species.** *Besnoitia* species are parasites of vertebrates, including lizards, opossums, rodents, rabbits, donkeys, cattle, goats, and wild ruminants, all of which serve as intermediate hosts, and cats, which are the definitive hosts of those species whose life cycles are known.[28] Cysts in intermediary hosts occur mainly in connective tissues, have very thick walls, and include host cell nuclei within the wall. In intermediate hosts, infections proceed through acute and chronic phases, the former characterized by weakness, fever, and swelling of lymph nodes; death may occur in severe infections. Chronic infections result in various skin problems and, in bulls, infertility.

The economically important *B. besnoiti* occurs in cattle in Africa, the Mediterranean countries, and Eurasia, but other *Besnoitia* species have been reported from North and South America and Australia.

Neospora caninum. *Neospora caninum* (Figs. 8.21 and 8.22) was recognized in the late 1980s as a cause of toxoplasma-like illness in dogs, resulting in paralysis in pups and early

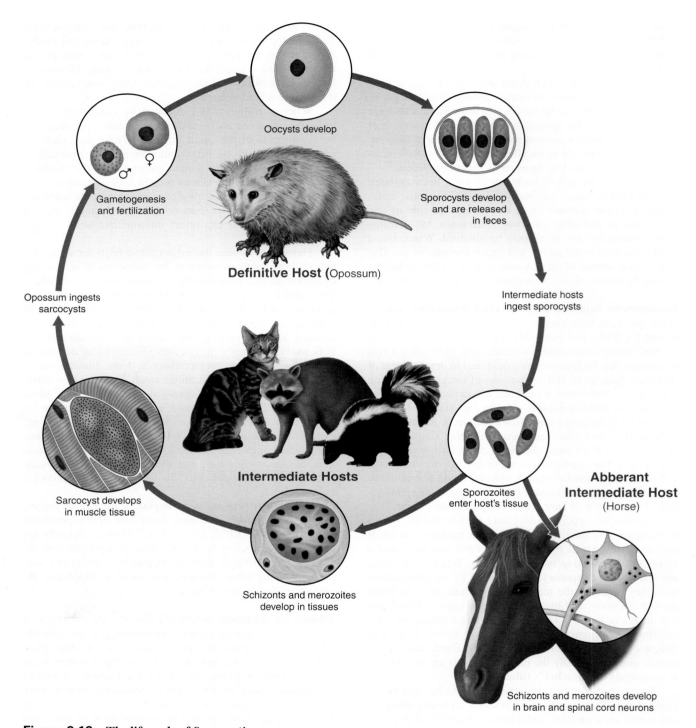

Oocysts develop

Gametogenesis
and fertilization

Sporocysts develop
and are released
in feces

Definitive Host (Opossum**)**

Opossum ingests
sarcocysts

Intermediate hosts
ingest sporocysts

Intermediate Hosts

Sporozoites
enter host's tissue

**Abberant
Intermediate Host**
(Horse)

Sarcocyst develops
in muscle tissue

Schizonts and merozoites
develop in tissues

Schizonts and merozoites develop
in brain and spinal cord neurons

Figure 8.18   The life cycle of *Sarcocystis neurona*.

Sarcocystis neurona causes a neurological disorder (equine protozoan myeloencephalitis, EPM) in horses, but the natural intermediate hosts evidently are a variety of carnivores, including mink, seals, and sea otters in addition to the species shown above. It took various scientists thirty years to finally work out the life cycle and demonstrate conclusively that *S. neurona* was a valid species and infected horses.[25a]

Drawing by Bill Ober and Claire Garrison.

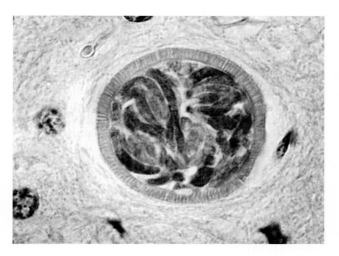

Figure 8.19   Cross section of zoitocyst of *Sarcocystis tenella* in muscle of experimentally infected sheep. (×6000)

From J. P. Dubey et al., "Development of sheep-canid cycle of *Sarcocystis tenella*," in *Canad. J. Zool.* 60:2464–2477. Copyright © 1982.

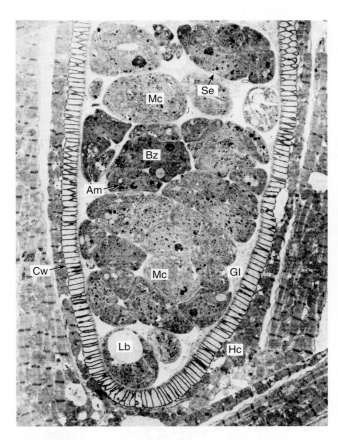

Figure 8.20   Transmission electron micrograph of *Sarcocystis tenella* sarcocyst.

Note fully formed wall composed of cytophaneres (*Cw*); indistinct granular septum (*Se*); fine granular layer of cytoplasm (*Gl*); bradyzoites (*Bz*); metrocytes (*Mc*); amylopectin (*Am*); and lipid bodies (*Lb*). *Hc* is host cell. (× 6000)

From J. P. Dubey et al., "Development of sheep-canid cycle of *Sarcocystis tenella*," in *Canad. J. Zool.* 60:2464–2477. Copyright © 1982.

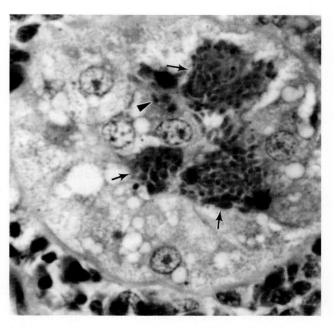

Figure 8.21   Cross section of a cat kidney tubule infected with *Neospora caninum*.

Tachyzoites are indicated by *arrows,* and the *arrowhead* points to a desquamated epithelial cell and tachyzoites in the lumen of the tubule.

From J. P. Dubey and D. S. Lindsay, "Transplacental *Neospora caninum* infection in cats," in *J. Parasitol.* 75:765–771. Copyright © 1989.

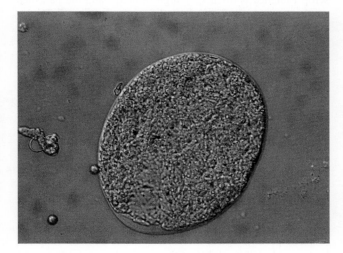

Figure 8.22   *Neospora caninum* tissue cyst isolated from the brain of an experimentally infected mouse.

The cyst was photographed using Nomarski illumination. It is packed with large numbers of bradyzoites.

From A. M. McGuire et al., "Separation and cryopreservation of *Neospora caninum* tissue cysts from murine brain," in *J. Parasitol.* 83:319–321. Copyright © 1997.

death, of generalized nervous system infection in kittens, and of fatal congenital infections or abortion in cattle and sheep.[25] Subsequent studies have shown that *N. caninum* is a major cause of abortion in dairy cattle and possibly other domestic livestock.[59] Dogs are one definitive host, although transplacental infection can occur in cats, dogs, cattle, and sheep.[25, 67]

Oocysts are spherical, 10 μm to 11 μm in diameter, and contain two sporocysts each with four sporozoites.[67] Tissue cysts containing tachyzoites are produced in intermediate hosts, although dogs may also have tissue infections and exhibit neurological symptoms.[25] Tachyzoites range from 3 μm to 7 μm long, depending on their stage of division, and multiply by endodyogeny. Many different cell types can be infected (see Fig. 8.21), with tachyzoites penetrating the host cell membrane and becoming enclosed in a parasitophorus vacuole. Cell death is evidently due to multiplication of tachyzoites. In dogs, severe infections occur in congenitally infected pups, which develop paralysis and hyperextension, especially of their hind legs, although older dogs may also develop dermatitis.

Neosporosis is distributed virtually worldwide in cattle, especially dairy cattle, and is now recognized as a leading cause of abortion. Individual cows may abort in succeeding pregnancies, and neosporosis is often endemic, with up to a third of the cows aborting either sporadically or in groups at almost any time during pregnancy. Inflammatory lesions are found throughout fetal tissues but especially in the central nervous system, heart, muscle, and liver, and congenitally infected calves may exhibit hind limb hyperextension. Congenital infection is the primary means of transmission in cattle, and, in some cases, endemic neosporosis has been traced to individual cows brought into a herd.

Diagnosis of *N. caninum* as the definitive cause of abortion in cattle requires a variety of techniques because seroprevalence is often very high in herds, and the parasites can be present in a fetus aborted for other reasons.[27] Thus a combination of serological, histological, and molecular studies must be used to conclusively establish a cause-and-effect relationship between parasites and fetal loss. Dubey and Schares give an excellent and detailed review of such diagnostic techniques appropriate for domestic livestock.[27]

Neospora caninum has been found in a wide variety of zoo animals and wild herbivores, especially whitetail deer, but antibodies have been detected in musk ox, bison, moose, and warthogs, as well as in rats and raccoons.[40] Wild canids ranging from Texas coyotes to Australian dingoes also have been shown to be seropositive, suggesting a global sylvatic cycle.[40]

A second species, *N. hughesi,* occurs in horses and there is evidence for transplacental transmission as with *N. caninum* in cattle, but the pathological effects in horses, including abortion, evidently are not as severe as with *N. caninum* in cattle.[77]

The parasitological literature contains an interesting debate regarding the validity of the species known as *Neospora caninum.* Heydorn and Mehlhorn contend that *N. caninum* is not distinguishable from the closely related species *Hammondia heydorni,* another coccidian of dogs.[43] Dubey and his coworkers, however, distinguish the species based on molecular and ultrastructural characters.[24] The cited papers make fascinating—and highly educational—reading for anyone who believes taxonomy to be irrelevant in the biotech age!

Pneumocystis carinii. *Pneumocystis carinii* is a parasite whose taxonomic position remained undetermined for nearly a century after its discovery. It is considered a fungus,[32] but it is mentioned here because of its importance as an "opportunistic parasite" that often causes severe pathology in

immunodeficient hosts. Although *P. carinii* has many fungal properties, it is sensitive to antiprotozoal agents such as pentamidine, trimethoprim-sulfamethoxazole, isethionate, pyrimethamine, and sulfadiazine.[39]

Pneumocystis carinii causes **interstitial plasma cell pneumonitis,** especially in immunosuppressed hosts. It occurs commonly in humans of all ages and is particularly important in the elderly and in children with primary disorders of immune deficiency. It is also a critical problem in patients receiving cytotoxic or immunosuppressive drugs for lymphoreticular cancers, organ transplantation, and a variety of other disorders. Persons with AIDS are particularly susceptible; 85% of such patients eventually present infections,[11] and pneumocystis pneumonia is a major cause of death in that segment of the population. In nature the organism is widespread in mammals. Many human infections may be acquired from pets.

- **Morphology and Biology.** In the lungs, *P. carinii* assumes three morphological forms: **trophozoite, precyst, and cyst.** Trophozoites are pleomorphic, 1 μm to 5 μm wide, and have small filopodia that form pockets in the membranes of interstitial cells. Precysts are oval, with few filopodia, and have a clump of mitochondria in their center. A precyst nucleus undergoes three divisions, after which it becomes "delimited" by membranes. A mature cyst (Fig. 8.23) is spherical, has a thick chitinous membrane, and contains eight "intracystic bodies," which are the infective young trophozoites.

All three stages live in interstitial tissues of the lungs and are not normally found in alveoli. In virulent cases parasites are abundant in pulmonary exudate; transmission probably is by aerosol droplets and direct contact.

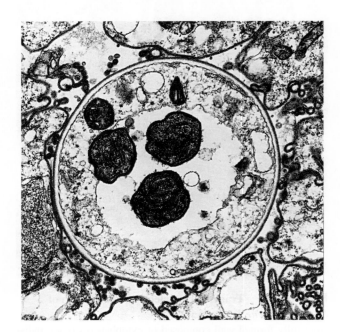

Figure 8.23 **Transmission electron micrograph of a** *Pneumocystis carinii* **cyst.**

Note the intracystic bodies.

From K. Yoneda et al., *"Pneumocystis carinii*: Freeze-fracture study of stages of the organism," in *Exp. Parasitol.* 53:68–76. Copyright © 1982.

Congenital infection is possible, as *P. carinii* has been found in stillborn infants, newborn germ-free rats, and three-day-old children.[66]

- **Pathogenesis.** In infected lungs the epithelium becomes desquamated and alveoli fill with foamy exudate containing parasites. The disease has a rapid onset associated with fever, cough, rapid breathing, and cyanosis (blue skin around the mouth and eyes). Death is caused by asphyxia. The mortality rate is virtually 100% in untreated patients. In some cases parasites disseminate to the spleen, lymph nodes, bone marrow, and even the eyes. Gutierrez[42] and Smulian[83] give excellent discussions of clinical manifestations, pathology, genetics, and cell biology of *P. carinii*.

- **Diagnosis.** Infections with *P. carinii* are suspected in any patient who presents clinical symptoms consistent with the disease, but positive diagnosis is possible only by demonstrating the organisms with special staining. Sputum examination is effective in about half the cases, but lung biopsy or bronchial lavage yields infected material most often. Toluidine blue or methenamine silver stains are apparently reliable, and the Gram-Weigert stain is accurate in demonstrating cysts. Although a number of other methods—such as ELISA, immunofluorescence assay, and DNA amplification techniques—are being developed, none as yet has proven quicker, easier, and more reliable than the classical staining.

- **Treatment.** Even with treatment, mortality is high in immunodeficient patients. The treatment of choice is a combination of trimethoprim-sulfamethoxazole. Pentamidine isethionate is equally effective, delivered as an inhalant spray, but pentamidine is toxic to the patient, too, and treatment must be monitored carefully for dangerous side effects.

Blastocystis hominis

Like *Pneumocystis carinii*, *Blastocystis hominis* is a parasite whose taxonomic status is unclear.[10] The life cycle includes vacuolar, amoeboid, precystic, and cyst stages.[85] Ameboid stages divide by binary fission and phagocytize bacteria. Two kinds of cysts are formed: thin walled and thick walled. The former evidently contain schizonts and are possibly autoinfective, whereas the latter are likely the means of external transmission. The parasite is included here primarily because of the molecular evidence linking it more closely to the alveolates than to the amebas.[10] Several species of *Blastocystis* have been described from ducks, geese, camels, and even koalas. *Blastocystis hominis* has been implicated in various intestinal disorders, including traveler's diarrhea and irritable bowel syndrome, but a clear link between infection and disease has yet to be established.[46]

Learning Outcomes

By the time a student has finished studying this chapter, he or she should be able to:

1. Diagram or label a diagram of the life cycle of *Monocystis lumbrici*.
2. Label a diagram of an apicomplexan sporozoite.
3. Label a diagram of an oocyst of the coccidian genus *Eimeria*.
4. Explain the life history of a typical gregarine parasite.
5. Write an extended paragraph on the life cycle of *Toxoplasma gondii*.
6. Tell the pathological changes that occur during infection of a chicken with *Eimeria tenella*.
7. Describe the distinguishing characteristics of different *Hepatozoon* species.
8. Describe the economic impacts of *Neospora caninum*.
9. Explain why apicomplexan taxonomy is such a challenging subdiscipline within parasitology.
10. Explain the transmission ecology of *Cryptosporidium* species and tell why members of genus *Cryptosporidium* are considered closely related to gregarines.

References

References for superscripts in the text can be found at the following Internet site: www.mhhe.com/robertsjanovynadler9e

Additional Readings

Chartier, C., and C. Paraud. 2012. Coccidiosis due to *Eimeria* in sheep and goats, a review. *Small Ruminant Res.* 103:84–92.

Contreras-Ochoa, C. O., A. Lagunas-Martinez, J. Belkind-Gerson, and D. Correa. 10121. *Toxoplasma gondii* invasion and replication in astrocyte primary cultures and astrocytoma cell lines: systematic review of the literature. *Parasitol. Res.* 110:2089–2094.

Desmonts, G., and J. Couveur. 1974. Congenital toxoplasmosis. *N. Eng. J. Med.* 290:1110–1116. A study of 378 pregnancies.

Feldman, H. A. 1974. Congenital toxoplasmosis, at long last. *N. Eng. J. Med.* 290:1138–1140. A short summary of the discovery of congenital toxoplasmosis.

Feustel, S. M., M. Meissner, and O. Liesenfeld. 20121. *Toxoplasma gondii* and the blood-brain barrier. *Virulence* 3:182-192.

Lillehoj, H. S., and E. P. Lillehoj. 2000. Avian coccidiosis. A review of acquired intestinal immunity and vaccination strategies. *Avian Dis.* 44:408–425.

Lindsay, D. S., and J. P. Dubey. 2011. *Toxoplasma gondii:* the changing paradigm of congenital toxoplasmosis. *Parasitol.* 138:1829–1831.

Long, P. L. (Ed.). 1982. *The biology of the coccidia.* Baltimore, MD: University Park Press.

Peng, H.-J., X.-G. Chen, and D. S. Lindsay. 2011. A review: competence, compromise, and contomitance-reaction of the host cell to *Toxoplasma gondii* infection and development. *J. Parasitol.* 97:620–628.

Shirley, M. W., and H. S. Lillehoj. 2012. The long view: a selective review of 40 years of coccidiosis research. *Avian Pathol.* 41:111–121.

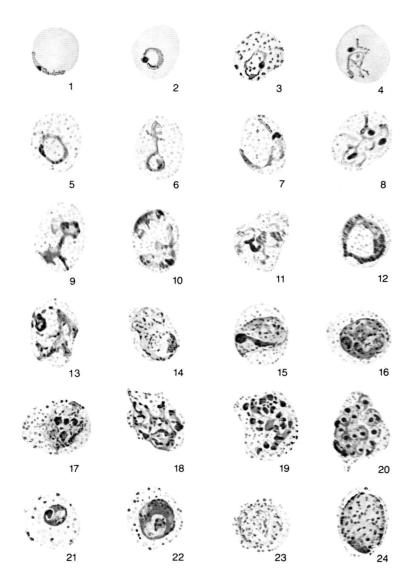

Plate 1

Plasmodium vivax. (*1*) Normal-sized red cell with marginal ring-form trophozoite. (*2*) Young signet-ring–form trophozoite in macrocyte. (*3*) Slightly older ring-form trophozoite in red cell showing basophilic stippling. (*4*) Polychromatophilic red cell containing young tertian parasite with pseudopodia. (*5*) Ring-form trophozoite showing pigment in cytoplasm, in enlarged cell containing Schüffner's stippling (*dots*). (Schüffner's stippling does not appear in all cells containing growing and older forms of *P. vivax*, as would be indicated by these pictures, but it can be found with any stage from fairly young ring form onward.) (*6, 7*) Very tenuous medium trophozoite forms. (*8*) Three ameboid trophozoites with fused cytoplasm. (*9, 11–13*) Older ameboid trophozoites in process of development. (*10*) Two ameboid trophozoites in one cell. (*14*) Mature trophozoite. (*15*) Mature trophozoite with chromatin apparently in process of division. (*16–19*) Schizonts showing progressive steps in division (presegmenting schizonts). (*20*) Mature schizont. (*21, 22*) Developing gametocytes. (*23*) Mature microgametocyte. (*24*) Mature macrogametocyte.

From A. Wilcox, *Manual for the Microscopial Diagnosis of Malaria in Man.* Public Health Service, Publication No. 796. U.S. Government Printing Office, Washington D.C., 1960.

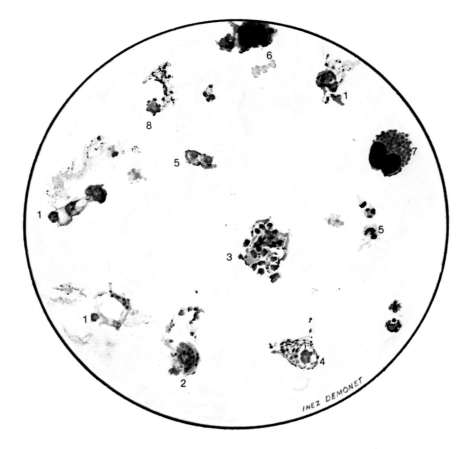

Plate 2

Plasmodium vivax in thick smear. (*1*) Ameboid trophozoites; (*2*) Schizont—two divisions of chromatin; (*3*) Mature schizont; (*4*) Microgametocyte; (*5*) Blood platelets; (*6*) Nucleus of neutrophil; (*7*) Eosinophil; (*8*) Blood platelet associated with cellular remains of young erythrocytes.

From A. Wilcox, U.S. Government Printing Office, Washington D.C., 1960.

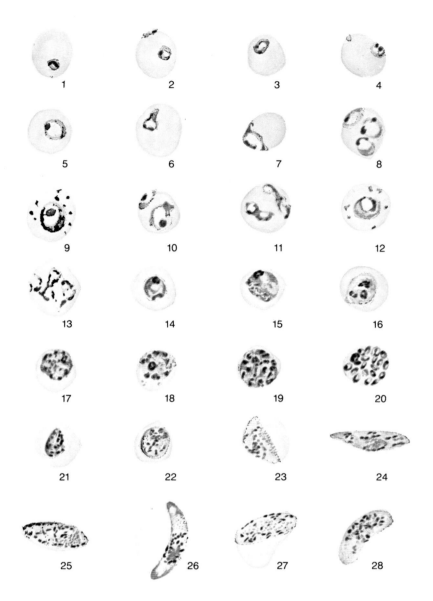

Plate 3

Plasmodium falciparum. (*1*) Very young ring-form trophozoite. (*2*) Double infection of single cell with young trophozoites—one a marginal form, the other a signet-ring form. (*3, 4*) Young trophozoites showing double chromatin dots. (*5–7*) Developing trophozoite forms. (*8*) Three medium trophozoites in one cell. (*9*) Trophozoite showing pigment, in cell containing Maurer's dots. (*10, 11*) Two trophozoites in each of two cells, showing variations of forms that parasites may assume. (*12*) Almost mature trophozoite showing haze of pigment throughout cytoplasm. Maurer's dots in cell. (*13*) Estivoautumnal slender forms. (*14*) Mature trophozoite showing clumped pigment. (*15*) Parasite in process of initial chromatin division. (*16–19*) Various phases of development of schizont (presegmenting schizonts). (*20*) Mature schizont. (*21–24*) Successive forms in development of gametocyte—usually not found in peripheral circulation. (*25*) Immature macrogametocyte. (*26*) Mature macrogametocyte. (*27*) Immature microgametocyte. (*28*) Mature microgametocyte.

From A. Wilcox, U.S. Government Printing Office, Washington D.C., 1960.

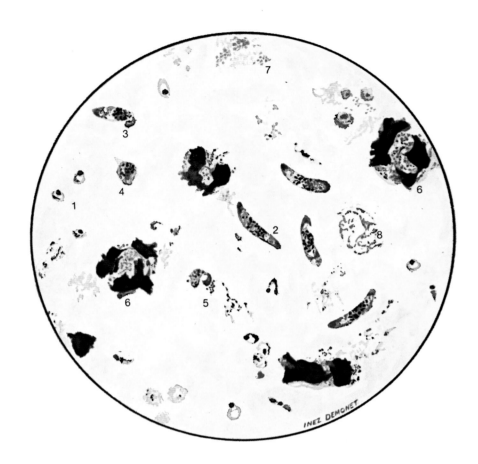

Plate 4

Plasmodium falciparum in thick film. (*1*) Small trophozoites; (*2*) Gametocytes—normal; (*3*) Slightly distorted gametocyte; (*4*) "Rounded-up" gametocyte; (*5*) Disintegrated gametocyte; (*6*) Nucleus of leukocyte; (*7*) Blood platelets; (*8*) Cellular remains of young erythrocyte.

From A. Wilcox, U.S. Government Printing Office, Washington D.C., 1960.

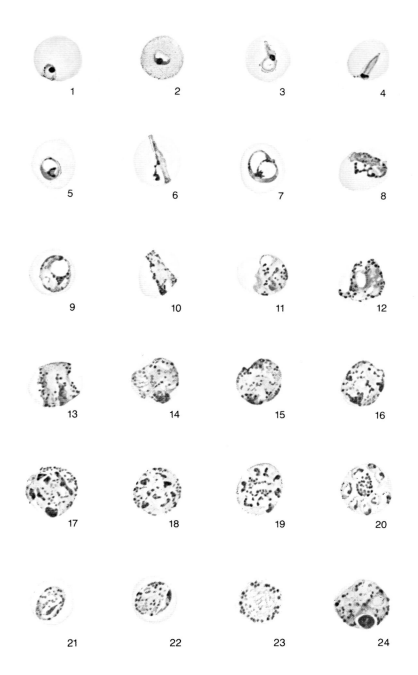

Plate 5

Plasmodium malariae. (*1*) Young ring-form trophozoite of quartan malaria. (*2–4*) Young trophozoite forms of parasite showing gradual increase of chromatin and cytoplasm. (*5*) Developing ring-form trophozoite showing pigment granule. (*6*) Early band-form trophozoite—elongate chromatin, some pigment apparent. (*7–12*) Some forms that developing trophozoite of quartan may take. (*13, 14*) Mature trophozoites—one a band form. (*15–19*) Phases in development of schizont (presegmenting schizonts). (*20*) Mature schizont. (*21*) Immature microgametocyte. (*22*) Immature macrogametocyte. (*23*) Mature microgametocyte. (*24*) Mature macrogametocyte.

From A. Wilcox, U.S. Government Printing Office, Washington D.C., 1960.

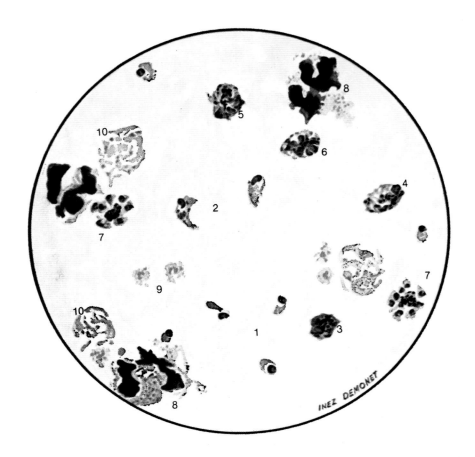

Plate 6

Plasmodium malariae in thick smear. (*1*) Small trophozoites; (*2*) growing trophozoites; (*3*) mature trophozoites; (*4–6*) schizonts (presegmenting) with varying numbers of divisions of chromatin; (*7*) mature schizonts; (*8*) nucleus of leukocyte; (*9*) blood platelets; (*10*) cellular remains of young erythrocytes.

From A. Wilcox, *Manual for the Microscopical Diagnosis of Malaria in Man*, Department of Health, Education, and Welfare Public Health Service. U.S. Government Printing Office, Washington D.C., 1960.

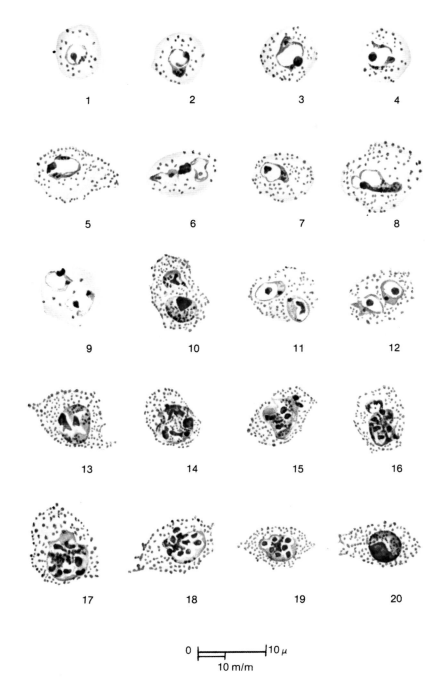

0 |———————|10 μ
10 m/m

Plate 7

Plasmodium ovale. (*1*) Young ring-shaped trophozoite; (*2–5*) older ring-shaped trophozoites; (*6–8*) older ameboid trophozoites; (*9, 11, 12*) doubly infected cells, trophozoites; (*10*) doubly infected cell, young gametocytes; (*13*) first stage of schizont; (*14–19*) schizonts, progressive stages; (*20*) mature gametocyte.

From A. Wilcox, *Manual for the Microscopical Diagnosis of Malaria in Man*, (2nd Ed.). National Institutes of Health Bulletin, No. 180 (Revised). U.S. Government Printing Office, Washington D.C., 1960.

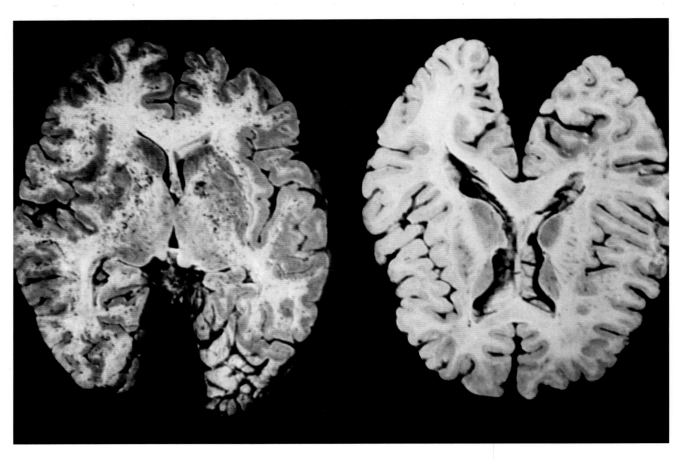

Plate 8

Cut section of brain from cerebral malaria victim (left) compared with normal brain (right). The cortex shows a slate-gray color from hemozoin, and tiny hemorrhages (petechiae) around blood vessels in white matter can be seen. The degree to which there is swelling from fluid in the nervous tissue (edema) can be observed by comparing with ventricles and sulci of the normal brain.

From Toro González, G., G. Román-Campos, and L. Navarro de Román. 1983. *Neurología Tropical: Aspectos Neuropatológicos de la Medicina Tropical*. Colombia: Editorial Printer Columbiana, Ltda.

Phylum Apicomplexa: Malaria Organisms and Piroplasms

Parasitic elements are found in the blood of patients who are ill with malaria. Up to now, these elements were thought incorrectly to be pigmented leukocytes. The presence of these parasites in the blood probably is the principal cause of malaria.

—Charles Louis Alphonse Laveran, 1880

Order Heamosporida contains family Plasmodiidae, including genera *Plasmodium*, *Haemoproteus*, and *Leucocytozoon*, species which cause malaria and malarialike diseases in humans, other mammals, birds, and reptiles. When in host cells, *Plasmodium* and *Haemoproteus* usually produce a pigment called **hemozoin** from host hemoglobin, distinguishing them from the closely related *Leucocytozoon,* which does not produce hemozoin. The ultrastructure of these parasites is basically similar to that of coccidia, except that these organisms lack conoids. Syzygy is absent, and the macrogametocyte and microgametocyte develop independently. Microgametocytes produce about eight flagellated gametes. Zygotes are motile and are called **ookinetes;** sporozoites are not enclosed within sporocysts. Haemosporideans are heteroxenous, with merozoites produced in a vertebrate host and sporozoites developing in an invertebrate host. It is possible that these parasites evolved from coccidia of vertebrates rather than of invertebrates, with mites or other bloodsuckers initiating the cycle in arthropods.

Although most species of Haemospororida are parasites of wild animals and appear to cause little harm in most cases, a few cause diseases that are among the worst scourges of humanity. Indeed, malaria has played an important part in the rise and fall of nations and has killed untold millions the world over.

Despite the combined efforts of 102 countries to eradicate malaria, it remains one of the most important diseases in the world today in terms of lives lost and economic burden. Progress has been made, however. In some countries, such as the United States, eradication of endemic malaria is complete. Between 1948 and 1965 the number of cases was cut from a worldwide total of 350 million to fewer than 100 million. However, more recent estimates put the worldwide prevalence as high as 659 million.[127] Development of resistance in the parasite to antimalarial drugs and in the vector to insecticides deserves much of the blame for the increase in prevalence.[80] In Africa, malaria is now a much worse problem than it was 30 years ago.[106]

Over 1.5 billion people live in malarious areas of the world. These areas lack the administrative, financial, and human resources necessary for control of this disease.

ORDER HAEMOSPORORIDA

Order Haemosporida contains family Plasmodiidae, including genera *Plasmodium, Haemoproteus,* and *Leucocytozoon,* species of which cause malaria and malaria-like diseases in humans, other mammals, birds, and reptiles.

Genus *Plasmodium*

Malaria has been known since antiquity; recognizable descriptions of the disease were recorded in various Egyptian papyri. Hippocrates studied medicine in Egypt and clearly described quotidian, tertian, and quartan fevers with splenomegaly. He believed that bile was the cause of the fevers. Greeks built beautiful city-states in the lowlands only to see them devastated by the disease, and wealthy Greeks and Romans traditionally summered in the highlands to escape the heat, mosquitoes, and mysterious fevers. Herodotus (c. 2500–2424 B.C.) states that Egyptian fishermen slept with their nets arranged around their beds so that mosquitoes could not reach them. Medieval England saw crusaders falter and fail as they encountered malaria. As had happened before and has happened since, malaria killed more warriors than did warfare. When Europeans imported slaves and returned their colonial armies to their continent, they brought malaria with them, increasing the concentration of the disease with devastating results.

Throughout history a connection between swamps and fevers has been recognized. It was commonly concluded that the disease was contracted by breathing "bad air" or *mal aria.* This belief flourished until near the end of the 19th century. Another name for the disease, *paludism* (marsh disease), is still in common use in the world.

There has been much speculation as to whether malaria existed in the Western Hemisphere before the Spanish conquest. It seems inconceivable that the great Olmec and Mayan civilizations could have developed in highly malarious regions. The Spanish conquistadors made no mention of fevers during the early years of the conquest, and in fact they holidayed in Guayaquil and the coastal area near Veracruz, regions that soon after became very unhealthy because of malaria. Balboa did not mention any encounters with malaria while traversing the Isthmus of Panama. It therefore seems likely that malaria was introduced into the New World by the Spaniards and their African slaves. However, some evidence that Africans reached South America during pre-Columbian times suggests that, while improbable, it is not impossible that malaria existed in localized areas of the continent before the Spanish conquest,[47] could have been brought from Oceania or from Asia by way of the Bering Strait, or could have been introduced by the Vikings.

The French army physician Charles Louis Alphonse Laveran accurately described the male and female gametes, the trophozoite, and the schizont while working with a poor, low-power microscope and unstained preparations. By 1890 several scientists in different parts of the world verified his findings.

The mode of transmission of malaria was, however, still unknown. Patrick Manson favored the hypothesis of transmission by mosquitoes; he was conditioned by proof of mosquitoes as vectors of filariasis, which gave him some insight. While on leave from the Indian Medical Service, Surgeon-Major Ronald Ross was 38 years old when he and Manson met for the first time. Finding in Ross a man who was interested in malaria and who could test his ideas for him, Manson lost no time in convincing Ross that malaria was caused by a protozoan parasite. For the next several years, in India, Ross worked during every spare minute, searching for the mosquito stages of malaria that he had become certain existed. Dissecting mosquitoes at random and also after allowing them to feed on malarious patients, he found many parasites, but none of them proved to be what he searched for.

Ross's first significant observation was that exflagellation normally occurs in the stomach of a mosquito, rather than in the blood as was then thought. At this time he was posted to Bangalore to help fight a cholera epidemic, the first in a series of frustrating interruptions by superiors who had no concept of the importance of the work Ross was doing in his spare time. After two years of work, which his superior officers ignored as harmless lunacy, he seemed to have reached an impasse. He was eligible for retirement soon but was determined to try "one more desperate effort to solve the Great Problem." He toiled far into the nights, dissecting mosquitoes in a hot little office. He could not use the overhead fan lest it blow his mosquitoes away, so while he worked swarms of gnats and mosquitoes avenged themselves "for the death of their friends." At last, late in the night of August 16, 1897, he dissected some "dapple-winged" mosquitoes (*Anopheles* spp.) that had fed on a malaria patient, and he found some pigmented, spherical bodies in the walls of the insects' stomachs. The next day he dissected his last remaining specimen and found the spheroid cells had grown. They were most certainly the malaria parasites!

He reported his discovery to Manson and immediately set about breeding the correct kind of mosquito in preparation for the first step of transmitting the disease from the insect to humans. Unfortunately, he was immediately posted to Bombay, where he could do no further research on human malaria.

Nevertheless, he found similar organisms (*Plasmodium relictum*) in birds. He repeated his feeding experiments with mosquitoes and found similar parasites when they fed on infected birds. He also found that the spheroid bodies ruptured, releasing thousands of tiny bodies that dispersed throughout the insect's body, including into the salivary glands.

Through Manson, Ross reported to the world how malaria was transmitted by mosquitoes. It remained only for a single experiment to prove the transmission to humans. Ross never did it. The authorities were so impressed with his work they ordered Ross to work out the biology of kala-azar in another part of India. This transfer seems to have broken his spirit, for he never really tried again to finish the study of malaria. The concentration had made him ill, his eyes were bothering him, and his microscope had rusted tight from his sweat. Anyway, he was a physician, not a zoologist, and he was most interested in learning how to prevent the disease, as opposed to determining the finer points of the parasite's biology. This he considered done, and he retired from the Army. He was awarded the Nobel Prize in Medicine in 1902 and was knighted in 1911. He died in 1932 after a distinguished postarmy career in education and research.

The history of malariology is tarnished by strife and bitterness. Several persons who were working on the life cycle of the parasite claimed credit for the discovery that pointed to the means of control for malaria. Italian, German, and American scientists all made important contributions to the solution of the problem of malaria transmission. Several of them, including Ross, spent a good portion of their lives quibbling about priorities in the discoveries. Manson-Bahr[70] and Harrison (in Additional Readings) give fascinating accounts of the personalities of the men who conquered the life cycle of malaria. Credit for completing study of the life cycle should go to Amigo Bignami and Giovanni Grassi, who experimentally transmitted the malaria parasite from mosquito to human in 1898.

Although medical scientists thought they knew the life cycle of malaria after Ross's work, they knew nothing of the stages in the liver. They thought that the cycle progressed from blood to mosquito and back to blood. Quinine was a well-known antimalarial drug, but it had effect only on the erythrocytic forms. Soldiers treated with the drug were apparently cured; that is, no parasites could be found in their blood. However, when treatment stopped and patients moved to a nonmalarious area, parasites returned to their blood at certain time intervals.

In 1938 S. P. James and P. Tate discovered the exoerythrocytic stages of *P. gallinaceum*. After this discovery large-scale work began to find the exoerythrocytic stages of human malaria parasites. Finally, in 1948 H. C. Shortt and P. C. C. Garnham demonstrated the exoerythrocytic stages of *P. cynomolgi* in monkeys and *P. vivax* in humans.[112]

These historical notes should not be concluded without mention of a man who applied these early discoveries for the immense benefit of his country and humanity: William C. Gorgas. Gorgas was the medical officer placed in charge of the Sanitation Department of the Canal Zone when the United States undertook construction of the Panama Canal. Were it not for his mosquito control measures, malaria and yellow fever would have defeated American attempts to build the Canal, just as they had defeated the French. During July 1906 the malaria rate in the Canal Zone was 1263 hospital admissions per 1000 population![104] Gorgas's work reduced the rate to 76 hospital admissions per 1000 in 1913, saving his country $80 million and the lives of 71,000 fellow humans. Gorgas was a hero: The president made him Surgeon General, Congress promoted him, Oxford University made him an honorary Doctor of Science, and the King of England knighted him. Sir William Osler said, "There is nothing to match the work of Gorgas in the history of human achievement." It is a sad commentary on our cultural memory that the name of Gorgas is now known by so few, while so many easily remember the names of generals and tyrants who caused great bloodshed.

For students interested in more details about humanity's fight against malaria, we highly recommend Harrison's book and those by Desowitz cited in Additional References.

Life Cycle and General Morphology

Following is a general account of development and structure of malaria parasites (Fig. 9.1), without reference to particular species. Specific morphological details for each species (except for *P. knowlesi*) are in Table 9.1. *Plasmodium* spp. require two types of hosts: an invertebrate (mosquito) and a vertebrate (reptile, bird, or mammal). Technically, the invertebrate is the definitive host because sexual reproduction occurs there. Asexual reproduction takes place in the tissues of a vertebrate, which thus is the intermediate host. *Plasmodium* spp. were probably derived from an ancestral coccidian whose asexual and sexual reproduction took place in the same (presumably vertebrate) host.

Vertebrate Phases. When an infected mosquito takes blood from a vertebrate, she injects saliva containing tiny, elongated sporozoites into the bloodstream. Sporozoites are similar in morphology to those of *Eimeria* and other coccidia. They are about 10 μm to 15 μm long by 1 μm in diameter and have a pellicle composed of a thin outer membrane, a doubled inner membrane, and a layer of subpellicular microtubules. There are three polar rings. The rhoptries are long, extending to the midportion of the organism, and much of the rest of the anterior cytoplasm is taken up by the micronemes. An apparently nonfunctional cytostome is present, and there is a mitochondrion in the posterior end of the sporozoite.[1]

After being injected into the bloodstream, sporozoites disappear from the circulating blood within an hour. Their immediate fate was a great mystery until the mid-1940s, when it was shown that within one or two days they enter the

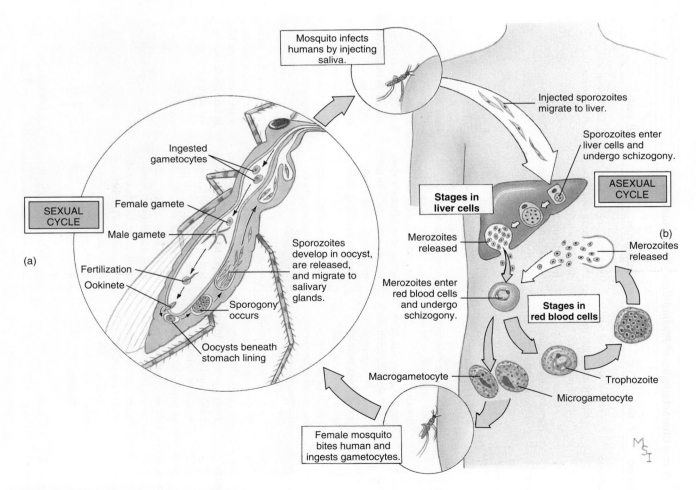

Figure 9.1   **Life cycle of *Plasmodium vivax*.**

(*a*) Sexual cycle produces sporozoites in body of mosquito. Meiosis occurs just after zygote formation (zygotic meiosis). (*b*) Sporozoites infect a human and reproduce asexually, first in liver cells and then in red blood cells.

Table 9.1 Some Characteristics of *Plasmodium* spp. in Humans

Stage or Period	*P. vivax*	*P. falciparum*	*P. ovale*	*P. malariae*	*P. knowlesi*
Early trophozoite	About ⅓ diameter of red cell; chromatin dot heavy; vacuole prominent	About ⅕ diameter of red cell; chromatin dot small; two dots frequent; marginal forms frequent	Like *P. vivax* and *P. malariae*	Single, heavy chromatin dot; cytoplasmic circle often smaller, thicker, heavier than in *P. vivax*; vacuole fills in early	
Growing trophozoite	Pseudopodia common; one or more food vacuoles	This stage not usually seen in circulating blood	Compact, little vacuolation	Cytoplasm usually compact; little or no vacuole; sometimes in a band form across the red cell	
Late trophozoite	Large mass of chromatin; fine brown hemozoin; almost fills red cell	This stage not usually seen in circulating blood		Chromatin often elongated, less definite in outline than in *P. vivax*; cytoplasm dense, rounded, oval, or band shaped; almost fills red cell	
Hemozoin	Short, delicate rods, irregularly scattered; yellowish brown	Granular; has tendency to coalesce; coarse in gametocytes	Hemozoin lighter than in *P. malariae*; similar to *P. vivax*	Granules rounded; larger, darker than in *P. vivax*; tendency to peripheral arrangement	
Appearance of erythrocyte	Larger than normal, often oddly shaped; Schüffner's dots at all stages but young rings; multiple infection occasional	Normal size; Maurer's spots common in cells with later trophozoites (not usually seen in circulating blood)	Schüffner's dots often present in ring and later stages; red cell larger than normal, oval, often with irregular edge	About normal or slightly smaller; stippling rarely seen; multiple infection rare	
Schizont	12 to 24 merozoites; hemozoin in one or two clumps; almost fills red cell	8 to 24 or more merozoites; rare in circulating blood	4 to 16 but usually 8 merozoites	6 to 12 but usually 8 or 10 merozoites in a rosette or cluster arrangement; often found in peripheral blood	
Microgametocytes (usually smaller and fewer than macrogametocytes)	Rounded or oval; almost fill red cell; dark hemozoin throughout cytoplasm; chromatin diffuse, in large mass, pink; small amount of light blue cytoplasm; no vacuoles	Crescent shaped; length about 1.5 times diameter of red cell; chromatin diffuse, pink; hemozoin granules in central portion; cytoplasm pale blue	Like *P. vivax* but somewhat smaller; mature macrogametocyte fills infected cell; microgametocytes smaller	Like *P. vivax*, but smaller; pigment more conspicuous	
Macrogametocytes	As in microgametocytes, except cytoplasm stains darker blue; chromatin more compact, dark red	Size and shape about as in microgametocytes; chromatin more compact, red; cytoplasm darker; hemozoin concentrated		Pigment abundant; round, dark brown granules; coarser than *P. vivax*	
EE cycle	8 days	5½ to 6 days	9 days	13 days	
Prepatent pd., minimum	11 to 13 days	9 to 10 days	10 to 14 days	15 to 16 days	
Schizogonic cycle	48 hours	36 to 48 hours, usually 48	about 48 hours	72 hours	24 hours
Development in mosquito	10 days at 25°C to 30°C	10 to 12 days at 27°C	14 days at 27°C	25 to 28 days at 22°C to 24°C	

parenchyma of the liver or other internal organ, depending on the species of *Plasmodium*. Where they are the first 24 hours still is unknown. A protein covering the surface of the sporozoite (**circumsporozoite protein**) bears a ligand (molecule that specifically and noncovalently binds to another molecule) that binds to receptors on the basolateral domain of the hepatocyte cell membrane.[13] That is why sporozoites enter liver cells and not other cells in the body. Extrusion of their contents from the rhoptries facilitates penetration into host cells.[96] Rhoptries' contents, function, and biogenesis make them most analogous to secretory lysosomal granules found in mammalian cells.[87] Entry into a hepatocyte initiates a series of asexual reproductions known as the **preerythrocytic cycle** or **primary exoerythrocytic schizogony,** often abbreviated as **PE** or **EE** stage. Once within a hepatic cell, the parasite metamorphoses into a feeding trophozoite. Organelles of the apical complex disappear, and trophozoites feed on the cytoplasm of the host cell by way of their cytostome and, in the species in mammals, by pinocytosis.

After about a week, depending on species, trophozoites are mature and begin schizogony. Numerous daughter nuclei are first formed, transforming the parasite into a schizont (Fig. 9.2), also known as a **cryptozoite.** During nuclear divisions nuclear membranes persist, and the microtubular spindle fibers are formed within the nucleus. The mitochondrion becomes larger during growth of a trophozoite, forms buds, and then breaks up into many mitochondria. Elements of the apical complex form subjacent to the outer membrane, and schizogony proceeds as previously described. Merozoites are much shorter than sporozoites—2.5 μm long by 1.5 μm

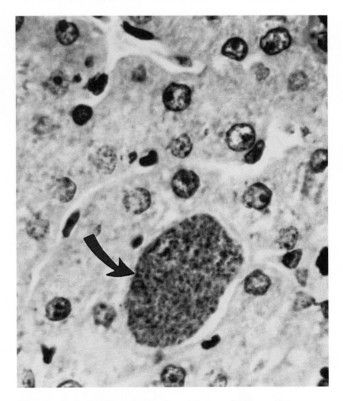

Figure 9.2   **Preerythrocytic schizont of *Plasmodium* (*arrow*) in liver tissue.**

Courtesy of Peter Diffley.

in diameter—and have small, teardrop-shaped rhoptries and small, oval micronemes.

What happens next has been a subject of lively debate. For many years it was believed that merozoites entered new hepatocytes to form new schizonts and then merozoites, at least in species of *Plasmodium* that are capable of causing a relapse.[112] However, as early as 1913 it was postulated that some sporozoites become dormant for an indefinite time after entering the body.[6] Such dormant cells, called **hypnozoites,** have now been demonstrated.[59] They are discussed under relapse in malaria (p. 154).

When merozoites leave liver cells to penetrate erythrocytes in the blood, they initiate an **erythrocytic cycle.** Some merozoites may be phagocytized by Kupffer cells in the liver, which may be an important host defense mechanism.[128] On entry into an erythrocyte, the merozoite again transforms into a trophozoite. Host cytoplasm ingested by a trophozoite forms a large food vacuole, giving the young *Plasmodium* the appearance of a ring of cytoplasm with the nucleus conspicuously displayed at one edge (Plate 1, *1* and *2*). Distinctiveness of the "signet-ring stage" is accentuated by Romanovsky stains: The parasite cytoplasm is blue, and the nucleus is red. As the trophozoite grows (see Plate 1, *3* to *15*), its food vacuoles become less noticeable by light microscopy, but pigment granules of **hemozoin** in the vacuoles become apparent. Hemozoin is an end product of the parasite's digestion of the host's hemoglobin. It is an insoluble polymer of heme (hematin, ferriprotoporphyrin-IX).[126]

The parasite rapidly develops into a **schizont** (or **meront**) (see Plate 1, *16* to *20*). The stage in the erythrocytic schizogony (also called *merogony*) at which the cytoplasm is coalescing around the individual nuclei, before cytokinesis, is called a **segmenter.** When development of merozoites is completed, the host cell ruptures, releasing parasite metabolic wastes and residual body, including hemozoin. Metabolic wastes thus released are one factor responsible for the characteristic symptoms of malaria. A great many of the merozoites are ingested and destroyed by reticuloendothelial cells and leukocytes, but, even so, the number of parasitized host cells may become astronomical because erythrocytic schizogony takes only from one to four days, depending on the species. Hemozoin has a toxic effect on macrophages, depressing their effectiveness as phagocytes.[133]

After an indeterminate number of asexual generations, some merozoites enter erythrocytes and become **macrogamonts (macrogametocytes)** and **microgamonts (microgametocytes)** (see Plate 1, *21* to *24*). The size and shape of these cells are characteristic for each species (see Table 9.1); they also contain hemozoin. Unless they are ingested by a mosquito, gametocytes soon die and are phagocytized by cells of the reticuloendothelial system.

Invertebrate Stages. When erythrocytes containing gametocytes are imbibed by an unsuitable mosquito, they are digested along with the blood. However, if a susceptible species of mosquito is the diner, gametocytes develop into gametes. Suitable hosts for the *Plasmodium* spp. of humans are a wide variety of *Anopheles* spp. (see chapter 39). After release from its enclosing erythrocyte, a macrogametocyte matures to a macrogamete in a process involving little obvious change other than a shift of its nucleus toward the periphery. In contrast, the microgametocyte displays a rather astonishing transformation, **exflagellation.** As a microgametocyte becomes extracellular, within 10 to 12 minutes its nucleus divides repeatedly to form six to eight daughter nuclei, each of which is associated with elements of a developing axoneme. The doubled outer membrane of the microgametocyte becomes interrupted; flagellar buds with their associated nuclei move peripherally between the interruptions and then continue outward covered by the outer membrane of the gametocyte. These break free and are then microgametes.

The stimulus for exflagellation is an increase in pH caused by escape of dissolved carbon dioxide from the blood.[88] The life span of microgametes is short since they contain little more than nuclear chromatin and a flagellum covered by a membrane. A microgamete swims about until it finds a macrogamete, which it penetrates and fertilizes. The resultant diploid zygote quickly elongates to become a motile ookinete. The ookinete is reminiscent of a sporozoite or merozoite in morphology. It is 10 μm to 12 μm in length and has polar rings and subpellicular microtubules but no rhoptries or micronemes.

The ookinete penetrates the peritrophic membrane in the mosquito's gut and migrates intracellularly and intercellularly[131] to the hemocoel side of the gut. There it begins its transformation into an oocyst. An oocyst (Fig. 9.3) is covered by an electron-dense capsule and soon extends out into the insect's hemocoel. The initial division of its nucleus is reductional; meiosis takes place immediately after zygote formation as in other coccideans.[113] The oocyst reorganizes internally into a number of haploid nucleated masses called **sporoblasts,** and the cytoplasm contains many ribosomes, an endoplasmic reticulum, mitochondria, and other inclusions. Sporoblasts, in turn, divide repeatedly to form thousands of sporozoites[101] (Fig. 9.4). These break out of the oocyst into the hemocoel and migrate throughout the mosquito's body. On contacting the salivary gland, sporozoites enter its channels and can be injected into a new host at the next feeding.

Sporozoite development takes from 10 days to two weeks, depending on the species of *Plasmodium* and temperature. Once infected, a mosquito remains infective for life, capable of transmitting malaria to every susceptible vertebrate it bites. *Anopheles* spp. that are good vectors for human malaria live long enough to feed on human blood repeatedly. Infection appears to stimulate mosquitoes to feed more frequently, thus increasing the chance of transmission.[83]

Plasmodium sometimes is transmitted by means other than the bite of a mosquito. The blood cycle may be initiated by blood transfusion, by syringe-passed infection among drug addicts, in laboratory accidents,[43] or, rarely, by congenital infection.

Classification of *Plasmodium*

Genus *Plasmodium* was divided by Garnham[31] into nine subgenera, of which three occur in mammals, four in birds, and two in lizards. Most *Plasmodium* spp. are parasites of birds; others occur in such animals as rodents, primates, and

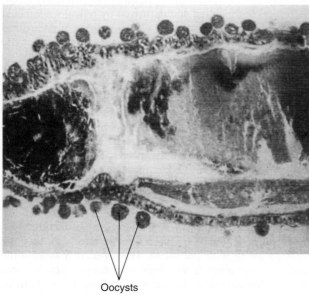
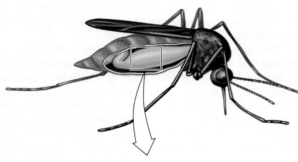

Oocysts

Figure 9.3 **Longitudinal section of a mosquito intestine with numerous oocysts of *Plasmodium* sp.**

Mosquito drawing by William Ober and Claire Garrison; photo from H. Zaiman (Ed.), A *pictorial presentation of parasites.*

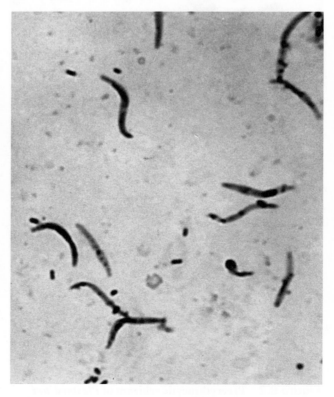

Figure 9.4 *Plasmodium* **sporozoites.**

Courtesy of Peter Diffley.

reptiles. Some species, such as the rodent parasite *P. berghei* and the chicken parasite *P. gallinaceum,* are very useful in laboratory studies of immunity, physiology, and so forth. Still other species normally parasitic in nonhuman primates occasionally infect humans as zoonoses or can be acquired by humans when infected experimentally.

There are five species of *Plasmodium* normally parasitic in humans: *P. falciparum, P. vivax, P. malariae, P. knowlesi,* and *P. ovale.* Molecular analysis of the small subunit rRNA gene of three of these species suggests they are members of separate phylogenetic lineages, each more closely related to *Plasmodium* of other animal parasites than to each other.[139] According to this analysis, *Plasmodium falciparum* is apparently most closely related to *P. gallinaceum* and *P. lophurae* from birds, *P. malariae* seems to form a lineage of its own, and *P. vivax* seems most closely related to several species from other primates. Insufficient data are available to place *P. ovale* in a lineage. On the other hand, studies on base sequences of the mitochondrial cytochrome *b* gene place *P. falciparum* in a clade with *P. reichenowi* (from chimpanzees), not related to *Plasmodium* spp. of birds or reptiles

and only distantly related to other plasmodia of mammals.[97] This study also supports placement of *Hepatocystis* spp. and *Haemoproteus* spp. (p. 159) in separate clades with different species of *Plasmodium,* thus making genus *Plasmodium* paraphyletic.

Plasmodium vivax. *Plasmodium vivax* (see Plates 1 and 2) causes **benign tertian malaria,** also known as **vivax malaria** or **tertian ague** (pronounced ag-yoo). When early Italian investigators noted the actively motile trophozoites of the organism within host corpuscles, they nicknamed it *vivace,* foreshadowing the Latin name *vivax,* which later was accepted as its epithet. It is called *tertian* because fever paroxysms typically recur every 48 hours (Fig. 9.5); the name is derived from the ancient Roman custom of calling the day of an event the first day, making 48 hours later the third day.

The species flourishes best in temperate zones, rarely as far north as Manchuria, Siberia, Norway, and Sweden and as far south as Argentina and South Africa. Because malaria eradication campaigns have been so successful in many of the temperate areas of the world, however, the disease has

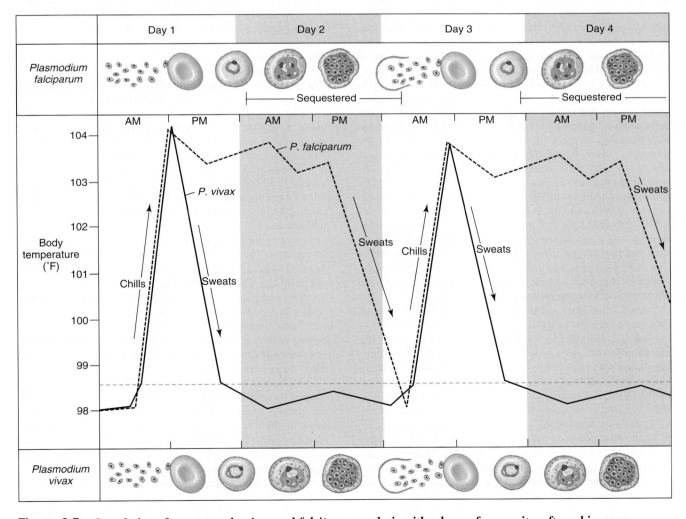

Figure 9.5   Correlation of paroxysms in *vivax* and *falciparum* malaria with release of merozoites after schizogony.
The periodicity of both is about 48 hours, but the hot phase in *P. falciparum* infection is more drawn out; a patient experiences little relief between paroxysms.

Redrawn from L. J. Bruce-Chwatt, *Essential malariology.* London: William Heinemann Medical Books Ltd., 1980.

practically disappeared from them. Most vivax malaria today is found in Asia; about 40% of malaria among U.S. military personnel in Vietnam resulted from *P. vivax.*[11] It is common in North Africa but drops off in tropical Africa to very low levels, partly because of a natural resistance of black people to infection with this species (see Duffy blood groups). About 43% of malaria in the world is caused by *P. vivax.*

Sporozoites that are 10 μm to 14 μm long invade cells of the liver parenchyma within one or two days of injection with a mosquito's saliva. By the seventh day the exoerythrocytic schizont is an oval body about 40 μm long that has blue-staining cytoplasm, a few large vacuoles, and lightly staining nuclei. On attaining maturity, the schizonts' vacuoles disappear, and about 10,000 merozoites are produced. The fate of these merozoites is a subject of debate. Certainly many of them are killed outright by host defenses. Others invade erythrocytes to initiate erythrocytic stages of development. Still others may remain in hepatic cells as hypnozoites (see "Relapse in Malarial Infections," p. 154).

Relapses up to eight years after initial infection are characteristic of vivax malaria. A patient is in normal health during intervening periods of latency. Relapses are believed to result from genetic differences in the original sporozoites; that is, some give rise to tissue schizonts that take much longer to mature.[18] However, occurrence of relapses may also be related to the immune state of the host (see discussion of immunity later in this chapter).

Plasmodium vivax merozoites invade only young erythrocytes, the reticulocytes, and apparently are unable to penetrate mature red cells because receptor sites change as the cells mature.[18] Merozoites can only penetrate erythrocytes with mediated receptor sites, such sites being genetically determined.[49] Known as **Duffy blood groups,** two codominant alleles, Fy^a and Fy^b, are recognized by their different antigens. A third allele, Fy, has no corresponding antigen. Fy/Fy genotype is common in West Africans and their descendants (40% or more) and rare in people of European or Asian descent[72] (about 0.1%). Fy^a and Fy^b proteins are receptors for *P. vivax* and *P. knowlesi;*[81] hence, Fy/Fy individuals have no such receptors on their red cells and are refractory to infection. Receptors (Fy^a and Fy^b) normally bind two chemotactic cytokines that mediate inflammation on leukocytes, but their physiological function on erythrocytes is unknown.[49]

Soon after invasion of erythrocytes and formation of ring stages, the parasites become actively ameboid, throwing out pseudopodia in all directions and fully justifying the name *vivax.* As a trophozoite grows, the red cell enlarges, loses its pink color, and develops a peculiar stippling known as **Schüffner's dots** (see Plate 1, 5). These dots are visible by light microscopy after Romanovsky staining. With electron microscopy they can be seen as small surface invaginations (**caveolae**) surrounded by small vesicles.[110] A trophozoite occupies about two-thirds of the red cell after 24 hours. Its vacuole disappears, it becomes more sluggish, and hemozoin granules accumulate as the trophozoite grows. By 36 to 42 hours after infection, nuclear division begins and is repeated several times, yielding 12 to 24 nuclei in mature schizonts. Once schizogony begins, hemozoin granules accumulate in two or three masses in the parasite, ultimately to be left in a residual body (see Fig. 4.7) and engulfed by the host's reticuloendothelial system. The rounded merozoites, about 1.5 μm in diameter, immediately attack new

erythrocytes. Erythrocytic schizogony takes somewhat less than 48 hours, although early in the disease there are usually two populations, each maturing on alternate days, resulting in a daily, or quotidian, periodicity (refer to discussion of pathology later in the chapter).

Some merozoites develop into gametocytes rather than into schizonts. Factors determining the fate of a given merozoite are not known, but since gametocytes have been found as early as the first day of parasitemia in rare instances, it may be possible for exoerythrocytic merozoites to produce gametocytes. The appearance of micro-and macrogametocytes differs (see Table 9.1). A mature macrogametocyte fills most of the enlarged erythrocyte and measures about 10 μm wide. Mature microgametocytes are smaller than macrogametocytes and usually do not fill an erythrocyte.

Gametocytes take four days to mature, twice the time required for schizont maturation. Macrogametocytes often outnumber microgametocytes two to one. A single host cell may contain both a gametocyte and a schizont.

Formation of zygote, ookinete, and oocyst are as described previously. Oocysts may reach a size of 50 μm and produce up to 10,000 sporozoites. If ambient temperature is too high or too low, the oocyst blackens with pigment and degenerates, a phenomenon noted by Ross. Too many developing oocysts kill a mosquito before sporozoites complete development.

Plasmodium falciparum. Malaria known as **malignant tertian, subtertian,** or **estivoautumnal (E-A)** is caused by *P. falciparum* (see Plates 3 and 4), the most virulent of *Plasmodium* spp. in humans. It was nearly cosmopolitan at one time, with a concentration in the tropics and subtropics. It still extends into the temperate zone in some areas, although it has been eradicated in the United States, the Balkans, and around the Mediterranean. Nevertheless, falciparum malaria reigns supreme as *the* greatest killer of humanity in the tropical zones of the world today, accounting for about 50% of all malaria cases.

Among the many cases studied by Laveran, persons suffering from "malignant tertian malaria" interested him the most. He had long noticed a distinct darkening of the gray matter of the brain and abundant pigment in other tissues of his deceased patients. When in 1880 he saw crescent-shaped bodies in blood and watched them exflagellate, he knew he had found living parasites. Confusion that surrounded the correct name for this species continued until 1954, when the International Commission of Zoological Nomenclature validated the epithet *falciparum.*

Malignant tertian malaria is usually blamed for the decline of the ancient Greek civilization, the halting of Alexander the Great's progress to the East, and the disintegration of some of the Crusades. In more modern times, the Macedonian campaign of World War I was destroyed by falciparum malaria, and this disease caused more deaths than did battles in some theaters of World War II.

As in other species, exoerythrocytic schizonts of *P. falciparum* grow in liver cells. They are more irregularly shaped than those of *P. vivax,* with projections extending in all directions by the fifth day. A schizont ruptures in about five and one-half days, releasing about 30,000 merozoites. True relapses do not occur; however, recrudescences of the disease may follow remissions of up to a year, occasionally up to two or three years, after initial infection, apparently because small populations of the parasites remain in red blood cells.

Merozoites of *P. falciparum* can invade erythrocytes by any of at least four different pathways, in contrast to *P. vivax*, which is limited to a single receptor.[19] Therefore, falciparum malaria usually has much higher levels of parasitemia than the other types. Soon after invasion of an erythrocyte, a trophozoite produces proteins that are deposited beneath and within the erythrocyte surface membrane in deformations called *knobs*.[25] One or more of these proteins bind to certain glycoproteins on postcapillary venular endothelium. This binding causes **sequestration** of infected erythrocytes along the venular endothelium. Gametocyte-infected erythrocytes have no knobs and do not stick to endothelium. Hence, one usually observes only early ring stages and/or gametocytes in blood smears from patients with falciparum malaria. If schizogony is well synchronized, parasites may be practically absent from peripheral blood toward the end of a 48-hour cycle. Infected red cells are antigenically distinct from uninfected erythrocytes, and sequestration may decrease clearance of infected cells by phagocytes in the spleen.[5, 21] Infected erythrocytes can also bind to uninfected red cells, forming rosettes, which may also play a role in clogging venules.[136] A number of receptors have been described on the surface of endothelial cells, and some of these can be upregulated by cytokines such as IL-1, TNF, and IFN-γ. Because of sequestration, parasites may be difficult to demonstrate in circulating blood, and examination of skin biopsies may be helpful in diagnosis.[85]

The early ring-stage trophozoite is the smallest of any *Plasmodium* spp. of humans: about 1.2 μm. There are other diagnostic aids for ring stages of *P. falciparum* (see Table 9.1 and Plate 3). The frequency of multiple infections in the same cell has led some parasitologists to believe that the ring stages divide and that the binucleate rings are division stages. As they grow, trophozoites extend wispy pseudopodia, but they are never as active as those of *P. vivax*. An infected erythrocyte develops irregular blotches known as **Maurer's clefts** (see Plate 3, *9*). These are much larger than the fine Schüffner's dots found in *P. vivax* infections. They are apparently associated with the **tubovesicular membrane network,** which is continuous with the parasitophorous vacuole and extends toward the erythrocyte membrane.[25, 134] This network probably has a significant role in transport of nutrient molecules to the parasite and export of *P. falciparum* molecules to the surface of the red cell.

Mature schizonts are less symmetrical than those of the other species infecting humans. They develop 8 to 32 merozoites, with 16 being the usual number. In contrast to the usual situation, schizonts may be fairly common in peripheral blood in some geographical areas. This may reflect strain differences. The erythrocytic cycle takes 48 hours, but periodicity is not as marked as in *P. vivax,* and it may vary considerably with the strain of parasite. Extremely high levels of parasitemia may occur, with more than 65% of erythrocytes containing parasites; a density of 25% is usually fatal. Two or three parasites per milliliter of blood may be sufficient to cause disease symptoms.

In *P. vivax* gametocytes may appear in peripheral blood almost at the same time as the trophozoites, but in *P. falciparum* sexual stages require nearly 10 days to develop, and then they appear in large numbers. They develop in blood spaces of spleen and bone marrow—first assuming bizarre, irregular shapes and then becoming round and finally changing into the crescent shape so distinctive of the species (see Plate 3, *26* and *27*). Hemozoin granules cluster around the nucleus in micro- and macrogametocytes. This distribution differs from that of *P. vivax,* in which pigment is diffuse throughout the cytoplasm (see Table 9.1).

Plasmodium malariae. Quartan malaria, with paroxysms every 72 hours, is caused by *P. malariae* (see Plates 5 and 6). It was recognized by early Greeks because the timing of fevers differed from that of the tertian malaria parasites. Although Laveran saw and even illustrated the characteristic schizonts of this parasite, he refused to believe it was different from *P. falciparum*. In 1885 Golgi differentiated the tertian and quartan fevers and gave an accurate description of what is now known as *P. malariae.*

Plasmodium malariae is a cosmopolitan parasite but does not have a continuous distribution anywhere. It is common in many regions of tropical Africa, Myanmar (formerly Burma), India, Sri Lanka, Malaya, Java, New Guinea, and Europe. It is also distributed in the New World, including Guadeloupe, Guyana, Brazil, Panama, and at one time the United States. The peculiar distribution of this parasite has never been satisfactorily explained. It may be the only species of human malaria organism that also regularly lives in wild animals. Chimpanzees are infected at about the same rate as humans but are unimportant as reservoirs, since they do not live side by side with people. Some workers believe that *P. brasilianum* is really *P. malariae* in New World monkeys.[61] As noted before, present molecular evidence suggests that *P. malariae* is alone in its evolutionary lineage. This species accounts for about 7% of malaria cases in the world.

Exoerythrocytic schizogony is completed in 13 to 16 days. Erythrocytic forms build up slowly in the blood; the characteristic symptoms of the disease may appear before it is possible to find the parasites in blood smears. The ring forms are less ameboid than those of *P. vivax,* and their cytoplasm is somewhat thicker. Rings often retain their shape for as long as 48 hours, finally transforming into an elongated "band form," which begins to collect pigment along one edge (see Plate 3, *6* and *10*). Their nucleus divides into 6 to 12 merozoites at 72 hours. The segmenter is strikingly symmetrical and is called a *rosette* or *daisy-head* (see Plate 5, *20*). Parasitemia levels are characteristically low, with one parasite per 20,000 red cells representing a high figure for this species. This low density is accounted for by the fact that merozoites apparently can invade only aging erythrocytes, which are soon to be removed from circulation by the normal process of blood destruction.

Gametocytes probably develop in internal organs, since immature forms are rare in peripheral blood. They are slow to develop in sporozoite-induced infections. Recrudescences of quartan malaria can occur up to 53 years after initial infection.[30] Because *P. malariae* can live in blood so long, it is the most important cause of transfusion malaria.

Plasmodium ovale. This species causes **ovale,** or **mild tertian, malaria** and is rarest of the four malaria parasites of humans. It is confined mainly to the tropics, although it has been reported from Europe and the United States. Although common on the west coast of Africa, which may be its original home, this species is scarce in central Africa and present but not abundant in eastern Africa. It is known also in India, the Philippine Islands, New Guinea, and Vietnam. *Plasmodium ovale* is difficult to diagnose because of its similarity to *P. vivax* (see Table 9.1). Use of acridine-orange staining and PCR-based methods have shown that *P. ovale* is much more widespread and prevalent in East Asia than thought previously.[147]

The youngest ring stage has a large, round nucleus and a rather small vacuole that disappears early. Mature schizonts are oval or spheroid and are about half the size of the host cell. Eight merozoites are usually formed, but there is a range of 4 to 16. Schüffner's dots appear early in infected blood cells. They are very numerous and larger than those in *P. vivax* infections and stain a brighter red color. As in *P. vivax,* Schüffner's dots are due to caveolae.

Gametocytes of *P. ovale* take longer to appear in blood than do those of other species. They are numerous enough three weeks after infection to infect mosquitoes regularly.

Plasmodium knowlesi. This species has been known as a parasite of macaque monkeys in southeast Asia. When it was found in humans it was, until recently, misdiagnosed as *Plasmodium malariae.* Now *P. knowlesi* has become recognized as an important, and sometimes fatal, parasite in humans in Malaysia, and also Thailand, Myanmar, Singapore, the Philippines, Vietnam, and Indonesia.[20, 114, 115] Microscopically, it closely resembles *P. malariae,* but *P. knowlesi* can be distinguished by a PCR assay. Clinically, *P. knowlesi* differs from *P. malariae* in a critical respect: *P. knowlesi* has a 24-hour schizogonic cycle, rather than a 72-hour cycle, which not only does not allow a sufferer a respite between fever days, but the population of malarial organisms builds more rapidly and can lead to death.

Malaria: The Disease. Certain disease aspects of *Plasmodium* spp. have been mentioned in the preceding pages; following is a brief consideration of the subject, particularly in relation to pathogenesis and public health. We urge you to consult other references for more complete treatment.[123, 142]

Diagnosis. Diagnosis depends to some extent on the clinical manifestations of the disease, but most important is demonstration of the parasites in stained smears of peripheral blood. Technical details can be found in many texts and laboratory manuals of medical parasitology. A number of criteria are useful for differential diagnosis (see Table 9.1).

Diagnosis by this conventional method requires training and time, and very low parasitemias are easily missed. A number of techniques using newer methods have been described. An inexpensive method of visualizing the parasites after staining with a fluorescent dye is simple, very sensitive, and adaptable for field conditions.[55] A DNA probe specific for the detection of *P. falciparum* is sensitive and suitable for field conditions.[4] Diagnostic methods based on polymerase chain reaction have been described.[92, 130] In chapter 3 (p. 35) we explained a dipstick method for detecting *P. falciparum* antigen[111] which compares favorably in accuracy and sensitivity to both microscopical diagnosis and PCR methods.[50] Available dipstick methods were reviewed by Wongsrichanalai.[144]

Pathogenesis. Most major clinical manifestations of malaria may be attributed to two general factors: (1) the host inflammatory response, which produces the characteristic chills and fever as well as other related phenomena, and (2) anemia, arising from the enormous destruction of red blood cells. Severity of the disease is related to the species producing it: Falciparum malaria is most serious, and quartan and ovale are the least dangerous.

Malaria is characterized by overproduction of proinflammatory cytokines of the innate immune system (p. 27).[86] The characteristic paroxysms of fever in malaria closely follow maturation of each generation of merozoites and rupture of red blood cells that contain them (schizont burst). Glycophosphatidylinositols (GPIs, p. 29) specific to the parasite are released along with cellular debris, host and parasite membranes, and hemozoin. GPIs are the dominant parasite-associated molecular pattern recognized by host monocytes and macrophages.[39] The most important receptors on macrophage surfaces are TLR2 dimerized with either TLR1 or TLR6. Activation requires MyD88 adaptor protein and initiates MAPK and NF-kB signaling pathways, which stimulate a burst of pro-inflammatory mediators. These include TNF, IL-6, IL-12, IL-1, IFN-γ, and nitric oxide synthase.[39] Some evidence suggests that hemozoin is also responsible for the burst of TNF.[99] TNF toxicity can account for most or all of the symptoms described in the next few paragraphs.[17]

A few days before the first paroxysm, a patient may feel malaise, muscle pain, headache, loss of appetite, and slight fever; or the first paroxysm may occur abruptly, without any prior symptoms. A typical attack of benign tertian or quartan malaria begins with a feeling of intense cold as the hypothalamus, the body's thermostat, is activated, and the temperature then rises rapidly to 104°F to 106°F. The teeth chatter, and the bed may rattle from the victim's shivering. Nausea and vomiting are usual. The hot stage begins within one half to one hour later, with intense headache and feeling of intense heat. Often a mild delirium stage lasts for several hours. As copious perspiration signals the end of the hot stage, the temperature drops back to normal within two to three hours, and the entire paroxysm is over within 8 to 12 hours. A person may sleep for a while after an episode and feel fairly well until the next paroxysm. Time periods for stages are usually somewhat shorter in quartan malaria, and paroxysms recur every 72 hours. In vivax malaria periodicity is often quotidian early in the infection because two populations of merozoites usually mature on alternate days. "Double" and "triple" quartan infections also are known. Only after one or more groups drop out does fever become tertian or quartan, and a patient experiences the classical good and bad days.

Because synchrony in falciparum malaria is much less marked, the onset is often more gradual, and the hot stage is extended (see Fig. 9.5). Fever episodes may be continuous or fluctuating, but a patient does not feel well between paroxysms, as in vivax and quartan malaria. In cases in which some synchrony develops, each episode lasts 20 to 36 hours, rather than 8 to 12, and is accompanied by much nausea, vomiting, and delirium. Concurrent infections with more than one species were formerly thought to occur in less than 2% of patients, but use of sensitive PCR techniques has shown they may be as frequent as 65%.[73] Occasionally, all four species may be present.

Falciparum malaria is always serious, and sometimes severe complications occur. Although severe malaria develops in only about 1% of patients, it causes around one million deaths in sub-Saharan Africa alone.[65, 120] Severe malaria traditionally has been understood as caused by two major syndromes: (1) severe anemia resulting from destruction of red blood cells, and (2) cerebral malaria, primarily a result of blockage of small blood vessels in the brain by sequestration of infected red blood cells (Fig. 9.6). However, in recent years there has been increased realization that severe malaria is a complex, multisystem disease (Table 9.2). Release of proinflammatory cytokines, such as TNF and IFNγ, cause serious metabolic changes.

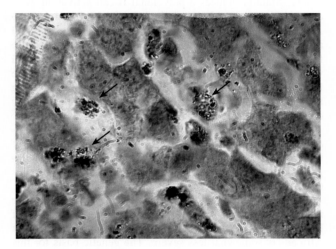

Figure 9.6   **Section of liver tissue with numerous deposits of malarial pigment (*arrows*).**

Photograph by Larry S. Roberts.

The main causes of the anemia are destruction of both parasitized and nonparasitized erythrocytes, inability of the body to recycle the iron bound in hemozoin, and an inadequate erythropoietic response of the bone marrow. Why such large numbers of nonparasitized red cells are destroyed is still not understood, but some evidence has indicated complement-mediated, autoimmune hemolysis. In acute malaria the spleen removes substantial numbers of unparasitized red cells from the blood, an effect that may persist beyond the time of parasite clearance.[14] Both the splenic removal of red cells and the defective bone marrow response may be due in part to TNF toxicity.[16, 77] Destruction of erythrocytes leads to an increase in blood bilirubin, a breakdown product of hemoglobin. When excretion cannot keep up with formation of bilirubin, jaundice yellows the skin. Hemozoin is taken up by circulating leukocytes and deposited in the reticuloendothelial system. In severe cases the viscera, especially the liver, spleen, and brain, become blackish or slaty as the result of pigment deposition (Fig. 9.7, see Plate 8). After ingesting hemozoin, macrophages suffer impairment in phagocytic ability.[133]

Hypoglycemia (reduced concentration of blood glucose) is a common symptom in falciparum malaria. It is usually found in women with uncomplicated or severe malaria who are pregnant or have recently delivered as well as in other cases of severe falciparum malaria.[145] Coma produced by hypoglycemia has commonly been misdiagnosed as cerebral malaria. This condition is usually associated with quinine treatment. Pancreatic islet cells are stimulated by quinine to increase insulin secretion, thus lowering blood glucose.[138] This effect may also be due to excessive TNF.[16]

Immunity and Resistance. Despite the fact that much of the disease results from inflammatory and immune responses of the host, host defenses are vital in limiting the infection.

Table 9.2   Clinical Features of Malaria and Disease Mechanisms*

Syndromes	Clinical Features	Disease Mechanisms
Severe anaemia	Shock; impaired consciousness; respiratory distress	Reduced RBC production (reduced erythropoietin activity, proinflammatory cytokines); increased RBC destruction (parasitemediated, erythrophagocytosis; antibody and complement-mediated lysis)
Cerebral complications (cerebral malaria)	Impaired consciousness; convulsions; long-term neurological deficits	Microvascular obstruction[26] (parasites, platelets, rosettes, microparticles); proinflammatory cytokines; parasite toxins (i.e., GPIs)
Metabolic acidosis	Respiratory distress, low blood oxygen, rapid breathing, high lactic acid in blood, reduced central venous pressure	Reduced circulation to tissues (low blood volume, reduced cardiac output, anemia); parasite products; proinflammatory cytokines; lung pathology (airway obstruction, reduced diffusion)
Other	Low blood sugar; disseminated intravascular coagulation	Parasite products and/or toxins; proinflammatory cytokines; cytoadherence
Malaria in pregnancy	Placental infection; low birth weight and fetal loss; maternal anemia	Premature delivery and fetal growth restriction; placental infiltration by leucocytes and inflammation; proinflammatory cytokines

*Modified from Mackintosh et al.[65] See[65] for references.

Abbreviations: RBC, red blood cell; GPIs, glycosylphosphatidylinositols (anchor malarial proteins to the cell membrane, but many have no associated protein; potent stimulators of innate immune system)[86]

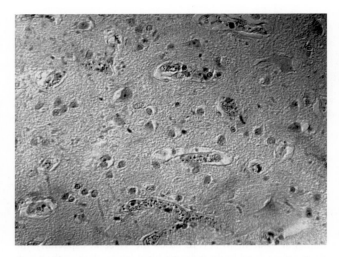

Figure 9.7 **Section of cerebral tissue, demonstrating capillaries filled with erythrocytes infected by *Plasmodium falciparum*.**
Infected red cells are marked by pigment; the parasites themselves are transparent.

Photograph by Larry S. Roberts.

One vivax segmenter producing 24 merozoites every 48 hours would give rise to 4.59 billion parasites within 14 days, and the host would soon be destroyed if the organisms continued reproducing unchecked.[61] Development of some protective immunity is evident in malaria, and we will consider only briefly some practical effects.

Relapses and recrudescences may be due to lowered antibody titers or increased ability of the parasite to deal with the antibody, but they also may depend on genetic variation of the parasites to evade host immune defenses.[79] Variant antigens in *P. falciparum* are encoded by a large family (about 50) of genes called *var*. Proteins encoded by *var* genes are incorporated into the erythrocyte membrane where they mediate cytoadhesion to vascular endothelium and to uninfected erythrocytes. Only one gene is expressed at a given time, and parasite switching to another *var* gene appears to be responsible for recrudescence.[94, 124]

Symptoms in a relapse are usually less severe than those in the primary attack, but the level of parasitemia is higher. After the primary attack and between relapses, a patient may have a *tolerance* to effects of the organisms, remaining asymptomatic while, in fact, having as high a circulating parasitemia level as during a primary attack. Such tolerant carriers are very important in epidemiology of the disease. Tolerance may be related to loss of reactivity to TNF; humans can become refractory to TNF on continued exposure.[16]

Protective immunity to malaria is primarily a premunition (p. 24)—that is, a resistance to superinfection, while the host's immune response controls numbers of parasites remaining in its body. Premunition is effective only as long as a residual population of parasites is present; if a person is completely cured, susceptibility returns.[93] Thus, in highly endemic areas, infants are protected by maternal antibodies, and young children are at greatest risk after weaning. Immunity of children who survive a first attack will be continuously stimulated by bites of infected mosquitoes as long as they live in the malarious area. Nonimmune adults are highly susceptible. Protective immunity apparently has some components that are species, strain, and variant specific, but there is now evidence that existing infection with *P. vivax* can provide some protection against infection with *P. falciparum* or, at least, prevent severe symptoms.[66]

Protective mechanisms by immune effectors against the parasites remain unclear. In vitro binding of specific antibodies to surface proteins of sporozoites and merozoites can prevent penetration of host cells, and there is some evidence for an antibody-dependent, cell-mediated cytotoxicity (ADCC).[76] At least part of sporozoite-induced immunity depends on the killing of infected liver cells by cytotoxic T lymphocytes.[24] However, the immune response is inefficient: Malaria induces a polyclonal B-cell activation, with dramatic synthesis of especially IgG and IgM, only 6% to 11% of which is specifically against malarial antigens.[66] IgG and IgE levels differ markedly in uncomplicated and severe falciparum malaria. Both total IgG and antiplasmodial IgG were higher in patients with uncomplicated malaria, while IgE was highest in the group with severe disease, suggesting that IgG may play a role in reducing severity, while IgE may contribute to pathogenesis.[98]

It is probable that an important mechanism for *Plasmodium* to evade the host defense system involves exposure of the host to a large repertoire of antigenic epitopes.[118] Analogous systems of immune evasion are shown in trypanosomes (p. 68) and schistosomes (p. 38).

West Africans and their descendants elsewhere are much less susceptible to vivax malaria than are people of European or Asian descent, and falciparum malaria in West Africans is somewhat less severe. The genetic basis for this phenomenon in the case of vivax malaria is explained by the inheritance of Duffy blood groups (p. 150). Other factors that can contribute to genetic resistance are certain heritable anemias (sickle cell, favism, and thalassemia) and several other heritable traits.[140] Although these conditions are of negative selective value in themselves, they have been selected for in certain populations because they confer resistance to falciparum malaria.

The most well known of these genetic conditions is **sickle-cell anemia.** In persons homozygous for this trait a glutamic acid residue in the amino acid sequence of hemoglobin is replaced by a valine, interfering with the conformation of the hemoglobin and oxygencarrying capacity of the erythrocytes. Individuals with sicklecell anemia usually die before the age of 30. In heterozygotes some of the hemoglobin is normal, and such people can live normal lives, but the presence of the abnormal hemoglobin confers 80% to 95% protection against severe malaria.[140] The selective pressure of malaria in Africa has led to maintenance of this otherwise undesirable gene in the population. This legacy has unfortunate consequences when the people are no longer threatened by malaria, as in the United States, where 1 in 10 Americans of African ancestry is heterozygous for the sickle-cell gene, and 1 in 500 is homozygous.[68]

Relapse in Malarial Infections. Since the discovery of an effective antimalarial drug (quinine) in the 16th century, it has been noted that some persons who have been treated and seemingly recovered relapse back into the disease weeks, months, or even years after the apparent cure.[69] Coatney contributed an interesting history of the phenomenon.[18] Malarial relapse engendered much speculation and research for many years. The discovery of preerythrocytic schizogony in the liver by Shortt and Garnham in 1948 seemed to have solved the mystery. It appeared most reasonable to assume that preerythrocytic merozoites simply reinfected other hepatocytes, with subsequent reinvasion of red blood cells. This would explain why relapse occurred after erythrocytic forms were eliminated by erythrocytic schizontocides, such as quinine and chloroquine.

However, not all species of *Plasmodium* cause relapse. Among the parasites of primates, only *P. vivax* and *P. ovale* of humans and *P. cynomolgi, P. fieldi,* and *P. simiovale* of simians cause true relapse. If preerythrocytic merozoites reinvade hepatocytes, then relapse should occur in all species.

In species that undergo relapse, there are two populations of exoerythrocytic forms.[59] One develops rapidly into schizonts, as previously described, but the other remains dormant as **hypnozoites** ("sleeping animalcules"). *Plasmodium vivax, P. ovale,* and *P. cynomolgi* have hypnozoites, but they have not been found in any species that does not cause relapse. How long hypnozoites can remain capable of initiating schizogony and what triggers them to do so are unknown. Primaquine is an effective hypnozoiticide.[36]

Malariologists long thought that *P. malariae,* a dangerous species in humans, also exhibited relapse, but we now know that this species can remain in blood for years, possibly for the lifetime of a host, without showing signs of disease and then suddenly initiate a clinical condition. This is more correctly known as a *recrudescence,* since preerythrocytic stages are not involved. The danger of transmission of this parasite through blood transfusion is evident. Because there are no hypnozoites, treatment of *P. malariae* with primaquine is unnecessary.

Epidemiology, Control, and Treatment. In light of the prevalence and seriousness of the disease, epidemiology and control are extremely important, and thorough consideration is far beyond the scope of this book. Some aspects of these subjects have been touched on in the preceding pages, and the following will give you additional insight into the problem involved (see also chapter 39, and Strickland,[123] and Bruce-Chwatt[10]).

In addition to natural or biological transmission, discussed next, human to human transmission can spread malaria. Accidental transmission can occur by blood transfusion and by the sharing of needles by IV drug users. Although rare, infection of the newborn from an infected mother also occurs.[10] Neurosyphilis was formerly treated by deliberate infection with malaria. (A great deal of knowledge about malaria was gained during these treatments, but we still do not understand why infection with malaria alleviated the symptoms of the terrible disease of neurosyphilis.)[15]

A variety of interrelated factors contributes to the level of natural transmission of the disease in a given area (Fig. 9.8). It is necessary to study and understand all these factors

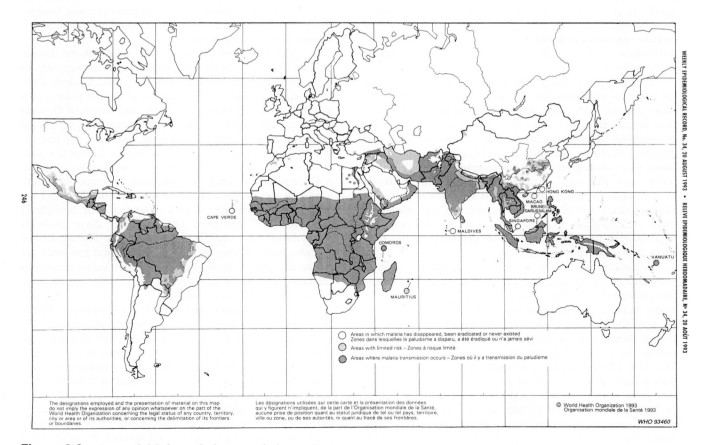

Figure 9.8 **Areas of risk for malaria transmission, 1991.**

Reproduced by permission of the World Health Organization, from World malaria situation in 1991, Parts I and II. *Weekly Epidemiological Record* 68 (245–252/253–260, 1993).

thoroughly before undertaking a malaria control program with any hope of success. Following (modified from Strickland[123]) are the most important factors:

- Reservoir—the prevalence of the infection in humans, including persons with symptomatic disease and tolerant individuals and, in some cases, other primates, with high enough levels of parasitemia to infect mosquitoes.
- Vector—suitability of the local anophelines as hosts; their breeding, flight, and resting behavior; feeding preferences; and abundance.
- New hosts—availability of nonimmune hosts.
- Local climatic conditions.
- Local geographical and hydrographical conditions and human activities that determine availability of and accessibility to mosquito breeding areas.

Abundance of appropriate vectors has crucial importance to endemicity. In many areas reproduction of mosquitoes, and thus transmission, is virtually constant, with abundant rainfall throughout the year and/or water available in irrigation ditches or ponds. Humans are subjected to a high entomologic inoculation rate and develop an immunity that can block transmission to mosquitoes, even while they may have a high circulating parasitemia.[8] Young children are at greatest risk. These conditions are described as **stable endemic malaria.**[123] In other areas transmission may be seasonal, being interrupted by a dry or cool period, or may vary from year to year. With a low entomologic inoculation rate, people do not become resistant, symptoms are usually much more serious, and epidemics often occur. These conditions produce **unstable malaria.** Climate change may affect the distribution of stable malaria in Africa, but probably not in the next few decades.[129]

Of the approximately 390 species of *Anopheles,* some are more suitable hosts for *Plasmodium* than are others. Of those that are good hosts, some prefer animal blood other than human; therefore, transmission may be influenced by the proximity in which humans live to other animals. The preferred breeding and resting places are very important. Some species breed only in fresh water; some prefer brackish water; some like standing water around human habitations, such as puddles or trash that collects water such as bottles and broken coconut shells. Water, vegetation, and amount of shade are important, as are whether a species enters dwellings and rests there after feeding and whether a species flies some distance from breeding areas.

Anopheles spp. show an astonishing variety of such preferences; two specific examples can be cited for illustration. *Anopheles darlingi* is the most dangerous vector in South America, extending from Venezuela to southern Brazil, breeding in shady fresh water among debris and vegetation. It invades houses and prefers human blood. *Anopheles bellator* is an important vector in cocoa-growing areas of Trinidad and coastal states of southern Brazil, breeding in partial shade in the "vases" of epiphytic bromeliads (plants that grow attached to trees and collect water in the center of their leaf rosettes). It prefers humans but enters dwellings only occasionally and then returns to the forest. The importance of thorough investigation of such factors is demonstrated by cases in which swamps have been flooded with seawater to destroy the breeding habitat of the species, only to create extensive breeding areas for a brackish water species that turned out to be just as effective a vector.

Valuable actions in mosquito control include destruction of breeding places when possible or practical, introduction of mosquito predators such as the mosquito-eating fish *Gambusia affinis,* and judicious use of insecticides. The efficacy and economy of DDT have been a boon to such efforts in underdeveloped countries. Although we now are aware of environmental dangers of DDT, these dangers may be preferable to and minor compared with the miseries of malaria. Unfortunately, reports of DDT-resistant strains of *Anopheles* are increasing, and this phase of the battle is becoming more difficult. For exterminating susceptible *Anopheles* spp. that enter dwellings and rest there after feeding, spraying insides of houses with residual insecticides can be effective and cheap, without incurring any environmental penalty. Unfortunately, some *Anopheles* rest in houses only briefly before or after feeding, and sufficient quantities of DDT are difficult to obtain on the world market.[22]

In the 1980s it was reported that use of bed nets treated with pyrethroid insecticides (insecticide-treated nets, ITNs) could significantly reduce mortality and morbidity of both *P. falciparum* and *P. virax* malaria. Indeed, some authorities assert that "mortality trials showed that ITNs are the most powerful malaria control tool to be developed since the advent of indoor residual spraying . . . and chloroquine, more than four decades earlier."[46] It is estimated that about 370,000 deaths could be avoided per year if all children in sub-Saharan Africa were protected by ITNs. While ITNs will be a key element in malaria control in years to come, troublesome problems remain, such as cost, distribution, and development of insecticide resistance.

Appropriate drug treatment of persons with the disease as well as prophylactic drug treatment of newcomers to malarious areas are integral parts of malaria control. Centuries ago the Chinese used extracts of certain plants, such as *chang shan* and *shun qi* (the roots and leaves of *Dichroa febrifuga,* family Saxifragaceae) and *qing hao* (the annual *Artemisia annua,* family Compositae, Fig. 9.9), that had antimalarial properties.[48, 57] In the meantime Europeans were medically powerless and depended on absurd and superstitious remedies. Extracts of bark from Peruvian trees were used with varying success to treat malaria, but alkaloids from the bark of certain species of *Cinchona* proved dependable and effective.[48] The most widely used of these alkaloids was quinine, discovered in the 16th century. The alkaloid of *D. febrifuga,* febrifugine, is now considered too toxic for human use, but the terpene from *A. annua,* called *qinghaosu* **(artemisinin),** which has recently been "rediscovered," and its derivatives are valuable drugs.

Only two synthetic antimalarials were discovered before World War II. Japanese capture of cinchona plantations early in the war created a severe quinine shortage in the United States, stimulating a burst of investigation that produced a number of effective drugs. The most important of these was **chloroquine.** Subsequently a number of valuable drugs have been developed, including primaquine, mefloquine, pyrimethamine, proguanil, sulfonamides such as sulfadoxine, and antibiotics such as tetracycline. Only primaquine is effective against all stages of all species; the others vary in efficacy according to stages and species, with the erythrocytic stages being most susceptible. Drugs of choice are

Figure 9.9 *Artemisia annua,* the source of the anti-malarial drug qinghaosu, growing in the herb garden of the College of Traditional Medicine, Guangzhou, China.
Photograph by Larry S. Roberts.

chloroquine and primaquine for *P. vivax* and *P. ovale* malarias and chloroquine alone for *P. malariae* infections. Chloroquine is still recommended for strains of *P. falciparum* sensitive to that drug.[32]

Resistance of *P. falciparum* to chloroquine has now spread through Asia, Africa, and South America,[23] and resistance to other drugs is often present. A combination of sulfadoxine and pyrimethamine (Fansidar) was used for chloroquine-resistant falciparum malaria, but Fansidar-resistant *P. falciparum* is now present in a number of areas. For multidrug-resistant *P. falciparum,* mefloquine (Lariam) is still effective, but resistance to mefloquine is established in several endemic areas.[82] Artemisinin and its derivatives are effective for drug-resistant *P. falciparum,* both in severe and uncomplicated malaria.[121] The artemisinin derivative is commonly given in combination with other drugs (artemisinbased combination therapies, ACTs).[27] Dihydroartemisin with piperaquine (Artekin) is an ACT with the important advantage that it is also low cost (U.S. $1 per adult treatment).[84] Resistance to artemisinin has not been reported in the field, but resistant *Plasmodium* strains have been produced in the laboratory.[51] For the most current recommendations on malaria chemotherapy and prophylaxis, consult the U.S. Centers for Disease Control and Prevention, Yellow Book,[12] or the CDC web page at http://www.cdc.gov/malaria/travel. Research continues to develop new drugs and combinations of drugs.[2, 30, 63]

Because of the ominous and dangerous multidrug resistance in various strains of *P. falciparum,* it is clear that the search for satisfactory malaria treatments must continue; perhaps the answer lies in the development of vaccines.

This area is the subject of intensive investigation, and much progress has been made.[41] The current thrust has been made possible in part by the development of methods whereby *P. falciparum* could be cultured in vitro,[132] thus making a large supply of organisms available. However, difficulties have been numerous. Different stages of the parasite have different antigens on their surface, and surface proteins on both sporozoites and merozoites are highly polymorphic.[37, 38] Sporozoites continually slough off their outer coat, restoring it with newly synthesized protein.[35] Thus, sporozoites evade the host defense by producing new binding sites and producing "decoys" in the form of sloughed molecules.

Because *P. vivax* is the second most prevalent species of *Plasmodium,* is responsible for significant morbidity, and is frequently co-endemic with *P. falciparum,* a multispecies vaccine would be very beneficial.[45] The paper by Higgs and Sina[45] is the introduction to an issue of the American Journal of Tropical Medicine and Hygiene that deals almost entirely to the search for a *P. vivax* vaccine. Of course, an effective malaria vaccine would be an enormous help, and strong efforts are currently underway for such a development. Significant progress has been made, although a vaccine that confers 100% protection has not been developed.[102, 116]

In the 1950s many parasitologists thought that an expenditure of effort and money could achieve the eradication of malaria from large areas of the globe: Its scourge would rest only in history. Such views were naively optimistic. Not only did we not anticipate insecticide-resistant *Anopheles* spp., drug-resistant *Plasmodium,* and animal reservoirs, but we took insufficient account of the enormous logistical problems of control in wilderness areas and we failed to consider the disruptive effects of wars and political upheavals on control programs. Malaria will be with us for a long time, probably as long as there are people.

Metabolism, Drug Action, and Drug Resistance

Energy Metabolism. *Plasmodium* spp. derive energy primarily by the degradation of glucose to lactate, even though oxygen is available. Genes encoding proteins necessary for a complete Krebs cycle have been identified, although this pathway may not be involved in energy production.[60] *Plasmodium* species from birds have cristate mitochondria, but they nevertheless depend heavily on glycolysis for energy, converting four to six molecules of glucose to lactate for every one they oxidize completely. Asexual stages of most mammalian species have acristate mitochondria, but sporogonic stages of these organisms in the mosquito possess prominent, cristate mitochondria.[60] This difference might reflect a developmental change in metabolic pattern analogous to that observed in trypanosomes (p. 67). Treatment of a host with qinghaosu leads to swelling of mitochondria in *P. inui* (a monkey parasite with prominent mitochondria) within two and one-half hours.[53] Host mitochondria are unaffected. Similar reactions have been observed after primaquine treatment, leading to the suggestion that these drugs act via inhibition of mitochondrial metabolic reactions.

Erythrocytic forms of *Plasmodium* act as facultative anaerobes, consuming oxygen when it is available. They probably use oxygen for biosynthetic purposes, especially synthesis of nucleic acids. Also, a branched electron transport system, analogous to that suggested for some helminths (p. 320), was proposed,[109] but a classical cytochrome system has not been demonstrated. A limiting factor may be the parasite's inability to synthesize coenzyme A, which it must obtain from its host; this cofactor is necessary to introduce two-carbon fragments into the Krebs cycle. Supplies of CoA in the mammalian erythrocyte may be even more limited and may impose restrictions on any CoA-dependent reaction.

Both bird and mammal plasmodia fix carbon dioxide into phosphoenolpyruvate, as do numerous other parasites (see Fig. 20.28). In plasmodia the carbon dioxide-fixation reaction can be catalyzed by either phosphoenolpyruvate carboxykinase or phosphoenolpyruvate carboxylase. Chloroquine and quinine inhibit both enzymes, possibly accounting for some antimalarial activity of these drugs. The significance of the carbon dioxide fixation is not clearly understood; it may be to reoxidize NADH produced in glycolysis, or its reactions may function to maintain levels of intermediates for use in other cycles.

Activity of the pentose phosphate pathway is low in plasmodia; however, plasmodia have a complete array of pentose pathway enzymes, including glucose 6-phosphate dehydrogenase (G6PDH), 6-phosphogluconate dehydrogenase (6PGDH), transaldolase, and transketolase. Because an important function of the pathway is to furnish reducing power in the form of NADH, it has been suggested that *Plasmodium* gets NADH from its host. Persons with a genetic deficiency in erythrocytic G6PDH are more resistant to malaria than are *G6PDH*[+] homozygotes. Ingestion of various substances such as the antimalarial drug primaquine or the broad bean *Vicia favia* can bring on a hemolytic crisis of varying severity.[68] Such genes are relatively frequent in black people and some Mediterranean whites. Over 5% of Southeast Asian refugees entering the United States have had a G6PDH deficiency.[105] Presence of the deficiency should be determined before treatment with primaquine to avoid a hemolytic crisis.

Protein Degradation. Some 25 different proteases have been described from various species of *Plasmodium*.[9] They are vital in maturation and release of merozoites from red cells, invasion of cells, and digestion of hemoglobin in the food vacuole.

Plasmodia depend heavily on host hemoglobin as a source of nutrition. They ingest a portion of host cytosol via the cytostome, and the vesicle thus formed migrates to and joins the central food vacuole, where the hemoglobin is rapidly degraded.[146] One of the products of hemoglobin digestion is ferriprotoporphyrin IX (FP, heme), but FP inhibits several of the plasmodial proteases and disrupts membranes.[28] Therefore, the parasites sequester FP as insoluble hemozoin. Chloroquine is a dibasic amine (a weak base) and increases pH in a food vacuole and prevents digestion of hemoglobin. It binds to FP and prevents sequestration of the FP into inert hemozoin. Chloroquine is not effective against *Plasmodium* stages that do not form hemozoin. Mefloquine also affects the food vacuoles,[52] and quinine may act by a similar mechanism.[64] The mechanism of action of artemisinin and its analogs apparently is inhibition of heme polymerization into hemozoin.[95]

In chloroquine-sensitive strains of *P. falciparum,* chloroquine accumulates in the food vacuole, but in chloroquine-resistant strains, the drug moves out again just as rapidly.

The rapid-efflux phenotype is apparently due to a mutation at a single genetic locus that spread rapidly from one or two foci in Southeast Asia and South America.[141] The mechanism of the efflux is still controversial. It is reversed (and chloroquine sensitivity is regained) by verapamil and other Ca^+ channel blockers.[34] Some scientists believe that interaction with a permease on the lysosomal membrane may be involved.[137] The mechanism of resistance of *P. falciparum* to mefloquine is distinct from that to chloroquine, but it is associated with overexpression of a multidrug resistance gene (*pfmdr 1*).[82] Given the multidrug resistance developed by *P. falciparum,* it is curious that the parasite remains susceptible to quinine after 350 years of use.[78]

Resistance to *P. falciparum* by persons homozygous and heterozygous for sickle-cell hemoglobin (HbS) may involve several mechanisms, partly involving feeding and digestion by the protozoa. The parasite develops normally in cells with HbS until those cells are sequestered in the tissues.[29] Kept in this low-oxygen environment for several hours, the cells have more of a tendency to sickle than do cells that pass through at a normal rate. When sickling occurs, HbS forms filamentous aggregates. The filamentous aggregates actually pierce the *Plasmodium*, apparently releasing digestive enzymes that lyse both parasite and host cell. K^+ leaks out of the sickled cell, depriving the parasite of this ion. Sickled cells also may block capillaries, further decreasing local oxygen concentration. Finally, sickling denatures hemoglobin and releases ferriprotoporphyrin IX, which has a membrane toxicity;[91] the effect of FP on plasmodial proteases was already mentioned.

Synthetic Metabolism. As specialized parasites, *Plasmodium* species depend on their host cells for a variety of molecules other than the strictly nutritional ones. Specific requirements for maintenance of the parasites free of host cells are pyruvate, malate, NAD, ATP, CoA, and folinic acid. Inability of the organisms to synthesize CoA has been mentioned. They are unable to synthesize the purine ring de novo, thus requiring an exogenous source of purines for DNA and RNA synthesis. Their purine source seems to be hypoxanthine salvaged from the normal purine catabolism of host cells.[67]

Although plasmodia have cytoplasmic ribosomes of the eukaryotic type, several antibiotics that specifically inhibit prokaryotic (and mitochondrial) protein synthesis, such as tetracycline and tetracycline derivatives, have a considerable antimalarial potency. Tetracycline inhibits protein synthesis in *P. falciparum* as well as growth in vitro.[7]

Tetrahydrofolate is a cofactor very important in transfer of one-carbon groups in various biosynthetic pathways in both prokaryotes and eukaryotes. Mammals require a precursor form, **folic acid,** as a vitamin, and dietary deficiency in this vitamin inhibits growth and produces various forms of anemia, particularly because of impaired synthesis of purines and the pyrimidine thymine. In contrast, *Plasmodium* species (in common with bacteria) synthesize tetrahydrofolate from simpler precursors, including *p*-aminobenzoic acid, glutamic acid, and a pteridine (Fig. 9.10); the organisms are apparently unable to assimilate folic acid. Analogs of *p*-aminobenzoic acid such as **sulfones** and **sulfonamides** block incorporation of the precursor, and some of these analogs (such as sulfadoxine and dapsone) are effective

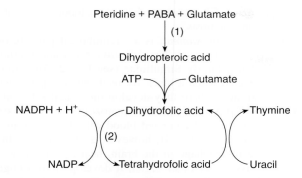

Figure 9.10   Metabolism of folate in *Plasmodium*.
(*1*) Site of action of PABA analogs, such as sulfadoxine, which inhibit the synthesis of dihydropteroic acid from PABA and pteridine. (*2*) Site of action of pyrimethamine, which inhibits synthesis of tetrahydrofolic acid from dihydrofolic acid, which prevents the synthesis of thymine required for DNA synthesis.

From D. L. Looker et al., *Chemotherapy of parasitic diseases.* New York: Plenum Press, 1986.

antimalarials. In both the mammalian pathway and the plasmodial-bacterial pathway, an intermediate product is dihydrofolate, which must be reduced to tetrahydrofolate by the enzyme **dihydrofolate reductase.** Also, this enzyme is necessary for tetrahydrofolate regeneration from dihydrofolate, which is produced in a vital reaction for which tetrahydrofolate is a cofactor: thymidylic acid synthesis. Thus, this enzyme is vital to both parasite and host, but fortunately the dihydrofolate reductases from the two sources vary in several respects. These differences include affinity for certain inhibitors.[64] A concentration of pyrimethamine and trimethoprim required to produce 50% inhibition of the mammalian enzyme is more than 1000 times that yielding 50% inhibition of the plasmodial one.

Resistance to pyrimethamine is due to point mutations, changing one or another of the amino acids in the parasite's dihydrofolate reductase. These changes reduce the binding of the protein with pyrimethamine, but they do not affect its enzymatic function.[141]

Apicoplast. In the 1960s an unusual membranous structure was noticed in electron micrographs of apicomplexans (p. 120). This structure was neglected until the 1990s, upon realization that it was a nonphotosynthetic plastid[89] (plastids are organelles in plants and algae that bear photosynthetic pigments, such as chloroplasts). Like mitochondria, plastids arose as a result of **endosymbiosis,** in which an ancient prokaryote was engulfed by a host cell and became a permanent resident. Most plastids are enveloped by two membranes, one apparently representing the plasma membrane of the ancestral prokaryote, the other being the phagosome lining of the host cell. Apicoplasts, however, have four membranes, indicating their origin as a **secondary endosymbiotic event.**[134] Also like mitochondria, plastids carry their own genome, although genes encoding many of their proteins have been transferred to the nuclear genome through the course of evolution and are imported to the organelles posttranslationally. Apicoplasts bear their genes in a 35 kb circle of DNA. Ribosomal and tRNA genes in the 35 kb circle are

sufficient to transcribe protein-encoding genes in the circle.[71] However, fully 10% of genes in the genome of *Plasmodium falciparum* encode proteins targeted to their apicoplast.[143]

Numerous possible functions of apicoplasts have been suggested, and the organelle is necessary for survival and transmission of *P. falciparum.*[125] Substantial experimental data support an essential role in lipid metabolism (fatty acids and isoprenoids such as sterols).[71] In all organisms studied so far, isoprenoids are synthesized by condensation of varying numbers of activated isoprene units (isopentenyl diphosphate), which are formed from acetate in mammals and fungi via the mevalonate pathway. A mevalonate-independent pathway has been found in some bacteria, algae, plants, and *P. falciparum.* Inhibitors of the nonmevalonate pathway have significant potential as antimalarial drugs.[54, 135]

Genus *Haemoproteus*
Protozoa belonging to the genus *Haemoproteus* are primarily parasites of birds and reptiles and have their sexual phases in insects other than mosquitoes. Exoerythrocytic schizogony occurs in endothelial cells; merozoites enter erythrocytes to become pigmented gametocytes in the circulating blood (Fig. 9.11).

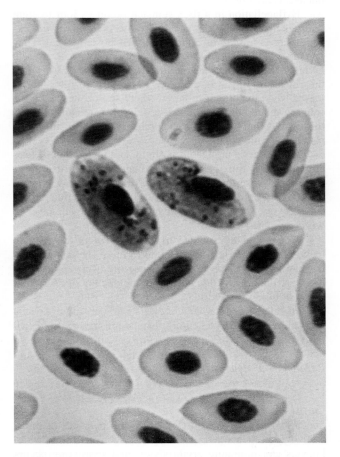

Figure 9.11   *Haemoproteus* gametocytes in blood of a mourning dove.
They are about 14 μm long.

Courtesy of Sherwin Desser.

Haemoproteus columbae is a cosmopolitan parasite of pigeons. The definitive hosts and vectors of this parasite are several species of ectoparasitic flies in the family Hippoboscidae (chapter 39), which inject sporozoites with their bite. Exoerythrocytic schizogony takes about 25 days in the endothelium of lung capillaries, producing thousands of merozoites from each schizont. Merozoites presumably can develop directly from a schizont, or a schizont can break into numerous multinucleate "cytomeres." In the latter case the host endothelial cell breaks down, releasing the cytomeres, which usually lodge in the capillary lumen, where they grow, become branched, and rupture, producing many thousands of merozoites. A few of these may attack other endothelial cells, but most enter erythrocytes and develop into gametocytes. At first they resemble ring stages of *Plasmodium,* but they grow into mature microgametocytes or macrogametocytes in five or six days. Multiple infections of young forms in a single red blood cell are common, but one rarely finds more than one mature parasite per cell.

Mature macrogametocytes are 14 μm long and grow in a curve around the host nucleus. Their granular cytoplasm stains a deep blue color and contains about 14 small, dark-brown pigment granules. The nucleus is small. Microgametocytes are 13 μm long, are less curved, have lighter-colored cytoplasm, and have six to eight pigment granules. Their nucleus is diffuse.

Exflagellation in a fly's stomach produces four to eight microgametes. Ookinetes are like those of *Plasmodium* except there is a mass of pigment at their posterior end. They penetrate intestinal epithelium and encyst between muscle layers. Oocysts grow to maturity within nine days, when they measure 40 μm in diameter. Myriad sporozoites are released when the oocyst ruptures. Many sporozoites reach the salivary glands by the following day. Flies remain infected throughout the winter and can transmit infection to young squabs the following spring.

Pathogenesis in pigeons is slight, and infected birds usually show no signs of disease. Exceptionally, birds appear restless and lose their appetite, and their lungs may become congested. Some anemia may result from loss of functioning erythrocytes, and spleen and liver may be enlarged and dark with pigment.

More than 80 species of *Haemoproteus* have been named from birds, mainly Columbiformes. The actual number may be much less than that, since life cycles of most of them are unknown. *Culicoides* spp. are vectors of *H. meleagridis* of turkeys in Florida.[62]

Of related genera, *Hepatocystis* spp. parasitize African and oriental monkeys, lemurs, bats, squirrels, and chevrotains; *Nycteria* and *Polychromophilus* are in bats; *Simondia* occurs in turtles; *Haemocystidium* lives in lizards; and *Parahaemoproteus* is common in a wide variety of birds.

Genus *Leucocytozoon*

Species of *Leucocytozoon* (Fig. 9.12) are parasites of birds. Schizogony is in fixed tissues, gametogony is in both leukocytes and immature erythrocytes, and sporogony occurs in insects other than mosquitoes. Pigment is absent from all phases of their life cycles. A related genus, *Akiba,* with only one species, occurs in chickens. *Leucocytozoon* has about 60 species in various birds. These are the most important blood protozoa of birds, since they are pathogenic in both domestic and wild hosts.

Leucocytozoon simondi is a circumboreal parasite of ducks, geese, and swans. The definitive hosts and vectors are black flies, family Simuliidae (see Fig. 29.7*b*). Sporozoites injected when a black fly feeds enter hepatocytes of the avian host where they develop into small schizonts (11 μm to 18 μm). They produce merozoites in four to six days. Merozoites that enter red blood cells become round gametocytes (Fig. 9.13). If, however, a merozoite is ingested by a macrophage in the brain, heart, liver, kidney, lymphoid tissues, or other organ, it develops into a huge megaloschizont 100 μm to 200 μm in diameter. The large form is more abundant than the small hepatic schizont.

The megaloschizont divides internally into primary cytomeres, which, in turn, multiply in the same manner. Successive cytomeres become smaller and finally multiply by schizogony into merozoites. Up to a million merozoites may be released from a single megaloschizont.

Merozoites penetrate leukocytes or developing erythrocytes to become elongated gametocytes (see Fig. 9.13). Gametocytes of both sexes are 12 μm to 14 μm in diameter in fixed smears and may reach 22 μm in living cells. Macrogametocytes have a discrete, red-staining nucleus. The male cell is pale staining and has a diffuse nucleus that takes up most of the space within the cell. The diffuse nucleus of macrogametocytes has large numbers of ribosomes.[58] As gametocytes mature, they cause their host cells to become elongated and spindle shaped (see Fig. 9.13).

Exflagellation begins only three minutes after the organism is eaten by a fly. A typical ookinete entering an intestinal cell becomes a mature oocyst within five days. Only 20 to 30 sporozoites form and slowly leave an oocyst. Rather than entering the salivary glands of a vector, they enter the proboscis directly and are transmitted by contamination or are washed in by saliva.

Leucocytozoon simondi is highly pathogenic for ducks and geese, especially young birds. The death rate in ducklings may reach 85%; older ducks are more resistant, and the disease runs a slower course in them, but they still may succumb. Anemia is a prominent symptom of leukocytozoonosis, as are elevated numbers of leukocytes. The liver enlarges and becomes necrotic, and the spleen may increase to as much as 20 times the normal size. *Leucocytozoon simondi* probably kills the host by destroying vital tissues, such as of brain and heart. An outstanding feature of an outbreak of leukocytozoonosis is suddenness of its onset. A flock of ducklings may appear normal in the morning, become ill in the afternoon, and be dead by the next morning. Birds that survive are prone to relapses but, as a result of premunition, are generally immune to reinfection.

Another species of importance is *L. smithi,* which can devastate domestic and wild turkey flocks. Its life cycle is similar to that of *L. simondi.*

ORDER PIROPLASMIDA

Members of order Piroplasmida are small parasites of ticks and mammals. They do not produce spores, flagella, cilia, or true pseudopodia; their locomotion, when necessary, is

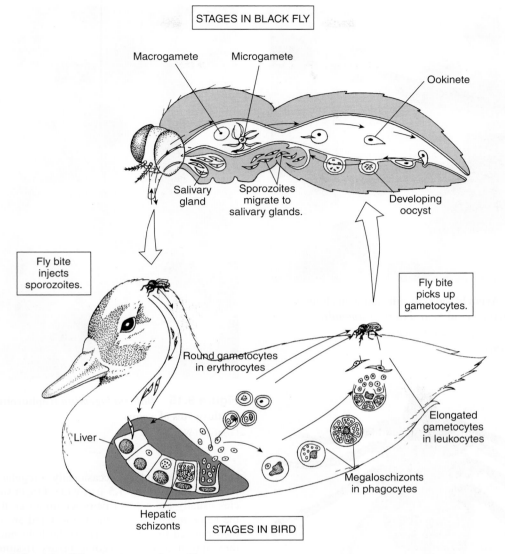

STAGES IN BLACK FLY

Macrogamete   Microgamete

Ookinete

Salivary gland

Sporozoites migrate to salivary glands.

Developing oocyst

Fly bite injects sporozoites.

Fly bite picks up gametocytes.

Round gametocytes in erythrocytes

Liver

Elongated gametocytes in leukocytes

Hepatic schizonts

Megaloschizonts in phagocytes

STAGES IN BIRD

Figure 9.12   Life cycle of *Leucocytozoon simondi*.

Drawing by William Ober and Claire Garrison.

accomplished by body flexion or gliding. No stages produce intracellular pigment. Asexual reproduction is in erythrocytes or other blood cells of mammals by binary fission or schizogony. Sexual reproduction occurs, at least in some species.[74] Components of the apical complex are reduced but warrant placement in the phylum Apicomplexa.

Piroplasmida contains the two families Babesiidae and Theileriidae, both of which are of considerable veterinary importance.

Family Babesiidae

Babesiids are usually described from their stages in the red blood cells of vertebrates. They are pyriform, round, or oval parasites of erythrocytes, lymphocytes, histiocytes, erythroblasts, or other blood cells of mammals and of various tissues of ticks. Their apical complex is reduced to a polar ring, rhoptries, micronemes, and subpellicular

microtubules. A cytostome is present in at least some species. Schizogony occurs in ticks. By far the most important species in America is *Babesia bigemina,* the causative agent of **babesiosis,** or **Texas red-water fever,** in cattle.

Babesia bigemina

By 1890 the entire southeastern United States was plagued by a disease of cattle, variously called *Texas cattle fever, redwater fever,* or *hemoglobinuria.* Infected cattle usually had red-colored urine resulting from massive destruction of erythrocytes, and they often died within a week after symptoms first appeared. The death rate was much lower in cattle that had been reared in an enzootic area than in northern animals that were brought south. Also, it was noticed that, when southern herds were driven or shipped north and penned with northern animals, the latter rapidly succumbed to the disease.

The cause of red-water fever and its mode of dissemination were a mystery when Theobald Smith and Frank Kilbourne began their investigations in the early 1880s. In a

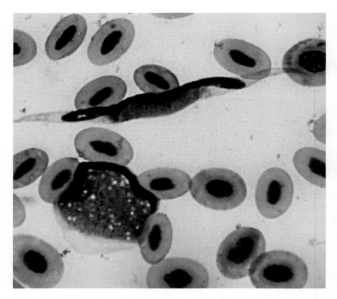

Figure 9.13   **Avian blood cells infected with elongate and round gametocytes of *Leucocytozoon simondi.***
The elongate form is up to 22 μm long.
Courtesy of Sherwin Desser.

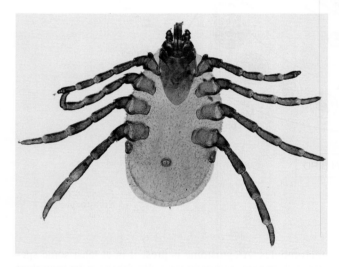

Figure 9.14   *Boophilus annulatus,* **the vector of *Babesia bigemina.***
Courtesy of Jay Georgi.

series of intelligent, painstaking experiments, they showed that the tick *Boophilus annulatus* (Fig. 9.14) was the vector and alternate host of a tiny protozoan parasite that inhabited red blood cells of cattle and killed these relatively immense animals.[119] Their investigations not only pointed the way to an effective means of control, but were also the first demonstrations that a protozoan parasite could develop in and be transmitted by an arthropod. The book in your hand is replete with other examples of this phenomenon, as we have already seen.

Babesia bigemina infects a wide variety of ruminants, such as deer, water buffalo, and zebu, in addition to cattle. When in an erythrocyte of a vertebrate host, the parasite is

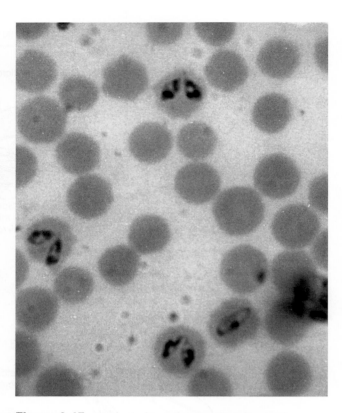

Figure 9.15   *Babesia bigemina* **trophozoites in the erythrocytes of a cow.**
Courtesy of Warren Buss.

pear shaped, round, or, occasionally, irregularly shaped, and it is 4.0 μm long by 1.5 μm wide. Organisms of this species usually are seen in pairs within an erythrocyte (hence the name *bigemina* for "the twins") and are often united at their pointed tips (Fig. 9.15). At the light microscope level, they appear to be undergoing binary fission, but electron microscopy has revealed that the process is a kind of binary schizogony, a budding analogous to that occurring in Haemosporida, with redifferentiation of the apical complex and merozoite formation.[1]

Biology. The infective stage of *Babesia* in ticks is a sporozoite. It is about 2 μm long and is pyriform, spherical, or ovoid. After completing development, sporozoites in tick salivary glands are injected with its bite. There is no exoerythrocytic schizogony in the vertebrate. Parasites immediately enter erythrocytes, where they become trophozoites and escape from the parasitophorous vacuole.[3] They undergo binary fission and ultimately kill their host cell. Merozoites attack other red blood cells, building up an immense population in a short time. This asexual cycle continues indefinitely or until the host succumbs. Erythrocytic phases are reduced or apparently absent in resistant hosts. Some of the intraerythrocytic parasites do not develop further and are destined to become gametocytes, called **ray bodies,** when ingested by a tick.[74]

Ticks of genus *Boophilus* transmit *Babesia bigemina*; thus, distribution of the tick limits distribution of babesiosis. *Boophilus annulatus* is the vector in the Americas. It is a one-host tick, feeding, maturing, and mating on a single host (chapter 41). After engorging and mating, a female tick drops to the ground, lays her eggs, and dies. The larval, six-legged ticks that hatch from eggs climb onto vegetation and attach to animals that brush by the plants.

One would think that a one-host tick would be a poor vector—if they do not feed on successive hosts, how can they transmit pathogens from one animal to another? This question was answered when it was discovered that the protozoan infects the developing eggs in the ovary of the tick, a phenomenon called **transovarial transmission.**

After ingestion by a feeding tick, the parasites are freed by digestion from their dead host cells, and they develop into ray bodies. These are bizarrely shaped stages that have

a thornlike process and several stiff, flagellalike protrusions (Fig. 9.16).[74] Fusion of two ray bodies forms a zygote, which becomes a primary **kinete.** The primary kinetes leave the intestine and penetrate various cells, such as hemocytes, muscles, Malpighian tubule cells, and ovarian cells including oocytes. They enlarge and become polymorphic, dividing by multiple fission into a number of **cytomeres,** which differentiate into new kinetes. Some secondary kinetes migrate to the salivary glands, penetrate gland cells, and become polymorphic. They stimulate host cells and nuclei to hypertrophy. When a host begins to feed, parasites rapidly undergo multiple fission to produce enormous numbers of sporozoites about 2 μm to 3 μm long by 1 μm to 2 μm wide. These

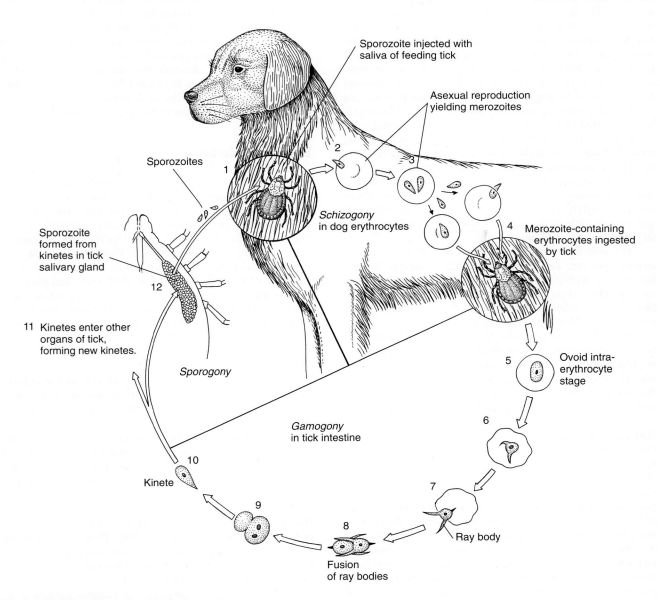

Figure 9.16   Life cycle of *Babesia canis.*
(*1*) Sporozoite injected with saliva of feeding tick. (*2, 3*) Asexual reproduction in red blood cells of vertebrate host by binary fission, yielding merozoites. (*4*) Merozoites in erythrocytes are ingested by tick. (*5, 6*) Gametocytes form protrusions after ingestion by tick and become ray bodies. (*7–9*) Two ray bodies fuse to form zygote. (*10*) Zygote becomes motile kinete. (*11*) Kinetes leave intestine, enter other cells, and form new kinetes. (*12*) Kinetes that enter cells of salivary gland give rise to thousands of small sporozoites.

Drawing by William Ober and Claire Garrison.

sporozoites enter the channels of the salivary glands and are injected into the vertebrate host by the feeding tick.[100]

Although this is the life cycle as it occurs in a one-host tick, two- and three-host ticks serve as hosts and vectors of *B. bigemina* in other parts of the world. With these ticks transovarial transmission is not required and may not occur. All instars of such ticks can transmit the disease.

Pathogenesis. *Babesia bigemina* is unusual in that the disease it causes is more severe in adult cattle than in calves. Calves less than a year old are seldom seriously affected, but the mortality rate in acute cases in untreated adult cattle is as high as 50% to 90%. The incubation period is 8 to 15 days, but an acutely ill animal may die only four to eight days after infection. The first symptom is a sudden rise in temperature to 106°F to 108°F; this may persist for a week or more. Infected animals rapidly become dull and listless and lose their appetite. Up to 75% of erythrocytes may be destroyed in fatal cases, but even in milder infections so many erythrocytes are destroyed that a severe anemia results. Mechanisms for clearance of hemoglobin and its breakdown products are overloaded, producing jaundice, and much excess hemoglobin is excreted by the kidneys, giving the urine the red color mentioned earlier. Chronically infected animals remain thin, weak, and out of condition for several weeks before recovering. Levine described damage to internal organs.[61]

Cattle that recover are usually immune for life with a sterile immunity or, more commonly, premunition. There are strain differences in the degree of immunity obtained; furthermore, little cross-reaction occurs between *B. bigemina* and other species of *Babesia*.

For unknown reasons drugs that are effective against trypanosomes are also effective against *Babesia* spp. A number of chemotherapeutic agents are available, some allowing recovery but leaving latent infection, others effecting a complete cure. It should be remembered that elimination of all parasites also eliminates premunition.

Infection can be prevented by tick control, the means by which red-water fever was eliminated from the United States. Regular dipping of cattle in a tickicide effectively eliminates vectors, especially if it is a one-host species. Another method that has been used is artificial premunizing of young animals with a mild strain before shipping them to enzootic areas.

Babesia microti

Prior to 1969 *Babesia* infections in humans were rare. There have been a few reports of infections caused by species normally parasitic in other animals. In several cases, three of which were fatal, patients had been splenectomized some time before infection, and it was believed that the disabling of the immune system by splenectomy rendered the humans susceptible. However, human infection with *Babesia* in a nonsplenectomized patient was reported from Nantucket Island off the coast of Massachusetts in 1969. Since then, hundreds of cases have been recorded, most in the northeast United States but some in Wisconsin, Washington, and California.[44] These have all been infections with *B. microti*, a parasite of meadow voles and other rodents that can also infect pets. The vector is *Ixodes scapularis*, whose adults feed on deer. Deer are refractory to infection with *B. microti*, and the infection is transmitted among rodents and to humans by nymphs of *I. scapularis* and among rodents by *I. muris,*

which does not feed on humans.[103] It is unclear why this formerly rare infection has now become almost common. However, as with Lyme disease (p. 613), the explanation probably lies in the increased contact of humans with ticks and the reservoir hosts.

Other Species of Babesiidae

Cattle seem particularly suitable as hosts to piroplasms. Other species of *Babesia* in cattle are *B. bovis* in Europe, Russia, and Africa; *B. berbera* in Russia, North Africa, and the Middle East; *B. divergens* in western and central Europe; *B. argentina* in South America, Central America, and Australia; and *B. major* in North Africa, Europe, and Russia. Several other species are known from deer, sheep, goats, dogs, cats, and other mammals as well as birds. Their biology, pathogenesis, and control are generally the same as for *B. bigemina.*

Babesia divergens occasionally occurs in splenectomized humans in Europe, and two such cases have been reported in the United States.[148] Another *Babesia* sp. occurs in rabbits (*Syvilagus floridanus*) in the United States. It is morphologically and genetically similar to *B. microti* and *B. divergens* but will not grow in bovine erythrocytes in vitro.[122]

Family Theileriidae

Like Babesiidae, members of this family lack a conoid. Rhoptries, micronemes, subpellicular tubules, and polar ring are well demonstrated in the tick stages. Theileriidae parasitize blood cells of mammals, and vectors are hard ticks of family Ixodidae. Gamogony occurs in the gut of nymphal ticks, resulting in the formation of kinetes, which are very similar to ookinetes of Haemosporida. Kinetes grow in the gut cells of a tick for a time and then leave and penetrate cells of the salivary glands, where sporogony takes place. Several members of this family infect cattle, sheep, and goats, causing **theileriosis,** which results in heavy losses in Africa, Asia, and southern Europe.

Theileria parva

Theilerosis due to *Theileria parva* is called **East Coast fever** in cattle, zebu, and Cape buffalo. It has been one of the most important diseases of cattle in southern, eastern, and central Africa, although it has been eliminated from most of southern Africa. After Romanovsky staining, forms within erythrocytes have blue cytoplasm and a red nucleus in one end. At least 80% of them are rod shaped, about 1.5 μm to 2.0 μm by 0.5 μm to 1.0 μm in size. Oval and ringor comma-shaped forms are also found.

Biology. East Coast fever, like red-water fever, is a disease of ticks and cattle, flourishing in both. Principal vectors are brown cattle ticks, *Rhipicephalus appendiculatus,* a three-host species. Other ticks, including oneand two-host ticks, can also serve as hosts for this parasite.

When a tick feeds, it injects sporozoites present in its salivary glands into the next host;[33] there they enter T and B lymphocytes, and only after entry do they discharge contents of their rhoptries and micronemes.[108] This causes the membrane surrounding their containing vacuole to disperse, so that they come to lie free in host-cell cytoplasm. They grow and undergo schizogony. Schizonts, called **Koch's blue bodies,** can be seen in circulating lymphocytes within three days of infection. Two types of schizonts are recognized. The first generation in lymph cells comprises **macroschizonts** and produces about 90 macromerozoites, each 2.0 μm to 2.5 μm in diameter. Some of these enter other lymph cells, especially in fixed tissues, and initiate further generations of macroschizonts. Others enter lymphocytes and become **microschizonts,** producing 80 to 90 micromerozoites, each 0.7 μm to 1.0 μm wide. Within lymphocytes, schizonts induce clonal expansion and blastogenesis in their host cells, imitating leukemia. Lymphoblast invasion of other tissues, such as of kidney and brain, contributes significantly to pathogenesis. Evidently, cellular transformation is by means of parasite induction of host-cell overexpression of a gene coding for **casein kinase II,** an important regulatory enzyme.[107]

If microschizonts rupture while in lymphoid tissues, micromerozoites enter new lymph cells, maintaining the lymphocytic infection. However, if they rupture in circulating blood, micromerozoites enter erythrocytes to become the "piroplasms" typical of the disease. Apparently, the parasites do not multiply in erythrocytes.

Ticks of all instars can acquire infection when they feed on blood containing piroplasms. However, because three-host ticks drop off the host to molt immediately after feeding, only nymphs and adults are infective to cattle. Transovarial transmission does not occur as it does in *Babesia* spp.

Ingested erythrocytes are digested, releasing piroplasms that undergo gamogony, differentiating into ray bodies. Fusion of ray bodies produces kinetes, as described before.[74]

Pathogenesis. As in babesiosis, calves are more resistant to *T. parva* than are adult cattle. Nevertheless, *T. parva* is highly pathogenic: Strains with low pathogenicity kill around 23% of infected cattle, whereas highly pathogenic strains kill 90% to 100%. Symptoms such as high fever first appear 8 to 15 days after infection. Other signs are nasal discharge, runny eyes, swollen lymph nodes, weakness, emaciation, and diarrhea. Hematuria and anemia are unusual, although blood is often present in feces.

Animals that recover from theileriosis are immune from further infection, without premunition. Immunity is cell mediated, including destruction of infected lymphocytes by activated cytotoxic T cells and natural killer cells.[42] Diagnosis depends on finding parasites in blood or lymph smears. No cheap and effective drug is currently available.[42] Control depends on tick control and quarantine rules.

Other species of *Theileria* are *T. annulata, T. mutans, T. hirei, T. ovis,* and *T. camelensis,* all parasites of ruminants. Other genera in the family are *Haematoxenus* in cattle and zebu and *Cytauxzoon* in antelope, both in Africa. *Cytauxzoon felis* parasitizes felines in the south central and southeastern United States.[75] Bobcats (*Lynx rufus*) are apparently the normal hosts, but infection of domestic cats is usually fatal.

Learning Outcomes

By the time a student has finished studying this chapter, he or she should be able to:

1. List the scientific names of the species of parasites that are recognized agents of malaria in humans.

2. Describe the life cycles of the parasites named in learning objective #1, above.

3. What is the practical (medical) significance of the time required for the schizogonic cycle to occur in each malaria species?

4. Describe the symptoms as related to phases of the schizogonic cycle in malaria, and explain why that is significant.

5. Name an important compound in host cells that serves as an important energy source for malaria parasites.

6. What is the parasite organelle through which the compound in learning objective #5 is ingested by malaria parasites?

7. What is an important basis for drug resistance in chloroquine-resistant *Plasmodium falciparum?*

8. Describe the results of filamentous aggregate formation in persons that are infected with *Plasmodium falciparum* and that have sickle-cell anemia.

9. Describe why many researchers think that apicoplasts originated through a secondary endosymbiotic event.

10. Name the primary hosts of *Haemoproteus* and *Leucocytozoon,* and their primary mode(s) of transmission.

11. Name the primary hosts of *Babesia bigemina* and explain its economic importance.

References

References for superscripts in the text can be found at the following Internet site: www.mhhe.com/robertsjanovynadler9e

Additional Readings

Desowitz, R. S. 1991. *The malaria capers.* New York: W. W. Norton & Company. This book includes a rather different portrait of Ronald Ross from that normally painted, such as the present chapter and in Hagan and Chauhan, below.

Desowitz, R. S. 1997. *Who gave pinta to the* Santa Maria? New York: W. W. Norton & Company. Another fascinating account by Desowitz, including chapters on malaria in the United States and England.

Garrett, L. 1995. *The coming plague: Newly emerging diseases in a world out of balance.* New York: Penguin Books.

Hagan, P., and V. Chauhan. 1997. Ronald Ross and the problem of malaria. *Parasitol. Today* 13:290–295.

Harrison, G. 1978. *Mosquitoes, malaria, and man: A history of the hostilities since 1880.* New York: E.P. Dutton & Co., Inc.

Honigsbaum, M. 2001. *The fever trail. In search of the cure for malaria.* New York: Farrar, Straus and Giroux. A fascinating account of the physical dangers and discomforts endured by the men who endeavored to recover cinchona trees and seed in South America.

Winstanley, P., and S. Ward. 2006. Malaria chemotherapy. In Molyneaux, D. H., ed., *Advances in parasitology* 61, London: Elsevier.

Chapter 10

Phylum Ciliophora: Ciliated Protistan Parasites

Nearly all the endoparasitic protozoa offer problems in the question of their transmission, which seem opposed to a simple solution.

— J. F. Mueller and H. J. Van Cleave, discussing the life cycle of *Trichodina renicola,* a parasite of fish kidneys[19]

Ciliophora possess simple cilia or compound ciliary organelles in at least one stage of their life cycle. A compound subpellicular infraciliature is universally present even when cilia are absent (see chapter 4). Most species have one or more macronuclei and micronuclei, and fission is homothetogenic. Some species exhibit sexual reproduction involving conjugation, autogamy, and cytogamy. Although each cilium has a kinetosome, centrioles functioning as such are absent. Most ciliates are free living, but many are commensals of vertebrates and invertebrates, and a few are parasitic.

Phylum Ciliophora has undergone extensive revision in the past two decades, with special focus on higher taxonomic criteria. Today ciliate taxonomy at virtually all levels depends on cortical structure, position and arrangement of kinetosomes, and ontogeny of ciliary distribution patterns during cell division, all in addition to molecular data. Investigative tools are various techniques for staining cortical structures with silver. As might be expected, in some cases phylogenetic relationships based on morphology are consistent with those revealed by molecular techniques, but in other cases they are not.[1, 4]

The following examples represent the most common and widely recognized ciliate commensals and parasites.

CLASS SPIROTRICHEA

Members of Spirotrichea have well-developed, conspicuous membranelles in and around their buccal cavity (adoral zone of membranelles, or AZM). Body ciliature may be reduced, or cilia may be joined into compound organelles called *cirri.*

Order Clevelandellida; Family Nyctotheridae

Members of Clevelandellida have unique nonmicrotubular fibrils associated with their somatic kinetids. Somatic ciliature may also be separated into defined areas by suture lines.

Buccal ciliature is conspicuous, with the AZM typically composed of one to many membranelles or undulating membranes that wind clockwise to the cytostome. Most species are quite large.

Nyctotheridae are robust parasites of the intestine of vertebrates and invertebrates. Their entire body has tiny cilia arranged in longitudinal rows. A single undulating membrane extends from the anterior end to deep within the cytopharynx.

Nyctotherus is the most common genus. These ciliates (Fig. 10.1) are ovoid to kidney shaped, with their cytostome on one side. The anterior half contains a massive macronucleus, with a small micronucleus nearby. *Nyctotherus* species are also known to contain hydrogenosomes, membrane-bound organelles of anaerobic unicellular eukaryotes that produce hydrogen and ATP, and that are thought to have evolved from mitochondria.[2] However,

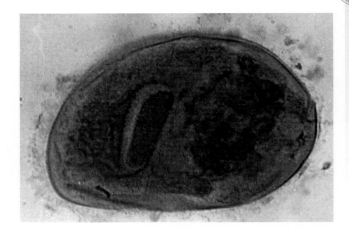

Figure 10.1 *Nyctotherus cordiformis* **trophozoite from the colon of a frog.**

These protozoa range from 60 μm to 200 μm in length.

Courtesy of Warren Buss.

167

unlike all hydrogenosomes studied previously (p. 96), hydrogenosomes contained in *Nyctotherus* spp. possess their own DNA.[16] Further research on these hydrogenosome may help bridge the evolutionary gap between mitochondria and hydrogenosomes.

CLASS LITOSTOMATEA

Order Vestibuliferida, Family Balantidiidae

Litostomatea have body monokinetids with tangential transverse microtubule ribbons and nonoverlapping, laterally directed, kinetodesmal fibrils (see chapter 4). Members of order Vestibuliferida have a densely ciliated vestibulum near the apex of the cell, and they have no polykinetids. The vestibulum is a depression or invaginated area that leads directly to the cytostome; it is lined with cilia predominantly somatic in nature and origin.

Balantidiidae has a single genus *Balantidium,* species of which are found in the intestine of crustaceans, insects, fish, amphibians, and mammals. A vestibulum leading into the cytostome is at the anterior end, and a cytopyge is present at the posterior tip (Fig. 10.2).

Balantidium coli

Balantidium coli (Fig. 10.3) is the largest protozoan parasite of humans. It is most common in tropical zones but is present throughout temperate climes as well. Epidemiology and effects on the host are similar to those of *Entamoeba histolytica.* The organism appears to be primarily a parasite of pigs, with strains adapted to various other hosts, including several species of primates.[21]

Morphology. *Balantidium coli* trophozoites are oblong, spheroid, or more slender, 30 μm to 150 μm long by 25 μm to 120 μm wide (Fig. 10.3). Encysted stages (Fig. 10.4), which are most commonly found in stools, are spheroid or ovoid, measuring 40 μm to 60 μm in diameter. The macronucleus is a large, sausage-shaped structure. The single micronucleus is small and often hidden from view by the macronucleus. There are two contractile vacuoles, one near the middle of the body and the other near the posterior end. The cytostome is at the anterior end. Food vacuoles contain erythrocytes, cell

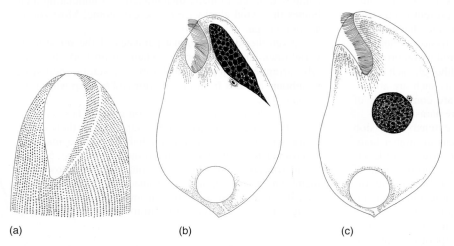

(a) (b) (c)

Figure 10.2 *Balantidium* species.

(*a*) Vestibule infraciliature of *Balantidium* spp. showing how vestibular infraciliature is actually a continuation of body kineties. (*b, c*) *Balantidium* species from cockroaches: (*b*) *B. praenucleatum* from *Blatta orientalis*: (*c*) *B. ovatum* from *Blatta americana.*

From E. Faure-Fremiet, "La position systematique du genre *Balantidium,*" in *Journal of Eukaryotic Microbiology* 2:54–58, vol. 2, no. 2. Copyright © 1955. Reprinted by permission of John Wiley & Sons.

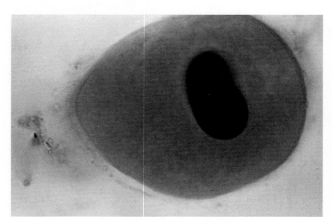

Figure 10.3 Trophozoite of *Balantidium coli.*
Trophozoites range from 30 μm to 150 μm long by 25 μm to 120 μm wide.
Courtesy of James Jensen.

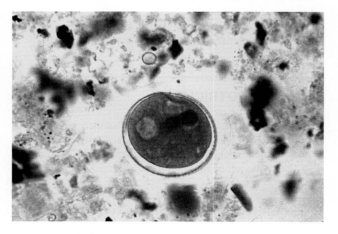

Figure 10.4 Encysted form of *Balantidium coli.*
Cysts are 40 μm to 60 μm in diameter.
Courtesy of James Jensen.

fragments, starch granules, and fecal and other debris. Living trophozoites and cysts are yellowish or greenish.

Biology. *Balantidium coli* lives in the cecum and colon of humans, pigs, guinea pigs, rats, and many other mammals. It is not readily transmissible from one species of host to another, because it seems to require a period of time to adjust to the symbiotic flora of a new host. However, when adapted to a host species, the protozoan flourishes and can become a serious pathogen, particularly in humans. In animals other than primates, *B. coli* is unable to initiate a lesion by itself, but it can become a secondary invader if the mucosa is breached by other means.

Trophozoites multiply by transverse fission. Conjugation has been observed in culture but may occur only rarely, if at all, in nature. Encystment is instigated by dehydration of feces as they pass posteriorly in the rectum. These protozoa can encyst after being passed in stools—an important factor in their epidemiology. Infection occurs when cysts are ingested, usually in contaminated food or water. Unencysted trophozoites may live up to 10 days and may possibly be infective if eaten, although this is unlikely under normal circumstances. Because *B. coli* is destroyed by a pH lower than 5, infection is most likely to occur in malnourished persons with low stomach acidity.

Pathogenesis. Under ordinary conditions trophozoites feed much like most other ciliates, ingesting particles through a cytostome. However, sometimes it appears that the organisms can produce proteolytic enzymes that digest away a host's intestinal epithelium. Production of hyaluronidase has been detected, and this enzyme could help enlarge an ulcer. Ulcers usually are flask shaped, like amebic ulcers, with a narrow neck leading into an undermining saclike cavity in the submucosa. Colonic ulceration produces lymphocytic infiltration with few polymorphonuclear leukocytes, and hemorrhage and secondary bacterial invasion may follow. Fulminating cases may produce necrosis and sloughing of the overlying mucosa and occasionally perforation of the large intestine or appendix, as in amebic dysentery. Death often follows at this stage. Secondary foci, such as liver or lung, may become infected.[9] Urogenital organs are sometimes attacked after contamination, and vaginal, uterine, and bladder infections have been discovered.

Epidemiology and Transmission Ecology. Balantidiasis in humans is most common in the Philippines but can be found almost anywhere in the world, especially among those who are in close contact with swine. Generally the disease is considered rare and occurs in less than 1% of the human population. Higher infection rates have been reported among institutionalized persons. However, in pigs the infection rate may be quite high; in a typical survey of pigs brought to slaughter in Japan, prevalence was 100%.[20] An interesting epidemiological situation evidently occurs in Iran. In contrast to most Middle Eastern countries, there is both a relatively high prevalence of balantidiasis and an increasing wild boar population.[25] Muslims consider pigs abhorrent, so boars are not hunted for religious reasons although they are important crop pests and thus can contaminate soil and water.[25]

Primates other than humans sometimes are infected and may represent a reservoir of infection to humans, although the reverse is probably more likely. The ciliates' ability to encyst after being passed increases the number of potential infections from a single reservoir host, and cysts can remain alive for weeks in pig feces, if the feces do not dry out. Pigs are probably the usual source of infection for humans, but the relationship is not clear. The protozoa in swine are essentially nonpathogenic and are considered by some a separate species, *B. suis.* There may be strains of *B. coli* that vary in their adaptability to humans. Infections often disappear spontaneously in healthy persons, or they can become symptomless, making infected persons carriers.

Treatment and Control. Several drugs are used to combat infections of *B. coli,* including carbarsone, diiodohydroxyquin, and tetracycline. Prevention and control measures are similar to those for *Entamoeba histolytica,* except that particular care should be taken by those who work with pigs. In one troop of free-ranging rhesus monkeys, eating of soil evidently functioned to virtually eliminate diarrhea from intestinal infections, including with *B. coli.* The soil contained kaolinitic clay with the same pharmaceutical properties as over-the-counter medicines used to treat human diarrhea.[13]

Other species of *Balantidium* are *B. praenuleatum,* common in the intestines of American and oriental cockroaches, *B. duodeni* in frogs, *B. caviae* in guinea pigs, and *B. procypri* and *B. zebrascopi* in fishes.

Order Entodiniomorphida

The curiously appearing entodiniomorphids have a generally firm pellicle and unique tufts of cilia on an otherwise naked body (Fig. 10.5). The order contains six families and

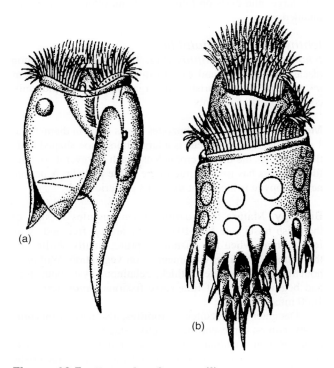

Figure 10.5 Examples of rumen ciliates.
(*a*) *Entodinium caudatum;* (*b*) *Ophryoscolex purkinjei.*

From Karl G. Grell, *Protozoology.* Copyright © 1973 Springer-Verlag, Heidelberg, Germany. Reprinted by permission.

17 genera of ciliates. All are commensals in mammalian herbivores, especially ruminants, where they occupy the rumen, although some species are found in the caecum and colon of horses, others are described from apes, and still others are described from marsupials.[8]

As many as ten entodiniomorphid genera can be present in individual ruminants, contributing up to half the total rumen microbial biomass.[11] Heavy "infections" tend to reduce the host's amino acid supply and increase methane production, but rumen ciliates also indirectly stimulate lysis of cellulose by bacteria.[11]

CLASS OLIGOHYMENOPHOREA

Members of this class have a buccal cavity bearing a well-defined but sometimes inconspicuous oral ciliary apparatus composed of only three or four specialized membranelles.

Subclass Hymenostomatia, Order Hymenostomatida, Family Ichthyophthiriidae

Ichthyophthiriidae contains one genus, *Ichthyophthirius*, with two species, *I. marinus* and *I. multifiliis;* the latter is a common pest in freshwater aquaria and in fish farming, causing much economic loss. In members of this subclass, body ciliature is often uniform and heavy, and conspicuous kinetodesmata are regularly present. Hymenostomatida have a well-defined buccal cavity on their ventral surface. Most species are small in size, but *Ichthyophthirius multifiliis* is very large, and cysts on infected fish are often visible to the unaided eye.

Ichthyophthirius multifiliis
Ichthyophthirius multifiliis (Fig. 10.6) causes a common disease in aquarium and wild freshwater fish. It is known as *ick* to many fish culturists. The organism attacks epidermis, cornea, and gill filaments.

Morphology. Adult trophozoites are as large in diameter as 1 mm. Their macronucleus is a large, horseshoe-shaped body that encircles the tiny micronucleus. Each of several contractile vacuoles has its own micropore in the pellicle. A permanent cytopyge is located at the cell's posterior end.

Biology. Mature trophozoites form pustules in skin of their fish hosts (see Fig. 10.6). They are set free and swim feebly about when the pustules rupture, finally settling on the bottom of their environment or on vegetation. Within an hour the ciliate secretes a thick, gelatinous cyst about itself and begins a series of transverse fissions, producing up to 1000 infective cells.

Daughter trophozoites, or **tomites,** also termed **theronts** or **swarmers,** represent the infective stages and can survive about 96 hours without a host. The tomite is about 40 μm by 15 μm. Its narrowed anterior end carries a characteristic long filament that emerges from a conical depression in the pellicle.[18] The parasite evidently burrows into the fish's skin with its pointed end and filament. There it becomes a trophozoite within three days, ingesting debris from host cells and forming a pustule that reaches over 1 mm in diameter. Although the life cycle of *I. multifiliis* is not typically shown with a sexual phase (see Fig. 10.6), there is some evidence that conjugation may occur between established parasites and newly entering theronts.[17]

Pathogenesis. Grayish pustules form wherever these parasites colonize skin (Fig. 10.7). Epidermal cells combat the irritation by producing much mucus, but many die and are sloughed. When many parasites attack gill filaments, they so interfere with gas exchange that the fish may die.

Research with carp has shown that the immune response is of a cellular nature, with macrophages accumulating at the site of epidermal infections in immunized fish, while in naive fish the cellular response to theronts consists of a diffuse infiltration of neutrophils.[3, 7] In one study, however, passively transferred antibodies caused the parasites to leave their hosts rapidly, suggesting not only a mechanism for the host to rid itself of parasites, but also a mechanism for the parasite to avoid host defenses.[5]

Species of fish, as well as populations of a single species, may differ significantly in their susceptibility to *I. multifiliis*. Susceptibility also can vary according to the time a species has been under domestication (from wild populations), the most recently isolated stocks being least resistant.[6]

Aquarium fish can be treated successfully with malachite green, methyline blue, or very dilute concentrations of formaldehyde. There are also commercial preparations, available in most pet stores, that usually are quite effective. Food with malachite green has been developed and been shown to be effective in the control of ick.[23] *Ichthyophthirius multifiliis* is an exceedingly common and widespread parasite in nature, but in the confines of an aquarium its populations can explode. One of the surest ways to infect an expensive carnivorous ornamental pet fish is to feed it wild caught minnows.

Subclass Peritrichia

Peritrichia contains two orders: Sessilida and Mobilida. Members of both orders have prominent oral ciliary fields with paroral and adoral membranelles. There is a temporary posterior circlet of locomotor cilia, and many are stalked and sessile. All possess an aboral **scopula,** a structure composed of a field of kinetosomes with immobile cilia and functioning either as a holdfast or in stalk formation. Molecular studies, however, suggest that these strucutal similarities between the two groups may be a result of evolutionary convergence.[10]

Order Sessilida

As the name implies, members of this order typically live attached to a substrate. Genera such as *Epistylis* (Fig. 10.8) and *Lagenophrys* are obligate ectocommensals that commonly occur on crustaceans, sometimes in large numbers, including species of economic importance.[14, 24] These protists may show site specificity, occuring most often on particular body regions of a host, and pathological effects have been reported.[24]

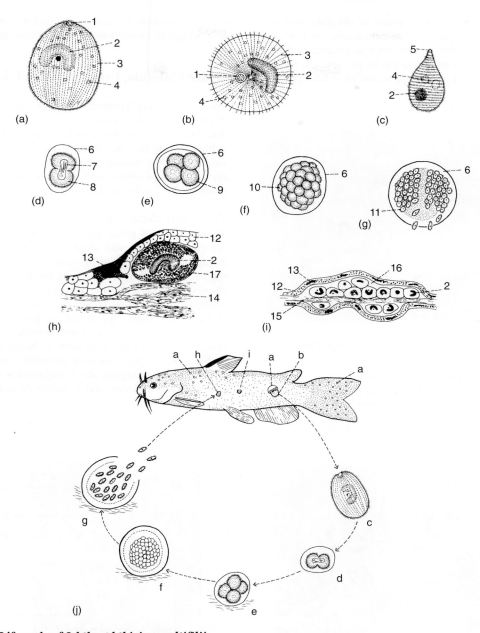

Figure 10.6   **Life cycle of *Ichthyophthirius multifiliis.***

(*A*) Fully developed trophozoite from pustule. (*B*) Anterior end of fully developed trophozoite. (*C*) Tomite from cyst. (*D, E*) First and second divisions of encysted trophozoite. (*F*) Later stage of cystic multiplication. (*G*) Cyst filled with tomites, some of which are escaping into water. (*H*) Section of skin of fish, showing full-grown trophozoite embedded in it. (*I*) Section of tail of carp, showing ciliates developing in pustule. (*J*) Infected bullhead (*Ameiurus melas*). (*1*) cytostome; (*2*) macronucleus with nearby micronucleus; (*3*) longitudinal rows of cilia; (*4*) contractile vacuoles; (*5*) boring or penetrating apparatus; (*6*) cyst; (*7*) dividing of macronucleus; (*8*) two daughter cells formed by first division; (*9*) four daughter cells formed by second division in cyst; (*10*) numerous daughter cells; (*11*) tomites; (*12*) epidermis of fish skin; (*13*) pigment cell in epidermis; (*14*) dermis; (*15*) cartilaginous skeleton of tail of carp; (*16*) pustule containing trophozoites; (*17*) trophozoite under skin; (*a*) pustules; (*b*) trophozoite escaping from pustule into water; (*c*) trophozoite free in water; (*d*) encysted trophozoite on bottom of pond in first division, showing two daughter cells; (*e*) cyst in second division with four daughter cells; (*f*) cyst with many daughter cells; (*g*) ruptured cyst liberating tomites; (*h*) tomite attached to skin; (*i*) tomite partially embedded in skin.

Scyphidia (Fig. 10.9) is a genus of obligate epibionts that occur widely, but not always commonly, on fish, amphibians, and aquatic invertebrates such as annelids and mollusks, including both marine and freshwater species. Species of a similar and related genus, *Apiosoma*, have been reported mainly from fish. Pathology, if present, has been attributed to blockage of gas exchanges due to heavy gill infestation.[22]

Order Mobilida, Family Trichodinidae

Species in family Trichodinidae lack stalks and are mobile. Their oral-aboral axis is shortened, with a prominent basal disc usually at the aboral pole. A protoplasmic fringe, or velum, lies on the margin of the basal disc, and a circle of strong cilia lies underneath. A second circle of cilia, above the disc, cannot always be found. The family contains seven genera, with *Trichodina* being a typical example.[26]

Trichodina Species

Members of this genus parasitize a wide variety of aquatic invertebrates, fish, and amphibians. A basal disc contains a ring of sclerotized "teeth" that aid the parasite in attaching to its host (Fig. 10.10). The number, arrangement, and shapes of these teeth are important taxonomic characters. A buccal ciliary spiral makes more than one but fewer than two complete turns. Species of *Trichodina* may cause some damage to fish gills, but most produce little pathogenic effect and are of interest only as beautiful examples of highly evolved protozoa with incredibly specialized organelles. Typical examples are *T. californica* on salmon gills, *T. pediculus* on *Hydra*, and *T. urinicola* in the urinary bladder of amphibians.

Figure 10.7 Sunfish infected with *Ichthyophthirius multifiliis*.

Note the light-colored pustules in the skin.

From G. Hoffman, "Ciliates of freshwater fishes," in J.P. Kreier (Ed.), *Parasitic protozoa,* vol. 2. © 1978. Academic Press, Inc. Reprinted by permission.

(a)

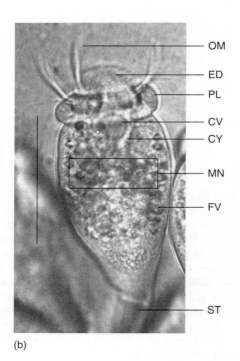

(b)

Figure 10.8 *Epistylis* sp., attached to the wing pad of a damselfly larva.

Epistylis species are colonial ciliates with a thick, non-contractile stalk. (*a*) A colony attached to the back of an aquatic insect, a damselfly larva; (*b*) higher magnification view of one individual ciliate (bar = 50 μm). CV, contractile vacuole; CY, cytopharynx (light, funnel-shaped area); ED, epistomal disc; FV, food vacuoles; MN, macronucleus (a C-shaped structure appearing as a dense area around the cytopharynx); OM; oral membranes used for feeding; PL, peristomal lip; ST, stalk.

Photographs by John Janovy, Jr.

Figure 10.9   *Scyphidia physarum* **from the body surface of a freshwater snail (*Physa fontinalis*).**
The figure shows individuals in various states of contraction. Bar = 20 μm.
Scanning electron micrograph courtesy of Alan Warren.

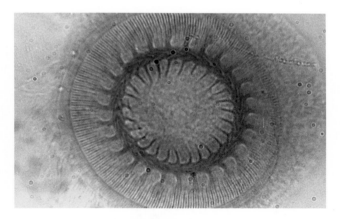

Figure 10.10   *Trichodina* **sp. from a fish gill.**
Trichodina are 35 μm to 60 μm in diameter, with a height of 25 μm to 55 μm.
Courtesy of Warren Buss.

Learning Outcomes

By the time a student has finished studying this chapter, he or she should be able to:

1. Describe the mode of transmission of *Balantidium coli* and describe the conditions under which this species could occur in multiple species of mammals.

2. Draw and label the life cycle of *Ichthyopthirius multifiliis* and describe the conditions under which this parasite could become fatal to fish.

3. Write an extended paragraph describing the various body types of, and infection sites utilized by, members of the subclass Peritrichia.

References

References for superscripts in the text can be found at the following Internet site: www.mhhe.com/robertsjanovynadler9e

Additional Readings

Bykhovskaya-Pavlovskaya, I. E. 1962. *Key to the parasites of freshwater fish* (Israel Program for Scientific Translations, Jerusalem [1964], Trans.). Moscow: Academy of Science. An outstanding reference to ciliate parasites.

Corliss, J. O. 1979. *The ciliated protozoa. Characterization, classification and guide to the literature*. Oxford: Pergamon Press. Advanced treatise but essential to serious students of ciliates.

Hoffman, G. L. 1999. *Parasites of North American freshwater fishes* (2d ed.). Ithaca, NY: Cornell University Press. Ciliates of North American fish are listed in this useful reference work.

Levine, N. D. 1973. *Protozoan parasites of domestic animals and of man* (2d ed.). Minneapolis: Burgess Publishing Co.

Chapter 11

Microsporidia and Myxozoa: Parasites with Polar Filaments

. . . an enigma wrapped in a puzzle.
— Arthur Koestler

Because they form spores, members of these two groups were formerly placed in a class (Cnidosporidea) of a subphylum Sporozoa (which also included the gregarines, coccidians, and malarial parasites). However, both Microspora and Myxozoa are quite different from apicomplexans and, indeed, bear little if any relationship to each other. Furthermore, Sporozoa is no longer considered a valid taxonomic group. Myxozoa are not protozoa at all but are now included in phylum Cnidaria (anemones, jellyfish, corals), although their position within that phylum is still a matter of discussion.[41]

Myxozoans occur mainly in fish. Microsporidians are mostly parasites of invertebrates, especially insects, but some are found in vertebrates, and a few are recognized as opportunistic parasites in humans, especially in immunodeficient patients. Life cycles have not been worked out for many of these parasites, but among those that are known, myxozoans and some microsporans require a second host.

Both groups possess **polar filaments,** which are tubelike and held coiled within the spores. When spores encounter the proper environment, typically a host's digestive system, the polar filaments are expelled. In Myxozoa the filaments lie within polar capsules and evidently serve an anchoring function after expulsion. In Microsporidia the polar filament is also called a **polar tube.** It pierces the intestinal epithelium of the host, and the amebalike sporoplasm passes through the tubular filament into the host cell. In both Myxozoa and Microsporidia, polar filaments can be stimulated to extrude artificially.

PHYLUM MICROSPORIDIA

Phylum Microsporidia includes about 1200 known species of intracellular parasites, and new ones are being described regularly.[6, 8] These species have been found in protozoa, platyhelminths, nematodes, bryozoa, rotifers, annelids, all classes of arthropods, fish, amphibians, reptiles, birds, and some mammals. Members of seven genera have been reported from humans.[49] A number are pathogenic, especially in immunodeficient patients, and several are of economic importance to agriculture.

Phylogenetic relationships of Microsporidia are still somewhat unsettled. Although molecular evidence, including a complete sequence of the *Encephalitozoon cuniculi* genome, suggests strong fungal affinities, the exact placement of these parasites relative to fungal and protist taxa remains to be resolved.[17, 26, 47]

Microsporidia formerly included a class Haplosporea. Haplosporideans are now placed in phylum Haplosporidia (see chapter 4).[42] Haplosporideans are parasites of invertebrates; *Haplosporidium* spp. (formerly *Minchinia* spp.) are pathogens in the economically important oyster *Crassostrea virginica.* We will not discuss them further.

Microsporidian taxonomy is in a state of flux because relationships revealed by molecular techniques do not necessarily match those postulated on morphological grounds.[8] For example, *Pleistophora,* a genus whose species infect a variety of fish, has been broken into several genera based on small subunit rDNA and RNA polymerase amino acid sequences.[38] Nuclear structure during development and presence or absence of a **sporophorous vesicle** (an envelope or membrane containing spores) within a host cell are both characters used for identification. Some species are diplokaryotic during merogony; that is, they have nuclei joined in pairs. However, others have single nuclei at this stage. Microsporidians that form a sporophorous vesicle usually have a characteristic number of spores within the vesicle.[8] Microsporidians lack Krebs cycle and some synthetic pathway enzymes, thus explaining their dependence on a host.[26]

Spores are the most conspicuous and morphologically distinctive stages in microsporidian life cycles; they are unicellular, contain a single sporoplasm, and are ovoid, spheroid, or cylindroid in shape. Spore walls are complete, without suture lines, pores, or other openings. They are trilaminar, consisting of an outer, dense exospore, an electron-lucent middle layer (endospore), and a thin membrane surrounding the cytoplasmic contents (Fig. 11.1). Some species have two to five layers of exospore. The wall is dense and refractile; its resistant properties contribute greatly to the survival of the spore. Spores have a simple or complex extrusion apparatus with polar tube and polar cap. A vacuolelike organelle, the **polaroplast,** may be located near the polar tube (see Fig. 11.1). Spores are usually about 3 μm to 6 μm in length, and little structure can be discerned under a light microscope other than an apparent vacuole at one or both ends. The smallest known spores are those of *Encephalitozoon* spp. from mammals (2.5 μm by 1.5 μm); the largest belong to *Mrazekia piscicola* from cod (20 μm by 6 μm).

There is no polar capsule in microsporidians, and neither is the polar filament formed by a separate capsulogenic cell (as it is in myxozoans). At the ultrastructural level we can see a small **polar cap** or **sac** covering the attached end of the filament (Fig. 11.1).

An ameboid sporoplasm surrounds the extrusion apparatus, with its nucleus and most of its cytoplasm lying within the filament coils. A **posterior vacuole** may be found at the end opposite the polaroplast. The sporoplasm has many free ribosomes and some endoplasmic reticulum but no mitochondria, peroxisomes, or typical Golgi membranes. The polar cap membrane and matrix are continuous, with a highly pleated membrane comprising the polaroplast. This, in turn, is continuous with the anchoring disc or polar filament base.[48]

When extrusion of the polar filament is stimulated in a host, a permeability change in the polar cap apparently allows water to enter the spore, and the filament is expelled explosively, simultaneously turning "inside out." The stacked membrane in the polaroplast unfolds as the filament discharges, and this membrane contributes to the expelled filament so that it is much longer than when it is coiled within the spore. The force with which the filament discharges causes it to penetrate any cell in its path, and the sporoplasm flows through the tubular filament, thereby gaining access to its host cell.[48] Within the host cell the filament's end expands to enclose the sporoplasm and becomes the parasite's new outer membrane.

The intracellular trophozoite's nuclei divide repeatedly, and the organism becomes a large, multinucleate plasmodium. Finally, cytokinesis takes place, and the process may then be repeated. In diplokaryotic species, nuclei are associated in pairs (**diplokarya**), but such association apparently is not involved with sexual reproduction. Trophozoite multiple fission (merogony) is usually regarded as schizogony, but the process may not be strictly analogous to schizogony found in Apicomplexa.

Sporogenesis occurs when nuclear divisions of monokaryotic or dikaryotic trophozoites give rise to nuclei destined to become spore nuclei. In a number of genera, not including *Nosema,* nuclear division preceding sporogony is meiotic (reductional), giving rise to haploid spores.[19] In these genera spores are not directly infective to new hosts, leading to the suggestion that there is an alternate (intermediate?) host in which restoration of diploidy occurs. For example, see Canning and Hollister[5] for *Amblyospora* in copepods and mosquitoes. Sexual reproduction seems to be restricted to plasmogamy, not karyogamy.

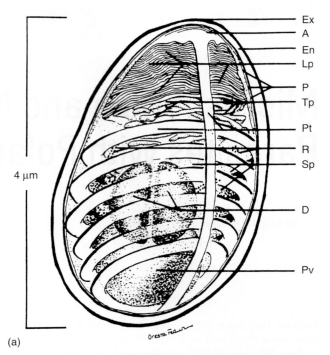

4 μm

(a)

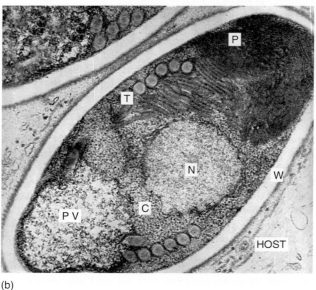

(b)

Figure 11.1 **Microsporidian spores.**

(*a*) Diagram of the internal structure of a microsporidian spore. *Ex*, electron dense exospore; *A*, filament anchoring disc; *Lp*, lamellar polaroplast; *Tp*, tubular polaroplast; *Pt*, polar tubule; *D*, diplokaryon nuclei; *Pv*, posterior vacuole; *En*, endospore; *P*, plasma membrane; *R*, ribosomes; *Sp*, sporoplasm, (*b*) *Nosema lophii* spore displaying polaroplast (*P*), nucleus (*N*), ribosome-rich cytoplasm (*C*), polar tube (*T*), posterior vacuole (*PV*), and wall (*W*).

(*a*) From A. Cali, "General microsporidian features and recent findings on AIDS isolates," in *J. Protozool.* 38:625–630, 1991. Copyright © 1991 The Society of Protozoologists. Reprinted by permission. (*b*) From E. Weidner, "Ultrastructural study of microsporidian invasion into cells," in *Z. Parasitenkd.* 40:227–242. Copyright © 1972.

During sporogony the organism becomes a multinucleate, sporogonial plasmodium. This change can occur either by internal segregation of cytoplasm around the nuclei to become sporont-determinate areas or by formation of an envelope at the sporont surface and subsequent separation from developing sporoblasts, leaving a vacuolar space.[30] Spores then differentiate and mature within the pansporoblast. In each sporoblast, there forms a mass of tubules, which becomes the polar tube and polaroplast.[30] Mitochondria are not present at any stage. A **xenoma** (combination parasite and hypertrophied cell) of considerable size develops in some species.[6]

Family Nosematidae

Genera of Nosematidae are separated on the basis of number of spores produced by each sporoblast mother cell during the life cycle (from 1 to 16 or more).

Nosema Species

Nosema apis is a common parasite of honey bees in many parts of the world, causing much loss annually to beekeepers. It infects epithelial cells in the insect's midgut. Infected bees lose strength, become listless, and die. Although a queen bee's ovaries are not directly infected, they degenerate when her intestinal epithelium is damaged, an example of parasitic castration. The disease is variously known as *nosema disease, spring dwindling, bee dysentery, bee sickness,* and *May sickness.*

Nosema apis spores are oval, measuring 4 μm to 6 μm long by 2 μm to 4 μm wide. The extended filament is 250 μm to 400 μm long. Infected bees defecate spores that are infective to other bees. Swallowed spores enter the midgut and lodge on the peritrophic membrane. Extruded filaments pierce the peritrophic membrane and intestinal epithelial cells, and sporoplasms enter epithelial cells. The entire process is accomplished within 30 minutes. Sporogony takes place in the second multiple fission generation, and spores rupture host cells to be passed with feces. The entire life history in bees is completed in four to seven days. Destruction of intestinal epithelium kills a host.

Nosema bombycis is a parasite of silk moth larvae, flourishing in the crowded conditions of silkworm culture. The parasite affects nearly all tissues of the insect's body, including intestinal epithelium. Parasitized larvae show brown or black spots on their bodies, giving them a peppered appearance. There is a high rate of mortality. Pasteur devoted considerable effort in 1870 to understanding and controlling this disease and is credited with saving the silk industry in the French colonies. *Nosema bombycis* also was one of the first "germs" proved to cause disease. Its life cycle is basically similar to that of *N. apis* and can be completed in four days.

Because of the pathological effects, microsporidians are also being studied as biological control agents. *Nosema algerae* infection, for example, reduces the number of malarial oocysts formed in *Anopheles* mosquitoes.[40] *Nosema whitei* is pathological to *Tribolium* (flour beetles) species, which, in addition to being favorites of experimental ecologists, are among the many stored grain pests that significantly reduce global food supplies.[2] Molecular studies have shown that an important and widely studied parasite of orthopterans (grasshoppers and allies), available as a commercial pesticide and previously known in the literature as *Nosema locustae,* is not closely related to other *Nosema* species after all, hence has been renamed *Antonospora locustae* (Fig. 11.2).[43]

Other Microsporidian Species

Species of *Glugea and Pleistophora,* as well as other genera, parasitize fish, including several economically important groups. Serious epizootics have been reported.

Encephalitozoon cuniculi is among the most extensively studied of all Microsporidia, occurring in laboratory mice and rabbits, monkeys, dogs, rats, birds, guinea pigs, and

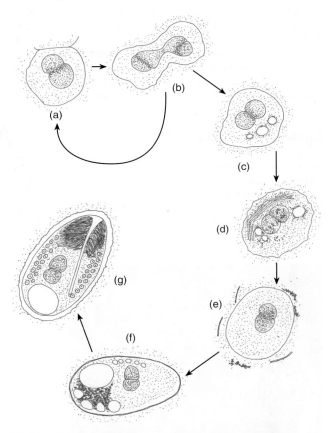

Figure 11.2   Development of *Antonospora locustae* in a grasshopper's fat body.

(a), (b) Merogonic cycle in which diplokaryotic meronts multiply by binary fission. *(c)–(e)* Transformation of meront into sporont, with vacuolization of the cytoplasm, buildup of rough endoplasmic reticulum, and accumulation of electron dense particles in the sporont cytoplasm. *(e)* Appearance of dense tubes and particles in the host cell cytoplasm; these materials will later adhere to the developing spore. *(f)* Following division of sporonts, cells become sporoblasts, recognizable by their elongate shape, thickening of wall, and accumulation of a vesicular-tubular material at their posterior end. *(g)* Mature spore.

Drawing by John Janovy Jr., based on electron micrographs from Y. Y. Sokolova and C. E. Lange, 2002. An ultrastructural study of *Nosema locustae* Canning (Microsporidia) from three species of Acrididae (Orthoptera) in *Acta Protozool.* 41:229–237, 2002.

other mammals, including humans, and at various early times it was thought to cause rabies and polio. It may be transmitted in body exudates or transplacentally. Although damage is usually minimal, an infection can be fatal, especially in AIDS patients.[37] In one such individual dying from a combination of opportunistic infections, molecular analysis revealed the *E. cuniculi* strain to be of dog origin; brain, heart, and adrenal glands were all infected.[37] High levels of anti–*E. cuniculi* antibodies are common in immunodeficient patients but low in uncompromised people.

Other microsporidian species have been isolated from AIDS patients and others who were unable to rally their immune defenses. Documented cases have been reported for species of genera *Pleistophora, Nosema, Enterocytozoon, Encephalitozoon, Vittaforma, Brachiola, Trachipleistophora,* and *"Microsporidium"* (a catchall genus for microsporidian parasites of unknown affinities).[4, 49] Although it is often difficult to pinpoint the exact pathological effect of one parasite in multiply infected hosts, one study showed that 44% of AIDS patients with diarrhea also had microsporidial infections, while only 2.3% of those without diarrhea had such infections.[10] There is no established treatment for humans, but polyamine analogs have been used to treat experimental infections in mice.[1]

Epidemiology and Zoonotic Potential

Microsporidian spores are exceedingly common in the environment, and consequently these parasites are candidates for opportunistic infections. *Enterocytozoon bieneusi,* which infects humans, also has been reported from a variety of wild animals.[45] Epidemiological studies show that people living under poor sanitary conditions and exposed to duck and chicken droppings are at a high risk of infection.[3] Aquatic birds in general can be carriers, and one study showed that "a single visit of a waterfowl flock can introduce into the surface water approximately 9.1×10^8 microsporidian spores of species known to infect humans."[44] Urban park pigeons may also be carriers, thus exposing elderly people and children who might not otherwise live in unsanitary circumstances.[18]

In one study, both supermarket and street vendor vegetables were found to be contaminated with spores of species infective to humans.[25]

MYXOZOA

Myxozoa are parasites both of invertebrates and vertebrates, the latter mostly fish; no myxozoans are known from birds or mammals. Two classes are recognized: Malacosporea, with species in freshwater bryozoans and fish, and Myxosporea, with species in annelids, sipunculids, fish, amphibians, and occasionally reptiles. Myxozoans whose life cycles are known have sexual-proliferative cycles in invertebrates and asexual-proliferative cycles in vertebrates. Some species are of economic importance because they are pathogenic to food and sport fish. Excellent reviews of the group can be found in Canning and Okamura,[7] and Lom and Dyková.[35]

Myxozoa are characterized by spores that are of multicellular origin and beautifully diverse structurally (Figs. 11.3, 11.4). The **myxospore** life-cycle phase occurs in vertebrate hosts, with spores typically arising in large

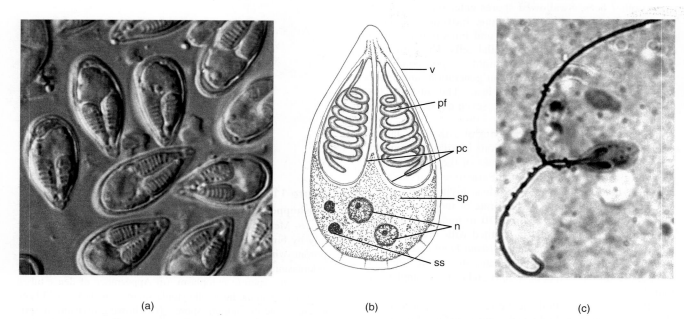

(a)   (b)   (c)

Figure 11.3 **General structure of myxozoan spores.**
(*a*) Myxospores from a *Myxobolus* species parasitizing a minnow, as seen in a fresh squash preparation of a cyst. (*b*) Drawing of a typical *Myxobolus* spore showing various internal structures and binucleate sporoplasm. Sporoplasmosomes are dense bodies of unknown function. (*c*) *Myxobolus* spore with polar filaments extruded by pressure. pc, polar capsules; pf, polar filament; n, nuclei; sp, sporoplasm; ss, sporoplasmosomes; v, valve.

(*a*) Courtesy of W. L. Current; (*b*) and (*c*) drawing and photograph by John Janovy, Jr.

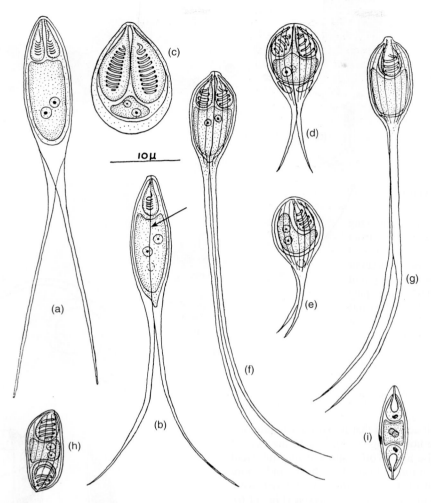

Figure 11.4   **Diverse spore structures among representative myxozoan species.**

(a) Henneguya umbri (frontal view); *(b) Henneguya umbri* (side view); *(c) Myxobolus eucalii* (frontal view); *(d) Myxobilatus noturii* (frontal view); *(e) Myxobilatus noturii* (side view); *(f) Myxobilatus cotti* (frontal view); *(g) Myxobilatus cotti* (side view); *(h) Myxidium* sp. (side view); *(i) Myxidium umbri* (side view). These species were all found in various tissues of the western mudminnow and tadpole madtom, freshwater fishes of Lake Michigan, and illustrate both intra- and interspecific diversity of spore morphology. Arrow indicates suture line between valves in *(b)*.

From H. G. Guilford, "New species of Myxosporidia from Green Bay (Lake Michigan)," in *Trans. Am. Micro. Soc.* 84:566–573. Copyright © 1965 Wiley-Blackwell. Reprinted by permission of Blackwell Publishing Ltd.

plasmodia called **pansporoblasts.** Myxospores contain one or more infective ameba-like **sporoplasms** and nematocyst-like **polar capsules,** all enclosed in up to seven **valves** joined along suture lines (see Fig. 11.4b). Sporoplasms may contain electron-dense bodies, called **sporoplasmosomes,** of unknown function. Spore components each arise from separate cells during development (see Fig. 11.11). Polar capsules contain coiled filaments that quickly discharge upon contact with hosts and aid in attachment and infection. Sexual reproduction occurs in invertebrate definitive hosts, with sporoplasms undergoing merogony to form gametes which then fuse and develop into **actinospores,** also called **triactinomyxons** in some species (see Fig. 11.6). Life-cycle details and terminology are provided by Lom and Dyková.[35]

More than 1300 species in 62 genera of Myxozoa have been described. Most are host and tissue specific. Molecular and ultrastructural studies show that Myxozoa are closely related to, if not members of, phylum Cnidara (corals, jellyfish, anemones).[27, 28] We continue to treat Myxozoa as phylum, however, because to date there are no publications that formally eliminate the phylum and establish a cnidarian taxon to contain these parasites.

Family Myxobolidae

Myxobolidae are parasites of fishes. They have two or four polar capsules in the spore stage, and their sporoplasm lacks iodinophilous vacuoles.[21] One species, *Myxobolus cerebralis,* is of circumboreal importance to salmonid fish, including trout.

Elucidation of myxozoan life histories has been one of the more interesting parasitological developments in recent

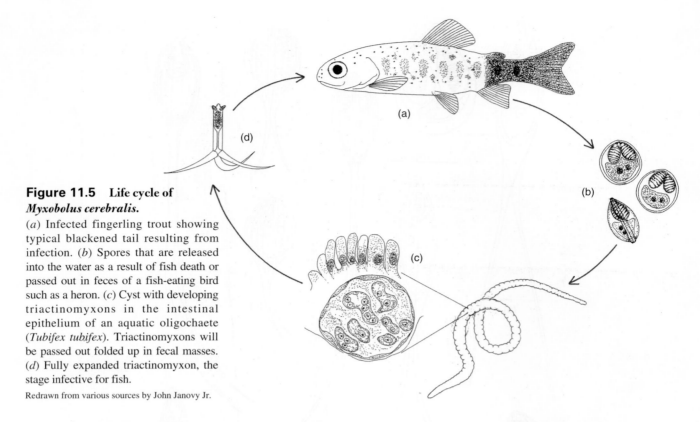

Figure 11.5 **Life cycle of**
Myxobolus cerebralis.

(*a*) Infected fingerling trout showing typical blackened tail resulting from infection. (*b*) Spores that are released into the water as a result of fish death or passed out in feces of a fish-eating bird such as a heron. (*c*) Cyst with developing triactinomyxons in the intestinal epithelium of an aquatic oligochaete (*Tubifex tubifex*). Triactinomyxons will be passed out folded up in fecal masses. (*d*) Fully expanded triactinomyxon, the stage infective for fish.

Redrawn from various sources by John Janovy Jr.

decades. It is now well accepted that myxozoans require annelids as intermediate hosts (Fig. 11.5).[16, 29] In the case of *M. cerebralis,* Markiw and Wolf[36] showed that tubificid oligochaetes were required intermediate hosts and, more remarkably, that the parasite stages infective for fish were identical to members of genus *Triactinomyxon* (Fig. 11.6), which had formerly been placed in a separate class (Actinosporea) of Myxozoa. Triactinomyxons liberated into water from worms were infective for fish, which developed typical *M. cerebralis* infections.[16]

The triactinomyxon, or actinospore, stage has its own complex life cycle in worms, involving sexual reproduction and production of actinospores with three valves (contrasted with two in spores that develop in fish).[34] Initial stages of spore development occur in cysts in intestinal epithelium. Sporogenesis begins when two cells envelop two others (Fig. 11.7). The enclosed cells then undergo a series of divisions and arrangements into their relative positions in the finished spore, with three flattened valve cells surrounding both a sporoplasm (Fig. 11.8) and capsulogenic cells (those that will develop into polar filaments). The actinospore shaft and long hooks are inflated to their final form after the spores' release from the worm in feces. In addition to *M. cerebralis,* a number of other myxozoan species have likewise been transmitted to various fishes by way of actinospores.[13, 14] The worm portion of the life cycle requires about three months for completion.

When exposed to a variety of stimuli (pressure, acid), myxozoan polar capsules shoot out their filaments in a manner analogous to eversion of the finger of a glove. This event presumably initiates the infection. *Myxobolus cerebralis* actinospores attach to trout fry quickly upon exposure, and the sporoplasm invades the epidermis within 15 minutes.[15]

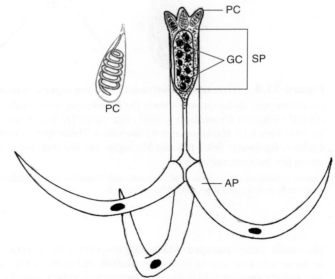

Figure 11.6 **A *Triactinomyxon* spore, typically released from aquatic oligochaetes.**

Sexual reproduction occurs during development of the spore, resulting in formation of infectious germ cells (GC) within the elongate style. AP, anchor-like processes; SP, sporoplasm containing germ cells; PC, polar capsules. Bar = 100 μm.

Drawn by John Janovy, Jr. from various sources.

After a few days, *M. cerebralis* sporoplasms, consisting of a primary cell containing up to several enveloped secondary cells, can be found in the central nervous system, although invasion of cartilage requires up to 80 days.[15]

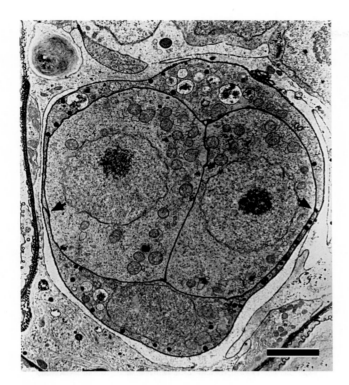

Figure 11.7   **Beginnings of spore formation in** *Triactinomyxon.*

Two outer cells (contact points indicated by *arrows*) envelop two inner cells. (Bar = 3 μm)

From J. Lom and I. Dyková, "Fine structure of *Triactinomyxon* early stages and sporogony: Myxosporean and actinosporean features compared," in *J. Protozool.* 39:16–27. Copyright © 1992 The Society of Protozoologists. Reprinted by permission.

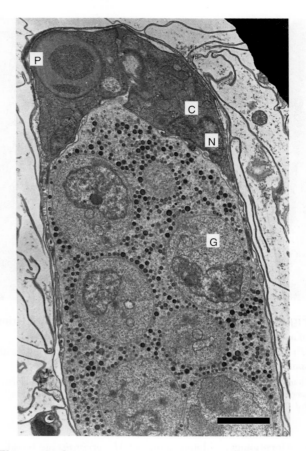

Figure 11.8   **Anterior part of** *Triactinomyxon.*

Note two capsulogenic cells (*C*) containing capsule primordium (*P*) and nucleus (*N*). (*G*) Germinal cells of the sporoplasm. (Bar = 2 μm)

From J. Lom and I. Dyková, "Fine structure of *Triacinomyxon* early stages and sporogony: Myxosporean and actinosporean features compared," in *J. Protozool.* 39:16–27. Copyright © 1992 The Society of Protozoologists. Reprinted by permission.

Some myxozoan species remain in the skin, while others eventually localize in other sites, such as gills, but the exact route of migration is not known in all cases. Once a sporoplasm reaches its characteristic infection site, it begins to grow, its nuclei dividing repeatedly.[12] A multinucleate trophozoite often grows until it is visible to an unaided eye (Fig. 11.9)—some species can reach a size of several millimeters—feeding from surrounding tissues by pinocytosis[11] (Fig. 11.10).

During growth and nuclear divisions, two types of nuclei can be distinguished, **generative** and **somatic** (Fig. 11.11). As development proceeds, a certain amount of cytoplasm becomes segregated around each generative nucleus to form a separate cell within the trophozoite. These cells will produce spores; hence, they are called **sporoblasts.** Because in most species each will give rise to more than one spore, they also are called **pansporoblasts.** Each pansporoblast in *M. cerebralis* will produce two spores. The generative nucleus for each spore divides four times, one daughter nucleus of each division remaining generative and the other becoming somatic. The first somatic daughter nucleus forms the spore's outer envelope; the second divides again to give rise to valvogenic cells; and the third nucleus divides to produce nuclei of polar capsule cells.[12] Thus, a myxozoan spore is of multicellular origin.

Genus *Myxobolus*

Myxobolus species (some of which were formerly placed in a now defunct genus *Myxosoma*) have ovoid or teardropshaped spores with a distinct sutural line and two polar capsules (see Fig. 11.3). A wide variety of fishes, especially minnows, are infected with *Myxobolus* spp. Infections occur in several tissues: skin, gills, and various internal organs.

Myxobolus cerebralis. In salmonids, *M. cerebralis* causes **whirling disease,** so called because fish with the disease swim in circles when disturbed or feeding. The parasite apparently was formerly endemic in the brown trout from central Europe to southeast Asia, and it causes no symptoms in that host. Whirling disease was first noticed in 1900 after introduction of rainbow trout to Europe. Since then it has spread to other localities in Europe, including Sweden and Scotland; to the United States; to South Africa; and to New Zealand.[20] The disease results in a high mortality rate in very young fish and causes corresponding economic loss, especially in hatchery-reared brook and rainbow trout. If a fish survives, damage to the cranium and vertebrae can cause crippling and malformation.

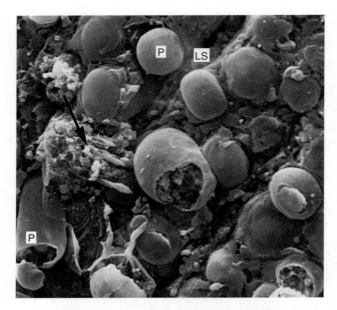

Figure 11.9 Scanning electron micrograph of the interior of a channel catfish gill infected with *Henneguya exilis.*

(*P*) Parasite cysts (plasmodia); *arrow* points to broken cyst with spores protruding; (*LS*) lamellar sinuses.

From W. L. Current and J. Janovy Jr, "Comparative study of ultrastructure of inter-lamellar and intralamellar types of *Henneguya exilis* Kudo from channel catfish," in *J. Protozool.* 25:56–65. Copyright © 1978 by the Society of Protozoologists.

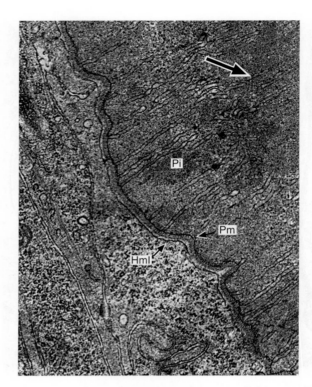

Figure 11.10 Transmission electron micrograph of the *Myxobolus (Myxosoma) funduli* cyst wall.

(*Pi*) Zone of pinocytic canals; (*Pm*) parasite cyst membrane; (*Hm*) host cell membranes; *large arrow,* pinocytic vesicle at end of canal.

From W. L. Current et al., "*Myxosoma funduli* Kudo (Myxosporida) in *Fundulus kansae:* Ultrastructure of the plasmodium wall and of sporogenesis," in *J. Protozool.* 26:574–583. Copyright © 1979 by the Society of Protozoologists.

- **Morphology.** Mature spores of *M. cerebralis* are broadly oval, with thick sutural ridges on the valve edges. They measure 7.4 µm to 9.7 µm long by 7 µm to 10 µm wide. Spores are covered with a mucoidlike envelope. There are two polar capsules at the anterior end, each with a filament twisted into five or six coils. During development each polar capsule lies within a polar cell that also contains a nucleus, and nuclei of the two valvogenic cells may be seen lying adjacent to the inner surface of each valve. The sporo-plasm contains two nuclei (presumably haploid), numerous ribosomes, mitochondria, and other typical organelles.[33]

- **Biology.** Following encounter with a fish, the triacti-nomyxon exsporulates, and the sporoplasm migrates to the spine and head cartilages; it begins growing, and its nuclei divide, as discussed previously, forming cavities in the surrounding cartilage tissue. These cavities within the cartilage become packed with trophozoites and spores by eight months after infection. Spores may live in fish for three or more years. Our understanding of how they escape into the water is speculative, but it seems reason-able to assume that, when the host is devoured by a larger fish or other piscivorous predator, such as a kingfisher or heron, spores are released by digestion of their former home. The parasite's crippling effect can make a fish es-pecially vulnerable to predation. *Myxobolus cerebralis* can be spread through feces of birds that have been fed fish.[46]

- **Pathogenesis.** The main pathogenic effects of this disease are damage to cartilage in the axial skeleton of young fish, consequent interference with function of adja-cent neural structures, and subsequent granuloma forma-tion in healing of the lesions. Invasion of the cartilaginous capsule of the auditory-equilibrium organ behind the eye interferes with coordinated swimming. Thus, when an infected fish is disturbed or tries to feed, it begins to whirl frantically, as if chasing its tail. It may become so exhausted by this futile activity that it sinks to the bot-tom and lies on its side until it regains strength. Predation most likely occurs at this stage.[22] Often the spine cartilage is invaded, especially posterior to the 26th vertebra. Func-tion of sympathetic nerves controlling melanocytes is impaired, and an infected fish's posterior part becomes very dark, producing the "black tail." If the fish survives, granulomatous tissue infiltration of the skeleton may produce permanent deformities: misshapen head, perma-nently open or twisted lower jaw, or severe spinal curva-ture (scoliosis; Fig. 11.12).

- **Epizootiology and Prevention.** It seems clear that in ponds in which infected fish are held, spores can accumu-late, whether by release from dead and decomposing fish, passage through predators, or some kind of escape from the tissue of infected living fish. Severity of an outbreak depends on the degree of contamination of a pond, and light infections cause little or no overt disease. Spores are resistant to drying and freezing, surviving for a long pe-riod of time, up to 18 days, at –20°C.[23]

No effective treatment for infected fish is known, and such fish should be destroyed by burial or incineration. Great care must be exercised to avoid transferring spores

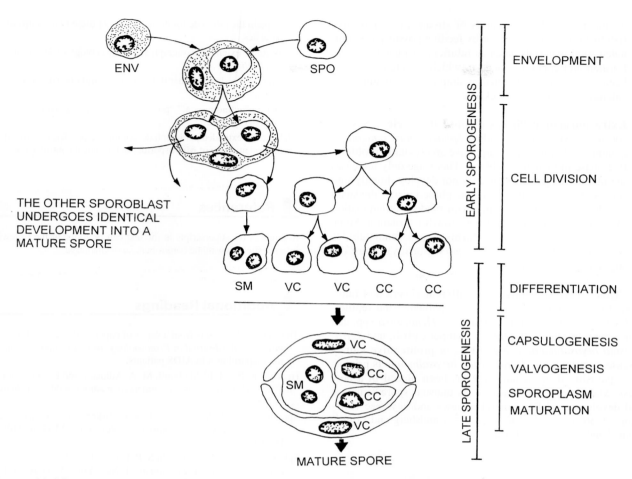

Figure 11.11   **Diagram of the development of a myxosporidian within a cyst in a vertebrate host.**

Initial stages of sporogenesis involve envelopment of a sporoblast cell (SPO) by an enveloping cell (ENV). While remaining inside the enveloping cell, sporoblasts subsequently divide into precursors of valves (valvogenic cells, VC), polar capsules (capsulogenic cells, CC), and binucleate infective sporoplasms (SM). In this particular case (*Henneguya exilis*), two myxospores are formed within a single enveloping cell. Differentiation into a mature spore involves deposition of valve proteins, acquisition of final shape, and formation of the polar filament with the capsulogenic cells.

Drawn by John Janovy, Jr., based on information from W. L. Current and J. Janovy, Jr. "Sporogenesis in Henneguya exilis infecting the channel catfish: an ultrastructural study," in Protistologica 13:157–167, 1977.

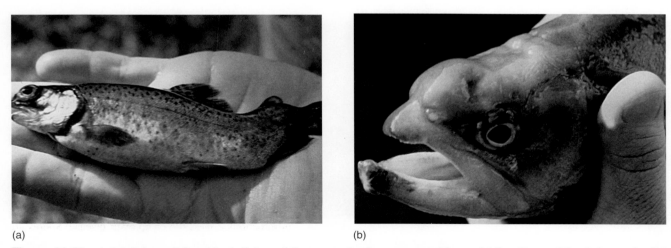

(a)   (b)

Figure 11.12   **Axial skeleton deformities in living rainbow trout that have recovered from whirling disease (*Myxobolus cerebralis*)**

(*a*) Note bulging eyes, shortened operculum, and both dorsoventral and lateral curvature of the spinal column (lordosis and scoliosis). (*b*) Note gaping, underslung jaw and grotesque cranial granuloma.

Photographs by Larry S. Roberts.

to uncontaminated hatcheries or streams, either by live fish that might be carriers or by feeding possibly contaminated food materials, such as tubificids, to hatchery fish. Earthen and concrete ponds in which infected fish have been held can be disinfected by draining and treating with calcium cyanamide or quicklime.

- **Extrasporogonic Phases of the Life Cycle.** Several species of Myxozoa, including *Sphaerospora renicola* in commercially important carp, have an asexually proliferative phase in their host's blood. This stage only increases the number of parasites; it does not develop directly into spores. A second extrasporogonic phase invades the swim bladder of carp fry, causing swim bladder inflammation that results in high mortality or growth retardation. Some small plasmodia (ameboid forms) reach renal tubules where they either produce spores (seasonally) or are destroyed by host reactions.[32]

Other species, belonging to different genera and families, also commonly occur in fish, amphibians, and reptiles, and some are of economic importance. *Henneguya* spp. can cause mass mortality of cultured channel catfish, and *Tetracapsula bryosalmonae* causes PKX, or proliferative kidney disease, in salmonids. *Tetracapsula bryosalmonae* is primarily a parasite of bryozoans and has been placed in a new class, Malacosporea, because of its unusual spore structure and development.[9] For general reviews and keys see Hoffman,[21] Hoffman et al.,[24] Kent et al.,[28] Landsberg and Lom,[31] Lom,[32] and Lom and Dykova.[35]

Learning Outcomes

By the time a student has finished studying this chapter, he or she should be able to:

1. Draw the structure of a microsporidian spore and diagram, with labels, the life cycle of a typical microsporidian.
2. Draw the structure of a myxozoan spore and diagram, with labels, the life cycle of a typical myxozoa.

3. Explain the rationale for classification of these two groups of parasites.
4. Write an extended paragraph about the ecology of *Myxobolus cerebralis.*
5. Draw some sketches that illustrate structural diversity among myxozoan spores.
6. Define "actinomyxon," "polar capsule," "sporoplasm," and "valvogenic cell."
7. Write an extended paragraph describing the problems surrounding the potential use of Microsporidia as biological control agents.

References

References for superscripts in the text can be found at the following Internet site: www.mhhe.com/robertsjanovynadler9e

Additional Readings

The Cali 4 reference is from a series of papers, all published in vol. 38 of the *Journal of Protozoology,* from a symposium on microsporidiosis in AIDS patients.

Hedrick, R. P., M. El-Matbouli, M. A. Adkison, and E. MacConnell. 1998. Whirling disease: Re-emergence among wild trout. *Immunol. Rev.* 166:365–376.

Sprague, V. 1982. Ascetospora. In S. P. Parker (Ed.), *Synopsis and classification of living organisms* 1. New York: McGraw-Hill Book Co., pp. 599–601.

Sprague, V. 1982. Myxozoa. In S. P. Parker (Ed.), *Synopsis and classification of living organisms* 1. New York: McGraw-Hill Book Co., pp. 595–597.

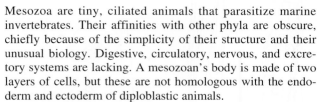

Chapter 12

The Mesozoa:
Pioneers or Degenerates?

It has proved to be a docile animal, easily anesthetized and with vascular and renal systems superbly available for preparative surgery. . . . catheters have been inserted wherever needed.

—A. W. Martin, describing the renal system of a 15 kg *Octopus dofleini*[15]

Mesozoa are tiny, ciliated animals that parasitize marine invertebrates. Their affinities with other phyla are obscure, chiefly because of the simplicity of their structure and their unusual biology. Digestive, circulatory, nervous, and excretory systems are lacking. A mesozoan's body is made of two layers of cells, but these are not homologous with the endoderm and ectoderm of diploblastic animals.

Two distinct groups were formerly placed in phylum Mesozoa: classes Rhombozoa and Orthonectida. However, these two groups are so different in morphology and life cycles that most current authors believe they should be placed in separate phyla; molecular studies support this position and we concur.[9, 11] The two mesozoan phyla are Dicyemida and Orthonectida.[3] Dicyemida are parasites of cephalopod molluscs exclusively, being reported from well over a hundred species of squid and octopus;[4] Orthonectida occur in several invertebrate groups, including annelids, bryozoans, echinoderms, and urochordates.[10]

Because of their presumed primitiveness, structural simplicity, limited cell numbers, and elaborate development, mesozoans have attracted researchers interested in cell-to-cell communication, chromosomal replication, responses to hormones, and mitochondrial differentiation.[1, 2, 7, 17] Such studies reveal a number of seemingly odd phenomena, especially ones involving chromosomal DNA replication, amplification, and ultimate reduction. And the cephalopod hosts are themselves fascinating and even captivating animals to study.

PHYLUM DICYEMIDA

Class Rhombozoa

Rhombozoans are parasites of renal organs of cephalopods, either lying free in the kidney sac or attached to renal appendages of the vena cava. Partial life cycles are known for a few species, but certain details are lacking in all

cases. Interesting histories of the group were presented by Stunkard.[22, 23]

Order Dicyemida

The most prominent developmental stages in cephalopods are **nematogens** and **rhombogens** (Figs. 12.1, 12.2, 12.3). Their bodies are composed of a **polar cap,** or **calotte,** and a trunk. The calotte is made up of two tiers of cells, usually with four or five cells in each. The anterior tier is called **propolar;** the posterior is called **metapolar.** Cells in the two tiers may be arranged opposite or alternate to each other, depending on the genus. The trunk comprises relatively large axial cells surrounded by a single layer of ciliated, somatic cells. Axial cells give rise to new individuals, as in the following description.

The earliest known stage in cephalopods is a ciliated larva, or **nematogen** (Figs. 12.2, 12.3a). Axial cells of a nematogen each contain a **vegetative nucleus** (AN in Fig. 12.2) and one or more **germinative nuclei** that develop into **agametes** (AG in Fig. 12.2), which in turn divide, becoming aggregates of cells, in a process much like the asexual internal reproduction of germ balls found in trematode larval stages within snails (see chapter 15). Within the axial cell, agametes develop into vermiform embryos that escape the nematogen's body and attach to host kidney tissues.

Agametes within an axial cell of a nematogen produce many generations of identical vermiform embryos that also develop into nematogens, building up a massive infection in the cephalopod. When a host becomes sexually mature, production of nematogens ceases. Instead, vermiform embryos form stages that become rhombogens, similar to nematogens in cell number and distribution but with a different method of reproduction and with lipoprotein- and glycogen-filled somatic cells. These cells may become so engorged that they swell out, and the animal appears lumpy.

Rhombogens (Fig. 12.3b) produce, in the axial cell, agametes that divide to become nonciliated **infusorigens** (Fig. 12.2). An infusorigen is a mass of reproductive cells

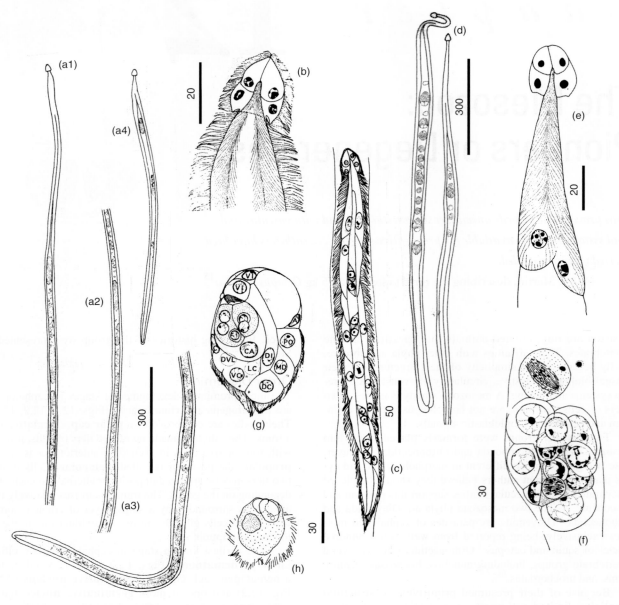

Figure 12.1   **Various life cycle stages of *Dicyemennea antarcticensis*.**
(*a*) Entire nematogens (*a1, a2,* and *a3* all sections of one; *a4* complete). (*b*) Anterior end of nematogen. (*c*) Vermiform embryo within axial cell of nematogen. (*d*) Rhombogens. (*e*) Anterior end of rhombogen. (*f*) Infusorigen. (*g–h*) Infusiform larvae. Cells in (*g*) are *A,* apical; *CA,* capsule; *DI,* dorsal interior; *LC,* lateral caudal; *MD,* median dorsal; *V1,* first ventral cells (see also Fig. 12.4). Scales are in micrometers; scale to the left of (*f*) also applies to (*g*).

From R. B. Short and F. G. Hochberg Jr., "A new species of *Dicyemenna* (Mesozoa: Dicyemidae) from near the Antarctic peninsula," in *J. Parasitol.* 56:517–522. Copyright © 1970. Reprinted with permission of the publisher.

that represents either a hermaphroditic sexual stage or a hermaphroditic gonad (see Fig. 12.1). It remains within the axial cell and produces male and female gametes, which fuse in fertilization. The zygotes detach from the infusorigen, and each then divides to become a hollow, ciliated ovoid stage called an **infusoriform larva.** This microscopic larva consists of a fixed number of cells of several different types. In some species that number is 37,[6, 7, 8] which is small enough for its complete lineages for all its cells to be described (Fig. 12.4).

The infusoriform larva escapes from the axial cell and parent rhombogen and leaves the host. It is the only stage

known that can survive in seawater. Subsequent to the larva leaving its host, its fate is unknown because attempts to infect new hosts with it have failed. It is possible that an alternate or intermediate host exists in the life cycle.

Order Heterocyemida
Although they are also parasites of cephalopods, heterocyemids differ in morphology from dicyemids. Nematogens of heterocyemids have no cilia or calotte and are covered by a syncytial external layer. Rhombogens are much like nematogens, and they produce infusorigens and infusiform larvae, as in dicyemids.

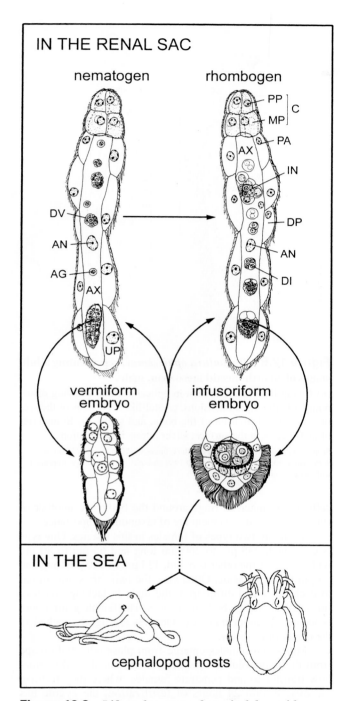

Figure 12.2   **Life cycle events of a typical dycemid mesozan (see text for details).**

The solid line represents known events and the dashed line indicates unknown ones involved in the infection of cephalopod molluscs. Two types of embryos are produced from worm-like organsims. Vermiform embryos are produced asexually from nematogen stages, and infusoriform embryos are produced sexually from hermaphroditic rhombogen stages. AG, agamete; AN, axial cell nucleus; AX, axial cell; C, calotte; DI, developing infusiform embryo; DP, diapolar cell; DV, developing vermiform embryo; IN, infusorigen; MP, metapolar cell; PA, parapolar cell; PP, propolar cell; UP, uropolar cell.

From H. Furuya and K. Tsuneki, "Biology of dicyemid mesozoans," in *Zoological Sci.* (Tokyo) 20:519–532. Copyright © 2003 Zoological Society of Japan. Reprinted by permission.

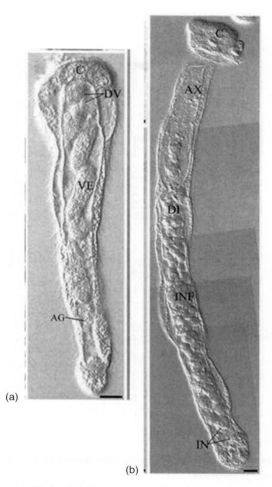

Figure 12.3   **Nematogen and rhombogen of *Dicyema japonicum* from *Octopus vulgaris*.**

Photomicrographs of developing stages in *D. japonicum;* (*a*) nematogen; (*b*) rhombogen. INF, infusorigen; VE, vermiform embryo; other abbreviations as in Fig. 12.2.

From Takahito Suzuki et al., "Phylogenetic analysis of dicyemid mesozoans (phylum Dicyemida) from innexin amino acid sequences: Dicyemids are not related to Platyhelminthes," in *J. Parasitol.* 96:614–625. Copyright © 2010. Reprinted with permission

PHYLUM ORTHONECTIDA

Class Orthonectida

Orthonectida are quite different from Rhombozoa in their biology and morphology. The 18 known species parasitize marine invertebrates, including brittle stars, nemerteans, annelids, turbellarians, and molluscs.[14] Complete life cycles are known for some.

Morphology and Biology

The best known orthonectid is *Rhopalura ophiocomae,* a parasite of brittle stars along the coast of Europe (Figs. 12.5 and 12.6). Both sexual and asexual stages exist in the life cycle.

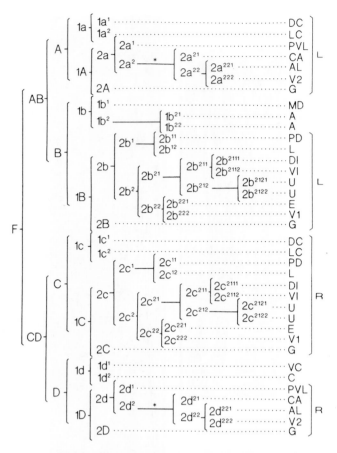

Figure 12.4   Cell lineage in the infusiform embryo of
Dicyema japonicum.

F, fertilized egg; *L*, left side of embryo; *R*, right side. Cells are
A, apical; *AL*, anterior lateral; *C*, couvercle; *CA*, capsule; *DC*,
dorsal caudal; *DI*, dorsal internal; *E*, enveloping; *G*, germinal;
L, lateral; *LC*, lateral caudal; *MD*, dorsal median; *PVL*, posterior
ventral lateral; *U*, urn; *V1, V2*, first and second ventral cells.

From Hidetaka Furuya et al., "Development of the infusoriform embryo of
Dicyema japonicum (Mesozoa: Dicyemidae)," in *Biol. Bull.* 183:248–257.
Copyright © 1992. Reprinted with permission of the author.

A **plasmodium stage** lives in tissues and spaces of
gonads and genitorespiratory bursae of the ophiuroid *Am-
phipholis squamata*. It may spread into the aboral side of the
central disc, around the digestive system, and into the arms.
Developing host ova degenerate, with ultimate castration,
but male gonads usually are unaffected.[13] The multinucleate plasmodia are usually male or female but are sometimes
hermaphroditic. Some nuclei are vegetative, whereas others
are agametes that divide to form balls of cells called **moru-
las.** Each morula differentiates into an adult male or female,
with a ciliated somatoderm of **jacket cells** and numerous
internal cells that become gametes. Monoecious plasmodia
that produce both male and female offspring may represent
the fusion of two separate, younger plasmodia. Male ciliated
forms are elongated and 90 µm to 130 µm long. Constrictions
around the body divide it into a conical cap, a middle portion, and a terminal portion. A genital pore, through which
sperm escape, is located in one of the constrictions. Jacket

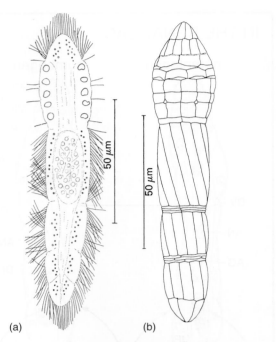

(a)   (b)

**Figure 12.5   *Rhopalura ophiocomae*, representing adult
stages of an orthonectid mesozoan, male.**

(*a*) Living individual, as seen in optical section, showing distribution of cilia, lipid inclusions, crystal-like inclusions of the second superficial division of the body, and testis. (*b*) Boundaries
of jacket cells, at the surface; silver nitrate impregnation.

From E. N. Kozloff, "Morphology of the orthonectid *Rhopalura ophiocomae*," in
J. Parasitol. 55:171–195. Copyright © 1969. *Journal of Parasitology.* Reprinted
with permission of the publisher.

cells are arranged in rings around the body; the number of
rings and their arrangement are of taxonomic importance.

There are two types of females in this species. One type
is elongated, 235 µm to 260 µm long and 65 µm to 80 µm
wide, whereas the other is ovoid, 125 µm to 140 µm long and
65 µm to 70 µm wide. Otherwise, the two forms are similar
to each other and differ from the male in lacking constrictions that divide the body into zones. The female genital pore
is located at about midbody. Oocytes are tightly packed in
the center of the body.

Males and females emerge from plasmodia and escape
from the ophiuroid into the sea. There, tailed sperm somehow transfer to and penetrate females, where they fertilize
the ova. Within 24 hours of fertilization, the zygote has
developed into a multicellular, ciliated larva that is born
through its mother's genital pore and enters the genital opening of a new host.

It is not known whether a plasmodium is derived from
an entire ciliated larva or from certain of its cells or whether
one larva can propagate more than one plasmodium.

PHYLOGENETIC POSITION

The phylogenetic position of Mesozoa has been a matter
of considerable debate.[24] The central issue is whether these
parasites are an early divergence from early metazoans, or

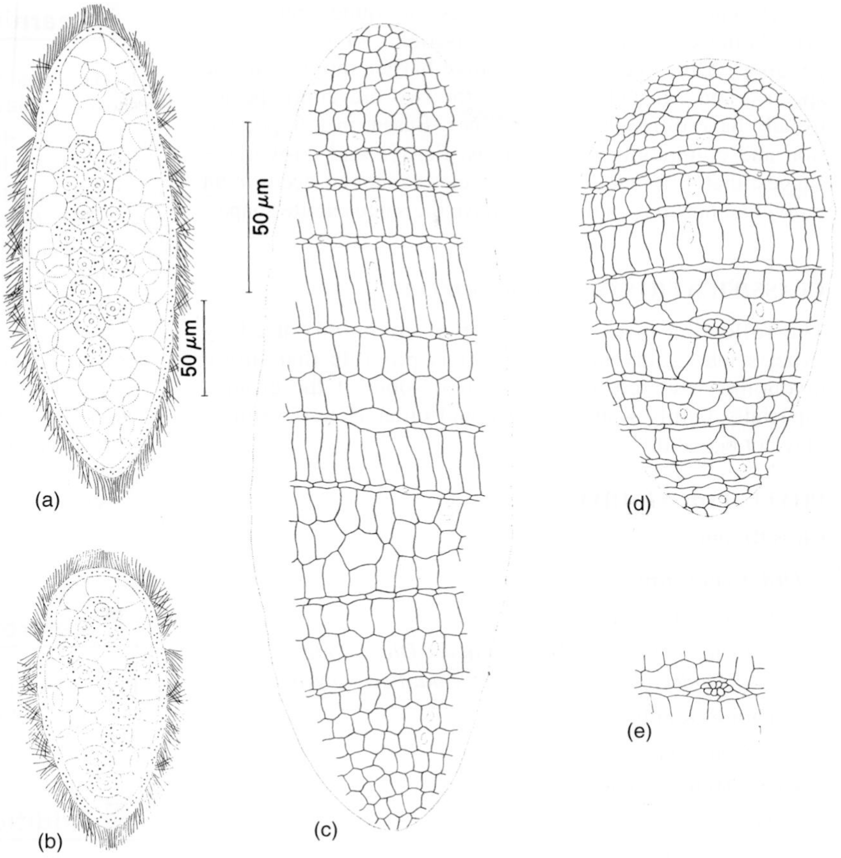

Figure 12.6   **Adult stages of *Rhopalura ophiocomae* (continued from Fig. 12.4), female.**

(*a*) Living specimen of elongated type, as seen in optical section. (*b*) Living specimen of ovoid type, as seen in optical section. (*c*) Boundaries of jacket cells of elongated type; silver nitrate impregnation. (The cells surrounding the genital pore have been omitted because they were not distinct; approximate proportions of nuclei of representative cells are based on specimens impregnated with Protargol.) (*d*) Cell boundaries of ovoid type; silver nitrate impregnation. (*e*) Genital pore of ovoid type; silver nitrate impregnation.

From E. N. Kozloff, "Morphology of the orthonectid *Rhopalura ophiocomae*," in *J. Parasitol.* 55:171–195. Copyright © 1969. *Journal of Parasitology.* Reprinted by permission.

are degenerate metazoans.[9] Early taxonomists placed them between protozoa and sponges because of their cilia, small size, and simple cellularity. Arguments also have been made for considering dicyemids to be primitive or degenerate platyhelminthes.

Molecular and ultrastructural studies provide strong evidence for the "degenerate" metazoan hypothesis. For example, dicyemids contain a peptide sequence characteristic of superphylum Lophotrochozoa (annelids, nemertenes, brachiopods, platyhelminths, and others) that is not found in superphylum Ecdysozoa (nematodes, arthropods, and others) or in superphylum Deuterostomia.[12] Other molecular research indicates that the sister group to dicyemids includes Echiura, Pogonophra, Mollusca, and Annelida, but not Platyhelminthes.[19] Furthermore, ultrastructural research reveals cell-to-cell junctions typical of complex metazoans and that are not found in Cnidaria.[7] We still do not know the ancestral group, although Furuya and co-workers suggest that dicyemids may be progenetic larval forms of parasites that once lived in now-extinct predatory marine vertebrates such as mosasaurs.[7]

HOST-PARASITE RELATIONSHIPS

What little is known of mesozoan physiology was reviewed by McConnaughey,[16] and most of that concerns dicyemids, based on the early observations of Nouvel.[18] Good ultrastructural studies of both dicyemids and orthonectids are available.[13, 20, 21]

Most dicyemids attach themselves loosely to the lining of a cephalopod kidney by their anterior cilia. They are easily dislodged and can swim about freely in their host's urine. The relationship appears to be a commensal one; no pathogenic consequences of the infection can be discerned. However, a few species have morphological adaptations for gripping renal cell surfaces, and, when the parasites are dislodged, renal tissues show an eroded appearance.

The ruffle membrane surface of nematogens and rhombogens evidently is an elaboration to facilitate uptake of nutrients. Ridley[20] showed that the membranes could fuse at various points and form endocytotic vesicles, and "transmembranosis" was suggested by uptake of ferritin. The peripheral cells of infusoriform larvae do not have ruffle membranes but do have microvilli.[21] Clearly, nematogens and rhombogens must derive their nutrients largely or entirely from their hosts' urine, whereas infusoriform larvae must live for a period on stored food molecules. Oxygen is very low or absent in a cephalopod's urine, so nematogens and rhombogens apparently are obligate anaerobes. The organisms live longer in vitro under nitrogen or even in the presence of cyanide than when maintained in urine under air or in the absence of cyanide. Infusoriforms can live anaerobically only until their glycogen supply is consumed. Adult orthonectids, on the other hand, require aerobic conditions.

Recent studies show that calottes vary quite a bit among taxa, being cone or cap-shaped, discoidal, or irregular depending on the species.[5] Hosts may harbor more than one species, and when that happens, the two dicyemid species usually have different calotte shapes. Based on infrapopulation and community observations, Furuya and co-workers suggest that different dicyemid species cannot co-occur in an individual host unless they both have the same calotte shape.[7]

CLASSIFICATION OF THE MESOZOA

The following classification is that commonly used by experts who study mesozoans. It is possible that future taxonomic research may reveal that some of the groups, especially family Kantharellidae and order Heterocyemida, may not be valid.[10]

PHYLUM DICYEMIDA

Class Rhombozoa

Order **Dicyemida**

Family Dicyemidae

Genera *Dicyema, Dicyemennea, Dicyemodeca, Dodecadicyema, Pleodicyema, Pseudicyema*

Family Kantharellidae

Genus *Kantharella*

Order **Heterocyemida**

Family Conocyemidae

Genera *Conocyema, Microcyema*

PHYLUM ORTHONECTIDA

Class Orthonectida

Family Rhopaluridae

Genera *Rhopalura, Intoshia, Stoecharthrum, Ciliacincta*

Family Pelmatosphaeridae

Genus *Pelmatosphaera*

Learning Outcomes

By the time a student has finished studying this chapter, he or she should be able to:

1. Draw the structure of a microsporidian spore and diagram, with labels, the life cycle of a typical microsporidian.
2. Draw the structure of a myxozoan spore and diagram, with labels, the life cycle of a typical myxozoa.
3. Explain the rationale for classification of these two groups of parasites.
4. Write an extended paragraph about the ecology of *Myxobolus cerebralis*.
5. Draw some sketches that illustrate structural diversity among myxozoan spores.
6. Define "actinomyxon," "polar capsule," "sporoplasm," and "valvogenic cell."
7. Write an extended paragraph describing the problems surrounding potential use of Microsporidia as biological control agents.

References

References for superscripts in the text can be found at the following Internet site: www.mhhe.com/robertsjanovynadler9e

Additional Readings

Grassé, P. P., and M. Caullery. 1961. Embranchement des mésozoaires. In P. P. Grassé (Ed.), *Traité de zoologie: Anatomie, systématique, biologie, vol. 4. Plathelminthes, Mésozoaires, Acanthocéphales, Némertiens.* Paris: Masson & Cie, pp. 693–729.

Chapter 13

Introduction to Phylum Platyhelminthes

There's no god dare wrong a worm.

—Ralph Waldo Emerson

Platyhelminthes, or flatworms, are so called because most are dorsoventrally flattened. They are usually leaf shaped or oval, but some, such as tapeworms and terrestrial planarians, are extremely elongated. Flatworms range in size from nearly microscopic to over 60 meters in length. These worms lack a coelom but do possess a well-developed mesoderm, which becomes parenchyma, reproductive organs, and musculature in adults. Traditionally, the phylum contained four classes. "Free-living" flatworms were included in class Turbellaria, which is no longer recognized, but we will use the term *turbellaria* as a common noun to refer to those generally free-living or ectocommensal platyhelminths that typically have ciliated epidermis as adults.

Platyhelminthes also are bilaterally symmetrical and thus have a definite anterior end, with associated sensory and motor nerve elements. This nervous system is surprisingly elaborate in many species and helps enable them to invade a wide variety of habitats, including lakes and streams, moist terrestrial environments, and ocean sediments from pole to pole. The bodies of other kinds of animals have proven quite hospitable to flatworms, and in fact most platyhelminths are parasitic. But flatworms can even serve as hosts for other flatworms; some cercariae (free-swimming transmission stages of trematodes) can and do penetrate planarians and encyst, becoming infective stages (metacercariae) for the next host in a complex life cycle.[11]

A peculiarity of platyhelminth physiology is their apparent inability to synthesize fatty acids and sterols de novo, which may help explain why flatworms are most often symbiotic with other organisms, either as commensals or parasites.[26] Free-living acoel turbellarians, sometimes considered illustrative of ancestral flatworms, also seem to lack this ability, indicating that the parasitic forms may not have lost it secondarily in their evolution. Being soft bodied, Platyhelminthes have left a relatively poor fossil record, but some evidence suggests they have been on Earth for eons. Fossil tracks from a slab of Permian siltstone have been interpreted as those of a land planarian.[1]

Tegument structure varies among the major taxonomic groups. Generally speaking, turbellarians and some free-living stages of Cestoidea and Trematoda have a ciliated epithelium, which in some cases is their primary mode of locomotion. This epithelium is very thin, being formed of a single layer of cells, and contains many glandular cells and ducts from subepithelial glands. Sensory nerve endings are abundant in the epithelium. In some flatworms, cells that produce adhesive secretions are paired with those that produce releasing secretions; the combination is known as a **duo-gland adhesive system.**

Trematoda and Cestoidea have lost their external cilia except in certain larval stages. During metamorphosis of these parasitic forms, the larval epidermis is replaced by a syncytial adult tegument, the nuclei of which are in cell bodies **(cytons)** located beneath a superficial muscle layer. Thus, the name *Neodermata* ("new skin") has been used in some classifications to distinguish such worms from free-living species that retain the ciliated epithelium as adults.

Embedded in the tegument in most free-living turbellarians and in members of trematode genus *Rhabdiopoeus* are numerous rodlike bodies called **rhabdites.** Their function is not always clear, but various authors have attributed lubrication, adhesion, and predator repellancy to them; they are generally absent from symbiotic turbellaria.

Most of a flatworm's body is made up of **parenchyma,** a loosely arranged mass of fibers and cells of several types. Some of these cells are secretory, others store food or waste products, and still others have huge mitochondria and function in regeneration. The internal organs are so intimately embedded in the parenchyma that dissecting them out is nearly impossible. The bulk of the parenchyma probably is composed of myocytons.

Muscle fibers course through the parenchyma. Contractile portions of muscle fibers are rarely striated and are usually arranged in one or two longitudinal layers near the body surface. Circular and dorsoventral fibers also occur.

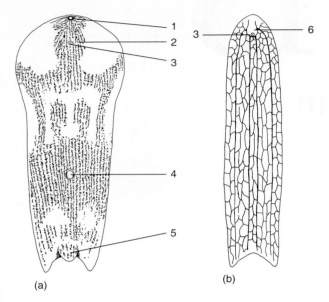

Figure 13.1 *Amphiscolops,* **an aceol from Bermuda.**
(*a*) Whole worm; (*b*) dorsal portion of the nervous system;
(*1*) frontal organ; (*2*) eyes; (*3*) statocyst; (*4*) mouth; (*5*) penis;
(*6*) brain.

From L. H. Hyman, *The Invertebrates* (vol. 2). McGraw-Hill Book Company, Inc.,
New York, NY, 1951. Courtesy of the American Museum of Natural History.

The **nervous system** of acoel turbellarians (Fig. 13.1)
includes central and peripheral components, the central nervous system consisting of ganglia around the statocyst, and
the peripheral portion consisting of networks supplying the
epithelium, muscles, and sensory structures.[2] In larger and
more structurally complex turbellarians and in trematodes
and cestodes, the nerve system is an **orthogon** (ladder) type,
with paired ganglia near the anterior end, nerves running
anteriorly toward sensory or holdfast organs, and longitudinal nerve trunks extending posteriorly to near the end of the
body. The number of trunks varies, but most trunks are lateral and are connected by transverse commissures.

Sensory elements are abundant, especially in turbellarians, and may be distributed in a variety of patterns, depending on the species. Tactile cells, chemoreceptors, eyespots,
and statocysts have been found. The nervous system of
turbellarians has attracted the interest of a number of recent
researchers, primarily because this system may hold clues
to the evolutionary origin of bilateral symmetry in animals.
Most of this work is done at the electron microscope level
and, as might be expected, has shown that diversity of neuron types and synaptic junctions is greater than that expected
of seemingly primitive animals.[2]

The **digestive system** is typically a blind sac, although
acoels and a few trematodes (*Anenterotrema* and *Austromi-
crophallus* spp.) have only a mouth but no permanent gut,
food being digested by individual cells of the parenchyma.
Most flatworms have a mouth near their anterior end, and
many turbellarians and most trematodes have a muscular
pharynx, behind their mouth, with which they suck in food.
In the familiar planarians as well as in some other freeliving
flatworms, the mouth is located midventrally and the pharynx
can be extended outward. The gut varies from a simple sac to

a highly branched tube, but only rarely does a flatworm have
an anus. Digestion is primarily extracellular, with phagocytosis by intestinal epithelium (gastrodermis), which may contain both secretory and phagocytic cells.[4] Undigested wastes
are eliminated through the mouth. A digestive system is completely absent from all life-cycle stages of cestodes.

The functional unit of most flatworms' **excretory system** is the **flame cell,** or **protonephridium** (see Fig. 20.15).
This is a single cell with a tuft of flagella that extends into
a delicate tubule, which may consist of another cell inter-
digitating with the first.[14, 32] As is the case with the nervous
system, ultrastructural studies aimed partly at uncovering
characters of evolutionary significance have shown that
platyhelminth excretory systems are far more complex than
originally thought. Rohde[32] showed that detailed structure
of the flame cell system is related more to evolutionary relationship than to the worms' habitat. Protonephridial systems
have at least three types of flame cells and as many kinds of
tubule cells.[32] Excess water, which may contain soluble nitrogenous wastes, is forced into the tubule, which joins with
other tubules, eventually to be eliminated through one or more
excretory pores. Filtration occurs through minute slits formed
by **rods,** or extensions of the cell, collectively called a **weir**
(Old English *wer:* a fence placed in a stream to catch fish). In
parasitic flatworms the weir is formed by rods from both the
terminal flagellated cell (the **cyrtocyte**) and a tubule cell and is
thus referred to as a **two-cell weir.** Because excreta are mainly
excess water, this system is often referred to as an **osmoregulatory system,** with excretion of other wastes considered a
secondary function. Some species have an excretory bladder
just inside the pore.

Reproductive systems follow a common basic pattern
in all Platyhelminthes. However, extreme variations of this
basic pattern are found among different groups. Most species
are monoecious, but a few are dioecious. Because reproductive organs are so important in identification of parasites and
therefore are considered in great detail for each group, we
will not discuss them here. Most hermaphrodites can fertilize
their own eggs, but cross-fertilization occurs in many. Some
turbellarians and cestodes practice **hypodermic impregnation,** which is sperm transfer through piercing the body wall
with a male organ, the cirrus, and injecting sperm into the
parenchyma of the recipient. How sperm find their way into
the female system is not known. Most worms, however, deposit sperm directly into the female tract. Young are usually
born within egg membranes, but a few species are viviparous
or ovoviviparous. In parasitic species and some turbellarians,
egg yolk is supplied by cells other than the ovum, and eggs
are thus **ectolecithal.** Asexual reproduction is also common
in trematodes and a few cestodes.

PLATYHELMINTH SYSTEMATICS

Phylogeny of Platyhelminthes is one of the most active areas of research in invertebrate biology, with many workers
attacking evolutionary problems from a variety of directions.
However, a great deal of useful information is found in older
literature organized according to traditional classifications.
Historically, the phylum included four classes: Turbellaria,
Monogenea, Trematoda (Digenea), and Cestoda, generally

corresponding to "free-living" flatworms, ectoparasitic single-host worms, endoparasitic flukes with two or more hosts (one a mollusc), and tapeworms, respectively.

Although *parasitic* is not necessarily a valid criterion for separating taxa, parasitic flatworms do form a monophyletic group, Neodermata, based on other characters. Neodermata shed their epidermis at the end of their larval life. Within Neodermata, cestodes and monogenes are sister taxa, as are trematodes and aspidobothreans.[20] Features such as nature of the egg yolk, spermiogenesis, body wall musculature, and structure of excretory organs, especially flame cells, are considered important morphologically and are useful in platyhelminth classification. Molecular characters that have been used include 18S and 28S ribosomal DNA sequences, genes for cytochrome oxidase, NADH dehydrogenase, and elongation factor 1-α, and immunochemistry of neurotransmitters.[23, 24, 28] Phylogenies based on molecular characters do not always agree with those based on morphology.[19]

In addition to many unresolved phylogenetic problems within the free-living turbellarians, there are three major issues that relate to Platyhelminthes as a whole. The first of

these is the position of Acoela, traditionally included in the phylum. A number of authors do not consider acoels to be flatworms at all, this conclusion being based largely on nervous system structure. Acoel flatworms, typically small species without an intestine (digesting food intracellularly and in temporary cavities), lack protonephridia. The remaining groups, including all the parasitic ones, have protonephridial flame cells with more than two and sometimes more than 100 flagella.[10] Acoels appear as the sister group to order Nemertodermatida in modern phylogenies, but in older literature, genera such as *Nemertoderma* were placed in Acoela.[14, 21] Anyone looking for pictures of all these enigmatic little worms should probably start with Hyman, volume 2.[15]

According to Ehlers,[10] Litvaitis and Rohde,[23] and Brooks and McLennan,[6] the subphylum Catenulida is the "basal" taxon of a platyhelminth cladogram; that is, the sister group of the "true" Platyhelminthes. Catenulida includes a number of delightful little worms whose ease of culture and asexual reproductive habits have made them favorite experimental animals for regeneration studies (Fig. 13.2). The major structural feature dividing catenulid

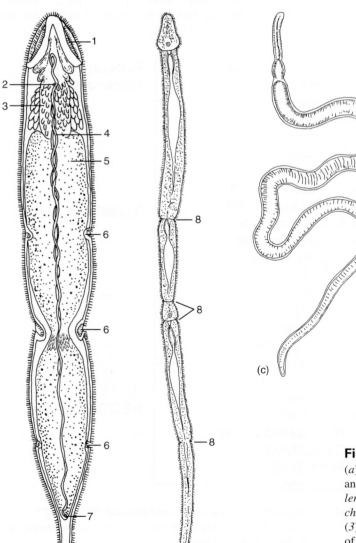

(a)   (b)   (c)

Figure 13.2   Some representative Catenulida.

(*a*) *Stenostomum tenuicauda* showing unpaired protonephridia and sites of asexual division (zooid ciliated pits). (*b*) *Catenula lemnae,* also in the process of sexual reproduction. (*c*) *Rhynchoscolex* sp. (*1*) Ciliated pits (not frontal organs); (*2*) mouth; (*3*) pharynx; (*4*) protonephridium; (*5*) intestine; (*6*) ciliated pits of zooids; (*7*) nephridiopore; (*8*) fission lines of zooid formation.

From L. H. Hyman, *The Invertebrates* (vol. 2). McGraw-Hill Book Company, Inc., New York, NY, 1951. Courtesy of the American Museum of Natural History.

platyhelminths from the rest is lack of a frontal organ, which is a terminal or subterminal pit with mucoid gland cells and sometimes cilia. Catenulids lack this organ, although some species have lateral pits. Some authors doubt that frontal organs are homologous among the taxa that possess them. Nevertheless, Catenulida appear as basal and as a sister taxon to all remaining Platyhelminthes (except Acoela and Nemertodermatida) in the consensus phylogeny of Littlewood et al.[21] The remaining platyhelminths (Euplatyhelminthes) also possess dense epidermal ciliature (three to six cilia per μm^2) compared to catenulids, which have about a tenth that many per unit area.

The classification that follows is based primarily on the phylogenies of Brooks and McLennan,[6] Ehlers,[10] Littlewood et al.,[21, 22] Litvaitis and Rohde,[23] and Rohde.[30, 31] Not all scientists agree upon taxon names or hierarchical levels. The Brooks and McLennan[6] taxonomy utilizes more levels than typically encountered in literature used by undergraduates. Thus, in that reference you will find superclasses, subsuperclasses, infraclasses, cohorts, and subcohorts in addition to familiar classes and orders. Newer phylogenies depend greatly on molecular data, but authors are not always willing to assign formal names and

hierarchical levels to their groupings. Rohde's[30, 31] phylogeny is based both on 18S ribosomal DNA and on reassessment of structural features, including newer information on spermiogenesis. This phylogeny differs from that of Brooks and McLennan[6] mainly in placement of Temnocephalidea and Udonellidea. Rohde[31] provides evidence that the ectocommensal Temnocephalidea are not the sister group to Neodermata, citing a number of structural features such as dual-gland adhesive systems and protonephridia in Temnocephalidea that are identical to those of free-living rhabdocoels. Rohde[31] also considers superclass Cercomeria invalid because of the inclusion of Temnocephalidea and because the doliiform pharynx and posterior adhesive organs evidently arose independently in several groups of Platyhelminthes. The Littlewood et al. phylogeny of Fig. 13.3 is a consensus one using all available data, both morphological and molecular.[20]

The classification of Platyhelminthes will likely undergo more changes based on newer phylogenies, but *Interrelationships of the Platyhelminthes,* edited by Littlewood and Bray, will be the standard reference on platyhelminth systematics for some years to come.[19] Whatever else it may accomplish, modern evolutionary biology has clearly demonstrated that

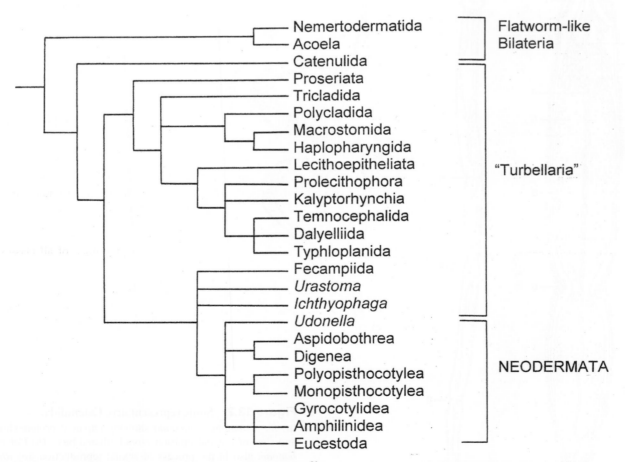

Figure 13.3   **A consensus tree obtained by Littlewood et al.[21] using both molecular and morphological characters.**
Free-living members of the phylum appear mostly at the base of the tree. A sister-group relationship between Digenea (Trematoda) and Aspidobothrea is supported, as is the basal position of catenulids and monophyly of the Neodermata.

Modified from D. T. J. Littlewood, K. Rohde and K. A. Clough, "The interrelationships of all major groups of Platyhelminthes: Phylogenetic evidence from morphology and molecules," in *Biol. J. Linnean Soc.* 66:75–114.

we still have a great deal to learn about animals we have been studying for a long time. The parasites involved do not give up their secrets easily.

CLASSIFICATION OF PHYLUM PLATYHELMINTHES (WITH EMPHASIS ON COMMENSAL AND PARASITIC REPRESENTATIVES)

SUBPHYLUM CATENULIDA

Lack a frontal organ and have monociliated epidermal cells.

SUBPHYLUM EUPLATYHELMINTHES

With a frontal organ, high density of epidermal cilia, and multiflagellated flame cells (when present).

SUPERCLASS ACOELOMORPHA

Reduction and loss of protonephridia and a modified (or missing) gut; tips of cilia with a distinct step. Haszprunar[13] considered Acoelomorpha to be the sister taxon to all Bilateria, not just Platyhelminthes. Littlewood et al.[21] agree that Acoelomorpha do not belong in Platyhelminthes, but their status as a separate phylum has not been established in the literature.

SUPERCLASS RHABDITOPHORA

With lamellated rhabdites, a duo-gland adhesive system, and multiflagellated flame cells.

Class Rhabdocoela

With a bulbous pharynx and simple intestine.

Order Dalyellioida

Suborder Temnocephalida

With cephalic tentacles.

Subsuperclass Neodermata

With ectolecithal eggs; loss of larval ciliated epidermis; acquisition of a syncytial adult epidermis. Neodermata is considered a monophyletic group.[21, 22, 23]

Class Trematoda

Posterior adhesive organ a sucker; male genital pore opening into an atrium; adults with pharynx near the oral sucker.

Subclass Aspidobothrea

With specialized microvilli on and microtubules in neodermis. Posterior sucker divided into compartments. Major order: Aspidobothriiformes.

Subclass Digenea

First larval stage a miracidium; life cycle with one or more sporocyst generations and cercarial stage; gut development paedomorphic. Important orders: Paramphistomiformes, Echinostomatiformes, Hemiuriformes, Strigeiformes, Opisthorchiformes, Plagiorchiformes.

Class Monogenoidea

Oncomiracidium (larva) with three ciliary bands; adults with single testis; all ectoparasitic. Molecular phylogenies suggest that Monogenoidea is paraphyletic, with polyopisthocotyleans evidently related to Aspidobothrea and Digenea and monopisthocotyleans, along with Udonellidea, being a sister group to remaining Neodermata.[23] *Udonella* species are ectoparasitic on crustaceans but feed on the crustaceans' fish hosts.[7, 30] Important orders: Dactylogyridea, Gyrodactylidea, Polystomatidea, Mazocraeidea, Diclybothriidea, Chimaericolidea.

Class Cestoidea

Intestine lacking; cercomer paedomorphic and somewhat reduced in size; oral sucker and pharynx vestigal; larval cercomer with 10 hooks.

Subclass Cestodaria

Order Gyrocotylidea

Rosette with funnel at posterior end; body margins crenulate.

Order Amphilinidea

Genital pores at posterior end; uterus *N*-shaped.

Subclass Eucestoda

Adults polyzoic; six-hooked larval cercomer lost during ontogeny; life cycles with more than one host. Orders: Pseudophyllidea, Caryophyllidea, Spathebothriidea, Cyclophyllidea, Proteocephalata, Rhinebothriidea,[13] Tetraphyllidea, Trypanorhyncha.

CLASSIFICATION OF PLATYHELMINTHES AS FOUND IN OLDER LITERATURE

Class Turbellaria

Mostly free-living worms in terrestrial, freshwater, and marine environments; some commensals or parasites of invertebrates, especially of echinoderms and molluscs.

Class Monogenea

All parasitic, mainly on the skin or gills of fish; although mostly ectoparasites, a few living within the stomodaeum, the proctodaeum, or their diverticula.

Class Trematoda

All parasitic, mainly in the digestive tract, of all classes of vertebrates.

Subclass Digenea

At least two hosts in life cycle, first almost always a mollusc; perhaps most diversification in bony marine fish, although many species in all other groups of vertebrates.

Subclass Aspidogastrea

Most with only one host, a mollusc; a few mature in turtles or fishes with mollusc or lobster intermediate host.

Subclass Didymozoidea

Tissue-dwelling parasites of fish; no complete life cycle known, but intermediate host may not be required.

Class Cestoidea

All parasitic, common in all classes of vertebrates except agnathan classes; intermediate host required for almost all species.

TURBELLARIANS

Most turbellarians are free-living predators, but several orders contain species that maintain varying degrees and types of symbiosis. Of these, most are symbionts of echinoderms, but others are found on or in sipunculids, arthropods, annelids, molluscs, coelenterates, other turbellarians, and fish. At least 27 families have symbiotic species. A considerable degree of host specificity is manifested by these worms. Most symbionts are commensals; a few are true parasites, and several degrees of these relationships are known.

Acoels

Acoels, which are entirely marine, are from one to several millimeters long. They exhibit several primitive characteristics, including absence of an excretory system, pharynx, and permanent gut, and many have no rhabdites. Most are free living, feeding on algae, protozoa, bacteria, and various other microscopic organisms. A temporary gut with a syncytial lining appears whenever food is ingested, and digestion occurs in vacuoles within it. After digestion is completed the gut disappears. Haszprunar[12] considered acoels the sister group to all other Bilateria. Ruiz-Trillo et al. agree with the conclusion that acoels are not flatworms at all.[33]

Some species have adopted a symbiotic existence, and it is difficult to decide which, if any, are true parasites. *Ectocotyla paguri* is the only ectocommensal known. It lives on hermit crabs, but nothing is known of its biology or feeding habits. Several species of acoels live in the intestines of Echinoidea and Holothuroidea. It is not known if any are parasites, but, because no apparent harm comes to their hosts, these acoels are usually considered endocommensals.

Rhabditophorans

Many orders of turbellarians contain mainly free-living species. Space limitations prevent presentation of a detailed taxonomy of these groups. However, Meglitsch and Schram[24] and Ehlers[10] provide informative reviews, Rohde[30, 31] and Litvaitis and Rohde[22] discuss some ongoing taxonomic problems associated with these worms, and several chapters in Littlewood and Bray[19] deal with free-living species. Meglitsch and Schram[25] give class status to Rhabditophora and subclass status to Macrostomida and Neoophora. They include orders Seriata, Typhloplanoida, and Dalyellioida within their Neoophora. Most symbiotic turbellarians belong to one of these orders. Again, most seem to be commensals, but a few are definitely parasitic. Dalyellioids are small, like acoels, but they have a permanent, straight gut and a complex, bulbous pharynx. Most are predators of small invertebrates. Of the four suborders in Dalyellioida, three have symbiotic species.

Fecampia erythrocephala (order Dalyellioida) lives in the hemocoel of decapod crustaceans. During development in a host, a young worm loses its eyes, mouth, and pharynx, and absorbs nutrients from the host's blood. When sexually mature it mates and leaves its host. After cementing itself to a substrate, the flatworm shrinks until all internal tissues vanish, leaving only a bottle-shaped cocoon made of degenerated epidermis. Each cocoon contains two eggs and several vitelline cells that produce two ciliated, motile juveniles. These swim about until contacting a crustacean.[3] Their mode of entry into a host is not known. The host is not killed by these parasites but does suffer adverse effects in its hepatopancreas and ovaries. Because of its fertility, *F. erythrocephala* presents a high risk for culture of prawns in Atlantic and Mediterranean marine areas.

Kronborgia amphipodicola is very unusual among turbellarians because it is dioecious.[8] Furthermore, there is pronounced sexual dimorphism: Males are 4 mm to 5 mm long, whereas females are 20 mm to 30 mm long and can stretch to 45 mm. Both sexes lack eyes and digestive systems at all stages of their life cycles. They mature in the hemocoel of a tube-dwelling amphipod *Amphiscela macrocephala,* with the male near the anterior end and the female filling the rest of the available space.

On reaching sexual maturity, the worms burrow out of the posterior end of the host, which becomes paralyzed and quickly dies; as if to add insult to injury, before the host is killed it is castrated. After emergence from the amphipod, a female worm quickly secretes a cocoon around herself and attaches the elongated cocoon to the burrow wall, from which it protrudes 2 cm to 3 cm into open water. A male enters the cocoon, crawls down to the female, and inseminates her. He then leaves the cocoon and dies. Each female produces thousands of capsules, each with two eggs and some vitelline cells, and then she also dies. A ciliated larva hatches from each egg and eventually encysts on the cuticle of another amphipod. While in the cyst, the larva bores a hole through the host's body wall and enters the hemocoel to begin its parasitic existence.

Tegumental ultrastructure of *Kronborgia amphipodicola* has been studied.[18] The lateral membranes of epidermal cells break down, and the epidermis thus becomes syncytial. Although short microvilli are not unusual on the outer surface of epithelial cells of free-living turbellaria, microvilli of *K. amphipodicola* are quite long and constitute an adaptation for increasing surface area to absorb nutrients. Subepidermal gland cells with long processes extending to the surface probably function in escape of the worm from its host and in construction of the cocoon.

Urastoma cyprinae (order Prolecithophora) is an ectoparasite on the gills of bivalve molluscs, including the giant clam *Tridacna gigas* in Australia and edible mussels and oysters in the Mediterranean.[9] Heavy infections result in damage to and ultimately necrosis of gill filaments; *U. cyprinae* is considered a threat to the mussel culture industry.[29]

Members of the dalyellioid family Umagillidae live in the digestive tract or coelom of Holothuroidea and Echinoidea. Crinoidea and Sipunculida also are infected. Traditionally considered harmless commensals, some species consume host intestinal cells as well as commensal ciliated protozoa.[17] For example, *Syndesmis franciscanus* and *S. dendrastrorum* ingest host intestinal tissue along with intestinal contents, whereas *S. echinorum* subsists entirely on host intestinal tissue.[34]

Syndesmis spp. (Fig. 13.4) and *Syndisyrinx* spp. are found in intestines of sea urchins and therefore are available to nearly any college laboratory with preserved or living sea urchins in its stock. Little is known of their biology, but they

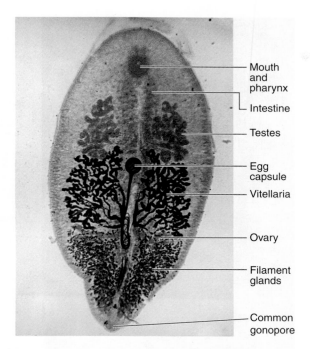

Madagascar, Sri Lanka, and India; a few are known from Europe. Species also occur on turtles, molluscs, and freshwater hydromedusae. Probably they are much more widespread than reported but have gone undiscovered or unrecognized because enough trained specialists simply have not looked for them extensively.

Temnocephalids are small and flattened, with tentacles at the anterior end and a weak, adhesive sucker at the posterior end (Fig. 13.5a). They have leechlike movements, alternately attaching with the tentacles and posterior sucker. Their tegument is syncytial with varying numbers of cilia or at least ciliated receptors, depending on the species.[37] Scanning electron microscope studies have shown that the tegument is structurally complex, with folds, microvilli, and occasionally scales (Fig. 13.5b).[16, 37] Rhabdites are located mainly at the anterior end, and mucous glands are most numerous around the posterior sucker.

The biology of temnocephalids is simple, as far as it is known. Eggs are laid in capsules and attached to a host's exoskeleton. Each hatches as an immature adult and matures with no further ado. What happens to those that are lost at ecdysis of the host is unknown; fate of the adults at that time is also unknown. It is possible that a free-living stage is present in these worms' life cycle, but one has yet to be found.

The pattern of nutrition apparently does not differ from that of free-living flatworms, with protozoa, bacteria, rotifers, nematodes, and other microscopic creatures serving as food. Cannibalism has been established. The host serves only as a substrate for attachment but the relationship is evidently an obligate one.

Mouth and pharynx

Intestine

Testes

Egg capsule

Vitellaria

Ovary

Filament glands

Common gonopore

Figure 13.4   *Syndesmis* **sp., a turbellarian from the intestine of a sea urchin.**

It is about 2.5 to 3.0 mm long.

Courtesy of Warren Buss.

appear to be excellent subjects for study. About 50 species have been described in this family.

Syndesmis franciscanus inhabits sea urchins of genus *Strongylocentrotus* along the northwest coast of North America. Cross-fertilization presumably occurs. Some species have a penis stylet so may practice hypodermic impregnation. The worms produce an egg capsule about every one and one-half days; these capsules are released one at a time into the host's intestine and pass to the outside with feces. Each capsule contains two to eight oocytes and several hundred vitelline cells. Embryogenesis requires about two months. The worms hatch when eaten by a suitable host and mature with no further migration.[35]

Molluscs also play host to turbellarians. Some bivalves have invaded the New World as a result of commerce and, as other animals are inclined to do when they travel, have brought along their symbionts. Thus, dalyellioid flatworms (along with a whole community of protozoans), interpreted as natural occupants of the Manila clam *(Tapes philippinarum),* have been found in the intestinal lumen of these bivalves accidently introduced into Canada.[5]

Temnocephalideans

Temnocephalidea were given class status by Brooks and McLennan[6] and considered the sister group to the parasitic platyhelminths, but Rohde[31] considered them members of Dalyellioida, which contains a number of free-living families. Most temnocephalids are ectocommensals on crustaceans in South and Central America, Australia, New Zealand,

Alloeocoels

Alloeocoels are turbellarians with an irregular gut. Most are marine, but a few inhabit brackish water or fresh water, and a few are terrestrial. Several are commensal on snails, clams, and crustaceans, but *Ichthyophaga subcutanea* is clearly a parasite of marine teleost fish. It lives in cysts under the skin in the branchial and anal regions of its host and apparently ingests blood. Morphologically it has nonparasite features, such as eyes and a ciliated epithelium.[36]

At least one *Monocelis* species lives within the valves of intertidal barnacles and snails during low tide but returns to the open water when the tide is in. This may illustrate a case of incipient endosymbiosis.

Tricladids

Tricladids are large worms, up to 50 cm in length, that occupy marine, freshwater, and terrestrial habitats. They are easily recognized by their tripartite intestine. Nearly all are free-living predators, feeding on small invertebrates and sucking the contents out of larger ones by means of their eversible pharynges.

Members of three genera, *Bdelloura, Syncoelidium,* and *Ectoplana,* live on the book gills of horseshoe crabs, *Limulus polyphemus.* Of these, *Bdelloura candida* (Fig. 13.6) is the most common. It has a large adhesive disc at its posterior end

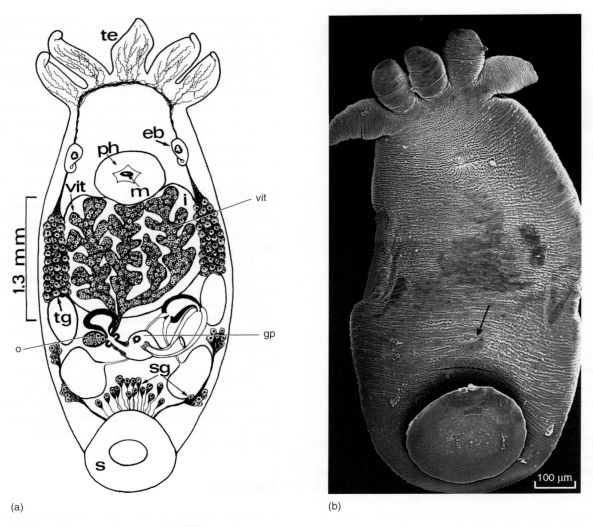

(a) (b)

Figure 13.5 **Structure of temnocephalidean worms.**

(*a*) *Temnocephala haswelli*, showing internal anatomy; (*b*) *Temnocephala dendyi*, showing body surface features; arrow indicates location of the gonopore. eb, excretory bladder; gp, gential pore; i, intestine; m, mouth; o, ovary; ph, pharynx; s, sucker; sg, sucker glands; t, testes; te, tentacles; tg, tentacular gland; vit, vitelline glands.

(*a*) From R. Ponce de León "Description of *Temnocephala hasewlli* n. sp. (Plathyhelminthes) from the mantle cavity of *Pomacea canaliculata* (Lamark)" in *J. Parasitol.* 75: 524–526. Copyright © 1989. Used by permission. (*b*) J. B. Williams, "Studies on the epidermis of *Temnocephala dendyi*," in *Austr. J. Zool.* 26:127–224. Copyright © 1978. Used by permission.

and well-developed eyespots. It evidently feeds on particles of food torn apart by its host's gnathobases and washed back to the gill area. No evidence of harm to its host has been detected. It lays its eggs in capsules on the book gill lamellae. These tricladids may migrate from one horseshoe crab to another during copulation of their hosts, in a sort of marine, verminous venereal disease! The biology and physiology of these worms would surely prove to be a rewarding area of research.

Polycladids

Polycladids have a complex gut with many radiating branches. Except for one freshwater species, they are marine. No parasites are known in this group, and the few reported "commensals" are suspect of even that degree of symbiosis.

Although some species are found with hermit crabs, they are also found in empty shells. Others, such as the "oyster leech," *Stylochus frontalis,* live between the valves of oysters and are predators on the original owner, devouring large pieces of it at a time.[27]

Truly parasitic turbellarians show structural changes expected with their specialized way of life: losses of ciliated epidermis, eyes, mucous glands, and rhabdites. Various commensals, however, show few or no specializations over their free-living brethren. The prevalence of rhabdocoels in echinoderms may simply be a result of the diverse fauna of ciliated, protozoan commensals in the latter, which offer rich pickings for the former. The origin of trematodes and cestodes from acoel-like ancestors, which became adapted to endocommensalism within molluscs and crustaceans, is not difficult to visualize.

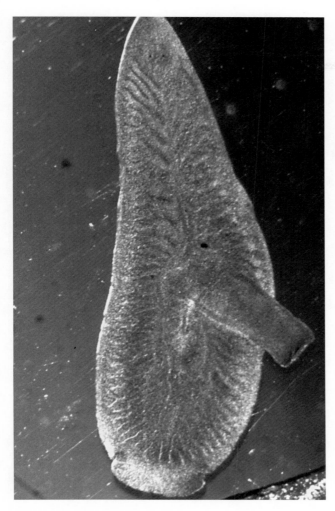

Figure 13.6   *Bdelloura candida,* **a triclad turbellarian from the gills of a horseshoe crab.**

Note the eyespots and the huge midventral pharynx. Overall length may reach 20 mm.

Courtesy of Warren Buss.

Learning Outcomes

By the time a student has finished studying this chapter, he or she should be able to:

1. Describe the general body plan of a platyhelminth.
2. Describe the level of organization of platyhelminth nervous systems, excretory system, and reproductive system.
3. Distinguish between Turbellaria and Neodermata.
4. Briefly describe the tegument, nervous system, excretory system, and digestive system of each class of platyhelminths.
5. Describe hypodermic impregnation briefly.
6. Briefly describe the difference in epidermal structure of Neodermata and that of other flatworms.

References

References for superscripts in the text can be found at the following Internet site: www.mhhe.com/robertsjanovynadler9e

Additional Readings

The entire volume 132 of the journal *Hydrobiologia* is devoted to information on phylogeny, development, reproduction, regeneration, and ecology; this publication entitled "Advances in the biology of turbellarians and related platyhelminthes" will be a primary reference for several years.

Baer, J. F. 1961. Classe des Temnocéphales. In P. P. Grassé (Ed.), *Traité de zoologie: Anatomie, systématique, biologie, vol. 4, part I. Plathelminthes, Mésozoaires, Acanthocéphales, Némertiens.* Paris: Masson & Cie, pp. 213–214.

Brooks, D. R. 1989. The phylogeny of the Cercomeria (Platyhelminthes: Rhabdocoela) and general evolutionary principles. *J. Parasitol.* 75:606–616.

Jennings, J. B. 1971. Parasitism and commensalism in the Turbellaria. In B. Dawes (Ed.), *Advances in parasitology 9.* New York: Academic Press, Inc, pp. 1–32. A most readable account of the subject. Recommended for all parasitologists.

Littlewood, D. T. J., and R. A. Bray (Eds.). 2001. *Interrelationships of the Platyhelminthes.* London: Taylor and Francis. For anyone with even passing interest in the evolution of Platyhelminthes, this volume is required reading. It is the printed version of a symposium hosted in July 1999 by The Linnean Society of London that brought together virtually all of the world's leading platyhelminth systematists to struggle with the major phylogenetic problems posed by this wonderful group of worms.

Chapter 14

Trematoda: Aspidobothrea

There's a sucker born every minute.

—Phineas Taylor Barnum (attributed)

Aspidobothrea constitute a small group of Digenea-like worms that, in most species, have established a loosely parasitic relationship with molluscs, but some are facultative or obligate parasites of fishes or turtles.[2, 6] Two other names have often been used for this group: Aspidocotylea and Aspidogastrea. Although most of the literature has accumulated under the name Aspidogastrea, Aspidobothrea undoubtedly has priority.

By any name, this group of parasites has attracted perhaps less than its fair share of attention because it contains no species of medical or known economic importance. Nevertheless, these innocuous little worms are of considerable biological interest: They seem to represent a step between free-living and parasitic organisms. And like many seemingly unimportant parasites, their structures and lives have been remarkable enough to capture the interest of many well-known parasitologists at some time in their careers. A comprehensive review of Aspidobothrea was presented by Rohde[7]; the "literature-cited" section of that review reads like a "Who's Who" of 20th-century helminthology, although few of these famous people spent very much time on Aspidobothrea.

FORM AND FUNCTION

Body Form

Externally, aspidobothreans exhibit three basic types of anatomy, corresponding to the three families that have been established for them. Members of family Aspidogastridae (Fig. 14.1) have a huge **ventral sucker,** extending most of their body length. This sucker (also known as an **opisthaptor** or **Baer's disc**) has muscular septa in longitudinal and transverse rows, dividing it into shallow depressions called **alveoli** or **loculi** (Fig. 14.2).[4] The number, shape, and arrangement of these loculi are of considerable taxonomic importance. Hooks or other sclerotized structures are never present. Between the marginal loculi usually are **marginal bodies,** which are

secretory organs, or short tentacles, also presumably secretory in nature. Exceptionally, both are absent.

Marginal bodies are round to oval organs and are connected to each other by fine ducts. They consist of gland cells, storage chambers (ampullae), and secretory ducts (Fig. 14.3). In some species, ampullae empty through a muscular papillae, and the terminal ducts can be protruded or retracted.[11] Although a sensory function has been suggested for marginal bodies, no indication exists that their function is other than secretory. The tentacles of *Lophotaspis interiora* (Fig. 14.4) are probably modified marginal bodies.

Members of family Stichocotylidae (see Fig. 14.13) have a longitudinal series of individual suckers instead of a single complex of loculi, whereas in Rugogastridae the ventral holdfast is made up of transverse ridges called **rugae** (see Schell[13]).

The **longitudinal septum** is a peculiar morphological characteristic of Aspidobothrea. It is a horizontal layer of connective tissue and muscle in the anterior part of the body, projecting like a shelf and dividing the body into dorsal and ventral compartments. The septum's function is not known, but it might be correlated with pressures exerted by contraction of the giant ventral sucker.

Tegument

Aspidobothrean tegument seems to be similar to that of other parasitic flatworms (see Fig. 15.8), although that conclusion is based largely on the study of one species (*Multicotyle purvisi*). The tegument is syncytial and has an outer stratum of **distal cytoplasm,** containing mitochondria and numerous vesicles of various types. Tegumental nuclei are in **cytons** internal to the superficial muscle layer and connected to the distal cytoplasm by internuncial processes. Cytons are rich in Golgi complexes. A mucoid layer of variable thickness is found on the outer surface membrane, and in some areas this surface membrane has riblike elevations to support the thick mucoid layer.

Figure 14.1 **Examples of family Aspidogastridae.**
(*a*) *Lobatostomum ringens;* (*b*) *Cotylogaster michaelis;*
(*c*) *Lophotaspis vallei;* (*d*) *Cotylaspis insignis.*

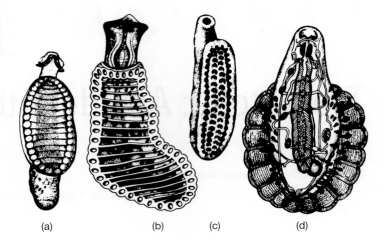

(a)   (b)   (c)   (d)

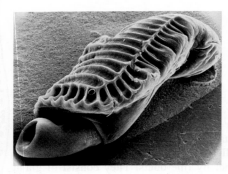

Figure 14.2 **Scanning electron micrograph: ventral view of *Cotylogaster occidentalis.***

The cup-shaped buccal funnel is at left, and the neck is semiretracted into the body by a fold. Note the smooth tegument and prominent alveoli in the ventral haptor.

From H. S. Ip, S. S. Desser, and I. Weller, "*Cotylogaster occidentalis* (Trematoda: Aspidogastrea): Scanning electron microscopic observations of sense organs and associated surface structures," in *Trans. Am. Microsc. Soc.* 101:253–261. Copyright © 1982.

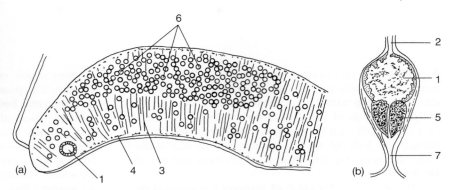

Figure 14.3 *Lobatostoma manteri.*
(*a*) Diagram of a section through a marginal alveolus. (*b*) Diagram through one of the marginal bodies. (*1*) Ampulla of marginal body; (*2*) duct of marginal gland; (*3*) dorsoventral muscles; (*4*) longitudinal muscles; (*5*) muscular papilla; (*6*) nuclei of the marginal gland; (*7*) terminal duct.

From K. Rohde and N. Watson, "Ultrastructure of the marginal glands of *Lobatostoma manteri* (Trematoda, Aspidogastrea)," in *Zool. Anz.* 223:301–310. Copyright © 1989. Reprinted by permission of Elsevier.

Digestive System

The digestive tract is simple. In some species the mouth is funnel-like, whereas in others it is surrounded by a muscular sucker or several muscular lobes. At the base of the mouth funnel is a spheroid **pharynx,** a powerful muscular pump. The **intestine,** or **cecum,** is a single, simple sac that usually extends to near the posterior end of the body (although in the Rugogastridae it is branched). Its epithelial cells bear a complex reticulum of lamellae on their luminal surface, presumably vastly increasing the absorptive surface. A layer of muscles, usually of both circular and longitudinal fibers, surrounds the cecum.

Osmoregulatory System

This system consists of numerous **flame cell protonephridia** connected to capillaries feeding into larger excretory ducts and eventually into an **excretory bladder** near the posterior end of the body. The flame cells are peculiar in that their flagellar membranes continue beyond the tips of the axonemes and anchor apically in the cytoplasm.[7] Lateral or nonterminal flagellar flames have been reported in a number of species. The small capillaries have numerous microvilli projecting into their lumina, and larger capillaries and excretory ducts are abundantly provided with lamellar projections of their surface membranes, thus suggesting secretory-absorptive function. The **excretory pore** is dorsosubterminal or terminal and usually single.

Nervous System

The aspidobothrean nervous system is very complex for a parasitic flatworm, reminiscent of a condition more typical of free-living forms. As in many turbellaria, there is a complex set of anterior nerves called a **cerebral commissure** and a modified ladder-type peripheral system. A wide variety of sensory receptors has been observed, mostly around the mouth and on

Tentacles

Figure 14.4 *Lophotaspis interiora* **from an alligator snapping turtle.**

From H. B. Ward and S. H. Hopkins, "A new North American aspidogastrid, *Lophotaspis interiora,*" in *J. Parasitol.* 18:69–78. Copyright © 1931.

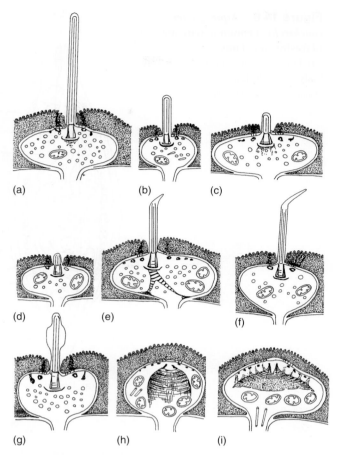

Figure 14.5   **Diagrams of various types of receptors of** *Lobatostoma manteri.*

(*a*) Uniciliate receptors with long cilium; (*b*) receptor with cilium of medium length; (*c*) short cilium receptor from posterior body surface; (*d*) short cilium receptor from anterior body surface; (*e*) receptor with ciliary rootlet; (*f*) cilium with bent tip; (*g*) inflated cilium receptor; (*h*) nonciliated receptor with large rootlet; (*i*) nonciliate disclike receptor.

From K. Rohde and N. Watson, "Sense receptors of *Lobatostoma manteri* (Trematoda, Aspidogastrea)," in *Int. J. Parasitol.* Copyright © 1992. Reprinted by permission.

the margins of the ventral disc. In a specimen of *Multicotyle purvisi* 6.1 mm long, Rohde[7] counted 360 dorsal and 260 ventral receptors in the prepharyngeal region and 140 in the oral cavity, not counting free nerve endings below the tegument. Three types of "ciliated sense organs" (sensilla) have been described on the body of *Cotylogaster occidentalis,*[3] and Rohde and Watson[12] distinguished nine types of receptors, most of them with cilia, in *Lobatostoma manteri,* a parasite of snails and fish of Australia's Great Barrier Reef. These receptors differed in the structure of their cilia and in the presence of a rootlet fiber (Fig. 14.5).

A complex system of connectives and commissures occurs in the ventral disc and alveolar walls, indicating a high degree of neuromuscular coordination. The septum, intestine, pharynx, prepharynx, cirrus pouch, uterus, and genital and excretory openings are all innervated by plexuses. Some cells in the nervous system are positive for paraldehyde-fuchsin stain, indicating possible neurosecretory function.

Reproductive Systems

The male reproductive system of Aspidobothrea is similar to that of Digenea (see Figs. 15.2 and 15.5). One, two, or many **testes** are present, located posterior to the ovary (Fig. 14.6). The **vas deferens** expands to form an **external seminal vesicle** before it enters the **cirrus pouch** to become an **ejaculatory duct.** A cirrus pouch is lacking in some species. The cirrus is unarmed and opens through a genital pore into a common genital atrium, located on the midventral surface just anterior to the leading margin of the ventral disc.

Spermiogenesis has been studied extensively in *Multicotyle purvisi,* and the general sequence of events appears to be widespread among parasitic Platyhelminthes.[5, 14] Spermatids develop as clusters of nuclei surrounding a central mass of cytoplasm (the **cytophore**). Two flagella, with striated fibers (rootlets) attached to their bases, grow out at right angles from a basal body (the **intercentriolar body, ICB**), located in an extension of the cytoplasm supported by pellicular microtubules (Fig. 14.7). A **median cytoplasmic process (MCP)** grows between the two flagella (Fig. 14.8). The nucleus and a mitochondrion migrate into the MCP, and the flagella bend at their bases, coming to lie parallel to the MCP and eventually fusing, starting with their bases. This process is known as *proximodistal fusion* (base to tip). Axonemes in the spermatozoan filament have a nine plus one structure, as is the case with other platyhelminth sperm (Fig. 14.9).

The female reproductive system consists of an ovary, vitelline cells, a uterus, and associated ducts. The **ovary**

Figure 14.6 *Aspidogaster conchicola,* **a common parasite of freshwater clams.**

(*a*) Dorsal view, showing general body form. (*b*) Lateral view.

Drawing by William Ober and Claire Garrison.

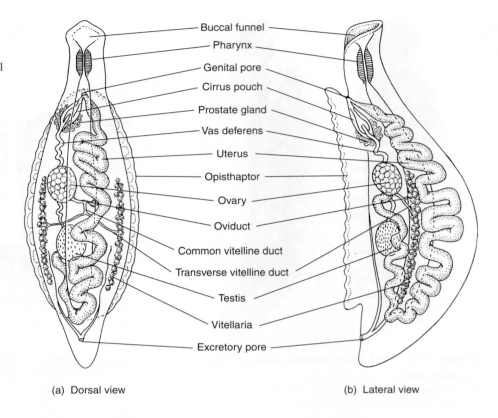

- Buccal funnel
- Pharynx
- Genital pore
- Cirrus pouch
- Prostate gland
- Vas deferens
- Uterus
- Opisthaptor
- Ovary
- Oviduct
- Common vitelline duct
- Transverse vitelline duct
- Testis
- Vitellaria
- Excretory pore

(a) Dorsal view   (b) Lateral view

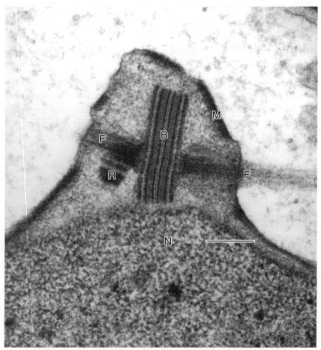

Figure 14.7 **A zone of differentiation associated with spermatid nucleus in *Microcotyle purvisi.***

B, Basal body; *F,* flagellar bases; *M,* pellicular microtubules; *N,* nucleus; *R,* beginning of striated fiber associated with flagellar bases (rootlet).

From N. A. Watson and K. Rohde, "Re-examination of spermatogenesis of *Multicotyle purvisi* (Platyhelminthes, Aspidogastrea)" in *Int. J. Parasitol.* 25:579–586. Copyright © 1995. Reprinted with permission of the publisher.

(see Fig. 14.6) is lobated or smooth and empties its products into an **oviduct.** The oviduct is peculiar among Platyhelminthes in that its lumen is divided into many tiny chambers by septa, and the lining along much of its length is ciliated[7] (Fig. 14.10). Each septum has a small hole in it through which the oocytes pass. The oviduct empties into an **ootype,** which is surrounded by Mehlis' gland cells (Fig. 14.11). A short tube leading from the ootype and ending blindly in the parenchyma or, in a few cases, connecting with the excretory canal is called **Laurer's canal** and probably represents a vestigial vagina.

Vitelline follicles occur in two lateral fields, each of which has a main **vitelline duct** that fuses with that from the other field to form a small **vitelline reservoir,** which in turn opens into the ootype. Finally, a **uterus** extends from the ootype to course toward the genital atrium, usually with a posterior loop and anterior, distal stem. The distal end of the uterus has powerful muscles in its walls and is called a **metraterm.** The metraterm propels eggs out of the system.

Some aspidobothreans are apparently self-fertilizing, with the cirrus depositing sperm in the terminal end of the uterus, which serves as a vagina. However, self-fertilization does not occur in *Lobatostoma manteri,* and unfertilized eggs do not develop beyond the blastula stage.[8]

DEVELOPMENT

As in other Platyhelminthes with separate vitellaria, eggs of aspidobothreans are ectolecithal; that is, most of the embryo's yolk supply is derived from separate cells packaged with the zygote inside the eggshell. Some species' eggs are completely embryonated when they pass from the parent, and

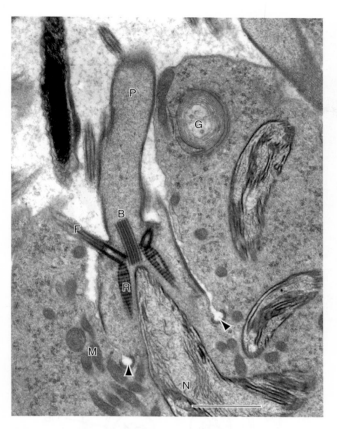

Figure 14.8   **Spermatid development in *Microcotyle purvisi*.**

Note intercentriolar body (*B*); free flagellum (*F*); Golgi body (*G*); mitochondria (*M*); nucleus (*N*); striated rootlet (*R*); and mediated cytoplasmic process (*P*). Arrowheads are "arching membranes" supported by microtubules and functioning to separate sperm. Bar = 1 μm.

From N. A. Watson and K. Rohde, "Re-examination of spermatogenesis of *Multicotyle purvisi* (Platyhelminthes, Aspidogastrea)." *Int. J. Parasitol.* 25:579–586. Copyright © 1995. Reprinted with permission of the publisher.

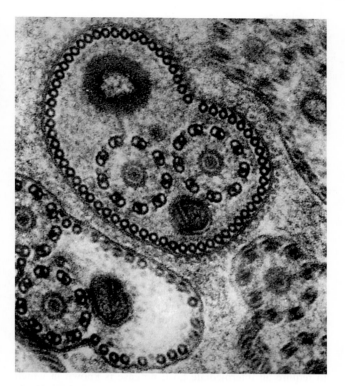

Figure 14.9   **Cross section of sperm filament of *Aspidogaster conchicola*.**

Courtesy of Ronald P. Hathaway.

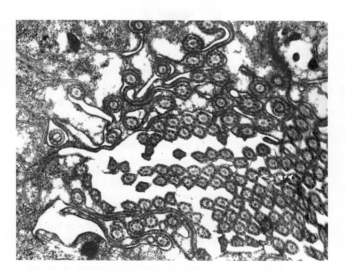

Figure 14.10   **Section of proximal oviduct of *Aspidogaster conchicola*, showing cilia that line much of its length.**

Courtesy of Ronald P. Hathaway.

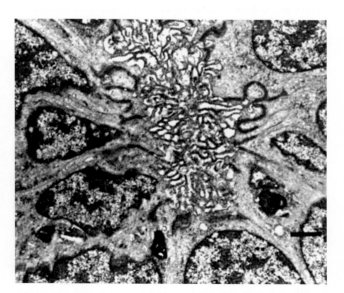

Figure 14.11   **Ootype of *Aspidogaster conchicola* surrounded by Mehlis' gland cells.**

Courtesy of Ronald P. Hathaway.

they hatch within a matter of hours, whereas others require three to four weeks of embryonation in the external environment. Larvae (**cotylocidia;** Fig. 14.12) hatching from eggs in most species have a number of ciliary tufts that are effective in swimming. These larvae possess a mouth, pharynx, simple gut, and prominent posterior-ventral disc without alveoli; there are no hooks. As the worm develops in its host, alveoli begin to form, tier by tier, in the anterior part of the ventral disc. The original cup of the disc remains apparent for some time behind the new ventral sucker and then disappears forever.

Larval ultrastructure has been studied in *M. purvisi* and in *C. occidentalis.*[1, 7] The tegument pattern is similar to that of an adult, with a distal cytoplasmic, syncytial layer at the surface and with internal cytons. Between the ciliary tufts and covering most of the body the tegument surface in *M. purvisi* bears unique filiform structures called **microfila.** These have one central filament and about 9 to 12 peripheral filaments, differing from microvilli in that they do not have a cytoplasmic core. Their function is unknown, but it has been suggested that they help the larva to float.[7] In contrast, the tegument of *C. occidentalis* bears short microvilli with an external glycocalyx coat.[1]

Most aspidobothreans have a direct life cycle, requiring no intermediate host. Those parasitic in vertebrates appear to require an intermediate host; no case is known in which the free larva is directly infective to vertebrates. Individuals can be removed from their definitive hosts and are capable of surviving for several days in water or saline, suggesting that they are rather generalized physiologically and not highly specialized for parasitism. Furthermore, if they are eaten by a fish or turtle, they can live for a considerable length of time in this new host. Therefore, it is not uncommon to find an aspidobothrean in a fish's intestine, although the worm normally parasitizes a mollusc. Some species have so little host specificity that they can mature both in clams and in fish, although those in fish may be larger and produce more progeny than when they develop in molluscs.[1] Others apparently will not mature in a mollusc and need a fish final host.[2] *Lobatostoma manteri* preadults develop in any of several species of snails, but they must reach the intestine of a snail-eating fish *(Trachinotus blochi)* to mature.[8, 9]

The following life cycles illustrate the biology of two aspidobothrean families.

Aspidogaster conchicola

This common representative of Aspidogastridae (see Fig. 14.6) is most often found in the pericardial cavity of freshwater clams in Europe, Africa, and North America, although it is known from other molluscs, fishes, and turtles. The adult is 2.5 mm to 3.0 mm long by 1.0 mm wide; it is oval and has a long, mobile "neck" with a buccal funnel at its end. Loculi on the ventral sucker are arrayed in four longitudinal rows, totaling 64 to 66.

When eggs hatch within a molluscan host, the young can develop without further migration. If an egg or cotylocidium leaves the mollusc and is drawn into the incurrent siphon of the same or another clam, it can reach the nephridiopore and migrate through the kidney into the pericardium.

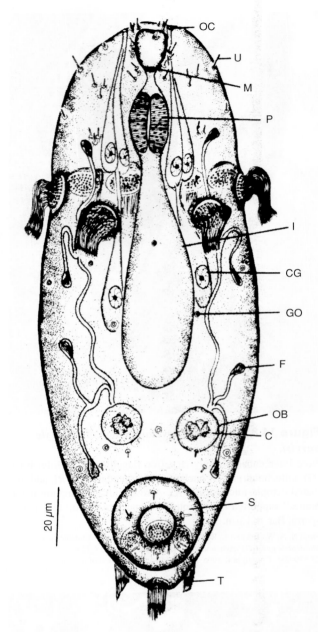

Figure 14.12 Composite drawing of cotylocidium of *Cotylogaster occidentalis.*

C, concretion; *CG,* cephalic gland; *GO,* opening of gobletlike gland cells; *F,* flame cell; *I,* intestine; *M,* mouth; *OB,* osmoregulatory bladder; *OC,* opening of cephalic gland; *P,* pharynx; *S,* sucker; *T,* tuft of cilia; *U,* uniciliated sensory structure.

From D. W. Fredericksen, "The fine structure and phylogenetic position of the cotylocidium larva of *Cotylogaster occidentalis* Nickerson 1902 (Trematoda: Aspidogastridae,)" in *J. Parasitol.* 64:961–976. Copyright © 1978 *Journal of Parasitology.* Reprinted by permission.

The cotylocidium is 130 μm to 170 μm long at hatching, lacks external cilia, and bears a simple posterior sucker without loculi. Growth and metamorphosis are rapid.

Lophotaspis vallei, also in Aspidogastridae, may use a marine snail as intermediate host. Mature forms have been found in marine turtles, but it is possible that they normally mature in molluscs.

Rugogaster hydrolagi

This species (Fig. 14.13) parasitizes rectal glands of ratfish (*Hydrolagus colliei*, a species of chimera, class Holecephali) in the Pacific Ocean. In 1973, Schell[13] erected a new family for these worms, based mainly on their branched gut and rugae, and only one additional species in the family has been described since then. The life cycle is unknown, but only gravid adults are found in ratfish, and eggs are embryonated, so there may be intermediate hosts involved.

Stichocotyle nephropsis

This parasite (Fig. 14.14) lives in bile ducts of rays in the Atlantic Ocean. It has been found in lobsters and other crustaceans and is thought to employ them as intermediate hosts. Adults are slender, are 115 mm long, and have 24 to 30 separate suckers along their ventral surface. This is the only species in Stichocotylidae and is the only aspidogastrean found in crustaceans. Crustaceans may not be the normal hosts in the life cycle of this parasite and the occurrence of *S. nephropsis* in these animals could be accidental. It is also possible that this worm does not belong in Aspidobothrea.

PHYLOGENETIC CONSIDERATIONS

Aspidobothrean anatomy is rather digenean, whereas their biology is suggestive of Monogenoidea. This tiny group of worms displays sufficient individuality to distinguish and separate it from both groups, although recent phylogenetic studies indicate that Aspidobothrea is both monophyletic and the sister taxon to Digenea (see chapter 13).[10, 15]

Aspidobothreans differ morphologically from Digenea in that the ventral sucker develops as a new structure,

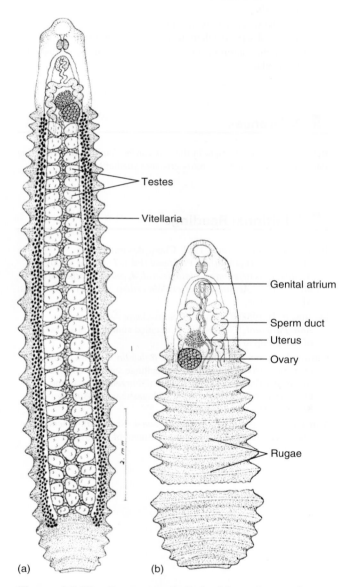

(a)   (b)

Figure 14.13   *Rugogaster hydrolagi* **from the rectal glands of the ratfish.**

(a) Dorsal view of a prepared specimen, showing the reproductive organs. *(b)* Interrupted ventral view showing the rugae or transverse ridges.

From S. S. Schell, "*Rugogaster hydrolagi gen.* et sp. n. (Trematoda: Aspidobothrea fam. n.) from the ratfish, *Hydrolagus colliei* (Lay and Bennett, 1839)," in *J. Parasitol.* 59:803–805. Copyright © 1973 Journal of Parasitology. Reprinted by permission.

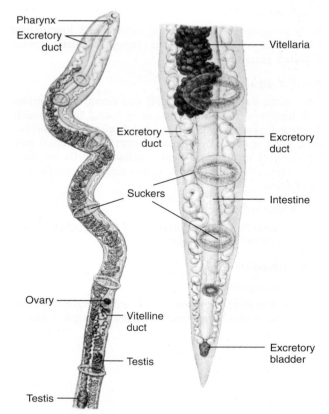

Figure 14.14   *Stichocotyle nephropsis* **from bile ducts of rays.**

From T. Odhner, "*Stichocotyle nephropis* J. T. Cunningham ein aberranter Trematode der Digenenfamilie Aspidogastridae," in *K. Svenska Vetensk. Acad. Handl.* 45:3–16, 1910.

unrelated to any homologue in larvae or adults of the other group. The frontal septum of aspidobothreans is not found in any other platyhelminth taxon; the septate, ciliated oviduct is not found in Digenea.[7] However, the predominance of mollusc hosts, presence of Laurer's canal, and a highly developed nervous system are suggestive of Digenea. The simple, saclike cecum and undemanding physiological requirements are more in keeping with free-living turbellaria. Rohde[8] believed that an organism such as *Lobatostoma manteri,* with its mollusc intermediate host and fish definitive host, probably lies close to the ancestral protodigenean stock.

CLASSIFICATION OF ASPIDOBOTHREA

Trematodes with single, large, ventral sucker subdivided by septa into numerous, shallow loculi or with one ventral row of individual suckers; no sclerotized armature on any species; mouth with or without sucker, sometimes lobated; pharynx well developed; intestine with a single or double median sac; testis single, double, or numerous; cirrus pouch present or absent; genital pores median, in front of sucker; ovary single, pretesticular; vagina absent, Laurer's canal sometimes present; vitellaria follicular, usually lateral but occasionally otherwise; eggs lacking polar prolongations; excretory pores on or near posterior end; development direct, without intermediate host; parasites of molluscs, fishes, and turtles.

Order Stichocotylida

Family Stichocotylidae

Body elongated, slender; ventral surface with longitudinal row of separate suckers; two testes present; vitellaria tubular, unpaired; parasites of Batoidea.

Family Rugogastridae

Body elongated; most of ventral and lateral body surface with transverse rugae; musculature of buccal funnel weakly developed; pharynx, prepharynx, and esophagus present; two ceca; testes multiple; ovary pretesticular; Laurer's canal present, seminal receptacle absent; vitellaria distributed along ceca; uterus ventral to testes; eggs operculate; parasites of Holocephali.

Family Multicalycidae

Body elongated; holdfast composed of fused suckers; otherwise similar to Rugogastridae.

Order Aspidobothriiformes

Family Aspidogastridae

Body oval or elongate; ventral sucker with numerous shallow loculi; one or two testes present; vitellaria follicular, lateral; parasites of molluscs, fishes, or turtles; cosmopolitan. Subfamilies: Aspidogasterinae, Cotylaspidinae, Rohdellinae.

Learning Outcomes

By the time a student has finished studying this chapter, he or she should be able to:

1. Describe the structural features of *Aspidogaster conchicola* and tell which one(s) of these features are common to Aspidobothrea and distinguish members of this class from members of Digenea and Cestoda.
2. Define the vocabulary words: marginal body, loculi (alveoli), longitudinal septum, ciliary receptors, and cotylocidium.
3. Describe the basis for distinguishing between the two orders of Aspidobothrea.

References

References for superscripts in the text can be found at the following Internet site: www.mhhe.com/robertsjanovynadler9e

Additional Readings

Baer, J. G., and C. Joyeux. 1961. Classe des trematodes (Trematoda Rudolphi). In P. Grassé (Ed.), *Traité de zoologie: Anatomie, systématique, biologie, vol. 4, part I. Plathelminthes, Mésozoaires, Acanthocéphales, Némertiens* (pp. 561–570). Paris: Masson & Cie.

Dollfus, R. P. 1958. Trematodes. Sous-classe Aspidogastrea. *Ann. Parasitol.* 33:305–395. A detailed summary of knowledge of this group to 1958.

Gibson, D. I., and S. Chinabut. 1984. *Rohdella siamensis* gen. et sp. nov. (Aspidogastridae: Rohdellinae subfam. nov.) from freshwater fishes in Thailand, with a reorganization of the classification of the subclass Aspidogastrea. *Parasitology* 88:383–393.

Yamaguti, S. 1963. *Systema Helminthum 4.* New York: Interscience Publishers. A useful taxonomic treatment of the group.

Chapter 15

Trematoda: Form, Function, and Classification of Digeneans

The life cycles of the great majority of digeneans display a remarkable and highly characteristic alternation of asexual and sexual reproductive phases....

—P. J. Whitfield[94]

Digenetic trematodes, or flukes, are among the most common and abundant of parasitic worms, second only to nematodes in their distribution. They are parasites of all classes of vertebrates, especially marine fishes, and nearly every organ of the vertebrate body can be parasitized by some kind of trematode, as adult or juvenile. Digenean development occurs in at least two hosts. The first is a mollusc or, very rarely, an annelid. Many species include a second and even a third intermediate host in their life cycles. Several species cause economic losses to society through infections of domestic animals, and others are medically important parasites of humans. Because of their importance, Digenea have stimulated vast amounts of research, and literature on the group is immense. This chapter will summarize digenean morphology and biology, illustrating them with some of the more important species.

Trematode development will be considered in detail later (p. 219), but a "typical" life cycle (Fig. 15.1) is as follows: A ciliated, free-swimming larva, a **miracidium,** hatches from its shell and penetrates a first intermediate host, usually a snail. At the time of penetration or soon after, the larva discards its ciliated epithelium and metamorphoses into a rather simple, saclike form, a **sporocyst.** Within the sporocyst a number of embryos develop asexually to become **rediae.** Rediae are somewhat more differentiated than sporocysts, possessing, for example, a pharynx and a gut, neither of which are present in a miracidium or sporocyst. Additional embryos develop within the redia, and these become **cercariae.** Cercariae emerge from the snail. They usually have a tail to aid in swimming. Although many species require further development as **metacercariae** before they are infective to a definitive host, cercariae are properly considered juveniles; they have organs that will develop into an adult digestive tract and suckers, and genital primordia are often present. A fully developed, encysted metacercaria is infective to a definitive host and develops there into an adult trematode. Many trematodes have a second intermediate host which bears their encysted metacercariae. Their vertebrate definitive hosts are then infected when they consume the second intermediate host.

FORM AND FUNCTION

Body Form

Flukes exhibit a great variety of shapes and sizes as well as variations in internal anatomy. They range from the tiny *Levinseniella minuta,* only 0.16 mm long, to the giant *Fascioloides magna,* which reaches 5.7 cm in length and 2.5 cm in width.

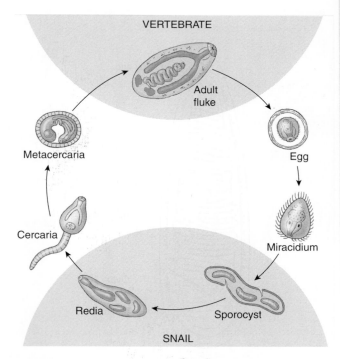

Figure 15.1 **Typical trematode life cycle.**
Many variations occur.

Drawing by William Ober and Claire Garrison.

209

Most flukes are dorsoventrally flattened and oval in shape, but some are as thick as they are wide. Some species are filiform, round, or even wider than they are long. Flukes usually possess a powerful oral sucker that surrounds the mouth, and most also have a midventral **acetabulum** or **ventral** sucker. The words **distome, monostome,** and **amphistome,** which formerly had taxonomic significance, are sometimes used as descriptive terms, and of course they refer to suckers, not mouths (Gr., *stoma:* mouth). If a worm has only an oral sucker, it is a monostome (Fig. 15.2); with an oral sucker and an acetabulum at the posterior end of the body, it is an amphistome (Fig. 15.3); and if the acetabulum is elsewhere on the ventral surface, the worm is a distome (Fig. 15.4). The oral sucker may have muscular lappets, as in genus *Bunodera* (Fig. 15.5), or there may be an anterior adhesive organ with tentacles, as in *Bucephalus* (Fig. 15.6). *Rhopalias* spp., parasites of American opossums, have a spiny, retractable proboscis on each side of the oral sucker (Fig. 15.7). In species of Hemiuridae the posterior part of the body telescopes into the anterior portion. Some workers use additional terms to describe body forms of digeneans, such as *holostome* (p. 236, Fig. 16.1), *schistosome* (p. 240), and *echinostome* (p. 254).

Tegument

The tegument of trematodes, like that of cestodes, formerly was considered a nonliving, secreted cuticle; but, as in cestodes, electron microscopy reveals that the body covering of trematodes is a living, complex tissue. In common with Monogenoidea and Cestoidea, digenetic trematodes have a "sunken" epidermis; that is, there is a distal, anucleate

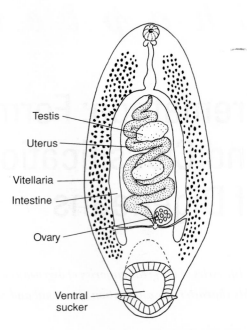

Figure 15.3 *Zygocotyle lunata,* **an amphistome fluke from ducks.**

Drawing by William Ober.

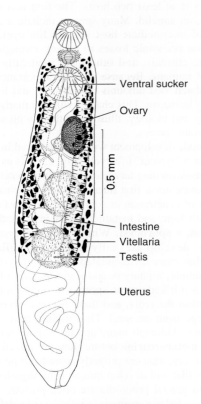

Figure 15.4 *Alloglossidium hirudicola,* **a distome trematode from leeches.**

From G. D. Schmidt and K. Chaloupka, "*Alloglossidium hirudicola* sp. n., a neotenic trematode (Plagiorchiidae) from leeches, *Haemopis* sp.," in *J. of Parasitol.* 55:1185–1186. Copyright © 1969. Reprinted by permission of the publisher.

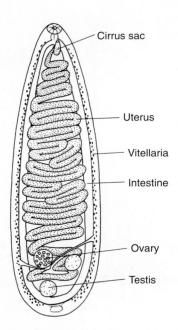

Figure 15.2 *Cyclocoelum lanceolatum,* **a common monostome fluke from the air sacs of shore birds.**

Drawing by William Ober.

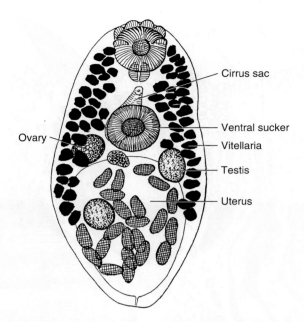

Figure 15.5 *Bunodera sacculata* **from yellow perch.**
Note the muscular lappets on the oral sucker.

From H. J. Van Cleave and J. F. Mueller, "Parasites of Oneida Lake Fishes, Part I. Descriptions of new genera and new species," in *Roosevelt Wildlife Annals* 3:9–71, 1932.

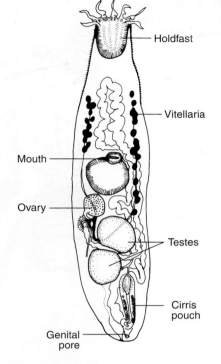

Figure 15.6 *Bucephalus polymorphus* **from European and Asian fishes.**

From S. Yamaguti, *Synopsis of digenetic trematodes of vertebrates.* Tokyo: Keigaku Publishing Company, 1971.

layer (**distal cytoplasm**). Cell bodies containing the nuclei (**cytons**) lie beneath a superficial layer of muscles, connected to the distal cytoplasm by way of channels (**internuncial processes,** Fig. 15.8). Because the distal cytoplasm is continuous, with no intervening cell membranes, tegument is **syncytial.** Although this is the same general organization found in cestodes, trematode tegument differs in many details, and striking differences in structure may occur in the same individual from one region of the body to another.

Ornamentation such as **spines** is often present in certain areas of a trematode's body and may be discernible with a light microscope. Oral and ventral suckers of *Schistosoma* spp. are densely beset with spines (Fig. 15.9). The tegumental surface of schistosomes bears ridges of various configurations, pits, and sensory papillae[42] (Fig. 15.10). Female schistosomes have many anteriorly directed spines on their posterior ends (Fig. 15.11). The spines consist of crystalline actin;[23] their bases lie above the basement membrane of the distal

cytoplasm, and their apices project above the surface, although generally they are covered by the outer plasma membrane. In some trematodes the spines are in the form of flattened plates with serrated edges.[2]

Distal cytoplasm usually contains vesicular inclusions, more or less dense, and tegument of the same worm sometimes may bear several recognizable types. The function of the vesicles is unclear, although in some cases they contribute to the outer surface. The surface membrane of *S. mansoni* is continuously renewed by multilaminar vesicles moving outward through the distal cytoplasm, perhaps to replace membrane damaged by host antibodies.[81] The outer layers with host antibody adsorbed to them are indeed shed by the worm.[56] In *Megalodiscus* contents of some vesicles seem to be emptied to the outside.[8]

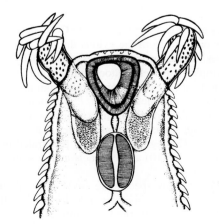

Figure 15.7 *Rhopalias caballeroi,* **a parasite of American opossums.**

Retractable proboscides are located on each side of the mouth.

From Caballero y C., E. 1946. An. Inst. Biol., Univ. Nac. Auton. Mex, Serie Zool. 1. 18:137–165 and Kifune, T., and N. Uyema. 1982. Med. Bull., Fukuoka Univ. 9:241–256.

Figure 15.8 **Diagram of the tegument of *Fasciola hepatica* at the ultrastructural level.**

Drawing by L. T. Threadgold.

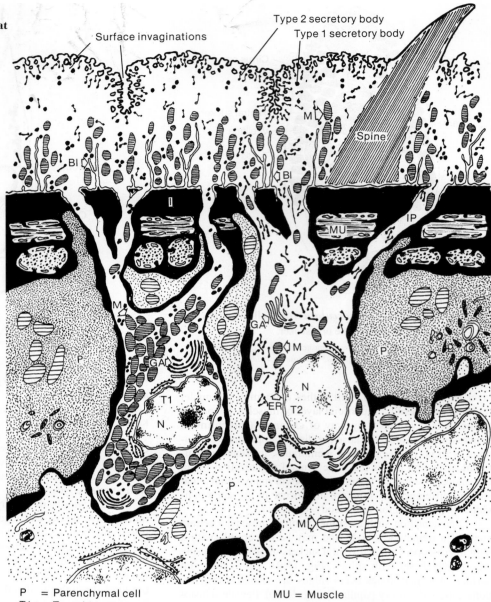

P = Parenchymal cell
T1 = Type 1 tegumentary cyton
T2 = Type 2 tegumentary cyton
GA = Golgi complex
I = Interstitial material (connective tissue)
IP = Internuncial process

MU = Muscle
BI = Basal invagination
N = Nucleus
ER = Granular endoplasmic reticulum
M = Mitochondria

Vesicles of the distal cytoplasm are Golgi derived. They usually pass outward from the cytons through the internuncial processes, although Golgi bodies occasionally occur in the distal cytoplasm as well. Mitochondria occur in the distal cytoplasm in most species examined, although not in *Megalodiscus* and some paramphistomes.[8, 10] The outer surface of adult trematodes does not bear microvilluslike microtriches, as in cestodes, but some structural features that increase surface area occur. The tegument is often penetrated by many deep pits (see Fig. 15.10) and channels and in some species bears short microvilli.[3, 81]

Miracidia of many species are covered by ciliated epithelial cells with nuclei, as is typical for such cells. The epithelial cells are interrupted by "intercellular ridges,"

extensions of cells whose cytons lie beneath the superficial muscle layer and bear no cilia, although some microvilli may be present (Fig. 15.12). On loss of the ciliated epithelium, metamorphosis to a sporocyst involves a spreading of the distal cytoplasm from the intercellular ridges over the worm's surface.[87] Well-developed microvilli are present on the surface of both sporocyst and redia. The luminal surface of tegumental cells in rediae may be thrown into a large number of flattened sheets that extend to other cells in the body wall and to cercarial embryos contained in the lumen. Nutritive molecules such as glucose can pass through the tegument to developing cercariae.[8]

Early embryos of cercariae are covered with a primary epidermis below which a definitive epithelium forms.

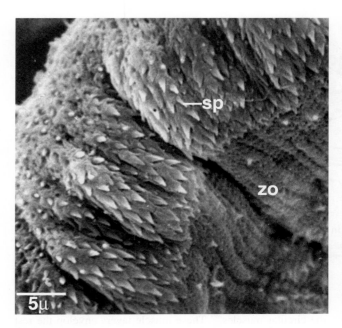

Figure 15.9   Scanning electron micrograph of male *Schistosoma japonicum.*

The rim of the ventral sucker shows increasingly larger spines (*sp*) toward the interior and a narrow zone (*zo*) with sensory papillae. (×2000)

From P. Sobhon and E. S. Upatham, *Snail hosts, life-cycle, and tegumental structure of Oriental schistosomes.* World Health Organization, Special Programme for Research and Training in Tropical Diseases, 1990.

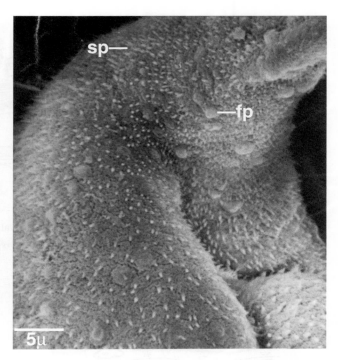

Figure 15.11   Scanning electron micrograph of the posterior end of a female *Schistosoma japonicum.*

Note numerous spines (*sp*) and sensory papillae (*fp*). (×1600)

From P. Sobhon and E. S. Upatham, *Snail hosts, life-cycle, and tegumental structure of Oriental schistosomes.* World Health Organization, Special Programme for Research and Training in Tropical Diseases, 1990.

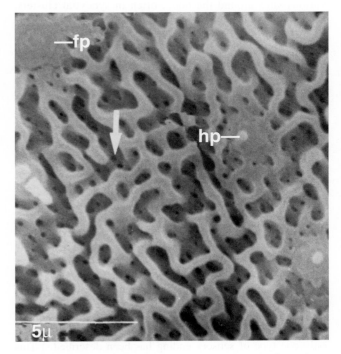

Figure 15.10   Scanning electron micrograph of a male *Schistosoma japonicum.*

Note ridges, pits (*arrow*), and sensory papillae with (*hp*) and without (*fp*) a short cilium. (×5100)

From P. Sobhon and E. S. Upatham, *Snail hosts, life-cycle, and tegumental structure of Oriental schistosomes.* World Health Organization, Special Programme for Research and Training in Tropical Diseases, 1990.

Nuclei of this secondary epithelium sink into the parenchyma, and in its final form cercarial tegument has an organization similar to that of adults. Cystogenic cells in the parenchyma begin to secrete cyst material, which passes into the distal cytoplasm of the tegument. The metacercarial cyst wall forms when the cercarial tegument sloughs off, and the cyst material it contains undergoes chemical and/or physical changes to envelop the worm in its cyst. The cystogenic cells in the parenchyma then flow toward the surface, their nuclei being retained beneath the superficial muscles, and a thin layer of cytoplasm spreads over the organism to become the adult tegument.

A differing mode of tegument formation occurs in cercariae of *S. mansoni.*[50] The definitive tegument, including its nuclei, forms beneath the primitive tegument and is syncytial. Subsequently nuclei of the primitive tegument degenerate and are lost, as processes from subtegumental cells grow outward and join the distal cytoplasmic layer (Fig. 15.13).

The tegument is variously interrupted by cytoplasmic projections of gland cells, by openings of excretory pores, and by nerve endings. Both miracidia and cercariae may have penetration glands that open at the anterior, and adults of some species have prominent glandular organs opening to the exterior.

Muscular System

Muscles that occur most consistently throughout Digenea are circular muscles lying just beneath the basal lamina of the tegument, with longitudinal and diagonal layers underlying

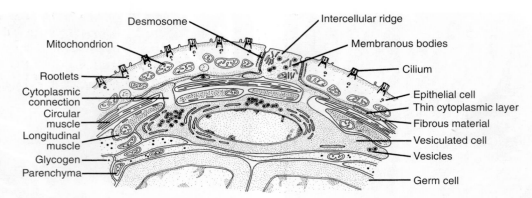

Figure 15.12 **Miracidium of** *Fasciola hepatica.*

The line drawing reconstructs a transverse segment of the body wall in the region of the germ cell cavity.

From R. A. Wilson, "Fine structure of the tegument of the miracidium of *Fasciola hepatica L.*," in *J. Parasitol.* 55:124–133. Copyright © 1969 *Journal of Parasitology.* Reprinted by permission.

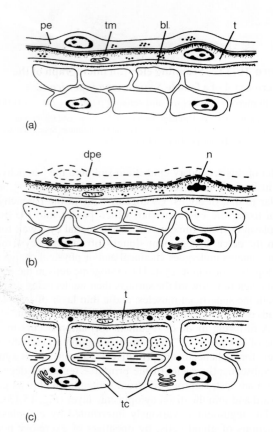

Figure 15.13 **Diagram summarizing three stages in the formation of the tegument during cercarial development.**

(*a*) Germ ball covered with a primitive epithelium (*pe*) and the tegument (*t*), which has a thickened outer membrane (*tm*) and an underlying basal lamina (*bl*). (*b*) Young cercaria with degenerating primitive epithelium (*dpe*) and degenerating tegumental nucleus (*n*). (*c*) Cercaria nearly ready to emerge from the sporocyst; the primitive epithelium has been lost, and the tegument (*t*) is connected to nucleated tegumental cytons (*tc*).

From D. J. Hockley, "Ultrastructure of the tegument of *Schistosoma*," in B. Dawes (Ed.), *Advances in parasitology*, vol. 11. London: Academic Press, LTD, 1973.

the circular muscles.[59] These muscle layers envelop the rest of the body like a sheath. The degree of muscularization varies considerably in the group, from rather feeble to robust and strong. The deep musculature found in cestodes is generally absent in trematodes. Muscles are often most prominent in the anterior parts of the body, and strands connecting dorsal to ventral superficial muscles are usually found in lateral areas. Fibers are smooth, and their nuclei occur in cytons called *myoblasts* connected to fiber bundles and located in various sites around the body, often in syncytial clusters. The suckers and pharynx are supplied with radial muscle fibers, often very strongly developed. A network of fibers may surround the intestinal ceca, helping these structures fill and empty.

Nervous System

Organization of the nervous system is the **orthogon** (ladderlike) type typical of Platyhelminthes, with longitudinal nerve cords connected at intervals by transverse ring commissures (connecting bands of nerve tissue, Fig. 15.14).[46] Several nerves issue anteriorly from the cerebral ganglion, and three main pairs of trunks—dorsal, lateral, and ventral—supply posterior parts of the body. Ventral nerves are usually the most developed. Branches provide motor and sensory endings to muscles and tegument. The anterior end, especially the oral sucker, is well supplied with sensory endings (see Fig. 15.14).

Sensory endings in Digenea are an interesting array of types, particularly in miracidia and cercariae. Adults require no orientation to such stimuli as light and gravity, and, in the few forms in which ultrastructure has been studied, only one type of sensory ending has been described.[51] This is a bulbous nerve ending in the tegument that has a short, modified cilium projecting from it, and the cilium is enclosed throughout its length by a thin layer of tegument. The general structure is similar to that of sense organs described for cestodes (see Fig. 20.13). Such structures generally have been regarded as **tangoreceptors** (receptors sensitive to touch) in trematodes.

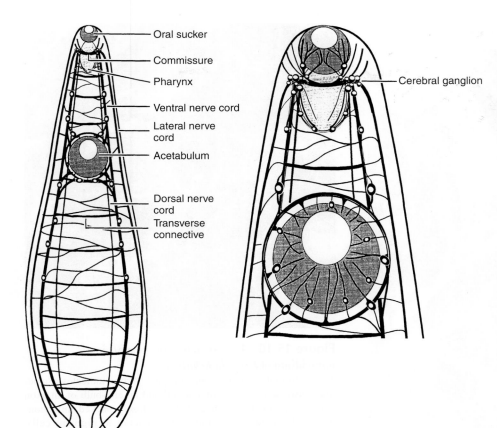

- Oral sucker
- Commissure
- Pharynx
- Ventral nerve cord
- Lateral nerve cord
- Acetabulum
- Dorsal nerve cord
- Transverse connective

- Cerebral ganglion

Figure 15.14   Generalized schematic pattern of the trematode nervous system.

From D. W. Halton and M. K. S. Gustafson, "Functional morphology of the platyhelminth nervous system," in *Parasitology* 113:547–572. Copyright © 1996. Cambridge University Press. Reprinted by permission of Cambridge University Press.

Cercariae and miracidia show more variety in sense organs than do adults, a condition doubtlessly related to the adaptive value of finding a host quickly.[11] Uniciliated bulbous endings are found on the anterior portion of cercariae of *S. mansoni,* similar to but smaller than those on adults. The tegumentary sheath opens at the ciliary apex. In addition, a bulbous type with a long (7 μm) unsheathed cilium is widely distributed over the body of cercariae, and its lateral areas bear small, flask-shaped endings containing five or six cilia and opening to the outside through a 0.2 μm pore (Fig. 15.15). This latter type is probably a chemosensory ending.

Another apparent chemoreceptor, this one described in miracidia of *Diplostomum spathaceum,* consists of two dorsal papillae between the first series of ciliated plates. Each papilla consists of a nerve ending and has radiating from it a number of modified cilia, which are parallel to the surface of the miracidium. These sensory endings are strikingly similar to olfactory receptors of vertebrate nasal epithelium! Thirteen morphologically distinct types of sensory endings can be recognized on the body and tail of *Diplostomum pseudospathaceum* cercariae.[29]

Eyespots are present in many species of miracidia and in some cercariae. Although also present in some adult trematodes, eyespots are apparently functionless in adults. The structure of eyespots in miracidia is generally similar to that of such organs found in turbellaria and some Annelida. Eyespots consist of one or two cup-shaped pigment cells surrounding parallel rhabdomeric microvilli of one or more retinular cells (Fig. 15.16). Mitochondria of the retinular cells

are packed in a mass near the rhabdomere. Because rhabdomeres are the photoreceptors, the cup shape of pigment cells allows the organism to distinguish light direction. Interestingly, some miracidia that do not have eyespots yet can nevertheless orient with respect to light. Some cells in miracidia of *D. spathaceum* and *S. mansoni* have large vacuoles; into these vacuoles project a number of cilia, each of which has a conspicuous membrane evagination. These membranes, which are stacked in a lamellar fashion, might be photoreceptors, thus providing, in the case of *S. mansoni,* a means of light sensitivity for a miracidium without eyespots.[11]

An important excitatory neurotransmitter is 5-hydroxytryptamine, and acetylcholine is apparently the major inhibitor of neuromuscular transmission.[46] A large number of neuropeptides have been found distributed through the nervous system of trematodes and other flatworms.[79] Although they probably serve as messenger systems that regulate and control a variety of bodily processes, their specific functions remain obscure. There is evidence that some neuropeptides help coordinate complex muscular activities involved in formation of eggs in the oogenotop (p. 219).[46]

Excretion and Osmoregulation

In her concise review, Hertel[48] paraphrased Beklemishev,[6] who said that excretion included (1) removal of waste products of metabolism; (2) regulation of internal osmotic pressure; (3) regulation of internal ionic composition; and

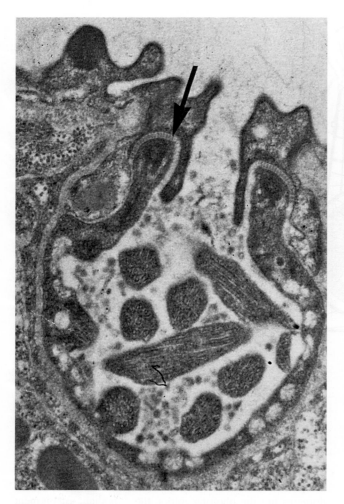

Figure 15.15 **Multiciliated pit in anterior body
tegument of cercaria of *Schistosoma mansoni.***
Arrow indicates septate desmosome. (×36,000)

From G. P. Morris, "The fine structure of the tegument and associated structures of
the cercaria of *Schistosoma mansoni*," in *Z. Parasitenkd.* 36:115–131. Copyright
© 1971.

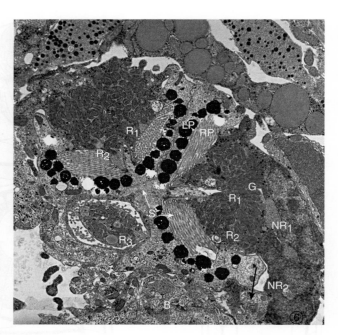

Figure 15.16 Ultrastructure of eyespots in a
miracidium of *Fasciola hepatica.*
Nearly frontal section in dorsal aspect. *Arrow* indicates a junction between a lateral retinular cell and an end-bulb of an axon from the brain. Retinular cells have closely packed mitochondria with rhabdomeric microvilli adjacent to pigment cells. *B*, brain; *G*, glycogen; *LP*, left pigment cell; *RP*, right pigment cell; NR_1, nucleus of anterior lateral retinular cell; NR_2, nucleus of posterior lateral retinular cell; R_1, anterior lateral retinular cell; R_2, posterior lateral retinular cell; R_3, median or posterior (5th) retinular cell; *S*, septum. (×7000)

From H. Isseroff, and R. M. Cable, "Fine structure of photoreceptors in larval
trematodes. A comparative study," in *Z. Zellforsch.* 86:511–534. Copyright © 1968.

(4) removal of unnecessary or harmful substances. Thus, by this definition excretion includes osmoregulation, and this is an important function in the so-called excretory systems of many flatworms. Removal of metabolic wastes takes place by diffusion across the tegument and epithelial lining of the gut and by exocytosis of vesicles, in addition to through the excretory system.

The excretory system of Digenea is based on flame bulb **protonephridia** (units of an excretory system closed at the proximal end and opening to the exterior at the distal end by way of a pore). A **flame bulb** (see Fig. 20.15), or cell, is flask shaped and contains a tuft of fused flagella to provide a motive force for fluid in the system. In Trematoda and Cercomeromorphae, flagella are surrounded by rodlike extensions of the flame cell, which extend between similar projections of the proximal tubule cell.[48] These interlacing rods form the latticelike weir (p. 309). The weir is a filtering apparatus. A thin membrane usually extends between the rods, and beating of the flagella creates a pressure gradient that draws fluid through the weir and into the collecting tubule. Fingerlike **leptotriches** sometimes extend from both the internal and external surfaces of the weir. They appear to increase filtration efficiency, possibly by keeping surrounding cells away from the weir and keeping the wall of the weir away from the tuft of flagella.

Ductules of flame cells join collecting ducts, those on each side eventually feeding into an excretory bladder in an adult that opens to the outside with a single pore. In digeneans the pore is almost always located near the posterior end of the worm. In some trematodes walls of collecting ducts are supplied with microvilli, indicating that some transfer of substances, absorption or secretion, is probably occurring.[8] That the system has osmoregulatory function may be inferred from the fact that, among free-living Platyhelminthes, freshwater forms have much better developed protonephridial systems than do marine flatworms. When the two free-swimming stages of digeneans occur in freshwater, they require an efficient water-pumping system.

The primary nitrogenous excretory product of trematodes is apparently ammonia, although excretion of uric acid and urea has been reported. What proportion of ammonia excretion takes place through the tegument, ceca, or excretory system is not known.

Acquisition of Nutrients and Digestion

Feeding and digestion in trematodes vary with nutrient type and habitat within their host. For example, two lung flukes of frogs, *Haematoloechus medioplexus* and *Haplometra cylindracea,* feed predominately on blood from the capillaries. Both species draw a plug of tissue into their oral sucker and then erode host tissue by a pumping action of their strong, muscular pharynx. Other trematode species characteristically found in the intestine, urinary bladder, rectum, and bile ducts feed more or less by the same mechanism, although their food may consist of less blood and more mucus and tissue from the wall of their habitat, and it may even include gut contents. In species without a pharynx that feed by this mechanism, the walls of their esophagus are quite muscular, and this apparently serves the function of a pharynx. In contrast *S. mansoni,* living in blood vessels of the hepatoportal system and immersed in its semifluid blood food, has no necessity to breach host tissues, and, not surprisingly, this species has neither pharynx nor muscular esophagus.

Digestion in most species studied is predominately extracellular in the ceca, but in *Fasciola hepatica* it occurs by a combination of intracellular and extracellular processes. A frog lung fluke, *Haplometra cylindracea,* has pear-shaped gland cells in its anterior end, and a nonspecific esterase is secreted from these cells through the tegument of the oral sucker, beginning the digestive process even before food is drawn into the ceca.

Those trematodes that feed on blood cope in various ways with the iron component of hemoglobin. In *F. hepatica,* in which final digestion of hemoglobin is intracellular, the iron is expelled through the excretory system and tegument. The fate of the iron in *H. cylindracea* is unclear, but apparently it is stored within the worm, tightly bound to protein. Extracellular digestion in *Haematoloechus medioplexus* and *S. mansoni* produces insoluble end products within the cecal lumen, and these wastes are periodically regurgitated.

In *S. mansoni* the end products are a heterogeneous population of molecules, but worms digest and incorporate some of both globin and heme moieties of hemoglobin.[38] Ceca of trematodes apparently do not bear any gland cells, but gastrodermal cells themselves may in certain species secrete some digestive enzymes: Proteases, dipeptidases, an aminopeptidase, lipases, acid phosphatase, and esterases have been detected.[39] Alkaline phosphatase has not been found in most trematodes. *Fasciola hepatica* secretes a dipeptidyl dipeptidase.[20]

There are several peptidases in the intestine of *S. mansoni,* of which the most abundant are cathepsins (a family of cysteine peptidases), especially an enzyme designated SmCB1 (for *S. mansoni* cathepsin B1).[19] *Trichobilharzia regenti* are schistosomatid parasites that live in nasal cavaties of ducks, where they feed on blood. As in other schistosomatids, their cercariae penetrate their host's skin, but juvenile *T. regenti* (schistosomula) follow an unusual route to the nasal cavity: via their host's nervous system. Cathepsins in their gut are inefficient in digesting hemoglobin but readily degrade myelin basic protein, the major protein component of nervous tissue.[35]

The gastrodermis of trematodes may be syncytial or cellular, according to species.[39] Cytoplasmic processes, which vary from short (1 μm to 15 μm) and irregular to long (10 μm to 20 μm), extend into the lumen (Fig. 15.17). Fujino[39]

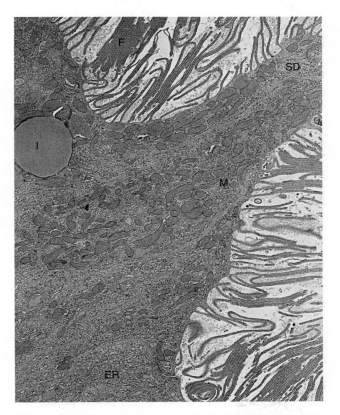

Figure 15.17   Apical portion of the cecal epithelium of *Paragonimus kellicotti.*

The apical surface has numerous folds (*F*) extending into the lumen. The cecal epithelial cells are joined by septate desmosomes (*SD*). The cytoplasm contains a well-developed granular endoplasmic reticulum (*ER*) and numerous mitochondria (*M*). An inclusion (*I*) is indicated.

From S. C. Dike, "Acid phosphatase activity and ferritin incorporation in the ceca of digenetic trematodes," in *J. Parasitol.* 55:111–123. Copyright © 1969.

distinguished three categories according to shape: (1) slender and ribbon shaped with narrow ends (for example, *Clonorchis sinensis, Eurytrema pancreaticum*); (2) broad and triangular with distal or marginal filamentous extensions (for example, *Fasciola hepatica, Echinostoma hortense*); and (3) broad, sheetlike, or triangular, with distal ends blunt or round (for example, *Haematoloechus lobatus, Schistosoma japonicum, Paragonimus* spp.). These processes greatly amplify the surface area of the gastrodermis for absorption of nutrients.

Within the gut cells of both *Gorgodera amplicava* and *Haematoloechus medioplexus* are abundant rough endoplasmic reticulum, many mitochondria, and frequent Golgi bodies and membrane-bound vesicular inclusions. High activity of acid phosphatase is found in vesicles of *H. medioplexus* and *Paragonimus kellicotti,* and after incubation in ferritin that material is found within the vesicles. No evidence of "transmembranosis" has been found, but the vesicles may be lysosomes that would function in degradation of nutritive materials after phagocytosis.

It is not surprising that trematodes can absorb small molecules through their tegument. Amino acids and hexoses can be absorbed, but various species differ in which molecules are absorbed by the tegument and which are absorbed by the gut.[39] *Schistosoma mansoni* takes in glucose only through its tegument.[70] Schistosomes absorb glucose both by diffusion and by a carrier-mediated system,[26] and even brief exposure to a glucose-free medium disrupts uptake of a variety of other small molecules.[27]

Megalodiscus temperatus cannot absorb glucose or galactose across its tegument,[78] and this species, as well as several other paramphistomes, has no mitochondria in its tegumental cytoplasm.[10, 34] This could be a reason the tegument of these trematodes may have little or no absorptive capacity.

Reproductive Systems

Most trematodes are hermaphroditic (important exceptions are schistosomes), and some are capable of self-fertilization. Others, however, require cross-fertilization to produce viable progeny. Some species inseminate themselves readily; others will do so if there is only one worm present, but they always seem to cross-inseminate when there are two or more in the host.[68] Some species will not inseminate themselves or even mature unless there is another adult worm present. Worms find each other by means of chemoattractants, and, except in schistosomes, the active compound appears to be cholesterol. A few instances are known in which adult trematodes can reproduce parthenogenetically.[68, 89, 95]

Male Reproductive System

Male reproductive systems (Fig. 15.18) usually include two testes, although the number varies with species from one testis to several dozen. Shape of the testes varies from round to highly branched, according to species. Each testis has a vas efferens that connects with others to form a vas deferens; this duct then courses toward the genital pore, which is usually found within a shallow genital atrium. The genital atrium is most often on the midventral surface, anterior to the acetabulum, but depending on the species it can be found nearly anywhere, including at the posterior end, beside the mouth, or even dorsal to the mouth in some species. Before reaching the genital pore, the vas deferens usually enters a muscular cirrus pouch where it may expand into an **internal seminal vesicle** (within the pouch) for sperm storage. Constricting again, the duct forms a thin ejaculatory duct, which extends the rest of the length of the cirrus pouch and forms at its distal end a muscular cirrus. The cirrus is the male copulatory organ. It can be invaginated into the cirrus pouch and evaginated for transfer of sperm to the female system. A cirrus may be naked or covered with spines of different sizes. The ejaculatory duct is usually surrounded by numerous unicellular **prostate gland cells.** A muscular dilation may form a **pars prostatica.**

Much variation in these terminal organs occurs among families, genera, and species. A cirrus pouch and prostate gland may be absent, with the vas deferens expanded into a powerful seminal vesicle that opens through the genital pore, as in *Clonorchis sinensis*. The vas deferens may expand into an **external seminal vesicle** before continuing into the cirrus pouch, and other more specialized modifications occur.

Female Reproductive System

The single ovary in the female reproductive tract (see Fig. 15.18) is usually round or oval, but it may be lobated or even branched. A short oviduct is provided with a proximal sphincter, or **ovicapt,** that controls passage of ova. The oviduct and most of the rest of the female ducts are ciliated. A seminal receptacle forms as an outpocketing of the wall of the oviduct. It may be large or small, but it is almost always present. At the base of the seminal receptacle there often arises a slender tube, **Laurer's canal,** which ends blindly in the parenchyma or opens through the tegument. Laurer's canal is probably a vestigial vagina that no longer functions as such (with a few possible exceptions), but it may serve to store sperm in some species.

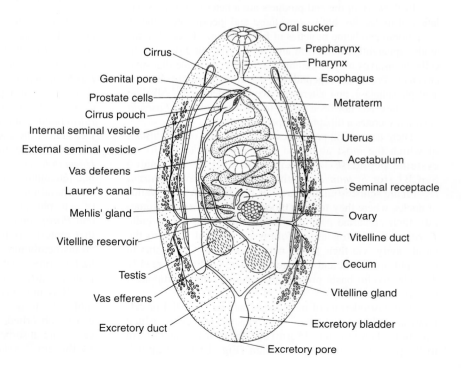

Figure 15.18 **Diagrammatic representation of a digenetic fluke.**
Note male and female reproductive systems.
Drawing by William Ober.

Unlike other animals but in common with Cercomeromorphae and some free-living flatworms, yolk is not stored in the female gamete (as in **endolecithal** eggs) but is contributed by separate cells called *vitelline cells.* Such a system is described as **ectolecithal.** Vitelline cells are produced in follicular vitelline glands, usually arranged in two lateral fields and connected by ductules to the main right and left vitelline ducts. These ducts carry vitelline cells to a single, median vitelline reservoir from which extends the common vitelline duct joining the oviduct. Anatomical distribution of vitelline glands tends to be constant within a species and so is an important taxonomic character. After the junction with the common vitelline duct, the oviduct expands slightly to form an **ootype.** Numerous unicellular **Mehlis' glands** surround the ootype and deposit their products into it by means of tiny ducts.

The structural complex just described (Fig. 15.19), including the upper uterus, is called the *egg-forming apparatus,* or **oogenotop.**[43] Beyond the ootype, a female duct expands to form a uterus, which extends to the female genital pore. The uterus may be short and fairly straight, or it may be long and coiled or folded. The distal end of the uterus is often quite muscular and is called a **metraterm.** The metraterm functions as ovijector and as a vagina. The female genital pore opens near the male pore, usually together with it in the genital atrium. In some species, such as in the Heterophyidae, the genital atrium is surrounded by a muscular sucker called a **gonotyl.**

At the time germ cells leave the ovary, they have not completed meiosis and thus are not ova at all but oocytes. Meiosis is completed after sperm penetration. The first meiotic division may reach pachytene or diplotene, at which point meiotic activity arrests, and the chromosomes return to a diffuse state. After sperm penetration the chromosomes quickly reappear as bivalents and proceed into the first meiotic metaphase. The two meiotic divisions occur, with extrusion of polar bodies, and only then do male and female pronuclei fuse.[17, 57]

As an oocyte leaves the ovary and proceeds down the oviduct, it becomes associated with several vitelline cells and a sperm emerging from the seminal receptacle. These all come together in the area of the ootype, and there are contributions from the cells of Mehlis' gland as well. It was long thought that Mehlis' gland contributed the shell material, and it was called a *shell gland* in older texts. However, we now know that the bulk of shell material is contributed by the vitelline cells, and the function of the Mehlis' gland remains obscure. In at least some species two distinct types of secretions are released—mucoid dense bodies and membranous bodies.[9] The mucoid dense bodies may serve as an adhesive mediating coalescence of vitelline globules to form the shell, or they may serve as a lubricant for the various components in the ootype. The membranous bodies aggregate to surround the oocyte, two or three vitelline cells, and some spermatozoa. Globules released from the vitelline cells coalesce against the membranous aggregate; the aggregate thus forms a kind of template for the shell material before stabilization. Additional evidence suggests that Mehlis' gland secretions serve as an eggshell template,[65] but they may have other functions as well.

"Stabilization" of structural proteins (e.g., sclerotin, keratin, and resilin) to impart qualities of physical strength

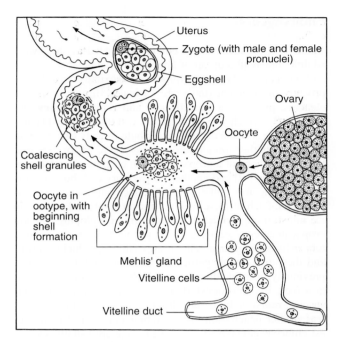

Figure 15.19   **Schematic representation of the oogenotop of a digenetic trematode.**

Drawing by William Ober and Claire Garrison.

and inertness occurs by crosslinkage to amino acid moieties in adjacent protein chains. Most trematode eggshells appear to be stabilized primarily by the quinone-tanning process of sclerotization (see Fig. 33.4).[24, 25] In some trematodes keratin or elastin may be the major structural protein.

DEVELOPMENT

At least two hosts serve in the life cycle of a typical digenetic fluke. One is a vertebrate (with a few exceptions) in which sexual reproduction occurs, and the other is usually a mollusc in which one or more generations are produced by an unusual type of asexual reproduction. A few species have asexual generations that develop in annelids.

This alternation of sexual and asexual generations in different hosts is one of the most striking biological phenomena. The variability and complexity of life cycles and ontogeny have stimulated the imaginations of zoologists for more than 100 years, creating a huge amount of literature on the subject. Even so, many mysteries remain, and research on questions of trematode life cycles remains active.

As many as six recognizably different body forms may develop during the life cycle of a single species of trematode (see p. 209 for a summary). In a given species certain stages may be repeated during ontogeny, and stages found in other species may be absent. So many variations occur that few generalizations are possible. Therefore, we will first examine each form separately and then consider a few examples.

Embryogenesis

Apart from the fact that the embryo produced by the sexual adult begins with a union of female and male gametes, early embryogenesis of progeny produced asexually and that of progeny produced sexually are basically similar. The first cleavage produces a **somatic cell** and a **propagatory cell (stem cell),** which are cytologically distinguishable. Daughter cells of a somatic cell will contribute to body tissues of an embryo, whether miracidium, sporocyst, redia, or cercaria. Further divisions of a propagatory cell each may produce another somatic cell and another propagatory cell, but, at some point, propagatory cell divisions produce only more propagatory cells. Each of these will become an additional embryo in the miracidium, sporocyst, or redia. In a developing cercaria propagatory cells become gonad primordia. Thus, propagatory cells are germinal cells in asexually reproducing forms, and they give rise to germ cells in sexual adults. As noted previously, a miracidium metamorphoses into a sporocyst; however, if a sporocyst stage is absent from a particular species, redial embryos develop in the miracidium to be released after penetration of an intermediate host. The youngest embryos developing in a given stage are usually seen in the posterior portions of its body and are often referred to as **germ balls.**

The nature of this asexual reproduction has long been controversial; different workers have argued that it represents alternatively **budding, polyembryony,** or **parthenogenesis.** The view of early zoologists—that it is an example of metagenesis (an alternation of generations in which the asexual generation reproduces by budding)—was discarded when it was realized that specific reproductive cells (propagatory cells) are kept segregated in the germinal sacs. The most widely held opinion has been that the process is one of sequential polyembryony;[28] that is, multiple embryos are produced from the same zygote with no intervening gamete production as, for example, in monozygotic twins in humans. Whitfield and Evans[95] reviewed the evidence for parthenogenesis and found it insubstantial. They felt that asexual reproduction in Digenea "most probably represents a budding process in which the development of the buds is initiated by the division of diploid totipotent (propagatory) cells."

Larval and Juvenile Development

Egg (Shelled Embryo)

The structure referred to as an *egg* of trematodes is not an ovum but a developing (or developed) embryo enclosed by its shell, or capsule. The egg capsule of most flukes has an **operculum** at one end, through which the larva eventually will escape. It is not clear how the operculum is formed, but it appears that the embryo presses pseudopodiumlike processes against the inner surface of the shell while it is being formed, thereby forming a circular groove. An operculum is absent from eggshells of blood flukes. Considerable variation exists in shape, size, thickness, and coloration of fluke capsules.

In many species an egg contains a fully developed miracidium by the time it leaves the parent; in others development has advanced to only a few cell divisions by that time. In *Heronimus mollis* miracidia hatch while still in their parent's uterus. For eggs that embryonate in the external environment, certain factors influence rate of development. Water is necessary, since eggs desiccate rapidly in dry conditions. High oxygen tension accelerates development, although eggs can remain viable for long periods under conditions of low oxygen. Eggs of *Fasciola hepatica* will not develop outside a pH range of 4.2 to 9.0.[72] Temperature is critical, as would be expected. Thus, *F. hepatica* requires 23 weeks to develop at 10°C, whereas it takes only eight days at 30°C. However, above 30°C development again slows, and it completely stops at 37°C. Eggs are killed rapidly at freezing. Light may be a factor influencing development in some species, but this has not been thoroughly investigated.

Eggs of many species hatch freely in water, whereas others hatch only when eaten by a suitable intermediate host. Light and osmotic pressure are important in stimulation of hatching for species that hatch in water, and osmotic pressure, carbon dioxide tension, and probably host enzymes initiate hatching in those that must be eaten. Time of hatching is correlated with the time the snail intermediate host is nearby. Light is also required for hatching of *Echinostoma caproni* eggs, which likewise show a circadian hatching pattern.[5]

The miracidium of *F. hepatica* within its capsule is surrounded by a thin **vitelline membrane,** which also encloses a padlike **viscous cushion** between the anterior end of the miracidium and the operculum.[96] Light stimulates hatching activity. Apparently, the miracidium releases some factor that alters the permeability of the membrane enclosing the viscous cushion. The latter structure contains a mucopolysaccharide that becomes hydrated and greatly expands the volume of the cushion. The considerable increase in pressure within the capsule causes the operculum to pop open, remaining attached at one point, and the miracidium rapidly escapes, propelled by its cilia. The nonoperculated capsules of *Schistosoma* spp. are fully embryonated when passed from the host, and they hatch spontaneously in freshwater. Miracidia release substantial quantities of leucine aminopeptidase, and this enzyme probably helps digest the capsule from the inside.[97] Unlike leucine aminopeptidases from other sources, the enzyme produced by schistosome miracidia is inhibited by NaCl, which prevents hatching while in the host's body.

Miracidium

A typical miracidium (Fig. 15.20) is a tiny, ciliated organism that could easily be mistaken for a protozoan by a casual observer. It is piriform, with a retractable **apical papilla** at the anterior end. The apical papilla has no cilia but bears five pairs of duct openings from glands and two pairs of sensory nerve endings (Fig. 15.21). The gland ducts connect with **penetration glands** inside the body. A prominent **apical gland** can be seen in the anterior third of the body. This gland probably secretes histolytic enzymes. An apical stylet is present on some species, and spines are found on others. Sensory nerve endings connect with nerve cell bodies that in turn communicate with a large ganglion. Miracidia have a variety of sensory organs and endings, including adaptations for photoreception, chemoreception, tangoreception, and statoreception.

The outer surface of a miracidium is covered by flat, ciliated epidermal cells, the number and shape of which are constant for a species. Underlying the surface are longitudinal and circular muscle fibers. Cilia are restricted to protruding **ciliated bars** in genus *Leucochloridiomorpha* (Brachylaimidae) and family Bucephalidae, and they are absent altogether from families Azygiidae and Hemiuridae. One or two pairs of protonephridia are connected to a pair of posterolateral excretory pores.

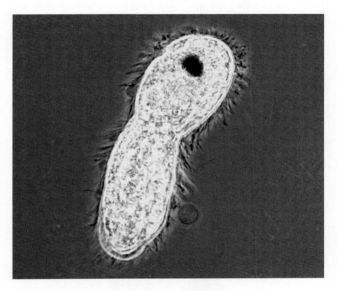

Figure 15.20 Miracidium of *Alaria* sp.

Courtesy of Jay Georgi.

In the posterior half of a miracidium are found propagatory cells, or germ balls (embryos), which will be carried into the sporocyst stage to initiate further individuals.

Free-swimming miracidia are very active, swimming at a rate of about 2 mm per second, and they must find a suitable molluscan host rapidly, since they can survive as free-living organisms for only a few hours. In many cases mucus produced by the snail is a powerful attractant for miracidia.[21]

On contacting an appropriate mollusc, the miracidium attaches to it with its apical papilla, which actively contracts and extends in an augerlike motion. Cytolysis of snail tissues can be seen as the miracidium embeds itself deeper and deeper. As penetration proceeds, the miracidium loses its ciliated epithelium, although this may be delayed until penetration is complete. A miracidium takes about 30 minutes to complete penetration. Miracidia of many species will not hatch until they are eaten by the appropriate snail, after which they penetrate the snail's gut.

Sporocyst

Often miracidia undergo metamorphosis near their site of penetration, such as foot, antenna, or gill, but they may migrate to any tissue, depending on the species, before beginning metamorphosis. Metamorphosis of a miracidium into a sporocyst involves extensive changes. In addition to loss of ciliated epithelial cells, there is formation of new tegument with its microvilli.[30] Sporocysts retain the subtegumental muscle layer and protonephridia of the miracidium, but all other miracidial structures generally disappear. A sporocyst has no mouth or digestive system; it absorbs nutrients from its host tissue, with which it is in intimate contact, and the entire structure serves only to nurture the developing embryos. The

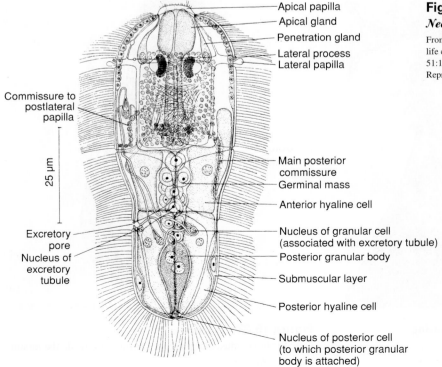

Figure 15.21 Miracidium of *Neodiplostomum intermedium*, dorsal view.

From J. C. Pearson, "Observations on the morphology and life cycle of *Neodiplostomum intermedium*," in *Parasitology* 51:133–172. Copyright © 1961 Cambridge University Press. Reprinted with the permission of Cambridge University Press.

Labels for Figure 15.21:
- Apical papilla
- Apical gland
- Penetration gland
- Lateral process
- Lateral papilla
- Commissure to postlateral papilla
- 25 μm
- Excretory pore
- Nucleus of excretory tubule
- Main posterior commissure
- Germinal mass
- Anterior hyaline cell
- Nucleus of granular cell (associated with excretory tubule)
- Posterior granular body
- Submuscular layer
- Posterior hyaline cell
- Nucleus of posterior cell (to which posterior granular body is attached)

sporocyst (or other stage with embryos developing within it—that is, the miracidium or redia) may be referred to as a **germinal sac.** Sometimes sporocysts may become very slender and extended or branched or highly ramified.

Embryos in a sporocyst may develop into another sporocyst generation (daughter sporocysts), into a different form of germinal sac (redia), or directly into cercariae (Fig. 15.22).

Leucochloridium paradoxum has a specialized sporocyst with a fascinating adaptation that evidently enhances transfer to its bird definitive host. The sporocyst is divided into three parts: a central body located in the snail's hepatopancreas, where the embryos are produced; a broodsac lying in the head-foot of the snail and entering its tentacles; and a tube connecting the broodsac to the central body. Embryos pass from the central body through the tube to the broodsac, where they mature into cercariae. The sporocyst within the snail's tentacles causes the tentacles to enlarge, become brightly colored, and pulsate rapidly. Although the effect seems analogous to a neon sign attracting the birds to "Dine here," evidence that this is actually the case has been elusive.[66]

Redia

Rediae (Fig. 15.23) burst their way out of the sporocyst or leave through a terminal birth pore and usually migrate to the hepatopancreas or gonad of their molluscan host. They are commonly elongated and blunt at the posterior end and may have one or more stumpy appendages called **procrusculi.** More active than most sporocysts, they crawl about within their host. They have a rudimentary but functional digestive system, consisting of a mouth, muscular pharynx, and short, unbranched gut. Rediae pump food into their gut by means of pharyngeal muscles, as previously described in adults. They not only feed on host tissue but also can prey on sporocysts of their own or other species.[62] The luminal surface of their gut is greatly amplified by flattened lamelloid or ribbonlike processes.[58] Gut cells are apparently capable of phagocytosis. The outer surface of their tegument also functions in absorption of food, and it is provided with microvilli or lamelloid processes.

Embryos in rediae develop into daughter rediae or into the next stage, cercariae, which emerge through a birth pore near the pharynx. The epithelial lining of the birth pore in *Cryptocotyle lingua* (and probably other species) is highly folded, making it able to withstand the extreme distortion produced by the exit of a cercaria.[53] It appears that rediae must reach a certain population density before they stop producing more rediae and begin producing cercariae: Young rediae have been transplanted from one snail to another through more than 40 generations without cercariae being

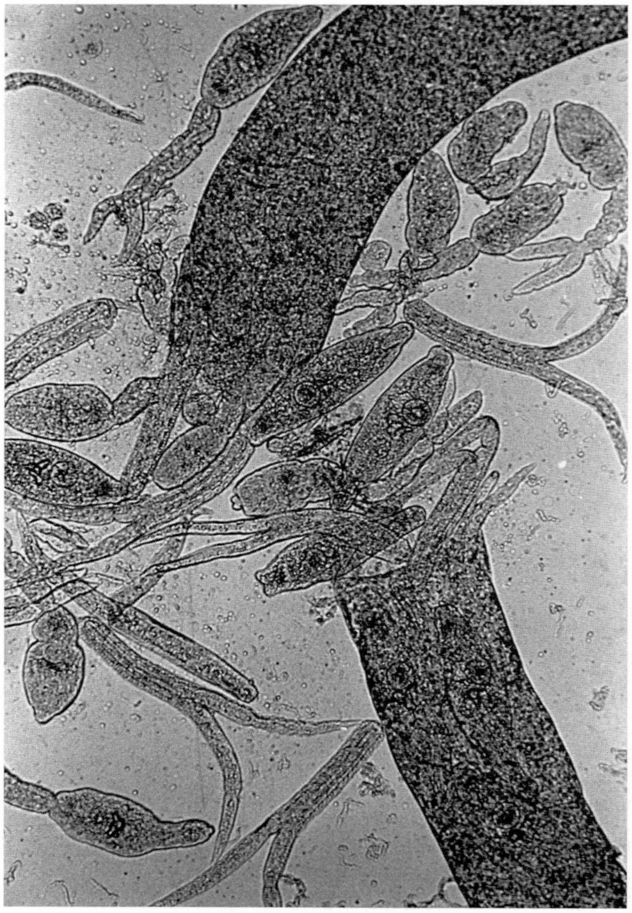

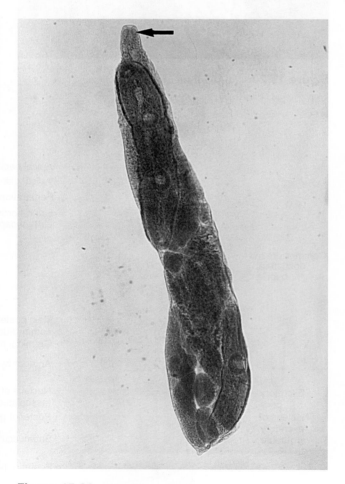

Figure 15.22 **Ruptured sporocyst releasing furcocercous cercariae.**

Courtesy of James Jenson.

Figure 15.23 **A redia.**

Note the large, muscular pharynx (*arrow*) just inside the mouth.

Courtesy of Warren Buss.

developed.[32] This type of regulation is an interesting parallel to certain free-living invertebrates that reproduce parthenogenetically only as long as certain environmental conditions are maintained.[33]

Hechinger and his colleagues described a surprising evolutionary development among digenetic trematodes: separate soldier and reproductive castes in their snail hosts. Reproductive individuals were much larger than soldiers and produced large numbers of cercariae asexually. Such caste formation is known in several arthropods and a few other invertebrates but is most unusual among Platyhelminthes. In this case the soldiers were much smaller in body size than reproductive individuals but had comparably large pharynges. They attacked and killed individuals of the same or other trematode species that invaded the same host snail.[47]

Cercaria

Cercariae represent a juvenile stage of the vertebrate-inhabiting adult. There are many varieties of cercariae, and most have specializations that enable them to survive a brief free-living existence and make themselves available to their definitive or second intermediate hosts (Fig. 15.24). Most have tails that aid them in swimming, but many have only rudimentary tails or none at all; these cercariae can only creep about, or they may remain within the sporocyst or redia that produced them until they are eaten by the next host.

Structure of a cercaria is easily studied, and cercarial morphology often has been considered a more reliable indication of phylogenetic relationships among families than has adult morphology. Cercariae are widely distributed, abundant, and easily found; hence, they have attracted much attention from zoologists. The name *Cercaria* can be used properly in a generic sense for a species in which the adult form is unknown, as is done with the term *Microfilaria* among some nematodes.

Most cercariae have a mouth near the anterior end, although it is midventral in Bucephalidae. The mouth is usually surrounded by an oral sucker, and a prepharynx, muscular pharynx, and forked intestine are normally present. Each branch of the intestine is simple, even those that are ramified in adults. Many cercariae have various glands opening near the anterior margin, often called *penetration glands* because of their assumed function. Cercariae of most trematodes probably have glands that serve several functions; schistosome cercariae have no fewer than four distinguishable types:[82]

1. **Escape glands.** They are so called because their contents are expelled during emergence of the cercaria from the snail, but their function is not known.
2. **Head gland.** The secretion is emitted into the matrix of the tegument and is thought to function in the postpenetration adjustment of the schistosomule.
3. **Postacetabular glands.** They produce mucus, help cercariae adhere to surfaces, and have other possible functions.
4. **Preacetabular glands.** The secretion contains calcium and a variety of enzymes including a protease. The function of these glands seems most important in actual penetration of host skin.

Secretory cystogenic cells are particularly prominent in cercariae that will encyst on vegetation or other objects.

Cercariae have many morphological variations that are constant within a species (or larger taxon); thus, certain descriptive terms are of value in categorizing the different varieties (see Fig. 15.24). Some of the more commonly used terms are **xiphidiocercaria** (with a stylet in the anterior margin of the oral sucker), **ophthalmocercaria** (with eyespots), **cercariaeum** (without a tail), **microcercous cercaria** (with a small, knoblike tail), and **furcocercous cercaria** (with a forked tail).

The excretory system is well developed in cercariae. In some cercariae the excretory vesicle empties through one or two pores in the tail. Because the number and arrangement of protonephridia are constant for a species, these are important taxonomic characters. Each flame cell has a tiny **capillary duct** that joins with others to form an **accessory duct.** The accessory ducts join the **anterior** or **posterior collecting ducts,** whose junction forms a **common collecting duct** on each side (Fig. 15.25). When the common collecting ducts extend to the region of the midbody and then fuse with the excretory vesicle, the cercaria is called **mesostomate.** If the tubules extend to near the anterior end and then pass posteriorly to join the vesicle, the cercaria is known as **stenostomate.** The number and arrangement of flame cells can be expressed conveniently by a **flame cell formula.** For example, $2[(3 + 3)(3 + 3)]$ means that both sides of the cercaria (2) have three flame cells on each of two accessory tubules $(3 + 3)$ on the anterior collecting tubule, plus the same arrangement on the posterior collecting tubule $(3 + 3)$. The flame cell formula for the cercaria in Figure 15.25 would be $2[(3 + 3 + 3)(3 + 3 + 3)]$.

Mature cercariae emerge from the mollusc and begin to seek their next host. Many remarkable adaptations facilitate host finding. Most cercariae are active swimmers, of course, and rely on chance to place them in contact with an appropriate organism. Some species are photopositive, dispersing themselves as they swim toward the surface of the water, but then become photonegative, returning to the bottom where the next host is. Some opisthorchiform cercariae remain quiescent on the bottom until a fish swims over them; the resulting shadow activates them to swim upward. Some plagiorchiform cercariae cease swimming when in a current; hence, when drawn over the gills of a crustacean host, they can attach and penetrate rather than swim on. Large, pigmented azygiids and bivesiculids are enticing to fish, which eat them and become infected. Some cercariae float; some unite in clusters; some creep at the bottom. Cercariae of *Schistosoma mansoni,* which directly penetrate a warm-blooded definitive host, concentrate in a thermal gradient near a heat source (34°C).[22] In certain cystophorous hemiurid cercariae, their body is withdrawn into the tail, which becomes a complex injection device.[64] The second intermediate host of the trematode is a copepod crustacean, which attempts to eat the caudal cyst bearing the cercaria. When the narrow end of the cyst is broken by the mandibles of the copepod, a delivery tube everts rapidly into the mouth of the copepod, piercing its midgut. The cercaria then slips through the delivery tube into the hemocoel of the crustacean! These and many more adaptations help the trematode reach its next host.

Mesocercaria

Species of strigeiform genus *Alaria* have a unique larval form called a mesocercaria, which is intermediate between a cercaria and metacercaria (see Fig. 16.3).

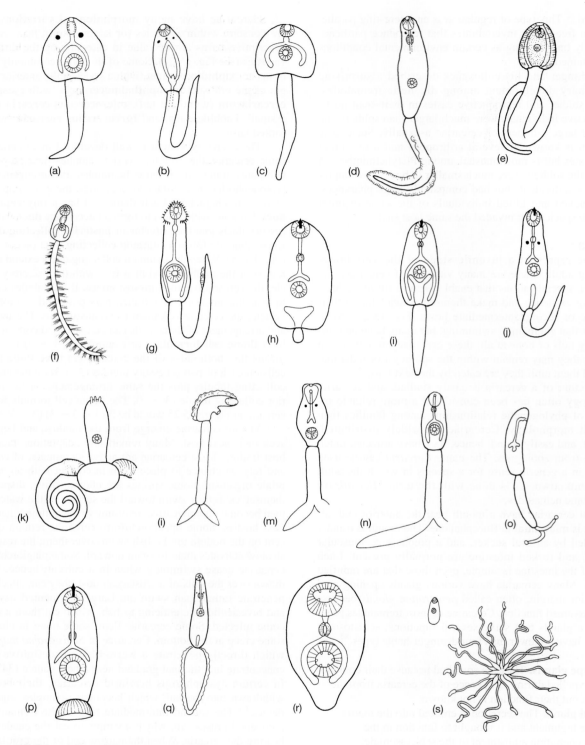

Figure 15.24 **A few of the many types of cercariae.**

(*a*) Amphistome cercaria; (*b*) monostome cercaria; (*c*) gymnocephalous cercaria; (*d*) gymnocephalous cercaria of pleurolophocercous type; (*e*) cystophorous cercaria; (*f*) trichocercous cercaria; (*g*) echinostome cercaria; (*h*) microcercous cercaria; (*i*) xiphidiocercaria; (*j*) ophthalmoxiphidiocercaria; (*k–o*) furcocercous types of cercariae: (*k*) gasterostome cercaria; (*l*) lophocercous cercaria; (*m*) apharyngeate furcocercous cercaria; (*n*) pharyngeate furcocercous cercaria; (*o*) apharyngeate monostome furcocercous cercaria without oral sucker; (*p*) cotylocercous cercaria; (*q*) rhopalocercous cercaria; (*r*) cercariae; (*s*) rattenkönig, or rat-king, cercariae.

From O. W. Olsen, *Animal parasites: Their life cycles and ecology.* Copyright © 1974 Dover Publications, Inc., New York, NY. Reprinted by permission.

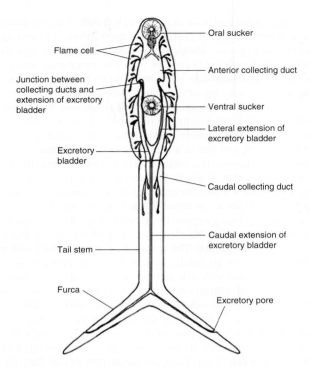

Flame cell

Oral sucker

Junction between
collecting ducts and
extension of excretory
bladder

Anterior collecting duct

Ventral sucker

Lateral extension of
excretory bladder

Excretory
bladder

Caudal collecting duct

Caudal extension of
excretory bladder

Tail stem

Furca

Excretory pore

Figure 15.25 **Diagrammatic representation of the excretory system of a fork-tailed cercaria.**
The caudal flame cells are absent from the tail in nonfurcate forms.

Drawn by John Janovy, Jr., based on D. A. Erasmus, *The biology of trematodes.*
Crane-Russak & Company, Taylor & Francis, Inc., 1972, figure 17–23. Reproduced with permission of the author.

Metacercaria

Between cercaria and adult is a quiescent stage, or metacercaria, although this stage is absent from blood flukes. Metacercariae are usually encysted, but in genera *Brachycoelium, Halipegus, Panopistus,* and others they are not. Most metacercariae are found in or on an intermediate host, but some (Fasciolidae, Notocotylidae, and Paramphistomidae) encyst on aquatic vegetation, sticks, and rocks or even freely in the water.

A cercaria's first step in encysting is to cast off its tail. Cyst formation is most elaborate in metacercariae encysting on inanimate objects or vegetation. The cystogenic cells of *Fasciola hepatica* are of four types, each with the precursors of a different cyst layer. Metacercariae encysting in intermediate hosts have thinner and simpler cyst walls, with some components contributed by the host.

The extent of development in metacercariae varies widely according to species, from those from which a metacercaria is absent (*Schistosoma* spp.) to those in which the gonads mature and viable eggs are produced (*Proterometra* spp.). Often some amount of development is necessary as a metacercaria before a trematode is infective for its definitive host. We can arrange metacercariae in three broad groups on this basis:[32]

1. Species such as *Fasciola* spp., whose metacercariae encyst in the open on vegetation and inanimate objects and that can infect a definitive host almost immediately after encystment, in some cases within only a few hours, with no growth occurring.

2. Species that do not grow in an intermediate host but that require at least several days of physiological development to infect a definitive host, such as those in family Echinostomatidae.
3. Species whose metacercariae undergo growth and metamorphosis before they enter their resting stage in a second intermediate host and that usually require a period of weeks for this development; examples are found in family Diplostomidae.

These developmental groups are correlated with longevity of the metacercariae: Those in group 1 must live on stored food and can survive the shortest time before reaching a definitive host, whereas those in groups 2 and 3 obtain some nutrients from their intermediate hosts and so can remain viable for the longest periods—in one case up to seven years. After the required development, metacercariae go into a quiescent stage and remain in readiness to excyst on reaching a definitive host. *Zoogonus lasius,* a typical example of group 2, has a high rate of metabolism for the first few days after infecting its second intermediate host, a nereid polychaete, and then drops to a low level, only to return to a high rate on excystation.[92] Metacercariae of *Bucephalus haimeanus* remain active in the liver of their fish host and increase threefold in size. They take up nutrients from degenerating liver cells, including large molecules, by pinocytosis.[49] Metacercariae of *Clinostomum marginatum* take up glucose both by facilitated diffusion and by active transport.[88]

The metacercarial stage has a high selective value for most trematodes. It can provide a means for transmission to a definitive host that does not feed on the first intermediate host or is not in the environment of the mollusc, and it can permit survival over unfavorable periods, such as a season when definitive hosts are absent.

Development in a Definitive Host

Once a cercaria or metacercaria has reached its definitive host, it matures in a variety of ways: either by penetration (if a cercaria) or by excystation (if a metacercaria) and then by migration, growth, and morphogenesis to reach gamete production. If the species does not have a metacercaria and the cercaria penetrates the definitive host directly, as in schistosomes, the most extensive growth, differentiation, and migration will be necessary. At the other extreme some species acquire adult characters as metacercariae, the gonads may be almost mature, or some eggs may even be present in the uterus; and little more than excystation is needed before the production of progeny (*Bucephalopsis, Coitocaecum, Transversotrema*). A very few species (*Proctoeces maculatus*[1] and *Proterometra dickermani*) reach sexual reproduction in the mollusc and apparently do not have a vertebrate definitive host. Others may mature in another invertebrate; for example, several species of Macroderoididae mature in leeches (see Fig. 15.3),[91] and *Allocorrigia filiformis* matures in the antennal gland of a crayfish.[84] These are probably examples of neoteny.

Normally, development in a definitive host begins with excystation of the metacercaria, and species with the heaviest, most complex cysts, such as those with cysts on vegetation (for example, *Fasciola hepatica*), seem to require the most complex stimuli for excystation. Digestive enzymes largely remove the outer cyst of *F. hepatica,* but escape from the inner cyst requires presence of a temperature of about 39°C, a low oxidation-reduction potential, carbon dioxide, and bile. Such conditions enhance excystment of a number of other species.[52] This combination of conditions is not likely to be present anywhere but in the intestine of an endothermic vertebrate; like the conditions required for the hatching or exsheathment of some nematodes, these requirements constitute an adaptation that avoids premature escape from protective coverings. This kind of adaptation is less important to metacercariae that are not subjected to the widely varying physical conditions of the external environment, such as those encysted within a second intermediate host. These have thinner cysts and excyst on treatment with digestive enzymes. A number of species require presence of a bile salt(s) or excyst more rapidly in its presence. Some metacercariae may release enzymes that assist in excystment.[52]

After excystation in the intestine, a more or less extensive migration is necessary if the final site is in some other organ. The main sites of such parasites are the liver, lungs, and circulatory system. Probably the most common way to reach the liver is by way of the bile duct (*Dicrocoelium dendriticum*), but *F. hepatica* burrows through the gut wall into the peritoneal cavity and finally, wandering through the tissues, reaches the liver. *Clonorchis sinensis* usually penetrates the gut wall and is carried to the liver by the hepatoportal system. *Paragonimus westermani* penetrates the gut wall, undergoes a developmental phase of about a week in the abdominal wall, and then reenters the abdominal cavity and makes its way through the diaphragm to the lungs.

Host hormones have significant effects on survival, growth, and maturation of schistosomes.[31] There is little evidence that such effects are mediated by control of gene expression via nuclear receptors, but a number of apparent nuclear receptors ("orphan" receptors) have been described in schistosomes for which a ligand is currently unknown.

Trematode Transitions

A remarkable physiological aspect of trematode life cycles is the sequence of totally different habitats in which the various stages must survive, with physiological adjustments that must often be made extremely rapidly. As an egg passes from a vertebrate, it must be able to withstand rigors of the external environment in freshwater or seawater, if only for a period of hours, before it can reach a haven in a mollusc. There conditions are quite different from those of both the water and the vertebrate. A trematode's physiological capacities must again be readjusted on escape from the intermediate host and again on reaching a second intermediate or definitive host. Environmental change may be somewhat less dramatic if the second intermediate host is a vertebrate, but often it is an invertebrate. Although adjustments must be extensive, the nature of these physiological adjustments made by trematodes during their life cycles has been little investigated, the most studied trematodes in this respect being *Schistosoma* spp.

Penetration of a definitive host is a hazardous phase of the life cycle of schistosomes, and it requires an enormous amount of energy. Hazards include a combination of dramatic changes in the physical environment, in physical and chemical nature of host skin through which a schistosome must penetrate, and in host defense mechanisms. Depending on the host species, losses at this barrier may be as high as 50%, and the glycogen content of newly penetrated schistosomules (**schistosomule** is the name given a young developing worm) is only 6% of that found in cercariae.

Among the most severe physical conditions the organism must survive is a sequence of changes in ambient osmotic pressure. Osmotic pressure of freshwater is considerably below that in a snail, and that in a vertebrate is twice as great as in a mollusc. Assuming that osmotic pressure of cercarial tissues approximates that in snails, the trematode must avoid taking up water after it leaves the snail and avoid a serious water loss after it penetrates a vertebrate. Aside from the possible role of the osmoregulatory organs (protonephridia), there appear to be major changes in the character and probably permeability of the cercarial surface. The cercarial surface is coated with a fibrillar layer, or glycocalyx, which is lost on penetration of a vertebrate, and with it is lost the ability to survive in freshwater; 90% of schistosomules recovered from mouse skin 30 minutes after penetration die rapidly if returned to freshwater.

Biochemical changes in the tegument occur after penetration: The schistosomule surface is much less easily dissolved by a number of chemical reagents, including 8 M urea, than is that of cercariae. Antigenic epitopes on the tegument are changed as well. When cercariae are incubated in immune serum, a thick envelope called the CHR (*cercarienhüllenreaktion*) forms around them, but schistosomules do not give this reaction.

In several cases cercarial attraction to the next host is mediated by substances different from those that stimulate actual penetration.[44] Schistosome cercariae are apparently attracted to host skin by the amino acid arginine, whereas the most important stimulus for actual penetration is the skin lipid film, specifically essential fatty acids, such as linoleic and linolenic acids, and certain nonessential fatty acids. Human skin surface lipid applied to walls of their glass container will cause cercariae to attempt to penetrate it, lose their tails, evacuate their preacetabular glands, and become intolerant to water. The presence of the penetration-stimulating substances causes loss of osmotic protection and a reduction of the CHR, even in cercariae free in the water.[45] Successful penetration and transformation have been correlated with cercarial production of eicosanoids, such as leukotrienes and prostaglandins (fatty acid derivatives with potent pharmacological activity).[73] These eicosanoids may enable schistosomules to evade host defenses by inhibiting superoxide production by neutrophils.[67]

After penetration the tegument of developing schistosomules undergoes a remarkable morphogenesis. Within 30 minutes numerous subtegumental cells connect with the distal cytoplasm to become the tegumental cytons. Abundant multilaminate vesicles pass from the cytons through the distal cytoplasm to fuse at the surface. These laminae coalesce to form two layers: an outer **membranocalyx** and an inner apical plasma membrane.[81] The old cercarial outer membrane, along with its remaining glycocalyx, is cast off. These changes are almost entirely complete within three hours after penetration. During the next two weeks the main changes in the tegument are a considerable increase in thickness and development of many invaginations and deep pits. The pits increase the surface area fourfold between 7 and 14 days after

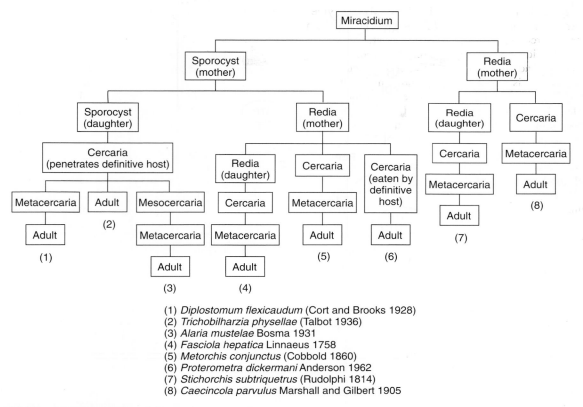

(1) *Diplostomum flexicaudum* (Cort and Brooks 1928)
(2) *Trichobilharzia physellae* (Talbot 1936)
(3) *Alaria mustelae* Bosma 1931
(4) *Fasciola hepatica* Linnaeus 1758
(5) *Metorchis conjunctus* (Cobbold 1860)
(6) *Proterometra dickermani* Anderson 1962
(7) *Stichorchis subtriquetrus* (Rudolphi 1814)
(8) *Caecincola parvulus* Marshall and Gilbert 1905

Figure 15.26   **Some life cycles of digenetic trematodes.**
A very few cases have been reported in which miracidia are produced by sporocysts.[4]

Redrawn from S. C. Schell, *How to know the trematodes.* Dubuque, IA.: Wm. C. Brown Publishers, Inc., 1970.

penetration. It is likely that this change represents an adaptation for nutrient absorption through the tegument.

Summary of Life Cycle

In summary, the basic pattern of a digenetic trematode's life cycle is egg → miracidium → sporocyst → redia → cercaria → metacercaria → adult. You should learn this pattern well because it is the theme on which to base all variations. The most common variations are (1) more than one generation of sporocysts or rediae, (2) deletion of either sporocyst or redial generations, and (3) deletion of metacercaria. Much less common are cases in which miracidia are produced by sporocysts and forms with adult morphology in the mollusc that produce cercariae (these in turn lose their tails and produce another generation of cercariae).[54] Figure 15.26 shows some possible life cycles.

METABOLISM

Energy Metabolism

Metabolism of trematodes has been accorded considerable study.[83] Compared with that of certain vertebrates, however, trematode metabolism is meagerly known, and metabolism of larval stages has received scant attention. Furthermore, because of size, availability, and/or medical importance, metabolism

of *Schistosoma* spp. and that of *Fasciola hepatica* have been much more thoroughly investigated than those of other species.

The overall scheme of nutrient catabolism is surprisingly similar in adult trematodes, cestodes, and even nematodes. Their main sources of energy are degradation of carbohydrate from glycogen and glucose. They are facultative anaerobes, and even in the presence of oxygen they excrete large amounts of short-chain acid end products (Fig. 15.27). In other words, the energy potential in the glucose molecule is far from completely harvested. Why the worms should excrete such reduced compounds, from which so much additional energy could be derived, is not at all obvious. In the past most investigators concluded that, since parasites have what is, for practical purposes, an inexhaustible food supply, they simply do not need to catabolize a glucose molecule completely. Other investigators, unsatisfied with that conclusion, have maintained that "there is a payoff somewhere."[13] Some possibilities follow.

The "usual" glycolytic pathway produces pyruvate, which in most aerobic organisms is then decarboxylated and condensed with coenzyme A to form acetyl CoA. Acetyl CoA enters the Krebs cycle, reactions that release carbon dioxide and electrons. Electrons are passed through a series of oxidation-reduction reactions (electron transport pathway) where energy is reaped in generation of ATP, and the final electron acceptor is oxygen. In the absence of oxygen, pyruvate becomes the final electron acceptor and is reduced to the end product of anaerobic glycolysis, lactate. Glycolysis yields only 1/18 the amount of energy resulting from full oxidation

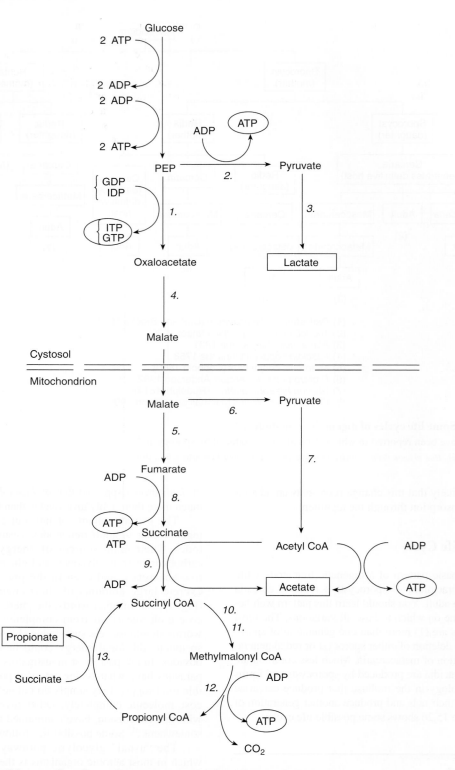

Figure 15.27 **Possible overall pathway for energy metabolism of *Fasciola hepatica*.**

Compounds in boxes represent end products; circled compounds are net energy derived in phosphate bonds. (*1*) PEP carboxykinase; (*2*) pyruvate kinase; (*3*) lactate dehydrogenase; (*4*) malate dehydrogenase; (*5*) fumarate hydratase; (*6*) malate dehydrogenase (decarboxylating); (*7*) pyruvate dehydrogenase; (*8*) fumarate reductase; (*9*) succinyl CoA synthetase; (*10*) methylmalonyl CoA mutase; (*11*) methylmalonyl CoA racemase; (*12*) propionyl CoA carboxylase; (*13*) acyl CoA transferase.

Redrawn from C. M. Lloyd, "Energy metabolism and its regulation in the adult liver fluke *Fasciola hepatica*," in *Parasitology* 93:217–248, 1986.

of the glucose molecule through the Krebs cycle and electron transport system. Although adults of many parasitic worms do excrete lactate (sometimes almost exclusively), many have a further series of reactions that derive some additional energy (see Fig. 15.27).[83] Carbon dioxide is fixed into phosphoenol-pyruvate (PEP), which is reduced to malate and then passed into the mitochondria where these reactions take place. End products are varying amounts of acetate, succinate, and propionate, which are excreted. (Additional reduced end products are excreted by some worms; see p. 371.)

Several trematode species can catabolize certain of the Krebs cycle intermediates, and they have various enzymes in that cycle. In fact, all enzymes necessary for a functional citric acid cycle are present in *F. hepatica,*[63] but the levels of aconitase and isocitrate dehydrogenase activities are quite low. There is also evidence for a functional Krebs cycle in schistosomes, albeit at a low level. Functional significance of the Krebs cycle in trematodes may lie in pathways other than energy derivation.[83]

A classical electron transport chain through cytochrome *c* oxidase is present in free-living and perhaps in larval stages of trematodes but is generally absent from adults.[83] In classical electron transport, electrons are passed from NADH to ubiquinone (coenzyme Q) and hence to cytochrome *c*. Electrons apparently also pass to ubiquinone from the oxidation of succinate catalyzed by succinate dehydrogenase. However, in anaerobic metabolism such as found in adult trematodes, succinate dehydrogenase functions in the opposite direction; that is, as a fumarate reductase (see Fig. 15.27). Electrons from oxidation of NADH are carried by rhodoquinone, which, although it resembles ubiquinone in chemical structure, has different electron transport characteristics. Fumarate reduction is coupled to phosphorylation of ADP to produce an ATP. Thus, trematodes must reorganize their metabolic machinery from a ubiquinone-mediated system in juveniles to a rhodoquinone system in adults.

Whether or not adult trematodes oxidize any glucose completely, they nevertheless excrete copious amounts of the short-chain acids. Bryant[13] speculated that, like some microorganisms, the worms might obtain an energetic advantage by pumping out acids. Microorganisms establish a proton gradient across their cell membranes and use the energy of that gradient to generate ATP. Could the mitochondria of parasitic helminths retain that ability? Alternatively, considering that "waste" molecules of the worm remain quite usable by the host, an explanation may lie in the continuing evolution of the host-parasite relationship and physiological interactions with the host.

The astonishing ability of trematodes to survive radical changes in environment requires important adjustments in their energy metabolism, such as the ubiquinone to rhodoquinone switch already mentioned. Clearly, an ability to derive every possible ATP from every glucose unit would be of great selective value to free-swimming miracidia or cercariae that do not feed. Thus, miracidia and cercariae are obligate aerobes in all species investigated so far, and they are killed by short exposures to anaerobiosis. Miracidia of *S. mansoni* may have a functional Krebs cycle even before hatching. Cercariae oxidize pyruvate rapidly and produce carbon dioxide from all three pyruvate carbons.

Sporocysts are facultative anaerobes, apparently adjusting their metabolism according to aerobic or anaerobic conditions prevailing in their snail host.[90] When oxygen is present, most glucose is degraded to CO_2, but sporocysts also produce some lactate. Under anaerobic conditions they produce lactate and succinate, the succinate resulting from fumarate reduction and electron transport by rhodoquinone.[90] Schistosome cercariae, however, have a functional Krebs cycle and a classical electron transport pathway. Immediately after penetration, schistosome energy metabolism undergoes a major adjustment. Their ability to use pyruvate drops dramatically, and schistosomules produce lactate aerobically.[23]

Developmental studies on metabolism of other trematodes would be very interesting, but few have been reported. Oxygen consumption of adult *Gynaecotyla adunca,* an intestinal parasite of fish and birds, drops sharply 24 hours after excystation and then even more after 48 and 72 hours.[92] Juvenile *F. hepatica,* living in the liver parenchyma, have a cyanide-sensitive respiration but are facultative anaerobes, so they seem to be in transition from the aerobic cercariae to the anaerobic adults.[84]

We have no evidence that lipids are used as energy sources or energy storage compounds, but requirement of lipid is probably high for certain functions, such as replacement of surface membrane. Trematodes cannot synthesize fatty acids de novo, but they can modify fatty acids obtained from the host.[83] Fatty acids are incorporated into phospholipids, triglycerides, and cholesterol esters. Sizable quantities may be excreted. *Fasciola hepatica* excretes about 2% of its net weight per day as polar and neutral lipids (including cholesterol and its esters), and excretion is mainly by way of its excretory system.[16]

Digeneans contain large amounts of stored glycogen: 9% to 30% of dry weight, according to species. Amounts in female *S. mansoni* are unusually low: only about 3.5% of dry weight. Although glycogen content of cestodes may range higher than 30%, it is still surprising that trematodes, even tissue-dwelling species, store so much because availability of their food should not be subject to vagaries of their host's feeding schedule, as it is with cestodes. In cases in which measurements have been performed, a large proportion of a trematode's glycogen is consumed under starvation conditions in vitro. In fact, maintenance of a high glycogen concentration in the worms may be of critical importance.

A pentose phosphate pathway may function in schistosomes, but critical enzymes for a glyoxylate cycle have not been found. In contrast, the pentose cycle in *Fasciola hepatica* appears to be minimal, but enzymes necessary for the glyoxylate path are all present.[63] Schistosomes evidently have all the necessary enzymes for gluconeogenesis, but no one has been able to demonstrate that the process occurs.[85]

Transamination ability appears limited, but the α-ketoglutarate–glutamate transaminase reaction is active.[93] Ammonia and urea are both important end products in degradation of nitrogenous compounds in *F. hepatica* and *Schistosoma* spp., and both worms excrete several amino acids as well. A full complement of enzymes necessary for the ornithine-urea cycle is not present in *F. hepatica;* the urea produced must be by other pathways.[55]

Effect of Drugs on Energy Metabolism

Niridazole, an antischistosomal drug, causes glycogen depletion in schistosomes, and its mode of action is very interesting. Glucose units in glycogen are mobilized for glycolysis by the action of glycogen phosphorylase, as in other systems, and extent of mobilization is controlled by how much enzyme is in the physiologically active *a* form. Niridazole inhibits conversion of phosphorylase *a* to the inactive *b* form; thus, the phosphorolysis of glycogen is uncontrolled, the worm's glycogen stores are depleted, and it is finally killed if the niridazole concentration is maintained.[15] As with any good chemotherapeutic agent, the corresponding host enzyme is not affected.

Organic trivalent antimonials, traditional antischistosomal drugs, inhibit a critical enzyme in glycolysis, phosphofructokinase (PFK). The PFK of the schistosomes is much more sensitive to the antimonials than is the corresponding host enzyme.[14] However, there is evidence that the action on PFK does not fully account for the effect of these drugs.[7] Antimonials have severe side effects on the host and have now been replaced by other compounds.

Synthetic Metabolism

Stimulated by the search for chemotherapeutic agents, researchers have studied purine and pyrimidine metabolism in schistosomes. *Schistosoma mansoni* cannot synthesize purines de novo, but they are capable of de novo pyrimidine synthesis.[37] However, the worms probably depend on salvage pathways for supplies of both types of bases. Kurelec[60] showed that *Fasciola hepatica* and *Paramphistomum cervi* could not synthesize carbamyl phosphate, and he concluded that they depended on their hosts for both pyrimidines and arginine. In light of the high arginine requirement of *S. mansoni,* it would seem probable that the situation is the same in that species.

The requirement of schistosomes for arginine is so high, in fact, that these parasites reduce the level of serum arginine to almost zero in mice with severe infections. The worms take up arginine more rapidly than they do histidine, tryptophan, or methionine. Both gut and tegument of male schistosomes absorb proline rapidly, but only a little is absorbed by the tegument of females.[76] Interestingly, proline consumed is concentrated in the ventral arms of the gynecophoral canal, the region of contact with the female. Glycogen concentrations in male and female schistosomes fluctuate in a parallel manner.[61]

Biochemistry of Trematode Tegument

Recognition that schistosomes' tegument represented their barrier of defense against the host led to much investigation of its structure and chemistry. This research has shown that the tegumental surface is active and complex. We have already mentioned the complex structure of the tegument in schistosomes (p. 226). Vesicles and granules in the distal cytoplasm appear to replace the membranocalyx continuously, and turnover is quite rapid.[81] A variety of carbohydrates, including mannose, glucose, galactose, *N*-acetyl glucosamine, *N*-acetyl galactosamine, and sialic acid, are exposed on the surface. There are receptors for both host antigens (for example, blood group antigens) and host antibodies, including IgG, IgA, and IgM. Extracts of worm tegument include all classes of host immunoglobulin (except IgD and IgE), albumin, and α-2-macroglobulin.[40] The worms synthesize and replace their surface glycans.[75]

Praziquantel and the Tegument

The aggregate of host molecules bound on their tegument plus the rapid turnover of tegumental membrane evidently shield schistosomes from their host's immune defenses. It would follow that chemicals that could disrupt integrity of the membrane could allow host immune effectors to recognize the worm. Such an action appears to be important in the effect of praziquantel, a drug that is highly effective against many flatworms.[81] In addition, praziquantel affects permeability to calcium ions, allowing a rapid influx and resulting in a muscular tetany. Some evidence suggests that both effects on Ca^{++} metabolism and on tegumental structure are necessary for the lethal effect of the drug.[7] Praziquantel is not effective against *Fasciola hepatica,* possibly because the tegument of *F. hepatica* is much thicker than that of *Schistosoma* spp. Also, tetanic contraction of *F. hepatica* in vitro requires 100 times greater concentration of praziquantel than that required for *Schistosoma* spp.

PHYLOGENY OF DIGENETIC TREMATODES

Numerous schemes have been suggested for origin of Digenea. Various authors have derived the ancestral form from Monogenoidea, Aspidobothrea, and even insects.[80] However, most authorities today believe that trematodes share a near common ancestor with some free-living flatworms, probably rhabdocoels.[36, 41] Whatever the ancestral digenean, any system of their phylogeny must rationalize the evolution of their complex life cycles in terms of natural selection, a most perplexing task.

Digeneans display much more host specificity to their molluscan hosts than to their vertebrate hosts. This suggests that they may have established themselves as parasites of molluscs first and then added a vertebrate host as a later adaptation. It is not difficult to imagine a small, rhabdocoel-like worm invading the mantle cavity of a mollusc and feeding on its tissues. In fact, the main hosts of known endocommensal rhabdocoels are molluscs and echinoderms. The stage of the protodigenean in the mollusc was probably a developmental one, with a free-living, sexually reproducing adult. There are several reasons this is a reasonable hypothesis. First, a free reproductive stage would be of selective value in dispersion and transferral to new hosts. Precisely this lifestyle is shown by *Fecampia erythrocephala,* a rhabdocoel symbiont of various marine crustaceans. Second, possession of a cercarial stage is ubiquitous among digeneans, and most of these are adapted for swimming. Those without tails show evidence that the structure has been secondarily lost.

If one grants that present adults represent ancestral adults that were free living, it is clear that additional (asexual) multiplication in molluscs would have been advantageous also, and alternation of the two reproductive generations could have been established. It is likely that such free-living adults would often be eaten by fish, and individuals in the population that could survive and maintain themselves in a fish's digestive tract for a period of time would have selective advantage in extending their reproductive life. *Fecampia erythrocephala,* for example, dies after depositing its eggs.

Further evidence that protodigeneans were originally free living as adults is demonstrated by flukes still having to leave the snail to infect their next host. With few exceptions, they are incapable of infecting a definitive host while still in their first intermediate host, even if eaten. In most cases when a life cycle requires that the fluke be eaten within a mollusc, it has left its first host and penetrated a second to become infective. Some workers, however, feel that the protodigenean adult was a parasite of molluscs, as in modern aspidobothreans.[36, 71]

It is likely that miracidia represent a larval form of the fluke's ancestor; all digeneans still have them, even though they are not now all free swimming.

With the basic two-host cycle and two reproducing generations established in protodigeneans, it is less difficult to visualize how further elaborations of the life cycle could have been selected for.

We can assume that a digenean adaptation to vertebrate hosts has occurred relatively recently. Digenetic trematodes are very common in members of all classes of vertebrates except Chondrichthyes; extremely few species of digeneans are found in sharks and rays. Urea in the tissue of most elasmobranchs, which plays such an important role in their osmoregulation, is quite toxic to the flukes on which it has been tested. It is supposed that elasmobranchs did not have digenean parasites when that particular osmoregulatory adaptation evolved and that urea has since proved a barrier to invasion of elasmobranch habitats by flukes. The situation is quite the opposite with cestodes; sharks and rays have a rich tapeworm fauna, and their cestodes either tolerate the urea or degrade it.[69]

Cladistic analysis yields the relationships among the groups of Digenea shown in Figure 15.28.

CLASSIFICATION OF SUBCLASS DIGENEA

Diagnoses are adapted from the synapomorphies and diagnoses listed by Brooks and McLennan.[12]

Subclass Digenea

With characteristics of Trematoda (p. 195) and with primitive character states; first larval stage a miracidium; miracidium with a single pair of flame cells; saclike sporocyst stage in snail host following miracidium; cercarial stage developing in snail host following mother sporocyst; cercariae with tail; cercariae with primary excretory pore at posterior end of tail; cercarial excretory ducts stenostomate; cercarial intestine bifurcate; gut development pedomorphic (gut does not appear until redial or cercarial stage).

Order Heronimiformes

With symmetrically branched sporocysts; eggs hatching in utero; ventral sucker degenerating in adults. Family Heronimidae.

Order Paramphistomiformes

Cercariae with two eyespots; redial stage with appendages; cercariae leaving snail and encysting in the open ("on" something, either animal, vegetable, or mineral); pharynx in adults at junction of esophagus and cecal bifurcation. Families: Gyliauchenidae, Paramphistomidae, Microscaphidiidae, Pronocephalidae, Notocotylidae.

Order Echinostomatiformes

Redial stage with appendages; cercariae encyst in the open; primary excretory pore in anterior half of cercarial tail; ventral sucker in cercariae midventral; secondary excretory pore terminal; acetabulum in adult midventral; adult body with spines; no eyespots in cercariae; rediae with collars; uterus extending from ovary to preacetabular. Families: Cyclocoelidae, Psilostomidae, Fasciolidae, Philophthalmidae, Echinostomidae, Rhopaliasidae.

Order Haploporiformes

Cercariae with two eyespots; cercariae encyst in the open; primary excretory pore in anterior half of cercarial tail; ventral sucker in cercariae midventral; secondary excretory pore terminal; acetabulum in adult midventral; adult body with spines; rediae without appendages; hermaphroditic duct present; uterus extending from ovary anteriorly to halfway between bifurcation and pharynx. Families: Haploporidae, Haplosplanchnidae, Megaperidae.

Order Transversotrematiformes

Cercariae with two eyespots; primary excretory pore in anterior half of cercarial tail; furcocercous cercariae; body transversely elongated; rediae with appendages. Family Transversotrematidae.

Order Hemiuriformes

Cercariae with two eyespots; primary excretory pore in anterior half of cercarial tail; ventral sucker in cercariae midventral; secondary excretory pore terminal; acetabulum in adult midventral; adult body with spines; rediae without appendages; furcocercous cercariae; cystophorous cercaria. Families: Vivesiculidae, Ptychogonimidae, Azygiidae, Hirudinellidae, Bathycotylidae, Hemiuridae, Accacoeliidae, Syncoeliidae.

Order Strigeiformes

Cercariae with two eyespots; acetabulum in adult midventral; adult body with spines; rediae without appendages; cercariae encyst in second intermediate host; mesostomate excretory system; two pairs of flame cells in miracidium; no secondary excretory pore in cercariae; brevifurcate cercariae; primary excretory pores at tips of furcae in cercarial tail; ovary between testes; genital pore midhindbody; uterus extending anteriorly from ovary to near acetabulum then posteriorly to genital pore. Families: Clinostomidae, Sanguinocolidae, Spirorchidae, Schistosomatidae, Gymnophallidae, Fellodistomidae, Brachylaemidae, Bucephalidae, Liolopidae, Cyathocotylidae, Proterodiplostomidae, Neodiplostomidae, Bolbophoridae, Diplostomidae, Strigeidae.

Order Opisthorchiformes

Cercariae with two eyespots; secondary excretory pore in cercariae terminal; acetabulum in adult midventral; adult body with spines; rediae without appendages; cercariae encyst in second intermediate host; mesostomate excretory system; cercarial tail not furcate; cercarial excretory bladder lined with epithelium; seminal receptacle present; primary finfold present on cercarial tail; eggs small, generally less than 40 μm; eggs ingested and hatch in molluscan host; no cirrus sac; no cirrus. Families: Opisthorchiidae, Cryptogonimidae, Heterophyidae.

Order Lepocreadiiformes

Cercariae with two eyespots; secondary excretory pore in cercariae terminal; acetabulum in adult midventral; adult

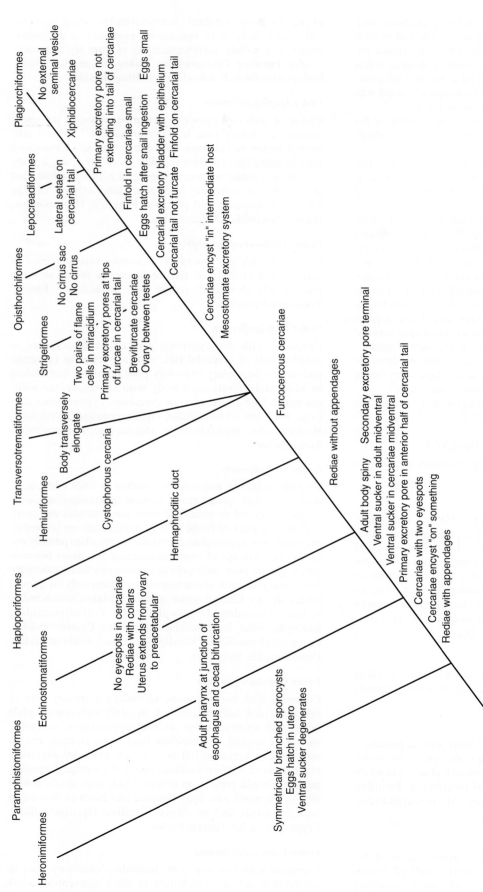

Figure 15.28   A hypothetical phylogenetic tree for the orders of Digenea.

Redrawn from Daniel R. Brooks and Deborah A. McLennan, *Parascript: Parasites and the language of evolution.* Copyright © 1993 by the Smithsonian Institution. Used by permission of the publisher.

body with spines; rediae without appendages, cercariae encyst in second intermediate host; mesostomate excretory system; cercarial tail not furcate; cercarial excretory bladder lined with epithelium; seminal receptacle present; primary excretory vesicle in cercariae extending a short distance into tail; dorsoventral finfold on cercarial tail; eggs small, generally less than 40 μm; eggs hatch in molluscan host; primary excretory pore not extending into tail of cercariae; finfold in cercariae small; lateral setae on cercarial tail. Families: Deropristidae, Homalometridae, Lepocreadiidae.

Order Plagiorchiformes

Secondary excretory pore in cercariae terminal; acetabulum in adult midventral; adult body with spines; rediae without appendages; cercariae encyst in second intermediate host; mesostomate excretory system; cercarial tail not furcate; cercarial excretory bladder lined with epithelium; seminal receptacle present; primary excretory vesicle in cercariae extending a short distance into tail; dorsoventral finfold present on cercarial tail; eggs small, generally less than 40 μm; primary excretory pore not extending into tail of cercariae; finfold in cercariae small; xiphidiocercariae; no external seminal vesicle. Families: Allocreadiidae, Acanthocolpidae, Campulidae, Troglotrematidae, Renicolidae, Macroderoididae, Opecoelidae, Zoogonidae, Lissorchiidae, Microphallidae, Lecithodendriidae, Prosthogonimidae, Plagiorchiidae, Dicrocoeliidae, Brachycoeliidae, Cephalogonimidae, Gorgoderidae, Auridistomidae, Rhytidodidae, Telorchiidae, Ochetosomatidae, Urotrematidae, Pleorchiidae, Pachypsolidae, Calycodidae, Haematoloechidae.

Learning Outcomes

By the time a student has finished studying this chapter, he or she should be able to:

1. Describe the differences between miracidia, sporocysts, rediae, cercariae, and metacercariae.
2. Describe the main differences between distome, amphistome, and monostome with respect to mouth location, and with respect to sucker location in digeneans.
3. Describe the organization of the tegument in trematodes, including cytons, distal cytoplasm, and internuncial processes.
4. Explain the difference between ectolecithal and endolecithal eggs.

5. Explain the differences between miracidium, sporocysts, redia, cercaria, and metacercaria.
6. Suggest a reason why digenean trematodes may have become parasites of vertebrates relatively recently.

References

References for superscripts in the text can be found at the following Internet site: www.mhhe.com/robertsjanovynadler9e

Additional Readings

Baer, J. G., and C. Joyeux. 1961. Classe des Trématodes (Trématoda Rudolphi). In P. P. Grassé (Ed.), *Traité de zoologie: Anatomie, systématique, biologie, vol. 4, part* I. Plathelminthes, *Mésozoaires, Acanthocéphales, Némertiens.* Paris: Masson & Cie, pp. 561–692. A well-illustrated overview of trematodes.

Barrett, J. 1981. *Biochemistry of parasitic helminths.* Baltimore, MD: University Park Press.

Cable, R. M. 1965. "Thereby hangs a tail." *J. Parasitòl.* 51:3–12. An interesting overview of the nature of the juvenile stages of trematodes.

Dawes, B. 1946. *The Trematoda, with special reference to British and other European forms.* Cambridge: Cambridge University Press. A classic reference work of value to all interested in trematodes.

Hyman, L. H. 1951. *The invertebrates, vol. 2. Platyhelminthes and Rhynchocoela. The acoelomate Bilateria.* New York: McGraw-Hill Book Co. A standard reference to all aspects of Trematoda.

Schell, S. C. 1985. Handbook of trematodes of North America North of Mexico. Moscow, ID: University Press of Idaho.

Sobhon, P., and E. S. Uptham. 1990. *Snail hosts, life-cycle, and tegumental structure of Oriental schistosomes.* Geneva, Switzerland: UNDP/World Bank/WHO.

Trager, W. 1986. *Living together. The biology of animal parasitism.* New York: Plenum Press.

Yamaguti, S. 1971. *Synopsis of digenetic trematodes of vertebrates 1.* Tokyo: Keigaku Publishing Co.

Yamaguti, S. 1975. *A synoptical review of life histories of digenetic trematodes of vertebrates : with special reference to the morphology of their larval forms.* Tokyo: Keigaku Publishing Company.

Chapter 16

Digeneans: Strigeiformes

. . . though fully appreciating professor Looss's vast erudition, we must not forget that without the complement of good judgment, it is quite easy to strain learning into absurdity.

—L. W. Sambon, on Looss's insistence that there was only one species of schistosome[11]

Of the several superfamilies in this order, only two, Strigeoidea and Schistosomatoidea, are of much economic or medical significance. The latter, however, contains some of the most important disease agents of humans.

SUPERFAMILY STRIGEOIDEA

Strigeoidea are bizarre in appearance, with their bodies divided into two portions (Fig. 16.1). The anterior portion usually is spoon or cup shaped, with accessory **pseudosuckers** on each side of the oral sucker. Behind the acetabulum is a spongy, padlike organ referred to as the **adhesive** or **tribocytic organ.** This structure secretes proteolytic enzymes that digest host mucosa, probably functioning both as an accessory holdfast and as a digestive-absorptive organ. The hindbody contains most of the reproductive organs, although vitelline follicles often extend into the forebody. The genital pore is located at the posterior end.

Most strigeoids are quite small and are found commonly in digestive tracts of fish-eating vertebrates. Their cercariae are easily recognized because they have both a pharynx and a forked tail. No adult strigeoids are known to parasitize humans, but they are so ubiquitous and their biology so interesting that we will briefly consider a few species.

Family Diplostomidae

Alaria americana

Genus *Alaria* contains several very similar species, all of which mature in the small intestines of carnivorous mammals. *Alaria americana* is found in various species of Canidae in northern North America. They are about 2.5 mm to 4.0 mm long, with the forebody longer than the hindbody. The forebody has a pair of ventral flaps that are narrowest at the anterior end (Fig. 16.2). A pointed process flanks each side of the oral sucker. The tribocytic organ is relatively large and elongated and has a ventral depression in its center.

Life cycles of *Alaria* spp. are remarkable in that the worms may require four hosts before they can develop to maturity (see Fig. 16.2). Eggs are unembryonated when laid, and they hatch in about two weeks. Miracidia swim actively and will attack and penetrate any of several species of planorbid snails.[47] Mother sporocysts develop in the renal veins and produce daughter sporocysts in about two weeks. Daughter sporocysts migrate to the digestive gland and need about a year to mature and begin producing cercariae. The furcocercous cercaria leaves the snail during daylight hours and swims to the surface, where it hangs upside down. Occasionally, it sinks a short distance and then returns to the surface. If a tadpole swims by, the resulting water currents stimulate the cercaria to swim after it. If it contacts a tadpole, the cercaria will quickly attack, drop its tail, penetrate the skin, and begin wandering within the amphibian.

The trematode remains viable if the tadpole undergoes metamorphosis. In about two weeks the cercaria has transformed into a **mesocercaria,** an unencysted form between a cercaria and a metacercaria. It is then infective to its next host, which may be the definitive host or a paratenic host. If a canid eats an infected tadpole or adult frog, the mesocercariae are freed by digestion, penetrate the coelom, and then move to the diaphragm and lungs. After about five weeks in the lungs, mesocercariae have transformed into **diplostomulum metacercariae** (see Fig. 16.2). Diplostomula migrate up the trachea and then to the intestine, where they mature in about a month.

Tadpoles, however, are not always available to terrestrial canids and, furthermore, are distasteful to all but the hungriest carnivores. This ecological barrier is overcome when a water snake eats the infected tadpole or frog and thereby becomes a paratenic host. A snake (or other animal) can accumulate large numbers of mesocercariae in its tissues, producing a heavy infection in a definitive host when eaten. Mesocercariae migrate, develop into diplostomula, and mature in the intestine, as do those from tadpoles. Life cycles of other species of *Alaria* are similar.

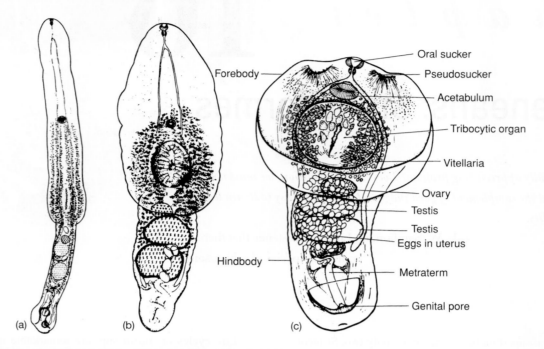

Figure 16.1 **Typical strigeoid trematodes, illustrating body forms.**

(*a*) *Mesodiplostomum gladiolum* Dubois 1936. (*b*) *Pseudoneodiplostomum thomasi* (Dollfus 1935). (*c*) *Proalarioides serpentis* (Yamaguti 1933).

From S. Yamaguti, *Synopsis of Digenetic Trematodes of Vertebrates.* Tokyo: Keigaku Publishing Company, 1971.

Mature *Alaria* spp. are quite pathogenic, causing severe enteritis that often kills the definitive hosts in severe infections. Also, the mesocercariae are pathogenic, especially when accumulated in large numbers. Figure 16.3 is from a fatal infection of mesocercariae in a human.

Shoop and Corkum[61] demonstrated that *A. marcianae* can be transmitted to a juvenile definitive host through the milk of its mother. In this species the parasite matures normally in the intestine of adult males and nonlactating females but remains as diplostomula disseminated throughout the tissues of lactating females. In an experimental infection a single cat infected 21 of her offspring via milk over the course of five litters and still harbored infective juveniles after three years. Primates also can transmit this worm by transmammary means.[62]

Uvulifer ambloplitis

Several species of strigeoid trematodes cause black spots in the skin of fish; one such species is *Uvulifer ambloplitis,* a parasite of kingfishers, fish-eating birds that are widely distributed across the United States. The spoon-shaped forebody of the parasite is separated from the longer hindbody by a slender constriction. Adults are 1.8 mm to 2.3 mm long.

Eggs, which are unembryonated when laid, hatch in about three weeks. Miracidia penetrate snails of genus *Helisoma* and transform into mother sporocysts that retain the eyespots of the first larva. Daughter sporocysts invade the digestive gland and produce cercariae in about six weeks. Cercariae escape from the tissues of the snail and rise to the surface of the water, where they are sensitive to the passing of fish. If they contact a centrarchid fish, they drop their tails and penetrate the skin. Once inside the dermis, the flukes metamorphose into **neascus meta-cercariae** and secrete a delicate, hyaline cyst wall around themselves. A neascus is similar to a diplostomulum except that the

forebody is spoon shaped, without anterolateral "points." The fish host responds to the neascus by deposition of melanin. The result is a conspicuous black spot indicating the presence of a metacercaria (Fig. 16.4). When fish are heavily infected, they are often discarded as diseased by people who catch them. Kingfishers become infected when they eat fish with metacercariae. The flukes mature in 27 to 30 days.

Other related flukes also cause black spots in a wide variety of fish and have similar life cycles (see Fig. 16.4). When a neascus larva is encountered for which the adult genus is unknown, it is proper to refer it to genus *Neascus.* This is also true for *Diplostomulum, Tetracotyle,* and *Cercaria.* In fact, new species can be named in these genera; of course, when its adult form becomes known, the species reverts to the proper genus.

Family Strigeidae

Cotylurus flabelliformis

This is a common parasite of wild and domestic ducks in North America. Adult flukes are 0.5 mm to 1.0 mm long. The forebody is cup shaped, with the acetabulum and tribocytic organ located at its depths. The hindbody is short and stout and is curved dorsally.

Adult worms live in the small intestine of ducks. Eggs passed in feces hatch in about three weeks, and miracidia attack snails of family Lymnaeidae *(Lymnaea, Stagnicola).* There are two sporocyst generations. After about six weeks sporocysts begin to release furcocercous cercariae. If cercariae contact a snail of family Lymnaeidae, they penetrate, migrate to the ovotestis, and transform into **tetracotyle meta-cercariae.** Tetracotyles are similar to diplostomula except that they have an extensive system of excretory canals that

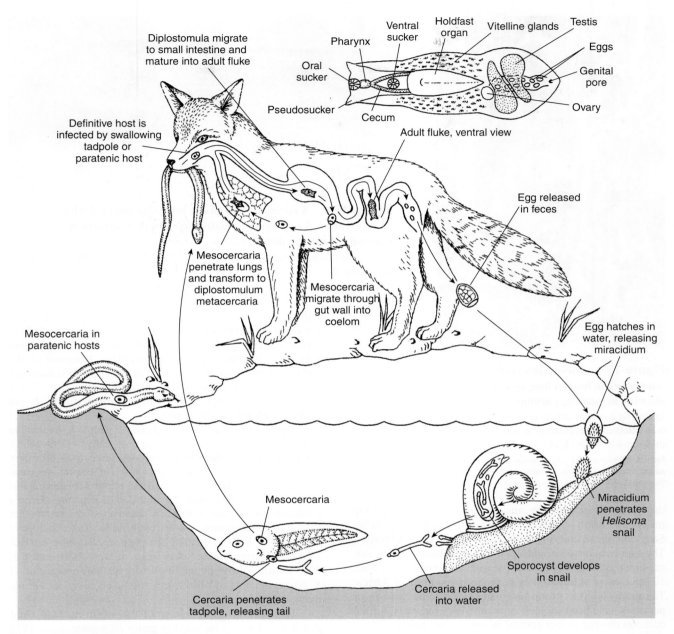

Figure 16.2   **Life cycle of *Alaria americana*.**

Drawing by William Ober and Claire Garrison.

often are filled with excretory products. The canals are called the *reserve bladder system*. When a duck eats the snail, the flukes excyst and mature in about one week.

If a cercaria enters a snail of families Planorbidae or Physidae, however, it will attack sporocysts or rediae of other species of flukes already present and will develop into a tetracotyle metacercaria within them. These too will mature in about a week if a duck eats them.

Strigeoid trematodes, then, exhibit complex life cycles that involve several unrelated hosts. Their adaptability seems amazing when one considers the differences in environments provided by snails, pond water, fish, amphibians, reptiles, and birds or mammals. It is also remarkable that a parasite that may require more than a year to complete its juvenile development can become sexually mature in its definitive

host in a week or less and die a few days later. Such is the pattern in life cycles of strigeoids.[47]

SUPERFAMILY SCHISTOSOMATOIDEA

Flukes of superfamily Schistosomatoidea are peculiar in that they have no second intermediate host in their life cycles and also in that they mature in the blood vascular system of their definitive hosts. Most species are dioecious.

Popiel[55] discusses the tantalizing puzzle of the possible selective advantage of being dioecious in a phylum of worms that is overwhelmingly monoecious. These flukes are parasites of fishes, turtles, birds, and mammals throughout

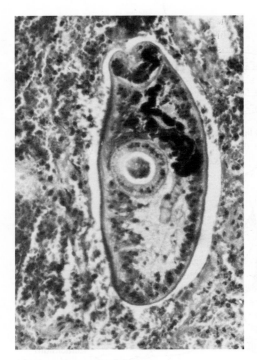

Figure 16.3 Mesocercaria of *Alaria americana* in human lung at autopsy showing hemorrhage around the worm.

In this fatal case nearly every organ of the body was infected, presumably as a result of eating undercooked frogs' legs.

From R. S. Freeman et al., "Fatal human infection with mesocercariae of the trematode *Alaria americana*," in *Am. J. Trop. Med. Hyg.* 25:803–807. Copyright © 1976.

Figure 16.4 A minnow, *Pimephales promelas*, infected with metacercariae of a strigeoid trematode species that matures in fish-eating birds, probably herons.

Arrows indicate typical "blackspot" resulting from cercarial penetration and encycstment.

Courtesy of Alaine Knipes.

the world. Several species are parasites of humans, causing misery and death wherever they are distributed. The families Sanguinicolidae and Spirorchidae parasitize fish and turtles. The family Schistosomatidae, however, includes species that are among the most dreaded parasites of humans. To date, 12 species of schistosomes have been reported in Africa. Taxonomy of this group and validity of some species of parasites have been subjects of controversy for years; a final decision on species recognition awaits further studies.

Family Schistosomatidae: *Schistosoma* Species and Schistosomiasis

Schistosomiasis is a major human disease with about 800 million people in 74 countries at risk.[13, 21] Some 200 million people are infected with schistosomes, of whom 120 million are symptomatic and 20 million suffer from severe disease.

A Fascinating History

Three species of schistosomes are of vast medical significance: *Schistosoma haematobium*, *S. mansoni*, and *S. japonicum*— all parasites of humans since antiquity.[35] Bloody urine was a wellrecognized disease symptom in northern Africa in ancient times. At least 50 references to this condition have been found in surviving Egyptian papyri, and calcified eggs of *S. haematobium* have been found in Egyptian mummies

dating from about 1200 B.C. Hulse[30] presented a well-reasoned hypothesis that the curse that Joshua placed on Jericho can be explained by an introduction of *S. haematobium* into the communal well by the invaders. Removal of the curse occurred after abandonment of Jericho and subsequent droughts eliminated the snail host, *Bulinus truncatus*. Today Jericho (Ariha, Jordan) is well-known for its fertile lands and healthy, well-nourished people.

The first Europeans to record contact with *S. haematobium* were surgeons with Napoleon's army in Egypt (1799–1801). They reported that **hematuria** (bloody urine) was prevalent among the troops, although the cause, of course, was unknown. Nothing further was learned about schistosomiasis haematobia for more than 50 years, until a young German parasitologist, Theodor Bilharz, discovered the worm that caused it. He announced his discovery in letters to his former teacher, Von Siebold, naming the parasite *Distomum haematobium*.[20] During the next few years, it was discovered that 30% to 40% of the population in Egypt bore infections of *S. haematobium*, and the worm was even found in an ape dying in London.

The peculiar morphology of the worm made it clear that it could not be included in genus *Distomum*, so in 1858 Weinland proposed the name *Schistosoma*. Three months later Cobbold named it *Bilharzia*, after its discoverer. This latter name became widely accepted throughout the world, and the parasite was even given the nickname "Bill Harris" by British soldiers serving in Europe during World War I. Today, however, the strict rules of zoological nomenclature decree that *Schistosoma* has priority and is thus the current name for the parasite. Even so, health officers in many parts of the world erect signs next to ponds and streams that warn prospective bathers of the dangers of "bilharzia." Nonetheless, *Schistosoma* is an apt name, referring to the "split body" (gynecophoral canal) of the male.

While information was accumulating on biology of *S. haematobium*, some investigators began to doubt whether it was a single species or whether two or more species were being confused. The problem was confounded by the observation in some patients of eggs with terminal spines in both urine and feces. Whenever eggs with lateral spines were noticed, they were ignored as "abnormal." In 1905 Sir Patrick Manson decided that intestinal and vesicular (urinary bladder) schistosomiasis usually were distinct diseases, caused by distinct species of worms. He reached this conclusion when he examined a man from the West Indies who had never been to Africa and who passed laterally spined eggs in his feces but none at all in his urine.[41]

Sambon[58] argued in favor of the two-species concept in 1907, and he named the parasites producing laterally spined eggs *Schistosoma mansoni.* (Japanese zoologists had already detected still another species by this time, but their reports were generally unknown to Europeans.) However, the eminent German parasitologist Looss disagreed and brought the full sway of his reputation and dialectic against the notion and even stated that he had seen a female worm with both kinds of eggs in its uterus. Sambon, undaunted, replied that, until Professor Looss could "show me an actual specimen, I am bound to place the worm capable of producing the two kinds of eggs with the phoenix, the chimaera, and other mythical monsters."[58]

The question was finally resolved by Leiper[37] in 1915. Working in Egypt, he discovered that cercariae emerging from the snail *Bulinus* spp. could infect the vesicular veins of various mammals, and they always produced eggs with terminal spines. Those emerging from a different snail, *Biomphalaria* spp., infected intestinal veins and produced laterally spined eggs. It was soon determined that *S. mansoni* had a broad distribution in the world, having been widely scattered by the slave trade. It is now widespread in Africa and the Middle East and is the only blood fluke of humans in the New World, with the possible exception of a small focus of *S. haematobium* in Suriname.[40] The original endemic area of *S. mansoni* was probably the Great Lakes region of central Africa.

While Cobbold, Weinland, Bancroft, Sambon, and others were wrestling with the problem of *S. haematobium* and *S. mansoni,* Japanese researchers were investigating a similar disease in their country. For years physicians in the provinces of Hiroshima, Saga, and Yamanachi had recognized an endemic disease characterized by an enlarged liver and spleen, ascites, and diarrhea. At autopsy they noted eggs of an unknown helminth in various organs, especially in the liver. In 1904 Professor Katsurada of Okayama recognized that larvae in these eggs resembled those of *S. haematobium.* Because he was unable to make a postmortem examination of an infected person, he began examining local dogs and cats, in hopes that they were reservoirs for the parasite. He soon found adult worms containing eggs identical to those from humans and named them *Schistosoma japonicum.* The experimental elucidation of the life cycle by various Japanese researchers was a milestone in the history of parasitology and formed the basis for Leiper's work on blood flukes in Egypt. The distribution of *S. japonicum* is limited to Japan, China, Taiwan, the Philippines, and Southeast Asia.

In more recent years other species of *Schistosoma* parasitic in humans have been distinguished (p. 249), and *Schistosoma* spp. seem to be evolving actively.[32]

Morphology

Although *Schistosoma* spp. are generally similar structurally, several differences in detail are listed in Table 16.1. Considerable sexual dimorphism exists in the genus, males being shorter and stouter than females (Fig. 16.5). The males have a ventral, longitudinal groove, the **gynecophoral canal,** where the female normally resides. The mouth is surrounded by a strong oral sucker, and the acetabulum is near the anterior end (Fig. 16.6). There is no pharynx. The paired intestinal ceca converge and fuse at about the midpoint of the worm and then continue as a single gut to the posterior end. Males possess five to nine testes, according to species, each of which has a delicate vas efferens, and these combine to form a vas deferens. The latter dilates to become a seminal

Table 16.1   Comparative Morphology of the Three Primary Species of Human Schistosomes

Characteristic	*S. haematobium*	*S. mansoni*	*S. japonicum*
Tegumental papillae	Small tubercles	Large papillae with spines	Smooth
Size			
Male			
Length	10–15 mm	10–15 mm	12–20 mm
Width	0.8–1.0 mm	0.8–1.0 mm	0.50–0.55 mm
Female			
Length	ca. 20 mm	ca. 20 mm	ca. 26 mm
Width	ca. 0.25 mm	ca. 0.25 mm	ca. 0.3 mm
Number of testes	4–5	6–9	7
Position of ovary	Near midbody	In anterior half	Posterior to midbody
Uterus	With 20–100 eggs at one time; average 50	Short; few eggs at one time	Long; may contain up to 300 eggs; average 50
Vitellaria	Few follicles, posterior to ovary	Few follicles, posterior to ovary	In lateral fields, posterior quarter of body
Egg	Elliptical, with sharp terminal spine; 112–170 μm × 40–70 μm	Elliptical, with sharp lateral spine; 114–175 μm × 45–70 μm	Oval to almost spherical; rudimentary lateral spine; 70–100 μm × 50–70 μm

vesicle, which opens ventrally through the genital pore immediately behind the ventral sucker. Cirrus pouch, cirrus, and prostate cells are absent.

The suckers of females are smaller and not so muscular as those of males, and tegumental tubercles (see Fig. 16.5), if any, are confined to the ends of females. The ovary is anterior or posterior to or at the middle of the body, and the uterus is correspondingly short or long, depending on species.

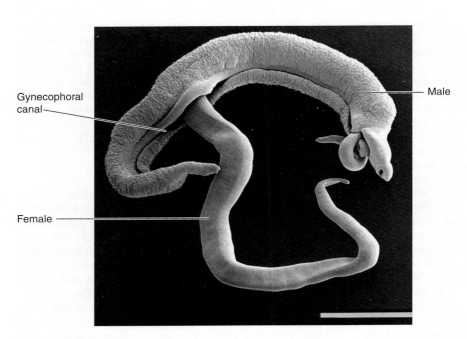

Figure 16.5 **Scanning electron micrograph of male and female *Schistosoma mansoni.***

The female is lying in the gynecophoral groove in the ventral surface of the male. (Bar = 2 mm)

Courtesy of D. W. Halton.

Biology

Adult worms live in veins that drain certain organs of their host's abdomen (Fig. 16.7), and the three main species have distinct preferences: *S. haematobium* lives principally in veins of the urinary bladder plexus; *S. mansoni* prefers the portal veins draining the large intestine; and *S. japonicum* is more concentrated in veins of the small intestine. Female worms are usually in the gynecophoral canal of male worms, where copulation takes place, and there are other physiological reasons for this location.[3] The more robust muscles of the males allow the paired worms to work their way "upstream" into smaller veins, where the female deposits eggs. The eggs (Fig. 16.8) must then traverse the wall of the venule, some intervening tissue, and gut or bladder mucosa before they are in a position to be expelled from the host. The mechanism by which this "escape" is achieved is not self-evident and has been the subject of much speculation. Spines on the eggs were traditionally credited with contributing to the expulsion, but this feat is also accomplished by *S. japonicum* and other Asian schistosomes that have only the most rudimentary spines. It appears that the worms enlist the aid of their host.

According to File,[19] endothelial cells lining the venule actively move over the schistosome eggs to exclude them from the lumen (Fig. 16.9). Damian[14] and Doenhoff et al.[17] postulated that the worm then exploits the host immune response to transport its egg to the lumen of the gut or the

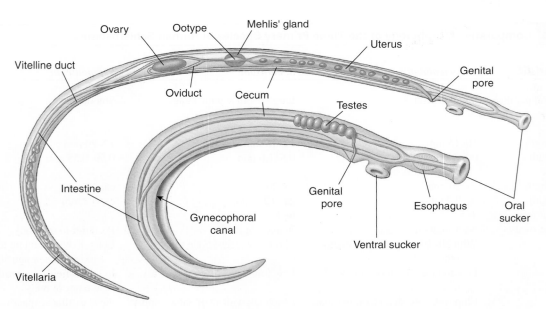

Figure 16.6 **Diagram of schistosome anatomy.**

Drawing by William Ober and Claire Garrison.

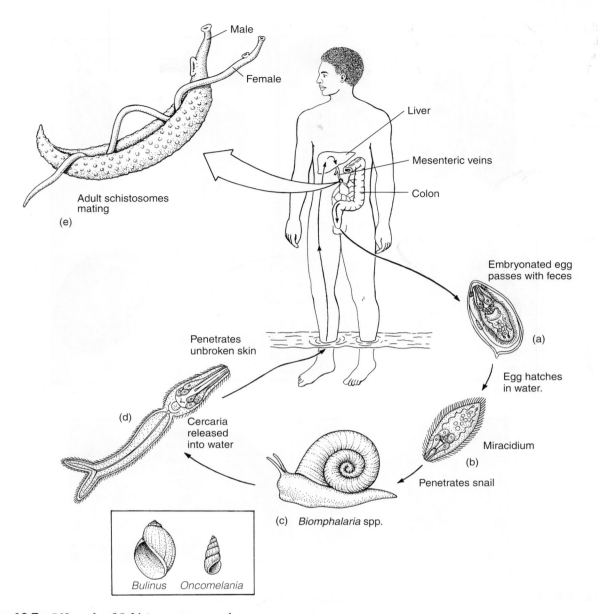

Male

Female

Adult schistosomes
mating

(e)

Liver

Mesenteric veins

Colon

Embryonated egg
passes with feces

(a)

Egg hatches
in water.

Penetrates
unbroken skin

(d)

Cercaria
released
into water

Miracidium

(b)

Penetrates snail

(c) *Biomphalaria* spp.

Bulinus   *Oncomelania*

Figure 16.7   **Life cycle of *Schistosoma mansoni.***
The main intermediate hosts of *S. mansoni* are species of *Biomphalaria.* Those of *S. haematobium* are *Bulinus* spp. (left, in box), and those of *S. japonicum* are *Oncomelania* spp. (right, in box). (*a*) Embryonated egg is passed in feces. (*b*) Miracidium hatches spontaneously and penetrates *Biomphalaria.* (*c*) Two sporocyst generations develop in snail. (*d*) Cercaria leaves snail and penetrates skin of definitive host. (*e*) Adult schistosomes mate in portal venules of intestine.

Drawing by William Ober and Claire Garrison.

bladder. The extravasated egg stimulates a granuloma to form around it (see Fig. 16.15). The granuloma, consisting of motile cells (such as eosinophils, plasma cells, and macrophages), then moves to the intestinal or bladder lumen, carrying the egg with it. Once in the lumen, cells of the granuloma disperse, and the egg is excreted with feces or urine.

In any case, about two-thirds of the eggs do not make it, and large numbers build up in the gut or bladder wall, particularly in chronic cases in which the wall is toughened by an extensive buildup of fibrous tissue. Of course, many eggs are never expelled from the venules but are swept away by

blood, eventually to lodge in liver or capillary beds of other organs. By the time eggs reach the outside by way of urine or feces, they are completely embryonated and hatch when exposed to the lower osmolarity of fresh water.

The mechanism of hatching is poorly understood. The first indication of hatching is activation of cilia on the miracidium. Ciliary activity increases until the miracidium is a veritable spinning ball. Then, suddenly, an osmotically induced vent opens on the side of the egg, and the miracidium emerges (Fig. 16.10). The miracidium usually contracts a few times, completely clearing the shell, after which it rapidly

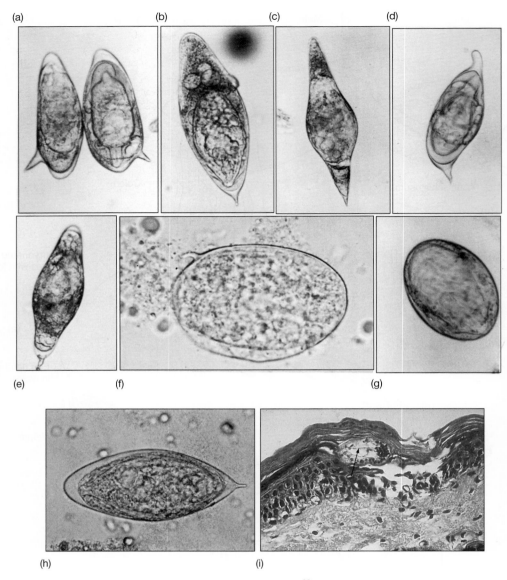

(a) (b) (c) (d)

(e) (f) (g)

(h) (i)

Figure 16.8 **Eggs of schistosome flukes (*a* to *h*; egg sizes from Loker[38] in μm).**

(*a*) *S. mansoni* (142 × 60); (*b*) *S. intercalatum* (175 × 62); (*c*) *S. bovis* (202 × 58); (*d*) *S. rodhaini* (149 × 55); (*e*) *S. mattheei* (173 × 53); (*f*) *S. japonicum* (81 × 63); (*g*) *Schistosomatium douthitti* (91 × 68); (*h*) *Schistosoma haematobium* (144 × 58). Skin section showing schistosomule (*arrow*) moments after a cercaria of *S. mansoni* penetrated skin of a parasitologist (*i*).

(*a* through *h*) Courtesy of Robert E. Kuntz and Jerry A. Moore. (*i*) Courtesy of William C. Campbell.

Figure 16.9 **Scanning electron micrograph of endothelial cells and eggs of *Schistosoma japonicum* in vitro.**

The eggs have just been expelled by a female worm, and the endothelial cells are moving over them.

From S. File, "Interaction of schistosome eggs with vascular endothelium," in *J. Parasitol.* 81:234–238. Copyright © 1995.

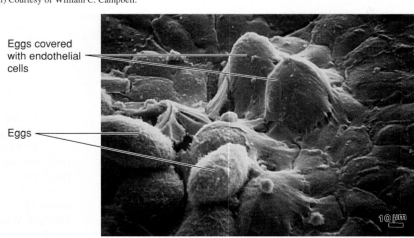

Eggs covered with endothelial cells

Eggs

10 μm

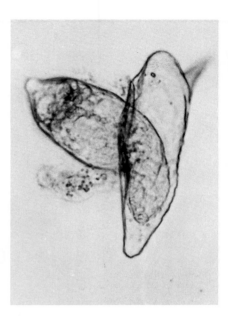

Figure 16.10   Miracidium of *Schistosoma mansoni* escaping from its eggshell.

Courtesy of Robert E. Kuntz.

swims away. However, some eggs do not hatch, no matter how active the miracidium becomes, and others hatch before the larva becomes activated.

Miracidia swim ceaselessly during their short free-living life. If hatching from an old egg, a miracidium will live only one to two hours; in optimal conditions it will survive for five to six hours. Although miracidia of schistosomes do not have eyespots, they evidently have photoreceptors, and they are positively phototropic.[8] When miracidia enter the vicinity of a snail host, they are stimulated to swim more rapidly and change direction much more frequently, thus increasing their chances of encountering the host. Following are the important snail vectors for the three most common *Schistosoma* species (see Fig. 16.7):

1. for *S. haematobium,* several species of *Bulinus* and *Physopsis,* possibly also *Planorbarius;*
2. for *S. mansoni, Biomphalaria alexandrina* in northern Africa, Saudi Arabia, and Yemen; *B. sudanica, B. rupellii, B. pfeifferi,* and others in the genus in other parts of Africa; *B. glabrata* in the Western Hemisphere; and *Tropicorbis centrimetralis* in Brazil; and
3. for *S. japonicum,* several species of *Oncomelania.*

After penetration of a snail a miracidium sheds its epithelium and begins development into a mother sporocyst, usually near its point of entrance. After about two weeks the mother sporocyst, which has four protonephridia, gives birth to daughter sporocysts, which usually migrate to other organs of the snail, if there is room. The mother sporocyst continues producing daughter sporocysts for up to six to seven weeks.[48] There is no redial stage.

The furcocercous cercariae (see Fig. 16.7) start to emerge from daughter sporocysts and the snail host about four weeks after initial penetration by the miracidium. Cercariae have a body 175 μm to 240 μm long by 55 μm to 100 μm wide and

a tail 175 μm to 250 μm long by 35 μm to 50 μm wide, bearing a pair of furci 60 μm to 100 μm long. An oral sucker is absent, being replaced by a head organ composed of penetration glands, and the ventral sucker is small and covered with minute spines. Four types of glands open through bundles of ducts at the anterior margin of the head (p. 223).

There is no second intermediate host in the life cycle. Cercariae alternately swim to the water surface and slowly sink toward the bottom, continuing to live this way for one to three days. If they come into contact with the skin of a prospective host, such as a human, they attach and creep about for a time as if seeking a suitable place to penetrate. They are attracted to secretions of the skin, showing a strongly positive response to the amino acid arginine. Upon stimulation by arginine cercariae begin to produce arginine themselves from postacetabular glands, thus attracting other cercariae in the neighborhood.[23]

Cercariae require only half an hour or less to completely penetrate the epidermis, and they can disappear through the surface in 10 to 30 seconds (see Fig. 16.8i). Penetration is accompanied by a vigorous wiggling, together with secretion of products from the head organ. The tail drops off in the process. Worms are somewhat smaller once the penetration glands have emptied their contents. Within 24 hours the schistosomules (little schistosomes) enter the peripheral circulation and are swept off to the heart. Some of the schistosomules may migrate through lymphatics to the thoracic duct and from there to the subclavian veins and heart.

Leaving the right side of the heart, the small worms wriggle their way through pulmonary capillaries to gain access to the left heart and systemic circulation. Migration through the lung capillaries is evidently a major obstacle; up to 70% may be eliminated there.[69] It appears that only schistosomules that enter the mesenteric arteries, traverse the intestinal capillary bed, and reach the liver by the hepatoportal system can continue to grow. After undergoing a period of about three weeks of development in the liver sinusoids, young worms pair up and then migrate to the gut or bladder wall (according to species) and begin producing eggs.[47] The entire prepatent period is about five to eight weeks. Adult schistosomes may live 20 to 30 years.[34]

Unpaired female worms do not become sexually mature and have the appearance of starving. Their esophageal musculature is weak and thin, they produce little of at least some digestive enzymes, and they ingest about one-fourth as many erythrocytes as paired females. A growth-stimulating function may result from the muscular action of the clasping male, which helps the immature female pump blood into her intestine.[3]

Surprisingly, normal development of schistosomes requires cues from their host's immune system.[15] TNF stimulates egg production, and growth is impaired in IL7-deficient mice. Parasite growth is stimulated by host IL-7 and thyroxin.[55] CD4+ lymphocytes are an important part of immune signals that are recognized by the parasites.

Epidemiology and Transmission Ecology

Human waste in water containing intermediate hosts is the single most important epidemiological factor in schistosomiasis, and availability of suitable snail species will determine endemicity of a particular species of *Schistosoma.* The latter is well illustrated by the fact that, although both *S. mansoni* and *S. haematobium* are widespread in Africa (Fig. 16.11), only *S. mansoni* became established in the New World by the

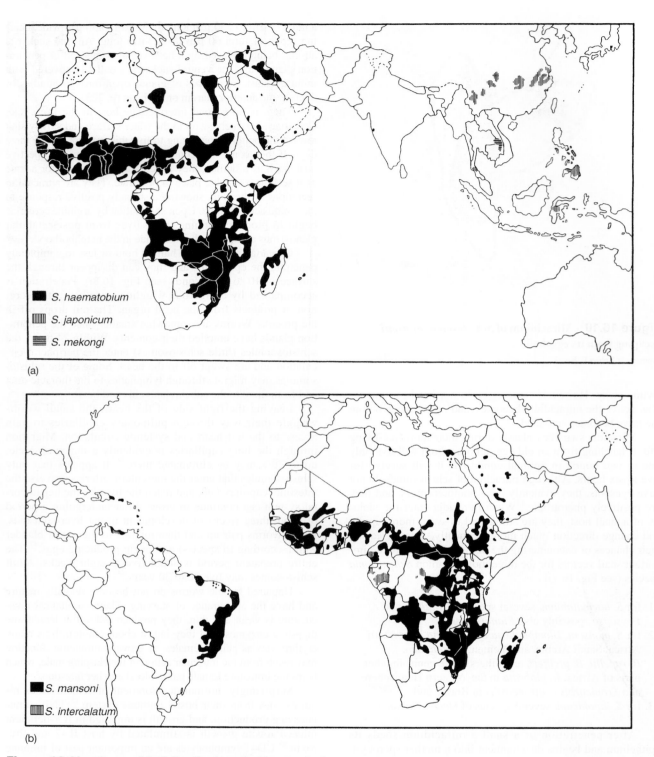

Figure 16.11 **Geographical distribution of schistosomiasis.**

(*a*) Distribution of *S. haematobium, S. japonicum,* and *S. mekongi.* (*b*) Distribution of *S. mansoni, S. intercalatum,* and *S. guineensis.* In this diagram, hatching for *S. intercalatum* represents both *S. intercalatum* and *S. guineensis.* Genuine *S. intercalatum* is found only in the Democratic Rupublic of Congo, and *S. guineensis* is found in Cameroon, Equatorial Guinea, Gabon, Nigeria, and São Tomé.

From World Health Organization, Geneva, Report of a WHO Expert Committee, WHO Technical Report Series, No. 830. Reprinted by permission.

slave trade, almost certainly because snails suitable for only that species were present there.[24, 70]

Survival of these parasites depends on human habits of polluting water with their own feces. Hygienic waste disposal is sufficient to eliminate schistosomiasis as a disease of humans (Fig. 16.12). Tradition, at once the salvation and the bane of culture, prompts people to use the local waterway (Figs. 16.13 and 16.14) for sewage disposal instead of foul-smelling

Figure 16.12   **Government-encouraged pit latrines in the Philippines serve as a means to prevent schistosomiasis.**

Courtesy of Robert E. Kuntz.

Figure 16.14   **Slow-running streams and protecting tropical vegetation provide ideal habitats for *Biomphalaria,* snail host for *S. mansoni* in Puerto Rico.**

Photograph by Larry S. Roberts.

Figure 16.13   **Typical native dwelling in a schistosomiasis area of the Philippines.**

Oncomelania, the snail host of *S. japonicum,* is found in the stream near the house, which is also likely to be contaminated by human feces.

Courtesy of Robert E. Kuntz.

outhouses. A bridge across a small stream becomes a convenient toilet; a grove of mango trees over a rivulet is a haven for children who bombard the area with their feces.

Especially vulnerable to infection are farmers who wade in their irrigation water, fishermen who wade in their lakes and streams, children who play in any contaminated body of water, and people who wash clothes in streams. A focus of infection in Brazil was a series of ditches in which watercress was grown for food. In some Muslim countries the religious requirement of ablution—that is, washing the anal or urethral orifices after urination or defecation, an act intended to achieve greater cleanliness—is an important factor in transmission. Not only is a convenient water source used to perform ablution likely to be a contaminated river or canal, but it is likely to be near the spot chosen for the deposition of additional feces and urine, ensuring further contamination.

Clearly, a population's economic and education levels both influence transmission of the disease, and age and sex are important factors as well. Males usually show the highest rates of infection and the most intense infections, and the most hazardous age is the second decade of life. This distribution of disease appears to reflect occupational and recreational differences, rather than sex or age resistance to infection. In Suriname, where both sexes work in the fields, the highest prevalence occurs in adults of both sexes. Certain other factors, such as immunity and cessation of egg release in chronic infections, must be considered when a population is surveyed and transmission is studied. Buildup of granulation tissue in the gut or bladder wall prevents release of eggs into feces or urine and will mask infection in the absence of immunodiagnostic or biopsy methods.

During the course of infection some protective immunity to superinfection is elicited, either by repeated exposures to cercariae or by the presence of adult worms, although the adult worms themselves are not affected by the immune response, an example of concomitant immunity[27] (p. 24). Resistance to reinfection is most apparent in adults, and children are least resistant.[26] Resistance becomes apparent during the period from 8 to 18 years of age. It is unclear whether the resistance is due to a slowly acquired immunity during the course of repeated exposure through childhood or to some other age-related factor, such as sexual maturation.[9]

It is of utmost importance to recognize that, by extending snail habitats, agricultural projects intended to increase food production in underdeveloped countries have, in many cases, created more misery than they have alleviated.[18] For example, a $10 million irrigation project in southern Zimbabwe had to be abadoned 10 years after it was started because of schistosomiasis.[49] The Aswan High Dam in Egypt has had many of its benefits canceled by the increase in disease prevalence it has caused. Restraint of the wide fluctuation in the water level of the Nile, although making possible four crops per year by perennial irrigation, has also created conditions vastly more congenial to snails.[13] Before the dam construction, perennial irrigation was already practiced in the Nile delta region, and the prevalence of schistosomiasis was about 60%; in the 500 miles of river valley between Cairo and Aswan, where the river was subject to annual floods, prevalence was only about 5%. Four years after the dam was completed prevalence of *S. haematobium* ranged from 19% to 75%, with an average of 35%, between Cairo and Aswan, or an average sevenfold increase! In the area above the dam, prevalence was very low before its construction; in 1972, 76% of the fishermen examined in the impounded area were infected. A 1982 study showd a continued increase in prevalence in six villages of upper Egypt.[36]

Damming the Yangtze River in China created the enormous Three Gorges Reservoir and substantially changed ecosystems in the entire area. Widespread increase in the prevalence of *Schistosoma japonicum* is likely.[73]

The roles of reservoir hosts and strains of parasites have some importance as epidemiological factors, depending on species. Members of no fewer than seven mammalian orders have successfully been infected experimentally with *S. mansoni;* however, certain monkeys and a variety of rodents are probably important natural reservoir hosts in Africa and tropical America. *Schistosoma haematobium* is more host specific than is *S. mansoni,* and it is thought that no natural reservoir hosts exist for it. The opposite is true of *S. japonicum,* which seems to be the least host specific. It can develop in dogs, cats, horses, swine, cattle, caribou, rodents, and deer; but there seems to be more than one race of this worm, and susceptibility of a given host varies. For example, *S. japonicum* is widely prevalent in rats in Taiwan, but it is rare in humans there. The descriptions of *S. mekongi* and *S. malayensis* as distinct from *S. japonicum* suggest that the traditional *S. japonicum* may be a complex of cryptic species.[12]

Pathogenesis

Schistosomiasis is unusual among parasitic infections in that pathogenesis is almost entirely due to eggs and not to adult worms. It was mentioned before that eggs traverse the gut or

bladder wall surrounded by a granuloma (Fig. 16.15) which disperses upon reaching the lumen. Clearly, the granulomas do not disperse from the eggs that do not reach the lumen, remaining in place and leaking antigens over a considerable length of time. Thus, the primary lesion in schistosomiasis is a delayed type hypersensitivity (DTH) reaction around eggs (Fig. 16.16). However, progress and outcome of the disease are a result of a complex interplay of immunopathology involving both the T_H1 and T_H2 arms of the immune response (see p. 32).

Schistosomiasis is often divided into three phases: migratory, acute, and chronic. The **migratory phase** encompasses the time from penetration until maturity and egg production; it is often symptomless. Penetration of cercariae may produce a dermatitis if a patient's immune system has been sensitized by earlier experiences of cercarial penetration. Schistosome dermatitis is usually most severe when

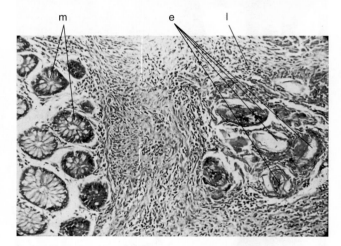

Figure 16.15 Eggs of *Schistosoma mansoni* in granuloma in intestinal wall. Eggs (e), mucosa (m), and leukocytic infiltration (l) can be seen.

Courtesy of H. Zaiman, from *Pictorial Presentation of Parasites.*

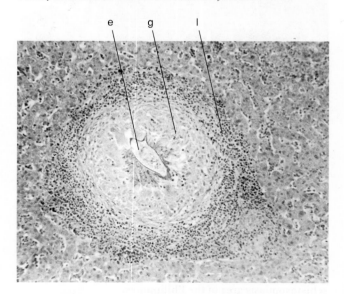

Figure 16.16 Egg of *Schistosoma mansoni* in granuloma. Egg (e), granuloma (g), leukocytic infiltration (l).

AFIP neg. no. 64-6532.

caused by bird schistosomes (p. 249), probably because cercariae are killed.

The **acute phase** is sometimes called **Katayama fever** (for the Katayama region of Japan, a former endemic area[67]) and occurs when the schistosomes begin producing eggs about 4 to 10 weeks after initial infection. By this time a host has had considerable exposure to various schistosome antigens, sufficient to mount a humoral response, but the advent of egg production substantially increases the amount of antigen release. The change in antigen-antibody ratio leads to formation of large immune complexes that must be cleared by cells of the RE system. The syndrome is marked by chills and fever, fatigue, headache, malaise, muscle aches, lymphadenopathy, and gastrointestinal discomfort.[46] There is a high eosinophilia, and granulomas around eggs contain large numbers of eosinophils, as well as neutrophils and macrophages.

Macrophages in early acute granulomas secrete predominantly IL-1, and then TNF increases after one to two weeks. Chronic granulomas are dominated by macrophages, lymphocytes, fibroblasts, and multinucleated giant cells.[64] These become small fibrous granulomas, or **pseudotubercles,** so called because of their resemblance to the localized nodules of tissue reaction (tubercles) in tuberculosis. Many eggs are carried by the hepatic portal circulation back up into the liver, where they stimulate granuloma formation (see Fig.16.16), and some may be carried to the lungs or other tissues.

Chronic phase patients indigenous to endemic areas are commonly asymptomatic,[64] or, with intestinal schistosomiasis, they may show mild, chronic, bloody diarrhea with mild abdominal pain and lethargy. With schistosomiasis haematobia, there may be pain on urination and blood in the urine. Affected people usually accept these conditions as normal and only seek medical assistance with heavy infections or when more serious complications develop. In most cases the patient's immune responses are modulated so that granulomatous reactions do not become too severe. For example, although macrophages secrete fibroblast growth factors, which mediate fibrosis, they also secrete collagenases that digest collagen fibers. A balance between collagen synthesis and degradation in most cases prevents progression to serious hepatic fibrosis.

In about 8% of cases of infection with *S. japonicum* and *S. mansoni,* development of egg granulomas and fibrosis in the liver seriously impedes portal blood flow. As the eggs accumulate and the fibrotic reactions in the liver continue, a periportal cirrhosis and portal hypertension ensue. A marked enlargement of the spleen (splenomegaly) occurs, partly because of eggs lodged in it and partly because of the chronic passive congestion of the liver. Ascites (accumulation of fluid in the abdominal cavity) is common at this stage (Fig. 16.17). Some eggs pass the liver, lodging in the lungs, nervous system, or other organs and produce pseudotubercles there.

Pathological changes due to *S. japonicum* tend to involve the small intestine more extensively than those due to *S. mansoni.* Frequently, fibrous nodules containing nests of eggs occur on the serosal and peritoneal surfaces. Eggs of *S. japonicum* reach the brain more often do those of the other species; 60% of all neurological disease in schistosomiasis and almost all brain lesions are due to *S. japonicum.*

Because adults of *S. haematobium* live in the venules of the urinary bladder, the chief symptoms are associated with the urinary system. Onset of bloody urine is usually gradual and becomes marked as the disease develops and the bladder

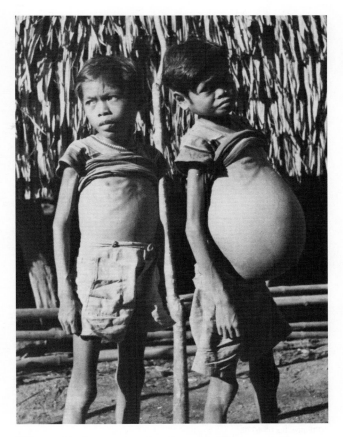

Figure 16.17   **Ascites in advanced schistosomiasis japonica, Leyte, Philippines** *(right).*
This is an example of dwarfing caused by schistosomiasis. The male on the left is 13 years old; the one on the right is 24 years old.
Courtesy of Robert E. Kuntz.

wall becomes more ulcerated. Changes in the bladder wall (Fig. 16.18) are associated with the DTH reactions around the eggs—that is, pseudotubercles, fibrous infiltration, thickening of the muscularis layer, and ulceration. Chronic heavy infections lead to genital, ureteral, and kidney involvement and to lesions in other parts of the body, as with other species. Major disease manifestations in chronic *S. haematobium* infections are urinary tract blockages, chronic urinary bacterial infections, bladder cancer, and bladder calcification.[64]

Invasion of the female reproductive tract is most common with *S. haematobium,* but it can also happen in *S. mansori, S. japonicum,* and *S. intercalatum* (p. 249) infections. Intricate vascular links between the rectal and the bladder venous plexus provide easy access of migrating worms to internal and external female genitalia, and up to 75% of women infected with *S. haematobium* have eggs in their genitals.[31]

Diagnosis and Treatment

A simple, cheap, sensitive, and specific technique for routine diagnosis of schistosomiasis is still not available.[16] As is the case with many other helminths, demonstration of eggs in excreta is the most straightforward mode of diagnosis. However, the number of eggs produced per female schistosome, even for *S. japonicum,* is far smaller than for most

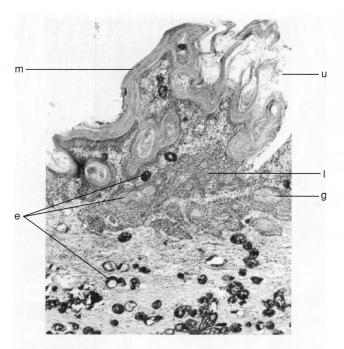

Figure 16.18   Schistosomiasis of the urinary bladder.
In this case, many eggs (*e*) of *S. haematobium* can be seen in all the layers of the bladder. Many of the eggs are calcified. The epithelium has undergone squamous metaplasia (*m*). Note also leukocytic infiltration (*l*), granulosis (*g*), and ulceration (*u*).

AFIP neg. no. 65-6779.

other helminth parasites of humans; therefore, direct smears must be augmented by concentration techniques and other diagnostic methods, such as biopsy and immunodiagnosis. With concentration techniques, such as gravity or centrifugal sedimentation, more than 90% of the coprologically demonstrable cases can be diagnosed. The Kato technique is a simple method for discovery and quantification of eggs in 20 mg to 50 mg stool samples.[52] However, few or no eggs may be passed, particularly in chronic cases. In such cases rectal, liver, or bladder biopsies may be of great value as can ultrasonography,[57] but these require services of specialists and availability of appropriate surgical facilities. Hence, substantial effort has been directed at finding sensitive, accurate, and reliable immunodiagnostic methods.

Serological tests based on detection of antibodies in the patient's blood have several inherent problems: (1) they only become positive some time after infection, (2) they only become negative some time after cure, and (3) they may cross-react with other helminth infections. Tests designed to detect schistosome antigens, however, offer potential solutions to these problems. They become positive as soon as antigens are present and become negative quickly after cure. Monoclonal antibodies to one of the best characterized schistosome antigens were prepared by Deelder and his coworkers.[65] In a modified ELISA test they detected the antibody in patients; sera at levels as low as 1 ng/ml, and the antigen concentration was highly correlated with egg output—that is, it provided estimation of worm burden. However, antigen-detection techniques appear to be no more sensitive than microscopic examination of excreta for eggs.[16]

Diagnosis of schistosome infections is an active area of research, with a number of molecular techniques being explored experimentally, including those that target parasite DNA.[21, 54, 72]

Difficulty in treating schistosomiasis has been a major factor contributing to the disease as a world health problem. Until recently the most effective drugs were organic trivalent antimonials, but these are quite toxic to humans and must be given carefully—in small doses over a period of two to six weeks, depending on the drug. Problems inherent in treating large numbers of people with such drugs over wide areas in developing countries are obvious.

Therefore, much effort was spent on investigating other, less toxic but effective drugs. The drug of choice is now praziquantel, which is effective against all species of schistosomes in humans. We described its mode of action in chapter 15 (p. 230). Praziquantel is less effective against schistosomules, but artemesin and its derivatives, which have proven valuable animalarials (p. 156), have antischistosomal activity against all stages.[71] Appropriate chemotherapy can lead to reversal and even resolution of much fibrosis and urinary pathology, but long-standing heavy infection can result in irreversible liver or bladder damage.[64]

A few years ago, it was reported that myrrh, a plant substance known since antiquity, was an effective antischistosomal drug.[60] However, further investigations have not confirmed that result.[2, 6]

Control

Control of schistosomiasis, as with many infectious diseases, must be a multifold effort, including (1) education of a population to undertake activities to prevent transmission, (2) curing of infected persons, (3) control of vectors, and (4) protective vaccination.

Education. Although potentially very effective, education is often exceedingly difficult, depending ultimately on the task of persuading masses of uneducated, poor people to change their customs and traditions. Some such efforts have reported little success,[63] but others have been highly successful.[29]

Control by Chemotherapy. In the past curing infected persons was not a practical strategy for control. Development of safe and effective schistosomacides has altered that circumstance. Effective strategies for schistosomiasis control with praziquantel have been developed.[18]

Vector Control. Control of snails was the major thrust of control efforts before the advent of safe and effective antihelminthics, and it is still important to support chemotherapy campaigns and to reduce reinfection.[39] Snail control may be undertaken by environmental management, by molluscicides, and by biological agents.

Although draining snail habitats is of value, we pointed out before that many more good habitats are being created by efforts to increase agricultural production. This need not be so. Environmental management measures, such as stream channelization, seepage control, canal lining, and canal relocation with deep burial of snails, can prevent increase in transmission.[39] Such actions helped prevent an increase in prevalence after construction of a lake and irrigation canals in Cameroon between 1979 and 1985.[1] This project included regular cleaning of vegetation from secondary and tertiary canals, use of only the concrete-lined primary canals for bathing and clothes washing, and provision of a clean source of drinking water.

Use of chemical molluscicides has met with some success; but problems involved include determination and application of the proper quantity in a given body of water, dilution, effects on other organisms in the environment, and errors in estimating the physical and chemical characteristics of water. Niclosamide is the only molluscicide currently available. Molluscicidal control of *S. japonicum* is virtually ineffective because *Oncomelania* spp. are amphibious and visit water only to lay eggs.

Some biological control efforts have been successful, mostly the introduction of potential predators and competitors of vector snails.[53] The best-studied are potential competitive species of snails: *Marisa cornuarietis, Helisoma duryi, Thiara granifera,* and *Melanoides tuberculata.* In Martinique, for example, *M. tuberculata* colonized rapidly after its introduction in January 1983, and by October 1984, *Biomphalaria glabrata* and *B. straminea* had disappeared and have not been found since. Introduction of *M. cornuarietis* has successfully controlled *B. glabrata* in several habitats in Puerto Rico; *M. cornuarietis* not only competes with *B. glabrata* for food, but also preys on the vector snails. A North American crayfish, *Procambarus clarkii,* was introduced into East Africa for aquaculture in about 1970 and has since then dispersed into all major drainages in Kenya.[28] *Procambarus clarkii* feeds on *Biomphalaria* spp., and under certain environmental circumstances, it can have a significant impact on schistosome transmission.[39]

Snail-eating fish have been cultured and released in infected waters with some success.

Vaccination. The development of an effective vaccine would have great potential value in the control of schistosomiasis, and this area of research is active. An advantage for investigators searching for an antischistosome vaccine is that the vaccine would not have to be more than 90% effective, as would a vaccine against a virus or bacteria. Because serious symptoms are shown only by patients with large numbers of worms, dramatic reduction in numbers of severe cases would result from only 50% or so reduction in worm burden.

Some protection can be conferred by vaccination with irradiated cercariae and/or schistosomules,[5] but for practical use, a long-acting, killed antigen would be more desirable. A number of parasite-derived antigens confer partial protection against reinfection when used to immunize mice, rats, and other animals,[4] and promising candidate vaccines are enzymes from *S. mansoni* (Sm28GST, for *Schistosoma mansoni* 28 kDa glutathione-*S*-transferase) and the same enzyme derived from *S. haematobium* (Sh28GST).[10] Vaccination with these antigens resulted in a partial protection against reinfection and, more importantly, resulted in a significant inhibition of female worm fecundity. Recombinant forms (rSh28GST) produced in yeast are being tested.[10]

Other *Schistosoma* spp.
In addition to *S. mansoni, S. haematobium,* and *S. japonicum,* several other species of schistosomes can infect humans, although their distribution and prevalence are much less than those of the "big three." In Africa hundreds of cases

are known in which terminal-spined eggs are recovered from stools only. The worms were named *Schistosoma intercalatum,* but they are morphologically indistinguishable from *S. mattheei,* a natural parasite of African ruminants. Evidently, however, *S. mattheei* can infect primates only in the presence of a coinfection with *S. haematobium* or *S. mansoni.*[32] In Southeast Asia, careful examination of biology, morphology, and molecular analysis has supported recognition of *S. mekongi* and *S. malayensis* as distinct from *S. japonicum,* which they closely resemble.[25]

A number of other species of *Schistosoma* exist, and these differ with respect to morphology, host specificity, distribution, and DNA sequences (Table 16.2), but some are quite difficult to distinguish from each other. *Schistosoma* spp. can be arranged in five groups: *S. mansoni* group (with lateral-spined eggs), *S. haematobium* group (with terminal-spined eggs; including the recently recognized *S. guineensis*[68]), *S. japonicum* group (rounded, minutely spined, or spineless eggs), and *S. indicum* group (an as yet poorly defined group from India and Southeast Asia), and a "long tail-stem lineage."[32, 68] All species except those in the *S. japonicum* group use pulmonate snails as intermediate hosts.

Characterization of groups on the basis of egg shape and spines is convenient but not entirely satisfactory. For example, *S. margrebowiei* in the *S. haematobium* group has round eggs with small spines, and *S. sinensium* in the *S. japonicum* group has lateral-spined eggs. These species nevertheless fit into their assigned groups on the basis of other criteria, such as specificity of intermediate hosts. Molecular analysis has so far supported the identity of the species recognized,[68] but within groups, some species are very closely related. *Schistostoma mansoni* and *S. rodhaini* can form hybrids that produce viable and fertile young in the laboratory.[7] *Schistostoma intercalatum* and *S. haemotobium* also produce hybrids in the laboratory, and naturally occurring hybrids have been found in certain areas in Africa.[56]

Molecular analysis has indicated that species in the *S. japonicum* group are basal to all other *Schistosoma* spp.[68]

Schistosome Cercarial Dermatitis ("Swimmer's Itch")
Several species of *Schistosoma* cause a severe rash when their cercariae penetrate skin of an unsuitable host; *S. spindale* and *S. bovis* are agents of dermatitis in humans throughout the range of these schistosomes. More importantly, several species of bird schistosomes are distributed throughout the world and cause "swimmer's itch" when their cercariae attack anyone on whose skin the organisms land. Species in genera *Trichobilharzia, Gigantobilharzia, Ornithobilharzia, Microbilharzia,* and *Heterobilharzia* are the guilty parties.

For the most part the skin reaction is a product of sensitization, with repeated infections causing increasingly severe reactions. When a cercaria penetrates skin and is unable to complete its migration, the host's defense responses rapidly kill it. At the same time the cercaria releases allergens that cause inflammation and, typically, a pus-filled pimple (Fig. 16.19). The reaction may also be general, with an itching rash produced over much of the body. The condition is not a serious threat to health but is a terrific annoyance, much like poison ivy, that interrupts a summer vacation, for instance, or decreases the income of someone who rents lakefront cottages to the summer crowd. In the United States

Table 16.2 Distribution and Host Specificity of *Schistosoma* spp. (modified from Johnston et al.)[32]

Species group	Distribution[a]	Snail host genera	Mammalian host[b]
Schistosoma haematobium			
S. haematobium	Af & ad	*Bulinus*	Pr
S. intercalatum	Af	*Bulinus*	Pr
S. guineensis	Af	*Bulinus*	Pr
S. mattheei	Af	*Bulinus*	Pr, Ar
S. bovis	Af & ad	*Bulinus, Planorbarius*	Ar
S. curassoni	Af	*Bulinus*	Ar
S. margrebowiei	Af	*Bulinus*	Ar
S. leiperi	Af	*Bulinus*	Ar
Schistosoma mansoni			
S. mansoni	Af, SA	*Biomphalaria*	Pr, R
S. rodhaini	Af	*Biomphalaria*	R, C
Schistosoma hippopotami			
S. hippopotami	Af	*Bulinus*[44]	Ar
S. edwardiense	Af	*Biomphalaria*[44]	Ar
Schistosoma indicum			
S. indicum	SEA, SWA	*Indoplanorbis*	Ar
S. spindale	SEA, SWA	*Indoplanorbis*	Ar
S. nasale	SWA	*Indoplanorbis*	Ar
S. incognitum	SEA, SWA	*Lymnea, Radix*	Ar, R, C
Schistosoma japonicum			
S. japonicum	SEA	*Oncomelania*	Pr, Ar, R, C, Pe
S. mekongi	SEA	*Neotricula*	Pr, C
S. sinensium	SEA	*Neotricula*	R
S. malayensis	SEA	*Robertsiella*	Pr, R

[a]Af = Africa; Af & ad = Africa and adjacent regions; SA = South America and Caribbean; SEA = Southeast Asia; SWA = Southwest Asia.
[b]Pr = Primates; Ar = Artiodactyla; R = Rodentia; C = Carnivora; Pe = Perissodactyla.

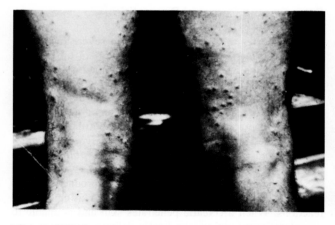
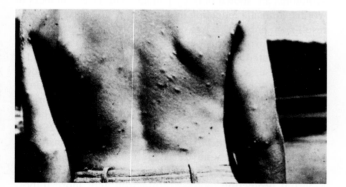

Figure 16.19 Cercarial dermatitis, or "swimmer's itch," caused by cercariae of avian blood flukes.
AFIP neg. no. 77203.

the problem is most serious in the Great Lakes area, but it has been reported from nearly all states.

Control depends mainly on molluscicides, but their usefulness is limited because they threaten sport fishing by poisoning fish and, of course, because of the other problems mentioned previously. Ocean beaches are occasionally infested with avian schistosome cercariae, for which no control has yet been devised.

Learning Outcomes

By the time a student has finished studying this chapter, he or she should be able to:

1. Diagram the life cycle of *Schistosoma mansoni* and tell how that cycle differs from those of *S. haematobium* and *S. japonicum*.

2. Distinguish between the three *Schistosoma* species that most commonly parasitize humans, using both structural and clinical observations.

3. Explain how cultural practices promote infections with schistosomes in human populations.

4. Explain the development of pathology in human schistotomiasis (all three species), with emphasis on the factors that contribute most to pathology.

5. Tell why developments such as dams have an impact on the distribution of blood fluke infections.

References

References for superscripts in the text can be found at the following Internet site: www.mhhe.com/robertsjanovynadler9e

Additional Readings

Ansari, N. (Ed.). 1973. *Epidemiology and control of schistosomiasis (bilharziasis)*. Basel: S. Karger. AG. An official publication of the World Health Organization, outlining advances in the area of schistosomiasis.

Berrie, A. D. 1970. Snail problems in African schistosomiasis. In B. Dawes (Ed.), *Advances in parasitology* 8. New York: Academic Press, Inc., pp. 43–96.

Brant, S. V., J. A. T. Morgan, G. M. Mkoji, S. D. Snyder, R. P. V. Jayanthe Rajapakse, and E. S. Loker. 2006. An approach to revealing blood fluke life cycles, taxonomy, and diversity: provision of key reference data including DNA sequence from single life cycle stages. *J. Parasitol.* 92:77–88.

Bruce, J. I., and S. Sornmani (Eds.). 1980. The Mekong schistosome. *Malacological Reviews,* Suppl. 2.

Jordan, P., G. Webbe, and R. F. Sturrock. 1993. *Human schistosomiasis*. Wallingford, Oxon, UK: CAB International.

Rollinson, D., and A. J. G. Simpson (Eds.). 1987. *The biology of schistosomes. From genes to latrines*. London: Academic Press.

Roueché, B. 1988. A swim in the nile. The medical detectives. New York: Truman Talley Books/Plume, pp. 124–137. An interesting tale of a malady unexpected in North America, schistosomiasis haematobium.

Sobhon, P., and E. S. Upatham. 1990. *Snail hosts, life-cycle, and tegumental structure of oriental schistosomes*. Geneva: UNDP/ WORLD BANK/WHO Special Program Research Training Tropical Diseases.

Chapter 17

Digeneans: Echinostomatiformes

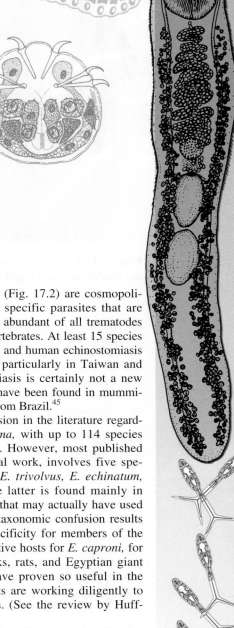

Die Frage nach der Lebensund Entwickelungsgeschichte des Distomum hepaticum *hat mich bereits seit vielen Jahren beschäftigt.*

—Dr. R. Leuckart, Leipzig, 1881[29]

Members of order Echinostomatiformes often show little resemblance to one another in their adult stages, but embryological studies have indicated common ancestries through developmental similarities. Often, though by no means always, the tegument bears well-developed scales or spines, particularly near the anterior end. Their acetabulum is near the oral sucker. In many cases a second intermediate host is absent, and cercariae encyst on underwater vegetation or debris or even return to the snail first intermediate host. Most species are parasitic in wild animals, but a few are important as agents of disease in humans, domestic animals, or both.

SUPERFAMILY ECHINOSTOMATOIDEA

Parasites of Echinostomatoidea infect all classes of vertebrates and are found in marine, freshwater, and terrestrial environments. Some are among the most common parasites encountered, and a few cause devastating losses to agriculture.

Family Echinostomatidae

Echinostomes are easily recognized by their circumoral collar of peglike spines; hence their name (Fig. 17.1). The spines are arranged either in a single, simple circle or in two circles, one slightly lower than and having spines alternating with the other. The collar is interrupted ventrally, and at each end it has a group of "corner spines." Size, number, and arrangement of these spines are of considerable taxonomic importance in both cercariae and adults. Echinostomes typically are slender worms with large preequatorial acetabula, pretesticular ovaries, and tandem testes, although exceptions occur. Vitellaria are voluminous and mainly postacetabular. These worms are parasites of the intestine or bile duct of reptiles, birds, and mammals, particularly those frequenting aquatic environments.

Genus *Echinostoma*

Members of genus *Echinostoma* (Fig. 17.2) are cosmopolitan, sometimes rather non–host specific parasites that are among the most widespread and abundant of all trematodes in warm-blooded, semiaquatic vertebrates. At least 15 species have been reported from humans, and human echinostomiasis is fairly common in the Orient, particularly in Taiwan and Indonesia.[24] Human echinostomiasis is certainly not a new phenomenon; echinostome eggs have been found in mummified bodies 600–1,200 years old from Brazil.[45]

There has been some confusion in the literature regarding the taxonomy of *Echinostoma,* with up to 114 species being described from the genus. However, most published research, especially experimental work, involves five species: *E. caproni, E. paraensei, E. trivolvus, E. echinatum,* and *E. revolutum,* although the latter is found mainly in older literature reporting studies that may actually have used one of the other species. Some taxonomic confusion results from the relatively low host specificity for members of the genus. "Preferred" natural definitive hosts for *E. caproni,* for example, include domestic ducks, rats, and Egyptian giant shrews. *Echinostoma* species have proven so useful in the lab, however, that parasitologists are working diligently to solve these taxonomic problems. (See the review by Huffman and Fried.[24])

Eggs hatch in water and miracidia penetrate a first intermediate host. Sporocysts develop from germinal cells in the miracidia, and mother rediae are produced from these sporocysts. *Echinostoma* species differ in the timing of redia production in snails.[4] Routine inspections of snails in common and widespread genera *Physa, Lymnaea, Helisoma, Paludina,* and *Segmentina* often reveal echinostome infections. Metacercariae occur in molluscs, planaria, fish, and tadpoles. Infection of a definitive host is accomplished when the definitive host eats one of these. Humans usually become infected by eating raw mussels or snails.

Morphologically, worms of genus *Echinostoma* are easily identified by their circumoral collar spines arranged in two rows. The number of spines ranges from 27 to 51, but

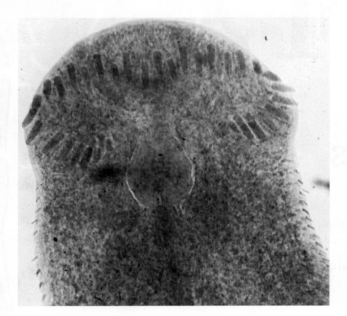

Figure 17.1 Anterior end of *Echinostoma* sp., showing the double crown of peglike spines on the circumoral collar. Courtesy of Warren Buss.

the common species used in experimental work belong to a group with 37. The operculate eggs are large, typically about 100 μm by 60 μm, and few of them occur in the uterus at any one time. The genital pore is median and preacetabular, and the cirrus pouch is large, passing dorsal to the voluminous acetabulum. The short uterus has an ascending limb only. Overall size varies considerably among species.

Echinostoma caproni. Huffman and Fried[24] considered many reports of *E. revolutum* actually to be studies of *E. caproni,* and Christensen et al.[10] considered *E. liei,* described by Jeyarasasingam et al.,[25] synonymous with *E. caproni;* molecular work supports this conclusion.[37] Snails of genus *Biomphalaria,* especially *B. glabrata* and *B. alexandrina,* are good first intermediate hosts. Daughter rediae migrate into the snail's gonad and digestive gland. Cercariae have a number of options, including migration up a snail's nephridiopore or penetration of a variety of second intermediate hosts such as clams, frogs, and sometimes fish. *Echinostoma caproni* adults survive longer in mice and hamsters than in chickens.

Echinostoma paraensei. *Echinostoma paraensei* is similar to the other species in many respects. Lie and Basch[30] reported that both *Biomphalaria glabrata* and *Physa rivalis*—that is, snails of two distinct families (Planorbidae and Physidae)—served as first intermediate hosts. Sporocysts develop in a snail's ventricle, but rediae migrate through tissues to a variety of organs. Cercariae emerge about 25 days after infection, live for around six hours, and in experimental situations encyst as metacercariae in the very snails they came from. Metacercariae accumulate in the tentacle tips of *B. glabrata* snails, and dead or dying snails typically have many cysts in their heads.[30] Although adult worms live for five months in hamsters, infections of over 100 parasites kill the hosts.

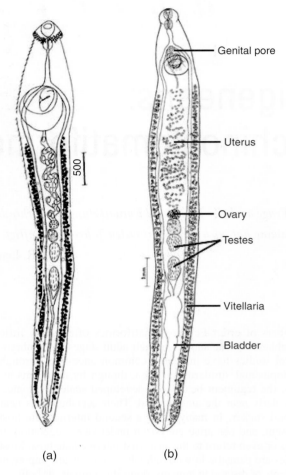

(a)   (b)

Figure 17.2 Representative species of genus *Echinostoma.*

(*a*) *Echinostoma revolutum* (= *E. audyi*); (*b*) *E. paraensi.* Figures from the original species descriptions. Note the elongate body, extensive vitellaria, tandem testes, and large excretory bladder.

(*a*) From K. J. Lie and T. Umatheva, "Studies on Echinostomatidea (Trematoda) in Malaya VIII. The life history of *Echinostoma audyi* sp. n." in *J. Parasitol.* 51:781–788. (*b*) From K. J. Lie and P. F. Basch, "The life history of *Echinostoma paraensei* sp. n. (Trematoda: Echinostomatidae)," in *J. Parasitol.* 53:1192–1199.

Echinostoma trivolvus. *Echinostoma trivolvus* is distinguished by a rather remarkable list of definitive hosts, including several species of ducks, geese, hawks, owls, doves, flamingos, dogs, cats, guinea pigs, rabbits, pigs, rats, and of course, mice.[24] Not all hosts are equal, however; both worm size and number of eggs produced differ, depending on the definitive host. Christensen et al.[10] stated that much of the early research on "*E. revolutum*" was actually done on *E. trivolvus.* For anyone who thinks all the world's systematic problems are solved or easily solvable, a journey through these discussions in the echinostome literature would be exceedingly educational.

Echinostoma revolutum (syn. E. audyi). There are numerous reports in older literature of trematodes identified as *E. revolutum,* but the name is now restricted to worms matching Frolich's 1802 description.

Huffman and Fried[24] considered *E. audyi* a synonym of *E. revolutum,* and the former's life-cycle description by Lie and Umathevy[31] is a fine illustration of the way these seemingly complex events are discovered and deciphered. In their experiments, the snail *Lymnaea rubiginosa* served as both first and second intermediate host, and both rediae and daughter rediae were produced. Cercariae encysted in *L. rubiginosa* as well as in species of other snail genera and, in some cases, within the redia itself. Lie and Umathevy[31] were able to infect pigeons, ducklings, and sparrows experimentally, but they could not determine the preferred host in nature. Egg production started eight days after infection in birds, but adult worms evidently died after eight weeks. Worms from pigeons were over 10 mm long, while those from ducklings were less than 7 mm, a difference that might lead a naive parasitologist to suspect that two species were represented.

Euparyphium ilocanum. This species was formerly placed in genus *Echinostoma,* but Huffman and Fried[24] considered it a member of a related genus *Euparyphium.* Eggs of *Euparyphium ilocanum* were first seen in the stool of a prisoner in Manila in 1907. The organism has since been found commonly throughout the East Indies and China. Tubangui[55] discovered that Norway rats were an important reservoir of infection. *Euparyphium ilocanum* has 49 to 51 spines, with five or six corner spines on each side. The double row of spines is continuous dorsally, and the testes are deeply lobate.

The biology of *Eu. ilocanum* is similar to that of other echinostomes, with metacercaria encysting in any freshwater mollusc. Infected snails are eaten raw by people, who thereby become infected. The worms cause inflammation at their sites of attachment within the small intestine. Intestinal pain and diarrhea may develop in severe cases.

Other Echinostomatid Species Reported from Humans. Several species of echinostomes in different genera that normally parasitize wild animals have been reported from humans. These include *Echinostoma lindoense* in Celebes and possibly Brazil (considered a synonym of *E. echinatum* by Christensen et al.[10]); *E. malayanum* from India (Fig. 17.3), Southeast Asia, and the East Indies; *E. cinetorchis* from Japan, Taiwan, and Java; *E. hortense* from Japan and Korea; *E. melis,* which is circumboreal; and *Hypoderaeum conoidum* in Thailand. Others are *Himasthla muehlensi* in New York; *Paryphostomum surfrartyfex* in Asia Minor; and *Echinochasmus perfoliatus* from eastern Europe and from Asia. *Acanthoparyphium tyosenense,* first described from a duck, has also been reported from humans in Korea; infections were acquired from raw or improperly cooked brackish water clams.[8]

These are only some of the echinostome species of wild animals that have found their ways into humans, but considering the lack of host specificity in this group, other species probably do so fairly often and remain undetected. Thus, when zoonotic infections are found, it immediately becomes necessary to identify the pathogen and determine how the person became infected. Faunal and systematic surveys, sometimes considered a less glamorous sort of science than experimental work, are the only ways to fulfill this need.

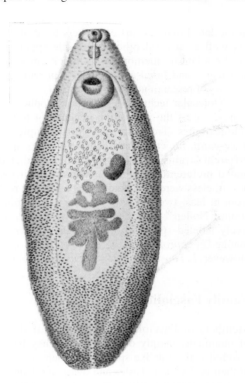

Figure 17.3 *Echinostoma malayanum,* **a parasite of humans in southern Asia.**

From T. Odhner, "Ein zweites Echinostomum aus dem Menschen in Ostasien (*Ech. malayanum* Leiper)," in *Zool. Anz.* 41:577–582, 1913.

Echinostomatids as Models in Experimental Parasitology

Echinostomes are exceptionally good models for use in experimental research.[18] *Echinostoma caproni,* in nature a parasite of a Madagascar falcon, has been particularly well studied.[19] For example, research with this species, as well as some other echinostomes, especially *E. paraensei,* has shown parasite species-specific pathological effects on the intestinal epithelium of their vertebrate hosts.[20] Because *E. caproni* can survive in so many different kinds of hosts, it has been used to explore host factors that influence course of infection. For example, mice, rats, and hamsters react somewhat differently to *E. caproni;* mice are excellent hosts, allow worm survival up to 29 weeks, although villi are eroded, and display humoral responses.[53] In hamsters, there is a strong inflammatory response to infection, and worm Excretory/Secretory (E/S) antigens evidently move across the mucosa into the blood. Rats are far less hospitable, however, and expel worms in 10 weeks.[53]

Other studies have revealed alterations in the fatty acid composition of infected snails and snail strain-specific responses to parasites that are correlated with resistance or susceptibility to important trematodes such as *Schistosoma mansoni.* Snail resistance to trematodes is a well-known phenomenon, but in the case of echinostomes experimental work shows that parasite excretory and secretory proteins inhibit a host snail's ability to mount a cellular immune response.[32] And these tractable generalist parasites do not always need a complete definitive host in order to complete their life cycles in the lab. Fried and his coworkers have grown *E. caproni* (as well as several other trematode species) to maturity on the allantoic membranes of chick embryos.[9] Fried and Huffman[19] and Fried and Graczyk[18] provide comprehensive reviews of research utilizing *E. caproni*.

Molecular techniques have been applied to some problems, such as the population genetics of trematodes, using echinostomes. Trouve et al.,[54] for example, using two- and three-worm infections and genetic markers specific to three different strains of *E. caproni*, showed that the parasites mated preferentially with members of their own strain but nevertheless were capable of donating to and receiving sperm from at least two different co-infecting worms of different strains. Nollen[39, 40] carried this work further, using radioactively labeled sperm to demonstrate one-way interspecific mating (*E. caproni* sperm to *E. trivolvus* but not the reverse). However, *E. caproni* and *E. paraensei* did not interbreed.[39]

Family Fasciolidae

Members of Fasciolidae are large, leaf-shaped parasites of mammals, mainly of herbivores. They have a tegument covered with scalelike spines, and their acetabulum is close to their oral sucker. Testes and ovary are dendritic, and vitellaria are extensive, filling most of the postacetabular space. There is no second intermediate host in the life cycle; metacercariae encyst on submerged objects or freely in the water. One important species lives in the intestinal lumen, but most parasitize the liver of mammals.

Fasciola hepatica. *Fasciola hepatica* (Fig. 17.4) has been known as an important parasite of sheep and cattle for hundreds of years. Because of its size and economic importance, it has been the subject of many scientific investigations and may be the best known of any trematode species. Jean de Brie published the first record of it in 1379. He was well acquainted with a disease of sheep called "liver rot," in which the liver of an animal is infected with large, flat worms. In 1668 the great pragmatist Francisco Redi was the first to illustrate this fluke, thereby stimulating other researchers to investigate its biology. Leeuwenhoek was interested in the organism but apparently was distracted by all of the other tiny wonders he found with his microscopes.

Cercariae and rediae of *F. hepatica* were described in 1737 by Jan Swammerdam, a man with a remarkable ability to see and understand microscopic objects with the use of a primitive microscope. Linnaeus gave the worm its name in 1758 but considered it a leech. Pallas first found it in a human in 1760. Professor C. L. Nitzsch, in 1816, was first to recognize the similarity of cercariae and adult liver flukes. Thus, the history of *F. hepatica* parallels the history of trematodology itself in that discoveries proved to be generally applicable to the biology of digeneans. In 1844 Johannes Steenstrup published a landmark book, *Alternation of Generations,* in which he postulated that trematodes have two generations, one adult and one not.

By the mid-1800s, circumstantial evidence indicated that molluscs were involved in transmission of *F. hepatica*. In 1880 George Rolleston, professor of anatomy and physiology at Oxford, was convinced that a common slug was the

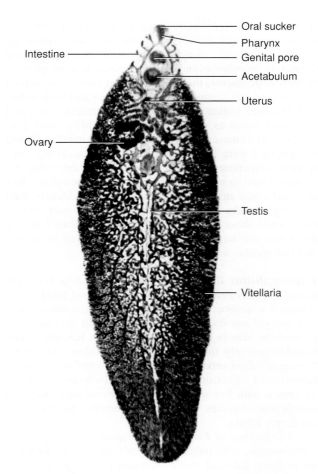

Figure 17.4   ***Fasciola hepatica,*** **the sheep liver fluke.**
Courtesy of Turtox/Cambosco.

intermediate host of *F. hepatica*. Although he was wrong in this assumption, he recommended that A. P. Thomas undertake an investigation to determine the life cycle of this parasite. Thomas was young—a 23-year-old demonstrator at the time—but he took on this formidable task with zeal. He soon found the snail *Lymnaea truncatula* infected with rediae and cercariae that were similar in many regards to *Fasciola*. Then he successfully infected this snail with miracidia and followed their development through the sporocyst, redia, and cercarial stages.

At the same time as this "lowly" Oxford demonstrator was investigating liver rot, fascioliasis was also engrossing the mind of the greatest parasitologist then living, Rudolph Leuckart.[29] After a series of false starts, Leuckart traced the development of *F. hepatica* through the same species of snail and, as a final irony, published his results 10 days before Thomas published his. Credit is given to both men equally, but one can scarcely refrain from lending sympathy to the young Englishman who elucidated the first trematode life cycle in a truly scientific manner, without the advantages of a large budget and long experience. Even today there are many trematode species—indeed, many parasite species—whose life cycles are yet to be fully described and might well yield to insight, creative experiments, and hard work, regardless of funds available for research.

Neither Thomas nor Leuckart determined the mode of infection of the definitive host. This was done by Adolph

Lutz, a Brazilian working in Hawaii, who demonstrated between 1892 and 1893 that ruminants become infected by eating juveniles encysted on vegetation. That Lutz was actually working with a different species, *F. gigantica,* is immaterial, because the biology of both species is the same.

- **Morphology.** *Fasciola hepatica* is one of the largest flukes of the world, reaching a length of 30 mm and a width of 13 mm. It is rather leaf shaped, pointed posteriorly and wide anteriorly, although the shape varies somewhat. The oral sucker is small but powerful and is located at the end of a cone-shaped projection at the anterior end. A marked widening of the body at the base of the socalled oral cone gives the worm the appearance of having shoulders. This combination of an oral cone and "shoulders" is an immediate means of identification. The acetabulum is somewhat larger than the oral sucker and is quite anterior, at about shoulder level. The tegument is covered with large, scalelike spines, reminding one of echinostomes, to which they are closely related. The intestinal ceca are highly dendritic (branched) and extend to near the posterior end of the body.

  The testes are large and greatly branched, arranged in tandem behind the ovary. The smaller, dendritic ovary lies on the right side, shortly behind the acetabulum, and the uterus is short, coiling between the ovary and the preacetabular cirrus pouch. Vitelline follicles are extensive, filling most of the lateral body and becoming confluent behind the testes. The operculate eggs are 130 μm to 150 μm by 63 μm to 90 μm.

- **Biology.** Adult *F. hepatica* (Fig. 17.5) live in bile passages of the liver of many kinds of mammals, especially ruminants. Humans are occasionally infected. In fact, fascioliasis is one of the major causes of hypereosinophilia in France.[12] The flukes feed on the lining of biliary ducts.

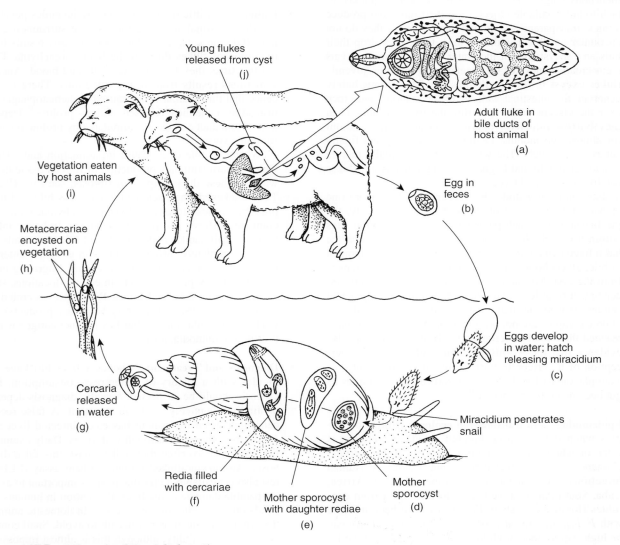

Young flukes released from cyst (j)

Adult fluke in bile ducts of host animal (a)

Vegetation eaten by host animals (i)

Egg in feces (b)

Metacercariae encysted on vegetation (h)

Eggs develop in water; hatch releasing miracidium (c)

Cercaria released in water (g)

Miracidium penetrates snail

Redia filled with cercariae (f)

Mother sporocyst with daughter rediae (e)

Mother sporocyst (d)

Figure 17.5   **Life cycle of *Fasciola hepatica.***

(*a*) Adult worm in bile duct of sheep or other mammal; (*b*) egg; (*c*) miracidium; (*d*) mother sporocyst; (*e*) mother sporocyst with developing rediae; (*f*) redia with developing cercariae; (*g*) free-swimming cercaria; (*h*) metacercaria, encysted on aquatic vegetation; (*i*) host animals eating vegetation; (*j*) flukes released from cyst.

Drawing by William Ober.

Their eggs are passed out of the liver with bile and into the intestine to be voided with feces. If they fall into water, eggs will complete their development into miracidia and hatch in 9 to 10 days during warm weather. Colder water retards their development. On hatching, miracidia have about 24 hours in which to find a suitable snail host, which can be any of several genera in the gastropod family Lymnaeidae, depending on location.

In other parts of the world different but related snails are important first intermediate hosts. Mother sporocysts produce first-generation rediae, which in turn produce daughter rediae that develop in a snail's digestive gland. Cercariae begin emerging five to seven weeks after infection. If the water in which the snails live dries up, the snails burrow into the mud and survive, still infected, for months. When water is again present, the snails emerge and rapidly shed many cercariae.

Cercariae have a simple, club-shaped tail about twice their body length. Once in the water, cercariae quickly attach to any available object, drop their tails, and produce a thick, transparent cyst around themselves. If they do not encounter an object within a short time, they drop their tails and encyst free in the water. When a mammal eats metacercariae encysted on vegetation or in water, juvenile flukes excyst in the small intestine. They immediately penetrate the intestinal wall, enter the coelom, and creep over the viscera until contacting the liver capsule. Then they burrow into liver parenchyma and wander about for almost two months, feeding and growing and finally entering bile ducts. The worms become sexually mature in another month and begin producing eggs. Adult flukes can live as long as 11 years.

Fascioloid trematodes have been used to explore one of parasitology's most obdurate mysteries—namely, the mechanisms by which parasites find their infection sites within hosts. Sukhdeo[48] showed that *Fasciola hepatica* has a fixed behavioral pattern cued by a single component of bile, glycocholic acid. This molecule elicits emergence from the cyst. Migration to the bile ducts is evidently independent of at least some brain function; developmental studies showed that major portions of their cerebral ganglionic complex are not present until after the worms have reached their final infection site.[48] However, there is also evidence that the worms' diet may influence their perception of a vertebrate host's internal environment. For example, only when *F. hepatica* starts eating liver does its gut become highly branched.[49]

- **Epidemiology and transmission ecology.** Infection begins when metacercaria-infected aquatic vegetation is eaten or when water containing metacercariae is drunk. Humans are often infected by eating watercress. Human infections occur in parts of Europe, northern Africa, Cuba, South America, and other locales. Data from coprolites (fossil feces) show Europeans have been infected with *F. hepatica* for at least 5000 years.[3] Prevalence can be high: Up to 38% of children, ages 5 to 19, may be infected in the Altiplano region of Bolivia.[16] Surprisingly, few cases are known in humans in the United States, although the worm is fairly common in parts of the South and West. Sheep, cattle, and rabbits are the most frequent reservoirs of infection.

Whether or not humans are infected, veterinary fascioliasis is a major economic problem. *Fasciola hepatica* is one of the most important disease agents of domestic stock throughout the world and shows promise of remaining so for years to come. Losses are enormous because of mortality, reduction of milk and meat production, secondary bacterial infection, expensive antihelmintic treatment, and especially condemnation of livers.[35] For example, out of the nearly 6 million bovines slaughtered for food in Mexico between 1979 and 1987, 424,000 livers were confiscated.[7] And in a later study in Montana, over 17% of the livers were infected, as determined by inspection during slaughter.[28] Regardless of whether you like liver, that is a lot of high-quality food lost to the human population. However, in one study in Greece, involving 10,277 cattle processed at an abattoir, it was determined that the cost of antihelmintic treatment "far exceeded the economic loss due to condemnation resulting from parasitic infection."[52]

- **Pathology.** Little damage is done by juveniles penetrating the intestinal wall and the capsule surrounding the liver (Glisson's capsule), but much necrosis results from migration of flukes through the liver parenchyma. During this time, they feed on liver cells and blood. Anemia sometimes results from heavy infections. There is evidence that this anemia is not caused by hematophagia but instead by a chemical released, perhaps proline.[46] Deposition of bile duct collagen is stimulated by proline released from the worms.[58]

Worms in bile ducts cause inflammation and edema, which in turn stimulate production of fibrous tissue in the walls of these ducts (pipestem fibrosis; see Fig. 18.26). Thus thickened, the ducts can handle less bile and back pressure causes atrophy of liver parenchyma, with concomitant cirrhosis and possibly jaundice. In heavy infections the gallbladder is damaged, and bile duct walls are eroded completely through, with worms then reentering the parenchyma, causing large abscesses. Migrating juveniles frequently produce ulcers in ectopic locations, such as eyes, brain, skin, and lungs. Proteinases secreted by these worms (E/S, or Excretory/Secretory products) degrade extracellular matrix, but E/S product antigen is also useful in immunodiagnosis.[5]

- **Diagnosis and Treatment.** Whenever liver blockage coincides with a history of watercress consumption, fascioliasis should be suspected. Specific diagnosis depends on finding eggs (Fig. 17.6) in the stool. A false record can result when the patient has eaten infected liver and *F. hepatica* eggs pass through with feces. Daily examination during a liverfree diet will unmask this false diagnosis. An enzyme-linked immunosorbent assay (ELISA) test also is available. Early diagnosis is important to avoid irreparable damage to the liver. Prevention in humans depends on eschewing raw watercress. In domestic animals the problem is much more difficult to avoid. Snail control is always a possibility, although this is almost impossible in areas of high precipitation. Reservoir hosts, particularly rabbits, can maintain infestation of a pasture when pasture rotation is attempted as a control measure.

Excretory/secretory (E/S) antigens are useful both in immunodiagnostic tests and as potential vaccines. Some

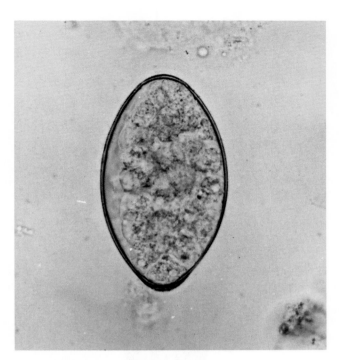

Figure 17.6   **Egg of *Fasciola hepatica*.**
Eggs of this species are 130 μm to 150 μm by 63 μm to 90 μm.
Courtesy of Jay Georgi.

Figure 17.7   *Fasciola gigantica,* a liver fluke.
They may be as large as 75 mm long.
Courtesy of Warren Buss.

monoclonal antibodies to E/S products cross-react with antigens from related helminths, but others are specific to *F. hepatica*, and antigens from bile and feces (copro-antigens) include a 26 kDa protein stable enough to use in diagnostic assays.[14] ELISA tests are available commercially and can detect anti-*F. hepatica* antibodies in serum and milk, but new ones, especially intended for use on fecal samples, are being developed, in part because of the expense of gathering serum samples from large herds. In one study, monoclonal antibodies to a 13–25 kDa *F. hepatica* E/S protein could detect single worms in sheep, and coproantigen concentration was correlated with known worm burden.[36] The ELISA tests also revealed infections up to five weeks earlier than did fecal egg counts.

Proteases secreted by *F. hepatica* also have been used experimentally in immunizing antigens. In studies using vaccines made from a combination of two such proteases, plus fluke hemoglobin, worm burdens and viability of eggs in surviving worms were both reduced.[11] In other studies, vaccines with vectors (plasmids) containing cDNA coding for *F. hepatica* cysteine proteases reduced worm numbers by up to 75% in rats.[57] In addition, *F. hepatica* possesses fatty acid binding proteins (FABP) and saposin-like lytic proteins, both of which have been used experimentally as vaccines.[15, 34] In some cases, these vaccines are crossreactive with *Schistosoma mansoni*.[2]

Experimental work on vaccines continues, especially with a focus on recombinant and DNA-based antigens, because if successful, vaccination reduces the need to locate, handle, and deliver medicines to free-ranging livestock. Recent studies with such vaccines show significant reductions of worm burdens in both cattle and sheep, not only in experimental infections, but also in pasture settings.[21, 33]

Several drugs are effective in chemotherapy of fascioliasis, both in humans and in domestic animals. One of these, rafoxanide, apparently acts by uncoupling oxidative phosphorylation in the fluke.[41] Triclabendazole is the current drug of choice; it binds to worm tubulin and thus interferes with processes involving microtubules.[17, 22] Praziquantel is ineffective against *F. hepatica*.

Other Fasciolid Trematodes

Fasciola gigantica (Fig. 17.7), a species that is longer and more slender than but otherwise very similar to *F. hepatica,* is found in Africa, Asia, and Hawaii, being relatively common in herbivorous mammals, especially cattle, in these areas. Morphology, biology, and pathology are nearly identical to those of *F. hepatica,* although different snail hosts are necessary. Strains of sheep differ in their susceptibility to *F. gigantica*. For example, Indonesian Thin Tail (ITT) sheep are relatively resistant because of a single gene that shows incomplete dominance.[43] Such sheep are not resistant to *F. hepatica,* however.[42] Cattle strains also differ in both susceptibility and in production loss due to infection.[56]

Fasciola jacksoni causes anemia, loss of weight, hypoproteinemia, abdominal and submandibular edema, and sometimes death in Asian elephants.[6]

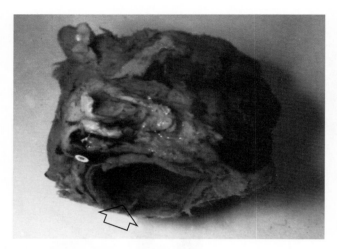

Figure 17.8 Large calcareous cyst from the liver of a steer (*arrow* points to its opening).

This cyst contained two *Fascioloides magna* and is nearly 8 cm wide.

Courtesy of Warren Buss.

Fascioloides magna is the giant in a family of large flukes, reaching nearly 7.5 cm in length and 2.5 cm in width. Formerly strictly an American species, it was first discovered in Europe in an American elk from an Italian zoo. Now the fluke has become established in Europe, mainly in game reserves. It is easily distinguished from *Fasciola* spp. by its large size and the absence of a cephalic cone and "shoulders." Its life cycle is similar to that of *Fasciola hepatica,* except that adults live in liver parenchyma rather than in bile ducts. Because of their large size, they cause extensive damage. Often, but not always, they become encased in a calcareous cyst of host origin (Fig. 17.8). Their excretory system produces great amounts of melanin, which fills their excretory canals and also the cyst containing them. Their normal hosts are probably elk and other Cervidae, but cattle are commonly infected in endemic areas. Human infections have not been found.

Fasciolopsis buski (Fig. 17.9) is a common parasite of humans and pigs in the Orient. Stoll[47] estimated 10 million human infections in 1947. The number may be greater today, with prevalences ranging up to near 60% in India and mainland China.[23] Although *F. buski* is a typical fasciolid, it is peculiar because it lives in the small intestine of its definitive host rather than in the liver. It is elongated and oval, reaching a length of 20 mm to 75 mm and a width of up to 20 mm. There is no cephalic cone or "shoulders." The acetabulum is larger than the oral sucker and is located close to it. Another difference from "typical" fasciolids is the presence of unbranched ceca. The dendritic testes are tandem in the posterior half of the worm. The ovary is also branched and lies in the midline anterior to the testes. Vitelline follicles are extensive, filling most of the lateral parenchyma all the way to the caudal end. The uterus is short, with an ascending limb only. Eggs are almost identical to those of *Fasciola hepatica.*

The life cycle of *F. buski* parallels that of *F. hepatica.* Each worm daily produces about 25,000 eggs (Fig. 17.10), which take up to seven weeks to mature and hatch at 27° C to 32° C. Several species of snails of genera *Segmentina* and *Hippeutis* (Planorbidae) serve as intermediate hosts.

Figure 17.9 *Fasciolopsis buski,* an intestinal fluke in the Fasciolidae.

It may reach 75 mm long.

Courtesy of Robert E. Kuntz.

Cercariae encyst on underwater vegetation, including cultivated water chestnut, water caltrop, lotus, bamboo, and other edible plants (Fig. 17.11). Metacercariae are swallowed when these plants are either eaten raw or peeled and cracked with the teeth before eating. The worms excyst in the small intestine, grow, and mature in about three months without further migration. Infection, then, depends on human or pig feces being introduced directly or indirectly into bodies of water in which edible plants grow.[44]

Disease conditions resulting from *F. buski* are immuno-pathologic, obstructive, and traumatic. Inflammation at the site of attachment provokes excess mucous secretion, which is a typical symptom of infection. Heavy infections block the passage of food and interfere with normal digestive juice secretions. Ulceration, hemorrhage, and abscess of the intestinal wall result from long-standing infections. Chronic diarrhea is symptomatic. Another aspect of disease is a sensitization caused by absorption of the worm's allergenic metabolites. This may eventually cause death of the patient. Treatment is usually effective in early or lightly infected cases. Late cases do not fare so well. Prevention is easy. Immersion of vegetables in boiling water for a few seconds will kill the metacercariae. Snail control should be attempted whenever it is impractical to prevent the use of nightsoil as a fertilizer. High prevalence is maintained, especially in school-age children, by a variety of social and economic factors, particularly poverty, lack of sanitation, and traditional dietary practices.[23]

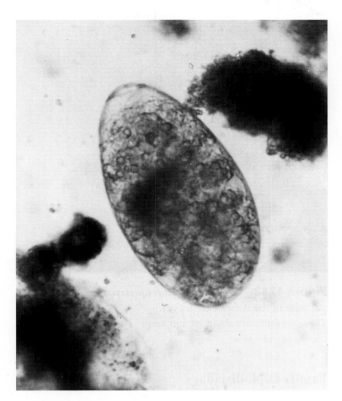

Figure 17.10 **Egg of *Fasciolopsis buski* in a human stool in Taiwan.**

Eggs of this species are 130 μm to 140 μm by 80 μm to 85 μm.

Courtesy of Robert E. Kuntz.

Figure 17.11 **Woman harvesting water caltrop.**

This serves as a medium for transport of metacercariae of *Fasciolopsis buski* to humans in Taiwan.

Courtesy of Robert E. Kuntz.

Family Cathaemasiidae

Ribeiroia ondatrae. *Ribeiroia ondatrae* is the species considered responsible for much of the recently observed deformity in frogs (Fig. 17.12*a*).[26] In some localities, high percentages of frogs have been observed to have obvious malformations—for example missing, extra, and misshapen limbs—particularly in

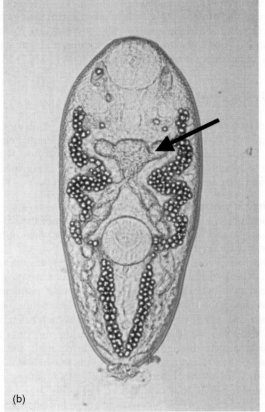

Figure 17.12 (*a*) Oregon spotted frog (*Rana pretiosa*) showing an extra but underdeveloped hind limb and missing toes. (*b*) Metacercaria of *Ribeiroia ondatrae* (approximately 0.7 mm long), showing esophageal diverticula (arrow) that are characteristic of *Ribeiroia* species.

Courtesy of Pieter Johnson.

their hind legs. There is considerable evidence that many if not most of these deformities are caused by *R. ondatrae*.[26]

First intermediate hosts are various planorbid snails; second intermediate hosts are fish and larval amphibians, including both frogs and salamanders. Tadpoles react to penetrating cercariae with rapid and angular swimming, but the junction between tail and body is a "dead water zone" where tadpoles have difficulty shedding the parasites.[51] Thus large numbers of metacercariae (Fig. 17.12*b*) can end up in the region where hind limb buds form. Larval amphibians are vulnerable to deformation for a short period during limb development, however, so that within a given pond, some individual amphibians are infected but not deformed. The actual mechanism by which deformation is produced has not been determined, but mechanical disruption of cells involved in limb bud formation may be sufficient to cause the deformities.[26] Definitive hosts for *R. ondatrae* are predators such as hawks, herons, and badgers, but muskrats, which are mostly vegetarian, can also serve in this role.

SUPERFAMILY PARAMPHISTOMOIDEA

Superfamily Paramphistomoidea contains "amphistomes," flukes in which the acetabulum is located at or near the posterior end. Usually they are thick, fleshy worms with their genital pore preequatorial and the ovary usually posttesticular. Species are found in fishes, amphibians, reptiles, birds, and mammals. Of several families in this group we will consider three.

Family Paramphistomidae

Members of family Paramphistomidae occur in mammals, especially herbivores. Several species in different genera of this family parasitize sheep, goats, cattle, cervids, water buffalo, elephants, and other important animals. One species was found once in humans.

Paramphistomum cervi. *Paramphistomum cervi* lives in the rumen of domestic animals throughout most of the world. Adults are almost conical in shape and are pink when living. The testes are slightly lobate.

The life cycle is similar to that of *Fasciola hepatica* and, in North America at least, these parasites develop in the same snail species. Cercariae are large and pigmented and have eyespots. There is no second intermediate host; metacercariae encyst on aquatic vegetation. When eaten, the worms excyst in the duodenum, penetrate the mucosa, and migrate anteriorly through the tissues. On reaching the abomasum, or true stomach, they return to the lumen and creep farther forward to the rumen. There they attach among villi and mature in two to four months.

Paramphistomum cervi is a particularly pathogenic species. Migrating juveniles cause severe enteritis and hemorrhage, often killing the host. Secondary bacterial infection often complicates the problem. No adequate prevention or treatment is known.

Three other *Paramphistomum* species occur in European cattle, the most common of these being *P. daubneyi*. Prevalence can range up to nearly 30%. Females are more often infected than males; the difference is attributed to the keeping of young males inside for rapid fattening and females being sent out to pasture early.[50]

Stichorchis subtriquetrus (Fig. 17.13) is a parasite of beavers, occurring throughout their range. Like *P. cervi*, metacercariae encyst on underwater objects, including sticks that beavers embed in the bottom of their ponds. When beavers later eat bark off these sticks, they swallow any attached metacercariae. A peculiarity of this worm's early embryogenesis is development of a mother redia within the miracidium. This redia is released in the snail immediately after penetration, and the remainder of the miracidium then disintegrates. Adult worms live in the stomach and reportedly have caused mortality in beavers in the former Soviet Union.

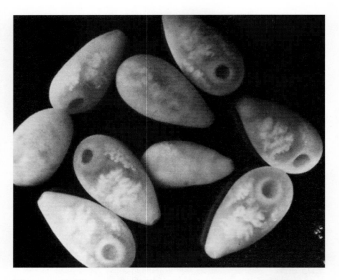

Figure 17.13 *Stichorchis subtriquetrus*, a stomach parasite of the American beaver.
These parasites are about 10 mm long.
Courtesy of Warren Buss.

Family Diplodiscidae

Flukes in family Diplodiscidae have a pair of posterior diverticula in the oral sucker.

Megalodiscus temperatus. *Megalodiscus temperatus* and other genera and species of amphistomes are common parasites of the rectum and urinary bladder of amphibians, especially frogs. They measure up to 6 mm long and 2 mm to 3 mm wide at the posterior end. Their posterior sucker is equal to about the greatest width of the body.

The life cycle of this species is similar to that of other amphistomes in that no second intermediate host is required. Miracidia hatch soon after the eggs reach water and penetrate snails of genus *Helisoma*. Cercariae have eyespots and swim toward lighted areas. If they contact a frog, they will encyst almost immediately on its skin, especially on its dark spots. Frogs molt the outer layers of their skin regularly and not infrequently will eat the sloughed skin. Metacercariae excyst in the rectum and mature in one to four months. If a tadpole eats a cercaria, the worm will encyst in the stomach and excyst when it reaches the rectum. At metamorphosis, when the amphibian's intestine shortens considerably, the flukes migrate anteriorly as far as the stomach and then to the rectum. These parasites are an easily obtained amphistome for general studies.

Family Gastrodiscidae

The morphology of family Gastrodiscidae is essentially similar to that of Paramphistomidae and perhaps should not be separate from it. One of its species is a common parasite of humans in restricted areas of the world.

Gastrodiscoides hominis This typical amphistome is cone shaped, fleshy, and pink. It is an important parasite of humans in India, southeastern Asia, and the Philippines, inhabiting the lower small intestine and upper colon. Rodents and primates are reservoirs.

Adult worms are 5 mm to 8 mm long by 5 mm to 14 mm wide at the ventral disc, which occupies about two-thirds of the ventral surface. There is a conspicuous posterior notch in the rim of the ventral sucker.

The complete life cycle of *G. hominis* is unknown, but the planorbid snail *Helicorbus coenosus* serves as an experimental host in India.[13] Presumably, humans are infected by eating uncooked aquatic plants. An adult worm draws a mass of mucosal tissue into the ventral sucker and remains attached for some time, causing a nipplelike projection on the intestinal lining.[1] The most common symptom is mucoid diarrhea. Treatment and prevention have not been well studied.

Learning Outcomes

By the time a student has finished studying this chapter, he or she should be able to:

1. Draw and label the life cycle of a typical member of genus *Echinostoma*.
2. Draw and label the life cycle of *Fasciola hepatica*.
3. Explain why *Fasciola hepatica* can be considered a zoonotic parasite and how human infections with this parasite typically occur.
4. Write an extended paragraph on the subject of control of fascioliasis in both humans and domestic animals.
5. Write an extended paragraph on the subject of parasite-induced morphological deformity in amphibians.

References

References for superscripts in the text can be found at the following Internet site: www.mhhe.com/robertsjanovynadler9e

Additional Readings

Boray, J. C. 1969. Experimental fascioliasis in Australia. In B. Dawes (Ed.), *Advances in parasitology 7*. New York: Academic Press, Inc., pp. 96–210. A very interesting general account of fascioliasis, with many experimental approaches. Accent is on special problems of Australia.

Connor, D. H., and R. C. Neafie. 1976. Fasciolopsiasis. In C. H. Binford and D. H. Connor (Eds.), *Pathology of tropical and extraordinary diseases* (vol. 2, sect. 10). Washington, D.C.: Armed Forces Institute of Pathology.

Horak, I. G. 1971. Paramphistomiasis of domestic ruminants. In B. Dawes (Ed.), *Advances in parasitology 9*. New York: Academic Press, Inc., pp. 33–72. A thorough discussion of paramphistomiasis, mainly outside North America.

Kendall, S. B. 1970. Relationships between the species of *Fasciola* and their molluscan hosts. In B. Dawes (Ed.), *Advances in parasitology 8*. New York: Academic Press, Inc., pp. 251–258. This short article brings the earlier one [1966] by this author up-to-date.

Reinhard, E. G. 1957. Landmarks of parasitology. I. The discovery of the life cycle of the liver fluke. *Exp. Parasitol.* 6:208–232.

Roberts, J. A., and Suhardono. 1996. Approaches to the control of fasciolosis in ruminants. *Int. J. Parasitol.* 26:971–981.

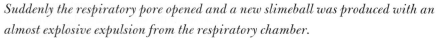

Chapter 18

Digeneans: Plagiorchiformes and Opisthorchiformes

Suddenly the respiratory pore opened and a new slimeball was produced with an almost explosive expulsion from the respiratory chamber.

—W. H. Krull and C. R. Mapes,[29] describing a snail infected with *Dicrocoelium dendriticum*

In this chapter we discuss two orders of trematodes, Plagiorchiformes and Opisthorchiformes. Members of these orders occur in a wide variety of vertebrate definitive hosts. Their cercariae all encyst in second intermediate hosts, in contrast to many Echinostomatiformes, that encyst on vegetation or other substrates.

ORDER PLAGIORCHIFORMES

In contrast to adults of some other orders, adults of Plagiorchiformes often show little resemblance to each other. However, they have many larval and juvenile similarities. The wall of the excretory bladder is epithelial. Cercariae have a simple tail with a dorsal finfold, the primary excretory vesicle extends a short distance into the tail, and an oral stylet is usually present (xiphidiocercariae). In most members of the order, eggs are small and must be eaten by a snail intermediate host before hatching.

Brooks and McLennan[7] include six suborders in Plagiorchiformes of which we will describe some examples from Plagiorchiata and Troglotrematata. Members of Troglotrematata retain the homoplasious character of free-swimming miracidia. Plagiorchiata is one of the most derived taxa of Digenea, and sequence analysis of its large-subunit ribosomal DNA indicates that Plagiorchiata may be paraphyletic.[48]

Suborder Plagiorchiata

Members of Plagiorchiata parasitize fishes, amphibians, reptiles, birds, and mammals and thereby are among the most commonly encountered flukes. They inhabit hosts in marine, freshwater, and terrestrial environments. Most are parasites of wild animals, but a few are important disease agents of humans and domestic animals. Cercariae possess an oral stylet, and metacercariae usually encyst in an invertebrate intermediate host. We discuss only a few of the many families in the suborder.

Family Dicrocoeliidae

Dicrocoeliidae is one of the three major families of liver flukes that we will consider (see also Fasciolidae and Opisthorchiidae). Some species parasitize the gallbladder, pancreas, or intestine. All are parasites of terrestrial or semiterrestrial vertebrates and use land snails as first intermediate hosts. All dicrocoeliids are medium sized and flattened, with a subterminal oral sucker and a powerful acetabulum in the anterior half of the body. The body is usually pointed at both ends. Ceca are simple. Testes are preequatorial, and the ovary is posttesticular. The voluminous uterus has both a descending and ascending limb, commonly filling most of the medullary parenchyma.

Dicrocoeliids are common among a wide variety of familiar vertebrates, but they rarely parasitize humans. One cosmopolitan species is an important parasite of domestic mammals and occasionally of humans.

Dicrocoelium dendriticum. *Dicrocoelium dendriticum* (Fig. 18.1) is common in the bile ducts of sheep, cattle, goats, pigs, and cervids. It is common throughout most of Europe and Asia and has foci in North America and Australia, where it was recently introduced. Ducommun and Pfister[15] found over 45% of cattle in Switzerland were infected with liver flukes, most of which were *D. dendriticum.* Its importance is often underestimated.[35] The trematode is commonly known as the *lancet fluke* because of its bladelike shape.

- **Morphology.** *Dicrocoelium dendriticum* is 6 mm to 10 mm long by 1.5 mm to 2.5 mm at its greatest width, near the middle. Both ends of its body are pointed. The ventral sucker is larger than the oral sucker and is located near it. The large, lobate testes lie almost in tandem directly behind the acetabulum, and the small ovary lies immediately

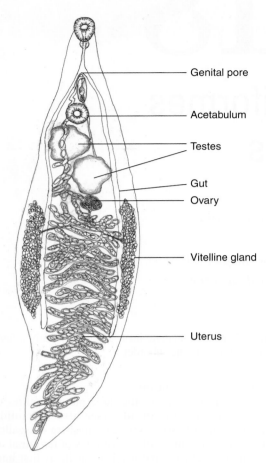

Genital pore

Acetabulum

Testes

Gut

Ovary

Vitelline gland

Uterus

Figure 18.1 *Dicrocoelium dendriticum,* **a liver fluke of mammals.**

Drawing by John Janovy, Jr.

behind them. Loops of the uterus fill most of the body behind the ovary. Vitellaria are lateral and restricted to the middle third of the body. The operculate eggs are 36 μm to 45 μm by 22 μm to 30 μm.

• **Biology.** *Dicrocoelium dendriticum* is a fascinating example of a trematode that has dispensed with any requirement for an aquatic environment at all stages of its life cycle. Adult *D. dendriticum* live in bile ducts within the liver, much like *Fasciola hepatica.* When laid, eggs contain miracidia and must be eaten by land snails before they will hatch (Fig. 18.2). *Cionella lubrica* is evidently the most important snail host in the United States, but *D. dendriticum* shows little specificity as to snail host. On hatching in the snail's intestine, a miracidium penetrates the gut wall and transforms into a mother sporocyst in the digestive gland. Mother sporocysts produce daughter sporocysts, which in turn produce xiphidiocercariae. The fact that these cercariae possess well-developed tails probably indicates a recent aquatic origin.

About three months after infection, cercariae accumulate in the "lung" (mantle cavity) of the snail or on its body surface. The snail surrounds this irritant with thick mucus and eventually deposits these cercariae-containing slimeballs as it crawls along. Each slimeball can be

expelled from the snail's pneumostome (mantle cavity opening) with some force. Slimeballs are most abundantly produced during a wet period immediately following a drought. Individual slimeballs may contain up to 500 cercariae each. Drying of the slimeball surface retards desiccation of the interior and thereby prolongs the lives of cercariae within.

Continued development of the fluke depends on its ingestion by an ant, which becomes the second intermediate host. The common brown ant *Formica fusca* is the arthropod host in North America. On eating the apparently delectable slimeballs or feeding them to their larvae, ants become host to metacercariae, most of which encyst in the hemocoel and are then infective to definitive hosts. Over 100 metacercariae may occur in a single ant. One or two, however, migrate to the subesophageal ganglion and encyst there.[31] These will become so-called brainworms. They are not infective, but they alter the ant's behavior in a most remarkable way. When the temperature drops in the evening, ants thus infected climb to the tops of grasses and other plants, and their mandibular muscles undergo tetanic spasm, firmly grasping the plant (Fig. 18.3).[31]

Infected ants remain attached until later the next day when they warm up and seemingly resume normal behavior. This behavioral pattern keeps infected ants near the tops of vegetation during active periods of grazing by ruminants in the evening and morning hours but allows them to retreat to cooler places during hot hours of the day. The parasite thus influences its intermediate host to behave in a manner that increases the probability of passage to the definitive host.

A similar phenomenon is found when metacercariae of the dicrocoeliid *Brachylecithum mosquensis,* a parasite of American robins, encyst near the supraesophageal ganglion of carpenter ants, *Campanotus* spp. Instead of retreating from brightly lighted areas, as is normal for these ants, infected individuals actually seek such places and wander aimlessly or in circles on exposed surfaces. This behavior makes them much more obvious to the bird definitive host than they would be otherwise.[10]

On being eaten by a definitive host, *D. dendriticum* excysts in the duodenum. Evidently it is attracted by bile and quickly migrates upstream to the common bile duct and thence into the liver. The flukes mature in sheep in six or seven weeks and begin producing eggs about a month later. Up to 50,000 *D. dendriticum* have been found in a single sheep.

Pathological conditions of dicrocoeliiasis are basically the same as those for fascioliasis, except that there is no trauma to the gut wall or liver parenchyma resulting from migrating juveniles. General biliary dysfunction, with several symptoms, such as bile duct inflammation and fibrosis and hepatocyte degeneration, is typical.

Numerous cases of *D. dendriticum* in humans have been reported. Most of these were false infections. That is, the eggs that were detected in the stool were actually part of a liver repast that the person had enjoyed a few hours earlier. There have been some genuine infections in humans, however, mainly in Russia, Europe, Asia, and Africa. One human case was reported in New Jersey.[14] Many cases of human infection with a related species, *D. hospes,* have been reported from Africa.[25]

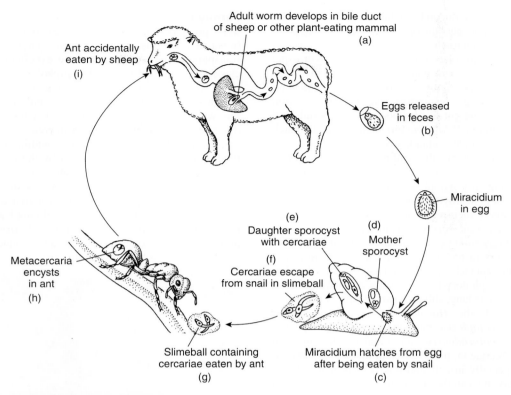

Figure 18.2   **Life cycle of *Dicrocoelium dendriticum*.**

(*a*) Adult, in bile duct of sheep or other plant-eating mammal. (*b*) Egg released in feces. (*c*) Miracidium hatching from egg after being eaten by snail. (*d*) Mother sporocyst. (*e*) Daughter sporocyst. (*f*) Cercariae escaping from snail in slimeball. (*g*) Slimeballs containing cercariae eaten by ant. (*h*) Metacercaria encysting in ant. (*i*) Ant accidentally eaten by sheep.

Drawing by William Ober and Claire Garrison.

Figure 18.3   **Ants *Formica rufibarbis*.**

The ants are infected with metacercariae of *Dicrocoelium dendriticum*. In the evening their mandibles lock on plants where they are available to infect grazing definitive hosts.

Courtesy of M. Y. Manga-González, from M. Y. Manga-González, C. González Lanza, E. Cabanas, and R. Campo, "Contributions to and review of dicrocoeliosis, with special reference to the intermediate hosts of *Dicrocoelium dendriticum*," in *Parasitology* 123:S91–S114, 2001. Reprinted with permission of Cambridge University Press.

Some benzimidazoles and praziquantel are effective against this trematode in domestic animals but at a high dose rate that may not be economically feasible.[6] Praziquantel is the drug of choice for humans.[8] Control promises to be difficult in the foreseeable future because of the ubiquity of land snails and ants.

Family Haematoloechidae

Flukes of family Haematoloechidae are parasitic in lungs of frogs and toads. They are of no economic or medical importance to humans, but, because of their large size and easy availability, they are often the first live parasites seen by biology students. Their transparent beauty, enigmatic location, and fascinating biology have led more than one biology student into a career in parasitology. In fact, the parasitology literature abounds with first papers by later famous scientists dealing with medically unimportant organisms. Indeed, it might be argued that these parasites are important in that they have served to attract serious scientists into the discipline.

Haematoloechus medioplexus (Fig. 18.4) is typical among the more than 40 known *Haematoloechus* species that are found in all parts of the world where amphibians occur. It is a flat, nonmuscular worm up to 8.0 mm long and 1.2 mm wide. The acetabulum is small and inconspicuous in this and related species. The uterus is voluminous, with a descending limb reaching to near the posterior end and then ascending with wide loops to the genital pore near the oral sucker. So many eggs fill the uterus that most internal organs are obscured. However, when living worms are placed in tap water, they will expel most eggs and thereby become transparent enough to study.

Adult flukes lay prodigious numbers of eggs, which are carried out of the frog's respiratory tract by ciliary action and thence through the gut to the outside. When swallowed by a scavenging *Planorbula armigera* snail, miracidia hatch and migrate to the hepatic gland, where they develop into sporocysts. Cercariae escape the snail by night and live a free life for up to 30 hours. When sucked into the rectal branchial chamber of a dragonfly nymph, cercariae penetrate the thin cuticle and encyst in nearby tissues.

Although many metacercariae are lost when the dragonfly larva molts, some survive metamorphosis. Frogs can thus get infected by eating either larval or adult dragonflies.[5] Excystation occurs in the frog's stomach. The little flukes creep through the stomach, up the esophagus, through the glottis, and into the respiratory tree. As many as 75 worms have been found in a single lung, although two or three is average.

Life cycles of other *Haematoloechus* spp. are similar to those of *H. medioplexus,* but cercariae of some species, for example *H. coloradensis,* can actually penetrate, and become metacercariae in, many different kinds of aquatic invertebrates, especially insects and microcrustaceans. These *Haematoloechus* spp. can thus infect frogs too small to eat an adult dragonfly.[4]

Family Prosthogonimidae

Most prosthogonimids are parasites living in the oviduct, bursa of Fabricius, or gut of birds. They are remarkably transparent, stain well, and make excellent examples for classroom studies of trematode morphology.

Prosthogonimus macrochis (Fig. 18.5) is about 8 mm long and is found in the oviduct of domestic fowl and various wild birds in North America. It causes considerable damage to an oviduct and can decrease or even prevent egg laying. Many have been found within eggs after being trapped in the membranes formed by the oviduct, presumably giving cooks an unexpected surprise.

The life cycle of *P. macrochis* is similar to that of *Haematoloechus* spp. When embryonated eggs pass into water, they sink to the bottom. They do not hatch until eaten by a snail (*Amnicola* spp.). Then they burrow into the digestive gland, become sporocysts, and produce short-tailed

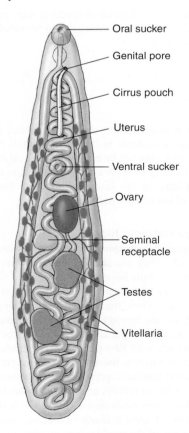

Figure 18.4 *Haematoloechus medioplexus,* common in the lungs of frogs.

Drawing by William Ober and Claire Garrison.

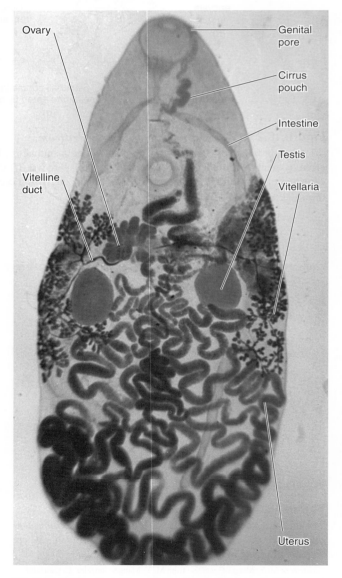

Figure 18.5 *Prosthogonimus macrochis,* an oviduct fluke of birds.

Courtesy of Warren Buss.

xiphidiocercariae. When these cercariae are sucked into the rectal branchial chamber of a dragonfly nymph, they attach and penetrate into the hemocoel and encyst in muscles of the body wall, remaining infective after the insect metamorphoses. When eaten by a bird, they excyst in the intestine, migrate downstream to the cloaca and into the bursa of Fabricius or oviduct, and mature in about a week. In male birds the infection is lost when the bursa atrophies. In the United States prevalence of *P. macrorchis* is declining in confined spaces and are thus no longer able to pursue dragonfly prey. More than 30 species of *Prosthogonimus* are known from various areas around the world.

Some Other Plagiorchiata

Another family of flukes often encountered in wild animals is Plagiorchiidae. They parasitize vertebrate classes from fishes through birds and mammals. All species use aquatic snails as first intermediate hosts and insects, such as mayflies and dragonflies, as usual second intermediate hosts. The literature on plagiorchiids is replete with variations of life cycles and descriptions of the many species of worms encountered. Evidently, this is an evolutionarily plastic group, adaptable to different situations. Representative species are *Plagiorchis muris* in dogs, rats, and a variety of birds (Fig. 18.6); *P. maculosus,* cosmopolitan in swallows; *P. nobeli* in American blackbirds; and *Neoglyphe soricis* in shrews. Several species of *Plagiorchis* have been reported from humans in Korea, Japan, Indonesia, Thailand, and the Philippines.[22]

Snakes that feed on amphibians, which bear the metacercariae, are often infected by members of Ochetosomatidae: species of *Ochetosoma* (= *Renifer*) and *Dasymetra* in mouth and esophagus and *Pneumatophilus* and *Lechriorchis* in trachea and lung.

Suborder Troglotrematata

Family Troglotrematidae

Troglotrematidae are oval, thick flukes with a spiny tegument and dense vitellaria. They are parasites of lungs, intestine, nasal passages, cranial cavities, and various ectopic locations in birds and mammals in many parts of the world. We will examine the biology of this interesting group with discussions of two species.

***Paragonimus* spp.** Paragonimiasis is an excellent example of a **zoonosis.** About 48 species and subspecies of *Paragonimus* have been described as parasites of carnivorous mammals, but not all of these may be valid. Seven species have been recorded from humans from three main foci: Asia and Oceania *(P. westermani, P. skrjabini, P. miyazakii,* and *P. heterotremus);* western, sub-Saharan Africa *(P. africanus* and *P. uterobilateralis).*[9] A total of nine *Paragonimus* species have been described from South and Central America, but there is a disagreement about the validity of several.[50] Several million people are infected in Asia.

Paragonimus westermani is the most widely prevalent species. It was first described from two Bengal tigers that died in zoos in Europe in 1878. During the next two years, infections by this worm in humans were found in Formosa. It was very quickly found in lungs, brain, and viscera of humans in Japan, Korea, and the Philippines. The life cycle was worked out by Kobayashi[26] and Yokagawa.[52]

- **Morphology.** Adult worms (Fig. 18.7) are 7.5 mm to 12.0 mm long and 4 mm to 6 mm at their greatest width. They are very thick, measuring 3.5 mm to 5.0 mm in the dorsoventral axis. In life they are reddish brown, lending them the overall size, shape, and color of coffee beans.

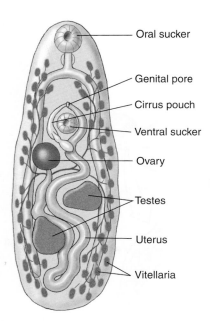

Figure 18.6 *Plagiorchis muris,* **a common parasite of swallows.**

Birds and some mammals are infected when they eat arthropods containing metacercariae.

Drawing by William Ober and Claire Garrison.

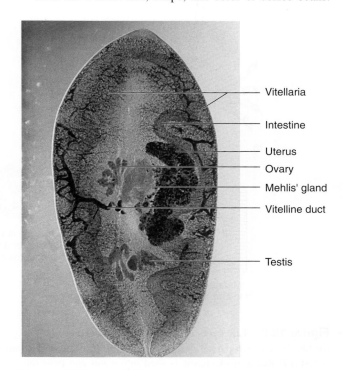

Figure 18.7 Adult *Paragonimus westermani.*

Courtesy of Robert E. Kuntz and Jerry A. Moore.

Their tegument is densely covered with scalelike spines. Oral and ventral suckers are about equal in size, with the latter placed slightly preequatorially. The excretory bladder extends from the posterior end to near the pharynx. The lobated testes are at the same level, located at the junction of the posterior fourth of the body. A cirrus and cirrus pouch are absent. Their genital pore is postacetabular.

The ovary is also lobated and is found to the left of midline, slightly postacetabular. The uterus is tightly coiled into a rosette at the right of the acetabulum and opens into a common genital atrium with the vas deferens. Vitelline follicles are extensive in lateral fields, from the level of the pharynx to the posterior end. Eggs are ovoid and have a rather flattened operculum set into a rim. They measure 80 μm to 118 μm by 48 μm to 60 μm.

Identification of the 30 or so species of *Paragonimus* is difficult, with much emphasis being placed on characters of metacercariae and shape of the tegumental spines.[32] Several nominal species are probably synonyms. There are several genetically distinct populations of *P. westermani* in east and Southeast Asia.[2]

- **Biology.** Adult *Paragonimus* spp. (Fig. 18.8) usually live in the lungs, encapsulated in pairs by layers of

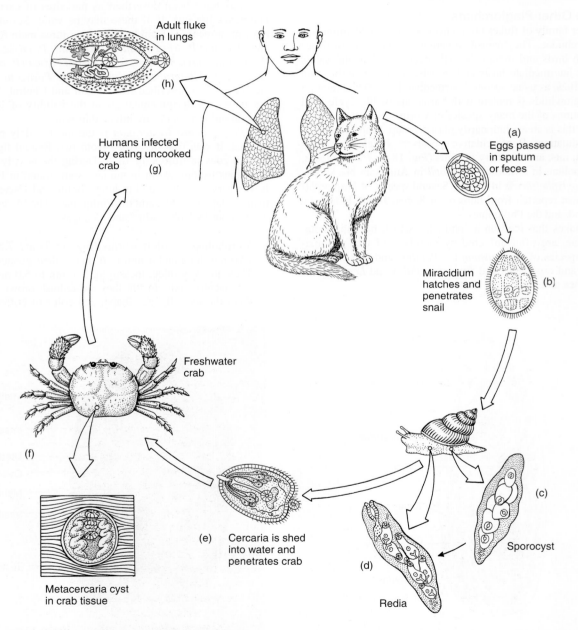

Figure 18.8   **Life cycle of *Paragonimus westermani.***
(*a*) Shelled embryo passed in feces or sputum. (*b*) After development miracidium hatches spontaneously and penetrates snail. (*c*) Sporocyst. (*d*) Redia. (*e*) Cercaria is shed into water and penetrates crab. (*f*) Metacercarial cyst in tissue of freshwater crab. (*g*) Cats or humans infected by eating uncooked crab. (*h*) Adult fluke in lungs.

Drawing by William Ober and Claire Garrison.

granuloma (Fig. 18.9). They sometimes occur in many other organs of the body, however (Fig. 18.10). Cross-fertilization normally occurs. Some *P. westermani* are triploid and tetraploid; triploid individuals are unable to form sperm, but reproduce parthenogenetically.[2, 49] The tetraploid condition evidently arises when a triploid oocyte is fertilized by a sperm from a diploid fluke. Eggs (Fig. 18.11) are often trapped in surrounding tissues and cannot leave the lungs, but those that escape into air passages are moved up and out by the ciliary epithelium. Most eggs escape before encapsulation is complete. Arriving at the pharynx, they are swallowed and passed through the alimentary canal to be voided with feces. Larvae within eggs require from 16 days to several weeks in water before development of miracidia is complete.

• Hatching is spontaneous, and miracidia must encounter a snail in family Thieridae if they are to survive. Because these snails usually live in swift-flowing streams, chances of survival of any miracidium are slight. This problem is offset by the numbers of eggs produced by an adult. On entering a snail, a miracidium forms a sporocyst that produces rediae, which in turn develop many cercariae. These cercariae (Fig. 18.12) are microcercous, with spined, knoblike tails and minute oral stylets.

After escaping from a snail, cercariae become quite active, creeping over rocks in inchworm fashion, and attack crabs and crayfish of at least 11 species, encysting in the viscera and muscles. A common second intermediate host in Taiwan is *Eriocheir japonicus* (Fig. 18.13). Some evidence suggests that crustaceans may become infected by eating infected snails.[33] Metacercariae (Fig. 18.14) are pearly white in life and can be identified to species by an expert. When the crustacean is eaten by an appropriate definitive host, the worms excyst in the duodenum; they produce cysteine protease, which apparently facilitates excystment.[12] After excystment, juveniles penetrate the intestine and embed themselves in the abdominal wall. Several days later they reenter the coelom and penetrate the diaphragm and pleura. They find their mate in the pleural spaces, and if a potential mate is not present, subadults can wait in there for some

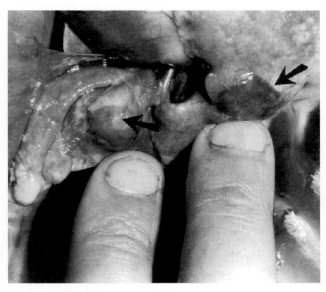

Figure 18.9   **Lung of a cat with two cysts containing adult *Paragonimus westermani* (arrows).**

Courtesy of Robert E. Kuntz.

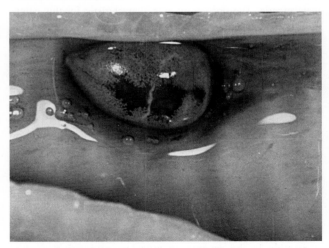

Figure 18.10   **Adult *Paragonimus westermani* in the trachea of an experimentally infected cat.**

Courtesy of Robert E. Kuntz.

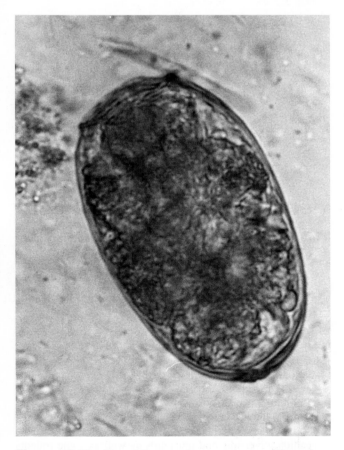

Figure 18.11   **Egg of *Paragonimus westermani* from the feces of a cat.**

Eggs average 87 μm by 50 μm.

Courtesy of Robert E. Kuntz.

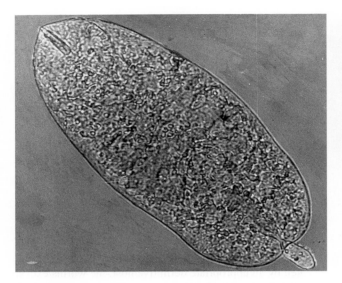

Figure 18.12   Microcercous cercaria of *Paragonimus westermani*.

It is about 500 μm long.

Courtesy of Robert E. Kuntz.

Figure 18.13   *Eriocheir japonicus,* second intermediate host for *Paragonimus westermani* in Taiwan.

Courtesy of Robert E. Kuntz.

(a)

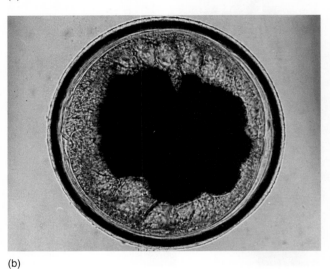

(b)

Figure 18.14   Metacercariae of *Paragonimus westermani*.

(*a*) Several metacercariae in a gill filament of a crab. (*b*) A single metacercaria. The opaque mass is characteristic for this genus. The size is 340 μm to 480 μm.

Courtesy of Robert E. Kuntz.

weeks until such time as another worm may arrive.[2] Finally, after penetrating a lung, the pair forms a cyst together. They normally mature in 8 to 12 weeks. Wandering juveniles may locate in ectopic locations, such as brain, mesentery, pleura, or skin.

- **Epidemiology and transmission ecology.**   Natural, definitive, and, therefore, reservoir hosts of *Paragonimus* spp. are several species of carnivores, including felids, canids, viverrids, and mustelids as well as some rodents and pigs. Humans are probably a lesser source of infective eggs than are other mammals, but, like the others, humans become infected when they eat raw or insufficiently cooked crustaceans. Crab collectors in some

countries distribute their catch miles from their source, effectively propagating paragonimiasis (Fig. 18.15). In addition, crabs exported internationally for consumption in sushi restaurants can also be a source of infection in people who have never traveled to Asia. In one unusual case, a raw crab in a martini was evidently the source of metacercariae.[3, 49]

Completely raw crabs and crayfish are not as commonly eaten in Southeast Asia as are those prepared by marination in brine, vinegar, or wine, which coagulates protein in the crustacean muscles, giving it a cooked appearance and taste but not affecting the metacercariae. Exposure commonly is effected by contamination of fingers or cooking utensils during food preparation

Figure 18.15   Crab collectors stringing crabs for a trip to market in Taiwan.

This practice distributes *Paragonimus* far from its source.

Courtesy of Robert E. Kuntz.

Figure 18.16   Children "cooking" fresh-caught crabs on an open fire.

Such practices contribute to the widespread prevalence of infection in eastern Asia.

Courtesy of Robert E. Kuntz.

(Fig. 18.16).[46] It is even possible that people accidentally become infected when they smash rice-eating crabs in the paddies, splashing themselves with juices that contain metacercariae. Another factor of possible epidemiological

significance in some ethnic groups is the medicinal use of juices strained from crushed crabs or crayfish.

A variety of mammals and some birds can serve as paratenic hosts, and ingestion of a paratenic host is probably the means by which large carnivores, such as tigers, become infected.[2] Paratenic hosts can serve as a source of infection for people who have never eaten freshwater crustaceans. Guinea pigs, considered a delicacy in the Andean region, are paratenic hosts in Ecuador and Peru.[21]

- **Pathology.**   The early, invasive stages of paragonimiasis cause few or no symptomatic pathological conditions. Once in a lung or an ectopic site, the worm stimulates an inflammatory response that eventually enshrouds it in a capsule of granulation tissue. Such capsules often ulcerate and heal slowly. Eggs in surrounding tissues themselves become centers of pseudotubercles. Worms in the spinal cord can cause paralysis, which sometimes is total. Fatal cases of *Paragonimus* spp. in the heart have been recorded. Cerebral cases have the same results as those of cerebral cysticercosis (see p. 333).[30] Pulmonary cases usually cause chest symptoms, with breathing difficulties, chronic cough, and sputum containing blood or brownish streaks (fluke eggs). Fatal cases are rare in pulmonary paragonimiasis.[9]

- **Diagnosis and Treatment.**   The only sure diagnosis, aside from surgical discovery of adult worms, is by finding the highly characteristic eggs in sputum, aspirated pleural fluid, feces, or matter from a *Paragonimus*-caused ulcer. Pulmonary infection is easily mistaken for tuberculosis, pneumonia, spirochaetosis, and other such illnesses; and X-ray examination may be incorrectly interpreted. Cerebral involvement requires differentiation from tumors, cysticercosis, hydatids, encephalitis, and others. Seroimmunological diagnosis is useful and particularly valuable in detecting ectopic infection. Intradermal tests are practiced for surveys but must be followed by other assays on people testing positive, because a dermal reaction persists for long periods after recovery from the disease. An assay that detects worm antigens using a monoclonal antibody is now available.[53]

The drug of choice is praziquantel.[9] Clinical symptoms decrease after five to six years of infection, but worms can live for 10 to 20 years. Infection can be avoided by cooking crustaceans before eating them and by avoiding contamination with their juices.

Paragonimus kellicotti closely resembles *P. westermani*. It has been found in a wide variety of mammals (cat, dog, raccoon, opossum, skunk, mink, muskrat, bobcat, pig, goat, red fox, coyote, weasel) in North America east of the Rocky Mountains, and many details of its life history and pathogenesis are known.[43, 45] The first intermediate host is *Pomatiopsis lapidaria*. Crayfish of the common genus *Cambarus* serve as second intermediate hosts, with metacercariae usually encysting on the heart. Like *P. westermani*, *P. kellicotti* in the definitive hosts is usually found in cysts occupied by pairs of worms in the lung. How the migrating worms find each other is still unknown, but encounter with another worm may be necessary for them both to mature.[43] One case of infection in a human has been reported.[9]

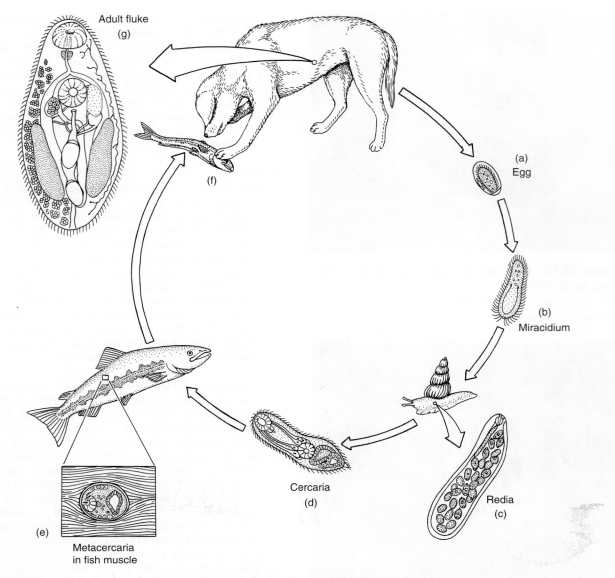

Figure 18.17   **Life cycle of *Nanophyetus salmincola* and of the neorickettsia it harbors.**

(*a*) Shelled embryo is passed in feces. (*b*) After development, miracidium hatches spontaneously. (*c*) Redia. (*d*) Cercaria leaves snail and penetrates fish. (*e*) Metacercaria in fish muscle. (*f*) Dog eats raw fish. (*g*) Adult fluke in small intestine of dog.

Drawing by William Ober and Claire Garrison after C. B. Philips, "Canine rickettsiosis in western United States and comparison with a similar disease in the Old World," in *Arch. Inst. Pasteur Tunis.* 36:505–603, 1959.

Nanophyetus salmincola. By 1814 people realized that, when dogs ate raw salmon in northwestern North America, they were prone to a disease so severe that scarcely 1 in 10 survived.[20] Over 100 years later an association with a minute fluke (Fig. 18.17) was found.[13] Metacercariae were common in the flesh and viscera of salmon. We now know that the disease itself is due to a rickettsia, *Neorickettsia helminthoeca,* which is transmitted to dogs by the flukes. Infected dogs can be treated effectively with sulfanilamides and antibiotics.

- **Morphology.**   Adult worms are 0.8 mm to 2.5 mm long and 0.3 mm to 0.5 mm wide. The oral sucker is slightly larger than the midventral acetabulum. Testes are side by side in the posterior third of the body. A cirrus pouch is present, but there is no cirrus. The small ovary is lateral to the acetabulum, and the uterus is short, containing only a few eggs at a time.

- **Biology.**   Adult *Nanophyetus salmincola* live deeply embedded in crypts in the wall of the small intestine of at least 32 species of mammals, including humans, as well as of fish-eating birds. They produce unembryonated eggs that hatch in water after 87 to 200 days. The snail host in the northwestern United States is *Oxytrema silicula,* an inhabitant of fast-moving streams. Experimental infection of snails in the laboratory has not been accomplished. Sporocysts have not been found, but rediae are well-known, occurring in nearly all tissues of the snail. The xiphidiocercaria is microcercous. It penetrates and encysts in at least 34 species of fish, but salmonid fish are more susceptible than are fish of other families. Metacercariae can be found in nearly any tissue of the fish, but they are most numerous in kidneys, muscles, and fins. Young fish have a high rate of mortality in heavy infection.[18] A variety of mammals and even two bird species (heron and merganser) can be infected with the trematode, but raccoons and spotted skunks are clearly the main definitive hosts in nature.[40] The worm has been reported from humans in North America on at least 10 occasions and infects up to 98% of the people in some villages in Siberia.[16]

- **Pathology.**   Adult flukes themselves cause surprisingly little disease. Philip[37] noted that inflammatory changes in the intestine of a dog carrying hundreds of *N. salmincola* were no more extensive than in animals infected with salmon poisoning disease by injection with lymph node suspensions. Salmon poisoning disease is restricted to dogs, coyotes, and other canids and does not affect humans. Its course in dogs is rapid and severe. After an incubation period of 6 to 10 days, the dog's temperature rises to 40°C to 42°C, often with edematous swelling of the face and discharge of pus from the eyes. An infected dog exhibits depression, loss of appetite, and increased thirst. Vomiting and diarrhea begin four to seven days after onset of symptoms. Fever usually lasts from four to seven days, and the dog usually dies about 10 days to two weeks after onset; however, those that recover are immune for the rest of their lives.

    Much remains to be learned about the biology of this fluke and the rickettsia it harbors. Dogs are extremely susceptible, and, when untreated, their mortality is about 90%. The disease can be transmitted experimentally by injection of lymph node preparations from other infected dogs or by injecting eggs (evidence of transovarial transmission in the fluke), metacercariae, or adult flukes and digestive glands from infected snails.[34] The geographical range of the disease coincides with distribution of the snail host of the fluke, and the proportion of salmonids within this range that are infected is extremely high. In light of these facts and the mortality in dogs, we can assume that there is some reservoir of the rickettsia, but identity of that reservoir is not at all clear. Raccoons do not seem to be susceptible; after fluke infection or injection with infected lymph nodes, they have a transitory, low-grade fever, but attempts to transmit the disease from them to dogs by way of lymph node preparations were unsuccessful.[37]

ORDER OPISTHORCHIFORMES

These are medium to small flukes, often spinose and with poorly developed musculature. Testes are at or near the posterior end, and a cirrus pouch and cirrus are absent. A seminal receptacle is present, and metraterm and ejaculatory ducts unite to form a common genital duct. Eggs are embryonated when passed, but hatching occurs only after ingestion by a suitable snail. Adults live in the intestine or biliary system of fishes, reptiles, birds, and mammals. Metacercariae are in fishes.

Family Opisthorchiidae

Opisthorchiids are delicate, leaf-shaped flukes with weakly developed suckers. Most are exceptionally transparent when prepared for study and so are popular subjects for parasitology classes. Adults are in the biliary system of reptiles, birds, and mammals. Three species, assigned to genera *Clonorchis* and *Opisthorchis,* are of substantial consequence to humans. Over 30 million people are infected with these flukes.[23]

Clonorchis sinensis.   *Clonorchis sinensis* was first discovered in bile passages of a Chinese carpenter in Calcutta in 1875. Other infections were quickly discovered in Hong Kong and Japan. Although some authors have used the name *Opisthorchis sinensis,* Looss erected the genus *Clonorchis* in 1907 on the basis of its branched testes in contrast to the lobed testes of *Opisthorchis.*[27] Although *Clonorchis* and *Opisthorchis* have been long considered separate genera, molecular genetic studies have shown that they are "extremely closely related to each other."[36]

Today we know that *C. sinensis* is widely distributed in Japan, Korea, China, Taiwan, and Vietnam, where it causes untold suffering and economic loss. Reports of this parasite outside eastern Asia involve infections people acquired while visiting there or by eating frozen, dried, or pickled fish imported from endemic areas. Prevalence of infection among 150 New York City immigrant Chinese was 26%.

- **Morphology.**   Adults (Fig. 18.18) measure 8 mm to 25 mm long by 1.5 mm to 5.0 mm wide. The tegument lacks spines, and musculature is weak. The oral sucker is slightly larger than the acetabulum, which is about a fourth of the way from the anterior end.

    The male reproductive system consists of two large, branched testes in tandem near the posterior end and a large, serpentine seminal vesicle leading to the genital pore. A cirrus and cirrus pouch are absent. The pretesticular ovary is relatively small and has three lobes. The seminal receptacle is large and transverse and is located just behind the ovary. The uterus ascends in broad, tightly packed loops and joins the ejaculatory duct to form a short, common genital duct. The genital pore is median, just anterior to the acetabulum. Vitelline follicles are small and dense and are confined to the level of the uterus. A Laurer's canal is conspicuous.

- **Biology.**   Human liver flukes (Fig. 18.19) mature in the bile ducts and produce up to 4000 eggs per day for at least six months. The mature egg (Fig. 18.20) is yellow-brown, 26 μm to 30 μm long and 15 μm to 17 μm wide. The operculum is large and fits into a broad rim of the eggshell. There is usually a small knob or curved spine on the abopercular end that helps distinguish eggs of this species. When passed, eggs contain a well-developed miracidium that is rather asymmetrical in its internal organization.

    Hatching of a miracidium will occur only after the egg is eaten by a suitable snail, of which *Parafossarulus manchouricusis* the most common and, therefore, most important first intermediate host throughout east Asia. A miracidium transforms into a sporocyst in the wall of the intestine or in other organs within four hours of infection. Sporocysts produce rediae within 17 days. Each redia

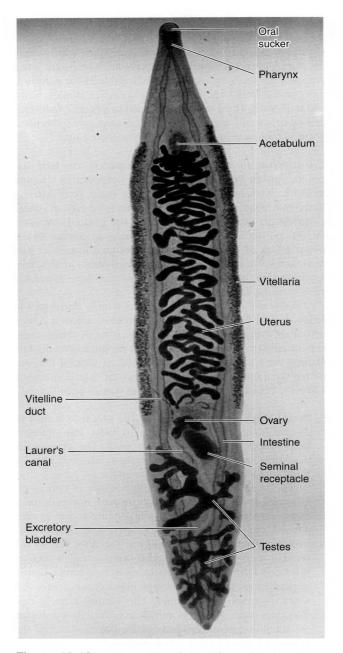

Figure 18.18   *Chinese liver fluke, Clonorchis sinensis.*
Adults measure 8 mm to 25 mm long by 1.5 mm to 5.0 mm wide.
Courtesy of Robert E. Kuntz.

produces from 5 to 50 cercariae. Cercariae have a pair of eyespots and are beset with delicate bristles and tiny spines. The entire cercaria is brownish. Its tail has dorsal and ventral fins (pleurolophocercous cercaria).

Cercariae hang upside down in the water and slowly sink to the bottom. When contacting any object, they rapidly swim upward toward the surface and again begin to sink. Even a slight current of water will also cause this reaction. Thus, when a fish swims by, a cercaria is stimulated to react in a way favoring its contact with its next host.

On touching the epithelium of a fish, a cercaria attaches with its suckers, casts off its tail, and bores through the skin, coming to rest and encysting under a scale or in a muscle (Fig. 18.21). Nearly a hundred species of fishes, mostly in Cyprinidae (Fig. 18.22), have been found naturally infected with metacercariae of *Clonorchis sinensis*, although some species are more susceptible than others. Thousands of metacercariae may accumulate in a single fish, but the number usually is much smaller. Metacercariae will also develop in species of crayfish genera *Caridina, Macrobrachium,* and *Palaemonetes;* and such metacercariae are infective, at least to guinea pigs and reportedly to humans.[27, 47] Definitive hosts are infected when eating raw or undercooked fish or crustaceans.

Mammals other than humans that have been found infected with adult *C. sinensis* are pigs, dogs, cats, rats, and camels.[28] Experimentally, rabbits and guinea pigs are highly susceptible. Perhaps any fish-eating mammal can become infected. Dogs and cats undoubtedly are important reservoir hosts. Birds are possibly infected.

Young flukes excyst in the duodenum. The route of migration to the liver is not clear; conflicting reports have been published. It is probable that juveniles migrate up the common bile duct to the liver. Young flukes have been found in the liver 10 to 40 hours after infection of experimental animals. The worms mature and begin producing eggs in about a month. The entire life cycle can be completed in three months under ideal conditions. Adult worms can live at least eight years in humans.

- **Epidemiology and Transmission Ecology.**   It is easy to see why clonorchiasis is common in countries in which raw fish is considered a delicacy (Fig. 18.23). In some areas the most heavily infected people are wealthy epicures who can afford beautifully cut and arranged slices of raw fish. But the poor are also afflicted because fish is often their only source of animal protein. Prevalence may range from an average of 14% in cities such as Hong Kong to 80% in some endemic rural areas. Although complete protection is achieved simply by cooking fish, it would be a futile exercise to try to get millions of people to change centuriesold eating habits. In addition, educating people to cook their fish would not change matters because fuel is a luxury that many cannot afford.

Fish farming is a mainstay of protein production throughout eastern Asia, in Europe, and increasingly in the United States. More protein in the form of fish can be harvested from an acre of pond than protein in the form of beef, beans, or corn from an acre of the finest farmland. The fastest growing fish are primary consumers of algae and other plants. Fish ponds are fertilized with human feces throughout much of Asia (Fig. 18.24), which increases the growth rate of water plants and thereby that of fish. In a study conducted in Guangdong Province, China, 40% of fish pond owners fed their fishes feces from domestic animals and humans.[54] Of course these practices abet the life cycle of *C. sinensis*. Where fish farming is not so important, dogs and cats serve as reservoirs of infection, contaminating streams and ponds with their feces.

Metacercariae will withstand certain types of preparation of fish, such as salting, pickling, drying, and smoking. Because of this resistance, people can become infected thousands of miles from an endemic area when they eat imported fish.

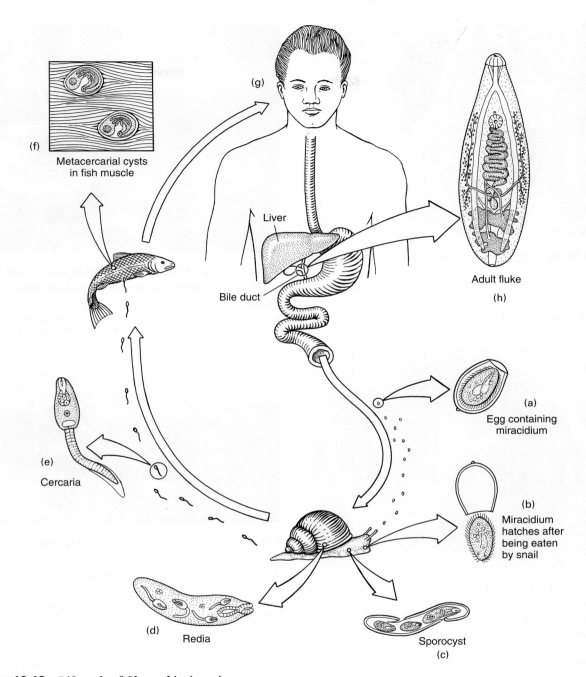

Figure 18.19   Life cycle of *Clonorchis sinensis.*
(*a*) Egg containing miracidium is passed in feces. (*b*) Miracidium hatches after being eaten by snail. (*c*) Sporocyst. (*d*) Redia.
(*e*) Cercaria leaves snail and penetrates fish. (*f*) Metacercarial cysts in fish muscle. (*g*) Human becomes infected by eating raw fish.
(*h*) Adult fluke in bile duct.

Drawing by William Ober and Claire Garrison.

• **Pathology.**   Pathogenesis of *C. sinensis* is similar to pathogenesis in infections with *Opisthorchis* spp.[8] Typically, flukes do not inhabit first-order bile ducts (smallest bile ducts); their size confines them to second-order ducts (larger bile ducts with their main branches). There they cause erosion of epithelium lining (Fig. 18.25), excess mucus production, and epithelial cell proliferation. Inflammatory changes become prominent, with heavy eosinophil and mononuclear infiltration around ducts,

periductal fibrosis, and necrosis and atrophy of surrounding liver cells. In experimental infections with *Opisthorchis viverrini,* parasite antigens were detected early in the infection, both in tissues surrounding worms and in first-order bile ducts where no worms were present.[44] Worm antigens were also located in damaged liver cells. Thus, host immune response plays a critical role in pathogenesis.

Eventual outcome in humans depends mainly on intensity and duration of infection; fortunately, worm burdens

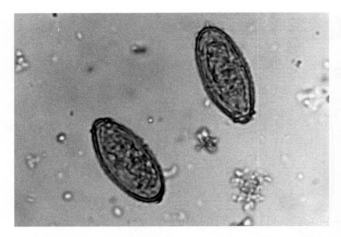

Figure 18.20 Eggs of *Clonorchis sinensis* from a human stool.

They are 26 μm to 30 μm long. Note the small knob on the abopercular end.

Courtesy of Robert E. Kuntz and Jerry A. Moore.

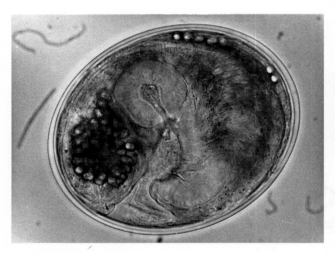

Figure 18.21 Encysted metacercaria of *Clonorchis sinensis* from fish muscle.

The oral and ventral suckers are clearly seen; the round bodies are excretory corpuscles.

From J. B. Gibson and T. Sun, in Marcial-Rojas (Ed.) *Pathology of protozoal and helminthic diseases with clinical correlation* (Baltimore, MD: Williams & Wilkins, 1971).

Figure 18.22 Grass carp, *Ctenopharyngodon idellus*, a common second intermediate host of *Clonorchis sinensis*.

This fish is widely cultivated in eastern Asia.

From J. B. Gibson and T. Sun, in Marcial-Rojas (Ed.), *Pathology of protozoal and helminthic diseases with clinical correlation* (Baltimore, MD: Williams & Wilkins, 1971).

Figure 18.23 "Yue-shan chuk," thin slices of raw carp with rice soup, vegetable garnishing, and soy sauce—a Cantonese delicacy.

From J. B. Gibson and T. Sun, in Marcial-Rojas (Ed.), *Pathology of protozoal and helminthic diseases with clinical correlation* (Baltimore, MD: Williams & Wilkins, 1971).

Figure 18.24 Privy over a fish-culture pond in Hong Kong.

The Chinese characters on the structure advertise a worm medicine.

From J. B. Gibson and T. Sun, in Marcial-Rojas (Ed.), *Pathology of protozoal and helminthic diseases with clinical correlation* (Baltimore, MD: Williams & Wilkins, 1971).

are usually small. Mean intensity of infection in most endemic areas is 20 to 200 flukes, but as many as 21,000 have been removed at a single autopsy. Chronic defoliation of biliary epithelium leads to gradual thickening and occlusion of the ducts (Fig. 18.26). Pockets form in ductal walls, and complete perforation into surrounding parenchyma may result. Infiltrating eggs become surrounded by granulomas, thereby interfering with liver function. Jaundice is found in a small percentage of cases and is probably caused by bile retention when ducts are obstructed. Eggs and sometimes entire worms often become nuclei of gallstones. Cancer of the bile ducts is commonly associated with advanced clonorchiasis and opisthorchiasis.[41]

- **Diagnosis and Treatment.** Diagnosis is based on recovery of characteristic eggs in feces. Liver abnormalities just described should suggest clonorchiasis in endemic areas, but care must be taken to exclude cancer, hydatid

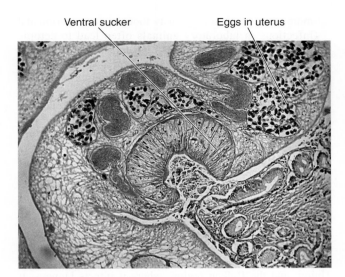

Ventral sucker   Eggs in uterus

Figure 18.25   **Adult *Clonorchis sinensis* attached by its ventral sucker to biliary epithelium in a human.**

From J. B. Gibson and T. Sun, in Marcial-Rojas (Ed.), *Pathology of protozoal and helminthic diseases with clinical correlation* (Baltimore, MD: Williams & Wilkins, 1971).

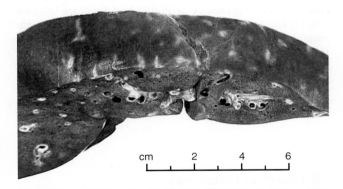

cm   2   4   6

Figure 18.26   **Severe clonorchiasis with "pipestem fibrosis" in a human.**

The dilated, thick-walled bile ducts are full of flukes.

From J. B. Gibson and T. Sun, in Marcial-Rojas (Ed.), *Pathology of protozoal and helminthic diseases with clinical correlation* (Baltimore, MD: Williams & Wilkins, 1971).

disease, beriberi, amebic abscess, and other types of hepatic disease. Praziquantel is the drug of choice.

***Opisthorchis* spp.** *Opisthorchis felineus* is very similar to *C. sinensis* but occurs in Europe as well as Asia. Originally described from a domestic cat in Russia, it is common throughout southern, central, and eastern Europe, Turkey, southern Russia, Vietnam, India, and Japan. Besides parasitizing cats and other carnivores, it parasitizes humans, probably infecting more than a million people within its range.

Seven million people in northeast Thailand are infected with *Opisthorchis viverrini*.[39] The pathology and epidemiology of this species, as well as those of *O. felineus,* are similar to those of *C. sinensis*. Morbidity can be resolved in mild and moderate infections after treatment with praziquantel.[39]

Chronic inflammation due to *Opisthorchis viverrini* infection is apparently mediated by a TLR2 pathway through NF-kB (page 26).[38] Such infection is highly correlated with

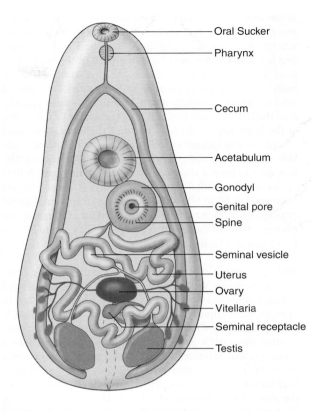

Oral Sucker

Pharynx

Cecum

Acetabulum

Gonodyl

Genital pore

Spine

Seminal vesicle

Uterus

Ovary

Vitellaria

Seminal receptacle

Testis

Figure 18.27   *Heterophyes heterophyes.*

Its size is 1.0 mm to 1.7 mm long.

Drawing by William Ober and Claire Garrison.

incidence of bile duct cancer, much more so than infection with *C. sinensis*.[21, 42] Although the causative mechanism remains unclear, there is strong evidence that infection with *O. viverrini* is responsible for the cancer.

Family Heterophyidae

Heterophyids are tiny, teardrop-shaped flukes, usually maturing in the small intestine of fish-eating birds and mammals. The distal portion of the vas deferens and uterus join to form a hermaphroditic duct, which opens into a genital sac. The genital sac may bear a muscular sucker, or **gonotyl,** which is greatly modified in different species. In some heterophyids the genital sinus encloses the acetabulum. A cirrus pouch is absent. The tegument is scaly, especially anteriorly. This is a large family with several subfamilies. Several species are important parasites of humans.

Heterophyes heterophyes *Heterophyes heterophyes* (Fig. 18.27) is a minute fluke that was first discovered in an Egyptian man in Cairo in 1851. It is common in northern Africa, Asia Minor, and the Far East, including Korea, China, Japan, Taiwan, and the Philippines. Because the eggs cannot be differentiated from those of related species, an accurate estimate of human infections cannot be made.

- **Morphology.**   Adults are 1.0 mm to 1.7 mm long and 0.3 mm to 0.4 mm at their greatest width. The entire body is covered with slender scales, most numerous near the anterior end. The oral sucker is only about 90 μm in diameter, whereas the acetabulum is around 230 μm wide and is located at the end of the first third of the body. Two oval testes lie side by side near the posterior end of the body. A vas deferens expands to form a sinuous seminal vesicle, which constricts again, becoming a short ejaculatory duct. The ovary is small, medioanterior to the testes, at the beginning of the last fourth of the body. A seminal receptacle and Laurer's canal are present. The uterus coils between the ceca and constricts before joining the ejaculatory duct to form a short common genital duct, which then opens into the genital sinus. The gonotyl is about 150 μm wide and has 60 to 90 toothed spines on its margin. Lateral vitelline follicles are few in number and are confined to the posterior third of the worm. The eggs are 28 μm to 30 μm by 15 μm to 17 μm.

- **Biology.** Adult worms live in the small intestine, burrowed between villi. Eggs contain a fully developed miracidium when laid but hatch only when eaten by an appropriate freshwater or brackish-water snail (*Pironella conica* in Egypt, *Cerithidia cingula* in Japan). After penetrating the snail's gut, the miracidium transforms into a sporocyst that produces rediae. A second-generation redia gives birth to cercariae with eyespots and finned tails (ophthalmolophocercous cercariae), which emerge from the snail. Like cercariae of *Clonorchis sinensis,* those of *H. heterophyes* swim toward the surface of the water and slowly drift downward. On contacting a fish, they penetrate the epithelium, creep beneath a scale, and encyst in muscle tissue. Metacercariae are most abundant in various species of mullet, which are exposed when they enter estuaries or brackish-water shorelines. Several thousand metacercariae have been found in a single, small fish. A definitive host becomes infected when it eats raw or undercooked fish.

- **Epidemiology.** For eggs to be available to estuarine and brackish-water snails, pollution must occur in these waters. Boatmen, fishermen, and others who live by or on the water are often the main sources of infection. Infected fish are distributed widely in fish markets. Other fish-eating mammals, such as cats, foxes, and dogs, serve as reservoirs of infection.

- **Pathology.** Each worm elicits a mild inflammatory reaction at its site of contact with the intestine. Heavy infections, which are common, cause damage to the mucosa and produce intestinal pain and mucous diarrhea. Perforation of the mucosa and submucosa sometimes occurs and allows eggs to enter the blood and lymph vascular systems and to be carried to ectopic sites in the body.[1] The heart is particularly affected, with tissue reactions in the valves and myocardium leading to heart failure. Kean and Breslau[24] reported that 14.6% of cardiac failure in the Philippines resulted from heterophyid myocarditis.

Eggs in the brain or spinal cord lead to neurological disorders that are sometimes fatal. Two bizarre cases are known in which adult *H. heterophyes* were found in the brains of humans, and in another case an adult worm was found in the myocardium.[1] Such infections are probably more common than previously thought, for experimental infections in laboratory animals often lead to ectopic flukes.[19] In such experiments immature flukes have been found inside lymphoid follicles and Peyer's patches. Young flukes had migrated from the sinuses in Peyer's patches via the lymphatics to mesenteric lymph glands, which became enlarged and hyperplastic and contained mature worms.

Diagnosis is difficult when adult worms are not available. Eggs closely resemble those of several other heterophyids and are not very different from those of *C. sinensis.* Praziquantel is effective in treatment.

Other Heterophyid Parasites of Humans

Heterophyes katsuradai is very similar to *H. heterophyes.* It has been found in humans near Kobe, Japan. Infection is acquired by eating raw mullet. *Metagonimus yokagawai* is a very common heterophyid in the Far East, the former Soviet Union, and the Balkan region, where it infects humans. It superficially resembles *H. heterophyes,* but its acetabulum is displaced to the left, where it is fused with the gonotyl. The biology of *M. yokagawai* is identical to that of *H. heterophyes,* except that a different snail host (*Semisulcospira* spp.) is required and the second intermediate hosts are freshwater fish of several species. A definitive host becomes infected when it eats uncooked fish. Various fish-eating mammals are natural reservoirs, and even pelicans have been implicated in this regard. Pathogenesis, diagnosis, and treatment are as for *H. heterophyes.*

Until proved otherwise, all species of Heterophyidae should be considered potential parasites of humans. More than 22 species have been found infective to date.[11]

Learning Outcomes

By the time a student has finished studying this chapter, he or she should be able to:

1. Draw and label the life cycles of *Dicrocoelium dendriticum* and *Paragonimus westermani,* and state how they differ.

2. Compare and contrast the social and cultural factors that promote human infection with *Paragonimus westermani* and *Clonorchis sinensis.*

3. Draw adult flukes of *Dicrocoelium dendriticum, Plagiorchis muris,* and *Clonorchis sinensis* and describe the structural features that distinguish them from one another.

4. Alternatively, label the structural features and life-cycle stages of *Paragonimus westermani, Clonorchis sinensis, Dicrocoelium dendriticum,* and *Plagiorchis muris.*

5. List the factors that may determine whether a frog becomes infected with different species of *Haematoloechus.*

References

References for superscripts in the text can be found at the following Internet site: www.mhhe.com/robertsjanovynadler9e

Additional Readings

Dooley, J. R., and R. C. Neafie. 1976. Clonorchiasis and opisthorchiasis. In C. H. Binford and D. H. Connor (Eds.), *Pathology of tropical and extraordinary diseases* (vol. 2, sect. 10). Washington, DC: Armed Forces Institute of Pathology.

Holmes, J. C., and W. M. Bethel. 1972. Modification of intermediate host behaviour by parasites. In E. U. Canning and C. A. Wright (Eds.), *Behavioural aspects of parasite transmission. Linnaean Society of London.* London: Academic Press, pp. 123–149. An outstanding summary of the subject. Should be required reading for all students of parasitology.

Kaewkes, S. 2003. Taxonomy and biology of liver flukes. *Acta. Trop.* 88:177–186. Despite its title, this review is not about all liver flukes, but covers only *Opisthorchis viverrini, O. felineus,* and *Clonorchis sinensis.*

Mairiang, E., and P. Mairiang. 2003. Clinical manifestation of opisthorchiasis and treatment. *Acta Trop.* 88:221–227.

McManus, D. P., T. H. Le, and D. Blair. 2004. Genomics of parasitic flatworms. *Int. J. Parasitol.* 34:153–158.

Meyers, W. M., and R. C. Neafie. 1976. Paragonimiasis. In C. H. Binford and D. H. Connor (Eds.), *Pathology of tropical and extraordinary diseases* (vol. 2, sect. 10). Washington, DC: Armed Forces Institute of Pathology.

Millemann, R. E., and S. E. Knapp. 1970. Biology of *Nanophyetus salmincola* and "salmon poisoning" disease. In B. Dawes (Ed.), *Advances in parasitology 8.* New York: Academic Press, Inc., pp. 1–41.

Sithithaworn, P., and M. Haswell-Elkins. 2003. Epidemiology of *Opisthorchis viverrini. Acta Tropica* 88:187–194.

Upatham, E. S., and V. Viyanant. 2003. *Opisthorchis viverrini* and opisthorchiasis: historical review and future perspective. *Acta Tropica* 88:171–176.

Wongratanacheewin, S., R. W. Sermswan, and S. Sirisinha. 2003. Immunology and molecular biology of *Opisthorchis viverrini* infection. *Acta Tropica* 88:195–207.

Yokagawa, M. 1965. *Paragonimus* and paragonimiasis. In B. Dawes (Ed.), *Advances in parasitology 3.* New York: Academic Press, Inc., pp. 99–158. A complete summary of paragonimiasis. Required reading for all who are interested in the subject.

Chapter 19

Monogenoidea

For oaths are straws, men's faiths are wafer-cakes, And hold-fast is the only dog, my duck.

—William Shakespeare (*Henry V*)

Monogenoidea are hermaphroditic flatworms that mainly are external parasites of vertebrates, particularly fish, and especially on gills and external surfaces. Some species, however, are found internally in diverticula of the stomodeum or proctodeum and also in ureters of fishes and bladders of turtles, frogs, salamanders, and caecilians. A few are external parasites of invertebrates, including crustaceans and squid.[25] A single species is known from mammals: *Oculotrema hippopotami* from the eye of the hippopotamus.[50] These worms are not usually regarded as hazardous to wild populations, although a few fish deaths have been attributed to monogenes in nature. However, like copepods and numerous other fish pathogens, monogenes can become a serious threat when fish are crowded together, as in hatcheries or farming operations.

Although monogenes were first distinguished from Digenea in 1858 by van Beneden, there is still nomenclatural debate over origin of the proper noun "Monogenea," used by many authors.[54] Boeger and Kritsky[5] make a strong case for using the name Monogenoidea to denote the class, a practice we follow here. However, the vernacular term "monogene" is so deeply embedded in both the invertebrate literature and professional conversation that we also use it as the common noun.

The first comprehensive overview of the group was by Braun[6] in 1889 and 1893. Fuhrmann,[16] in 1928, helped establish Monogenoidea as a category separate from, although closely allied with, Digenea. Bychowsky,[8] in 1937, was evidently the first to elevate Monogenoidea to class level, apart from and equal to Digenea.[54] Some relatively modern literature still treats the group as a subclass of Trematoda,[47] but phylogenetic analysis of parasitic flatworms reveals that Monogenoidea are more closely related to tapeworms than to trematodes (see chapter 13), and there is no question that they belong in their own class.

Ecological research on monogenes has focused on population and community dynamics, with special attention to site specificity on aquatic vertebrate hosts, intra- and interspecific competition, and habitat factors that influence prevalence and abundance. Monogenes are often very particular about both species of host and site where they live on that host, restricting themselves to extremely narrow niches in some cases. Thus one species may live only at the base of a gill filament, whereas another is found only at its tip.[45] It is possible that such niche specificity is related to structure of the highly specialized, posterior attachment organ, the haptor. A similar phenomenon occurs with the scolex of tetraphyllidean cestodes of elasmobranchs (p. 346). Although some monogenes are highly site specific, others are less so and may be found distributed generally across gill arches, especially in the case of freshwater hosts.[21, 41]

Some monogenes remain fixed to their original site of attachment and cannot relocate later. Others, especially those on the skin, move about actively, leechlike, relocating at will. Infection site specificity may also vary with intraspecific competition. For example, *Gyrodactylus anisopharynx,* a parasite of Brazilian freshwater fishes, evidently responds to infrapopulation size, preferentially occupying the head, dorsal fin, and caudal fin when worm numbers are low but spreading to other parts of the body with increasing infrapopulations.[38]

Certain species are found only on young fish, whereas others occur only on mature hosts. Interspecific competition evidently is not as important as host habitat in maintaining richness of monogene communities, at least in fish,[33] with factors such as water current and dissolved oxygen evidently playing major roles.[41] Nutritional requirements of these parasites may also help determine host specificity because in some cases the free-swimming larvae are particularly attracted by mucus produced by the epidermis of their host species.[22, 56]

The life span of monogenes varies from a few days to several years. Many are incapable of living more than a short time after death of their host.

FORM AND FUNCTION

Body Form

Monogenoidea are bilaterally symmetrical, but with partial asymmetry, particularly involving the haptor (opisthaptor of older literature), superimposed on a few species. The body can be subdivided roughly into the following regions: **cephalic region** (anterior to pharynx), **trunk** (body proper), **peduncle** (portion of body tapered posteriorly), and **haptor** (Fig. 19.1).

Most monogenes are quite small, but a few are large; their sizes range from 0.03 mm to 20.00 mm long. Marine forms are usually larger than those from freshwater hosts. All are capable of stretching and compressing their bodies, so, unless the worms are properly relaxed before fixing, a permanent slide preparation may give a false impression of the true morphology. The dorsal surface is usually convex, while the ventral side is concave. The body is colorless or gray, but eggs, internal organs, or ingested food may cause it to be red, pink, brown, yellow, or black.

The anterior end bears various adhesive and feeding organs, collectively called a **prohaptor,** which sometimes is associated with compound sense organs.[42] There are two main types of prohaptor: those that are not connected with the mouth funnel and those that are. In the first case (Fig. 19.2), the head end usually is truncated, lobed, or broadly rounded. These worms have **anterior attachment organs** consisting of glands and specialized tegument, sometimes opening on small lobes or in sacs, and typically occurring in groups of three on each side.[55] These areas usually bear dense, long microvilli on the tegument, in contrast to the short, scattered microvilli on the remainder of the body. These microvilli may function to spread and mix secretions of the different types of head glands. Many genera have two different types of anterior gland cells, distinguished by their inclusions. Sticky substances are released by these cells through individual ducts or groups of ducts. The utility of substances produced by anterior attachment organs for adhesion is clear to anyone who has watched a monogene with no anterior sucker progress down a fish gill filament in an inchwormlike manner, alternately attaching and releasing its anterior and posterior ends. Some species in this group have shallow, muscular bothria, which serve as suckers, in conjunction with the head gland secretions. Most species have two bothria, but some species have four.

The second type of prohaptor (Fig. 19.3) involves specializations of the mouth and buccal funnel. The simplest types have an **oral sucker** that surrounds the mouth. This structure may be a slightly muscular anterior rim of the mouth or a powerful circumoral sucker. In members of order Mazocraeidea, two **buccal organs (buccal suckers)** are embedded within the walls of the buccal funnel. Ultrastructural studies of buccal organs have shown muscular, glandular, and sensory components; and they appear to play some role in the worm's feeding on host blood.[46]

The posterior end of all monogenes also bears a highly characteristic organ, the haptor (see Figs. 19.1 and 19.4). It is clear that, in a group whose primary habitat is the surface and gills of fish, great adaptive value will accrue from

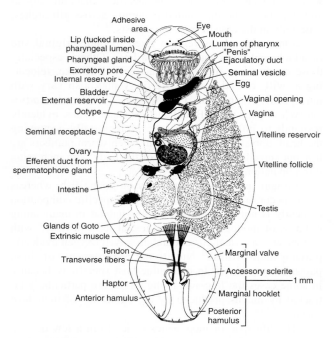

Figure 19.1   **Anatomy of an adult specimen of** *Entobdella soleae* **(ventral view).**

From G. C. Kearn, *Ecology and Physiology of Parasites,* edited by M. Fallis. Copyright © 1971 University of Toronto Press, Toronto. Reprinted by permission.

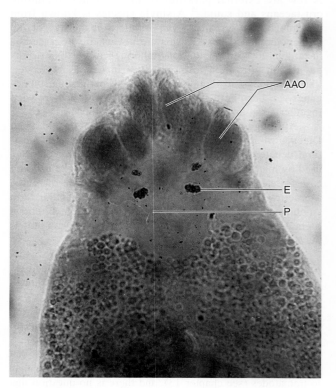

Figure 19.2   **A prohaptor not connected to a mouth funnel.**

Anterior attachment organs (*AAO*), eyespots (*E*), and pharynx (*P*) are clearly visible.

Courtesy of Warren Buss.

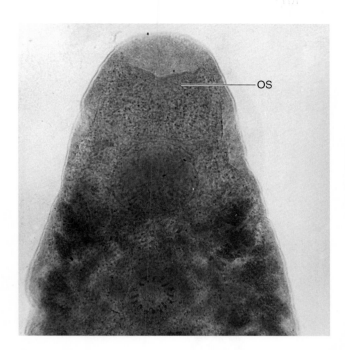

Figure 19.3   **A monogene in which the prohaptor has an oral sucker (*OS*).**

Courtesy of Warren Buss.

an efficient attachment organ that prevents dislodgment by strong water currents, particularly one that will allow the mouth end to "hang downstream" and graze at will. The haptor is such an organ; unsurprisingly, it exhibits great variation within the group, and many forms have been interpreted as adaptations to particular hosts and infection sites.

The haptor may extend for a considerable distance anteriorly along the worm's trunk or may be confined to the posterior extremity. It may be sharply delineated from the body by a peduncle or may be merely a broad continuation of it. Haptors develop into one or two basic types during ontogeny. The larva that hatches from an egg always has a tiny haptor armed with sclerotized hooks or spines. This structure is retained in adults of most species and either expands into a definitive haptor or remains juvenile, while the adult organ develops from other sources near or surrounding it. In this first basic type the muscles expand into a large disc that often has shallow **loculi** or well-developed **suckers,** as well as large hooks called **anchors** or **hamuli.** Anchors occur in one to three pairs, usually in the center of the haptor, although they may be displaced to the side or posterior margin of the disc. Hamuli often have **connecting bars,** or **accessory sclerites,** supporting them (Fig. 19.5). The homologies of central hooks and their supporting bars are not always clear, at least at higher taxonomic levels. The protein in anchors and marginal hooks is evidently keratin.

The tiny larval hooklets, and their persistent forms in adults, are termed **hooks,** as opposed to anchors (see Fig.19.5). Haptor hooks are characterized as being either "marginal" or "central." **Marginal hooks,** which are not always strictly marginal, are usually the tiny hooklets of the larval haptor, some of which may be missing. It sometimes takes careful and skillful microscopy to find these.

Rarely **supplementary discs,** or **compensating discs,** are developed near the base of the haptor (family Diplectanidae). These discs consist of a series of sclerotized lamellae or spines and are not technically a part of the larval or adult haptor. Two to eight suckers are found on the ventral surface of the haptors of many species.

Complex **clamps** are found on many species; on some the clamp is muscular, whereas on others it is mainly sclerotized. It functions as a pinching mechanism, aiding in adherence to a host. Although many variations of structure occur, all are based on a single, basic type of clamp (Fig. 19.6). Identity of the material of which clamps are constructed is enigmatic; it is not keratin, chitin, quinone-tanned protein, or collagen.[27] The number of clamps varies from two to many, distributed symmetrically in some species and asymmetrically in others. The combinations of hooks, suckers, and clamps vary among several families (see Fig. 19.4).

Because it has undergone such evolutionary diversification and varies considerably among species, the haptor is an important taxonomic character. Consequently, specialists studying monogenes rely heavily on the sclerotized hamuli (anchors), hooks, bars, clamps, and so on, which are relatively easy to study. However, in a single species, sizes of sclerotized parts may vary with the season, primarily due to development of worms at different temperatures.[32] But although sizes vary, hamulus shape and proportions remain stable, suggesting that these features are influenced more by genes than by the environment.[15] Sclerotized portions of the male reproductive system also vary between species, and consequently they have been used extensively in taxonomy and phylogenetic analysis.[3, 34]

Tegument

As in digeneans and cestodes, the tegument of monogenes traditionally was called a cuticle because light microscopists could discern little structure within it. However, by use of electron microscopy, the cuticle has now been recognized as a living tissue, the **tegument.** Its fundamental structure is similar to that of digenean and cestode tegument, with some noteworthy differences. The surface layer is, as in cestodes and digeneans, a syncytial stratum laden with vesicles of various types and mitochondria. This layer is bounded externally by a plasma membrane and glycocalyx (fine filamentous layer on surface) and internally by a membrane and basal lamina. This stratum is the **distal cytoplasm,** and it is connected by trabeculae (**internuncial processes**) to the cell bodies, or **cytons (perikarya),** located internal to a superficial muscle layer. Often the tegument of monogenes is supplied with short, scattered microvilli; in some species microvilli are absent, and shallow pits occur.

A curious condition has been reported in certain species. Some areas of the body are without a tegument, and large pieces of tegument are only loosely connected to the surface. In these areas the basal lamina constitutes the external covering. Rohde[44] contends that these cases might offer a clue to the adaptive value of the tegumental arrangement of monogenes, digeneans, and cestodes; that is, syncytial distal cytoplasm with internal perikarya. Because the superficial tegument may be subjected to such damaging influences as host secretions, it then easily can be replaced from the internal cell machinery that is still intact.

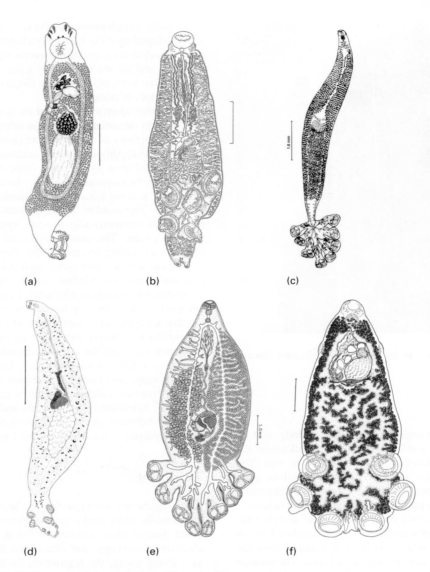

(a) (b) (c)

(d) (e) (f)

Figure 19.4   **A variety of monogenes, showing diverse body shapes and haptor morphology.**
(*a*) *Demidospermus peruvianus* from an Amazonian catfish, bar = 100 μm; (*b*) *Branchotenthes robinoverstreeti* from an Indian Ocean guitarfish (*Rhina ancylostoma*), bar = 1 mm; (*c*) *Choricotyle scapularis* from the gills of *Anisotremus scapularis,* a Chilean marine fish, bar = 1.6 mm; (*d*) *Mazocraes australis* from an Argentine anchovy (*Engraulis anchoita*), bar = 0.4 mm;   (*e*) *Diclidophora embiotocae* from the gills of *Hyperprosopon ellipticum,* a marine fish from the Oregon coast, bar = 1 mm; (*f*) *Polystoma cuvieri* from urinary bladder of a frog (*Physalaemus cuvieri*) in Paraguay, bar = 0.5 mm.

All figures are from the *Journal of Parasitology* and are reprinted by permission. (*a*) C. A. Mendoza-Palmero and T. Scholz, "New species of *Demidospermus* (Monogenea: Dactylogyridae) of pimelodid catfish (Siluriformes) from Peruvian Amazonia and the reassignment of *Urocleidoides lebedevi* Kritsky and Thatcher, 1976," 97:586–592, © 2011; (*b*) S. A. Bullard and S. M. Dippenaar, "*Branchotenthes robinoverstreeti* n. gen. and n. sp. (Monogenea: Hexabothriidae) from gill filaments of the bowmouth guitarfish, *Rhina ancylostoma* (Rhynchobatidae), in the Indian Ocean," 89:595–601, © 2003; (*c*) M. E. Oliva, M. T. González, P. M. Ruz, and J. L. Luque, "Two new species of *Choricotyle* van Beneden & Hesse (Monogenea: Diclidophoridae), parasites from *Anisotremus scapularis* and *Isacia conceptionis* (Haemulidae) from northern Chilean coast," 95:1108–1111, © 2009; (*d*) J. T. Timi,. H. Sardella, and J. A. Etchegoin, "Mazocraeid monogeneans parasitic on engraulid fishes in the southwest Atlantic," 85:28–32, © 1999; (*e*) A. W. Hanson, "*Diclidophora embiotocae* sp. n. (Monogenea, Diclidophoridae), a parasite of embiotocid fishes of Oregon," 65:457–459, © 1979; (*f*) C. Vaucher, "*Polystoma cuvieri* n. sp. (Monogenea: Polystomatidae), a parasite of the urinary bladder of the leptodactylid frog *Physalaemus cuvieri* in Paraguay," 76:501–504, © 1990.

Muscular and Nervous Systems

The main musculature, other than that in haptors, consists of **superficial muscles** immediately below the distal cytoplasm of the tegument, arranged in circular, diagonal, and longitudinal layers. Muscles of the haptor, in suckers or inserted on hooks and accessory sclerites, are clearly important in adhesion. We understand the mechanics of their operation in several species, an example of which is *Entobdella soleae* (Fig. 19.7). This species lives on the skin of soles, and its haptor anchors the parasite firmly in its relatively smooth and exposed site on its host.[22] The disc-shaped haptor forms an effective suction cup. Prominent muscles in the peduncle insert on a tendon that passes down to near the ventral surface of the

Figure 19.5   Examples of haptor hooks, anchors, and sclerotized portions of the male reproductive system, as typically seen in papers describing new species from small fishes (not all drawn to the same scale).

(*a*) *Dactylogyrus magnus* from a silver chub; (*b*) *D. manicatus* from a striped shiner; (*c*) *Gyrodactylus tennesseensis* from a redbelly dace; (*d*) *G. illigatus* from a silverband shiner; *an,* anchor; *ap,* accessory piece; *b,* bar(s) and shields; *h,* hook(s); *p,* penis. Note that bar shape, anchor (hamulus) shape, hook shape, and structure of the sclerotized parts of the male system all vary among species. Terminal portions of the male reproductive system, as seen at the worm's surface, are to the right of the anchors in (*c*) and (*d*).

(*a, b*) From W. A. Rogers, "Studies on *Dactylogyrus* (Monogenea) with descriptions of 24 new species of *Dactylogyrus,* 5 new species of *Pellucidhaptor,* and the proposal of *Aplodiscus* gen.n," in *Journal of Parasitology* 53:501–524. Copyright © 1967 Journal of Parasitology. Reprinted by permission. (*c, d*) From W. A. Rogers, "Eight new species of *Gyrodactylus* (Monogenea) from the southeastern U.S. with redescriptions of *G. fairporti* Van Cleave, 1921, and *G. cyprini* Diarova, 1964," in *Journal of Parasitology* 54:490–495. Copyright © 1968 Journal of Parasitology. Reprinted by permission.

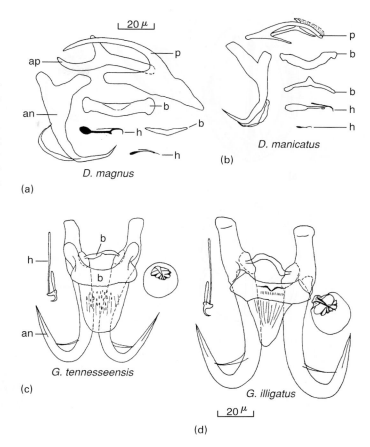

disc, up over a notch in the accessory sclerites, and then to the proximal end of the large anchors. Contraction of these muscles erects the accessory sclerites so that their distal ends push down against the fish's skin and their proximal ends serve as a prop toward which the proximal ends of the anchors are pulled. This action tends to lift the center area of the haptor, thus reducing pressure and creating suction, at the same time that the distal, pointed ends of the anchors are forced downward to penetrate the host's epidermis.

The nervous system is a typical flatworm ladder (orthogon) type with **cerebral ganglia** in the anterior and several nerve trunks coursing posteriorly from them. Nerve trunks connect by ladder commissures, and additional nerves emanate from the cerebral ganglia to connect with a pharyngeal commissure. As would be expected, adhesive organs of the haptor are well innervated. Cholinesterase was found in the nervous system of *Diclidophora merlangi;* therefore, at least some fibers are probably cholinergic.[20] Neuropeptides were demonstrated in *Gyrodactylus salaris,* and both peptidergic and serotoninergic innervation were demonstrated in *Eudiplozoon nipponicum,* all by use of immunocytochemical staining.[43, 59] Monogenes, like other flatworms, have a chemically diverse nervous system.

Monogenes also have a variety of sense organs. Most have pigmented eyes in the free-swimming larval stage, and in many species adults have eyes as well. Oncomiracidia of subclass Polyonchoinea usually have four eyespots, which persist in adults, the two larval eyespots of infrasubclass Polystomatoinea, on the other hand, disappear during maturation. These are rhabdomeric eyes similar to those found in turbellaria and some larval Digenea. In addition, what appears to be a nonpigmented ciliary photoreceptor has been found in larvae of *Entobdella soleae,* with counterparts in structures described in larval Digenea.

Several different types of ciliary sense organs (**sensillae**) occur in the tegument and anterior attachment organs. These include single receptors (one modified cilium in a single nerve ending) and compound receptors (consisting both of several associated nerve endings, each with a single cilium, and of one or a few nerves, each with many cilia).[28] Sensillae can be stained with silver nitrate, and their distribution on the body may be useful in distinguishing taxa.[48]

Finally, a very interesting, nonciliated sense organ occurs on the haptor of *Entobdella soleae.* The disc surface of the haptor is covered with more than 800 small papillae (Fig. 19.8), and beneath the tegument of each papilla are packed nerve endings that double over and pile on top of one another. The function of these peculiar organs is probably mechanoreception, perhaps to sense contact with a host or detect local tensions in the haptor. It is not known whether similar organs occur in other monogenes.[29]

Osmoregulatory System

The excretory system has not been used as a tool for systematics in this group in quite the way it has in Digenea, although *Gyrodactylus* has been divided into subgenera based

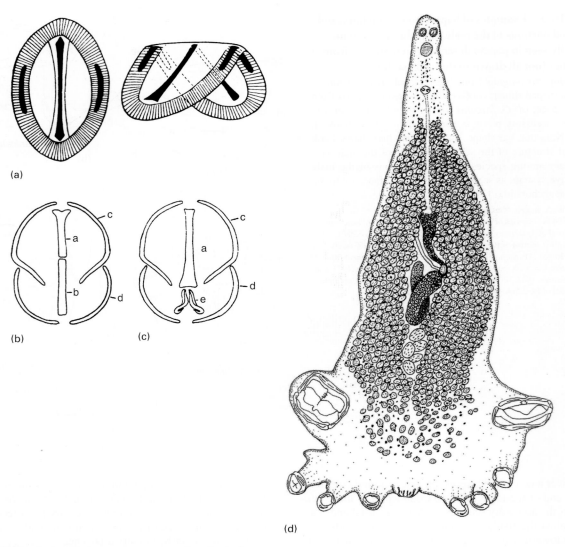

(a)

(b)   (c)

(d)

Figure 19.6   **Monogene clamps and associated sclerites.**

(*a*) Diagram of an attaching clamp. On the left, it is fully open; on the right, it is partially closed. The sclerotized parts are black; the musculature is crosshatched. (*b, c*) Types of clamp sclerites from members of the order Mazocraeidea; *a,* midsclerite or anterior midsclerite; *b,* posterior midsclerite; *c,* anterolateral sclerite; *d,* posterolateral sclerite; *e,* accessory sclerite.

Figure 19.7   **A diagrammatic parasagittal section through the adhesive organ of *Entobdella soleae.***

The arrows show the direction of movement of the tendon when the extrinsic muscle contracts. The posterior hamulus has been omitted.

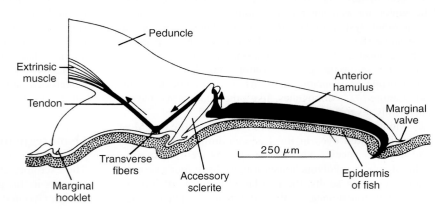

Peduncle

Extrinsic muscle

Tendon

Anterior hamulus

Marginal valve

Transverse fibers

Accessory sclerite

250 μm

Epidermis of fish

Marginal hooklet

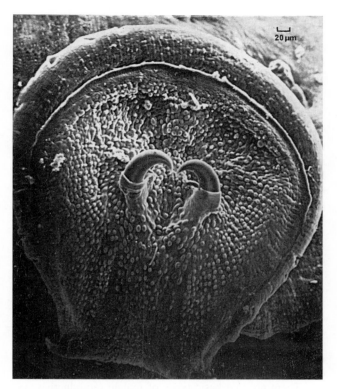

20 µm

Figure 19.8   **Scanning electron micrograph of the opisthaptor of *Entobdella soleae.***

Note the probable sensory papillae.

From K. M. Lyons, "Scanning and transmission electron microscope studies on the sensory sucker papillae of the fish parasite *Entobdella soleae* (Monogenea)," in *Zeitschr. Zellforsch.* 137:471–480. Copyright © 1973. Springer-Verlag.

on structure of the excretory system, and molecular studies support this division.[14, 57]

Typical of Platyhelminthes, excretory units are **flame cell protonephridia.** There are several patterns of distribution in the body, and individual worms add flame cells, especially to the haptor, during development.[14] A thin-walled capillary leads from this unit to fuse with a succession of ducts leading to two lateral **excretory pores** near the worm's anterior end. Each terminal duct often has a contractile bladder at its distal end.

Fine structure of the excretory system has been studied in two species of *Polystomoides,* and it is generally similar to that of Digenea and Cestoda with minor differences.[44] The internal surface area of the tubules is increased in a manner differing from that in either of the other groups; that is, by strongly reticulated walls. Lateral or nonterminal flames are frequent.

Acquisition of Nutrients

The mouth and buccal funnel often have associated suckers. Behind the buccal funnel a short **prepharynx** is followed by a muscular and glandular **pharynx.** This powerful sucking apparatus draws food into the system. In *Entobdella soleae* the pharynx can be everted and the pharyngeal lips closely applied to a host's skin.[22] Pharyngeal glands secrete a strong

protease that erodes host epidermis, and the worm sucks up the lysed products. Fortunately for the fish, its epidermis is capable of rapid migration and regeneration to close the wound left by the parasite's feeding. Posterior to the pharynx may be an **esophagus,** although it is absent from many species. The esophagus may be simple or have lateral branches and may have unicellular digestive glands opening into it.

In most monogenes the **intestine** divides into two lateral **crura,** which are often highly branched and may even connect along their length. If the crura join near the posterior end of the body, it is common for a single tube to continue posteriorly for some distance. There is no anus.

Feeding habits among Monogenoidea vary along taxonomic lines, with members of subclass Polyonchoinea (including families such as Gyrodactylidae and Dactylogyridae) feeding mainly on epidermal cells and secretions, and members of subclass Heteronchoinea typically feeding on blood.[23] In Polyonchoinea, the gut epithelium usually consists of a single layer of one cell type, although it is syncytial in *Gyrodactylus* species.[23] In blood-feeding heteronchoineans, host hemoglobin is taken up by endocytosis, which is initiated by contact between hemolysed host blood and host protein-specific receptors on the parasite's cells. As a result of digestion, hematin accumulates in gut epithelial cells, from which it is eventually released and regurgitated.[18]

It was believed formerly that the unusable breakdown product of hemoglobin digestion, hematin, was eliminated in the gut by sloughing off gut cells containing it so that parts of the cecal wall were denuded. However, ultrastructural studies on *Diclidophora merlangi* have shown that cecal epithelium is not discontinuous but that hematin-containing cells are interspersed with a different kind of cell type called *connecting cells.*[19] Both hematin cells and connecting cells have their luminal surfaces increased by long, thin lamellae. Digestion of hemoglobin in *D. merlangi,* at least, is evidently mostly or entirely intracellular: The protein is taken into a cell by pinocytosis and digested within an extensive, intracellular reticular space, and hematin is subsequently extruded by temporary connections between the reticular system and the gut lumen. Indigestible particles are eliminated through the mouth in all monogenes. Finally, *D. merlangi* can absorb neutral amino acids through its tegument, suggesting the possibility that direct absorption of low–molecular weight organic compounds could supplement its blood diet.[17]

Male Reproductive System

Monogenes are hermaphroditic with cross-fertilization usually taking place (see Fig. 19.1). **Testes** usually are round or ovoid, but they may be lobed. Most species have only one testis, but the number varies according to species, and one species has more than 200 per individual. Each testis has a **vas efferens,** which expands or fuses with others to become a **vas deferens;** this structure may in turn lead into an ejaculatory duct. There is no trace of a cirrus pouch or eversible cirrus in the sense of those in cestodes or trematodes. In some cases the ejaculatory duct is simple and terminates within a shallow, sometimes suckerlike **genital atrium,** which propels sperm into the female system at copulation. In many species tissues surrounding the terminal ejaculatory duct are thick and muscular, forming a papillalike copulatory organ or penis. Hooks of consistent size and form for each species commonly arm the distal end of the copulatory organ. In many the lining of the distal ejaculatory duct is sclerotized, sometimes for a considerable portion of its length. (We use the term *sclerotized,* although the chemical nature of the stabilized protein is unknown.) A simple, saclike seminal vesicle is present in some species. **Prostatic glands** are usually present.

In several families still another type of copulatory organ exists: a complex **sclerotized copulatory apparatus** that joins with the ejaculatory duct. This apparatus commonly consists of a penis and **accessory piece** (Fig. 19.9). These

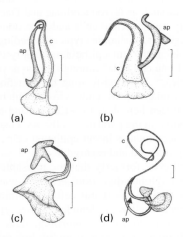

Figure 19.9 **Examples of male copulatory complexes from Dactylogyridea.**

(*a*) *Anchoradiscus triangularis.* (*b*) *Actinocleidus bifidus.* (*c*) *Actinocleidus georgiensis.* (*d*) *Crinicleidus crinicirrus. c,* Copulatory organ; *ap,* accessory piece. (Bar = 10 μm.)

From M. Beverley-Burton, "The taxonomic status of *Actinocleidus* Mueller, 1937; *Anchoradiscus* Mizelle, 1941; *Clavunculus* Mizelle et al., 1956; *Anchoradiscoides* Rogers, 1967; *Syncleithruim* Price, 1967; and *Crinicleidus* n. gen.: North American ancyrocephalids (Monogenea) with articulating haptoral bars," in *J. Parasitol.* 72:22–44. Copyright © 1986.

components vary widely in structure among species but are similar within a species and so are important taxonomic characters.[3] These structures are contained in a membranous sac and are controlled by muscles.

Female Reproductive System

The single **ovary (germarium)** of all species is usually anterior to the testes (Fig. 19.10; see also Fig. 19.1). Among species it varies in shape from round or oval to elongated or lobed. The **oviduct** leaves the ovary and courses toward the ootype, receiving vitelline, vaginal, and genitointestinal ducts along the way. More specifically, the oviduct extends from the ovary to a confluence with the vitelline duct (canal); the remainder is often referred to as the **female sex duct.** A seminal receptacle is present, either as a simple swelling of the oviduct or as a special sac with a separate duct to the oviduct.

Vitellaria are abundant, usually extending throughout the parenchyma and often even into the haptor. Despite their many ramifications, vitellaria consist of left and right groups. Each has an efferent duct; these ducts fuse midventrally near the oviduct, forming a small **vitelline reservoir.** Each vitelline follicle consists of a few cells surrounded by a thin, muscular membrane. Vitelline ducts are lined with ciliated epithelium.

There are two basic types of female reproductive systems in monogenes,[4] distinguished by connections of the vagina(s) and presence (in Heteronchoinea) or absence (from Polyonchoinea) of a curious structure called the **genitointestinal canal** connecting the female system to the gut (see Fig. 19.10). A vagina may be present or absent, and when present it may be doubled. Vaginal openings are dorsal, ventral, or lateral. The terminal portion is sclerotized in some species; in others the vaginal pore is multiple or surrounded by spines. In those species with a "true" vagina, the vaginal opening and duct lead directly to the oviduct. The second basic type of system, typical of heteronchoineans, is one in which there is a "ductus vaginalis" connecting the vagina to vitelline canals (see Fig. 19.10).

Figure 19.10 **Basic types of female reproductive systems in monogeneans.**

(*a*) Vagina connecting to oviduct ("true" vagina), with genitointestinal canal absent; (*b*) vagina connecting to vitelline ducts ("ductus vaginalis"), with genitointestinal canal present; *dv,* "ductus vaginalis"; *g,* gut; *gi,* genitointestinal canal; *o,* germarium; *oo,* ootype; *ov,* oviduct; *u,* uterus; *v,* "true" vagina; *vc,* vitelline canal.

From W. A. Boeger and D. C. Kritsky, "Phylogeny and a revised classification of the Monogenoidea Bychowsky, 1937 (Platyhelminthes)," in *Systematic Parasitol.* 26:1–32. Copyright © 1993 The Natural History Museum, London. Reprinted by permission.

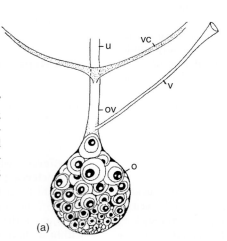

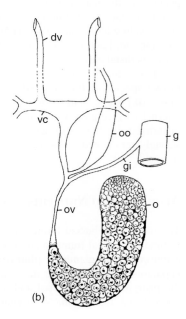

In some species, such as *Entobdella soleae* (see Fig. 19.1), the vagina may be much smaller than the penis, and sperm transfer is achieved by deposition of a spermatophore adjacent to the vagina of the mating partner rather than by direct copulation. *Diclidophora merlangi,* which does not have a vagina, practices a kind of hypodermic impregnation.[30] The suckerlike penis of an individual attaches at a ventrolateral position posterior to the genital openings of its partner, draws up a papilla of tegument into the penis, and then breaches the tegument with spines in the penis. Sperm enter the partner and make their way between cells to a seminal receptacle, a distance of 1 mm to 2 mm.

The function, if any, of the genitointestinal canal is unknown. Sometimes yolk granules and sperm are observed in the gut, presumably having arrived there through the genitointestinal canal. One hypothesis is that the canal represents a vestige of a mechanism by which eggs were passed into the intestine to be expelled through the mouth. Such a canal occurs in many turbellarians, especially polyclads, but the homology of this canal in various platyhelminth groups as well as its function in polyclads is obscure.

After an oocyte is fertilized in the oviduct or ovary itself, the zygote and attendant vitelline cells pass into the ootype, a muscular expansion of the female duct. In the species studied, the **Mehlis' gland** around the ootype comprises two cell types, mucous and serous. (The ootype epithelium may also be secretory.) The function of Mehlis' gland is not known. It was formerly thought to contribute shell material, but in Monogenoidea, as in Digenea and Cestoda, shell material evidently comes from vitelline cells. Egg shape evidently is determined by the walls of the ootype. In *Entobdella soleae* the tetrahedral egg shape is imparted by four pads in the ootype walls.[22] The eggs of many monogenes have a filament at one or both ends, also characteristic of a given species. This filament may have an adhesive property that serves to attach the egg to the host or substrate on which it falls after release into open water.

It is generally believed that protein in the eggshell is stabilized by a process of quinone tanning to form sclerotin. Some work, however, indicates stabilization may not be by quinone tanning but by means of dityrosine and disulfide links as in resilin and keratin (see Fig. 33.4).[40]

Although many eggs may be produced (*Polystoma* species may shed one to three eggs every 10 to 15 seconds), they are passed out of the worm fairly rapidly; therefore, not many may be present within the parent at one time. Some species may store a few eggs in the ootype and then pass them to the outside directly through a pore, but in most species eggs pass from the ootype into a uterus, which courses anteriorly to open into the genital atrium together with the ejaculatory duct. Hence, the uterus, at least in most cases, does not function as a vagina as it does in digeneans.

DEVELOPMENT

Life cycles of a few species have been well studied, but little or nothing is known about most. With the exception of the mostly viviparous Gyrodactylidae, monogenes usually have a single-host life cycle involving an egg, oncomiracidium, and adult. Some evidence suggests that two species

of gastrocotylids that parasitize predatory fish do not infect their definitive hosts directly but undergo a period of development on fish preyed on by the parasite's definitive hosts.[24]

Oncomiracidium

The oncomiracidium (Figs. 19.11 and 19.12) hatches from an egg and resembles a ciliate protozoan in size and shape. It is elongated and bears three zones of cilia: one in the middle and one at each end. The zones of ciliated epidermal cells are separated by an interciliary, nonnucleate syncytium. It has been shown in *Entobdella soleae* that nuclei of the interciliary regions are actually extruded during embryogenesis. Subsequently cytons of the "presumptive adult" tegument, which are located within the superficial muscle layer, extend processes out to underlay the ciliated cells and join the syncytial interciliary regions. The larva rapidly sheds its ciliated cells on attachment to a host; stimulus for this shedding in *E. soleae* is mucus from the host epidermis. The process takes only 30 seconds.

Oncomiracidia have cephalic glands with efferent ducts opening on the anterior margin and, as previously noted, have one or two pairs of eyes. The digestive tract is well differentiated, and excretory pores are already formed. The posterior end always is developed into an attachment organ that bears hook sclerites, and these sclerites usually are retained

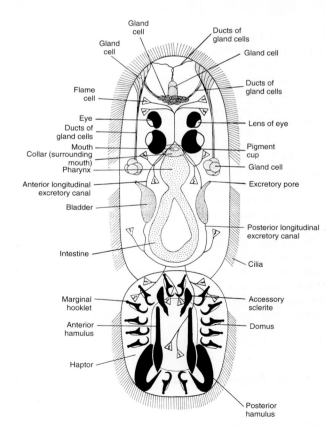

Figure 19.11   **The oncomiracidium of *Entobdella soleae* (ventral view).**

From G. C. Kearn, *Ecology and physiology of parasites,* edited by M. Fallis. Copyright © 1971 University of Toronto Press, Toronto. Reprinted by permission.

Figure 19.12 Oncomiracidia of *Callorhynchicola multitesticulatus* from the elephant fish, *Callorinchus milli.*
The larva on the left is shown in dorsal view, with the anterior end at the bottom; the larva on the right is in ventral view, with the posterior end at the bottom.

Courtesy of M. Beverley-Burton, F. R. Allison, and J. McKenzie, University of Guelph, Guelph, Ontario, and University of Canterbury, Christchurch, New Zealand.

in adults. Larvae swim about until they contact a host; then they attach, lose their ciliated cells, and develop into adults.

The free-swimming life of an oncomiracidium is short, and its potential hosts are widely dispersed most of the time. In addition, potential hosts may not even be present in an aquatic habitat except during breeding season. Thus, it is of selective value for the worm's egg production to be closely related to its host's reproduction—to coincide, for instance, with a time when the host is concentrated in spawning areas. Other features of a host's habits also may enhance chances for infection. Such correlation has been shown in several species, and similar adaptations are probably widely prevalent.[26] Some of these adaptations will be illustrated in the discussion of life cycles that follows.

Subclass Polyonchoinea

Order Dactylogyridea
Members of Dactylogyridea (Fig. 19.13) usually have two pairs of anchors. Species in this order occur on both marine and freshwater fishes, and they are among the most commonly encountered monogenes on familiar fishes such as minnows and sunfish, often being found on the gills.

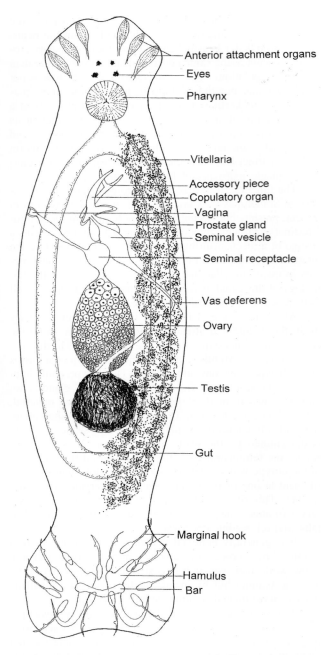

Figure 19.13 Generalized anatomy of *Dactylogyrus* sp., ventral view.
Eyespots are visible through the worm but are actually on the dorsal surface. Hamuli emerge on the dorsal surface, but the haptor is flattened as it would be with a living worm under a cover glass. Only the left side of the vitellaria is shown; the glands would occupy much of the worm's right side.

Drawn by John Janovy Jr. from various sources.

***Dactylogyrus* Species** A large number of species in this genus have been described, and some of them, such as *Dactylogyrus vastator, D. anchoratus,* and *D. extensus,* are of great economic importance as pathogens of hatchery fish. *Dactylogyrus* species typically have large anchors on their haptor and live on gill filaments of their host. Heavy infections cause loss of blood, erosion of epithelium, and access for secondary bacterial or fungal infections. Irritation to the gills stimulates increased mucus production, which often smothers the fish. Massive die-offs due to monogene infection are common in the crowded situations of fish culture ponds.

The life cycles and factors influencing the economically important species are reasonably well-known and correspond to the preceding general outline.[9] *Dactylogyrus vastator* on carp shows marked seasonal fluctuations correlated with temperature. Each worm deposits 4 to 10 eggs per 24 hours during the summer, and this rate increases with increasing temperature. The eggs require four to five days with temperatures between 20°C and 28°C for embryonation, but this rate slows as temperatures drop to 4°C, at which point development is completely suppressed. Adult worms are adversely affected by lower temperatures so that the number of parasites on a fish decreases greatly during winter. The net effect is that the parasite population builds up over the summer, but eggs deposited toward the end of the season winter over and result in a mass emergence in spring to infest young-of-the-year fish.

Order Gyrodactylidea

Members of this order have two seminal vesicles, large vitelline follicles, and hinged hooks (Fig. 19.14). The group includes both oviparous and viviparous species, but most are viviparous. Species in this order are common on freshwater fishes and may be very active when observed on fins of a live host.

***Gyrodactylus* Species.** Despite the (unfortunate) similarity in name with *Dactylogyrus, Gyrodactylus* spp. are very different organisms. They are also economically significant as important pests, particularly of trout, bluegills, and goldfish in fish ponds.[13] Family Gyrodactylidae has viviparity in many members, exhibiting a type of sequential polyembryony. Instead of having a discrete ovary, viviparous *Gyrodactylus* species posses a chamber posterior to the uterus, called an **egg cell formation region (ECFR),** in which oocytes develop and sperm are stored.[10] The ovary consists of a layer of cells in this chamber. Both the ECFR and uterus are lined with syncytium, which is thickened in a uterus with an early embryo and presumably functions in transport of nutrients. Vesicles in the embryonic tegument suggest exchange of materials with the parent by means of endocytosis.[11] Young are retained in the uterus until they develop into functional subadults. Inside such a developing juvenile, one can often see a second juvenile developing, with a third juvenile inside of it and a fourth inside the third; that is, several generations in a single worm! After birth the young worm begins feeding on its host and gives birth to the juvenile "remaining" inside. Only then can an egg from its own ovary be fertilized to repeat the sequence. Because only a day or so is required for a worm to mature after birth and give birth to another worm already developing within it, massive infections can build up quickly.

Not having an oncomiracidium, gyrodactylids must depend on transmission of the adult or subadult from one host to another. Because these forms appear unable to swim, it is clear that a prospective host must be quite close to the worm's current host for transfer to take place. However, infections seem to spread easily in hatcheries, so that lack

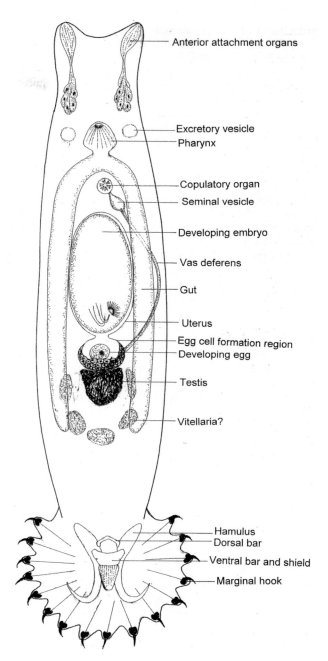

Figure 19.14   Generalized anatomy of a viviparous *Gyrodactylus* sp., ventral view.

Cell masses labeled "vitellaria?" have been identified as such in the older literature but evidently have unknown functions; viviparous species of this genus do not have vitellaria in the same sense as most other flatworms with ectolecithal eggs.

Drawn by John Janovy Jr. from various sources.

of a swimming stage is not a serious barrier to transmission. These worms also sometimes leave a dead fish and remain active for several hours. Experiments have shown that *G. salaris* can leave a dead salmon and infect a live eel. The same studies have shown that, when infected salmon were removed from a tank, an uninfected eel placed in the same tank the next day became infected.[2]

Subclass Polystomatoinea

Family Polystomatidae

Polystoma integerrimum (Fig. 19.15) is parasitic in the urinary bladder of Old World frogs. It is of particular interest because the worm's reproductive cycle is synchronized with that of its host by means of host hormones, a mechanism that provides a ready supply of hosts to hatching oncomiracidia. Furthermore, two different types of adults develop: normal and neotenic.

Worms are dormant during the winter while the frogs hibernate, but they become active in spring along with their hosts. When the frogs' gonads begin to swell and produce gametes, worms begin to copulate and produce eggs that are released into the urine. These eggs are then voided into water in the frogs' spawning area. Depending on temperature, oncomiracidia hatch in 20 to 50 days. By then the frog eggs have developed into tadpoles that will be the next host generation. Tadpoles have gills and ventilate by sucking water into their mouths, then pushing it over the gills and out through slits. An oncomiracidium contacting a gill attaches, metamorphoses, and begins producing eggs in 20 to 25 days.

The gill form of *P. integerrimum* has a narrower body than the bladder form, and the opisthaptor is not as sharply set off from the body. The gill form's intestine has fewer lateral branches, and the ovary is a different shape from that of bladder worms. Furthermore, there is no uterus or genitointestinal canal. Some authors have considered the gill stage neotenic.[1]

When the water is warm, eggs of *P. integerrimum* hatch in 15 to 20 days. These larvae also attach to tadpole gills, but migrate to the bladder when tadpoles begin their

metamorphosis, including resorption of gills. This migration occurs over the tadpole's ventral skin and takes only about a minute.[24] Once in the bladder, the worms develop slowly, requiring four to five years to mature and begin egg production.

A similar species, *P. nearcticum*, occurs in tree frogs in the United States. It also has a gill form but no slowly developing form that migrates to the bladder. Instead, oncomiracidia enter the host's cloaca directly when they contact metamorphosing tadpoles. Larvae of *Protopolystoma xenopi* also directly enter the cloaca of their host, *Xenopus laevis*. Interestingly, *X. laevis* remains in water all year, and reproduction of *P. xenopi* continues correspondingly.

At the other extreme are *Pseudodiplorchis americanus* and *Neodiplorchis scaphiopodis*, parasites in the urinary bladder of spadefoot toads (*Scaphiopus* spp.) in Arizona. These toads live in one of the hottest and driest areas in the United States and spend about 10 months per year beneath the soil in a state of torpor. They breed during only one to three nights per year in temporary pools formed by brief desert rains. These conditions present an extraordinary challenge to a parasite that depends on an aqueous environment for transmission, and transmission must be exquisitely coordinated with activities of their hosts. During the time that toads are in hibernation, the encapsulated oncomiracidia become fully developed in the uterus of adult worms.[51] When toads enter water, larvae are deposited, pass out with urine, hatch within seconds, and are fully infective for the next host. The oncomiracidia are much larger than most other oncomiracidia (up to 600 μm), have a much longer free-swimming life (up to 48 hours), and are more resistant to drying (up to one hour). Thus, if they do not reach another host during the first night when toads are spawning, they have a good chance of succeeding the next night. When oncomiracidia contact another spadefoot toad as it is floating with the end of its nose above water, they crawl on the toad's skin up out of the water and enter its nares. Then they migrate to the lungs, undergo a period of development, and finally migrate to the urinary bladder by passing through the stomach and intestine.[52]

Obviously, these worms pass through many relatively hostile environments in a fairly short length of time before reaching their final home in the urinary bladder. Unique vesicles found in their tegument allow such adventurous migrations.[12] Contents of these vesicles are released during migration through the gut, producing a thick, presumably protective glycocalyx and mucin layer over the worm. The toads feed gluttonously during their brief annual stay above ground, consuming nearly 30% of their body weight in a single feed and replenishing the fat stores needed to sustain hibernation in just two nights of eating. This massive food intake may stimulate the worms to migrate from the lungs and into the bladder. The vesicles that protect them are an adaptation for running the digestive system gauntlet, an event that consumes about 30 minutes of their year-long lives.[12]

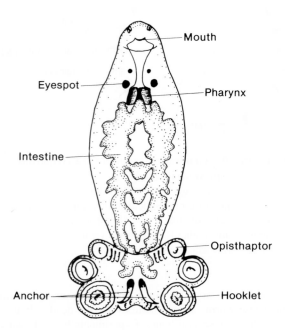

Figure 19.15 *Polystoma integerrimum,* a parasite of Old World frogs.

From E. Zeller, "Untersuchungen über die Entwicklung und den Bau des *Polystomum integerrimum* Rud," in *Zeitschr. Wissensch. Zool.* 22:1–28, 1872.

Subclass Oligonchoinea

Diplozoon paradoxum

Diplozoon paradoxum (Fig. 19.16) is a common parasite on the gills of European cyprinid fishes. Like *Dactylogyrus*

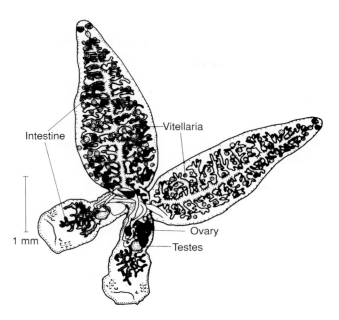

Figure 19.16   *Diplozoon paradoxum*, a parasite of freshwater fishes in Europe and Asia.

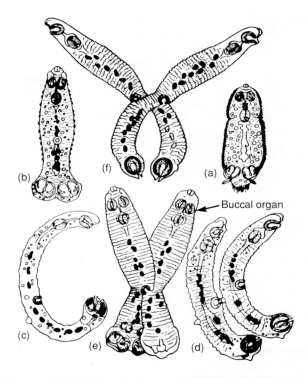

Figure 19.17   Development of *Diplozoon paradoxum.*
(*a*) Freshly hatched, free-swimming juvenile. (*b*) Diporpa juvenile. (*c–f*) Diporpa juveniles attaching themselves to one another.

spp., *Diplozoon paradoxum* exhibits a strong seasonal variation in its reproductive activity. Virtually no gametes are produced during winter, but gonads begin to function during the spring, reaching a peak during May to June and continuing through the summer. Eggs, which have a long, coiled filament at their ends, can hatch about 10 days after deposition; light intensity and water turbulence, as might be caused by host feeding or spawning activity, stimulate hatching. Oncomiracidia bear two clamps on their haptor with which they attach to a gill filament; they lose their cilia almost immediately. Worms feed and begin to grow, adding another pair of clamps to the haptor. A small sucker also appears on the ventral surface and a tiny papilla appears on the dorsal surface, slightly more posterior than the sucker. When this stage (Fig. 19.17) was first discovered, it was thought to represent a new genus and was named *Diporpa.* When *Diporpa* was recognized as a juvenile stage of *Diplozoon,* the term **diporpa** was applied to the stage. Curiously, in contrast to *Dactylogyrus* spp., *Diplozoon paradoxum* (even its diporpae) rarely infects young-of-the year fish.

A diporpa juvenile can live for several months, but it cannot develop further until encountering another diporpa; unless this happens, the diporpa usually perishes by winter. When one diporpa finds another, each attaches its sucker to the dorsal papilla of the other. Thus begins one of the most intimate associations of two individuals in the animal kingdom. The two worms fuse completely, with no trace of partitions separating them. The fusion stimulates maturation. Gonads appear; the male genital duct of one terminates near the female genital duct of the other, permitting crossfertilization. Two more pairs of clamps develop in the opisthaptor of each. Adults apparently can live in this state for several years.

PHYLOGENY

Monogenes are excellent material for studies of phylogeny and host specificity because of the taxonomic diversity of both worms and their fish hosts. For example, several hundred species of *Gyrodactylus* have been described and placed in six subgenera based on excretory system characteristics.[31] Molecular analysis of these worms shows that subgenus *Limnonephrotus,* common on freshwater fishes including trout and salmon, probably originated on minnows, but its species evidently colonized other fish families at least eight times, subsequently undergoing rapid adaptive radiation.[58] Morphological and molecular studies using worms from another family, Dactylogyridae, on cichlid fishes, also revealed host switching and previously unrecognized sister-species relationships among worms thought to be a single species.[39] Monogenes are useful in studies of more ancient evolutionary events, too; molecular phylogenies of Polystomatoinea (p. 294) show the group evidently originated in lungfish, but lineages in amphibians and turtles then diverged in the late Devonian or early Carboniferous.[53]

Although some modern authors have persisted in allying Monogenea with Digenea and making Aspidobothrea an order or subclass within Trematoda,[47] Monogenoidea has historically been considered the sister group to Cestodaria, an infraclass that includes cohorts Gyrocotylidea, Amphilinidea, and Eucestoda (see chapter 13, Boeger and Kritsky,[4, 5] and

Brooks and McLennan[7]). The primary structural feature uniting all these taxa is a posterior adhesive organ with hooks. Recent studies, however, suggest monogenes form a sister group to a clade composed of trematodes and cestodes.[36]

Major systematic issues are placement of the group within Neodermata, monophyly of Monogenoidea and relationships among families.[5, 36] Some molecular studies support the hypothesis of monophyly, although selection of an outgroup may influence this conclusion, and more recent work provides evidence that Monogenoidea is polyphyletic.[35, 37] Subclasses as listed next are now well accepted, as is the division of Heteronchoinea into two infrasubclasses, again supported by molecular studies. Relationships between families are not so clear, however, and some families are considered polyphyletic.[5]

An extensive and detailed revision of the group was published by Boeger and Kritsky, and the following classification is based on their work.[4, 5] These authors provided a taxonomic scheme that reflects evolutionary relationships as we currently understand them. The characteristics given are major synapomorphies that distinguish groups in the Boeger and Kritsky[4] cladogram.

CLASSIFICATION OF CLASS MONOGENOIDEA

Hermaphroditic, dorsoventrally flattened, elongated or oval worms with a syncytial tegument; conspicuous posterior adhesive organ present (haptor) that is muscular, sometimes divided into loculi, usually with sclerotized anchors, hooks, and/or clamps, often subdivided into individual suckers or clamps without sclerites; an anterior adhesive organ (prohaptor) usually present, consisting of one or two suckers, grooves, glands, or expanded ducts from deeper glands; eyes, when present, usually of two pairs; mouth near anterior, pharynx usually present; gut usually with two simple or branched stems often anastomosing posteriorly, rarely a single median tube or sac; male genital pore usually in atrium common with female pore; genitointestinal duct present or absent; reproduction oviparous or viviparous; vitelline follicles extensive, usually lateral; two lateral osmoregulatory canals present, each with an expanded vesicle opening dorsally near anterior end; parasites on or in aquatic vertebrates, especially fishes, or rarely on aquatic invertebrates; cosmopolitan.

Subclass Polyonchoinea

Mouth ventral; sperm microtubules lying along one-fourth of cell periphery; male copulatory organ sclerotized, muscular, elongated, with spines absent; 14 marginal and two central hooks on the oncomiracidium.

Order Dactylogyridea

Two pairs of ventral anchors; sperm with one axoneme.

Suborder Dactylogyrinea

Twelve marginal and two central hooks on the oncomiracidium; eight marginal, two central, and four dorsal hooks on the adult. Families: Dactylogyridae, Pseudomurraytrematidae, Diplectanidae.

Suborder Tetraonchinea

Gut single. Families: Tetraonchidae, Neotetraonchidae.

Suborder Amphibdellatinea

Eyes absent from the oncomiracidium. Family: Amphibdellatidae.

Suborder Neodactylodiscinea

Bars absent; two lateral "true" vaginas. Family: Neodactylodiscidae.

Suborder Calceostomatinea

Twelve marginal and two central hooks on the oncomiracidium; 12 marginal and two central hooks on the adult. Family: Calceostomatidae.

Order Gyrodactylidea

Two ventral bars and 16 marginal hooks on the oncomiracidium; 16 marginal hooks on the adult; hooks hinged. Families: Gyrodactylidae, Anoplodiscidae, Bothitrematidae, Tetraonchoididae.

Order Lagarocotylidea

Gut diverticula absent; ventral mouth; oral sucker absent, germarium intercecal; ventrolateral vagina; 16 hooks. Family: Lagarocotylidae.

Order Montchadskyellidea

Gut diverticula present; germarium/oviduct looping around right caecum; one midventral "true" vagina. Family: Montchadskyellidae.

Order Capsalidea

Single genital aperture marginal. Families: Acanthocotylidae, Capsalidae, Dionchidae.

Order Monocotylidea

Sperm with two axonemes, one reduced; 14 marginal hooks on both oncomiracidium and adult. Families: Monocotylidae, Loimoidae.

Subclass Heteronchoinea

Two ventrolateral ducti vaginalis; genitointestinal canal present; four haptoral suckers associated with hooks; spermatozoa with lateral microtubules.[5]

Infrasubclass Polystomatoinea

More than two testes; gastrointestinal canal present; haptoral suckers, with associated hook, present in the adult; three pairs of haptoral suckers; two lateral "ducti vaginalis."

Order Polystomatidea

Egg filaments absent. Families: Polystomatidae, Sphyranuridae.

Infrasubclass Oligonchoinea

One pair of haptoral suckers; eyes absent from the oncomiracidium; "crochet en fleau" hooklike; one pair of lateral sclerites; four pairs of haptoral suckers; gut diverticula present; midsclerite terminates in hook.

Order Mazocraeidea

Two oral suckers present; germarium elongated, inverted, *U*-shaped; egg with two filaments; midsclerite flared or truncated; one pair of eyes fused on the oncomiracidium; eyes absent from the adult; two pairs of lateral sclerites.

Suborder Microcotylinea

Germarium elongated, double inverted, *U*-shaped. Families: Axinidae, Diplasiocotylidae, Heteraxinidae, Microcotylidae, Allopyragraphioridae, Diclidophoridae, Pterinotrematidae, Rhinecotylidae, Pyragraphoridae, Heteromicrocotylidae.

Suborder Hexostomatinea

Spines of male copulatory organ absent; two pairs of eyes in oncomiradicium. Family: Hexostomatidae.

Suborder Discocotylinea

Haptoral suckers present on oncomiracidium; six marginal hooks on oncomiracidium; anchor absent from all developmental stages. Families: Discocotylidae, Diplozoidae, Octomacridae.

Suborder Gastrocotylinea

Accessory sclerite parallel to midsclerite (see Fig. 19.6). Families: Anthocotylidae, Pseudodiclidophoridae, Protomocrocotylidae, Allodiscocotylidae, Pseudomazocreaidae, Chauhaneidae, Bychowskycotylidae, Gastrocotylidae, Neothoracocotylidae, Gotocotylidae.

Suborder Mazocraeinea

Posterior midsclerite platelike; two pairs of lateral sclerites, posterior pair broken. Families: Plectanocotylidae, Mazoplectidae, Mazocraeidae.

Order Diclybothriidea

Haptoral suckers present in appendix; lateral sclerites absent. Families: Diclybothriidae, Hexabothriidae.

Order Chimaericolidea

Fourteen marginal hooks on the oncomiracidium and adult; germarium lobed; eyes absent from the oncomiracidium. Family: Chimaericolidae.

Learning Outcomes

By the time a student has finished studying this chapter, he or she should be able to:

1. Label diagrams of both a *Dactylogyrus* and a *Gyrodactylus* species.
2. Draw or label diagrams of several different types of haptors.
3. Explain the criteria used for classification of Monogenoidea.
4. Write an extended paragraph on the subject of monogene life cycles.
5. Write an extended paragraph on the subject of monogene infection site specificity.

References

References for superscripts in the text can be found at the following Internet site: www.mhhe.com/robertsjanovynadler9e

Additional Readings

Bakke, T. A., P. D. Harris, and J. Cable. 2002. Host specificity dynamics: Observations on gyrodactylid monogeneans. *Int. J. Parasitol.* 32:281–308.

Buchmann, K., and T. Lindenstrøm. 2002. Interactions between monogenean parasites and their fish hosts. *Int. J. Parasitol.* 32:309–319.

Gelnar, M., I. D. Whittington, and L. A. Chisholm. 1998. Preface. *Int. J. Parasitol.* 28:1479–1480. This special issue of the journal is entirely devoted to the Third International Symposium on Monogenea, held in Brno, Czech Republic, 25 to 30 August 1997.

Gusev, A. V. 1995. Some pathways and factors of monogene microevolution. *Can. J. Fish. Aquatic Sci.* 52(suppl):52–56.

Hargis Jr., W. J., A. R. Lawler, R. Morales-Alamo, and D. E. Zwerner. 1969. *Bibliography of the monogenetic trematode literature of the world. Special Scientific Report* 55:1–95. Gloucester Point, VA: Virginia Institute of Marine Science.

Hendrix, S. 1994. *Marine flora and fauna of the eastern United States.* Platyhelminthes: Monogenea. NOAA Tech. Rept. NMFS 121.

Kearn, G. C. 1999. The survival of monogenean (platyhelminth) parasites on fish skin. *Parasitol.* 119(suppl.):S57–S88.

Llewellyn, J. 1963. Larvae and larval development of monogeneans. In B. Dawes (Ed.), *Advances in parasitology* 1. New York: Academic Press, Inc., pp. 287–326.

Yamaguti, S. 1963. *Systema Helminthum* 4. New York: Interscience Publishers.

20

Cestoidea: Form, Function, and Classification of Tapeworms

He was as fitted to survive in this modern world as a tapeworm in an intestine.

—William Golding *(Free Fall)*

Fear and superstition still abound among laypersons, who generally view tapeworms as the lowliest and most degenerate of creatures (Fig. 20.1). Most of the repugnance with which people regard these animals derives from the fact that tapeworms live in the intestine and are only seen when they are passed with the host's feces. Furthermore, tapeworms seem to be generated spontaneously, and mystery is nearly always accompanied by fear. Finally, in a few instances their presence initiates disease conditions that traditionally have been difficult to cure. A scientific approach to cestodology, however, has increased understanding of tapeworms and shown that they are one of the most fascinating groups of organisms in the animal kingdom. Their complex life cycles and intricate host-parasite relationships are rivaled by few known organisms.

In ancient times Hippocrates, Aristotle, and Galen appreciated the animal nature of tapeworms.[35] The Arabs suggested that segments passed with the feces were a separate species of parasite from tapeworms; they called these segments the *cucurbitini,* after their similarity to cucumber seeds.[47] Andry, in 1718, was the first to illustrate the scolex of a tapeworm from a human. Three common species in humans, *Taenia saginata, Taenia solium,* and *Diphyllobothrium latum,* were confused by all scientists until the brilliant efforts of Küchenmeister, Leuckart, Mehlis, Siebold, and others in the 19th century determined both the external and internal anatomy of these and other common species. These researchers also showed conclusively that bladder worms, hydatids, and coenuri were juvenile tapeworms and not separate species or degenerate forms in improper hosts. Although these organisms have been removed from the realms of ignorance and superstition within the past 150 years, much misconception persists.

Sexually mature tapeworms live in the intestine or its diverticula (rarely in the coelom) of all classes of vertebrates. Two forms are known that mature in invertebrates: *Archigetes* spp. (order Caryophyllidea) in the coelom of a freshwater oligochaete and *Cyathocephalus truncatus* (order Spathebothriidea) in the hemocoel of an amphipod.[1]

FORM AND FUNCTION

Strobila

The strobila (Fig. 20.2) of cestodes is a structure unique among Metazoa. Typically it consists of a linear series of sets of reproductive organs of both sexes; each set is referred to as a **genitalium,** and the area around it is a **proglottid** or **proglottis.** Cestodes with multiple proglottids are described as **polyzoic,** but of Eucestoda members of order Caryophyllidea have only one genitalium and so are **monozoic** (see Fig. 21.11). Some workers advise avoiding use of the terms *polyzoic* and *monozoic* because such usage implies that polyzoic tapeworms are chains of zoids (individuals), an hyphothesis that is no longer tenable.[72] Nonetheless, in the absence of a better term, we shall retain *polyzoic* to describe tapeworms with multiple genitalia. Polyzoic cestodes may have only a few proglottids, but others may have thousands. Usually there are constrictions between proglottids, and the worms are said to be segmented. However, tissues such as tegument and muscle are continuous between proglottids, and no membranes separate them. Some workers have contended, therefore, that the word *segment* is a misnomer.[71] Because some polyzoic cestodes lack constrictions of any kind between proglottids (order Spathebothriidea), the words *segment* and *proglottid* are not synonymous, although they are often used as such by parasitologists. Finally, segments of tapeworms should not be confused with segments or metameres of metameric animals such as annelids and arthropods (p. 489).

In many polyzoic species new proglottids (and segments) are continuously differentiated near the anterior end in a process called **strobilation.** Each segment moves toward the posterior end as a new one takes its place and during the process becomes sexually mature. By the time they approach the posterior end of the strobila, genitalia have copulated and produced eggs. A proglottid can copulate with itself, with others in its strobila, or with those

in other worms, depending on the species. After a segment contains fully developed eggs or shelled embryos, it is said to be **gravid.**

When a segment reaches the end of its strobila, it often detaches and passes intact out of the host with feces, as in *Taenia* spp., or disintegrates en route, releasing eggs, as in *Hymenolepis* spp. This process is called **apolysis.** In some species eggs are released from a gravid segment through a uterine pore, such as in *Diphyllobothrium* spp., or through tears or slits in the segment, as in Trypanorhyncha; a segment only detaches when it is senile or exhausted (**pseudoapolysis** or **anapolysis**). In some forms segments may be shed while immature and lead an independent existence in the gut while developing to maturity (**hyperapolysis),** as in some Tetraphyllidea. If the posterior margin of a segment overlaps the anterior of the following one, the strobila is said to be **craspedote;** if not, it is **acraspedote** (Fig. 20.3).

Scolex

Most tapeworms bear a "head," or **scolex** (plural **scolices**), at the anterior end that may be equipped with a variety of holdfast organs that function to maintain the position of the animal in the gut (Fig. 20.4). A scolex may bear suckers, grooves, hooks, spines, glands, tentacles, or combinations of these (Fig. 20.5). However, a scolex can be simple or absent altogether. In some forms the holdfast function of the scolex is lost early in life, and the anterior end of the strobila becomes distorted into a **pseudoscolex** to function as a holdfast (Fig. 20.6). Some species penetrate the gut wall of the host to a considerable distance, with the scolex and a portion of the strobila then encapsulated by reacting host tissues.

Suckerlike organs on scolices of tapeworms can be divided into three types: **acetabula, bothria,** and **bothridia.** An acetabulum is more or less cup shaped, circular or oval in outline, with a heavy muscular wall. There are normally four acetabula on a

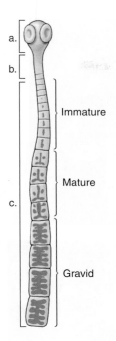

Figure 20.2   Generalized diagram of a tapeworm.

Note scolex (*a*), neck (*b*), and strobila (*c*).

Drawing by William Ober and Claire Garrison.

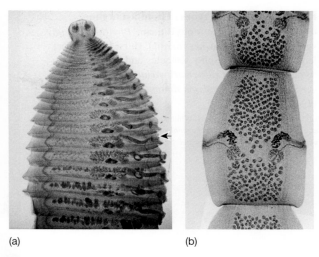

(a)   (b)

Figure 20.3   Proglottids.

(*a*) Scolex and proglottids of *Paranoplocephala mamillana,* a craspedote cestode. Mature proglottids of *P. mamillana (arrow)* are about 5 mm wide. (*b*) A proglottid of *Dipylidium caninum,* an acraspedote species whose proglottids are about 2 mm wide.

Courtesy of Jay Georgi.

scolex, spaced equally around it. Bothridia usually are in groups of four; are quite muscular, projecting sharply from the scolex; and can have highly mobile, leaflike margins. Bothria are usually two in number, although as many as six may occur and take the form of shallow pits or longer grooves. They are arranged either in lateral or dorsoventral pairs. Accessory suckers sometimes occur, and most cestodes have a variety of proteinaceous hooks for anchoring the scolex to the host gut.

In acetabulate worms hooks often are arranged in one or more circles anterior to the suckers and are borne on a protrusible, dome-shaped area on the apex of the scolex called a **rostellum.** Both presence or absence and shape and arrangements of hooks are of great taxonomic value. If a rostellum is armed with hooks, it is supplied internally with a heavy muscular pad, which becomes flat and disc shaped when the hooks attach to a host's gut wall. Retraction of the central area of the pad allows withdrawal of the hooks. In some species the rostellum can be withdrawn into the end of the scolex. Members of order Cathetocephalidea have a very unusual scolex (Fig. 20.5*m*). Cathetocephalideans have so far only been found in carcharhinid sharks. Their scolex bears no suckers or hooks. It is a single, transversely extended organ with a fleshy pad, band of papillae, and wrinkled base.[15]

Various kinds of gland cells occur in the scolices of a variety of tapeworms, but their function(s) remain(s) enigmatic.[62, 63] In some Pseudophyllidea and Caryophyllidea, secretions of the glands may aid in adhesion of the scolex to the host gut mucosa.[40, 44] Contents of one type of gland in the pseudophyllidean *Diphyllobothrium dendriticum* are expelled within three days of infection of the definitive host; another gland type remains active and is associated with the nervous system in the scolex.[40]

Hymenolepis diminuta, a cyclophyllidean with an unarmed rostellum, has an invagination of the apical tegument termed an **apical organ** or **anterior canal** (see Fig. 20.4).[109] Modified tegumentary cytons (see pp. 212 and 305) secrete material into the apical organ and through the surrounding rostellar tegument.[110] It is possible that these materials play some regulatory role in the development of the worms; there is circumstantial evidence that they are antigenic.[28, 50, 80] Apical organs are found in many other cestodes, but they may not be homologous or even structurally similar to that of *H. diminuta.* Apparently, similar and homologous structures are found in the Proteocephalata where, at least in certain cases, their secretion has proteolytic activity and probably functions in penetration (p. 345).[19]

The scolex contains the chief neural ganglia of the worm (as we will examine further), and it bears numerous sensory endings on its anterior surface, probably detecting both physical and chemical stimuli. Such sensory input may allow optimal placement of the scolex and entire strobila with respect to the gut surface and physicochemical gradients within the intestinal milieu.

Commonly, between the scolex and the strobila lies a relatively undifferentiated zone called the **neck,** which may be long or short. It contains stem cells that evidently are responsible for giving rise to new proglottids. In the absence of a neck, similar cells may be present in the posterior portion of the scolex. Praziquantel, a chemotherapeutic agent highly effective against cestodes, preferentially damages tegument of the neck region and leaves tegument of segments farther down the strobila unaffected.[6]

Tegument

Cestodes lack any trace of a digestive tract and therefore must absorb all required substances through their external

Figure 20.4 **Two types of holdfast organs.**
(*a*) Bothridea of a tetraphyllidean, *Phyllobothrium* sp., with accessory sucker *(arrow).* (First bar in upper left = 10 μm). (*b*) Spiny tentacles and bothridea of a trypanorhynchan, *Callitetrarhynchus gracilis* (×80). (*c*) Scolex of *Hymenolepis diminuta.* Note the apical organ *(arrow).* (×200)

(*a*), (*b*) Courtesy of Frederick H. Whittaker. (*c*) From J. E. Ubelaker et al., "Surface topography of *Hymenolepis diminuta* by scanning electron microscopy," in *J. Parasitol.* 59:667–671. Copyright © 1973.

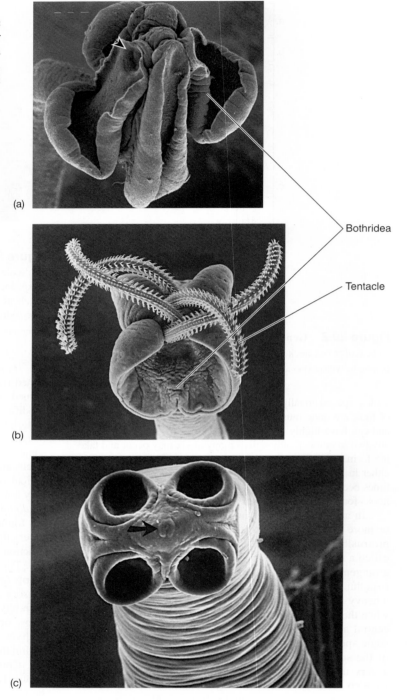

(a)

Bothridea

Tentacle

(b)

(c)

covering. Thus, structure and function of the body covering have been of great interest to parasitologists, who have used electron microscopy and radioactive tracers to contribute much to this area of cestodology. Before 1960 the body covering of cestodes and trematodes was commonly referred to as a *cuticle,* but it is now known that it is a living tissue with high metabolic activity, and most parasitologists prefer the more noncommital term **tegument.**

Tegumental structure is generally similar in all cestodes studied, differing in details according to species. The basic plan is similar to that of Trematoda. One major difference is

that the outer limiting membrane of cestode tegument projects out toward the host as numerous finger-shaped tubes called **microtriches** (singular **microthrix**) (Fig. 20.7–Fig. 20.9). Microtriches are similar in some respects to microvilli found on gut mucosal cells and other vertebrate and invertebrate transport epithelia, and they completely cover the worm's surface, including its suckers. They have a dense distal portion set off from the base by a multilaminar plate (see Fig. 20.9). The distal portion can take one of a variety of forms in different cestode groups. The cytoplasm of the base is continuous with that of the rest of the tegument, and the

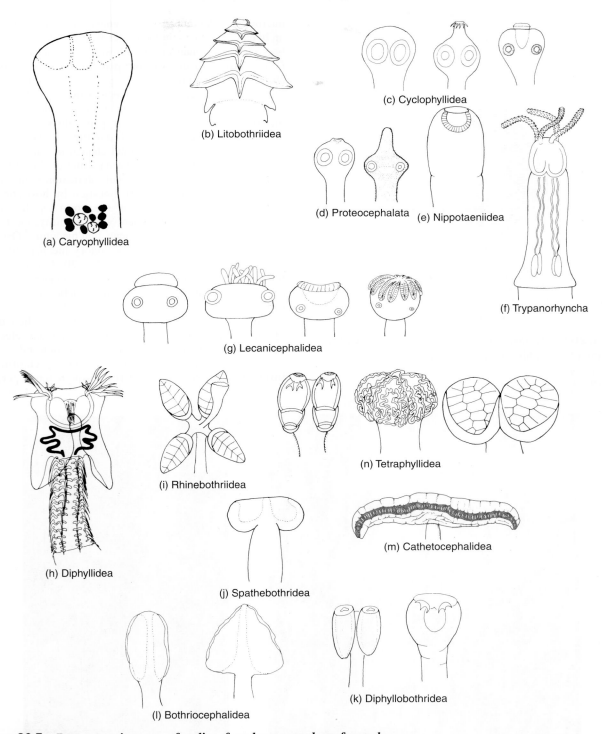

Figure 20.5   **Representative types of scolices found among orders of cestodes.**
(*a*) Caryophyllidea, (*b*) Litobothriidea, (*c*) Cyclophyllidea, (*d*) Proteocephalata, (*e*) Nippotaeniidea, (*f*) Trypanorhyncha, (*g*) Lecanicephalidea, (*h*) Diphyllidea, (*i*) Rhinebothriidea, (*j*) Spathebothriidea, (*k*) Diphyllobothriidea, (*l*) Bothriocephalidea, (*m*) Cathetocephalidea.

Mostly from G. D. Schmidt, *How to know the tapeworms*, 1970. Wm. C. Brown Publishers, Inc. Reprinted by permission.

entire structure is covered by a plasma membrane. Micro-triches serve to increase absorptive area of the tegument. They can vary dramatically from the form just described. Microtriches on many tetraphyllideans and trypanorhynchs (p. 346) are surprisingly ornate.[16] Trypanorhynchs may have four kinds of microtriches (palmate, filamentous, hairlike, and cilialike) on their scolex,[80] and the palmate microtriches may be 3, 5, or 6 fingered (see Fig. 20.7*b*).

A layer of carbohydrate-containing macromolecules, or a **glycocalyx,** is found on the surface membrane of

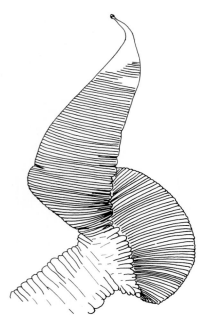

Figure 20.6 *Fimbriaria fasciolaris,* a tapeworm with a pseudoscolex in addition to a tiny true scolex.

Source: H. O. Mönnig, *Veterinary helminthology and entomology,* 1934, London: William Wood.

microtriches. A number of phenomena, apparently depending on interaction of certain molecules with the glycocalyx, have been reported: enhancement of host amylase activity; inhibition of host trypsin, chymotrypsin, and pancreatic lipase; absorption of cations; and adsorption of bile salts. Several of these phenomena seem to depend on adsorption of the molecules to the glycocalyx, but present evidence suggests that this is not the case with trypsin inhibition.[103] When incubated in the presence of *H. diminuta,* trypsin seems to undergo a subtle conformational change that decreases its proteolytic activity. The functional value of such phenomena to the worm is uncertain, but interaction with nutrient absorption, protection against digestion by host enzymes, and maintenance of integrity of the worm's surface membrane may be involved. Whatever its identity, the trypsin inhibitor is liberated into the incubation medium.[84] Oaks and Holy[77] described two distinct secretory mechanisms from cestode tegument, one for vesicles (Fig. 20.10) and the other for endogenous macromolecules. There is evidence for presence of a *G* protein–linked signal transduction system on the surface membrane.[120]

Beneath the microtriches lies a layer called **distal cytoplasm** that contains abundant vesicles and electron-dense bodies as well as numerous mitochondria. The distal cytoplasm is connected to cytons by **channels** or **internuncial processes** that run through the superficial

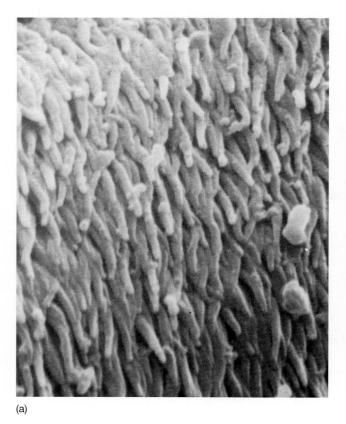

(a)

(b)

Figure 20.7 **Microtriches.**

(*a*) Posteriorly directed microtriches on the surface of a proglottid of *Hymenolepis diminuta.* (×44,925) (*b*) Five-to six-fingered palmate microtriches on bothridial surface of a juvenile trypanorhynch (plerocercus, p. 326). (*Bar* = 1 μm)

(*a*) From J. E. Ubelaker et al., "Surface topography of *Hymenolepis diminuta* by scanning electron microscopy," in *J. Parasitol.* 59:667–671. Copyright © 1973. (*b*) From H. W. Palm and R. M. Overstreet, "*Otobothrium cysticum* (Cestoda: Trypanorhyncha) from the muscle of butterfishes (Stromateidae)," in *Parasitol. Res.* 86:41–53. Copyright © 2000.

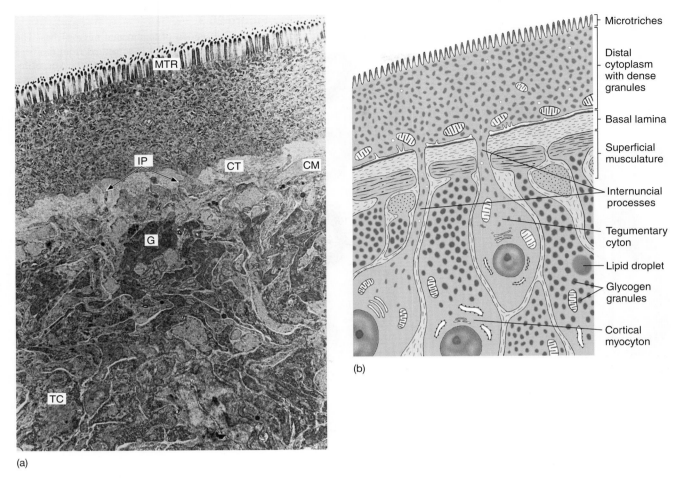

(a)

(b)

Figure 20.8   **Longitudinal section through immature proglottid of** *Hymenolepis diminuta,* **showing nature of tegumentary cortical region.**

Basal tegumentary cytons (perikarya, *TC*) are surrounded by glycogen-filled processes (*G*) of cortical myocytons. Internuncial processes (trabeculae, *IP*) from tegumentary cytons extend through longitudinal and circular (*CM*) muscles as well as a connective tissue (*CT*) layer (basement lamina), before joining syncytial distal cytoplasm. Microtriches (*MTR*) line free surface of syncytial layer, and discoidal vesicles occupy distal cytoplasm. (×5900)

(a) From R. D. Lumsden and R. D. Specian, in H. P. Arai, editor, *Biology of the Rat Tapeworm,* Hymenolepis diminuta, Academic Press, Inc. *(b)* Drawing by William Ober and Claire Garrison.

muscle layer (see Fig. 20.8). Nuclei lie in the cytons, not the distal cytoplasm. Vesicles are secreted in the cytons, passed to the distal cytoplasm through the internuncial processes, and at least some of them contribute to microthrix and hook formation.[76, 95] Although each cyton contains just one nucleus, the distal cytoplasm is continuous, with no intervening cell membranes; therefore, the tegument of cestodes is a **syncytium.** However, there is immunocytochemical evidence that different subpopulations of cytons exist along the strobila, which may reflect functional differences.[49]

Calcareous Corpuscles

Tissues of most cestodes contain curious structures termed **calcareous corpuscles.** They also are found in excretory canals of some trematodes.[112] They are secreted within the nucleus or in cytoplasm of differentiated calcareous corpuscle cells, which are themselves destroyed in the process,

or they may be secreted within excretory canals, as in trematodes (Fig. 20.11).[117] Corpuscles are from 12 μm to 32 μm in diameter, depending on species, and consist of inorganic components, principally calcium and magnesium carbonates, along with a hydrated form of calcium phosphate embedded in an organic matrix.[105] The organic matrix is organized into concentric rings and a double outer envelope; the matrix contains protein, lipid, glycogen, mucopolysaccharides, alkaline phosphates, RNA, and DNA. They always contain a series of minor inorganic elements, and these, as well as the amount of phosphate, are affected by diet of the host.

Possible functions of calcareous corpuscles have been a subject of much speculation. For example, mobilization of the inorganic compounds might buffer the tissues of the worm against the large amounts of organic acids produced in its energy metabolism (p. 319). Another suggestion has been that they provide depots of ions or carbon dioxide for use when such substances are present in insufficient quantity in the environment, such as on initial establishment in a host

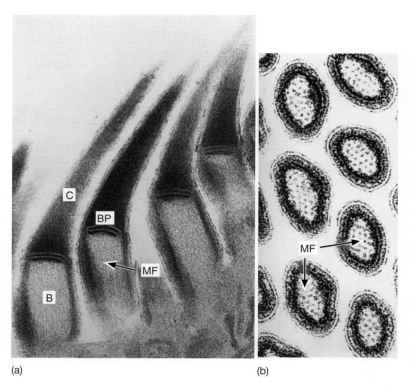

(a)   (b)

Figure 20.9

(*a*) Sagittal section of tegumental microtriches. The electron-opaque cap (*C*) is separated from the base (*B*) by a multilaminar base plate (*BP*). Microfilaments (*MF*) are regularly arranged within the base. Tegumental plasmalemma extends over the entire length of each microthrix. (×71,000) (*b*) Cross section through bases of tegumental microtriches revealing an orderly array of microfilaments (*MF*) surrounded by an accumulation of electron-dense material. (×120,000)

From R. D. Lumsden and R. D. Specian, in H. P. Arai, editor, *Biology of the Rat Tapeworm,* Hymenolepis diminuta, Academic Press, Inc.

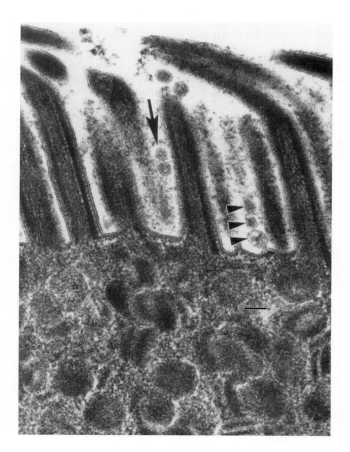

Figure 20.10   Electron micrograph showing one of the two mechanisms for secretion of materials from the tegument of *Hymenolepis diminuta.*

A chain of microvesicles *(arrowheads)* is attached to the plasma membrane at the base of the microtriches. Note the external filamentous material on the microtriches *(arrow).* (Scale bar = 0.1 μm.)

From J. A. Oaks and J. M. Holy, "*Hymenolepis diminuta:* two morphologically distinct tegumental secretory mechanisms are present in the cestode," in *Exp. Parasitol.* 79:292–300. Copyright © 1994. Reprinted with permission of the publisher.

gut. Still another idea is that they are an excretory product.[29] Confirmation of these proposals or discovery of the true function of calcareous corpuscles will require the application of a creative scientific mind.

Muscular System

Muscle cells of *Hymenolepis diminuta* consist of two portions: a contractile myofibril and a noncontractile myocyton (see Fig. 20.8).[67] The contractile portion contains actin and myosin fibrils, and, like muscles of other platyhelminths, it is nonstriated and lacks transverse sarcolemmal tubules (T tubules)[66] as might be expected of muscles with slow contraction. Myocytons comprise the bulk of the worm's parenchyma, and they are often referred to as *parenchymal*

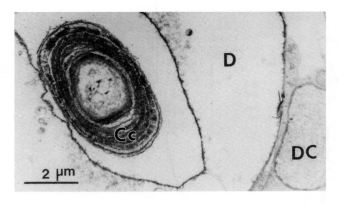

Figure 20.11

Calcareous corpuscle (*Cc*) formed in an excretory canal of *Taenia solium* cysticercus, showing typical matrix of granular concentric layers. (*D*), protonephridial duct. (*DC*), duct cell.

From L. Vargas-Parada, M. T. Merchant, K. Willms, and J. P. Laclette, "Formation of calcareous corpuscles in the lumen of excretory canals of *Taenia solium* cysticerci," in *Parasitol. Res.* 85:88–92. Copyright © 1999. Reprinted with permission of the publisher.

cells. Myocytons contain a nucleus, rough endoplasmic reticulum, free ribosomes, a vesicular Golgi apparatus, few mitochondria, and abundant glycogen. Lipid is stored here as well. Although these cytological details are best known for *H. diminuta,* it is highly likely that they pertain to all other cestodes and even trematodes.

Contractile portions of muscle cells are arranged in discrete bundles in specific regions of the worms. Just internal to tegumental distal cytoplasm are bundles of longitudinal and circular fibers. More powerful musculature lies below the superficial muscles. Longitudinal bundles are usually arranged around a central parenchymal area, which itself is largely free of contractile elements. There may be a zone of cortical parenchyma, also free of longitudinal fibers. There are numbers of dorsoventral and transverse fibers and sometimes radial fibers as well. The pattern and relative development of muscle bundles are highly variable in Cestoidea but constant within a species; therefore, they are often valuable taxonomic characters.

Internal musculature of the scolex is complex, making the scolex extraordinarily mobile. Three distinct muscle types have been found in scolices of trypanorhynchs (p. 346): peripheral myofibers similar to those previously described, tentacle retractor muscles, and tentacle bulb muscles. The bulb muscles are obliquely striated and have numerous motor end plates; motor innervation of the peripheral muscles and retractor muscles has not been found.[123]

Nervous System

The main nerve center of a cestode is in its scolex, and the complexity of ganglia, commissures, and motor and sensory innervation there depends on number and complexity of other structures on the scolex. Among the simplest are bothriate cestodes such as *Bothriocephalus* spp., which have only a pair of lateral cerebral ganglia united by a single ring and a transverse commissure. Arising from the cerebral ganglia is a pair of anterior nerves, supplying the apical region of the scolex; four short posterior nerves; and a pair of lateral nerves that continue posteriorly through the strobila. The bothria are innervated by small branches from the lateral nerves.

In contrast, worms with bothridia or acetabula and hooks, a rostellum, and so on may have a substantially more complex system of commissures and connectives in the scolex, with five pairs of longitudinal nerves running posteriorly from the cerebral ganglia through the strobila (Fig. 20.12). In addition to the motor innervation of the scolex, there may be many sensory endings, particularly at the apex of the tegument. Stretch receptors have been described.[94]

The nervous system of cestodes again illustrates the orthogon plan typical of Platyhelminthes,[43] made even more striking by the intraproglottidal commissures connecting the longitudinal nerves in every proglottid. Smaller nerves emanate from the commissures to supply the general body musculature and sensory endings. The cirrus and vagina are richly innervated, and sensory endings around the genital pore are more abundant than in other areas of the strobilar tegument.[71] Such an arrangement has obvious value.

Study of the neuroanatomy of cestodes formerly was difficult because the nerves are unmyelinated and do not stain well with conventional histological stains. However, histochemical techniques that show sites of acetylcholinesterase activity and immunocytological techniques to demonstrate neuropeptides and 5-hydroxytryptamine (serotonin, 5-HT) have permitted elegant studies of tapeworm nervous systems. Serotonin is an important excitatory neurotransmitter, and acetylcholine seems to be the main inhibitory neurotransmitter.[107] Some 20 different neuropeptides occur in cestodes. There are cholinergic, serotoninergic, and peptidergic components throughout the central and peripheral nervous systems in *Moniezia expansa,*[71] and the "classical" neurotransmitters and peptides may coexist in certain populations of neurons in flatworms.[42] The presence of a nitric oxide/guanosine 3′, 5′-cyclic monophosphate (NO/cGMP) signalling pathway has been described in tapeworm nervous systems, but its function(s) remain(s) unclear.[39] Functions of the plethora of neuropeptides are even more obscure in cestodes than in trematodes (p. 214).

Sensory function probably includes both tactoreception and chemoreception, and as many as seven morphologically distinct types of sensory endings have been described in some species.[13] Several types have a modified cilium projecting as a terminal process (Fig. 20.13), as is common in such cells in invertebrates.

Excretion and Osmoregulation

In many families of cestodes the main excretory canals run the length of the strobila from the scolex to the posterior end. These are usually in two pairs, one ventrolateral and the other dorsolateral on each side (Fig. 20.14). Most often the dorsal pair is smaller in diameter than is the ventral pair, a useful criterion for determining the dorsal and ventral sides of a tapeworm. The canals may branch and rejoin throughout the strobila or may be independent. Usually a transverse canal joins the ventral canals at the posterior margin of each proglottid. The dorsal and ventral canals unite in the scolex, often with some degree of branching. Posteriorly, the two pairs of canals merge into an excretory bladder with a

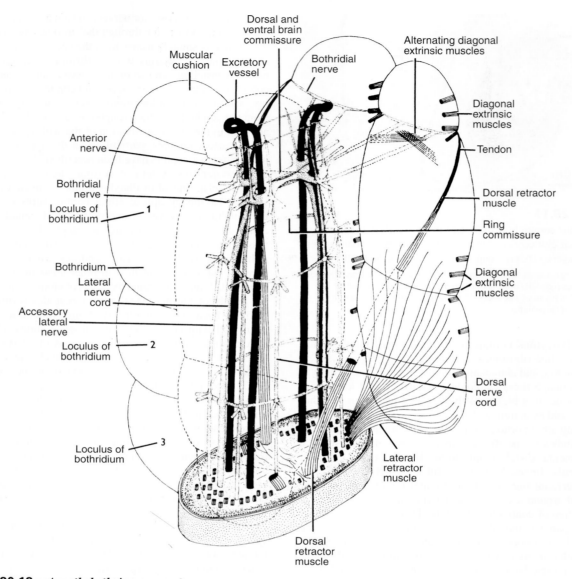

Figure 20.12 *Acanthobothrium coronatum.*
Reconstruction of nervous system of scolex. Excretory vessels and some muscles are included.

Source: G. Rees and H. H. Williams, "The functional morphology of the scolex and the genitalia of *Acanthobothrium coronatum* (Rud.) (Cestoda: Tetraphyllidea)" in *Parasitology*, 55:617–651. Copyright © 1965.

single pore to the outside. When the terminal proglottid of a polyzoic species detaches, the canals empty independently at the end of the strobila. Rarely the major canals also empty through short, lateral ducts. In some orders, such as Pseudophyllidea, canals form a network that lacks major dorsal and ventral ducts.

Embedded throughout the parenchyma are flame cell protonephridia (Fig. 20.15), p. 309, whose ductules feed into the main canals. The flagella of a flame cell provide motive force to the fluid in the system. Protonephridia of tapeworms show the weir construction already described (p. 192). In contrast to trematodes, however, the distal tubules of tapeworms are not formed by a single cell but may be syncytial.[46] Furthermore, the excretory ducts are lined with microvilli (Fig. 20.16), thus suggesting that the duct linings serve a transport function. Therefore,

functions of the system might include active transport of excretory wastes and ionic regulation of the excretory fluid. Fluid from the excretory canals of *H. diminuta* contains glucose, soluble proteins, lactic acid, urea, and ammonia but no lipid.[125]

The principal end products of cestode energy metabolism, short-chain organic acids, are probably excreted through the tegument, either by diffusion or perhaps by one of the mechanisms demonstrated by Oaks and Holy.[77]

Osmoregulation is another function of the tegumental surface. Although cestodes have been regarded as osmoconformers, with little ability to regulate their body volume in media of differing osmotic concentrations, *Hymenolepis diminuta* can osmoregulate between 210 and 335 mOsm/L in a balanced salt solution if 5 mM glucose is present.[116] The worms rapidly lose water at pH 7.4 and 300 mOsm/L

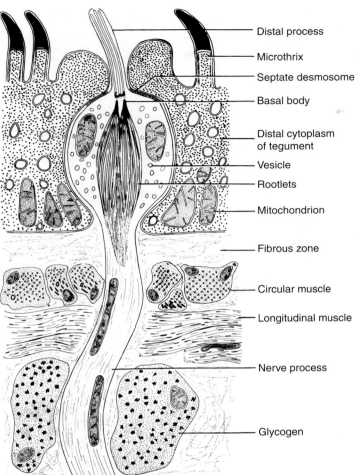

- Distal process
- Microthrix
- Septate desmosome
- Basal body
- Distal cytoplasm of tegument
- Vesicle
- Rootlets
- Mitochondrion
- Fibrous zone
- Circular muscle
- Longitudinal muscle
- Nerve process
- Glycogen

Figure 20.13   **Schematic drawing of a longitudinal section through a sensory ending in the tegument of *Echinococcus granulosus*.**

From D. J. Morseth, "Observations on the fine structure of the nervous system of *Echinococcus granulosus*" in *J. Parasitol.* 53:492–500. Copyright © 1967 Journal of Parasitology. Reprinted by permission.

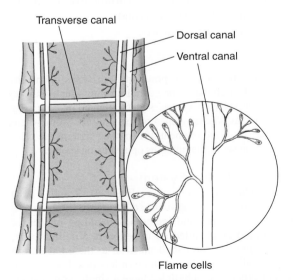

- Transverse canal
- Dorsal canal
- Ventral canal
- Flame cells

Figure 20.14   **Diagram showing the typical arrangement of dorsal (*d*) and ventral (*v*) osmoregulatory canals.**

Drawing by William Ober and Claire Garrison.

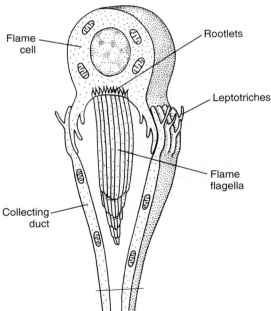

- Flame cell
- Rootlets
- Leptotriches
- Flame flagella
- Collecting duct

Figure 20.15   **Diagram of terminal organ of flame cell protonephridium in *Hymenolepis diminuta*.**

The flame is composed of approximately 50 flagella. The collecting duct is syncytial.

Redrawn by William C. Ober from M. B. Hildreth, R. D. Lumsden, and R. D. Specian, *Biology of the Rat Tapeworm,* Hymenolepis diminuta, edited by H. P. Arai. Copyright © 1980 Academic Press, Inc., New York, NY. Reprinted by permission.

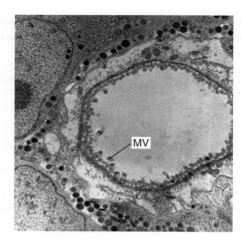

Figure 20.16   **Low-magnification electron micrograph of excretory duct of *Hymenolepis diminuta* showing beadlike microvilli (*MV*).**
Level of section is indicated by position of line in Fig. 20.15.
Courtesy of H. E. Potswald.

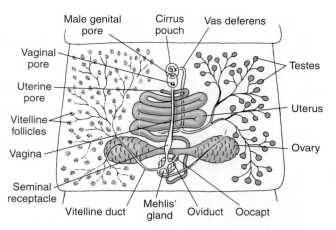

Figure 20.17   **Diagram showing reproductive system of *Diphyllobothrium latum* (infracohort Pseudophylla).**
Note that the testes have been drawn on one side of the proglottid and the vitellaria on the other.
Drawing by William Ober and Claire Garrison.

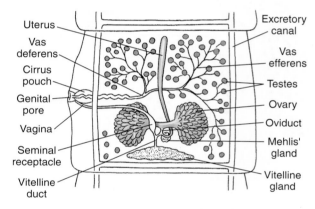

Figure 20.18   **Diagram showing reproductive system of *Taenia* spp. (infracohort Saccouterina).**
Drawing by William Ober and Claire Garrison.

without glucose. Uglem[116] concluded that water balance in *H. diminuta* is closely related to excretory acid concentration and pH of the medium.

Reproductive Systems

Tapeworms are monoecious with the exception of a few species from water birds and two from a stingray that are dioecious. Usually each segment has one complete set of both male and female systems, but some genera have two sets of genitalia per segment (see Fig. 20.3*b*), and a few species in water birds have one male and two female systems in each segment.

As a segment is pushed toward the posterior end of the strobila, the reproductive systems mature, sperm are transferred, and oocytes are fertilized. Usually male organs mature first and produce sperm that are stored until maturation of the ovary; this is called **protandry** or **androgyny**. In a few species the ovary matures first; this is called **protogyny** or **gynandry**. Such asynchronous development may be an adaptation that prevents self-fertilization of the same proglottid. Many variations occur in structure, arrangement, and distribution of reproductive organs in tapeworms. These variations are useful at all levels of taxonomy.

Male Reproductive System
Male reproductive systems (Figs. 20.17 and 20.18) consist of one to many testes, each of which has a very thin vas efferens. Vasa efferentia unite into a common vas deferens that channels sperm toward the genital pore. The vas deferens may be a simple duct, or it may have sperm storage capacity in convolutions or in a spheroid external seminal vesicle. As the vas deferens leads into the cirrus pouch, which is a muscular sheath containing terminal organs of the male system, it may form a convoluted ejaculatory duct or dilate into an internal seminal vesicle. The male copulatory organ is a muscular cirrus, which may or may not bear spines. It can invaginate into the cirrus pouch and evaginate through the cirrus pore.

Commonly reproductive pores of both sexes open into a common sunken chamber, or genital atrium, which may be simple or equipped with spines, stylets, glands, or accessory pockets. The cirrus pore may open on the margin or somewhere on the flat surface of the segment. If two male systems are present, they open on segmental margins opposite one another.

Female Reproductive System
Female reproductive systems (see Figs. 20.17 and 20.18) consist of an ovary and associated structures, which are variable in size, shape, and location, depending on genus. The entire complex is called an **oogenotop.** Vitelline cells, which contribute yolk and shell material to the embryo, are scattered as follicles in most cestode orders (see Fig. 20.17). Members of order Cyclophyllidea have a single, compact vitelline gland (see Fig. 20.18). As oocytes mature, they leave the ovary through a single oviduct, which often has a controlling sphincter, or **oocapt** (see Fig. 20.17).

Oocytes leave the ovary arrested in prophase of meiosis I.[22] Sperm penetration occurs in the proximal oviduct and stimulates resumption and completion of meiotic divisions. One or more cells from the vitelline glands pass through a common vitelline duct, sometimes equipped with a small vitelline reservoir, and join with the zygote. Together they pass into an area of the oviduct known as an *ootype.* This zone is surrounded by unicellular Mehlis' glands, which appear to secrete a thin membrane around the zygote and its associated vitelline cells.

Shell formation is then completed from within by the vitelline cells and in many cases cells of the embryo. Eggs of pseudophyllidean tapeworms are covered by a thick **capsule** of sclerotin. These capsules are apparently homologous with trematode eggshells and are formed in a similar manner (Fig. 20.19). Some shelled embryos develop in water after passing from the host and usually hatch to release a free-swimming larval stage that is eaten by an aquatic intermediate host.

Shell formation in cestodes in the infracohort Saccouterina (p. 322) is complicated, with several layers being contributed by embryonic cells.[21, 115] These layers include a **coat, embryophore,** and **oncospheral membrane;** a capsule is thin or lacking. Three different types are distinguished (see Fig. 20.19): (1) *Dipylidium* type with a thin capsule and an embryophore (as in cyclophyllidean genera *Dipylidium, Moniezia,* and *Hymenolepis* and orders Proteocephalata and Tetraphyllidea); (2) *Taenia* type with a very thin capsule but a thick embryophore (as in *Taenia* and *Echinococcus* spp.); and (3) *Stilesia* type, formed by species with no distinct vitellaria, with cellular covering apparently laid down by the uterine wall.[107] In contrast with the pseudophyllidean type, in *Dipylidium* and *Taenia* types only one or a few vitelline cells associate with a zygote. During early embryogenesis some cells become segregated from the rest of the embryo, fuse, surround the embryo, and form an **outer envelope (OE)** (Fig. 20.20). Other cells become an **inner envelope (IE).** The vitelline cell contributes to the OE. A coat forms within the OE and adds to or replaces the capsule. An embryophore and oncospheral membrane are formed by the IE.

As the zygote and vitelline cells pass through the ootype, secretions of Mehlis' glands are added. These secretions may

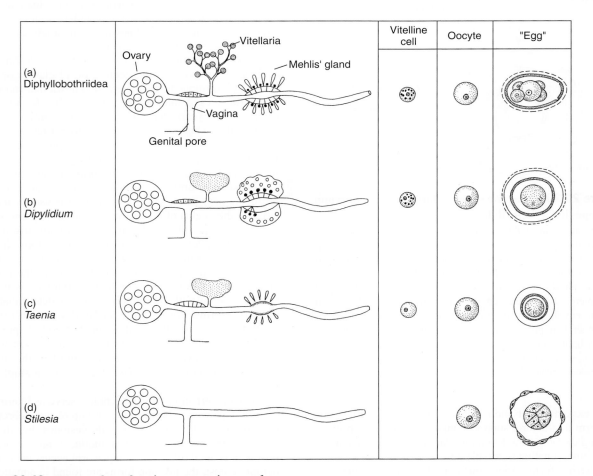

Figure 20.19   Types of egg-forming systems in cestodes.

(*a*) Diphyllobothriidean type, found also in the Caryophyllidea. A relatively thick capsule of sclerotin is formed from material from the vitelline cells. (*b–d*) "Oligolecithal" types found in the infracohort Saccouterina. (*b*) *Dipylidium* type, found in some Cyclophyllidea, Tetraphyllidea, and Proteocephalata. A thin coat from the outer envelope is added to the thin capsule. In Tetraphyllidea, two to three vitelline cells contribute to the capsule. (*c*) *Taenia* type with thick embryophore and very thin capsule. (*d*) *Stilesia* type, found in cestodes with no distinct vitellaria in which the cellular covering is apparently laid down by the uterine wall.

Source: Drawing by William Ober and Claire Garrison after E. Löser, "Die Eibildung bei Cestoden," in *Z. Parasitenkd.,* 25:556–580, 1965.

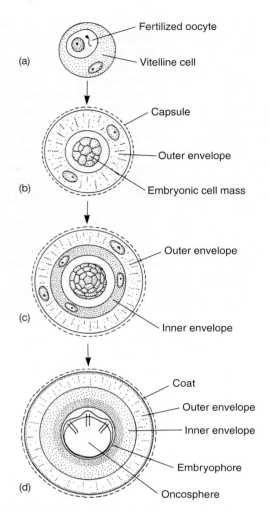

Figure 20.20   **Diagram showing formation of embryonic envelopes in Cyclophyllidea.**

The organization of the envelopes is similar in other cestodes. (*a*) Fertilized oocyte surrounded by vitelline cell and capsule. (*b*) Early phase of development showing formation of outer envelope from vitelline cell and embryonic blastomeres. (*c*) Later phase showing formation of inner envelope from other blastomeres. (*d*) Mature oncosphere with fully developed embryonic envelopes: coat, embryophore of inner envelope, and oncospheral membrane (the oncospheral membrane cannot be seen by light microscopy).

Redrawn by William C. Ober from K. Rybicka, "The embryonic envelopes in cyclophyllidean cestodes" in *Acta Parasitol. Pol.* 13:25–34, 1965.

cause exocytosis of shell material from the vitelline cells and form a structural support component for the capsule.[115] Leaving the ootype, a developing larva passes into the uterus where embryonation is completed.

Uterine form varies considerably among groups. The uterus may be reticulated, lobulated, or circular; it may be a simple sac or a simple or convoluted tube; or it may be replaced by other structures. In some tapeworms the uterus disappears, and eggs, either singly or in groups, are enclosed within hyaline **egg capsules** embedded within the parenchyma. In some species one or more fibromuscular structures called **paruterine organs** form, attached to the uterus. In these species eggs pass from the uterus into the paruterine organ, which assumes the functions of a uterus. The uterus then usually disintegrates.

DEVELOPMENT

Nearly every life cycle known for tapeworms requires two hosts for its completion. One notable exception is *Hymenolepis nana,* a cyclophyllidean parasite of mice and humans, which can complete its juvenile stages within its definitive host (p. 340). (*Hymenolepis nana* has been called *Vampirolepis nana* because the type species of *Hymenolepis, H. diminuta,* has an unarmed rostellum, and the rostellar hooks of *H. nana* were viewed as the basis for putting it in a separate genus. However, due to the case of *Taenia saginatus* vs. *Taeniarhynchus saginatus* [see p. 332], we can no longer justify presence or absence of rostellar hooks as a generic distinction.) Complete life cycles are known for only a comparatively few species of tapeworms. In fact, there are some orders in which not a single life cycle has been determined. Among life cycles that are known, much variety exists in juvenile forms and patterns of development.

Sexually mature tapeworms live in the intestine or its diverticula or rarely in the coelom of all classes of vertebrates. As mentioned, two genera are known that can mature in invertebrates. A mature tapeworm may live for a few days or up to many years, depending on species. During its reproductive life a single worm produces from a few to millions of eggs, each with the potential of developing into an adult. Because of the great hazards obstructing the course of transmission and development of each worm, mortality is high.

Most tapeworms are hermaphroditic and are capable of fertilizing their own eggs. Sperm transfer is usually from the cirrus to the vagina of another proglottid in the same strobila or between adjacent strobilas, if the opportunity affords.

A few species of tapeworms are dioecious. In these it is not clear what determines the gender of a given strobila, because it appears that each strobila has the potential of maturing as either male or female. Interaction between two or more strobilas is important in sex determination of dioecious forms. For example, in *Shipleya inermis* (Cyclophyllidea, Dioecocestidae), if a single strobila is present in its shorebird host, it is usually female; if two are present, one is nearly always a male. In fact, most of the time the host intestine contains a single female and a single male worm.[104]

Both invertebrates and vertebrates serve as intermediate hosts of tapeworms. Nearly every group of invertebrates has been discovered harboring juvenile cestodes, but the most common are crustaceans, insects, molluscs, mites, and annelids. As a general rule, when a tapeworm occurs in an aquatic definitive host, the juvenile forms are found in aquatic intermediate hosts. A similar assumption can be made for terrestrial hosts.

Vertebrate intermediate and paratenic hosts are found among fishes, amphibians, reptiles, and mammals. Tapeworms found in these hosts normally mature within predators whose diets include the intermediary.

Larval and Juvenile Development

Among life histories that are known, much variety exists in juvenile forms and details of development, but there seems to be a single basic theme:[37] (1) embryogenesis within the egg to result in a larva, the **oncosphere**; (2) hatching of the oncosphere after or before being eaten by the next host, where it penetrates to a **parenteral (extraintestinal)** site; (3) metamorphosis of the larva in the parenteral site into a juvenile **(metacestode)** usually with a scolex; and (4) development of an adult from the metacestode in the intestine **(enteral** site) of the same or another host. Oncospheres of all Eucestoda have three pairs of hooks (Fig. 20.21) and thus also are referred to as **hexacanths.** Free-swimming oncospheres hatching from an egg of some Diphyllobothriidea and a few Tetraphyllidea have a ciliated inner envelope (IE) and are called **coracidia** (Fig. 20.22).[115] Larvae of gyrocotylideans and amphilinideans have 10 hooks (hence, they are **decacanths**), are also ciliated, and are called **lycophoras.**

In cestodes with free-swimming larvae a coracidium must be eaten by an intermediate host, usually an arthropod, within a short time. A coracidium sheds its ciliated IE and actively uses its six hooks to penetrate the gut of its host. In the hemocoel it metamorphoses into a **procercoid** (Fig. 20.23). During this reorganization the oncospheral hooks are relegated to the posterior end in a structure known as a **cercomer.** A procercoid is defined as the stage in which larval hooks are still present but the definitive holdfast has not developed. It is regarded by some authors as a differentiating metacestode.[37] When the first intermediate host is consumed by a second intermediate host—often a fish—the procercoid penetrates the host gut into the peritoneal cavity and mesenteries and then commonly into skeletal muscles. Development of a scolex characterizes a plerocercoid (see Fig. 20.23), and there is commonly strobila formation at this stage, with or without concomitant proglottid formation.

In the pseudophyllideans *Ligula* and *Schistocephalus,* development as plerocercoids proceeds so far that little growth occurs when these worms reach a definitive host, and the gonads mature within 72 hours and start producing eggs within 36 hours thereafter.[67, 79] Proteocephalata develop a first-stage plerocercoid in an arthropod intermediate host, with no intervening procercoid, and a second-stage plerocercoid in a parenteral site in a second intermediate host. In some species of this order, metacestode development (plerocercoid II) may be completed in the gut of a definitive host, or the metacestodes may develop through a sequence of sites: parenterally in an intermediate host, then parenterally in a definitive host, and finally enterally in the definitive host.[33, 34, 127] Coracidia, procercoids, and plerocercoids of diphyllobothriideans and plerocercoids of proteocephalatans are all plentifully supplied with penetration glands that aid in penetration of, and migration in, host tissues.[19, 63]

Life cycles of cyclophyllideans differ in that there is neither a procercoid nor a plerocercoid. Larvae are fully developed and infective when they pass from their definitive host, but they do not hatch until eaten by an intermediate host. The oncosphere penetrates the gut of its intermediate host to reach a parenteral site and metamorphoses to a

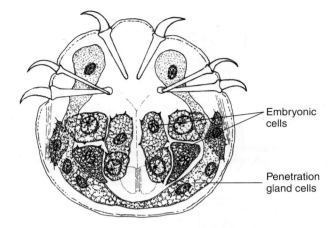

Figure 20.21   **Diagram of oncosphere of *Hymenolepis diminuta,* showing hooks, U-shaped penetration gland, and embryonic cells.**

Drawing by John Janovy, Jr., from various sources.

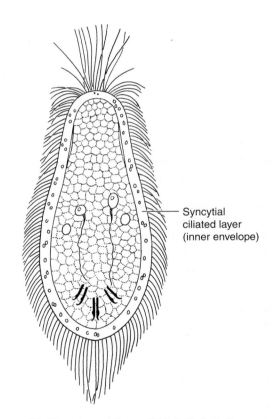

Figure 20.22   **Coracidium of *Diphyllobothrium erinacei.***

From: Neveu Lemaie, Traité d'helminthologie médicale et veterinaire, 1936, in R. A. Wardle and J. A. McLeod, *The Zoology of Tapeworms,* 1952 University of Minnesota Press.

cysticercoid or to a **cysticercus** type of metacestode. Cysticercoids (Fig. 20.24 and see Fig. 20.23) are solid-bodied organisms with a fully developed scolex invaginated into their body. It is surrounded by cystic layers, and the cercomer,

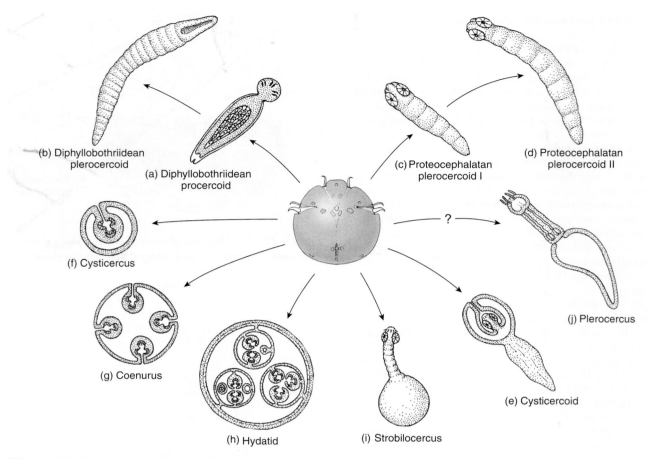

Figure 20.23 **Types of cestode metacestodes.**

(*a*) The procercoid can be regarded as a differentiating plerocercoid of diphyllobothriideans (*b*). According to some authors, the differentiating plerocercoid of proteocephalatans is a plerocercoid I (*c*), and the infective stage develops into plerocercoid II (*d*). Cysticercoids and cysticerci are metacestodes in the Cyclophyllidea. Cysticercoids (*e*) have an invaginated scolex and a solid body, and the scolices of cysticerci (*f–i*) are both invaginated and introverted into a fluid-filled bladder. The plerocercus (*j*), found in some Trypanorhyncha, is like a plerocercoid with a posterior bladder.

Drawings by William Ober and Claire Garrison.

which contains the larval hooks, is outside the cyst. If not displaced mechanically, the cercomer will be digested away, along with parts of the cyst, in the gut of the definitive host. A few cysticercoids have been described that undergo asexual reproduction by budding.

Members of cyclophyllidean family Taeniidae form a cysticercus metacestode (see Figs. 20.23, 21.12, and 21.16), which differs from a cysticercoid in that the scolex is introverted as well as invaginated, and the scolex forms on a germinative membrane enclosing a fluid-filled bladder. Several variations from the simple cysticercus in the Taeniidae undergo asexual reproduction by budding (as will be discussed further). These juvenile stages are of considerable medical and veterinary importance.

Numerous other kinds of metacestodes can be distinguished from the typical forms just described, but they are, for the most part, simply modifications of the basic types:

1. **Sparganum**—a term originally proposed for any pseudophyllidean plerocercoid of unknown species but now usually used for some plerocercoids of genus *Diphyllobothrium* (formerly *Spirometra*).

2. **Plerocercus**—a modified plerocercoid found in some Trypanorhyncha, in which the posterior forms a bladder, the **blastocyst,** into which the rest of the body can withdraw (as in *Gilquinia* spp.). Also applied to plerocercoids of proteocephalatans with an invaginated scolex.

3. **Strobilocercoid**—a cysticercoid that undergoes some strobilation; found only in *Schistotaenia* spp.

4. **Tetrathyridium**—a fairly large, solid-bodied juvenile that can be regarded as a modified cysticercoid, developing in vertebrates that have ingested the cysticercoid encysted in the invertebrate host. It is known only in the atypical cyclophyllidean *Mesocestoides*.

5. Variations on cysticercus.

   a. **Strobilocercus** (Fig. 20.25)—a simple cysticercus in which some strobilation occurs within the cyst (for example, *Taenia taeniaeformis*).

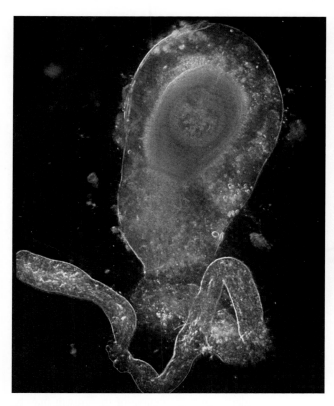

Figure 20.24   Fully developed cysticercoid of *Hymenolepis diminuta*.

From M. Voge, in G. D. Schmidt, editor, *Problems in Systematics of Parasites,*
© 1969 University Park Press.

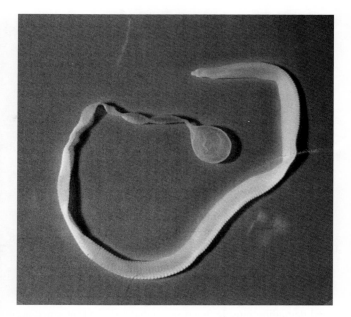

Figure 20.25   Strobilocercus from the liver of a rat.
Note the small bladder at the posterior end.

Courtesy of James Jensen.

b. **Coenurus** (Fig. 20.26 and see Fig. 21.20)—budding of a few to many scolices (called *protoscolices*) from the germinative membrane of the cyst, each on a simple stalk invaginated into the common bladder (as in *Taenia multiceps*).

c. **Unilocular hydatid** (see Fig. 21.23)—with up to several million protoscolices present; there are occasional sterile specimens. Usually there is *inner,* or *endogenous,* budding of **brood cysts,** each with many protoscolices inside. Exogenous budding rarely occurs, resulting in two more hydatids called **daughter cysts.** Unilocular hydatids may grow very large, sometimes containing several quarts of fluid. Occasionally many protoscolices break free and sink to the bottom of the cyst, forming **hydatid sand** (see Fig. 21.26), but this is probably rare in the living, normal cyst. This metacestode form is known only for the cyclophyllidean genus *Echinococcus.*

d. **Multilocular** or **alveolar hydatid** (see Fig. 21.27)—known only for *Echinococcus multilocularis,* exhibiting extensive exogenous budding, resulting in an infiltration of host tissues by numerous cysts. It forms a single mass with many little pockets that contain protoscolices when in a normal host.

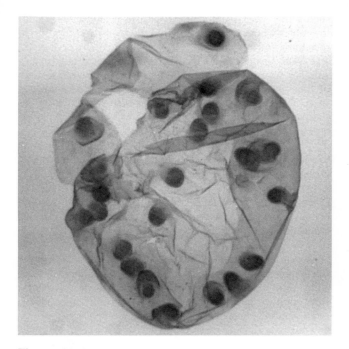

Figure 20.26   Coenurus.
Each round body in the bladder is an independent scolex.

Courtesy of Warren Buss.

EFFECTS OF METACESTODES ON HOSTS

Tapeworms present many examples of a phenomenon called *p*arasite *i*nduced *t*rophic *t*ransmission (PITT), in which parasite infection causes changes in the behavior, physiology, or morphology of one host that facilitates transmission to the next host.[64] Because increased transmission generally increases parasite fitness, a selective value for a parasite from such host manipulation seems clear. We have space here only for a few examples of PITT shown in cestodes; many more, as well as discussion of their evolution, can be found in Holmes and Bethel,[48] Hurd,[53] Lafferty,[64] and Moore.[75]

Life cycles of tapeworms in order Diphyllobothriidea commonly include a procercoid stage in a crustacean first intermediate host and a plerocercoid in a second intermediate host (usually a fish but sometimes another vertebrate). Copepods infected with procercoids of *Triaenophorus crassus* swim near the water surface, where they are more likely to be preyed on by fish, their second intermediate host, whereas uninfected copepods remain near the bottom.[74] Furthermore, infected copepods only show this behavior beginning 10 days after infection, when the procercoids have become mature enough to survive and continue development in a fish. Infected copepods are less motile and have reduced escape responses.[85]

Plerocercoids often develop in the skeletal muscle of second intermediate hosts, diverting energy from the muscle and degrading muscle function, so that a second intermediate host can be captured more easily by a predator. Plerocercoids such as those of *Ligula intestinalis* develop in the abdominal cavity of their fish host, growing rapidly and greatly distending the host belly.[12, 65] Not only is the swimming ability of such a fish seriously impaired, but also it now prefers shallow water where it can be captured more easily by a piscivorous bird.

Many instances are known in which cyclophyllideans enhance transmission by affecting behavior or physiology of their intermediate hosts. Hydatids (p. 337) and coenuri (p. 336) directly disable hosts and facilitate predation by definitive hosts.[48] Infections of mice with cysticerci of *Taenia crassiceps* can prevent adult males from becoming behaviorally dominant by influencing host endocrine function, leading to exclusion of an infected host from its group and increasing chance of predation.[38] Infections of beetles (*Tenebrio molitor*) with cysticercoids of *Hymenolepis diminuta* extend the life of female beetles by reducing host fecundity and thus exposing beetles to more predation.[52] These beetles become infected when they consume shelled larvae of *H. diminuta* in feces of infected rodents. Feces of infected hosts release a volatile attractant that causes hungry beetles (*Tribolium confusum*) to feed preferentially on feces containing larvae rather than feces from uninfected hosts.[30] Whether the attractant originates from adult worms in the definitive host or from some induced modification in host physiology is not known.

Development in Definitive Hosts

As in many other areas of parasitology, generalizations regarding this phase of development may be ill advised because detailed studies of only relatively few species are available. However, substantial data have accumulated for the cestode species that have been examined.[99]

When a juvenile tapeworm reaches the small intestine of its definitive host, certain stimuli cause it to excyst, evaginate, or both and begin growth and sexual maturation. In encysted forms action of digestive enzymes in the host's gut may be necessary to at least partially free the organism from

its cyst. In *Hymenolepis diminuta* most of the cyst wall may be removed by treatment with pepsin and then with trypsin, but few worms will evaginate and emerge from the cyst unless bile salts are present.[101]

In some diphyllobothriideans with a well-developed strobila in the plerocercoid (for example, species of *Ligula* and *Schistocephalus*), an increase in temperature to that of their definitive host is all that is required for them to mature.[2] The temperature "activation" of such plerocercoids is accompanied by a great increase in the rate of carbohydrate catabolism, excretion of organic acids, and levels of tricarboxylic acid cycle intermediates.[7, 59] A burst of neurosecretory activity occurs during activation of *Diphyllobothrium dendriticum* plerocercoids.[41] Contact of the rostellum with a suitable protein substrate is necessary to induce strobilar growth in the cyclophyllidean *Echinococcus granulosus*.[108]

As strobilar development begins, subsequent events are influenced by a variety of conditions, including size of the infecting juvenile, species of the worm and host, size and diet of the host, presence of other worms, and the immune and/or inflammatory state of the host intestine.[52] Under optimal conditions certain species have a burst of growth that must surely rival growth rates found anywhere in the animal kingdom. *Hymenolepis diminuta* can increase its weight by up to 1.8 million times within 15 to 16 days.[98] Such rapid growth, accompanied by strictly organized differentiation, makes this worm a fascinating system for developmental studies, particularly since the course of the growth may be altered experimentally.

Worm growth is especially sensitive to composition of the host diet with respect to carbohydrates. The situation is best known for *H. diminuta,* but the findings can be extended to other tapeworms, to some extent at least. *Hymenolepis diminuta* apparently has a high carbohydrate requirement, but it can only absorb glucose and to a lesser degree galactose across its tegument. This is true for other cestodes tested, although some can absorb a limited number of other monosaccharides and disaccharides.[10] For optimal growth carbohydrate must be supplied in the host diet in the form of a polysaccharide so that glucose will be released as digestion proceeds in the host gut. If glucose per se—or a disaccharide containing glucose, such as sucrose—is furnished in the host diet, worms are placed at a competitive disadvantage for glucose with respect to the gut mucosa, physiological conditions in the gut are altered, or both, so that the worm's growth is dramatically restrained.

Another important condition affecting worm growth is the increased presence of other tapeworms in the gut, the so-called *crowding effect*. This is an interesting adaptation by which parasite biomass adjusts to carrying capacity of a host. Again evidence exists that, although best known in *H. diminuta,* the crowding effect occurs in at least several other species.[93] Within certain limits, the weight of individual worms in a given infection is, on average, inversely proportional to number of worms present. In consequence, total worm biomass and number of eggs produced are the same and are maximal for that host, regardless of number of worms present.

The operational mechanism of the crowding effect is of considerable biological interest as a mode of developmental control. One view has been that the individual worms compete for available host dietary carbohydrate. However, the means by which such competition might be translated into lower rates of cell division and cell growth have not been elucidated, and the worms apparently secrete "crowding factors" that influence the development of other worms in the population.[55, 98]

As a worm approaches maximal size, growth rate decreases, and production of new proglottids is only sufficient to replace those lost by apolysis. Although some species, such as *H. nana,* characteristically become senescent and pass out of the host after a period, others may be limited only by length of their host's life. *Taenia saginata* may live in a human for more than 30 years, and *H. diminuta* may live as long as the rat it inhabits. In fact, Read[92] reported an "immortal" worm that he kept alive for 14 years by periodically removing it from its host, severing the strobila in the region of the germinative area, and then surgically reimplanting the scolex in another rat.

Some tapeworms manage a surprising degree of mobility within their host's intestine. Cestodes may establish initially in one part of the gut and then move to another as they grow. *Hymenolepis diminuta* actually undergoes a diurnal migration in the rat's gut (Fig. 20.27). This migration correlates with the nocturnal feeding habits of rats and can be reversed by giving food to the rat only in daytime. In fact, migration of the worms is apparently mediated by vagal nerve stimulation of gastrointestinal function rather than by the presence of food itself.[73]

Wang and McKay found that *Hymenolepis diminuta* could modulate their host's immune response.[122] Worm extracts and soluble products released by the cestodes suppressed immune cell proliferation and influenced cytokine production. IL-2 and IL-4 secretion was inhibited, and a cytokine with properties of IL-2 was stimulated.

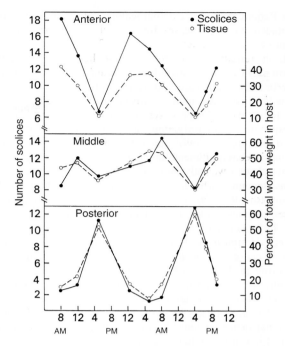

Figure 20.27   Distribution of scolices and of wet tissue of *Hymenolepis diminuta* in the host intestine at various times of the day.

Anterior refers to the first 10 in., *middle* to the second 10 in., and *posterior* to the remainder of the small intestine. Each point is the mean of determinations from four host animals, representing 110 to 120 worms.

From C. P. Read, "Some physiological and biochemical aspects of host-parasite relations" in *J. of Parasitol.* 56:643–652. Copyright © 1970 Journal of Parasitology. Reprinted by permission.

METABOLISM

Acquisition of Nutrients

All nutrient molecules must be absorbed across the tegument. Mechanisms of absorption include active transport, mediated diffusion, and simple diffusion.[83] Whether pinocytosis is possible at the cestode surface has been the subject of some dispute,[68] but plerocercoids of *Schistocephalus solidus* and *Ligula intestinalis* are capable of this process.[51, 114] Cysticerci of *Taenia crassiceps* are capable of pinocytosis, and the process is stimulated by presence of glucose, yeast extract, or bovine serum albumin in the medium.[112, 113]

Glucose is the most important nutrient molecule to fuel energy processes in tapeworms. As noted before, the only carbohydrates that most cestodes can absorb are glucose and galactose, and although some tapeworms can absorb other monosaccharides and disaccharides, we know of none other than glucose and galactose that can actually be metabolized. The primary fate of galactose seems to be incorporation into membranes or other structural components, such as glycocalyx.[78] Galactose can be incorporated into glycogen but does not support net glycogen synthesis.[58] Both glucose and galactose are actively transported and accumulate in the worm against a concentration gradient. Of the two sugars, glucose has been studied more extensively. Glucose influx in

a number of species couples to a sodium pump mechanism; that is, the system of maintenance of a sodium concentration difference across the membrane. Accumulation of glucose, in *H. diminuta* at least, is also sodium dependent. At least two transport sites for glucose are kinetically distinct in the tegument of *H. diminuta,* and the relative proportion of these sites changes during development.[96, 111]

Fully developed larvae of *H. diminuta* with intact shells absorb very little glucose, but when the shell is removed, as it would be when eaten by a beetle intermediate host, they absorb much larger amounts.[81, 82]

Amino acids are also actively transported and accumulated, although less is known about them than about glucose. However, presence of other amino acids in the ambient medium stimulates efflux of amino acids from the worm; therefore, the worm pool of amino acids rapidly comes to equilibrium with amino acids in the intestinal milieu.

Purines and pyrimidines are absorbed by facilitated diffusion, and the transport locus is distinct from the amino acid and glucose loci.[69]

The actual mechanism of lipid absorption has not been investigated, but it is likely to be a form of diffusion. Fatty acids, monoglycerides, and sterols are absorbed at a considerably greater rate when they are in a micellar solution with bile salts.[4] *Hymenolepis diminuta* has a specific transporter for cholesterol.[57]

Requirements for external supplies of vitamins are substantiated in only two cases. Investigations of vitamin requirements are difficult, as they often are in parasites, because of limitations in in vitro cultivation techniques, because the worm may be less sensitive than its host to a vitamin-deficient diet, or both. In any case, pathogenesis of a vitamin deficiency in its host may have indirect effects on the worm. The necessity for an external supply of a vitamin has been demonstrated unequivocally in only one case—that of pyridoxine in *H. diminuta*.[89, 100] We can infer that *Diphyllobothrium latum* has a requirement for vitamin B_{12} because the worm accumulates unusually large amounts of it.[9] In some cases, *D. latum* can compete so successfully with its host for the vitamin that the worm can cause pernicious anemia in persons genetically susceptible to its effects (chapter 21).

In a phenomenon possibly related to acquisition of nutrients, *H. diminuta* slows down intestinal transit.[24, 25] The worms cause myoelectrical alterations in the host intestine resulting in decreased lumenal transit and increased nonpropulsive contractility. Myoelectrical activity returns to normal when worms are expelled by drugs,[25] and introduction of worm extracts by cannula mimics actual infection.[26] A signal factor responsible for the myoelectrical alterations in host intestine is one of the molecules suggested as crowding factors, cyclic GMP.[20] Interestingly, altered myoelectrical activity occurs only in ileum, not jejunum, and only at times when tapeworms are in the ileum, not when they are in the jejunum.[126] Further experiments suggested that these observations may be related to a requirement of the worms for host bile salts.[27]

Energy Metabolism

Glycolysis

The patterns of energy metabolism in cestodes are much like those already described in trematodes (p. 227). In brief, adult cestodes are facultative anaerobes that derive energy from catabolism of glucose and glycogen, but they only oxidize a glucose molecule in part, and they excrete highly reduced end products, such as short chain organic acids (Fig. 20.28).[97]

Because cestodes have very limited ability to degrade fatty and amino acids, their processes of carbohydrate storage and catabolism assume critical importance for energy production. Indeed, juvenile and adult cestodes characteristically store enormous amounts of glycogen, ranging from about 20% to more than 50% of dry weight. Whereas tissue-dwelling juveniles are exposed to a reasonably constant glucose concentration maintained by homeostatic mechanisms of the host, adults must survive between host feeding periods. The large amount of stored glycogen serves at these times as an effective cushion. *Hymenolepis diminuta* consumes 60% of its glycogen during 24 hours of host starvation and another 20% during the next 24 hours. When glucose is again available, glycogen stores are rapidly replenished.[31]

As in trematodes, glucose from glycogen or absorbed directly from the host intestine is degraded by classical glycolysis as far as phosphoenolpyruvate (PEP), but at this point there is a branch in the pathway (see Fig. 20.28). Either lactate is produced by dephosphorylation of PEP and reduction of pyruvate, or malate is produced by fixation of carbon

dioxide to form oxaloacetate, which is then reduced to malate.[97] Both branches thus far are functionally equal because each generates a high-energy phosphate bond and reoxidizes the NADH formed in glycolysis; therefore, cytoplasmic redox balance is preserved.

However, additional energy is obtained when malate enters the mitochondria, where part of the malate is metabolized and excreted as acetate or is transaminated and excreted as alanine. The other half of the malate is metabolized and reduced to succinate (see Fig. 20.28). Reducing equivalents for reduction of fumarate are provided by oxidation of malate. However, in *H. diminuta* and *H. microstoma*, oxidative decarboxylation of malate is NADP dependent, whereas fumarate reduction is NAD dependent. Therefore, a hydride ion must be transferred from NADPH to NAD, and this is accomplished by an NADPH:NAD transhydrogenase.[119] Excretion of succinate and acetate produces two more ATPs than if the glucose carbon were excreted solely as lactate. In some cestodes propionate is formed by decarboxylation of succinate, generating additional ATP.[88] An advantage of alanine excretion would be that it is less acidic than lactate.

Despite the energetic advantage of the mitochondrial reactions and excretion of acetate and succinate, the value of these reactions to the worms remains quite unclear. Some strains of *Hymenolepis diminuta* excrete mostly lactate and little acetate and succinate, and others excrete predominately acetate and succinate.[14] Moreover, within the same strain the proportions of these acids excreted varies according to worm development, part of the strobila, and even immune status of the host. The acids excreted by the tapeworms can be catabolized by the host to CO_2 and water, and the explanation may lie in still obscure aspects of the host-parasite interaction.[14]

Krebs Cycle

The tricarboxylic acid cycle is of little or no importance in adult cestodes, but a substantial amount of glucose carbon may flow through the Krebs cycle in certain metacestodes. As much as 40% of carbohydrate utilized by protoscolices of *Echinococcus multilocularis* and the sheep strain of *E. granulosus* may be channeled into the Krebs cycle, and only 22% of glycogen catabolized by plerocercoids of *Schistocephalus solidus* is accounted for by excreted acids.[59] Activity of the Krebs cycle increases in *S. solidus* when the plerocercoids are activated by an increase in the ambient temperature.[7] Tegumental mitochondria of *Taenia crassiceps* cysticerci are aerobic.[23]

Electron Transport

Cestodes take up oxygen when it is available, but oxygen probably does not function as a terminal electron acceptor in an energy-producing series of reactions (for example, oxidative phosphorylation via the "classical" cytochrome system). Although earlier research indicated that some cytochromes might be present in some cestodes, later research failed to confirm that a cytochrome system was operating, and the function of such cytochromes as were present was a mystery. Use of more sensitive techniques now has provided evidence that a classical mammalian type of electron transport system is present in at least some cestodes, that the classical chain is probably of minor importance (Fig. 20.29), and that the major cytochrome system is a so-called *o*-type, similar to

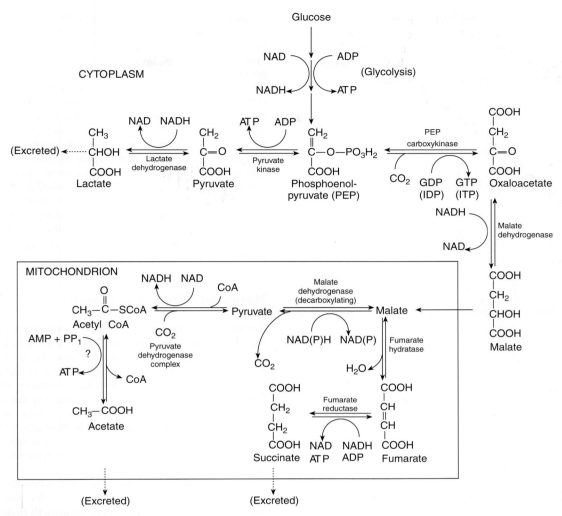

Figure 20.28   **Reactions forming the major end products of energy metabolism from phosphoenolpyruvate in** *Hymenolepis diminuta* **(adapted and proposed from various sources).**

These reactions yield additional ATP above that from classical glycolysis, with a balanced cytoplasmic oxidation-reduction and a balanced mitochondrial oxidation-reduction (ratio of succinate to acetate excreted approximately 2:1). Fumarate reductase usually is referred to as succinate dehydrogenase, but it acts in an opposite direction from mammalian systems; that is, as a fumarate reductase (Watts and Fairbairn).[124] In *Hymenolepis,* the malate dehydrogenase (decarboxylating) reaction is NADP dependent, and the hydride ion is transferred to NAD for the fumarate reductase reaction by a transhydrogenase, thus maintaining the redox balance in the mitochondrion.[32]

that reported in many bacteria.[18] This system may be an adaptation to facultative anaerobiosis. The terminal oxidase can transfer electrons to either fumarate or oxygen, depending on whether conditions are aerobic or anaerobic, and the products are either succinate or hydrogen peroxide, respectively. A peroxidase destroys the hydrogen peroxide before it reaches toxic levels. Under anaerobic conditions, succinate formed in this pathway would be excreted.

Oxygen consumption increases by 40% when cysticerci of *Taenia solium* are stimulated to evaginate by treatment with trypsin, but evagination is not affected by respiratory poisons such as cyanide.[17] These metacestodes apparently also have a branched electron transport system.

Tapeworms probably do not derive any energy from degradation of lipids or proteins. *Hymenolepis diminuta* has

only a modest capacity for carrying out transaminations and can degrade only four amino acids.[118] They can convert cystine to cysteine and catabolize the latter by the oxidative but not the transaminative pathway.[121]

The function served by much of the lipid in cestodes remains a mystery, since no one has been able to show that lipids are depleted at all during starvation, even though they may comprise up to 20% of total worm dry weight or more than 30% in parenchyma of gravid proglottids. *Schistocephalus solidus* has all the enzymes necessary for the b oxidation sequence of lipids; nevertheless, it appears unable to catabolize them.[5] Lipids in cestodes may represent metabolic end products, since they are relatively nontoxic to store, and parenchyma of gravid proglottids is discarded during apolysis.

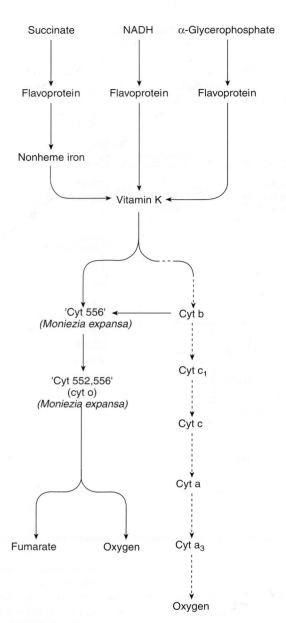

Figure 20.29 **Branched chain electron transport system with cytochrome *o,* facultatively transporting electrons to fumarate or oxygen.**

Evidence exists that a similar system operates in species of *Moniezia, Taenia,* and probably other cestodes. *Solid line,* major pathway; *dotted line,* minor pathway.

Source: C. Bryant, "Electron transport in parasitic helminths and protozoa" in *Advances in Parasitology,* Vol. 8, edited by B. Dawes, 1970, Academic Press, Inc., New York, NY.

Nitrogenous end products excreted by *H. diminuta* include considerable quantities of ammonia, α-amino nitrogen, and urea.[31]

Synthetic Metabolism

Little need be said of protein and nucleic acid synthetic abilities of cestodes. It is clear that they can absorb amino acids, purines, pyrimidines, and nucleosides from the intestinal milieu and synthesize their own proteins and nucleic acids (see, for example, Bolla and Roberts[8]). *Moniezia expansa* cannot synthesize carbamyl phosphate; therefore, it depends on its host for both pyrimidine and arginine.

In contrast, capacity for synthesis of lipids appears minimal. The worm can neither synthesize fatty acids de novo from acetyl-CoA nor introduce double bonds into the fatty acids it absorbs.[56] *Hymenolepis diminuta* rapidly hydrolyzes monoglycerides after absorption, and it can then resynthesize triglycerides. It can lengthen a chain of fatty acids provided that the acid already contains 16 or more carbons. Similar observations have been reported for *Diphyllobothrium mansonoides.*[74] There is some evidence that *Hymenolepis microstoma* might be able to synthesize fatty acids de novo.[91]

Finally, *H. diminuta* cannot synthesize cholesterol, a biosynthesis that requires molecular oxygen in other systems.[36]

Hormonal Effects of Metabolites

Certain substances produced by cestodes have effects on their hosts that mimic the host's own hormones. For example, fish infected with the plerocercoids of *Ligula intestinalis* are unable to reproduce: Their gonads do not develop, and there is an apparent suppression of the presumed gonadotropin-producing cells in their pituitary glands. On the possibility that a sex steroid was produced by the worms and that the steroid interfered with gonadotropin production and hence gonad development, the presence of such compounds was investigated.[3] None was found, and the mechanism of the effect on the host remains enigmatic.

A more surprising case is that of a substance produced by plerocercoids of *Diphyllobothrium* (= *Spirometra*) *mansonoides.* Referred to as **plerocercoid growth factor (PGF),** it acts as a growth hormone in several mammalian species (Fig. 20.30).[86] It is definitely not growth hormone (GH), although the pituitary recognizes it as such and decreases its own production of GH. Under normal circumstances administration of human GH can have anti-insulin and diabetogenic activities; these properties do not accompany administration of PGF. Growth hormones from other vertebrates (except primates) are not active in humans because of the strict binding specificity of the receptor molecules in humans. However, PGF has the same binding specificity as human GH, and monoclonal antibodies raised against the unique epitope of human GH crossreact with PGF. Nevertheless, when the gene for PGF was sequenced, there was *no homology* with human GH or any other hormone.[86] Surprisingly, the gene for PGF shared 40% to 50% homology with known cysteine proteinases! Subsequent studies showed that PGF did in fact have cysteine proteinase activity, and the substrate most extensively hydrolyzed was collagen.[87] This observation, plus localization of PGF/proteinase to the plerocercoid surface, led Phares to suggest that the main function of PGF/proteinase is to facilitate migration of the worm in host tissues.[86] Furthermore, because one function of mammalian GH is stimulation of the immune system, an action not shared by PGF, and because PGF suppresses host production of GH, PGF may mediate evasion of host immune response by the worm.

Figure 20.30 Illustration of growth hormonelike action of *Diphyllobothrium mansonoides* plerocercoids.

All rats were hypophysectomized when they weighed 90 g, but the two larger rats received 20 juvenile scolices of *D. mansonoides* approximately one month after the operation. The photograph was taken six months later, and the experimental animals outweigh the controls by three or four times.

From J. F. Mueller, "The biology of *Spirometra*," in *J. Parasitol.* 60:3–14. Copyright © 1974.

CLASSIFICATION OF CLASS CESTOIDEA

Classification of tapeworms is in a state of flux as phylogenetic systematists strive to construct a system that avoids paraphyly and polyphyly. We are adopting what we believe is the most acceptable system for the higher classification of the Cestoidea currently available.[11, 61] Thus, cohort Cestoidea belongs to subclass Cercomeromorphae (possessing a cercomer with hooks) and infraclass Cestodaria (intestine lacking, cercomer reduced). Figure 20.31 is a cladogram showing relationships among Cestodaria. Keep in mind that each character listed in the diagnoses and cladograms is apomorphic for the groups but that all members of a group do not necessarily have that character. Just as snakes are tetrapods with no legs, caryophyllaeids are eucestodans with but one proglottid.

Mariaux[70] provided a phylogeny based on base sequence analysis of 18S rDNA. His analysis largely supported current hypotheses based on morphology, ontogeny, and ultrastructure. Caryophyllidea was basal to and the sister group of other eucestodes; thus, the monozoic condition is plesiomorphic. Tetraphyllidea was paraphyletic, as was Pseudophyllidea (if it included Diphyllobothriidae) and Cyclophyllidea (if it includes Tetrabothriidae and Mesocestoididae). Within Cyclophyllidea, Taeniidae is basal.

Order Cathetocephalidea, established by Schmidt and Beveridge,[102] has not been adopted widely, but the highly unusual scolex of these worms and the molecular evidence provide ample reasons for recognition of this order.[15]

Cohort Gyrocotylidea

Rosette at posterior end of body; funnel connecting with rosette; anterolateral genital notch present; body margins crenulate; body spines small over most of body, large at pharyngeal level; metraterm; vitellaria encircling entire body, extending along entire body length; no nuclei in larval epidermis; no multiciliary nerve receptors; copulatory papilla present.

Cohort Cestoidea

Male genital pore and vagina proximate; cercomer totally invaginated during ontogeny; hooks on larval cercomer in two size classes (six large and four small); protonephridial ducts lined with microvilli; protonephridia in larvae in posterior end of body; genital pores marginal.

Subcohort Amphilinidea

Uterine pore and genital pores not proximate; male pore and vaginal pore at posterior end; uterus *N*-shaped; uterine pore proximal to vestigial pharynx; inner longitudinal muscle layer weakly developed; adults parasitic in body cavity of fishes and turtles.

Subcohort Eucestoda

Adults polyzoic; cercomer lost during ontogeny; hooks on larval cercomer reduced to six; medullary portion of proglottids restricted; tegument covered with microtriches; sperm lacking mitochondria.

Infracohort Pseudophylla

Bilaterally symmetrical, bipartite scolices ("difossate" condition), with modifications for attachment consisting of longitudinal flaps (bothria) and their modifications; polylecithal eggs (a large component of vitelline material forming a true shell that is quinone tanned); one embryonic membrane (no embryophore); hexacanth larva hatches from egg, is ingested in water, and is followed by procercoid and plerocercoid stages; oncospheres with unicellular protonephridium.

Order Caryophyllidea

Scolex unspecialized or with shallow grooves or loculi or shallow bothria; monozoic; genital pores midventral; testes numerous; ovary posterior; vitellaria follicular, either scattered or lateral; uterus a coiled median tube, opening, often together with vagina, near male pore; parasites of teleost fishes and aquatic annelids. Families: Caryophyllaeidae, Balanotaeniidae, Lytocestidae, Capingentidae.

Order Spathebothriidea

Scolex feebly developed, either undifferentiated or with funnelshaped apical organ or one or two hollow, cuplike organs; constrictions between proglottids absent; proglottids

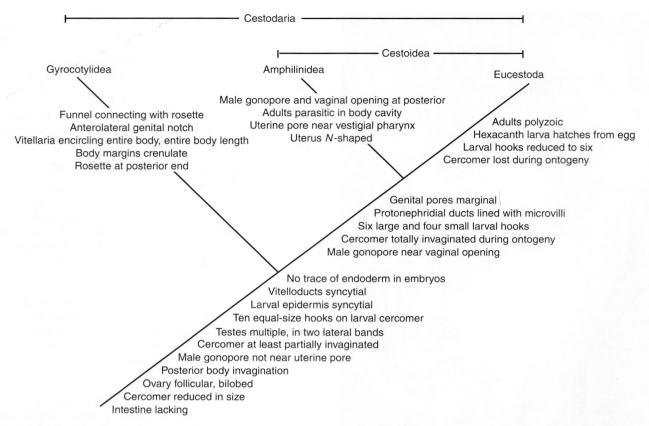

Figure 20.31   Cladogram showing hypothetical relationships of groups within the infraclass Cestodaria.
Gyrocotylidea and Amphilinidea will be described in chapter 21. The name Cestodaria is commonly applied to the Gyrocotylidea and Amphilinidea, but it appears that the Amphilinidea and Eucestoda share a common ancestor and form a clade with the Gyrocotylidea as sister group. One primitive character state of the Cestodaria is 10 larval hooks, of which four become reduced in the Cestoidea and disappear in the Eucestoda.

Source: Based on D. R. Brooks and D. A. McLennan, *Parascript, Parasites & the Language of Evolution.* Copyright © 1993 Smithsonian Institution Press, Washington, DC.

distinguished internally; genital pores and uterine pore ventral or alternating dorsal and ventral; testes in two lateral bands; ovary dendritic; vitellaria follicular, either lateral or scattered; uterus coiled; parasites of teleost fishes. Families: Cyathocephalidae, Spathebothriidae, Bothrimonidae.

Order Bothriocephalidea

Genital pore on dorsal, dorsolateral, or lateral aspects and posterior to the ventral uterine pore; uterine sac present; definitive hosts mainly teleost fishes, never homoiothermic vertebrates.

Order Diphyllobothriidea

Genital pore anterior to uterine pore; uterine sac absent; definitive hosts homoiothermic tetrapods, mainly mammals.

Infracohort Saccouterina

Bilateral saccate uteri lacking permanent pores; oncospheral flame cells absent; oncospheres not ciliated; generally "oligolecithal" eggs (minimal vitelline component, with shell formed by embryo); two embryonic membranes; oncosphere matures in utero; indeterminate growth (continuous budding from a growth zone); generally apolytic.

Order Nippotaeniidea

Scolex with single sucker at apex, otherwise simple; neck short or absent; strobila small; proglottids each with single set of reproductive organs; genital pores lateral; testes anterior; ovary posterior; vitelline gland compact, single, between testes and ovary; osmoregulatory canals reticular; parasites of teleost fishes. Family Nippotaeniidae.

Order Lecanicephalidea

Scolex divided into anterior and posterior regions by transverse groove; anterior portion cushionlike or with unarmed tentacles, capable of being withdrawn into posterior portion, forming a large suckerlike organ; posterior portion usually with four suckers; neck present or absent; testes numerous; ovary posterior; vitellaria follicular, either lateral or encircling proglottid; uterine pore usually present; parasites of elasmobranchs. Families: Adelobothriidae, Disculicepitidae, Lecanicephalidae.

Order Trypanorhyncha

Scolex elongated, with two or four bothridia and four eversible (rarely atrophied) tentacles armed with hooks; each tentacle invaginating into internal sheath provided with muscular bulb; neck present or absent; strobila apolytic, anapolytic, or hyperapolytic; genital pores lateral, rarely ventral; testes numerous; ovary posterior; vitellaria as in Pseudophyllidea; uterine pore present or absent; parasites of elasmobranchs. Families: Dasyrhynchidae, Eutetrarhynchidae, Gilquiniidae, Gymnorhynchidae, Hepatoxylidae, Hornelliellidae, Lacistorhynchidae, Mustelicolidae, Otobothriidae, Paranybeliniidae, Pterobothriidae, Sphyriocephalidae, Tentaculariidae, Mixodigmatidae, Rhinoptericolidae.

Order Aporidea

Scolex with simple suckers or grooves and armed rostellum; constrictions between proglottids absent; proglottids distinguished internally or separate proglottids not evident; genital ducts and pores, cirrus, ootype, and Mehlis' gland absent; hermaphroditic, rarely dioecious; vitelline cells mixed with ovarian cells; parasites of anseriform birds. Family Nematoparataeniidae.

Order Tetraphyllidea

Scolex with highly variable bothridia, sometimes also with hooks, spines, or suckers; myzorhynchus present or absent; proglottids commonly hyperapolytic; hermaphroditic, rarely dioecious; genital pores lateral, rarely posterior; testes numerous; ovary posterior; vitellaria follicular, usually medullary in lateral fields; uterine pore present or absent; vagina crossing vas deferens; parasites of elasmobranchs. Families: Onchobothriidae, Phyllobothriidae, Triloculariidae, Dioecotaeniidae.

Order Rhinebothriidea

Characteristics generally as in Tetraphyllidea, except bothridea borne on stalks.[45]

Order Cathetocephalidea

Scolex a single transversely expanded, fleshy organ lacking bothridia, suckers, or armature consisting of an apical pad, and a rugose base.[15] Genitalia essentially as in Tetraphyllidea. Family Cathetocephalidae.

Order Diphyllidea

Scolex with armed or unarmed peduncle; two spoon-shaped bothridia present, lined with minute spines, sometimes divided by median, longitudinal ridge; apex of scolex with insignificant apical organ or with large rostellum bearing dorsal and ventral groups of *T*-shaped hooks; strobila cylindrical, acraspedote; genital pores posterior, midventral; testes numerous, anterior; ovary posterior; vitellaria follicular, either lateral or surrounding other organs; uterine pore absent; uterus tubular or saccular; parasites of elasmobranchs. Families: Ditrachybothridiidae, Echinobothriidae.

Order Litobothridea

Scolex a single, well-developed apical sucker; anterior proglottids modified, cruciform in cross section; neck absent; strobila dorsoventrally flattened, with numerous segments, each with single set of medullary genitalia; proglottids laciniated and craspedote, apolytic or anapolytic; testes numerous, preovarian; genital pores lateral; ovary with two or four lobes, posterior; vitellaria follicular, encircling medullary parenchyma; parasites of elasmobranchs. Family Litobothridae.

Order ....

Scolex with............; occasionally ....; genital pores la.............; vitelline glands foll............; or medullary; uterine p............; fishes, amphibians, and rep............; Monticellidae.

Order Cyclophyllidea

Scolex usually with four suckers; rostellum armed or unarmed; neck present or absent; s....... with distinct segments, monoecious or rarely dioec.... pores lateral (ventral in Mesocestoididae); vitelline gla.... pact, single (double in Mesocestoididae), posterior to o.... (anterior or beneath ovary in Tetrabothriidae); uterine pore absent; parasites of amphibians, reptiles, birds, and mammals. Families: Amabiliidae, Anoplocephalidae, Catenotaeniidae, Davaineidae, Dilepididae, Dioecocestidae, Diploposthidae, Hymenolepididae, Mesocestoididae, Nematotaeniidae, Progynotaeniidae, Taeniidae, Tetrabothriidae, Triplotaeniidae.

Learning Outcomes

By the time a student has finished studying this chapter, he or she should be able to:

1. Explain the general structure and function of tapeworm scolices.
2. Identify the structure of the different organs on the various types of scolices.
3. Identify the structure of tapeworm tegument, including microtriches, distal cytoplasm, internuncial processes, and perikarya.
4. Identify the structures in proglottids, including reproductive, excretory, and neural structures.
5. Explain the means by which tapeworms derive energy by glycolysis and electron transport.
6. Describe the organization of tapeworm eggs, including inner and outer envelopes, embryophore, and oncosphere.
7. Explain the differences in various types of metacestodes.
8. Explain the differences between the various types of cysticerci.
9. Name the order of tapeworms that contains the great majority of cestode parasites of amphibians, reptiles, birds, and mammals.

References

References for superscripts in the text can be found at the following Internet site: www.mhhe.com/robertsjanovynadler9e

Additional Readings

Arme, C., and P. W. Pappas (Eds.). 1983. *The biology of the Eucestoda* (2 vols.). London: Academic Press. Summary of cestodology; covers evolution and systematics, ecology, morphology and fine structure, development, biochemistry and physiology, pathology, immunology, and chemotherapy.

Barrett, J. 1981. *Biochemistry of parasitic helminths.* Baltimore, MD: University Park Press.

...ptation and loss of genetic
...es. *Biol. Rev.* 45:29–72.

... *invertebrates 2.* New York: McGraw-Hill
...omplete summary of knowledge of cestodes up

...ll, L. F., and A. Jones (Eds.). 1994. *Keys to the cestode parasites of vertebrates.* Wallingford, Oxon, England: CAB International.

Pax, R. A., and J. L. Bennett. 1992. Neurobiology of parasitic flatworms: How much "neuro" in the biology? *J. Parasitol.* 78:194–205.

Read, C. P. 1959. The role of carbohydrates in the biology of cestodes. VIII. *Exp. Parasitol.* 8:365–382.

Schmidt, G. D. 1986. *Handbook of tapeworm identification.* Boca Raton, FL: CRC Press.

Wardle, R. A., and J. A. McLeod. 1952. *The zoology of tapeworms.* New York: Hafner Publishing Co. This monograph is the classic in its field. No student of tapeworms should be without it.

Yamaguti, S. 1959. *Systema helminthum, vol. 2. The cestodes of vertebrates.* New York: Interscience.

Chapter 21

Tapeworms

. . . we should all brush up on tapeworms from time to time. . . .

—Dave Barry *(Bad Habits)*

Although most species of cestodes are parasites of wild animals, a few infect humans or domestic animals and so are of particular interest. All tapeworms that parasitize humans belong to orders Diphyllobothriidea and Cyclophyllidea.

Many tapeworms cause no medical or economic problem. Still, they are interesting in their own right and deserve at least an introduction. Their diversity of morphology is astonishing, and study of their many varieties of life cycles is a science in itself. Many opportunities are available for research on these worms. For example, many cestodes exist for which not a single life cycle is known. Following discussions of Diphyllobothriidea and Cyclophyllidea and brief descriptions of several other orders, we give some very brief accounts of the sister group of subcohort Eucestoda, Amphilinidea, and of Gyrocotylidea, the sister group of cohort Cestoidea.

ORDER DIPHYLLOBOTHRIIDEA

A diphyllobothriidean cestode typically has a scolex with two longitudinal bothria. The bothria may be deep or shallow and smooth or fimbriated, and in some cases they are fused along all or part of their length, forming longitudinal tubes. Proteinaceous hooks accompany the bothria in some species. Genital pores may be lateral or medial, depending on the species. The vitellaria are always follicular and scattered throughout the segment. Testes are numerous. Generally life cycles of diphyllobothriideans involve crustacean first intermediate hosts and fish second intermediate hosts.

Some species are fairly small, but the largest tapeworms known are in Diphyllobothriidea. For example, *Hexagonoporus physeteris* from sperm whales measures more than 30 meters long. In addition, each segment has 4 to 14 complete sets of genitalia. One worm has up to 45,000 segments. The reproductive capacity of such an animal is staggering.

FAMILY DIPHYLLOBOTHRIIDAE

Diphyllobothrium Species

Species of *Diphyllobothrium* are difficult to distinguish from one another morphologically, and they typically exhibit little host specificity.[2] Usually called *broad fish tapeworms,* they have commonly been designated *D. latum* and have been reported from many canines, felines, mustelids, pinnipeds, bears, and humans. Many of these records, however, are misidentifications. Humans seem to be quite suitable as hosts for *Diphyllobothrium* spp., with at least 13 distinct species having been reported. The most prevalent are *D. dendriticum* and *D. latum.*[2] An estimated 9 million people are infected worldwide.[48] *Diphyllobothrium dendriticum* is most common and occurs throughout the Northern Hemisphere. Less widespread is *D. latum,* with major endemicity in Scandinavia, the Baltic states, and western Russia. It has been introduced in other parts of the world, including the Great Lakes area of the United States. There are numerous reports of *Diphyllobothrium* sp. from the West Coast of North America but species involved are not clear.[88] Although more recently reported for the first time from Argentina,[84] *Diphyllobothrium* spp. apparently parasitized native South Americans well before the "discovery" of the New World by Columbus.[51]

- **Morphology.** Adult *D. latum* (Fig. 21.1) may attain a length of 10 m and shed up to a million eggs a day. *Diphyllobothrium dendriticum* attains a length of 1 m. These species are anapolytic and characteristically release long chains of spent segments, usually the first indication that the infected person has a secret guest.

  Their scolex (Fig. 21.2) is finger shaped and has dorsal and ventral bothria. Proglottids (Fig. 21.3; see also Fig. 20.17) usually are wider than long. There are numerous testes and vitelline follicles scattered throughout each

325

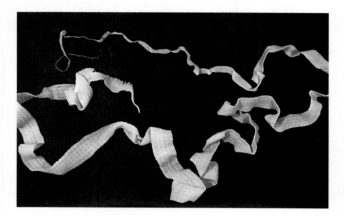

Figure 21.1 *Diphyllobothrium latum.*
The scolex is at the tip of the threadlike end at upper left.
Courtesy of Warren Buss.

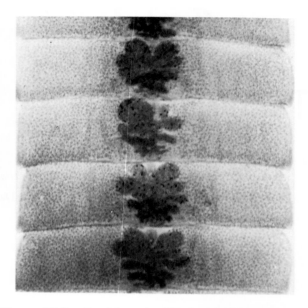

Figure 21.3 Gravid proglottids of *Diphyllobothrium* sp.
Note the characteristic rosette-shaped uterus.
Courtesy of Larry Jensen.

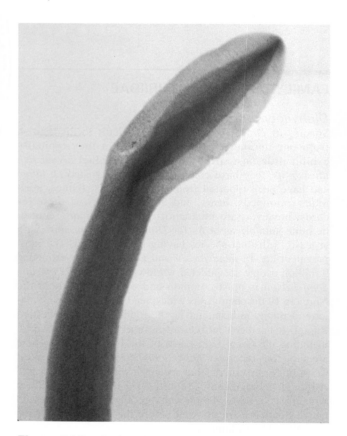

Figure 21.2 Scolex of *Diphyllobothrium* sp.
It is about 1 mm long.

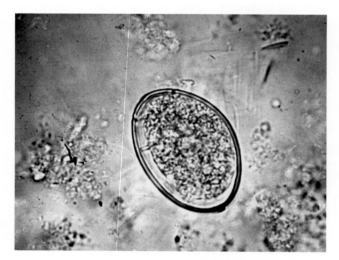

Figure 21.4 Egg of *Diphyllobothrium* sp. in a human stool.
Note the operculum at the upper end and the small knob at the opposite end. It is 40 μm to 60 μm long.
Courtesy of David Oetinger.

proglottid, except for a narrow zone in the center. Male and female genital pores open midventrally. The bilobed ovary is near the rear of the segment. The uterus consists of short loops and extends from the ovary to a midventral uterine pore.

- **Biology.** The ovoid eggs measure about 60 μm by 40 μm and have a lidlike operculum at one end and a small knob on the other (Fig. 21.4). When released through the

uterine pore, the shelled embryo is at an early stage of development, and it must reach water for development to continue (Fig. 21.5). Completion of development to coracidium takes from eight days to several weeks, depending on the temperature. Emerging through the operculum, the ciliated coracidium swims randomly about, where it may attract the attention of predaceous copepods such as species of *Diaptomus* and related genera. Soon after being eaten, the coracidium loses its ciliated epithelium and immediately begins to attack the midgut wall with its six tiny hooks. Once through the intestine and into the crustacean's hemocoel, it becomes parasitic, absorbing nourishment from the surrounding hemolymph. Infection with

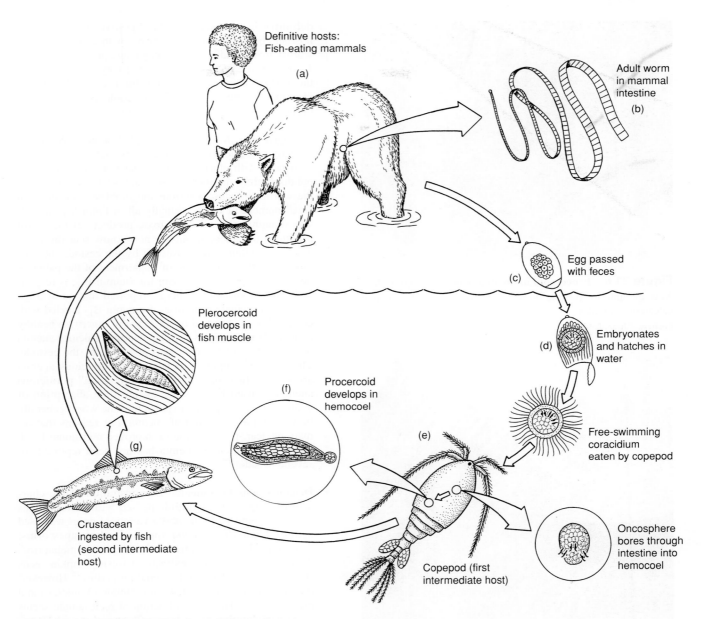

Figure 21.5   Life cycle of *Diphyllobothrium latum*.

(*a*) Definitive hosts are any of a number of fish-eating mammals. (*b*) Adult worm is in mammal small intestine. (*c*) Shelled embryo passes in feces at an early stage of development. (*d*) Embryogenesis continues in water, and free-swimming coracidium hatches. (*e*) Coracidium eaten by copepod, and oncosphere penetrates intestine into hemocoel. (*f*) Procercoid develops in hemocoel. (*g*) Copepod eaten by fish, where procercoid penetrates into muscle and develops into pleroceroid.

Drawing by William Ober and Claire Garrison.

Diphyllobothrium spp. impairs motility of copepods, thus rendering them more vulnerable to predation.[78]

In about three weeks the worm increases its length to around 500 μm, becoming an elongated, undifferentiated mass of parenchyma with a cercomer at the posterior end. It is now a procercoid (Fig. 21.6), incapable of further development until it is eaten by a suitable second intermediate host—any of several species of freshwater fishes, especially pike and related fishes, or any of the salmon family. The cercomer may be lost while still in the copepod or soon after the procercoid enters a fish. Large, predaceous fish eat comparatively few microcrustaceans but can still become infected by eating smaller fish containing plerocercoids, which then migrate into the new host. The larger fish is thus a paratenic host.

When a fish eats an infected copepod, the procercoid is released and bores its way through the intestinal wall and into the body muscles. Here it absorbs nutrients and grows rapidly into a plerocercoid. Mature plerocercoids vary in length from a few millimeters to several centimeters. They are still mainly undifferentiated, but there may be evidence of shallow bothria at the anterior end. Usually plerocercoids are found unencysted and coiled up in the musculature, although they may be encysted in the

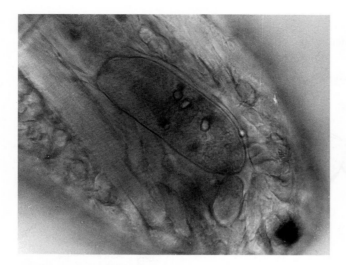

Figure 21.6 **Procercoid in the hemocoel of a copepod.**

Note the anterior pit, the posterior cercomer, and the internal calcareous granules.

Courtesy of Justus F. Mueller.

Figure 21.7 **Two plerocercoids in the flesh of a perch.**

From R. Vik, in Marcial-Rojas (Ed.), *Pathology of protozoal and helminthic diseases, with clinical correlation,* © 1971. (Baltimore, MD: Williams & Wilkins).

viscera. They are easily seen as white masses in uncooked fish (Fig. 21.7), but when the flesh is cooked, worms are seldom noticed. They are also killed and thus rendered noninfective.

Plerocercoids of other diphyllobothriideans, as well as those of proteocephalatans and trypanorhynchans, are also found in fish and are often mistaken for those of *Diphyllobothrium*. When a plerocercoid is ingested by a suitable

definitive host, it survives the digestive fate of its late host and begins a close relationship with a new one. The worms grow rapidly and may begin egg production in 7 to 14 days. Little of this initial growth may be attributed to the production of new proglottids but is caused by growth in primordia already in the plerocercoid. As much as 70% of the strobila may mature on the same day.[7]

- **Pathogenesis.** Many cases of diphyllobothriasis are apparently asymptomatic or have poorly defined symptoms associated with other tapeworms, such as vague abdominal discomfort, diarrhea, nausea, and weakness. However, the worm can cause a serious megaloblastic anemia in a small number of cases, virtually all in Finnish people. It was thought originally that toxic products of the worm produced the anemia, but we now know that the large amount of vitamin B_{12} absorbed by the cestode, in conjunction with some degree of impairment of the patient's normal absorptive mechanism for vitamin B_{12}, is responsible for the disease. Nyberg[74] reported that an average of 44% of a single oral dose of vitamin B_{12} labeled with cobalt 60 was absorbed by *D. latum* in otherwise healthy patients, but in patients with tapeworm pernicious anemia 80% to 100% of the dose was absorbed by the cestode. The clinical symptoms of tapeworm pernicious anemia are similar in many respects to "classical" pernicious anemia (caused by a failure in intestinal absorption of vitamin B_{12}), except that expulsion of the worm generally brings a rapid remission of anemia. For reasons that remain unclear, possibly due to improved nutritional level, tapeworm pernicious anemia has not been reported for several decades.[90]

- **Diagnosis and Treatment.** Demonstration of the characteristic eggs or proglottids passed with a stool gives positive diagnosis. In the past a variety of drugs was used against *Diphyllobothrium* spp. and other tapeworms; aspidium oleoresin (extract of male fern), mepacrine, dichlorophen, and even extracts of fresh pumpkin seeds (*Cucurbita* spp.) have anticestodal properties.[25] However, the drugs of choice are now niclosamide (Yomesan) and praziquantel.[89] The mode of action of niclosamide seems to be an inhibition of an inorganic phosphate—ATP exchange reaction associated with the worm's anaerobic electron transport system. We described the action of praziquantel on p. 230.

- **Epidemiology.** Obviously, persons become infected when they eat raw or undercooked fish. Hence, infection rates are highest in countries where raw fish is eaten as a matter of course. Communities that dispose of sewage by draining it into lakes or rivers without proper treatment create an opportunity for a massive buildup of *D. latum* or *D. dendriticum* in local fish. These fish may be harvested for local consumption or shipped thousands of miles by refrigerated freight to distant markets. An unsuspecting customer may thus gain infection in a restaurant or at home by tasting such dishes as gefilte fish during preparation. The higher prevalence of *Diphyllobothrium* in women is probably due to the higher prevalence of women among the ranks of cooks. The fad in the United States of eating raw salmon as sushimi has led to infections.[88]

Other Diphyllobothriideans Found in Humans

Several other species of *Diphyllobothrium* have been reported from humans in different parts of the world. These include *D. cordatum, D. pacificum, D. cameroni, D. hians,* and *D. lanceolatum,* parasites of pinnipeds, and *D. ursi* of bears. At least some infections of humans on the West Coast of North America and Hawaii apparently are due to *D. dendriticum,*[5] but some are likely to be from one or more species from pinnipeds with a marine life cycle. *Diphyllobothrium nihonkaeiense* is the dominant species in Japan, although other species occur there.[3, 65]

Digramma brauni and *Ligula intestinalis* have also been reported from humans, but such occurrences must be rare. *Diplogonoporus grandis (D. balaenopterae)* has been reported numerous times from humans in Japan.[2] A parasite of whales, its plerocercoid occurs in marine fish, the mainstay of the Japanese protein diet.

Sparganosis

With the exception of forms with scolex armature, species of plerocercoids found in humans are impossible to distinguish by examining their morphology. When procercoids of some species are ingested accidentally, usually when a person swallows an infected copepod in drinking water, they can migrate from the gut and develop into plerocercoids, sometimes reaching a length of 35 cm. This infection is called *sparganosis* and may have severe pathological consequences. Cases have been reported from most countries of the world but are most common in eastern Asia. Yamane, Okada, and Takihara[103] reported a living sparganum that had infected a woman's breast for at least 30 years.

Another means of infection is by ingestion of insufficiently cooked amphibians, reptiles, birds, or even mammals such as pigs.[24] Plerocercoids present in these animals may then infect a person indulging in such delicacies. Many Chinese are infected in this way because of their tradition of eating raw snake to cure a panoply of ills.[54]

A third method of infection results from the east Asian treatment of skin ulcers, inflamed vagina, or inflamed eye (Fig. 21.8) by poulticing the area with a split frog or flesh of a vertebrate that may be infected with spargana. The active worm then crawls into the orbit, vagina, or ulcer and establishes itself. Most cases of sparganosis in eastern Asia are probably caused by *Diphyllobothrium erinacei,* a parasite of carnivores.

In North America most spargana are probably *Diphyllobothrium mansonoides,* a parasite of cats.[71] It usually does not proliferate, except by occasionally breaking transversely, and may live up to 10 years in a human.[95] The current public awareness of the symptoms of cancer has led to an increase in reported cases of sparganosis in this country. Subdermal lumps are no longer ignored by an average person, and more than one physician has been shocked to find a gleaming, white worm in a lanced nodule. Wild vertebrates are commonly infected with spargana (Fig. 21.9).

Rarely a sparganum will be proliferative, splitting longitudinally and budding profusely. Such cases are very serious, since many thousands of worms can result, with the infected organs becoming honeycombed.[69]

Treatment of sparganosis is usually by surgery, but supplementary treatment with praziquantel may be advisable.[88]

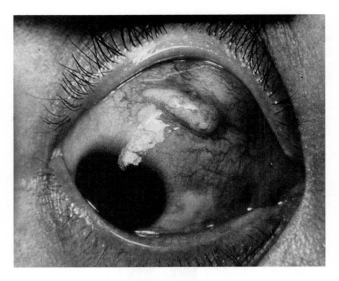

Figure 21.8   Right eye of patient with sparganosis.
Note the protruding mass in the upper conjunctiva.

From L. T. Wang and J. H. Cross, "Human sparganosis on Taiwan. A report of two cases," in *J. Formosan Med. Assoc.* 73:173–177. Copyright © 1974.

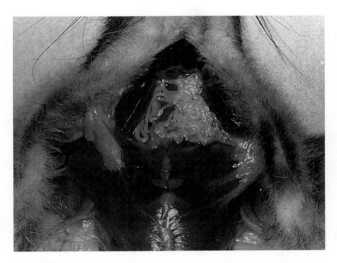

Figure 21.9   Spargana in subcutaneous connective tissues of a wild rat in Taiwan.

Courtesy of Robert E. Kuntz.

ORDER CARYOPHYLLIDEA

Caryophyllideans are intestinal parasites of freshwater fishes, except for a few that mature in the coelom of freshwater oligochaete annelids.[59, 62] All are monozoic, showing no trace of internal proglottisation or external segmentation (Fig. 21.10). The scolex is never armed. Usually it is quite simple, bearing shallow depressions (loculi), or it is frilled or entirely smooth. Some species seem to lack a scolex altogether. The anterior end of the worm is very motile, however, and functions well as a holdfast. Some species induce a pocket in the wall of the host's intestine in which one or more worms remain.

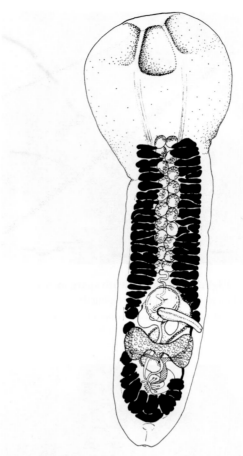

Figure 21.10 *Penarchigetes oklensis*, a typical caryophyllidean cestode, from a spotted sucker.

From J. S. Mackiewicz, "*Penarchigetes oklensis* gen. et sp. n. and *Biacetabulum carpiodi* sp. n. (Cestoidea: Caryophyllaeidae) from catostomid fish in North America," in *Proc. Helm. Soc. Wash.* 37:110–118, 1969.

Each worm has a single set of male and female reproductive organs. In most the ovary is near the posterior end. Testes fill the median field of the body, and vitelline follicles are mainly lateral. Male and female genital pores open near each other on the midventral surface.

Catfishes, true minnows (Cyprinidae), and suckers are the most common hosts of Caryophyllidea. *Glaridacris* spp., predominantly *G. catostomi*, are found abundantly in suckers (*Catostomus* spp.) in North America.[59] Intermediate hosts are aquatic annelids. After an oligochaete eats an egg, the oncosphere hatches and penetrates the intestine into the coelom. There it grows into a procercoid with a prominent cercomer, similar to that of *Diphyllobothrium* spp. When eaten by a fish, the procercoid loses its cercomer and grows directly into an adult.

It has been suggested that segmented adults once existed but became extinct with their hosts, probably aquatic reptiles. However, this did not happen before plerocercoids developed neotenically in fish second intermediate hosts. If this hypothesis is true, extant caryophyllaeid species actually are neotenic plerocercoids. Support is lent to this idea by the existence of several species of *Archigetes* that become sexually mature while in annelids. Reproductive adults retain their

cercomer and infect no additional host, although they can live for some time if eaten by a fish. *Archigetes* spp., then, appear to be neotenic procercoids. However, this hypothesis is not supported by molecular evidence, which suggests that the monozoic condition is plesiomorphic.[64]

ORDER SPATHEBOTHRIIDEA

These are peculiar parasites of marine and freshwater teleost fishes. Their most striking characteristic is a complete absence of segmentation, with possession of a typically linear series of internal proglottids. The scolex always lacks armature. It may be totally undifferentiated, as in *Spathebothrium simplex;* it may be a shallow funnel-shaped organ, as in *Cyathocephalus truncatus;* or it may consist of one or two powerful cuplike organs (see Fig. 20.5*j*). Genital pores are ventral, testes are in two lateral bands, the ovary is dendritic, and vitellaria are follicular and lateral or scattered. The uterus is rosettelike and opens ventrally, usually near the vaginal pore.

No life cycles are known. Although these worms are of no known economic importance, they remain an interesting zoological group that should be studied further. *Bothrimonus,* a common genus in North America, has been investigated more fully;[19] *B. sturionis* parasitizes a wide variety of marine and freshwater fish, from sturgeons to salmonids to flounders.

ORDER CYCLOPHYLLIDEA

Cyclophyllidea and Proteocephalata both have scolices with four acetabula. In these orders, parenchyma is divided into highly distinct medullary and relatively extensive cortical regions defined by the longitudinal musculature. Brooks and McLennan[17] combined these two orders into the Proteocephaliformes, but we are retaining the traditional separation for the present.

The most obvious morphological feature that characterizes Cyclophyllidea is a single compact, postovarian vitelline gland (see Fig. 20.18). A rostellum, which usually bears an armature of hooks, is commonly present. Genital pores are lateral in all except family Mesocestoididae, in which they are midventral. The number of testes varies from one to several hundred, depending on the species. Most species are rather small, although some are giants of more than 10 m in length. Most tapeworms of birds and mammals belong to this order.

Family Taeniidae

The largest cyclophyllideans are in family Taeniidae, as are the most medically important tapeworms of humans. A remarkable morphological similarity occurs among species in the family; striking exceptions are *Echinococcus* spp., which are much smaller than cestodes of other taeniid genera. An armed rostellum is present on most species and when present is not retractable. Testes are numerous, and the ovary is a bilobed mass near the posterior margin of the

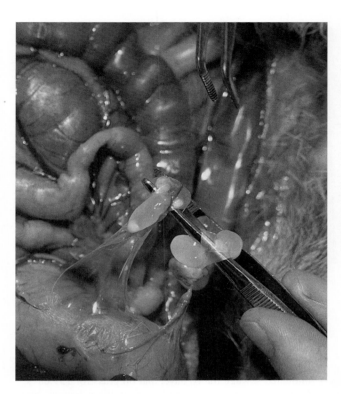

Figure 21.11   **Cysticerci of *Taenia pisiformis* in the mesenteries of a rabbit.**

Courtesy of John Mackiewicz.

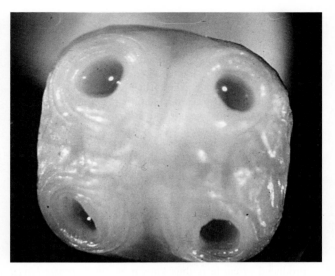

Figure 21.12   **En face view of the scolex of *Taenia saginata*.**

Note the absence of a rostellum or armature.

AFIP neg. no. 65-12073-2.

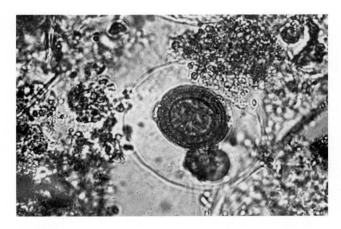

Figure 21.13   *Taenia* **sp. egg in human feces.**

The thin outer membrane is often lost at this stage.

Courtesy of David Oetinger.

proglottid. Metacestodes are various types of **bladderworms** (Fig. 21.11), and mammals serve as their intermediate hosts.

Taenia saginata. Among Taeniidae, *Taenia saginata* is by far the most common in humans, occurring in nearly all countries where beef is eaten. The beef tapeworm, as it is usually known, lacks a rostellum or any scolex armature (Fig. 21.12). Individuals of this exceptionally large species may attain a length of over 20 m, but 3 m to 5 m is much more common. Even the smaller specimens may consist of as many as 2000 segments.

In earlier editions of this text, we considered the un-armed scolex of this species sufficient reason to consign it to a separate genus, *Taeniarhynchus*.[91] However, nucleotide-sequence data support placement of *T. saginata* with other, well-recognized species of *Taenia*.[75] Subsequent cladistic studies with both morphological and molecular data clearly show that retention of *Taeniarhynchus* would render genus *Taenia* paraphyletic.[39, 47, 73, 80]

- **Morphology.**   The scolex, with its four powerful suckers, is followed by a long, slender neck. Mature segments are slightly wider than long, whereas gravid ones are much longer than wide. Usually it is a gravid segment passed in the feces that is first noticed and taken to a physician for diagnosis. Because eggs of this species cannot be differentiated from those of *Taenia solium,* the next most common taeniid of humans, accurate diagnosis depends on other criteria. Of course, if an entire worm is passed, the unarmed scolex leads to an unmistakable diagnosis.

The spherical eggs are characteristic of Taeniidae (Fig. 21.13). A thin, hyaline, outer membrane is usually lost by the time the egg is voided with the feces. The embryophore is very thick and riddled with numerous tiny pores, giving it a striated appearance in optical section. Unfortunately, the egg sizes of several taeniids in humans overlap, making diagnosis of species impossible on this character alone.

- **Biology.**   When gravid, segments detach and either pass out with feces or migrate out of the anus. Each segment behaves like an individual worm, crawling actively about, as if searching for something. Segments are easily mistaken for trematodes or even nematodes at this stage. As a segment begins to dry up, a rupture occurs along the midventral body wall, allowing eggs to escape. Larvae are fully developed and infective to their intermediate host at this time; they remain viable for many weeks. Cattle are the usual intermediate hosts, although cysticerci have also been reported from llamas, goats, sheep, giraffes, and even reindeer (perhaps incorrectly).

When eaten by a suitable intermediate host, eggs hatch in the duodenum under the influence of gastric and intestinal secretions. Hexacanths quickly penetrate the mucosa and enter intestinal venules, to be carried throughout the body. Typically they leave a capillary between muscle cells and enter a muscle fiber, developing into infective cysticerci in about two months. These metacestodes are white, pearly, and up to 10 mm in diameter and contain a single, invaginated scolex. Humans are probably unsuitable intermediate hosts, and the few records of *Taenia saginata* cysticerci in humans are most likely misidentifications. Before the beef cysticercus was known to be a juvenile form of *T. saginata,* it was placed in a separate genus under the name of *Cysticercus bovis.* The disease produced in cattle is thus known as *cysticercosis bovis,* and flesh riddled with the juveniles is called *measly beef.*

A person who eats infected beef, cooked insufficiently to kill the juveniles, becomes infected. The invaginated scolex and neck of cysticerci evaginate in response to bile salts. The bladder is digested by the host or absorbed by the scolex, and budding of proglottids begins. Within 2 to 12 weeks the worm begins shedding gravid proglottids.

- **Pathogenesis.** Disease characteristics of *T. saginata* infection are similar to those of infection by any large tapeworm, except that the avitaminosis B_{12} found in association with *D. latum* is unknown. Most people infected with *T. saginata* are asymptomatic or have mild to moderate symptoms of dizziness, abdominal pain, diarrhea, headache, localized sensitivity to touch, and nausea. Delirium is rare but does occur. Intestinal obstruction with need for surgical intervention sometimes occurs. Hunger pains, universally accepted by lay people as a symptom of tapeworm infection, are not common, and loss of appetite is frequent. Worms release antigens, which sometimes result in allergic reactions. In addition, it is difficult to estimate the psychological effects on an infected person of observing continued migration of proglottids out of the anus.

- **Diagnosis and Treatment.** Identification of taeniid eggs according to species is impossible. Therefore, accurate diagnosis depends on examination of a scolex or perhaps a gravid segment. The latter is characterized by 15 to 20 branches on each side (contrasted with 7 to 13 in *T. solium*). Because these branches tend to fuse in deteriorating segments, freshly passed specimens must be obtained for reliable results; furthermore, several investigators have reported overlapping numbers.[86] A test for worm antigens passed in the feces (coproantigens) using a variant of the ELISA has been described, and an improved polymerase chain reaction-restriction fragment length polymorphism assay suitable for field samples is available.[4, 86]

Numerous taeniicides have been used in the past. Today niclosamide and praziquantel are the drugs of choice.

- **Epidemiology.** Human infection is highest in areas of the world where beef is a major food and sanitation is deficient. Thus, in several developing nations of Africa and South America, for instance, ample opportunity exists for cattle to eat tapeworm eggs and for people to eat infected flesh. Many people are content to eat a chunk of meat that is cooked in a campfire, charred on the outside and raw on the inside.

Local custom may have profound effects on infection rates. Hence, in India there may be a high rate of infection among Moslems, whereas Hindus, who do not eat beef, are unaffected. In the United States, federal meat inspection laws and a high degree of sanitation combine to keep the incidence of infection low. However, not all cattle slaughtered in the United States are federally inspected, and standard inspection procedures fail to detect one-fourth of infected cattle.[28] One wonders if backyard cookery and the popularity of rare steaks and undercooked hamburgers might not contribute to prevalence of taeniiasis, although well-publicized cases of bacterial (*Escherichia coli*) contamination of ground beef may have the reverse effect.

Despite a high level of sanitation in any country, it still is possible for cattle to be exposed to eggs of this parasite. One infected person who defecates in a pasture or cattle-feeding area can quickly infect an entire herd. The use of human feces as fertilizer can have the same effect. Shelled larvae can remain viable in liquid manure for 71 days, in untreated sewage for 16 days, and on grass for 159 days.[51] Cattle are coprophagous and often will eat human dung, wherever they find it. In India, where cattle roam at will, it is common for a cow to follow a person into the woods, in anticipation of obtaining a fecal meal.[22]

Prevention of human infection is easy; when meat is cooked until it is no longer pink in the center, it is safe to eat, since cysticerci are killed at 56° C. Furthermore, meat is also rendered safe by freezing at −5° C for at least a week.

Taenia asiatica. A taeniid very similar morphologically to *Taenia saginata* has been distinguished in Southeast Asia and China.[32] Many authors referred to this form as "Asian *Taenia*" because of uncertainty that it was a separate species, but *T. asiatica* can be distinguished from other *Taenia* species on biological, morphological, and molecular bases.[33] A striking biological difference from *T. saginata* is that cysticerci of *T. asiatica* develop in pigs, primarily in liver and other viscera and not in muscles. (Many rural Asians relish raw pig viscera.) Cysticerci of *T. asiatica* may have small hooklets on their scolex. *Taenia asiatica* is genetically and immunologically much closer to *T. saginata* than to other species of taeniids.[38, 79] They are sister species and only distantly related to *T. solium.*[47]

Taenia solium. The most dangerous adult tapeworm of humans is the pork tapeworm, *Taenia solium,* because humans can also serve as intermediate hosts; that is, infection with eggs results in development of cysticerci (**cysticercosis**) in humans. Thus, a person can become infected through contamination of food or fingers with eggs. Likewise, it is possible to infect others in the same household by the same means, often with grave results.

- **Morphology.** The scolex of an adult (Fig. 21.14) bears a typical, nonretractable taeniid rostellum armed with two

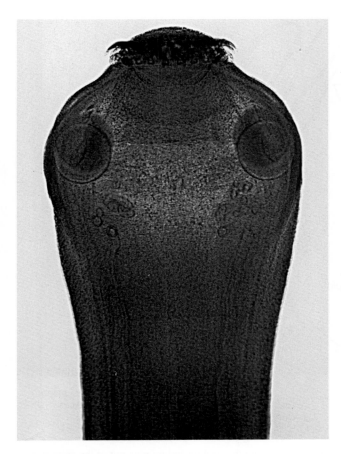

Figure 21.14   Scolex of *Taenia solium*.
Note the large rostellum with two circles of hooks.
Courtesy of David Oetinger.

circles of 22 to 32 hooks measuring 130 μm to 180 μm long. Whereas the scolex of *Taenia saginata* is cuboidal and up to 2 mm in diameter, that of *T. solium* is spheroid and only half as large. There are reports of strobilas as long as 10 m, but 2 to 3 m is much more common. Mature segments are wider than long and are nearly identical to those of *T. saginata*, differing in number of testes (150 to 200 in *T. solium*, 300 to 400 in *T. saginata*). Gravid segments are longer than wide and have the typical taeniid uterus, a medial stem with 7 to 13 lateral branches.

- **Biology.**   The life cycle of *T. solium* (Fig. 21.15) is in most regards like that of *T. saginata*, except that its normal intermediate hosts are pigs instead of cattle. Gravid proglottids passed in feces are laden with eggs infective to swine. When eaten, oncospheres develop into cysticerci *(Cysticercus cellulosae)* in muscles and other organs. Blowflies can carry eggs from infected feces to uninfected meat, which is readily eaten by pigs,[55] and pigs feed on human feces where it is available.[37] Infection of black bears in California with *T. solium* cysticerci has been confirmed.[99]

A person easily becomes infected when eating a bladderworm along with insufficiently cooked pork. Dogs and cats also can serve as intermediate hosts and so can serve as a source of infection for people where those animals are eaten.[90] Evaginating by the same process as in *T. saginata*, the worm attaches to the mucosa of the small intestine and matures in 5 to 12 weeks. Specimens of *T. solium* can live for as long as 25 years. Pathogenesis caused by adult worms is similar to that in taeniiasis saginata.

- **Cysticercosis.**   Unlike those of most other species of *Taenia*, cysticerci of *T. solium* develop readily in humans. Infection occurs when shelled larvae pass through the stomach and hatch in the intestine. People who are infected by adult worms may contaminate their households or food with eggs they or others accidentally eat. Possibly, a gravid proglottid may migrate from the lower intestine to the stomach or duodenum, or it may be carried there by reverse peristalsis. Subsequent release and hatching of many eggs at the same time results in a massive infection by cysticerci.

Virtually every organ and tissue of the body may harbor cysticerci. Most commonly they are found in the subcutaneous connective tissues. The second most common site is the eye, followed by the brain (Fig. 21.16), muscles, heart, liver, lungs, and coelom. A fibrous capsule of host origin surrounds the metacestode, except when it develops in the chambers of the eye. The effect of any cysticercus on its host depends on where it is located. In skeletal muscle, skin, or liver, little noticeable pathogenesis usually results, except in massive infection. Ocular cysticercosis may cause irreparable damage to the retina, iris, or choroid. A developing cysticercus in the retina may be mistaken for a malignant tumor, resulting in the unnecessary surgical removal of the eye. Removal of the cysticercus by fairly simple surgery is usually successful.

Cysticerci occur rarely in the spinal cord but commonly in the brain.[13] Symptoms of infection are vague and rarely diagnosed except at autopsy. Pressure necrosis may cause severe central nervous system malfunction, blindness, paralysis, disequilibrium, obstructive hydrocephalus, or disorientation. Perhaps the most common symptom is epilepsy of sudden onset. When this occurs in an adult with no family or childhood history of epilepsy, cysticercosis should be suspected. Praziquantel is the drug of choice for neurocysticercosis, but it is not used against cysticerci in the brain ventricles or the eye.[37] Praziquantel can also be used against porcine cysticercosis, thus avoiding the need for condemnation of the carcass.[99]

Cysticerci apparently evade a host's defenses by down-modulating its immune response,[44, 50] but when a cysticercus dies, it elicits a rather severe inflammatory response. Many of them may be rapidly fatal to the host, particularly if the worms are located in the brain. This was observed frequently in former British soldiers; a high proportion of those who served in India became infected. Other types of cellular reaction also occur, usually resulting in eventual calcification of the parasite (Fig. 21.17). If this occurs in the eye, there is little chance of corrective surgery.

Cysticerci occur in three distinct morphological types, of which the most common is the ordinary "cellulose" cysticercus, with an invaginated scolex and a fluid-filled bladder about 0.5 cm to 1.5 cm in diameter. The "intermediate" form (with a scolex) and the "racemose" (with no evident scolex) are much larger and more dangerous.

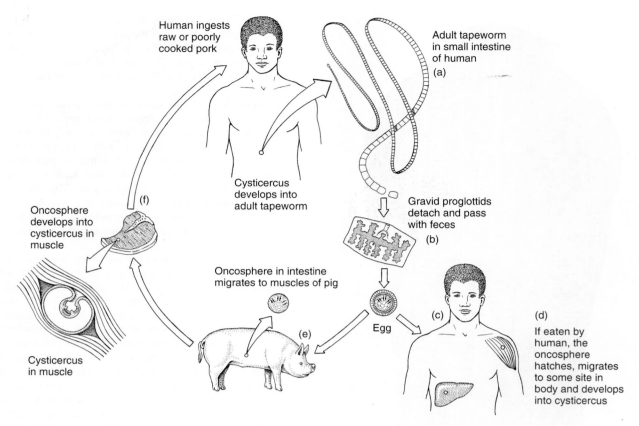

Figure 21.15   Life cycle of *Taenia solium*.

(*a*) Adult tapeworm in the small intestine of a human. (*b*) Gravid proglottids detach from the strobila and migrate out of the anus or pass with feces. (*c*) Shelled oncosphere. (*d*) If eaten by a human, the oncosphere hatches, migrates to some site in the body, and develops into a cysticercus. (*e*) Cysticerci will also develop if the eggs are eaten by a pig. (*f*) The life cycle is completed when a person eats pork containing live cysticerci.

Drawing by William Ober and Claire Garrison.

Figure 21.16   **Human brain containing numerous cysticerci of *Taenia solium*.**

From A. Flisser, "Neurocysticercosis in Mexico," in *Parasitol. Today* 4:131–137. Copyright © 1988.

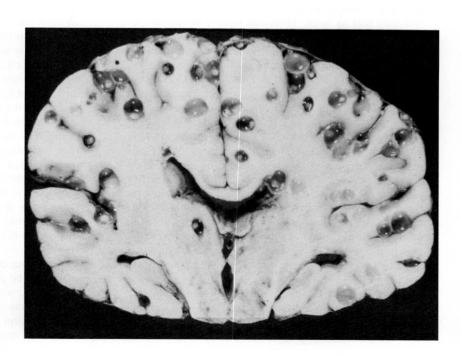

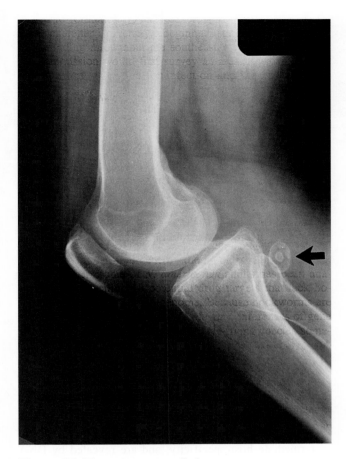

Figure 21.17   *Cysticercus cellulosae.*

Partially calcified cyst *(arrow)* found in a routine X-ray examination of a human leg.

From R. L. Roudabush and G. A. Ide, "*Cysticercus cellulosae* on X-ray," in *J. Parasitol.* 61:512. Copyright © 1975.

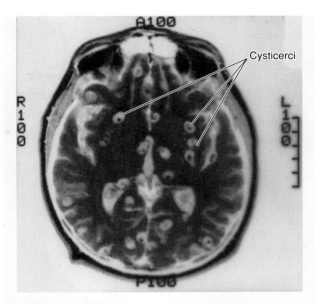

Figure 21.18   **Neurocysticercosis.**

Computerized axial tomography scan (CAT scan) of a patient with multiple *T. solium* cysticerci in the brain.

Courtesy Herman Zaiman.

They can measure up to 20 cm and contain 60 ml of fluid. Up to 13% of patients may have all three types in the brain.[81]

Prevention of cysticercosis depends on early detection and elimination of the adult tapeworm and a high level of personal hygiene. Fecal contamination of food and water must be avoided and the use of untreated sewage on vegetable gardens eschewed. The majority of cases apparently originate from such sources, including contamination by infected food handlers.[31] ELISA-based tests for *T. solium* antigens in feces have been described that are more sensitive than microscopical diagnosis.[4]

Although neurocysticercosis has been considered uncommon, improved brain imaging (through computerized axial tomography and magnetic resonance imaging Fig. 21.18) has demonstrated a higher frequency in the United States than once thought.[20] Of 138 cases reported from Los Angeles County, California, from 1988 to 1990, most were in immigrants from Mexico,[94] where cysticercosis is a major public health problem;[36] however, 9 were travel-associated cases, and 10 were infections acquired in the United States.

A very sensitive and specific, enzyme-linked immunoelectrotransfer blot (EITB) test for serum antibodies has been described.[69, 100] It can detect 98% of parasitologically proven cases with two or more cysts and 60% to 80% of patients with only one lesion. The EITB showed that 1.3% of the people in an Orthodox Jewish community in New York were infected, apparently having become so by domestic employees who were immigrants from countries endemic for *T. solium.*[66]

Cysticercosis due to *T. solium* is highly endemic in Mexico, Central and much of South America, much of sub-Saharan Africa, India, China, and other parts of eastern Asia.[24] Some observers believe that it was deliberately introduced into Irian Jaya by Indonesians as a biological weapon against certain primitive peoples that opposed annexation.[49] This worm remains one of the most serious parasitic diseases in Irian Jaya, Papua Indonesia, and Papua New Guinea.[8, 92]

Other Taeniids of Medical Importance

Taenia multiceps, T. glomeratus, T. brauni, and *T. serialis* are all characterized by a coenurus type of bladderworm (Fig. 21.19). This juvenile type is similar to an ordinary cysticercus but has many rather than one scolex. Such coenuri occasionally occur in humans, particularly in the brain, eye, muscles, or subcutaneous connective tissue, where they often grow to be longer than 40 mm. The resulting pathogenesis is similar to that of cysticercosis. Adults are parasites of carnivores, particularly dogs, with herbivorous mammals serving as intermediate hosts. Accidental infection of humans occurs when eggs are ingested. Coenuriasis of sheep, caused by *T. multiceps,* causes a characteristic vertigo called *gid,* or *staggers.*

Echinococcus granulosus. Genus *Echinococcus* contains the smallest tapeworms in Taeniidae. However, their juveniles often form huge cysts and are capable of infecting

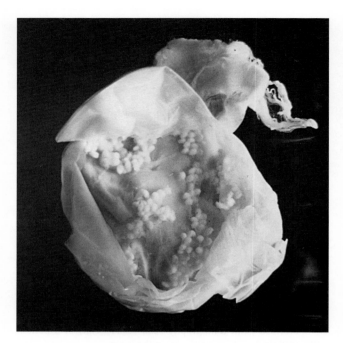

Figure 21.19 **Coenurus metacestode of *Taenia serialis*.**
This metacestode is from the muscle of a rabbit that has been
opened to show the numerous scolices arising from the germinal
epithelium. The cyst is about 4 in. wide.

Courtesy of James Jensen.

humans, resulting in a very serious disease in many parts of
the world.

Echinococcus granulosus causes **cystic echinococcosis.**
It uses carnivores, particularly dogs and other canines, as
definitive hosts (Fig. 21.20). Many mammals may serve as
intermediate hosts, but herbivorous species are most likely
to become infected by eating eggs on contaminated herbage.

Adults (Fig. 21.21) live in the small intestine of their definitive host. They measure 3 mm to 6 mm long when mature
and consist of a typically taeniid scolex, a short neck, and
usually only three proglottids. The nonretractable rostellum
bears a double crown of 28 to 50 (usually 30 to 36) hooks.
The anteriormost segment is immature; the middle one is
usually mature; and the terminal one is gravid. The gravid
uterus is an irregular longitudinal sac. The eggs cannot be
differentiated from those of other taeniids. Ripe segments
detach and develop a rupture in their wall, releasing the eggs,
which are fully capable of infecting an intermediate host.

Hatching and migration of oncospheres are the same as
previously described for *Taenia saginata*, except that liver
and lungs are the most frequent sites of development. By a
very slow process of growth, an oncosphere metamorphoses into a type of bladderworm called a **unilocular hydatid**
(Fig. 21.22). In about five months the hydatid develops a
thick outer, laminated, noncellular layer and an inner, thin,
nucleated germinal layer. The inner layer eventually produces brood capsules, within which develop protoscolices
(Fig. 21.23) that are infective to definitive hosts. Brood
capsules are small cysts, containing 10 to 30 protoscolices,
which usually are attached to the germinal layer by a slender stalk; they may break free and float within the hydatid
fluid. Similarly, individual scolices may break free from

the germinal layer of the capsule (Fig. 21.24). Rarely germinal cells penetrate the laminated layer and form daughter
capsules. When a carnivore eats the hydatid, the cyst wall
is digested away, freeing the protoscolices, which evaginate
and attach among the villi of the small intestine. A small percentage of hydatids lack protoscolices and are sterile, being
unable to infect a definitive host. The worm matures in about
56 days and may live for 5 to 20 months.

- **Epidemiology**. The life cycle of *Echinococcus granulosus* in wild animals may involve a wolf-moose,
  wolf-reindeer, dingo-wallaby, lion-warthog, or other
  carnivore-herbivore relationship, which is known as
  sylvatic echinococcosis. Humans are seldom involved
  as accidental intermediate hosts in these cases. However,
  ample opportunities exist for human infection in situations
  in which domestic herbivores are raised in association
  with dogs. For example, hydatid disease is a very serious
  problem in sheep-raising areas of Australia, New Zealand,
  North and South America, Europe, Asia, and Africa. Similarly, goats, camels, reindeer, and pigs, together with dogs,
  maintain the cycle in various parts of the world. Dogs are
  infected when they feed on the offal of butchered animals,
  and herbivores are infected when they eat herbage contaminated with dog dung. Humans are infected with hydatids when they accidentally ingest *Echinococcus* spp. eggs,
  usually as a result of fondling dogs.

  The species *E. granulosus* is composed of a number
  of genetically differing strains.[56, 104] Strain differences
  include morphology, development, metabolism, and intermediate host specificity, and are revealed by DNA
  hybridization and restriction site analysis. Worms of one
  strain are adapted to one species of intermediate host—for
  example, cattle, horses, sheep, or pigs—and they do not
  develop well in other species. The strains have considerable epidemiological significance for humans: Horse and
  pig strains in Europe probably do not infect humans, but
  sheep and cattle strains do. Molecular evidence suggests
  that several strains of *E. granulosus* may represent distinct species and that *E. granulosus* is paraphyletic.[16, 58]

  Local traditions may contribute to massive infections.
  Some tribes of Kenya, for instance, are said to relish
  dog intestine roasted on a stick over a campfire. Because
  cleaning of the intestine may involve nothing more than
  squeezing out its contents, and cooking may entail nothing more than external scorching, these people probably
  have the highest rate of infection with hydatids in the
  world.[71] A further complication lies in the lack of burial
  of the dead by the Turkana people of Kenya. When the
  corpses are eaten by carnivores, humans become true intermediate hosts of *E. granulosus*.[61]

  A different set of circumstances leads to infection in
  tanners in Lebanon, where dog feces are used as an ingredient of a solution for tanning leather. Scats picked off
  the street are added to the vats, and any eggs present may
  contaminate their handler.[91]

  Sheepherders in the United States and elsewhere risk
  infection by living closely with their dogs. Surveys of cattle, hogs, and sheep in abattoirs reveal that *E. granulosus*
  occurs throughout most of the United States, with greatest
  concentrations in the deep South and far West. Recent
  outbreaks have been diagnosed in California and Utah.

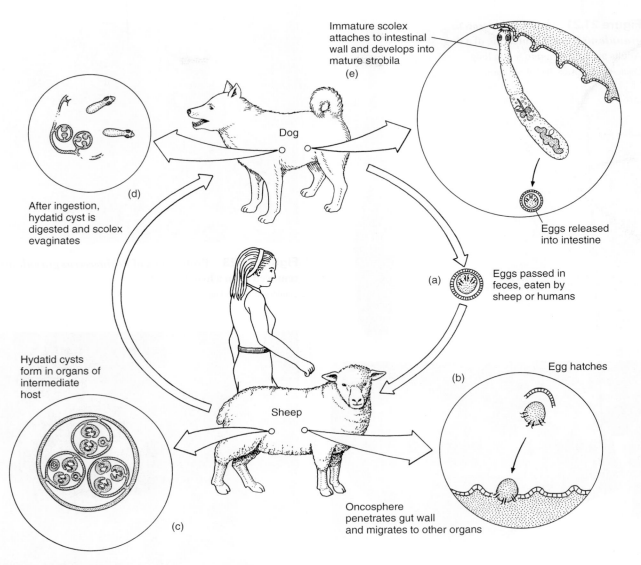

Immature scolex
attaches to intestinal
wall and develops into
mature strobila
(e)

Eggs released
into intestine

After ingestion,
hydatid cyst is
digested and scolex
evaginates
(d)

(a)   Eggs passed in
feces, eaten by
sheep or humans

Dog

Egg hatches
(b)

Hydatid cysts
form in organs of
intermediate
host

Sheep

(c)

Oncosphere
penetrates gut wall
and migrates to other organs

Figure 21.20   **Life cycle of *Echinococcus granulosus*.**

(*a*) Shelled oncosphere is passed in feces and eaten by hooved animals or humans. (*b*) Oncosphere penetrates gut wall and is carried to liver and other sites by circulation. (*c*) Hydatid cysts form in organs of intermediate host. (*d*) After ingestion by canid, hydatid cyst is digested and scolex evaginates. (*e*) Scolex attaches to intestinal wall and develops into strobila.

Drawing by William Ober and Claire Garrison.

Nevertheless, 98% of hydatid infections diagnosed in the United States are imported, the most frequent contributor of cases being Italy.[17]

This disease can be eliminated from an endemic area only by interrupting the life cycle by denying access of dogs to offal, by destroying stray dogs, and by a general education program.[39, 65]

• **Pathogenesis.**   Effects of a hydatid may not become apparent for many years after infection because of its usual slow growth. Up to 20 years may elapse between infection and overt pathogenesis. If infection occurs early in life, the parasite may be almost as old as its host.[12]

Cases have been reported in which liver, lungs, and brain simultaneously bore hydatids,[103] and cysts may occur in almost any organ.[68] Thus, the type and extent of

pathological conditions depend on the location of the cyst in the host. As the size of a hydatid increases, it crowds adjacent host tissues and interferes with their normal functions (Fig. 21.25). The results may be very serious. If the parasite is lodged in the nervous system, clinical effects may be manifested relatively early in the infection before much growth occurs. When bone marrow is affected, the growth of the hydatid is restricted by lack of space. Chronic internal pressure caused by the parasite usually causes necrosis of the bone, which becomes thin and fragile; characteristically, the first sign of such an infection is a spontaneous fracture of an arm or leg. When a hydatid grows in an unrestricted location, it may become enormous, containing more than 15 quarts of fluid and millions of protoscolices. Even if it does not occlude a vital organ, it can still cause sudden death if it ruptures.

Figure 21.21 **Adult *Echinococcus granulosus* from the intestine of a dog.**
Adults are only 3 mm to 8 mm long.
Courtesy of Ann Arbor Biological Center.

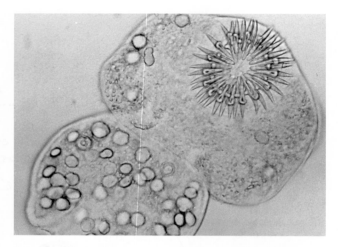

Figure 21.23 **Protoscolex of *Echinococcus granulosus*, removed from a hydatid cyst.**
Courtesy of Sharon File.

Figure 21.22 **Several unilocular hydatids in the lung of a sheep.**
Each hydatid contains many protoscolices.
Courtesy of James Jensen.

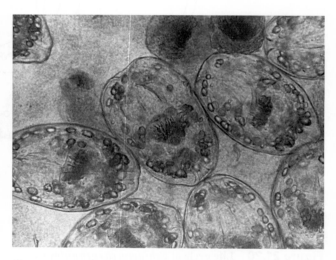

Figure 21.24 **Unattached protoscolices of *Echinococcus granulosus* from a hydatid cyst.**
Courtesy of Robert E. Kuntz.

The host is sensitized to *Echinococcus* antigens during its long-standing infection, and sudden release of massive amounts in the hydatid fluid induces an adverse host reaction called *anaphylactic shock*. Unconsciousness and death are nearly instantaneous in such instances.

- **Diagnosis and Treatment.**   When hydatids are found, it is often during X-radiography, ultrasonography, or CAT scans. Several immunodiagnostic techniques are available, but these are generally less sensitive than imagery.[10]

Surgery remains the only routine method of treatment and then only when the hydatid is located in an unrestricted location; for treatment of an inoperable hydatid, albendazole is recommended.[68] The typical surgical procedure involves incising the surrounding adventitia until the capsule is encountered and aspirating the hydatid fluid with a large syringe. Considerable delicacy is required at this point, since fluid spilled into a body cavity can

quickly cause fatal anaphylactic shock. After aspiration of the cyst contents, 10% formalin is injected into the hydatid to kill the germinal layer. This fluid is withdrawn after five minutes, and the entire cyst is then excised. A high rate of surgical success is obtained on ocular hydatidosis.

Echinococcus multilocularis. *Echinococcus multilocularis* causes **alveolar echinococcosis**. It is primarily boreal in its distribution, being widespread in Eurasia (central Europe, most of Russia, northern and western China, and Tibet, with southernmost records from Kashmir); Hokkaido Island in Japan; and two cases diagnosed from North Africa. In North America its range extends from the tundra zone in Alaska and northwest Canada through the provinces of Alberta, Saskatchewan, and Manitoba in Canada and in the United States from western Montana east to Ohio and south to Missouri. Adults are mainly parasites of foxes, but dogs, cats, and coyotes may also serve as definitive hosts. Several

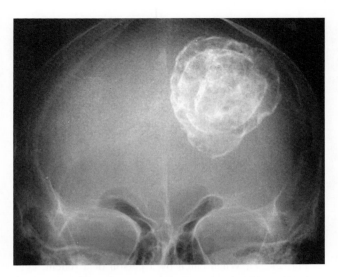

Figure 21.25   **Partially calcified hydatid cyst in the brain.**
AFIP neg. no. 68-2740.

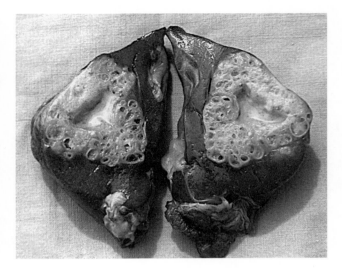

Figure 21.26   **Alveolar metacestode in the liver of a rhesus monkey infected experimentally 15 months earlier.**
The central cavity is a result of necrosis.
Courtesy of Robert Rausch.

species of small rodents such as voles, lemmings, and mice are intermediate hosts.

- **Morphology.**   Adults are very similar to those of *E. granulosus,* differing from them in the following characteristics: (1) *E. granulosus* is 3 mm to 6 mm long, whereas *E. multilocularis* is only 1.2 mm to 3.7 mm long; (2) genital pores of *E. granulosus* are about equatorial, but they are preequatorial in *E. multilocularis;* and (3) *E. granulosus* has 45 to 65 testes with a few located anterior to the cirrus pouch, but *E. multilocularis* has 15 to 30 testes, all located posterior to the cirrus pouch.

- **Biology.**   Metacestodes (Fig. 21.26) differ in several respects from those of *E. granulosus.* Instead of developing a thick, laminated layer and growing into large, single cysts, this parasite has a thin outer wall that grows and infiltrates processes into surrounding host tissues like a cancer. Following transformation of the oncosphere, extensions of germinal tissue invade surrounding hepatic tissue of the host, followed by formation of small chambers, each surrounded by a thin layer of laminated tissue lined by germinal tissue, from which a brood capsule arises. In rodents, each chamber contains a few protoscolices. In humans and other unnatural hosts the pockets typically lack protoscolices. In natural intermediate hosts the cyst is more regular. In humans, pieces of the cyst sometimes break off and metastasize to other parts of the body.[40] Because of its type of construction, this metacestode form is called an **alveolar** or **multilocular hydatid.**

  Human infection with alveolar echinococcosis is unusual, although more intensive investigation in recent years has shown a much higher prevalence than previously realized. Humans do not seem to be very good hosts because protoscolices may not develop in humans, but the germinal membrane is still viable.[82] Dogs that catch and eat wild rodents seem to be the main source of infection for humans.[31] Studies conducted in Alaska showed that human infection was highly correlated with a close human-dog association, and trapping or skinning foxes

was not associated with greater infection risk.[31] There is some evidence that there are strains that differ in virulence.[55] Some human infections seem to disappear spontaneously, while others seem to march inexorably toward death. Such differences may be explained by differing immune responses.[41]

- **Diagnosis and Treatment.**   Diagnosis of an alveolar echinococcosis is difficult, particularly because the protoscolices may not be found. Even at necropsy cysts may be mistaken for malignant tumors. As a result of the difficulties of liver surgery, excision is usually practical only when the hydatid is localized near the tip of a lobe of the liver; infections of the hilar area are inoperable. The infiltrative nature of the cyst and its slow rate of growth may advance the disease to an inoperable state before its presence is detected. Praziquantel, the drug that is so effective for most flatworm parasites, may actually enhance growth of alveolar hydatids.[62] Albendazole may be effective in some patients; encouraging results have been obtained in mice treated with albendazole-loaded nanoparticles.[84]

  Alveolar echinococcosis can be prevented only by avoiding dogs and their feces in endemic regions, by carefully washing all strawberries, cranberries, and the like that may be contaminated by dung, and by regularly worming dogs that may be liable to infection.

Echinococcus vogeli* and *E. oligarthrus. *Echinococcus vogeli* causes polycystic echinococcosis in humans in tropical America and is responsible for significant morbidity and mortality in several countries, while infections with *E. oligarthrus* are apparently quite rare.[81] Normal definitive hosts of *E. vogeli* are bush dogs, *Speothus venaticus,* while those of *E. oligarthrus* are several species of wild cats. The most important sources of infection for humans are domestic dogs and possibly domestic cats. Cysts of these species are relatively large, fluid-filled vesicles with numerous protoscolices. Their natural intermediate hosts are pacas and agoutis (rodents).[25]

Family Hymenolepididae

The huge family Hymenolepididae consists of numerous genera with species in birds and mammals. Only two species, *Hymenolepis nana* and *H. diminuta,* can infect humans. The family offers considerable taxonomic difficulties because of the large number of species and the immense and far-flung literature that has accumulated. However, the family's morphology is relatively simple compared with, for example, Pseudophyllidea, and most species are small, transparent, and easy to study.

The most characteristic morphological feature of the group is the small number of testes: usually one to four. The combination of few testes, usually unilateral genital pores, and a large external seminal vesicle permits easy recognition of the family. All except *H. nana* require arthropod intermediate hosts.

Hymenolepis nana. Commonly called the *dwarf tapeworm, Hymenolepis nana* (Fig. 21.27), also known as *Vampirolepis nana,* is a cosmopolitan species that is one of the most common cestodes of humans in the world, especially among children. Rates of infection run from 1% in the southern United States to 9% in Argentina and to 97.3% in Moscow.[51]

- **Morphology.** As its name implies (Gr. *nanos* = dwarf), this is a small species, seldom exceeding 40 mm long and 1 mm wide. The scolex bears a retractable rostellum armed with a single circle of 20 to 30 hooks. The neck is long and slender, and the segments are wider than long. Genital pores are unilateral, and each mature segment contains three testes. After apolysis gravid segments disintegrate, releasing eggs, which measure 30 μm to 47 μm in diameter. The oncosphere (Fig. 21.28) is covered with a thin, hyaline, outer membrane and an inner, thick membrane with polar thickenings that bear several filaments. The heavy embryophores that give taeniid eggs their characteristic striated appearance are lacking in this and the other families of tapeworms infecting humans.

- **Biology.** The life cycle of *H. nana* is unique among tapeworms in that an intermediate host is optional (see Fig. 21.27). When eaten by a person or a rodent, eggs hatch in the duodenum, releasing oncospheres, which penetrate the mucosa and come to lie in lymph channels of the villi. Here each develops into a cysticercoid (Fig. 21.29). In five to six days cysticercoids emerge into the lumen of the small intestine, where they attach and mature.

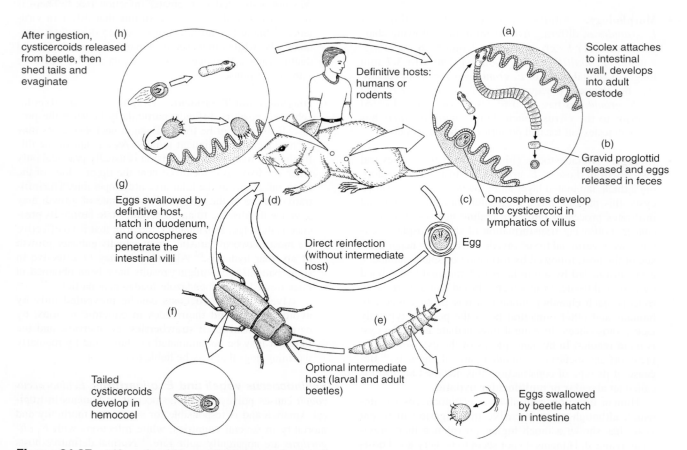

Figure 21.27 Life cycle of *Hymenolepis nana,* the dwarf tapeworm.

(*a*) Adult attached to intestinal wall releases gravid proglottids (*b*) and shelled oncospheres in feces (*c*). (*d*) Direct reinfection when definitive host swallows shelled oncosphere. (*e*) Larval and adult beetles are optional intermediate hosts. (*f*) Cysticercoids develop in hemocoel of beetle. (*g*) When shelled oncospheres are ingested by definitive host, they hatch in the duodenum and penetrate the intestinal villi. (*h*) Ingested cysticercoids shed tails and evaginate, and scolices attach to intestinal wall.

Drawing by William Ober and Claire Garrison.

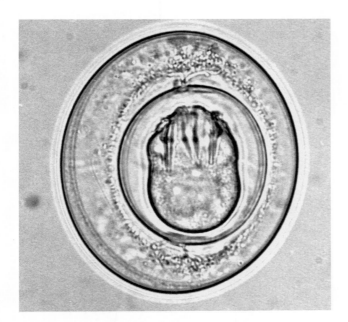

Figure 21.28 Egg of *Hymenolepis nana*.

Note the polar filaments on the inner membrane and the well developed oncosphere. Its size is 30 μm to 47 μm.

Courtesy of Jay Georgi.

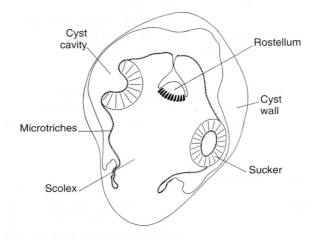

Figure 21.29 *Hymenolepis nana*.

Diagrammatic representation of a longitudinal section through a cysticercoid from a mouse villus.

From J. Caley, "A comparative study of two alternative larval forms of *Hymenolepis nana*, the dwarf tapeworm, with special reference to the process of encystment," in *Z. Parasitenkd*. 47:218–228, 1975. Copyright © 1975 Springer-Verlag, New York. Reprinted by permission.

This direct life cycle is doubtless a recent modification of the ancestral two-host cycle, found in other species of hymenolepidids, because cysticercoids of *H. nana* can still develop normally within larval fleas and beetles (Fig. 21.30). One reason for the facultative nature of the life cycle is that *H. nana* cysticercoids can develop at higher temperatures than can those of other hymenolepidids. Direct contaminative infection by eggs is probably the most common route in human cases, but accidental ingestion of an infected grain beetle or flea cannot be ruled out.

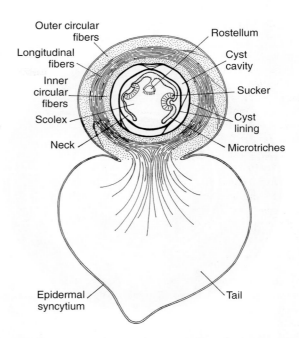

Figure 21.30 Diagrammatic representation of a longitudinal section through a cysticercoid of *Hymenolepis nana* from the insect host.

From J. Caley, "A comparative study of two alternative larval forms of *Hymenolepis nana*, the dwarf tapeworm, with special reference to the process of encystment," in *Z. Parasitenkd*. 47:218–228, 1975. Copyright © 1975 Springer-Verlag, New York. Reprinted by permission.

Besides humans, domestic mice and rats also serve as suitable hosts for *H. nana*.[35] Some authors contend that two subspecies exist: *H. nana nana* in humans and *H. nana fraterna* in murine rodents. Differences do seem to exist in the physiological host-parasite relationships of these two subspecies, because higher rates of infection result from eggs obtained from the same host species than from the other.[76] This is probably an example of allopatric speciation in action.

Pathological results of infection by *H. nana* are rare and usually occur only in massive infections. Heavy infections can occur through autoinfection,[45] and the symptoms are similar to those already described for *Taenia saginata* infection. Praziquantel acts very rapidly against *H. nana* and *H. diminuta*.[5, 13] In vitro the drug produces vacuolization and disruption of the tegument in the neck of the worms but not in more posterior portions of the strobila.

Hymenolepis diminuta. *Hymenolepis diminuta* is a cosmopolitan worm that is primarily a parasite of rats (*Rattus* spp.), but human infections are not uncommon. It is a much larger species than *H. nana* (up to 90 cm) and differs from *H. nana* in lacking hooks on the rostellum. Typical of the genus, it has unilateral genital pores and three testes per proglottid. Eggs (Fig. 21.31) are easily differentiated from those of *H. nana*, since they are larger and have no polar filaments. It has been demonstrated experimentally that more than 90 species of arthropods can serve as suitable intermediate hosts. Stored-grain beetles (*Tribolium* spp.) are probably

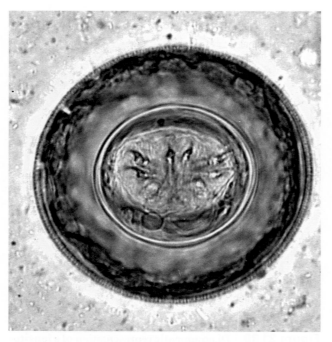

Figure 21.31 **Egg of *Hymenolepis diminuta*.**
It is 40 μm to 50 μm wide.
Courtesy of Jay Georgi.

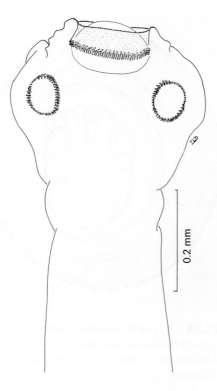

Figure 21.32 **Scolex of *Raillietina*.**
The suckers are weak and have a double circle of spines, and the massive rostellum has many hammer-shaped hooks.
Drawing by Thomas Deardorff.

most commonly involved in infections of both rats and humans. A household shared with rats is also likely to have its cereal foods infested with beetles. Treatment is as recommended for *H. nana*.

The ease with which this parasite is maintained in laboratory rats and beetles makes it an ideal model for many types of experimental studies; its physiology, metabolism, development, genetics, and nutrient uptake have been more thoroughly examined than those of any other tapeworm.[6] Research on *H. diminuta* has contributed enormously to our concepts of the world of tapeworms, including our understanding of how they survive, reproduce, and develop and of their adaptations for parasitism.

Family Davaineidae

Raillietina Species

The following species of *Raillietina* have been reported from humans: *R. siriraji*, *R. asiatica*, *R. garrisoni*, *R. celebensis*, and *R. demarariensis*. All normally parasitize domestic rats and possibly represent no more than two actual species. The genus is easily recognized by its large rostellum with hundreds of tiny, hammer-shaped hooks and by its spiny suckers (Fig. 21.32). Life cycles of these species are unknown, but they probably use insects as intermediate hosts; their epidemiology is probably similar to that of *H. diminuta*.

Raillietina cesticillus is one of the most common poultry cestodes in North America, and a wide variety of grain, dung, and ground beetles serve as intermediate hosts. The genus is very large, with species in many birds and mammals. The closely related genus *Davainea* is nearly identical to *Raillietina*, except that its strobila is very short, consisting of only a few proglottids. *Davainea* spp. are found in galliform birds and use terrestrial molluscs as intermediate hosts. Other genera infect a wide variety of hosts, from passeriform birds to scaly anteaters.

Family Dilepididae

***Dipylidium caninum*.** A cosmopolitan, common parasite of domestic dogs and cats, *Dipylidium caninum* often occurs in children.[67] It is easily recognized because each segment has two sets of male and female reproductive systems and a genital pore on each side (Fig. 21.33). The scolex has a retractable, rather pointed rostellum with several circles of rose thorn-shaped hooks. Its uterus disappears early in its development and is replaced by hyaline, noncellular egg capsules, each containing 8 to 15 eggs. Gravid proglottids detach and either wander out of the anus or are passed with feces. They are very active at this stage and are the approximate size and shape of cucumber seeds. As detached segments begin to desiccate, egg capsules are released.

Fleas are the usual intermediate hosts, although chewing lice have also been implicated. Unlike adults, larval fleas have simple, chewing mouthparts and feed on organic matter, which may include *D. caninum* egg capsules. The resulting cysticercoids survive their host's metamorphosis into the parasitic adult stage, when fleas may be nipped or licked out of the fur of a dog or cat, thereby completing the life cycle. This, by the way, is an example of hyperparasitism, since the flea is itself a parasite.

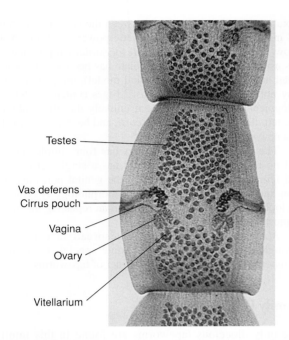

Testes

Vas deferens

Cirrus pouch

Vagina

Ovary

Vitellarium

Figure 21.33   **Mature segment of *Dipylidium caninum*, the "double-pored tapeworm" of dogs and cats.**
The two vitelline glands are directly behind the larger ovaries. The smaller spheres are testes.

Courtesy of Ann Arbor Biological Center.

Nearly every reported case of infection of humans has involved a child. Adult humans may be more resistant, or else children may have increased chances of accidentally swallowing a flea. The symptoms and treatment are the same as for *Hymenolepis nana.*

The only feature separating this family from Hymenolepididae is a larger number of testes, usually more than 12. This family, too, consists of hundreds of species that parasitize birds and mammals. Taxonomic difficulties also attend this family.

Family Anoplocephalidae

Moniezia Species

The numerous species of *Moniezia* use hoofed animals as definitive hosts. The most frequently encountered species in domestic animals are *M. expansa* and *M. benedeni* in sheep, cattle, and goats. These are large tapeworms, up to 6 m, and their proglottids are much wider than long. They have two sets of reproductive organs in each proglottid, and their scolex is unarmed. They have curious interproglottidal glands that open along the junctions between proglottids; the function of the glands is unknown.[93] Eggs of *M. benedeni* are rather square in shape, and those of *M. expansa* are more triangular. Oncospheres of both are borne within an oddly shaped pyriform apparatus, an embryophore with long hook or hornlike extensions.

Moniezia expansa was the first anoplocephalid for which a life cycle was discovered. Parasitologists had been mystified for many years as to how herbivorous animals could be infected with tapeworms, and small arthropods that might be ingested along with herbage were diligently investigated as potential intermediate hosts. Finally in 1937 Horace Stunkard announced that the intermediate host of *M. expansa* was a minute, free-living mite in the family Oribatidae.[95]

Ba and coworkers[9] sorted *M. expansa* and *M. benedeni* from France and Senegal, Africa, on the basis of patterns of their interproglottidal glands and then performed isoenzyme analysis on them. They found that French worms were very similar genetically to some African worms, and these were identified as *M. expansa.* They came from sheep and goats but not from cattle. Other African worms, putatively assigned to either *M. expansa* or *M. benedeni,* seemed to represent at least four distinct species. Thus, species diversity of *Moniezia* spp. in domesticated ruminants may be greater than previously believed.

Bertiella studeri. Normally a parasite of Old World primates, *Bertiella studeri* has been reported many times from humans, especially in southern Asia, the East Indies, and the Philippines. The scolex is unarmed, and proglottids are much wider than they are long, with the ovary located between the middle of the segment and the cirrus pouch. The egg is characteristic: 45 μm to 50 μm in diameter, with a bicornuate pyriform apparatus on the inner shell.

Ripe segments are shed in chains of about a dozen at a time. Intermediate hosts are various species of oribatid mites. Accidental ingestion of mites infected with cysticercoids completes the life cycle within primates. No disease has been ascribed to infection by *B. studeri.* Treatment is as for *Hymenolepis nana.*

Bertiella mucronata is similar to *B. studeri* and also has been reported from humans.[75] It appears to be a parasite of New World monkeys, and children may become infected when living with a pet monkey and the ubiquitous oribatid mite. Distinguishing this species from *B. studeri* normally requires a specialist.

Inermicapsifer madagascariensis. *Inermicapsifer madagascariensis* is normally parasitic in African rodents, but it has been reported repeatedly in humans in several parts of the world, including South America and Cuba. Baer[11] concluded that humans are the only definitive host outside Africa.

The scolex is unarmed. The strobila is up to 42 cm long. Mature proglottids are somewhat wider than long. The uterus becomes replaced by egg capsules in ripe segments, each capsule containing 6 to 10 eggs, which do not possess a pyriform apparatus.

The life cycle of this parasite is unknown but undoubtedly involves an arthropod intermediate host. Clinical pathological conditions have not been studied, and treatment is similar to that described for other species.

Family Mesocestoididae

Mesocestoides Species

Unidentified specimens of the genus *Mesocestoides,* whose definitive hosts are normally various birds and mammals, have occasionally been reported from humans in Denmark, Africa, the United States, Japan, and Korea.[48] The ventromedial location of the genital pores is clearly diagnostic of the genus (Fig. 21.34). The complete life cycle is not known for

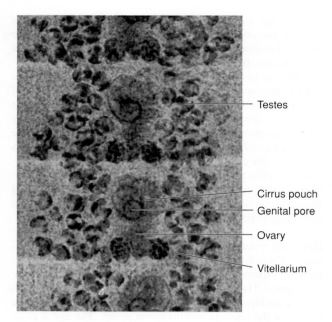

Figure 21.34 *Mesocestoides* **sp., a cyclophyllidean cestode with a midventral genital pore and a bilobed vitellarium.**

Courtesy of Larry Shults.

Figure 21.35 **Tetrathyridial metacestodes of** *Mesocestoides* **sp. in the mesenteries of a baboon,** *Papio cyanocephalus.*

Courtesy of Robert E. Kuntz.

any species in this difficult family, but many have a rodent or reptile intermediate host, in which a cysticercoid type of larva known as a *tetrathyridium* (Fig. 21.35) develops. Neither mammals nor reptiles can be infected directly by eggs, so a first host must be involved. As yet, such a host has not been identified (Fig. 21.36). Pathological conditions and treatment of humans have not been studied.

Mesocestoides sp. is very curious in that it may undergo asexual multiplication in the definitive host (see Fig. 21.36)—not by budding, as in coenuri and hydatids, but by longitudinal fission of the scolex! An inwardly directed

protuberance of the tegument between the suckers, the "apical massif," has morphocytogenetic power.[44] Although we know that this one isolate of *Mesocestoides* reproduces in this astonishing way (and the isolate has been widely propagated and used as an experimental model), the phenomenon may not be typical of the genus.[23] Etges contended that developmental differences made it unlikely that the proliferative tetrathyridium could be *M. corti,* and he thus described it as a new species, *M. vogae.*[34]

The scolex of *Mesocestoides* has four simple suckers and no rostellum. Each proglottid has a single set of male and female reproductive systems; the genital pores are median and ventral. Otherwise, their morphology is typically cyclophyllidean, with a paruterine organ replacing the uterus in most species.

Mesocestoides spp. are widespread in carnivores throughout most of the world. Some specialists consider members of this family a distinct order of tapeworms.

Family Dioecocestidae

The only dioecious tapeworms are found in this family, except for species in the genus *Dioecotaenia,* which form a family of tetraphyllideans in rays. All dioecocestids are parasites of shorebirds, grebes, or herons. Some species are completely dioecious, whereas others are only regionally so. There are wide variations in scolex types, although all have four suckers. In some species, such as *Shipleya inermis* in dowitchers, both sexes have secondary sex organs of the opposite sex. Hence, the male has a uterus and the female a cirrus and cirrus pouch, but each has only an ovary or testes.[90]

ORDER PROTEOCEPHALATA

Proteocephalatans are all parasites of freshwater fishes, amphibians, or reptiles. Scolices (see Fig. 20.5*d*) are much like those occurring in cyclophyllideans, bearing four simple suckers and occasionally armature or a rostellum. Proglottids, however, are much more like those of Tetraphyllidea. Genital pores are lateral. The ovary is posterior, and numerous testes fill most of the region anterior to it. Vitellaria are follicular and are restricted to lateral margins of the proglottid.

We know complete life cycles for several species. All involve a cyclopoid crustacean intermediate host, in which the worm develops into a procercoid (the plerocercoid I, according to some authors; see p. 314). This metacestode has a well-developed scolex and a cercomer at the posterior end. In some species the procercoid is directly infective to a definitive host; in others it burrows into the viscera for a time before reemerging and maturing in the lumen of the gut. Paratenic hosts are common in proteocephalid life cycles. The boring action of the plerocercoid (as it is now called, since it loses the cercomer as it penetrates the intestinal wall; it is also known as plerocercoid II) may be highly pathogenic to its host. For example, *Proteocephalus ambloplitis* in bass in North America sometimes castrates its fish host.

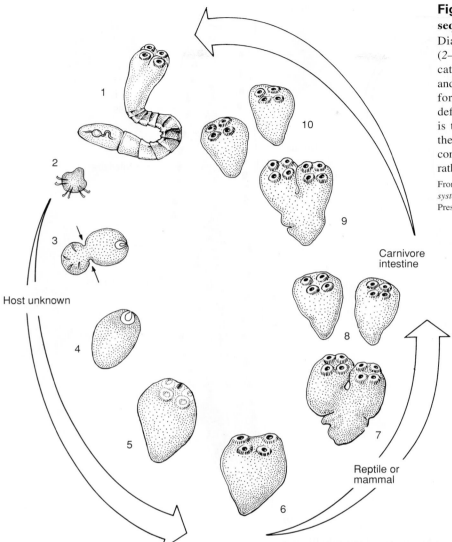

Figure 21.36   **Developmental sequence of *Mesocestoides vogae*.**
Diagram illustrates developmental stages (*2–5*), tetrathyridium and asexual multiplication in second intermediate host (*6–8*), and asexual multiplication with subsequent formation of adult worms in intestine of definitive host (*9–10*). Not illustrated here is the potential reinvasion of tissues from the intestinal lumen of the carnivore with continuing asexual multiplication of the tetrathyridial stage.

From M. Voge, in G. D. Schmidt (Ed.), *Problems in systemics of parasites*. Baltimore, MD: University Park Press, 1969.

ORDER TETRAPHYLLIDEA

Tetraphyllideans are notable for their astonishing variety of scolex forms (see Figs. 20.4*a* and 20.5*i*). Basically, there are four bothridia, which may be stalked or sessile, smooth or crenate, or subdivided into loculi or major units. Often there are accessory suckers (see Fig. 20.4*a*) and/or hooks (Fig. 21.37) or spines. An apical, stalked, suckerlike organ, the myzorhynchus, is present on some. A neck is present or absent. The strobila and proglottids are essentially identical to those of Lecanicephalidea and Trypanorhyncha, and, like members of those orders, adult tetraphyllideans are all parasites of the spiral intestine of elasmobranchs.

As far as is known, life cycles are also similar. No complete cycle has been discovered, but infective plerocercoids are common in molluscs, crustaceans, and fishes. Fishes undoubtedly are paratenic hosts, as may be some molluscs and crustaceans. In vitro cultivation of *Acanthobothrium* sp. plerocercoids in the presence of urea, a substance they encounter in their definitive host, causes the scolex to differentiate into the adult condition.[42]

ORDER TRYPANORHYNCHA

Scolices (see Fig. 20.5*f*) of trypanorhynchans are extraordinary organs. They usually are elongated with two or four shallow bothridia, which may be covered with minute microtriches. Four eversible tentacles (atrophied in *Aporhynchus norvegicum*) emerge from the apex of the scolex. The tentacles are armed with an astonishing array of hooks and spines (Fig. 21.38), shaped and arranged differently in each species. Interpretation of the hook arrangement is difficult but must be accomplished before species identification is possible. Each tentacle invaginates into an internal tentacle sheath, provided at its base with a muscular bulb. A retractor muscle originates at the base or front end of the bulb, courses through

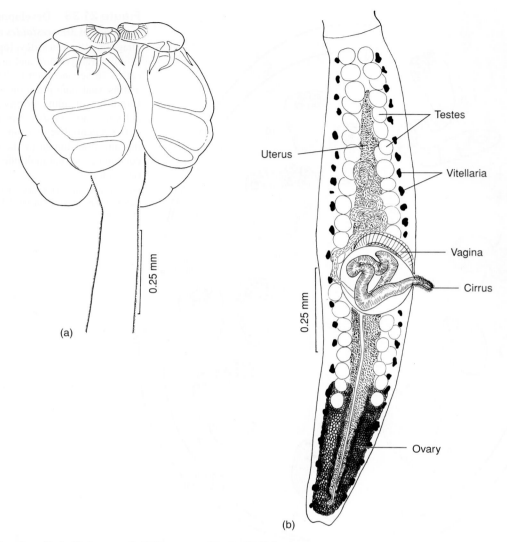

0.25 mm

(a)

Testes

Uterus

Vitellaria

Vagina

Cirrus

0.25 mm

Ovary

(b)

Figure 21.37 *Acanthobothrium urolophi,* **an armed tetraphyllidean.**
(*a*) Scolex; (*b*) mature proglottid.

From G. D. Schmidt, "*Acanthobothrium urolophi* sp. n., a tetraphyllidean cestode (Oncobothriidae) from an Australian stingaree," in *Proc. Helm. Soc. Wash.* 40:91–93. Copyright © 1973. Reprinted with permission of the publisher.

the tentacle sheath, and inserts inside the tip of the tentacle. When the retractor muscle contracts, it invaginates the tentacle, detaching it from host tissues. When the bulb contracts, it hydraulically evaginates the tentacle, driving it deep into the host's intestinal wall. This process is very similar to that which manipulates the proboscis of an acanthocephalan (chapter 32).

A neck is present or absent; the strobila varies from hyperapolytic to anapolytic. Proglottids of trypanorhynchans are morphologically very consistent with those of tetraphyllideans. The single ovary is basically bilobed and posterior. Vitellaria are follicular, cortical, and lateral or circummedullary. The uterus is a simple sac, usually in the anterior two-thirds of a gravid segment. Testes are few to many and medullary, and cirrus pouch and cirrus often are huge relative to the proglottid. All genital pores are lateral.

Adult trypanorhynchans are all parasites of the spiral intestine of sharks and rays. Infective metacestodes are common in marine molluscs,[19] crustaceans, and fishes. Sakanari

and Moser[89] reported experimental infection of copepods with coracidia of *Lacistorhynchus tenuis* which developed into procercoids. These grew into plerocercoids after being eaten by mosquito fish, which produced immature adults after being fed to leopard sharks. This life cycle is similar to that of another trypanorhynchan, *Grillotia erinaceus,* but many other members of this order do not have operculated eggs that release ciliated coracidia.[37]

The plerocercoid, sometimes called a *plerocercus,* may or may not bear a posterior sac, or blastocyst, into which the scolex is inverted. Plerocerci may be so plentiful in the flesh of certain fish or shrimps as to make them unpalatable and thereby unsalable. This is one known economic importance of trypanorhynchans. They have never been reported from humans. However, they remain among the most enigmatic and challenging invertebrates for taxonomists. Dollfus published classical reviews,[28, 29] and Schmidt provided a key to families and genera.[90]

Figure 21.38   *Eutetrarhynchus thalassius,*
a typical trypanorhynch.

(*a*) Scolex; (*b*) proglottid.

From K. J. Kovacs and G. D. Schmidt, "Two new species of
cestode (Trypanorhyncha, Eutetrarhynchidae) from the yellow-
spotted stingray, *Urolophus jamaicensis,*" in *Proc. Helm. Soc.
Wash.* 47:10–14. Copyright © 1980. Reprinted with permission
of the publisher.

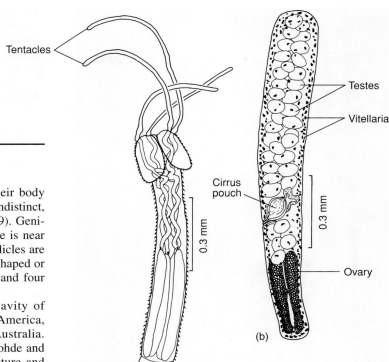

SUBCOHORT AMPHILINIDEA

Amphilinidea is the sister group of Eucestoda. Their body
is monozoic and dorsoventrally flattened, with an indistinct,
proboscislike holdfast at the anterior end (Fig. 21.39). Genital pores are near the posterior end; the uterine pore is near
the anterior end. The ovary is posterior, vitelline follicles are
bilateral, and testes are preovarian. The uterus is *N*-shaped or
looped. The oncosphere is provided with six large and four
small hooks.

Amphilinideans are parasites of the body cavity of
fishes and turtles in Asia, Japan, Europe, North America,
Sri Lanka, Brazil, Africa, the East Indies, and Australia.
They are of no medical or economic importance. Rohde and
Georgi[86] presented an excellent study on the structure and
life history of *Austramphilina elongata.*

COHORT GYROCOTYLIDEA

This is the sister group of Cestoidea. They are also monozoic, and their anterior end is provided with a small, in-
versible holdfast organ. The posterior end is a frilled,
rosette-like organ, and the lateral margins may be frilled
(*Gyrocotyle* spp., Fig. 21.40), or it is a long, simple cylinder, and the lateral margins are smooth (*Gyrocotyloides
nybelini).* The ovary is posterior; the uterus has extensive
lateral loops, terminating in a midventral pore in the anterior half. Testes are anterior. Genital pores are near the
anterior end. Whether or not the structures on their tegumental surface were indeed microtriches (p. 304) has been
controversial, but evidence appears strong that they are
indeed microtriches.[78] Gyrocotylideans have a larva with
10 equal-sized hooks. They are parasites of the spiral intestine of Holocephali.

Gyrocotylidea have traditionally been placed with
Amphilinidea in subclass Cestodaria of Class Cestoidea.
Present opinion places them as the sister group of cohort
Cestoidea in infraclass Cestodaria, and makes Cesto-
daria the sister group of infraclass Monogenea in subclass
Cercomeromorphae.[16]

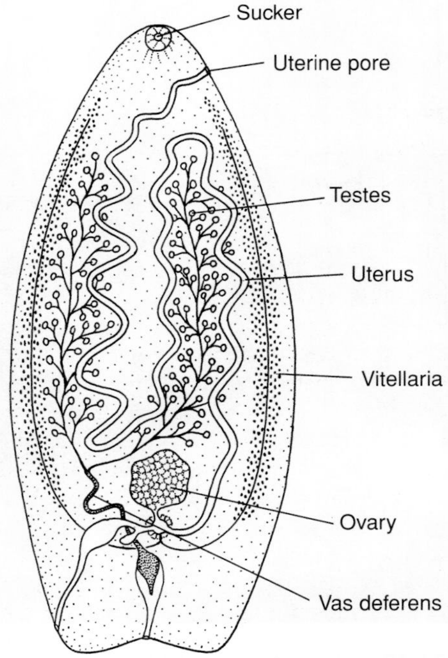

Learning Outcomes

By the time a student has finished studying this chapter, he or she
should be able to:

1. Explain the differences among the various types of cestode
   larvae.

Figure 21.39   **The monozoic tapeworm** *Amphilina
foliacea.*

Redrawn from R. A. Wardle and J. A. McLeod, *The Zoology of Tapeworms,* 1952,
Hafner Publishing Co., New York, NY.

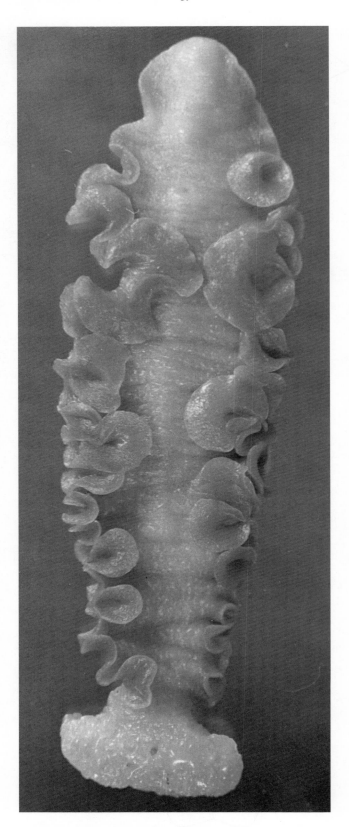

Figure 21.40 *Gyrocotyle parvispinosa* **from the ratfish** *Hydrolagus colliei.*

Courtesy of Warren Buss.

2. Explain the epidemiological significance of different kinds of cestode larvae, with emphasis on cysticerci and cysticercoids.

3. Describe the difference between a cysticercus and a cysticercoid.

4. Name a tapeworm with a cysticercus in its life cycle and one with a cysticercoid. What are implications for the epidemiology of each?

5. Understand and diagram the life cycle of *Diphyllobothrium latum, Dipylidium caninum, Taenia saginata, Taenia solium, Taenia asiatica, Echinococcus granulosus, Hymenolepis diminuta, H. nana, Bertiella studeri,* and *Bertiella mucronata,* including the source(s) of human infection for each.

6. Tell which of the aforementioned has been the most thoroughly studied tapeworm in laboratories. Why?

7. Which of the aforementioned is the most common cause of sparganosis in humans? Of cysticercosis?

References

References for superscripts in the text can be found at the following Internet site: www.mhhe.com/robertsjanovynadler9e

Additional Readings

Andersen, F. L., J. Chai, and F. Liu (Eds.) 1993. *Compendium on cystic echinococcosis with special reference to the Xinjiang Uygur Autonomous Region, The People's Republic of China.* Provo, UT: Brigham Young University. Contains much basic information, in addition to reports on the Xinjiang Uygur Autonomous Region.

Arai, H. P. (Ed.). 1980. *Biology of the tapeworm* Hymenolepis diminuta. New York: Academic Press, Inc.

Arme, C., and P. W. Pappas (Eds.). 1983. *Biology of the Eucestoda* (2 vols.). London: Academic Press.

Binford, C. H., and D. H. Connor (Eds.). 1976. *Pathology of tropical and extraordinary diseases, sect. 11. Disease caused by cestodes.* Washington, DC: Armed Forces Institute of Pathology.

Hoffman, G. L. 1999. *Parasites of North American freshwater fishes,* 2d ed. Ithaca, NY: Cornell University Press. Keys to all genera of tapeworms of North American fishes with lists of species and hosts.

Soulsby, E. J. L. 1965. *Textbook of veterinary clinical parasitology 1.* Philadelphia: F. A. Davis Co. A useful reference to tapeworms of veterinary significance.

Southwell, T. 1925. *A monograph on the Tetraphyllidea.* Liverpool: Liverpool University Press, Liverpool School of Tropical Medicine Memoir n. s. 2:1–368. An old but useful monograph on this order.

Spasskaya, L. P. 1966. *Cestodes of birds of SSSR.* Moscow: Akademii Nauk. SSSR. A useful illustrated survey of tapeworms of northern birds.

Stunkard, H. W. 1962. The organization, ontogeny and orientation of the cestodes. *Q. Rev. Biol.* 37:23–34.

Thompson, R. C. A. (Ed.). 1986. *The biology of* Echinococcus *and hydatid disease.* London: George Allen and Unwin.

22

Phylum Nematoda: Form, Function, and Classification

If all the matter in the universe except the nematodes were swept away, . . . we should find [our world's] mountains, hills, vales, rivers, lakes and oceans represented by a thin film of nematodes.

—N. A. Cobb (1914, Yearbook of the United States
Department of Agriculture)

Nematodes are the most abundant multicellular animals on earth. Ninety thousand nematodes were once found in a single rotting apple; 1074 individuals, representing 236 species, were counted in 6.7 ml of coastal mud; and up to 9 billion per acre may be found in good farmland.[84] Of course more species of insects have been formally described and named, but when one realizes that many kinds of insects harbor at least one species of parasitic nematode and when one further calculates the number of kinds of nematodes parasitic in the rest of the animal kingdom, there is no contest. Also, about 2000 species of nematodes that parasitize plants have been described.[4] Finally, it is estimated that 75% of all nematode species are free-living in marine, freshwater, and soil habitats.[4]

Most nematodes are microscopic, inconspicuous, and seemingly unimportant to humans and therefore attract the attention only of specialists. However, many free-living nematodes are vital to ecosystem services, such as nitrogen mineralization in soils. A few, however, cause diseases of great importance to humans, as well as domestic and wild plants and animals. In addition, some nematode pathogens of insects have been used for biological control of insect pests.

Obviously, these animals have a lot going for them. Roundworms are studied by parasitologists and nematologists, with the latter emphasizing free-living, predatory, and plant-parasitic species. The bacterial-feeding nematode *Caenorhabditis elegans* is an extremely important model organism that has contributed to fundamental understanding of eukaryotic life, particularly molecular, cell, and developmental biology. This and subsequent chapters should give you some appreciation of the diversity of nematodes, particularly in parasitic habitats.

HISTORICAL ASPECTS

Ancient people were probably familiar with the larger nematodes, which they encountered when they slew game or passed worms in their own feces. Some ancient records mention these worms or contain recognizable allusions to them. Aristotle discussed the worm we now call *Ascaris lumbricoides,* and the Ebers Papyrus of 1550 B.C. Egypt described clinical hookworm disease, as did Hippocrates, Lucretius, and the ancient Chinese. Moses wrote of a scourge that probably was caused by guinea worms. Eggs of *Ascaris lumbricoides* and *Trichuris trichiura* were found in the intestine of a 2300-year-old body preserved in a peat bog in the Orkney Islands.[24] Avicenna and Avenzoar, Arabians who kept parasitology (and much other science) alive during the Dark Ages in Europe, studied elephantiasis, differentiating it from leprosy.[45]

Linnaeus, in 1758, placed roundworms in his class Vermes, along with all other worms and wormlike animals. Goeze, Zeder, and Rudolphi made great advances in recognition of various nematodes, although they still believed the worms arose by spontaneous generation. Further work by some great 18th- and 19th-century zoologists such as Gegenbauer, Huxley, Hatschek, Leuckart, Beneden, Diesing, Linstow, Looss, Railliet, and Stossich established nematodes as a distinct and important group of animals.

The name Nematoda is a modification of Rudolphi's Nematoidea and was applied to worms placed in Nemathelminthes, itself first considered a class in phylum Vermes by Gegenbauer in 1859 and then elevated to phylum status. The taxon Aschelminthes was proposed by Grobben in 1910 as a superphylum to contain several divergent groups of wormlike animals that had in common a pseudocoelomic body cavity. Hyman resurrected the name Aschelminthes for use as a phylum containing classes Nematoda, Rotifera, Priapulida, Gastrotricha, Nematomorpha, and Kinorhyncha.[58] Today many authorities consider each of these groups separate phyla. Some authorities prefer the name Nemata to Nematoda, but Nematoda remains widely accepted as designating this phylum. Finally, certain molecular phylogenetic hypotheses place Nematoda with arthropods, kinorhynchs, nematomorphs, and other animals that molt in a superphylum Ecdysozoa. Several other of the remaining phyla with pseudocoelomic body plans are placed in the large proto-stome superphylum Lophotrochozoa.[53]

Curiously, studies of nematode parasites developed along two separate lines, with parasitologists claiming parasites of vertebrates and nematologists accounting for plant- and invertebrate-parasitic roundworms. This division of labor is a result of parasitologists' historical concern for parasites of medical and veterinary importance. Exceptions occur, of course; a few individuals, such as Chitwood, Bird, Crites, Inglis, Mawson, and Schuurmanns-Steckhoven, have made significant contributions to both fields. Nevertheless, each discipline publishes in its own journals, uses unique terminology and taxonomic formulas, and to a large extent employs different techniques for handling and preparing specimens for study. This historical fragmentation of specialists studying nematodes has resulted in discipline-specific taxonomic classifications, and an emphasis on parasites without connection to the nonparasitic taxa that compose the majority of nematodes.[4] Some trends suggest that the two disciplines are beginning to merge, aided in part by the broad-based evolutionary context provided by molecular phylogenies for Nematoda.[4, 18] If so, a comprehensive and integrated view of Nematoda should result.

FORM AND FUNCTION

Typical nematodes are bilaterally symmetrical, elongated, and tapered at both ends, and they possess a pseudocoel, a body cavity derived from the embryonic blastocoel. There are variations on this basic theme, however (Fig. 22.1). The digestive system is usually complete, with a mouth at the extreme anterior end and an anus (or cloaca in males) near the posterior tip (Fig. 22.2). The lumen of the pharynx (or esophagus, as *pharynx* and *esophagus* are synonyms) is characteristically **triradiate** (see Fig. 22.19). The body is covered with a noncellular cuticle that is secreted by an underlying hypodermis (or epidermis) and is shed four times during ontogeny. Muscles of the body wall are only one layer thick and are distinguished by all being longitudinally arranged with no circular layer. The secretory-excretory (SE) systems of nematodes are multifunctional, involving both osmoregulation and secretion of compounds with different functions, depending on species. The structure of the SE systems differ substantially between members of the classes Enoplea and Chromadorea. Except for some sensory endings of modified cilia, neither cilia nor flagella are present, even in male gametes. Most nematodes are dioecious and show considerable sexual dimorphism: Females are usually larger, and the tail of males is often curled. Some species are hermaphroditic, and others are parthenogenetic. Most are oviparous, but some are ovoviviparous. The female reproductive system opens through a ventral vulva; the male system opens into a cloaca, together with the digestive system. Adult nematodes vary in size from less than 0.5 mm, as in the genus *Halicephalobus* to more than 10 meters, as in *Placentanema gigantisma.*

A considerable body of knowledge has accumulated on the function and structure (both at the light and electron microscope levels) of nematodes, far beyond our ability to review it within the confines of this chapter. Many reviews and literature references are available.[32, 71, 97]

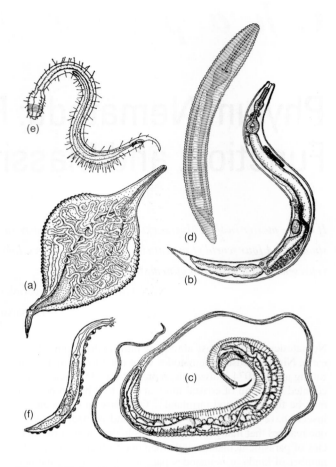

Figure 22.1   **Variety of form in nematodes.**
Variety is illustrated in the following genera: (*a*) *Tetrameres;* (*b*) *Rhabditis;* (*c*) *Trichuris;* (*d*) *Criconemoides;* (*e*) *Draconema;* (*f*) *Bunonema.*

Partly from H. D. Crofton, *Nematodes.* Copyright © 1966. Hutchinson University Library, London. Reprinted with permission.

Body Wall

The nematode body wall consists of cuticle, hypodermis, and body wall musculature. The outermost covering is cuticle, a complex structure of great functional significance to the animals. Cuticle also lines structures that open to the exterior, including the buccal cavity (synonymous with stoma), esophagus, anus, cloaca, excretory pore, and vagina. Overlying the cuticle in free-living and parasitic nematodes is a glycoprotein or mucinlike protein surface coat, 5 nm to 20 nm in thickness.[17] The surface coat is released onto the cuticle surface from the SE system, pharyngeal glands, or amphid openings. It is readily shed *in vitro,* and is important in parasite-host interactions, including evasion of the host immune response.[72]

In many species the cuticle consists of three main layers—**cortical, median,** and **basal zones** (Fig. 22.3) with the outer cortex layer enclosed by a thin lipid **epicuticle.** The exact boundaries of each zone may be difficult to distinguish. The lipoprotein epicuticle appears in electron micrographs as a trilaminar membrane, but it is not a cell membrane; the outermost cell membrane in nematodes occurs where the

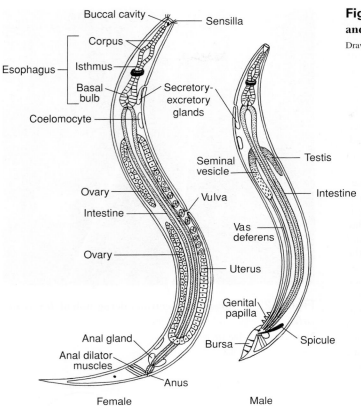

Figure 22.2   Morphology of a typical nematode male and female.

Drawing by William Ober and Claire Garrison.

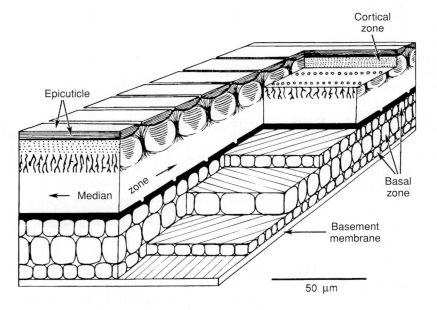

Figure 22.3   Diagram of the cuticle of *Ascaris suum*.

The basal zone is composed of a trellislike arrangement of crossed fiber layers. Strands of each fiber layer run at an angle of about 75° to the longitudinal axis of the worm, and strands of the middle layer run about 135° from those of the inner and outer layers.

From Bird and Bird, *The Structure of Nematodes*, 2d ed. Copyright © 1991. Academic Press, Orlando, FL. Reprinted by permission.

hypodermis interfaces with the basal zone of the cuticle.[75] Just beneath the epicuticle, a cortical zone contains a highly resistant noncollagen protein called **cuticulin,** that probably strengthens structures of the outer cuticle, such as annulations.[43, 101] The inner cortical zone as well as the other zones in the cuticle are primarily collagens, a protein type also abundant in vertebrate connective tissue. The median zone shows much structural variation among nematodes, but many vertebrate parasites have a homogeneous gel or fluid-filled median layer.

The basal zone of larger nematodes is composed of two or three **fibrous layers,** each of parallel strands of collagen, running at an angle of about 75 degrees to the longitudinal axis of the worm. Strands of the second fibrous layer run at an angle of about 135 degrees to those of the first (and third, if present) layer, thus forming a latticelike arrangement. The fibrous layers are important components of the hydrostatic skeleton in larger nematodes, providing strength to resist the high internal pressure.[73] The strands themselves are not extensible, but they do allow longitudinal stretching and compression of the overlying cuticle by changes in the angles between the layers. This arrangement also permits flexibility, which is important for serpentine, spiraling, and coiling movements. Beneath the basal zone is a **basement membrane,** a layer of fine fibrils that merges with the underlying hypodermis. Ringlike depressions of the cortical zone called **annules** enhance flexibility of the animal. These annules are more prominent in some species than in others. The cuticle of the parasitic juveniles of mermithids (described later) is quite different in structure from that just described, and some small free-living nematodes lack fibrous strands in the basal zone.[9]

Cuticular markings and ornamentations of many types occur in nematodes. These structural features include shallow **punctations,** deeper **pores,** and **spines** of varying complexity. Lateral or sublateral cuticular thickenings called **alae** are present in many parasites. Cervical alae (Fig. 22.4) are found on the anterior part of the body; caudal alae are on the tail ends of some males; and longitudinal alae, when present, extend the entire length of both sexes. Longitudinal alae may be of value to the animal when it is swimming, or they may lend stability on solid substrate, when the nematode is crawling on its side by dorsoventral undulations, as do juveniles of *Nippostrongylus brasiliensis*[69] (Fig. 22.5). Longitudinal ridges that encircle the body occur in many adult trichostrongylids. In *N. brasiliensis* ridges are supported by a series of struts or skeletal rods in the middle layer of cuticle[68] (Fig. 22.6). The struts are held erect by collagenous fibers inserted in the cortical and basal zones, but the median zone itself is fluid filled and contains hemoglobin. Longitudinal ridges in trichostrongylids aid in locomotion, as the worm moves between intestinal villi with a corkscrew motion (Fig. 22.7), and the ridges abrade the microvillar surface, perhaps helping the animal obtain food. The pattern of cuticular ridges is called the **synlophe;** it is typically observed in cross sections of the animal and used for species diagnosis in some genera.

The **epidermis** (or **hypodermis**) lies just beneath the basement membrane of cuticle and is usually syncytial in adult worms; nuclei lie in four thickened portions (six to eight in mermithids), known as **epidermal cords,** that project into the pseudocoel. Epidermal (or hypodermal) cords run longitudinally and divide the somatic musculature into four quadrants. On large nematodes cords may be discernible with the unaided eye as pale lines. Dorsal and ventral cords contain longitudinal nerve trunks, whereas lateral cords contain canals of the SE system in chromadorean species. An important function of epidermis is secretion of cuticle, as described in the section on development.

Specialized areas of epidermis, **bacillary bands,** occur in at least three enoplean genera, *Trichuris, Trichinella* and *Capillaria.* These bands include unicellular epidermal gland cells that open that through pores lateral to the esophagus in *Trichuris* spp. and extend the length of the body in *Capillaria* spp. Dendritic processes of adjacent nerve cells project into the gland cells.[14] The bacillary band cells appear to have a secretory function, but the nature of the secretions is unknown.

Epidermal cords of enoplean nematodes bear structures that apparently function as proprioceptors (receptors sensitive to stimuli from within the body).[56] These structures were named **metanemes** by Lorenzen.[79]

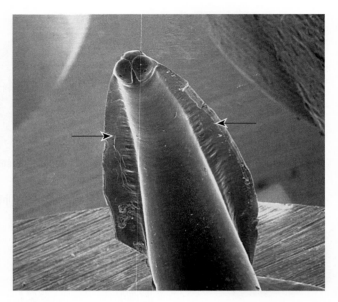

Figure 22.4 **Scanning electron micrograph of *Toxocara cati.***

Note cervical alae (*arrows*).

Courtesy of John Ubelaker.

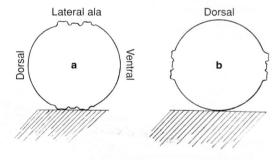

Figure 22.5 **Outline of a transverse section through a third-stage juvenile.**

Note the more stable position, when it lies on a lateral side (*a*) when moving by two-dimensional undulatory propulsion, and the less stable position, when it lies on its ventral surface (*b*).

From D. L. Lee, "*Nippostrongylus brasiliensis:* Some aspects of the fine structure and biology of the infective larva and the adult," in A. E. R. Taylor (Ed.), *Nippostrongylus* and *Toxoplasma.* Copyright © 1969 Blackwell Science, Ltd., Oxford, UK. Reprinted by permission.

Musculature

The somatic musculature is technically a part of the body wall, but it is convenient to consider its function along with that of the pseudocoel. Indeed, somatic musculature, pseudocoel and the fluid it contains, and cuticle function together as a **hydrostatic skeleton.**

Nematode muscles commonly have a contractile portion and a noncontractile "cell body" or **myocyton. Platymyarian muscle** cells are rather wide and shallow, with their contractile portion lying close to the hypodermis (Fig. 22.8*b*). The myocyton contains the nucleus, large mitochondria with numerous cristae, ribosomes, endoplasmic reticulum, glycogen, and lipid. **Coelomyarian** cells

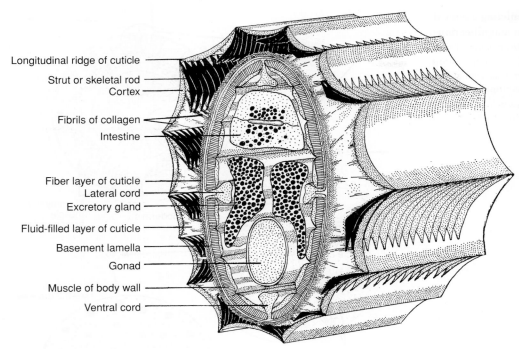

Figure 22.6 **Stereogram of a thick section taken from the middle region of an adult *Nippostrongylus brasiliensis.***

Note arrangement of the various layers of cuticle and other internal organs.

From D. L. Lee, "The cuticle of adult *Nippostrongylus brasiliensis,*" in *Parasitology* 55:173–181. Copyright © 1965 Cambridge University Press. Reprinted with the permission of Cambridge University Press.

Longitudinal ridge of cuticle
Strut or skeletal rod
Cortex
Fibrils of collagen
Intestine
Fiber layer of cuticle
Lateral cord
Excretory gland
Fluid-filled layer of cuticle
Basement lamella
Gonad
Muscle of body wall
Ventral cord

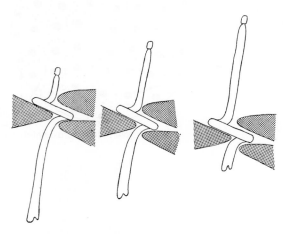

Figure 22.7 **Locomotion of an adult *Nippostrongylus* removed from the intestine of a host and placed among moist sand grains.**

It is probable that similar movements are performed by the nematode among villi of a host intestine.

From D. L. Lee, *The Physiology of Nematodes.* Copyright © 1965 Oliver & Boyd Ltd., London. Reprinted by permission of Addison Wesley Longman.

are spindle shaped, with their contractile portion in the shape of a narrow *U* (Fig. 22.9). The distal end of the *U* is placed against the hypodermis, contractile fibrils extending up along its sides, and the space in the middle is tightly packed with mitochondria. In some cases the elongated contractile portion does not sandwich the mitochondria, but these organelles are concentrated in the distal portion of the myocyton close to contractile fibrils.[134] Coelomyarian myocytons bulge medially into the pseudocoel; they contain a nucleus, some mitochondria, endoplasmic reticulum, a Golgi body, and a large amount of glycogen. An important function of these cytons is glycogen storage. In **circomyarian**

cells, contractile fibrils at the periphery entirely encircle the myocyton.

Myofilaments seem to be essentially similar in all nematode muscle types. Contraction apparently occurs in a manner similar to the Hanson-Huxley model for vertebrate striated muscle, with thick filaments containing myosin and thin filaments actin. The actin filaments slide past myosin filaments in contraction. The *A, H,* and *I* bands typical of striated muscle can be distinguished, but *Z* lines are absent. Thus, the structure of nematode muscle is similar to those of vertebrate striated muscle and insect flight muscle, except that rows of myofilaments are offset, a condition referred to as *obliquely striated*[107] (see Fig. 22.9). In addition, nematode muscles contain additional proteins, such as paramyosin and twitchin in thick filaments.

In body-wall muscles, conduction of nerve impulses is via an innervation process that runs from myocytons to nerves in the epidermal cords (Fig. 22.10). This pattern of innervation is unusual, but it is also known in platyhelminths and some other invertebrate groups.[132] Nematode muscle cells have frequent muscle-muscle connectives, at least in coelomyarian types.[135] Such connectives occur most often in anterior regions of the worms and between innervation processes of muscle cells, although they may be between cytons. Transmission of nerve impulses between muscle cells that are so connected may increase the degree of coordination.

Pseudocoel and Hydrostatic Skeleton

The somatic musculature and the rest of the body wall enclose a fluid-filled cavity, the **pseudocoelom,** or **pseudocoel** (Fig. 22.11). A pseudocoel is derived embryonically from the blastocoel, rather than being a cavity within the endomesoderm; thus, a pseudocoel differs from a true coelom in that it has no peritoneal (mesodermal) lining. A nematode's pseudocoel functions as a hydrostatic skeleton.

Figure 22.8 **Diagrams depicting a typical body-wall muscle at different magnifications.**
(*a*) Whole transverse section; (*b*) portion of muscle cells on either side of dorsal nerve cord; (*c*) two types of muscle filaments as seen at high resolution with the aid of an electron microscope.

From A. F. Bird, *The Structure of Nematodes.* Copyright © 1971 Academic Press, Orlando, FL. Reprinted by permission.

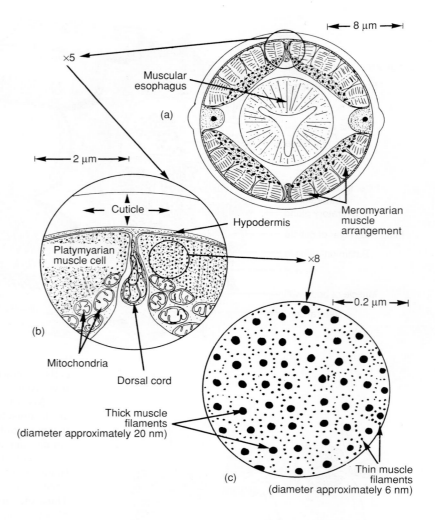

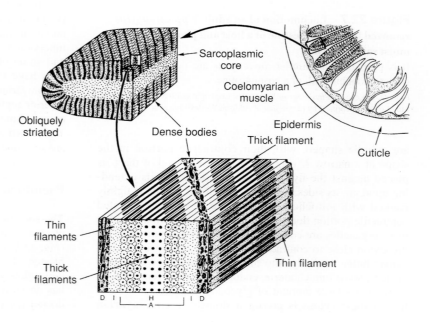

Figure 22.9 **Diagram of somatic musculature of *Ascaris suum*.**

Note the fine structure of coelomyarian oblique striated muscle.

From Bird and Bird, *The Structure of Nematodes,* 2nd edition. Copyright © 1991 Academic Press, Orlando, FL. Reprinted by permission.

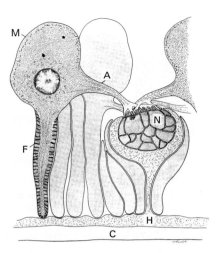

Figure 22.10   Diagram of muscle cells and myoneural junctions in transverse section.

The myocyton (*M*), containing the nucleus of a muscle cell, is continuous with the core of the striated fiber (*F*) and with its innervation process (*A*). The process subdivides as it approaches the nerve cord (*N*). Individual axons comprising the nerve cord are embedded in a troughlike extension of the epidermis (*H*), which underlies the animal's cuticle (*C*).

From J. Rosenbluth, "Ultrastructure of somatic cells in *Ascaris lumbricoides.* II Intermuscular junctions, neuromuscular junctions, and glycogen stores," in *J. Cell Biol.* 26:579–591. Copyright © 1965. The Rockefeller University Press, New York.

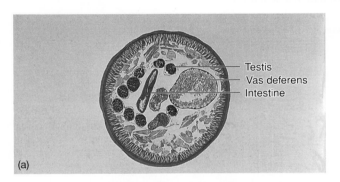

(a)

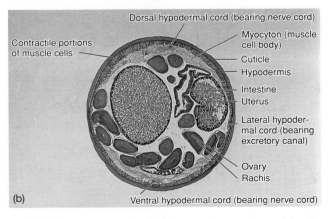

(b)

Figure 22.11   Cross sections of male and female *Ascaris suum.*

The "white" space between organs is pseudocoel. (*a*) Male; (*b*) female.

© Carolina Biological Supp/PHOTOTAKE.

Hydrostatic skeletons are widespread in invertebrates. Skeletal function depends on enclosure of a volume of noncompressible fluid, ability of muscle contraction to apply pressure to that fluid, and transmission of the pressure in all directions in the fluid as the result of its incompressibility. Thus, simultaneous contraction of circular muscles and relaxation of longitudinal muscles cause an animal to become thinner and longer, whereas relaxation of circular muscles and contraction of longitudinal muscles make an animal shorter and thicker. However, in nematodes somatic musculature is entirely longitudinal, and muscles act not against other antagonistic muscles but against stretching and compression of cuticle.[51]

The mechanism of body movement can be summarized as follows: As muscles on the ventral or dorsal side of the body contract, they compress cuticle on that side, and the force of the contraction is transmitted (by fluid in the pseudocoel) to the other side of the nematode, stretching cuticle on the opposite side. Compression and stretching of cuticle serve to antagonize the muscle and are the forces that return the body to resting position when muscles relax. Alternation of contraction and relaxation in dorsal and ventral muscles impels the body into a series of curves in a single, dorsoventral plane, producing the characteristic serpentine motion seen in nematode locomotion.[120] An increase in efficiency of this system can only be achieved by an increase in hydrostatic pressure, and hydrostatic pressure in the pseudocoel of nematodes is extraordinarily high. In *Ascaris suum* the pressure can average from 70 mm Hg to 120 mm Hg and vary up to 210 mm Hg,[50] an order of magnitude higher than the pressure in body fluids of animals with hydrostatic skeletons in other phyla. Limitations imposed by this high internal pressure determine many features of nematode morphology and physiology, such as how they eat, defecate, copulate, and lay eggs.

Pseudocoelomic fluid is a clear, almost cell-free, complex solution. In some nematode parasites of vertebrates, nematode hemoglobin gives the fluid a pink hue. Aside from its structural significance, this fluid certainly is important in transport of solutes from one tissue to another. These solutes include a variety of electrolytes, proteins, fats, and carbohydrates. Curiously, the fluid has far less chloride than would be required to balance the cations present, and the anion deficiency is made up mostly of volatile and nonvolatile organic acids.[40]

A peculiar and unique cell type found in the pseudocoel is the **coelomocyte.** Usually two, four, or six such cells, ovoid or with many branches, lie in the pseudocoel, attached to surrounding tissues. Although often small, in some species these cells are enormous; in *Ascaris suum* coelomocytes are 5 mm by 3 mm by up to 1 mm thick. Their function is still obscure, although they may have a role in the accumulation and storage of vitamin B_{12} and in protein synthetic and secretory function.[21]

Nervous System

Morphology

The nervous system of nematodes is relatively simple. There are two main concentrations of nerve elements in nematodes, one in the esophageal region and one in the anal area, connected by longitudinal nerve trunks. The most prominent

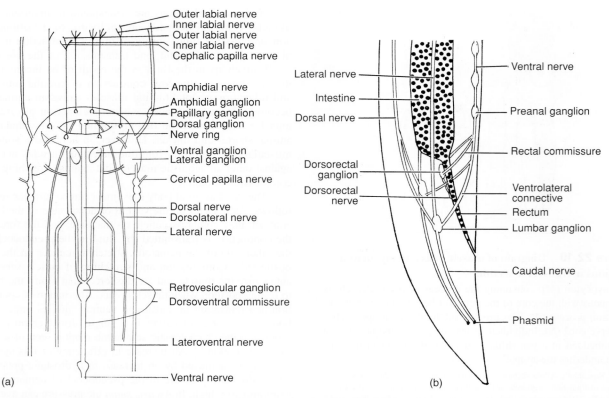

Figure 22.12 **Diagrammatic representation of the nervous system of a nematode.**
(*a*) Anterior end; (*b*) posterior end.

From H. D. Crofton, *Nematodes.* Copyright © 1966 Hutchinson University Library, London. Reprinted with permission.

feature of the anterior concentration is the **nerve ring,** or **circumesophageal commissure.** In *Ascaris suum* the nerve ring comprises eight cells, four of which are nerve cells and four of which are supporting, or **glial,** cells. The ring lies close to the outer wall of the esophagus, but can be difficult to observe unless special stains or high-contrast light microscopy is used. The ring serves as a commissure for **ventral, lateral,** and **dorsal cephalic ganglia** (Fig. 22.12*a*), which are usually paired. Emanating from each ganglion posteriorly are **longitudinal nerve trunks,** which become embedded in the epidermal cords; the ventral nerve is largest. Proceeding anteriorly from the lateral ganglia are two **amphidial nerves,** which innervate the amphids (explained next). Six **papillary nerves,** which are derived directly from the nerve ring, innervate the cephalic sensory papillae surrounding the mouth.

The ventral nerve trunk runs posteriorly as a chain of ganglia, the last of which is the **preanal ganglion.** The preanal ganglion gives rise to two branches that proceed dorsally into the pseudocoel to encircle the rectum, thus forming the **rectal commissure,** or **posterior nerve ring.** Other posterior nerves and ganglia are depicted in Figure 22.12*b*. The peripheral nervous system consists of a latticework of nerves that interconnect with fine commissures and supply nerves to sensory endings within cuticle.

Nematodes have a variety of **sensilla** (small sense organs), the most prominent of which are cephalic and caudal papillae, amphids, phasmids (in Chromadorea), and, in certain free-living species, ocelli. The patterns of sensory papillae on the surface of a nematode are an important taxonomic

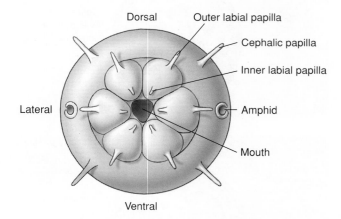

Figure 22.13 **Diagram of the anterior end of a hypothetical ancestral nematode showing the arrangement of the sense organs or sensilla.**

Redrawn by William Ober and Claire Garrison, from L. A. P. de Connick, "Les relations de symetrie, regissant la distribution des organes sensibles anterieurs chez les nematodes," in *Ann. Soc. Roy. Zool. Belgique* 81:25–31, 1950.

character. The ancestral bauplan of lips surrounding the mouth of nematodes probably was two lateral, two dorsolateral, and two ventrolateral, each of which was supplied with sensory papillae (Fig. 22.13). In addition to 12 papillae forming the **inner** and **outer labial circles,** there were four **cephalic papillae,** one located behind the lips in each of the dorsolateral and ventrolateral quadrants. Most parasitic nematodes are modified from this basic form. Labial papillae are often lost or fused together, and cephalic papillae usually are very small and difficult to resolve with light microscopy. Some papillae are found on all species, and careful study will reveal all 16 nerve endings on most species, even those that have lost all semblances of lips. The pattern of lips and papillae on large nematodes is studied by slicing the anterior tip from the worm with a sharp blade and orienting the end for an en face view on a microscope slide. For small nematodes, lips and associated papillae are studied by scanning electron microscopy. The sensory endings of the papillae are modified cilia.[14] Papillae on the head of nematodes include both chemosensory and tactile receptors.

Amphids are a pair of rather complex sensilla that open laterally on each side of the head at about the same level as the cephalic circle of papillae. The ampid openings are most conspicuous in marine, free-living forms in the Enoplea and usually are reduced in animal parasites. The amphidial opening leads into a deep, cuticular pit, which contains several nerve processes (Fig. 22.14). Sensory endings of neurons, dendrites, are modified cilia, and there are up to 23 in one amphid, in contrast to the one to three per papilla. Until modified cilia were discovered in sensilla of nematodes, it was thought that these worms had no cilia. Of course, their structure is rather different from ordinary motile cilia. They have no kinetosomes, and microtubules usually diverge from the normal 9 + 2 pattern (p. 45); for example, to 9 + 4, 8 + 4, or 1 + 11 + 4. Amphids are chemoreceptors in many nematodes.[101, 121] In several species, amphidial neurons function as thermoreceptors and mediate thermotaxis.[13, 77, 78] The amphid sheath cell (Fig. 22.14) may have a secretory function in some species;[103, 119] extracts of hookworm amphids inhibit clotting of vertebrate blood.[124] In what is their most remarkable function yet discovered, amphidial neurons control the life cycle of *Strongyloides stercoralis,* specifically whether juveniles (p. 393) produce free-living or parasitic adults.[3]

Most parasitic nematodes have a pair of cuticular papillae, known as **deirids,** or **cervical papillae,** at about the level of the nerve ring, and other sensory papillae are at different levels along the body of many species. **Caudal papillae** (Fig. 22.15) are more numerous in males, aiding in tactile responses related to copulation. Distribution of caudal papillae is an important taxonomic character, but such diagnostic features are typically found only in males. These papillae reach maximal development in superfamily Strongyloidea, where along with other supporting structures they form a complex copulatory bursa (chapter 25).

Near the posterior end of many chromadorean nematodes is a bilateral pair of cuticle-lined organs called **phasmids** (Fig. 22.16). Phasmids are similar in structure to amphids except that they have fewer neural endings, and the gland, if present, is smaller.[73] Presence or absence of phasmids was once used as a character to separate classes in nematode taxonomy. However, more recent analyses have suggested that the group of nematodes lacking phasmids is not monophyletic. Phasmids are recognized by their cuticle-lined ducts (Fig. 22.17) that open at the apices of papillae in the lateral field near the tail. They are difficult to recognize in many small species, and thus are of limited practical utility for diagnostics.

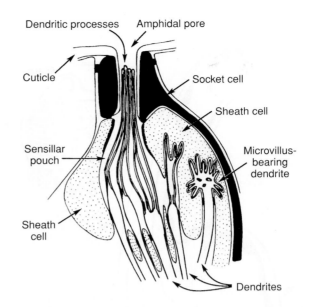

Figure 22.14 **Diagram of an amphid in *Caenorhabditis elegans.***

From K. A. Wright, "Nematode Sense Organs," in B. M. Zuckerman (Ed.), *Nematodes as Biological Models, Vol. 2, Aging and Other Model Systems.* Copyright © 1980. Academic Press, New York. Reprinted by permission.

Figure 22.15 **Ventral view of male *Toxascaris* sp., showing caudal papillae.**

Courtesy of Jay Georgi.

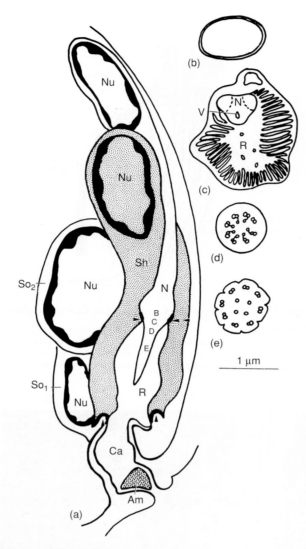

Figure 22.16   **Reconstruction of a phasmid in adult male *Meloidodera floridensis*, a plant-parasitic nematode.**
(*a*) Phasmid from dorsoventral view. Sensory ending is a modified cilium extending into a receptor cavity (*R*). The cavity opens to the outside at the ampulla (*Am*), which is often plugged by an electron-dense material. The neuron (*N*) is surrounded by cytoplasm of a sheath cell (*Sh*), which in turn is partially or completely surrounded by socket cells (*So*). *Ca,* canal; *Nu,* nucleus; *arrows,* intercellular junctions. (*b–e*) Cross sections at levels indicated by small letters inside neuron in (*a*). *V,* vesicle.

From L. K. Carta and J. G. Baldwin, "Ultrastructure of phasmid development in *Meloidodera floridensis* and *M. charis* (Heteroderinae)," in *J. Nematol.* 22:362–385. Copyright © 1990 Journal of Nematology. Reprinted by permission.

Neurotransmission

The predominant excitatory neurotransmitter in nematodes is acetylcholine.[67] Muscle cells undergo spontaneous depolarization in the innervation arm and then generate action potentials in a repeated or oscillatory manner, somewhat similar to that of vertebrate cardiac muscle, which consists of a spontaneous, rhythmic spike production. Rate of firing increases with lowered resting potential and decreases with higher resting potential. Nerve fibers play primarily a

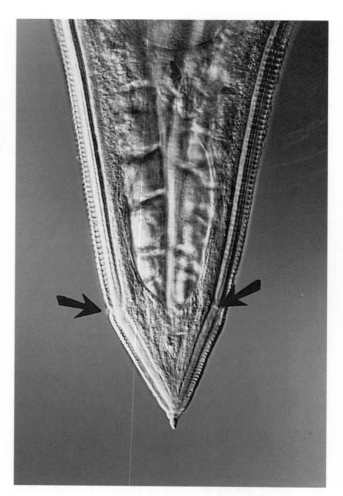

Figure 22.17   **Ventral view of female *Toxascaris* sp.**
Note ducts (*arrows*) leading to phasmids.
Courtesy of Jay Georgi.

modulating role, with both excitatory and inhibitory fibers. Stimulation of excitatory fibers releases acetylcholine at neuromuscular junctions, depolarizes muscle membranes, and increases the rate of spikes (Fig. 22.18). Inhibitory fibers release gamma-aminobutyric acid (**GABA**), hyperpolarize muscle, and decrease the rate of action potentials.

Although nematode nervous systems are somewhat simpler in organization than those of platyhelminths, they nevertheless display an "astounding level of neurochemical diversity."[83] The full complement of neuropeptides in *Caenorhabditis elegans* alone has been estimated at 400–500.[116] The biological functions of most of these are unknown, but some apparently are excitatory, mediating contraction of body-wall or ovijector muscles.[87] Along with serotonin and acetylcholine, they may function in control and modulation of feeding activities by their effects on muscles of the esophagus.[25]

Effects of Drugs

The foremost reason that neurobiology of parasitic nematodes has elicited so much research interest is that several nematicidal drugs interfere with neural function. For example, piperazine is a GABA-agonist that opens chloride-ion channels

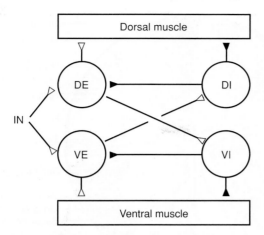

Figure 22.18   **Diagram illustrating synaptic relationships of excitatory and inhibitory motor neurons in nematodes.**

Open triangles are excitatory, and closed triangles are inhibitory synapses. Stimulation of dorsal excitatory motor neuron (*DE*) by interneuron (*IN*) causes dorsal muscle contraction and stimulation of ventral inhibitory motor neuron (*VI*). *VI* inhibits ventral excitatory motor neuron (*VE*) and hyperpolarizes the ventral muscle, thus relaxing it. Stimulation of the *VE* neuron has the opposite effect, causing the ventral muscle to contract and the dorsal muscle to relax (*DI*, dorsal inhibitory neuron). (Direct recordings have not been made from *VE* neurons, but their action is inferred from recordings made from *DE* neurons.)

From A. Stretton et al., "Motor behavior and motor nervous system function in the nematode *Ascaris suum*," in *J. Parasitol.* 78:206–214, 1992.

and thereby hyperpolarizes the muscle membrane, effectively paralyzing the worms, so that they pass out of the host.[47] Levamisole and pyrantel mimic effects of acetylcholine, depolarizing the muscle membrane, resulting in paralysis.

Earlier results suggested that the action of ivermectin was due to stimulation of GABA release and enhancement of its binding to postsynaptic receptors. Subsequent research has indicated that ivermectin and other macrocyclic lactones irreversibly open glutamate-gated chloride ion channels. Macrocyclic lactones appear to have different primary effects in different nematode species, but include inhibition of pharyngeal pumping and interference with secretion and maintenance of hydrostatic pressure.[26]

The benzimidazoles (mebendazole, parbendazole, fenbendazole) appear to have two modes of action.[110] They inhibit mitochondrial electron transport, especially the fumarate reductase system in species with this pathway (see p. 319), thus inhibiting energy metabolism; and they also inhibit the polymerization of beta-tubulin, which interferes with microtubule-dependent processes such as acetylcholinesterase secretion, and paralyze the worms. Tubulin in intestinal cells of *A. suum* binds three times more mebendazole than body-wall muscles, suggesting that interference with intestinal function may be a mechanism of action.[28]

Digestive System and Acquisition of Nutrients

The digestive system is complete in most nematodes, with mouth, buccal cavity or stoma, esophagus, intestine, and anus, although in mermithids and a few filariids the anus is atrophied. Cuticle lines the stomodeum (buccal cavity and esophagus) and proctodeum (rectum), and nematodes shed the cuticular lining of these cavities along with the exterior cuticle when they molt.

Mouth and Esophagus

The mouth is usually a circular opening surrounded by a maximum of six lips. Few parasitic nematodes possess six lips; in some nematodes lips fuse in pairs to form three. In many species lips are indistinct and appear absent, whereas in others two lateral lips develop as new structures derived from the inner margin of the mouth. However, the pattern of sensilla associated with the lips and head region of nematodes is generally conserved.

A buccal cavity or stoma lies between the mouth and esophagus of nematodes. The size and shape of this area vary extensively among species and are important taxonomic characters. In some species the cuticular lining is quite thick, forming a rigid structure known as a **buccal capsule.** The cavity may be elongated, reduced, or absent altogether, with a mouth that opens almost immediately into the lumen of the esophagus. Buccal armament such as teeth, denticles, or spears is often present in parasitic and predaceous nematodes. The elements arise from modifications of the cuticle lining the cavity.

Food ingested by a nematode moves into a muscular region of the digestive tract known as the **esophagus,** or **pharynx.** This is a pumping organ that sucks food into the alimentary canal and forces it into the intestine. Such an arrangement is necessary because of high pressure in the surrounding pseudocoel. The esophagus assumes a variety of shapes, depending on the order and species of nematode, and for this reason it is an important taxonomic character. It is highly muscular and cylindrical and often has one or more enlargements **(bulbs).** In some free-living and parasitic nematodes, esophageal bulbs contain structures for macerating ingested microorganisms. The lumen of the esophagus is lined with cuticle, and much of it is triradiate in cross section, with one radius directed ventrally and the other two pointed laterodorsally (Fig. 22.19). Radial muscles insert on cuticular lining in interradii and run the length of the esophagus.

Interspersed among muscles of the esophagus are three or more glands, one in each of the interradial zones. The dorsal gland is usually more extensive than are the ventrolaterals. The products of the glands are released into the esophageal lumen. In Chromadorea the dorsal gland commonly opens near the buccal cavity, whereas ducts of the subventral glands open near the esophageal base. Secretions produced by these glands include digestive enzymes such as amylase, proteases, pectinases, chitinases, and cellulases.[73] In hookworms the secretions have anticoagulant properties.[124] In some sedentary endoparasites of plant roots, esophageal secretions induce the formation of nurse cells in the plants, which provide nutrition for the nematode. In some species the glands fuse together near the posterior end of the esophagus, and in some nematodes, such as certain spiruromorpha, the posterior portion of the esophagus is mostly glandular. In some species the glands are so extensive that much of their mass lies outside the esophagus proper (Fig. 22.20). In class Enoplea there are five or more esophageal glands; in the specialized esophagus of Trichinellida and Mermithida the anterior portion is a thin-walled, muscular tube, whereas

Figure 22.19 **Diagram of a transverse section through the posterior part of the esophagus of *Nippostrongylus brasiliensis.***

Note arrangement of various cells, cell membranes, and cellular organelles. Reconstructed from several electron micrographs.

From D. L. Lee, "The ultrastructure of the alimentary tract of the skin-penetrating larva of *Nippostrongylus brasiliensis* (Nematoda)," in *J. Zool.* 154:9–18. Copyright © 1968 Oxford University Press. Reprinted by permission of Oxford University Press.

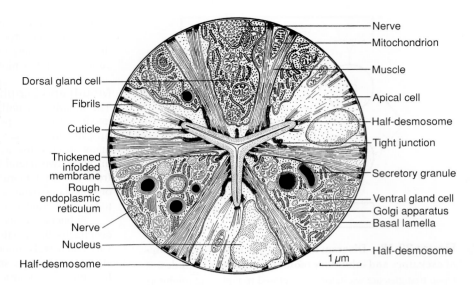

Figure 22.20 *Syphacia*, **a rodent pinworm with enlarged esophageal glands (*g*).**

From G. D. Schmidt and R. E. Kuntz, "Nematode parasites of Oceanica. IV. Oxyurids of mammals of Palawan, P.I., with descriptions of four new species of *Syphacia*," in *Parasitology* 58:845–854. Copyright © 1968 Cambridge University Press. Reprinted with the permission of Cambridge University Press.

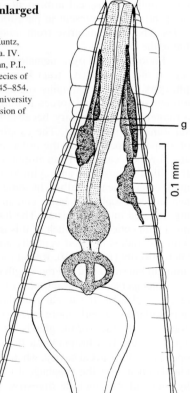

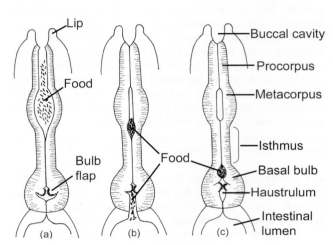

Figure 22.21 **Diagrams to show the structure and function of the *Rhabditis*-type esophagus during feeding.**

Food particles, small enough to pass through the buccal cavity, are drawn into the lumen of the metacorpus by sudden dilation of the procorpus and metacorpus (*a*). Closure of the lumen of the esophagus in these regions expels excess water (*b*), and the mass of food particles is passed backward along the isthmus (*b, c*). Food is drawn between the bulb flaps of the posterior bulb by dilation of the haustrulum, which inverts the bulb flaps (*a*) and is passed to the intestine by closure of the haustrulum and by dilation, followed by closure of the pharyngeal-intestinal valve (*b*). Bulb flaps contribute to the closure of the valve in the basal bulb and, when they invert (*a*), also crush food particles.

Drawn by John Janovy, Jr., from various sources.

the posterior portion is a very thin tube surrounded by a column of single glandular cells called **stichocytes,** the entire structure being referred to as a **stichosome.** Stichocytes communicate with esophageal lumen by small ducts.[117] The stichosome may be homologous to esophageal glands of other nematodes, perhaps derived by multiplication of the number of glands.[1]

Rapid contraction of the buccal muscles and anterior esophageal muscles opens the mouth and dilates the anterior end of the esophagus, sucking in food (Fig. 22.21). The high hydrostatic pressure in the pseudocoel surrounding the esophagus closes the mouth and esophageal lumen when the muscles relax. Food passes down the esophagus by a posteriorly progressing wave of muscle contraction opening the lumen for it until it reaches the intestine. A posterior bulb in many species seems to function as a one-way, nonregurgitation valve for food in the intestine. Thus, the mechanism is a kind of peristalsis in which the force moving food is not contraction of circular muscles but closure of the esophageal

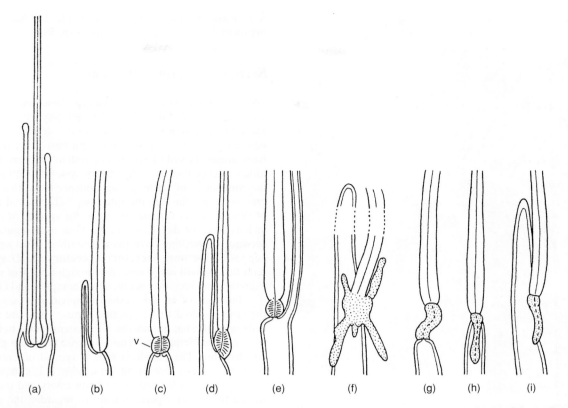

(a)   (b)   (c)   (d)   (e)   (f)   (g)   (h)   (i)

Figure 22.22   **Variations in the anterior alimentary tract in genera of ascaridoid nematodes showing esophageal and intestinal diverticuli.**

Nematodes shown are of genera (*a*) *Crossophorus,* (*b*) *Angusticaecum,* (*c*) *Toxocara,* (*d*) *Porrocaecum,* (*e*) *Paradujardinia,* (*f*) *Multicaecum,* (*g*) *Anisakis,* (*h*) *Raphidascaris,* (*i*) *Contracaecum. v,* ventriculus.

From G. Hartwich, *CIH Keys to the Nematode Parasites of Vertebrates,* no. 2. Farnham Royal, Bucks, England: Commonwealth Agricultural Bureaux, 1974.

lumen by hydrostatic pressure behind the food. Frequency of pumping has been recorded as 2 to 24 per second.[34]

In a few ascaridomorphs (species of *Contracaecum, Multicaecum,* and others) one to five posteriorly directed esophageal ceca originate from a short, glandular ventriculus between the body of the esophagus and the intestine (Fig. 22.22). The generic names (*Contracaecum, Multicaecum*) refer to these structural features.

Intestine

The intestine is a simple, tubelike structure extending from esophagus to proctodeum and is constructed of a single layer of intestinal cells.

In females a short, terminal, cuticle-lined rectum runs between anus and intestine. In males the rectum receives products of the reproductive system into its terminal portion and, therefore, is a cloaca. The dorsal wall of the cloaca is usually invaginated into two pouches, or spicule sheaths, that contain **copulatory spicules;** these will be described along with the reproductive system. The vas deferens opens into the ventral wall of the cloaca.

The intestine is nonmuscular. Its contents are forced posteriorly by action of the esophagus as it adds more food to the front end of the system and perhaps by locomotor activity of the worm. Internal pressure in the pseudocoel flattens the intestine when empty. Between the dorsal wall of the cloaca and the body wall is a powerful muscle bundle called **depressor ani.** This is a misnomer because, when it contracts, the anus

is opened; it is therefore a dilator rather than a depressor. Hydrostatic pressure surrounding the intestine causes defecation when the anus is opened. Hydrostatic pressure expels feces with some force: When removed from a saline solution, *Ascaris suum* can project its feces 60 cm.[34]

The intestine consists of tall, simple columnar cells with prominent brush borders of microvilli (Fig. 22.23). Although several digestive enzymes have been identified in intestinal lumen, intestinal digestion is probably of minor importance in most species because of rapid rate of food movement through the intestine.

Numbers of intestinal cells vary from about 30 in some free-living species to more than a million in larger parasitic forms. These cells rest on a basement membrane, which is attached to extensions of body wall musculature. It is probable that the intestine serves as a primary means of excretion of nitrogenous waste products in addition to functioning in nutrient absorption. Crofton[34] maintained that the intestine of *Ascaris lumbricoides* is emptied by defecation every three minutes under experimental conditions. Such a rapid turnover of materials must surely limit the amount of enzymatic action possible in intestinal lumen but favors excretion of water-soluble waste products.

Food

Food of nematodes parasitic in animals includes blood, tissue cells and fluids, intestinal contents, bacteria, or some combination of these. Some species parasitic in the intestine

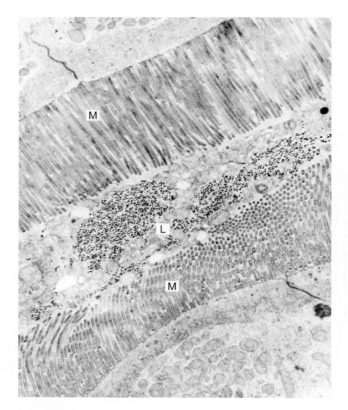

Figure 22.23 **Cross section of intestine showing microvilli (*M*) of dorsal and ventral sides.**
Cellular debris fills the lumen (*L*). (×10,800)

From H. G. Sheffield, Electron microscope studies on the intestinal epithelium of *Ascaris suum*, in *J. Parasitol.* 50:365–379. Copyright ©1964.

feed only on tissue and not on blood or host ingesta.[5] Adult hookworms, which feed solely on blood, accumulate granules of zinc sulfide in their intestinal cells, apparently as a waste product.[48] Some parasites of vertebrates, including hookworms, feed on bacteria as juveniles, but host tissues as adults. These differences in feeding choices are reflected by changes in esophagus structure in the different stages. Other parasites such as the whipworm *Trichuris,* have a very small stoma that is only suited for uptake of fluid. Since the anterior end of *Trichuris* is located within mucosa tissue, nutrient-containing fluids must be derived from the host cells.

Feeding in Mermithids

Members of order Mermithida are unusual in that adults are free living but juveniles are parasitic in invertebrates, primarily insects. Adults do not feed, and at no stage is there a functional gut. The body wall in adults and first-stage juveniles has a structure typical of other nematodes, described previously, but the body wall in parasitic juveniles is greatly modified for absorption of nutrients.[9] The cuticle is very thin, and the hypodermis is thick and metabolically active, with microvilli underlying the cuticle. The hypodermis is connected by cytoplasmic bridges to a food storage organ, or **trophosome.**[10] During the sometimes long, nonfeeding adult life, worms apparently live on nutrients stored in the trophosome. Because mermithids almost always kill their host, they have potential as biological control agents of insect pests;

Romanomermis culicivorax, for example, has been used experimentally in control of mosquitoes (p. 584).[98, 100]

Secretory-Excretory System

So-called excretory systems have been observed in all nematodes except Trichinellida and Dioctophymatida,[31] but an excretory function was assigned to the systems in various nematodes solely on a morphological basis; that is, the systems simply looked like excretory systems. However, there is little evidence that these "excretory systems" are involved in the elimination of waste; strong evidence exists that most excretion occurs through the intestine.[111] The actual functions of these systems vary according to the species of nematode and its stage of development, but both osmoregulatory and secretory functions have been described. Bird and Bird[14] suggested the term **secretory-excretory (S-E) system,** a term that we will adopt here. A thorough review of nematode excretion and secretion is given by Thompson and Geary.[123]

Presence of an S-E system is apparently ancestral and probably evolved first in marine forms. There are no flame cells or nephridia; in fact, the S-E system seems to be a neoformation of Nematoda.[1] The two basic types are **glandular** and **tubular.** The glandular type is typical of Enoplea and is found in many free-living nematodes. It is composed of a single gland cell located in the body cavity and that is connected to a ventral pore. Adamson[1] regarded the glandular type as plesiomorphic. We will confine our discussion to the tubular type, which is characteristic of most Chromadorea.

Several varieties of tubular excretory systems occur (Fig. 22.24). Each type includes a pore cell, duct cell, a SE cell, and a gland cell. Two long canals in the lateral hypodermis connect to each other by a transverse canal near the anterior end. This transverse canal opens to the exterior by means of a median ventral duct and pore, the **excretory pore.** This pore usually is conspicuous; its location is fairly constant within a species and therefore is a useful taxonomic character. Derived states in form of excretory system correlate with evolutionary group in Ascaridomorpha.[90, 91]

Osmoregulation and Secretion

Ability to osmoregulate varies greatly among nematodes and corresponds generally with the requirements of their habitats. Body fluids of species parasitic in animals may differ somewhat in osmotic pressure from the tissues they inhabit but not dramatically so. For example, *Ascaris suum* hemolymph is about 80% to 90% of the osmotic pressure of pig intestinal contents.[47] *Ascaris suum* clearly can control its electrolyte concentrations to some degree: Chloride ion concentration of host intestinal contents varies between 34 mM and 102 mM, but *A. suum* hemolymph is fairly constant at around 52 mM. Adults of most parasitic species cannot tolerate media much different in osmotic pressure from their hemolymph; when placed in tap water, they will burst, sometimes within minutes, from addition of imbibed water to the already high internal pressure. Of course, freshwater and terrestrial nematodes, including juveniles of many parasitic species, must withstand extremely hypotonic conditions (and regulate their internal osmotic constitution accordingly).

Details of water and ion excretion are poorly known. Contractions of S-E canals and the ampulla near the pore have

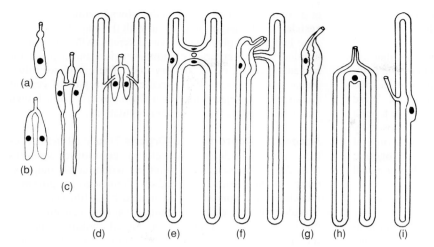

Figure 22.24   Excretory systems.

(*a*) Single renette in a dorylaimid; (*b*) two celled renette in *Rhabdias* spp.; (*c*) larval *Ancylostoma* spp.; (*d*) rhabditoid type; (*e*) oxyuroid type; (*f*) *Ascaris* spp.; (*g*) *Anisakis* spp.; (*h*) *Cephalobus* spp.; (*i*) *Tylenchus* spp.

From H. D. Crofton, *Nematodes.* Copyright © 1966 Hutchinson University Library, London. Reprinted with permission.

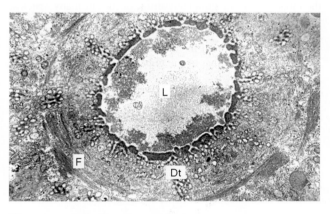

Figure 22.25   Transverse section through main excretory canal of *Anisakis* sp.

Note round canal with interrupted dense material lining the lumen (*L*), ramifying drainage tubules (*Dt*), and congregated vesicles surrounding the main canal and drainage tubules. Filaments (*F*) appear in a circular arrangement around the main canal. (×10,000)

From H.-F. Lee et al., "Ultrastructure of the excretory system of *Anisakis* larva (Nematoda: Anisakidae)," in *J. Parasitol.* 59:289–298. Copyright © 1973.

been observed in several species. Rates of contraction of the ampulla in free-living, third-stage juveniles of *Ancylostoma* sp. and *Nippostrongylus brasiliensis* are inversely proportional to the salt concentration of the solution in which they are maintained. Some evidence suggests that osmoregulation by the ampulla is part of a homeostatic mechanism to maintain constant volume so that locomotor activity is not impaired. Nelson and Riddle[92] ablated different portions of the S-E system of the free-living nematode *Caenorhabditis elegans* with a laser microbeam. Destruction of the pore cell, duct cell, or excretory cell led to accumulation of water and death of the worm, but ablation of the gland cell had no effect.

Ultrastructure of the S-E system strongly suggests that the system functions in osmoregulation and secretion as well.[76] Surface area of the peripheral cell membrane may be greatly increased by numerous bulbular invaginations, and on its interior the lumen is perforated by drainage ductules, or canaliculi (Fig. 22.25). Filaments that are presumably contractile may surround the lumen of the duct. Hydrostatic

pressure in the pseudocoel may provide pressure for movement of substances through the canals embedded in the hypodermal cords.

Ultrastructure of the gland cells clearly suggests secretory function. Enzymes responsible for exsheathment (shedding the old cuticle at ecdysis) are produced there by various strongyle juveniles. A variety of nematodes excrete substances antigenic for their hosts through the S-E pores. Lee[69, 70] suggested that digestive enzymes were secreted by adult *Nippostrongylus brasiliensis* to act in conjunction with the abrading action of the cuticle. Other types of secretory gland cells are also present in some nematodes. For example, rectal gland cells in root-knot nematodes secrete the gelatinous matrix that is part of the external egg mass of this plant parasite.

Excretion

The major nitrogenous waste product of nematodes is ammonia. In normal saline, *A. suum* excretes 69% of the total nitrogen excreted as ammonia and 7% as urea. Under conditions of osmotic stress, these proportions can be changed to 27% ammonia and 52% urea. Amino acids, peptides, and amines may be excreted by nematodes. Other excretory products include carbon dioxide and a variety of organic acids. These organic acids are end products of energy metabolism (p. 371). The role of the S-E system in elimination of the foregoing substances is not well established. Juvenile *Nippostrongylus brasiliensis* excrete several primary aliphatic amines through their S-E pore. A large proportion of nitrogenous waste products can be excreted via the intestine and anus by *A. suum,*[111] and it would seem that epidermis and cuticle must play a major role in ammonia excretion in most nematodes.

Reproduction

Most nematodes are dioecious, but a few monoecious species are known. Parthenogenesis also exists in some. Sexual dimorphism usually attends dioecious forms, with females growing larger than males. Furthermore, males have a more coiled tail than do females and often have associated external features, such as bursae, caudal alae, and papillae. Such dimorphism achieves the ultimate in Tetrameridae and the plant-parasitic Heteroderidae, in which males have typical vermiform anatomy, but mature females are little more than swollen bags of uteri.

Gonads of nematodes are solid cords of cells that are continuous with ducts that lead to the external environment. This allows their reproductive systems to function despite the high pressure of the surrounding pseudocoelom. Were the cords not continuous, an oocyte would not be able to gain access to the oviduct, which would be squeezed closed by surrounding pressure. When germ cells proliferate only at the inner end of a gonad, the gonad is described as **telogonic;** but if germ cells proliferate throughout the length of a gonad (as in Trichinellida), the gonad is **hologonic.**

Male Reproductive System

Testes are generally paired in Enoplea and some Chromadorea, and the paired condition is probably plesiomorphic. This organ may be relatively short and uncoiled, but in larger animal parasites it appears as a long, thread-like structure that is coiled around the intestine and itself at various levels of the body. There are usually three zones in telogonic testes: a **germinal zone,** incorporating the blind end and in which spermatogonial divisions take place; a **maturation** or **growth zone** (Fig. 22.26); and a **storage zone** or **seminal vesicle,** which merges on one end with the end of the growth zone and on the other with a vas deferens, which is usually divided into an anterior glandular region and a posterior muscular region, or ejaculatory duct. The ejaculatory duct opens into a cloaca. Some species have a pair of cement glands near the ejaculatory duct that secrete a hard, brown material that plugs the vulva after copulation.

Accessory Reproductive Organs Nearly all nematodes have a pair of sclerotized, acellular, **copulatory spicules** (Fig. 22.27). These spicules originate within dorsal outpocketings of the cloacal wall and are controlled by proximal muscles. Each spicule is surrounded by a fibrous sheath. Spicule structure varies substantially among nematode taxa but is fairly constant among individuals within a species, making size and morphology of spicules two of the most important taxonomic characters. A dorsal sclerotization of the cloacal wall, or **gubernaculum,** occurs in many species; this structure guides exsertion of the spicules from the cloaca at copulation. In several strongyloid genera an additional ventral sclerotization of the cloaca, or **telamon,** has the same general function as that of the gubernaculum. Both structures

Figure 22.26 **Diagram of a nematode testis and spermatogenesis in *Brugia malayi.***

Spermatids are activated to become mature sperm with a pseudopodium by mating.

Redrawn by William Ober and Claire Garrison, from A. L. Scott, "Nematode Sperm," in *Parasitol. Today,* 12:425–430. Copyright © 1996 Elsevier Science.

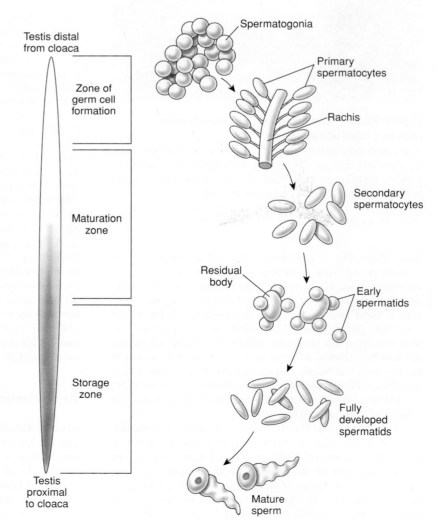

are important taxonomic characters. Spicules are inserted into the vulva at copulation. They are not true intromittent organs, since they do not conduct sperm, but they are another adaptation to cope with high internal hydrostatic pressure. Spicules must hold the vulva open while ejaculatory muscles overcome hydrostatic pressure in the female and inject sperm into her reproductive tract.

Spermatogenesis. Spermatogonia divide mitotically in the germinal zone and then attach by cytoplasmic bridges to a supporting structure called a **rachis** (see Fig. 22.26). On the rachis these cells enter prophase of meiosis I.[114] As the primary spermatocytes move into the maturation zone, they detach from the rachis and continue meiosis I through meiosis II with no intervening cytokinesis. The four haploid

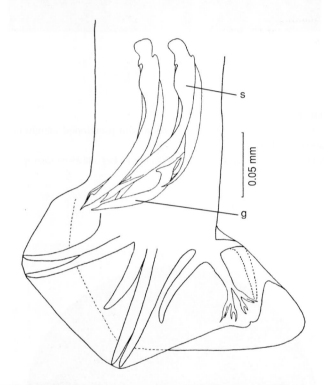

Figure 22.27   **Bursate nematode *Molineus mustelae.***
Note complex spicules (*s*) and a gubernaculum (*g*).

From G. D. Schmidt, "*Molineus mustelae* sp. n. (Nematoda: Trichostrongylidae) from the long-tailed weasel in Montana and *M. chabaudi* nom. n. with a key to the species of *Molineus*," in *J. Parasitol.* 51:164–168. Copyright © 1965. Reprinted by permission.

nuclei move to the margin of the cell and, along with certain organelles, bud off to become round spermatids. The remaining **residual body** contains all biosynthetic organelles, such as endoplasmic reticulum, ribosomes, and Golgi apparatus, which means that all molecules necessary for subsequent sperm differentiation and function must have been synthesized earlier.

Spermatids remain in the storage zone, where they are activated upon mating. Their nuclear material is highly condensed into two or three discrete bodies that are not membrane bound (Fig. 22.28). Nematode spermatozoa are unusual because they lack a flagellum and acrosome and depend on pseudopodial locomotion. Upon activation, a spermatid becomes a mature spermatozoon and protrudes a pseudopodium that bears numerous processes called **villipodia** (Fig. 22.28). The pseudopodium contains no organelles nor any actin or myosin.[114] Locomotion is due to action of a fibrous material composed of a sperm-specific substance known as **major sperm protein (MSP).** MSP is highly conserved among nematodes but has *no sequence homology* with molecules such as actin, myosin, or tubulin that are associated with movement in other organisms. For pseudopodial movement, MSP forms fibrous complexes anchored to dense material inside the cell membrane at locations of villipodia (see Figs. 22.28 and 22.29*b*). Movement depends on MSP assembly into fibers, disassembly, and reassembly as villipodia treadmill rearward. MSP assembly and disassembly apparently are controlled by a precise gradient of intracellular pH.[114]

Female Reproductive System

Most female nematodes have two ovaries, although some have one and others more than six. The general pattern of structure of the reproductive system in females is similar to that in males, except the gonopore is independent of the digestive system. The pattern is a linear series of structures, with the gonad at the internal or end distal to the gonopore, followed by developmental, storage, and ejective areas. Female reproductive tracts of most nematodes are telogonic, but some are hologonic.

Ovaries and Oviducts. Ovaries are solid cords of cells that produce gametes and move them distally into the terminal portion of the system. The proximal end of a telogonic ovary is the germinal zone, which produces oogonia; oogonia become oocytes and move into the growth zone of the ovary, toward the oviduct. In large ascarids oocytes are attached to a rachis. In *Ascaris* spp. the germinal zone is very short, and

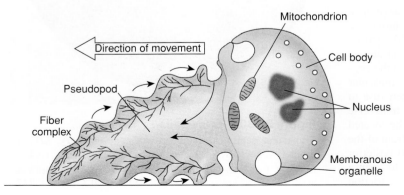

Figure 22.28   **Diagram of mature sperm from *Brugia malayi*, consisting of a cell body and a pseudopod.**

The cell body contains the nucleus, mitochondria, and membranous organelles, and the pseudopod has fibrous complexes and villipodia. *Arrows* indicate direction of movement of the villipodia and proposed movement of the disassembled and reprocessed fiber constituents from the tip of the pseudopod.

Drawing by William Ober and Claire Garrison.

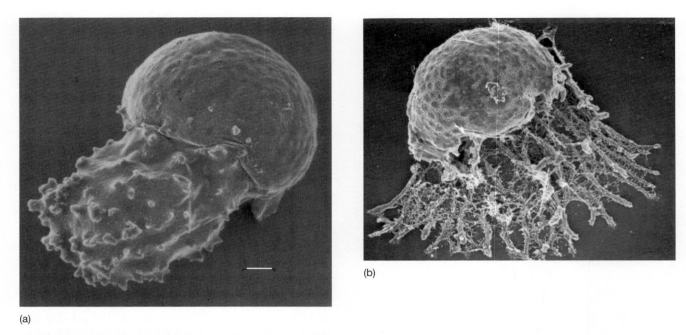

(a)

(b)

Figure 22.29 **Scanning electron micrograph of *A. suum* spermatozoa.**
(*a*) Seventeen minutes after in vitro activation on glass. (Scale bar = 1 μm) (*b*) Activated spermatozoon with pseudopod membrane removed with Triton X-100, exposing the MSP-rich fiber complexes. (about ×4000)

(*a*) From S. Sepsenwol et al., "A unique cytoskeleton associated with crawling in the amoeboid sperm of the nematode, *Ascaris suum.*," in *J. Cell Biol.* 108:55–66. Copyright © 1989. (*b*) Courtesy S. Sepsenwol.

most of the 200 cm to 250 cm length of the ovary consists of oocytes attached to the rachis in a radial manner by cytoplasmic bridges[44] (Fig. 22.30). Oocytes increase in size as they move down the rachis, becoming detached from it when they reach a point about 3 cm to 5 cm from the oviduct. In some nematodes the rachis ends at the beginning of the growth zone, and oocytes pass single file down the growth zone, increasing greatly in volume.[81]

The proximal end of the oviduct in most nematodes is a distinct **spermatheca,** or sperm storage area. In some species the spermatheca is offset from the path of the main gonad tube. As oocytes enter the oviduct (spermathecal area), sperm penetrate them; only then do the oocytes proceed with meiosis. A polar body is extruded at each of the two meiotic divisions. Concurrent with these events, shell formation occurs, described in the section on development.

Uterus and Vulva. The uterine wall has well-developed circular and diagonal muscle fibers, and these move the developing embryos ("eggs") distally by peristaltic action. The shape of eggs may be molded by the uterus, and uterine secretory cells may contribute additional material to the eggshells. The distal end of the uterus is usually quite muscular and constitutes an **ovijector.** Ovijectors of the uteri fuse to form a short vagina that opens through a ventral, transverse slit in the body wall, the **vulva.** The vulva may be located anywhere from near the mouth to immediately in front of the anus, depending on species. The vulva never opens posterior to the anus and only very rarely into the rectum to form a cloaca. Muscles of the vulva act as dilators, and constriction

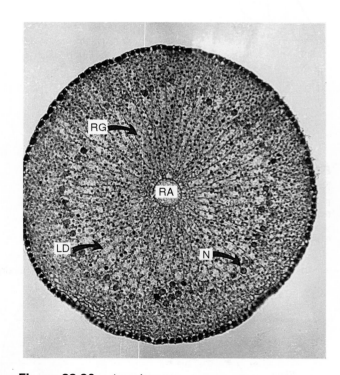

Figure 22.30 *Ascaris suum.*
Transverse section through growth zone of ovary. *LD,* lipid droplet; *N,* nucleus; *RA,* rachis; *RG,* refringent granule. (×440)

From W. E. Foor, "Ultrastructural aspects of oocyte development and shell formation in *Ascaris lumbricoides,*" in *J. Parasitol.* 53:1245–1261. Copyright © 1967.

of the circular muscles of the ovijectors both expels mature eggs and restrains more proximal, undeveloped eggs from being expelled due to hydrostatic pressure.

Mating Behavior

Clearly, adult worms of opposite sexes must find each other and copulate for reproduction to occur. Both chemotactic and thigmotactic mechanisms operate in these processes.

Pheromone sex attractants have now been shown for about 40 species of nematodes,[52, 80] usually by means of an in vitro assay. Pheromones may thus have some potential for use as biological control agents. For example, male *N. brasiliensis* migrate toward a source of medium in which females have been incubated or that contains an aqueous extract of females. There also may be a "medley" of attractants, more complex than hitherto realized.

After a male and female nematode have found each other, thigmotactic responses mediated by papillae facilitate copulation. Females in some species seek the coiled posterior end of males, which they enter. Caudal papillae of a male detect the vulva; this encounter excites a probing response of the spicules, leading to sperm transfer. Spicules have neurons running up their center that terminate distally in sensory endings, probably allowing a spicule to "feel" its way into a female's reproductive tract without damaging her tissues.[114] Females of some species have vulvar papillae. Curiously, if no males are present within a host, females of some species tend to wander, seeking a constriction to squeeze through. This may result in dire consequences to the host if a bile duct, for example, is selected for exploration. Other unexpected results of this behavior have been recorded (Fig. 22.31). Female *Ascaris suum* cease producing eggs if they are transferred to a host without any male worms, and they readily resume when a male is transferred to join them.[60]

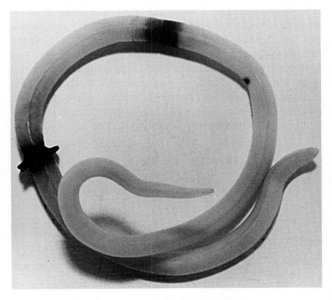

Figure 22.31   **Female *Ascaris lumbricoides* strangled by a shoe eyelet.**
This illustrates the tropism of female nematodes to seek the coiled tail of males.

From P. C. Beaver, "*Ascaris* strangled in a shoe-eyelet," in *Am. J. Trop. Med. Hyg.* 13:295–296. Copyright © 1964.

DEVELOPMENT

Studies on development of nematodes have led to fundamental discoveries in zoology. For example, in 1883 van Beneden[12] was first to elucidate the meiotic process and realize that equal amounts of nuclear material were contributed by sperm and egg after fertilization. Boveri[22] (1899) first demonstrated genetic continuity of chromosomes and determinate cleavage; that is, the fate of blastomeres is determined very early in embryogenesis. Boveri was also among the first to observe the reduction in chromosome number in gametes that reflects meiosis. Both men based their insights on studies of nematode material (from a species of *Parascaris* in horses). *Caenorhabditis elegans* is the only metazoan for which complete development has been described at the cellular level. The transparency of certain nematode eggs and their embryos makes them excellent organisms for developmental studies.

Not surprisingly, in such a large and diverse phylum as Nematoda, details of development and life history differ greatly among various groups. However, the general pattern is remarkably conserved. Four juvenile stages and an adult stage are each separated from the one preceding by an ecdysis, or molting of cuticle. Animal parasitologists traditionally (but incorrectly) refer to juvenile stages as *larvae* (see Fig. 22.34). The first-stage juvenile is often quite similar in body form to the adult and no real metamorphosis occurs during ontogeny.

Eggshell Formation

Penetration of an oocyte by a sperm initiates the process by which protective layers are produced around the zygote and developing embryo. A fully formed shell in most nematodes consists of three layers: (1) an outer **vitelline layer,** often not detectable by light microscopy; (2) a **chitinous layer;** and (3) an innermost **lipid layer.** A fourth, **proteinaceous layer,** which consists of an acid mucopolysaccharide-tanned protein complex, is contributed by uterine cell secretions in some nematodes (*Ascaris* spp., *Thelastoma* spp., *Meloidogyne* spp.). Formation of shell layers is best known in *Ascaris suum,* but it seems likely that the process is similar in other nematodes.

Immediately after sperm penetration, a new plasma membrane forms beneath the original; the old plasma membrane becomes the vitelline layer and separates from peripheral cytoplasm. Then the cytoplasm shrinks back, leaving a clear space within which the chitinous layer forms[44, 74] (Fig. 22.32). Refringent bodies, previously dispersed throughout the cytoplasm, migrate to the periphery and expel their contents, the fusion of which forms the lipid layer.

The so-called chitinous layer is probably supportive or structural in function and also contains protein; the proportion of chitin present varies among groups from great (ascaridomorphs, oxyuridomorphs) to very small (strongyloids). The lipid layer confers resistance to desiccation and to penetration of water-soluble substances. At least in ascarids, the lipid layer is composed of 25% protein and 75% **ascarosides.** Ascarosides are very interesting and unique glycosides (compounds with a sugar and an alcohol joined by a glycosidic bond). In ascarosides the sugar is ascarylose (3,6-dideoxy-L-arabinohexose), and the alcohols are a series of secondary monols and diols containing 22 to 37 carbon atoms.[59] Ascarosides render the eggshell virtually impermeable to substances other than gases and lipid solvents; chapter 26 further explores resistance of *Ascaris* spp. eggs. We do not know whether ascarosides are present in lipid layers of nematode

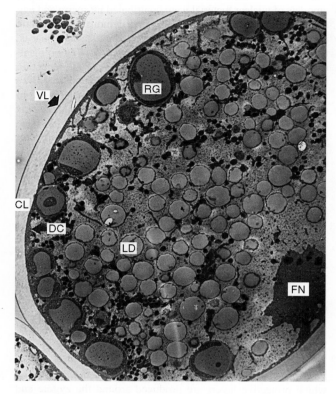

Figure 22.32 **Low magnification electron micrograph of a newly fertilized egg.**
Note vitelline layer (*VL*); incipient chitinous layer (*CL*); dense, particulate, cortical cytoplasm (*DC*); and numerous lipid droplets (*LD*). Female nucleus (*FN*) lies near the surface, and refringent granules (*RG*) have migrated to position just beneath the cortical cytoplasm. After extrusion, contents of refringent granules will become the lipid layer. (×3800)

From W. E. Foor, "Ultrastructural aspects of oocyte development and shell formation in *Ascaris lumbricoides*," in *J. Parasitol.* 53:1245–1261. Copyright © 1967.

eggs other than ascaridoids. Some water can pass across the lipid layer of *A. suum* and at least some other nematodes, but embryos can continue to develop despite water stress.[118, 131]

In oxyuridomorphs, eggshell formation is similar to that just described, but there are two uterine layers, and the lipid layer in some (*Syphacia* spp., for example) is thin and of dubious protective value.[129, 130] Oxyuridomorphs and some other nematodes also have an operculum, which is a specialized area on the egg that facilitates hatching of juveniles.[120] Trichurid eggs have an opercular plug at each end.

Embryogenesis

The most detailed information on nematode development and embryogenesis comes from studies of *C. elegans*.[55] However, early embryonic development and the cell lineage have now been characterized for many species, including several parasites. Certain variable aspects of early development appear consistent with clades in molecular phylogenetic trees, such as specification of the axis polarity of the embryo.[49] Although post-embryonic development is more readily studied in small, transparent nematodes like *C. elegans*, embryonic development has a long history of study within parasites. *Parascaris equorum* was the focus of many classical investigations in the early 1900s, and the cell lineage nomenclature developed for *P. equorum* formed the basis for subsequent work in other species.

The determinate cleavage of nematode embryos has been considered the clearest and best documented example of germinal lineage in the animal kingdom,[34] however, recent studies reveal that not all development is determinate, but that consistent adult form can result by reproducible intercellular communication and indeterminate development. Because of early determination of the fate of each cell (blastomere) in a cleaving embryo, names or letter designations can be given to each blastomere, and tissues that will develop from each are known (Fig. 22.33). At the first cleavage, a zygote produces one cell that will give rise to somatic tissues and one cell whose progeny will comprise more somatic cells and the germinal cells. Early cleavages of some nematodes (in ascaridoids, but also some copepods, ciliates, and insects) are marked by a very curious phenomenon called **chromatin diminution.** The chromosomes

Figure 22.33 **Cell lineage of nematodes.**
The two cells produced at the first cleavage of the zygote are P_1 and S_1. The diagram indicates the progeny of these cells and the tissues to which they give rise.

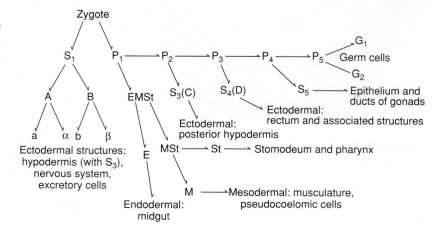

fragment, and only the middle portions are retained, the heterochromatic ends being extruded to the cytoplasm to degenerate.[88] Oddly, during this diminution there is an approximate doubling of the histone:DNA ratio. The explanation for this observation is unclear; the distance between the nucleosomes does not decrease, nor is there a detectable free histone pool.[36]

Since chromatin diminution occurs only in somatic cells, the germ line can be recognized by its full chromatin complement. The only cells left with complete chromosomes at completion of embryogenesis are G_1 and G_2, which will give rise to gonads. Interestingly, further nuclear differentiation in the various tissues seems to go in both directions with respect to chromatin content. Some, such as muscle and ganglia nuclei, further diminish whereas others, particularly those with high levels of protein synthetic activity such as excretory and pharyngeal glands and uterine cells, exhibit polyploidy with respect to DNA content.[122] Clearly, there must be a great redundancy of the genes left after chromatin diminution. Biological and evolutionary significance of chromatin diminution remain unclear, but the phenomenon may have arisen as a gene-silencing mechanism in somatic cells.[88]

Rather typical morula and blastula stages are formed. Gastrulation is by invagination and also by epiboly (movement of the micromeres down over macromeres).

Studies of embryonic and post-embryonic development in *C. elegans* have led to the common belief that individuals of a nematode species have the same number of cells, or **eutely.** Even in *C. elegans,* some cell lineages show variation in cell number, such as the intestine. Individual nematode species commonly vary in epidermal cell number, and this variation appears to increase with cell lineage complexity.[35] Further investigation is required to determine if certain tissues have cell number constancy, but clearly some post-embryonic growth results from cell enlargement or increase in the mass of syncytia rather than cell divison.

Timing, site, and physical requirements for embryogenesis vary greatly among species. In some, cleavage will not begin until an egg reaches the external environment and oxygen is available. Others begin (or even complete) embryogenesis before an egg passes from its host, whereas, in some, juveniles complete development and hatch within the female nematode (ovoviviparity).

Embryonic Metabolism

Embryonation of *A. suum* eggs demonstrates a most fascinating sequence of biochemical epigenetic adaptation: adaptive appearance and disappearance of biochemical pathways through ontogeny, based on gene regulation.[42]

Energy metabolism of adult *A. suum* is anaerobic, but that of its embryonating egg is obligately aerobic. Dependence on pathways such as glycolysis would not only be wasteful of the limited stored nutrient in the embryos, but it would also soon result in a toxic concentration of acidic end products because the eggshell is not permeable to such compounds. Eggs survive temporary anaerobiosis, but they do not develop unless oxygen is present. They are completely embryonated and infective after 20 days at 30°C, and throughout this time a Krebs cycle and

cytochrome *c*-cytochrome oxidase electron transport system are active.

As in many parasitic nematodes, the infective stage of *A. suum* is the third-stage juvenile, the worms having undergone two molts in the egg.[39] Eggs hatch in the intestine of the pig host. Then juveniles go through a tissue migration. They emerge into lung alveoli, travel up the trachea, and then after being swallowed gain access to the intestine, where they become adults (see Fig. 26.5). Cytochrome oxidase is still present in juveniles recovered from lungs, and they require oxygen for motility. Oxidase activity disappears from fourth-stage juveniles in the intestine and remains repressed through adult life.

A similar phenomenon happens to enzymes of the glyoxylate cycle.[8] *Ascaris suum* embryos consume both lipid and carbohydrate reserves during the first 10 days of development and then resynthesize carbohydrate (glycogen and trehalose) from fat.[95] Other nematodes, including the free-living model organism *C. elegans* use the glyoxylate cycle in the first juvenile stage(s) and then shift to aerobic respiration in subsequent juvenile stages. Finally, all activity of the two critical enzymes in the cycle (isocitrate lyase and malate synthase) seems to be repressed in muscle of *A. suum* adults.

Hatching

In nematodes whose juveniles are free-living before becoming parasitic, hatching occurs spontaneously.[96] Some plant-parasitic species hatching is induced in the presence of substances from their prospective hosts.[67] Eggs of many species parasitic in animals, however, will hatch only after being swallowed by a prospective host. On reaching the infective stage, such eggs remain dormant until the proper stimulus is applied, and this requirement has the obvious adaptive value of preventing premature hatching. Ascarid eggs require a combination of conditions: a temperature of about 37°C, a moderately low oxidation-reduction potential (presence of an oxidizing agent reversibly inhibits hatching[57]), a high carbon dioxide concentration, and a pH of about 7. These conditions are present in the gut of many warm-blooded vertebrates, and indeed *A. suum* will hatch in a wide variety of mammals and even in some birds, but all four conditions are unlikely to be present simultaneously in the external environment.

The lipid layer of the egg is impermeable to water, but fluid around a juvenile has a trehalose concentration of about 0.2 M, resulting in a high osmotic pressure surrounding the juvenile.[33] The first change detectable on application of hatching stimuli is a rapid change in permeability; trehalose from the perivitelline fluid leaks from the eggs. Increase in water concentration apparently activates juveniles. The lipid layer also becomes permeable to enzymes secreted by the juvenile, such as chitinase, esterases, and proteinases, and these enzymes attack the hard shell, digesting it sufficiently for the worm to force a hole in it and escape.[40]

First-stage juveniles of some nematodes, such as *Trichuris* spp., possess a stylet or spear on their anterior end. When juveniles are activated by a hatching stimulus, they penetrate the operculum (polar plug) with their stylet and emerge from the eggshell.[94]

Growth and Ecdysis

Molting

There is growth in body dimensions of nematodes between molts of their cuticle (Fig. 22.34). After the fourth molt in large nematodes such as *A. suum,* there is considerable increase in size, and the cuticle itself continues to grow after the last ecdysis. The molting process has been studied in several species. First the epidermis detaches from the basement membrane of the old cuticle and starts to secrete a new one, beginning with the cortical zone. This process may continue until the new cuticle is substantially folded under the old cuticle, to be stretched out later after ecdysis. In some cases old cuticle up to the cortical zone dissolves and new cuticle absorbs the resulting solutes. This conservation of resources is particularly important when materials and space are limited, such as in the first molt of *A. suum,* but less so when there is plenty of food and the old cuticle is very complex in structure, as in the fourth molt of *Nippostrongylus brasiliensis.*[70] Escape from the old cuticle is facilitated by several enzymes, such as a collagenase-like enzyme that attacks it.[105]

Developmental Arrest

A common adaptation in many nematodes is a resting stage at one or more points in their development (**developmental arrest** or **hypobiosis**), enabling them to survive adverse conditions while awaiting return of more congenial circumstances. A great deal has been learned about genetic control of development in nematodes using a free-living species, *Caenorhabditis elegans.*[23, 29] This species can undergo developmental arrest as third-stage (J_3, **dauer juveniles**). Specific neurons in their amphids sense environmental signals, such as ambient temperature, food supply; these signals interact with a *C. elegans* pheromone that regulates dauer arrest in an insulin-like signaling pathway that involves nuclear receptor genes.[38] All the *C. elegans* dauer signaling pathways also occur in parasitic *Strongyloides stercoralis.*[82] Depending on conditions, *S. stercoralis* may have free-living or parasitic adults (see chapter 24).

Numerous examples in which developmental arrest is of survival value among free-living nematodes could be cited. A particularly interesting one was described by Hominick and Aston.[54] Dauer juveniles of *Pelodera strongyloides* attach to mice, upon which they enter hair follicles on abdominal skin and molt to fourth-stage juveniles. They will develop no further at the body temperature of the mouse, and a mouse may accumulate hundreds or even thousands of nematode juveniles during its life. When the mouse dies and its body cools, nematodes rapidly emerge and, in the presence of a food source, molt to the adult stage. The mouse seems little inconvenienced by its passengers.

Many parasitic nematodes produce infective third-stage juveniles comparable to dauer juveniles. They develop no further until a new host is available, some remaining ensheathed in their second-stage cuticle. They live on stored food reserves and usually exhibit behavior patterns that enhance the likelihood of reaching a new host. For example, third-stage juveniles of *Haemonchus* and *Trichostrongylus* species migrate out of a fecal mass and onto vegetation that

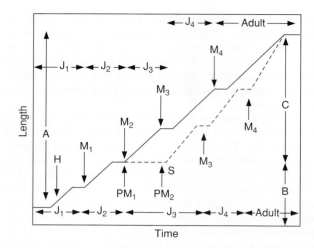

Figure 22.34   Idealized form of the basic life cycle of nematodes.

The life cycle of a free-living nematode is represented by a solid line. Hatching (*H*) is "spontaneous," and there are four molts (M_1–M_4). The broken line represents a life cycle in which a change in environment is necessary to stimulate (*S*) the completion of the second molt (PM_2). (*A–C*) are different environments. (J_1–J_4 are the juvenile stages.)

From W. P. Rogers and R. I. Sommerville, "The infective stage of nematode parasites and its significance in parasitism" in *Advances in Parasitology,* Vol. 1, edited by B. Dawes, 1963, page 112. Copyright © 1963 by permission of Academic Press.

is eaten by the host. Third-stage juveniles of species that penetrate host skin, such as hookworms and *Nippostrongylus brasiliensis,* migrate onto small objects (sand grains, leaves, and others) and move their anterior ends freely back and forth, in the same manner as some dauer juveniles (Fig. 22.35).

In both dauer juveniles and infective juveniles, a more or less specific stimulus is required for resumption of development and completion of ecdysis of the second-stage cuticle. Those that penetrate skin usually exsheath in the processes of penetration, but stimuli for exsheathment of swallowed juveniles (*Haemonchus* spp., *Trichostrongylus* spp., and others) are very similar to those required for hatching of *A. suum* eggs, including carbon dioxide, temperature, redox potential, and pH. In terms of developmental function, infective eggs (shelled juveniles) are fundamentally the same as dauer juveniles and infective juveniles.[106] For most nematodes tested, carbon dioxide seems to be the most important stimulus for hatching or exsheathing.[99]

Nematodes with intermediate hosts normally undergo hypobiosis at the third stage and remain dormant until they reach a definitive host. Some species are astonishingly plastic in their capacities to sustain more than one developmental arrest if necessary. For example, if some species of hookworms and ascarids infect an unsuitable host, they enter another developmental arrest and lie dormant in host tissues until they receive another stimulus to migrate.[85] In several of these, an older animal of the proper species is treated as an unsuitable host, and the worms lie dormant until they are stimulated by hormones of host pregnancy. They then

Figure 22.35   **Third-stage infective juveniles of**
Nippostrongylus brasiliensis.

This illustrates the typical behavior of crawling up on pebbles,
blades of grass, or the like and waving anterior ends to and fro.
In this photograph of living worms, juveniles have mounted
granules of charcoal and even each other. This waving behavior
is termed "nictation" or "questing."

Photograph by Larry S. Roberts.

migrate to the uterus or mammary glands and infect the progeny by way of the placenta in utero or the milk after birth.[30]
Some species, such as *Strongyloides ratti,* may not undergo
a second developmental arrest at this stage, but if a lactating
female is infected, the juveniles are somehow diverted from
completing their migration to the adult's intestine and migrate instead to the mammary glands and infect the suckling
young.[133]

Females of many species of filaroids are parasitic in tissues of birds and mammals. Rather than eggs, they release
relatively undeveloped J_1s, termed *microfilariae*. Microfilariae do not develop further unless consumed by their invertebrate (commonly an insect) intermediate host, in which they
undergo two molts and enter developmental arrest at J_3. That
arrest is broken when a definitive host, commonly a bird or
mammal, is infected.

Thus in many species initiation or cessation of developmental arrest is triggered by significant ambient temperature
change, either up or down. Regulation of genes and signaling pathways involved are complex and beyond the scope
of this book, but in several cases initiation of such pathways
includes *heat shock factor* (HSF) and one or more *heat shock
proteins* (HSPs).[38]

METABOLISM

Energy Metabolism

Adult Nematodes

Probably more is known about nematode metabolism
than about metabolism of any other group of parasitic helminths,[27, 113] but most of what we know has been derived
from studies on *Ascaris suum*. This species was one of the
first organisms in which cytochrome was demonstrated.[62]
Nevertheless, many questions await resolution.

The overall scheme of energy metabolism of adult *A. suum*—
and other such nematode parasites of animals that have been
examined—is basically similar to that of adult trematodes
and cestodes (see Figs. 15.27 and 20.28). Many nematode
parasites that dwell in the intestinal lumen live in the presence of reduced quantities of oxygen, and they do not completely oxidize all of their nutrient molecules. They degrade
glucose to phosphoenolpyruvate and fix CO_2 to form oxaloacetate, which is reduced to malate, and the malate enters the
mitochondrion to undergo further reactions, however, these
mitochondrial pathways are quite distinct from the normal
TCA cycle. Two additional ATP are produced from each
glucose molecule in the mitochondrial reactions, which is
far less than what is produced in organisms with oxidative
phosphorylation. Reduced acid end products include lactate,
acetate, and succinate. Some succinate is decarboxylated to
form propionate. Further reactions may occur in cytoplasm
to produce a variety of other unusual end products, such as
α-methylbutyrate and α-methylvalerate (Fig. 22.36).

As with trematodes and cestodes, the adaptive value of
this seemingly wasteful scheme is not at all obvious. We can
speculate that it may involve some as yet unclear aspects of
the host-parasite relationship.[27]

Other electron transport reactions are present in *Ascaris*
spp., but their importance and sequence are still not known
with certainty. Oxygen in moderate concentrations is toxic
to this nematode. Succinate and malate are oxidized in mitochondrial preparations with hydrogen peroxide as an end
product, and toxicity of hydrogen peroxide probably accounts
for the nematode's intolerance of oxygen. In the absence of
classical catalase activity, how the worms can avoid hydrogen

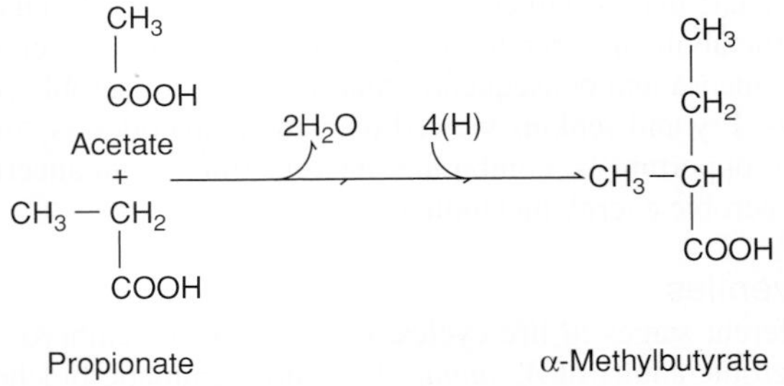

$$CH_3 \qquad\qquad\qquad\qquad CH_3$$
$$|\qquad\qquad\qquad\qquad\qquad\qquad |$$
$$COOH \qquad\qquad\qquad\qquad CH_2$$
$$Acetate \quad\; 2H_2O \quad\; 4(H) \qquad |$$
$$+ \qquad\qquad\qquad\qquad\qquad\longrightarrow CH_3-CH$$
$$CH_3-CH_2 \qquad\qquad\qquad\qquad |$$
$$|\qquad\qquad\qquad\qquad\qquad\qquad COOH$$
$$COOH$$

Propionate   α-Methylbutyrate

Figure 22.36   **Mode of methylbutyrate formation in**
Ascaris **muscle.**

Methylvalerate is formed in a similar manner except that the carbon of one propionate unit condenses with the carboxyl carbon
of another propionate unit.

peroxide poisoning under normal conditions in their host gut is a fascinating question. This vital function apparently is mediated by a form of hemoglobin in *Ascaris* hemolymph. *Ascaris* hemoglobin binds oxygen some 25,000 times more avidly than does human hemoglobin, so the worm's hemoglobin cannot function in oxygen transport. Evidence now suggests that *Ascaris* hemoglobin acts as a pseudoperoxidase, detoxifying hydrogen peroxide enzymatically using nitric oxide as a co-substrate.[6, 86] Hemoglobins have been known from nematodes for over 100 years, and they well may have other functions in nematodes other than *Ascaris* spp.[15]

Other nematodes have been studied that survive and metabolize carbohydrates in the absence of oxygen for extended periods; examples are *Heterakis gallinarum* and *Trichuris vulpis*. These organisms serve as particularly good examples of how some parasites have solved the metabolic problem of reoxidizing NADH (chapter 4) in the absence of oxygen as a terminal electron acceptor. Some adults (*Haemonchus contortus*) apparently have a tricarboxylic acid cycle that operates when oxygen is present and an *A. suum* type of system operative in the absence of oxygen.[128] *Rhabdias bufonis*, a parasite in the lungs of frogs, also apparently has alternative systems,[2] despite the unlikelihood of its being subjected to anaerobiosis.

Nevertheless, other adult nematodes seem to be obligate aerobes with respect to their energy metabolism, requiring presence of at least low concentrations of oxygen for survival and motility. Even so, glucose is not oxidized completely to carbon dioxide and water, and substantial quantities of various reduced end products are excreted. Some species apparently have a classical cytochrome system, or the electrons may be transported by a flavoprotein and terminal flavin oxidase to oxygen, producing hydrogen peroxide.

Nippostrongylus brasiliensis and *Litosomoides carinii* can survive short periods of anaerobiosis but are killed by longer periods (a few hours).[104] *Nippostrongylus brasiliensis* has a fluid-filled layer in its cuticle (see Fig. 22.6) that contains hemoglobin, and the hemoglobin loads and unloads oxygen in the living animal.[115] Thus, the worm can exploit areas in the intestine that are quite hypoxic.[112] *Brugia pahangi* has an active Krebs cycle, but the contribution of aerobic metabolism to its energy metabolism appears limited.[7]

Some evidence suggests that the extent of dependence on aerobic pathways in nematodes is correlated with body diameter; that is, larger nematodes have a relatively anaerobic metabolism, whereas smaller ones can get to oxygen near the mucosa and consequently tend to use more aerobic pathways. Fry and Jenkins viewed parasitic nematodes as "metabolic opportunists, combining the versatility of an anaerobic and aerobic energy metabolism."[46]

Juveniles

Different stages in life cycles, such as aerobic embryos and anaerobic adults of *A. suum*, often show dramatic biochemical adaptations in energy metabolism. Developing juveniles change over to anaerobic metabolism during the J_3 stage.[64] Although metabolism of the *A. suum* testis/seminal vesicle is anaerobic, *A. suum* mitochondria have elevated levels of Krebs cycle enzymes, possibly for use later during embryonic development.[63] Juveniles of *Trichinella spiralis* in muscles of their host have more aerobic metabolism than do

adult *A. suum*, but some of their energy is apparently generated anaerobically.[19]

Strongyloides spp. are another interesting example of biochemical epigenetic adaptation.[65] *Strongyloides* spp. have a complex life cycle with free-living adults (males and females) and parasitic adults (parthenogenetic females only). The first three juvenile stages of both types are free living, but those destined to become parasitic undergo developmental arrest at the third stage until penetration of a host occurs (chapter 24). All free-living stages are subjected to a selective pressure common to other free-living animals—to derive as much energetic value from their nutrient molecules as they can—and they have a complete Krebs cycle and probably a cytochrome system. In contrast, parasitic females have neither a complete Krebs cycle nor a cytochrome system, as in many other intestinal helminths. Juveniles of several other parasitic species have apparently functional Krebs cycles,[20, 127] although in some species the significance of the cycle may lie in regulation of the supply of fourcarbon intermediates rather than in energy production. Oddly, the first- and second-stage juveniles of *Ancylostoma tubaeforme* and *Haemonchus contortus* are apparently anaerobic, and the infective third stage is aerobic.[93]

The normal pathway of fatty acid oxidation is called β-*oxidation* because the β-carbon of the fatty acetyl-CoA is oxidized, and the two-carbon fragment, acetyl-CoA, is cleaved off to enter the Krebs cycle. It would be expected, therefore, that the presence of β-oxidation enzymes would be correlated with a functional Krebs cycle (though not always), and in the few cases investigated this was confirmed. β-oxidation of fatty acids has been found in developing *Ascaris suum* embryos and in free-living *Strongyloides ratti* juveniles and adults.[65, 126] Tissue lipids gradually disappear from infective eggs or juveniles of several species. Interestingly, neither muscle from adult *A. suum* nor parasitic females of *Strongyloides ratti* can carry out β-oxidation.

Synthetic Metabolism

Proteins and Nucleic Acids

Synthetic metabolism of nematodes has not been as intensively studied as has energy metabolism, probably because energy pathways of helminths, in contrast to those of prokaryotes, usually offer the better sites for chemotherapeutic attack. However, there are several points of interest. In light of the enormous number of progeny produced by organisms such as *Ascaris* spp., protein and nucleic acid synthetic ability must be correspondingly great. In this connection RNA metabolism of fertilized *A. suum* eggs deserves further comment (see earlier discussion of embryogenesis). Young oocytes have nucleoli and large amounts of cytoplasmic RNA, and these presumably are responsible for the very large amount of yolk protein synthesized in a developing oocyte. By the time an oocyte matures, the nucleoli and most of the cytoplasmic RNA have disappeared.[61] At the same time, sperm contain little or no RNA. Immediately after fertilization there is a massive ribosomal RNA synthesis in male pronuclei, along with a smaller amount of messenger RNA, while female pronuclei are going through their maturation divisions. Kaulenas and Fairbairn suggested, therefore, that the female genome is responsible for the high rate of oocyte production and yolk synthesis, whereas ribosomes provided by the male genome largely support shell formation and cleavage.[61]

Presumably, much of the amino acid supply for protein synthesis in oocytes is furnished by intestinal absorption nearby, but some amino acids are synthesized in the ovaries as well. The ovaries contain active transaminases, which form amino acids from corresponding α-keto acids derived from carbohydrate metabolism. In addition, ovaries can condense pyruvate with ammonia to form the amino acid alanine. Some free-living nematodes can synthesize a wide variety of amino acids from a simple substrate, such as acetate. When incubated in a medium containing glycine, glucose, and acetate, *Caenorhabditis briggsae* synthesizes an array of "nonessential" and "essential" amino acids; this species was the first metazoan known that could synthesize "essential" amino acids.[109]

As noted in the discussion of the body wall, much collagen is found in cuticle of *Ascaris* spp. Muscle, intestine, and reproductive organs also contain collagens. Such stabilized proteins are important factors in resistance and strength of a nematode's cuticle. Collagens are stabilized by bonds between lysine residues in subunits, and they are unusual in that they contain around 12% proline and 9% hydroxyproline; hydroxyproline is an amino acid rarely found in other proteins. Normal collagen precursor is a polypeptide called *protocollagen,* and proline in protocollagen is hydroxylated to hydroxyproline by protocollagen proline hydroxylase (PPH, or proline monoxygenase). Among other cosubstrates, this enzyme requires molecular oxygen to carry out hydroxylation of proline. Here is an example of a biosynthetic reaction that requires oxygen in an organism that is anaerobic with respect to its energy metabolism. Oxygen concentration greater than 5% even inhibits PPH from *A. suum* muscle but not the enzyme from its embryos.[31]

Lipids

At least some nematodes can synthesize polyunsaturated fatty acids de novo but apparently are unable to synthesize sterols de novo.[108] *Ascaris suum* incorporates acetate into long-chain fatty acids, probably by the malonyl-CoA pathway as found in vertebrates.[11] Nonsugar parts of ascarosides (the alcohols) in *Ascaris* spp. are synthesized from long-chain fatty acids. These reactions involve a condensation in which the carboxyl carbon of one fatty acid condenses with carbon number 2 of another, with the elimination of a molecule of carbon dioxide.[41] Ascarylose is freely synthesized by *A. suum* ovaries from glucose or glucose-1-phosphate, and the end product of the synthesis is probably ascarylose-dinucleotide phosphate, which then condenses with the nonsugar moiety to give the ascaroside. *Dirofilaria immitis* can synthesize all classes of complex lipids, including cholesterol.[125]

CLASSIFICATION OF PHYLUM NEMATODA

For higher classification of Nematoda that follows, we are using the system proposed by Blaxter and by De Ley and Blaxter, with our own minor modifications based on more recently published molecular phylogenetic hypotheses for nematode parasites.[16, 37, 66, 89] The original system is based on phylogenetic analysis of small subunit ribosomal RNA genes from more than 300 species. To fit this classification into the traditional ranks of taxa, De Ley and Blaxter found it necessary to "downgrade" the ranks of several groups previously regarded as orders. Similarly, De Ley and Blaxter reduced in rank many superfamilies to families, however, this usage has rarely been adopted in taxonomic papers, and the older ranks and usage (superfamilies) are mainly retained in the chapters that follow. Figure 22.37 includes free-living nematodes as well as plant and animal parasites. In the remarks to follow, we will confine ourselves to taxa containing animal parasites. As shown in the figure, evidence indicates that nematode parasitism of animals evolved at least eight separate times and parasitism of plants at least three times. This illustrates that free-living nematodes are preadapted to parasitism.

Order Muspiceida does not appear in Fig. 2 of Blaxter[16] (or here in Fig. 22.37), but De Ley and Blaxter[37] considered it an order of subclass Dorylaimia, and we are thus inserting it in that position on p. 375.

Class Chromadorea

Subclass Chromadoria

Orders Plectida, Araeolaimida, Monhysterida, Desmodorida, Chromadorida
No known parasites in these groups.

Order Rhabditida (= Phasmidea; = Secernentea)

Amphids generally poorly developed, with small, simple pores near or on the lips; caudal and hypodermal glands absent; phasmids present; excretory system with one or two lateral canals, with or without associated glandular cells; deirids commonly present; free living in soil or freshwater or parasitic in plants or animals.

Suborder Rhabditina

Infraorder Bunonematomorpha

Free living nematodes.

Infraorder Diplogasteromorpha

Free living and insect associates; few insect parasites.

Infraorder Rhabditomorpha

De Ley and Blaxter[35] reallocated some long-established orders to lower ranks (superfamilies) on the basis of relationships shown by sequence analysis.

Superfamily Mesorhabditoidea

Families Mesorhabditidae, Peloderidae. Insect associates.

Superfamily Strongyloidea

Commonly long, slender worms. Esophagus usually swollen posteriorly but lacking definite bulb. Male with well-developed copulatory bursa supported by sensory rays. Ovijector complex, with well-developed sphincter. Excretory system with *H*-shaped tubular arrangement and two subventral glands. First-, second-, and beginning of third-stage juveniles free living or parasitic in invertebrates. Usually oviparous. Eggs thick shelled, rarely developed beyond morula when laid. Parasites of all classes of vertebrates (rare in fishes). Families Diaphanocephalidae, Ancylostomatidae, Uncinariidae,

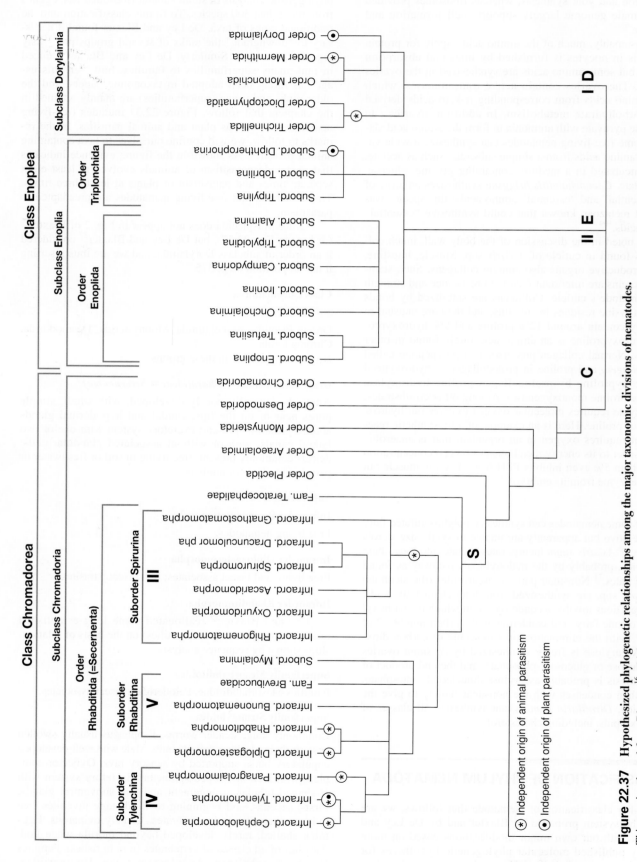

Figure 22.37 **Hypothesized phylogenetic relationships among the major taxonomic divisions of nematodes.**
This tree is adapted from Blaxter,[16] which was derived from analysis of over 300 nematode small-subunit rRNA gene sequences, complemented by consideration of morphological characters. Major clades of nematodes: C (Chromadoria, called C&S in Blaxter et al.[18], E (Enoplia, Clade I of Blaxter et al.[18], **D** (Dorylaimia, Clade II, and clades III (Spirurina), IV (Tylenchina), and V (Rhabditina).

Globocephalidae, Strongylidae, Cloacinidae, Syngamidae, Trichostrongylidae, Amidostomatidae, Strongylacanthidae, Heligmosomidae, Ollulanidae, Dictyocaulidae, Metastrongylidae, Angiostrongylidae, Heterorhabditidae.

Suborder Tylenchina

Lips variable but almost always hexaradiate. Amphids usually pore-like and located on the lips. Hollow stylet in esophagus, not evident in some adult insect parasites and some male plant parasites. Orifice of dorsal esophageal gland in the procorpus, usually near base of stylet. Female with one or two ovaries. Males never with more than one pair of phasmids. Caudal alae present, may be reduced. Free living or parasitic in plants, insects, molluscs, annelids, or vertebrates.

Infraorder Panagrolaimomorpha

Free living and parasitic in insects, and vertebrates.

Superfamily Strongyloidoidea

Two ovaries. Zooparasitic. Discrete dauer stage in life cycle. Families Strongyloididae, Rhabdiasidae, Steinernematidae.

Infraorder Cephalobomorpha

Mainly free living, but some parasites of molluscs or annelids.

Infraorder Tylenchomorpha

Parasitic in plants and insects.

Suborder Myolaimina: All known species free living.

Suborder Spirurina

Infraorder Ascaridomorpha

Usually with three prominent lips. Buccal capsule weakly cuticularized and surrounded by esophageal tissue. Esophagus consisting of corpus and posterior ventriculus that may be muscular or glandular. Eggs thick shelled. Life cycle direct or indirect with invertebrate or vertebrate intermediate hosts. Families Ascarididae, Anisakidae, Cosmocercidae, Atractidae, Kathlaniidae, Heterakidae, Aspidoderidae, Ascaridiidae, Cucullanidae, Quimperiidae, Chitwoodchabaudiidae, Schneidernematidae, Raphidascarididae, Subuluridae, Maupasinidae, Heterocheilidae.

Infraorder Spiruromorpha

Mouth surrounded by six lips, lips lost, or lateral pseudolabia present. Buccal capsule often well developed. Esophagus divided into anterior muscular and posterior glandular portion, never with bulb. Development from first- to third-stage juvenile in arthropod intermediate host. Parasites in intestine and deeper tissues of all vertebrate classes. Families Physalopteridae, Thelaziidae, Rhabdochonidae, Pneumospiruridae, Gongylonematidae, Spiruridae, Spirocercidae, Hatertiidae, Hedruridae, Habronematidae, Tetrameridae, Cystidocolidae, Acuariidae, Filariidae, Onchocercidae, Aproctidae, Desmidocercidae, Diplotriaenidae, Oswaldofiliariidae.

Infraorder Rhigonematimorpha

Complex cuticular modifications present at base of buccal cavity. Vagina long and muscular. Parasites in posterior gut of Diplopoda. Families Rhigonematidae, Ichthyocephalidae, Ransomnematidae, Carnoyidae, Hethidae.

Infraorder Oxyuridomorpha

Medium to small worms often with sharply pointed tails. Esophagus with prominent posterior bulb with valve. Excretory system X-shaped with prominent sinus and vesiculateduct. Males with single (or no) spicule and reduced number of caudal papillae. Sperm comet shaped. Eggs often flattened on one side. Haplodiploid (male haploid derived from unfertilized egg, female diploid derived from fertilized egg). Parasites of colon or rectum of Arthropoda (rarely annelids) and vertebrates; life cycles direct. Families Thelastomatidae, Travassosinematidae, Hystrignathidae, Protrelloididae, Oxyuridae, Pharyngodonidae, Heteroxynematidae.

Infraorder Gnathostomatomorpha

Head with two large lateral lips. Anterior end swollen, separated from rest of body by a constriction. Four glandular cervical sacs hang into pseudocoel from attachments near anterior end of esophagus. Head bulb divided internally into four hollow areas. Parasites of reptiles, elasmobranchs, fish, and mammals. Families Gnathostomatidae, Anguillicolidae, Seuratidae.

Infraorder Dracunculomorpha

Buccal capsule reduced. Anterior muscular and posterior glandular portions of esophagus not separated into distinct compartments. Female often highly enlarged, filled with first-stage juveniles. Parasites of various tissue sites of vertebrates (mostly fish). Families Camallanidae, Dracunculidae, Philometridae, Phlyctainophoridae, Skrjabillanidae.

CLASS ENOPLEA

Subclass Enoplia

Orders Enoplida, Triplonchida. Free living nematodes.

Subclass Dorylaimia

Order Trichinellida

Anterior end more slender than posterior. Lips and buccal capsule absent or much reduced. Esophagus a very slender capillary-like tube, embedded within one or more rows of large, glandular cells (stichocytes) along posterior portion. Bacillary band present. Both sexes with a single gonad. Males with one spicule or none. Eggs with polar plugs (opercula) except in *Trichinella* spp. Histiotrophic parasites of nearly all organs of all classes of vertebrates. Families Anatrichosomatidae, Capillariidae, Cytoopsidae, Trichinellidae, Trichosomoididae, Trichuridae.

Order Dioctophymatida

Stout worms, often very large. Esophageal glands highly developed, multinucleate. Lips and buccal capsule reduced, replaced by muscular oral sucker, Soboliphymatidae. Esophagus cylindrical. Nerve ring far anterior. Anus at posterior end in both sexes. Male with bell-shaped muscular copulatory bursa without rays. Boths sexes with a single gonad. Males with a single spicule. Eggs deeply sculptured or pitted. Histiotrophic parasites of birds and mammals. Families Dioctophymatidae. Eustrongylidae, Soboliphymatidae.

Order Muspiceida

Alimentary tract reduced or vestigial. Male unknown. Parasites of skin and deeper tissues of rodents, deer, bats, marsupials, and crows. Families Muspiceidae, Robertdollfusiidae.

Orders Mononchida, Dorylaimida. Free living and plant parasitic nematodes.

Order Mermithidia

Six or eight hypodermal cords visible in cross section. Juvenile stages parasitic in body cavity of various invertebrates. Families Mermithidae, Marimeremithidae, Echinodermellidae, Tetradonematidae.

Learning Outcomes

By the time a student has finished studying this chapter, he or she should be able to:

1. List structural differences and similarities between nematodes and trematodes.
2. Describe how the high internal pressure in the pseudocoelom of nematodes impacts on their biology.
3. Differentiate among the different sensilla, or sensory structures found in nematodes.
4. Explain differences between the metabolism of *Ascaris suum* adults and their vertebrate host.
5. List steps in the general development or life cycle of a nematode, from egg to adult.
6. Describe the biological functions of different nematode structures, organs, and systems relative to movement, feeding, and reproduction.

References

References for superscripts in the text can be found at the following Internet site: www.mhhe.com/robertsjanovynadler9e

Additional Readings

Anderson, R. C., 1993. *Nematode parasites of vertebrates. Their transmission and development.* Wallingford, Oxon, England: CAB International. A complete compilation of what is known about nematode life cycles in vertebrates.

Anderson, R. C., A. G. Chabaud, and S. Willmott. 1974–1985. *CIH keys to the nematode parasites of vertebrates.* Bucks, England: Commonwealth Agricultural Bureaux, Farnham Royal. A 10-part up-to-date series of keys for the identification of nematode parasites.

Barrett, J. 1976. Bioenergetics in helminths. In H. Van den Bossche (Ed.), *Biochemistry of parasites and host-parasite relationships.* Amsterdam: Elsevier/North Holland Biomedical Press.

Bryant, C. 1978. The regulation of respiratory metabolism in parasitic helminths. In W. H. R. Lumsden, R. Muller, and J. R. Baker (Eds.), *Advances in parasitology 16.* New York: Academic Press.

Crofton, H. D. 1996. *Nematodes.* London: Hutchinson University Library. A very useful summary of nematode characteristics.

Croll, N. A., and B. E. Matthews. 1977. *Biology of nematodes.* New York: John Wiley & Sons.

Grassé, P. P. 1965. *Traité de zoologie: Anatomie, systématique, biologie, vol. 4, parts 2 and 3 Némathelminthes (Nématodes-Gordiacés), rotifères-gastrotriches, kinorinques.* Paris: Masson & Cie. An indispensable reference for serious students of nematodes.

Lee, D. L. (Ed.). 2002. *The biology of nematodes.* London: Taylor and Francis.

Lee, D. L., and H. J. Atkinson. 1977. *Physiology of nematodes,* 2d ed. New York: Columbia University Press.

Levine, N. D., 1980. *Nematode parasites of domestic animals and of man,* 2d ed. Minneapolis: Burgess Publishing. An excellent general reference.

Nematodes: Trichinellida and Dioctophymatida, Enoplean Parasites

Trichinella spiralis: the worm that would be virus.

—D. D. Despommier[23]

Some of the most dreaded, disfiguring, and debilitating diseases of humans are caused by nematodes. Of the 13 main neglected tropical diseases of mankind, six are caused by nematodes.[35] In addition, agriculture suffers mightily from attacks by these animals, due to parasitism of both production animals and crops. Nematodes normally parasitic in wild animals can occasionally infect humans and domestic animals, causing mystifying diseases. Furthermore, free-living nematodes may accidentally find their way into a vertebrate and occasionally become a short-lived but pathogenic parasite. Many thousands of nematodes are known to parasitize vertebrates; many still are unknown. A few examples are presented here and in the following chapters, selected for their interest as parasites of humans and as illustrations of parasitism by nematodes.

Although most nematodes remain undescribed and unknown to science, current knowledge suggests that most species are free-living rather than parasitic. However, the two nematode classes appear to have different proportions of parasitic species, with more free-living taxa represented within Enoplea. This chapter is devoted to the two orders of Class Enoplea that are parasites of animals.

ORDER TRICHINELLIDA

The Trichinellida contains, among others, three genera of medical importance: *Trichuris, Capillaria,* and *Trichinella.*

Family Trichuridae

Whipworms, members of family Trichuridae (Gr. *trichos* = a hair + *oura* = the tail), are so called because they are threadlike along most of their body, and then they abruptly become thick at the posterior end, reminiscent of a whip with a handle (Fig. 23.1). The name *Trichocephalus* (Gr. *trichos* = a hair + *kephalē* = head), in widespread use in some countries, was coined when it was realized that the "hair" was the anterior end rather than the tail, but the name *Trichuris* has priority by nomenclatural rules. There are many species in a wide variety of mammalian hosts, and *Trichuris trichiura* is a very important parasite of humans.

Eggs of *T. trichiura* have been found in a glacier mummy more than 5000 years old,[5] and the worm has likely been with us for much longer; it probably coevolved with us as a parasite of our nonhuman ancestors.[12] *Trichuris* sp. eggs recovered from archaeological sites have been confirmed as *T. trichiura* by gene sequence.[60]

Trichuris trichiura

- **Morphology.** *Trichuris trichiura* measures 30 mm to 50 mm long, with males being somewhat smaller than females. The mouth is a simple opening, lacking lips. The buccal cavity is tiny and has a minute spear. The esophagus of *Trichuris* spp. and other Trichinellida is quite different in comparison to those of typical chromadorean species. The esophagus is very long, occupying about two-thirds of the body length, and consists of a thin-walled tube surrounded by large, unicellular glands, or **stichocytes.** The entire structure is referred to as a **stichosome.** The anterior end of the esophagus is somewhat muscular and lacks stichocytes. Both sexes have a single gonad, and the anus is near the tip of the tail. Males have a single spicule that is surrounded by a spiny spicule sheath. The ejaculatory duct joins the intestine anterior to the cloaca. The male tail is typically coiled. In females the vulva is near the junction of esophagus and intestine. The uterus contains many unembryonated, lemon-shaped eggs, each with prominent opercular plugs at each end (Fig. 23.2).

  The ventral surface along the esophageal region bears a wide band of minute pores, leading to underlying glandular and nonglandular cells.[73, 86] This **bacillary band** is typical of the order. Although the function of the cells in the bacillary band is unknown, their ultrastructure

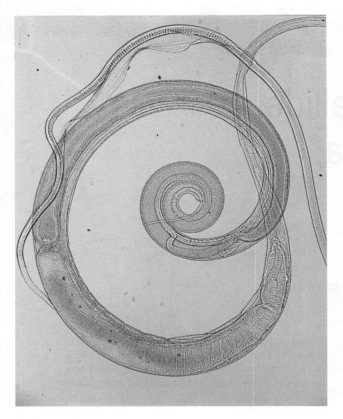

Figure 23.1 Male *Trichuris* sp.

Note the slender anterior end and the stout posterior end with a single, terminal spicule.

Courtesy of Jay Georgi.

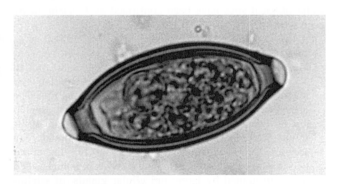

Figure 23.2 Egg of *Trichuris trichiura*.

It measures 50 μm to 54 μm by 22 μm to 23 μm.

Courtesy of Robert E. Kuntz.

suggests that the gland cells may have roles in osmotic regulation and secretion.

- **Biology.** Each female worm produces from 3000 to 20,000 eggs per day.[12] Embryonation is completed in about 21 days in soil, which must be moist and shady. When swallowed, eggs containing infective first-stage juveniles[43] hatch in the small intestine and subsequently enter the crypts of Lieberkühn in the large intestine. After penetration of cells in the base of the crypts, worms begin to grow and tunnel within the epithelium back toward the luminal

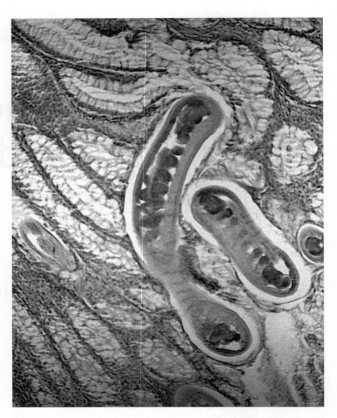

Figure 23.3 Section of large intestine with sections of *Trichuris trichiura* embedded in the mucosa.

Individual stichocytes are evident.

Courtesy of Robert E. Kuntz.

surface. They can penetrate the gut mucosa in many places, but, according to Bundy and Cooper, there is no evidence that worms entering cells other than in the large intestine develop further, nor is there evidence of a duodenal phase and later migration down the gut.[12] As worms approach maturity, the enlarging posterior portion breaks out of the epithelium and protrudes into the intestinal lumen. The prepatent period is about 2.5 months. The slender anterior ends remain embedded in the gut mucosa and induce the formation of a host epithelial syncytium, which appears to be the parasite's food source (Fig. 23.3). Adults live for up to four years, so large numbers may accumulate in a person, even in areas in which the rate of new infection is low.

- **Epidemiology.** The two requirements for *T. trichiura* to become a serious health problem are poor standards of sanitation, in which human feces are deposited on the soil, and conditions favoring egg survival and development: a warm climate, high rainfall and humidity, moisture-retaining soil, and dense shade. Although generally coextensive in distribution with *Ascaris lumbricoides*, *T. trichiura* is more sensitive to desiccation and direct sunlight. Appropriate physical conditions exist in much of the world, including small foci within the southeastern United States, where prevalence of infection may be high in small children. A 1987 survey based on samples submitted to U.S. state public health labs reported an overall prevalence of 1.2%,[40] however, such samples are not expected to be representative for the U.S. population. A recent estimate puts the world prevalence at 795 million.[26] Often both *A. lumbricoides* and *T. trichiura* infections are present concurrently.[10]

Infective eggs are acquired from contaminated soil, and geophagy (eating soil) contributes to transmission in some localities.[12] Use of nightsoil as fertilizer for vegetables can be an important source of infection, and house flies can serve as mechanical vectors of eggs. Chickens and pigs can probably serve as transport hosts.[61] *Trichuris trichiura* can apparently survive and reproduce in dogs,[79] although the role of canids in the epidemiology of human infections remains uncertain.

Infections with *T. trichiura* in human populations are characteristically aggregated.[15] Furthermore, after some individuals are treated for the infection, these same people tend to be predisposed to picking up large numbers of worms, as with *A. suum* (p. 414). Recent studies of human populations indicate that genetic factors explain more of the variation in *T. trichiura* fecal egg counts than household environmental effects.[81, 82] Experimental studies of *Trichuris suis* in pigs of known pedigree also demonstrate a genetic (heritable) component to fecal egg counts and reveals a strong genetic component to infection resistance, consistent with aggregation.[58]

- **Pathology.** Fewer than 100 worms rarely cause clinical symptoms, and the majority of infections are symptomless. A higher infection intensity may result in a variety of conditions, occasionally resulting in severe mucosal hemorrhage that can result in death. Small children are particularly prone to heavy infections, which may involve 200 to more than 1000 worms.[18] In intense trichuriasis, diarrhea, abdominal pain, blood-streaked stools, tenesmus, anemia, and growth retardation are very common, and severe tenesmus may lead to rectal prolapse (Fig. 23.4). Moderate to heavy infections adversely affect cognitive function in children,[59] and the accompanying anemia and malnutrition favors **pica,** which may result in increased infection intensity.

    With their anterior ends buried in mucosa, worms feed on cells and blood, although blood loss by this mechanism is negligible. Trauma to intestinal epithelium and underlying submucosa, however, can cause a chronic hemorrhage that results in anemia.[70] Tissue inflammation is highly localized, and there is a striking absence of normal markers of cell-mediated immunopathology.[19] Nonetheless, there is an increase in macrophages in the colonic lamina propria and increased TNF (p. 25) concentration both in the mucosa and systemic circulation. There is an increase in degranulating mast cells and a 10-fold increase in the proportion of lamina propria cells with surface IgE. This and other evidence suggests that inflammation in *Trichuris* colitis may be considered a local tissue anaphylactic response.[19]

- **Diagnosis and Treatment.** Specific diagnosis depends on demonstrating a worm or eggs in the stool. Worms can be demonstrated dramatically by colonoscopy.[12] Eggs are 50 μm to 56 μm by 22 μm to 23 μm and have smooth outer shells with distinctive bipolar plugs. Their structure and formation have been reported by Preston and Jenkins.[69] Clinical symptoms may be confused with those of hookworm, amebiasis, or acute appendicitis.

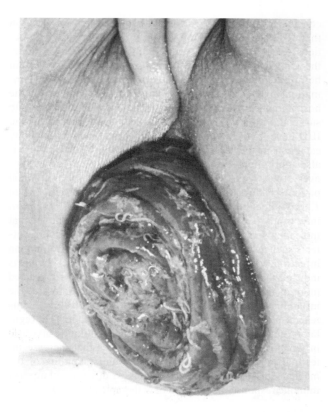

Figure 23.4   Prolapse of rectum caused by whipworm infection.

Courtesy of Herman Zaiman; also, courtesy of University of Miami School of Medicine. From J. W. Beck and J. E. Davies, *Medical parasitology,* 3d. ed. St. Louis, The C. V. Mosby Co., 1981.

Mebendazole and albendazole are often used for treatment, but have lower efficacy against *T. trichiura* than other intestinal nematodes such as *A. lumbricoides*. Combination therapy using either albendazole or mebendazole along with ivermectin has a higher cure rate, and is more promising for control programs.[17, 44] Nitazoxanide is also effective, as is tribendimidine, a new drug developed in China.[28, 74] Training of children and adults in sanitary disposal of feces and in washing of hands is necessary to prevent reinfection.

Other *Trichuris* Species and Therapeutic Human Infections

Some 60 to 70 other species of *Trichuris* have been described from a wide variety of mammals. Several species occur in wild and domestic ruminants, of which *T. ovis* is the most important in domestic sheep, goats, and cattle. *Trichuris suis* is found in swine and although very morphologically similar to *T. trichiura,* is distinct based on ribosomal sequence.[21] *T. vulpis* is found in the cecum of dogs, foxes, and coyotes and is common in the United States except in drier areas. This species has been reported to infect humans.[42] *Trichuris muris* occurs often in rats and mice.

There has been an increased interest in the use of immunomodulatory effects of *Trichuris* sp. infection as a treatment for human inflammatory bowel disease (IBD), including ulcerative colitis and Crohn's disease. One rationale for considering intestinal worms as therapy is the observation that IBD is rare in regions with endemic helminth infections (e.g., Asia, Africa), but common in Northern Europe, the United Kingdom, and North America. Perhaps certain helminth infections protect against the intestinal inflammation in IBD as a result of their down-regulation of the host immune system; the latter is believed to favor parasite survival within the host. Most clinical treatments of human IBD have used *T. suis,* a parasite that does not permanently establish in the large intestine. In randomized clinical trials, approximately half of the patients given *T. suis* eggs every two weeks showed improvement.[77] Some individual patients have shown complete remission from IBD when *T. suis* infection was maintained, or in induced *T. trichiura* infection.[11] However, double-blind trials have shown that individuals ingesting *T. suis* eggs suffer some adverse effects such as diarrhea and abdominal pain.[7] Studies of human infection with *T. trichiura* and the mouse model *T. muris* indicate that following infection, intestinal tissues produced more of the mucosal-healing cytokine IL-22, and less of the inflammatory cytokine IL-17 characteristic of IBD;[11, 29] this response is thought to be related to an increase in TH2 regulatory cells, which are known to release cytokines involved in wound healing and induce cell changes that increase intestinal mucus production. Increased understanding of the mechanisms used by these parasites to down-regulate host immune response may lead to new therapies for IBD.

Family Capillariidae

Members of genus *Capillaria* are small nematodes that share certain structural characteristics with *Trichuris* spp., including a stichosome with stichocytes, and bacillary bands. The body can be subdivided into two regions corresponding to the stichosome versus the portion containing the intestine and reproductive organs, however, in adult *Capillaria* sp. there is no narrowing in diameter like for *Trichuris* spp. A large genus, *Capillaria* includes species that are parasitic in nearly all organs and tissues of all classes of vertebrates.

Capillaria hepatica

- **Biology.** *Capillaria hepatica* is a parasite of the liver, mainly of rodents, but it has been found in more than 140 mammal species, including humans, and in over 50 countries worldwide.[33] Females deposit eggs in liver parenchyma, where they have no means of egress until they are eaten by a predator or until the liver decomposes after death. Eggs cannot embryonate while in the liver, so a new host cannot be infected when it eats an egg-laden liver. The eggs merely pass through the digestive tract of the predator with feces. Embryonation occurs in soil within 30 days, and new infection is by contamination. After hatching in the small intestine, juveniles use the portal vein to reach the liver, where they mature.

- **Epidemiology.** As with *T. trichiura,* infection occurs when contaminated objects or food is ingested. Choe and coworkers, who reported the first case from Korea, believed that geophagy is especially important in transmission.[16] In support of this hypothesis, 60% of diagnosed human cases have been in children less than eight years

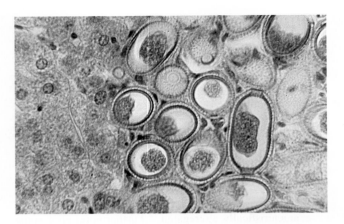

Figure 23.5 Eggs of *Capillaria hepatica* in liver.
Note the extensive damage to hepatic parenchyma.
Courtesy of Warren Buss.

old.[33] Unlike with whipworm, however, human feces are not the source of contamination; more likely, feces of carnivores or flesh-eating rodents are involved. Eggs of *C. hepatica* have been found in several species of earthworms. These transport hosts may facilitate infections in normal definitive hosts.[72]

This nematode has some potential as a biological control agent for rodent populations.[8, 75]

- **Pathology.** Wandering of adult *C. hepatica* through the host liver causes loss of liver cells and thereby loss of normal function. Large areas of parenchyma may be replaced by masses of eggs and granulomas (Fig. 23.5). Rarely eggs will be carried to the lungs or other organs by the bloodstream.

  Hepatomegaly can become severe, and eggs become encased in granulomatous tissue, with heavy infiltration of eosinophils and other leukocytes.[16] Other symptoms include persistent fever, eosinophilia and increases in certain blood enzymes, including some markers of liver function (e.g., alanine aminotransferase).

- **Diagnosis and Treatment.** There are only 85 reported cases of this parasite in humans,[33] partly because of difficulties of diagnosis. Most cases have been determined after death, but liver biopsy, serology, and ultrasonography have uncovered others.[33] The mortality rate in untreated cases is approximately 50%.[33] Clinical symptoms resemble numerous liver disorders, especially hepatitis with eosinophilia, and other parasitic diseases, including visceral larva migrans. Definitive diagnosis depends on demonstrating eggs or adult worms; the eggs, resemble those of *T. trichiura* except the polar plugs protrude much less and they measure 40 μm to 67 μm by 27 μm to 36 μm and have striated shells. Treatment for this disease has not been investigated adequately, but albendazole is currently the drug of choice; effective drug therapy requires several weeks of treatment.[17]

  Discovery of *C. hepatica* eggs in human feces may indicate the presence of a spurious infection caused by eating an infected liver. More than 70 spurious cases have been documented.[33]

Capillaria philippinensis. *Capillaria philippinensis* was discovered in 1963 as a parasite of humans in the Philippines. In contrast to *C. hepatica*, *C. philippinensis* is an intestinal parasite. Its appearance as a human pathogen was sudden and unexpected. One or two isolated cases were followed by an epidemic in Luzon with over 1400 infections and 95 deaths.[20, 27] It has now been reported from Thailand, Iran, Japan, Indonesia, and Egypt.

It is probably a zoonotic disease, but the original host remains unknown. *Capillaria philippinensis* has been transmitted experimentally to monkeys, gerbils, *Rattus* spp., and several species of migratory fish-eating birds.[20] Some female worms bear living juveniles, and eggs, juveniles, and adults pass from the definitive host in feces. When feces reach water, eggs embryonate and are eaten by small fishes. After hatching in a fish's intestine, juveniles develop for a few weeks until they become infective for a definitive host. Juveniles released by females in a definitive host's intestine are autoinfective, and massive populations can accumulate, causing severe pathology.

- **Morphology.**   This parasite is small; males measure 2.3 mm to 3.2 mm, and females measure 2.5 mm to 4.3 mm long. The male has small caudal alae and a spineless spicule sheath. The esophagus of females is about half as long as the body. Females produce *Capillaria*-type eggs measuring 36–45 μm long and approximately 21 μm wide.

- **Epidemiology.**   Intensive surveys of Philippine fauna have so far failed to identify any reservoir host, but fish-eating birds are prime suspects. Migratory birds are probably the means by which the infection has spread to other Asian countries and even to the Middle East. Because infective juveniles are in fish intestines, any region where people savor small, whole, raw fish may experience new cases.

- **Pathology.**   Worms repeatedly penetrate mucosa of the small intestine and reenter the lumen, especially jejunum, leading to progressive degeneration of mucosa and submucosa; although the villi atrophy, the small intestine thickens. Infected people usually experience diarrhea and abdominal pain, progressing to weight loss, weakness, malaise, anorexia, and emaciation.[20] Protein and electrolytes, especially potassium, are lost, and there is malabsorption of fats and sugars. Patients die from loss of electrolytes, heart failure, and sometimes secondary bacterial infection. Mortality rates can reach 20%.

- **Diagnosis and Treatment.**   Both adults and eggs, as well as juveniles, are abundant in feces of heavily infected people, and at least one of them is necessary for specific diagnosis. Eggs of *C. philippinensis* must be distinguished from those of *T. trichiura*; *C. philippinensis* eggs have nonprotruding polar plugs and are slightly smaller in size. The only other intestinal nematode in humans in which juveniles normally pass in feces is *Strongyloides stercoralis* (p. 393), and presence of the easily distinguished eggs and adults of *C. philippinensis* differentiates these two.

Prolonged adminiinistration of mebendazole or albendazole is effective in curing this disease. Control consists of persuading people to refrain from eating small raw fish whole.

Other *Capillaria* Species

Several species of *Capillaria* are important parasites of domestic animals. *Capillaria aerophila* is a lung parasite of cats, dogs, and other carnivores and has been reported several times from humans. It is probably the most destructive parasite of commercial fox farms.

Capillaria annulata and *C. caudinflata* infect the esophagus or crop and intestine, respectively, of chickens, turkeys, and several other species of birds. Unlike most species in the genus, these species require an intermediate host (earthworm) in the life cycle. *Capillaria* spp. can be highly pathogenic in domestic fowl.

Family Anatrichosomatidae

Anatrichosoma Species

Species of *Anatrichosoma* are very similar to *Capillaria*, except that they lack a spicule and spicule sheath. They have been reported from tissues of a wide variety of Asian and African monkeys and gerbils and from the North American opossum. *Anatrichosoma ocularis* (Fig. 23.6) lives in the corneal epithelium of tree shrews, *Tupaia glis*. Eggs of this genus (Fig. 23.7) have the polar plugs characteristic of the order. A handful of infections have been reported from humans, including one in North America.[31]

Family Trichinellidae

Trichinella Species

Curiously, the smallest nematode parasite of humans, which exhibits the most unusual life cycle, is one of the most widespread and clinically important parasites in the world. This parasite was described in 1835 and since 1895 has been known as *Trichinella spiralis*. For many decades the genus *Trichinella* was considered to include only a single species

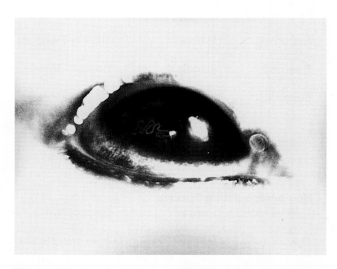

Figure 23.6   *Anatrichosoma ocularis* **in the eye of a tree shrew, *Tupaia glis*.**

From S. K. File, "*Anatrichosoma ocularis* sp. n. (Nematoda: Trichosomoididae) from the eye of the common tree shrew, *Tupaia glis*," in *J. Parasitol.* 60:985–988. Copyright © 1974.

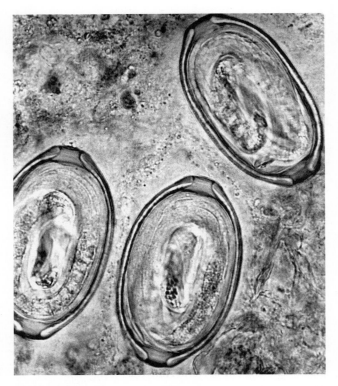

Figure 23.7   **Eggs of *Anatrichosoma ocularis* from eye secretions.**

Courtesy of Sharon K. File.

(*T. spiralis*), but eight cryptic species (Table 23.1) and four additional genotypes of uncertain taxonomic status are currently recognized.[65] There are, however, no morphological differences among the different *Trichinella* spp., although differences in host response to infection differentiates encapsulated versus unencapsulated muscle juveniles; this division also reflects distinct phylogenetic lineages as inferred from molecular data.[65] Species and genotypes can be distinguished by disease produced in humans,[67] recognition of antigens by monoclonal antibodies,[41] and various molecular tools including multiplex PCR and PCR-RFLP analysis.[65] Male and female first-stage juveniles can be distinguished so that potential hybridization of putative species can be tested.[47] In the discussion to follow, we will be referring to *T. spiralis* in the strict sense, except where noted otherwise.

These parasites are responsible for the disease variously known as *trichinosis, trichiniasis,* or *trichinellosis.* It is common in carnivorous and omnivorous mammals, including rodents and humans; human infections have been reported from 55 countries. *Trichinella* sp. infections are cosmopolitan, but certain species have restricted geographic distributions. Incidence of infection is always higher than suspected because of the vagueness of symptoms, which usually suggest other conditions; more than 50 different diseases have been diagnosed incorrectly as trichinellosis.

• **Morphology.**   Males (Fig. 23.8) measure 1.4 mm to 1.6 mm long and are more slender anteriorly than posteriorly. The anus is nearly terminal and has a large copulatory

Table 23.1   **Biological and Distributional Characteristics of *Trichinella* spp.**

Species	Cycle	Distribution	Main hosts	Biological characters
Encapsulating Species				
T. spiralis	Domestic, sylvatic	Cosmopolitan	Mammals (swine, carnivores, rats)	• Juvenile production per 72 hr in vitro > 90 (others < 60) • No freezing resistance • High RCI* in rats and pigs
T. native	Sylvatic	Arctic and subarctic	Mammals (carnivores)	• Low RCI in rats and pigs • High freezing resistance
T. nelsoni	Sylvatic	Tropical Africa	Mammals (carnivores)	• Low RCI in pigs and rats • No freezing resistance
T. britovi	Sylvatic	Temperate zone, Palaearctic region, West and North Africa	Carnivorous mammals, pigs, game, horses	• Moderate freezing resistance • Low RCI in rats and pigs
T. murrelli	Sylvatic	United States	Carvivorous mammals, horses	• Low freezing resistance • Low RCI in Swiss mice and pigs • High RCI in wild mice
Non-encapsulating Species				
T. pseudospiralis	Sylvatic	Cosmopolitan	Mammals, birds	• Very low RCI in pigs; high in rats • No freezing resistance
T. papuae	Sylvatic	Papua New Guinea	Reptiles, mammals	• Moderate RCI in mice • No freezing resistance
T. zimbabwensis	Sylvatic	Africa	Reptiles, exptl. mammals	• No freezing resistance • Low RCI in mice

Data from Pozio et al.,[64] Pozio and Murrell,[66] and Pozio and Zarlenga[68]
*RCI = reproductive capacity index

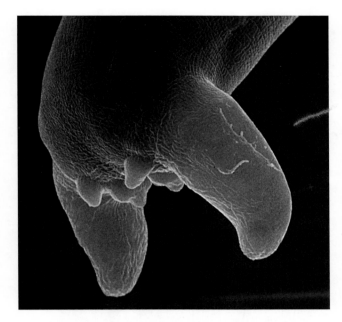

Figure 23.9   **Scanning electron micrograph of male** *Trichinella nativa.*

Note the posterior end including copulatory pseudobursae and papillae.

From J. R. Lichtenfels et al., "Comparison of three subspecies of *Trichinella spiralis* by scanning electron microscopy," in *J. Parasitol.* 69:1131–1140. Copyright © 1983.

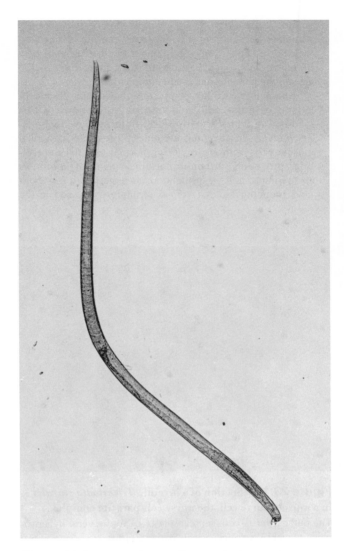

Figure 23.8   Male *Trichinella spiralis* (1.4 mm to 1.6 mm in length) from the intestine of a rat.

Courtesy of Jay Georgi.

pseudobursa on each side of it (Fig. 23.9). There is no copulatory spicule. As in other members of order Trichinellida, stichocytes are arranged in a row following a short muscular esophagus. Females are about twice the size of males and also taper anteriorly. The anus is nearly terminal. The vulva opens near the middle of the esophagus, which is about a third the length of the body from the anterior end. The single uterus is filled with developing eggs in its posterior portion, whereas the anterior portion contains fully developed, hatching juveniles.

• **Biology.**   The biology of this organism is unusual in that the same individual animal serves as both definitive and intermediate host, with juveniles and adults located in different tissues. Even more interesting, however, is another unique character: juveniles are the world's largest intracellular parasites! Adults are parasites of crypts within the small intestine, and individual juveniles reside in nurse cells, whose formation they themselves induce, within skeletal muscle cells.

Juveniles within nurse cells have an anaerobic (or facultatively anaerobic) metabolism (p. 372). When infective juveniles are swallowed and reach the stomach of the host, they are released from their nurse cells and become "activated": They shift to the aerobic metabolism characteristic of adults.[36] After passing into their host's small intestine, they quickly grow and undergo four molts. Copulation occurs within the mucosal epithelium within 30 to 32 hours of infection.[23] The worms thread through a serial row of intestinal cells (Fig. 23.10). During her sojourn through intestinal epithelium, a female gives birth to between hundreds and thousands of juveniles over a period of 4 to 16 weeks. Eventually the female dies and passes out of the host. Males can copulate several times but then die shortly after.

Most juveniles are transported by the hepatoportal system through the liver and then to the heart, lungs, and arterial system, which distributes them throughout the body. During this migration, they may be found in literally every kind of tissue and space in the body. When they reach skeletal muscle, they penetrate individual fibers. Then, in a strategy reminiscent of viruses, they subvert and redirect host cell activities to their own survival. They alter gene expression of the host cell from that of a contractile fiber to that of a nurse cell, a cell that functions in nourishing the worm. After the nematode enters it, the fiber loses its myofilaments, but its nuclei enlarge (hypertrophy) and smooth endoplasmic reticulum increases.[85] Preexisting mitochondria degenerate consistent with apoptosis, and are replaced by new smaller mitochondria. Eventually the entire unit becomes encapsulated with collagen secreted by the nurse cell[25] (Fig. 23.11). For unknown reasons, neither *T. pseudospiralis, T. papuae*, nor *T. zimbabwensis* stimulate formation of a collagenous capsule to surround their nurse cells.[63] Angiogenesis, or the formation of new blood vessels, occurs (a **circulatory rete,** Fig. 23.12) around the parasite-nurse-cell complex.[23, 25]

Although we do not completely understand the mechanisms of nurse cell induction, investigations of this fascinating transformation are yielding new insights. *Trichinella* spp. juveniles are able to penetrate individual muscle cells without triggering normal suicidal intracellular processes, although certain aspects of apoptosis are evident such as mitochondrial degeneration. Apoptosis inducing factor (AIF) is strongly expressed in the

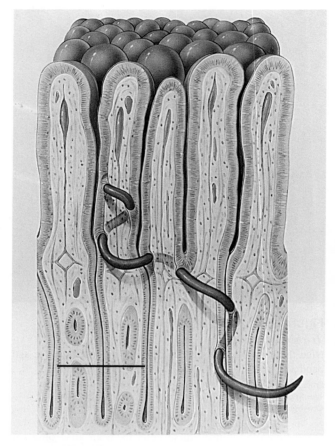

Figure 23.10 **Conceptual illustration of an adult *T. spiralis* threading its way through the intestinal epithelium.**
(Scale bar = 400 μm.)

Illustration by J. Karapelou, from D. D. Despommier, "*Trichinella spiralis* and the concept of niche," in *J. Parasitol.* 79:472–482. Copyright © 1993 Journal of Parasitology. Reprinted by permission.

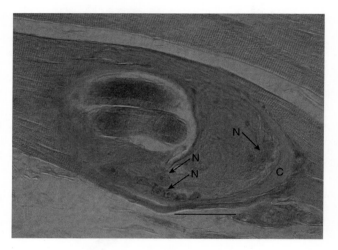

Figure 23.11 **Section of a juvenile *Trichinella spiralis* in a muscle nurse cell, the nurse cell-parasite complex.**
The outer layer is collagen capsule (*C*). Note several hypertrophied nuclei (*N, arrows*). (Scale bar = 50 μm)

Courtesy of Steve Nadler.

Figure 23.12 **Schematic drawing of an intact nurse cell-parasite complex, showing the surrounding circulatory rete.**

From D. D. Despommier, "*Trichinella spiralis;* the worm that would be a virus," in *Parasitol. Today* 6:193–196. Copyright © 1990 Elsevier Trends Journals Cambridge, England. Reprinted by permission.

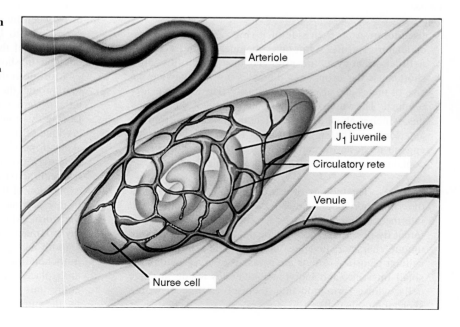

sacroplasm and nucleus of infected muscle cells in the earliest stages of infection. Muscle cell chromatin condenses, and is eventually fragmented, presumably from a pathway involving AIF.[14] Later during infection, AIF expression is greatly reduced. This suggests that juveniles use excretory-secretory products to up- or down-regulate apoptosis during nurse cell formation and maintenance.[6] The parasites seize control of their host's gene expression, downregulating genes for muscle-specific proteins and upregulating genes for vascular endothelial growth factor (VEGF) and collagen synthesis. Expression of VEGF and another factor that induces it (Thymosin Beta 4) is increased by 10 days post-infection and remains high for six weeks, the time required for nurse cell complex maturation.[38] A central role for materials secreted from stichocytes is strongly suspected in alteration of host cell gene expression.[48] Stichocytes of juveniles produce about 40 different proteins joined to an unusual, very antigenic sugar called *tyvelose* (3,6-dideoxy arabinohexose). The antigen is present in stichocytes of six-day-old juveniles but not in nurse-cell nuclei at that time. However, worm peptides, including four tyvelosylated antigens, can be detected in nurse cell nuclei by eight days, when nuclear hypertrophy is greatest.[25] These peptides remain in nurse-cell nuclei for the life of the parasite.[25] Injection into muscle of secretory-excretory material collected from juveniles cultured in vitro can produce changes in muscle fibers similar to those seen in nurse cells.[45] Another effect on host cell nuclei is their repositioning in the cell cycle.[37] Normal muscle nuclei are arrested in the G_0/G_1 phase of the cell cycle, the phase to which muscle gene expression is usually restricted. Nuclei of nurse cells are induced to undergo DNA synthesis, become 4N, and then stop in the G_2/M phase. Therefore, nurse cells are arrested at a stage at which they cannot express muscle genes but only genes normally characteristic of the proliferative phase. Expression patterns of thymidylate synthase and two other enzymes indicate that an unusual cell cycle regulation mechanism is found in arrested muscle juveniles of *Trichinella* spp.,[22, 71] a pattern shared with dauer juveniles of *C. elegans*.[83] Normally, high levels of these enzymes are associated with cell proliferation or regeneration and not arrest. Interestingly, *T. spiralis* juveniles and their extracts have been shown to have anti-tumor effects in vitro and in vivo, involving both apoptosis and cell cycle arrest.[80] Perhaps increased understanding of these parasites will lead to development of new antitumor therapeutics.

Some muscles are much more heavily invaded than are others, but we do not know why. Most susceptible are eye, tongue, and masticatory muscles, then the diaphragm and intercostals, and finally the heavy muscles of arms and legs. Juveniles absorb their nutrients from their enclosing nurse cell and increase in length to about 1 mm in four to eight weeks, at which time they are infective to their next host. Heavy *T. spiralis* infection induces hypoglycemia during the period of muscle juvenile growth and is correlated with the accumulation of glycogen within infected muscle cells. Genes of the insulin signaling pathway are up-regulated in infected cells during this phase, leading to increased uptake of glucose.[87]

Nurse cells become gradually larger during this time, and finally they achieve a length of 0.25 mm to 0.50 mm.

Juveniles enter a developmental arrest and can live for months or years while encased. Gradually, host reactions begin to calcify the nurse cell and eventually the worms themselves. However, living juveniles in muscle up to 39 years after first infection have been documented.[24]

It was the presence of such calcified granules in human cadavers that led to the discovery of *Trichinella* sp. in 1835 by James Paget, a medical student in London. Noticing that his subject had gritty particles in its muscles that tended to dull his scalpels, he studied the particles and demonstrated their wormlike nature to his fellow students. He then showed them to the eminent anatomist Richard Owen, who reported on them further and gave them their scientific name. It was another 25 years before it was determined that these minute animals cause disease.

- **Epidemiology.**   Trichinosis today is best considered a zoonotic disease, because humans can scarcely be important in the life cycle of the parasite. Unless an infected person is eaten by a carnivorous predator or becomes a cannibal's supper, both unlikely events these days, the parasites are at a dead end in a human.

We have traditionally considered the life cycle of *Trichinella spiralis* as two epidemiologically distinct types: a domestic cycle (involving pigs and rats, around human habitation) and a sylvatic cycle (involving wild animals). Campbell[13] provided a different framework, taking account of host preferences of different species of *Trichinella*. He recognized four distinct epidemiological cycles: **domestic, sylvatic-temperate zone, sylvatic-tropical variant,** and **sylvatic-arctic variant** (Fig. 23.13).

Domestic trichinosis involves *Trichinella spiralis* only (see Table 23.1). It is epidemiologically most important to humans because of the close relationship among rats, pigs, and people. Infected pork is our most common source of infection. Pigs become infected by eating offal or trichinous meat in garbage or by eating rats, which are ubiquitous in pig farms. Garbage containing raw pork scraps is probably the usual source of infection for pigs, but pigs will greedily devour dead or even live rats when they can catch them. Rats likely maintain their infections by cannibalism, although rat and mouse feces can contain juveniles capable of infecting rats, pigs, or humans.[49, 62] Juveniles in muscles of mice can alter host behavior, making them more vulnerable to predation.[54, 88] Infection can spread from pig to pig when they nip off and eat each other's tails, a common practice in crowded piggeries.[76] Piggish cannibalism also is involved. The worm can survive and remain infective after anaerobic digestion of sewage sludge,[32] but freezing at $-15°$ C for 20 days destroys all *T. spiralis*.

The importance of cooking pork (and other meats) properly before it is eaten cannot be overstated. A roast or other piece of solid meat is safe when it has been thoroughly heated to $71°$ C ($160°$ F), usually eliminating all traces of pink. Many people are careful about cooking roast pork but become careless when cooking sausage, which is equally dangerous. Raw sausage is a delicacy among many peoples of the world, particularly in areas where trichinellosis is a chronic health problem. Even a

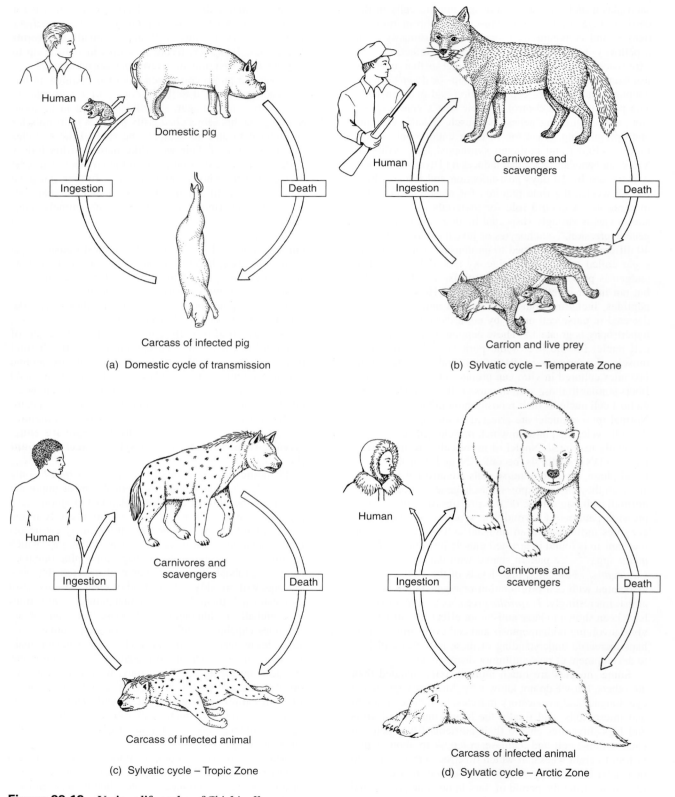

(a) Domestic cycle of transmission

(b) Sylvatic cycle – Temperate Zone

(c) Sylvatic cycle – Tropic Zone

(d) Sylvatic cycle – Arctic Zone

Figure 23.13   Various life cycles of *Trichinella* spp.

These are all essentially zoonoses, with humans becoming infected incidentally and not playing an essential role in the life cycle. (*a*) Domestic cycle, primarily *T. spiralis*. (*b*) Temperate zone sylvatic cycle, *T. spiralis, T. britovi,* and *T. murrelli*. (*c*) Tropic zone sylvatic cycle, *T. nelsoni*. (*d*) Arctic zone sylvatic cycle, *T. nativa*.

Drawing by William Ober and Claire Garrison after W. C. Campbell, "Trichinosis revisited—another look at modes of transmission," in *Parasitol. Today* 4:83–86, 1988.

casual taste to determine proper seasoning can be fatal. Particularly important in transmission is meat processed by "backyard butchers" (individual who slaughter their own stock).

Reported cases (which are likely a tiny fraction of the true number) in the United States declined from over 400 per year in the 1940s to about 30 to 40 year from 1987 to 1989. A similar decline was observed in Europe. During the 1990s there was a resurgence of trichinellosis, and it is again becoming a threat in developed and developing regions.[57] Dramatic increases occurred in a number of countries, such as Romania (17-fold), Argentina (7-fold), and Lithuania (9-fold). Murrell and Pozio[57] attributed the increase to several factors, including mass-marketing of meat that spreads infection from endemic to nonendemic areas and an increase in importance of sylvatic trichinellosis, both directly from game meat and through spillover of sylvatic species to domestic animals, such as sheep, goats, cattle, and horses. A recent outbreak in Izmir, Turkey, involved 1098 documented cases;[2] this is unexpected in a Muslim country where consumption of pork is against Islamic law. The source of the infection was an illegal mixture of ground beef and pork, distributed by a butcher to restaurants, but represented as beef.

Sylvatic trichinellosis in the temperate zone is due to both *T. spiralis* and *T. britovi* in Europe and Asia and to *T. spiralis* and *T. murrelli* in North America.[57] The range of *T. britovi* extends from western Europe to Japan.[67] *Trichinella britovi* has some resistance to freezing but not nearly so much as *T. nativa,* which is responsible for sylvatic trichinosis in the circumpolar Arctic.[39] Because sylvatic trichinosis usually involves wild mammals, humans are infected when they interject themselves into the sylvatic food chain or possibly when domestic animals have been fed infected meat from game.

Native Americans and others who rely on wild carnivores for food and urban dwellers who return home with the spoils of the hunt are all subject to infection with *Trichinella* spp. Fatal cases of trichinellosis occur among those who eat undercooked or underfrozen bear, wild pig, cat, dog, or walrus meat. Theoretically, any wild mammal may be a source of infection, but of course most rarely find their way to the dinner table. In Alaska, polar bears, black bears, and walrus are common sources of *T. nativa.*[51] Arctic explorers have been killed by *Trichinella* acquired from uncooked polar bear meat. The cause of death of the three members of the ill-fated André polar expedition of 1897 was determined 50 years later when *Trichinella* juveniles were found in museum specimens of the polar bear meat that the men had been eating before they died.[78]

The cause of sylvatic trichinosis in tropical Africa, *T. nelsoni,* has a low infectivity for pigs and rats, but it can cause intensive infections and even death in humans.[13] Carrion-eating habits of hyenas suggest that they are important in transmission, but any mammal feeding on a carcass of another mammal could transmit the infection. Humans in this cycle would not be dead-end hosts if they were fed upon by wild predators.

Epidemiological importance of the three unencapsulated species, *T. pseudospiralis, T. papuae,* and *T. zimbabwensis,* is largely unassessed, although several human infections with *T. pseudospiralis* have been reported.[57] *Trichinella*

pseudospiralis has a high reproductive capacity in rats but a low capacity in pigs, and it may be primarily a parasite of birds. However, there are increasing reports of this species in domestic and sylvatic swine, indicating it may become a greater human health concern.[68] *Trichinella papuae* apparently cannot infect birds. All three species have no freezing resistance.

Although it is unclear how they acquire infection, herbivores such as horses and cattle can carry *Trichinella* spp.[55] It is possible that they acquire infection from uncooked table scraps.[57] Consumption of horsemeat has been responsible for several outbreaks in France and Italy, including several deaths, and 4% of a sample of horses in Mexico was found infected.[4]

Survivors of trichinellosis have varying degrees of immunity to further infection, and this immunity in mice can be passed by a mother to her young by suckling.[30] Suckling rats rapidly expel *T. spiralis* if the mother is immune. Her serum antibodies are passed through the milk to the pups.[3]

- **Pathogenesis.** Pathogenesis of trichinellosis can be considered in three successive stages: the enteric phase when adult worms reside in the small intestine, migration of juveniles, and penetration and nurse cell formation.

First symptoms may appear between 12 hours and two days after ingestion of infected meat. Commonly, this phase is clinically inapparent because of low-grade infection, or it is misdiagnosed because of vagueness of symptoms. Worms migrating in intestinal epithelium cause traumatic damage to host tissues; a host begins to react to their waste products. Inflammation causes symptoms such as nausea, vomiting, abdominal pain, and diarrhea. The enteric phase usually lasts about one week before the migration phase ensues.

During migration newborn juveniles damage blood vessels, resulting in localized edema, particularly in the face and hands. Wandering juveniles may also cause pneumonia, pleurisy, encephalitis, meningitis, nephritis, deafness, peritonitis, brain or eye damage, and subconjunctival or sublingual hemorrhage. Death resulting from myocarditis (inflammation of heart muscle) may occur at this stage. Although juveniles do not stay in the heart, they migrate through its muscle, causing local areas of necrosis and infiltration of leukocytes. The migration phase is accompanied by fever and myalgia.

By the 10th day after appearance of first symptoms, juveniles begin penetration of muscle fibers. Attendant symptoms are again varied and vague: intense muscular pain, difficulty in breathing or swallowing, swelling of masseter muscles (occasionally leading to a misdiagnosis of mumps), weakening of pulse and blood pressure, heart damage, and various nervous disorders, including hallucination. Heart muscle damage is accompanied by changes in electrocardiograms and creatine phosphokinase levels. Heavy infection significantly suppresses muscle contractibility[1] and reduces mechanical stress, work, power output, and fatigue resistance of the diaphragm.[34, 84] Extreme eosinophilia and high fever (40° C) are common but may not be present even in severe cases. Death is usually caused by heart failure, respiratory complications, or kidney malfunction.

- **Diagnosis and Treatment.** Most cases of trichinellosis, particularly subclinical cases, go undetected. Although muscle biopsy is infrequently employed, it remains an accurate diagnostic if trichinellosis is suspected. Pressing the tissue between glass plates and examining it by low-power microscopy is useful, although digestion of the muscle in artificial gastric enzymes for several hours provides a much more reliable diagnostic technique. Detection of *Trichinella*-specific DNA in muscle biopsy by PCR is also very sensitive. Several immunodiagnostic tests are available.[56]

  No entirely satisfactory treatment for trichinellosis is known. The diagnosis, if made, usually happens after most juveniles have been produced, so targeting adults with anthelminthics is usually not an option. In addition, the rapid death of juveniles due to drug therapy can lead to adverse host inflammatory reactions, worsening the disease outcome. Treatment is basically that of relieving the symptoms by use of analgesics and corticosteroids. Purges during the initial symptoms may dislodge females that have not yet begun penetrating intestinal epithelium. Thiabendazole has been shown effective in experimental animals, but results in clinical cases have been variable.

  Despite immense research, trichinosis remains an important disease of humans, one that has the potential of striking anyone, anywhere.

ORDER DIOCTOPHYMATIDA

The few members of order Dioctophymatida are parasites of aquatic birds and terrestrial mammals. There are two families in the order: Soboliphymatidae, which has a single genus *Soboliphyme,* parasites of shrews and mustelid carnivores (Fig. 23.14); Dioctophymatidae, which has the genera *Eustrongylides,*[61] *Hystrichis,* both parasites of birds, and *Dioctophyme.* Molecular phylogenetic analysis supports the monophyly of each family and the superfamily that includes them, Dioctophymatoidea.[46]

Family Dioctophymatidae

The genus *Dioctophyme* appears to include a single valid species, *D. renale.* This species has been reported from 28 species of mammals, including humans, but is primarily reported from mustelids and canids in many parts of the world. The genus *Eustrongylides* contains approximately 11 described species, although fewer may be valid.[52] Infection of humans by juvenile *Eustrongylides* sp. has been reported; these infections can be highly pathogenic. Species of *Hystrichis* are found within tumors in the proventriculus of certain waterfowl and wading birds.

Dioctophyme renale
- **Morphology.** *Dioctophyme renale* is truly a giant among nematodes, with males up to 20 cm long and 6 mm wide and females up to 100 cm long and 12 mm wide. In terms of length and body mass, only *Placentonema gigantissima* Gubanov, 1951 at eight meters long is larger! They

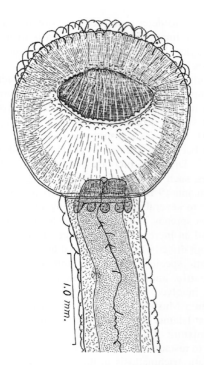

Figure 23.14 *Soboliphyme jamesoni* **from a shrew.**
Note the swollen mouth capsule typical of this genus.

From C. P. Read, "*Soboliphyme jamesoni* n. sp., a curious nematode parasite of California shrews," in *J. Parasitol.,* 38:203–206. Copyright © 1952. Reprinted by permission.

are blood red and rather blunt at the ends. Males have a conspicuous, bell-shaped copulatory bursa that lacks any supporting rays or papillae (Fig. 23.15). A single, simple spicule is 5 mm to 6 mm long. The vulva is near the anterior end. Eggs (Fig. 23.16) are lemon shaped, with deep pits in the shells, except at the poles.

- **Biology.** Mace and Anderson described the life cycle of this parasite.[50] The thick-shelled eggs require about 35 days in water at 20° C to embryonate. First-stage juveniles, which bear an oral spear, hatch when eaten by the aquatic oligochaete annelid *Lumbriculus variegatus.* They penetrate the ventral blood vessel, where they develop into third-stage juveniles. When the small annelid is swallowed, juveniles migrate to a kidney of the new host, where they mature. If the annelid is eaten by any of several species of fish, frogs, or toads, juveniles will encyst in the muscle or viscera, using the animal as a paratenic host.[53] When swallowed by a definitive host, juveniles penetrate the stomach wall. After about five days they migrate to the liver and remain in the liver parenchyma for about 50 days; then they migrate to the kidney. Usually the right kidney is invaded, perhaps because of the proximity of the stomach to right lobes of the liver, from which the worms can migrate directly to the kidney.[50] Worms mature in the kidney, and eggs are voided from the host in its urine. In the rare human infections worms usually are in a kidney, but one third-stage juvenile was found in a subcutaneous nodule.[9]

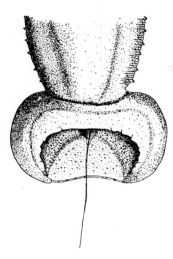

Figure 23.15 **Posterior end of a male *Dioctophyme renale*.**

Note the single spicule and the copulatory bursa.

From W. Stefanski, "Quelques précisions sur les caractères spécifiques du strongle géant du chien," in *Ann. Parasitol. Hum. Comp.* 6:93–100. Copyright © 1928. Reprinted by permission.

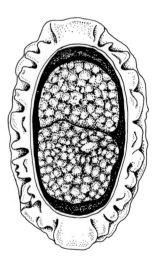

Figure 23.16 **Egg of *Dioctophyme renale*, showing the corrugated shell and the two-cell embryo.**

This stage is released by female worms.

From T. F. Mace and R. C. Anderson, "Development of the giant kidney worm, *Dioctophyma renale* (Goeze, 1782 Nematoda: Dioctophymatoidea)," in *Can. J. Zool.* 53:1552–1568. Copyright © 1975 National Research Council of Canada. World rights reserved.

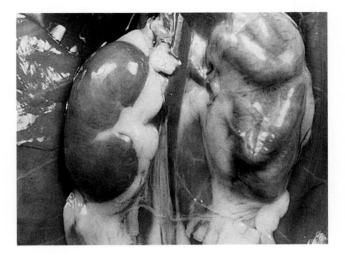

Figure 23.17 **Ferret dissection.**

The kidney on the left is normal, whereas that on the right is distended by an adult *Dioctophyme renale*. (Anterior of animal is toward bottom.)

Arthur E. Woodhead. Courtesy Ann Arbor Biological Center.

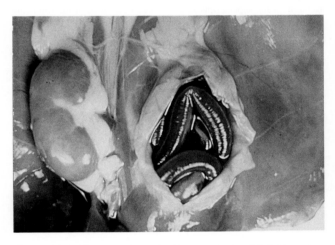

Figure 23.18 **Specimen in Figure 23.17 with infected kidney opened to reveal the worm.**

The organ is reduced to a hollow shell.

Arthur E. Woodhead. Courtesy Ann Arbor Biological Center.

- **Epidemiology.** Probably any species of large mammal can serve as a definitive host. Because of their fish diets, mustelids, canids, procyonids, and bears are particularly susceptible, as are humans. However, even such non-fish-eating mammals as cows, horses, and pigs can become infected when they accidentally ingest an infected oligochaete. Thorough cooking of fish and drinking of only pure water will prevent infection in people.

- **Pathology.** Pressure necrosis caused by growing worms and their feeding activities reduce the infected kidney to

a thin-walled, ineffective organ (Figs. 23.17 and 23.18). Loss of kidney function is compounded by uremic poisoning. Infection symptoms include hematuria, nephritis, and intermittent pain in the kidney region. Worms will sometimes develop to adults in the coelomic cavity.

- **Diagnosis and Treatment.** The rarity of this parasite makes physicians and veterinarians unlikely to suspect its presence. Demonstration of the characteristic eggs in urine is the definitive means of diagnosis, aside from surgical discovery of the worm itself. Imaging techniques including radiology and ultrasonagraphy are also useful diagnostic approaches. Surgical removal of the worm is the only treatment known. Unfortunately, most diagnoses are made postmortem.

Learning Outcomes

By the time a student has finished studying this chapter, he or she should be able to:

1. Compare the differences in biology and life cycles of parasitic nematodes representing the Enoplea.

2. Assess the epidemiological differences between nematode species with a direct life cycle involving fecal-oral transmission of infective eggs versus species that are transmitted via consumption of prey items that contain infective juveniles.

3. Explain why understanding parasite biogeography is important for control of human trichinellosis.

4. Analyze strategies that enoplean parasites use to increase their fitness by manipulation of host biology.

5. Describe how specific manipulations of the host employed by enoplean parasites might be used advantageously for human medicine.

6. List human cultural practices that can alter the risk of infection by *Capillaria* spp. and *Trichinella* spp.

References

References for superscripts in the text can be found at the following Internet site: www.mhhe.com/robertsjanovynadler9e

Additional Readings

Bundy, D. A. P., and E. S. Cooper. 1989. *Trichuris* and trichuriasis in humans. In J. R. Baker and R. Muller (Eds.), *Advances in parasitology 28*. London: Academic Press, pp. 107–173

Campbell, W. C. 1983. Trichinella *and trichinellosis.* New York: Plenum Press.

Gajadhar, A. A., E. Pozio, H. R. Gamble, K. Nöckler et al. 2009. *Trichinella* diagnostics and control: Mandatory and best practices for ensuring food safety. *Vet. Parasitol.* 159:197–205.

Reddy, A., and B. Fried. 2007. The use of *Trichuris suis* and other helminth therapies to treat Crohn's disease. *Parasitol. Res.* 100:921–927.

The *Trichinella* page, http://www.trichinella.org/.

Chapter 24

Nematodes: Tylenchina, a Functionally Diverse Clade

. . . occasionally parasitic females and males are passed in diarrheic stools . . . . The rare parasitic males are practically indistinguishable from the free-living males of the indirect cycle.

—E. C. Faust[12]

Although Faust continued to describe the parasitic male in his textbook as late as 1970, no other investigators were able to find such worms and the concept fell into disrepute.

—D. I. Grove, describing the controversy over the existence of a parasitic male *Strongyloides stercoralis*[18]

The tiny worms in suborder Tylenchina include many species and a diverse spectrum of feeding mechanisms and life histories, ranging from free-living **microbivores** to parasites of plants and animals. This suborder includes the most economically important plant parasites, including root-knot (*Meloidogyne* spp.), cyst (*Heterodera* spp.), and lesion (*Pratylenchus* spp.) nematodes. These and other species of plant parasites cause annual losses of agricultural crops (food and fiber) totaling approximately 12%.[2] Several species of Tylenchina that parasitize vertebrates are considered unusual because they alternate between free-living and parasitic generations. Many of the entirely free-living species live in habitats rich in bacteria, including decaying organic matter, soils, aquatic sediments, decaying fruit, and so on. Due to abundance, these species may find their way into the bodies of animals; the digestive, reproductive, respiratory, and excretory tracts are particularly susceptible, as are open wounds. Some species may become facultatively parasitic for a time whereas most others simply pass through the body. The eggs of such transient nematodes, including certain plant parasites, are sometimes present in the feces of animals including humans, presenting a challenge to the diagnostician. Rarely, facultative Tylenchina cause serious disease in humans and other animals.[13] For example, *Halicephalobus gingivalis* is a very rare, but typically fatal pathogen of horses and humans.[29]

Some behavioral and developmental adaptations of free-living species in this suborder may serve as preadaptations to parasitism. Dauer juveniles (chapter 22) and anhydrobiosis are mechanisms used by free-living nematodes to persist until new food resources and conditions are favorable for growth. Other species living in organic matter rich in biological activity have become adapted to high temperature. It is not difficult to visualize how adaptations that enhance persistence and thermal tolerance, and reinitiate development in the presence of bacterial food sources may also serve as preadaptations for gut parasitism of animals. The diversity in life cycles of parasites within Tylenchina further illustrates this evolutionary opportunism. Thus, within the suborder, one can observe completely free-living species, facultative parasites, obligate parasites, and even species that produce free-living or parasitic populations, depending on environmental conditions (Rhabdiasidae, Strongyloididae).

Based on their analysis of small subunit ribosomal DNA (SSU rDNA) sequences, Blaxter and his coworkers concluded that order Rhabditida is paraphyletic, its members being distributed in two distinct clades along with members of several other orders.[5] In their classification De Ley and Blaxter avoided paraphyly for Rhabditida by making the order much more inclusive.[10] Almost all the nematodes we have covered in this chapter in previous editions are included in De Ley and Blaxter's suborder Tylenchina, infraorder Panagrolaimomorpha.

FAMILY STEINERNEMATIDAE

Species in the family Steinernematidae are one type of entomopathogenic nematode, so called because when they successfully infect an insect host they result in its eventual death in a manner similar to parasitoids (p. 599). Initial SSU rDNA phylogenies[10] indicated that steinernematids belong to Tylenchina, but more recent and comprehensive analyses have revealed uncertainty about *Steinernema* relationships.[46] Nevertheless, this family is covered here along with a more distantly related one, Heterorhabditidae (suborder Rhabditina), which has a remarkably similar life cycle that appears to have evolved through convergent evolution. Steinernematid and heterorhabditid nematodes are widely used as biological control agents of insect pests in commercial agriculture,[21] and may have wider residential application as potential biocontrol agents of fleas, ticks, and termites.[38]

The life cycles of these nematodes are quite remarkable as they involve an obligate bacterial symbiosis. Third-stage dauer juveniles (J3) are the infective stage for the insect host. Infective juveniles can live a considerable period in soil, depending on species, temperature, and moisture. For *Steinernema,* different species have different host-finding behavior. Some species are mobile cruisers, actively seeking sedentary insects such as grubs deep in the soil. Other species are sedentary and remain near the soil surface, ambushing mobile insects. *Steinernema* sp. and *Heterorhabditis* sp. (Heterorhabditidae) dauers enter the mouth, anus, or spiracles of a host and penetrate to the hemocoel. There they release symbiotic bacteria from their gut (*Xenorhabdus* spp. for *Steinernema* spp., *Photorhabdus* spp. for *Heterorhabditis* spp.). These bacteria are essential to establish a successful infection because they produce toxins that kill the host, enzymes that digest the insect tissues, and antimicrobials that inhibit growth of other microorganisms, reducing putrefaction. These antimicrobials may even prove useful against drug-resistant human bacterial pathogens.[52] The bacteria themselves are a major food source supporting the growth and reproduction of the nematodes. Nematode reproduction continues for several generations until supplies of nutrients become limiting, then dauer J3s are produced, and take up symbiotic bacteria before entering the soil to seek new hosts. The relationship between the nematodes and bacteria is a true mutualism. The bacteria require nematodes for transmission between insect hosts (food sources), and the worms need the bacteria for several functions that are critical to the nematode life cycle.

FAMILY RHABDIASIDAE

Phylogenetic analysis of SSU rDNA sequences reveals that Rhabdiasidae are closely related to the Strongyloididae, and this is reflected by similarities between members of these two families. *Rhabdias bufonis* and *R. ranae,* common parasites in lungs of toads and frogs, have very curious life cycles. The parasitic adult is a **protandrous hermaphrodite;** that is, an individual that is a functional male before it becomes a female. Sperm (chromosome number [N] = 5 or 6)[36] are produced in an early male phase and stored in a seminal receptacle. Then the gonad produces functional ova (N = 6) that are fertilized by the stored sperm. Resulting shelled zygotes (2N = 11 or 12) pass up the trachea of the host and then are swallowed, embryonating along the way. Juveniles hatch in the intestine of the frog, and first-stage juveniles (J1s) accumulate in the cloaca to be voided with feces. The J1s often are referred to as **rhabditiform** because the posterior end of their esophagus has a prominent **basal bulb** that

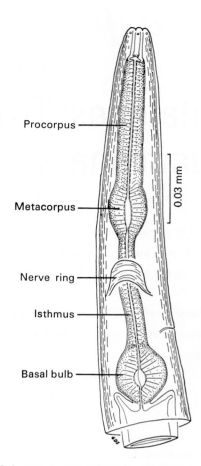

Figure 24.1 Typical rhabditiform esophagus.

From G. D. Schmidt and Robert E. Kuntz, "Nematode parasites of Oceania. XVIII, *Caenorhabditis avicoloa* sp. n. (Rhabditidae) found in a bird from Taiwan," in *Proc. Helm. Soc. Wash.* 39:189–191. Copyright © 1972. Reprinted by permission.

is separated from the anterior portion (**corpus**) by a narrower region (**isthmus**) (Fig. 24.1).

These juveniles undergo four molts to produce a dioecious generation of free-living males (2N = 11) and females (2N = 12). This nonparasitic generation feeds on bacteria and other inhabitants of humusy soil. Its progeny hatch in utero and proceed to consume the internal organs of their mother, destroying her (matricidal hatching). As Shakespeare wrote in *King Lear,* "How sharper than a serpent's tooth it is to have a thankless child!" By the time they escape from the female's body they have become infective J3s. They are referred to as *filariform* at this stage because the esophagus has no apparent basal bulb or isthmus. Filariform juveniles undergo developmental arrest unless they penetrate a toad's or frog's skin. After penetration *Rhabidas* larvae develop in the fascia before penetrating into the host's body cavity, where they molt to become adults. These adults penetrate tissues to reach the lungs where they feed on blood and reproduce.[25] This type of life cycle is **heterogonic;** that is, a free-living generation is interspersed between parasitic generations.

Rhabdias fuscovenosa is a common parasite in lungs of some kinds of aquatic snakes (Natricinae). Its life cycle differs somewhat from those of *R. bufonis* and *R. ranae* in that most eggs from parasitic forms yield filariform juveniles; these juveniles are unable to penetrate the skin of snakes. Instead they enter esophageal tissue following ingestion.

Labels on figure: Procorpus, Metacorpus, Nerve ring, Isthmus, Basal bulb, 0.03 mm

Snake lungworms are also able to establish infections using juveniles in transport hosts, whereas this mode of infection appears to be used infrequently for *Rhabdias* spp. from frogs and toads.[25] Few free-living adults are found;[7] therefore, the life cycle in this species is predominately **homogonic,** and the worm is an obligate parasite.

In the field of conservation biology there is much interest in biological invasions and the impact of nonnative species on native ones. The impact of parasites and pathogenic microorganisms from introduced hosts on native species has also generated recent attention. For example, laboratory studies have demonstrated that *Rhabdias pseudosphaerocephala* from introduced Cane toads in Australia can greatly decrease the viability of metamorphs of an endemic tree frog (*Litoria splendida*), whereas parasitism by this lungworm has no apparent effect on another closely related endemic species, *L. caerulea*.[35] Such differences in response between closely related host species illustrate the difficulty of predicting the ecological impact of invading parasite species. An interesting twist concerning such host-parasite interactions involves the potential use of parasites for the biological control of introduced species. The Puerto Rican coqui frog was accidentally introduced to Hawaii and has significantly impacted tourism due to the extremely loud mating calls of this frog (about 80 dB) in combination with their high densities. In Hawaii, coqui frogs lack the lungworm (*R. elegans*) that parasitizes this host in Puerto Rico. However, laboratory studies designed to examine the effects of *R. elegans* on coquis showed little negative impact,[27] suggesting that this lungworm species is unlikely to be effective as a biocontrol agent of coquis in Hawaii, and that other candidate *Rhabdias* species should be considered.

FAMILY STRONGYLOIDIDAE

Strongyloides Species

Some species of this family are among the smallest nematode parasites of humans, males being even smaller than *Trichinella* spp. More than 50 *Strongyloides* species have been described, mainly from mammals.[47] *Strongyloides stercoralis* is the most common, widespread species infecting humans; *S. fuelleborni* also frequently infects humans in parts of Africa.[19] *Strongyloides stercoralis* infects other primates besides humans as well as dogs, cats, and some other mammals, and strains isolated from humans vary in infectivity for different hosts. Similarly, *Strongyloides fuelleborni* has been reported to infect a variety of primates and dogs.[17] Molecular identifications have revealed that *S. fuelleborni* infects both wild and captive orangutans, and infections of humans with this species in Borneo may be zoonotic.[24] Molecular evidence also suggests that some highly pathogenic human infections in New Guinea may be caused by a previously unrecognized species rather than a subspecies of *S. fuelleborni*.[11] Other species include parasites of other mammals, such as *S. ratti* and *S. venezuelensis* in rats, *S. ransomi* in swine, and *S. papillosus* in sheep. Fewer species have been described from birds, amphibians, and reptiles, but infections can be common in these hosts. Species of *Strongyloides* are remarkable in their ability, at least in some cases, to maintain homogonic, parasitic life cycles or to

repeat free-living generations indefinitely, depending on conditions. The parasitic generation consists only of parthenogenetic females according to most authorities; sperm have not been found in the seminal receptacle of females.[6] In contrast, *Parastrongyloides,* the sister group of *Strongyloides*[11] has both parasitic males and females. The freeliving generation of both *Strongyloides* spp. and *Parastrongyloides* spp. consists of both males and females. Parasites with free-living generations offer the potential for classical genetic investigations, and indeed *Strongyloides ratti* is the first animal-parasitic nematode for which a genetic map has been developed.[30] The following description pertains to *S. stercoralis* (Fig. 24.2).

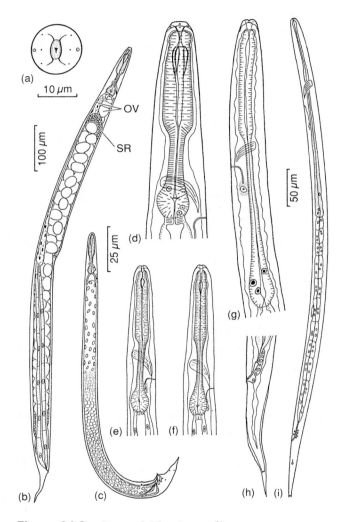

Figure 24.2 *Strongyloides stercoralis.*

(*a*) Free-living female, en face view. (*b*) Free-living female, lateral view (*OV*, ovary; *SR*, seminal receptacle containing sperm). (*c*) Free-living male. (*d*) Anterior end of free-living female, showing details of esophagus. (*e*) Newly hatched J$_1$ obtained by duodenal aspiration from human. (*f*) J$_1$ from freshly passed feces of same patient as was juvenile in (*e*). (*g*) J$_2$ developing to the filariform (J$_3$) stage; cuticle is separating at anterior end. (*h*) Tail of same juvenile as shown in (*g*); notched tail can be observed within J$_2$ cuticle that is separating near tip. (*i*) Filariform J$_3$.

From M. D. Little, "Comparative morphology of six species of *Strongyloides* (Nematoda) and redefinition of the genus," in *J. Parasitol.* 52:69–84. Copyright © 1966. Reprinted by permission.

- **Morphology.** Parthenogenetic females reach a length of 2.0 mm to 2.5 mm. The buccal capsule of both sexes is small, and the animals possess a long, cylindrical esophagus that lacks a basal bulb. The vulva is in the posterior third of the body; uteri contain only a few eggs at a time. Both sexes of free-living adults have a rhabditiform esophagus. Males are up to 0.9 mm long and 40 μm to 50 μm wide. Males have two simple spicules and a gubernaculum; their pointed tail is curved ventrally. Females are stout and have a vulva that is located at the midbody; uteri generally contain more eggs than do those of parasitic females.

- **Biology.** Parasitic females (Fig. 24.3) burrow their anterior ends into submucosa of the small intestine. They are found occasionally in the respiratory, biliary,

or pancreatic system. They produce several dozen, thin-shelled, partially embryonated eggs a day and release them into gut lumen or submucosa. Eggs measure 50 μm to 58 μm by 30 μm to 34 μm. They hatch during passage through the gut or within submucosa, and juveniles escape to the lumen. These J_1s are 300 μm to 380 μm long, and are usually passed with feces. Juveniles either develop into free-living adults or become infective, filariform J_3s with a developmental arrest; these are 490 μm to 630 μm long. Filariform juveniles develop no further unless they gain access to a new host by skin penetration or ingestion. Cues for filariform juvenile attraction include non-specific substances such as carbon dioxide and sodium chloride, whereas urocanic acid in mammalian skin is a specific attractant for *S. stercoralis*.[37]

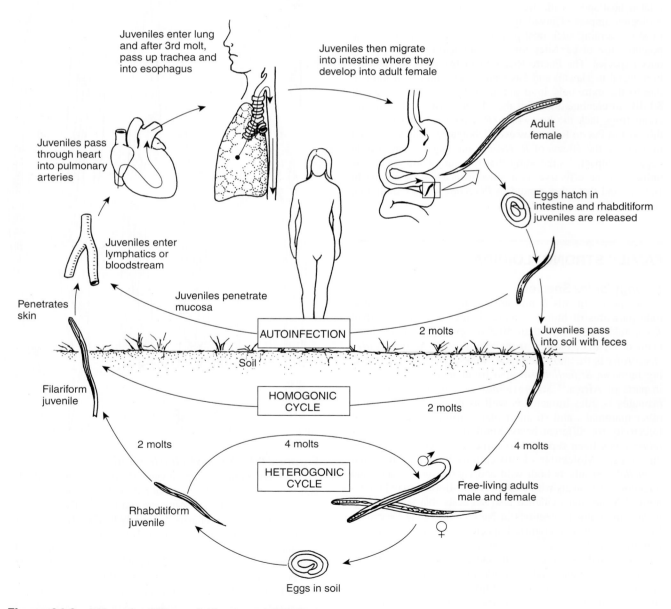

Figure 24.3 **Life cycle of *Strongyloides stercoralis* in humans.**

Drawing by William Ober and Claire Garrison after *Medical Protozoology and Helminthology*. Bethesda, MD: U.S. Naval Medical School, 1959.

If infection is by skin penetration, then juveniles must undergo a tissue migration to reach the small intestine. It has been widely believed that juveniles were carried by blood to the lungs, where they emerged into alveoli and gained access to the digestive system after being carried up the tracheae. However, the migration route of many *Strongyloides* spp. juveniles does not include the lungs.[23, 41] In dogs *S. stercoralis* juveniles seem to "scramble" from infection site to intestine by any route they can.[26] *Strongyloides ratti* accumulates in the nasofrontal region and cerebrospinal fluid of rats, apparently moving subcutaneously.[23] After a period of time, they move through olfactory bulbs and nasal mucosa, reaching the small intestine by 40 to 50 hours after infection. Given the lung symptoms sometimes seen in humans infected with *S. stercoralis,*[14] which can be quite severe,[22] significant numbers of juveniles must migrate through lung tissue.

Strongyloides stercoralis has a single free-living adult generation, whereas *S. ratti* and *S. planiceps* can complete multiple free-living cycles. Free-living adults can produce successive generations of free-living adults. Both parasitic and free-living females can produce juveniles that become filariform, infective J_3s and juveniles that mature into free-living adults. In other words, in some species the homogonic and heterogonic life cycles seem to be mixed. The mechanism that determines whether a given embryo will become a free-living female or a parasitic female is, at least in part, ambient temperature in which juveniles develop. If the temperature is 34°C or higher (temperatures unfavorable for freeliving stages), they become J_3 infective juveniles and will not develop further unless they encounter a potential host.[31] If the ambient temperature is less than 34°C, they tend to molt to J_4 and become free-living females. The developmental switch occurs during J_1 and is mediated by a single pair of amphidial neurons. Other environmental sex determination mechanisms in *Strongyloides* spp. involve host immunity; increased host immune response against the parasite results in more males being produced by parasitic females.[48] Potential advantages of the facultative free-living generation is that sexual reproduction can yield new genetic variants through recombination, and the free-living females are more fecund.

Still other permutations of the life cycle occur. If juveniles have time to molt twice during their transit down the digestive tract, they may penetrate lower gut mucosa or perianal skin, complete migration, and mature. This process is called **autoinfection.** It appears that the host immune response slows down development of juveniles in the gut because there is evidence for an "autoinfective burst" early in infection, before the immune response becomes effective.[42] The potential for autoinfection means that *S. stercoralis* is one of the few multicellular parasites that can multiply within its definitive host. Normally autoinfection is kept in check by the host immune response, but, in immunocompromised hosts, hyperinfection can become overwhelming and life threatening.[9, 14] Nevertheless, disseminated strongyloidiasis is uncommon in organ transplant patients, probably because the drug used to suppress their immune response (cyclosporin) has an anthelmintic effect, and, for reasons that remain unclear,

incidence of disseminated strongyloidiasis is lower than would be expected in AIDS patients.[19] Treatment of humans with corticosteroids greatly aggravates hyperinfection, resulting in life-threatening, disseminated strongyloidiasis.[50] This could result from binding of steroids with a nuclear steroid-hormone receptor that upregulates molting rate.[45]

Nevertheless, whether by normal reinfection, by autoinfection, or by both, cases are known in which patients have had *S. stercoralis* infections for up to 65 years.[19] In 1984 Pelletier[34] reported on 142 American ex-prisoners of war who had worked on the Burma-Thailand Railroad in World War II. Of these, 52 had symptomatic, previously unrecognized strongyloidiasis. Such extreme longevity also has been reported in British and Australian ex-prisoners of war.

- **Epidemiology.**   People typically become infected with *S. stercoralis* by contacting juveniles in contaminated soil or water. Transmammary infection can occur in dogs, and it presumably can also occur in humans.[43] This infection primarily occurs in humans within tropical regions but extends well into temperate zones on several continents. Like most filthborne diseases, strongyloidiasis is most prevalent under conditions of low sanitation standards. Although traditionally a disease of poor, uneducated people in depressed areas of the world, it is often found wherever conditions of filth prevail, such as some mental institutions. It could also become important among the affluent with "vacation hideways" with inadequate sewage disposal facilities, such as in mountain environments. Surveys based solely on stool examination underestimate the extent of infection, particularly when based on a single sample. In a study in the Peruvian Amazon, only 8.7% of participants were found infected by stool examination, but 72% were positive by ELISA.[53] Prevalence of human infection shows an increase with age,[3] with those over 60 having approximately twice the occurrence compared to other age groups in certain countries.[33]

- **Pathology.** Effects of strongyloidiasis may be described in three stages: invasive, pulmonary, and intestinal.

  Penetration of skin by filariform juveniles results in slight hemorrhage and swelling, with intense itching at the site of entry. If pathogenic bacteria are introduced with juveniles, inflammation may result as well.

  During migration, damage to lung tissues results in occasional pulmonary eosinophilic lung infiltrates or wheezing. A burning sensation in the chest, a nonproductive cough, and other symptoms of bronchial pneumonia may accompany this phase. The lung phase can be mistaken for asthma, which, if treated with corticosteroids, can result in rampant autoinfection, as already noted.

  After juvenile females enter crypts of intestinal mucosa, they rapidly mature and invade tissues. They rarely penetrate deeper than the muscularis mucosae, but some cases of deeper penetration have been reported. Worms migrate through mucosa, depositing eggs, and repeatedly burrowing into and exiting the epithelial layer. Esophageal glands secrete adhesion molecules, which form tunnels through which the worms move.[28] An intense, localized burning sensation or aching pain in the abdomen usually is felt at this time. Destruction of tissues by adult worms and juveniles results in sloughing of patches of mucosa, with fibrotic changes in chronic cases, sometimes with death resulting from septicemia (bacterial infection of blood) following intestinal ulceration. Sometimes edema may lead to obstruction of the small intestine and failure of peristalsis.[32]

Infection with *S. stercoralis* is most commonly asymptomatic, but cases are known in which hosts were asymptomatic carriers for years and then developed serious disease. Chronic strongyloidiasis can result in relapsing colitis.[4] Fulminant, fatal hyperinfection can occur in immunocompromised patients. Rarely, renal transplant patients have been infected by receiving donor-infected tissues.[20] Immunosuppression to control organ rejection places such recipients at high risk of hyperinfection.

Diagnosis and Treatment

Demonstration of rhabditiform (or occasionally filariform) juveniles in freshly passed stools is a sure means of diagnosis. A direct fecal smear is often effective in cases of massive infections, but, in more moderate infections, finding juveniles may be very difficult.[39] Special isolation or concentration techniques for juveniles increase chances, but a technique of culturing fecal samples on nutrient agar is most effective.[15] Serodiagnosis by ELISA[40] or detection of *S. stercoralis* antigens in patient serum[44] may be useful. ELISA methods based on coproantigens and detection of *Strongyloides* DNA in feces by real-time PCR also appear promising.[16, 49, 51] Rarely, after purgation or in severe diarrhea, embryonating eggs may be seen in the stool. These resemble hookworm eggs but are more rounded.

Difficulties in diagnosis arise because of day-to-day variability in numbers of juveniles in feces. Furthermore, once autoinfection or hyperinfection becomes established, numbers of juveniles passed in feces may decrease. Duodenal aspiration is a very accurate technique but only applies to duodenal infections: Juveniles farther down the intestine cannot be obtained.

Once juveniles are obtained, problems of identification can occur. First-stage juveniles are similar to rhabditiform hookworm juveniles, which may be present if the stool was constipated or remained at room temperature long enough for hookworm eggs to hatch. Two morphological features can be useful to separate the two: *S. stercoralis* has a short buccal cavity and a large genital primordium, whereas hookworm juveniles have a long buccal cavity and a tiny genital primordium (p. 398).

If a stool has been exposed to soil or water, species of free-living microbivores may invade it, compounding the confusion. Furthermore, filariform juveniles appear in cases of constipation or autoinfection. These juveniles, however, are easily recognized by their notched tails.

Ivermectin is considered to be the drug of choice, but thiabendazole and albendazole can also be effective. These drugs are not effective against juveniles, and therefore multiple treatments may be required.[1, 8] In some individuals the infection may be difficult to eliminate, despite multiple rounds of drug treatment. No entirely satisfactory drug for strongyloidiasis is currently available. Ivermectin shows greatest promise.[1, 6]

Learning Outcomes

By the time a student has finished studying this chapter, he or she should be able to:

1. List differences between a parasite and a parasitoid.
2. Describe how microbivorous soil nematodes are preadapted for endoparasitism.
3. Identify differences between the life cycle of *Strongyloides stercoralis* and other nematodes you have studied.
4. Explain the advantages and disadvantages of sexual versus asexual reproduction in the life cycle of *Strongyloides sterocoralis*.
5. Discuss the process of autoinfection in *Strongyloides sterocoralis* relative to potential benefits to the parasite versus potential harm to the host.
6. Hypothesize concerning the potential adaptive value of differences in the life cycle characteristics of different nematode species.

References

References for superscripts in the text can be found at the following Internet site: www.mhhe.com/robertsjanovynadler9e

Additional Readings

Borgonie, G., A. García-Moyano, D. Litthauer, W. Bert, E. van Heerden, C. Möller, M. Erasmus and T. C. Onstott. 2011. Nematoda from the terrestrial deep subsurface of South Africa. *Nature* 474:79–82.

Krichbaum, K., C. G. Mahan, M. Steele, G. Turner, and P. J. Hudson. 2010. The potential role of *Strongyloides robustus* on parasite-mediated competition between two species of flying squirrels (*Glaucomys*). *J. Wild. Dis.* 46:229–235.

Lewis, E. E. and D. Clarke. 2012. Nematode Parasites and Entomopathogens. In: *Insect Pathology* (Vega and Kaya, eds.). China: Academic Press, pp. 395–424.

Little, M. D. 1966. Comparative morphology of six species of *Strongyloides* (Nematoda) and redefinition of the genus. *J. Parasitol.* 52:69–84.

Viney, M. E. 2006. The biology and genomics of *Strongyloides*. *Med. Microbiol. Immunol.* 195:49–54.

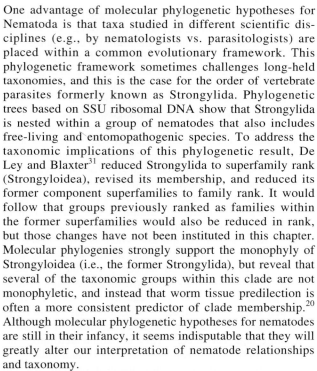

Chapter 25

Nematodes: Rhabditomorpha, Bursate Roundworms

The disease induced by the hookworm . . . was never suspected to be a disease at all. The people who had it were merely supposed to be lazy, and were therefore despised and made fun of, when they should have been pitied.

—Mark Twain *(Letters from the Earth)*

One advantage of molecular phylogenetic hypotheses for Nematoda is that taxa studied in different scientific disciplines (e.g., by nematologists vs. parasitologists) are placed within a common evolutionary framework. This phylogenetic framework sometimes challenges long-held taxonomies, and this is the case for the order of vertebrate parasites formerly known as Strongylida. Phylogenetic trees based on SSU ribosomal DNA show that Strongylida is nested within a group of nematodes that also includes free-living and entomopathogenic species. To address the taxonomic implications of this phylogenetic result, De Ley and Blaxter[31] reduced Strongylida to superfamily rank (Strongyloidea), revised its membership, and reduced its former component superfamilies to family rank. It would follow that groups previously ranked as families within the former superfamilies would also be reduced in rank, but those changes have not been instituted in this chapter. Molecular phylogenies strongly support the monophyly of Strongyloidea (i.e., the former Strongylida), but reveal that several of the taxonomic groups within this clade are not monophyletic, and instead that worm tissue predilection is often a more consistent predictor of clade membership.[20] Although molecular phylogenetic hypotheses for nematodes are still in their infancy, it seems indisputable that they will greatly alter our interpretation of nematode relationships and taxonomy.

A feature of most Strongyloidea is a broad copulatory bursa in males that is supported by rays. Most but not all species of Strongyloidea that are parasites in the intestine have a direct life cycle, requiring no intermediate hosts. Certain species that infect other sites such as lungs or muscles have indirect life cycles, and some use molluscs as intermediate hosts. This chapter will illustrate the parasitological importance of this superfamily chiefly using examples of importance to humans.

FAMILY ANCYLOSTOMATIDAE

Members of this family are commonly known as *hookworms.* They live in their host's intestine, attaching to the mucosa and feeding on blood and tissue fluids sucked from it. A recent review of hookworm infection in humans was provided by Brooker et al.[14]

- **Morphology.** Much similarity of morphology and biology exists among the numerous species in this family, so they will first be given a general consideration. However, hookworm species living in marine mammals have remarkable differences in life cycles, as necessitated by the aquatic life history of their hosts.[64] Most species are rather stout, and the anterior end is curved dorsally, giving the worm a hooklike appearance. The buccal capsule is large and heavily sclerotized and usually is armed with cutting plates, teeth, lancets, or a dorsal cone (Fig. 25.1). Lips are reduced.

  The esophagus is robust, with a swollen posterior end, giving it a club shape. It is mainly muscular, corresponding to its action as a powerful pump. Esophageal glands are extremely large and are mainly outside the esophagus, extending posteriorly into the body cavity. Cervical papillae are present near the rear level of the nerve ring.

  Males have a conspicuous copulatory bursa, consisting of two broad **lateral lobes** and a smaller **dorsal lobe,** all supported by fleshy rays (Fig. 25.2). These rays follow a common pattern in all species, varying only in relative size and point of origin; consequently, they are important taxonomic characters. The number and general pattern of rays is also common to other male nematodes in Rhabditomorpha, including free-living species. Spicules

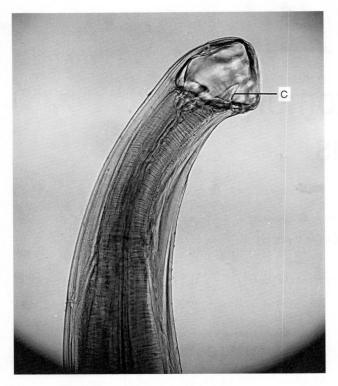

Figure 25.1 **Lateral view of the anterior of *Bunostomum* sp., a hookworm of ruminants.**

Note the large buccal capsule and dorsal flexure, typical of hookworms. The dorsal cone (*C*) bears the duct of the dorsal esophageal gland, similar to that in *Necator americanus*.

Courtesy of Jay Georgi.

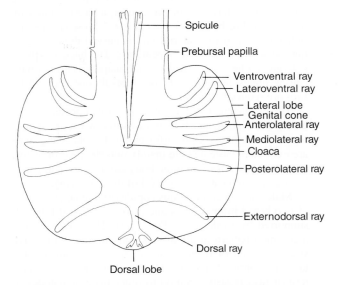

Figure 25.2 **Ventral view of a typical strongyloid copulatory bursa.**

The basic pattern is found in all Strongyloidea. Rays and papillae are also referenced by a numerical labeling system that is used in taxonomic descriptions.

are simple, needlelike, and similar. A gubernaculum is present.

Females have a simple, conical tail. The vulva is postequatorial, and two ovaries are present. About 5% of the daily output of eggs is found in the uteri at any one time; the total production is several thousand per day for as long as nine years.

- **Biology.** Hookworms mature and mate in the small intestine of their host (Fig. 25.3). Embryos develop into two-, four-, or several-cell stages by the time they are passed with feces (Fig. 25.4). Species infecting humans cannot be diagnosed by egg structure or size. Eggs require warmth, shade, and moisture for continued development. Coprophagous insects may mix the feces with soil and air, thus hastening embryogenesis, which is completed within 24 to 48 hours in ideal conditions (Fig. 25.5). Newly hatched J_1s have a rhabditiform esophagus (p. 392) with its characteristic constriction at the level of the nerve ring and a basal bulb with valve, such as that occurring in rhabditiform J_1s of *Strongyloides* spp. In fact, differentiation of hookworm juveniles from those of *Strongyloides* spp. is difficult for a beginner.

Juveniles live in the feces, feeding on fecal bacteria, and molt their cuticle in two to three days. Secondstage juveniles, which also have a rhabditiform esophagus, continue to feed and grow and, after about five days, molt to the third stage, which is infective to a host. Second-stage cuticle may be retained as a loosefitting sheath until penetration of a new host, or it may be lost earlier. Filariform J_3s have a **strongyliform esophagus;** that is, with a reduced basal bulb that is not separated from the corpus by an isthmus. Their intestine is filled with stored nutrients that sustain them through the nonfeeding J_3 period. Hookworm J_3s are similar to filariform J_3s of *Strongyloides* spp. but can be distinguished by the tail tip, which is pointed in hookworms and notched in *Strongyloides* spp. (Fig. 24.2*i*).

Living in the upper few millimeters of soil, J_3s remain in the water film surrounding soil particles. Freezing or desiccation kills them quickly. There is a short, vertical migration in the soil, depending on the weather or time of day. When the ground surface is dry, they migrate a short distance into the soil, following the retreating water. Under ideal conditions, they can live for several weeks. When the ground surface is wet, after rain or morning dew, they move to the surface, remaining in a resting posture until activated.[38] They are stimulated into sinusoidal motion by a variety of environmental cues, such as touch, vibration, water currents, heat, or light. Warmth and moisture stimulate them to stand upright on their tail, waving to and fro in a searching behavior termed *questing* or *nictation* (see Fig. 22.35). Warmth and fatty acids in skin induces penetration behavior.[38]

Infection occurs when J_3s contact a host's skin and burrow into it, and they resume feeding at about this time.[40] They usually shed second-stage cuticle as they penetrate, but presence of cuticle does not preclude resumption of feeding.[50] Activation and resumption of feeding in J_3s are mediated by parallel signalling pathways involving cyclic guanosine 3',5'-monophosphate (cGMP), transforming growth factor beta (TGF-β), and insulin.[13, 39] Juveniles can penetrate any epidermis, although parts most

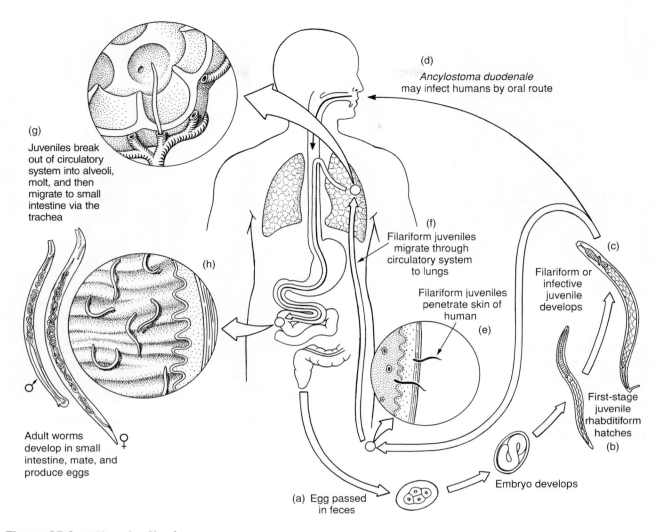

(d)

Ancylostoma duodenale
may infect humans by oral route

(g)
Juveniles break
out of circulatory
system into alveoli,
molt, and then
migrate to small
intestine via the
trachea

(f)
Filariform juveniles
migrate through
circulatory system
to lungs

(h)

Filariform juveniles
penetrate skin of
human

(e)

Filariform or
infective
juvenile
develops

(c)

First-stage
juvenile
rhabditiform
hatches

(b)

Adult worms
develop in small
intestine, mate, and
produce eggs

(a) Egg passed
in feces

Embryo develops

Figure 25.3   Life cycle of hookworms.

(*a*) Embryonated egg passed in feces. (*b*) First-stage juvenile (rhabditiform) hatches. (*c*) Two molts ensue and then infective third-stage juvenile (filariform) enters developmental arrest until it reaches a new host. (*d*) *Ancylostoma duodenale* may infect humans by oral route. In such cases, the juveniles develop to adults without migration to the lungs. (*e*) Filariform juveniles penetrate skin of humans. (*f*) Juveniles migrate through circulatory system to lungs. (*g*) Juveniles break out of circulatory system into alveoli where they molt to the fourth stage; juveniles then migrate to small intestine via the trachea. (*h*) Fourth stage juveniles molt to adults in the small intestine, mate, and produce eggs.

Drawing by William Ober and Claire Garrison.

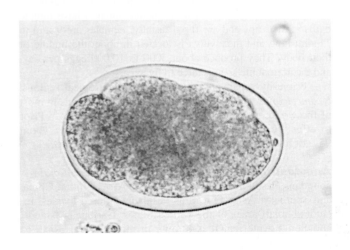

Figure 25.4   *Necator americanus* egg in early cleavage stage, as normally passed in feces.

Size is 50 μm to 80 μm by 36 μm to 42 μm. Eggs of *Ancylostoma duodenale* are of similar size and are not distinguishable.

Photograph by Gerald D. Schmidt.

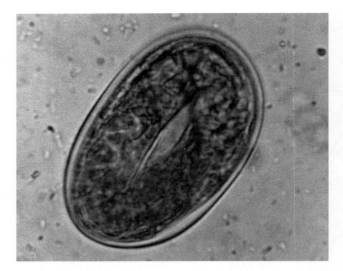

Figure 25.5 Hookworm egg containing fully developed J₁.

Courtesy of Robert E. Kuntz and J. Moore.

often in contact with the soil, such as hands, feet, and buttocks, are most often attacked. *Necator americanus* (and probably other skin-penetrating nematodes) secretes a variety of enzymes that hydrolyze skin macromolecules.[15] This species must penetrate the skin to infect humans, but *Ancylostoma duodenale* can infect by ingestion, skin penetration, in a mother's milk (transmammary infection), and probably transplacentally.[25, 90]

After gaining entry to a blood or lymph vessel, juveniles are carried to the heart and then to the lungs. They break into the air spaces of alveoli where they molt to the fourth stage, which has an enlarged buccal capsule. They are carried by ciliary action up the respiratory tree to the glottis where they are swallowed and finally arrive in the small intestine. There they attach to the mucosa, grow, and molt to the adult stage. After further growth, worms become sexually mature. The worms feed heavily on blood, for which they have a multiprotease cascade to digest host hemoglobin.[88]

At least five weeks are required from the time of infection to the beginning of egg production. However, *A. duodenale* can undergo developmental arrest for up to 38 weeks, its maturation coinciding with seasonal return of environmental conditions favorable to transmission.[10] *Ancylostoma caninum,* a widespread hookworm of dogs and other carnivores, arrests during its tissue migration and then is reactivated in female dogs during lactation, resulting in transmammary transmission.[6] Reactivation is not due directly to hormones of pregnancy, but rather to up-regulation of transforming growth factor-β in uterine and mammary tissues by estrogen and prolactin.[6]

Several genera and many species of hookworm plague humans and domestic and wild mammals. The following two species infect approximately 700 million people worldwide and are responsible for 65,000 deaths each year.[44]

Necator americanus. *Necator americanus,* the "American killer," was first discovered in Brazil and then Texas, but it was later found indigenous in Africa, India, Southeast Asia, China, and the southwest Pacific islands. It probably came

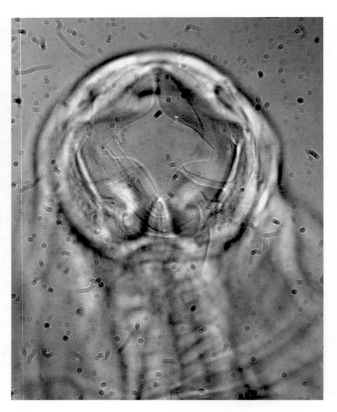

Figure 25.6 En face view of the mouth of *Necator americanus.*

Note the two broad cutting plates in the ventrolateral margins *(top).*

Photograph by Larry S. Roberts.

to the New World with the slave trade. The worm has had an important impact on the economic and cultural development of the southern United States, as well as of other regions of the world in which it occurs.

Necator americanus has a pair of dorsal and a pair of ventral cutting plates surrounding the anterior margin of the buccal capsule (Fig. 25.6). In addition, a pair of subdorsal and a pair of subventral teeth are near the rear of the buccal capsule. The duct of the dorsal esophageal gland opens on a conspicuous cone that projects into the buccal cavity (see Fig. 25.1).

Males are 5 mm to 9 mm long and have a bursa diagnostic for the genus (Fig. 25.7). The needlelike spicules have minute barbs at their tips and are fused distally. Females are 9 mm to 11 mm long and their vulva is located in about the middle of their body. They produce about 5000 to 10,000 eggs per day, and the normal life span is three to five years.[45]

Primarily a parasite of tropical and subtropical regions, *N. americanus* is the most common species in humans in most of the world, accounting for about 85% of infections.[44] Prior to effective hookworm control in the United States, about 95% of hookworms in the southern states were this species.

Ancylostoma duodenale. *Ancylostoma duodenale* is abundant in southern Europe, northern Africa, India, China, and Southeast Asia, as well as in other scattered locales, including small areas of the United States, Caribbean Islands, and South America. It is known in mines as far north as

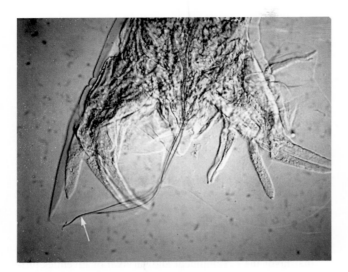

Figure 25.7   **Copulatory bursa and spicules of *Necator americanus.***

The spicules are fused at their distal ends *(arrow)* and form a characteristic hook.

Photograph by Larry S. Roberts.

England and Belgium; since Lucretius, in the first century, it was known to cause a serious anemia in miners. Mines offer an ideal habitat for egg and juvenile development because of their constancy in temperature and humidity. The problem is apt to occur whenever miners are promiscuous in defecation habits.

The anterior margin of the buccal capsule has two ventral plates, each with two large teeth that are fused at their bases (Fig. 25.8). A pair of small teeth is found in the depths of the capsule. The duct of the dorsal esophageal gland runs in a ridge in the dorsal wall of the buccal capsule and opens at the vertex of a deep notch on the dorsal margin of the capsule.

Adult males are 8 mm to 11 mm long and have a bursa characteristic for the species (Fig. 25.9). The needlelike spicules have simple tips and are never fused distally. Females are 10 mm to 13 mm long, with the vulva located about a third of the body length from the posterior end. A single female can lay from 10,000 to 30,000 eggs per day, and the normal life span is one year.[45]

This is the first hookworm for which a life cycle was elucidated. In 1896 Arthur Looss, working in Egypt, was dropping cultures of *A. duodenale* juveniles into the mouths of guinea pigs when he spilled some of the culture onto his hand. He noticed that it produced an itching and redness, and he wondered if infection would occur this way. He began examining his feces at intervals and, after a few weeks, found that he was passing hookworm eggs. He next placed some juveniles on the leg of an Egyptian boy who was to have his leg amputated within an hour. Subsequent microscopic sections showed juveniles penetrating the skin. Looss's monograph on the morphology and life cycle of *A. duodenale* remains one of the most elegant of all works on helminthology.[58]

It is possible for swallowed juveniles to develop to adults without migration through the lungs, but this is probably a fairly rare means of infection.

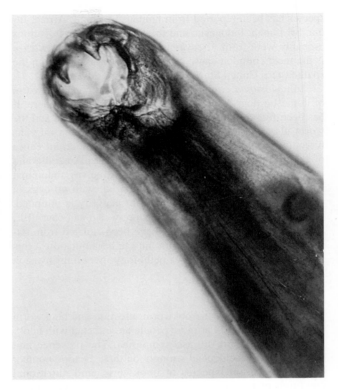

Figure 25.8   *Ancylostoma duodenale,* **dorsal view.**

Notice the powerful ventral teeth.

AFIP neg. no. N-41730-2.

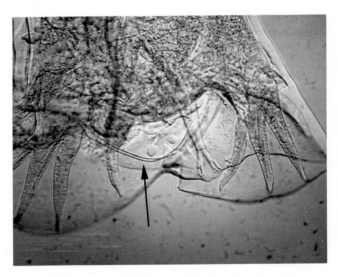

Figure 25.9   **Copulatory bursa and spicules *(arrow)* of *Ancylostoma duodenale.***

The tips of the spicules are not fused into a hook, as in *Necator americanus.*

Photograph by Larry S. Roberts.

Other Hookworms Reported from Humans

Ancylostoma ceylanicum is normally a parasite of carnivores in Sri Lanka, Southeast Asia, and the East Indies, but it has been reported from humans in the Philippines.[85] A morphologically similar species, *A. braziliense,* is found in domestic and wild carnivores in most of the tropics. Although this species has been reported from humans in Brazil, Africa, India, Sri Lanka, Indonesia, and the Philippines, the infections probably were from *A. ceylanicum. Ancylostoma braziliense* is the most common cause of cutaneous larva migrans in the southeastern United States and the New World Tropics.[71]

Ancylostoma caninum is the most common hookworm of domestic dogs, especially in the Northern Hemisphere. It has been found in humans on at least five occasions, and the worm also is a common cause of cutaneous larva migrans (creeping eruption). This hookworm is an important cause of eosinophilic enteritis (EE) in northeastern Australia and now reported in the United States.[23] EE involves abdominal pain with peripheral blood eosinophilia but with no eggs in the feces. Apparently, development to maturity is inhibited, but the presence of even one immature worm can cause EE. *Ancylostoma caninum* juveniles have been isolated from human muscle and associated with muscle inflammation;[57] this species is implicated in other pathology involving invasion of human tissues.[12]

Hookworm Disease

The distinction between hookworm infection and hookworm disease is important. Far more people are infected with hookworms than exhibit disease symptoms. The presence and severity of disease depend strongly on three factors: number of worms present, species of hookworm, and nutritional condition of the infected person. In general fewer than 25 *N. americanus* in a person will cause no symptoms, 25 to 100 worms lead to light symptoms, 100 to 500 produce considerable damage and moderate symptoms, 500 to 1000 result in severe symptoms and grave damage, and more than 1000 worms causes very grave damage that may be accompanied by drastic and often fatal consequences. Because *Ancylostoma* spp. suck more blood than *N. americanus,* fewer worms cause greater disease; for example, 100 worms may cause severe symptoms. However, the clinical disease is intensified by a nutritional condition, corresponding impairment of host's immune response, and other considerations.

The human immune response to hookworm infection is complex, but it is clear that hookworms have evolved strategies that modulate the host's defense system. Survival of hookworms appears to depend upon a balance between immune responses that protect the parasite, such as the Th1 arm, and responses that protect the host such as the Th2 arm. The Th2 response to hookworms is associated with increased levels of worm-specific (and total) IgE with activation of effector cells such as eosinophils and basophils. When attached to the host mucosa, mature hookworms seem to be protected from the host's immune response. The hookworm-induced immunoregulation is believed to be caused by IL-10-producing Th1 cells and other regulatory cells. In contrast to established adults, newly recruited worms appear to cause a strong eosinophilic response that expels them from the small intestine.[24] Several potential mechanisms for evading the host's defensive systems have been discovered. For example, *Ancylostoma* spp. secrete a neutrophil inhibition factor that interferes with activation of neutrophils.[67] *Necator americanus* secretes acetyl cholinesterase, which inhibits gut peristalsis and possibly is an anti-inflammatory factor. It also secretes glutathione-S-transferase and superoxide dismutase, substances that interfere with antibody-dependent, cell-mediated cytotoxicity

(ADCC, p. 29). Nine genes in *N. americanus* code for proteins similar to neutrophil inhibitory factor.[28] Immunomodulation by hookworms appears to protect against asthma more than any other parasite, and intentional infection with small numbers of worms may have benefits for asthma patients.[62] Alternatively, the hookworm molecules responsible for this effect, once characterized, might be developed as a therapeutic. The protection is correlated with higher levels of IL-10 production to hookworm antigens, whereas production of interferon (IFN-gamma) is suppressed.

- **Epidemiology.** From the discussion on biology of hookworms, it is obvious that a combination of poor sanitation and appropriate environmental conditions is necessary for high endemicity.

- **Environmental Conditions.** Environmental conditions that favor the development and survival of juveniles promote transmission. The disease is restricted to warmer parts of the world (and to specialized habitats such as mines in more severe climates) because juveniles will not develop to maturity at less than 17°C, with 23°C to 30°C being optimal. Frost kills eggs and juveniles. Oxygen is necessary for hatching of eggs and juvenile development because their metabolism is aerobic. Thus, juveniles will not develop in undiluted feces or in waterlogged soil. Therefore, a loose, humusy soil that has reasonable drainage and aeration is favorable. Both heavy clay and coarse sandy soils are unfavorable for the parasite, the latter because juveniles are also sensitive to desiccation. Alternate drying and moistening are particularly damaging to juveniles; hence, very sandy soils become noninfective after brief periods of frequent rainfall. However, juveniles live in the film of water surrounding soil particles, and even apparently dry soil may have enough moisture to enable survival, particularly below the surface.

  Juveniles are quite sensitive to direct sunlight and survive best in shady locations, such as coffee, banana, or sugarcane plantations. Humans working in such plantations often have preferred defecation sites, not out in the open where juveniles would be killed by sun, of course, but in shady, cool, secluded spots beneficial for juvenile development. Repeated return of people to the defecation site exposes them to continual reinfection. Furthermore, use of preferred defecation sites makes it possible for hookworms to be endemic in otherwise quite arid areas.

  Juveniles develop best near a neutral pH, and acid or alkaline soils inhibit development, as does the acid pH of undiluted feces (pH 4.8 to 5.0). Chemical factors have an influence. Urine mixed with feces is fatal to eggs, and several strong chemicals that may be added to feces as disinfectants of fertilizers are lethal to free-living stages. Salt in the water or soil inhibits hatching and is fatal to juveniles.

- **Longevity of Juveniles and Adults.** Longevity of the worms is important in transmission to new hosts, continuity of infection in a locality, and introduction to new areas. Juveniles can survive in reasonably good environmental conditions for about three weeks; in protected sites like mines, they can last for a year. There is some dispute about the life span of adults, but a good estimate is 5 to 15 years. A person who moves from an endemic area loses the infection in about that time.

- **Degree of Soil Contamination.**   Obviously, a higher average number of worms per individual will seed the soil with more eggs. Promiscuous defecation, associated with poverty and ignorance, keeps soil contamination high. Use of nightsoil as fertilizer for crops is an especially important factor in eastern Asia.

- **Soil Contact with Skin.**   Because worms penetrate skin, a habit of going barefoot in tropical countries is an elemental contribution to transmission. The role of skin penetration presumably accounts for a general lack of high correlation of hookworm with *Ascaris lumbricoides* and *Trichuris trichiura* infections, which must be acquired by ingestion.[11] This finding has important implications for control strategies.

- **Genetics.**   When hookworm was endemic in the southern United States, it was found that among children (ages 6–16), whites had significantly higher infection intensities than blacks.[75] This suggests a differential susceptibility to infection, and generated much speculation concerning potential contributing factors. Experimental infections of *N. americanus* in human volunteers revealed that infection intensity may vary between individual hosts due to differences in immune response that regulate each host's population size of hookworms.[24] Host genetics are one important factor in susceptibility to and pathogenicity of hookworms, as documented for species from nonhuman hosts. For example, in California sea lions, lower average genetic heterozygosity is associated with higher hookworm intensity, and individuals homozygous for one (as yet uncharacterized) genetic marker are predisposed to greater hookworm-induced anemia.[1]

- **Paratenic Hosts.**   A new dimension in the epidemiology of hookworm disease was the discovery by Schad and coworkers[72] that juveniles will survive in muscles of paratenic hosts. Thus, *A. duodenale,* at least, can be transmitted through ingestion of undercooked meat, including rabbit, lamb, beef, and pork. Pigs can serve as transport hosts for *N. americanus.*[77] Similarly, dogs can be infected with the canid hookworm, *A. caninum,* by ingestion of juveniles in mice, cockroaches, and possibly other paratenic hosts that might be consumed through predation.

- **Coinfection with Other Helminths.**   Higher egg counts have been reported in instances of hookworm coinfection with *Ascaris lumbricoides* (p. 411), suggesting a synergistic effect.[32]

- **Pathogenesis.**   Hookworm disease manifests three main phases of pathogenesis: the cutaneous or invasion period, the migration or pulmonary phase, and the intestinal phase. When a juvenile enters an unsuitable host, it generates another pathogenic condition, which will be discussed separately.

- **Cutaneous Phase.**   The cutaneous phase begins when juveniles penetrate skin. They do little damage to superficial layers, since they seem to slip through tiny cracks between skin scales or penetrate hair follicles. Juveniles are stimulated to penetrate by host fatty acids; they must remain in a water film for successful penetration.[38] Proteases released by J_3 cause an increase in cellular permeability and disruption of vascular endothelial cell junctions. Once in the dermis, however, their attack on blood vessels initiates a tissue reaction that may isolate and kill the worms. If, as usually happens, pyogenic bacteria are introduced into skin with the invading juvenile, a urticarial reaction will result, causing a condition known as **ground itch.**

- **Pulmonary Phase.**   The pulmonary phase occurs when juveniles break out of the lung capillary bed into alveoli and progress up bronchi to the throat. Small hemorrhages occur in the alveoli, and in some cases juveniles induce eosinophilic pneumonia, or Loffler's syndrome. The pulmonary phase is usually asymptomatic, although there may be some dry coughing and sore throat.

- **Intestinal Phase.**   The intestinal phase is the most important period of pathogenesis. On reaching the small intestine, young worms attach to the mucosa with their strong buccal capsule and teeth, and they begin to feed on blood (Fig. 25.10). In human volunteers, initiation of this phase is accompanied by painful eosinophilic enteritis. In heavy

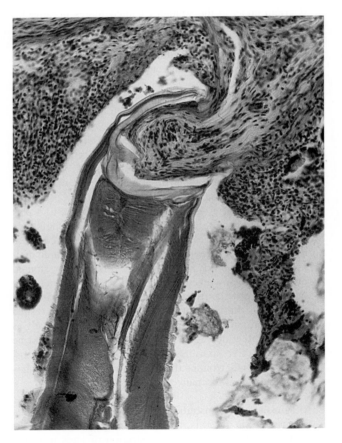

Figure 25.10   Hookworm attached to intestinal mucosa.
Notice how the ventral tooth in the depth of the buccal capsule lacerates the host tissue.

AFIP neg. no. N-33818.

infections, worms are found from the pyloric stomach to the ascending colon, but usually they are restricted to the anterior third of the small intestine. Worms move from place to place, and blood loss is exacerbated by bleeding at sites of former attachment.[37] Hookworms produce proteins that inhibit host blood clotting factors,[34] and these molecules may contribute to bleeding at former feeding sites. Ironically, such anticoagulants may have beneficial medical applications.[33] Worms pass substantially more blood through their digestive tracts than would appear necessary for their nutrition alone, but the reason for this is unknown. Blood loss per worm is about 0.03 ml per day for *N. americanus* and around 0.26 ml per day for *A. duodenale.*

Patients with heavy infections may lose up to 200 ml of blood per day, but around 40% of the iron may be reabsorbed before it leaves the intestine.[56] Nevertheless, a moderate hookworm infection will gradually produce an iron-deficiency anemia as body reserves of iron are used up. Severity of anemia depends on worm load and dietary iron intake of a patient. Anemia during pregnancy can cause serious complications, putting both mother and child at risk. In hookworm endemic regions, iron deficiency resulting from hookworm infection during pregnancy is common.[9] Slight, intermittent abdominal pain, loss of normal appetite, and desire to eat soil (geophagy) are common symptoms of moderate hookworm disease. (Certain areas in the southern United States became locally famous for the quality of their clay soil, and people traveled for miles to eat it. In the early 1920s an enterprising person began a mail-order business, shipping clay to hookworm sufferers throughout the country!)

- **Heavy Infections.** In very heavy infections, patients suffer severe protein deficiency, with dry skin and hair, edema, and potbelly in children and with delayed puberty, mental dullness, heart failure, and death. Intestinal malabsorption is not a marked feature of infection with hookworms, but hookworm disease is usually manifested in the presence of malnutrition and is often complicated by infection with other worms and/or malaria.

The drain of protein and iron is catastrophic to a person subsisting on a minimal diet. In addition, the staple foods of many developing countries, such as cassava, rice, and corn, are poor sources of iron. Such chronic malnutrition, particularly in the young, often causes stunted growth and below-average intelligence, but treatment for the worms can significantly increase fitness, appetite, and growth.[54, 78] Impairment in ability to produce IgG results in lowered antibody response to hookworms as well as to other infectious agents. No living organism could be expected to live up to its potential under such conditions, and it is small wonder that economic development has been so difficult for many tropical countries.

- **Diagnosis.** Demonstration of hookworm eggs or worms themselves in feces is, as usual, the only definitive diagnosis of the disease. Demonstration of eggs in direct smears may be difficult, however, even in clinical cases, and one of the several concentration techniques should be used. If estimation of worm burden is necessary, techniques are available that give reliable data on egg

counts.[27] It is not possible to distinguish hookworm eggs from those of *Oesophagostomum bifurcum* or *Ternidens deminutus*, and this is important in the areas of Africa where *O. bifurcum* and *T. deminutus* are widely prevalent in humans. PCR methods have been described for these identifications, and a multiplex real-time PCR test based on DNA from a 200-ul fecal sample can diagnose species in mixed infections, and provide quantitative results that correlate with egg counts.[83] However, in many developing countries, implementing advanced molecular diagnostics for routine testing is often not practical.[73, 84, 30]

It is not necessary or possible to distinguish *N. americanus* eggs from those of *Ancylostoma* ssp., but care should be taken to differentiate *Strongyloides stercoralis* infections. This is not a problem unless some hours pass between time of defecation and time of examination of feces. Then hookworm eggs may have hatched, and juveniles must be distinguished from those of *S. stercoralis.*

It is desirable, nevertheless, to distinguish *N. americanus* and *Ancylostoma* spp. in studies on the efficacy of various drugs or chemotherapeutic regimens because the two species are not equally sensitive to particular drugs. *Necator americanus,* for example, has low sensitivity to ivermectin, in contrast with *Ancylostoma* spp.[69] Differentiation can be accomplished by recovery of adults after anthelmintic treatment, culturing juveniles from feces, or molecular identification based on single eggs.

- **Treatment.** Mebendazole or albendazole are commonly used for treatment, as they remove both species of hookworm and also any concurrent infection with *Ascaris lumbricoides.* Single-dose therapy is inexpensive and convenient, but reports of drug failure and decreased efficacy for mebendazole suggest that albendazole is now the drug of choice.[3, 44] There is also evidence for resistance of *N. americanus* to mebendazole in Africa.[29] Similarly, the dog hookworm *A. caninum* shows evidence of resistance to the anthelminthic pyrantel.[49] Routine treatment of pregnant women in areas of high hookworm prevalence significantly decreases incidence of infants with very low birthweight.[53]

Treatment for hookworm disease should always include dietary supplementation. In many cases, provision of an adequate diet alleviates symptoms of the disease without worm removal, but treatment for the infection should be instituted, if only for public health reasons.

- **Control.** Control of hookworm disease depends on lowering worm burdens in a population to an extent that remaining worms, if any, can be sustained within nutritional limitations of people without causing symptoms. Mass treatment campaigns do not eradicate the worms but certainly lower the "seeding" capacity of their hosts. Education and persuasion of a population in sanitary disposal of feces are also vital. Economic dependence on nightsoil in family gardens remains one of the most persistent of all problems in parasitology.

Recognizing these factors, the American zoologist Charles W. Stiles persuaded John D. Rockefeller to donate $1 million in 1909 to establish the Rockefeller Sanitary Commission for the Eradication of Hookworm Disease.[2] (The activities of the Commission eventually led to formation of the Rockefeller Foundation and then Rockefeller University.) Beginning state by state and then extending throughout the southeastern United States, the Commission would first survey an area. Residents of the area were examined for infection and then treated with anthelmintics. Thousands of latrines were provided with instructions on how to use and maintain them. It says something about human nature that many people at first refused to use latrines and ultimately were persuaded only with great difficulty. As a result of efforts of this and other similar hygiene commissions, hookworm prevalence is now much lower in some areas of the world. Nevertheless, worldwide prevalence of hookworms is still high; about one-tenth of the Earth's population remains infected.[18, 44]

- **New approaches.**   New tools in molecular biology hold much promise for advances in understanding hookworm biology and implementing their control. For example, the transcriptome of *N. americanus* adults has been analyzed,[22] revealing 18 potential drug targets that lack homologues in the human genome. Because hookworms are relatively closely related to *C. elegans,* inferences of gene orthology in comparison to this free-living species sometimes permit functional predictions for *N. americanus* proteins. The wealth of *C. elegans* research has benefited understanding of hookworm in other cases. For example, the well-characterized signaling pathway that controls dauer-stage juveniles of *C. elegans* is also present in hookworms, where it similarly governs development of the infective third stage.[87]

Efforts to produce a vaccine to combat hookworms have been advanced by new knowledge concerning their biology, particularly molecular aspects of blood feeding.[44] In the intestine, hookworms ingest and digest red cells and serum proteins, absorbing the peptides and amino acids in their intestines. An ordered pathway of hemoglobin-degrading proteases is used by *N. americanus* for digestion. The first in this series of proteins, APR1 (an aspartic protease) and a second protein (GST1) that is involved in detoxification of released heme are being used as antigens for a hookworm vaccine that is under development. In experimental studies, antibodies to these proteins have reduced worm burdens and egg output.[65, 89, 91]

Creeping Eruption

Also known as **cutaneous larva migrans,** creeping eruption is caused by invasive juvenile hookworms of species normally maturing in animals other than humans. Juveniles manage to penetrate the skin of humans but are incapable of successfully completing migration to the intestine. However, before they are overcome by immune effectors, they produce distressing but rarely serious complications of the skin (Fig. 25.11). Species of hookworms from cats, dogs, and other domestic animals are most likely to come into contact with people. *Ancylostoma braziliense,* a common hookworm of dogs and cats, appears to be the most common agent throughout its geographic range.[71] Vacationers to tropical resorts who acquire this infection by sunbathing on beaches may encounter difficulty obtaining a correct diagnosis and medication upon returning home where cooler weather prevails.[82]

After entering the top layers of epithelium, juveniles are usually incapable of penetrating the basal layer (stratum

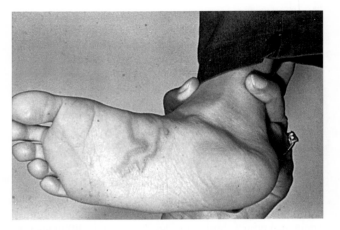

Figure 25.11   Creeping eruption caused by infection with *Ancylostoma* sp. juvenile.

H. Zaiman, Ed., *A pictorial presentation of parasites.* Photo courtesy of F. Battistine.

germinativum), so they begin an aimless wandering. As they tunnel through skin, they leave a red, itchy wound that usually becomes infected by pyogenic bacteria. Juveniles may live for weeks or months. It is known that some can enter muscle fibers and become dormant.[57] Juveniles can attack skin anywhere on the body, but people's feet and hands are more in contact with the ground and so are most often affected. Thiabendazole has revolutionized treatment of creeping eruption, and topical application of a thiabendazole ointment has supplanted all other forms of treatment.

FAMILY STRONGYLIDAE

Members of Strongylidae occur in a variety of mammals, especially herbivores such as horses, in which they are a serious veterinary problem. They are commonly recognized as large strongyles (several species of *Strongylus,* of which *S. vulgaris* is the most important) and small strongyles (mostly the numerous species of *Cyathostomum*).[41] Adults of both are found in the large intestine of equines. Eggs pass out in feces, hatch as J1s, and develop in soil into infective J3s; the latter retain the cuticle of the J_2 as a close-fitting sheath. These crawl onto vegetation and are eaten by grazing hosts. All undergo a migration and period of development in various tissues, the details of which vary with species.

Developing juveniles of *S. vulgaris* migrate into the arteries, especially the anterior mesenteric artery, where they cause thrombosis and arteritis. After three to four months in the arteries, young adults migrate to the intestine where they eventually enter the lumen and reach maturity. Formerly arterial stages of *S. vulgaris* showed 90% to 100% prevalence in horses in the United States, and it was the most feared equine parasite.[41] *S. vulgaris* remains sensitive to benzimidazole and ivermectin anthelmintics, but cyathostomes are relatively resistant. As a result, *S. vulgaris* is almost wiped out, and small strongyles such as *Cyathostomum* spp. are a much bigger problem. A quantitative real-time PCR test has been developed for *S. vulgaris*.[63]

Oesophagostomum spp. are called *nodular* worms; they are parasites of primates, rodents, ruminants, and pigs. Adults live in the large intestine and developing juveniles form nodules in walls of the small and large intestine. Infections are normally acquired by ingestion of J_3. Infections in humans are considered zoonoses. However, *O. bifurcum* is highly prevalent in humans and nonhuman primates in one small area in Africa (northern Togo and Ghana); individuals with hookworm infection have a higher likelihood of being infected with *O. bifurcum*.[92] Approximately 250,000 people are infected in these regions of Africa. Human infection typically presents as a painful abdominal mass that sometimes requires surgical intervention. Eggs of *O. bifurcum* are indistinguishable morphologically from hookworm, but J_3s obtained after fecal culture show clear differences. Ultrasound and DNA analysis can also be used to diagnose *O. bifurcum*.[79, 83, 84] Although morphologically indistinguishable, *O. bifurcum* from humans and three nonhuman primate hosts show levels of genetic differentiation. This observation is consistent with low levels of gene flow between these host-associated populations.[35]

FAMILY SYNGAMIDAE

Syngamus trachea is the gapeworm of poultry, so called because adults live in the trachea of their hosts, causing gasping and gaping. The fowl coughs up eggs, swallows them, and then passes them in feces. Juveniles molt twice in the egg to become infective J_3s. Eggs may or may not hatch in soil, and a variety of terrestrial molluscs, earthworms, and arthropods can serve as paratenic hosts. These worms can survive several years in earthworms, and numerous wild bird species serve as reservoirs. Definitive hosts become infected when they swallow embryonated eggs or juveniles. Infective juveniles penetrate the gut wall, are carried by blood to the lungs where they break out into alveoli, and then proceed up to the trachea. Males remain attached to a female via their copulatory bursa. Young birds are most severely affected and may die with a heavy infection.

FAMILY TRICHOSTRONGYLIDAE

Many genera and an enormous number of species comprise family Trichostrongylidae. They are primarily parasites of the stomach or small intestine of all classes of vertebrates, causing great economic losses in domestic animals, especially ruminants, and in a few cases causing disease in humans.

Trichostrongylids (Fig. 25.12) are small, very slender worms, with a rudimentary buccal cavity in most cases. Lips are reduced or absent, and teeth rarely are present. The cuticle of the head may be inflated. Males have a well-developed bursa, and spicules vary from simple to complex, depending on species. Females are considerably larger than males. The vulva is located anywhere from preequatorial to near the anus, according to species. Worms lay thin-shelled eggs that are in the morula stage.

Life cycles are similar in all species. No intermediate host is required; eggs hatch in soil or water and develop directly into infective J_3s. Some infections may occur through skin, but as a rule juveniles must be swallowed

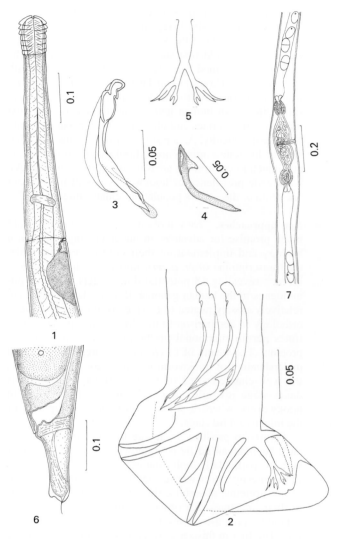

Figure 25.12 *Molineus mustelae,* showing characters typical of Trichostrongylidae.

(*1*) Anterior end, lateral view. (*2*) Posterior end of male. (*3*) Complex spicules, lateral view. (*4*) Gubernaculum, lateral view. (*5*) Dorsal ray of bursa. (*6*) Posterior end of female, lateral view. (*7*) Midregion of female, showing ovijectors. (All scales are in millimeters.)

From G. D. Schmidt, "*Molineus mustelae* sp. n. (Nematoda: Trichostrongylidae) from the long-tailed weasel in Montana and *M. chabaudi* nom. n., with a key to the species of *Molineus*," in *J. Parasitol.* 51:164–168. Copyright © 1965. Reprinted by permission.

with contaminated food or water. Many trichostrongylids undergo exsheathment, where J_3 escape the J_2 cuticle during initial infection. The host stimuli that induce production of exsheathing fluid by the J_3 has been extensively investigated. Enormous numbers of juveniles may accumulate on heavily grazed pastures, causing serious or even fatal infections in ruminants and other grazers. A given host usually is infected with several species since their life cycles are similar, and severe pathogenesis results from the cumulative effects of all the worms. Cost to the sheep industry in Australia was estimated recently at $222 million annually.[59]

Drug resistance in nematodes of livestock has been reported for every class of anthelminthic, and multidrug resistance (MDR) was reported in worms of sheep and goats in the 1980s.[48, 74] MDR in trichostrongylids infecting small ruminants threatens production throughout the world, but particularly in South America, South Africa, Malaysia, and the United States. Resistance by trichostrongyles to benzimidazole drugs (for example, albendazole, mebendazole, and thiabendazole) is increasing and quite ominious.[21, 36]

Haemonchus contortus. *Haemonchus contortus* lives in the "fourth stomach," or abomasum, of sheep, cattle, goats, and many wild ruminants. The species has been reported in humans in Brazil and Australia. It is one of the most important nematodes of domestic animals, causing a severe anemia in heavy infections.

The small buccal cavity contains a single well-developed tooth that pierces a host's mucosa. The blood this species sucks from this wound gives the transparent worms a reddish color. The large females have white ovaries wrapped around the red intestine, lending it a characteristic red and white appearance and leading to its common names: *twisted stomach worm* and *barber-pole worm.* Prominent cervical papillae are found near the anterior end. The male's bursa is powerfully developed, with an asymmetrical dorsal ray (Fig. 25.13). Spicules are 450 μm to 500 μm long, each with a terminal barb. The vulva has a conspicuous anterior flap in many individuals but not in all. Frequency of occurrence of the vulvar flap seems to vary according to strain.

Infection occurs when J₃s, still wearing the loosely fitting second-stage cuticle, are eaten with forage. Exsheathment takes place in the forestomachs. Arriving in the abomasum or upper duodenum, worms molt within 48 hours, becoming J₄s with a small buccal capsule. They feed on blood, which forms a clot around the anterior end of the worms. The worms molt for a final time in three days and begin egg production

about 15 days later. Fourth-stage juveniles can undergo developmental arrest, typically in fall with maturation to adults in spring. Arrest is considered a mechanism promoting survival and transmission in temperate climates, leading to the spring rise in eggs passed in feces of sheep.

Anemia, emaciation, edema, and intestinal disturbances caused by these parasites result principally from loss of blood and injection of hemolytic proteins into the host's system. A host often dies with heavy infections, but those that survive usually effect a self-cure, a result of inflammatory responses in the intestinal mucosa.

Ostertagia Species

Ostertagia spp. are similar to *H. contortus* in host and location, but they differ in color, being a dirty brown—hence, their common name, *brown stomach worm.* The buccal capsule is rudimentary and lacks a tooth. Cervical papillae are present. The male bursa is symmetrical. The vulva has a large anterior flap, and the tip of the female's tail bears several cuticular rings.

Their life cycle is similar to that of *H. contortus* except that J₃s invade gastric glands and elicit nodules. J₃s molt before returning to the lumen, where they feed, molt, and begin producing eggs about 17 days after infection. *Ostertagia* spp. suck blood but not as much as does *Haemonchus contortus.* Species of *Ostertagia* often undergo developmental arrest as J₄.

Some common species of *Ostertagia* are *O. circumcincta* in sheep, *O. ostertagi* in cattle and sheep, and *O. trifurcata* in sheep and goats. Economic losses in the cattle industry due to *O. ostertagi* and other nematodes probably exceed $600 million per year in the United States alone.[76]

Trichostrongylus Species

Trichostrongylus spp. are the smallest members of the family, seldom exceeding 7 mm in length. Many species parasitize the small intestine of ruminants, rodents, pigs, horses, birds, and humans.

They are colorless, lack cervical papillae, and have a rudimentary, unarmed buccal cavity. The male's bursa is symmetrical, with a poorly developed dorsal lobe. Spicules are brown and distinctive in size and shape in each species. The vulva lacks an anterior flap.

Their life cycle is similar to that of *Haemonchus* spp. except that J₃s burrow into mucosa of the anterior small intestine, where they molt. After returning to the lumen, they bury their heads in mucosa and feed, grow, and molt for the last time. Egg production begins about 17 days after infection.

Common species of *Trichostrongylus* are *T. colubriformis* in sheep, goats, cattle, and deer; *T. tenuis* in galliform birds such as grouse, pheasant, chickens, and turkeys; *T. capricola, T. falcatus,* and *T. rugatus* in ruminants; *T. retortaeformis* and *T. calcaratus* in rabbits; and *T. axei* in a wide variety of mammals. Hudson and coworkers showed that the periodic crashes in populations of British red grouse (*Lagopus lagopus scoticus*) were due to negative impact on fecundity caused by build-up of *Trichostrongylus tenuis.*[17, 46]

Approximately 10 species of *Trichostrongylus* have been reported in humans, with records from nearly every country of the world: There are nine species in Iran alone.[66] Prevalence varies from very low to as high as 69% in southwest Iran[70] and 70% in a village in Egypt.[55]

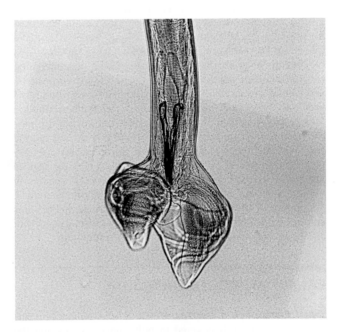

Figure 25.13 *Haemonchus contortus,* **ventral view of male.**

Note asymmetrical copulatory bursa.

Courtesy of Jay Georgi.

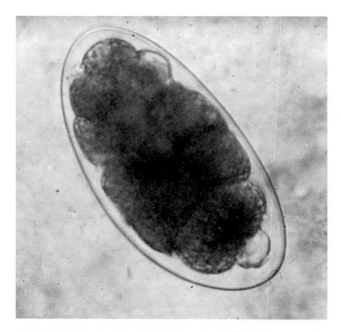

Figure 25.14 *Trichostrongylus* **egg found in a human stool.**

Eggs of *Trichostrongylus* spp. resemble those of hookworms but are usually larger, about 81 μm to 104 μm by 40 μm to 48 μm.

Courtesy of David Oetinger.

Pathological conditions are identical in humans and other infected animals. Traumatic damage to intestinal epithelium may be produced by burrowing juveniles and feeding adults. Systemic poisoning by metabolic wastes of the parasites and hemorrhage, emaciation, and mild anemia may develop in severe infections.

Diagnosis can be made by finding characteristic eggs (Fig. 25.14) in feces or by culturing juveniles in powdered charcoal. Juveniles are very similar to those of hookworms and *Strongyloides* spp., and careful differential diagnosis is required. Molecular diagnostics are available for the common trichostrongylid species from ruminants.[81]

Treatment with thiabendazole or with pyrantel pamoate has proven effective. Cooking vegetables adequately will prevent many infections in humans.

FAMILY DICTYOCAULIDAE

Species in this genus are medium-sized nematodes that as adults parasitize the bronchi and trachea, and are associated with bronchitis in their hosts. *Dictyocaulus filaria* is an important parasite of sheep and goats, but also infects wild antelope and deer. Adults live in bronchi and bronchioles, where females produce embryonated eggs. Eggs hatch while being carried toward the trachea by ciliary action. First-stage juveniles appear in feces and develop to J_3s in contaminated soil, without feeding. Cuticles of both first and second stages are retained by the third stage until the worm is eaten by a definitive host; then cuticles of all these stages are shed together. J_3s penetrate the musosa of the small intestine and enter mesenteric lymph nodes. There they undergo two molts to become small adults (about 500 μm long), enter the circulation by way of the thoracic duct, and parasitize the trachea and bronchi. They commonly cause death of their host.

Fully grown adults are slender and long, with males reaching 80 mm and females 100 mm. The bursa is small and symmetrical; spicules are short and boot shaped in lateral view. The uterus is near the middle of the body. Other species in horses and cattle are similar to *D. filaria* in morphology and biology.

Other Trichostrongyles

In addition to species from ruminants already mentioned, *Cooperia curticei* (Trichostrongylidae), *Nematodirus spathiger*, and *N. filicollis* (Molineidae) should be included among trichostrongyles that often occur in the same host and cause so much damage. *Hyostrongylus rubidus* (Trichostrongylidae) is a serious pathogen of swine and can cause death when present in large numbers. *Heligmosomoides polygyrus* (Heligmosomidae, *H. polygyrus = Nematospiroides dubius*) in mice and *Nippostrongylus brasiliensis* (Heligmonellidae) in rats are easily kept in the laboratory, and they serve as important tools for research on nematode biochemistry, immunology, life cycles, and other topics.

METASTRONGYLES

Metastrongyles are worms in several families formerly placed in superfamily Metastrongyloidea, but recently reclassified as a family by DeLey and Blaxter.[31]

Metastrongyles, along with dictyocaulids, are both nematode groups commonly referred to as lungworms. However, metastrongyles and dictyocaulids are each monophyletic groups related as sister clades according to molecular phylogenies.[16, 20] The common name "lungworm" is really a misnomer for metastrongyles because adults of different species occupy a variety of other vertebrate tissue sites, including skeletal muscles, central nervous system, circulatory system, and frontal sinuses. This is a good example of why biologists prefer to use formal taxonomic names rather than common ones. Most species for which life cycles are known use gastropod intermediate hosts, although earthworms and marine fish serve this role for certain species. Some species also employ a vertebrate or invertebrate transport host. Most species mature in terrestrial mammals, although several species in numerous genera are important parasites of marine mammals. Taxonomy of the group is unsettled. Molecular phylogenies conflict with traditional taxonomies for metastrongyles,[16] but additional refinement of these hypotheses is required.

FAMILY ANGIOSTRONGYLIDAE

Angiostrongylus cantonensis. *Angiostrongylus cantonensis* was first discovered in pulmonary arteries and the heart of domestic rats in China in 1935. Later the worm was found in many species of rats and bandicoots, and it may mature in other mammals throughout Southeast Asia, the East Indies, Madagascar, and Oceanica, with infection rates as high as 88%. As a parasite of rats, it attracted little attention, but 10 years after its initial discovery it was found in the spinal fluid of a 15-year-old boy in Taiwan. It has been discovered since in humans in Hawaii, Tahiti, the Marshall Islands, New Caledonia, Thailand, Vanuatu, the Loyalty Islands, and other places in the Eastern Hemisphere. It is now known to exist in Louisiana, the West Indies, and in the Bahamas.[68] This is another illustration of the value of basic research in parasitology to medicine, because when the medical importance of this parasite was realized, the reservoir of infection in rats already was known. Surveys of parasites endemic to wild fauna of the world remain the first step in understanding epidemiology of zoonotic diseases.

- **Morphology.** *Angiostrongylus cantonensis* is a delicate, slender worm with a simple mouth and no lips or buccal cavity. Males are 15.5 mm to 25 mm long, whereas females attain 19 mm to 34 mm. The bursa is small and lacks a dorsal lobe. Spicules are long, slender, and about equal in length and form. An inconspicuous gubernaculum is present. In females the intertwining of intestine and uterine tubules gives the worm a conspicuous barber-pole appearance. The vulva is about 0.2 mm in front of the anus. Eggs are thin shelled and unembryonated when laid. Eggs are not produced in human infections.

- **Biology.** Eggs are laid in the pulmonary arteries, are carried to capillaries, and break into air spaces, where they hatch. Juveniles migrate up the trachea, are swallowed, and are expelled with feces.

  Many types of molluscs serve as intermediate hosts, including slugs and aquatic and terrestrial snails. Terrestrial planarians, freshwater shrimp, land crabs, and coconut crabs serve as paratenic hosts. Frogs have been found naturally infected with infective juveniles.[7] Experimentally, Cheng[19] infected American oysters and clams, and Wallace and Rosen[86] succeeded in infecting crabs. All juveniles thus produced were infective to rats.

  When eaten by a definitive host, J₃s undergo an obligatory migration to the brain, which they leave four weeks later as subadults. In rats, the time from infection to egg appearance in feces is about six weeks.

- **Epidemiology.** Humans or other mammals become infected when they ingest J₃s. There may be several avenues of human infection, depending on the food habits of particular peoples.[5, 26] In Tahiti it is a common practice to catch and eat freshwater shrimp raw or to make sauce out of their raw juices. It is also possible to eat slugs or snails accidentally with raw vegetables or fruit. In Thailand and Taiwan, raw snails are often considered a delicacy. Infective juveniles escape from slugs and can be left behind in their mucous trail on vegetables over which they crawl.[42] Such juveniles were found on lettuce sold in a public market in Malaya. Fish can serve as paratenic hosts in some circumstances. Thus, although the epidemiology of angiostrongyliasis is not completely known, ample opportunities for infection exist.

- **Pathology.** For many years a disease of unknown cause was recognized in tropical Pacific islands and was named **eosinophilic meningoencephalitis.** Patients with this condition have high eosinophil counts in peripheral blood and spinal fluid in about 75% of cases and increased lymphocytes in cerebrospinal fluid. Neural disorders commonly accompany these symptoms, particularly cranial nerve involvement. We now know that *A. cantonensis* is at least one cause of this condition.

  The presence of worms in blood vessels of the brain and meninges, as well as that of free-wandering worms in brain tissue , or subdural and subarachnoid spaces, results in serious damage. Some effects of such infection are severe headache, fever in some cases, muscle paralysis and speech impairment, stiff neck, coma, and death. The clinical symptoms mimic migraine, brain tumor, and psychoneurosis. In nonsusceptible hosts such as mice and guinea pigs, IL-5 activates eosinophils that kill the worms.[80]

- **Diagnosis and Treatment.** When the symptoms described appear in a patient in areas of the world where *A. cantonensis* exists, angiostrongyliasis should be suspected. It should be kept in mind that many of these symptoms can be produced by hydatids, cysticerci, flukes, *Strongyloides* spp., *Trichinella* spp., various juvenile ascarids, and possibly other lungworms. Alicata[4] and Ash[8] differentiated the juveniles of several species of metastrongylids that could be confused with *A. cantonensis.*

  Thiabendazole and ivermectin show promise in treating infection, but no anthelminthic appears reliably therapeutic. Dead worms in blood vessels and the central nervous system may be more dangerous than live ones. A spinal tap to relieve headache may be recommended.

Other Metastrongyles

Angiostrongylus costaricensis parasitizes mesenteric arteries of many species of rodents in Central and South America, southern North America, and Cuba.[60] Cases in humans have been diagnosed from countries in North, Central, and South America and several Caribbean islands. Worms mature in mesenteric arteries and their branches. In humans, most damage is to the wall of the intestine, especially cecum and appendix, which become thickened and necrotic, with massive eosinophilic infiltration. Abdominal pain and high fever are the most evident symptoms. Evidently, these intestinal disorders are caused by pathogenic changes that affect blood vessels, or pseudo-neoplastic tissue thickening. No symptoms of meningoencephalitis, typical of *A. cantonensis,* are noted.

Angiostrongylus vasorum is a serious, emerging disease of dogs.[61] It has been reported from many countries in Europe, North and South America, and Africa. Adults localize in the right ventricle and pulmonary arteries of dogs and other canids and causes labored breathing, exercise intolerance, weight loss, abdominal and lumbar pain, heart failure, and sudden death. Snails and slugs can serve as experimental intermediate hosts, and frogs as transport hosts. However, the role of different infection sources for wild and domestic canids remains undetermined.[61] Genetic studies indicate that transmission occurs between wild and domestic canids.[47]

Protostrongylus rufescens (family Protostrongylidae) parasitizes bronchioles of ruminants in many parts of the world. Its intermediate hosts are terrestrial snails, in which it develops to the third stage. The definitive host is infected when it eats the snail along with forage. Mountain sheep in America are seriously threatened by this and related species, which take a high toll of lambs every spring. Hibler and co-workers[43] demonstrated transplacental transmission of *Proto-strongylus* spp. in bighorn sheep.

Umingmakstrongylus pallikuukensis (family Protostron-gylidae) is a parasite in lungs of muskoxen in the Canadian Arctic. It has a snail intermediate host, and its transmission dynamics are being radically altered by global warming.[51, 52]

Learning Outcomes

By the time a student has finished studying this chapter, he or she should be able to:

1. List the sequential steps in the life cycle of a human hookworm, from egg to adult.

2. Explain aspects of the hookworm life cycle that appear most vulnerable as targets for hookworm control.

3. Hypothesize about why some hookworm species can success-fully complete their life cycles in humans, whereas other species only cause cutaneous larva migrans.

4. Describe the difference between hookworm infection and hookworm disease in humans.

5. Discuss how use of anthelminthic drugs on trichostrongylid parasites of farm animals might contribute to the development or spread of anthelminthic resistance.

6. Identify examples of human parasites from this chapter that represent zoonotic diseases.

References

References for superscripts in the text can be found at the following Internet site: www.mhhe.com/robertsjanovynadler9e

Additional Readings

Dooley, J. R., and R. C. Neafie. 1976. Angiostrongyliasis: *Angio-strongylus cantonensis* infections. In C. H. Binford and D. H. Connor (Eds.), *Pathology of tropical and extraordinary diseases,* vol. 2, sect. 9. Washington, DC: Armed Forces Institute of Pathology.

Frenkel, J. K. 1976. Angiostrongyliasis: *Angiostrongylus costari-censis* infections. In C. H. Binford and D. H. Connor (Eds.), *Pathology of tropical and extraordinary diseases,* vol. 2, sect. 9. Washington, DC: Armed Forces Institute of Pathology.

Hotez, P. J., and D. I. Pritchard. 1995. (June). Hookworm infection. *Sci. Am.* 272(6):68–74.

Meyers, W. M., and R. C. Neafie. 1976. Creeping eruption. In C. H. Binford and D. H. Connor (Eds.), *Pathology of tropical and extraordinary diseases,* vol. 2, sect. 9. Washington, DC: Armed Forces Institute of Pathology.

Meyers, W. M., R. C. Neafie, and D. H. Connor. 1976. Ancylosto-miasis. In C. H. Binford and D. H. Connor (Eds.), *Pathology of tropical and extraordinary diseases,* vol. 2, sect. 9. Washington, DC: Armed Forces Institute of Pathology.

Pawlowski, Z. S., G. A. Schad, and G. J. Stott. 1991. *Hookworm infection and anaemia: Approaches to prevention and control.* Geneva: World Health Organization.

Schad, G. A., and K. S. Warren (Eds.). 1990. *Hookworm disease. Current status and new directions.* London: Taylor and Francis.

Stoll, N. R. 1972. The osmosis of research: Example of the Cort hookworm investigations. *Bull. N.Y. Acad. Med.* 48:1321–1329.

Chapter 26

Nematodes: Ascaridomorpha, Intestinal Large Roundworms

There are two things for which animals are to be envied: They know nothing of future evils, or of what people say about them.

—Voltaire (1739, correspondence)

Ascaridomorpha includes a diverse group of parasites that live in the alimentary tract of their definitive hosts, and includes species that are of veterinary, medical, and economic importance. The life cycles of these parasites are quite variable, ranging from species with simple direct patterns involving the ingestion of eggs containing infective juveniles, to others that use invertebrates or vertebrates as intermediate or paratenic hosts. Species of Ascaridomorpha are familiar to biologists and laypersons alike as the large intestinal roundworms that infect pet dogs and cats; however, a much wider range of vertebrates serves as definitive hosts, including elasmobranchs, teleost fishes, amphibians, reptiles, birds, and mammals. Ascaridomorpha parasitizing mammals are typically large, stout nematodes with three large lips; however, there is substantial variation in body size and morphological characteristics among genera and species, even though different taxa are superficially similar in structure. Phylogenetic analysis of SSU rDNA sequences has shown that the infraorder itself is not monophyletic,[58] whereas certain families and subfamilies with the infraorder are strongly supported as clades by molecular data.[59] Of several families in this infraorder, this chapter will emphasize Ascarididae (superfamily Ascaridoidea), which includes many species of medical importance. We will discuss representative members of certain other superfamilies briefly.

SUPERFAMILY ASCARIDOIDEA

Family Ascarididae

Ascaridids are among the largest of nematodes, some species achieving a length of 45 cm or more. Three large rounded or trapezoidal lips are present. Cervical, lateral, and caudal alae may be present. Spicules are equal and rodlike or alate. This family contains a cosmopolitan associate of humankind: *Ascaris lumbricoides,* the large intestinal roundworm.[18]

Ascaris lumbricoides and *Ascaris suum.*

Because of their great size and high prevalence, these nematodes may well have been the first parasites known to humans. Certainly, the ancient Greeks and the Romans were familiar with them, and they were mentioned in the Ebers Papyrus. It is probable that *A. lumbricoides* was originally a parasite of pigs that adapted to humans when swine were domesticated and began to live in close association with humans—or perhaps it was a human parasite that we gave to pigs (the physiologies of people and swine are remarkably similar). Populations of *Ascaris* spp. exist in both humans and pigs, but the extent of genetic isolation between these putative species (*A. lumbricoides* and *A. suum,* respectively) has been the subject of much recent research.

The two forms are so close morphologically that they were long considered the same species. Slight differences in the tiny denticles (small "teeth") on the inner edge of the lips were described between these species,[78] but were later found to reflect age-related wear rather than reliable taxonomic characters.[49] None of the genetic markers examined to date consistently discriminates between pig and human-source *Ascaris* spp. Experimental cross-transmission studies show that both putative species can reach maturity in humans and pigs. Genetic studies based on microsatellite markers reveal that there is a low level of hybridization between these species that occurs during co-infection. The distribution of maternally inherited mitochondrial DNA (mtDNA) haplotypes also reveals patterns that are consistent with low levels of cross-infection, but this interpretation is complicated by the possible retention of ancestral mtDNA polymorphisms between these very recently diverged taxa.[3, 17] This seems to be a good example of evolution in action, perhaps with each of these host-associated lineages (*A. suum* and *A. lumbricoides*) continuing to diverge with time now that there may be more barriers to gene flow between parasites from each host, and pigs no longer enjoy the homes of their masters.[56, 57] The following remarks on morphology and biology apply to both species equally, except where otherwise noted.

- **Morphology.** These species are characterized by, in addition to their great size (Fig. 26.1), having three prominent lips, each with a dentigerous ridge, and no alae. Lateral hypodermal cords are visible with the unaided eye.

  Males are 15 cm to 31 cm long and 2 mm to 4 mm in diameter at greatest width. The posterior end is curved ventrally, and the tail tip is blunt. Spicules are simple, nearly equal, and measure 2.0 mm to 3.5 mm long. No gubernaculum is present.

  Females are 20 cm to 49 cm long and 3 mm to 6 mm in diameter. The vulva is about one-third the body length from the anterior end. The ovaries are extensive, and uteri may contain up to 27 million eggs, with 200,000 being laid per day. When transferred to parasite-naive pigs,

female *A. suum* cease producing eggs after two to three weeks.[37] They resume egg production when male worms are transferred.

Fertilized eggs (Fig. 26.2) are oval to round, 45 μm to 75 μm long by 35 μm to 50 μm wide, with a thick, lumpy outer shell (mammillated, uterine, or proteinaceous layer) that is contributed by the uterine wall. When eggs are passed in feces, the mammillated layer is bile-stained a golden brown. The embryos within are usually uncleaved when eggs are passed. An uninseminated female, or one in early stages of oviposition, commonly deposits unfertilized eggs (Fig. 26.3) that are longer and narrower

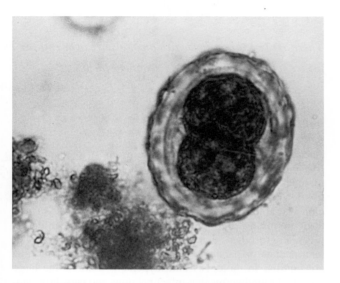

Figure 26.2 Fertilized egg of *Ascaris lumbricoides* from a human stool.

Eggs of this species are 45 μm to 75 μm long.

Courtesy of Robert E. Kuntz.

Figure 26.1 *Ascaris suum,* males (*right*) and females (*left*).

Females are up to 49 cm long.

Courtesy of Ann Arbor Biological Center.

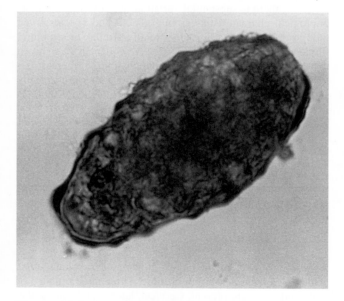

Figure 26.3 Unfertilized egg of *Ascaris lumbricoides* from a human stool.

Such eggs are 88 μm to 94 μm long.

Courtesy of Robert E. Kuntz.

than fertilized ones, measuring 88 μm to 94 μm long by 44 μm wide. Only the proteinaceous layer can be distinguished in unfertilized eggs because the vitelline, chitinous, and lipid layers of the eggshell are formed only after sperm penetration of the oocyte (p. 367).

- **Biology.** A period of 9 to 13 days is the minimal time required for embryos to develop into active J_3s. Embryos are extremely resistant to low temperature, desiccation, and strong chemicals. Sunlight and high temperatures are lethal in a short time (e.g., 2 days at 47°C). Human ascariasis does not occur where average land temperatures exceed 37°–40°C. Clearly, changes in climate variables may change the distribution of ascariasis and other parasitic diseases.[87] Juveniles must molt to the third stage[28] to be infective.

  Infection occurs when unhatched juveniles are swallowed with contaminated food and water. They hatch in the duodenum through an indistinct operculum (Fig. 26.4), where juveniles penetrate the mucosa and submucosa and enter lymphatics or venules (Fig. 26.5). After passing through the right heart, they enter the pulmonary circulation and break out of capillaries into air spaces. Many worms get lost during this migration and accumulate in almost every organ of the body, causing acute tissue reactions. In contrast to this classical pattern, Murrell and his coworkers[55] reported that juvenile *A. suum* did not penetrate the mucosa immediately after hatching but rather rapidly transited the small intestine and penetrated the mucosa of the cecum and upper colon. Then juveniles accumulated in the liver for up to 48 hours. This research on the pig parasite, *A. suum,* strongly suggests that the actual migration pattern of *A. lumbricoides* in humans involves the liver, rather than

the "classical" one determined by experiments with abnormal hosts such as guinea pigs and rats.[18]

While migrating in tissues, juveniles molt to the fourth stage and during a period of about 10 days, grow to a length of 1.4 mm to 1.8 mm. They then move up the respiratory tree to the pharynx, where they are swallowed. Many juveniles make this last step of their migration before molting to the fourth stage, but these J_3s cannot survive gastric juices in the stomach. Fourth-stage juveniles are resistant to such a hostile environment and pass through the stomach to the small intestine, where they molt again and mature. Within 60 to 65 days of being swallowed, they begin producing eggs. Genetic markers show that females may use more than one sire in producing offspring.[90]

It seems curious that these worms embark on such a hazardous migration only to end up where they began. One hypothesis to account for it suggests that migration simulates an intermediate host, which normally would be required during juvenile development for species with indirect life cycles. Indeed, molecular phylogenetic hypotheses confirm that indirect life cycles are ancestral for ascaridoids, and that the direct (one-host) life cycle of *Ascaris* sp. and *Parascaris* sp. is a derived condition.[59] After comparing many nematode taxa having tissue migration with closely related taxa that remain in the gut, Read and Skorping[65] concluded that tissue migration enables faster growth and larger size, thus increasing reproductive capacity.

- **Epidemiology.** The dynamics of *Ascaris* spp. infection are similar to those of *Trichuris trichiura* (p. 378). Indiscriminate defecation, particularly near habitations, "seeds" the soil with eggs that may remain viable for years. Resistance of *Ascaris* spp. eggs to chemicals is almost legendary. They can embryonate successfully in 2% formalin, in potassium dichromate, and in 50% solutions of hydrochloric, nitric, acetic, and sulfuric acid, among other similar inhospitable substances.[74] Eggs can survive in anaerobic sewage lagoon sludge for more than 10 years.[70] This extraordinary chemical resistance is a result of the lipid layer of their eggshell, which contains ascarosides (p. 367).

  Longevity of *Ascaris* spp. eggs also contributes to success of the parasite. Brudastov and coworkers[9] infected themselves with eggs kept for 10 years in soil at Samarkand, Russia. Of these eggs, 30% to 53% were still infective. Because of such longevity, it is impossible to prevent reinfection when yards have been liberally seeded with eggs, even when proper sanitation habits are initiated later.

  Contamination, then, is the typical means of infection. Children are the most likely to become infected (or reinfected) by eating dirt or placing soiled fingers and toys in their mouths. Chickens can serve as paratenic hosts.[62] In regions in which nightsoil is used as fertilizer, principally eastern Asia, Germany, and certain Mediterranean countries, uncooked vegetables become important mechanical vectors of *A. lumbricoides* eggs.[88] Experimental support for this hypothesis came from Mueller,[54] who seeded a strawberry plot with eggs; he and volunteers ate unwashed strawberries from this plot every year for six years and became infected each year. Cockroaches can

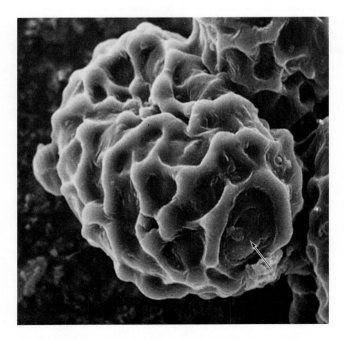

Figure 26.4   Scanning electron micrograph showing an egg of *Ascaris lumbricoides.*

An operculum (*arrow*) is visible at one end.

Courtesy of John Ubelaker.

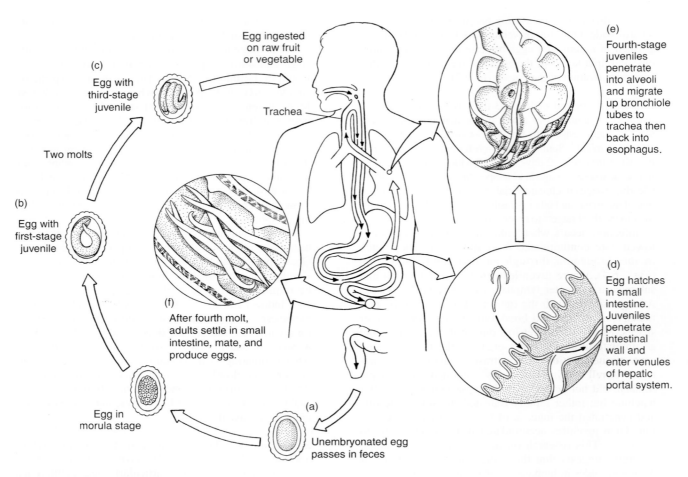

Figure 26.5 **Life cycle of *Ascaris lumbricoides*.**

(*a*) Embryo within shell passes in feces. (*b*) Egg containing J_1. (*c*) Egg containing J_3 ingested on raw fruit or vegetable. (*d*) Egg hatches in small intestine, and juvenile penetrates intestinal wall and enters venules of hepatic portal system. Juveniles undergo further development during migration. (*e*) J_4 emerges into alveoli and migrates in bronchioles to trachea and then up trachea and into esophagus. (*f*) Worms reach small intestine again and develop into adults.

Drawing by William Ober and Claire Garrison.

carry and disseminate *A. lumbricoides* eggs.[11] Similarly, in some areas dogs acquire *A. lumbricoides* eggs by coprophagy and spread viable eggs in their feces.[81] Even windborne dust can carry eggs when conditions permit. Bogojawlenski and Demidowa[6] found *A. lumbricoides* eggs in nasal mucus of 3.2% of schoolchildren examined in the Soviet Union. From the nasal mucosa to the small intestine is a short trip in children. Dold and Themme[23] found *A. lumbricoides* eggs on 20 German banknotes in actual circulation.

Worldwide, 1.27 billion persons, about one-quarter of the world population, are infected.[13] Most infections occur in east Asia, China, sub-Saharan Africa, and South and Central America.[5] Morbidity as assessed by disability-adjusted life years (DALYs) totals ~10.5 million.[13] Severe morbidity occurs in >100 million cases each year;[13] intestinal obstruction, mainly in children, occurs in roughly 1 out of 1000 infections.

Worms are commonly aggregated in local populations, with a small number of people harboring infections of high intensity. These individuals seem to be predisposed

to infection; when they are cured, they tend to become reinfected with large numbers of worms. The reasons for predisposition may be social, behavioral, environmental, and genetic, either alone or in combination. Members of a household tend to have similar infection intensities (household clustering), and individual household risk factors account for much of the variation in household worm counts.[86]

- **Pathogenesis.** Little damage is caused by penetration of intestinal mucosa by newly hatched worms. Juveniles that become lost and wander and die in anomalous locations, such as spleen, liver, lymph nodes, or brain, often elicit an inflammatory response. Symptoms may be vague and difficult to diagnose and may be confused with those of other diseases. Transplacental migration into a developing fetus is also known. Allergy and immunopathology of ascariasis was reviewed by Coles.[14] The polyprotein allergens (lipid binding proteins) of *Ascaris* spp. are known to elicit IgE antibody responses and appear to be a contributing factor in *Ascaris* pneumonitis (Loeffler's pneumonia, see below).

When juveniles break out of lung capillaries into the respiratory system, they cause a small hemorrhage at each site. Heavy infections will cause small pools of blood to accumulate, which then initiate edema (swelling) with resultant clogging of air spaces. Accumulations of eosinophils and dead epithelium add to the congestion, which is known as *Ascaris* pneumonitis. Large areas of lung can become diseased, and, if bacterial infections become superimposed, death can result. Once, an unbalanced student vented his ire on his roommates by "seeding" their breakfast with embryonated *A. suum* eggs. One roommate almost died before his malady was diagnosed.[4, 63]

- **Pathogenesis from "Normal Worm Activities."**   The main food of *Ascaris* spp. is liquid contents of the small intestinal lumen. In moderate and heavy infections the resulting theft of nourishment can cause malnutrition, underdevelopment, and cognitive impairment in small children.[18, 43] Abdominal pains and sensitization phenomena—including rashes, eye pain, asthma, insomnia, and restlessness—often result as allergic responses to metabolites produced by the worms.

   A massive infection can cause fatal intestinal blockage[6] (Fig. 26.6). Why in one case do large numbers of worms cause no apparent problem, whereas in another worms knot together to form a mass that completely blocks the intestine? The drug tetrachloroethylene, which was formerly used to treat hookworm, can cause *A. lumbricoides* to knot up, but other factors remain unknown. Penetration of the intestine or appendix is not uncommon. The resulting peritonitis is usually quickly fatal. According to Louw,[45] 35.5% of all deaths in acute abdominal emergencies of children in Capetown were caused by *A. lumbricoides*.

- **Wandering Worms.**   Wandering adult worms cause various conditions, some serious, some bizarre, all unpleasant. Overcrowding in high-intensity infections may lead to wandering. Downstream wandering leads to the appendix, which can become inflamed or penetrated, or to the anus, with an attendant surprise for an unsuspecting host. Upstream wandering leads to pancreatic and bile ducts, possibly occluding them with grave results. Multiple liver abscesses have resulted from such invasion.[69] Worms reaching the stomach are aggravated by the acidity and writhe about, often causing nausea. The psychological trauma induced in one who vomits a 45-cm ascarid is difficult to quantify. Aspiration of a vomited worm can result in death.[20] Worms that reach the esophagus, usually while the host is asleep, may crawl into the trachea, causing suffocation or lung damage; they may crawl into eustachian tubes and middle ears, causing extensive damage; or they may simply exit through the nose or mouth, causing understandable consternation.

- **Diagnosis and Treatment.**   Accurate diagnosis of migrating juveniles is impossible at this time. Demonstration of juveniles in sputum is definitive, provided a technician can identify them. Most diagnoses are made by identifying the characteristic, mammillated eggs in feces or by an appearance of the worm itself. Adults can also be diagnosed by ultrasound and other noninvasive radiographic methods.[30] So many eggs are laid each day by one worm that direct fecal smears are usually sufficient to demonstrate eggs. *Ascaris lumbricoides* should be suspected when any of the previously listed pathogenic conditions are noted. Most light infections are asymptomatic, and such infections are typically diagnosed only following spontaneous elimination of adults from the anus.

   Benzimidazole-based drugs (e.g., mebendazole, albendazole) are often effective in a single dose. Benzimidazoles bind to tubulin in the worm's intestinal cells and body wall muscles.[10] Emodepside, a novel anthelminthic so far licensed in combination with praziquantel for use in cats, causes relaxation of body-wall muscle of *Ascaris* and inhibits contraction.[89] Nitazoxanide and ivermectin are also effective.[24, 53] In regions endemic for many different soil-transmitted nematodes, certain drugs may be preferable to others due to their broader spectrum of efficacy in cases of multiple-species infections.

Toxocara canis. This species is a cosmopolitan intestinal parasite of domestic dogs and other canids, and it is the chief cause of visceral larva migrans in humans, discussed later.

   As a result of prenatal infections, even puppies in well-cared-for kennels are typically infected at birth, and require anthelminthic treatment. It is not uncommon for 100% of puppies to be infected. The casual owner of a new puppy is likely to be startled by the pet's vomiting up several large, active worms. Puppies tend to have the highest infection prevalence. The infective dose of eggs has a large impact on the success of infection in adult dogs where protective immunity may have a larger role in the fate of juveniles; a smaller number of eggs administered is more likely to lead to patent infection.[25]

   Adults resemble *Ascaris* spp., only are much smaller. Three lips are present. Unlike *Ascaris* spp., however, *Toxocara canis* has cervical alae in both sexes. Males are 4 cm to 6 cm and females are 6.5 cm to more than 15.0 cm long.

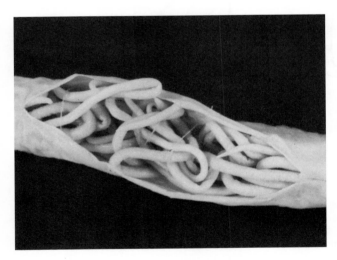

Figure 26.6   **Small intestine of a pig, nearly completely blocked by *Ascaris suum* (threads were inserted to hold worms in place).**

Such heavy infections are also fairly common with *A. lumbricoides* in humans.

Photograph by Larry S. Roberts.

The brownish eggs are almost spherical and roughly 75 μm × 85 μm, with surface pits, and are unembryonated when laid.

- **Biology.** Adult worms live in the small intestine of their host, producing prodigious numbers of eggs, which are passed with the host's feces (Fig. 26.7). Development of J_3s within eggs takes nine days under optimal conditions. The fate of ingested J_3s depends on host age and immunity. If a puppy is young and has had no prior infection, worms hatch and migrate through the portal system and lungs and back to the intestine, as in *A. lumbricoides*. If the host is an older dog, J_3 fate is variable. Most J_3s will not complete the tracheal migration to become adults, but instead will enter the capillaries and undergo a somatic migration, eventually entering developmental arrest, with most individuals residing in the skeletal muscles.

If a bitch becomes pregnant, arrested juveniles apparently are reactivated late in the pregnancy and reenter the circulatory system, where they are carried to the placentas. There they penetrate through to the fetal bloodstream, and migrate to the liver where the reside until birth. Juveniles begin migration to the lungs within 30 minutes following birth, and then undergo a tracheal migration. Thus, a puppy can be born with an infection of *T. canis*, even though the dam has shown no sign of patent infection. The puppy may become infected by the transmammary route, in the mother's milk, but this is probably less common than the transplacental route.[29] If a lactating bitch ingests infective juveniles, they can complete migration to the intestine and produce a patent infection. The molecular basis of *T. canis* juvenile reactivation has not been clearly established.

Another option in the life cycle of *T. canis* is offered when a rodent or other animal ingests embryonated eggs. In this host the juvenile begins to migrate but then becomes dormant and continues its developmental arrest. If the rodent is eaten by a dog, the worms promptly migrate through the lungs to the intestine or into tissues to continue their wait, depending on the the dog's age. Thus, rodents are paratenic hosts. Although this adaptability favors survival of the

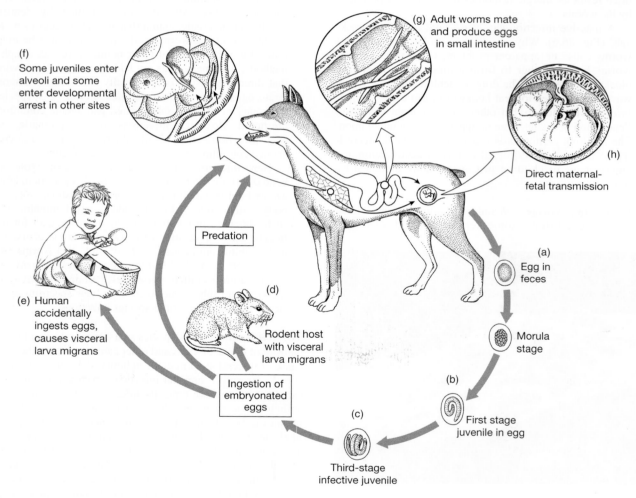

Figure 26.7 **Life cycle of *Toxocara canis*.**

(*a*) Shelled embryo passed in feces. (*b*) J_1 in egg. (*c*) Infective J_3 in egg. (*d*) Eggs hatch in rodent host, and juveniles enter developmental arrest in viscera. (*e*) Eggs hatch in human, and juveniles cause visceral larva migrans. (*f*) After penetration of intestinal wall, some juveniles break out into alveoli, ascend trachea, and finally mature in small intestine. Other juveniles (especially in mature dogs) enter developmental arrest in other sites. (*g*) Adult worms mate and produce eggs in small intestine. (*h*) Direct maternal-fetal transmission.

Drawing by William Ober and Claire Garrison.

parasite, it bodes ill for paratenic hosts, which may undergo behavioral changes as a result of infection that increases their risk of predation.[31]

Such versatility in timing and nature of developmental arrest is clearly adaptive, but it raises questions about how worms can evade host defenses for long periods. Maizels and coworkers found certain novel gene sequences expressed in arrested *T. canis* that had no resemblance to those encoded by the free-living nematode *Caenorhabditis elegans.*[51] These sequences accounted for almost 20% of the cDNA isolated from arrested juveniles and were expressed at high levels. These parasite-specific genes may include proteins that suppress host immune response.

Visceral Larva Migrans. When nematode juveniles gain access to the wrong host species they do not complete the normal migration but undergo developmental arrest and begin an extended, random wandering through various organs and soft tissues of the body. The resulting disease entity is known as **visceral larva migrans (VLM),** in contrast to cutaneous larva migrans (p. 405). Visceral larva migrans can be caused by a variety of spirurid, strongylid, and other nematodes in addition to ascaridoids. However, we will focus on *Toxocara canis,* which is the most common species causing VLM in humans.

- **Epidemiology.** Many years ago it was assumed that dog and cat ascaridoids could not infect humans or were not dangerous to them. In the early 1950s it was discovered that this assumption is not true, particularly for nematodes such as *T. canis.* About 2.2% of adult dogs and 98% of puppies in the United States are infected with *T. canis;* with the population of pet dogs in the United States, more than 1 million dogs are shedding *T. canis* eggs. Thus, risk of exposure to infective eggs is very high. However, most human infections are covert, and even overt symptoms may go unrecognized and unreported.

  Development of a specific immunodiagnostic test, an ELISA using secretory-excretory antigens collected from cultured juveniles, has been a boon to epidemiological studies of visceral larva migrans.[73] This test can distinguish between *Ascaris lumbricoides* and *T. canis,* but does not distinguish *T. canis* from *T. cati.*[47] In the United States, a recent extensive survey showed an overall seroprevalence of 13.9% (in people >6 years), but was much higher for non-Hispanic blacks (21.2%). Other risk factors included low socioeconomic status, living in rural areas, and geographic region.[33, 60] A seroprevalence of 34% has been found among Irish schoolchildren, and 31% in children from Croatia with eosinophilia.[34, 80] Seroprevalence among children in developing tropical countries has been much higher, from 50% to 80%. VLM is predicted to have a substantial impact on individuals living in poverty, both in the United States and abroad.

  Dogs and cats defecating on the ground seed an area with eggs, which embryonate and become infective to any mammal or bird ingesting them, including children. Small mammals are important paratenic hosts; infected mice undergo behavioral changes that increase their risk of predation, which increases the chance to complete the life cycle.[15, 16, 23, 31] Considering that the crawling-walking age of small children is a time when virtually every

available object goes into the mouth for a taste, it is not surprising that the disease is common in children between one and three years old. In an urban setting, dog owners look upon the city park as the perfect place to walk a dog, while parents bring young children there to play on egg-seeded grass. An especially unhappy fact in epidemiology of larva migrans is the high risk to children by exposure to the environment of puppies.[73] Finally a factor to contemplate in light of the foregoing is the durability and longevity of *T. canis* eggs, which are comparable to those of *Ascaris* spp. (discussed previously).

- **Pathogenesis.** Juveniles provoke a delayed-type hypersensitivity reaction in paratenic hosts, and degree and timing of the reaction depend on the infecting dose.[73] In experimental hosts most juveniles eventually end up in the brain; it is unclear whether this is because juveniles have a predilection for the brain or because they are destroyed in other sites but remain in the brain. In sites other than the brain, juveniles are encapsulated by a granulomatous reaction (Fig. 26.8).

  Characteristic symptoms of visceral larva migrans include fever, pulmonary symptoms, hepatomegaly, and

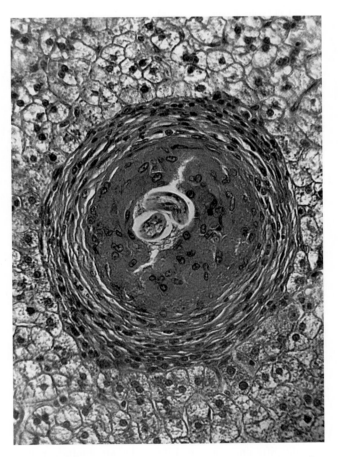

Figure 26.8 *Toxocara canis* **juvenile section in liver of a monkey at nine months' infection.**
The juvenile rests in a matrix of epithelioid cells surrounded by a fibrous capsule lacking intense inflammatory reaction.

From P. C. Beaver, "The nature of visceral larva migrans," in *J. Parasitol.* 55:3–12. Copyright © 1969.

eosinophilia. The extent of damage usually is related to numbers of juveniles present and their ultimate homestead in the body. Various neurological symptoms have been reported, and deaths have occurred when juveniles were especially abundant in the brain. Presence of juveniles in the spinal cord can lead to inflammatory lesions and sensory or motor dysfunction; treatment with albendazole can yield neurologic improvement in such patients.[35] There is little doubt that most cases result in rather minor, transient symptoms such as abdominal pain, headache, and cough. This condition is known as covert or common toxocariasis. The most common site of juvenile residence is the liver (see Fig. 26.8), but no organ is exempt. Rarely, juveniles of *T. canis* cause eosinophilic meningoencephalitis.[84]

Juveniles in an eye cause chronic inflammation of the inner chambers or retina or provoke dangerous granulomas of the retina. These reactions can lead to blindness in the affected eye. The frequency of ocular toxocariasis (OT) in the United States is difficult to assess. OT was diagnosed in 1% of patients examined for vision loss in Alabama eye clinics in 1987,[50] and a noncomprehensive survey conducted by the CDC revealed 68 new U.S. cases in a single year.[46] Other lesions destroy lung, liver, kidney, muscle, and nervous tissues. Generally ocular damage is a result of invasion of only a single juvenile.[73] It may be that, because heavy infections stimulate a much stronger immune response, juveniles survive longer in light infections, giving them more time to wander into an eye.

- **Diagnosis and Treatment.** An ELISA using secretory-excretory antigens has facilitated clinical diagnosis enormously. This test is more sensitive for detecting covert toxocariasis and VLM than ocular disease. A high eosinophilia is suggestive, especially if the possibility of other parasitic infections can be eliminated.

Usually only patients with severe symptoms are treated.[29] Diethylcarbamzine and mebendazole appear to be effective.[52, 77] Control consists of periodic worming of household pets, especially young animals, and proper disposal of the animals' feces. Thus, for toxocariasis, veterinary medicine practices are important to mitigate disease transmission to humans. Some anthelminthics have been reported to be effective against all stages in dogs, including juveniles in arrested development,[1] which presents new options for reducing transmission among dogs. Dogs and cats should be restrained, if possible, from eating available transport hosts. Sandpits in public parks can be protected from contamination by covering them with vinyl sheets when not in use.[82, 83]

Other *Toxocara* Species. *Toxocara cati* is widely prevalent among domestic cats and other felids (Fig. 26.9). The cervical alae (Fig. 26.10) of *T. cati* are shorter and broader than those of *T. canis,* and the eggs of the two species have a slight difference in size. Life cycles are similar, including the use of paratenic hosts, but kittens are infected with *T. cati* only by the transmammary route if queens are infected during late gestation.[29] *Toxocara cati* may be an important cause of visceral larva migrans but it is difficult to determine the relative importance of each species because the current ELISA test for human infection does not

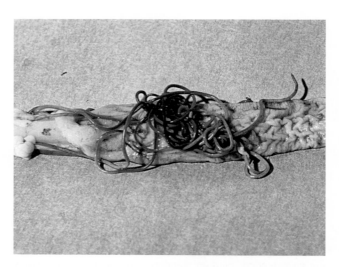

Figure 26.9 **Intestine of a domestic cat, opened to show numerous *Toxocara cati*.**
Courtesy of Robert E. Kuntz.

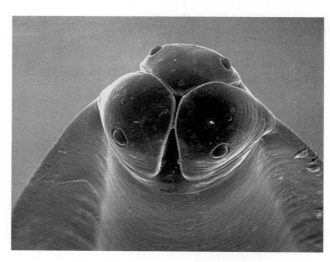

Figure 26.10 **Scanning electron micrograph of *Toxocara cati*. En face view.**
Note the three lips with sensory papillae and broad cervical alae on each side.
Courtesy of John Ubelaker.

distinguish between them. Adult *T. cati* have occasionally been reported from humans.[27]

Toxocara vitulorum is the only ascaridid that occurs in cattle. Its life cycle is similar to that of *T. cati,* with the young being infected by their mother's milk.[66] Adult hosts are refractory to intestinal infection. Young calves may succumb to verminous pneumonia during migratory stages of the parasites. Diarrhea or colic results in economic losses to the owner.

Parascaris equorums. This large nematode and its congener *P. univalens* are the only ascaridoids found in horses and other equids. *Parascaris equorum* is a cosmopolitan species.

It is very similar in gross appearance to *A. lumbricoides* but is easily differentiated by its huge lips, which give it the appearance of having a large, round head. In addition, *Parascaris* spp. individuals are white, whereas fresh *Ascaris* spp. specimens have a reddish color due to their characteristic muscle hemoglobin.

The life cycle is similar to that of *A. lumbricoides,* involving a lung migration. Foals are often infected soon after birth; there is no evidence of prenatal or transmammary transmission. Resulting pathogenesis is especially important in young animals, with pneumonia, bronchial hemorrhage, colic, and intestinal disturbances resulting in unthriftiness and morbidity. Intestinal perforation or obstruction is common. Prevalence and intensity of infection decreases with horse age, presumably due to acquired immunity. In several regions *P. equorum* shows strong resistance to the drug ivermectin, although certain other compounds remain viable alternatives for treatment.[48] The development and spread of drug resistance in nematodes is of concern because relatively few new anthelminthic drugs are being developed.

Baylisascaris procyonis.
This is a very common intestinal parasite of raccoons in North America. Other, related species in this genus occur in bears, skunks, badgers, and other carnivores. When embryonated eggs are ingested by a young raccoon, they will hatch in the small intestine, burrow in the intestinal wall, and mature. Older raccoons are typically infected by juveniles in tissues of rodent paratenic hosts. Other common paratenic hosts include birds and lagomorphs; more than 90 species of birds and mammals have been reported infected.[40] In these animals parasite juveniles wander, often invading the central nervous system, resulting in neurological damage and debilitation, or death. This makes infected hosts vulnerable to predation or scavenging by raccoons, in which juveniles are freed by digestion to grow to maturity. Unfortunately, juveniles affect humans in the same way. Neural larva (juvenile) migrans (NLM) caused by *B. procyonis* occurs almost exclusively in children less than two years old; risk factors for egg ingestion include **geophagia** and **pica.** A substantial fraction of NLM cases are fatal. Ocular larva migrans may occur in association with NLM, or independently when it occurs in adult humans. Serological diagnosis of infection in humans has been difficult, but a new ELISA method based on a recombinant DNA antigen appears promising.[19]

An important epidemiological factor is close contact between humans and raccoons or raccoon feces. Scavenging raccoons are bold animals that prowl near human dwellings and outbuildings. Their preferred communal defecation sites are dangerous sources of infection to humans and other animals.[61] Infected raccoons shed approximately 25,000 eggs per gram of feces, and communal raccoon latrines almost always contain infective eggs. Eggs can remain infective for years under ideal conditions, so once an area is contaminated it is nearly impossible to decontaminate using chemical treatments. Methods using heat such as steam generators or a propane flame gun can be effective for small areas[41] because juveniles within eggs are killed at 62C.[39, 76] Pet kinkajous (another procyonid) sold in the United States have been reported infected with *B. proyconis,*[41] revealing the potential hazards of exotic pets, including raccoons and skunks. Domestic dogs can also serve as hosts of adult *B. procyonis,* and

if such infections were to become common, this could alter factors influencing human infection.

Other species of *Baylisascaris* may have similar pathogenicity, but most hosts are not as likely to come in close contact with humans. Skunks infected with *B. columnaris* are potential hazards, however.

Toxascaris leonina.
Toxascaris leonina is a cosmopolitan parasite of dogs and cats and related canids and felids. It is similar in appearance to *Toxocara* spp., being recognized in the following ways: (1) the body tends to flex dorsally in *T. leonina* and ventrally in *Toxocara* spp.; (2) alae of *T. cati* are short and wide, whereas they are long and narrow in *T. canis* and *T. leonina* (Fig. 26.11); (3) the egg surface is smooth in *T. leonina* but pitted in *Toxocara* spp.; and (4) the tail of male *Toxocara* spp. constricts abruptly behind the cloaca, whereas it gradually tapers to the tail tip in *T. leonina.*

The life cycle of *T. leonina* is simple. Ingested eggs hatch in the small intestine, where juveniles penetrate the mucosa. After a period of growth they molt and return directly to the intestinal lumen, where they mature. Alternatively, juveniles in intermediate hosts such as rodents can infect definitive hosts following predation.

Like for *Toxocara* spp., pathogenicity of *T. leonina* for the definitive host depends on infection intensity, and in

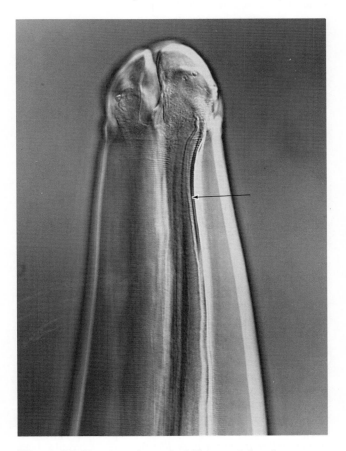

Figure 26.11   Anterior end of *Toxascaris leonina*, an intestinal parasite of dogs, cats, and other canids and felids. Note the narrow cervical alae (*arrow*) as compared with the broad alae of *Toxocara cati.*

Courtesy of Jay Georgi.

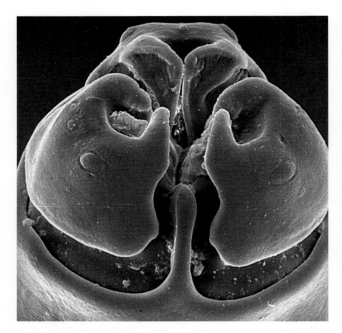

Figure 26.12 *Lagochilascaris turgida.*
Note the prominent cleft in the tip of each lip, typical of the genus (Gr.: *lagos,* hare + *cheilos,* lip).
Courtesy of John Sprent.

severe cases can involve intestinal obstruction or rupture of the intestine. Visceral larva migrans involving *T. leonina* has been implicated as a possible cause of human eosinophilia on St. Lawrence Island (Bering Sea), where this nematode commonly infects arctic fox, working dogs, and vole (rodent) paratenic hosts.[64]

***Lagochilascaris* species.** Relatively little is known about the natural definitive host ranges of the five described species in *Lagochilascaris,* a genus mainly reported from North, Central, and South America. The genus name is derived from the prominent cleft on the inner margin of each lip (Fig. 26.12). These nematodes normally mature in the gastrointestinal tract but seem to have a tendency to develop in abscesses outside the gut. The life cycle is indirect, with juveniles developing to the infective stage within rodent intermediate hosts that ingest the eggs. Embryonated eggs are not directly infective for definitive hosts. *Lagochilascaris minor* and *L. major* have often been reported from domesticated cats; *L. minor* is typically found in subcutaneous abscesses in the head or neck of such hosts whereas it localizes in the stomach, esophagus, and trachea of wild cats in South America and the Caribbean. Domestic cats have been experimentally infected with the third-stage juveniles of *L. minor* from mice.[7] Experimental infections were patent, suggesting that domestic cats may serve as a reservoir for zoonotic infection. The pharynx of domestic cats appears to be the preferred site for *L. major;* this species has also been reported from wild and domestic canids, and the raccoon. Wild cats are believed to represent the natural definitive host for both *L. minor* and *L. major* in South America, but host records are few. In North America, *L. sprenti* uses opossums as its definitive host.

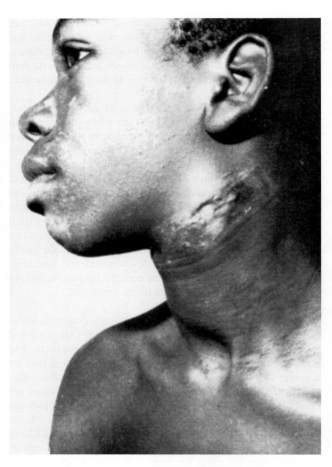

Figure 26.13 **Abscess in the neck of a 15-year-old native of Surinam.**
It contained numerous adults, juveniles, and eggs of *Lagochilascaris* sp. After treatment with thiabendazole, the fistula closed, and the abscess healed, leaving only a small scar.

From B. F. J. Oostburg, "Thiabendazole therapy of *Lagochilascaris minor* infection in Surinam. Report of a case," in *Am. J. Trop. Med. Hyg.* 20:580–583. Copyright © 1971.

Lagochilascaris minor has been reported in humans at least eight times, usually in tonsils, nose, or neck.[79, 85] A fatal brain infection has been reported.[68] When present, worms cause abscesses that may contain from 1 to more than 900 individuals (Fig. 26.13). Juveniles can mature in these locations, and they produce pitted eggs, much like those of *Toxocara* spp. Human infections may last many years, or kill infected people rapidly. How humans become infected is unknown. Humans are unnatural, accidental hosts for this zoonotic infection.

Family Anisakidae

The many species in the family Anisakidae are stomach parasites of fish-eating birds and marine mammals. In the genus *Anisakis,* the life cycle involves passage of eggs in feces of their definitive hosts, embryogenesis and hatching of J_3s, ingestion of J_3s by a crustacean, development in the hemocoel of the crustacean, and then either (1) ingestion by a definitive

Figure 26.14   **Scanning electron micrograph of**
***Terranova* sp. juveniles (family Anisakidae) penetrating**
the stomach of a rat on day three postinfection.
Arrows indicate acute lesions caused by juveniles. (Scale bar
= 1 mm)

From T. L. Deardorff et al., "Histopathology induced by larval *Terranova*
(Type HA) (Nematoda: Anisakinae) in experimentally infected rats," in *J. Parasitol.*
69:191–195. Copyright © 1983.

host or (2) ingestion by a fish paratenic host, which is ultimately consumed by a definitive host.[21, 71] Definitive hosts of
Anisakis spp. are marine mammals. Two aspects of this life
history are important to humans: aesthetics and public health.
The first relates to the disgust experienced by persons who
find large, stout juvenile anisakids in the flesh of the meal
they are preparing or eating. Many a finnan haddie has ended
up in the garbage pail when *Anisakis* sp. juveniles were discovered in it.

More importantly, living *Anisakis* spp. juveniles can
produce pathological conditions in humans who eat them in
raw, salted, marinated, or pickled fish. Such conditions may
be asymptomatic, mild, or severe.[8] Symptoms generally com-
mence when juveniles begin to penetrate the stomach lining
or intestinal mucosa (Fig. 26.14), and this may happen from
1 to 12 hours after ingestion of infected seafood (gastric)
or after up to 14 days in the case of intestinal penetration.
Symptoms may include severe epigastric pain, nausea, vomiting, diarrhea, and hives, but the disease may be confused
with other disorders, such as peptic ulcers. Sometimes severe
IgE-mediated hypersensitivity reactions occur, and because
the allergens may be heat resistant, cooking fish may not render them harmless.[12]

Diagnosis of gastric anisakiasis by endoscope and removal of worms with biopsy forceps is effective, although
catching lively worms with a forceps may be challenging.[21]
In intestinal anisakiasis, or cases in which the worm has fully
penetrated into submucosa or migrated beyond the gastrointestinal tract, diagnosis is more problematic, and symptoms
can mimic a number of other, more common conditions. In
such cases serodiagnosis can be very helpful, and recombinant antigens have made detection of IgE antibodies highly
specific.[2, 72]

Most cases have been reported from Japan, Korea,
Spain, and Scandinavia, where raw or marinated fish is rel-
ished. Approximately 2000 cases per year are reported from

Japan, where it is a major foodborne disease, and the number
of cases reported from the United States is increasing.[21, 38]
Fatalities due to peritonitis have been recorded.[8]

Anisakis spp. juveniles are the most frequent cause
of anisakiasis, but the name of this disease is a misnomer
because other anisakid genera, and even species from other
families (such as Raphidascarididae), can be responsible. A
common feature of the causative organisms is that they are
transmitted through aquatic food chains that involve invertebrates and most typically fish paratenic hosts; these paratenic
hosts can be infective for humans.

Cooking kills juveniles, but continued popularity of
rawfish dishes, such as sushi, sashimi, ceviche, and lomi-
lomi, ensures a continued risk of human infection. Commercial blast freezing causes little change in the texture or taste
of fish, and the process kills *Anisakis* juveniles.[22]

SUPERFAMILY HETERAKOIDEA

This superfamily includes gastrointestinal parasites of amphibians, reptiles, birds, and some mammals. The life cycle
of heterakoid nematodes is simple; eggs containing the infective third-stage juvenile are ingested by the definitive host,
although for some species, paratenic hosts can be involved.
Phylogenetic analysis of SSU rDNA sequences reveals that
as currently defined, this superfamily is not monophyletic
and requires taxonomic revision.[58] Two genera, *Ascaridia*
and *Heterakis,* include important parasites of birds, and both
impact on rearing of commercial poultry. In some countries,
regulatory bans on keeping laying hens in metal cages have
led to husbandry conditions that increase transmission of
these nematodes, providing new challenges to their control.[36]

Family Ascaridiidae

Ascaridia galli is a cosmopolitan parasite of the small intestine of domestic fowl and game birds. Males reach a length
of 77 mm, and females reach 115 mm.

Juveniles within eggs hatch after they are ingested with
contaminated food or water. The life cycle does not involve
extensive tissue migration. Instead, eight or nine days after
infection, juveniles molt to the third stage and begin to burrow into the mucosa, where they generally remain with their
tails still in the intestinal lumen. After molting to J_4 at about
18 days, they return to the lumen, where they undergo their
final molt. Probably a majority of worms complete their two
molts and attain maturity without ever leaving the lumen.
However, some juveniles burrow their anterior ends into the
intestinal mucosa where they remain for up to two months
before molting and returning to the lumen to complete development to the adult stage. Those that attack the mucosa
cause extensive damage, and *A. galli* causes production
losses in chickens. High-intensity infections can obstruct the
small intestine and cause death. In addition, adult *A. galli*
are sometimes found in chicken eggs destined for human
consumption. This is obviously of concern to egg producers.
Improved management practices to control infection through
sanitation are important because in some countries few anthelminthics are approved for use in poultry.

Family Heterakidae

Heterakis gallinarum is cosmopolitan in domestic chickens and turkeys. It was probably brought to the United States in imported ring-necked pheasants. The worms live in the cecum, where they feed on its contents. *Heterakis gallinarum* is unusual because in chickens it serves as a vector of the parasitic protozoan, *Histomonas meleagridis,* the causative agent of blackhead. Hence, we encounter the curious phenomenon of one parasite acting as an intermediate host and vector of another.

Three large lips and an esophageal basal bulb as well as lateral alae are found in this genus. Males are as long as 13 mm and possess wide caudal alae supported usually by 12 pairs of papillae (Fig. 26.15). Their tail is sharply pointed, and there is a prominent preanal sucker. Spicules are strong and dissimilar, and a gubernaculum is absent. Females have the vulva near the middle of their body and a long, pointed tail.

Several species of *Heterakis* are known from birds, particularly in ground feeders, and one species, *H. spumosa,* is cosmopolitan in rodents.

- **Biology.** Eggs of *H. gallinarum* contain a developing zygote when laid. They develop into the infective stage in 12 to 14 days at 22°C and can remain infective for four years in soil. Infection is contaminative: When embryonated eggs are eaten, third-stage juveniles hatch in the

Figure 26.15 **Posterior end of male** *Heterakis variabilis,* **a parasite of pheasants that is similar to** *H. gallinarum.*

Note the conspicuous preanal sucker and copulatory bursa with rays (ventral view).

From W. G. Inglis, et al., "Nematode parasites of Oceanica. XII. A review of *Heterakis* species, particularly from birds of Taiwan and Palawan," in *Rec. S. Aust. Mus.* 16:1–14. Copyright © 1971.

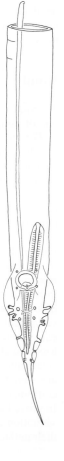

gizzard or duodenum and pass down to the ceca. Most complete their development in the lumen, but some penetrate the mucosa, where they remain for two to five days without further development. Returning to the lumen they mature about 14 days after infection.

If eaten by an earthworm, a juvenile may hatch and become dormant in the worm's tissues, remaining infective to chickens for at least a year. Since these nematodes do not develop further until eaten by a bird, an earthworm is a paratenic host. Grasshoppers, flies, and sowbugs can also serve as mechanical vectors of eggs.

- **Epidemiology.** As a result of the longevity of the eggs, it is difficult to eliminate *H. gallinarum* from a domestic flock. The many different mechanisms for persistent contamination of poultry farms by eggs remains a challenge to implementing sanitation methods, such as cleaning and disinfection, without concurrent use of strict hygiene barriers. In addition, wild birds may also serve as sources of infection. Furthermore, as earthworms feed in contaminated soil, they accumulate large numbers of juveniles, which in turn cause massive infections in unlucky birds that eat them.

- **Pathogenesis.** Generally speaking, *H. gallinarum* is not highly pathogenic in itself. Chickens typically have only minor histopathological lesions, but show localized cellular immune effects, particularly a Th2-dominated response at the site of infection.[75] However, the protozoan, *Histomonas meleagridis,* is transmitted between birds within eggs of *Heterakis gallinarum.*[44] This protozoan is the etiological agent of histomoniasis (blackhead), a particularly serious disease in turkeys where mortality in captive flocks can exceed 85%. Unlike in chickens, blackhead can be directly transmitted between turkeys by fecal contamination. Typically the protozoan is eaten by the nematode and multiplies in the worm's intestinal cells, ovaries, and finally the embryo within the egg (p. 98). Hatching of the worm within a new host releases *Histomonas meleagridis.* In chickens co-infected with *H. gallinarum* and *H. meleagridis,* severe ulceration of the cecal mucosa may occur. The protozoan infection elicits a different, Th1-dominated immune response and a higher T-cell infiltration rate than with infection of *H. gallinarum* alone.[75]

- **Diagnosis and Treatment.** *Heterakis gallinarum* can be diagnosed by finding eggs in feces of its host. Birds allowed to roam a barnyard usually are infected. Worms are effectively eliminated with mebendazole. Usually a flock of birds routinely gets this or other drugs in its feed or water. Other benzimidazole drugs that are effective against juvenile stages, such as albendazole and febendazole, have been shown to be useful for preventing establishment of *H. meleagridis* by preventing nematode infection.[32] Unfortunately, drugs directly effective against *H. meleagridis* were found to be carcinogenic, and are no longer registered for use in poultry. Without effective drugs or a vaccine, control of blackhead disease currently relies on management practices, including prophylaxis by regular deworming.

Learning Outcomes

By the time a student has finished studying this chapter, he or she should be able to:

1. Differentiate among the different types of pathology that can result from infection of humans by different species of ascaridomorph nematodes.

2. Explain how human food containing noninfectious nematode juveniles can pose a hazard for human consumption.

3. Discuss how methods to reduce the transmission of human parasites may differ between zoonotic diseases and parasite species that use only humans as definitive hosts.

4. Assess how raising hosts in a "natural environment" can cause more parasitism than raising hosts in an artificial environment, such as certain modern animal production systems.

5. List the sequential steps in the life cycle of *Ascaris lumbricoides*, beginning with the fertilized egg passed in human feces.

References

References for superscripts in the text can be found at the following Internet site: www.mhhe.com/robertsjanovynadler9e

Additional Readings

Chabaud, A. G. 1974. Keys to subclasses, orders and superfamilies. In R. C. Anderson, A. G. Chabaud, and S. Willmott (Eds.), *CIH keys to the nematode parasites of vertebrates.* Bucks, England: Commonwealth Agricultural Bureaux, Farnham Royal.

Criscione, C. D., J. D. Anderson, D. Sudimack, J. Subedi, R. P. Upadhayay, et al. 2010. Landscape genetics reveals focal transmission of a human macroparasite. *PLoS Negl. Trop. Dis.* 4: e665. doi:10.1371/journal.pntd.0000665

Gavin, P. J., K. R. Kazacos, and S. T. Shulman. 2005. Baylisascariasis. *Clin. Micro. Rev.* 18:703–718.

Little, S. E., E. M. Johnson, D. Lewis, R. P. Jaklitsch, M. E. Payton, B. L. Blagburn, D. D. Bowman, S. Moroff, T. Tams, L. Rich, and D. Aucoin. 2009. Prevalence of intestinal parasites in pet dogs in the United States. *Vet. Parasitol.* 166:144–152.

McDougald, L. R. 2005. Blackhead disease (Histomoniasis) in poultry: A critical review. *Avian Dis.* 49:462–476.

Chapter 27

Nematodes: Oxyuridomorpha, Pinworms

One can be fooled by appearances, which happens only too frequently, whether one uses a microscope or not.

—François-Marie Arouet (Voltaire) in Micromégas (1752)

Members of Oxyuridomorpha are called pinworms because they, especially females, have slender, sharp-pointed tails. As a group, adult oxyurids parasitize a greater taxonomic range of hosts than any other nematode group; definitive hosts include both invertebrate and vertebrate animals. In the context of the Nematoda, Oxyuridomorpha is a member of a larger clade of entirely parasitic taxa (clade III) that includes Ascaridomorpha, Spiruromorpha, and Rhigonematomorpha.[6] Members of Oxyuridomorpha were originally grouped together because they parasitize the posterior gut of animals (both vertebrates and invertebrates) and have a posterior pharyngeal bulb (Fig. 27.1). Phylogenetic analysis of morphological characters supports pinworm monophyly[1] and molecular phylogenetic analysis based on SSU sequences strongly supports Oxyuridomorpha as a clade.[25] Within the pinworms, SSU phylogenies support a sister-group relationship between the two recognized oxyurid superfamilies, the clade representing parasites of arthropods (Thelastomatoidea) and one including parasites of vertebrates (Oxyuroidea).[25] This molecular phylogenetic hypothesis[25] does not lend support to hypotheses depicting the evolution of pinworms of vertebrates from ancestors parasitizing arthropods.[4, 9] The sister group of Oxyuridomorpha is not unequivocally resolved based on SSU sequence data.

Oxyuridomorpha are the only known endoparasites with haplodiploidy. In haplodiploidy, males are haploid and develop parthenogenetically from unfertilized eggs, and females are diploid, developing from fertilized eggs. Haplodiploidy occurs among some rotifers, insects (for example, most Hymenoptera), and Acari (but not ticks). It has important implications for population dynamics and genetic relatedness of individuals. For example, haplodiploid species tend to be divided into small, semi-isolated subpopulations of related individuals.[1, 2, 3] There is a high level of inbreeding, which may be tolerable because deleterious recessives are exposed to selection in the haploid males.

Pinworms of the superfamily Oxyuroidea are common in mammals, birds, reptiles, and amphibians but are rare in fish. Most domestic birds and mammals harbor pinworms, but curiously pinworms are absent from dogs and cats. One species, *Enterobius vermicularis,* is among the most common nematodes of humans. Pinworms of the superfamily Thelastomatoidea parasitize invertebrates, and are very common in terrestrial arthropods such as insects and millipedes. Some species of cockroaches may host 10 different species of pinworms.[18] Thelastomatids are represented by five different families, and four of these appear to have relatively high levels of host specificity, infecting particular families of arthropod hosts. In contrast, the family Thelastomatidae (Figs. 27.2, 27.3) has been reported from many different host groups, and the same pinworm species have been reported to be shared among different arthropod species from the same habitat,[17] suggesting lower host specificity.

FAMILY OXYURIDAE

Enterobius vermicularis. Pinworms have probably infected *Homo sapiens* since the time of our species' origin in Africa.[12, 19] In some ways they are paradoxical among nematode parasites of humans. Their prevalence appears to be greater in temperate regions than in subtropical and tropical areas. Furthermore, pinworms often are found in families at high socioeconomic levels, where, after introduction into the premises by one member, they rapidly become a "family affair." Pinworms are well-adapted for transmission among groups of individuals living in close proximity, such as families, schoolchildren, and in certain institutional settings including day care centers. The greatest pinworm problems are among institutionalized persons, such as those in orphanages and mental hospitals, where conditions facilitate transmission and reinfection. For example, 45% of 148 children in a pediatric hospital in South Africa were infected.[23]

That these worms inhabit at least 400 million people[15] is perhaps less surprising than the fact that practically nothing

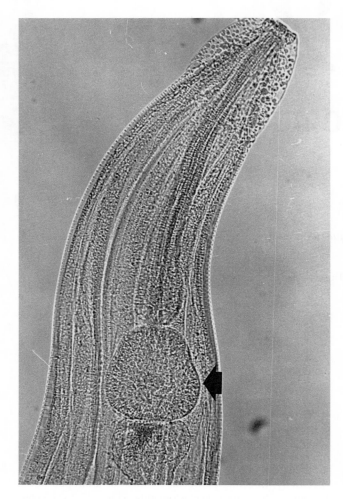

Figure 27.1 **Anterior end of a pinworm *Enterobius vermicularis*.**

Note the large basal esophageal bulb (*arrow*) and swollen cuticle at the head end, typical of this genus.

Courtesy of Warren Buss.

Figure 27.2 **Anterior end of *Cranifera cranifera* female, a parasite of cockroaches, showing marked annulation of the cuticle.**

From R. A. Carreno and L. Tuhela, "Thelastomatid Nematodes (Oxyurida: Thelastomatoidea) from the Peppered Cockroach, *Archimandrita tesselata* (Insecta: Blattaria) in Costa Rica," in *Comp. Parasitol.* 78:39–55, Copyright © 2011.

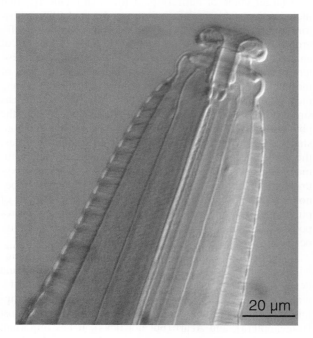

Figure 27.3 **Anterior end of *Cranifera cranifera* female as revealed in optical section by differential interference contrast microscopy.**

Note tubular-shaped stoma (buccal cavity) joining the esophagus to the oral opening. Compare cuticle annules with perspective provided by scanning electron microscopy (Fig. 27.2).

From R. A. Carreno and L. Tuhela, "Thelastomatid Nematodes (Oxyurida: Thelastomatoidea) from the Peppered Cockroach, *Archimandrita tesselata* (Insecta: Blattaria) in Costa Rica," in *Comp. Parasitol.* 78:39–55, Copyright © 2011.

is being done to reduce this infection. One reason is simple and practical: Pinworms cause no obvious debilitating or disfiguring effects. Their presence is considered an embarrassment and an irritation, like dandruff. Resources of countries could scarcely be expected to be used for such an apparently innocuous foe, particularly when there is so much difficulty mounting efforts against more disabling infectious agents.

And, yet, is enterobiasis so unimportant after all? Certainly, it is important to the millions of persons who suffer the discomforts of infection. Furthermore, a great deal of money is spent in efforts to be rid of pinworms. The frantic efforts to rid one's household of the tiny worms often lead to what has been called a "pinworm neurosis." Mental stress suffered by families who know they harbor parasites are unmeasurable but very real consequences of infection. Finally, the pathogenesis of these worms may be greatly underrated.[31] For example, experimental studies using a rodent pinworm system have shown that the infection causes immunological changes that can impact the host allergic response, in addition to gross pathological effects.[22]

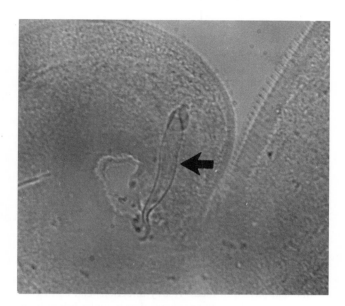

Figure 27.4   **Posterior end of a male *Enterobius vermicularis*, illustrating the single spicule (*arrow*).**

Courtesy of Warren Buss.

- **Morphology.**   In addition to *Enterobius vermicularis*, another species, *E. gregorii*, has been described from humans, but *E. gregorii* is likely only a morphological variant or developmental phase of *E. vermicularis*.[13, 14] Synonymy of *E. gregorii* and *E. vermicularis* is supported by molecular evidence.[26] Both sexes of *E. vermicularis* have three lips surrounding the mouth, followed by a cuticular inflation of the head (see Fig. 27.1). Males have a single spicule, which is 100 μm to 141 μm long (Fig. 27.4). Males are 1 mm to 4 mm long and have their posterior ends strongly curved ventrally. Conspicuous caudal alae are supported by papillae.

    Females measure 8 mm to 13 mm long and have the posterior end extended into a long, slender point (Fig. 27.5), giving pinworms their name. The vulva opens between the first and second thirds of the body. When gravid, the two uteri contain thousands of eggs, which are elongated-oval and flattened on one side with a thin shell (Fig. 27.5), measuring 50 μm to 60 μm by 20 μm to 30 μm.

- **Biology.**   Adult worms congregate mainly in the ileocecal region of the intestine, but developing juveniles and young adults may be found throughout the small intestine. They attach themselves to the mucosa where they presumably feed on epithelial cells and bacteria. Gravid females migrate within the lumen of the intestine, commonly passing out of the anus onto perianal skin. As they crawl about, both within the bowel and on outer skin, they leave a trail of eggs. These eggs do not normally hatch when they are still inside the intestine. One worm may deposit from 4600 to 16,000 eggs. Females die soon after oviposition, and males die soon after copulation. Consequently, many more females than males are recovered from hosts.

    When laid, each egg contains a partially developed juvenile, which can develop to infectivity within six hours at body temperature.[16] Most oxyurids have been reported to have two molts during development of juveniles within

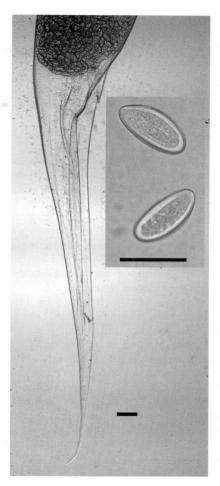

Figure 27.5   **Posterior end of a gravid female *Enterobius vermicularis* (scale bar 100 μm). Inset photo, *E. vermicularis* eggs (scale bar 50 μm).**

The long pointed tail lends this species its name *pinworm*.

Photographs by S. Nadler.

eggs, although in some arthropod pinworms the fully developed juvenile is the second stage.[11, 32] *Enterobius vermicularis* is presumed to have two molts within the egg. Juveniles within shells are resistant to putrefaction and disinfectants but succumb to dehydration in dry air within a day.

Infection occurs by two routes. Most often the eggs, containing J_3s, are swallowed, and they hatch in the duodenum. They slowly move down the small intestine, molting twice to become adults by the time they arrive at the ileocecal junction. The total time from ingestion of the eggs to sexual maturity of the worms is 15 to 43 days.

Some authorities have indicated that if the perianal folds are unclean for long periods, attached eggs may hatch and juveniles may enter the anus and hence to the intestine in a process known as **retrofection** (or **retroinfection**). However, with respect to maintaining an infection in a single host, the relative contributions of retrofection versus self-contamination with infective eggs is unknown.

- **Epidemiology.**   Clothing and bedding rapidly become seeded with eggs when an infection occurs. Even curtains, walls, and carpets become sources of subsequent infection (or reinfections). Eggs have been found in the dust of schoolrooms and school cafeterias,[7] providing a source of infection not only for children but also for teachers and other school personnel. The microscopic eggs are very light and are wafted about by the slightest air currents, depositing them throughout a building. Eggs remain viable in cool, moist conditions for up to a week.

The most common means of infection is through insertion of soiled fingers or other objects in the mouth or through use of contaminated bedding, towels, and other such objects (fomites). Obviously, it becomes next to impossible to avoid contamination when eggs are abundant. Furthermore, it remains impossible to avoid reinfection when retrofection occurs.

Humans can inhale and subsequently swallow airborne eggs, or eggs may remain in the nose until they hatch. This, together with nose picking, accounts for rare cases of pinworm in the nose. Contrary to popular belief, pinworms cannot be transmitted by dogs and cats because these animals are free of pinworms. But their fur can become contaminated with *E. vermicularis* eggs from their environment, and thereby serve as another potential source of infection. *Enterobius vermicularis* and *E. gregorii* have been reported from captive chimpanzees.[13] Oddly, infections of captive chimpanzees with human pinworms can be highly pathogenic, or even fatal[24] with dissemination of worms outside the intestinal lumen.

According to some authorities, 20% to 30% of elementary school children in the United States are infected.[8] In surveys of Southern California elementary schools conducted between 1960–1982[33] pinworm prevalence ranged from 7–43%, with an average of 21% in 1982. Similar average prevalence has been reported in schoolchildren from other U.S. states. Remm[29] reported a prevalence of 24.4% in 954 nursery school children in southeast Estonia, whereas an average prevalence of only 5.7% was reported in nursery schools in Khorrambad (western) Iran.[27]

• **Pathogenesis.** About one-third of infections are completely asymptomatic, and, in many more, clinical symptoms are negligible. Nevertheless, very large numbers of worms may be present and lead to more serious consequences. Pathogenesis has three aspects: damage caused by worms within the intestine, damage caused by extraintestinal migration of adult worms, and damage resulting from egg deposition around the anus. Minute ulcerations of intestinal mucosa from attachment of adults may lead to mild inflammation and bacterial infection.[30] Movements of females out of the anus to deposit eggs, especially when the patient is asleep, lead to a tickling sensation of the perianus, causing the patient to scratch. Eggs and dead female pinworms present on the perianal skin can also cause dermatitis and itching. The subsequent vicious circle of bleeding, bacterial infection, and intensified itching can lead to a nightmare of discomfort, but the resultant scratching is conducive for transmission of pinworms.

Although *E. vermicularis* infections are usually asymptomatic, migration of adult worms outside the intestine can lead to serious complications. Cases have been reported of pinworms wandering up the vagina, uterus, and oviducts into the coelom, to become encased by granulomatous

tissue in the peritoneum. They have been known to become encapsulated within an ovarian follicle,[4, 5] and cause inflammatory tissue reactions within fallopian tubes that lead to infertility.[20] Granulomas caused by *E. vermicularis* have been reported in perianal tissue and even in the vulva.[21, 31] There is a higher incidence of pinworm infection in individuals with chronic appendicitis, but no evidence of a relationship in cases of acute appendicitis.[20]

A variety of other symptoms have been ascribed to heavy pinworm infection in children, but establishment of a causal relationship is lacking in most instances. These symptoms include: nervousness, restlessness, irritability, loss of appetite, nightmares, insomnia, bed wetting, grinding of the teeth, perianal pain, nausea, and vomiting.

• **Diagnosis and Treatment.** Positive diagnosis can be made only by finding eggs or worms on or in the patient. Ordinary fecal examinations often give false negatives because relatively few eggs are deposited within the intestine and passed in feces. Heavy infections can be discovered by examining the perianus closely under bright light, during the night or early morning. Wandering worms glisten and can be seen easily.

When adults cannot be found, eggs often can be, as they are left behind on the perianal skin. A short piece of cellophane tape, held against a flat, wooden applicator or similar instrument, sticky side out, is pressed against the junction of the anal canal and the perianus. The tape is then reversed and stuck to a microscope slide for observation. If a drop of xylene or toluene is placed on the slide before the tape, it will dissolve the glue on the tape and clear away bubbles, simplifying the search for the characteristic, flat-sided eggs (see Fig. 27.5). A physician can teach an infected child's parent how to prepare the slide, since it should be done just after awakening in the morning, certainly before bathing the child for a trip to the doctor's office.

The preferred drugs are mebendazole or albendazole, given as a single dose. Treatment should be repeated after about 10 days to kill worms acquired after the first dose. All members of a household should be treated simultaneously, regardless of whether the infection has been diagnosed in all.

Although diagnosis and cure of enterobiasis are easy, preventing reinfection is more difficult. Personal hygiene is most important. Completely sterilizing the household is a difficult activity, albeit gratifying, but it is of limited usefulness. Nevertheless, at the time of treatment, all bed linens, towels, and undergarments should be washed in hot water to lower the prevalence of infective eggs in the environment. If all people in a household are undergoing chemotherapy while reasonable care is being taken to avoid reinfection, family infection can be eradicated—until the next time a child brings it home from school.

Rodent Pinworms

Pinworms of the genera *Syphacia* and *Aspiculuris* are commonly encountered in their natural hosts: wild and domestic rodents. A large number of *Syphacia* species have been described, and many seem to be host specific based on available survey data. Laboratory rats and mice are frequently infected by *S. muris* (in rats), and *S. obvelata* and *A. tetraptera* (in mice). These pinworms are known to have effects on rodent behavior, growth, intestinal physiology and immune response. In the BALB/c mouse strain, *S. obvelata* infection elicits a Th2-type immune response with elevated interlukin 4, 5, and 13 cytokine production, plus parasite-specific IgG. Mice strains deficient in IL-13 (or IL-4/13) have hundredfold greater parasite intensity and experience chronic disease. Pinworm infection also increases the anaphylactic response to a nonparasite dietary antigen.[22] Given the widespread use of rodents as experimental organisms, the need to control pinworm infections in laboratory rodent colonies should be clear. Chow medicated with febendazole appears to be the most effective current treatment, and this drug has been reported to have ovacidal activity.[10] However, direct transmission and contamination of food, water, and bedding with eggs makes these pinworms difficult to control in laboratory rodents.

Male *Syphacia* spp. are easily recognized by their mamelons, two or three ventral, serrated projections (Fig. 27.6). Females are typical pinworms, with long, pointed tails. Eggs are operculated. The life cycle is direct, and worms mature in the cecum or large intestine.[28] No migration within the host is known.

Learning Outcomes

By the time a student has finished studying this chapter, he or she should be able to:

1. Differentiate between the predicted outcomes of association by descent (host-parasite cophylogeny) and colonization (host-switching) for pinworms infecting humans and the living great apes.

2. Draw phylogenetic trees for a hypothetical group of eight pinworm species and their eight hypothetical host species that illustrates strict cospeciation.

3. Consider two different pinworm species, only one of which can undergo retroinfection. Predict how these two species might differ with respect to the genetic composition of infrapopulations.

4. For the hypothetical pinworm species in learning outcome 3, predict which differences in life history features (e.g., fecundity, longevity) will be favored by natural selection in these two species (one with, and the other without retroinfection).

References

References for superscripts in the text can be found at the following Internet site: www.mhhe.com/robertsjanovynadler9e

Additional Readings

Petter, A. J., and J.-C. Quentin. 1976. *CIH keys to the nematode parasites of vertebrates, no. 4 Keys to genera of the Oxyuroidea.* Bucks, England: Commonwealth Agricultural Bureaux, Farnham Royal.

Skrjabin, K. I., N. P. Schikhobolova, and E. A. Lagodovskaya. 1960–1967. *Essentials of nematodology 8, 10, 13, 15,* and *18. Oxyurata.* Moscow: Akademii Nauk SSSR. Indispensable reference works for the oxyurid taxonomist.

Skrjabin, K. I., N. P. Schikhobolova, and A. A. Mosgovoi. 1951. *Key to parasitic nematodes 2. Oxyurata and Ascaridata.* Moscow: Akademii Nauk SSSR. A useful key to genera, with lists of species.

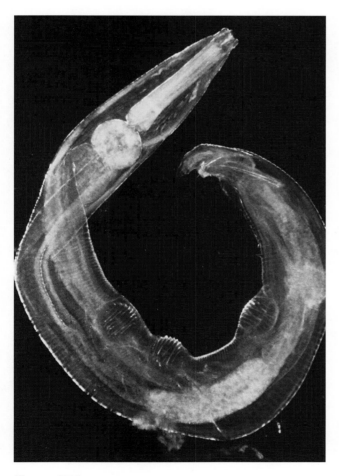

Figure 27.6   *Syphacia* sp., a pinworm of rodents.

Note the three corrugated mamelons on the male's ventral surface.

Courtesy of Warren Buss.

Chapter 28

Nematodes: Gnathostomatomorpha and Spiruromorpha, a Potpourri

Where the telescope ends, the microscope begins. Which of the two has the grander view?

—Victor Hugo *(Les Miserables)*

De Ley and Blaxter[7] elevated gnasthostomatids to infraorder rank (Gnasthostomatomorpha), and they established infraorder Spiruromorpha to contain ten superfamilies. Some of these superfamilies such as Camallanoidea are now considered more closely related to Dracunculomorpha[21] and are discussed in Chapter 30. Other superfamilies are part of a monophyletic Spiruromorpha as inferred from SSU sequence data[21] including Acuarioidea, Diplotriaenoidea, Filarioidea, Habronematoidea, Physalopteroidea, Spiruroidea, and Thelazoidea. In this chapter we cover representatives of several of these spiruromorph superfamilies, with the exception of Filarioidea, which is treated separately in Chapter 29. Considering Gnathostomatomorpha and Spiruromorpha together within a single chapter is done for convenience; indeed this grouping is a phylogenetic potpourri because these infraorders do not form an exclusive monophyletic group.

GNATHOSTOMATOMORPHA

FAMILY GNATHOSTOMATIDAE

Family Gnathostomatidae contains genera *Tanqua* from reptiles, *Spiroxys* from turtles and frogs, *Echinocephalus* from elasmobranchs, and *Gnathostoma* from stomachs of carnivorous mammals (Fig. 28.1). These distinctive nematodes have two pseudolabia, followed by a swollen "head," or cephalic inflation, which is separated from the rest of the body by a constriction. Internally, four peculiar, glandular cervical sacs, reminiscent of acanthocephalan lemnisci (p. 476), hang into pseudocoel from their attachments near the anterior end of the esophagus. The cephalic inflations are divided internally into four hollow areas called **ballonets.** Each cervical sac has a central canal, which is continuous with a ballonet. The functions of these organs are unknown.

Gnathostoma spp. are particularly interesting because of their widespread distribution and peculiar biology. In the United States *G. procyonis* is common in the stomach of raccoons, *G. turgidum* is common in opossums; *G. spinigerum* has been reported from a wide variety of carnivores in Asia. Juvenile stages of *Gnathostoma* sp. have been reported in humans from Mexico, Ecuador, Japan, and Africa,[12] and it was responsible for an outbreak in Myanmar.[5] *Gnathostoma doloresi* is common in pigs in Asia (Fig. 28.2). Of the 20 or so species that have been described, *G. spinigerum* has been demonstrated as a cause of disease in humans in Southeast Asia, whereas *G. binculeatum* causes human disease in parts of North and South America.

Gnathostoma spinigerum. In 1836 Richard Owen, the famous British anatomist who established our basic definitions of *homology* and *analogy,* discovered *G. spinigerum* in the stomach wall of a tiger that had died in the London Zoo. Since then, the species has been found in many kinds of mammals in several countries, although it is most common in Southeast Asia.

- **Morphology.** The body is stout and pink in life. The swollen cephalic inflations are covered with four circles of stout spines. The anterior half of the body is covered with transverse rows of flat, toothed spines, followed by a bare portion. The posterior tip of the body has numerous tiny cuticular spines.

  Males are 11 mm to 31 mm long and have a bluntly rounded posterior end. The cloaca is surrounded by four pairs of stumpy papillae. Spicules are 1.1 mm and 0.4 mm long and are simple with blunt tips.

  Females are 11 mm to 54 mm long and also have a blunt posterior end. The vulva is slightly postequatorial in position. Eggs are unembryonated when laid, 65 μm to 70 μm by 38 μm to 40 μm in size, and have a polar cap at only one end. The outer shell is pitted.

431

Figure 28.1 Morphological comparison among six species of female *Gnathostoma.*

S, *G. spinigerum;* H, *G. hispidum;* T, *G. turgidum;* D, *G. doloresi;* N, *G. nipponicum;* P, *G. procyonis.* This figure indicates the arrangement and shape of the cuticular spines and fresh fertilized uterine eggs, which at times may show various developmental stages when preserved.

From I. Miyazaki, "*Gnathostoma* and gnathostomiasis in Japan," in K. Morishita et al. (Eds.), *Progress of medical parasitology in Japan* 3:529–586. Copyright © 1966 Meguro Parasitological Museum, Tokyo. Reprinted by permission.

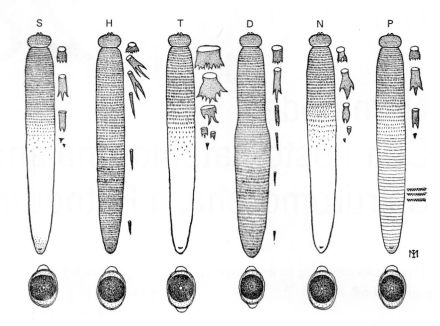

Figure 28.2 *Gnathostoma doloresi* attached to the stomach mucosa of a pig.

Courtesy of Robert E. Kuntz.

- **Biology.** Eggs complete embryonation and the J_1 molts to the J_2 stage and then hatch in about a week at 27°C to 31°C.[19] Actively swimming J_2s are eaten by cyclopoid copepods in which they penetrate into the hemocoel and develop further into J_3s in 7 to 10 days (Fig. 28.3). Third-stage juveniles already have a swollen head bulb covered with four transverse rows of spines.

  When an infected crustacean is eaten by a vertebrate host, J_3s penetrate the intestine of their new host and migrate to muscle or connective tissue. Third-stage juveniles are infective to a definitive host. If they are eaten by the wrong host, they may wander in that animal's tissues. More than 35 species of paratenic hosts are known, among them crustaceans, freshwater fishes, amphibians, reptiles, birds, and mammals, including humans.[20] It has been suggested that advanced J_3s may molt only

once to become adults, but this idea merits additional investigation.[8, 14]

In the definitive host, ingested juvenile worms migrate to the liver and mature before migrating to the stomach. In *G. turgidum,* there are marked seasonal changes in the presence of adult and juvenile worms in opossum definitive hosts, with peak prevalence and intensity of adult worms immediately prior to the rainy season in Mexico.[23] Adult worms are found embedded in tumorlike growths in the stomach wall of the definitive host. They begin producing eggs about 100 days after infection.

- **Epidemiology.** Human infection results from eating a raw or undercooked intermediate or paratenic host containing J_3s. In Japan, this is most often fish whereas, in Thailand, domestic ducks and chickens are probably the most important vectors.[6] In Mexico, a large survey of vertebrates found that paratenic hosts of *Gnathostoma* spp. were predominantly fish. However, any amphibian, reptile, or bird may harbor juveniles and thereby contribute an infection if eaten raw. Human infection with *G. hispidum, G. doloresi,* and *G. nipponicum* also have been reported in Japan.[28, 29] Gnathostomiasis due to *G. binucleatum* has emerged as a serious public health problem in northwestern Mexico and in Ecuador.[4] In the Mexican state of Nayarit, more than 6000 cases were reported between 1995–2005.

- **Pathology.** In humans J_3s usually migrate to superficial layers of the skin, causing creeping eruption (cutaneous larva migrans), pain, pruritis, and erythema. Juveniles may become dormant in abscessed pockets in the skin, or they may wander, leaving swollen red trails in the skin behind them. Untreated infections can persist for more than 10 years!

  If worms remain in the skin with little wandering, they cause relatively little disease. Often they erupt out of the skin spontaneously. However, erratic migration may take them into an eye, the brain, or the spinal cord,

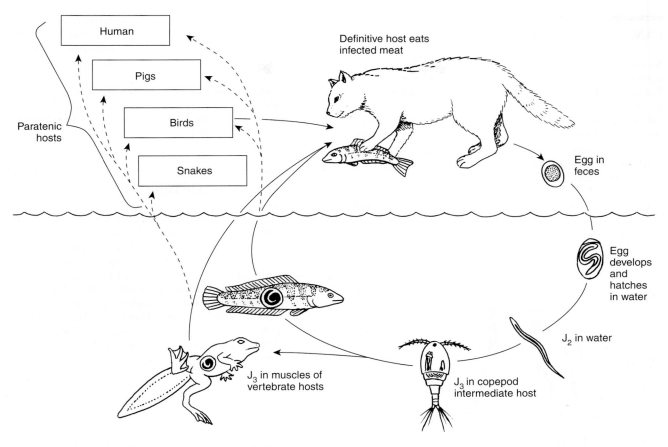

Figure 28.3   Life history of *Gnathostoma spinigerum*.

Drawing by William Ober and Claire Garrison after I. Miyazaki, "*Gnathostoma* and gnathostomiasis in Japan," in K. Morishita et al. (Eds.), *Progress of medical parasitology in Japan* 3:529–586. Tokyo: Megura Parasitological Museum, 1966.

with serious results that even may cause death; ocular gnathostomiasis has been reported in Mexico.[3] Although not neurotropic, juveniles can migrate along peripheral or cranial nerves, causing inflammation, pain, weakness, and paralysis.[27]

- **Diagnosis and Treatment.**   Morphologically based diagnosis is often difficult, particularly when the nematodes are not in the superficial skin. An intradermal test, using an antigen prepared from *G. spinigerum,* has been employed with success in Japan. An ELISA using a crude extract of *G. doloresi* has been used in Mexico.[4] Tests based on cloned specific antigens may offer higher sensitivity and specificity.[18] Gnathostomiasis should be suspected in an endemic area when a localized edema is accompanied by leukocytosis with a high percentage of eosinophils.

    For dermatologic infection, a 21-day treatment with oral ivermectin or albendazole is effective.[24] In cases of neurological gnathostomiasis, the benefits of anthelminthics and steroids have not been established, and treatment is supportive.[27]

    Prevention is the most realistic means of controlling this disease. In regions where ritualistic consumption of raw fish is an important tradition, the fish should be well frozen before preparation. Consumption of raw, previously unfrozen fish in any area of the world is dangerous for a variety of parasitological reasons.

SPIRUROMORPHA

Members of Spiruromorpha are parasitic in all classes of vertebrates and employ an intermediate host, usually an arthropod, in their development. They are a very large, heterogeneous group with many species. Phylogenies based on nuclear ribosomal RNA sequences are providing a new framework for evolutionary and taxonomic conceptions of the infraorder.[17, 21] The numerous variations of morphology in this infraorder make generalizations difficult, but spiruromorphs often have two lateral lips, called **pseudolabia,** and an esophagus that is divided into anterior muscular and posterior glandular portions. Pseudolabia do not represent the fusion of primitive lips but are evolutionarily new structures that originate in anterior shifting of tissues from within the buccal walls. Although the esophagus usually has both muscular and glandular portions, there are species whose esophagus is primarily muscular and others in which it is mainly glandular. Some of these, however, are of uncertain taxonomic position. Spicules are usually dissimilar in size and shape.

FAMILY ACUARIIDAE

Nematodes of this family, all parasites of birds, exhibit very peculiar morphological structures at their head ends. Some have four grooves or ridges, called **cordons,** which begin two dorsally and two ventrally at the junctions of the lateral lips and proceed posteriorly for varying distances (Fig. 28.4). Cordons may be straight, sinuous, recurving, or even anastomosing in pairs. Other acuariids do not have cordons but instead possess four extravagant cuticular projections, sometimes simple, sometimes serrated or even feather like (Fig. 28.5). Both specializations, cordons and cuticular projections of the head, seem to correlate with the parasite's location within the host, the stomach. Most acuariids mature under the koilon, or gizzard lining, where

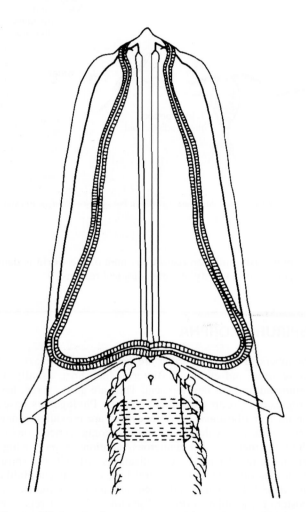

Figure 28.4 *Cordonema venusta,* **from the stomach of an aquatic bird (dipper).**
Note the helmetlike inflation of the cuticle, which bears two cordons on each side. The cordons of each side join at their posterior ends in this genus.

P. L. Wong and R. C. Anderson, Revision of the genera *Cordonema* Schmidt and Kuntz, 1972 and *Skrjabinoclava* Sobolev, 1943 (Nematoda: Acuarioidea)," in *Canadian Journal of Zoology* 61:339–348. Copyright 1983. Reprinted by permission.

they cause considerable damage to underlying epithelium. How these anterior modifications aid the parasites is not known. Acuariid life cycles involve arthropod intermediate hosts, and in some cases vertebrate paratenic hosts such as fish.

Members of only one genus, *Echinuria* spp., in ducks, geese, and swans, are of economic importance; however, acuariids represent an interesting example of nematode morphological diversity.

FAMILY PHYSALOPTERIDAE

Members of family Physalopteridae are mostly rather large, stout worms that live in the stomach or intestine of all classes of vertebrates. All have two large, lateral pseudolabia, usually armed with teeth that are used to attach to mucosa. The head papillae are on the pseudolabia. Cuticle at the base of the lips is swollen into a "collar" in some genera. Caudal alae are well developed in males. Spicules are equal or unequal, and a gubernaculum is absent. This family has a tendency toward **polydelphy,** or having many ovaries and uteri. Of the three subfamilies and several genera in this family, we will briefly consider *Physaloptera.*

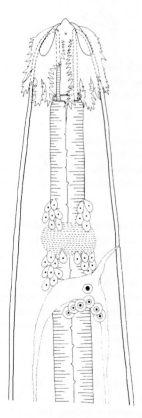

Figure 28.5 *Sobolevicephalus chalcyonis* **from under the gizzard lining of a kingfisher.**
The head cuticle has four "feathered" projections.

P. L. Wong and M. W. Lankester, "Revision of the genus *Sobolevicephalus* Parukhin, 1964 (Nematoda: Acuaridoidea)," in *Canadian Journal of Zoology* 63:1576–1581. Copyright 1985. Reprinted by permission.

Physaloptera Species

Members of genus *Physaloptera* have a conspicuous cephalic collar, and their pseudolabia are triangular and armed with varying numbers of teeth. Males have numerous pedunculated caudal papillae and caudal alae that join anterior to the cloaca. In a few species cuticle of the posterior end is inflated into a prepucelike sheath, which encloses the tail. Three species have been described in Amphibia, around 45 species in reptiles, 24 in birds, and nearly 90 in mammals. Physalopterids detach from the stomach mucosa to feed on food ingested by their host.

Physaloptera praeputialis (Fig. 28.6) lives in the stomach of domestic and wild dogs and cats throughout the world except Europe. It is common in dogs, cats, coyotes, and foxes in the United States. A flap of cuticle covers the posterior ends of both sexes. Its life cycle is incompletely known, but field crickets appear to be an important intermediate host, and insectivorous vertebrates appear to serve as paratenic hosts. This nematode can cause significant illness in dogs and cats, and pathological findings in cats include thickening of the stomach mucosa, leucocyte infiltration of tissues, and inflammation.[22]

Physaloptera rara is a common physalopterid of coyotes and certain other carnivores in North America, to which it is apparently restricted. It is similar to *P. praeputialis* but lacks the posterior cuticular flap. The life cycle involves an insect intermediate host, usually a field cricket, in which it develops into the J$_3$. A paratenic host, such as a rattlesnake, is commonly necessary in the life cycle because of the feeding habits of definitive hosts.

Physaloptera caucasica is the only species recorded from humans.[32] It is normally parasitic in African monkeys. Most recorded cases in humans were from Africa, although several records, some based only on eggs found in patients' feces, have been reported from South and Central America, India, and the Middle East. It is possible that some of these eggs were misidentified or represent a different *Physaloptera* species.

The life cycle is unknown but most likely involves insect intermediate hosts and a vertebrate paratenic host.

Symptoms in humans include vomiting, stomach pains, malaise and eosinophilia. Tentative diagnosis can be made by demonstrating eggs in a fecal sample or by obtaining an adult specimen for accurate identification.

FAMILY TETRAMERIDAE

Members of family Tetrameridae have extreme sexual dimorphism, similar in many respects to certain plant parasites, such as cyst nematodes. Whereas males exhibit typical nematoid shape and appearance, females are greatly swollen and often colored bright red. Females remain stationary within the host, again similar to certain plant-parasitic nematodes. The three genera in this family are all parasitic in the stomach of birds. Genus *Geopetitia* is represented by five rare species that live in cysts on the outside of the proventriculus or gizzard of birds, where they communicate with the enteric lumen through a tiny pore. The other two genera in the family are very common parasites that live in the branched secretory glands of the proventriculus, although males can be found wandering throughout the organ.

Tetrameres (Fig. 28.7) is a large genus of about 50 species, mainly parasites of aquatic birds. A well-developed, sclerotized buccal capsule is present in both sexes. Males are typically nematoid in form, lacking caudal alae and possessing spicules that are vastly dissimilar in size. Lateral, longitudinal rows of spines are present on many species. Females, however, are greatly swollen, with only the head and tail ends retaining a nematode appearance. The midbody of the female is divided into four equal sectors by longitudinal grooves in the cuticle, thus the genus name. In addition, females are blood-red with a black, saclike intestine. They are easily seen as reddish spots in the proventricular wall, where they mature with the tail end near the lumen of that organ.

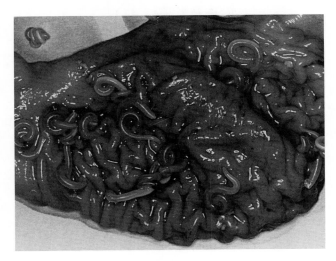

Figure 28.6   *Physaloptera praeputialis* **in the stomach of a domestic cat.**

Courtesy of Robert E. Kuntz.

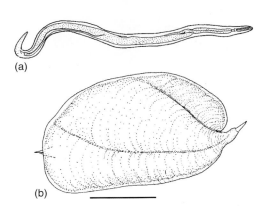

(a)

(b)

Figure 28.7   *Tetrameres megaphasmidiata* **from sandpipers. Bar = 0.5 mm.**

(*a*) male; (*b*) female.

From F. Cremonte et al., "*Tetrameres (Tetrameres) megaphasmidiata* n. sp. (Nematoda: Tetrameridae), a parasite of the two-banded plover, *Charadrius falklandicus*, and white-rumped sandpiper, *Calidris fuscicollis*, from Patagonia, Argentina," in *J. Parasitology* 87:148–151. Copyright 2001. Reprinted by permission.

The vulva is near the anus and thus is available to males who find it. Eggs are embryonated when laid. Juveniles in intermediate hosts, including insects and crustaceans, require about one month to become infective. Definitive hosts are water birds, chickens, owls, and hawks.

Approximately 40 species of *Microtetrameres* (Fig. 28.8) have been described, all of which live in proventricular glands of insectivorous birds. Females of this genus are also swollen and, in addition, are twisted into a spiral. Morphological and biological characteristics are otherwise similar to those of *Tetrameres* spp., with terrestrial insects serving as intermediate hosts.

The quaint little worms in this family are familiar to all who survey parasites of birds, since they are common in many species of hosts. Even though a hundred or more females are commonly embedded in the proventriculus of a single duck, for example, they seem to have little effect on the overall health of their host. In contrast, *T. americana* infection in free-range chickens is associated with anemia, weight loss, and even death.[11]

FAMILY GONGYLONEMATIDAE

This family contains the single genus *Gongylonema,* which has several species that inhabit the mucosa and submucosa of the upper digestive tracts of birds and mammals. Morphologically, they resemble several spiruridans, except that cuticle of the anterior end is covered with large bosses, or irregular scutes, arranged in eight longitudinal rows (Fig. 28.9). Cervical alae are present, as are cervical papillae. The posterior end of males bears wide caudal alae, which are supported by numerous pedunculated papillae.

Of the 25 or so species in this genus, *Gongylonema pulchrum* is probably the best known. Primarily a parasite of ruminants and swine, the worm also has been reported from monkeys, hedgehogs, bears, and humans.[9, 31] The life cycle involves an insect intermediate host; at least 16 genera of beetles from four families are suitable intermediate hosts, as are cockroaches. Although these would appear rather unpalatable fare for people, more than 50 cases of human

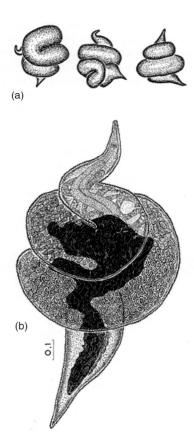

(a)

(b)

Figure 28.8 *Microtetrameres centuri,* **a parasite of the proventricular glands of the western meadowlark.**

(*a*) Variation in the shape of females; (*b*) gravid female; scale for female in millimeters.

From C. J. Ellis, "Life history of *Microtetrameres centuri* Barus, 1966 (Nematoda: Tetrameridae). II. Adults," in *J. Parasitol.* 55:713–719. Copyright ©1953. Reprinted by permission.

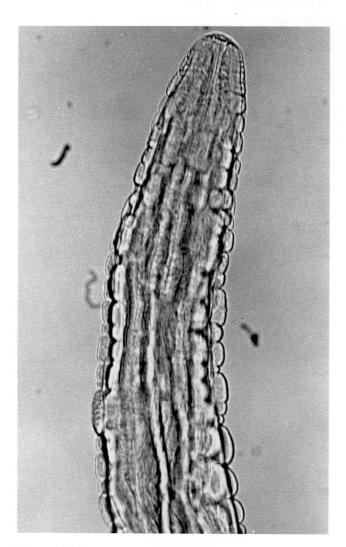

Figure 28.9 **Anterior end of** *Gongylonema,* **demonstrating the cuticular bosses typical of the genus.**

Courtesy of Warren Buss.

gongylonemiasis have been reported worldwide, perhaps due to accidental ingestion of infected beetles or when third-stage juveniles contaminate water sources, such as wells.[13] In normal hosts worms invade the esophageal epithelium, where they burrow stitchlike in shallow tunnels. In an abnormal host, such as humans, they behave similarly but do not mature and are typically found in epithelium of the tongue, gums, or buccal cavity. Their active movements often produce a sensation of worms moving in the mouth. Irritation and bleeding are usual. Patients can sometimes remove worms from their mouth with their fingers.[9] Treatment is surgical removal of worms that can be seen. In experimental studies of *G. pulchrum* in rabbits, levamisole appeared to have high efficacy against adult nematodes.[16] Mebendazole and albendazole have also been shown to be effective in naturally infected primates, and human cases. *Gongylonema* sp. eggs have sometimes been found in human feces, but these likely result from ingestion of infected meat containing eggs rather than from patent infections in humans.

FAMILY SPIROCERCIDAE

With the exception of one genus that occurs in birds, species in the family Spirocercidae are parasites of mammals. Of these, *Spirocerca lupi* is the most interesting because of its complex life cycle and its relationship to esophageal cancer in dogs.[1, 2]

Spirocerca lupi. This stout worm is bright pink to red when alive. Its mouth is surrounded by six rudimentary lips, and the buccal capsule is well developed, with thick walls. A short muscular portion of the esophagus is followed by a longer glandular portion. Males are 30 mm to 54 mm long, with a left spicule 2.45 mm to 2.80 mm long and a right spicule 475 μm to 750 μm long. Females are 50 mm to 80 mm long, with the vulva 2 mm to 4 mm from the anterior end. Eggs are small and cylindrical, approximately 35 by 15 μm, and embryonated when laid.

- **Biology.** Adults normally are found in clusters, entwined within nodules in the esophageal submucosa of dogs, and several other carnivorous hosts. Hounds seem to be the breeds most frequently infected in the United States, and worldwide there is a greater prevalence in large versus small dog breeds. This is probably related to their opportunities for exposure rather than to breed susceptibility.

  Juveniles within their eggshells pass out of the host with its feces. Any of several species of scarabaeid dung beetles can serve as intermediate host. A wide variety of paratenic hosts are known, including birds, reptiles, and other mammals. Dogs can become infected by eating dung beetles or infected paratenic hosts. Domestic dogs are probably most often infected by eating offal of chickens or gamebirds that have J$_3$s encysted in their crops.

  Once in the stomach of a definitive host, juveniles penetrate its wall and enter the wall of the gastric artery, migrating up to the dorsal aorta and forward to the area between diaphragm and aortic arch. They remain in the wall of the aorta for two and one-half to three months, after which they molt to J$_4$s and migrate to the nearby esophagus, which they penetrate and establish a nodule. After establishing a passage into the lumen of the

esophagus, they move back into the submucosa or muscularis where they complete their development about five to six months after infection. Eggs pass into esophageal lumen through the tiny passage formed by the worm.

Many worms get lost during migration and may be found in abnormal locations, including lung, mediastinum, subcutaneous tissue, trachea, urinary bladder, and kidney.

- **Epidemiology.** *Spirocerca lupi* is most common in warm climates but has been found in Manchuria and northern regions of the former Soviet Union. Many questions are still unanswered: Why is the parasite distributed so sporadically throughout the United States and the world? What factors influence the change in prevalence of infection in a given area? The attractiveness of dog feces to susceptible beetle species is a factor, as is the application of pesticides in an endemic area. Certainly, the successful transfer of J$_3$s from one paratenic host to another increases the parasite's chances for survival.

- **Pathology.** When J$_3$s penetrate mucosa of the stomach, they cause a small hemorrhage in the area. This irritation commonly causes a dog to vomit. Lesions in the aorta caused by migrating worms are often severe, with hemorrhage accounting for death in some dogs with heavy infections. Destruction of tissues in the wall of the aorta with subsequent scarring is typical of this disease and may lead to numerous aneurysms.

  Worms that leave the aorta and migrate upward come in contact with tissues surrounding the vertebral column, where they frequently cause, by a mechanism not yet elucidated, a condition known as **spondylitis.** This deformation may be so severe as to cause adjacent vertebrae to fuse. Hypertrophic pulmonary osteopathy, which involves growth of bone tissue in abnormal locations, leading to inflammation and swelling, is a common sequel to this disease (Fig. 28.10).

  The most striking lesions associated with spirocercosis are in the esophageal wall, where the worms mature. Here their presence stimulates the formation of a **reactive granuloma,** made up of fibroblasts. These granulomas are

Figure 28.10   Hound with severe hypertrophic pulmonary osteoarthropathy associated with esophageal sarcoma.

From W. S. Bailey, "Parasites and cancer: Sarcoma with *Spirocerca lupi*," in *Ann. N.Y. Acad. Sci.* 108(Art 3):890–923. Copyright © 1963.

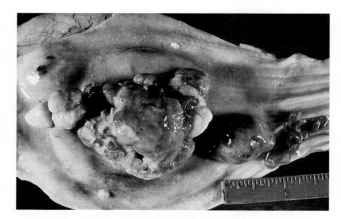

Figure 28.11 **Esophageal sarcoma associated with** *Spirocerca lupi* **infection.**

Pedunculated masses protrude into the lumen; adult *S. lupi* are partially embedded in the neoplasm.

From W. S. Bailey, "*Spirocerca lupi:* A continuing inquiry," in *J. Parasitol.* 58:3–22. Copyright © 1972. Photo courtesy of the Department of Pathology and Parasitology, Auburn University School of Veterinary Medicine, Auburn, AL.

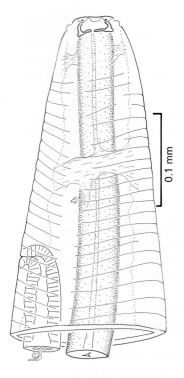

Figure 28.12 *Thelazia digiticauda* **from under the nictitating membrane of the eye of a kingfisher.**

Female, lateral view of anterior end.

From G. D. Schmidt and R. E. Kuntz, "Nematode parasites of Oceanica. XVIII. Subuluridae, Thelaziidae, and Acuariidae of birds," in *Parasitology* 63:91–99. Copyright © 1971 Cambridge University Press. Reprinted with the permission of Cambridge University Press.

rather more loosely organized than is usually the case in granulomatous reactions, and cell structure is characteristic of incipient **neoplasia** (cancer). Some granulomas change to **sarcomas,** true cancerous growths (Fig. 28.11); in some regions 20% of infected dogs have sarcomas. Worms may continue to live inside these tumors for some time, or they may be extruded or compressed and killed by rapidly growing tissue. These tumors may metastasize to other locations, including lungs. The precise oncogenic factor responsible for stimulation of neoplasia is still unknown. Although several other helminths have been associated with malignancy, in no other instance is there as strong evidence for a cause-effect relationship as in canine spirocercosis.

• **Diagnosis and Treatment.**   Diagnosis in dogs is by demonstration of the characteristic eggs in feces, or by radiological or other advanced imaging techniques that reveal caudal esophageal nodules, aortic abnormalities due to aneurysms, and spondylitis. Diagnosis in the early stage of the infection when juveniles are developing is difficult, whereas the presence of diagnostic adults is often accompanied by advanced disease. At necropsy, aortic scarring, aneurysms, and esophageal granuloma or sarcoma are considered diagnostic, even if worms are no longer present.

Doramectin is effective against adult *S. lupi*,[33] and milbemycin oxime prevents establishment of juveniles in the esophagus and may have value as a prophylatic drug.[15] However, if extensive aneurysms or a sarcoma have developed, anthelminthic drug treatment will not affect these conditions.

FAMILY THELAZIIDAE

Members of family Thelaziidae live on the surface of the eye in birds and mammals, usually remaining in lacrimal ducts or conjunctival sacs or under the nictitating membrane. Most

are parasites of wild animals, but two species of *Thelazia* have been reported from humans.[30]

These worms lack lips but show evidence of having a hexagonal mouth. The buccal capsule is well developed, with thick walls. Alae and cuticular ornamentations are absent, except for conspicuous annulations near the anterior end (see Fig. 28.12). These are deep, and their overlapping edges ostensibly aid movements across the smooth surface of the cornea.

Thelazia callipaeda is a parasite of dogs (Fig. 28.13) and other mammals in Southeast Asia, China, Europe, and Korea. In Italy, this species has been reported from red foxes, wolves, martens, brown hares, and wild cats[26] and *T. californiensis* parasitizes deer, canids, felids and other mammals in western North America. Both species have been reported from humans several times.[30]

Female worms release J_1s onto the surface of the host eye, where they are picked up by face flies, such as *Musca autumnalis,* feeding on lacrimal secretions. Some *Thelazia* spp. are vectored by drosophilid midges such as *Amiota* spp. and *Phortica* spp. A peculiarity in the transmission of *T. callipaeda* is that for one of its vectors, *P. variegata,* only male flies are zoophilic and vector the nematode.[26] After development in the fly, J_3s emerge from the fly's labellum when it is feeding at the eye of another host.[25] Sixteen species of *Thelazia* have been described, and six are of veterinary or medical importance. In the United States, *T. skrjabini* and *T. gulosa* most commonly occur in the eyes of cattle, and *T. lacrymalis*

Figure 28.13   **Eye of a dog with massive *Thelazia callipaeda* infection.**

From Ontranto, D., R. P. Lia, C. Cantacessi, G. Testini, A. Troccoli, J. L. Shen, and Z. X. Wang. 2005. Nematode biology and larval development of *Thelazia callipaeda* (Spirurida, Thelaziidae) in the drosophilid intermediate host in Europe and China. *Parasitology* 131:847–855. Copyright © 1971 Cambridge University Press. Reprinted with the permission of Cambridge University Press. (Photo courtesy of D. Ontranto and R. P. Lia, University of Bari, Italy.)

is found in the eyes of horses. Several anthelminthics have proven effective for treatment of infected dogs, including milbemycin oxime, which also works as a prophylactic at doses used for heartworm prevention.[10]

Learning Outcomes

By the time a student has finished studying this chapter, he or she should be able to:

1. Compare the life cycles of *Gnathostoma spinigerum* and *Toxocara canis* (chapter 26), analyzing their similarities and differences.

2. Certain nematodes have remarkable structural modifications of the cephalic (head) region. Explain why scientists often lack detailed information on the biological significance of such structures.

3. List examples of how particular parasite species extend the life-span of one or more of their life cycle stages.

4. Females of *Tetrameres* species have a rotund body shape and are sedentary parasites, whereas males of the same species are vermiform and mobile. Some plant-parasitic nematode species (e.g., *Heterodera* spp.) that live in root tissues have similar morphological differences between sedentary females versus mobile males. Explain how knowledge of phylogenetic relationships could be used to assess be if this shared (female) morphology is a shared-derived characteristic.

5. *Spirocerca lupi* frequently causes malignant tumors in infected dogs that result in death of the host. Explain one situation wherein there would not be strong (host) natural selection against death due to this parasite.

6. Benjamin Franklin stated, "An ounce of prevention is worth a pound of cure." Discuss how this quotation pertains to *Spirocerca lupi* infection of dogs.

References

References for superscripts in the text can be found at the following Internet site: www.mhhe.com/robertsjanovynadler9e

Additional Readings

Anderson, R. C., and O. Bain. 1976. *Keys to genera of the order Spirurida, part 3. Diplotriaenoidea, Aproctoidea, and Filaroidea.* In R. C. Anderson, A. G. Chabaud, and W. Willmott (Eds.), *CIH keys to the nematode parasites of vertebrates*, no. 3. Bucks, England: Commonwealth Agricultural Bureaux, Farnham Royal.

Chabaud, A. G. 1954. Valeur des caracteres biologiques pour la systématique des nématodes spirurides. *Vie et Millieu* 5:299–309.

Chabaud, A. G. 1975. Keys to the genera of the order Spirurida, part 1. Camallanoidea, Dracunculoidea, Gnathostomatoidea, Physalopteroidea, Rictularioidea, and Thelazioidea. In R. C. Anderson, A. G. Chabaud, and W. Willmott (Eds.), *CIH keys to the nematode parasites of vertebrates*, no. 3. Bucks, England: Commonwealth Agricultural Bureaux, Farnham Royal.

Chabaud, A. G. 1975. Keys to the order Spirurida, part 2. Spiruroidea, Habronematoidea and Acuarioidea. In R. C. Anderson, A. G. Chabaud, and S. Willmott (Eds.), *CIH keys to the nematode parasites of vertebrates* (no. 3). Bucks, England: Commonwealth Agricultural Bureaux, Farnham Royal.

Otranto, D., and D. Traversa. 2005. *Thelazia* eyeworm: an original endo and ecto parasitic nematode. *Trends Parasit.* 21:1–4.

Skrjabin, K. I., A. A. Sobolev, and V. M. Ivaskin. 1963–1967. *Essentials of nematodology, 11, 12, 14, 16, and 19. Spirurata of animals and man and the diseases they cause.* Moscow: Akademii Nauk SSSR. The most complete monographs on the subject.

Chapter 29

Nematodes: Filarioidea: Filarial Worms

Our eye-beams twisted, and did thread our eyes, upon one double string...

—John Donne *(The Extasy)*

Superfamily Filaroidea of infraorder Spiruromorpha include tissue-dwelling parasites of all vertebrate classes, except fish. All species employ arthropods as intermediate hosts, most of which are hematophagous and deposit J_3s on host skin with their bite (Fig. 29.1). Generally speaking, they are slender worms with reduced lips and buccal capsule. Most are parasites of wild animals, particularly of birds, but several are very important disease organisms of humans and domestic animals. The majority of these belong to the large family Onchocercidae.

It has been known for over 25 years that filarioid nematodes are infected with intracellular, rickettsial bacteria.[83, 91] Members of genus *Wolbachia,* they are related to congeners that infect at least 20% of all insects.[83] Unlike insect infection with *Wolbachia* sp., however, infection with these bacteria is often essential to the continued good health and reproduction of their nematode hosts. This mutualistic relationship implies a long period of coevolution of nematodes with their bacteria. These mutualistic bacteria are not present in filarioids of amphibians and reptiles, and their presence in species of Onchocercidae is variable; presence even varies among individuals within species.[32] Tissue localization of *Wolbachia* sp. also varies; hypodermal cords and female germ cells are often infected, but intestinal cells and somatic cells of the female reproductive tract can also harbor these bacteria.[32] Several early generalizations about filarioid *Wolbachia* sp. have been overturned by recent studies, and much remains to be learned. However, their presence in certain filarioid pathogens of humans has recently been exploited for new approaches to treatment. We will include interactions with *Wolbachia,* where known, in discussions to follow.

FAMILY ONCHOCERCIDAE

Members of family Onchocercidae live in tissues of amphibians, reptiles, birds, and mammals. Most are of no known medical or economic importance, but a few cause some of the most tragic, disfiguring, and debilitating diseases of

humankind. Of these, species of *Wuchereria, Brugia, Onchocerca,* and *Loa* will be considered in some detail. Short mention will be made of others.

Wuchereria bancrofti

Perhaps the most striking disease of humans is a clinical entity known as **elephantiasis** (Fig. 29.2). Horrible swelling of parts of the body afflicted with this condition has been experienced since antiquity. Ancient Greek and Roman writers likened the thickened and fissured skin of infected persons to that of elephants, although they also confused leprosy with this condition. Actually, *elephantiasis* is a nonsense word, since literally translated it means "a disease caused by elephants." The word is so deeply entrenched, however, that it is not likely ever to be abandoned. Classical elephantiasis is a relatively rare outcome of infection by *Wuchereria bancrofti* and by at least two other species of filarioids.

Infection of the lymphatic system by filarial worms is best referred to as **lymphatic filariasis** and is much more common than elephantiasis; recent estimates put the global prevalence at 120 million cases.[63] **Bancroftian filariasis** *(W. bancrofti)* is responsible for 91% of lymphatic filariasis,[45] extending throughout central Africa, the Nile Delta, Turkey, India and Southeast Asia, the East Indies, the Philippine and Oceanic Islands, Australia, New Guinea,[46] Latin America, and the Caribbean, and parts of South America (in short, across a broad equatorial belt). It was probably brought to the New World by the slave trade.[53] A nidus of infection remained in the vicinity of Charleston, South Carolina, until it died out spontaneously in the 1920s.[19] The remaining 9% of lympatic filariasis is caused by *Brugia malayi* or *B. timori,* which are distributed in Southern and Southeast Asia.[10]

In sub-Saharan Africa (SSA), where 73% of the population lives on less than $2 per day, there are approximately 50 million cases of lymphatic filariasis. The highest burden of disease is in Southeast Asia, with 859 million people at risk of infection. The Global Program to Eliminate Lymphatic

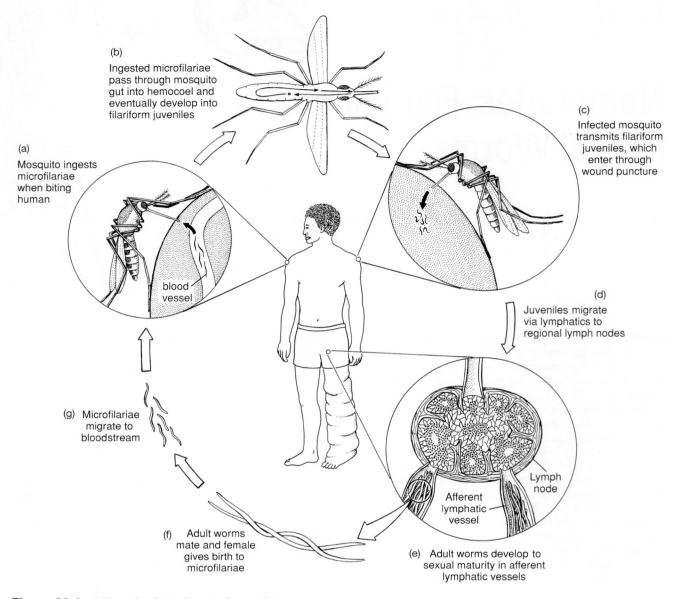

Figure 29.1 **Life cycle of *Wuchereria bancrofti*.**

(*a*) Mosquito ingests microfilariae when biting human. (*b*) Microfilariae pass through mosquito gut and develop to filariform J$_3$s. (*c*) Filariform juveniles escape from mosquito's proboscis when the insect is feeding and then penetrate wound. (*d*) Juveniles migrate via lymphatics to regional lymph nodes. (*e*) Worms develop to sexual maturity in afferent lymphatic vessels. (*f*) Adult worms mate, and female gives birth to microfilariae. (*g*) Microfilariae enter blood circulation.

Drawing by William Ober and Claire Garrison.

Filariasis was begun in 2000, and 2 billion drug control treatments have been provided to people living in 48 of 83 endemic countries. All endemic countries in Southeast Asia have participated in mass drug administration programs, with coverage reaching 546 million people in 2007. In contrast, only 15 of 39 endemic SSA countries implemented mass drug administration in 2007, and only 47 million of the 382 million people at risk were treated. All 39 endemic countries in SSA are underdeveloped economically, and this poses significant challenges to the success of global filariasis eradication.[11]

In the Pacific theater in World War II, the potential disfiguring effects of lymphatic filariasis was a cause of great psychological concern to American armed forces, who visualized themselves returning home carrying their scrotum in a wheelbarrow. Although thousands of cases of filariasis were in fact contracted by American servicemen, no single case of classical elephantiasis resulted. Some persons experienced symptoms for as long as 16 years.[95] The evolution of our knowledge of this disease and its cause remains one of the classics of medical history.[33]

• **Morphology.** Adult worms are long and slender with a annulated cuticle and bluntly rounded ends. Their head is slightly swollen and bears two circles of well-defined papillae. Their mouth is small; the buccal capsule is reduced.

   Males are about 40 mm long and 125 μm wide. Their spicules are unequal in length. Females are 6 cm to 10 cm long and 200 μm wide. Their vulva is near the level of the

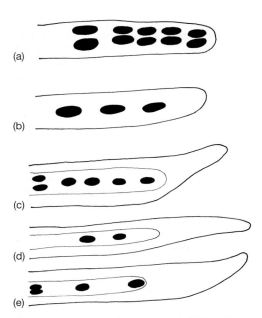

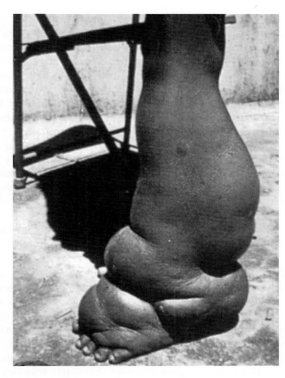

Figure 29.2   Elephantiasis of the leg caused by infection with *Wuchereria bancrofti* in India.

Courtesy of E. L. Schiller, AFIP neg. no. 74-6426-2.

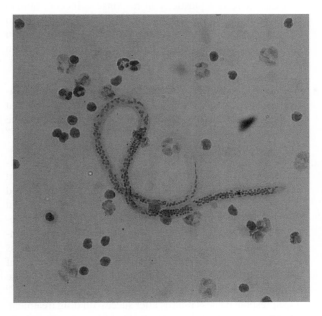

Figure 29.3   Microfilaria of *Wuchereria bancrofti*. A sheath is present, and the discrete nuclei do not reach the tip of the tail.

Photograph by Larry S. Roberts.

Figure 29.4   Presence or absence of a sheath and the arrangement of nuclei in the tail are useful criteria in identifying microfilariae.

(*a*) *Mansonella perstans;* (*b*) *Mansonella ozzardi;* (*c*) *Loa loa;* (*d*) *Wuchereria bancrofti;* (*e*) *Brugia malayi.*

near major lymph glands in the lower half of the body. Adults also inhabit lymphatics of the testes, epididymis, and scrotum. Females are ovoviviparous, producing thousands of juveniles known as **microfilariae** (Fig. 29.3). Microfilariae have a specialized anatomy. They lack some features found in most nematode J_1s such as a complete alimentary system, but have a secretory-excretory system. Microfilariae of *W. bancrofti* retain the egg membrane as a "sheath" (not to be confused with the sheath of some strongyle J_3s, which is the second-stage cuticle). The sheath is rather delicate and close fitting but can be detected where it projects at the anterior and posterior ends of a microfilaria. When stained, several internal nuclei and primordia of organs can be seen in microfilariae. Location of these and presence or absence of a sheath help to identify the eight common species of microfilariae found in humans (Fig. 29.4).

Females release microfilariae into lymph. Some may wander into adjacent tissues, but most are swept into the blood through the thoracic duct. Throughout much of the geographical distribution of this parasite, there is a marked periodicity of microfilariae in peripheral blood; that is, they are more abundant at certain times of day, whereas at other times they virtually disappear from peripheral circulation. The maximal number usually can be found between 10 P.M. and 2 A.M. For this reason, night-feeding mosquitoes are primary vectors of *W. bancrofti* in areas where microfilarial periodicity occurs. During the day, microfilariae are concentrated in blood vessels of deep tissues of the body, predominately in pulmonary capillaries.[56]

Causes of periodicity remain obscure, but do not result from daily release of a new generation of progeny by females. Stimuli such as arterial oxygen tension and body temperature probably are involved. Administration of pure oxygen to a patient during peak **microfilaremia** (condition of having microfilariae in the blood) can cause

middle of their esophagus. The two uteri occupy much of the body's length.

- **Biology.**   Adult *W. bancrofti* live in lymphatic vessels of humans. They are normally found in afferent lymph channels microfilariae to localize in deep tissues. Reversal of a patient's sleep schedule causes reversal of periodicity so that microfilaremia becomes diurnal. The adaptive value of the periodicity is difficult to explain. Although it is clearly advantageous for microfilariae to be present in peripheral blood when a vector is likely to be feeding, what value is there in being absent when vectors are not feeding? Periodicity is unimportant clinically, but it has significant diagnostic and epidemiological implications.

In certain areas of the South Pacific, including Fiji, Samoa, the Philippines, and Tahiti, a strain of *W. bancrofti* that lacks periodicity or shows diurnal periodicity is common. It is described as **subperiodic.** Daytime-feeding mosquitoes are major vectors of the subperiodic strain. Adult morphology for this strain is identical to that of other *W. bancrofti* adults producing periodic microfilariae. Most investigators believe that only one species is involved, although the subperiodic type was designated as a separate species named *W. pacifica;* this name is now considered a synonym of *W. bancrofti.*

Mosquitoes ingest microfilariae along with their blood meal. Microfilariae may or may not lose their sheath before penetration of the insect's gut. This usually occurs within two hours, after which the parasites reach their host's thoracic muscles. Here they develop as J$_1$s for about eight days before molting to J$_2$s. Second-stage juveniles are short, sausage-shaped worms (sausage stage; see Fig. 29.8) in which most of the organ systems are present, but they retain an anal plug and thus lack a functional anus. However, they do feed at this stage and so cause some damage to their host. Mosquitoes fed on microfilaremic volunteers die at a rate 11 to 15 times greater than mosquitoes fed on amicrofilaremics.[50]

After two to four days, development is complete, and J$_2$s molt to become elongated, slender filariform J$_3$s (see Fig. 29.9), and development ceases. Filariform juveniles are 1.4 mm to 2.0 mm long and are infective to a definitive host. They migrate throughout the hemocoel, eventually reaching the labium, or proboscis sheath, from which they escape when the mosquito is feeding. They enter the skin through the wound made by the mosquito. After migrating through the peripheral lymphatics, the worms settle in larger lymph vessels, where they mature. Worms require 6–12 months before females begin releasing microfilariae. Microfilariae can be released for up to 10 years in the absence of reinfection.[35]

- **Epidemiology.** Many mosquito vectors of *Wuchereria* have a preference for human blood and often breed near human habitation. At least 77 species and subspecies of mosquitoes in genera *Anopheles, Aedes, Culex, Ochlerotatus,* and *Mansonia* can support development of the parasite. This broad diversity of intermediate hosts is unusual for vector-borne parasites, however, transmission usually involves far fewer species at particular localities. Susceptibility is under strict genetic control and varies within mosquito populations. In areas in which the periodic strain of *W. bancrofti* is found, mosquito vectors are primarily night feeders. Periodicity has practical epidemiological significance because it determines which mosquito species must be controlled and consequently what control measures must be applied.

Suitable breeding sites for mosquitoes abound in tropical areas. Some sites are difficult or impossible to control, such as tree holes and hollows at bases of palm fronds, whereas others can be controlled with a degree of effort. Hollow coconuts, killed while still green by rats gnawing holes in them, fall, fill with rainwater, and become havens for developing mosquito larvae. These can be collected and burned. Even dugout canoes that are unused for a few days can partially fill with rainwater and become mosquito nurseries. Conditions for transmission of *W. bancrofti* vary from locality to locality and country to country. An epidemiologist must consider each case independently within the framework of the biology of a particular vector, putting economic and technical resources that are available to best advantage.

Humans seem to be the only animals naturally infected with *W. bancrofti;* there is no evidence of a reservoir.[49]

- **Pathogenesis.** Pathogenesis in lymphatic filariasis depends heavily on inflammatory and immune responses and involves responses to adult worms and *Wolbachia* sp. endosymbionts. Effects of infection with *W. bancrofti* display a wide spectrum, from clinically silent infections with no apparent inflammation or parasite damage, to mild-to-intense nongranulomatous chronic lymphatic inflammation, to a variety of granulomatous obstructive reactions.[45] Some investigators contend that these states represent a progression from initial infection through inflammation to obstructive disease.[15] Others maintain that progression from one clinical form to another is not inevitable and that there is a plasticity in response.[74] Indeed, human pedigree analysis shows that inflammatory responses to *W. bancrofti* are highly heritable;[23] likewise, **lymphedema** tends to be clustered within families, and individuals with pathology show stronger Th1 and Th2 responses.

- **Asymptomatic Phase.** Individuals with asymptomatic infections usually have high microfilaremias. It appears that in such people the Th1 inflammatory arm of the immune response is downregulated, and the Th2 arm is stimulated.[59] The cytokine IL-4, which suppresses activation of Th1 cells, is elevated, and IFN-γ is depressed (see p. 30). Downregulation of Toll-like receptors (TLRs) on T cells and antigen-presenting cells seems to occur during the period of microfilaremia, whereas expression of TLRs is increased during chronic lymphatic pathology.[7] Eventually, after a period of perhaps several years, this tolerance (or hyporesponsiveness) often breaks down and inflammatory reactions rise. Thus, lymphatic filariasis often involves a period of hyporesponsiveness followed by chronic lymphatic disease initiated by parasite, *Wolbachia,* and immune factors.[60] T-cell responsiveness returns after successful drug treatment.

Some individuals in endemic areas show neither symptoms nor microfilaremia; these are known as "endemic normals." However, lack of symptoms and microfilaremia does not mean uninfected. Worm antigen can be detected in blood of many and, in one study, about one in four eventually developed **hydrocele** (forcing of lymph into the tunica vaginalis of the testis or spermatic cord).[86]

Casual visitors to endemic areas may become infected with the parasite and suffer from acute lymphatic inflammation, but they usually have no microfilaremia.[73] This condition may persist for many years, subsiding and re-curring from time to time.[95]

One of the most perplexing problems of the disease has been determining why a person first infected as an adult so rarely shows a microfilaremia. In World War II, 10,431 U.S. naval personnel were infected with *W. bancrofti*, yet only 20 showed a microfilaremia.[9]

We now know that in endemic areas, fetuses in pregnant women are exposed to substantial amounts of filarial antigens.[29, 40] Exposure to antigens at this stage of development promotes tolerance by the immune system. Consequently, microfilariae may not be recognized as foreign. A child born to a microfilaremic mother is 2.4 to 2.9 times more likely to become microfilaremic than is a child born to an amicrofilaremic mother, although some data do not seem to support this result.[51]

- **Inflammatory (Acute) Phase.**   Inflammatory responses are due to antigens from adult worms; little or no disease is caused by microfilariae. Molecules derived from *Wolbachia* sp. released from dead or degenerating adult worms[68] are important inducers of cytokines TLR2, TLR4, and TLR6, resulting in inflammation.[7] Proteomic analysis of adult secretory-excretory products reveals that *Wolbachia* sp. proteins are not secreted from "healthy" parasites.[39] Inflammation also occurs in response to invasion of bacteria from the skin surface.[71] A fascinating aspect of host response is that parasites use host immune status as a cue for altering parasite development. Specifically, in a laboratory filarial model system (*Litomosoides sigmodontis*) infective juveniles accelerate their development in response to a stronger host eosinophil activation, leading to earlier reproduction and increased microfilaremia.[6] Such plasticity in parasite reproduction in response to host immunity has implications for antiparasite vaccines.

Adult worms in lymph channels cause dilation of the channels and interfere with lymph flow, resulting in lymphedema. Immune responses to *Wolbachia* sp. molecules cause activation of vascular endothelial growth factors, leading to excessive growth of cells lining the lymph channels.[80] Patients with lymphedema have periodic attacks of **adenolymphangitis** (inflammation of lymph channels) and **lymphadenitis** (inflammation of lymph nodes). Attacks are marked by chills and fever; acutely swollen, warm and tender skin of the lymphedematous extremity; tenderness along superficial lymphatics; and painful lymph nodes. The attack tends to subside after 5 to 7 days; this period can be shortened by administration of antibiotics. During an attack, bacteria normally present on skin surface can be cultured from tissue fluid, lymph, lymph node biopsies, and blood in a high proportion of patients.[71]

Additional common symptoms in the acute stage of filariasis include **orchitis** (inflammation of the testes, usually with sudden enlargement and considerable pain), hydrocele and **epididymitis** (inflammation of the spermatic cord). On the histological level, extensive proliferation of lining cells occurs in the lymphatics, with much inflammatory cell infiltration, especially of polymorphonuclear leukocytes and eosinophils, around lymphatics and adjacent veins. The most prominent cells in this infiltration become lymphocytes, plasma cells, and eosinophils, as the most acute phase subsides. Abscesses around dead worms may develop, with accompanying bacterial infection. Microfilariae may disappear from peripheral blood during febrile episodes, perhaps because the lymphatic vessel containing the female becomes blocked.[71]

- **Obstructive Phase.**   The obstructive phase is marked by **lymph varices, lymph scrotum,** hydrocele, **chyluria,** and elephantiasis. Lymph varices are "varicose" lymph ducts, caused when lymph return is obstructed and lymph "piles up," greatly dilating the affected duct. This can causes chyluria, or lymph in the urine, if collateral lympatic channels form and connect with the urinary tract. Chyle (lymph with emulsified fats) gives urine a milky appearance, and some blood is often present. A feature of chronic obstructive phase is progressive infiltration of the affected areas with fibrous connective tissue, or "scar" formation, after inflammatory episodes. However, dead worms are sometimes calcified instead of absorbed, usually causing little further difficulty.

In a certain proportion of cases, thought to be associated with repeated attacks of acute lymphatic inflammation, the condition known as *elephantiasis* gradually develops. This is a chronic lymphedema with much fibrous infiltration and thickening of skin. In men the organs most commonly afflicted with elephantiasis are the scrotum, legs, and arms; in women the legs and arms are usually afflicted, with the vulva and breasts being affected more rarely. Elephantoid organs are composed mainly of fibrous connective tissues, granulomatous tissue, and fat. Skin becomes thickened and cracked, and invasive bacteria and fungi further complicate the matter, leading to adenolymphangitis of bacterial origin. Microfilariae usually are not present.

Elephantiasis is thus a result of complex immune responses of long duration. Repeated superinfections over many years are usually necessary for elephantiasis to occur. Treatment of lymphedema and elephantiasis can be effective, but surgical intervention is often required when damage to extremities is severe. Such surgery is often not available to individuals suffering from elephantiasis in developing countries.

- **Social and Psychological Impact.**   An estimated 40 million people suffer from chronic disfiguring results of lymphatic filariasis, including 27 million men with testicular hydrocele, lymph scrotum or elephantiasis of the scrotum.[25] Some 13 million people, mostly women, have lymphedema or elephantiasis of the leg, arm, or breast. Approximately 12% of lymphatic filariasis infections result in lymphedema and 21% in hydrocele.[42] Negative social consequences of these afflictions are severe, including sexual disability. In one clinic where caregivers gained sufficient trust that their patients were willing to confide such personal details, patients told of marriages devoid of physical and sexual intimacy and a "conspiracy of silence" that included both patient and partner. Men with lymph scrotum, which causes leaking of lymph through scrotal skin and soiled clothing, were profoundly ashamed and entertained thoughts of suicide.[25] Surgical removal of hydrocele (hydrocelectomy) results in major improvements in work capacity, sexual performance, and self esteem.[2]

- **Diagnosis and Treatment.** Demonstration of microfilariae in blood involves a simple and fairly accurate diagnostic technique,[41] provided that thick blood smears are made during the period when juveniles are in peripheral blood. However, the need to take blood samples late at night is "a situation not well tolerated by community members or health workers."[64] In addition, technicians must be able to distinguish different species of microfilariae that could be present. A technique based on the polymerase chain reaction can detect as little as 1 pg of filarial DNA (just 1% of the DNA in one microfilaria).[101] In areas with mass drug administration for filariasis control, the night blood microfilariae test is not useful due to suppression of microfilarial release from females. Because many infected people are amicrofilaremic, techniques to detect antigens from adult worms, or circulating filarial antigens (CFAs), are considered the best available for *W. bancrofti.* CFA detection works on virtually all patients with microfilaremia, plus many individuals that are amicrofilaremic. An immunochromatographic CFA "card" test is easy to use in the field.[70, 96] The vigorous movements of adults can often be detected by ultrasonography, a pattern of noises referred to as a "filaria dance sign."[3] X-ray examinations can detect dead, calcified worms. Tests based on antibodies to the Bm-14 antigen detect both *W. bancrofti* and *B. malayi.*[100] Similarly, filaria-specific antibodies can be detected in urine.[43] Antibody-based tests are most useful in disease surveillance for control programs.

  The drug of choice for the past 40 years has been diethylcarbamazine (DEC, Hetrazan), which eliminates microfilariae from the blood and causes approximately 40% mortality of adult worms; it is not known if repeated doses will kill all adults.[28, 35] The standard treatment has been 6 mg/kg doses given over a period (usually 7 to 12 days) to a cumulative total of 72 mg/kg, which has had some significant disadvantages. For example, side effects have often been marked, and it has been difficult to persuade patients in the field to return repeatedly for treatments.[48] In recent years it was discovered that good microfilariae control could be achieved with a single dose of 6 mg/kg given annually or semiannually; side effects are fewer, and logistics are much easier. DEC is contraindicated for lymphatic filariais control in areas where *O. volvulus* co-occurs; in such regions, ivermectin is often substituted. However, ivermectin is contraindicated where *L. loa* co-occurs, and sometimes causes adverse reactions in patients with onchocerciasis. Doxycycline therapy eliminates *Wolbachia* sp. symbionts and is promising where *L. loa* is present because this species lacks *Wolbachia* sp.[35] Treatment based on anti-*Wolbachia* compounds, including doxycycline, may yield significant improvements in lymphedema and hydrocoele.[10] A drawback of current anti-*Wolbachia* agents is that they require prolonged administration. Simultaneous treatment with albendazole and ivermectin or DEC is frequently employed, but the benefits of this combined drug therapy for control of lymphatic filariasis remains controversial.[35] Other health benefits of including albendazole, such as control of intestinal nematodes, is noncontroversial. Given that drugs for control programs have been donated, the total cost of mass drug administration is only about 46 cents per person.[21] DEC-fortified table salt is effective, and its use was important for eradication of filariasis in China.[12, 62, 96]

  Edematous limbs are sometimes successfully treated by applying pressure bandages, which force lymph out of the swollen area. This treatment may gradually reduce the size of the member to nearly normal. Any connective tissue proliferation that might have developed will not be affected, however.

  Prevention primarily involves protection against mosquito bites in endemic areas. People temporarily visiting such places should use insect repellent, mosquito netting, and other preventive measures rigorously. Long-term protection requires mosquito control and mass chemotherapy of indigenous people to eliminate microfilariae from the circulating blood, where they are available to mosquitoes. To be successful, such measures require some education of people in endemic areas. For example, Eberhard and coworkers[30] found that *less than half* of an affected population had heard of filariasis and only 6% knew it was transmitted by mosquitoes. In the *same population,* 25% were microfilaremic, 5% of women had elephantiasis, and 30% of men had hydrocele.

Brugia malayi

It was first noticed in 1927 that a microfilaria, different from that of *W. bancrofti,* occurred in the blood of people native to Celebes. It was not until 1940 that the adult form was found in India; a year later it was discovered in Indonesia. We now know that *Brugia malayi* parasitizes humans in Southeast Asia, Indonesia, Malaysia, Thailand, India, Sri Lanka, the East Indies, and the Philippines. Much of its distribution overlaps that of *W. bancrofti,* but unlike *W. bancrofti,* *B. malayi* has not spread to Africa and the New World.[38, 53] *Brugia malayi* has both nocturnally periodic and subperiodic types; the latter includes nonhuman reservoir hosts, including domestic dogs and cats, and monkeys.[4] These reservoirs complicate human filariasis control in some regions.

The morphology of this parasite is very similar to that of *W. bancrofti,* although males are shorter and more slender. The number of anal papillae of males differs slightly between the two species, and the left spicule of *B. malayi* is a little more complex than that of *W. bancrofti.* Phylogenetic analysis of molecular sequence data places the genus *Brugia* as the sister genus to *Wuchereria.*[66]

- **Morphology.** Males are 13.5 mm to 25 mm long and 70 μm to 90 μm wide. The tail is curved ventrally and bears three or four pairs of adanal and three or four pairs of postanal papillae. Spicules are unequal and dissimilar, and a small gubernaculum is present.

  Females are 50 mm to 100 mm long by 130 μm to 170 μm wide. The cuticle is annulated and the tail covered with minute cuticular bosses. The vulva is near the level of the middle of the esophagus.

- **Biology and Pathology.** The life cycle of *B. malayi* is nearly identical to that of *W. bancrofti.* Mosquitoes of genera *Mansonia, Aedes,* and *Culex* are intermediate hosts. Adults live in lymphatics and cause the same disease symptoms as *W. bancrofti,* although elephantiasis, when it occurs, usually does not extend beyond the knees or elbows.[49]

Microfilariae are somewhat similar to those of *W. bancrofti* but can be differentiated by the presence of nuclei in the tail tip. Sequences of 5S rRNA, amplified by PCR from microfilariae or adults is also diagnostic for *B. malayi.*[18]

Diagnosis and treatment are as for *W. bancrofti.* Control is also primarily by mass drug administration and mosquito eradication. *Mansonia* spp. are major vectors in many areas. Their larvae pierce stems of aquatic vegetation to tap air, so they do not need to reach water surface regularly. For this reason, herbicides can be put to good use in mosquito control.

Another species of *Brugia, B. timori,* was first known from its distinctive microfilariae, and since then adults have been described.[76] It has been found only from the Lesser Sunda Islands of southeast Indonesia and can cause severe disease in affected populations. It shows nocturnal periodicity and is transmitted by *Anopheles barbirostris;* there is no known animal reservoir.[82]

Brugia malayi was the first parasitic nematode to have its genome completely sequenced. In addition, the genome of the *Wolbachia* sp. mutualist from *B. malayi* has also been sequenced. These genome sequences have been important in beginning to characterize the interaction between the bacterial symbiont and its host, and to understand unique aspects of the nematode proteome that might be new targets for chemotherapy.[52] A disadvantage of parasitic nematode models is that they are frequently difficult for classical genetics and establishment of gene function. An alternative is using a comparative approach, developing hypotheses of gene homology by sequence comparison to *C. elegans,* and exploiting the rich functional genomic resources of this free-living nematode to identify essential *B. malayi* genes. This approach appears useful for identifying potential drug targets.[52]

Comparison of the genomes of *B. malayi* and *C. elegans,* and consideration of the *Wolbachia* sp. genome of *B. malayi,* suggests that coevolution between the parasite and bacterium is reflected in the *Brugia* genome.[88] *Brugia malayi* lacks most of the enzymes for de novo synthesis of purines, whereas *Wolbachia* sp. from the parasite has all these enzymes. Other critical biosynthetic pathways have genes missing from *B. malayi,* but many are present in the *Wolbachia* symbiont. These observations may explain why eliminating *Wolbachia* sp. from *B. malayi* causes reductions in function for the nematode. Additional functional studies are needed to more fully understand the interactions between these mutualists. Likewise, comparative genomic analysis of filarial parasites lacking *Wolbachia* sp. symbionts should be most instructive.

Brugia pahangi is a very important laboratory model for the study of lymphatic filariasis. It is easily manipulated and maintained in small rodents (Mongolian jird or gerbil, *Meriones unguiculatus*) and so lends itself to far more experimentation than *W. bancrofti* and *B. malayi.* However, *B. malayi*—but not *W. bancrofti*—will survive and develop from J3s to adults in athymic and SCID (severe combined immunodeficiency) mice.[54]

Onchocerca volvulus

Onchocerciasis is a disease caused by this large filaroid worm in areas of Africa (where more than 18 million people are infected despite control programs), Arabia, Guatemala, Mexico, Venezuela, Colombia, and Ecuador. In the Americas, more than 500,000 people are at risk of infection. **River blindness,** as the condition is also known, is not a fatal disease. However, it does cause disfigurement and blindness in many cases; in some small communities in Africa and Central America, most people middle aged and older have significant visual impairment. Infection with *Onchocerca volvulus* also causes severe skin disease, and severe itching is responsible for roughly half of the disability impact of onchocerciasis.[75, 84]

Eradication of this disease would not result in the "parasitologist's dilemma," since it would not increase the birthrate or increase chances for infant survival. It would, instead, free hundreds of thousands of persons from a debilitating disease and thereby remove this economic burden from developing nations.

- **Morphology.** Worms characteristically are knotted together in pairs or groups in subcutaneous tissues (Fig. 29.5). Adults are long, thin worms. Lips and the buccal capsule are greatly reduced and two circles of four papillae each surround the mouth. The esophagus is not conspicuously divided. Females are 230 mm to 500 mm long and 250 μm to 450 μm wide, with the vulva opening just behind the posterior end of the esophagus. Males are 16 mm to 42 mm long and 125 μm to 200 μm wide. The posterior end of the male has two spicules, is curled ventrally and lacks alae. Cuticle annulation patterns differ between the sexes. Microfilariae are unsheathed and measure 220 μm to 360 μm long and 5 μm to 9 μm wide. In the tail end of microfilaria there are several elongated nuclei followed by a clear space in the tail tip.

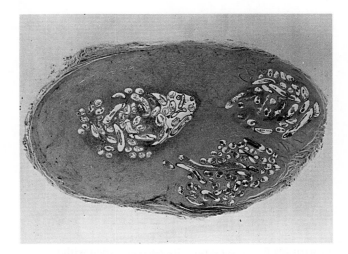

Figure 29.5   **Cross section of a fibrous nodule (onchocercoma) removed from the chest of an African patient.**

It contained several worms bound together in a mass.

From D. H. Connor et al., "Onchocercal dermatitis, lymphadenitis, and elephantiasis in the Ubangi Territory," in *Human Pathol.* 1:553–579. Copyright © 1970. AFIP neg. no. 69-3625.

- **Biology.**   Adult worms locate under the skin, where they become encapsulated by host fibrous tissue, mainly composed of collagens. Blood vessels are found throughout the nodules. The adult worms are in close proximity to lymphatic vessels and tissues,[58] a tissue association similar to species causing lymphatic filariasis. If the encapsulation is over a bone, such as at a joint or over the skull, a prominent nodule appears (Fig. 29.6). Location of these nodules differs according to geographical area. In Africa most infections are below the waist, whereas in Central America they are usually above the waist. These distributions correspond to biting preferences of insect vectors in the two areas. As a consequence, microfilariae are concentrated at sites where the insects prefer to bite.

    Unsheathed microfilariae remain in skin, where they can be ingested by black fly intermediate hosts *Simulium* spp. (Fig. 29.7). These ubiquitous pests become infected when they take a blood meal. Their mouthparts are not adapted for deep piercing, so much of their food consists of tissue juices, which contain numerous microfilariae in infected persons. After ingestion by a black fly, microfilariae are attracted to the thoracic muscles, to which they migrate.[55] There they develop to the sausage stage (Fig. 29.8) to complete J₁ development, molt to the second stage, and then molt again to infective, filariform J₃s

(Fig. 29.9). Infective juveniles move to the fly's labium and can infect a new host when the insect next feeds. Mature worms appear in skin in less than a year.

Onchocerca volvulus was probably introduced to the Americas with African slaves. It became established in Central America and has since changed sufficiently to cause its own distinct clinical symptoms in its definitive host and to differ from African strains in its infectivity to various vectors and laboratory animals. That the species has done this within about 400 years is an indication of the microevolutionary potential of dioecious parasites with high reproductive capacity. In common with

Figure 29.7   A black fly, *Simulium damnosum,* biting the arm of a human.

This insect is a major vector of *Onchocerca volvulus* in Africa.

From D. H. Connor et al., "Onchocercal dermatitis, lymphadenitis, and elephantiasis in the Ubangi Territory," in *Human Pathol.* 1:553–579. Copyright © 1970. AFIP neg. no. 68-2763-1.

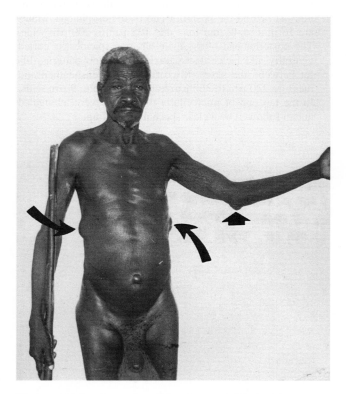

Figure 29.6   Several nodules *(arrows)* filled with *Onchocerca volvulus* are found in the skin of this man.

Note also the elephantoid scrotum and the depigmentation and wrinkling of the skin of the upper arms, also symptoms of onchocerciasis.

From D. H. Connor et al., "Onchocercal dermatitis, lymphadenitis, and elephantiasis in the Ubangi Territory," in *Human Pathol.* 1:553–579. Copyright © 1970. AFIP neg. no. 68-10071-3.

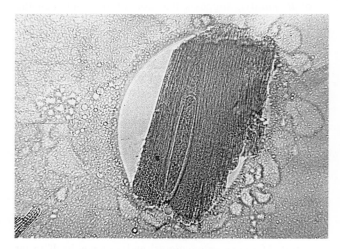

Figure 29.8   Piece of thoracic muscle from *Simulium damnosum.*

Note the "sausage-stage," juvenile of *Onchocerca volvulus.*

Courtesy of John Davies.

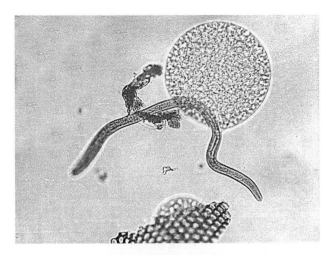

Figure 29.9   **Third-stage, or filariform, juvenile of** *Onchocerca volvulus.*

It was dissected from the head of a *Simulium damnosum.*

Courtesy of John Davies.

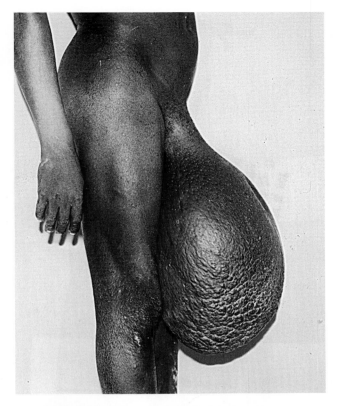

Figure 29.10   **Severe elephantoid scrotum on a native resident of Ubangi territory.**

It was removed surgically and a good cosmetic result was obtained. The scrotum weighed 20 kg and, when viewed microscopically, appeared as an edematous mass of interlacing collagen and smooth muscle fibers.

From D. H. Connor et al., "Onchocercal dermatitis, lymphadenitis, and elephantiasis in the Ubangi Territory," in *Human Pathol.* 1:553–579. Copyright © 1970. AFIP neg. no. 68-8582-9.

other insects, *Simulium* spp. have a variety of defense mechanisms against infections, and several species and/or strains of black flies are not susceptible.[38] We do not understand why. Humans appear to be the only natural definitive host for *O. volvulus.*

- **Epidemiology.**   Generally speaking, onchocerciasis is a model system for landscape epidemiologists. Larval stages of *Simulium* spp. only develop in well-oxygenated, fast running streams. Adult flies survive only where there is high humidity and plenty of streamside vegetation. Some African indigenous people have long known that the disease is associated with rivers (and even with black flies, although this fact was not officially "discovered" until 1926), and they gave it the name *river blindness.* Anyone who intrudes into such a river area is viciously attacked by these insects. Wild-caught black flies are often infected with a variety of species of filarioids, most of which are still unidentified, but in areas endemic with *O. volvulus,* juveniles can often be recognized as that species. In parts of the Amazon region, *O. volvulus* is sympatric with *Mansonella ozzardi* and *M. perstans;* a PCR test has been developed to distinguish these microfilariae, whether obtained from humans or insects.

  Surprisingly, foci of onchocerciasis occur in arid savannas of West Africa and desert areas along the Nile near the Egypt-Sudan border. Epidemiology in these areas has not been thoroughly studied, but it may depend on adaptations for survival of the black fly vectors or strain of *O. volvulus* or both.

- **Pathogenesis.**   Two different elements contribute to the pathogenesis of onchocerciasis: adult worms and microfilariae, both through consequences of the host immune response. Adults are the least pathogenic, often causing no symptoms whatever and, at worst, stimulating the growth of palpable subcutaneous nodules called **onchocercomas** (see Fig. 29.6), especially over bony prominences. In African strains these nodules are most frequent in the

pelvic area, with a few along the spine, chest, and knees. The Venezuelan strain is much like African ones, but in Central America nodules are mostly above the waist, especially on the neck and head. These nodules are relatively benign, causing some disfigurement but no pain or ill health. Number of nodules may vary from 1 to well over 100. Adult worms may live for 10 to 15 years. Rarely the nodule will degenerate to form an abscess, or worms will become calcified.

True elephantiasis sometimes ensues (Fig. 29.10), and another condition, known as *hanging groin,* is common in some areas of Africa. Dermatological changes in response to migrating microfilariae are numerous, and include a loss of skin elasticity that causes a sagging of the groin into pendulous sacs, often containing lymph nodes. Testes and scrotum are not affected, and hydrocele does not accompany the condition. Females are similarly affected (Fig. 29.11). Onchocerciasis frequently causes hernias, especially femoral hernia, in Africa.

In hyperendemic areas more than 90% of people can be microfilaremic.[89] Live microfilariae elicit little inflammatory response,[47] but degenerating juveniles in the skin often result in a severe **dermatitis.** The dermatitis appears

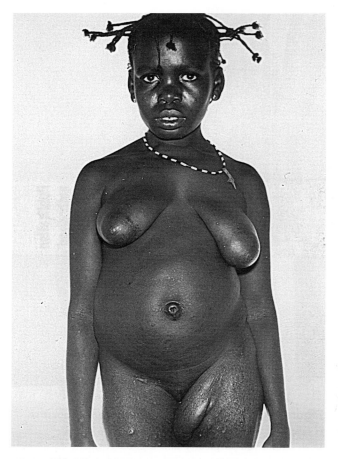

Figure 29.11 "Hanging groin," or adenolymphocele.
The tissue was excised and contained a group of lymph nodes embedded in subcutaneous tissue. The nodes contained many microfilariae of *Onchocerca volvulus.*

From D. H. Connor et al., "Onchocercal dermatitis, lymphadenitis, and elephantiasis in the Ubangi Territory," in *Human Pathol.* 1:553–579. Copyright © 1970. AFIP neg. no. 68-10066-1.

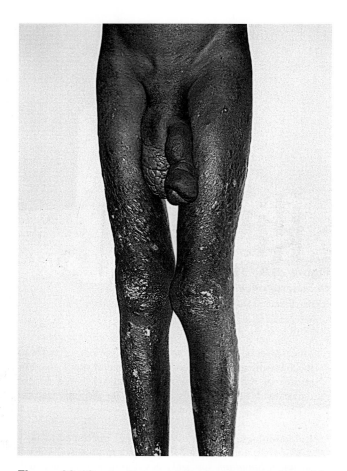

Figure 29.12 An 11-year-old boy with severe dermatitis characterized by depigmentation, wrinkling, and thickening of the skin.
He also has elephantoid changes of the penis and scrotum and onchocercomas over the knees.

From D. H. Connor et al., "Onchocercal dermatitis, lymphadenitis, and elephantiasis in the Ubangi Territory," in *Human Pathol.* 1:553–579. Copyright © 1970. AFIP neg. no. 68-7912-1.

due to inflammation caused by release of *Wolbachia* bacteria from dead juveniles. Treatment with antibiotics such as doxycycline to deplete *Wolbachia* dramatically improves skin lesions.[98] The first dermatitis symptom is intense itching, which may lead to secondary bacterial infection, often accompanied by abnormal pigmentation of the skin in small or extensive areas. This is followed by a thickening, discoloration, and cracking of the skin **(lichenification).** The last stage of the skin lesion is characterized by a loss of elasticity, which gives the patient a look of premature aging. Severe lichenified onchodermatitis is known in some areas as sowda.[67] Loss of pigment may be accentuated and extend over large areas, especially of the legs (Fig. 29.12). Patients at this stage may be misdiagnosed as leprous. Compared to the disabling effects of blindness, skin disease in onchocerciasis is often neglected, but onchodermatitis places a grave psychosocial burden on victims and their communities,[84] and has been shown to have marked socioeconomic impacts on families. Microfilariae in advanced cases often are located in the deeper part of the dermis and are not detected by skin-snip biopsy.

The most dreadful complications of onchocerciasis involve the eyes. The rate of impaired vision may reach 30% in some communities of Africa, where blindness affects in excess of 10% of the adult population. In these areas and in some areas of Guatemala, it is not unusual to see a child with good vision leading a string of blind adults to the local market. Ocular complications are less common in the rain forest areas of Africa but are frequent in the savanna, a difference probably related to the differences among strains of *O. volvulus.*[79, 84]

Eye lesions take many years to develop; most affected persons are over 40 years old. However, in Central America, with more worms concentrated on the head, young adults also show symptoms.

Live microfilariae invade many parts of the eye, but here again they cause little reaction; their death leads to lesions. Neutrophil infiltration of the eye leads to development of corneal and stromal haze; this infiltration is in response to *Wolbachia* sp. bacteria, and involves TLR2 responses and chemokine production that recruits and activates neutrophils.[37] Chronic inflammatory cells with eosinophils and neutrophils surround dead worms, followed by fibroblast proliferation and chronic inflammatory infiltrates.[47] The most important cause of blindness is **sclerosing** (scarring) **keratitis**,[79] a hardening inflammation of the cornea. The predominant immune response is Th2 (p. 30), with T cells producing IL-4 and IL-5. It appears that degranulating eosinophils, in response to worm antigens, disrupt fibril arrangement in the cornea.[79]

In some cases much visual loss is due to lesions of the choroid and retina, or **chorioretinopathy**.[22] In early stages, microfilariae and microfilarial debris can be seen in areas of choroidal inflammation. In advanced disease, photoreceptor and ganglion cell layers degenerate, nerve fiber and ganglion layers atrophy, and choriocapillaries and neuroretina are obliterated. New lesions are not initiated after treatment with DEC or ivermectin (when microfilariae are controlled; see following text), but established disease continues progressing after treatment.[22]

Microfilariae in chambers of the eye are easily demonstrated in onchocerciasis. In fact, often it is an ophthalmologist who first diagnoses the disease during routine ocular inspection.

- **Diagnosis and Treatment.**   The simplest method of diagnosis is demonstration of microfilariae in bloodless skin snips. A small bit of skin is raised with a needle and sliced off with a razor or scissors. The bit of skin is then placed in saline on a slide or in a microtiter plate and observed with a microscope for emerging microfilariae (Fig. 29.13). These must be differentiated from those of other species that might be present. Skin-snip biopsies can be taken anywhere, but if only one sample is taken, it should be from the buttock. If the snip is so deep as to draw blood, the sample might be contaminated with other species of filarioids. On the other hand, in old infections microfilariae may be so deep as to elude the snip. Nodules may be aspirated to sample for microfilariae. A given case may show other, overt symptoms that obviate need for demonstrating microfilariae.

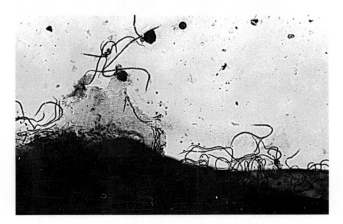

Figure 29.13   Skin snip from a patient with onchocerciasis.
Note emerging microfilariae.
Courtesy of Warren Buss.

Diagnostic methods based on microfilariae can fail in circumstances that are important to monitoring success of control programs. The 9 to 15 month prepatent period contributes to the failure to detect new infections using skin snips. Methods based on specific PCR amplification of microfilaria DNA[90] will also fail in such circumstances. Antibody-based tests offer significant advantages for early detection of prepatent infections for monitoring transmission but cannot discriminate between current and past infections. Several tests based on antibodies to *O. volvulus* recombinant antigens have been developed and field-tested.[57, 85] These antibody "card-style" tests are less invasive, requiring only a drop of blood from a finger prick, and require less equipment and training than using a skin snip. These tests have great potential for monitoring the success of control programs, particularly the frequency of new infections in children.

Surgical excision of nodules, especially those around the head, may be effective in lowering both the rate of eye damage and the number of new infections within a population. Chemotherapy with ivermectin has been a major advance and a modern success story. Merck & Co., Inc. developed and marketed the drug initially for veterinary use. After its effectiveness in onchocerciasis became known, the company generously donated the drug for mass treatment in developing countries; during the last 20 years, more than 570 million doses have been donated. Previously there had been no satisfactory drug available; the serious side effects of DEC and suramin precluded their use for mass treatment. Ivermectin is well tolerated, and annual or biannual treatment significantly decreases incidence of infection in populations.[92] A single dose of ivermectin greatly reduces skin microfilariae, typically to 1% of pre-treatment levels. Ivermectin also has an embryostatic effect on adult females reducing release of microfilariae for many months[87] thus interrupting transmission, and significantly improving skin disease.[16] Biannual doses of ivermectin suppress development of J_3s in the blackfly vector,[24] providing an independent effect for control. Although usually considered only a microfilaricide, ivermectin in repeated doses slowly kills adult worms.[27] Ivermectin does not have a cumulative embryostatic effect,[13] and onchocerciasis control relies upon continued community treatment. Administering ivermectin every 3 months causes greater lethality for adults, both through macrofilaricidal effects and by increasing the incidence of a lethal ovarian neoplasm in the nematodes.[26]

Prevention by elimination of vectors has been, over the years, very difficult. Application of DDT to swift-running streams destroys all simuliids,[34] but undesirable environmental effects of DDT are well known. Other more biodegradable insecticides are more expensive. Nevertheless, vector control and chemotherapy campaigns have prevented 125,000 to 200,000 people from going blind, protected 40 million people from ocular and skin lesions, and ensured that between 1974 and 1995 10 million children were born with no risk of blindness in the 11 African countries participating in the Onchocerciasis Control Programme (OCP). The net present value (equivalent discounted benefits minus discounted costs) of the OCP in West Africa has been estimated at $485 million for the program over a 39-year period.[99] The OCP ended in 2002 and was replaced by the African Program for Onchocerciasis Control (APOC), which operates in 19 countries in Central and East Africa and employs village-based volunteers to implement annual ivermectin distribution. The Onchocerciasis Elimination Program for the Americas (OEPA), founded in 1992, has employed biannual ivermectin treatment and succeeded in interrupting transmission in 7 of 13 foci, and has set a target date of elimination of 2015.

Loa loa

Loa loa is the "eye worm" of Africa and produces loiasis or fugitive or **Calabar swellings.** It is distributed in rain forest areas of Central and West Africa and equatorial Sudan where 3–13 million people are infected. In some rural areas of Africa, loiasis is the third most common cause for seeking medical attention. In some African villages the prevalence may reach 40%.[102] Although it was established for a short time in the West Indies, where it was first discovered during slavery, it no longer exists there.

The morphology of *L. loa* is typical of the family: a simple head with no lips and eight cephalic papillae; a long, slender body; and a blunt tail. The cuticle is covered with irregular, small bosses, except at the head and tail. Males are 20 mm to 34 mm long by 350 μm to 430 μm wide. The three pairs of preanal and five pairs of postanal papillae are often asymmetrical. Copulatory spicules are unequal and dissimilar, 123 μm and 88 μm long. Females are 20 mm to 70 mm long and about 425 μm wide. The vulva is about 2.5 mm from the anterior end, and the tail is about 265 μm to 300 μm long. *Wolbachia* have not been reported in *L. loa*.[91]

- **Biology.** Adults live in subcutaneous and intermuscular connective tissues (Fig. 29.14), including the back, chest, axilla, groin, penis, scalp, and eyes in humans. Infections of deep tissues, including fatal encephalitis,[69] are also known. Sheathed microfilariae (see Fig. 29.4) appear in peripheral blood in maximal numbers during daylight hours and concentrate in the lungs at night. Intermediate hosts belong to any of several species of deer fly, genus *Chrysops,* which feed by slicing the skin and imbibing blood as it wells into the wound. Worms develop into filariform J_3s in the fly's abdominal fat body, after which they develop in the hemocoel and then migrate to the mouthparts. The prepatent period in humans is about a year, and adult worms may live 15 years or longer. There are probably no important reservoir hosts of the *L. loa* strain found in humans, although several species of nonhuman primates have been experimentally infected.[81]

- **Pathogenesis.** These worms have a tendency to wander through subcutaneous connective tissues, provoking inflammatory responses as they go. When they remain in one spot for a short time, the host reaction results in localized Calabar swellings, especially in the wrists and ankles, that disappear when the worm moves on. Occasionally adult worms migrate through the conjunctiva and cornea (Fig. 29.15), with swelling of the orbit and psychological results to the host. Eosinophilia of up to 80%,

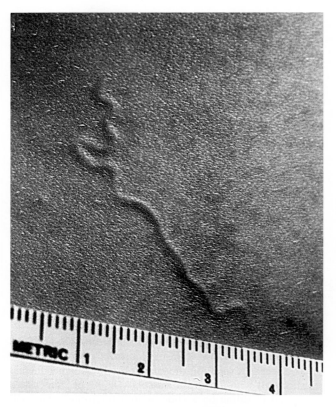

Figure 29.14   Adult female *Loa loa* visible under the skin of a patient.

From D. L. Price and H. C. Hopps, in R. A. Marcial-Rojas, editor, *Pathology of protozoal and helminthic diseases, with clinical correlation,* © 1971 The Williams & Wilkins, Co. AFIP neg. no. 67-5366.

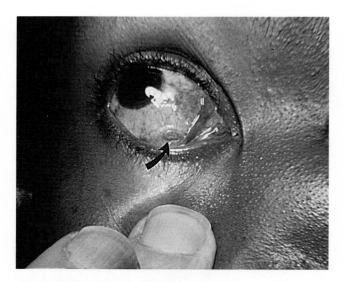

Figure 29.15   Adult female *Loa loa* coiled under the conjunctival epithelium *(arrow)* of the eye of a patient from the Congo.

From D. L. Price and H. C Hopps, in R. A. Marcial-Rojas, editor, *Pathology of protozoal and helminthic diseases, with clinical correlation,* © 1971 The Williams & Wilkins, Co. AFIP neg. no. 67-5368-1.

and marked monocytosis are common. Intense pruritis (itching), arthralgia (joint pain), and fatigue are common, and serious complications can occur.[81] The pathogenicity of loaisis may be generally underestimated.[1]

As in *Wuchereria bancrofti,* many people indigenous to endemic areas can be asymptomatic but still have a high microfilaremia, whereas newcomers may be amicrofilaremic but clinically and immunologically hyperresponsive.[73]

- **Diagnosis and Treatment.** Demonstration of typical microfilariae in the blood (see Fig. 29.4) is ample proof of loiasis. Visual observation of a worm in the cornea or over the bridge of the nose also indicates this species. Finally, transient skin swellings are suspect, although sparganosis or onchocerciasis may be confused with loiasis. Surgical removal is simple and effective, providing the worm is properly located, but most worms are inapparent.

Drug treatment of loiasis is complicated by severe side effects and fatalities that may occur when microfilaremia is high. In regions with co-infections of *L. loa* and *O. volvulus,* adverse reactions to ivermectin complicates drug-based control programs, which require special precautions in such regions.[20] Quantitative and specific PCR assays for *L. loa* microfilaria have the potential to help detect and mitigate such risks. Conventional treatment with DEC has troublesome side effects and can lead to encephalitis, sometimes fatal.[17] Ivermectin seems to be effective,[61] but neither it nor DEC affects adults.[81] Drug treatment based on combined doxycylcine and ivermectin treatment appears effective and well-tolerated with moderate *L. loa* infection intensities.[97] The antibiotic doxycyline has an anti-*Wolbachia* effect in filarioids with these symbionts (not including *L. loa*) but also appears to have a direct effect on adults.[97] Control of deer flies, which breed in swampy areas of the forest, is extremely difficult.

Other Filaroids Found in Humans

Mansonella ozzardi is a filaroid parasite of the New World, with distribution that encompasses northern Argentina, the Amazon drainage, the northern coast of South America, Central America, and several islands of the West Indies. It has never been found in the Old World. Adults live in the body cavity, threaded among mesenteries and peritoneum and in subcutaneous tissues. Symptoms can mimic those of bancroftian filariasis, with polylymphadenitis, lymphedema, elephantiasis, and hepatomegaly.[44] It is infected with *Wolbachia.*[91] Intermediate hosts are species of *Culicoides* and *Simulium.*[94]

Mansonella perstans (formerly *Dipetalonema perstans*) exists in people in tropical Africa and South America. Several primates have been incriminated as reservoir hosts. Adult worms live in the coelom and produce unsheathed microfilariae (Fig. 29.16). Intermediate hosts are species of biting midges of genus *Culicoides.* The worms appear to cause little pathological effect.

Mansonella streptocerca (formerly *Dipetalonema streptocerca*) is a common parasite in the skin of humans in many of the rain forests of Africa.[65] It probably is a parasite of chimpanzees in nature.

Dirofilaria repens is a subcutaneous filarial parasite of cats and dogs in Europe, sub-Saharan Africa, and

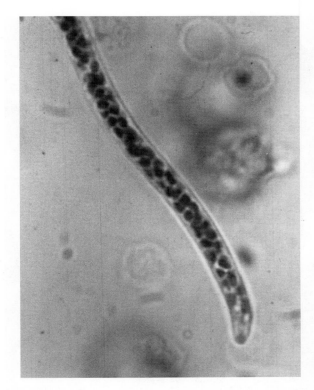

Figure 29.16 Tail end of microfilaria of *Mansonella perstans.*
The infection was acquired in Nigeria.
From H. Zaiman, editor, *A pictorial presentation of parasites.* Photo by M. G. Schultz.

South Asia. The distribution of this parasite is changing in Europe, moving north into regions that were formerly free of *Dirofilaria* spp., a result predicted based on models of regional climate change and recorded temperature increases.[36] It may be asymptomatic or it may cause pruritis or eczematous eruptions. In humans, *D. repens* is usually found in nodules or cysts in or around the eye.[72] It is the main agent of human dirofilariasis in the Old World and an emerging disease in southern Europe, with 492 known cases, most reported from Italy.[72] Juvenile stages can be distinguished from *Dirofilaria immitis* by polymerase chain reaction techniques and sequencing of mitochondrial DNA;[31] such distinction is particularly important in Italy, where the two species coexist. Microfilaria can be controlled in dogs using dermal application of macrocyclic lactones. Resolution of human cases normally involves surgical removal of adults.[72]

Dirofilaria immitis

Adult heartworms, *Dirofilaria immitis* (Fig. 29.17) are parasitic in the pulmonary arteries, caval veins and right side of the heart, and is frequently found in domesticated dogs and more rarely cats. Heartworm has been reported from many wild mammals, but appears most common in various canids, felids, and mustelids. Approximately 26 other species of *Dirofilaria* have been described, and some reports of *D. immitis* from uncommon hosts may be misdiagnoses. *Dirofilaria immitis* has been found in humans several times, including in the

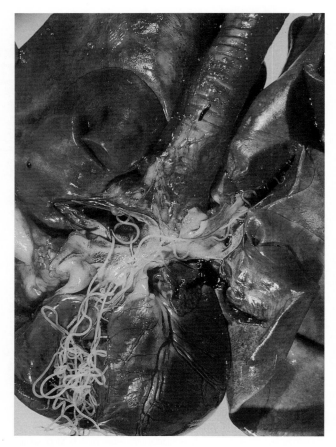

Figure 29.17 *Dirofilaria immitis* **in right ventricle of an eight-year-old Irish setter.**
It is extending up into right and left pulmonary arteries.

United States, but they apparently do not mature in humans, and therefore their microfilariae are absent from human blood.[68]

- **Biology.** Mosquitoes ingest microfilariae with their blood meal, and filariform J_3s develop in the Malpighian tubules. J_3s then migrate through the thorax and to the mosquito's mouthparts, to be deposited onto the host's skin during the mosquito's next meal. These larvae develop in the dog's tissues for about 3 months before entering the pulmonary vasculature. After about six months mature female worms are capable of producing microfilariae. *Dirofilaria immitis* is infected with *Wolbachia,* and when worms are "cured" of their bacterial infection with tetracycline, their reproduction and embryogenesis are inhibited.[8]

- **Epidemiology.** This worm is transmitted by many species of mosquitoes. Development to the infective J_3 within mosquito intermediate hosts is faster at higher ambient temperature; the daily minimum must exceed 57°F for continued development,[5] and infectivity generally requires temperatures exceeding 64°F for a month. At 80°F, J_3 development can be completed in 10–14 days. The highest prevalence in the United States corresponds to regions with temperatures conducive for rapid nematode development and high mosquito density. In the southeastern United States, including the Gulf Coast, prevalence is highest, averaging 3.9%.[14] The northeastern states have the lowest average prevalence (0.6%). However, even within regions of low average prevalence, "hotspots" of transmission can occur. For example, the average prevalence in California is 1.6%, but some counties have a prevalence exceeding 10%.[14] Human heartworm infection is reported most frequently in areas of high canine prevalence. Prevalence is much lower in the western United States but can range to 5% in some areas of California and Oregon. Heartworm has been diagnosed in domestic dogs from all 50 U.S. states. Heartworm positive dogs from Alaska have typically had a history of recent travel or residence in areas of endemic transmission outside the state, and heartworm transmission has not yet been reported there.

- **Pathogenesis.** This worm is a dangerous pathogen for dogs, and although prevalence and worm load are usually lower in cats, even a few adults can cause serious disease in cats.[62] Clinical signs include respiratory insufficiency, vomiting, chronic cough, and exercise intolerance.[78] Obviously, large worms extending through the openings of the tricuspid and semilunar valves will prevent efficient operation of the heart. Pulmonary arteries show thickening and inflammation of their inner walls. Adult worms can precipitate pulmonary embolism, leading to coughing up blood and even worms from ruptured vessels. Death may occur from cardiopulmonary failure.

  In humans symptoms are vague and unpredictable. They may include chest pain, cough, coughing up blood, fever, and malaise.[68, 93]

- **Diagnosis and Treatment.** Routine diagnosis in canines is by detection of a heartworm antigen from female worms in a blood sample. The antigen test is sensitive and specific to *D. immitis,* but cannot be used during the first 5 months post-infection. Antigen-based detection works for "occult" infections lacking circulating microfilariae, which may represent 20% of infections in dogs. Examination of microfilariae in blood smears can also be diagnostic, but *D. immitis* must be distinguished from other species such as *Acanthocheilonema reconditum* in the United States,[77] and *D. repens* in Europe.

  Dogs and other susceptible animals should be given prophylactic doses of macrocyclic lactone drugs such as ivermectin or milbemycin once a month as prescribed by a veterinarian. These drugs kill J_3 and early J_4 stages, but at prophylactic doses have minimal effect on adult worms. Even with chemoprophylaxis, dogs should be tested for heartworm on a periodic basis because protection from infection is not absolute. Treatment of adults requires a different drug regime. The death of adult worms during treatment can block blood vessels, induce emboli, and become life-threatening in severe cases. Animals require diligent care during treatment. No responsible pet owner should fail to provide chemoprophylaxis during periods of transmission.

Learning Outcomes

By the time a student has finished studying this chapter, he or she should be able to:

1. List the sequential steps in the life cycle of a nematode from the superfamily Filarioidea.
2. Explain how natural selection might favor nocturnal periodicity of microfilariae.
3. Discuss how the pathology of onchocerciasis differs from lymphatic filariasis.
4. Compare the benefits of community-based drug treatment programs for lymphatic filariasis versus drug treatment focused on individual patients.
5. Explain the potential hazards associated with using ivermectin or diethylcarbamazine (DEC) for community-based control programs of filarial diseases.

References

References for superscripts in the text can be found at the following Internet site: www.mhhe.com/robertsjanovynadler9e

Additional Readings

Chabaud, A., and R. C. Anderson. 1959. Nouvel essai de classification des filaries (Superfamille des Filarioidea) II, 1959. *Ann. Parasitol.* 34:64–87. An accurate, easy-to-use key to the genera of Filaroidea.

Chernin, E. 1983. Sir Patrick Manson's studies on the transmission and biology of filariasis. *Rev. Infect. Dis.* 5:148–166.

Chernin, E. 1983. Sir Patrick Manson: An annotated bibliography and a note on a collected set of his writings. *Rev. Infect. Dis.* 5:353–386.

Duke, B. O. L. 1971. The ecology of onchocerciasis in man and animals. In A. M. Fallis (Ed.), *Ecology and physiology of parasites.* Toronto: University of Toronto Press, pp. 213–222.

Hoerauf, A., and N. Brattig. 2002. Resistance and susceptibility in human onchocerciasis—beyond Th1 vs Th2. *Trends Parasitol.* 18:25–31.

Hoerauf, A., L. Volkmann, K. Nissen-Paehle, C. Schmetz, I. Autenrieth, D. W. Büttner, and B. Fleischer. 2000. Targeting of *Wolbachia endobacteria* in *Litosomoides sigmodontis:* Comparison of tetracyclines with chloramphenicol, macrolides and ciprofloxacin. *Trop. Med. Int. Health.* 5:275–279. Tetracyclines can deplete *Wolbachia* and thereby inhibit female worm development and early embryogenesis. Other antibiotics were not effective.

Khanna, N. N., and G. K. Joshi. 1971. Elephantiasis of female genitalia. A case report. *Plast. Reconstr. Surg.* 48:374–381.

Meyers, W. M. et al. 1976. Diseases caused by filarial nematodes. In C. H. Binford and D. H. Connor (Eds.), *Pathology of tropical and extraordinary diseases,* vol. 2, sect. 8. Washington, DC: Armed Forces Institute of Pathology.

Nelson, G. S. 1970. Onchocerciasis. In B. Dawes (Ed.), *Advances in parasitology 8.* New York: Academic Press, Inc., pp. 173–224.

Sasa, M. 1976. *Human filariasis. A global survey of epidemiology and control.* Baltimore, MD: University Park Press.

Zimmer, C. 2001. *Wolbachia:* A tale of sex and survival. *Science* 292:1093–1095.

Chapter 30

Nematodes: Dracunculomorpha, Guinea Worms, and Others

And they journeyed from Mount Hor by the way of the Red Sea, to compass the land of Edom. . . .
And the Lord sent fiery serpents among the people, and they bit the people; and much people of Israel died. . . .
And the Lord said unto Moses, "Make thee a fiery serpent and set it upon a pole; and it shall come to pass that everyone that is bitten, when he looketh upon it, shall live."

—Bible, Numbers 21:4–8

In previous editions of this book, the taxonomic coverage of this and other chapters has varied, reflecting differences in prevailing phylogenetic hypotheses. This edition is no exception; updated molecular phylogenetic hypotheses for nematodes have provided a broader and more resolved framework for dracunculoid nematodes and their relatives.[11, 15, 22, 23] At one time, members of Philometridae, Anguillicolidae, Skyrjabillanidae, and Dracunculidae were considered to be part of a monophyletic Dracunculoidea. However, phylogenies based on SSU rDNA conclusively show that anguillicolids are members of a different clade, and are most closely related to Gnathostomatomorpha.[11, 15] In addition, molecular phylogenies strongly support Camallanoidea as the sister group to Dracunculoidea (absent Anguillicolidae).[15] Therefore, given the most current molecular phylogenetic framework, this chapter includes information on nematodes within both of these superfamilies. In keeping with the infraorder status of clades identified within suborder Spirurina, these sister groups are grouped under Dracunculomorpha. Greatest emphasis is given to the family Dracunculidae, which includes one of the most fascinating nematodes of mankind, *Dracunculus medinensis.*

DRACUNCULOMORPHA

Family Philometridae

Members of this family are tissue parasites of fishes. Two common genera in family Philometridae are *Philometra,* with a smooth cuticle, and *Philometroides,* which has a cuticle covered with bosses. The mouth is small, there is no sclerotized buccal capsule, and their esophagus is short. Males of many species are unknown. Gravid females live under the skin, in the swim bladder, or in the coelom of fishes, where they release first-stage juveniles. After reaching the external environment, juveniles develop further if they are eaten by a cyclopoid copepod. If the worm is to survive, a definitive host must eat the microcrustacean containing J₃s.

Development into adults is not well known. Males and females mate in deep tissues of the body, and males die soon after. Females then migrate to their definitive site, where they release young. *Philometra onchorhynchi,* a parasite of salmon in the western United States and Canada, apparently passes

Figure 30.1 *Philometroides* **sp. in the skin of a fin of a white sucker,** *Catostomus commersoni.*
Courtesy of John S. Mackiewicz.

out with the fish's eggs when its host spawns. Then females burst in the fresh water and thus release their juveniles.[18] *Philometroides nodulosa,* under the skin of the head and fins of suckers (Catostomidae), is a familiar sight to those who work with these fish in the United States (Fig. 30.1).

Family Dracunculidae

Members of family Dracunculidae are tissue dwellers of reptiles, birds, and mammals. Morphological characteristics of the several genera and species are very similar, with, for example, only small differences occuring between *Dracunculus* species in reptile hosts and those in mammals.

Seven species of *Dracunculus* are known from snakes or turtles; one is common in snapping turtles in the United States. Species of genus *Micropleura* are found in crocodilians and turtles in South America and India, whereas *Avioserpens* has species in aquatic birds.

Four species of the genus *Dracunculus* are known from mammals. In the Americas *D. insignis* appears to be a generalist, infecting multiple host species including raccoons, otters, mink, fishers, muskrats, and opossums. Another North American species, *D. lutrae* appears to have a more narrow host range, infecting only otter. *Dracunculus medinensis* was once prevalent in circumscribed areas of Africa, India, and the Middle East, but appears to be on the verge of eradication from human populations. It has been reported from humans in the United States several times, but these cases may have been caused by *D. insignis,* which is infective for rhesus monkeys.[1] However, until recently, *D. medinensis* has not been scarce in humans in many countries of the world, so we will examine it in greater detail.

Dracunculus medinensis. *Dracunculus medinensis* has been known since antiquity, particularly in the Middle East and Africa. It is estimated that in 1986 there were 3.5 million cases occurring in 20 countries throughout Africa and Asia, with 120 million people at risk of infection. A concerted effort at eradication has been underway since then, and by 2010, there were only 1,793 cases, with transmission in 2011 limited to Ethiopia, Mali, and South Sudan.[9] This is arguably

the single most remarkable example of progress in parasite control, particularly because it was accomplished without use of a vaccine or anthelminthic drug. Because of the worm's large size and the conspicuous effects of infection, it is not surprising that the parasite was mentioned by classical authors. The Greek Agatharchidas of Cnidus, who was tutor to one of the sons of Ptolemy VII in the second century B.C., gave a lucid description of the disease: "The people who live near the Red Sea are tormented by an extraordinary and hitherto unheard of disease. Small worms issue from their bodies in the form of serpents which gnaw their arms and legs; when these creatures are touched they withdraw themselves and insinuating themselves between the muscles give rise to horrible sufferings."[7] The Greek and Roman writers Paulus Aegineta, Soranus, Aetius, Actuarus, Pliny, and Galen all described the disease, although most of them probably never saw an actual case. The Spanish and Arabian scholars Avicenna (Abu Ali alHusein ibn Sina), Avenzoar, Rhazes, and Albucasis also discussed this parasite, probably from firsthand observations. In 1674 Velschius described winding the worm out on a stick as a cure. European parasitologists remained ignorant of it until about the beginning of the 19th century, when British army medical officers began serving in India. Information about *D. medinensis* slowly accumulated, but it remained for a young Russian traveler and scientist, Aleksej Fedchenko, to give between 1869 and 1870 the first detailed account of the worm's morphology and life cycle.[6] His discovery that humans become infected by swallowing infected copepods pointed the way to a means of prevention of dracunculiasis.

- **Morphology.** *Dracunculus medinensis* is one of the largest nematodes known. Adult females have been recorded up to 800 mm long, although males do not exceed 40 mm. The mouth is small and triangular and is surrounded by a quadrangular, sclerotized plate. Lips are reduced. Cephalic papillae are arranged in an outer circle of four double papillae at about the same level as the amphids and an inner circle of two double papillae, which are peculiar in that they are dorsal and ventral. Their esophagus has a large glandular portion that protrudes and lies alongside the thin muscular portion.

  In females the vulva is about equatorial in young worms; it is atrophied and nonfunctional in adults. The gravid uterus has an anterior and a posterior branch, each of which is filled with hundreds of thousands of embryos. The intestine becomes squashed and nonfunctional as a result of uterine pressure in gravid worms.

  A major difficulty in taxonomy of dracunculids is the sparsity of discovered males. The few specimens known of *D. medinensis* range from 12 mm to 40 mm long; spicules are unequal and 490 μm to 730 μm long. The gubernaculum ranges from 115 μm to 130 μm in length. Genital papillae vary considerably in published descriptions. In fact, in monkeys, at least, males taken from a single animal have varying numbers of papillae. Because of the technical difficulties of obtaining male specimens from natural infections, morphological characters of females in combination with molecular systematic methods will be needed to resolve systematic, taxonomic, and epidemiological problems within the genus *Dracunculus.* Molecular approaches seem promising for this purpose as gene sequences can be used to distinguish among three species from mammals, *D. insignis, D. lutrae* and *D. medinensis*, and also *D. oesophageus,* a parasite of a colubrid snake.[2, 5, 23]

• **Biology.**   The development of *D. medinensis* has been studied in dogs, cats, and rhesus monkeys. When the ovoviviparous female is gravid, embryos in the uteri cause a high internal pressure. At this stage the female has migrated to the skin of the host. The great majority of infections involve a lower limb, but gravid females can emerge from other parts of the body, including the head, torso, upper extremities, buttocks, and genitalia.[20] Internal pressure and progressive degeneration cause the body wall and uterus of the parasite to burst, forcing a loop of the uterus through the cuticle, freeing many juveniles. An undefined chemical in the uterine fluid causes a blister in the skin of the host (Fig. 30.2). The blister eventually ruptures, forming an exit for young worms, which trickle out onto the skin surface. Sometimes, instead of the body wall rupturing, the uterus forces itself out of the worm's mouth. Muscular contractions of the body wall force juveniles out in periodic spurts, with more than half a million ejected at a single time. These contractions are instigated by cool water, which causes the worm and its uterus to protrude through the wound. As portions of the uterus empty, they disintegrate, and adjacent portions move into the ulcer. Eventually all of the worm is "used up," and the wound heals.

After leaving their mother and host, J_1s must enter water directly to survive. They can live for four to seven days but able to infect an intermediate host for only three days. To develop further they must be eaten by a cyclopoid crustacean. Once in the intestine of their new host, juveniles penetrate into the hemocoel, especially dorsally in the gut, where they develop into infective J_3s in 12 to 14 days at 25°C (Fig. 30.3). Some species of copepods suffer high mortality as a result of infection, thus certain species are more efficient intermediate hosts than others.[2]

Definitive hosts are infected by swallowing infected copepods with drinking water. However, with *D. insignis,* tadpoles and adult frogs can serve as paratenic hosts.[4] Released juveniles penetrate the duodenum, cross the abdominal mesenteries, pierce abdominal muscles, and enter subcutaneous connective tissues, where they migrate to axillary and inguinal regions. A third molt occurs about 20 days after infection, and the final one at about 43 days. Females are fertilized by the third month after infection. Gravid females migrate to the skin between the eighth and tenth months, by which time embryos are fully formed. Males die by seven months postinfection and become encapsulated in tissues. Between 10 and 14 months after initial infection, a female causes a blister in the skin. The blister causes intense burning pain, which is somewhat alleviated by immersion in water.

Little is known about the physiology of this parasite, but the gut is often filled with a dark-brown material, suggesting that worms feed on blood. Glycogen is stored in several tissues of mature females. Glucose utilization and the rate of formation of lactic acid are not affected by presence or absence of oxygen.[3]

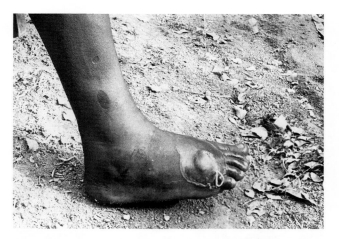

Figure 30.2   Blister, caused by a female *Dracunculus medinensis,* in the process of bursting.
There has been an unusually severe tissue reaction resulting in a very large blister. A loop of the worm can be seen protruding through the skin. Note swelling of foot and ankle.

From R. Muller, in B. Dawes, editor, *Advances in parasitology,* vol. 9, © 1971. London: Academic Press, Inc., Ltd.

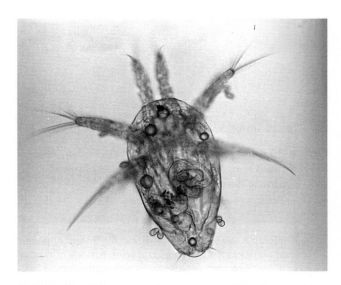

Figure 30.3   Living nauplius of *Cyclops vernalis* with a juvenile of *Dracunculus medinensis* in its hemocoel.
Courtesy of Ralph Muller.

• **Epidemiology.**   To become infected, a person must swallow a copepod that was exposed to juveniles previously released from the skin of a definitive host. Thus, three conditions must be met before the parasite's life cycle can be completed: (1) The skin of an infected individual must come in contact with water, (2) the water must contain the appropriate species of microcrustacean, and (3) the water must be used for drinking. There is circumstantial evidence that human infection can be acquired by eating a fish paratenic host.[10]

It is ironic that a parasite life cycle that is so dependent on water is most successfully completed under conditions

Figure 30.4 **Pond in the Mabauu area of Sudan, in the Sahel savanna zone. Conditions such as these favor the transmission of the guinea worm.**

From R. Muller, in B. Dawes, editor, *Advances in parasitology,* vol. 9, © 1971. London: Academic Press, Inc., Ltd. Photo by J. Bloss.

Figure 30.5 **Step well at Kantarvos, near Kherwara, India, infected with *Dracunculus medinensis.***

From R. Muller, in B. Dawes, editor, *Advances in parasitology,* vol. 9, © 1971. London: Academic Press, Inc., Ltd. Photo by A. Banks.

of drought. In some areas of Africa, for instance, people depend on rivers for their water. During periods of normal river flow, few or no new cases of dracunculiasis occur. During the dry season, however, rivers are reduced to mere trickles with occasional deep pools, which are sometimes enlarged and deepened by those who depend on them as a water source (Fig. 30.4). Planktonic organisms flourish in this warm, semistagnant water, and a cyclopean population explosion occurs. At the same time any bathing, washing, and water drawing bring infected persons in contact with water, into which juveniles are shed. When such water is drunk, many infected copepods may be downed at a quaff.

In areas of India step wells (Fig. 30.5) are a time-honored method of exposing groundwater. These wells, often centuries old, have steps the water bearers descend to enter the wells to fill their jars and, incidentally, release juveniles into the water at the same time.

In many desert areas the populace may depend on deep wells, which are crustacean free, during the dry season. Most villages also have one or more ponds that fill during

the rainy season and become sources of infection with *D. medinensis.* Most villagers prefer pond water because they have to pay for well water, and, moreover, well water is usually saline.

Given these examples, it is no wonder that a parasite with an aquatic life cycle should thrive in a desert environment because all animals, humans and beasts alike, depend on isolated waterholes for their existence. So does *D. medinensis.* On the other hand, the parasite's dependence on isolated waterholes exposes a weakness in its life cycle. The guinea worm is the only helminthic parasite of humans transmitted solely through drinking water.[17]

- **Pathogenesis.** Dracunculiasis may result in three major disease conditions: emergent adult worms, secondary bacterial infection, and nonemergent worms. There is no apparent pathogenesis in response to the worms normal development to adults within the host, prior to migration of females to the skin.

  At onset of migration to the skin, female worms release as yet undefined chemical toxins that may produce a rash, nausea, diarrhea, dizziness, and localized edema in the host. Worms remain just under the skin for a few days before a reddish papule develops. This rapidly becomes a blister. Local itching accompanies blister formation, often with an intense burning pain. On rupture of the blister, host reactions usually subside. However, allergic sensitivity can develop to worm secretions and rupture of worms can cause serious systemic allergic responses. The site of the blister becomes abscessed, but this lesion heals rapidly if serious secondary bacterial infections do not occur. A tiny hole remains, through which the worm protrudes. When the worm is removed or is expelled, healing is completed. An infected human may harbor many female worms, but typically only one emerges at a time. Infection, however, does not confer immunity, and a person may be reinfected many times. Secondary bacterial infections of ulcers are common and can become more serious if bacteria are drawn under the skin by retreating worms. In parts of Africa this is the third most common mode of entry of tetanus spores.[12] Some other complications are abscesses, synovitis, arthritis, and bubo. Bacterial infections cause death in about 1% of cases.[16]

Ulcers and abcesses are very painful and, depending on location and complications, can be disabling for 3 to 10 weeks. Pain may become so serious as to preclude sufferers from leaving their dwelling or doing any work. Application of antibacterial agents to ulcers is important in case management.

Worms that fail to reach the skin may cause complications, such as arthritis when they calcify in or alongside a joint. More serious symptoms, such as paraplegia, result from a worm in the central nervous system. Worms that rupture before emerging can cause an acute inflammatory response, leading to large aseptic abscesses and cystic swellings.

- **Diagnosis and Treatment.** Currently there is no dependable method for diagnosis during the long asymptomatic, prepatent period. Serologic tests lack sensitivity or specificity, or are simply not widely available.[19]

The appearance of an itchy, red papule that rapidly transforms into a blister is the first strong symptom of dracunculiasis. Patients who have endured previous infections become aware of the worm and its position days or even weeks before it emerges.[19] After the blister ruptures, juveniles can be obtained by placing cold water on the wound; when mounted on a slide, they can be seen actively moving about using a low-power microscope.

When a part of the worm emerges, diagnosis is fairly simple, although the drying, disintegrating worm does not show typical morphology of a nematode. An occasional sparganum or *Onchocerca volvulus* may be mistaken for dracunculiasis.

Pulling out guinea worms by winding them on a stick is a treatment used successfully since antiquity (Fig. 30.6). The biblical excerpt at the beginning of this chapter is a pretty fair account of dracunculiasis and its treatment. The burning pain of the blister could well be interpreted as the bite of a fiery serpent, and the serpent on a pole could easily represent the worm on a stick. Moses and his people were, at the time, near the Gulf of Akaba, where *D. medinensis* was then endemic. Also, the Israelites had for some time been in a drought area, existing on water where they could find it. This is consistent with the epidemiology of dracunculiasis.

Figure 30.6   Ancient woodcut showing removal of guinea worm by winding it on a stick.

From G. H. Velschius, *Exercitatio de Vena Medinensi, ad mentem Ebnisinae sive de dracunculi veterum. Specimens exhibens novae versionis ex Arabico, cum commentarion uberiori, cui accedit altera de vermiculus capillaribus infantium, Theophili Goebelli, Augustae Vindelicorum,* 1674.

Figure 30.7   Seal of the American Medical Association and the double-serpent caduceus of the military medical profession.

Might the serpent on a staff originally have depicted the removal of a guinea worm?

Courtesy of the AMA.

The staff with serpent carried by Aesculapius, the Roman god of medicine, adopted today as the official symbol of medicine (and the double-serpent caduceus of the military), may also depict removal of *D. medinensis* (Fig. 30.7). This form of cure is still used (Fig. 30.8). If cold water is applied to a worm, she will expel enough

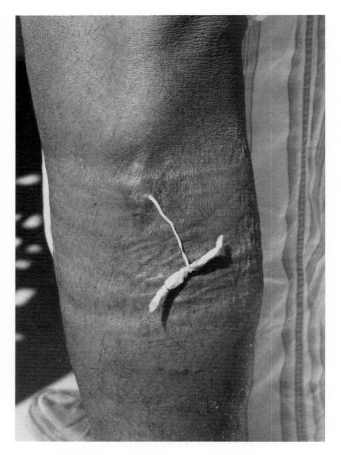

Figure 30.8   **Uncomplicated case of dracunculiasis.**
The worm is being pulled out through a small hole left after the ulcer is mostly healed.

From R. Muller, in B. Dawes, editor, *Advances in parasitology,* vol. 9, © 1971. London: Academic Press, Inc., Ltd.

juveniles to allow about 5 cm of her body to be pulled out. This procedure is repeated once a day, complete removal requiring about three weeks. In some areas of India worms are said to be sucked out by indigenous doctors using crude aspirators!

An alternate method is removal of a complete worm by surgery. This procedure is often successful when an entire worm is near the skin and also in the case of deep abscesses containing worms that have failed to reach the skin. However, if the worm is threaded through a tendon or deep fascia or is broken into several pieces, it may be impossible to remove completely. Worms can be removed much more easily and quickly if surgery is performed before ulcers develop and consequent bacterial contamination occurs.[19]

Several drugs have been used for the treatment of dracunculiasis, but evidence for their effectiveness is dubious.[14]

- **Eradication Efforts.** This is perhaps the only parasite covered in this book that is potentially eradicable at the present time, and the World Health Assembly declared in 1991 its goal of eradicating dracunculiasis by 1995; the goal date was subsequently changed to 2009, but this was also not met. Although eradication has not yet been

achieved, there has been a rapid and sustained reduction in all endemic countries, with the exception of a limited small outbreak of unknown origin in Chad,[9] where the program has been complicated by civil strife.[8] Civil unrest in South Sudan and Mali represent the most serious remaining challenges to eradication; they represent two of the three countries that still have indigenous dracunculiasis. India, Pakistan, and many countries in Africa have eradicated or are on the threshold of eradicating the disease.

The most important strategies for the campaign have been as follows:[8, 17, 20]

1. *Supply of safe drinking water.* Tube wells and hand pumps preclude infected persons from entering the water of ponds and stepwells.
2. *Health education.* This, combined with provisions of cloth filters, prevents consumption of unsafe water. More than a million cloth filters for household use have been distributed. Field workers persist in drinking from small ponds, but more than 7.8 million pipe filters (short lengths of plastic tubing with filter cloth closing one end, worn around the neck on a string, that workers can use as a drinking straw) have been distributed.[8]
3. *Early case containment.* This includes treatment and bandaging of lesions and preventing sufferers from contacting water sources.
4. *Vector control.* Temephos (Abate; American Cyanamid) is a chemical that has low toxicity to mammals and fish, and it kills copepods for four to five weeks at a concentration of one part per million. The manufacturer has donated an amount of Temephos estimated for total eradication requirements in Africa.

CAMALLANOMORPHA

Family Camallanidae

Camallanomorpha differ in important ways from other Spirurina. They have conspicuous phasmids with broad cavities and prominent pores. Esophageal glands are usually uninucleate, and intermediate hosts are copepods. Included in family Camallanidae are several structurally similar genera that inhabit intestines of fishes, amphibians, and reptiles. Their most conspicuous character is their head, in which the buccal capsule has been modified with a pair of large, bilateral sclerotized valves (Fig. 30.9). The complex ornamentation of these valves (Fig. 30.10) is a useful taxonomic character.

Genus *Camallanus* is common in freshwater fishes and turtles in the United States. *Camallanus oxycephalus* is often seen as a bright red worm extending from the anus of crappie (*Pomoxis* spp.) or other warm-water panfish. Life cycles of most species involve a cyclopoid copepod crustacean as intermediate host. Development proceeds to maturity in the intestine of the vertebrate host with no tissue migration. However, *C. cotti,* a parasite of various ornamental fishes from Asia, can dispense with its copepod intermediate host when it is not available;[12] this species also infects tropical fish kept in home aquaria. Monoxenous worms have significantly lower fitness compared with heteroxenous *C. cotti.*

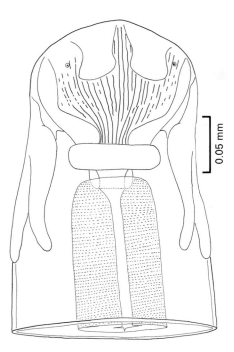

Figure 30.9   Head of *Camallanus marinus*, lateral view.
In this genus, the stoma is modified into large, sclerotized valves with various markings.

From G. D. Schmidt and R. E. Kuntz, "Nematode parasites of Oceanica. V. Four new species from fishes of Palawan, P. I., with a proposal for *Oceanicucullanus* gen. nov.," in *Parasitology* 59:389–396. Copyright © 1968 Cambridge University Press. Reprinted with permission of Cambridge University Press.

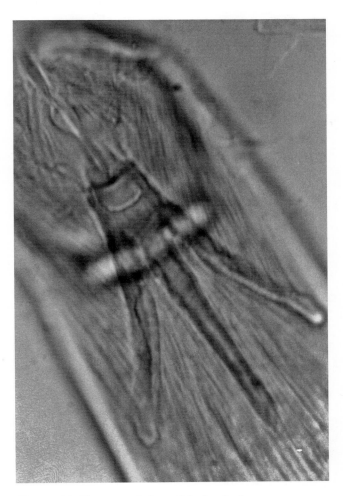

Figure 30.10   Dorsal view of the head of *Camallanus marinus*.
Note the large, sclerotized trident characteristic of this genus.

Photograph by Gerald D. Schmidt.

▍ Learning Outcomes

By the time a student has finished studying this chapter, he or she should be able to:

1. Some *Dracunculus* species appear to be generalists with respect to definitive host use, whereas others appear to use only one host species. Compare advantages and disadvantages of these two host usage patterns from the standpoint of the parasite species.

2. This chapter reveals that human knowledge of some parasites traces back to antiquity, long before development of the "germ theory of disease." Formulate explanations about how people in ancient cultures might have interpreted their infections with large, visible nematodes.

3. Scientists often debate the contributions of "nature versus nurture" in explaining differences in human behavior. Discuss how these two factors might influence efforts to control or eradicate a disease such as dracunculiasis.

4. Assessing parasites discussed in this chapter, and others you have studied, analyze the factors that make it more difficult to control or eradicate a parasitic disease.

5. Outbreaks of dracunculiasis have occurred within countries previously certified as free of this disease. List two different ways such human outbreaks could occur.

▍ References

References for superscripts in the text can be found at the following Internet site: www.mhhe.com/robertsjanovynadler9e

▍ Additional Readings

Muller, R. 1971. *Dracunculus* and dracunculiasis. In B. Dawes (Ed.), *Advances in parasitology* 9. New York: Academic Press, Inc., pp. 73–151.

Neafie, R. C., D. H. Connor, and W. M. Meyers. 1976. Dracunculiasis. In C. H. Binford and D. H. Connor (Eds.), *Pathology of tropical and extraordinary diseases,* vol. 2, sect. 9. Washington, DC: Armed Forces Institute of Pathology.

Chapter **31**

Phylum Nematomorpha, Hairworms

Who says that fictions only and false hair become a verse?

—George Herbert *(The Temple, Jordan)*

Nematomorpha are parasites of arthropods, primarily beetles and crickets. They are highly elongated, active worms that emerge, typically in dramatic fashion, usually from the rear ends of their hosts when the latter fall into water (Fig. 31.1). Hairworms are notorious for occurring in highly tangled masses, resembling the classical Gordian knot, which according to legend was tied by King Gordius of Phrygia and cut through by Alexander the Great. Thus, one of their common names is *Gordian worms.* Folklore has them originating from horse tail hairs that had fallen into watering troughs, explaining yet another of their common names: *horsehair worms.*[15] We now know that their occurrence in water is a natural part of the life cycle and that their tangled masses are actually mating events.

The phylum contains about 300 known species, but new ones are being described regularly from all continents (at least 50 since 1990).[11] Europe seems to have the most diverse fauna in terms of described species, but it is entirely possible that this distinction is based on the number and efforts of parasitologists. For a complete review of the phylum Nematomorpha, including all aspects of the biology of these enigmatic worms, see Hanelt et al.[11]

Traditionally, nematomorphs have been placed in two subordinate groups, treated as free-standing orders by some and referred to simply as "subordinate taxa" by others.[1, 3, 16] These groups are Nectonematida (= Nectonematoida) and Gordiida (= Gordioida). As is the case with some other invertebrate phyla, endings of taxonomic names higher than family level can vary with author; we adopt the spelling and nomenclature of Schmidt-Rhaesa.[24] Nectonematida contains a single genus, *Nectonema,* with five species all of which are marine and are found in crustaceans; gordiids, however, occur in freshwater, semiterrestrial, and terrestrial hosts such as crickets, cockroaches, mantids, millipedes, grasshoppers, and beetles.

Adults are not encountered very often in nature and are sometimes considered a curiosity, but ecological studies of larvae in addition to adults suggest that nematomorphs actually may be exceedingly common and widespread.[4, 10]

Juveniles occur as developing forms in a definitive host; neither their ecology nor the interactions between juveniles within a single host has been studied. There are a few strange cases of hairworms reported from humans, but these can probably be accounted for by the swallowing of infected arthropods or of the worms themselves in drinking water.[12, 24]

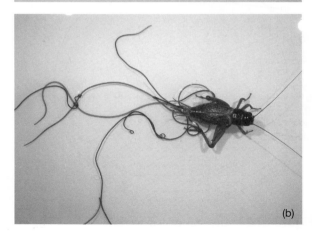

Figure 31.1   Gordian worms *(Paragordius varius)* emerging from two experimentally infected crickets.
(*a*) Emergence of a single worm within seconds. (*b*) Several worms emerging within a minute.

Photographs courtesy of Ben Hanelt.

465

FORM AND FUNCTION

Morphology

Hairworms are pseudocoelomates; as adults they are very long, cylindrical, and filamentous—ranging from a few centimeters to 3 meters in length (Fig. 31.2*a*).[24] Mature worms range from pure white to almost black, but many if not most species are brown. There is a lighter colored area **(calotte)** at the anterior end, and behind that, a pigmented ring (Fig. 31.2*b*). Females are generally larger than males. *Nectonema* species have dorsal and ventral rows of bristles that may function to aid in swimming (Fig. 31.2*c*).

The body wall consists of a thick cuticle with separate outer homogeneous and inner fibrous zones, the latter consisting of several layers (see Figs. 31.2*d*, and 31.3). Fibers are laid down in a crisscross pattern, and as a result, the body wall is very sturdy. The cuticle surface often is minutely ornate with surface features known as **areoles** (Fig. 31.4); depending on the species, areoles may include raised polygon-shaped areas, papillae, and bristles.[14] Beneath the cuticle lies a cellular epidermis that is thickened into a ventral cord (gordiids) or both dorsal and ventral cords (nectonematids). The nervous system consists of a cerebral ganglion and a midventral nerve cord either embedded in or loosely attached to the ventral epidermal cord. Both the calotte region and posterior end are extensively innervated, and histological evidence suggests the calotte may be photosensitive (Fig. 31.5).[14]

The digestive system is greatly reduced in adults; many species have no mouth opening. Histological similarities between nematomorph gut epithelium and that of insect Malpighian tubules have led to speculation that the gut serves as an excretory organ.[14] Posteriorly, the gut joins with reproductive ducts to form a cloaca lined with cuticle. Only longitudinal muscles are present, and locomotion consists mainly of coiling movements. Gordiids and nectonematids differ in muscle cell structure; myofibrils enclose the cell body in gordiids, whereas nectonematids' cells are coelomyarian as in certain nematodes (see Fig. 22.9). Individual worms differ in the extent to which their body cavities are filled with mesenchymal cells, gonads, or fluid, depending on their developmental stage.

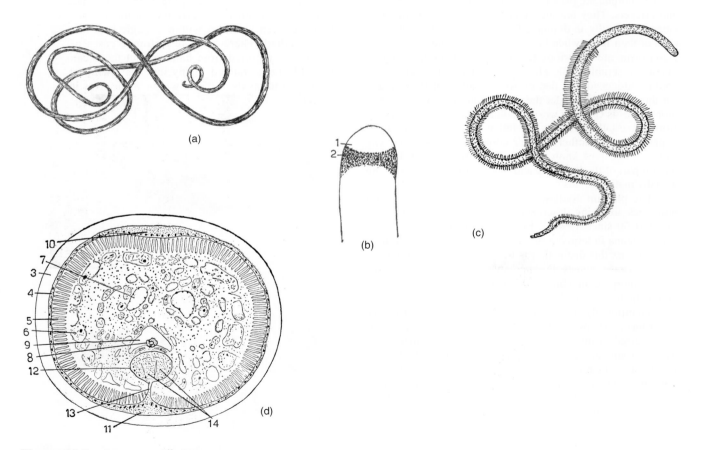

Figure 31.2   Anatomy of hairworms.

(*a*) Whole worm, illustrating almost filamentous condition. (*b*) Anterior end of gordiid, showing calotte (1) and pigmented band (2). (*c*) *Nectonema* sp., with dorsal and ventral rows of bristles. (*d*) Cross section of male gordiid, labeled as follows: 3, cuticle; 4, epidermis; 5, muscle layer; 6, mesenchyme; 7, testis; 8, intestine; 9, pseudocoel around intestine; 10, dorsal epidermal thickening (cord); 11, ventral cord; 12, nerve cord; 13, lamella connecting nerve cord and ventral cord; 14, tracts of the nerve cord separated by connective tissue (glial) partitions.

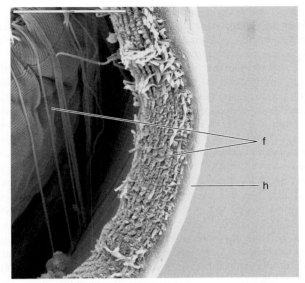

Figure 31.3 **Partial cross section (scanning electron micrograph) of nematomorph *(Gordius difficilis).***
Note homogeneous layer of cuticle (*h*) and fibrous layer (*f*). The fibrous layer is separated so that sheets made of crisscrossing parallel fibers are visible.
Courtesy of Matthew Bolek.

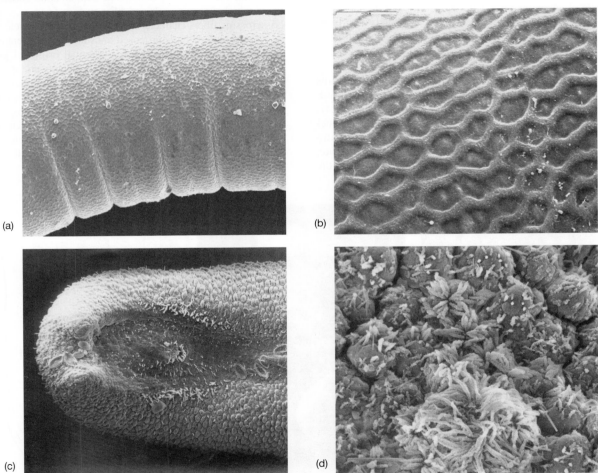

(a)   (b)

(c)   (d)

Figure 31.4 **Cuticular surface of nematomorphs, showing various surface features.**
(*a*) Low magnification SEM of female *Gordius difficilis,* showing a broad area of surface. (*b*) Higher magnification SEM, showing details of *G. difficilis* surface sculpturing. (*c*) Posterior end of male *Chordodes festar,* showing rough and textured surface (*d*) Higher magnification SEM of *C. festar,* showing areoles with minute blunt bristles.

(*a*), (*b*) From M. G. Bolek and J. R. Coggins. "Seasonal occurrence, morphology, and observations on the life history of *Gordius difficilis* (Nematomorpha: Gordioidea) from southeastern Wisconsin, United States," in *J. Parasitol.* 88:287–294. Copyright © 2002 *Journal of Parasitology.* Reprinted by permission. (*c*), (*d*) From C. de Villalobos and F. Zanca. "Scanning electron microscopy and intraspecific variation of *Chordodes festae* Camerano, 1897 and *C. peraccae* (Camerano, 1894) (Nematomorpha: Gordioidea)," in *Systematic Parasitol.* 50:117–125. Reprinted by permission.

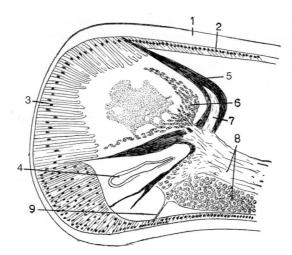

Figure 31.5 **Sagittal section of anterior end of**
Paragordius varius.

1, cuticle; *2*, epidermis; *3*, altered epidermis forming an "eye";
4, pharynx; *5*, capsule of "eye"; *6*, cells presumed to be light-
sensitive; *7*, dorsal nerve; *8*, cerebral ganglion; *9*, ventral nerve.

Sexes are separate, and most of the body cavity of adults
is filled with gonads. In gordiids the gonads are paired;
nectonematids possess a single, unpaired gonad. Each testis
joins the cloaca through a duct. As they mature, ovaries become highly lobed with many lateral diverticula that eventually fill up the pseudocoelom. The female cloaca is elongate,
with a glandular area at the anterior end where it joins the
oviduct and a seminal receptacle extending as a diverticulum
from the oviduct-cloaca juncture. Males and females differ
in the structure of their posterior ends (Fig. 31.6); the male's
lobes are used in clasping a female during copulation.

Anatomy of gordiid larvae has been well studied by both
light and electron microscopy.[7, 29, 30] Larvae are small, about
100 μm long, and quite different structurally from adults. The
larval body is divided into two parts, a preseptum and a postseptum (Fig. 31.7). Each division is superficially annulated
(ringed), similar to nematode cuticle. At the anterior end, the
preseptum contains an eversible proboscis bearing three rows
or crowns of hooks. This set of hooks is used to bore through
the gut of a paratenic host. The postseptum contains a large,
granular mass, the pseudointestine, which is thought to be used
in formation of the cyst wall.[5, 20] Nectonematid larvae have
been observed only rarely, but both their structure and behavior, including their eversible proboscis and hooks and probing
action, evidently are similar to those of gordiids.[13] Depending on development stage, juvenile *Nectonema munidae* may
have both a larval cuticle and, inside that, an adult cuticle. The
larval cuticle has no areoles or other surface features, whereas
crisscrossing fibers characterize developing adult cuticle.[22]

Some authors have proposed unusual development events
for nematomorphs, such as a hypothetical origin of gonads
from muscle. Ultrastructural studies of *Nectonema* sp., however, suggest that gonads originate from a "gono-parenchyme"
tissue that fills most of the pseudocoel and is distributed
among muscle cells toward the ends of female worms.[23] With
maturation, the body cavity becomes filled with parenchymal
cells and oocytes. In gordiids, female gonads develop from
well-formed cords of tissue that appear as rows of cell masses,
giving an impression of metamerism.[23]

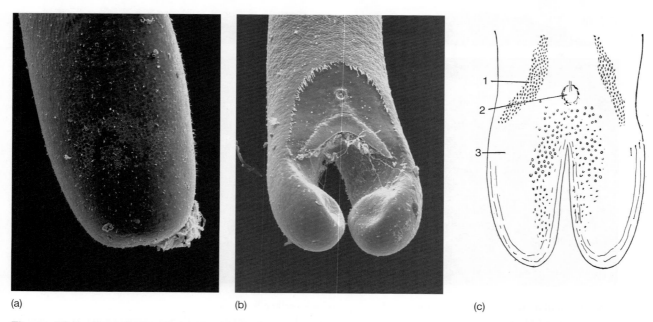

(a)   (b)   (c)

Figure 31.6 **Posterior ends of nematomorphs.**

Female (*a*) and male (*b*) *Gordius difficilis;* (*c*) male *Gordionus* sp. The male has a bilobed posterior end with a crescent-shaped cavity
and a curved row of spines, all presumably functioning to detect and clasp a female's posterior end during copulation. *1*, preanal tracts
of spines; *2*, anus; *3*, caudal lobe.

(*a*) Courtesy Matthew Bolek. (*b*) From M. G. Bolek and J. R. Coggins, "Seasonal occurrence, morphology, and observations on the life history of *Gordius difficilis* (Nematomorpha: Gordioidea) from southeastern Wisconsin, United States," in *J. Parasitol.* 88:287–294. Reprinted by permission.

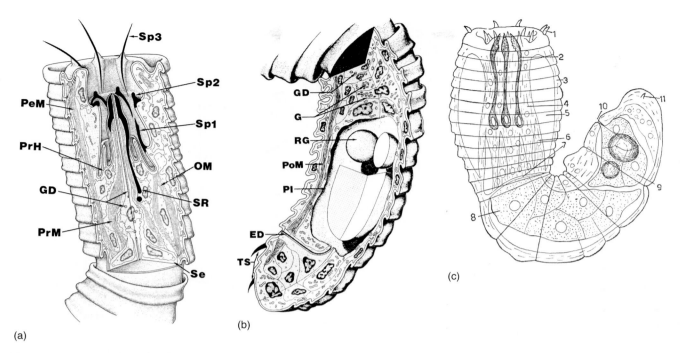

Figure 31.7   Structure of larval nematomorphs.

(*a*) and (*b*), Preseptum and postseptum, respectively, of *Paragordius varius.* (*c*) Larva of *Parachordodes* sp. *1,* spines; *2,* stylets; *3,* body wall; *4,* proboscis; *5,* pseudocoel; *6,* retractor muscles; *7,* septum; *8,* gland; *9,* intestine; *10;* anus; *11,* terminal spine. **ED,** exit duct; **G,** post-septal gland; **GD,** gland duct; **OM,** oblique muscles; **PI,** pseudointestine; **PeM,** preseptal parietal muscle; **Sp1–Sp3,** spines; **PoM,** postseptal muscle; **PrH,** proboscial hypodermis; **PrM,** proboscial muscle; **RG,** refringent granules; **SE,** septum; **SR,** support rod; **TS,** tail spine.

(*a*) and (*b*) from J. E. Zapotosky. 1974. "Fine structure of the larval stage of *Paragordius varius* (Leidy, 1851) (Gordioidea: Paragordidae). I. The preseptum." *Proc. Helm. Soc. Wash.* 41:209–221; and J. E. Zapotosky. 1974. "Fine structure of the larval stage of *Paragordius varius* (Leidy, 1851) (Gordioidea: Paragordidae. II. The postseptum." *Proc. Helm. Soc. Wash.* 42:103–111.

Physiology

Almost nothing is known about the physiology of gordian worms, a situation that may be explained in part by the inability of researchers to rear worms in the laboratory. This lack of knowledge may change in the future because of research on *Paragordius varius,* including laboratory maintenance of all developmental stages.[9] Adults probably do not feed but are able to survive and remain active for days, and sometimes weeks, in cool water, so whatever energy they require must be stored during development. Nonspecific esterases and alkaline phosphatases have been detected in both the intestinal epithelium and the body wall of juvenile *Nectonema* sp., and radioactive leucine is taken up across both surfaces.[21, 26] Thus, it is possible that nutrients are taken up across both surfaces, perhaps actively.

NATURAL HISTORY

Life Cycle

Nematomorphs occur as juveniles in arthropods but develop to near maturity before leaving their host and making the transition to adult very quickly upon encountering water (Fig. 31.8). There is some evidence from studies conducted near a French swimming pool that infected insects are more inclined to go swimming than are uninfected ones, provided they actually get close to the water.[28]

In freshwater forms, mating begins almost immediately upon emergence, with males actively entangling females, sometimes in large numbers, to produce the writhing Gordian knot or living ball of worms. Females produce astounding numbers of eggs, estimated to be in the millions, which they release as whitish strings sometimes fixed to vegetation or other substrates. Species differ in form of their egg strings and manner in which these are applied to substrates (Fig. 31.9). Eggs undergo holoblastic cleavage and then develop into active larvae within 14 to 30 days, depending on the species.

These larvae burrow through the gut wall and encyst when consumed by any of a variety of aquatic invertebrates, including small crustaceans, aquatic insects, and snails. It has also been suggested that encystment may occur on aquatic vegetation, explaining parasitism of some herbivorous hosts.[25] Encystment takes place within five to seven days, and cysts can occur in almost any part of a host's body.[6, 8] The thick cyst wall is presumably of parasite origin. Upon consumption by a second invertebrate, typically a cricket, grasshopper, or beetle, larvae excyst and burrow through the gut and into the hemocoel. Development into an adult requires one to three months, depending on species. The life cycle is completed when a parasitized terrestrial arthropod enters water and worms emerge, usually in rapid and fairly dramatic fashion, to begin mating.

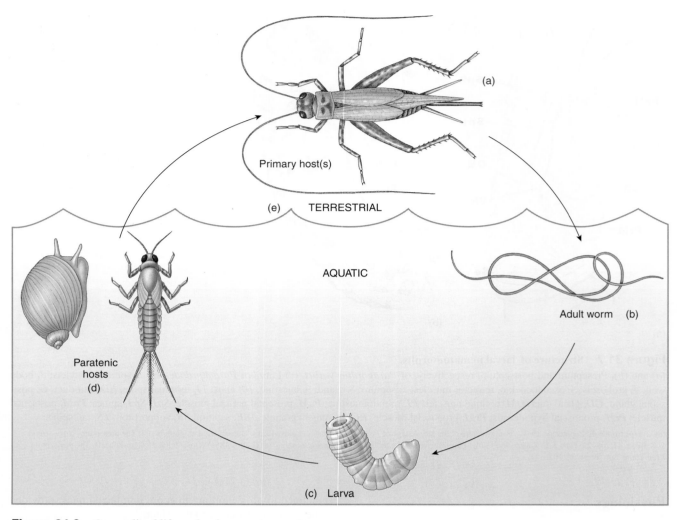

Figure 31.8 **Generalized life cycle of a nematomorph.**
(*a*) Infected cricket containing an adult worm. (*b*) Adult worm after emerging from cricket in water. (*c*) Larva developing from eggs. (*d*) Aquatic invertebrates infected by eating hairworm larvae. (*e*) Example of the movement of encysted larvae from the aquatic to the terrestrial environment upon being eaten by a cricket.

Drawing by Bill Ober based on figure from B. Hanelt and J. Janovy Jr. 2004. "Untying a Gordian knot: The domestication and laboratory maintenance of a Gordian worm, *Paragordius varius* (Nematomorpha: Gordiida)," in *J. Nat. History* 38:939–950.

Aquatic invertebrates in which nematomorph larvae encyst are considered transport or paratenic hosts because, other than acquisition of a cyst wall, there is no obvious structural development of larvae beyond their free-living stage.

Very little is known about the life history of marine nematomorphs. Their eggs are papillated and develop spines upon exposure to seawater.[13] Presumably, when a life cycle becomes known, we will find it similar in its general features to that of gordioids because marine crabs are often opportunistic scavengers, not unlike crickets. However, no nectonematid cysts ever have been reported.

Ecology

Only in the past decade have we begun to understand the ecology of hairworms, and as a result, many mysteries surrounding these parasites' distribution and movement through the environment have been explained.[8] Adults are observed sporadically, often by people who have thrown a cricket into the toilet or after a heavy summer rain when pools of water remain on the driveway and worms emerge there from crickets or grasshoppers. In general, therefore, adults are considered to develop from late-stage juveniles that emerge from terrestrial hosts, even though eggs are shed into water and larvae must develop in water.

In nature, individual hosts rarely have more than two or three worms, although the number can be much higher in laboratory infections. Temporal changes in sex ratios have been observed, with populations becoming male or female-biased at different seasons, but the ecological significance of this observation is not known.[2] Some species attach their egg strings to substrates or bury them in the sand or gravel of a stream, obviously influencing their access to certain kinds of paratenic hosts. Snails and oligochaetes are relatively long-lived and may serve as reservoirs for parasites in terrestrial

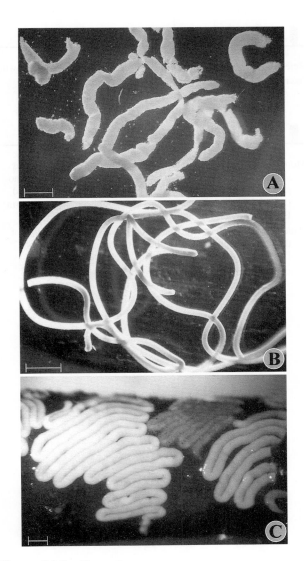

Figure 31.9   Egg strings.

(*a*) *Gordius robustus.* (*b*) *Paragordius varius.* (*c*) *Chordodes morgani.* (Scale bar = 1 mm.) Eggs of *C. morgani* are attached to a stick.

From B. Hanelt and J. Janovy Jr. "Morphometric analysis of nonadult characters of common species of American gordiids (Nematomorpha: Gordioidea), in *J. Parasitol.* 88:557–562. Copyright © 2002 *Journal of Parasitology.* Reprinted by permission.

insects. Dietary studies of carabid beetles show that they are quite omnivorous, with much of their diet consisting of invertebrates demonstrated to sustain encysted nematomorph larvae for long periods.[2]

We now know that nematomorph larvae can infect a number of transport or paratenic hosts, including aquatic oligochaetes, crustaceans, snails, immature stages of aquatic insects such as mayflies and midges, and even small fish.[2, 8] Encysted hairworm larvae can survive metamorphosis of their aquatic insect hosts and thus be carried to distant localities by terrestrial adults of such groups as mayflies and midges. Such transport helps explain their movement through ecosystems and their regular occurrence in seemingly odd places, such as the communal washhouse showers in one biological field station. Pulmonate snails are particularly good paratenic hosts because encysted hairworm larvae

can survive in them for months, even overwintering in snail tissues.[10] Terrestrial insects that feed on recently dead or dying snails, small freshwater crustaceans, and mayfly nymphs trapped at the edges of drying ponds or streams can become infected easily.

PHYLOGENY AND CLASSIFICATION

Nematomorphs have been found emerging from a cockroach trapped in amber somewhere between 15 and 45 million years old but obviously originated long before the Cenozoic.[18] They are usually considered the sister group to nematodes and are currently placed within superphylum Ecdysozoa, although similarity of larval structures, especially the introvert, suggests some possible and intriguing relationships to Kinorhyncha and Loricifera.[1, 24, 31] The strong superficial resemblance between nematomorphs and mermithid nematodes (see p. 362), both morphologically and in terms of life cycle, is considered convergence.[19, 24]

Members of phylum Nematomorpha lack cephalic papillae of any kind, lateral epidermal cords, secretory-excretory systems, amphids, and copulatory spicules (see chapter 22), all structures possessed by nematodes. Female genital openings are at their posterior end instead of near the middle of the body as in many nematodes. Hairworms have a greatly reduced digestive system, whereas nematodes have a well-developed and functional gut. No nematode has the kind of fibrous body wall characteristic of hairworms. Nematomorphs also have a "true" larval stage that undergoes considerable morphological change (metamorphosis) during development, in contrast to the juvenile developmental stages of nematodes.

Nematomorph phylogeny, taxonomy and nomenclature is far from settled. Some molecular phylogenetic trees based on nuclear genes depict a sister-group relationship between the phyla Loricifera and Nematomorpha,[17, 27] but statistical support for this relationship is weak. More broadly, molecular trees indicate other candidates as potential close relatives of nematomorphs, including the phyla Nematoda, Tardigrada, Onychophora, and Loricifera.[27] Hyman[14] treated hairworms as a class in now defunct phylum Aschelminthes. Nematomorpha is considered monophyletic; we treat the two subordinate taxa as orders, and recent molecular work shows they are not only monophyletic but also sister groups.[1] Linnaean classification schemes require names for all levels in the hierarchy, so we adopt a class name of Nematomorphidea.

Class Nematomorphidea

With the characters of the phylum.

Order Nectonematida

All marine; unpaired gonads; both dorsal and ventral thickened epidermal cords and coelomyarian muscle cells. Family Nectonematidae (single genus *Nectonema*).

Order Gordiida

All freshwater or semiterrestrial; paired gonads; only a ventral thickened epidermal cord; contractile fibers surround the muscle cell body. Families: Gordiidae (important genera: *Gordius, Paragordius*) and Chordodidae (important genus *Chordodes*).

Learning Outcomes

By the time a student has finished studying this chapter, he or she should be able to:

1. Describe the general body plan of a typical nematomorph worm.

2. Compare the internal anatomy of nematomorphs with nematodes and cestodes.

3. List the sequential steps in the life cycle of a nematomorph worm.

4. Explain how you might use information about the anatomy of nematomorph worms, both as adults and larvae, to develop hypotheses of phylogenetic relationships among all ecdysozoan animals.

5. A challenge to understanding the distribution and species diversity of nematomorphs is the rarity of adults. Describe how molecular tools can be used to increase understanding of their biodiversity, host range, and geographic range without acquiring adults.

References

References for superscripts in the text can be found at the following Internet site: www.mhhe.com/robertsjanovynadler9e

Additional Readings

Montgomery, T. H. 1903. The adult organization of *Paragordius varius* (Leidy). *Zool. Anz.* 97:377–470, plus plates. This monograph is a remarkable example of the detailed anatomical and histological work found in many older publications. Hairworm Biodiversity Survey. http://www.nematomorpha.net/.

Hanelt, B., F. Thomas, and A. Schmidt-Rhaesa. 2005. Biology of the phylum Nematomorpha. In: J. R. Baker, R. Muller, and D. Rollinson (Eds.) *Advances in Parasitology* 59:243–305.

Hairworm Biodiversity Survey http://www.nematomorpha.net/

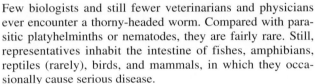

32

Phylum Acanthocephala: Thorny-Headed Worms

Every noble crown is, and on earth will forever be, a crown of thorns.

—Thomas Carlyle (1843, *Past and Present*)

Few biologists and still fewer veterinarians and physicians ever encounter a thorny-headed worm. Compared with parasitic platyhelminths or nematodes, they are fairly rare. Still, representatives inhabit the intestine of fishes, amphibians, reptiles (rarely), birds, and mammals, in which they occasionally cause serious disease.

The first recognizable description of an acanthocephalan in the literature is that of Redi, who, in 1684, reported white worms with hooked, retractable proboscides in eel intestines. From the time of Linnaeus to the end of the 19th century, all acanthocephalan species were placed in the collective genus *Echinorhynchus* Zoega in Mueller, 1776, although Koelreuther is credited with naming the genus *Acanthocephalus* in 1771. Hamann[23] divided *Echinorhynchus*, which by then had become large and unwieldy, into *Gigantorhynchus, Neorhynchus,* and *Echinorhynchus,* thereby beginning modern classification of Acanthocephala.

Lankester[32] proposed elevating order Acanthocephala, proposed by Rudolphi in 1808, to the level of phylum. This suggestion was not widely accepted until Van Cleave[67] convincingly argued in its favor. Today Acanthocephala is universally accepted as a separate phylum, although molecular biologists have found some intriguing relationships with other phyla, including Rotifera (see Phylogenetic Relationships, p. 506).

FORM AND FUNCTION

Acanthocephalan morphology reflects extensive adaptation to their parasitic mode of life and enteric habitat. There is evidently an evolutionary reduction in muscular, nervous, circulatory, and excretory systems and a complete loss, or at least absence, of a digestive system. The remaining animal seems little more than a pseudocoelomate bag of reproductive organs with a spiny holdfast at one end. The worms range in size from the tiny *Octospiniferoides chandleri,* only 0.92 mm to 2.40 mm long, to *Oligacanthorhynchus longissimus,* exceeding a meter in length.

General Body Structure

Superficially, the acanthocephalan body consists of an anterior proboscis, a neck, and a trunk (Fig. 32.1). The proboscis varies in shape, from spherical to cylindrical, depending on species (Fig. 32.2). It is covered by a tegument and has a thin, muscular wall within which are embedded the roots of recurved, sclerotized hooks. Sizes, shapes, and numbers of these hooks are important taxonomic characters. The proboscis is hollow and fluid filled. Attached to its inner apex is a pair of muscles, called **proboscis retractor muscles,** which extend the length of the proboscis and neck and insert in the wall of a muscular sac called the **proboscis receptacle** (see Figs. 32.1 and 32.3).

Proboscis receptacle morphology varies somewhat depending on family, but in general the receptacle consists of one or two layers of muscle fibers attached to the inner wall of the proboscis. When proboscis retractor muscles contract, the proboscis invaginates into its receptacle, with hooks completely inside. When the proboscis receptacle contracts, it forces the proboscis to evaginate by hydraulic pressure.[24] A nerve ganglion called the **brain** or **cerebral ganglion** is located within the receptacle.

The proboscis and its receptacle are sometimes referred to as the **presoma.** The neck is a smooth, unspined zone between the most posterior proboscis hooks and an infolding of the body wall. In some species, **neck retractor muscles** attach this infolding of the body wall to the inner surface of the trunk. When proboscis retractor and neck retractor muscles contract, the entire anterior end is withdrawn into the trunk. Some species have a sensory pit on each side of the neck, and two similar pits are found on the proboscis tip of many species.

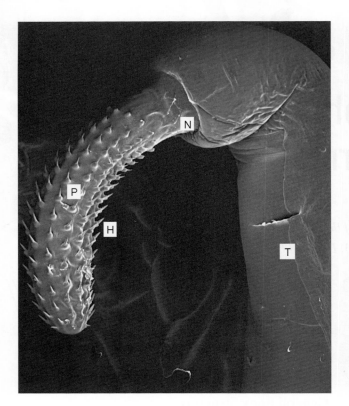
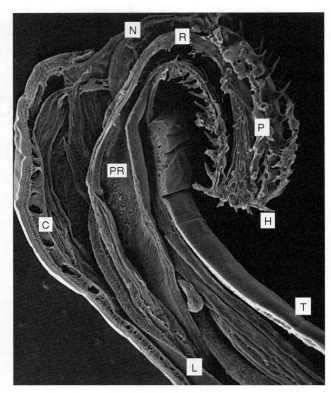

Figure 32.1 **Scanning electron micrographs of *Leptorhynchoides thecatus* from a sunfish.**

Note some of the major anatomical features of acanthocephalans. *P*, proboscis; *H*, hook; *N*, neck; *T*, trunk; *PR*, proboscis receptacle; *R*, proboscis retractor muscle; *L*, lemniscus; *C*, canals of the lacunar system.

Courtesy of Peter D. Olson.

The rest of the body, posterior to the neck, is called the **trunk,** or **metasoma.** Like proboscis and neck, it is covered by a tegument and has muscular internal layers. Many species have simple, sclerotized spines embedded in the trunk wall that maintain close contact with a host's intestinal mucosa. The trunk contains the reproductive system (see Fig. 32.3) and also functions in absorbing and distributing nutrients from the host's intestinal contents. In living worms the trunk is bilaterally flattened, usually with numerous transverse wrinkles, but when placed in a hypotonic solution, such as tap water, worms swell and become turgid. This procedure is important for identification because it places the internal organs in constant relationship with each other, and it usually forces the introverted proboscis to evaginate, allowing hooks to be counted and measured.

Body Wall

The body wall is a complex syncytium containing nuclear elements and a series of internal, interconnecting canals called a **lacunar system** (see Figs. 32.4, 32.5, and 32.6). In some species, nuclei are gigantic but few in number. In others, nuclei fragment during larval development and are widely distributed throughout the trunk wall. When entire nuclei are present, their number is constant for each species, demonstrating the principle of **eutely,** or nuclear constancy (see p. 368). Development of the wall was described by Butterworth.[7]

Tegument

The tegument has several regions differing in their construction. These are, beginning with the outermost, the (1) surface coat, (2) striped zone, (3) felt zone, (4) radial fiber zone, and (5) basement membrane (Fig. 32.4). Inside the tegument is a layer of irregular connective tissue, followed by circular and longitudinal muscle layers. Tegument is syncytial, but, unlike that of trematodes and cestodes, nuclei are in its basal region, not in cytons separated from distal cytoplasm.

The **surface coat,** or **glycocalyx,** a filamentous material, was formerly known as an *epicuticle.* It is about 0.5 μm thick, such as on *Moniliformis moniliformis,* an acanthocephalan of rats and the species most commonly investigated in the laboratory. The surface coat is a glycocalyx composed of acid mucopolysaccharides and neutral polysaccharides and/or glycoproteins.[47, 73] The stabilized system of filaments in the surface coat constitutes an extensive surface for molecular interactions, including those involved in transport functions and enzyme-substrate interactions.

Immediately beneath the surface coat and limited by a trilaminar outer membrane is the **striped zone.** This zone is 4 μm to 6 μm thick and is punctuated by a large number of crypts about 2 μm to 4 μm deep that open to the surface by pores.[8] These crypts give this zone a striped appearance *(Streifenzone)* under the light microscope. Crypts increase the worm's surface area by 44 times the area of a smooth surface. A filamentous molecular sieve is seen in the necks of these crypts, but particles of less than about 8.5 nm can gain

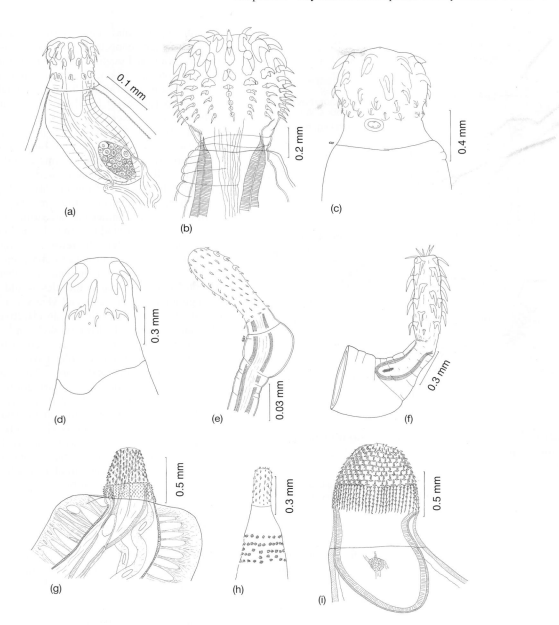

Figure 32.2   Examples of different types of acanthocephalan proboscides.

(*a*) *Octospiniferoides australis;* (*b*) *Sphaerechinorhynchus serpenticola;* (*c*) *Oncicola spirula;* (*d*) *Acanthosentis acanthuri;* (*e*) *Pomphorhynchus yamagutii;* (*f*) *Paracanthocephalus rauschi;* (*g*) *Mediorhynchus wardae;* (*h*) *Palliolisentis polyonca;* (*i*) *Owilfordia olseni.*

access to crypts and undergo pinocytosis by crypt membrane. The importance of pinocytosis in acquisition of nutrients is unknown. In deeper regions of the striped zone are found numerous lipid droplets, mitochondria, Golgi complexes, and lysosomes.

The striped zone grades into a region of numerous, closely packed, randomly arranged fibrils known as the **felt-fiber zone.** Mitochondria, numerous glycogen particles, vesicles, and occasionally lipid droplets and lysosomes also are found in the felt-fiber zone. The **radial fiber zone** is just within the felt-fiber zone and makes up about 80% of body wall thickness. It contains large bundles of filaments that course radially through the cytoplasm, large lipid droplets, and nuclei of the body wall. Here, too, are many glycogen

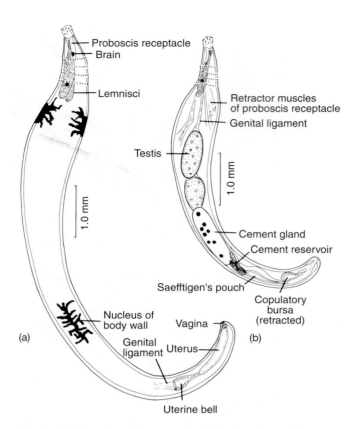

Figure 32.3 *Quadrigyrus nickoli,* **illustrating basic acanthocephalan morphology.**

(*a*) Female; (*b*) male.

From G. D. Schmidt and E. H. Hugghins, "Acanthocephala of South American fishes. Part I, Eoacanthocephala," in *J. Parasitol.* 59:829–835. Copyright © 1973 *Journal of Parasitology.* Reprinted by permission.

particles, mitochondria, Golgi complexes, and lysosomes. Rough endoplasmic reticulum is found in the perinuclear cytoplasm. The nuclei have numerous nucleoli. Lacunar canals course through the radial fiber zone.

Proboscis-wall structure is similar to that of the trunk, except it has fewer crypts and a thinner radial zone, and it lacks a felt zone.

Lacunar System and Muscles

A network of fluid-filled channels in the body wall called the *lacunar system* has been long known, but its function has remained enigmatic. A fascinating picture of the relationship of this curious system to functioning of body wall muscles emerged, largely as a result of the efforts of Miller and Dunagan and their coworkers.[36, 37, 38, 40]

The lacunar system is present in two parts apparently unconnected to each other: that in the proboscis and neck and that in the trunk. The presomal lacunar system has channels that run into two structures called **lemnisci,** extensions of the radial fiber zone, that grow from the base of the neck into the pseudocoelom. Each lemniscus has a central canal that is continuous with the presomal lacunar system. The function of the lemnisci is unknown, although they may contribute to hydraulics of the proboscis mechanism.

The metasomal lacunar system consists of a complicated network of interconnecting canals. In most species there are two main longitudinal canals, either dorsal and ventral or lateral. These are connected by numerous irregular or regular transverse canals. Location and arrangement of lacuni are used as taxonomic characters. In addition, at least in some species, there is a pair of medial longitudinal channels, each connected periodically by short radial canals to circular canals coursing between the dorsal and ventral longitudinal channels[36] (Fig. 32.5). The medial longitudinal channels lie on the pseudocoel side of the body-wall muscles, and radial canals pierce the muscle layers to intercept the ring canals.

Body-wall muscles are composed of a longitudinal layer surrounded by a circular muscle layer (Fig. 32.6). These muscles have a very curious structure. They are hollow, with tubelike cores and numerous, anastomosing interconnectives.[37] It has been found that muscle lumina are continuous with the lacunar system; therefore, circulation of lacunar fluid may well bring nutrients to and remove wastes from muscles. Although there is no heart or other circulatory organ, contraction of circular muscles would force fluid into the longitudinal components and vice versa. Thus, the lacunar system seems to function as an effective fluid transport system and possibly a hydrostatic skeleton.[40]

Acanthocephalan muscles are peculiar in other respects. They are electrically inexcitable, have low membrane potentials, and are slow conductors.[27] They are characterized by rhythmic, spontaneous depolarizations. Although the muscles appear to be stimulated by acetylcholine, nervous control of contraction is at present unclear. It is believed that nerves initiate contractions via a **rete system,** which is a highly branched, anastomosing network of thin-walled tubules lying on the medial surface of longitudinal muscles or between longitudinal and circular muscle layers. The rete system itself seems to be modified muscle cells.

Reproductive System

Acanthocephalans are dioecious and usually demonstrate some degree of sexual dimorphism in size, with females being larger (see Fig. 32.3). In both sexes one or two thin **ligament sacs** are attached to the posterior end of the proboscis receptacle and extend to near the distal genital pore. Within these sacs are gonads and some accessory organs of the reproductive systems. In some species ligament sacs are permanent; in others they break down as worms mature.

Male Reproductive System

Two testes normally occur in all species, and their location and size are somewhat constant for each species (see Fig. 32.3). Spermiogenesis has been described.[72] Each testis has a vas efferens through which mature spermatozoa, which appear as slender, headless threads, travel to a common vas deferens and/or to a small penis. Several accessory organs also are present, the most obvious of which are the **cement glands.** These syncytial organs, numbering from one to eight, contain one or more giant nuclei or several nuclear fragments and in many species are joined in places by slender bridges. They secrete a **copulatory cement** of tanned protein, which in some species is stored in a **cement reservoir** until copulation occurs. At that time the cement plugs the vagina after sperm transfer and rapidly hardens to form a **copulatory cap.** This cap remains attached to the female's

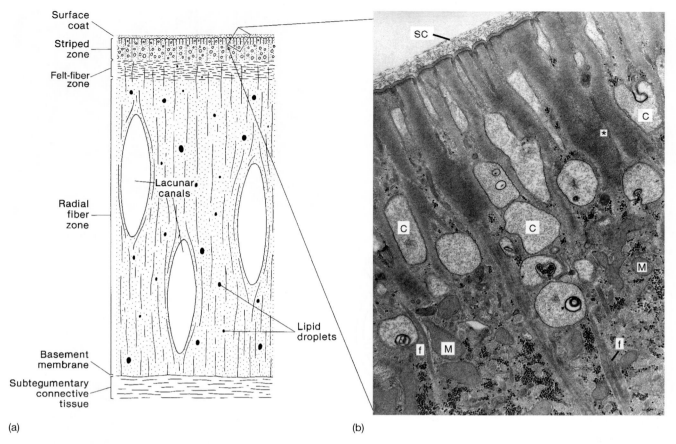

(a)   (b)

Figure 32.4   Tegument of *Moniliformis moniliformis.*

(*a*) Diagram of transverse section to show layers. The felt-fiber zone contains many vesicles and mitochondria with poorly developed cristae. Lacunar canals are in the radial fiber zone. (*b*) Electron micrograph showing the major features of the striped zone. The worm is coated with a finely filamentous surface coat (*SC*). Numerous surface crypts (*C*) appear as large scattered vesicular structures with elements occasionally appearing to course to the surface of the helminth. The crypts are separated by patches of moderately electronopaque material (***), giving the zone its striped appearance under the light microscope. Mitochondria (*M*), glycogen particles, microtubules, and other cytoplasmic details are evident in the inner portion of the striped zone. Bundles of fine cytoplasmic filaments (*f*) extend between this region and the deeper cytoplasm of the body wall. (×42,000)

(*a*) Drawing by William Ober. (*b*) From J. E. Byram and F. M. Fisher Jr., "The absorptive surface of *Moniliformis dubius* (Acanthocephala). I. Fine structure," in *Tissue and Cell* 5:553–579. Copyright © 1974.

posterior end during subsequent development of embryos within her body but eventually disintegrates.

Another male accessory sex organ is the **copulatory bursa** (Fig. 32.7), a bell-shaped specialization of the distal body wall that is invaginated into the posterior end of the body cavity except during copulation. A muscular sac, **Saefftigen's pouch,** is attached to the base of the bursa. When it contracts, fluid is forced into the lacunar system of the bursa, and it is everted by hydrostatic pressure. Many sensory papillae line the bursa; when it contacts the posterior end of a female, it clasps the female by muscular contraction, and sperm transfer is effected with a small penis.

Female Reproductive System

The ovary is peculiar in that it fragments into **ovarian balls** early in the life cycle, often while the worm is still a juvenile in an intermediate host. These balls of oogonia float freely within the ligament sac, increasing slightly in size before insemination occurs. The posterior end of the ligament sac

is attached to a muscular **uterine bell** (Fig. 32.8). This organ allows mature eggs to pass through into the uterus and vagina and out the genital pore, while returning immature eggs to the ligament sac.

After copulation, spermatozoa migrate from the vagina, through the uterus and uterine bell, and into the ligament sac. There they begin fertilizing oocytes of the ovarian balls. After the first few cleavages embryos detach from ovarian balls and float freely in pseudocoelomic fluid, exposing underlying oocytes for fertilization. Thus, several stages of early embryogenesis may be found in a single female. Eventually, from one copulation, many thousands or even millions of embryonated eggs are produced and released by each female and then pass from the host in its feces.

As shelled embryos are pushed into the uterine bell by peristaltic action, two possible routes are available. They may pass back into the pseudocoelom through slits in the bell or move on into the uterus. Fully developed embryos are slightly longer than immature ones and therefore

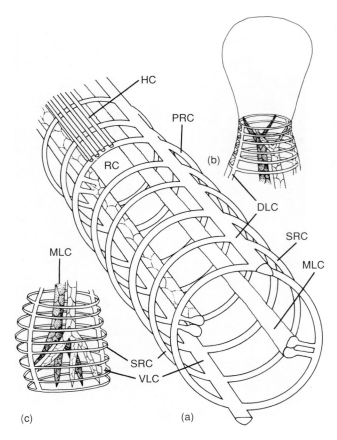

Figure 32.5 **Organization of lacunar system in** *Macracanthorhynchus hirudinaceus.*

(*a*) Midmetasomal region; (*b*) region near neck, with presomal lacunar system not indicated; (*c*) near posterior end of metasoma. *DLC,* dorsal longitudinal channel; *HC,* hypodermal canal (in radial fiber zone); *MLC,* medial longitudinal channel; *PRC,* primary ring canal; *RC,* radial canal; *SRC,* secondary ring canal; *VLC,* ventral longitudinal channel.

From D. M. Miller and T. T. Dunagan, "Body wall organization of the acanthocephalan, *Macracanthorhynchus hirudinaceus:* A reexamination of the lacunar system," in *Proc. Helm. Soc. Wash.,* 43:99–106. Copyright © 1976. Reprinted by permission.

cannot pass through the bell slits;[70] hence, they are passed on into the uterus. Immature eggs, however, are retained for further maturation. The efficiency of the sorting is quite high, and apparently no immature forms are passed into the uterus.

Excretory System

Excretion in most species appears to be effected by diffusion through the body wall. However, members of Oligacanthorhynchidae, one family in class Archiacanthocephala, are unique in possessing two **protonephridial excretory organs.** Each comprises many anucleate flame bulbs with tufts of flagella and may or may not be encapsulated, depending on species.[14] In males these organs are attached to the vas deferens and empty through it; in females they are attached to the uterine bell and empty into the uterus.

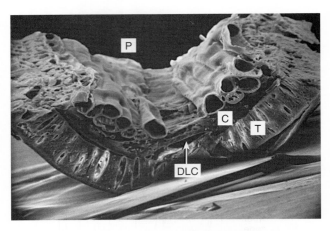

Figure 32.6 **Scanning electron micrograph of body wall** of *Oligacanthorhynchus tortuosa.*

C, circular muscle; *P,* pseudocoel; *T,* tegument, showing hypodermal lacunar canals; *DLC,* dorsal longitudinal channel.

Courtesy of D. M. Miller and T. T. Dunagan.

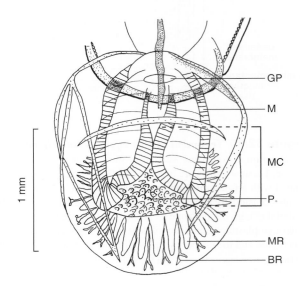

Figure 32.7 **Extended copulatory bursa of *Owilfordia olseni.*

Note the numerous sensory papillae (*P*) in a muscular cap (*MC*). The bursa is supported by major rays (*MR*) and smaller branched rays (*BR*). *GP,* genital pore; *M,* muscles.

From G. D. Schmidt and R. E. Kuntz, "Revision of the Porrorchinae (Acanthocephala: Plagiorhynchidae) with descriptions of two new genera and three new species," in *J. Parasitol.* 53:130–141. Copyright © 1967 *Journal of Parasitology.* Reprinted by permission.

Acanthocephalans show little ability to osmoregulate, swelling in hypotonic, balanced saline or sucrose solutions and becoming flaccid in hypertonic solutions. The osmotic pressure of their pseudocoelomic fluid is close to or somewhat above that of the intestinal contents. They take up sodium and potassium, swelling in hypertonic solutions of sodium chloride or potassium chloride at 37°C. In balanced saline they lose sodium and accumulate potassium against a concentration gradient. Their hexose transport mechanism is not sodium coupled.

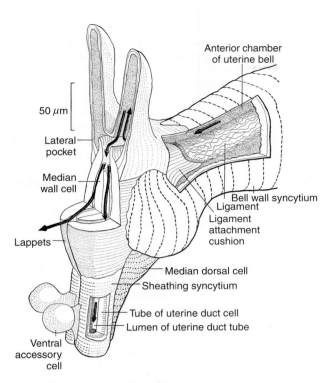

Figure 32.8   Stereogram of mature uterine bell, cut away to reveal complex internal luminal system.

Heavy arrows show possible routes for egg translocation.

From J. P. Whitfield, "A histological description of the uterine bell of *Polymorphus minutus* (Acanthocephala), in *Parasitology* 58:671–682. Copyright © 1968 Cambridge University Press. Reprinted with the permission of Cambridge University Press.

Nervous System

The nervous system of acanthocephalans is simple. The cerebral ganglion consists of only 54 to 88 cells in the species studied; it lies in the proboscis receptacle.[13, 41] Relatively few nerves issue from the ganglion, the largest of which are the anterior proboscis nerve and lateral posterior nerves.[13] Nerves supply two lateral sense organs and an apical sense organ, if present. A large multinucleate cell referred to as a **support cell** is located ventral and slightly anterior to the cerebral ganglion.[39] Processes from the support cell lead to lateral and apical sensory organs, but these processes are not nerves. Their function is unknown, but they may be secretory and help explain a host's inflammatory reaction to the worm's proboscis.

ACQUISITION AND USE OF NUTRIENTS

Because of the availability of a good laboratory subject (*Moniliformis moniliformis* in rats), investigators have been able to accumulate some knowledge of acanthocephalan metabolism. However, the problem of assessing the general applicability of observations on *M. moniliformis* and the few others reported is acute. (*M. dubius* is a junior synonym of *M. moniliformis,* and much literature accumulated on

physiology of the organism under that name.) Hatchery-reared rainbow trout are also good hosts for experimental work, and some physiological research has been done on *Neoechinorhynchus rutili* in that host. Extensive reviews of acanthocephalan physiology have been published.[10, 46]

Uptake

Because Cestoda and Acanthocephala are both groups that must obtain all nutrient molecules through their body surfaces, comparisons between the two are quite interesting, particularly in light of their structural differences.

Acanthocephalans can absorb at least some triglycerides, amino acids, nucleotides, and sugars. The presoma is the site of triglyceride uptake. In experiments using ^{3}H-labelled glyceroltrioleate and seven species of acanthocephalans, Taraschewski and Mackenstedt[63] showed that uptake started in the anterior half of the proboscis, but that the radioisotopes eventually accumulated in lemnisci. Amino acids are absorbed, at least partially, by stereospecific membrane transport systems in *M. moniliformis* and *Macracanthorhynchus hirudinaceus*.[66] The surface of *Moniliformis moniliformis* contains peptidases, which can cleave several dipeptides, and the amino acid products are then absorbed by the worm.[65] In several other species, lysine is absorbed across the metasomal tegument, especially the anterior portion, and accumulates in nuclei and the outer muscle belt.[62] Absorbed thymidine is incorporated into DNA in the perilacunar regions and into the nuclei of the ovarian balls and testes. Nuclei in the body wall are not labeled by radioactive thymidine; therefore, it is assumed that the DNA synthesized there is mitochondrial.

Like the tapeworm *Hymenolepis diminuta, M. moniliformis* has an absolute dependence on host dietary carbohydrate for growth and energy metabolism as an adult.[57, 58] The worm can absorb glucose, mannose, fructose, and galactose, as well as several glucose analogs. In contrast to *H. diminuta, M. moniliformis* can grow and mature in the host fed a diet containing fructose as the sole carbohydrate source.[11, 30] Absorption of glucose is through a single transport locus, whereas transport of mannose, fructose, and galactose is mediated both by the glucose locus and another site referred to by Starling and Fisher[58] as the **fructose site.** Maltose and glucose-6-phosphate (G6P) are absorbed also, but first they are hydrolyzed to glucose by enzymes in or on the tegumental surface.

In sharp contrast to tapeworm and other glucose transport systems, glucose absorption by *M. moniliformis* is not coupled to co-transport of sodium. In *M. moniliformis,* glucose is rapidly phosphorylated, removing free glucose from the vicinity of the tegumental transport loci, thus theoretically forming a metabolic sink for the flow of additional hexose down its concentration gradient. However, substantial amounts of free glucose are found in the body wall.[59] Evidence suggests that the free glucose pool is not derived directly from absorbed glucose but instead from the nonreducing disaccharide trehalose. If so, glucose may be deposited, perhaps by intervention of a membrane-bound trehalase, in an internal membranous compartment that by some means can resist efflux of glucose it contains. The scheme offers an interesting possible metabolic role for trehalose, perhaps similar to that in insects.

Acanthocephalans also accumulate a variety of nonorganic molecules, including heavy metals. In experimental studies, *Moniliformis moniliformis* took up more lead and cadmium than their rat hosts, concentrating both mainly in female worms, the lead especially in eggs.[52, 61] Acanthocephalan species parasitizing fish take up so much heavy metal—up to 200 times as much as their hosts—that the worms are potentially useful as bioindicators of pollution.[15] At least 16 different elements are taken up. In some cases worms compete with one another and with their fish hosts for these substances.[60]

Metabolism

As in the other helminth parasites, in acanthocephalans energy metabolism is adapted for facultative anaerobiosis. *Moniliformis moniliformis* can ferment hexoses it absorbs. The tricarboxylic acid cycle apparently does not operate in *M. moniliformis* or *Macracanthorhynchus hirudinaceus,* although there is evidence for it in *Echinorhynchus gadi,* a parasite of cod. *Moniliformis moniliformis* fixes carbon dioxide, and the principal enzyme of carbon dioxide fixation is phosphoenolpyruvate carboxykinase. Lactate and succinate are the main end products of glucose degradation in *Polymorphus minutus.* Interestingly, the main end products of glycolysis in *M. moniliformis* are ethanol and carbon dioxide with a small amount of lactate and only traces of succinate, acetate, and butyrate. Even though PEP carboxykinase activity in *M. moniliformis* is high,[31] it must be regulated in such a way that the major end products are ethanol and lactate, rather than succinate.

Lipids apparently are not used as energy sources. Körting and Fairbairn[31] found that endogenous lipids are not metabolized during in vitro incubation of *M. dubius (moniliformis).* Enzymes necessary for beta-oxidation of lipids are also low in activity, and one of them seems to be completely absent.

Electron transport in acanthocephalans has been studied very little. Oxidation of both succinate and NADH leads to reduction of cytochrome *b.*[6] Two pathways for reoxidation of this compound have been postulated, the major one independent of cytochrome *c* and cytochrome oxidase. This pathway is somewhat similar to the branched-chain electron transport postulated for the cestode *Moniezia expansa.*

DEVELOPMENT AND LIFE CYCLES

Each species of Acanthocephala uses at least two hosts in its life cycle (Fig. 32.9). The first must be an insect, crustacean, or myriapod, and the arthropod must eat an egg that was voided with feces of a definitive host. Development proceeds through a series of stages until the juvenile is infective to a definitive host. Many species, when eaten by a vertebrate that is an unsuitable definitive host, can penetrate the gut and encyst in some location where they survive without further development. This unsuitable vertebrate thus becomes a paratenic host, and, if it is eaten by the proper definitive host, the parasite excysts, attaches to the intestinal mucosa, and matures. Such adaptability has survival value. For example, ecological gaps exist in the food chain between

microcrustacea and large predaceous fishes. A paratenic host is one member of a food chain that bridges such a gap and incidentally ensures survival of a parasite; an example is the small fish in Figure 32.9.

The manner of early embryogenesis is an unusual characteristic of this group. Early cleavage is spiral, although this pattern is somewhat distorted by the spindle shape of the eggshell. At about the 4- to 34-cell stage, cell boundaries begin to disappear, and the entire organism becomes syncytial. Gastrulation occurs by migration of nuclei to the interior of the embryo.[54] Nuclei continue to divide but become smaller until they form a dense core of tiny nuclei, the **inner nuclear mass.** These nuclei give rise to all internal organ systems. In some species the uncondensed nuclei remaining in the peripheral area give rise to tegument; in some the tegument is derived from a nucleus that separates from the inner mass; in others there are contributions from both sources.

A fully embryonated larva that is infective to an arthropod intermediate host is called an **acanthor.** Acanthors are elongated organisms that are usually armed at their anterior end with six or eight bladelike hooks that aid in penetration of an intermediate host's gut (Fig. 32.10).[71] Hooks may be replaced by smaller spines in some species. Hooks or spines with their muscles are called an **aclid organ** or **rostellum.** The acanthor is a resting, resistant stage and will undergo no further development until it reaches an intermediate host. Under normal environmental conditions, acanthors may remain viable for months or longer. Acanthors of *Macracanthorhynchus hirudinaceus* can withstand subzero temperatures and desiccation and can remain viable for up to three and one half-years in soil.

Acanthors of some species completely penetrate the gut, coming to lie in a host's hemocoel, whereas others stop just under the serosa. In both cases the worm then becomes parasitic on the arthropod, absorbing nutrients and enlarging, thus initiating a developmental stage known as an **acanthella.** The end of the acanthor that bears the aclid organ apparently becomes the anterior end of the adult in some species, whereas others exhibit a curious 90-degree change in polarity, in which the anterior end of the adult develops from the side of the acanthor. During the acanthella stage, the organ systems develop from the central nuclear mass and hypodermal nuclei of the acanthor.

At termination of this development, the juvenile is an infective stage called a **cystacanth.** In most species the anterior and posterior ends invaginate, and the entire cystacanth becomes encased in a hyaline envelope. The parasite then must be eaten by a definitive host before it can fulfill its potential. Obviously, mortality is very high, because only a tiny fraction of the immense number of eggs produced may survive numerous hazards involved in completion of a life cycle.

Development of juveniles somewhat depends on their sex. In some species, female cystacanth size is correlated with intermediate host size, suggesting juvenile parasite growth is limited by resources. Female cystacanths also survive longer than males when both are incubated in salt solutions, indicating greater reserve energy storage in females than in males.[4]

Postcyclic transmission (adult worms in prey surviving and establishing in a predator) is known for some species, especially those found in fish. For example, *Acanthocephalus*

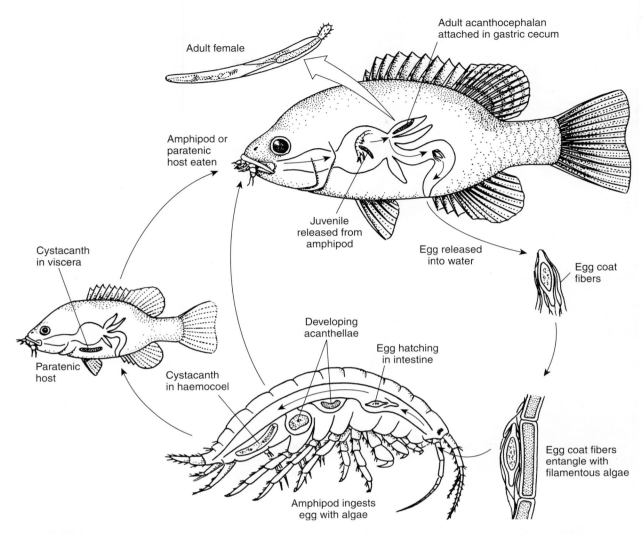

Figure 32.9   **Life cycle of a representative acanthocephalan, *Leptorhynchoides thecatus*, from the green sunfish, *Lepomis cyanellus.***

L. thecatus eggs have a fibrous coat that unravels and functions to entangle eggs in vegetation eaten by the amphipod intermediate host. Small fishes can serve as paratenic hosts in this system.

Drawing by William Ober and Claire Garrison.

tumescens in the gut of naturally infected inanga (a freshwater fish widely distributed in the Southern Hemisphere) survived for a month in cultured rainbow trout that ate inanga.[49] At least eight acanthocephalan species have been shown to be capable of postcyclic transmission, and such transmission may contribute to high acanthocephalan prevalence seen in predatory fish such as largemouth bass.[50]

Complete life cycles are known for only about 20 species in the phylum, although we have partial information on several more. The following examples illustrate the pattern followed in life histories of the three major groups.

Class Eoacanthocephala

Neoechinorhynchus saginatus is an eoacanthocephalan parasite of various species of suckers and of the creek chub, *Semotilus atromaculatus,* fish distributed in North America from Maine to Montana[64] (Fig. 32.11). When eggs are eaten

by a common ostracod crustacean *Cypridopsis vidua,* they hatch within an hour and begin penetrating the gut within 36 hours. After penetration, the unattached larva begins to enlarge and rearranges its nuclei, initiating the formation of internal organs. By 16 days after infection, the acanthella has developed into an infective cystacanth. Other eoacanthocephalan life cycles are similar, although paratenic hosts are known for several species, including *N. cylindratus* and *N. emydis.*[28, 68]

Class Palaeacanthocephala

Plagiorhynchus cylindraceus is a palaeacanthocephalan that is common in robins and other passerine birds in North America. Its life cycle and embryology were described by Schmidt and Olsen[55] (Fig. 32.12). When eggs are eaten by a terrestrial isopod crustacean *Armadillidium vulgare,* they hatch in the midgut within 15 minutes to two hours. Active

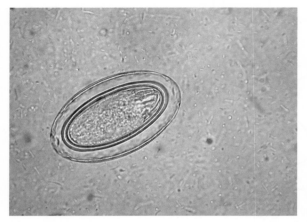

(a)

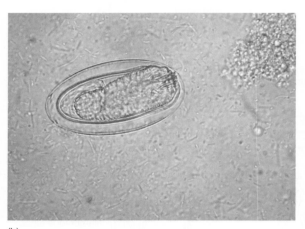

(b)

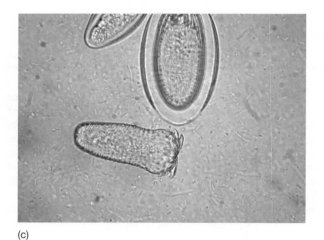

(c)

Figure 32.10 Hatching of a *Moniliformis moniliformis* egg.

(*a*) *In vitro* in a sodium bicarbonate-sodium chloride solution. (*b*) Acanthor cutting its way out of the egg. (*c*) Free acanthor. The time necessary for hatching under these circumstances is 10 to 30 minutes.

Photographs courtesy of Terry Miller.

entrance of an acanthor into the gut wall occurs within 1 to 12 hours, and the acanthor lies within tissues of the gut wall. After 15 to 25 days of apparent dormancy, it migrates to the outside of the gut, where it clings loosely to the serosa. Progressive changes follow in which overall size increases, and organs of the mature worm are delineated. Cystacanths (see Fig. 32.12) appear fully developed in 30 to 40 days but are not infective to the definitive host until 60 to 65 days. On ingestion of an infected isopod by a definitive host, the cystacanth proboscis evaginates, pierces the cyst, and attaches to the gut wall, where the worm develops to maturity.

Nickol and Oetinger[48] found encapsulated *P. cylindraceus* in the mesenteries of a shrew. This observation illustrates how interjection of a paratenic host into a life cycle may doom a parasite rather than serve it, because it is unlikely, although possible, that a robin would eat a shrew. (However, the parasite fauna of the American robin reveals the bird's diet to be more varied than most people realize.) But chances of a worm invading a hawk or owl are improved by such paratenesis.

Class Archiacanthocephala

Macracanthorhynchus hirudinaceus is a cosmopolitan parasite of pigs. Its life cycle has been known since 1868 but was described in detail by Kates.[29] When eggs are eaten by white grubs (larvae of beetle family Scarabaeidae), they hatch in the midgut within an hour and penetrate its lining. Within 5 to 20 days of infection, developing acanthellas are found free in the hemocoel or attached to the outer surface of the serosa. By 60 to 90 days after infection, cystacanths are infective to definitive hosts. Pigs are infected by eating grubs or adult beetles that have emerged from pupae with their parasites intact.

Most archiacanthocephalans are parasites of predaceous birds and mammals, so paratenic hosts often are involved in life cycles within this class.

EFFECTS OF ACANTHOCEPHALANS ON THEIR HOSTS

Behavior of an intermediate host is sometimes altered by infection, evidently increasing the probability of its being eaten by a definitive host.[43] Cockroaches (*Periplaneta americana* and *Blatella germanica*) infected with *Moniliformis moniliformis* move more slowly, travel shorter distances, and spend more time on horizontal surfaces than do uninfected controls.[21] However, a third cockroach species, *Supella longipalpa*, spends more time in the shade when infected with the same parasite.[45] Yet infection does not alter the behavior of a fourth cockroach species, *Diploptera punctata*.[1] If these behavioral changes do indeed increase the probability of transmission, then *M. moniliformis* may be one of the few species that has found a way to increase the utility of a cockroach.

Infection with *Polymorphus paradoxus,* a parasite of muskrats and surface-feeding ducks, changes phototaxis from negative to positive in its crustacean intermediate host, *Gammarus lacustris*.[35] Parasitized amphipods stay at the water surface instead of diving in response to disturbance and also

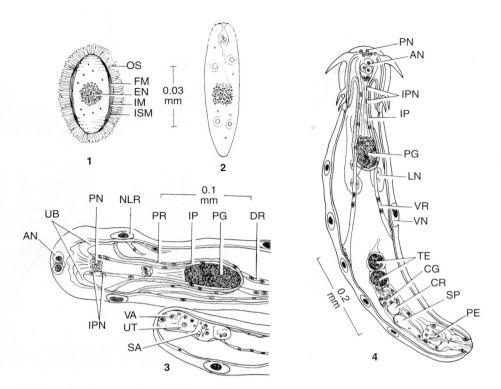

Figure 32.11   **Stages in development of *Neoechinorhynchus saginatus*.**

(*1*) Shelled acanthor from body cavity of adult female; (*2*) acanthor from gut of ostracod, one hour after feeding; (*3*) female acanthella, age 12 days; (*4*) late male acanthella, age 14 days (neck retractors omitted); *AN*, apical nuclei; *CG*, cement gland; *CR*, cement reservoir; *DR*, dorsal retractor of proboscis receptacle; *EN*, condensed nuclear mass; *FM*, fertilization membrane; *IM*, inner membrane; *IP*, proboscis inverter; *IPN*, proboscis inverter nuclei; *ISM*, inner shell membrane; *LN*, lemniscal nucleus; *NLR*, lemniscal ring nuclei; *OS*, outer shell; *PE*, penis; *PG*, brain anlage; *PN*, proboscis nuclear ring; *PR*, proboscis receptacle muscle sheath; *SA*, selector apparatus; *SP*, Saefftigen's pouch; *TE*, testes; *UB*, uncinogenous bands; *UT*, uterus; *VA*, vagina; *VN*, giant nucleus of ventral trunk wall; *VR*, ventral retractor of proboscis receptacle.

From G. L. Uglem and O. R. Larson, "The life history and larval development of *Neoechinorhynchus saginatus* Van Cleave and Bangham, 1949 (Acanthocephala: Neoechinorhynchidae)" in *J. Parasitol.* 55:1212–1217. Copyright ©1969 *Journal of Parasitology*. Reprinted by permission.

tend to cling to floating objects, thus becoming more vulnerable to predation. Injection of serotonin into unparasitized *G. lacustris* results in similar behavioral changes,[26] and nerve cords of parasitized crustacea contain more sites immunoreactive to antiserotonin antiserum, along nerve fibers, than do those unparasitized hosts, suggesting the fibers have increased numbers of neurotransmitter release points.[35] The parasite may thus be producing a serotonin-mediated change in intermediate host nervous system function. However, the congeneric *Polymorphus marilis,* a parasite of diving ducks, does not produce altered escape behavior in *G. lacustris.* A number of other acanthocephalan species produce altered behavior in crustaceans that seems to promote transmission to definitive hosts.[47]

Infection can have other effects on hosts that seem to have nothing to do with transmission. For example, male amphipods infected with *Pomphorhynchus laevis* are less successful at mating than are their uninfected competitors.[5] Studies also have shown that *Profilicollis antarcticus,* a parasite of gulls, increases the metabolic rate and activity of its crab host.[25] Although crabs infected with an unidentified *Profilicollis* species were less successful at mating, infection did not make the crabs more conspicuous, nor was their behavior different from uninfected crabs in the presence of a bird predator.[33]

A complete review of acanthocephalans', as well as other parasites', effects on host behavior can be found in the excellent book by Moore.[44]

In definitive hosts the nature of damage to intestinal mucosa is primarily traumatic, caused by penetration of the proboscis, and is compounded by the worms' tendency to release their hold occasionally and reattach at another place. Complete perforation of the gut sometimes occurs, and in mammals, at least, the results are often rapidly fatal (Fig. 32.13). Great pain accompanies this phase: Infected monkeys show evident distress, and Grassi and Calandruccio[22] recorded symptoms of pain and delirium experienced by Calandruccio after he voluntarily infected himself with cystacanths of *Moniliformis moniliformis.* It is suspected that secondary bacterial infection is responsible for localized and generalized peritonitis, hemorrhage, pericarditis, myocarditis, arteritis, cholangiolitis, and other complications.[3]

In view of the invasive nature of acanthocephalans, it is surprising they elicit so little inflammatory response in many cases. Host reaction seems mainly a result of the traumatic damage, with granulomatous infiltration and sometimes collagenous encapsulation around the proboscis (Fig. 32.14). Some species show evidence that antigens are released from the proboscis (as in *M. hirudinaceus*),

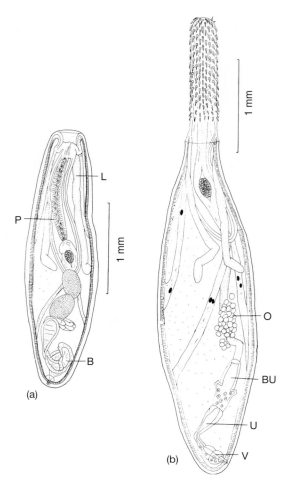

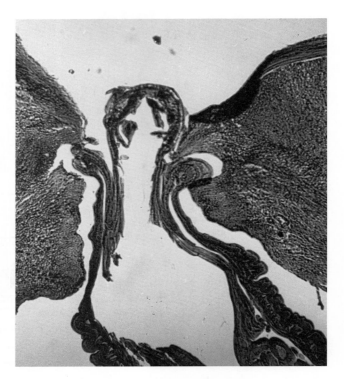

Figure 32.13 Complete perforation of the large intestine of a squirrel monkey by *Prosthenorchis elegans.*

From G. D. Schmidt, in R. N. T. W. Fiennes, editor, *Pathology of simian primates, vol. 2.* Basel: Karger AG, 1972.

Figure 32.12 Cystacanths of *Plagiorhynchus cylindraceus.*

(*a*) Thirty-seven days after infection of the pillbug intermediate hosts; (*b*) 60 days after infection; *B,* bursa; *L,* lemniscus; *P,* proboscis; *BU,* uterine bell; *O,* ovarian balls; *U,* uterus; *V,* vagina.

From G. D. Schmidt and O. W. Olsen, "Life cycle and development of *Prosthorynchus formosus* (Van Cleave 1918) Travassos, 1926, an acanthocephalan parasite of birds" in *J. Parasitol.* 50:721–730. Copyright ©1964 Journal of Parasitology. Reprinted by permission.

followed by an intense inflammatory response. It is clear that pathogenesis caused by acanthocephalans can be severe, but little consideration usually is given to the role this group of parasites could play as a factor controlling wildlife populations.

Little chemotherapy has been developed for acanthocephalans. Various authors have proposed chenopodium and castor oil, calomel and santonin, carbon tetrachloride, and tetrachloroethylene for primates and pigs, with varying results. Oleoresin of aspidium has been used successfully in human cases but is not recommended for children. Mebendazole was used successfully in a 12-month-old child, but in one study evidently produced hepatic dysfunction in rats and half the experimental group died.[20, 51] In the latter study, two doses of thiabendazole administered over a two-week period reduced worm burden almost 60%.[51]

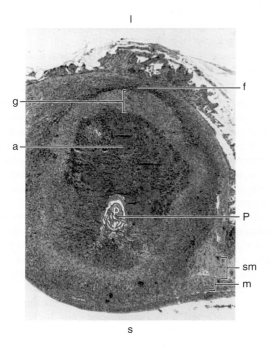

Figure 32.14 Cross section of a lesion produced by *Oligacanthorhynchus tortuosa* in an opossum.

P, Proboscis; *a,* necrotic abscess; *g,* granuloma; *f,* region of active fibrocyte proliferation; *m,* muscularis; *sm,* submucosa; *s,* serosal side of the intestinal wall; *l,* lumenal side.

Courtesy of Dennis Richardson.

ACANTHOCEPHALA IN HUMANS

Records of Acanthocephala in humans are few, no doubt because of the nature of intermediate and paratenic hosts involved in the parasites' life cycles. Few people eat such animals as insects, microcrustaceans, toads, or lizards, at least without cooking them first. However, human infections with seven different species have been reported.[53] *Macracanthorhynchus hirudinaceus* occasionally has been recognized as a parasite of humans from 1859 to the present. Nine *M. ingens* were recovered from a one-year-old child in Austin, Texas, in 1983.[12] *Moniliformis moniliformis* has been found repeatedly in people. *Bolbosoma* sp. has been reported twice from the body cavity of humans in Japan.[3] *Acanthocephalus rauschi* is known only from specimens taken from the peritoneum of an Alaskan Eskimo, an obvious case of accidental parasitism, since the proper host is undoubtedly a fish. Native Americans in the area often eat their fish raw, which contributes to such zoonotic infections. *Corynosoma strumosum*, a common seal parasite, also has been found in humans. More puzzling is a case of *Acanthocephalus bufonis*, a toad parasite, in an Indonesian. In this instance it is probable that the man ate a raw paratenic host.[53]

Thus, it seems that the Acanthocephala do not pose much of a threat to human health. They are much more important as parasites of wild and captive animals, in which sudden epizootics have been known to kill a great number of individuals in a short time.

PHYLOGENETIC RELATIONSHIPS

Origin of thorny-headed worms is one of the true parasitological mysteries,[9] and, although they have been included within phylum Aschelminthes, this phylum is considered neither monophyletic nor a valid taxon. Although a close association of Acanthocephala with Rotifera would seem unlikely, Lorenzen[34] provided morphological evidence for such a relationship based on cuticle structure and presence of lemnisci in certain rotifers. Molecular studies using 18S rDNA also concluded that Acanthocephala "share most recent common ancestry with rotifers of the class Bdelloidea."[19] Those same studies, as well as later research, strongly support the monophyly of Acanthocephala.[18]

More recently, Welch claimed on the basis of molecular evidence that acanthocephalans are a "highly derived class of Rotifera,"[69] and other authors have formally proposed inclusion of Acanthocephala within Rotifera.[56] This is one case in which molecular phylogenetics produced an increasingly accepted picture of relationships among seemingly quite dissimilar animals. We retain phylum status for acanthocephalans because we believe their suite of structural characters qualifies as a basic body plan.

Evolutionary relationships within Acanthocephala have been analyzed by Monks using morphological characters and Rotifera as an outgroup (Fig. 32.15).[42] Twenty-two species were chosen, representing all three classes. His results indicate that classes Palaeacanthocephala and Eoacanthocephala are monophyletic sister taxa but that Archiacanthocephala is not monophyletic. This last observation conflicts with results

of some molecular studies, but in both cases the small number of archiacanthocephalans included in the analysis may have contributed to the results.[17, 42] Molecular work has, however, supported the erection of class Polyacanthocephala by Amin in 1987 to include a single family (Polyacanthorhynchidae) with a single genus (*Polyacanthorhynchus*, not included in Monks' analysis).[2, 16]

CLASSIFICATION OF PHYLUM ACANTHOCEPHALA

Class Polyacanthocephala

Trunk spinose; tegumental nuclei many and small; main longitudinal lacunar canals dorsal and ventral; many hooks, in longitudinal rows; proboscis receptacle single walled; cement glands elongated, with giant nuclei; protonephridia absent; parasites of fishes and (?) crocodilians.

Order Polyacanthorhynchida

With characters of the class. Family Polyacanthorhynchidae.

Class Archiacanthocephala

Main longitudinal lacunar canals dorsal and ventral or just dorsal; tegumental nuclei few; giant nuclei present in lemnisci and cement glands; two ligament sacs persist in females; protonephridia present in one family; cement glands separate, pyriform; eggs oval, usually thick shelled; parasites of birds and mammals; intermediate hosts insects or myriapods.

Order Apororhynchida

Trunk short, conical; may be curved ventrally; proboscis large, globular, with tiny spinelike hooks (which may not pierce the surface of the proboscis) arranged in several spiral rows; proboscis not retractable; neck absent or reduced; protonephridial organs absent. Family Apororhynchidae.

Order Gigantorhynchida

Trunk occasionally pseudosegmented; proboscis a truncated cone, with approximately longitudinal rows of rooted hooks on the anterior portion and rootless spines on the basal portion; sensory pits present on apex of proboscis and each side of neck; proboscis receptacle single walled with numerous accessory muscles, complex, thickest dorsally; proboscis retractor muscles pierce ventral wall of receptacle; brain near ventral, middle surface of receptacle; protonephridial organs absent. Family Gigantorhynchidae.

Order Moniliformida

Trunk usually pseudosegmented; proboscis cylindrical, with long, approximately straight rows of hooks; sensory papillae present; proboscis receptacle double walled, outer wall with muscle fibers usually arranged spirally; proboscis retractor muscles pierce posterior end of receptacle or are somewhat ventral; brain near posterior end or near middle of receptacle; protonephridial organs absent. Family Moniliformidae.

Order Oligacanthorhynchida

Trunk may be wrinkled but not pseudosegmented; proboscis subspherical, with short, approximately longitudinal

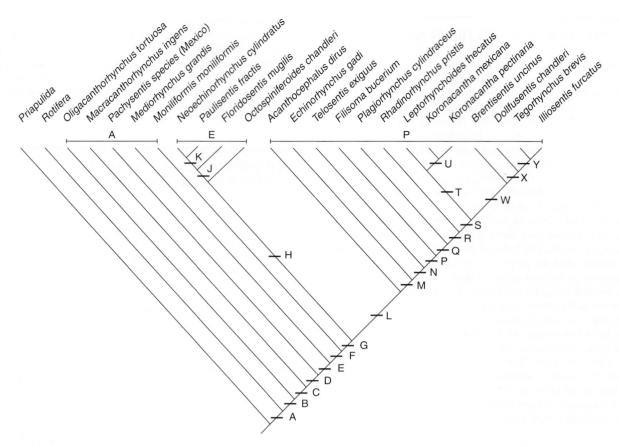

Figure 32.15 **Cladogram resulting from analysis of 22 acanthocephalan taxa using 138 morphological characters.**
A, E, and *P* on the horizontal bars above the clades represent Archiacanthocephala, Eoacanthocephala, and Palaeacanthocephala, respectively. Black crossbars with letters represent synapomorphies. *H* = five characters such as hook roots with anterior and posterior processes, lemnisci with one and two nuclei, respectively, syncytial cement glands, and cement reservoir present. *L* = 11 characters such as elongate proboscis, two muscle layers in receptacle wall, and lemnisci hanging free in pseudocoel.

From S. Monks, "Phylogeny of the Acanthocephala based on morphological characters," in *Syst. Parasitol.* 48:81–116. Reprinted by permission.

rows of few hooks each; sensory papillae present on apex of proboscis and each side of neck; proboscis receptacle single walled, complex, thickest dorsally; proboscis retractor muscle pierces dorsal wall of receptacle; brain near ventral, middle surface of receptacle; protonephridial organs present. Family Oligacanthorhynchidae.

Class Palaeacanthocephala

Main longitudinal lacunar canals lateral; tegumental nuclei fragmented, numerous, occasionally restricted to anterior half of trunk; nuclei of lemnisci and cement glands fragmented; spines present on trunk of some species; single ligament sac of female not persistent throughout life; protonephridia absent; cement glands separate, tubular to spheroid; eggs oval to elongated, sometimes with polar thickenings of second membrane; parasites of fishes, amphibians, reptiles, birds, and mammals.

Order Echinorhynchida

Trunk never pseudosegmented; proboscis cylindrical to spheroid, with longitudinal, regularly alternating rows of hooks; sensory papillae present or absent; proboscis receptacle double walled; proboscis retractor muscles pierce posterior end

of receptacle; brain near middle or posterior end of receptacle; parasites of fishes and amphibians. Families Diplosentidae, Echinorhynchidae, Fessisentidae, Heteracanthocephalidae, Heterosentidae, Hypoechinorhynchidae, Illiosentidae, Pomporhynchidae, Rhadinorhynchidae, Cavisomidae, Arythmacanthidae.

Order Polymorphida

Proboscis spheroid to cylindrical, armed with numerous hooks in alternating longitudinal rows; proboscis receptacle double walled, with brain near center; parasites of reptiles, birds, and mammals. Families Centrorhynchidae, Plagiorhynchidae, Polymorphidae.

Class Eoacanthocephala

Main longitudinal lacunar canals dorsal and ventral, often no larger in diameter than irregular transverse commissures; hypodermal nuclei few, giant, sometimes ameboid; proboscis receptacle single walled; proboscis retractor muscle pierces posterior end of receptacle; brain near anterior or middle of receptacle; nuclei of lemnisci few, giant; two persistent ligament sacs in female; protonephridia absent; cement gland single, syncytial, with several nuclei, with cement reservoir appended; eggs variously shaped; parasites of fish, amphibians, and reptiles.

Order Gyracanthocephalida

Trunk small or medium size, spined; proboscis small, spheroid, with a few spiral rows of hooks. Family Quadrigyridae.

Order Neoechinorhynchida

Trunk small to large, unarmed; proboscis spheroid to elongated, with hooks arranged variously. Families Neoechinorhynchidae, Tenuisentidae, Dendronucleatidae.

◼ Learning Outcomes

By the time a student has finished studying this chapter, he or she should be able to:

1. Draw a typical acanthocephalan male, and a female, and label the major structural features.

2. Explain why scientists believe that acanthocephalans and rotifers have a close phylogenetic relationship even though the two groups are structurally quite distinct.

3. Cite an example in which infection with an acanthocephalan species alters behavior of one species of an intermediate host, and speculate on why that same parasite species might not influence the behavior of another potential intermediate host species.

4. Sketch a life cycle that illustrates both paratenesis and postcyclic transmission and explain the possible results of these processes on acanthocephalan prevalence and infrapopulation in the definitive host.

◼ References

References for superscripts in the text can be found at the following Internet site: www.mhhe.com/robertsjanovynadler9e

◼ Additional Readings

Bullock, W. L. 1969. Morphological features as tools and as pitfalls in acanthocephalan systematics. In G. D. Schmidt (Ed.), *Problems in systematics of parasites.* Baltimore, MD: University Park Press, pp. 9–43. A useful, philosophical discussion of the subject, with recommended techniques for study.

Crompton, D. W. T. 1975. *Relationships between Acanthocephala and their hosts. Symposium of the Society for Experimental Biology, vol. 29. Symbiosis.* Cambridge: Cambridge University Press, pp. 467–504.

Crompton, D. W. T., and B. B. Nickol (Eds.). 1985. *Biology of the Acanthocephala.* Cambridge: Cambridge University Press.

Golvan, Y. J. 1969. Systématiques des Acanthocéphales (Acanthocéphala Rudolphi 1801). L'ordre des Palaeacanthocephala Meyer 1931. La super-famille des Echinorhynchoidea (Cobbold 1876) Golvan et Houin 1963. *Mem. Mus. Nat. Hist. Nat.* 47:1–373. An excellent account of this important superfamily. Besides descriptions of each species, it contains a key to genera and a host list.

Petrochenko, V. I. 1971. *Acanthocephala of domestic and wild animals, 1 and 2* (Israel Program for Scientific Translations, Trans.). Moscow: Akad. Nauk SSSR. (Original work published 1956, 1958.) An indispensable resource for students of the phylum. Descriptions are given for nearly every species known at the time of writing.

Schmidt, G. D. 1972. Revision of the class Archiacanthocephala Meyer, 1931 (Phylum Acanthocephala), with emphasis on Oligacanthorhynchidae Southwell and MacFie, 1925. *J. Parasitol.* 58:290–297.

Sures, B. 2001. The use of fish parasites as bioindicators of heavy metals in aquatic ecosystems: A review. *Aquatic Ecology* 353:245–255.

Yamaguti, S. 1963. *Systema helminthum, vol. 5, Acanthocephala.* New York: Interscience. In most regards a practical key to genera of Acanthocephala known to 1963. Lists of species and their hosts are included.

Chapter *33*

Phylum Arthropoda: Form, Function, and Classification

*Marvels indeed they are, and a feast to the eye and intellect of anyone interested
in polymorphisms, local races, rare aberrations, teratological specimens,
gynandromorphs, intersexes, and mosaics . . . and every aspect of Mendelian
genetics.*

—Cyril Clarke (commenting on Walter Rothschild's butterfly collection;
in M. Rothschild, *Dear Lord Rothschild*)

Phylum Arthropoda includes an enormous assemblage of both fossil and extant species that far outnumbers all other known animals put together. Nearly a million species of insects have been described, and more than a quarter of these are beetles. There are over 50,000 species of arachnids and another 30,000 of crustaceans. Yet analysis of Cambrian fossils reveals a number of arthropod body plans not represented among living groups, leading some authors to conclude that diversity of arthropods today may be lower than it was half a billion years ago.[17]

Arthropods are well represented in the geological record, revealing that all extant classes appeared during the Paleozoic. Chelicerate and crustacean fossils are present in Cambrian-age rocks. Most of the so-called "key innovations" leading to insect success, including internal fertilization, mandibles, and wings, all occurred before the end of the Devonian, and complete metamorphosis was present in early Carboniferous species.[22] Several modern orders, including Hymenoptera, Diptera, and Coleoptera, were present by the end of the Paleozoic, and roaches (order Dictyoptera) go back at least to the middle Carboniferous. Much current evolutionary research focuses on the origin of arthropod diversity and includes efforts to establish the evolutionary role of hormones, especially juvenile hormone (JH) because of its effect on postembryonic development. Similarly, the action of homeobox genes is now an active area of evolutionary research, especially because of the known influence these genes have on segmental development (see p. 506 for a more detailed discussion of arthropod phylogeny).[19, 22, 32]

Two structural features contribute significantly to arthropod success: relatively small size and a chitinous exoskeleton. Although some species such as lobsters and king crabs are quite large as adults, the vast majority of arthropods are less than 1 cm in length. The planet provides many places for small organisms to occupy: spaces between sand grains, for example, or cracks in tree bark and, of course, the bodies of other animals. As a general rule, complex environments support relatively diverse faunas and floras, and earth is an exceedingly complex environment. Small species, especially parasitic ones, therefore have a rich supply of potential ecological niches. In a number of previous chapters arthropods were discussed mainly in their roles as intermediate hosts and vectors; in the following chapters they will be discussed as parasites, with the understanding that the parasitic lifestyle is what makes some of them important transmitters of infectious disease.

Arthropods are involved in virtually every kind of parasitic relationship. They serve as both definitive and intermediate hosts for protozoans, flatworms, nematodes, and even other arthropods. They also function as vectors, transmitting infective stages of parasites to vertebrates, including humans and domestic animals. Many arthropods, such as fleas, ticks, and some crustaceans, are highly adapted parasites in their own right. Arthropod life cycles are as varied as their structures, and thus they present us with a seemingly never-ending series of challenges in our attempts to control parasitic infections. Large populations and rapid reproductive rate contribute to arthropods' ability to evolve genetic resistance to pesticides. And there is plenty of evidence that arthropod-borne diseases have been a deciding factor in numerous military operations, thus accomplishing defeat or ensuring victory for which human commanders subsequently received blame or claimed credit.[54]

This chapter is intended to be an easily accessible source of reference material on basic arthropod biology, a source you can quickly turn to if needed as you explore the following chapters on crustaceans, insects, ticks, and mites.

GENERAL FORM AND FUNCTION

Arthropod Metamerism

Segmentation, or metamerism, is the overriding structural feature of arthropods. In the ancestral condition each segment probably bore a pair of jointed appendages. This basic plan is manifested in many arthropod embryos, which pass through a stage consisting of a linear series of relatively similar tissue blocks, each with generalized and similar appendage buds (Fig. 33.1). Most living arthropods, however, have lost some of these appendages, particularly those on the abdomen, during their evolutionary history. In older literature, each segment (**metamere** or **somite**) of an arthropod's body was considered **homologous** to every other segment; that is, one segment could be thought of as a repeated expression of genes expressed during construction of other segments. We now know that arthropod metamerism is not so simple a phenomenon and that some segments may be secondarily derived early in development.[2] Molecular studies of development, however, indicate that, at least in insects and crustaceans, a set of neurons arises "in a repeated pattern in each segment," and the expression of certain genes in each segment suggest that metameres are indeed homologous units.[34, 44]

In different arthropod groups, metameres and appendages that appear to correspond in relative position and embryological origin also have been considered homologous. Again, however, modern molecular work suggests a more complex picture. For example, genes responsible for particular appendages may not be affiliated with corresponding segments in different classes.[1] Nevertheless, much of the evolutionary history of arthropods is written in the variation, in the divergence in form and function, of apparently corresponding segments and appendages.

Most adult arthropods are "modified" in the sense that segments are grouped into body regions, a feature known as **tagmatism** or **tagmatization.** The different body regions are called **tagmata** (singular **tagma**). As a result of grouping and subsequent development, individual skeletal plates may be shifted out of their embryonic positions or even fused together. Thus, the fundamental arthropod body architecture is not always obvious from adults, especially from their exterior (see Figs. 33.9 and 33.11).

Exoskeleton

The **cuticular exoskeleton** consists mostly of tanned proteins and chitin; is usually hard, virtually indigestible, and insoluble; may be impregnated with calcium salts or covered with wax; and provides physical as well as physiological protection. The exoskeleton also provides a place for muscle attachment, and, in the case of insects, wings (Fig. 33.2). The body skeleton is constructed of plates, or **sclerites,** laid down as dorsal **tergites,** ventral **sternites,** and lateral **pleurites.** Appendage segments are basically cylinders but, like body plates, may be so modified as to obscure their fundamental nature. Appendage joints are either hinges or pivots made from **condyles** and **sockets.** Skeletal pieces are joined by **articular membranes** where the cuticle is very thin; therefore, articular membranes are flexible, allowing movement of body and limbs (Fig. 33.3). Muscles are attached to inner skeletal ridges (**apophyses**) or spines (**apodemes**). Arthropod locomotion is influenced greatly by size, shape, and location of skeletal plates as well as by origin and insertion of muscles, and by the angles at which limbs are attached to the body.

The cuticle is made up of several layers containing protein, lipid, and polysaccharides, all secreted by the underlying **epidermis** (also called **hypodermis;** see Fig. 33.2). Much of this protein is stabilized (rendered relatively inert chemically) by processes of **sclerotization,** which results in crosslinking of amino acid chains in adjacent polypeptides (Fig. 33.4). Sclerotization reactions involve a number of compounds, particularly N-β-alanyldopamine and N-acetyldopamine, a tyrosine derivative.[3, 48] In **quinone tanning,** compounds such as N-acetyldopamine are first oxidized by phenoloxidase into quinones, which then spontaneously react with thiols and amine groups in adjacent polypeptides to form linkages between chains. In β-**sclerotization** the β-carbon of N-acetyldopamine is activated by an enzyme that forms covalent bonds between the β-carbon and adjacent polypeptides (see Fig. 33.4). So far, β-sclerotization is known to occur only in insects.[37] Protein stabilization by the formation of dityrosine and trityrosine crosslinks also has been reported. Some workers believe that sclerotization does not occur by covalent crosslinking of polypeptide chains but by controlled dehydration driven by quinones and other chemicals secreted into the cuticle.[51] In any case, when the protein is stabilized, it is virtually insoluble except by vigorous chemical treatment.

The main carbohydrate component of cuticle is the polysaccharide **chitin,** a polymer of N-acetylglucosamine linked by 1,4-α-glycosidic bonds into long, unbranched molecules

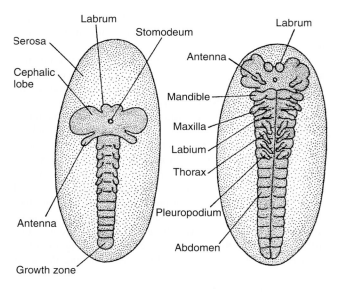

Figure 33.1   **Two stages in the embryonic development of a grain beetle,** *Tenebrio* **sp.**
Note undifferentiated mouth appendages and legs shown as relatively similar structures.

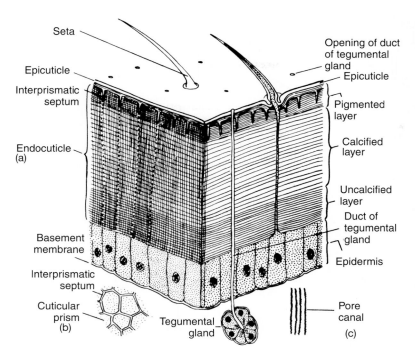

Figure 33.2   Cuticle.

(*a*) Diagram showing structure of crustacean cuticle. All layers are secreted by the hypodermis (epidermis). The thin epicuticle is of sclerotized protein, and procuticle (endocuticle) contains protein, chitin, and mineral salts. Protein in the uncalcified layer is unsclerotized. The procuticle of insects and arachnids is divided into highly sclerotized exocuticle and less sclerotized endocuticle. (*b*) Horizontal section through pigmented layer of endocuticle. (*c*) Pore canals as they appear in vertical sections.

From R. Dennell, in T. H. Waterman, editor, *The physiology of Crustacea,* vol. 1. Copyright © 1960 Academic Press, Inc., Orlando, FL. Reprinted by permission.

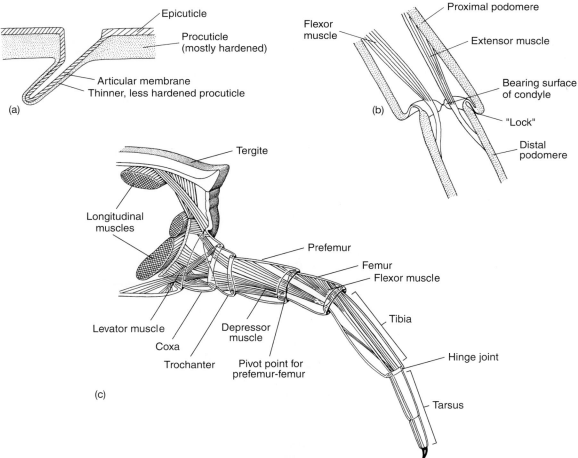

Figure 33.3   Diagram of articulation and musculature of arthropod trunk and limbs.

(*a*) Flexibility is provided by thinner cuticle between sclerites. (*b*) Movement of a joint is by contraction of muscles inserted on opposite sides of the articulation, and the bearing surfaces of joints are the condyles. (*c*) Muscles of a centipede leg. Levator muscles raise the leg; depressor muscles lower the leg; flexors bend the leg toward the body.

(*a, b*) Drawing by William Ober. (*c*) From S. Manton, *The Arthropoda: Habits, functional morphology, and evolution.* Copyright © 1977 Clarendon Press, Oxford, England.

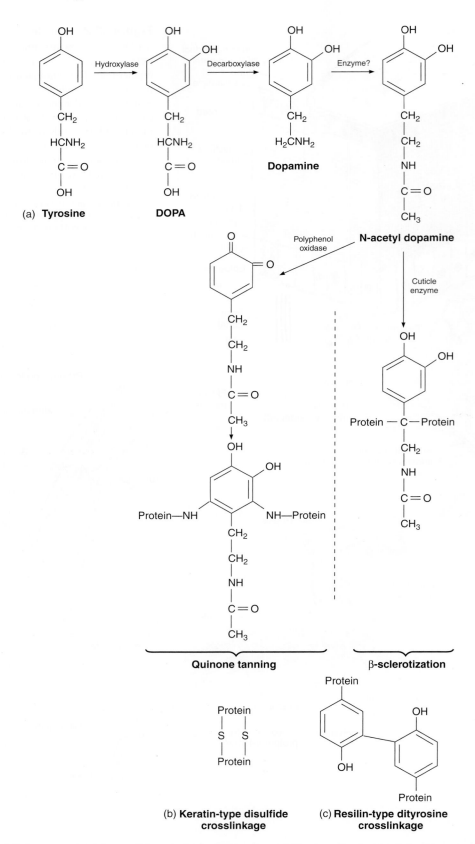

Figure 33.4 (*a*) Substrate and products of the reaction sequence for stabilization of sclerotins by quinone tanning in insects and other arthropods and by β-sclerotization in insects (based on Richards[37] and on Riddiford and Truman[39]). (*b*) Disulfide cross-linkages of adjacent amino acid chains, as in keratin; probably also occurs in arthropod cuticle.[3] (*c*) The highly elastic protein, resilin, stabilized by dityrosine and trityrosine cross-links, is sometimes found in insect cuticle, either mixed with other proteins or in almost pure form.[3]

of high molecular weight. Chitin is flexible and contributes little to the rigidity of an arthropod's skeleton; hardness is conferred by proteins and by deposition of inorganic salts. Evidence also exists that chitin is bonded to protein, although the structural significance of this linkage is unclear.[37] The origins of these reactions were important evolutionary events that allowed arthropods to achieve their long history of success, including their present and relatively recent success as distributors of human misery.

The outermost layer of cuticle, the **epicuticle,** is thin and contains stabilized protein, sometimes called **cuticulin,** but no chitin. Insects and arachnids usually have a lipoidal layer covering, or perhaps lipid interspersed in cuticulin, that functions to prevent water loss. Over the lipid is a "varnish" that protects the wax from abrasion. Beneath the epicuticle lies a thicker **procuticle,** which in insects and arachnids is further divided into **exocuticle** and **endocuticle,** endocuticle being much less sclerotized than exocuticle.

In Crustacea the waxy and varnish layers are absent, but the procuticle is often impregnated with calcium carbonate, calcium phosphate, and other inorganic salts. The entire procuticle of crustaceans is also called an *endocuticle,* which means that the term does not apply to the same structure as it does when used to refer to insects. In Crustacea the hardened layers containing salts and sclerotized proteins are the **pigmented** and **unpigmented** calcified layers. The unpigmented layers also contain chitin and protein, but the protein is unsclerotized and this layer is membranous and flexible.

Insertion of muscles in the unyielding exoskeleton, coupled with fulcra provided by condyles in the flexible joints, makes very fine control of movements mechanically possible. Increase in complexity of movements meant corresponding evolution of nervous elements to coordinate the movements. Furthermore, small changes in a given sclerite could substantially increase efficiency of a particular body part for a given function. Arthropods have many sclerites; the cuticle has evidently given evolutionary forces a great deal of raw material with which to work.

Molting

Although arthropod cuticle confers many evolutionary opportunities, it presents some problems as well, most important of which is growth of an animal enclosed in a nonexpansible covering. The solution to this problem is a series of **molts,** or **ecdyses,** through which all arthropods go during their development. Much of the physiological activity of any arthropod is related to its molting cycle, and this relationship must have held true since the evolutionary origin of the phylum; a sizeable fraction of trilobite fossils, for example, consists of cast exoskeletons.

Growth in tissue mass occurs during an interecdysial period, and dimensional increase occurs immediately after molting, while the new cuticle is still soft. Stages between each molt are referred to as **instars.** Length of an intermolt phase depends on the species involved, the animal's age and stage of development, and any interceding diapause (discussed later). The number of instars also varies with species and in some cases is influenced by nutritional state, especially in acarines and insects.

In those groups in which molting has been best studied—insects and malacostracan crustaceans—the process is controlled by hormones. However, there is evidence that ecdysteroid hormones induce molting in many if not all taxa now included in superphylum Ecdysozoa (arthropods, onychophorans, nematodes, nematomorphs, tardigrades).[22] In the malacostracan order Decapoda, which includes familiar lobsters and crayfish, a neuropeptide **molt-inhibiting hormone (MIH)** is produced in the **X-organ,** which is composed of neurosecretory cells in the eyestalk; **molting hormone (MH)** is produced in the **Y-organs,** a pair of glands near the mandibular adductor muscles. Structures comparable to X-organs are found within the head of sessile-eyed Crustacea (those species whose eyes are not on stalks).

In insects such as tobacco hornworms and silkworms, at least three hormones are directly involved in the molting process: **prothoracicotropic hormone** (**PTTH,** or **brain hormone**), **20-OH-ecdysone** (a steroid also known as 20E), and **bursicon.** PTTH is secreted by neurosecretory cells in the brain and released by organs called **corpora cardiaca** (Fig. 33.5). This hormone is carried in hemolymph to the **prothoracic glands (molting glands)** which are stimulated to produce 20E. The latter, also called *molting hormone*[22] stimulates molting and is analogous, perhaps homologous, to MH of crustaceans. In at least some species, an **eclosion hormone (EH),** released by protocerebral neurons, regulates behavior associated with ecdysis.[35] Bursicon is secreted from organs associated with ventral nerve ganglia and regulates postecdysial hardening of the cuticle. This summary of hormones involved in molting is by necessity an abbreviated one; Heming[22] provides a detailed description of the numerous central nervous system sites where neurosecretory cells occur in insects and of the various hormones, their origins, and activities.

As the level of MIH decreases and that of MH increases in crustaceans or as the level of 20-OH-ecdysone increases in insects, the organisms undergo a series of changes preparatory for a molt, the **preecdysial** period. DNA synthesis in hypodermal cells is stimulated, and then RNA and protein synthesis proceed. The next effect of ecdysial hormones is to cause the hypodermis to detach from the old procuticle (**apolysis**) and start secreting a new epicuticle (Fig. 33.6). At the same time, enzymes (including chitinases and proteinases) begin to dissolve old procuticle. As solution proceeds, products of the reactions, including amino acids, N-acetylglucosamine, and calcium and other ions, are resorbed into the animal's body. These materials are thus salvaged and later are incorporated into new cuticle.

Almost immediately after apolysis new epicuticle becomes limited in permeability. The new procuticle is thus protected from enzymes dissolving old cuticle above.[50, 53] Old cuticle is not completely dissolved; in insects the epicuticle and sclerotized exocuticle remain, and in crustaceans the epicuticle and calcified regions remain, although some decalcification of these layers occurs. At the time of ecdysis, the old cuticle splits, normally along particular lines of weakness, or dehiscence, and the organism climbs out of its old clothes.

The old cuticle splits, and the new one stretches, by expansion of the body. Insects accomplish this feat by inhaling air, and crustaceans do it by rapidly imbibing water, a process aided by increased osmotic pressure in the tissues

Figure 33.5 Generalized central nervous system of an insect, showing sources of hormones in the head and prothorax.

Neurosecretory hormones are produced by the median (*mnc*), lateral (*lnc*), and subesophageal (*snc*) neurosecretory cells and perhaps also by the corpora pedunculata (*cp*). Hormones from the neurosecretory centers (*NCCI, NCCII*) in the protocerebrum pass in two paired nerves to be stored in the corpora cardiaca (*CC*). Hormones are also secreted by the corpora allata (*CA*) and prothoracic or ecdysial glands. (*Hy*), hypocerebral ganglion.

Redrawn by William Ober and Claire Garrison from P. M. Jenkins, *Animal hormones: A comparative survey,* part 1. Oxford, UK: Pergamon Press Ltd., 1962.

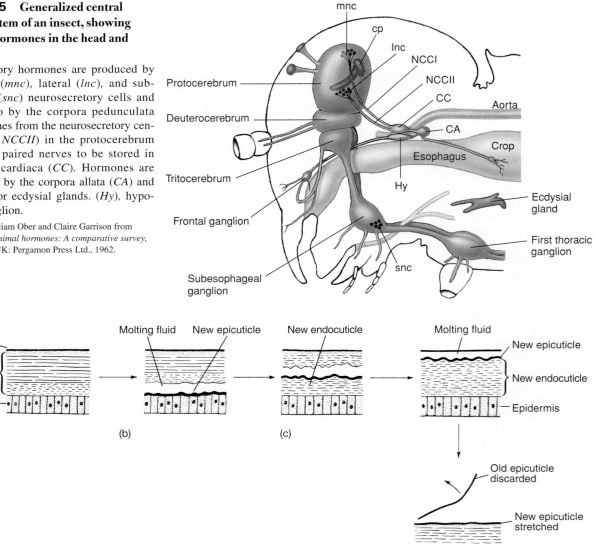

Figure 33.6 Cuticle secretion and resorption in preecdysis.

(*a*) Interecdysis condition. (*b*) Old endocuticle separates from epidermis, which secretes new epicuticle. (*c*) As new endocuticle is secreted, molting fluid dissolves old endocuticle, and the solution products are resorbed. (*d*) At ecdysis, little more than the old epicuticle is left to discard. In postecdysis, new cuticle is stretched and unfolded, and more endocuticle is secreted.

Drawing by William Ober.

and hemolymph due to mobilization of calcium ions from the cuticle prior to molt.[40] Increased hemolymph and tissue volume causes the small wrinkles in the still soft cuticle to smooth out, increasing body dimensions, and the cuticle begins to harden again. In this postecdysial period, sclerotization of protein and redeposition of calcium salts in the procuticle occur, and more procuticle is secreted. As in many biological phenomena, molting has both a benefit and a cost. The benefit is the growth allowed by a molt; the cost is vulnerability to predation that comes with a temporarily soft cuticle.

Molt Cycles

Length of time spent as a given instar depends on the species involved, its age and stage of development, the season or annual cycle, and sometimes the species' nutritional state.[47, 49] Decapod crustaceans that molt on an annual cycle and have a long intermolt period are said to be **anecdysic.**[25] Species in which one ecdysial cycle grades rapidly into another are **diecdysic.** Some crabs reach maximal size and stop molting, undergoing terminal anecdysis. In crustaceans other than malacostracans, little is known of the hormonal control of molting. Barnacles, at least, seem to be in a permanent diecdysis.[6] Furthermore, many copepod parasites of fish cease molting when they reach the adult stage, although they continue to grow actively. For example, females of genus *Lernaeocera* are about 2 mm long after their last molt, but they may attain an ultimate size of up to 60 mm *without molting*. We do not know what changes occur in the cuticle when the copepod reaches sexual maturity, but it is clear that the changes permit continuous growth.

More is known about mechanisms allowing such expansion in some parasitic insects and ticks than in crustaceans. The fourth-stage nymph of the bug *Rhodnius prolixus* takes a large blood meal that necessitates stretching its abdominal cuticle threefold.[52] Before this blood meal, its cuticle is stiff and inextensible, but this condition changes as the bug feeds. Change is mediated by neurosecretory axons running to the hypodermis, apparently stimulating an enzyme discharge that affects the cuticle. After feeding, additional cuticle material is deposited, probably to provide protection and a template for cuticle of the next instar.[23] Female *Boophilus microplus,* the cattle tick, ingest 150 times their body weight in blood after molting to adult, increasing in length from 2.5 mm to 11.0 mm. Filshie[15] reported that, before the last molt, epicuticle is laid down as a highly folded layer. Subsequent expansion is accommodated through unfolding of the inexpansible epicuticle and stretching and growth of underlying procuticle.

Early Development and Embryology

In arthropods, sexes are separate and fertilization is internal. However, mating behaviors, methods of sperm transfer, and subsequent treatment of eggs are all as varied as the phylum is structurally. Early embryology is discussed in fascinating detail, with wonderful pictures, in the book by D. T. Anderson.[4] Biologists have long used embryological development as structural evidence for evolutionary relationships, and Anderson provides a fairly accessible explanation of for such use. The molecular events described next are summarized from Nation[32] and Heming.[22]

Our current picture of arthropod early development is derived mostly from study of a fruit fly species, *Drosophila melanogaster,* and includes action of genes involved in pattern formation. These genes are **maternal** and **zygotic;** the former are present in ovarian nurse cells that contribute RNA to developing oocytes; the latter start to function after fertilization. Zygotic genes include both **segmentation genes** and **homeotic genes.**[32] Eggs become polarized into anterior and posterior ends very early in development due to secretion of RNA transcripts known as **morphogens** (proteins that influence an embryo's production of a particular structure) by nurse cells. Other morphogens, resulting from a **cascade** of gene expression (a series of specific proteins synthesized in response to a stimulus), are produced at each end and diffuse toward the opposite pole, interacting to establish regional gradients that, in turn, influence the fate of a region, such as development into head, thorax, or abdomen.[32]

In most Crustacea and Insecta, initial cleavage is **intralecithal,** with nuclei undergoing several divisions within the yolk mass and then migrating to the periphery to become **blastoderm.** Yolk is concentrated in the interior of the embryo (**centrolecithal**), and differentiation proceeds in the superficial areas. Blastoderm is also formed in chelicerates, but initial cleavages are sometimes complete (**holoblastic).**

Arthropod embryos exhibit a varying number of postoral or trunk segments (somites), each with a pair of jointed appendages. Segments are initially laid down as sometimes hollow mesoderm blocks; their cavities are the segmented coelom.

Segmentation genes are zygotic, and they consist of genes that divide the embryo into regions and specify structural arrangements with segments.[32] These genes interact with one another through their products to divide an embryo and direct the fate of tissues within segments. There is evidence for three anterior "naupliar segments" typical of all or most arthropods; a set of 10 posterior segments, also perhaps found in all arthropods; and a final group of segments—whose number is characteristic of a class—produced by secondary division of the posterior ones.[2]

Homeotic genes regulate sequence of segment development and provide each segment with its particular features.[22, 32] In *D. melanogaster,* for example, such genes control formation of antennae, legs, and bristles. The **homeobox** is a DNA sequence, about 180 base pairs long, found within homeotic genes and specifying their binding sequences. Portions of homeobox gene sequences are highly conserved, being found in all animals including vertebrates.[22] Because homeotic genes often code for transcription factors, controlling expression of other genes, variations in nonconserved homeobox sequences are ultimately responsible for specific differences in the final form of segments.[32] At the structural level comparative embryology provides evidence for the fate of corresponding appendages in various groups of arthropods. Thus, we claim that walking legs of spiders correspond to mouthparts of crayfish by reference to embryological development of both animals.

Postembryonic Development

Arthropods differ in their postembryonic development. In most species embryos develop into **larvae.** A larva is a life cycle stage that is structurally distinct from the adult, normally occupies an ecological niche separate from the adult, is sexually immature, and must undergo a structural reorganization (metamorphosis) before becoming an adult.

Crustaceans
The typical larva that hatches from a crustacean egg is called a **nauplius** (pl. **nauplii;** Fig. 33.7). Nauplii have only three pairs of appendages: antennules, antennae, and mandibles. These have locomotor function and are different in form from the adult appendages. Nauplii undergo several ecdyses, usually adding somites and appendages at each molt. Typically there are several instars, and later-stage larvae may be referred to as **metanauplii.** Metamorphosis may be gradual, occurring over several instars, or more abrupt, occurring from one instar to the next. But if a distinguishable larval stage occurs, development is said to be **indirect. Direct** development is that in which a juvenile, rather than a larva, hatches with segmentation and appendages complete. Juveniles, however, are sexually immature. Crustacea may vary widely in their development patterns, even within the same class (see chapter 34).

Insects
Of Insecta, only Thysanura have direct development. The winged orders, or Pterygota, are all metamorphic. In several orders (Dermaptera, Dictyoptera, Mallophaga, Anoplura, and

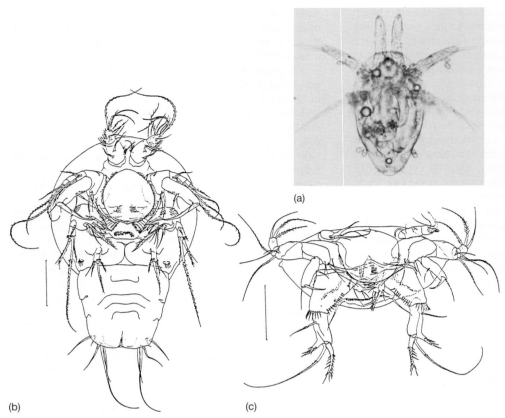

Figure 33.7 **Examples of nauplius larvae.**

(*a*) Copepod nauplius. This particular nauplius also has a parasitic nematode in it, as well as some ectocommensal peritrich ciliates attached to the surface. (*b*) Late nauplius of *Tisbe cucumariae.* (*c*) An early nauplius of *Stenhelia palustris. Tisbe cucumariae* and *Stenhelia palustris* are both harpacticoid copepods.

(*a*) Courtesy of Ralph Muller.
(*b, c*) From Hans-Uwe Dahms, "Pictorial keys for the identification of crustacean nauplii from the marine meiobenthos," in *Journal of Crustacean Biology* 13:609–616. Copyright © 1993 *Journal of Crustacean Biology.* Reprinted by permission.

Hemiptera) larval instars are called **nymphs,** and they become gradually more like an adult **(imago)** with each ecdysis (**gradual metamorphosis** or **hemimetabolous** development) (see Fig. 36.5). Wing buds develop externally **(exopterygote)** and can be observed readily in nymphal instars (except for bedbugs, which are wingless!). Nymphs of hemimetabolic insects have well-developed appendages, compound eyes, and rudiments of external genitalia, and their habits are generally similar to those of adults, although their microenvironments may differ slightly or, as in the case of dragonflies, for example, significantly. Nymphal dragonflies are aquatic and adults are terrestrial, but both stages are active predators.

In other orders (Neuroptera, Coleoptera, Strepsiptera, Siphonaptera, Diptera, Lepidoptera, and Hymenoptera) larvae bear little resemblance to adults. After several larval instars, these insects enter a nonfeeding period, the **pupa** stage, in which the animal is completely reorganized (see Fig. 33.8) and wing buds grow internally **(endopterygote).** This represents **complete metamorphosis** or **holometabolic** development. Nearly 90% of all insects undergo complete metamorphosis, and this type of development is considered a major factor in their evolutionary success.[22]

Larvae of holometabolic insects are extremely diverse and typically occupy ecological niches quite distinct from those of adults. Consider mosquitoes, for example (see Fig. 39.3); their larvae are aquatic and filter feed on microorganisms, whereas adult females of most species are terrestrial and feed on plant juices or blood. Holometabolic larvae occur in a variety of forms and exhibit a range of behaviors. Some are active predators with well-defined heads and thoracic appendages (**campodeiform** or **oligopod,** such as many beetles); some lack heads and appendages altogether (**vermiform** or **protopod,** such as Diptera and Hymenoptera); others are **polypod,** with abdominal **prolegs** (Lepidoptera); and still others are grublike, with swollen abdomens and lightly sclerotized body cuticle (**scarabaeiform,** such as some coleopteran families) (Fig. 33.8).

Ticks and Mites

Postembryonic development of acarines (ticks and mites, chapter 41) is characteristically direct, with a sexually immature **nymph,** a tiny facsimile of the adult, hatching from the egg. Some arachnids undergo one or two "larval" molts while still within the egg. Numbers of nymphal instars vary depending on the type of arachnid. Sixlegged larvae hatch from eggs in members of order Acari and become eight-legged nymphs at their first molt. Most mites have three nymphal instars: **protonymph, deuteronymph,** and **tritonymph.** However, hard ticks (family Ixodidae) have only one nymphal stage, and soft ticks (family Argasidae) may have as many as eight.

Endocrine Control of Development

Endocrine function during development is best known for insects, and it is essentially similar in both holometabolous and hemimetabolous forms. **Juvenile hormone (JH)** is produced by the **corpora allata** (see Fig. 33.5). Although three different chemical forms of the hormone are present in varying proportions in different insects, all three forms produce similar effects.[18, 39] The best known and documented of such

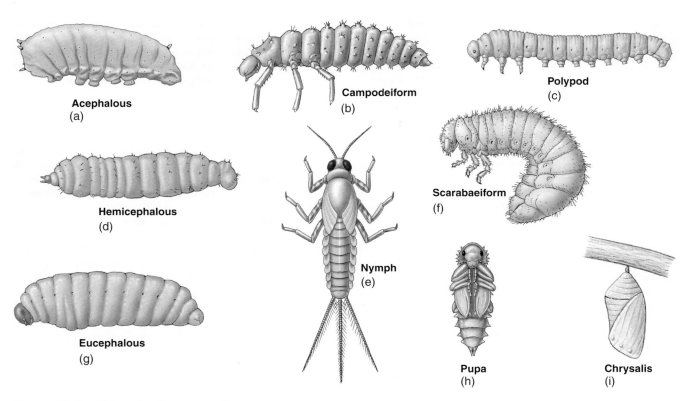

Figure 33.8   **Types of endopterygote larvae.**

(*a*) Acephalous type, found in some Diptera and Hymenoptera. (*b*) Campodeiform type, characterized by a prognathous head and well-developed thoracic legs, and resembling the thysanuran genus *Campodea*. (*c*) Polypod or eruciform type, found in some Hymenoptera, Diptera, and Coleoptera; this type usually has access to abundant food. (*d*) Hemicephalous type, apodous (lacking appendages), with some sclerotized mouthparts, found in some Diptera and Hymenoptera. (*e*) A nymph, resembling an adult except for reduced wings and lack of fully developed genitalia; found in Odonata, Ephemeroptera, Blatteria, and some other orders. (*f*) Scarabaeiform type, found in some Coleoptera. (*g*) Eucephalous type, apodous larva with a well-developed head, found in some Coleoptera, Hymenoptera, and Diptera. (*h*) A pupa, typical of holometabolous insects. (*i*) Chrysalis, lepidopteran pupa enclosed in a cocoon and attached by a silk thread.

Redrawn from various sources by William Ober and Claire Garrison.

actions is the so-called *status quo effect* in which JH acts on insect tissues to maintain larval or nymphal characters (impedance of maturation). The level (titer) of JH in an insect's hemolymph decreases as development proceeds through juvenile instars. Consequently, tissues and organs become progressively more adultlike. Finally, the titer drops to an undetectable level at about the beginning of the last nymphal instar in hemimetabolous insects, and the adult emerges at the next ecdysis. Disappearance of JH from the hemolymph of holometabolous insects usually occurs about midway through the last larval instar; the next molt produces a pupa. It is not clear, however, that absence of JH is the only cause of pupation.

The status quo action of JH is believed to result from its direction of the kinds of RNA produced after ecdysteroid stimulation, but the mechanism by which JH does this is still not well understood.[32] If a high titer of JH is present, the RNA produced will lead to larval proteins; if little or no JH is present in the hemolymph, RNA for synthesis of proteins with adult characteristics will be produced. Interestingly, although a shutdown of JH production by the corpora allata is necessary for maturation to the adult form, corpora allata are reactivated in adults of many insects and again secrete JH. The hormone is necessary for egg maturation in females

and proper development of sex accessory glands in males. Furthermore, adults of almost all insects do not molt again, but steroid hormones are once more secreted and also play a role in reproductive function.

Diapause

Without a doubt, another factor contributing greatly to arthropods' success has been their ability to withstand adverse environmental conditions, such as subfreezing temperatures or extreme dryness, in which normal physiological function would be impossible. Many arthropods have evolved the ability to enter a period of developmental arrest known as **diapause,** during which their metabolic rate is reduced and their chemical makeup is altered to provide resistance to seasonal fluctuations in temperature or moisture.[33] When diapause is an obligate stage in the life cycle, it occurs at a genetically determined and species-specific point in that cycle. Facultative diapause, however, is usually induced by environmental conditions.[33] Although diapause occurs in some crustaceans and arachnids, much more is known about it in insects than in those other groups. Knowledge of the role of diapause in the life of a particular species is sometimes of vital importance to our understanding of the biology of arthropod vectors of parasitic diseases.

In temperate regions diapause functions most often as an overwintering mechanism, whereas in the tropics it typically functions to allow survival through dry seasons.[33] Diapause in insects may occur in the egg, larva, pupa, or adult, depending on the species, but physiological changes accompanying quiescence usually begin before the onset of unfavorable conditions and are stimulated by cues such as day length. In immature stages diapause is characterized by a cessation of development and prolongation of that stage. In adults reproduction is inhibited.

In all insects that have been studied, the mechanisms of diapause initiation and termination is hormonal, but different hormones are involved, depending on the life cycle stage involved. In silkworms, for example, control of diapause in eggs largely depends on secretion of diapause hormone, an oligopeptide, from a female moth's subesophageal ganglion.[33] Larval diapause in other species is produced by a cascade of events resulting in elevated juvenile hormone levels, whereas in adults reproduction ceases, ovaries regress, and flight muscle is converted to fat, all in response to lowered JH levels.[32] Certain behaviors such as negative phototropism, digging, and even migration are also often associated with adult diapause. Diapause is clearly a complex phenomenon whose origin is to be found in the evolutionary history of those species that exhibit it.

EXTERNAL MORPHOLOGY

Phylum Arthropoda is so large and diverse that a detailed description of morphology and related physiology is far beyond the scope of this book. However, parasitism is inextricably intertwined with the lives of arthropods. From their roles as intermediate hosts, vectors of important human and veterinary diseases, and definitive hosts in their own right, to their often extraordinary adaptations to parasitic life, arthropods constantly present us with evidence of the many forms that parasitism can take. Thus, a parasitology student needs to know a modicum of structure to identify parasitic arthropods and to understand host/parasite relationships. In addition to the brief discussion that follows, more specialized details of structure and biology are given in chapters to come.

Form of Crustacea

In Crustacea the head is usually not clearly set off from the trunk. One or more thoracic somites are commonly fused with the head, but some thoracic segments are distinct (Fig. 33.9). Thus, typical tagmatization of crustaceans is **cephalothorax** (head, plus any thoracic segments fused with head), free **thorax,** and **abdomen,** although the degree of prominence and of fusion of tagmata varies greatly from group to group. The head seems to have been formed by fusion of five somites.[43] The cephalothorax and sometimes even the entire body may be covered by a **carapace,** which arises as a fold from the posterior margin of the head (see Fig. 33.9). Two types of eyes are found in crustaceans: **median eyes** and **compound eyes.** Median eyes, consisting of three or four pigment-cup ocelli, are also called **nauplius eyes** because they are present in that larval stage. Nauplius eyes sometimes persist into the adult, as in copepods. Adults of most species have a pair of compound eyes; these may be sessile, or they may be mounted on stalks and be very convex, with an angle of vision of 180 degrees or more.

The anteriormost head appendages are the **antennules** (first antennae) followed by **antennae** (second antennae). Crustaceans are the only arthropods with two pairs of antennae. Antennules and antennae are usually sensory, but in some forms they may be adapted for locomotion or grasping on to a host (prehension). Feeding appendages on the head are **mandibles, maxillules** (first maxillae), and **maxillae** (second maxillae) (see Fig. 33.9). One or more pairs of thoracic appendages may be incorporated into the mouthparts and are then called **maxillipeds.** Other thoracic appendages are **pereiopods,** and abdominal appendages are **pleopods.** Pereiopods and pleopods may be variously modified for

Figure 33.9 **Lateral view of a generalized malacostracan crustacean.**

Drawing by William Ober.

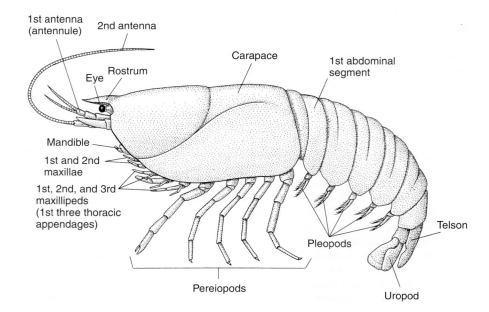

walking, swimming, or copulation. The numbers of pereio-pods and pleopods vary from group to group, and from some groups pleopods are absent. The abdomen ends in a **telson,** which may be flanked by the posteriormost pleopods, called **uropods.**

Appendages of Crustacea are primitively **biramous** (having two branches) (Fig. 33.10), and this condition prevails in at least some appendages of all living species during their lives. The terminology applied by various workers to crustacean appendages has not been blessed with uniformity. At least two systems are currently in wide use; and we are giving the alternative term for each structure in parentheses.

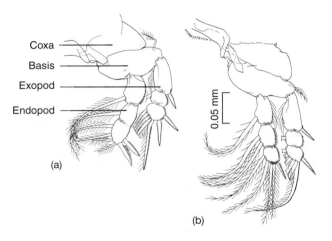

(a)

(b)

Figure 33.10   First (*a*) and second (*b*) thoracic appendages of *Ergasilus megaceros* (Copepoda), a parasite of the sucker *Catostomus commersoni.*
The terminal segments of the first endopod are fused, the ancestral condition being indicated by the presence of vestigial condyles. The medial side of the coxa may be modified for food handling in some Crustacea and is called a *gnathobase.*

From L. S. Roberts, "*Ergasilus* (Copepoda: Cyclopoida): Revision and key to species in North America," in *Trans. Am. Microsc. Soc.* 89:134–161. Copyright © 1970. Reprinted by permission.

The lateral branch is the **exopod** (exopodite), and the medial one is the **endopod** (endopodite). Each of these branches may contain several segments, varying by appendage and according to species. The endopod and exopod are borne on a **basis** (basipodite), and the basis, in turn, is attached to a **coxa** (coxopodite); together they are referred to as a **protopod.** Processes from the protopod are termed **endites** and **exites;** exites may be called **epipods** (epipodites).

Form of Pterygote (Winged) Insects

In all members of class Insecta the tagmata are **head, thorax,** and **abdomen** (see below). The head is made up of six fused metameres, four of which bear appendages in modern insects. Bases of the freely movable, sensory **antennae** are above or between the eyes (Fig. 33.11). **Mandibles** are usually the primary feeding appendages and are borne ventrally, lateral to the mouth. Immediately posterior to the mandibles are **maxillae** and following these, the **labium,** interpreted as a fused pair of appendages (Fig. 33.12). Maxillae and labium also may have **palps** with food-handling and sensory functions. Anteriorly the mouth is covered by a **labrum,** or upper lip. A tonguelike lobe, the **hypopharynx** arises from the floor of the mouth in some insects, and a similar extension, the **epipharynx,** may emerge from the roof of the mouth in others. The labrum, epipharynx, and hypopharynx are not considered appendages, but they do function in feeding. Depending on the insect group, these mouthparts may be highly modified or secondarily lost altogether. In addition to antennae and mouthparts, most insects' heads have a pair of **compound eyes** and one or more simple eyes, or **ocelli.**

The insect thorax consists of three segments—**prothorax, mesothorax,** and **metathorax**—each of which bears a pair of legs. Each leg usually is divided into five segments, or **podomeres** (see Fig. 33.11). The basal segment, or **coxa,** articulates with both the body and the **trochanter;** the latter is also fixed to the **femur,** the largest of the podomeres, which, in turn, articulates with the more slender **tibia.** Distal to the tibia is the **tarsus,** which is subdivided into two to five

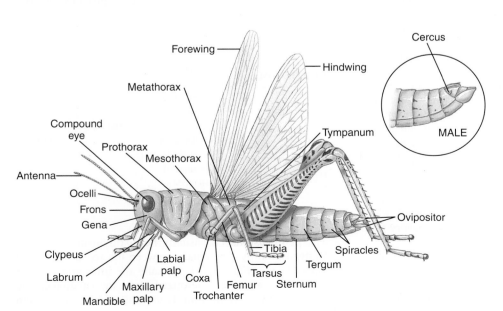

Figure 33.11   External features of a relatively generalized insect, the grasshopper *Romalea.*
The terminal segment of a male with external genitalia is shown in inset.

Figure 33.12 **Anterior view of the mouthparts of a grasshopper.**

Drawing by William Ober.

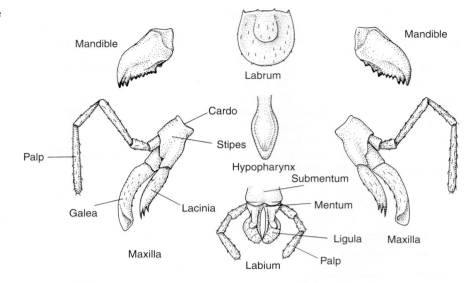

segments. The **pretarsus** consists of claws or other structures attached to the terminal tarsal segment.

Adult pterygote insects characteristically have wings, although some, such as fleas, lice, worker ants, and termites, have lost their wings during the course of evolution. Both the mesothorax and metathorax bear a pair of wings, but, in order Diptera, which includes mosquitoes, tsetse flies, sand flies, and black flies, all vectors of parasitic diseases, the metathoracic wings are reduced to balancing organs called **halteres** (see Fig. 39.2). In male Strepsiptera, mesothoracic wings are reduced to halteres, whereas females are highly modified parasites with no wings at all. Wings develop from evaginations of thoracic epidermis and thus consist of a double layer of epidermis. These layers are penetrated by canals, called **lacunae,** and lacunae contain nerves, tracheae, and hemolymph.

Epidermal cells atrophy as the wing approaches full development; thus, a wing consists of two thin layers of cuticle secreted by epidermis supported by more heavily sclerotized **veins,** which are the remains of lacunae. The pattern of wing venation is constant within a species and therefore is often of value in taxonomy. Entomologists have adopted a standard nomenclature for wing venation (Fig. 33.13), and anyone who regularly identifies insects using taxonomic keys quickly memorizes these terms.

An adult insect's abdomen consists of 11 somites plus a terminal **telson,** but all these segments usually can be discerned only in an embryo. Abdominal segment appendages also occur in embryos, but except for those on genital segments, these appendages are lost during transformation into an adult. Appendages of genital segments are called **external genitalia;** these consist of the **penis** or **aedeagus,** on the ninth segment of males, and the **ovipositor,** formed of the eighth and ninth segment appendages of females. The eleventh segment may also have appendages, the **cerci,** which range from vestigal in size to longer and have sensory function (see Figs. 33.11 and 39.8).

Most insects have **spiracles,** or openings into the respiratory system (see Fig. 33.11). Adult insects typically have eight abdominal pairs of spiracles. Spiracles usually have closing mechanisms that function to reduce water loss.

Although the vast majority of insects have only two pairs of thoracic spiracles, the mesothoracic and metathoracic, the former often migrates forward during embryological development and thus appears to be on the prothorax. Members of Diplura, primitively wingless forms typically found in leaf litter and rotting wood, are the only hexapods with true prothoracic spiracles.

Form of Acari

The primary tagmata in class Arachnida are a cephalothorax **(prosoma)** and abdomen **(opisthosoma).** Somites of these tagmata are fused to a greater or lesser degree, depending on the order. In spiders (order Araneae) fusion is complete in almost all species, but the prosoma and opisthosoma are distinct. In subclass Acari, even these tagmata are fused, and the opisthosoma is defined rather arbitrarily as a region posterior to the legs. This situation has given rise to a special nomenclature for the body regions of only Acari, given by Savory[41] as follows:

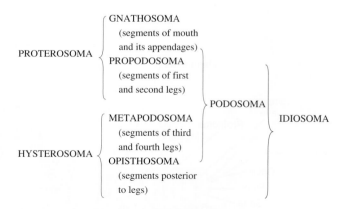

The **proterosoma** can usually be distinguished from the **hysterosoma** by a boundary between the second and third pairs of legs (Fig. 33.14). Dorsally the **idiosoma** is often covered by a single, sclerotized plate, the **carapace.** The **gnathosoma,** or **capitulum,** is usually sharply set off from

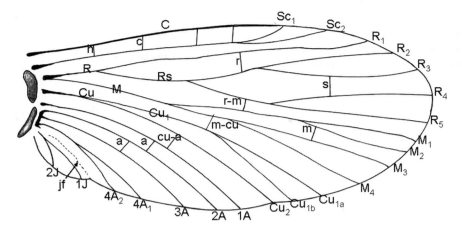

Figure 33.13   Diagram of typical venation of wing in modern insects.
Note standardized abbreviations for the veins: *C* (*costa*) is the unbranched thickened anterior margin of the wing. *Sc* (*subcosta*) is typically branched. *R* (*radius*) is the second major vein, connecting to the base of a sclerite. The radius usually branches, and the cells formed by these branches are important taxonomic characters. *M* (*media*) also branches into marginal cells which, like those of the radius, are numbered anterior to posterior. *Cu* (*cubitus*) articulates with a sclerite and is posterior to the media. *1A–4A* (anal veins) form a set. *jf* (jugal furrow) is a crease separating the anal region or fold from the jugal fold, which is the small area at the basal posterior corner of the wing. *Crossveins* are named according to the veins they connect.

From H. H. Ross, *Textbook of entomology,* 3d ed. Copyright © 1965 John Wiley & Sons, Inc. Reprinted with permission.

the idiosoma, and it carries the feeding appendages. These appendages are **chelicerae,** usually with three podomeres, and **pedipalps,** whose free segments may vary from one to five in different groups. Chelicerae may be **chelate** (pincer-like) in scavenging and predatory mites, but in parasitic mites they are usually modified to form stylets or bear teeth for piercing. Bases of pedipalps are lateral and just posterior to bases of chelicerae. Pedipalps may be leglike or chelate or reduced in size and serving as sense organs. Fused coxae of the pedipalps extend forward ventrally to form the **hypostome,** which, together with a labrum, makes up the **buccal cone** (Fig. 33.15). The dorsal part of the capitulum projects forward over the chelicerae as a **rostrum,** or **tectum.**

Acarines usually have four pairs of legs, as in other arachnids, but only one to three pairs may be present. Podomeres vary from two to seven, but six is the usual number, and these are **coxa, trochanter, femur, patella, tibia,** and **tarsus.** Each tarsus of most acarines has a pair of claws. Spiracles may or may not be present, and the position and existence of spiracles are important criteria for distinguishing suborders. The anus is near the posterior end of the body, but gonopore location is variable, being as far forward as the first legs in some forms. Some male mites have an intromittent organ, or **aedeagus.** The gonopore commonly opens through a more heavily sclerotized area, the **genital plate.** Other plates or shields are found on the idiosoma; their location and form are of taxonomic value.

The body and legs of most ticks and mites are well supplied with sensory (tactile) setae, which may be simple and hairlike, plumose, or leaflike; movement of a seta stimulates nerve cells at its base. One or two pairs of simple eyes are found laterally on the propodosoma in members of most suborders. Some mites have paired **Claparedé organs,** or **urstigmata,** between coxae of the first and second legs. Urstigmata are evidently humidity receptors. Ticks have a depression in their first tarsi called **Haller's organ,** which

bears four different kinds of sensory setae.[5] Haller's organ is a humidity and olfactory receptor and is of considerable value to the tick in finding hosts.[27, 46] More detailed anatomical information on ticks and mites is found in chapter 41.

Internal Structure

Body Cavity and Circulation of Fluids
The arthropod coelom is greatly reduced, its remnants being found in excretory organ or gonad spaces. The main body cavity of arthropods is thus a secondary space—the **hemocoel**—filled with fluid (**hemolymph**) containing a variety of cell types. Muscles, sometimes very large ones, are bathed in this fluid, which is circulated through an open circulatory system by means of a dorsal tubular heart. Hemolymph enters the heart from the surrounding **pericardial sinus** through pairs of lateral openings, the **ostia.** Ostia are one-way valves; when the heart contracts, ostia close, forcing hemolymph anteriorly into the arteries and finally into a system of tissue spaces, or sinuses. Hemolymph works its way back to the heart through these sinuses, often aided by body movements (Fig. 33.16).

Formed elements of hemolymph are mostly **amebocytes.** Parasites, especially larval stages, may penetrate the gut and come to lie in the hemocoel, as in the case of acanthocephalan or tapeworm larvae. And malarial sporozoites escape from their oocyst on the gut and migrate through the hemocoel to the mosquito vector's salivary glands (chapter 9).

Respiratory System
Gas exchange takes place directly through the body wall in very small arthropods that may lack specialized respiratory organs and even a heart. Larger Crustacea have **gills,** which are extensive folds of the epidermis, covered with thin cuticle, through which hemolymph circulates. Most insects, as well as many Acari, have a **tracheal system,** a branching

Figure 33.14 **A representative mite,** *Mycoptes neotomae* **(female, ventral view), parasitizing wood rats and white-footed mice.**

From A. Fain et al., "Two new Myocoptidae (Acari, Astigmata) from North American rodents," in *J. Parasitol.* 70:126–130. Copyright © 1970 *Journal of Parasitology.* Reprinted by permission.

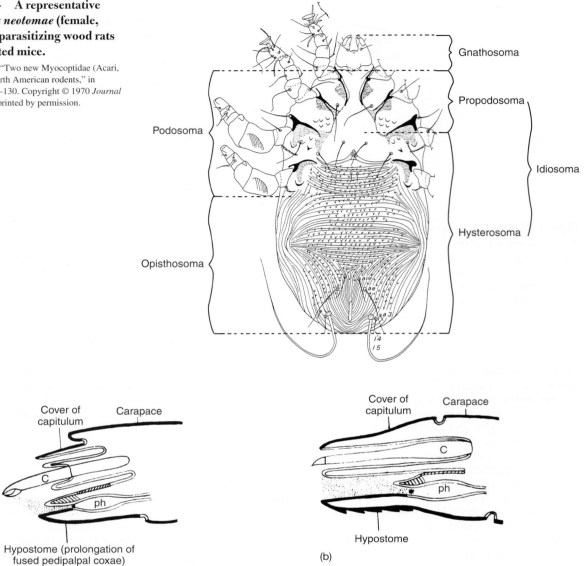

Figure 33.15 **Diagrammatic longitudinal section through the capitulum of acarines.**

(*a*) Mite; (*b*) hard tick. The hypostome and labrum (*crosshatched*) form the buccal cone, the anterior of which surrounds the preoral food canal (*shaded*). The mouth, designated by an *asterisk,* leads into the muscular pharynx (*ph*). Chelicerae (*C*) of ticks lie in a sheath and can be protracted and retracted.

From K. R. Snow, *The Arachnids: An introduction.* Copyright © 1970 Columbia University Press. Reprinted with permission of the publisher.

network of tubes. The tracheal system opens at spiracles and ramifies through the body into a large number of very fine **tracheoles** (Figs. 33.17 and 33.18). The cuticle of tracheae but not that of tracheoles is shed at ecdysis. Ventilation of the tracheal system is accomplished by pressure of body muscles on the walls of elastic tracheae, on tracheal air sacs, or both. Arachnid tracheal systems are thought to have evolved from **book lungs,** membranous folds inside a chamber that opens through a slit or spiracle. Book lungs occur in several arachnid orders but not in Acari.

Nervous System

The arthropod central nervous system consists of a dorsal ganglionic mass, the **brain,** lying above the **stomodaeum** (anteriormost part of the digestive system); nerves that supply cephalic sense organs; nerve trunks or **commissures** surrounding the esophagus and connecting the brain to a **subesophageal ganglion;** and a ventral nerve trunk that lies beneath the digestive tract. The ventral trunk consists of a double cord connecting **segmental ganglia.** However, in many if not most arthropods this fundamental structure is modified by postembryonic compression and shortening of the nerve trunk, fusion of ganglia, and lengthening of fibers to the posterior part of the animal.

The brain itself consists of three major regions: **protocerebrum, deuterocerebrum,** which in crustacea supplies nerves to the first antennae; and **tritocerebrum** (see Fig. 33.5). On the basis of evidence from comparative

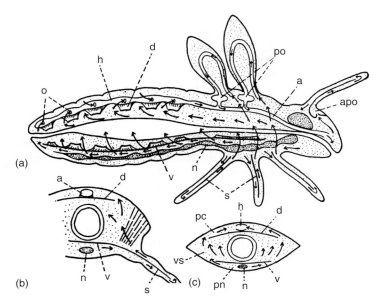

Figure 33.16   Insect circulatory system showing route of hemolymph circulation.

Although the hemolymph flows from the arteries into the open hemocoel, its circulation through the body is assured by partitions. (*a*) Schematic of insect with fully developed circulatory system; (*b*) transverse section of (*a*); (*c*) transverse section of abdomen; *arrows*, course of circulation; *a*, aorta; *apo*, accessory pulsatile organ of antenna; *d*, dorsal diaphragm with aliform muscles; *h*, heart; *n*, nerve cord; *o*, ostia; *pc*, pericardial sinus; *pn*, perineural sinus; *po*, mesothoracic and metathoracic pulsatile organs; *s*, septa dividing appendages; *v*, ventral diaphragm; *vs*, visceral sinus.

From V. B. Wigglesworth, *Principles of insect physiology*, 7th ed., figure 34.20. Copyright © 1972 Kluwer Academic Publishers, The Netherlands. Reprinted with permission.

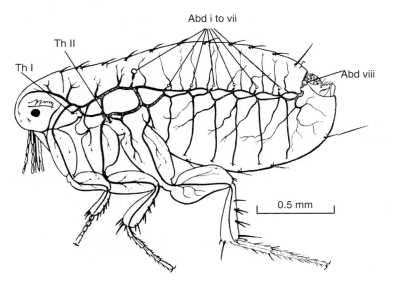

Figure 33.17   Half of the tracheal system of the flea *Xenopsylla* sp.

Main tracheae and locations of the spiracles are shown. *Th I*, *Th II*, thoracic spiracles; *Abd i–viii*, abdominal spiracles.

From V. B. Wigglesworth, *Principles of insect physiology*, 7th ed., figure 34.21. Copyright © 1972 Kluwer Academic Publishers, The Netherlands. Reprinted with kind permission from Kluwer Academic Publishers.

anatomy and embryological studies, the tritocerebrum consists of segmental ganglia incorporated by fusion into the brain. Evidence for homology of arthropod anterior appendages is found in the fact that nerve centers of crustacean second antennae, the chelicerae of chelicerates, and the antennae of insects are all located in the tritocerebrum.[45]

The peripheral nervous system includes axons that innervate muscles and glands and bi- or multipolar **neurocytes,** their distal processes, and axons. Sensory neurocytes are connected to a variety of sense organs, including tactile hairs and bristles and chemoreceptors.

Digestive System

In crustaceans the digestive tract consists of a **foregut, midgut,** and **hindgut.** Part of the foregut may be enlarged into a **triturating stomach,** bearing calcareous ossicles, chitinous ridges, or denticles on its walls and functioning to grind up food. The midgut is often enlarged to form a stomach, and it usually bears one or more pairs of **ceca.** One pair

of ceca may be modified to form a **digestive gland,** or **hepatopancreas,** which produces digestive enzymes. Absorption is confined to the midgut and tubules of the digestive gland.

The digestive system of insects is also divided into foregut, midgut, and hindgut. Distinct regions of the foregut are **esophagus, crop,** and **proventriculus** (Fig. 33.19). In insects that suck fluid meals from their hosts, the esophagus is a muscular **pharynx.** The crop is a storage chamber. The form and function of the proventriculus correspond to the type of food. In insects that eat solid food, the proventriculus is a gizzard, and in sucking insects, it is a valve regulating passage of food into the midgut. A pair of salivary glands usually lies beneath the midgut and opens into the buccal cavity by a common duct. Salivary secretions contain digestive enzymes and a variety of other substances, such as anticoagulants in bloodsucking species. The midgut is the principal site of digestive and absorptive function. In many insects the midgut secretes a thin, chitinous layer, the **peritrophic membrane,** which encloses the food mass. The peritrophic

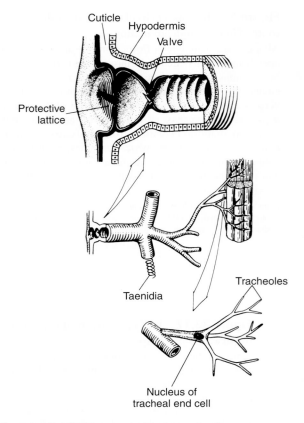

Figure 33.18 **Diagram of trachea of an insect.**
Tracheae are virtually impermeable to liquids, but the finely branching tracheoles, leading into tissues, are freely permeable, and their tips normally contain fluid. Oxygen primarily diffuses through the tracheolar walls, and elimination of carbon dioxide takes place more generally through the tracheal walls and body surface. Taenidia are chitinous bands that strengthen tracheae.

membrane is permeable to both enzymes and products of digestion, and it probably protects the midgut's delicate epithelial lining. Insects that live on liquid diets do not secrete a peritrophic membrane.

Gastric ceca, which increase the absorptive area, are found near the anterior end of the midgut of most insects. The hindgut, divided into **intestine** and **rectum,** functions not only in the elimination of wastes but also in regulation of water and ions.

In Acari the mouth leads into a muscular, sucking **pharynx,** which lies partly in the buccal cone. A slender esophagus proceeds posteriorly through the brain to the stomach, or **ventriculus.** A large pair of **salivary glands** lies above the ventriculus and esophagus and opens by means of ducts into the **salivarium** in the buccal cone, over the labrum (Fig. 33.20). In bloodsucking forms, the salivary secretions contain anticoagulants and histolytic components. The ventriculus has up to five pairs of ceca, which contain secretory and absorptive cells. The hindgut may be a short tube leading from the midgut to the anus, or an enlarged portion, the **rectal sac,** may precede the anus. In ticks, chiggers, water mites, feather mites, and many parasitic forms, the ventriculus has lost its connection with the midgut and ends blindly. From some of these acarines the indigestible food residues are removed by a remarkable process called **schizeckenosy.** The residues are stored in gut cells that detach from the epithelial lining and move into the posterodorsal gut lobes. When one of the lobes fills with waste-laden cells, it breaks free from the ventriculus and is extruded through a split in the dorsal cuticle.

Excretory System

Crustacean excretory organs are pairs of **antennal** and **maxillary glands** opening to the outside on or near the bases of antennae or maxillae, respectively. Both pairs are often present in larvae; adults normally retain only one or the other. The principal nitrogenous excretory products are ammonia with some amines and small amounts of urea and uric acid. Considerable excretion of ammonia also takes place across the gills.

Almost all insects have **Malpighian tubules,** ranging in number from 4 to over 100 (see Fig. 33.19). These thin-walled tubules are closed at their distal ends but open into the midgut near its junction with the hindgut. Uric acid is excreted, usually as an ammonium, potassium, or sodium salt. Water in the urine is reabsorbed by the proximal Malpighian tubules

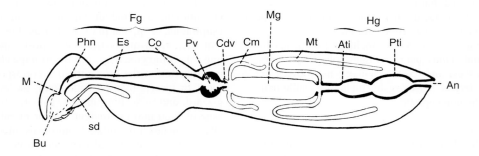

Figure 33.19 **Diagram of the digestive system of an insect.**
An, anus; *Ati,* anterior intestine; *Bu,* buccal cavity; *Cdv,* cardia valve; *Cm,* cecum; *Co,* crop; *Es,* esophagus; *Fg,* foregut; *Hg,* hindgut; *M,* mouth; *Mg,* midgut; *Mt,* Malpighian tubules; *Phn,* pharynx; *Pti,* posterior intestine; *Pv,* proventriculus; *sd,* salivary duct.

From R. M. Fox and J. W. Fox, *Introduction to comparative entomology.* New York: Reinhold Publishing Co., 1966.

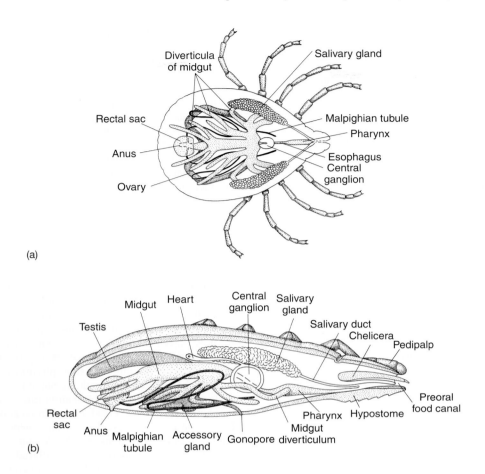

Figure 33.20 Internal anatomy of a hard tick.
(*a*) Dorsal view of female; (*b*) lateral view of male.

Drawings by William Ober.

or by the rectal wall; sodium and potassium are resorbed as bicarbonates, leaving virtually insoluble free uric acid as a precipitate. Thus, water and cations are recycled, as part of the overall water conservation mechanism of insects. Bloodsucking forms, however, produce large amounts of fluid urine after a meal, an event that rids the animal of excess water.

Excretory **coxal glands** are found in some mites and other arachnids. These glands open to the outside at the bases of one or more pairs of appendages. Most ticks and mites also have Malpighian tubules (see Fig. 33.20). Waste from the hemocoel is taken up by tubule walls and excreted into the lumen as guanine, the main excretory product. In those Prostigmata and Metastigmata whose ventriculus does not connect with the hindgut, an anteriorly directed excretory canal is joined to the hindgut, and guanine is excreted by this organ through the "anus" (uropore).

Reproductive Systems

Most Crustacea are dioecious; their gonopores open on a sternite or at the base of a trunk appendage. Males may have a penis, or appendages may be modified for copulation. Many crustaceans have nonflagellated, nonmobile sperm. In some groups the male places a packet of sperm (**spermatophore**) in a seminal receptacle or on the female's body surface. Many Crustacea retain fertilized eggs during embryonation, either in a brood chamber, attached to certain appendages, or within a sac formed during extrusion of the eggs.

Insects have a pair of testes. Vasa deferentia lead to a common, median **ejaculatory duct** that opens to the outside by the **aedeagus** (Fig. 33.21). **Accessory glands** join this ejaculatory duct and in many cases provide material comprising the spermatophores. In females the paired ovaries are subdivided into **ovarioles** (see Fig. 33.21). Each ovary usually has four to eight ovarioles, but some insects have more than 200; viviparous Diptera, such as tsetse flies, however, have only one. The upper end of an ovariole produces oocytes and nurse cells. Developing oocytes become larger by accumulation of yolk produced by nurse cells and surrounding follicular cells. The **common oviduct** enlarges into a vagina, which opens to the exterior behind the eighth or ninth abdominal sternite. The **seminal receptacle** connects to the oviduct or vagina by a slender **spermathecal duct.** Accessory glands (**colleterial glands**) also open into the common oviduct or vagina, and these glands may produce a substance that cements eggs together or to a substrate when they are laid or when they produce material for an egg capsule (**ootheca**).

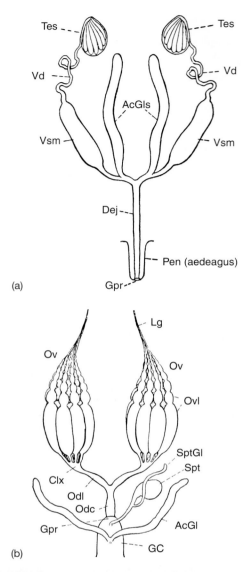

Figure 33.21   **General structure of insect reproductive organs.**

(*a*) Male: *AcGls,* accessory glands; *Dej,* ejaculatory duct; *Pen,* penis, or aedeagus; *Gpr,* gonopore; *Tes,* testis; *Vd,* vasa deferentia; *Vsm,* seminal vesicle. (*b*) Female: *Lg,* ovarian ligament; *Ov,* ovary; *Ovl,* ovariole; *Clx,* calyx; *Odl,* lateral oviduct; *Odc,* common oviduct; *Gpr,* gonopore; *GC,* genital chamber; *AcGl,* accessory gland; *Spt,* spermatheca; *SptGl,* spermathecal gland.

From R. E. Snodgrass, *Principles of insect morphology.* Copyright © 1993 by Cornell University. Used by permission of Cornell University Press.

ARTHROPOD PHYLOGENY

As might be expected from arthropods' long evolutionary history and extreme diversity, establishment of evolutionary relationships, especially among the more inclusive taxa, is a challenge that has occupied many scientists ever since Darwin's time. Exceptionally detailed comparative anatomical research, especially that focused on homology between body segments and formation of the head through segment fusion, was the standard approach through much of this period.[4, 21, 29] Embryological development also has been an important part of phylogenetic research on arthropods ever since Darwin claimed, based on nauplius structure, that barnacles were crustaceans rather than mollusks.[13] In recent years, such work has been complemented by molecular techniques, especially those allowing scientists to study the role of homeobox (Hox) genes. Consequently, our ideas about who is most closely related to whom have undergone, and may continue to undergo, rather substantial adjustment.

Traditionally, arthropods have been included in a single phylum of metameric, coelomate animals. Arthropods share many features with annelids, such as metamerism and a nervous system consisting of supraesophageal ganglia, nerves encircling the esophagus, and a ventral series of segmental ganglia. Such similarities led to claims that the two phyla are related and that arthropods likely evolved from annelidlike ancestors, but this idea is not supported by current research.[7, 9, 12] Annelids, molluscs, and several minor phyla, all sharing similar larval types and protostome development, are now placed in superphylum Lophotrochozoa. Based on 18S ribosomal RNA sequence data, phylum Arthropoda is considered a member of superphylum Ecdysozoa, i.e., animals that molt, along with Nematoda and other smaller phyla such as Nematomorpha (hairworms; chapter 31), Onychophora (wormlike tropical and subtropical organisms), and Tardigrada (water bears). Recent phylogenies using "nearly complete 28S + 18S ribosomal RNA gene sequences" show Arthropoda, Onychophora, and Tardigrada as a monophyletic clade with the other ecdysozoans comprising the sister group.[28]

Within Arthropoda, relationships are now proposed based on a combination of embryological, morphological, and molecular evidence to establish homologies between body parts, especially head segments. Although some authors use molecular data to argue that the mandibulate Myriapoda (centipedes and millipedes) are a sister group to the non-mandibulate Chelicerata (horseshoe crabs, ticks, mites, etc.), combining them into a taxon called Paradoxopoda or Myriochelata,[28] we accept the evidence as outlined by Scholtz and Edgecombe[42] that Mandibulata, which includes Hexapoda (insects), Crustacea, and Myriapoda, is monophyletic. Crustacea and Hexapoda, somtimes grouped in a clade named Pancrustacea, are now considered sister taxa within Mandibulata.[26]

Embryological evidence, especially that involving brain development and eye structure, suggests Hexapoda actually are most closely related to Malacostraca (p. 530) within Crustacea,[14] but 28S + 18S ribosomal RNA sequence data do not necessarily support this relationship.[28] Molecular sequence data support the derivation of insects from crustaceans.[20]

The position of myriapods remains unresolved, partly becausse they have not been studied to the same extent as insects and crustaceans.[14] Finally, Hexapoda does not include some six-legged arthropods, namely, members of classes Collembola (the exceedingly abundant springtails), Protura, and Diplura; members of all these classes occur mainly in or on damp soil or litter.

CLASSIFICATION OF ARTHROPODAN TAXA WITH SYMBIOTIC MEMBERS

This classification of Crustacea relies heavily on Kabata,[24] Marcotte,[30] Bowman and Abele,[10] and Martin and Davis.[31] Classification of Arachnida is according to Savory,[41] and diagnoses of the orders of pterygotes mainly follow Borrer et al.,[8] Gillott,[19] and Richards and Davies.[38] Subphylum Uniramia as traditionally constituted may not be a valid or monophyletic taxon.

CRUSTACEA

Head appendages consisting of two pairs of antennae, one pair of mandibles, and two pairs of maxillae; mostly aquatic; respiration usually with gills, sometimes through general body surface; head usually not clearly defined from trunk; cephalothorax usually with dorsal carapace; appendages, except first antennae (antennules), primitively biramous; sexes usually separate; development primitively with nauplius stage.

Class Maxillopoda

Typically with five cephalic, six thoracic, and four abdominal somites plus a telson, but reductions common; no typical appendages on abdomen; naupliar eye (when present) of unique structure and referred to as *maxillopodan eye*.

Subclass Copepoda

Typically with elongated, segmented body consisting of head, thorax, and abdomen; thorax with seven somites, of which first and sometimes second are fused with head to form cephalothorax; thoracic appendages biramous, but maxillipeds and often fifth swimming legs uniramous; no appendages on abdomen except pair of rami on telson; no carapace; compound eyes absent, but median nauplius eye often present; gonopores on "genital segment," usually considered last somite of thorax; parasitic forms may not fit some or much of foregoing diagnosis and may be highly modified as adults and sometimes as juveniles.

Order Calanoida

No symbiotic members but included because of ecological importance; antennules very long with 16 to 26 articles; buccal cavity open; antennae, mandibles, and maxillules biramous; mandibles gnathostomous; maxillae and maxilliped uniramous; first thoracic legs biramous, multiarticular, with plumose setae for swimming; last thoracic leg uniramous, modified, or missing; heart present in many; large order of important marine and freshwater planktonic organisms, never symbiotic.

Order Monstrilloida

Nauplii free swimming and then saclike endoparasites of marine polychaetes, prosobranch gastropods, and occasionally echinoderms; adults planktonic, without antennae, mouthparts, or functional gut; adult thoracic legs biramous for swimming.

Order Siphonostomatoida

Adult segmentation often reduced or lost; antennules reduced or elongated and multiarticulated; antennules may end in single massive claws for attachment to host; labrum and labium prolonged into siphon or tube, sometimes with some fusion; mandibles enclosed in buccal siphon, uniramous; maxillules ancestrally biramous, modified or reduced in derived forms; maxillae subchelate or brachiform (like human arm) for attachment to host; maxilliped subchelate or absent, sometimes absent in female only; adult thoracic limbs may be normal swimming appendages in some, variously modified and reduced in majority; adults ectoparasitic or endoparasitic on freshwater and marine fishes and on various invertebrates. Representative families: Caligidae, Cecropidae, Dichelesthiidae, Lernaeopodidae, Pandaridae, Pennellidae, Sphyriidae.

Order Cyclopoida

Antennules short with 10 to 16 articles; buccal cavity open; antennae uniramous; mandibles and maxillules usually biramous; mandibles gnathostomous; free living planktonic and benthic; commensal and ectoparasitic. Families: Ascidicolidae, Enterocolidae, Lernaeidae, Notodelphyidae.

Order Poecilostomatoida

Adult segmentation often lost with copepodid metamorphosis; antennules often insignificant in size; buccal cavity slitlike; antennae often end in many small claws for attaching to host; mandibles with falcate (*falcatus* = sickle-shaped) gnathobase, rami missing; maxillules much reduced; maxillae reduced with denticulate inward-pointing claw or slender, armed grasping claws; maxillipeds subchelate in males, often missing in females; adult thoracic limbs variously modified and reduced; adults parasitic on mostly marine invertebrates and fishes. Representative families: Bomolochidae, Chondracanthidae, Clausiidae, Ergasilidae, Lichomolgidae, Philichthyidae, Sarcotacidae, Tuccidae.

Order Harpacticoida

Antennules short with fewer than 10 articles; buccal cavity open; antennae and mandibles biramous; mandibles gnathostomous; maxillules usually biramous; various degrees of fusion, reduction, and loss of rami in cephalic and thoracic appendages; heart absent; mostly free living, benthic, epibenthic, planktonic.

Subclass Tantulocarida

No recognizable cephalic appendages; solid median cephalic stylet; six free thoracic somites, each with pair of appendages, anterior five biramous; six abdominal somites; anterior five thoracic appendages with well-developed protopod and large endite arising from base of protopod; class of minute, copepodlike ectoparasites of other deep-sea benthic crustaceans; described in 1983,[11] examples: spp. of *Basipodella, Deoterthron*.

Subclass Branchiura

Body with head, thorax, and abdomen; head with flattened, bilobed, cephalic fold incompletely fused to first thoracic somite; thorax with four pairs of appendages, biramous, and with proximal extension of exopod of first and second legs; abdomen without appendages, unsegmented, bilobed; eyes compound; both pairs of antennae reduced; claws on antennules; maxillules often forming pair of suctorial discs; maxillae uniramous; gonopore at base of fourth leg; ectoparasites of marine and freshwater fishes and occasionally of amphibians.

Order Argulidea

Families: Argulidae, Dipteropeltidae.

Subclass Pentastomida

Two pairs of sclerotized hooks near mouth; mouth held permanently open by sclerotized cadre; body superficially annulated; continuous production of protective protein-phospholipid secretion; all parasitic, mostly in vertebrate lungs.

Order Cephalobaenida

Mouth anterior to hooks; hooks lacking fulcrum; vulva at anterior end of abdomen.

Family Cephalobaenidae

Parasites of snakes, lizards, and amphibians. Genera: *Cephalobaena, Raillietiella.*

Family Reighardiidae

Parasites of marine birds. Genus *Rieghardia.*

Order Porocephalida

Mouth between or below level of anterior hooks; hooks with fulcrum; vulva near posterior end of body.

Family Sebekidae

Parasites of crocodilians and chelonians. Genera: *Sebekia, Alofia, Leiperia.*

Family Porocephalidae

Parasites of snakes. Genera: *Porocephalus, Kiricephalus.*

Family Subtriquetridae

Parasites of crocodilians. Genus *Subtriquetra.*

Family Sambonidae

Parasites of monitor lizards and snakes. Genera: *Sambonia, Elenia, Waddycephalus, Parasambonia.*

Family Diesingidae

Parasites of chelonians. Genus *Diesingia.*

Family Armilliferidae

Parasites of snakes. Genera: *Armillifer, Cubirea, Gigliolella.*

Family Linguatulidae

Parasites of mammals. Genus *Linguatula.*

Subclass Cirripedia

Sessile or parasitic as adults; head reduced and abdomen rudimentary; paired, compound eyes absent; body segmentation indistinct; usually hermaphroditic; in nonsymbiotic and epizoic forms, carapace becoming mantle, which secretes calcareous plates; antennules becoming organs of attachment; antennae disappearing; young hatching as nauplius and developing to bivalved cypris larva; all marine.

Order Thoracica

With six pairs of thoracic appendages; alimentary canal; usually nonsymbiotic, although some epizoic and commensal on whales, fishes, sea turtles, and crabs; examples: spp. of *Chelonibia, Conchoderma, Coronula, Xenobalanus.*

Order Acrothoracica

Bore into mollusc shells or coral; females usually with four pairs of thoracic appendages; gut present, no abdomen; dioecious; males very small, without gut and appendages except antennules, parasitic on outside of mantle of female.

Order Ascothoracica

With segmented or unsegmented abdomen; usually six pairs of thoracic appendages; gut present; parasitic on echinoderms and soft corals; example: genus *Trypetesa.*

Order Rhizocephala

Adults with no segmentation, gut, or appendages; with root-like absorptive processes through tissue of host; common parasites of decapod crustaceans. Families: Lernaeodiscidae, Peltogastridae, Sacculinidae.

Class Ostracoda

Body entirely enclosed in bivalve carapace; body unsegmented or indistinctly segmented; no more than two pairs of trunk appendages; at least one species parasitic on gills of a shark. Family Cypridinidae.

Class Malacostraca

Distinctly segmented bodies, typically with eight somites in thorax and six somites plus telson in abdomen (except seven in Nebaliacea); all segments with appendages; antennules often biramous; first one to three thoracic appendages often maxillipeds; carapace covering head and part or all of thorax, a primitive character, but carapace lost in some orders; gills usually thoracic epipods; female gonopores on sixth thoracic segment; male gonopores on eighth thoracic segment; largest subclass marine and freshwater, few terrestrial; many free living, but parasitic members relatively few, found in only three of the 10 to 12 extant orders commonly recognized.

Superorder Peracarida

Without carapace or with carapace, leaving at least four free thoracic somites; first thoracic somite fused with head; brood pouch in female (typically formed from modified thoracic epipods, the oostegites); several small, marine orders; the two large orders have parasitic members.

Order Amphipoda

No carapace; ventral brood pouch of oostegites; antennules often biramous; eyes usually sessile; gills on thoracic coxae; first thoracic limbs maxillipeds, second and third pairs usually prehensile (gnathopods); usually bilaterally compressed body form; marine, freshwater, and terrestrial; free living and symbiotic.

Suborder Hyperiidea

Head and eyes very large; only one thoracic somite fused with head; pelagic or symbiotic in medusae or tunicates. Families: Hyperiidae, Phronimidae.

Suborder Caprellidea

So-called skeleton shrimp and whale lice; two thoracic somites fused with head; abdomen much reduced, with vestigial appendages. Families: Caprellidae, Cyamidae.

Order Isopoda

No carapace; ventral brood pouch of oostegites; antennules usually uniramous, sometimes vestigial; eyes sessile; gills on abdominal appendages; second and third appendages usually not prehensile; body usually dorsoventrally flattened.

Suborder Gnathiidea

Thorax much wider than abdomen; first and seventh thoracic somites reduced, seventh without appendages; larvae parasitic on marine fishes. Family Gnathiidae.

Suborder Flabellifera

Flattened body, with ventral coxal plates sometimes joined to body; telson fused with next abdominal somite, and other abdominal somites sometimes fused; uropods flattened, forming tail fan; marine, free living, and ectoparasitic on fishes. Families with parasitic members: Aegidae, Corallanidae, Cymothoidae.

Suborder Epicaridea

Females greatly modified for parasitism; somites and appendages fused, reduced, or absent; mouthparts modified for sucking and mandible modified for piercing; maxillae reduced or absent; males small but less modified; marine parasites of Crustacea. Families: Bopyridae, Cryptoniscidae, Dajidae, Entoniscidae, Phryxidae.

Superorder Eucarida

All thoracic segments fused with and covered by carapace; no oostegites or brood pouch; eyes on stalks; usually with zoea larval stage.

Order Decapoda

First three pairs of thoracic appendages modified to maxillipeds (therefore, appendages on remaining five thoracic somites equal 10 [Gr.: *deka* = ten + *podos* = foot]); includes crabs, lobsters, and shrimp.

Suborder Pleocyemata

Eggs carried by female and brooded on pleopods, hatch as zoeae.

Infraorder Brachyura

Carapace broad; abdomen reduced and tightly flexed beneath cephalothorax; first legs in form of heavy chelipeds; typical crabs. Families with symbiotic members: Parthenopidae, Pinnotheridae.

"UNIRAMIA" (Insects)

All appendages uniramous; head appendages consisting of one pair of antennae, one pair of mandibles, and one or two pairs of maxillae.

Class Collembola

Springtails; tiny, primitively wingless; compound eyes lacking; six-segmented abdomen with furcula; abundant in soil; some species inquilines in termite and ant colonies.

Class Protura

Primitively wingless; minute; blind, lacking antennae; anamorphosis leads to 12 segments in abdomen; none parasitic or commensal.

Class Diplura (Entotrophi)

Primitively wingless; mouthparts withdrawn into head; with two cerci; none parasitic or commensal.

Class Hexapoda

Body with distinct head, thorax, and abdomen; one pair of antennae; thorax of three somites; abdomen with variable number, usually 11, of somites; thorax usually with two pairs of wings (sometimes one pair or none) and three pairs of jointed legs; separate sexes; usually oviparous; gradual or abrupt metamorphosis, few with direct development.

Subclass Apterygota

Primitively wingless insects; development direct or through slight metamorphosis.

Order Thysanura

Silverfish; flattened and elongate with three posterior filamentous appendages, body covered with scales, some occurring in termite colonies or ant nests.

Subclass Pterygota

Insects with wings (some secondarily wingless); all metamorphic; includes 97% of all insects; although members of all orders serve as hosts, what follow are only those with some medical or veterinary importance, in addition to orders that have appreciable numbers of symbiotic members.

Order Dermaptera

Earwigs; forewings represented by small tegmina; hindwings large, membranous, and complexly folded; mouthparts for biting; ligula bilobed; body terminated by forceps; few ectoparasites of mammals (*Arixenia, Hemimerus* spp.); some intermediate hosts of nematodes.

Order Dictyoptera

Cockroaches and mantids; antennae nearly always filiform with many segments; mouthparts for biting; legs similar to each other or forelegs raptorial; tarsi with five segments; forewings more or less thickened into tegmina with marginal costal vein; many cerci segmented; ovipositor reduced and concealed; eggs contained in an ootheca; none symbiotic but some implicated in mechanical transmission of human pathogens; some intermediate hosts of Acanthocephala; examples: spp. of *Blatta, Blatella, Periplaneta, Supella*.

Order Phthiraptera

Lice; wingless; metamorphosis slight; cerci absent; ectoparasitic on birds or mammals in all stages.

Suborder Amblycera

Chewing lice (in part); antennae club-shaped, partially to entirely beneath head; with maxillary palps; meso- and metathorax distinctly separate; parasitic on birds and mammals; examples: spp. of *Aotiella, Menacanthus, Menopon, Pseudomenopon*.

Suborder Ischnocera

Chewing lice (in part); antennae filiform and exposed; maxillary palps lacking; meso- and metathorax fused; parasitic on birds and mammals; examples: spp. of *Anaticola, Bovicola, Columbicola, Felicola, Philopterus, Trichodectes.*

Suborder Rhynchophthirina

Chewing lice (in part); head prolonged into snout with mandibles at tip; parasitic on elephants and some African pig species; single genus, *Haematomyzus.*

Suborder Anoplura

Sucking lice; head narrower than prothorax; mouthparts modified for piercing and sucking; retracted when not in use; thoracic segments fused; claws single; tarsi unisegmented; all stages ectoparasitic on mammals; examples: spp. of *Haematopinus, Pediculus, Phthirus.*

Order Hemiptera

True bugs, aphids, scale insects, etc.; wings variably developed with reduced or greatly reduced venation; forewings often more or less corneous; wingless forms frequent; mouthparts for piercing and sucking with mandibles and maxillae styletlike and lying in the projecting grooved labium, palps never evident; metamorphosis gradual with an incipient pupal instar sometimes present; many free living, some ectoparasites of birds and mammals; examples: spp. of *Cimex, Leptocimex, Rhodnius, Triatoma.*

Order Neuroptera

Alder flies, lacewings, ant lions, etc.; small to large, soft-bodied insects with two pairs of membranous wings without anal lobes; venation generally with many accessory branches and numerous costal veinlets; mouthparts for biting; antennae well developed; cerci absent; complete metamorphosis; campodeiform larvae with biting or suctorial mouthparts; few parasites of freshwater sponges and of spiders' egg cocoons. Families: Mantispidae, Sisyridae (*Climacia, Sisyra* spp.).

Order Coleoptera

Beetles; minute to large insects whose forewings are modified to form elytra and abut down line of dorsum; hindwings membranous, folded beneath elytra, or absent; prothorax large; mouthparts for biting; metamorphosis complete; larvae of diverse types but never typically polypod; largest order of animals (more than 330,000 species); 1.5% protelean parasites (immature stages parasitic) of insects; few ectosymbionts of mammals. Families: Leptinidae, Meloidae (some), Platypsyllidae, Rhipiphoridae, Staphylinidae (some).

Order Strepsiptera

Minute; males with branched antennae and degenerated biting mouthparts; forewings modified into small clublike processes; hindwings very large, plicately folded; females almost always extensively modified as internal parasites of other insects; larviform and devoid of wings, legs, eyes, and antennae; all protelean parasites of insects; examples: spp. of *Corioxenos, Elenchus, Eoxenos, Stylops.*

Order Siphonaptera

Fleas; very small; wingless; laterally compressed body; mouthparts for piercing and sucking; complete metamorphosis with vermiform larvae; pupation in silk cocoons; adults all parasitic on warm-blooded animals; examples: spp. of *Pulex, Ctenocephalides, Xenopsylla, Tunga.*

Order Diptera

Flies and mosquitoes; moderate size to very small; single pair of membranous wings (forewings), hindwings modified into halteres; mouthparts for sucking or for piercing and usually forming a proboscis; complete metamorphosis with vermiform larvae; many species of invertebrates and vertebrate protelean parasites; vertebrate and insect ectoparasites; examples: spp. of *Aedes, Anopheles, Bombylius, Chrysops, Conops, Culex, Glossina, Hippobosca, Melophagus, Phlebotomus, Simulium, Stomoxys, Stylogaster, Tabanus.*

Order Lepidoptera

Butterflies and moths; small to very large insects clothed with scales; mouthparts with galeae usually modified into a spirally coiled suctorial proboscis; mandibles rarely present; complete metamorphosis with larvae phytophagous, polypodous; large order with mostly free-living members; few insect protelean parasites and mammal ectoparasites; examples: spp. of *Bradypodicola, Calpe, Cyclotorna, Fulgoraecia.*

Order Hymenoptera

Sawflies, ants, bees, wasps, ichneumon flies, etc.; minute to moderate size; membranous wings, hindwings smaller and connected with forewings by hooklets, venation specialized by reduction; mouthparts for biting and licking; abdomen with first segment fused with thorax; sawing or piercing ovipositor present; complete metamorphosis with usually polypodous or apodous larvae; enormous insect order, about half of which are protelean parasites, mainly of other insects. Superfamilies: Bethyloidea, Chalcidoidea (many), Cynipoidea (some), Evanioidea, Ichneumonoidea, Orussoidea, Proctorupoidea (Serphoidea), Trigonaloidea, Vespoidea (some).

SUBPHYLUM CHELICERATA

Mostly terrestrial; respiration by gills, book lungs, or tracheae or through general body surface; first pair of appendages modified to form chelicerae; pair of pedipalps and usually four pairs of legs in adults; no antennae; tagmatization of prosoma (cephalothorax) and opisthosoma (abdomen), usually unsegmented.

Class Arachnida

Adult body fundamentally composed of 18 somites, divisible into 6-unit prosoma and 12-unit opisthosoma, but segmentation often obscured in either or both of these tagmata; eyes, if present, simple (ocelli), not more than 12; chelicerae of two or three podomeres, either chelate or unchelate; pedipalps of six podomeres, either chelate or leglike, often with gnathobases; respiration through general body surface or by book lungs or tracheae (or both); sexes separate, with orifices on lower side of second opisthosomatic somite. Alternate ordinal names (or subordinal, depending on author) used in the literature also follow.

Subclass Latigastra

Prosoma and opisthocoma joined across their whole breadth. (Some authors include Acari in this group.)[41]

Order Pseudoscorpiones

Prosoma undivided; opisthosoma with 12 distinguishable somites; chelicerae of two articles, chelate; pedipalps large, with six articles, chelate; no pedicel; no telson; several pseudoscorpions symbiotic on mammals; prey on ectoparasites (lice, mites); examples: spp. of *Lasiochernes, Megachernes, Chiridiochernes.*

Other orders: Opiliones, Scorpiones

Subclass Acari

Highly specialized arachnids, in which modifications of segmentation divide body into proterosoma and hysterosoma, usually distinguishable as boundary between second and third pairs of legs; segments of mouth and its appendages borne on gnathosoma (capitulum), more or less sharply set off from rest of body (idiosoma); typically four pairs of legs but sometimes three, two, or one pair; often six podomeres in legs but may vary from two to seven; position of respiratory and genital openings variable; includes free-living suborders Notostigmata, Tetrastigmata.

Order Ixodida (Metastigmata)

Large acarines (ticks); hypostome with recurved teeth, used as holdfast organ; sensory organ (Haller's organ) on tarsus of first leg with olfactory and hygroreceptor setae; single pair of spiracular openings close to coxae of fourth legs except in larvae; all parasitic.

Family Ixodidae

Capitulum terminal; anterodorsal sclerite (scutum) present; pedipalps rigid. Important genera: *Amblyomma, Boophilus, Dermacentor, Ixodes, Rhipicephalus.*

Family Argasidae

Capitulum subterminal; scutum absent; pedipalps leglike and articulating. Important genera: *Argas, Ornithodoros, Otobius.*

Order Mesostigmata (Gamasida)

Several sclerotized plates on dorsal and ventral surfaces; single pair of spiracular openings between second and fourth coxae; large group, many free living; parasitic examples: spp. of *Dermanyssus, Ornithonyssus, Sternostoma.*

Order Prostigmata (Trombidiformes)

Spiracular openings, when present, paired and located either between chelicerae or on dorsum of anterior portion of hysterosoma; usually weakly sclerotized; chelicerae vary from strongly chelate to reduced; pedipalps simple, fanglike, or clawed; terrestrial and aquatic, free-living, phytophagous, and parasitic forms; examples: spp. of *Demodex, Trombicula.*

Order Oribatida (Cryptostigmata)

Oribatid or beetle mites (so called because of superficial resemblance to beetles); spiracles absent, although some with trachea associated with paired dorsal pseudostigmata and with bases of first and third legs; free living, but some (*Galumna, Oppia* spp.) are vectors of tapeworms.

Order Astigmata (Sarcoptiformes)

Mostly slow moving and weakly sclerotized; no spiracles; respire through body surface; free-living and parasitic forms; examples: spp. of *Megninia, Otodectes, Sarcoptes.*

Other subclass Caulogastra

Orders: Palpigradi, Uropygi, Schizomida, Amblypygi, Araneida, Solfugae, Ricinulei.[41]

Learning Outcomes

By the time a student has finished studying this chapter, he or she should be able to:

1. Describe features of arthropod segmentation and metamerism.
2. Distinguish between quinone tanning and beta-sclerotization.
3. Tell what chitin is.
4. Explain why molting is necessary in arthropods.
5. Identify the stages of development in crustaceans, insects, and ticks and mites.
6. Explain the importance of diapause.
7. List the head appendages of crustaceans compared with those of insects.
8. Tell an important morphological difference between most body appendages of crustaceans compared with those of insects.
9. List important reasons why arthropods have been so successful.
10. Explain the endocrine control of development in insects.
11. Explain diapause and its importance in arthropods.
12. Name the main body regions in insects, crustaceans, and Acari.
13. Distinguish Crustacea, Uniramia, and Chelicerata.

References

References for superscripts in the text can be found at the following Internet site: www.mhhe.com/robertsjanovynadler9e

Additional Readings

Berenbaum, M. R. 1995. *Bugs in the system. Insects and their impact on human affairs.* Reading, MA: Addison-Wesley Publishing Co. An enjoyable account of the many ways in which insects are important to humans.

Brusca, R. C., and G. J. Brusca. 2003. *Invertebrates.* 2d ed. Sunderland, MA: Sinauer Associates, Inc., Publishers.

Burgess, N. R. H., and G. O. Cowan. 1993. *A colour atlas of medical entomology.* New York: Chapman and Hall.

Busvine, J. R. 1975. *Arthropod vectors of disease.* London: Edward Arnold.

Cloudsley-Thompson, J. L. 1976. *Insects and history.* London: Weidenfield and Nicolson.

Goddard, J. 1996. *Physician's guide to arthropods of medical importance,* 2d ed. Boca Raton, FL: CRC Press.

Gupta, A. P. (Ed.). 1990. *Morphogenetic hormones of arthropods: Discoveries, synthesis, metabolism, evolution, modes of action, and techniques.* New Brunswick, NJ: Rutgers University Press.

Harwood, R. F., and M. T. James. 1979. *Entomology in human and animal health,* 7th ed. New York: Macmillan Publishing Co., Inc. One of the best general texts available in medical entomology.

Service, M. W. 1986. *Blood-sucking insects: Vectors of disease.* London: Edward Arnold.

Snodgrass, R. E. 1935. *Principles of insect morphology.* New York: McGraw-Hill Book Co. This and the following are classics that remain valuable references on arthropod structure.

Snodgrass, R. E. 1965. *A textbook of arthropod anatomy.* (Facsimile of the 1952 edition.) New York: Hafner Publishing Co.

Strickland, G. T. (Ed.). 2000. *Hunter's tropical medicine and emerging infectious diseases,* 8th ed. Philadelphia: W. B. Saunders Co.

U.S. Department of Health, Education, and Welfare. 1960 (1979 revision). *Introduction to arthropods of public health importance.* HEW Pub. No. (CDC) 79-8139. Washington, DC: U.S. Government Printing Office. A short, concise introduction to the subject, with a key to some common classes and orders of public health importance.

Wigglesworth, V. B. 1972. *Principles of insect physiology,* 7th ed. London: Chapman & Hall Ltd.

Chapter 34

Parasitic Crustaceans

Morphological details of small animals are often ignored by observers just because they are small. . . . [I] venture to suggest that, had the copepod been the size of a cow, the tip of its first antenna would have become a topic for exhaustive studies. One tends to forget that the dimensional scale does not influence the biological importance.

—Z. Kabata[39]

A fascinating array of adaptations for symbiosis can be found among crustaceans. In addition to being of academic interest, many parasitic crustaceans are of substantial economic importance. Nonetheless, they are often neglected in both parasitology and invertebrate zoology courses.

Some crustacean parasites have been known since antiquity, although their identity as crustaceans or even arthropods was not recognized until the early 19th century. Aristotle and Pliny recorded the affliction of tunny and swordfish by large parasites we would now recognize as penellid copepods. In 1554 Rondelet[71] drew a figure of a tunny with a copepod in place near the pectoral fin. In 1746 Linnaeus first established genus *Lernaea*,[46] and in his 1758 edition of *Systema Naturae,* he called the species (from European carp) *Lernaea cyprinacea.*[45] Various other highly modified copepods were described in the latter half of the 18th and early 19th centuries, but they had so few obvious arthropod features that they were variously classified as worms, gastropod molluscs, cephalopod molluscs, and annelids. Finally, Oken[57] associated these animals with other parasitic copepods that could be recognized as such. Based on Surriray's important observation that their young resemble those of *Cyclops,* de Blainville[9] firmly established these animals as crustaceans. Copepoda is not the only crustacean group with members so modified for parasitism as to be superficially unrecognizable as arthropods; we will discuss some of these other groups as well.

The status of higher taxa in Crustacea, even the status of the taxon "Crustacea" itself, continues in a state of flux. Here we will accord Crustacea the rank of subphylum and follow the classification of Martin and Davis.[49]

CLASS MAXILLOPODA

Class Maxillopoda includes a number of crustacean groups traditionally considered classes unto themselves. There is evidence that these groups descended from a common ancestor and thus form a clade within Crustacea. They basically have five cephalic, six thoracic, and usually four abdominal somites plus a telson, but reductions are common. There are no typical appendages on the abdomen. The eye of nauplii (when present) has a unique structure and is referred to as a **maxillopodan eye.**

Members of subclass Pentastomida were long considered a separate phylum. They possess none of the foregoing characteristics, but other evidence supports their placement here.[49] We will discuss pentastomes in a separate chapter (chapter 35).

Subclass Copepoda

Copepods are extremely important both as free-living organisms and as parasites. They display enormous evolutionary versatility in exploitation of symbiotic niches and in their spectrum of adaptations to symbiosis, ranging from the slight to the extreme. Although penellids are so bizarre that 18th-century biologists did not recognize them as arthropods, many parasitic and commensal copepods are comparatively much less highly modified. In fact, one can arrange examples of the various groups in an arbitrary series to demonstrate the diversity from little to very high specialization.[38]

We shall cite but a few examples to illustrate trends in adaptation to parasitism in copepods. Some of these trends are (1) reduction in locomotor appendages; (2) development of adaptations for adhesion, both by modification of appendages and by development of new structures; (3) increase in size and change in body proportions, resulting from disproportionate growth of genital or reproductive regions; (4) fusion of body somites and loss of external evidence of segmentation; (5) reduction of sense organs; and (6) reduction in numbers of instars that are free living, both through passing of more stages before hatching and through larval instars becoming parasitic. "Typical" or primitive copepod development is gradual metamorphosis with a series of **copepodid** (subadult) instars succeeding the naupliar instars. Copepodid

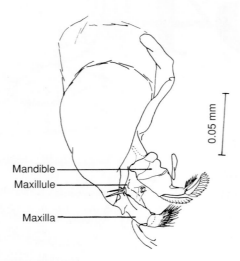

Figure 34.1 **Poecilostome mouthparts (one side drawn) of *Ergasilus cerastes,* a parasite of catfish (*Ictalurus* spp.).**

From L. S. Roberts, "*Ergasilus cerastes* sp. n. (Copepoda: Cyclopoida) from North American catfishes," in *J. Parasitol.* 55:1266–1270. Copyright © 1969 *Journal of Parasitology.* Reprinted by permission.

juveniles bear considerable similarity to adults except in dramatically metamorphic families such as Lernaeopodidae and Pennellidae.

The taxonomy of Copepoda at ordinal level is based partly on the morphology of mouthparts:[39] **gnathostome, poecilostome,** and **siphonostome.** Gnathostomous mandibles are fairly short, broad, biting structures with teeth at their ends, and buccal cavities are large and widely open. This is apparently the ancestral condition and is possessed by several copepod orders. Poecilostome mouths are rather similar, except that they are somewhat slitlike and have falcate (sickleshaped) mandibles (Fig. 34.1). The siphonostome condition is characterized by a more or less elongated, conical, siphonlike mouth formed by the labrum and labium (Fig. 34.2*a*), with mandibles that are styletlike and enclosed within the siphon (see Fig. 34.2*b*). Possession of poecilostome and of siphonostome mouths forms the basis for recognition of the orders Poecilostomatoida and Siphonostomatoida, respectively. According to Ho,[29] these two orders are sister groups.

Order Cyclopoida

Order Cyclopoida is a large group of copepods, most species of which are free living. Free-living cyclopoids occupy important niches as primary consumers in many aquatic habitats, particularly fresh water. Ho's[30] cladistic analysis suggested that parasitism has arisen twice in the evolution of cyclopoids. One line gave rise to two families parasitic in ascidian tunicates and a family whose members inhabit the mantle cavity of marine bivalve molluscs. The other line split into two groups, one of which produced yet another family of ascidian parasites and the other of which invaded fresh water. Descendants of the latter group produced a family that lives in blood of a freshwater snail and family Lernaeidae, an important and highly specialized family of fish parasites.

Family Lernaeidae. This is a relatively small family that parasitizes freshwater teleosts (bony fishes). Often quite

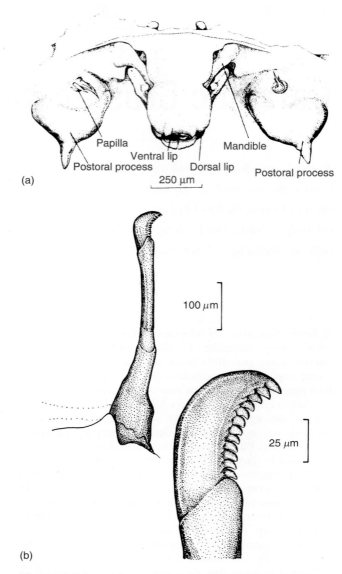

Figure 34.2 **Siphonostome mouthparts of *Caligus curtus,* which parasitizes a variety of marine fishes.**
(*a*) The base of the mandible can be seen as it extends into a tube formed by dorsal and ventral lips. (*b, c*) The mandible is a long, flat blade with teeth at its distal tip.

From R. R. Parker et al., "A review and description of *Caligus curtus* Miller, 1785 (Caligidae: Copepoda), type species of its genus," in *J. Fish. Res. B. Can.* 25:1923–1969. Copyright © 1968 Journal Fisheries Research Board of Canada. Reprinted with permission.

large in size and conspicuous, some species, especially *Lernaea cyprinacea,* are serious pests of economically important fishes. Therefore, they are among the best known parasitic copepods. *Lernaea cyprinacea* can infect a variety of fish hosts and even frog tadpoles.[78] The anterior of the parasite is embedded in its host's flesh and is anchored there by large processes that arise from the parasite's cephalothorax and thorax—hence, the common name *anchor worm* (Fig. 34.3). It causes damage to scales, skin, and underlying muscle tissue. There may be considerable inflammation, ulceration, and secondary bacterial and fungous infection.[8] Fishes that

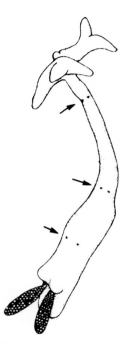

Figure 34.3 *Lernaea cyprinacea,* the "anchor worm," is a serious pest of a variety of fishes, including several of economic importance.

The anterior holdfast ("horns") is embedded in the host's flesh, and the posterior part of the body projects to the exterior. The swimming legs *(arrows)* do not participate in the rapid, final growth of the adult female (up to 16 mm long) and so remain proportionately very tiny.

From Z. Kabata, "Crustacea as enemies of fishes," in S. F. Snieszko and H. R. Axelrod, eds., *Diseases of fish,* book 1. Neptune City, NJ: T. F. H. Publications, Inc., 1970.

are small relative to the parasite can easily be killed by infection with several individuals. A fully developed *L. cyprinacea* may be more than 12 mm long. Epizootics of this pest occur in wild fish populations, and it is a serious threat wherever fishes are reared in hatcheries.[65]

Lernaeids are among the most highly specialized copepods. Once a sexually mature female is fertilized, she embeds her anterior end beneath a scale, near a fin base or in the buccal cavity. At that point the parasite is less than 1.5 mm long and is superficially quite similar to *Cyclops* spp. or other unspecialized cyclopoids. The female begins to grow rapidly, reaching "normal" size in little more than a week. The largest specimen recorded was 15.9 mm (22.0 mm including the egg sacs).[26] Interestingly, swimming legs and mouthparts remain in place but do not take part in this growth, so they quickly become inconspicuous. At the same time, the large anchoring processes, two ventral and two dorsal (see Fig. 34.3), grow into the fish's muscle. Body segmentation becomes blurred, being recognized only by location of the tiny legs. The result is an embryo-producing machine that bears practically no resemblance to an arthropod and that has its head permanently anchored in its food source. It is little wonder that early taxonomists had such trouble correctly placing *L. cyprinacea* in their system.

Nevertheless, larvae can be recognized clearly as crustacean and are typical nauplii. The primitive series of naupliar instars has been shortened to three. When nauplii hatch, they contain enough yolk material within their bodies to eliminate the need for feeding in any of the three naupliar stages. The third nauplius molts to give rise to the first copepodid, and this marks the end of the free-living life of *L. cyprinacea.* Thus, the length of time spent as a free-living organism has been markedly shortened, compared with the primitive condition, and the free-living instars do not even feed.

Order Poecilostomatoida

Members of this order range from little specialized parasites (Ergasilidae) to some highly modified and bizarre forms (Philichthyidae, Sarcotacidae). Poecilostomes have been successful as symbionts of several phyla, particularly Cnidaria. Of 1475 species of symbiotic copepods known from invertebrates, 416 belong to this order, and 373 species are associated with cnidarians.[32]

Family Ergasilidae. Ergasilids are among the most common copepod parasites of fishes. They have been a "thorn in the flesh for many valuable fisheries in the Old World" for a long time[38] and often frequent the gills of a variety of fishes in North America.[69] *Ergasilus* spp. are primarily parasites of freshwater hosts but are common on several marine fishes.

Ergasilidae show primitive morphological characteristics reminiscent of free-living copepods, with few but effective adaptations for parasitism. Their antennules are sensory, but the antennae have become modified into powerful organs of prehension (Fig. 34.4). *Ergasilus* spp. females usually are found clinging by their antennae to one of the fish's gill filaments. Each antenna ends in a sharp claw. The third segment and claw are opposable with the second (subchelate) (Fig. 34.5). Rather than depending on muscle and heavy sclerotization of antennae, the antennal tips may be fused or locked so that a gill filament is completely encircled *(E. amplectens, E. tenax)* (Fig. 34.6).

When removed from their position on a gill, most *Ergasilus* spp. can swim reasonably well; their pereiopods retain the flat copepod form, with setae and hairs well adapted for swimming. Their first legs, however, show adaptation for their feeding habit. These appendages are supplied with heavy, bladelike spines; in some species the second and third endopodal segments are fused, presumably lending greater rigidity to the leg. Such modifications increase the animal's ability to rasp off mucus and tissue from the gill to which it is clinging (Fig. 34.7). The first legs dislodge epithelial and underlying cells in this manner and sweep them forward to the mouth[20] (Fig. 34.8). It is easy to see that a heavy infestation with *Ergasilus* spp. could severely damage gill tissue, interfere with respiration, open the way to secondary infection, and lead to death. Epizootics of *Ergasilus* spp. on mullet *(Mugil* spp.) were recorded in Israel; in one case up to 50% of the stock in some ponds was lost, and hundreds of dead mullet were found daily.[72] Rogers and Hawke[70] found large numbers of *E. clupeidarum* infesting skin lesions of gizzard shad *(Dorosoma cepedianum)* in Tennessee; they believed the copepods were the primary cause of the moribund condition of the fish.

Ergasilus spp. have three naupliar and five copepodid stages, all free living.[80] Adult males are planktonic as well, and

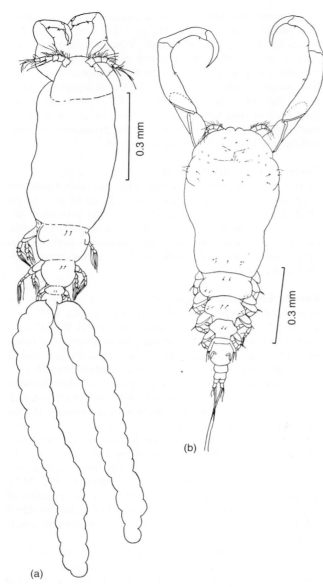

Figure 34.5 **Antenna of *E. centrarchidarum*, a common parasite of members of sunfish family Centrarchidae.**

Antennae of *Ergasilus* are usually modified into a powerful organ used to grasp their host's gill filament, with the third and fourth joints opposable with the second.

From L. S. Roberts, "*Ergasilus* (Copepoda: Cyclopoida): Revision and key to species in North America," in *Trans. Am. Microsc. Soc.* 89:134–161. Copyright © 1970. Reprinted by permission.

Figure 34.4 **Examples of *Ergasilus* spp., a common parasite of freshwater and some marine fishes.**

(*a*) *Ergasilus celestis,* from eels (*Anguilla rostrata*) and burbot (*Lota lota*), bearing egg sacs. (*b*) *Ergasilus arthrosis,* reported from several species of freshwater hosts; it is nonovigerous.

From L. S. Roberts, "*Ergasilus arthrosis* n. sp. (Copepoda: Cyclopoida) and the taxonomic status of *Ergasilus versicolor* Wilson, 1911, *Ergasilus elegans* Wilson, 1916, and *Ergasilus celestis* Mueller, 1936, from North American fishes," in *J. Fish. Res. B. Can.* 26:997–1011. Copyright © 1969 Journal of the Fisheries Research Board of Canada. Reprinted by permission.

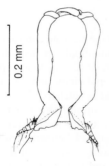

Figure 34.6 **Tips of antennae of *Ergasilus tenax* "lock" together, completely encircling the host's gill filament.**

From L. S. Roberts, "*Ergasilus tenax* sp. n. (Copepoda: Cyclopoda) from white crappie, *Pomoxis annularis* Rafinesque," in *J. Parasitol.* 51:987–989. Copyright © 1965 *Journal of Parasitology*. Reprinted by permission.

females are fertilized before attaching to a fish host. Only adult females have been found as parasites. In one species even females are planktonic as adults (*E. chautauquaensis,* which may be the only nonparasitic species in the genus), although females of several other species are sometimes encountered in plankton.[11]

Family Lichomolgidae. Lichomolgids are symbionts with a wide variety of marine animals, including serpulid polychaetes, alcyonarian and madreporarian corals, ascidians, sea anemones, nudibranchs, holothurians, starfish, bivalves, and sea urchins. It is evident that many species are involved, and many are yet to be discovered. Lichomolgidae (family here broadly accepted) were divided by Humes and Stock[33] into five families, embracing 76 genera and 324 species.

Lichomolgids are generally cyclopoid in body form, retaining segmentation and swimming legs (Fig. 34.9). Segments of the antennae are reduced to three or four, and they often end in one to three terminal claws. Antennae are

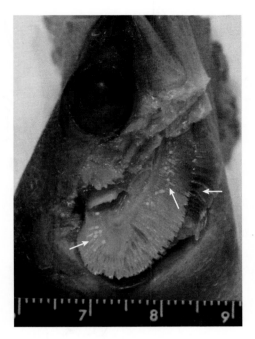

Figure 34.7   *Ergasilus labracis* **in situ on gills of striped bass.** *Morone saxatilis* **(two specimens are indicated by** *arrows***).**

The gill operculum has been removed. Note also that the fish is infected by an isopod, *Lironeca ovalis,* partly hidden under gill (*right arrow*).

Photograph by Larry S. Roberts.

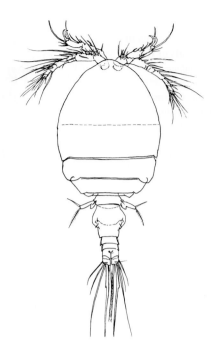

Figure 34.9   **Typical lichomolgid,** *Ascidioxynus jamaicensis,* **from the branchial sac of an ascidian,** *Ascidia atra* **(dorsal view of female).**

From *Smithsonian Contributions to Zoology,* no. 127, from the chapter entitled "A Revision of the Family Lichomolgidae Kossman, 1877." Smithsonian Institution, Washington, DC.

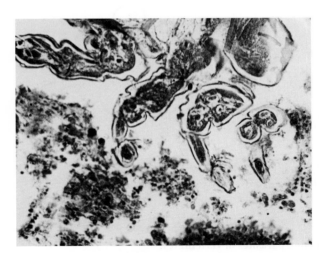

Figure 34.8   **Section of** *Ergasilus sieboldi* **in situ showing damage to gills inflicted by thoracic appendages.**

Tissue is rasped off, and the parasite feeds on detached epithelial, mucous, and blood cells. The first legs (*top left*) are particularly important in directing dislodged tissue anteriorly toward the mouth. (×200)

From T. Einszporn, "Nutrition of *Ergasilus sieboldi* Nordmann. II. The uptake of food and the food material," in *Acta Parasitol. Polon.* 13:373–380. Copyright © 1965.

apparently adapted for prehension in much the same manner as those of ergasilids. Greater specialization is shown in some species in which one or more swimming legs may be reduced or vestigial. Copepodids are often found parasitic on the same hosts as are adults, and relatively little time is apparently spent in free-living naupliar stages.

Families Philichthyidae, Sarcotacidae. Little is known of these families, but they deserve at least brief mention because of their extreme specialization for parasitism.

The general appearance of philichthyids is startling; unlikely looking processes emanate from their bodies (Fig. 34.10). This is a small group, completely endoparasitic in subdermal canals of teleosts and elasmobranchs; that is, in frontal mucous passages and sinuses and the lateral line canal. Some species retain external evidence of segmentation, but in others it is less apparent. Organs of attachment are reduced. Males are much smaller than females and are less highly modified.

Sarcotacids are also endoparasitic copepods and are probably the most highly specialized of any copepod parasite of a vertebrate (Fig. 34.11). They live in cysts in the muscle or abdominal cavity of their fish hosts. Their appendages are vestigial, and they appear to feed on blood from the vascular wall of the cyst. Adult females are little more than reproductive bags within the cysts and may reach several centimeters in size. Males are much smaller, and one lives in each cyst, mashed between the wall of the cyst and the huge body of its mate.

Nothing is known of the development and many other aspects of the biology of sarcotacids and philichthyids.

Figure 34.10

Philichthyids, parasites in subdermal canals of fish.

a) *Philichthys xiphiae;*
(*b*) *Sphaerifer leydigi;*
(*c*) *Colobomatus sciaenae;*
(*d*) *Lerneascus nematoxys;*
(*e*) *Colobomatus muraenae.*

From Z. Kabata, "Crustacea as enemies of fishes," in S. F. Snieszko and H. R. Axelrod, eds., *Diseases of fish,* book I. Neptune City, NJ: T. F. H. Publications, Inc., 1970.

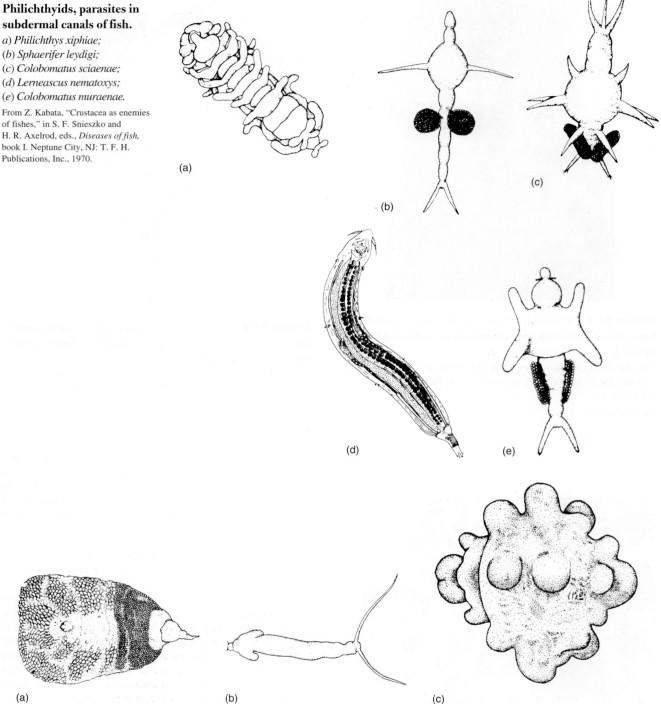

(a)

(b)

(c)

(d)

(e)

(a)

(b)

(c)

Figure 34.11 **Sarcotacids may be the most highly specialized copepod parasites of vertebrates.**

(*a*) *Sarcotaces* sp., female; (*b*) *Sarcotaces* sp., male; (*c*) *Ichthyotaces pteroisicola.*

From Z. Kabata, "Crustacea as enemies of fishes," in S. F. Snieszko and H. R. Axelrod, eds., *Diseases of Fish,* book I. Neptune City, NJ: T. F. H. Publications, Inc., 1970.

Family Xenocoelomatidae. *Xenocoeloma* spp. adults are essentially gonads grafted into the body wall of their hosts, polychaete annelids. Long thought hermaphroditic, they apparently are an example of cryptogonochorism (p. 528); all that remains of the male is a testis producing spermatozoa.[66] Their placement among copepods was formerly not established, but they are now considered poecilostomes.[49]

Order Siphonostomatoida

Members of this large group are mostly parasites of fishes. Only 31 species have been recorded from cnidarians, but there are probably many more.[32] Although even the least evolved siphonostomes show some adaptations to parasitism, like poecilostomes they show an array from generalized to extremely modified and bizarre.

Most siphonostomes are parasites of marine fishes, and, with increased aquaculture of marine fishes, these parasites have had an increasing economic impact. *Lepeophtheirus salmonis* and *Caligus elongatus,* known as *sealice,* are major pathogens of farmed salmon in Norway, Scotland, and Ireland, and damage caused by *L. salmonis* is the more severe.[62] They browse on and breach epidermal tissues. Adults feed on blood, and extensive damage to epidermis causes serious osmoregulatory problems for the fish. Economic impact is high: Costs due to sealice were £15 to £30 million in Scotland alone in 1998, Nk 500 million in Norway in 1997, and $20 million in Canada in 1995.[16, 64] €305 million, and U.S. $480 million (at 2006 exchange rates). A variety of chemical controls for sealice are available, and several wrasse species (Labridae, smaller species of which eat ectoparasites of other fish) are useful biological controls for fish in pens.

Species of salmonids vary in their susceptibility to sea lice. Rainbow trout (*Onchorhynchus mykiss*) and Atlantic salmon (*Salmo salar*) are susceptible species, while coho salmon (*O. kisutch*) are resistant. Incubation of *Lepeophtheirus salmonis* in mucus from the susceptible species causes the copepods to release a variety of low molecular weight proteases, but incubation with mucus from coho salmon did not stimulate such release.[21]

Family Caligidae. Adult caligids are obviously arthropods, although they have departed from a "typical" free-living copepod plan. They have, at least, some adaptations for prehension, tend to be larger than most free-living groups, and have some dorsoventral flattening for closer adhesion to their host's surface. Some tend to be more sedentary, being mostly confined to a fish's branchial chamber, but adults of many species can move rapidly over a host's fins, gills, and mouth. Adults can swim and change hosts.

The usual caligid body form shows a fusion of ancestral body somites: a large, flat cephalothorax followed by one to three free thoracic segments, a large genital segment, and a smaller unsegmented abdomen. Three segments between the cephalothorax and genital somite, as in *Dissonus nudiventris* (Fig. 34.12), is a primitive character, and reduction to a single segment, as in *Caligus* spp. (Fig. 34.13), is derived.[58] The principal appendages for prehension of *Caligus curtus* are antennae and maxillipeds. They have two **lunules** on the anterior margin of the cephalothorax that function as accessory organs of adhesion. Their cephalothorax is roughly disc shaped and bears a flexible, membranous margin. The posterior portion of the disc is not formed by the cephalothorax itself but by the greatly enlarged, fused protopod of the third thoracic legs (Fig. 34.14). Membranes on the margin of the protopods match those on the cephalothorax, and the arrangement forms an efficient suction disc when the cephalothorax is applied to a fish's surface and then arched.

The feeding apparatus of *C. curtus* is a good example of the tubular mouth type of siphonostomes. The mouth tube is carried in a folded position parallel to the body axis, but it can be erected so that its tip can be applied directly to a host surface. The tip of the tube bears flexible membranes analogous to those on the margin of the cephalothorax, again

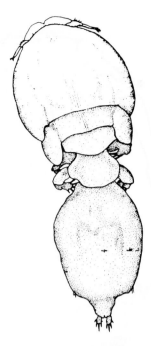

Figure 34.12   *Dissonus nudiventris,* **a caligid with the primitive character of three segments between cephalothorax and genital somite.**

From Z. Kabata, "Crustacea as enemies of fishes," in S. F. Snieszko and H. R. Axelrod, eds., *Diseases of Fish,* book I, edited by Neptune City, NJ: T. F. H. Publications, Inc., 1970.

increasing efficiency of the organ as a suction device. Bases of the mandibles are lateral to and outside the mouth tube (see Fig. 34.2). They enter the buccal cavity through longitudinal canals so that their tips lie within the opening of the cone. The mandibular tips bear a sharp cutting blade on one side and a row of teeth on the other. Thus, the mandibles can work back and forth like little pistons in their canals, piercing and tearing off bits of host tissue to be sucked up by muscular action of the mouth tube.

Caligus spp. and *Lepeophtheirus* spp. have only two naupliar stages, which evidently do not feed.[35, 45] The second nauplius molts to produce the first copepodid. The first copepodid must find a host or perish. If it finds its host, the copepodid clings to the fish with its prehensile antennae and molts to produce a specialized type of copepodid called a **chalimus** (Fig. 34.15). Three more chalimus instars follow, all of them attached to the host by the frontal filament. The actual attachment process of caligids is unknown, but it is probably similar to that of lernaeopodids. The chalimus backs off from its point of attachment, thus pulling more filament out of the frontal organ, while stroking the filament with its maxillae.[40] Four chalimus instars are followed by two preadult stages in all caligids. Preadults are not attached by a frontal filament, and they, as well as adults, have capacity for free movement over a host's body. Males are parasitic and not much smaller than females.

Family Trebiidae. Of the two genera in this family, all *Kabataia* spp. are parasites of teleost fishes, whereas *Trebius* spp. infect only elasmobranchs. *Trebius shiinoi* was found

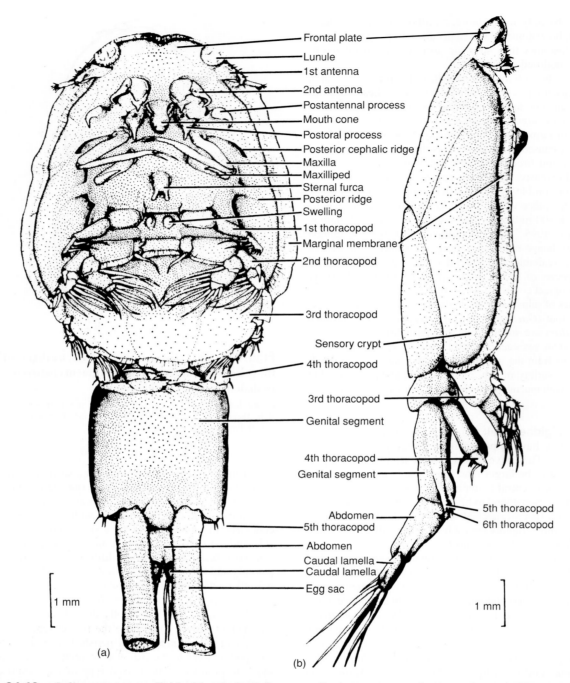

Frontal plate
Lunule
1st antenna
2nd antenna
Postantennal process
Mouth cone
Postoral process
Posterior cephalic ridge
Maxilla
Maxilliped
Sternal furca
Posterior ridge
Swelling
1st thoracopod
Marginal membrane
2nd thoracopod
3rd thoracopod
Sensory crypt
4th thoracopod
3rd thoracopod
Genital segment
4th thoracopod
Genital segment
Abdomen
5th thoracopod
Abdomen
Caudal lamella
Caudal lamella
Egg sac
5th thoracopod
6th thoracopod

1 mm

1 mm

(a)

(b)

Figure 34.13 *Caligus curtus,* **a caligid with a derived character of only one segment between the cephalothorax and genital somite.**

a) Female, ventral view; (*b*) male, lateral view.

From R. R. Parker et al., "A review and description of *Caligus curtus* Miller, 1785 (Caligidae: Copepoda), type species of its genus," in *J. Fish. Res. B. Can,* 25:1923–1969. Copyright © 1968 Journal Fisheries Research Board of Canada. Reprinted by permission.

on the uterine lining of two pregnant Japanese angelsharks, *Squatina nebulosa,* as well as on surfaces of their embryos (Fig. 34.16).[55] Adult females of this species have an extraordinarily long, two-segmented abdomen. These copepods presumably gain access to their host's uterus through its cloaca. Individuals on surfaces of the fetuses have the distinction of being endosymbiotic ectoparasites. Nagasawa and his coworkers describe them as "yet another example of Copepoda's amazing biological flexibility."[55]

Family Lernaeopodidae. Lernaeopodids are common, widespread parasites of marine and freshwater fish. Some species have done great damage to hatchery fish.[38] *Ommatokoita spp.* (Fig. 34.17) attach to the cornea of Greenland

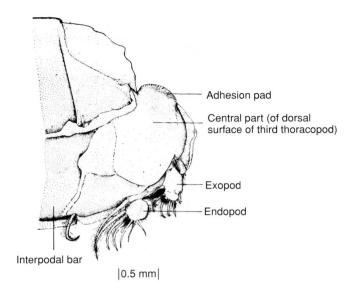

Figure 34.14   Third thoracic leg of *Caligus curtus*.
Note greatly enlarged, fused protopod with flexible marginal membrane.

From R. R. Parker et al., "A review and description of *Caligus curtus* Miller, 1785 (Caligidae: Copepoda), type species of its genus," in *J. Fish. Res. B. Can.* 25:1923–1969. Copyright © 1968 Journal Fisheries Research Board of Canada. Reprinted by permission.

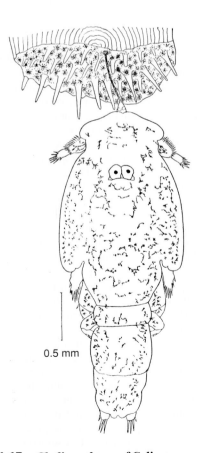

Figure 34.15   Chalimus larva of *Caligus rapax*.

From C. B. Wilson, "North American parasitic copepods belonging to the family Caligidae. Part 1, The Caligidae," in *Proc. U. S. Nat. Mus.*, 28:549, 1905.

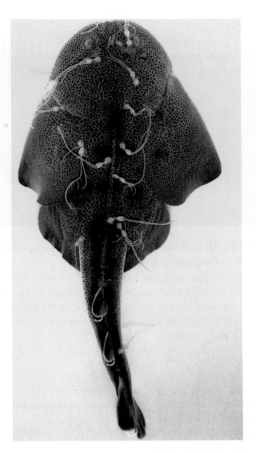

Figure 34.16   Embryo Japanese angelshark *Squatina japonica* infected with *Trebius shiinoi*.
Copepods with long abdomens are adult females and qualify as endosymbiotic (within their host's uterus) ectoparasites (on the surface of the embryo).

From K. Nagasawa et al., "*Trebius shiinoi* n.sp. (Trebiidae: Siphonosomatoida: Copepoda) from uteri and embryos of the Japanese angelshark (*Squatina japonica*) and the clouded angelshark (*Squatina nebulosa*), and redescription of *Trebius longicaudatus*," in *J. Parasitol.* 84:1218–1230. Copyright © 1998 *Journal of Parasitology*. Reprinted by permission.

Figure 34.17   *Ommatokoita elongata* on the eye of a Greenland shark, *Somniosus microcephalus*.
Note the opacity of the cornea.

Courtesy of G. Benz.

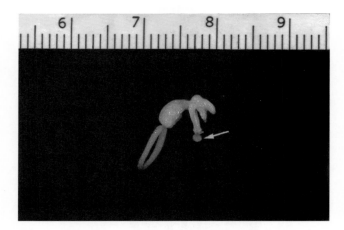

Figure 34.18   *Salmincola inermis,* a lernaeopodid parasite of whitefish, *Coregonus* spp.

The huge maxillae are fused to the bulla (*arrow*), which is embedded in the host's flesh, anchoring the female to that site. The powerful maxillipeds can be seen anterior to the maxillae, and the mouth is at the tip of the anteriormost conelike projection.

Photograph by Larry S. Roberts.

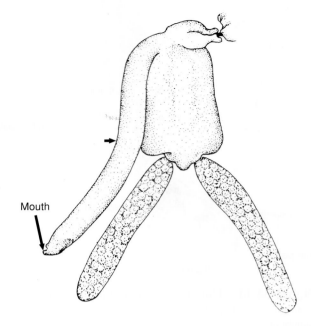

Figure 34.19   The maxillae are very short in *Clavella* spp., but the cephalothorax (*upper arrow*) is long and mobile providing an extended "grazing range."

From Z. Kabata, "Crustacea as enemies of fishes," in S. F. Snieszko and H. R. Axelrod, eds., *Diseases of fish,* book I. Neptune City, NJ: T. F. H. Publications, Inc., 1970.

sharks and Pacific sleeper sharks (*Somniosus microcephalus* and *S. pacificus*)—usually on both eyes—functionally blinding the sharks.[7, 39]

Lernaeopodids are substantially more modified away from the ancestral copepod form than are caligids and trebiids. Virtually all external signs of segmentation have disappeared in adults (see Figs. 34.17 and 34.18), as is the case with Lernaeidae and Pennellidae. Similarly, adult females are permanently anchored in one place on their host. However, in contrast to these families, lernaeopodid females are attached almost completely outside the host; an anchor, or **bulla,** is nonliving and is formed from head and maxillary gland secretions. Maxillae themselves are fused to the bulla, and they are often huge. In some genera maxillae are very short, as in *Clavella* spp. (Fig. 34.19); however, in these cases a very long, mobile cephalothorax provides a "grazing range" similar in extent to that possible with longer maxillae.

Maxillipeds are modified to form powerful grasping structures; although primitively they were posterior to the maxillae, in most species they are now located and function more anteriorly. The bases of the maxillae mark the approximate posterior limit of the cephalothorax, and the rest of the body is trunk, or fused thoracic and genital segments. Abdomen and swimming legs are absent or vestigial. There is extreme sexual dimorphism. Males are pygmies and are free to move around in search of females after the last chalimus stage. Both maxillae and maxillipeds of males are used as powerful grasping organs. Males do not use a bulla to anchor themselves, however.

Kabata and Cousens[40] gave a fascinating account of lernaeopodid development (Fig. 34.20). *Salmincola californiensis* hatches from the egg as a nauplius and molts simultaneously into a copepodid. After its cuticle hardens, a copepodid must find a host within about 24 hours or it dies. It attaches to the host with prehensile hooks on its antennae

and the powerful claws of its maxillae, and then it must find a suitable position on the fish for placement of the frontal filament. It wanders over its host's skin until it finds a solid structure, such as a bone or fin ray, close to the surface. Its maxillipeds excavate a small cavity at that position and press the anterior end of the cephalothorax into the cavity. The terminal plug of the frontal filament detaches and is fixed to the underlying host structure by a rapidly hardening cement produced by the frontal gland. The copepodid moves backward, pulling the filament out of the frontal gland, and, if the attachment site is favorable and the copepodid has not been too much damaged by detachment of the frontal filament, it soon molts to the first chalimus stage.

These hazards destroy many copepodids, but, even after a copepodid is safely attached to a host, it must pass through four chalimus instars. Each chalimus molt involves a complicated series of maneuvers in which the frontal filament is detached by the maxillae and then reattached when the molt is completed (Fig. 34.21). The fourth chalimus finally breaks free of the frontal filament.

After this chalimus is free, it must find a suitable location for its permanent residence. With its antennae and mouth appendages, it rasps out a site for the bulla, now developing in the frontal organ. Maxillipeds cannot be used because they are the principal means of prehension. After molting, the bulla is everted, placed in the excavation, and detached from the anterior end. These processes are again dangerous for the parasite, which loses considerable body fluid, causing substantial mortality. Finally the linked tips of the maxillae must find the opening in the implanted bulla, where they connect with small ducts and secrete cement

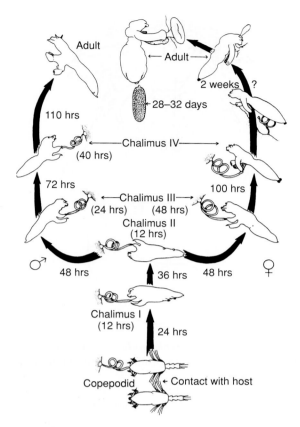

Figure 34.20   Life cycle of *Salmincola californiensis.*

Time periods in parentheses refer to duration of each stage, whereas those without parentheses denote time from first contact with host.

From Z. Kabata and B. Cousens, "Life cycle of *Salmincola californiensis* (Dana, 1852) (Copepoda: Lernaeopodidae)," in *J. Fish. Res. Bd. Can.* 30:881–903. Copyright © 1973.

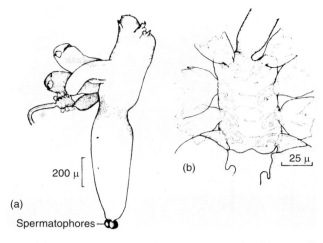

Figure 34.21   Early fourth chalimus of female *Salmincola californiensis.*

(*a*) Maxillae embedded in frontal filament. (*b*) Enlarged end of frontal filament, showing tips of maxillae embedded in it (at bottom), along with the molted cuticle of maxillae tips from earlier chalimus stages.

From Z. Kabata and B. Cousens, "Life cycle of *Salmincola californiensis* (Dana, 1852) (Copepoda: Lernaeopodidae)," in *J. Fish Res. Bd. Can.* 30:881–903. Copyright © 1973.

from the maxillary glands. If and when this last maneuver is successful, the parasite is permanently attached to its host and can graze at will on the surface epithelium. It is no surprise that many copepods fail in this complicated series of developmental events; it is amazing that so many succeed.

Family Pennellidae. Pennellids (formerly Lernaeoceridae) are widespread and conspicuous parasites of marine fish and mammals. They carry the evolutionary tendencies mentioned earlier to the extreme. Even small ones are usually large by free-living standards, and large ones are mammoths of the copepod world. *Pennella balaenopterae* from whales may be more than 30 cm long! Their loss of external segmentation, obscuration of swimming appendages in adults, and invasion of host tissue by their anterior ends are reminiscent of the cyclopoid family Lernaeidae. However, pennellids tend to be more invasive of the circulatory system, sense organs, and viscera than are lernaeids. Each species usually has a characteristic site into which the anterior end grows and feeds.

Several species, including all *Lernaeocera* spp., invade particular parts of the circulatory system, normally a large blood vessel. (The large trunk, bearing the reproductive organs and ovisacs, is external to the fish surface.) Common sites are heart, branchial vessels, and ventral aorta.

On Atlantic cod *Gadus morhua, L. branchialis* (Fig. 34.22) invades the bulbus arteriosus of its host. The parasite generally attaches in the branchial area, and the cephalothorax may have to grow into and follow the ventral aorta for some distance. The associated pathogenesis is severe and probably impacts commercial fisheries. Two or more mature parasites on haddock *(Melanogrammus aeglefinus)* can cause the fish to be as much as 29% underweight, have less than half the normal amount of liver fat, and lose half the hemoglobin content of its blood.[37, 42] Concurrent infections with a trypanosome can compound the damage.[41]

The form of adult females is wonderfully grotesque. Anchoring processes, sometimes referred to as *antlers,* emanate from the anterior end. These are often more elaborate than those found in lernaeids. The greatest development of antlers seems to be in *Phrixocephalus* spp., in which many branches are found (Fig. 34.23). Branches are structurally complex and may be involved in the exchange of molecules between parasite and host.[61] *Lernaeolophus* spp. and *Pennella* spp. have curious, branched outgrowths at the posterior part of the trunk, the function of which is unknown. As in lernaeids, limb appendages do not participate in metamorphosis undergone by the rest of the female body; they are so small compared with the rest of the body as to be hardly discernible.

Life cycles of pennellids are unique among copepods in that they often require an intermediate host, usually another species of fish but in some cases an invertebrate. *Lernaeocera branchialis* apparently has only one naupliar stage, which leads a brief pelagic existence. Copepodids infect any of several different species of fishes[39] and undergo several chalimus instars (Fig. 34.24). Females are fertilized as late chalimi while on their intermediate host and then detach from the frontal filament. They undergo another pelagic phase to search out a definitive host, usually a species of

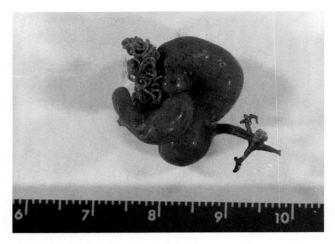

Figure 34.22 *Lernaeocera branchialis* **from an Atlantic cod,** *Gadus morhua.*

The voluminous trunk of the organism, containing the reproductive organs along with the coiled egg sacs, protrudes externally from the host in the region of the gills. The anterior end (*right*) extends into the flesh of the host, and the antlers are embedded in the wall of the bulbus arteriosus, which is severely damaged. Antlers rarely penetrate the lumen of the bulbus, since this would lead to thrombus formation and death of both parasite and host.

Photograph by Larry S. Roberts.

gadid (cod family). The copepod attaches in the gill cavity; the anterior end burrows into the host tissue, aided by strong antennae; and the dramatic metamorphosis begins. At the time she leaves the intermediate host, a female is only 2 mm to 3 mm long and is copepodan in appearance. In her metamorphosis she loses all semblance of external segmentation and grows to 40 mm or more.

Adult females of *Cardiodectes medusaeus* are found on lanternfishes, with their anterior ends embedded in the bulbus arteriosus of the heart.[59] Intermediate hosts are not fish but are thecosomate gastropods. Both *Lernaeocera* spp. and *C. medusaeus* feed on blood. *Cardiodectes medusaeus* completely digests hemoglobin and stores waste iron as ferritin crystals; the manner in which *Lernaeocera* spp. dispose of excess iron is unknown.[60]

Phrixocephalus cincinnatus invades the eye of Pacific sand dabs (lefteye flounders), *Citharichthyes sordidus.*[62] Its cephalic processes ramify throughout the choroid, where the parasite feeds on blood leaking from blood vessels, and its trunk elongates and breaks back out of the cornea. Extracts of the parasites show both cysteine and serine protease activities, which are no doubt important in digestion of host blood but also may act as anticoagulants and may aid in invasion and establishment.[62]

Order Monstrilloida

Monstrilloida have the distinction of being parasitic only during their immature stages. Adults are the free-living dispersal stage. They have the further distinction among Crustacea of having only one pair of antennae, and neither monstrilloid juveniles nor adults have a mouth or functional gut. A nauplius penetrates its host, which is either a polychaete or a prosobranch gastropod, depending on species. It

Figure 34.23 *Phrixocephalus longicollum,* **a lernaeocerid whose antlers proliferate into a luxuriant, intertwining growth.**

From Z. Kabata, "Crustacea as enemies of fishes," in S. F. Snieszko and H. R. Axelrod, eds., *Diseases of fish,* book I, Neptune City, NJ: T. F. H. Publications, Inc., 1970.

molts to become a rather undifferentiated larva with one to three pairs of apparently absorptive appendages (Fig. 34.25). Progressive differentiation and copepodid stages ensue, and adults finally break out of the host to reproduce. Thus, only adults and nauplii are free living, and intermediate stages absorb food in a manner analogous to that of a tapeworm. Adults may be found in plankton, and hosts of juvenile stages are often unknown.[28, 75, 76]

Members of family Thaumatopsyllidae formerly were regarded as monstrilloids because they have one pair of antennae, no mouthparts, and parasitic juvenile stages with planktonic adults, but they are now considered cyclopoids.[49, 77]

Forms Not Assignable to Orders

There are a number of strange copepods whose adults are so modified and of whose developmental stages we are so ignorant that we cannot even place them in an order.

Males and females of *Coelotrophus nudus* in the coelomic cavity of a sipunculan do not have a mouth.[31] Females have no appendages, and males have only one pair, with which they grasp the female.

Female *Ischnochitonika* spp. live in the branchial (pallial) groove of chitons, with long (presumably absorptive) processes extending into the host viscera. They have no appendages or segmentation, and several pygmy males live on the body of each female.[22, 51]

Pectenophilus ornatus females attach to a gill of a scallop, where they consume host blood. They lack segmentation and appendages, and pygmy males live in a special chamber in the female adjacent to the brood pouch.[52] They are a serious pest in the Japanese scallop industry.[54]

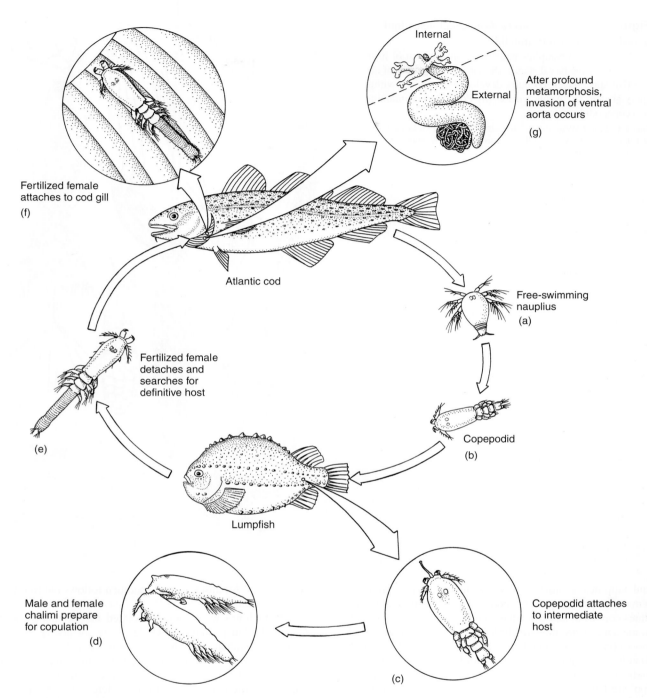

Figure 34.24   **Life cycle of *Lernaeocera branchialis*.**

(*a*) Free-swimming nauplius. (*b*) Copepodid. (*c*) Copepodid attaches to intermediate host—in this example, the lumpfish (*Cyclopterus lumpus*). Flounders (Pleuronectidae) and sculpins (Cottidae) can also serve. (*d*) Male and female chalimi preparing for copulation. (*e*) Fertilized female detaches and searches for a definitive host. (*f*) Fertilized female attaches to gill of a cod or other Gadidae (cod family). (*g*) After profound metamorphosis, invasion of ventral aorta occurs.

Drawing by William Ober and Claire Garrison after R. A. Khan et al., "*Lernaeocera branchialis:* A potential pathogen to cod ranching," in *J. Parasitol.* 76:913–917.

Tarificola bulbosus adult females have an elongate, wormlike body not differentiated into regions and with no recognizable appendages or segmentation.[48] They are parasites of compound ascidians *Polycitor crystallinus,* one copepod per zooid.

Subclass Branchiura

Subclass Branchiura is relatively small in numbers of species but great in its destructive potential in fish culture. All species are ectoparasites of fishes, although some can use frogs

Figure 34.25 *Haemocera danae,* **a monstrilloid parasite of polychaete annelids.**

(*a*) Nauplius. (*b*) Nauplius penetrating integument of host. (*c–e*) Successive larval stages, showing development of absorptive appendages. (*f*) Fully developed copepodid within spiny sheath. (*g*) Adult female. (*h*) Polychaete containing two copepodids in coelom.

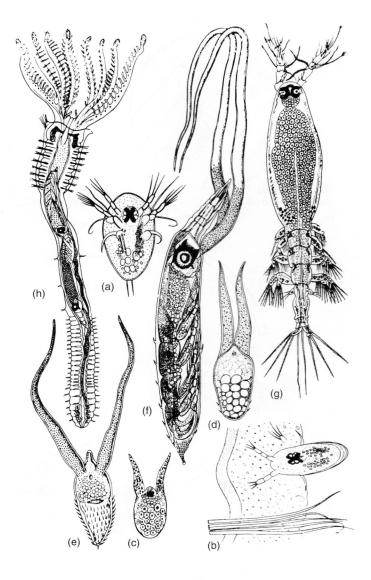

and tadpoles as hosts. They are dorsoventrally flattened, reminiscent of caligid copepods with which they are sometimes confused, and can adhere closely to the host's surface. Some species are moderately large, up to 12 mm or so. The most common cosmopolitan genus is *Argulus* (Fig. 34.26). *Argulus* spp. can swim well as adults; females must leave their hosts to deposit eggs on the substrate. Many *Argulus* spp. are not host specific and so have been recorded from a large number of fish species.

A branchiuran's carapace expands laterally to form respiratory alae. The parasites have two pairs of antennae. Homologies of the remaining head appendages have been disputed, but best evidence suggests that the only appendages in the suctorial proboscis, or mouth tube, are mandibles.[50] The large, prominent sucking discs are modified maxillules. Immediately posterior to the maxillular discs are large maxillae, apparently used to maintain the animal's position on its host and to clean other appendages. *Argulus* spp. have four pairs of thoracic swimming legs of typically crustacean biramous form. Exopods of the first two pairs often bear an odd, recurved process, or **flabellum,** thought by some to indicate affinities with the subclass Branchiopoda

(Fig. 34.27). An unsegmented abdomen follows the four thoracic segments.

Branchiura were long associated taxonomically with Copepoda, but present knowledge does not justify this association. Branchiurans have, among other characteristics that differ from copepods, a carapace, compound eyes, and an unsegmented abdomen behind the genital apertures, and no thoracic segments are completely fused with the head. Another feature present in most branchiurans but not in other Crustacea is a spiercing stylet, or "sting." It is located on the midventral line, just posterior to the antennae (Fig. 34.28). The function of this curious organ is unknown.

Development of *Argulus japonicus* and *Chonopeltis brevis* and various developmental stages of other species have been described,[23] illustrating further differences between Branchiura and Copepoda. Whereas development of copepods is metamorphic, that of most branchiurans is usually direct, although in some the first instar may be a modified nauplius.[24] As noted, eggs are laid on the substrate (no ovisacs, as in copepods), and the first instar is usually a juvenile.

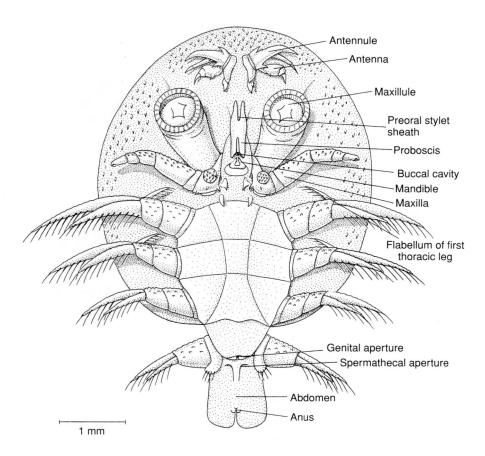

Figure 34.26   **Ventral view of** *Argulus viridis,* **female.**
Note suctorial proboscis, modification of maxillules into sucking discs, and lateral expansion of carapace into alae.

Drawing by William Ober and Claire Garrison after M. F. Martin, "On the morphology and classification of *Argulus* (Crustacea)," in *Proc. Zool. Soc. London* 1932:771–806.

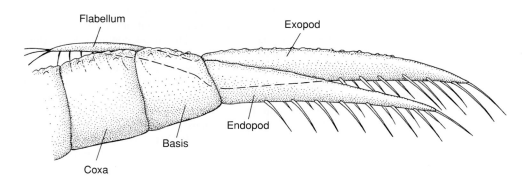

Figure 34.27   **First thoracic appendage of** *Argulus viridis,* **showing flabellum.**

Drawing by William Ober and Claire Garrison after M. F. Martin, "On the morphology and classification of *Argulus* (Crustacea)," in *Proc. Zool. Soc. London* 1932:771–806.

Subclass Thecostraca

Thecostracans most familiar to humans belong to superorder Thoracica, barnacles. They are important members of littoral and sublittoral benthic fauna and are economically important as fouling organisms. Some members of Thoracica are commonly found growing on other animals. Interestingly, *Conchoderma virgatum* is often found on copepods of genus *Pennella* spp., a good example of hyperparasitism. (Numerous species of parasitic copepods frequently have epizooic suctorians, hydroids, algae, and so on growing on them, encouraged by the fact that the copepod is in terminal anecdysis; that is, it does not molt further.)

Other groups of thecostracans contain some fascinating organisms that are among the most highly specialized parasites known. These are parasites of other invertebrates, and space will permit consideration only of the most important superorder, Rhizocephala.

Superorder Rhizocephala, Order Kentrogonida

Members of Kentrogonida are highly specialized parasites of decapod malacostracans. Decapods include animals most of us know as crabs, crayfish, lobsters, and shrimp. Sacculinidae are primarily parasites of a variety of brachyurans ("true" crabs), Peltogastridae are found on hermit crabs (anomurans), and Lernaeodiscidae prefer anomuran families Galatheidae and Porcellanidae as hosts.

As adults, kentrogonidans resemble arthropods even less than do lernaeids and pennellids. They have no gut or appendages, not even reduced ones, but get nutrients by means

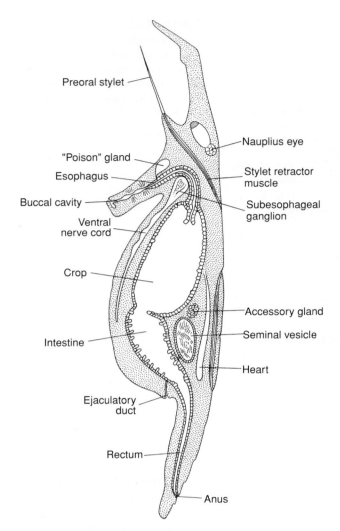

Figure 34.28 **Median longitudinal section of male**
Argulus viridis; **semidiagrammatic with preoral stylet**
extruded.

Drawing by William Ober and Claire Garrison after M. F. Martin, "On the morphology and classification of *Argulus* (Crustacea)," in *Proc. Zool. Soc.
London* 1932:771–806.

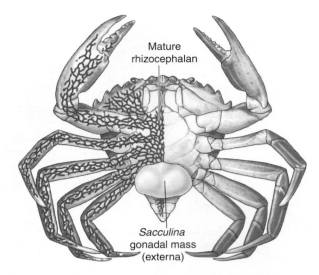

Figure 34.29 **Shore crab,** *Carcinus,* **infected with a**
mature rhizocephalan, *Sacculina.*

Drawing by William Ober and Claire Garrison from L. A. Borradaile et al.,
(editors), *The Invertebrata: A Manual for Students,* 2nd edition. Copyright © 1956
Cambridge University Press. Reprinted by permission of Cambridge University
Press.

of rootlike processes ramifying through tissues of their crab
host (Fig. 34.29). They start life much as do many other crustaceans, with a nauplius larva, but the nauplius has no mouth
or gut. Nauplii undergo four molts, and the fifth larval instar
is referred to as a **cyprid** or **cypris** (Fig. 34.30) because of its
resemblance to a free-living ostracod genus *Cypris.*

Thus far, the kentrogonidan life cycle is not unlike that
of a normal, thoracican barnacle. A barnacle cypris, however,
attaches to a suitable spot on the substrate by its antennules
and metamorphoses to the adult form. The halves of the carapace become mantle and secrete calcareous covering plates.
However, a female cypris of *Sacculina* spp. and related kentrogonidans attaches to a decapod with its antennules.

Most differentiated structures, including swimming legs
and their muscles, are shed from between the two valves
of the carapace. After these divestitures, the remaining peculiar larva from which the order gets its name is called a

kentrogon. It contains only an undifferentiated cell mass.
A kentrogon may be likened to a living hypodermic syringe.
The cell mass within is actually injected into the hemocoel
of the crab at the base of a seta or other vulnerable spot
where the cuticle is thin. The mass of cells (in some species
only a single cell) migrates to a site just dorsal to the ventral
nerve cord and begins to grow (see Fig. 34.30).

As absorptive processes grow out into the crab's tissue,
the central mass enlarges. This mass contains developing
female gonads. As it grows larger, it appears to press against
the host hypodermis in the ventral cephalothorax and thereby
prevents cuticle secretion. Finally the weakened cuticle
overlying the parasite breaks open, and the gonadal mass of
Sacculina spp. becomes external (now called an **externa**).[17]
After a sacculinid becomes externalized, it inhibits molting by its host;[54] therefore, further development of the host
essentially ceases. The externa attracts one or more male
cyprids. These extrude a mass of cells that become a spine-
covered larva called a **trichogon** and come to lie within male
cell receptacles in the female and undergo spermatogenesis.

A number of *Sacculina* species can produce multiple
externae on the same host crab. Some of these are accounted
for by fertilization from separate male cyprids, but some
appear to arise from budding of developing parasites.[25] Molecular studies support erection of a new genus, *Polyascus,*
for these asexually reproducing sacculinids.

Formerly these cells from the male cyprids in the females were interpreted as testes, and rhizocephalans were
considered hermaphroditic. However, we now know that
sexuality in kentrogonidans is an example of **cryptogono-**
chorism (Gr. *kryptos* = hidden + NL *gonas* = reproductive
organ + Gr. *choris* = apart; gonochoristic, gonochorism =
dioecious, sexes in separate individuals).[62] Life histories of
peltogastrids and lernaeodiscids seem to be similar to those
of sacculinids in many respects, although in the former further ecdyses of the host are not hindered.[56]

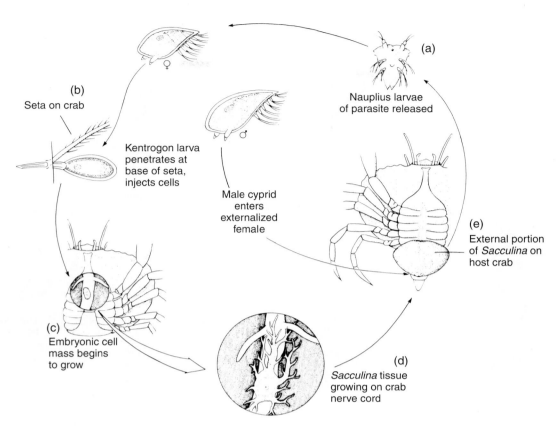

Figure 34.30   Life cycle of *Sacculina*, a rhizocephalan parasite of crabs.
(*a*) Nauplius larvae of parasite released. (*b*) Kentrogon larva penetrates at base of seta and injects cells. (*c*) Embryonic cell mass attaches and begins to grow. (*d*) *Sacculina* tissue growing on crab midgut. (*e*) External portion of *Sacculina* on host crab.

In light of the invasiveness of rhizocephalans, it is not surprising that there is a range of pathogenic effects on their hosts, including damage to hepatic, blood, and connective tissues and to the thoracic nerve ganglion of infected crabs.[67] However, some of the most interesting effects are on the hormonal and reproductive processes of hosts, through so-called **parasitic castration.**

Parasitic Castration. Crabs exhibit sexual dimorphism, and morphological differences between the sexes are especially pronounced in Brachyura. In a normal sequence of ecdyses, secondary sexual characteristics of the respective sexes become increasingly apparent as a crab approaches maturity. When a young male crab is infected with a species of *Sacculina,* various degrees of "feminization" appear in subsequent instars. Manifestations vary, depending on species of host and its degree of development when infected, but they may include a broader, more completely segmented abdomen and alteration of pleopods toward the female type. In female crabs effects seem to be more complex, involving some aspects of both hyperfeminization and hypofeminization. Somewhat similar effects of parasitic castration have been reported in hosts of peltogastrids and lernaeodiscids. The mechanism of host castration has yet to be explained satisfactorily.

Whatever the mechanism of host castration, however, the combined results of parasite structure and castration lead to an astonishing diversion of host behavior to promote parasite survival. The externa of the parasite (in many species) is in the same position and is the same size as the egg mass of the crab, as it would be carried on the crab's abdomen.[68] Reacting as though the parasite were in fact its own egg mass, the crab protects, grooms, and ventilates the parasite. If the grooming legs of the crab are removed artificially, the externa of the parasite soon becomes fouled and necrotic. At the time the parasite begins to release its larvae, the crab performs spawning behavior. The parasites apparently release pheromones that elicit larval-release behavior in the host crab.[19] And a final note: Because they are castrated and feminized, male crabs also display appropriate maternal behavior! The host departs from its normal hiding place, stands high on its legs, and waves its abdomen back and forth. Thus, nauplii of the parasite are released into the current created by its host.

Subclass Tantulocarida

This subclass was discovered and described relatively recently.[10] Its members are bizarre, minute ectoparasites of other crustaceans. They are most "uncrustacean" crustaceans, with no molts and no cephalic appendages except one pair of antennae on sexual females (Fig. 34.31). These are characters apparently unique to this group, and they apparently undergo a parthenogenetic phase, which is unusual in Crustacea.[34] The odd little larva, called a **tantulus,** attaches directly to a host by its median cephalic stylet. Immediately behind

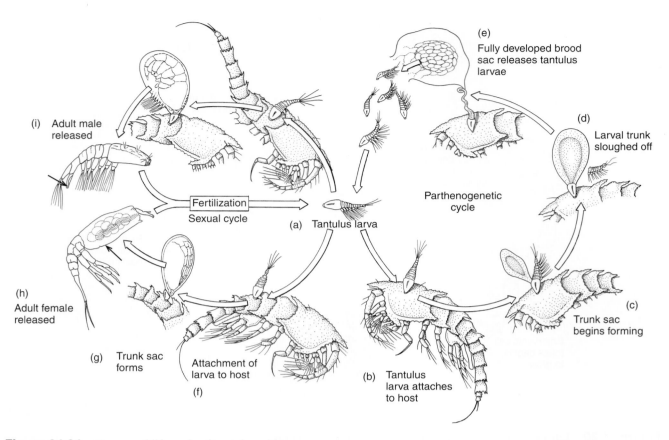

Figure 34.31 **Presumed life cycle of tantulocarids.**
Positions of genital apertures of free-swimming sexual stages marked with *small arrows.* Parthenogenetic cycle: (*a*) tantulus larva; (*b*) tantulus larva attaches to host; (*c*) trunk sac begins forming; (*d*) larval trunk is sloughed off; (*e*) fully developed brood sac releases tantulus larvae. Sexual cycle: (*f*) tantulus larva attaches to host; (*g*) trunk sac forms; (*h*) adult female is released; (*i*) adult male is released.

Drawing by William Ober and Claire Garrison after R. Huys et al., "The tantulocaridan life cycle: The circle closed?" in *J. Crust. Biol.* 13:432–442.

the larval head, a saclike trunk forms, and the larval trunk is sloughed off. Apparently, within the expanded trunk sac either (1) a brood of tantulus larvae develops parthenogenetically; (2) an adult male differentiates and breaks out of the trunk to swim free; or (3) a sexual female differentiates to be fertilized by the male during her free-swimming existence. Neither the free-swimming male nor the female feeds after liberation from the trunk sac.

CLASS OSTRACODA

Ostracods are enclosed in their bivalve carapace and resemble tiny clams. They have a long and diverse fossil record and are still widespread in both marine and freshwater habitats, benthic and planktonic. They show a variety of feeding habits, but the few reports of ostracods as parasites have been controversial. *Sheina orri* is a parasite on gills of sharks, but its morphology differs little from that of its free-living cypridinid relatives.[6] Claws on mandibles and maxillules of *S. orri* are adapted for attachment and cause some damage to host tissues.

CLASS MALACOSTRACA

Although malacostracans constitute the largest class of Crustacea, with members widespread and abundant in marine and freshwater habitats, comparatively few are symbiotic. Those that are, by and large, are confined to peracaridan orders Amphipoda and Isopoda. Isopods have been particularly diverse in this regard, and some of them have become highly modified for parasitism. Eucaridan order Decapoda is the largest order of crustaceans, but few of its members are symbiotic.

Order Amphipoda

Free-living amphipods are widely prevalent aquatic organisms, often abundant along seashores. Not many symbiotic species have been described, but some are quite common. Some Hyperiidae are frequent parasites of jellyfish (*Aurelia aurita, Cyanea* spp.), and Phronimidae are found in the tunic of planktonic ascidians (*Salpa* spp.), apparently killing the tunicate itself and taking over its gelatinous case.[43]

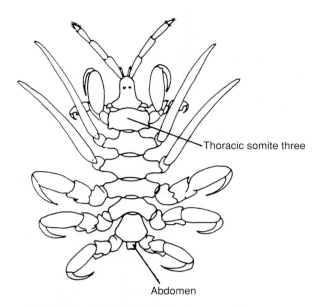

Figure 34.32   *Paracyamus,* **an amphipod parasite of whales.**

Cyamids are ectoparasites and are dorsoventrally flattened with several pairs of legs modified for clinging to their hosts.

From G. O. Sars, from W. T. Calman, "Crustacea," 1909.

Laphystius spp. (suborder Gammaridea) are relatively unmodified amphipods, parasitizing a variety of marine fishes.[27]

The most interesting symbiotic and most unlikely looking amphipods are among the Caprellidea. Cyamidae are curious ectoparasites of whales (Fig. 34.32). Their abdomen is vestigial. In contrast with most amphipods, cyamids are dorsoventrally flattened, a clearly adaptive characteristic in their ectoparasitic habitat. The second and the fifth through seventh legs are strongly modified adhesive organs.

Order Isopoda

Members of order Isopoda have exploited terrestrial environments more than any other group of crustaceans has, as limited as that exploration may be, and they are abundant in a variety of marine and freshwater habitats. Furthermore, they have invaded parasitic niches more than have any other malacostracans.

Flabelliferan families parasitic on marine fishes have relatively few modifications.[13] *Gnathia* spp. are parasites only in their larval stage, known as a **praniza,** which was originally described as a separate genus before its true identity was recognized. A praniza stage (Fig. 34.33) attaches to a fish and feeds on blood until its gut is hugely distended. It then leaves its host and molts to become an adult. Adults are benthic and do not feed. Smit and Davies[74] provided an interesting review of the parasitic stages of gnathiids.

Some cymothoids are of economic importance as fish parasites. The young of *Lironeca amurensis* and some other species burrow under a scale on their host. As the isopod grows, the underlying skin stretches to accommodate it; finally the enveloped crustacean communicates to the exterior

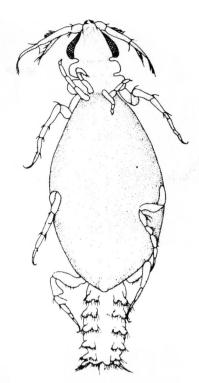

Figure 34.33   **Praniza larva of the isopod** *Gnathia* **sp.**

The praniza is the only parasitic stage of this isopod. The gut becomes greatly distended with blood from its fish host.

From Z. Kabata, "Crustacea as enemies of fishes," in S. F. Snieszko and H. R. Axelrod, eds., *Diseases of fish,* book I. Neptune City, NJ: T. F. H. Publications, Inc., 1970.

Figure 34.34   *Lironeca ovalis* **is a common parasite of fish along the Atlantic coast of the United States.**

Here it is found on a *Lepomis gibbosus* (operculum removed) from Chesapeake Bay.

Photograph by Larry S. Roberts.

only by a small hole. *Lironeca ovalis* (Fig. 34.34) is a common parasite of a variety of teleosts in the Atlantic Ocean and has been reported along the United States coast from Texas to Massachusetts.[73] The parasite is usually found beneath the gill operculum, where, on small host individuals, it

Figure 34.35 *Lironeca ovalis,* the same specimen as in Figure 34.34, removed from its site on the gills.

The gills show pressure atrophy and traumatic damage.

Photograph by Larry S. Roberts.

Figure 34.36 A cymothoid isopod (*Anilocra* sp.) on a coney (*Cephalopholis fulvus*) inhabiting a Caribbean coral reef.

Photograph by Larry S. Roberts.

causes a marked pressure atrophy of adjacent gills (Fig. 34.35). Juveniles of *L. vulgaris* locate their host by slowing their swimming activity in the presence of fish mucus, and white color (either paper or fish skin) plus mucus induces a settling response.[51] Adult females of *Anilocra* spp. are commonly found on the head of various coral reef fishes (Fig. 34.36). These isopods can significantly depress growth, reproduction, and survivorship of parasitized fish.[1] Species of *Ichthyoxenus* burrow through the body wall and live in a membranous sac within the body cavity of their host.[79] One male and one female usually occupy the sac, which opens to the exterior through an orifice near the base of their host's pectoral fin. The orifice provides a channel for release of young, gas exchange, and expulsion of wastes.

Epicaridean isopods are highly specialized parasites of other Crustacea. Adult females of some species are comparable to the most specialized copepods and to rhizocephalans in their loss of external segmentation and appendages. Portions of appendages that are not lost are oostegites that form the brood pouch. These may become enormously developed, whereas most of the other appendages disappear or become vestigial.

Life histories of epicarideans are very interesting. **Epicaridium** larvae hatch from eggs, and they are quite isopod-like in appearance. They have bloodsucking mouthparts and attach to free-swimming, calanoid copepods. Thus attached, they feed, grow rapidly,[2] and molt to become **microniscus** larvae in three to four days.[17] In a few more days microniscus larvae molt several times and develop into **cryptoniscus** larvae, which again are free swimming and must find a suitable definitive host. In Bopyridae, Entoniscidae, and Dajidae, the first individual isopod to infect a given host becomes a female, and subsequent cryptoniscus larvae become small males, sometimes living as parasites within the female brood sac. It seems clear that sex determination in a number of epicarideans is epigamic, depending on circumstances other than the chromosomal complement of the gametes.

A cryptoniscus enters the crab's hemocoel, apparently through the gills. In the gonads or alongside skeletal apodemes, it molts to an apodous juvenile.[44] Juveniles become invested with a cellular covering produced by the host. Female isopods become quite large and occupy the space normally filled by a host ovary, but males remain small and live on the surface of females. The female's oostegites are produced into extensive, thin lamellae that, together with the host-produced sheath, form a brood chamber. Extending from the sides of her abdomen are highly vascularized pleural lamellae (Fig. 34.37), apparently of respiratory function, but pleopods are vestigial. The short esophagus leads into a peculiar "cephalogaster," a contractile organ for sucking blood that apparently has absorptive function as well.[3] Upon hatching, larvae remain in the brood chamber until they exit through a pore formed between the sheath and the thin cuticle of the host's gill cavity.

In Cryptoniscidae, morphological modification appears even more extreme. After infection of a definitive host, female *Ancyroniscus bonnieri* feed heavily, and gorged isopods begin to produce eggs. During this process most of the internal organs, including those of the digestive and nervous system, disappear, and the animal becomes increasingly distended with eggs. Finally all that is left is a large, pulsatile sac of eggs that ruptures, freeing the eggs.[14]

Effects of epicarideans on their hosts are similar to those described for rhizocephalans, including parasitic castration. Secondary sexual characteristics of a male host are lost, the host becomes feminized, and gonads of both males and females are suppressed or atrophied.[5] If the parasite dies (perhaps killed by host defense mechanisms), reproductive capacity of the crab may return.[44] Feminization of brachyuran males by epicarideans is not as striking as that produced by rhizocephalans.[67] It is interesting that some rhizocephalans are themselves hyperparasitized by epicarideans, and these, in turn, induce castration of their rhizocephalan hosts!

Order Decapoda

Decapoda contains a relatively small number of symbiotic species, although some of them are quite common. They are of interest because of the slight but definite modifications for parasitism that they illustrate. Pinnotherids are frequent commensals with polychaetes, in their tubes or burrows, or

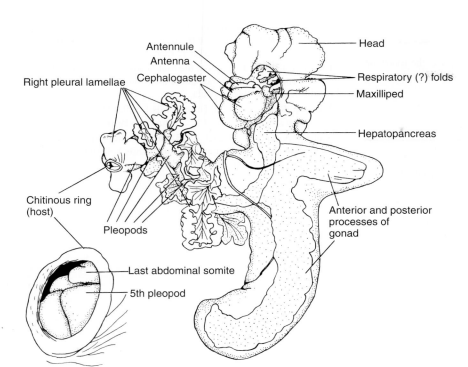

Antennule
Antenna
Cephalogaster
Right pleural lamellae
Head
Respiratory (?) folds
Maxilliped
Hepatopancreas
Chitinous ring (host)
Pleopods
Anterior and posterior processes of gonad
Last abdominal somite
5th pleopod

Figure 34.37   Young female *Pinnotherion vermiforme,* an entoniscid isopod parasite of a crab, *Pinnotheres pisum.*

The parasite develops in a closely invest-ing, thin layer of host origin that com-municates with the branchial chamber by a small opening. The opening into the host's branchial chamber is surrounded by a somewhat thickened ring of cuticle (*enlarged at left*). Vascularized pleural lamellae extending from the abdomen are prominent. Note the vestigial nature of appendages and presence of the peculiar contractile "cephalogaster."

Drawing by Larry S. Roberts after D. Atkins, "*Pinnotherion vermiforme* Giard and Bonnier, an entonscid infecting *Pinnotheres pisum,*" in *Proc. Zool. Soc. London* 1933:319–363.

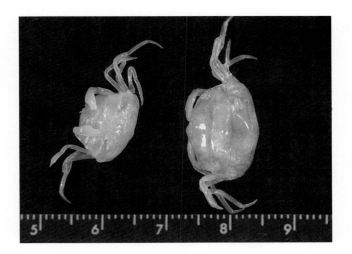

Figure 34.38   *Pinnotheres ostreum* damages gills of its host, the commercial oyster (*Crassostrea virginica*).

Its carapace is soft, and eyes and chelae are reduced.

Photograph by Larry S. Roberts.

are parasites in the mantle cavity of bivalves. *Pinnotheres pisum* is a parasite of mussels, *Mytilus edulis,* and *P. ostreum* is found in commercially important oysters, *Crassostrea vir-ginica* (Fig. 34.38). *Pinnotheres ostreum* interferes with the feeding of its host and damages its gills sufficiently to cause female oysters to become males.[4] A similar sex change can be produced experimentally by starving the oysters.

Pinnotherids are modified relatively little from typi-cal, free-living brachyurans. Adult females tend to be white or cream colored with thin, soft cuticle in the carapace and with reduced eyes and chelae. Younger stages and males

have hard carapaces and more well-developed eyes and chelae.[15] Symbiotic crabs are attracted to their host by chemical cues.[12]

A few species of symbiotic decapods belonging to other infraorders, such as Caridea and Anomura, are known. They live in such places as mantle cavities of clams, tubes of polychaetes, and stems of sea pens (Octocorallia) and are modified for symbiosis to about the same extent as are pin-notherids. The supposed rarity of these forms is probably a result of the failure of collectors to look for them.[36] A num-ber of different decapods are involved in cleaning associa-tions (chapter 1).

Learning Outcomes

By the time a student has finished studying this chapter, he or she should be able to:

1. Distinguish gnathostome, poecilostome, and siphonostome mouthparts in copepods.

2. List examples of copepods with each of the foregoing types of mouthparts.

3. Identify a copepod whose antennae have evolved into powerful holdfast organs.

4. Identify a copepod that has evolved to become essentially a gonad grafted onto (or into) the body wall of its host.

5. Describe the life cycle of *Lerneocera branchialis* and how this crustacean has become a highly adapted parasite.

6. Describe how *Haemocera danae* has become a highly special-ized parasite of polychaete annelids.

7. Describe how *Sacculina* changes the behavior of its host to enhance its own reproduction.

References

References for superscripts in the text can be found at the following Internet site: www.mhhe.com/robertsjanovynadler9e

Additional Readings

Bliss, D. E. (Series Ed.). 1982–1985. *The biology of Crustacea 1–10.* New York: Academic Press, Inc. This series is a standard reference for all aspects of crustacean biology.

Boxshall, G. A. 2004. *An introduction to copepod diversity.* London: Ray Society, 966 pp. (For a review, see G. W. Benz, 2005. *J. Parasitol.* 91:1512–1513.)

Gould, S. J. 1996. Triumph of the root-heads. *Nat. Hist.* 105:10–17. A wonderful Stephen Jay Gould essay on lessons of evolution illustrated by Rhizocephala.

Harrison, R. W., and A. G. Humes (Eds.). 1992. *Microscopic anatomy of invertebrates, vol. 9 Crustacea.* New York: Wiley-Liss.

Raibaut, A., and J. P. Trilles. 1993. The sexuality of parasitic crustaceans. In J. R. Baker and R. Muller (Eds.), *Advances in parasitology 32.* London: Academic Press.

Chapter 35

Pentastomida: Tongue Worms

Pentastome work warms up all the time. . . . The day after Labor Day I take off for Africa and ultimately will visit Nigeria, Kenya, Bombay, Bangkok, Kuala Lumpur, Taiwan, and Okinawa returning home the 28th of September.

—J. T. Self (July 7, 1969, correspondence with one of the authors)

Pentastomids, or tongue worms, are wormlike parasites of the respiratory systems of vertebrates. About 100 species are known. As adults, most live in the respiratory system of amphibians and reptiles, but one species lives in air sacs of sea birds, and another inhabits the nasopharynx of canines and felines. The latter species is occasionally found as transient nymphs in the nasopharynx of humans; other species, in their nymphal stages, also parasitize humans. Thus, pentastomids are certainly of zoological interest and also are of some medical importance.[29]

Prior to development of electron microscopy and molecular techniques, phylogenetic relationships of pentastomids were obscured by their seemingly aberrant adult morphology. Certain similarities with Annelida have been pointed out, but modern taxonomists align them with Arthropoda.[20, 26]

Recent phylogenetic studies using molecular techniques indicate that pentastomids arose in the Cambrian, confirming the fossil evidence (Fig. 35.1).[24] It is possible that the group reached its zenith in the Mesozoic age of reptiles and that today's few species are relicts derived from those ancestors. Wingstrand[36] proposed that Pentastomida be regarded as an order of crustacean class Branchiura (chapter 34). Wingstrand's[36] conclusion was based on demonstration that spermatozoa of the two groups are almost identical with regard to structure and development and that this type of spermatozoon represents a type of its own, not encountered in other animals. Each major crustacean group is characterized by its own type of spermatozoon, and if Pentastomida and Branchiura were unrelated, their sperm structure and development would represent a most extraordinary example of convergence in detail.

Subsequently Storch and Jamieson[31] confirmed Wingstrand's findings, as did Riley and his coworkers using embryogenesis, tegumental structure, and gametogenesis, further concluding that pentastomes should be regarded as a subclass of Crustacea, closely allied to Branchiura.[21] Abele and others[1] came to the same conclusion using 18S ribosomal RNA nucleotide sequences. Martin and Davis[14] formalized this proposal, and so Pentastomida is now "officially" a subclass of

class Maxillopoda. Orders and families within subclass Pentastomida are given in chapter 33 (p. 508); grouping within the subclass agrees with that of Riley.[20] Only infrequently do great parasitological mysteries get solved, but this one—the taxonomic home of the pentastomids—evidently is settled.

Phylogenetic analysis of Pentastomida, using morphological characters, generally support the traditional classification as listed by Riley[20] and Martin and Davis.[14] Almeida and Christoffersen[3] show that Sebekidae and Porocephalidae are sister taxa, as are Linguatulidae and Subtriquetridae; all are families of order Porocephalida. Relationships within order Cephalobaenida are not so clearly resolved, but Almeida and Christoffersen found no evidence for polyphyletic or paraphyletic groupings.[3] Surprisingly, specimens considered Cambrian fossil pentastomes were included in their analysis. These well-preserved fossils have several features in common with extant pentastome larvae, including two pairs of short appendages on their heads and pores on the inside of these appendages. If these specimens are indeed pentastomes, then they provide evidence that their parasitic lifestyle and characteristic morphology were present on Earth prior to their current terrestrial vertebrate hosts (Fig. 35.1).[34, 35]

MORPHOLOGY

The pentastomid body (Figs. 35.2, 35.3) is elongated, usually tapering toward its posterior end and often showing segmentation, forming numerous **annuli.** It is indistinctly divided into an anterior **forebody** and a posterior **hindbody,** which is bifurcated at its tip in some species.

Their exoskeleton contains chitin,[33] which is sclerotized around the mouth opening and accessory genitalia. A striking characteristic of all adult pentastomids is the presence of two pairs of sclerotized hooks in the mouth region (Fig. 35.4).

535

Figure 35.1 A fossil pentastome.

An unidentified "hammerhead" type specimen from the Orsten deposits, late Cambrian (Furongian) age, Kinnekulle, Västergötland, Sweden. A number of fossil genera have been described from these deposits, and all are interpreted as Pentastomida based on morphological and developmental similarities with extant forms, especially segment constancy. Bar is 100 μm.

Courtesy of Dieter Walobek, Ulm, Germany.

These may be located at the ends of stumpy stalks or may be nearly flush with the surface of the cephalothorax; in either case they can be withdrawn into cuticular pockets. The hooks are single in some species and appear to be double in others, but the apparently double hooks are actually single, with an accessory hooklike protrusion of the cuticle. In some species hooks articulate against a basal fulcrum. Hooks are manipulated by powerful muscles and are used to tear and embed the mouth region into host tissues.

The body cuticle in some species also has circular rows of simple spines; annuli may overlap enough to make the abdomen look serrated. There usually are transverse rows of cuticular glands, with conspicuous pores, which evidently function in regulation of water and mineral balance in hemolymph.[5]

The cuticle is similar to that of arthropods, although it is thin.[33] Muscles, too, are arthropodan in nature, being striated and segmentally arranged. The only sensory structures so far recognized are papillae, especially on the cephalothorax. The digestive system is simple and complete, with the anus opening at the posterior end of the abdomen. The mouth is permanently held open by its sclerotized lining, the **cadre,**

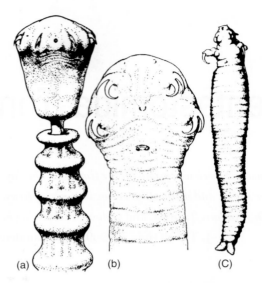

Figure 35.2 Examples of pentastome body types.

(*a*) Anterior end of *Armillifer annulatus;* (*b*) head of *Leiperia gracilis;* (*c*) entire specimen of *Taillietiella mabuiae.*

From J. G. Baer, *Ecology of animal parasites.* Copyright © 1952 by Board of Trustees of the University of Illinois. Used with permission of the University of Illinois Press.

which may be circular, oval, or *U* shaped and is an important taxonomic character. The nervous system is typical of arthropods and has been described by Doucet.[8]

Reproductive Anatomy

Pentastomids are dioecious and show sexual dimorphism in that males are usually smaller than females. Males have a single, tubular testis (two in *Linguatula* spp.), which occupies one-third to one-half of the body cavity (Fig. 35.5*a*). The testis is continuous with a seminal vesicle, which in turn connects to a pair of ejaculatory organs. These each have a duct extending to a terminal penis that fits into a **dilator organ,** which is usually sclerotized, serving as an intromittent organ in some species and as a dilator and guide for the penis in others. The male genital pore is midventral on the anterior abdominal segment, near the mouth.

In females a single ovary extends nearly the length of the body cavity (Fig. 35.5*b*). It may bifurcate at its distal end to become two oviducts. These unite to form a uterus. Oviducts and uterus usually are extensively coiled within the body. One or more uterine diverticulae serve as seminal receptacles. The uterus terminates as a short vagina that opens through the female gonopore, at the anterior end of the abdomen in order Cephalobaenida or at the posterior end in order Porocephalida.[11] Females mate once; males may be polygamous.

BIOLOGY

Adult pentastomids feed on host tissue fluids and blood cells. They appear to stimulate a strong immune response, but they also are long-lived, thus, to survive they must evade their

Figure 35.3   Female of *Linguatula arctica*, the reindeer sinus worm.

Female worm; dark streak down the center of the uterus filled with eggs. Area in rectangle is enlarged in Figure 35.4(2). Scale bar = 1 cm.

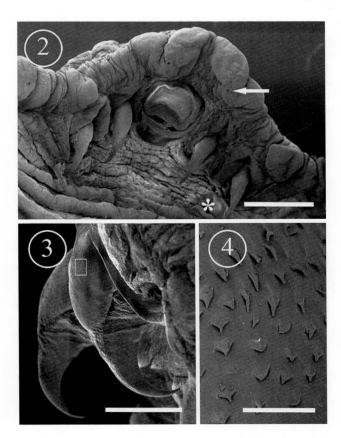

Figure 35.4   Anterior ventral view of *Linguatula arctica* from reindeer.

Figure numbers in circles are those from Nikander and Saari.[15] (2) Oral opening with hooks on either side. Papilla is marked by an asterisk(*); frontal gland opening is marked by an arrow. Scale bar = 500 μm. (3) Higher magnification view of hooks. Scale bar = 200 μm. (4) Enlargement of area in rectangle in (3), showing fine cuticular spines. Scale bar = 20 μm.

host's defenses.[22] Pentastome frontal and subparietal glands elaborate a lamellate secretion (secretory-excretory products, also known as S-E or ES), which is poured over the entire cuticular surface and evidently protects the parasites from antibody action.[22, 23] Host cells get caught up in this covering of S-E products, evidently reducing the inflammatory response that occurs each time a parasite molts. Vertebrate lungs produce a complex protein and phospholipid substance that acts as a surfactant, coating the alveoli and regulating activity of macrophages. In at least one pentastome species, S-E composition is quite similar to that of its host's own lung surfactant.[23] In experiments with *Porocephalus crotali* in mice, sustained cytotoxic responses killed parasites whose S-E coating had been removed.[4]

Development

Females, depending on size, may produce several million fully embryonated eggs, which pass up their host's trachea, are swallowed, and then pass out with feces. Intact eggs appear to be surrounded by two shell membranes—an outer, thin membrane and an inner, thick one.[10] The inner layer consists of three distinct layers. A characteristic of pentastomid eggs is the **facette,** a permanent, funnel-shaped opening through the inner membrane complex, with an inner opening extending toward the larva. A consistent feature of pentastomid embryos is a gland called a **dorsal organ, embryonic gland,** or **glandula embryonalis.** It consists of a number of gland cells surrounding a central hollow vesicle.[30] This vesicle opens through the cuticle by a dorsal pore. The dorsal organ secretes a mucoid substance that pours through the facette and ruptures the original outer membrane, which is lost. Mucoid material then flows over the inner membrane to form a new outer membrane, which is sticky when wet.[16] The viscid eggs cling together, sometimes resulting in massive infections in an intermediate host. Eggs can withstand drying for at least two weeks in the case of *Porocephalus crotali,* and they can remain viable in water at refrigerator temperatures for about six months.[10]

The larva that hatches from an egg is an oval, tailed creature with four stumpy legs, each with one or two retractable claws. The claws are manipulated by a combination of muscle fibers and an inner hydraulic mechanism. A **penetration organ** is located at the anterior end of the body. This is composed of a median spear and two lateral, pointed forks; together with the clawed legs, these structures can tear through tissues of an intermediate host. A duct opens on each side of the median spear. Accessory spinelets are present around the penetration organ of many species.

Pentastomes evidently have a developmental feature that is unique among crustaceans, namely segment constancy.[6] In *Reighardia sternae,* larvae hatch with only two head segments that bear appendages, and no more than three trunk segments. In contrast to all other crustaceans, there is no terminal budding zone and segments are not added during growth, so that the apparent segmentation of an adult is "pseudo-metamerism of its post-metameric region."[34]

The gut undergoes rather substantial morphological changes during development; these changes are probably associated with different larval and adult habitats.[32] In embryos the esophagus does not reach the midgut, but that connection is made between the first and fourth (infective) larval stages. In juveniles the mouth armature is inside the mouth, the foregut-midgut connection is a cardiac valve, a ringlike fold is present around the tongue, and sucking musculature is well developed. These anatomical features have been interpreted as aids to breaking lung capillaries in their definitive host and to sucking blood.[32]

Life Cycles

Complete life cycles are known for few species, but partial information is available for several more. With the exceptions of *Reighardia sternae* in birds and a few species of *Linguatula* in mammals, all pentastomids mature in reptiles. Intermediate hosts are various fishes, amphibians, reptiles, insects, and mammals. Typically, after ingestion by a poikilothermous (ectothermic) vertebrate, larvae hatch and penetrate the intestine and migrate randomly in the body, finally becoming quiescent and metamorphosing into nymphs (Fig. 35.6). Nymphs are infective to a definitive host; when eaten by the latter, they penetrate the host's intestine and bore into its lung, where they mature (Fig. 35.7). Each developmental stage undergoes one to several molts. Nymphal

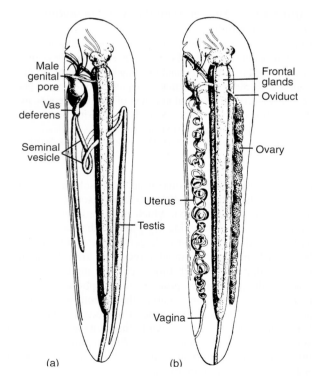

Figure 35.5   **Reproductive system of the pentastome *Waddycephalus teretiuscules.***

(*a*) Male; (*b*) female.

From J. G. Baer, *Ecology of animal parasites.* Copyright © 1952 by Board of Trustees of the University of Illinois. Used with permission of the University of Illinois Press.

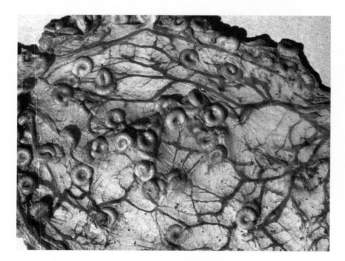

Figure 35.6   **Nymphs of *Porocephalus* sp. in the mesenteries of a vervet. *Cercopithicus aethops.***

From J. T. Self, "Pentastomiasis: Host responses to larval and nymphal infections." in *Trans. Am. Microsc. Soc.* 91:2–8. Copyright © 1972. Photograph by Robert E. Kuntz.

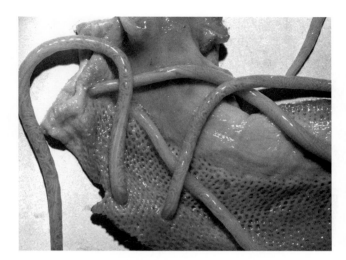

Figure 35.7 *Kiricephalus pattoni* in the lung of an Oriental rat snake, *Ptyas mucosus.*

Courtesy of Robert E. Kuntz.

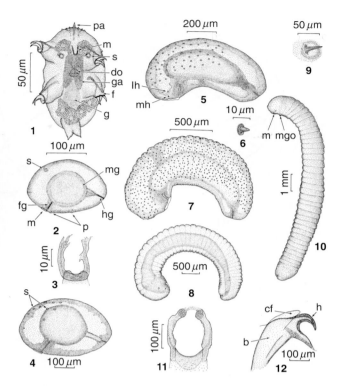

Figure 35.8 Developmental stages of *Porocephalus crotali* in experimental intermediate hosts (camera lucida drawings made from living specimens).

(*1*) primary larva (ventral view) after release from egg; (*2*) first nymphal stage (nymph I) in left lateral view (all succeeding nymphs identically oriented); (*3*) mouth ring of nymph I (en face view with anterior margin uppermost); (*4*) nymph II; (*5*) nymph III; (*6*) lateral mouth hook of nymph III; (*7*) nymph IV; (*8*) nymph V (individual stigmata not shown); (*9*) lateral mouth hook of nymph V; (*10*) nymph VI (infective stage), male, removed from enveloping cuticle of nymph V; (*11*) mouth ring of nymph VI; (*12*) lateral mouth hook of nymph VI; *b*, base of mouth hook; *cf*, cuticular fold or auxiliary hook; *do*, dorsal organ; *f*, foot or leg; *fg*, foregut; *g*, gut; *ga*, ganglion; *h*, external clawlike portion of mouth hook; *hg*, hindgut; *lh*, lateral mouth hook; *m*, mouth ring; *mg*, midgut; *mgo*, male genital opening; *mh*, medial mouth hook; *p*, papilla; *pa*, penetrating apparatus; *s*, stigma.

From J. H. Esslinger, "Development of *Porocephalus crotali* (Humboldt, 1808) (Pentastomida) in experimental intermediate hosts," in *J. Parasitol.* 48:452–456. Copyright © 1962 *Journal of Parasitology.* Reprinted by permission.

instars are difficult to differentiate. Some species even become sexually mature before completing their final ecdysis.

A definitive host that eats an egg can also serve as intermediate host, similar to the case of *Trichinella* spp. The parasites, however, probably cannot migrate to the lung and mature. Whereas vertebrates are intermediate hosts for Porocephalida, cockroaches are used by some species of *Raillietiella*,[2] and *Reighardia sternae* in gulls has a direct life cycle. *Subtriquetra subtriquetra,* a parasite of the nasopharynx of South American crocodilians, is unique in having a free-living larva, which somehow finds its fish intermediate host. Pentastomid reproductive biology was reviewed by Riley.[19]

Porocephalus crotali. The life cycle of *P. crotali* of crotalid snakes was experimentally demonstrated by Esslinger[9] (Fig. 35.8), using white mice as intermediate hosts; further elaboration was provided by Riley.[18] On hatching, larvae penetrate the duodenal mucosa and work their way to the abdominal cavity. Complete penetration can be accomplished within an hour after an egg is swallowed. After wandering about for seven or eight days, larvae molt and become lightly encapsulated in host tissue. Subsequent nymphal stages are devoid of larval characteristics, having lost their legs, penetration apparatus, and tail. During the next 80 days or so, *P. crotali* molts five more times, gradually increasing in size and becoming segmented. Mouth hooks appear during the fourth nymphal instar and increase in size through subsequent ecdyses. Sexes can be differentiated after the fifth molt. After the sixth molt, nymphs become heavily encapsulated and dormant. When eaten by a snake, nymphs are activated, quickly penetrate the snake's intestinal wall, and usually pass directly into its lung. They bury their forebody in lung tissues, feed on blood and tissue fluids, and mature.

Linguatula Species. *Linguatula serrata* (Fig. 35.9) is unusual among Pentastomida in that adults live in the nasopharyngeal region of mammals. Cats, dogs, foxes, and other carnivores are normal hosts of this cosmopolitan parasite. Almost any mammal is a potential intermediate host, and

reindeer are definitive hosts for *L. arctica* (Figs. 35.3, 35.4), a large species that evidently has a direct life cycle.[15]

Adult *L. serrata* embed their forebody into the nasopharyngeal mucosa, feeding on blood and fluids. Females live at least two years and produce millions of eggs.[12] Eggs are about 90 μm by 70 μm, with an outer shell that wrinkles when dry. Eggs exit the host in nasal secretions or, if swallowed, with feces. When swallowed by an intermediate host, the four-legged larvae hatch in the small intestine, penetrate the intestinal wall, and lodge in tissues, particularly in lungs, liver, and lymph nodes. There the nymphal instars develop, with infective stages becoming surrounded by host tissues. When eaten by a definitive host, infective nymphs either

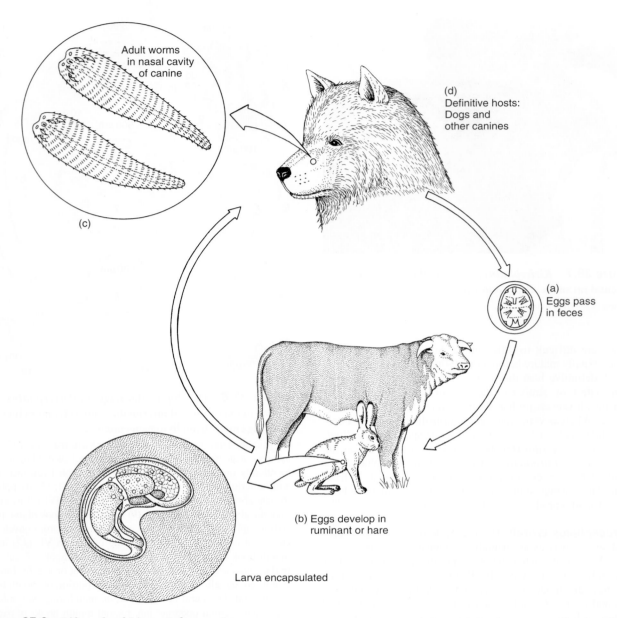

Adult worms in nasal cavity of canine (c)

(d) Definitive hosts: Dogs and other canines

(a) Eggs pass in feces

(b) Eggs develop in ruminant or hare

Larva encapsulated

Figure 35.9 **Life cycle of *Linguatula serrata*.**
Humans are dead-end hosts in this case. (*a*) Eggs pass in feces. (*b*) Larvae develop in ruminant or hare. (*c*) Adult worms in nasal cavity of canine. (*d*) Definitive hosts: dogs and other canines.
Drawing by William Ober and Claire Garrison.

attach in the upper digestive tract or quickly travel there from the stomach, eventually reaching the nasopharynx. Females begin egg production in about six months.

PATHOGENESIS

There are two types of **pentastomiasis** in humans. **Visceral pentastomiasis** results when eggs are eaten and nymphs develop in various internal organs, and **nasopharyngeal pentastomiasis** results when nymphs that are eaten locate in the nasopharynx. Both types are rather common in some parts of the world.

Visceral Pentastomiasis

Several species of pentastomids have been found encysted in humans. Probably the most commonly involved species is *Armillifer armillatus,* which has been reported from liver, spleen, lungs, eyes, and mesenteries of people in, among other places, Africa, Malaysia, the Philippines, Java, and China.[7, 28, 29] Other reported species are *A. moniliformis, Pentastoma najae, L. serrata,* and *Porocephalus* sp.

Most infections cause few if any symptoms and therefore go undetected. In fact, most recorded cases are found at autopsy, after death from other causes. However, infection of the spleen, liver, or other organs causes some tissue destruction. Ocular involvement may cause vision damage.[17] Prior visceral infection may sensitize a person, resulting in an allergy to subsequent infection.[13, 27] Host response to nymphs is often highly inflammatory, although little pathological response is elicited in definitive reptilian hosts. Dead nymphs are often calcified and are sometimes detected in X-ray films. Others begin a slow deterioration, causing a mononuclear cell response, with a subsequent abscess and granuloma formation. Experimentally produced heavy infections in rodents may kill them, indicating that visceral pentastomiasis possibly may be more important in human medicine than usually is thought.[27]

Nasopharyngeal Pentastomiasis

When nymphs of *L. serrata* invade the nasopharyngeal spaces of humans, they cause a condition usually called **halzoun** but also known as **marrara** or **nasopharyngeal linguatulosis.** According to Schacher and coworkers,[25] halzoun has been a clinically well-recognized but etiologically obscure disease since its original description in 1905. In Sudan it is known as *marrara syndrome*. In Lebanon the disease is linked in the popular mind with eating of raw or undercooked sheep or goat liver or lymph nodes; in Sudan it is linked with ingestion of various raw visceral organs of sheep, goats, cattle, or camels. A few minutes to half an hour or more after eating, there is discomfort and a prickling sensation deep in the throat; pain may later extend to the ears. Oedematous congestion of the fauces, tonsils, larynx, eustachian tubes, nasal passages, conjunctiva, and lips is sometimes marked. Nasal and lachrymal discharges, episodic sneezing and coughing, dyspnoea, dysphagia, dysphonia, and frontal headache are common. Complications may include abscesses in the auditory canals, facial swelling or paralysis, and sometimes asphyxiation and death.

At various times this condition was suspected to be caused by the trematodes *Fasciola hepatica, Clinostomum complanatum,* and *Dicrocoelium dendriticum* and also by leeches. However, recovery of *L. serrata* nymphs from the nasal passages and throats of patients in India, Turkey, Greece, Morocco, and Lebanon indicates that this species is the main cause of the condition in these areas. It is possible that the parasites can become mature if not removed or lost initially.

Epidemiology of this condition depends on cultural food patterns, in which nymphs are ingested in raw or undercooked visceral organs, primarily liver or mesenteric lymph nodes of domestic herbivores.[25]

Learning Outcomes

By the time a student has finished studying this chapter, he/she should be able to:

1. Diagram the life cycle of *Linguatula serrata*.
2. Draw a picture of a typical pentastome adult and label the important features.
3. Explain why the classification of pentastomes remained such a problem for so many decades and how this problem was finally solved.
4. Describe the causes and symptoms of halzoun.

References

References for superscripts in the text can be found at the following Internet site: www.mhhe.com/robertsjanovynadler9e

Additional Readings

Yao, M. H., F. Wu, and L. F. Tang, Lan Fang. 2008. Human pentastomiasis in China: Case report and literature review. *J. Parasitol.* 94:1295–1298.

Parasitic Insects: Phthiraptera, Chewing and Sucking Lice

Lice consultant. Heads checked and cleaned. Total cleanouts, usually in less than an hour. Satisfaction guaranteed.

—business card of Abigail Rosenfeld,
Brooklyn, New York, 2002.

Until recently in human history, lice and fleas were such common companions of *Homo sapiens* that they were considered one of life's inevitable nuisances for rich and poor, royalty and beggar alike. Evidence for our long-term relationship comes from a variety of sources, including ancient texts and mummies, such as on a 10,000-year-old human head hair from northeastern Brazil.[3] Not too many decades ago, the education of a French princess included instructions that "it was bad manners to scratch when one did it by habit and not by necessity, and that it was improper to take lice or fleas or other vermin by the neck to kill them in company, except in the most intimate circles."[58]

Lice remain widely prevalent, mostly among the very poor and in developing countries, and they obviously have contributed to the recent emergence of epidemic typhus in refugees from some troubled regions of sub-Saharan Africa.[48] Nevertheless, we in modern, industrialized countries tend to think of them as pests of the past. Hence, many upperand middle-class parents in the United States today are astonished to receive notes from a school nurse that their offspring must seek treatment for head lice. Astonishment usually evolves quickly into strong negative emotional reactions; in one psychological study of kindergarten drawings of infested people, unhappy faces, sometimes without mouths, were thought to reflect the universal negative reaction of parents and physicians, regardless of the cleanliness of a child's hair.[48]

Lice are part of our cultural heritage, giving rise to regularly used words and phrases, the origins of which most people are unaware. Do students, when complaining that their professor is a "lousy" teacher realize that they are literally accusing the august personage of being infested with lice? Few people appreciate the original meanings of the phrases "nit-picking" and "going over with a fine-toothed comb,"
which refer to the removal of louse eggs (nits) from a companion's hair. "Getting down to the nitty-gritty" and "nitwit" may assume new dimensions for some readers. And Robert Burns's poem "To a Louse" is the source of an insightful, louse-related quotation: "Oh wad some power the giftie gie us/To see oursels as ithers see us!"

Lice were traditionally assigned to two orders, Mallophaga (chewing lice) and Anoplura (sucking lice), but phylogenetic research using both morphological and molecular techniques shows that Mallophaga as previously construed was paraphyletic. Barker et al.[6] make a convincing case for inclusion of all lice within a single order, and we adopt that position. The word "mallophaga" is firmly entrenched in the literature, however, and students needing access to information on chewing lice should remember that a vast body of published research can be recovered using that word as a search term. Phthiraptera contains four suborders: Amblycera, Ischnocera, Rhynchophthirina, and Anoplura. The first three of these suborders formerly comprised Mallophaga.

Lice are wingless, are dorsoventrally flattened, and have reduced eyes or none; their tarsal claws are often enlarged, an adaptation for clinging to hair and feathers. Their development is hemimetabolous. Eggs are typically cemented to their host's feathers or hairs (Fig. 36.1), and development proceeds through three nymphal instars. The most critical difference between the suborders is the structure of their mouthparts, modified for chewing in Amblycera, Ischnocera, and Rhynchophthirina and for sucking in Anoplura.

All lice are highly adapted for parasitism: They have no free-living stages and soon die when separated from their host, so one would expect lice to have coevolved with their hosts. Molecular studies show that this expectation is met, at least in some cases. For example, pocket gophers have a

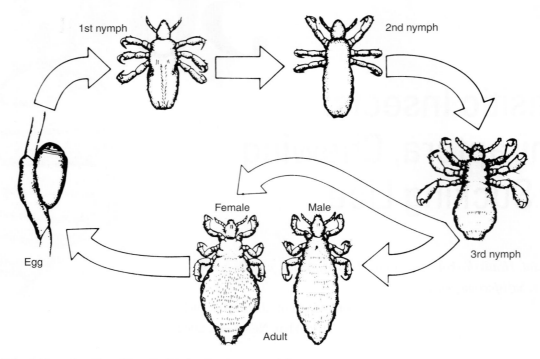

Figure 36.1 **Life cycle of head lice,** *Pediculus humanus capitis.*

Eggs (nits) are cemented to hairs and require 5 to 10 days to hatch. The life cycle requires about 21 days from egg to egg.

From H. D. Pratt and K. S. Littig, *Lice of public health importance and their control.* U.S. Government Printing Office, Washington, DC: Department of Health, Education, and Welfare, Pub. No. (CDC) 77-8265, 1973.

rich louse fauna; five species of lice from Central American pocket gophers of the genus *Orthogeomys* form a clade distinct from those on North American gopher genera (*Geomys, Thomomys,* and *Cratogeomys*). In this case enzyme studies generally confirmed earlier conclusions based on morphology.[35] Evidence from mitochondrial DNA also supports the cospeciation hypothesis.[39] On the other hand, similar studies of the lice of wallabies revealed much host switching and evident replacement of some parasite species by others, suggesting more convoluted evolutionary histories.[5] In the case of some bird lice, phoresis on parasitic but highly mobile hippoboscid flies (p. 590) evidently produces enough opportunity for colonization so that host and parasite phylogenies are not particularly congruent.[22] There is additional evidence from birds that lice can evolve faster than their hosts,[41] but human head and body lice also may represent evolutionary divergences on a single host, in this case related to loss of body hair during human evolution.[8]

CHEWING LICE

About 4400 species of chewing lice parasitize various birds and mammals. None is of direct medical importance, but some species are vectors of filarial nematodes, and others may become significant pests on domestic animals. They feed primarily on feathers and hair, but some eat sebaceous secretions, mucus, and sloughed epidermis; one study of cleared museum specimens' gut contents revealed cannibalism of eggs and nymphs, as well as predation on mites.[38] Most will eat blood, if available, such as that resulting from scratching by the host. *Menacanthus*

stramineus, a louse of chickens and turkeys, chews into developing quills to feed on blood, and some species on small birds pierce the skin to do so.

Morphology

Most lice are small, from one to a few millimeters in length. *Laemobothrion circi* is virtually a giant among lice at almost a centimeter long.[4] The head of chewing lice usually is broader than the prothorax and lacks ocelli. The short antennae have three to five segments; tarsi have one or two segments. Chewing louse mouthparts are most similar to the primitive chewing apparatus of free-living forms (see Figs. 33.12, 36.2). Mandibles are the most conspicuous of these appendages, whereas the maxillae and labium are reduced. Mandibles cut off pieces of feather or hair, and the labrum pushes them into the mouth.[4]

Three suborders of chewing lice are recognized: Amblycera, Ischnocera, and Rhynchophthirina. Amblycera are the most generalized and least host specific. They have maxillary palps, which have been lost in the other two suborders, and their antennae are carried in grooves in the head (Fig. 36.2*a*). The filiform, easily seen antennae of Ischnocera (Fig. 36.2*b*) distinguish them from Amblycera. Ischnocera are more specialized, more host specific, and more limited in food preferences; that is, their food is more confined to hairs and feathers (keratin) than is that of Amblycera. Ischnocera of birds are so specialized that they are usually limited to a particular region of their host's body or to a certain part of a feather. Rhynchophthirina is a much smaller suborder than the other two, comprising only three species: *Haematomyzus elephantis* on African and Indian elephants (see Fig. 36.9), and *H. hopkinsi*

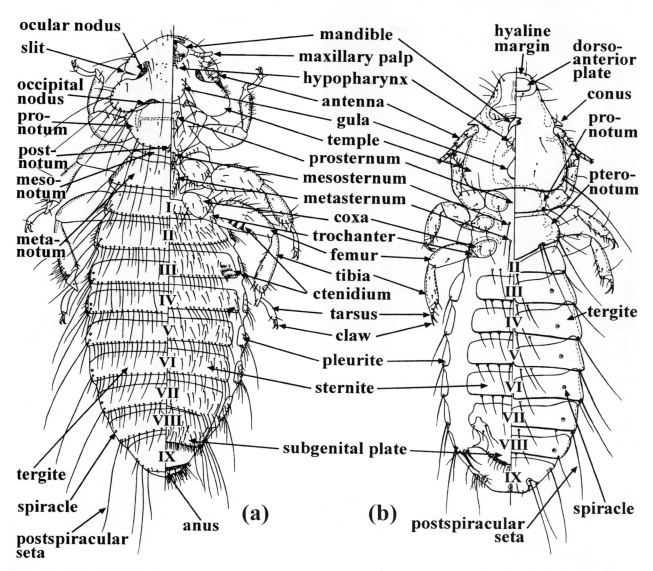

Figure 36.2   Anatomy of chewing lice.
The external characters shown are those needed to identify chewing lice using most keys.[45] A, dorsal and ventral views of female Amblycera. B, dorsal and ventral views of female Ischnocera.

From Price, R. D., R. A. Hellenthal, R. L. Palma, K. P. Johnson, and D. Clayton. 2003. *The chewing lice: world checklist and biological overview.* Il. Natural Hist. Survey Special Publ. 24., Champaign, IL. Copyright © 2003 Illinois Natural History Survey, reprinted by permission.

and *H. porci* on warthogs. Although of the chewing type, mouthparts of these lice are carried at the end of a projecting structure, and the insects feed on blood.

Biology of Some Representative Species

Because most chewing lice are parasites of birds, it is not surprising that domestic fowls host a number of species. The most common amblycerans are *Menopon gallinae,* the **shaft louse** of fowl, and *Menacanthus stramineus,* the **yellow body louse** of chickens and turkeys. Shaft lice are about 2 mm in length and usually cause little economic loss. *Menacanthus stramineus,* however, is about 3 mm long and may occur in large numbers, up to 35,000 per bird.[19] It frequents lightly feathered areas such as breast, thigh, and around the anus, gnawing through skin to reach quills of pinfeathers. This irritation can cause restlessness and disrupt a bird's feeding. Birds often become unhealthy, with reduced egg production and retarded development.

Important ischnoceran parasites of fowl include *Goniocotes gallinae,* called the **fluff louse** because it is found in fluff at

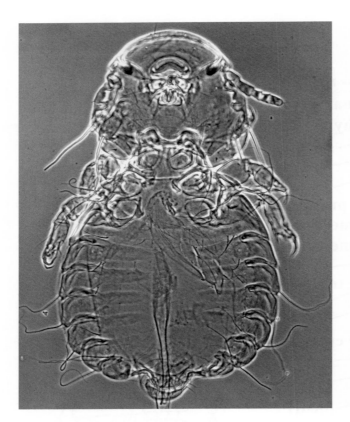

Figure 36.3 *Goniocotes gallinae* (Ischnocera), the fluff louse of fowl.

Antennae are clearly visible and do not lie in grooves on the head.

Courtesy of Jay Georgi.

the base of feathers (see Fig. 36.3); *Goniodes dissimilis,* the **brown chicken louse;** *Lipeurus caponis,* the **wing louse;** *Cuclotogaster heterographus,* the **chicken head louse;** *Chelopistes meleagridis,* the **large turkey louse;** *Oxylipeurus polytrapzius,* the **slender turkey louse;** *Columbicola columbae,* the **slender pigeon louse** (Fig. 36.4); and *Anaticola crassicornis* and *A. anseris,* **duck lice.**

Amblyceran parasites of mammals include *Gyropus ovalis* and *Gliricola porcelli* (Fig. 36.5) of guinea pigs and *Heterodoxus spiniger* on dogs. The guinea pig parasites are important to humans because of the wide use of their hosts in laboratory experiments. *Heterodoxus spiniger* is common on dogs in warmer parts of the world. Several ischnocerans are pests of other domestic mammals, including *Bovicola bovis* on cattle; *B. equi* on horses, mules, and donkeys; *B. ovis* on sheep; *B. caprae* on goats; *Trichodectes canis* (Fig. 36.6) on dogs; and *Felicola subrostratus* on cats. *Bovicola* species, when abundant, cause considerable irritation to their hosts, although the lice are only 1.5 mm to 1.8 mm long. *Bovicola bovis* causes cattle to rub against solid objects and bite at their skin in an attempt to alleviate the irritation, with consequent abrasions and hair loss. The reddish-brown color of *B. bovis* distinguishes it from sucking lice commonly found on cattle. Irritation caused by *Trichodectes canis* can become severe, especially on puppies. *Trichodectes canis* is an important intermediate host, along with dog and cat fleas, of the

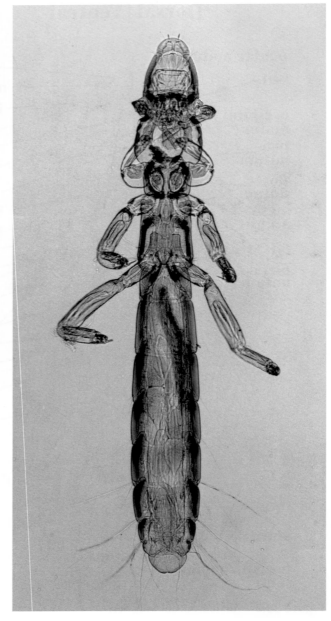

Figure 36.4 *Columbicola columbae* (Ischnocera), the slender pigeon louse.

Courtesy of Jay Georgi.

tapeworm *Dipylidium caninum,* which also can develop in humans, especially young children, who accidentally ingest the insects while playing with or petting their pets (p. 342).

Chewing lice have been implicated as intermediate hosts for several other endoparasites.[19] Several bird species, especially coots, grebes, and parrots, have filarial nematodes in their legs and ankles; the worms are transmitted by the lice, which can pick up microfilaria through chewing on skin and eating blood from minor wounds. Bartlett and Anderson[7] believe louse host specificity is responsible for evolutionary isolation of the worms in the parasites' respective host species. Even thick skin is no protection from lice; elephants may suffer a severe dermatitis caused by the rhynchophthirinan *Haematomyzus elephantis* (Fig. 36.9).[46]

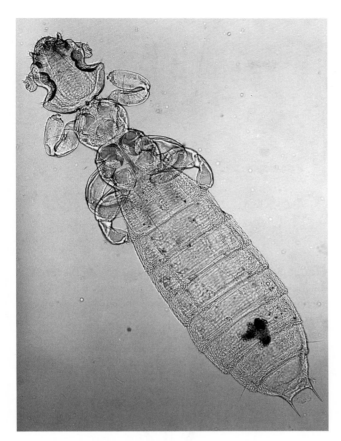

Figure 36.5   *Gliricola porcelli* **(Amblycera), a chewing louse of guinea pigs.**
Antennae are normally held in the deep grooves on the sides of the head.
Courtesy of Jay Georgi.

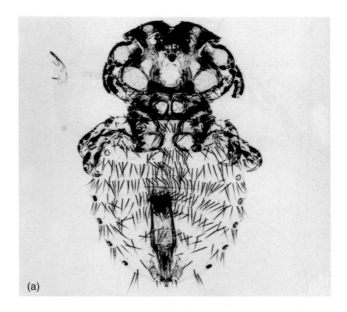

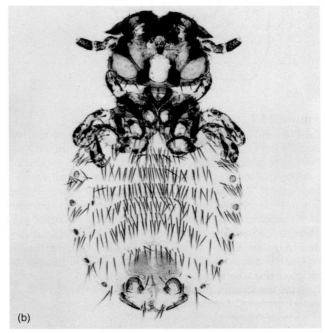

Figure 36.6   *Trichodectes canis* **(Ischnocera), the chewing louse of dogs.**
(*a*) Male; (*b*) female.
Courtesy of Jay Georgi.

In recent years, chewing lice also have been the subject of extensive, innovative, studies on host-parasite relationships and coevolution. In an ingenious series of experiments involving transfer of various louse species, Bush and Clayton[10] showed that host body size was a major factor in establishing louse host specificity. Lice could not survive on bird species smaller than their natural hosts unless the host was prevented from preening effectively. In reciprocal transfers, louse species from smaller birds did not survive on larger host species regardless of preening ability. The results suggested that in the case of both body and wing lice on birds, host switching (see chapter 2) is most likely to occur between host species of similar sizes.[10]

SUCKING LICE (SUBORDER ANOPLURA)

With fewer than 500 species,[23] sucking lice are a much smaller group than are chewing lice, parasitizing only mammals. Morphologically they are more specialized than members of the other suborder, but medically their importance and impact on human history are infinitely greater. Two species parasitize humans, *Pediculus humanus* (Fig. 36.10) and

Phthirus pubis (Fig. 36.12), of which *P. humanus* is the more important. The several species on domestic mammals are of considerable veterinary significance.

Morphology

Anoplurans superficially resemble chewing lice, with their small, wingless, flattened bodies, but anopluran heads are narrower than the prothorax. The sucking mouthparts are retracted

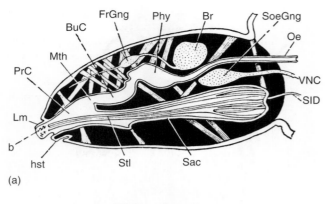

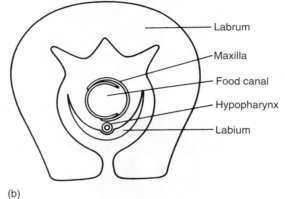

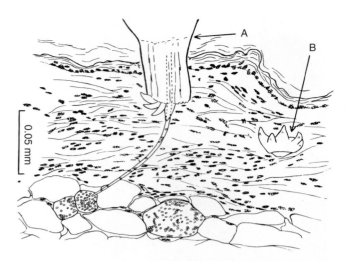

Figure 36.8 **Schematic diagram illustrating the feeding mechanism of *Haemotopinus suis*, drawn from a histologic section of the insect's mouthparts embedded in a mouse's skin.**

The labrum (*a*) is anchored in the dermis by the everted buccal teeth, and the stylets are inserted in a venule. (*b*) Lateral view of everted buccal teeth.

From M. M. J. Lavoipierre, "Feeding mechanism of *Haematopinus suis*, on the transilluminated mouse ear," in *Exp. Parasitol.* 20:303–311. Copyright © 1967.

Figure 36.7 **Sucking apparatus.**

(*a*) Piercing and sucking apparatus of Anoplura. At rest, the buccal teeth (*b*) are within the labrum (*Lm*), but when the louse bites its host, the labrum is everted, and buccal teeth cut into the epidermis of the host. *PrC,* preoral cavity, or "buccal funnel"; *Mth,* mouth; *BuC,* buccal cavity, first chamber of the sucking pump; *FrGng,* frontal ganglion; *Phy,* pharynx, second chamber of the sucking pump; *Br,* brain; *SoeGng,* subesophageal ganglion; *Oe,* esophagus; *VNC,* ventral nerve cord; *SID,* salivary duct; *Sac,* inverted sac holding the fascicle; *Stl,* stylet bundle, or fascicle; *hst,* hypostome. (*b*) Transverse section through the mouthparts of a sucking louse.

(*a*) From R. E. Snodgrass, *Principles of insect morphology.* New York: McGraw-Hill Book Co., 1935. (*b*) From R. R. Askew, *Parasitic insects.* New York: American Elsevier Publishing Co., 1971.

into the head when the animal is not feeding (Fig. 36.7*a*). Each leg has a single tarsal segment with a large claw, an adaptation for clinging to host hairs. The first legs, with their terminal claws, are often smaller than the other legs, and the third legs and their claws are usually largest. Eyes, if present, are small, and there are no ocelli. Antennae are short, clearly visible, and composed of a scape, a pedicel, and a flagellum that is divided into three subsegments. All three flagellar subsegments bear tactile hairs, and subsegments two and three bear chemoreceptors.[50]

Mode of Feeding

Lavoipierre[24] distinguished two distinct feeding methods used by bloodsucking arthropods. One of these he termed

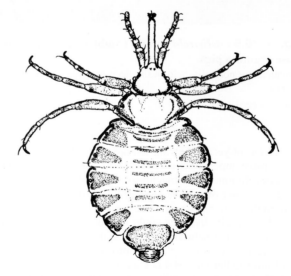

Figure 36.9 ***Haematomyzus elephantis* (Rhynchophthirina), a parasite of Indian and African elephants.**

The chewing mouthparts are at the end of a long proboscis.

From G. Lapage, *Veterinary parasitology.* Copyright © 1956 Oliver & Boyd, Ltd., London. Reprinted by permission of Addison Wesley Longman.

solenophage (Gr. for pipe + eating) for arthropods that introduce their mouthparts directly into a blood vessel to withdraw blood; the other he called telmophage (Gr. for pool + eating) for those whose mouthparts cut through the skin and vessels to produce and feed from a small pool of blood. Anoplurans are true solenophages.[25] Their proboscis is formed from the maxillae, hypopharynx, and labium, which are produced into long, thin stylets (see Fig. 36.7*b*). The maxillae

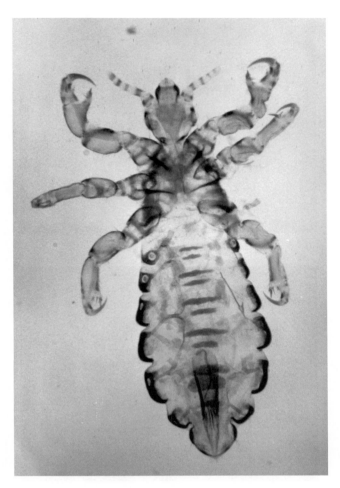

Figure 36.10 *Pediculus humanus* (Anoplura), the head and body louse of humans.

Courtesy of Warren Buss.

are flattened and rolled transversely to form a food canal, and a salivary duct passes down the hypopharynx. The third member of the stylet bundle, or **fascicle,** is the labium, which bears three serrated lobes at its tip. Mandibles are absent from most adult anoplurans.

Lavoipierre[25] observed feeding of *Haematopinus suis;* the process in other anoplurans is likely to be quite similar. The anterior tip of the head is formed by the labrum, which bears in its interior several strong, recurved teeth (Fig. 36.8). When a louse begins to feed, it places the tip of its labrum on the host's skin and begins to evert the structure, including the buccal teeth. The teeth serve to cut through the horny outer layer of the skin, and, when the labrum is fully everted, they are oriented so that their cutting edges point away from the central axis of the labrum (see Fig. 36.8). During feeding, the everted labrum and its teeth anchor the louse in place. The stylets evert and probe the tissues until they penetrate a blood vessel, usually a venule, whereupon the louse begins to suck blood. The louse sucks by means of a two-chambered pump in its head, the first chamber comprising the buccal cavity and the second the pharynx (see Fig. 36.7a). Contraction of muscles inserted on the walls of these structures and on the inner surface of the head cuticle serves to dilate the chambers. Salivary anticoagulants help keep blood flowing.[31]

How can the louse determine, when it is probing the tissue with its fascicle, that it has penetrated a venule? The stimulus for a number of hematophagous insects, presumably including lice, is detection of adenine nucleotides, particularly ADP and ATP, by chemoreceptors.[17] Nucleotides are released by platelets that aggregate in the bitten region as a result of damage done to the blood vessel by the probing fascicle.

The ability of lice (and fleas) to transmit prokaryotic pathogens may be due to the way in which they digest blood meals. In contrast to mosquitoes, lice hemolyze erythrocytes rapidly, their blood meals remain liquid, and they lack peritrophic membranes. Clotting and peritrophic membranes presumably inhibit prokaryote transmission but allow eukaryotic parasite (e.g., *Plasmodium* spp.) transmission.[55]

Pediculus humanus. Two distinct forms of *P. humanus* (see Fig. 36.10) parasitize humans: **body lice** (*P. humanus humanus*) and **head lice** (*P. humanus capitis*). Body lice also have been called *P. humanus corporis* and *P. humanus vestimenti,* and laypersons may know them by such common names as *cooties, graybacks,* or *mechanized dandruff* or by even more vulgar appellations. The two subspecies are difficult to distinguish morphologically, although they have slight differences. The subspecies will interbreed and are only slightly interfertile.[4]

A very convincing argument for separate species status for head lice (*Pediculus capitis*) and body lice (*Pediculus humanus*) was given by Busvine.[11] It seems likely that body lice descended from ancestral head lice after humans began wearing clothes. Body lice are much more common in cooler than in warmer parts of the world; in tropical areas people who wear few clothes usually have only head lice.[40] This difference makes typhus (discussed later) a disease of cooler climates because only body lice are vectors. Curiously, however, head lice can serve as hosts for the typhus-causing rickettsia and have a high potential for transmitting it.[34] Body lice are extremely unusual among Anoplura in that they spend most of their time in their host's clothing, visiting the host's body only during feeding. They nevertheless stay close to the body and are most commonly found in areas where clothing is in close contact.

Eggs (nits) of body lice are cemented to fibers in clothes and have a cap at one end that admits air and facilitates hatching. Eggs hatch in about a week, and the combined three nymphal stages usually require eight to nine days to mature when they are close to a host's body. Lower temperature lengthens the time of a complete cycle; for example, if clothing is removed at night, the life cycle will require two to four weeks. If clothing is not worn for several days, the lice will die. A female can lay 9 or 10 eggs per day, up to a total of about 300 eggs in her life; therefore, she has a high reproductive potential. Fortunately, this potential is usually not realized. It is typical to find no more than 10 lice per host, although as many as a thousand have been removed from the clothes of one person.[44]

Body lice normally do not leave their host voluntarily, but their temperature preferences are rather strict. They will depart when a host's body cools after death or if the person has a high fever. Nevertheless, they travel from one host to another fairly easily, and one can acquire them by contact with infested people in crowded locations such as buses and trains. Of course, they also may be acquired easily by donning infested clothing or occupying bedding recently vacated by a person with lice. Potential for transmission is highest when people are in crowded, institutionalized conditions or in war or prison camps, where sanitation is bad and clothing cannot be changed often.

Head lice tend to be somewhat smaller than body lice:1.0 mm to 1.5 mm for males and 1.8 mm to 2.0 mm for females, contrasted with 2 mm to 3 mm and 2 mm to 4 mm for male and female body lice, respectively.[44] Nits of both are about 0.8 mm by 0.3 mm. Head lice nits are cemented to hairs. Lice are usually most prevalent on the back of the neck and behind the ears, and they do not infest eyebrows and eyelashes. They are easily transmitted by physical contact and stray hairs, even under good sanitary conditions. Accordingly, they occur surprisingly often among schoolchildren. As in the case of body lice, however, the heaviest infestations are associated with crowded conditions and poor sanitation.[27]

Infestation with lice (**pediculosis**) is not life threatening, unless the lice carry a disease organism, but it can subject a host to considerable discomfort. The bites cause a red papule to develop, and it may continue to exude lymph. Intense pruritis induces scratching, which frequently leads to dermatitis and secondary infection. Symptoms may persist for many days in sensitized persons. Years of infestation lead to a darkened, thickened skin, a condition called **vagabond's disease.** In untreated cases of head lice the hair becomes matted together from exudate, a fungus grows, and the mass develops a fetid odor. This condition is known as **plica polonica.** Large numbers of lice are found under the mat of hair.

Pediculus humanus carries symbiotic bacteria, including *Wolbachia* species;[16] some endosymbionts occur in mycetomes (see chapter 37), and others have been used in coevolutionary studies of primates and their lice.[1]

Phthirus pubis. Origin of the common name of this insect, **crab louse** or, more popularly, **crabs,** is evident from its appearance (see Fig. 36.12). These lice are 1.5 mm to 2.0 mm long and nearly as broad as long, and the grasping tarsi on the two larger pairs of their legs are reminiscent of crabs' pincers (Fig. 36.13). *Phthirus pubis* dwells primarily in the pubic region, but it may also be found in armpits and rarely in beards, mustaches, eyebrows, and eyelashes. *Phthirus pubis* is less active than are *Pediculus* spp., and it may remain in the same position for some time with its mouthparts inserted in the skin. Bites can cause an intense pruritis but fortunately do not seem to transmit disease organisms.

Nits are cemented to hair (Figs. 36.11, 36.14), and the complete life cycle requires less than a month. A female deposits only about 30 eggs during her life. Infection can occur through contact with bedding or other objects, especially in crowded situations, but transmission is characteristically venereal and often a surprise to the new host. A few years ago one of the authors received an envelope addressed to the "Department of Microscopic Analysis, University of Massachusetts," along with the rather urgent instruction, "Please microscope these specimens immediately. They were taken from a living organism." The distress and embarrassment of the sender were apparent. On a more serious note, pubic lice can be considered a predictor of sexually transmitted diseases such as chlamydia and gonorrhea.

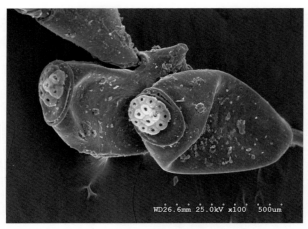

(a)

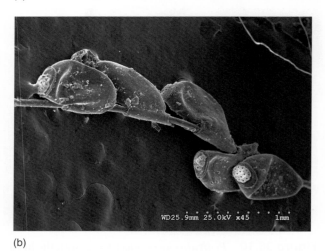

(b)

Figure 36.11   **Nits from a mummy.**

(*a*) High magnification view of head louse eggs from a South American mummy, A.D. 900–1200. Opercula are intact and the pores can be seen. (*b*) Several nits on a single hair from the same individual as (*a*).

Photographs courtesy of Nicole Searcey.

Other Anoplurans of Note

Like many chewing lice, Anoplura tend to be host specific. Interestingly, *Pediculus humanus* can also live and breed on pigs,[4] and *Haematopinus suis* of swine will readily feed on humans when they are hungry.[19] Principal effects on livestock are irritation, weight loss, and anemia in heavy infestations. The USDA estimated that the combined effects of chewing and sucking lice amounted to a $47 million loss in each of the cattle and sheep industries in 1965.[52] More recent estimates suggest that louse infestations may cost the sheep industry as much as $169 million annually in Australia alone, including $44 million spent on labor in control efforts and $55 million in wool loss.[29] In the United Kingdom, recent estimates for losses due only to lesions in hides range from $22–$30 million annually.[9]

Haematopinus suis (Fig. 36.15) on swine is considered their most serious infection after hog cholera.[52] Other

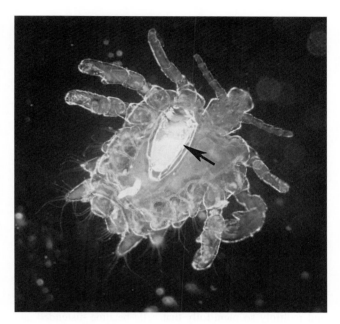

Figure 36.12   *Phthirus pubis* (Anoplura), the pubic, or crab, louse of humans.
Arrow, developing egg.
Courtesy of Warren Buss.

Figure 36.14   Nit of *Phthirus pubis* cemented to a hair.
Nits of *Pediculus* spp. are essentially similar. Note the operculum with pores.
Courtesy of John Ubelaker.

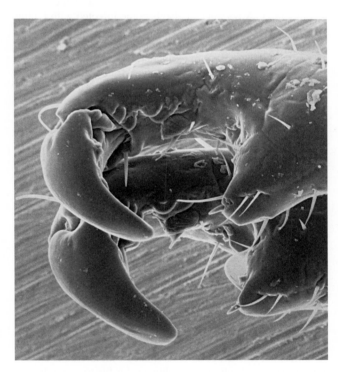

Figure 36.13   Scanning electron micrograph of the tarsi of the second and third legs of *Phthirus pubis,* with the terminal claw.
This nicely illustrates the adaptation for grasping hairs of its host.
Courtesy of John Ubelaker.

Figure 36.15   *Haematopinus suis,* the pig louse.
Courtesy of Warren Buss.

Haematopinus species infest cattle: *H. eurysternus,* the **short-nosed cattle louse;** *H. quadripertusus,* the **cattle tail louse;** and *H. tuberculatus,* primarily a parasite of water buffalo. The most serious economic losses resulting from lice on cattle are caused by *H. eurysternus. Haematopinus asini* is a parasite of horses, mules, and donkeys. *Haematopinus* spp. are large, blind lice; female *H. suis* are as long as 6 mm.

Louse infestations can be surprisingly common: one survey in Scotland showed 80% of the farms had cattle with lice, and in a sample of Norwegian herds, 94% had *Bovicola bovis,* with 27% of the animals infested.[15] Up to four species of lice were involved: *B. bovis, Linognathus vituli, Haematopinus eurysterms, Solenoptes capillatus,* with the first two most prevalent. Over 75% of hides in another Norwegian study showed lice damage.[15]

Like *Pediculus* spp., different members of *Linognathus* may specialize on different regions of the body: *L. pedalis* is found on the legs of sheep, whereas *L. ovillus* predominates on the head. *Pediculus mjobergi* is found on New World monkeys, sometimes becoming a problem in zoos. *Polyplax spinulosa* (Fig. 36.16) of *Rattus* spp. can transmit *Rickettsia typhi,* the causative agent of murine typhus, carried also by fleas (chapter 38).

LICE AS VECTORS OF HUMAN DISEASE

Three important human diseases are transmitted by *Pediculus humanus humanus:* epidemic, or louse-borne, typhus; trench fever; and relapsing fever.

Epidemic, or Louse-Borne, Typhus

Typhus is caused by a rickettsial organism, *Rickettsia prowazekii.* Rickettsias are bacteria that usually are obligate intracellular parasites. Various species can infect vertebrate and/or invertebrate hosts with effects ranging from symptomless to severe. Epidemic typhus has had an enormous impact on human history, detailed in Zinsser's classic book *Rats, Lice and History.*[58] Typhus epidemics tend to coincide with conditions favoring heavy and widely prevalent infestations of body lice, such as preand postwar situations and crowding, stress, poverty, and mass migration. Mortality rates during epidemics may approach 100%.

It is not certain which or how many of the great epidemics of earlier human history were caused by typhus, but in historical accounts of the decimation of the Christian and Moorish armies in Spain during 1489 and 1490 the role of typhus is clear. In 1528 typhus reduced the French army besieging Naples from 25,000 to 4000, leading to its defeat, to the crowning of Charles V of Spain as Holy Roman Emperor, and to the dominance of Spain among European powers for more than a century. The Thirty Years' War can be divided epidemiologically into two periods: 1618 to 1630, when the chief scourge was typhus, and 1630 to 1648, when the major epidemic was plague (see chapter 38). Zinsser contends that between 1917 and 1921, there "were no less and probably more than twenty-five million cases of typhus in the territories controlled by the Soviet Republic, with from two and one-half to three million deaths."[58]

Typhus starts with a high fever (39.5°C to 40.0°C), which continues for about two weeks, and backache, intense headache, and often bronchitis and bronchopneumonia. There is malaise, vertigo, and loss of appetite, and the face becomes flushed. A petechial rash appears by the fifth or sixth day, first in the armpits and on the flanks and then extending to the chest, abdomen, back, and extremities. The palms, soles, and face are rarely affected.[37] After about the second week, fever drops, and profuse sweating begins. At this point, stupor ends with clearing consciousness, which is followed either by convalescence or by an increased involvement of the central nervous system and death. The rash often remains after death, and subdermal hemorrhagic areas frequently appear. The disease can be treated effectively with broad-spectrum antibiotics of the tetracycline group and chloramphenicol. Also, although prior vaccination with killed *R. prowazekii* does not result in complete protection, severity of disease is greatly ameliorated in persons who have been vaccinated.

Curiously, typhus is a fatal disease for lice. When a louse picks up rickettsiae along with blood from a human host, the organisms invade the louse's gut epithelial cells and multiply so plentifully that cells become distended and rupture. After about 10 days so much damage has been done to the insect's gut that the louse dies. For several days before its demise, however, the louse's feces contain large numbers of rickettsiae. Scratching louse bites or when crushing the offending creature, a human is inoculated with typhus organisms from louse feces. A louse's strong preference for normal body temperature causes it to leave a febrile patient and search for a new host, thus facilitating spread of the disease in epidemics. A person can also become infected with typhus by inhaling dried louse feces or getting them in the eye. *Rickettsia prowazekii* can remain viable in dried louse feces for as long as 60 days at room temperature.[19]

Because infection is fatal to lice, transovarial transmission cannot occur, and humans are an important reservoir host. After surviving the acute phase of the disease, humans can be asymptomatic but capable of infecting lice for many years. The disease can recrudesce and produce a mild form known as **Brill-Zinsser disease.** Flying squirrels (*Glaucomys volans*) also can be a reservoir host, with the infection transmitted by lice (*Neohaematopinus sciuropteri*) and fleas (*Orchopeas howardii*).[51] Some cases in the United States were probably caused by contact with such animals.[28] Human and possibly the animal reservoirs could provide the source for a new epidemic in the event of a war, famine, or other disaster. As Harwood and James[19] point out, "Current standards of living in well-developed countries have largely eliminated the disease there, but its cause lies smoldering, ready to erupt quickly and violently under conditions favorable to it."

Both Ricketts and Prowazek, the pioneers of typhus research, became infected with typhus and died in the course of their work. Interestingly, Howard Ricketts was a football player in college who went to medical school, where he encountered an influential teacher, became fascinated with microbial disease transmission, and subsequently devoted his life to research.

Trench Fever

Trench fever is a nonfatal but very debilitating disease caused by another rickettsia, *Bartonella quintana,* transmitted by

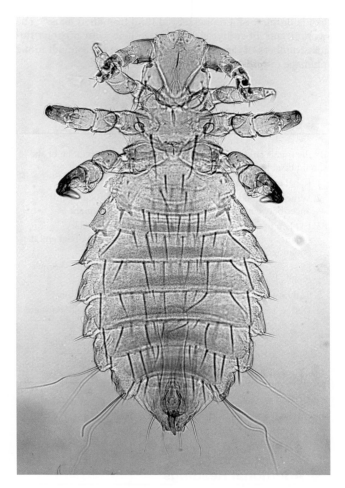

Figure 36.16   *Polyplax spinulosa* **(Anoplura), parasitic on brown and black rats.**

This louse transmits murine typhus from rat to rat, although not from rats to humans. Another species of *Polyplax, P. serrata,* parasitizes mice.

Courtesy of Jay Georgi.

Pediculus humanus humanus. Epidemics occurred in Europe during World Wars I and II, and foci have since been discovered in Egypt, Algeria, Ethiopia, Burundi, Japan, China, Mexico, and Bolivia. In lice the rickettsia multiplies in the gut lumen. Infection of humans occurs by contamination of abraded skin with louse feces or a crushed louse or by inhalation of louse feces. The organism is not pathogenic for lice; thus, the vector remains infective for the duration of its life.

A latent infection period lasts about 10 to 30 days, toward the end of which a person may experience headache, body pain, and malaise. Temperature then rises rapidly to 39.5°C to 40.0°C, accompanied by headache, pain in the back and legs (especially in the shins), dizziness, and postorbital pain in movement of the eyes. A typhuslike rash appears, usually early in the attack, on the chest, back, and abdomen, but it disappears within 24 hours. Fever continues for as long as a week and occasionally for several weeks. Convalescence is often slow, and an initial attack is followed in about half the cases by a regularly or irregularly relapsing fever. Tetracyclines are effective in treatment.

Bartonella quintana has no known nonhuman reservoirs.[17]

Relapsing Fever

The third important disease of humans transmitted by body lice is epidemic relapsing fever, which is caused by a spirochete, *Borrelia recurrentis.* Mortality is usually low, but the fatality rate can reach more than 50% in groups of undernourished people.[44] Lice pick up bacteria along with their blood meal, and spirochetes penetrate the insect's gut to reach the hemocoel. They multiply in hemolymph but do not invade salivary glands, gonads, or Malpighian tubules. Therefore, transmission is accomplished only when a louse is crushed by host scratching, which releases the spirochetes. Hence, infectious organisms gain entrance through abraded skin, but evidence also indicates that they can penetrate unbroken skin.[12] Louse-borne relapsing fever apparently has disappeared from the United States, but scattered foci are in South America, Europe, Africa, and Asia. Ethiopia had 4700 cases and 29 deaths in 1971.[19] Frequent epidemics occurred in Europe during the 18th and 19th centuries, and major epidemics befell Russia, central Europe, and North Africa during and after World Wars I and II. During the war in Vietnam an epidemic occurred in the Democratic People's Republic of Vietnam.[40]

Clinically, louse-borne relapsing fever is indistinguishable from the tick-borne relapsing fevers that are caused by other species of *Borrelia* (chapter 41). After an incubation period of 2 to 10 days, the victim is struck rather suddenly by headache, dizziness, muscle pain, and a fever that develops rapidly. Transitory rash is common, especially around the neck and shoulders and then extending to the chest and abdomen. The patient is severely ill for four to five days, when the temperature suddenly falls, accompanied by profuse sweating. Considerable improvement is seen for 3 to 10 days, and then another acute attack occurs. The cycle may be repeated several times in untreated cases. Antibiotic treatment is effective but complicated in this disease by serious systemic reactions to the drugs.

Humans are the only reservoirs, and epidemics are associated with the same kind of conditions connected with louse-borne typhus epidemics. The diseases often occur together.

CONTROL OF LICE

For detailed information on control of lice on humans, consult Pratt and Littig[44] and Wendel and Rompalo.[57] A variety of commercial preparations containing insecticides effective against lice are available. Within the last few years no fewer than six brands were found on the shelves of a supermarket in Homestead, Florida. Insecticides (permethrin) may also be incorporated into hair care products. In one study of 38,160 patients who used a permethrin creme rinse for 47,578 treatments, the delousing product proved both safe and effective.[2] But in a similar study in Israel 14 different antilouse shampoos varied in their ability to kill both lice and eggs.[33]

An extensive literature review revealed 1% permethrin creme rinse to be the only chemical treatment virtually guaranteeing at least a 90% cure rate.[53] However, permethrin resistance has been reported, and there is growing interest in use of naturally occurring phytochemicals, which may be lethal to a variety of arthropod vectors, including lice.[32, 43]

Hot air also kills head lice and nits, and in one study a single 30-minute treatment at temperatures "slightly cooler than a standard hair dryer" eradicated the parasites.[18] Extensive combing and picking helps with head lice (see epigraph quote from Abigail Rosenfeld's business card). Good personal hygiene with ordinary laundering of garments, including dry cleaning of woolens, will control body lice. Devices for large-scale treatment of civilian populations, troops, and prisoners of war by blowing insecticide dust into clothing are effective and have controlled or prevented typhus epidemics.

Lice on pets and domestic animals can be controlled by insecticidal dusts and dips. Ear tags impregnated with cypermethrin (a synthetic pyrethroid)[21] and slow-release moxidectin injected subcutaneously[56] have both been used on livestock. However, acquired resistance to cypermethrin has been demonstrated in laboratory studies.[26] Several commercially available endectocides (primarily ivermectin, doramectin, and avermectin formulations) also are effective, depending on the dose and delivery method.[13] As mentioned elsewhere, especially in the nematode chapters (for example, see p. 358), development of these macrocyclic lactone compounds has had a major impact on the treatment of various parasitic infections and infestations. Experimental work with mice also has shown that it is possible for a host to be partially immune to lice and injection of soluble antigens resulted in some protection.[47] Immunization of rabbits against *Pediculus humanus* has also been reported, with polyclonal antibodies being directed mainly at midgut proteins.[36]

Normal, healthy mammals and birds usually apply some natural louse control by grooming and preening themselves. Poorly nourished or sick animals that do not exhibit normal grooming behavior often are heavily infested with lice. Many species of passerine birds show an interesting behavior known as **anting** that may represent another natural method of louse control. The bird settles on the ground near a colony of ants, allowing the ants to crawl into its plumage, or it picks up ants and applies them to the feathers. The bird uses only ant species whose workers exude or spray toxic substances in attack and defense but do not sting. Ants in two subfamilies of Formicidae either spray formic acid or exude droplets of a repugnatorial fluid from their anus.[49] The worker ants liberally anoint the feathers with noxious fluids. Numbers of dead and dying lice have been found in the plumage of birds immediately after anting.

Learning Outcomes

By the time a student has finished studying this chapter, he or she should be able to:

1. Label an anatomical drawing of a sucking louse.
2. Write an extended paragraph describing the structural differences between Amblycera, Ischnocera, and Anoplura.
3. Explain the various ways that lice can be of veterinary importance, being certain to mention the hosts involved and the pathological conditions caused by lice.
4. Describe the factors involved in transmission of louse-borne typhus.
5. Tell how the mouthparts of an anopluran louse function.
6. Define "solenophage," "telmophage," and "vagabond's disease."

References

References for superscripts in the text can be found at the following Internet site: www.mhhe.com/robertsjanovynadler9e

Additional Readings

1. Kim, K. C., H. D. Pratt, and C. J. Stojanovich. 1986. *The sucking lice of North America: An illustrated manual for identification.* University Park, PA: Pennsylvania State University Press.
2. Rothschild, M., and T. Clay. 1952. *Fleas, flukes and cuckoos.* New York: Philosophical Library.

Chapter 37

Parasitic Insects: Hemiptera, Bugs

Snug as a bug in a rug.

> —Benjamin Franklin

Bug off!

> —familiar slang

Although to a layperson all insects, and sometimes even bacteria and viruses (including computer viruses), are bugs, to zoologists the only true bugs are members of order Hemiptera. It is one of the larger insect orders, containing more than 55,000 species, but relatively few (about 100 species) are ectoparasites of mammals and birds.[3] Two suborders are commonly recognized: Homoptera and Heteroptera. Homoptera (considered a separate order by some authors) includes such insects as aphids, scale insects, leaf hoppers, and cicadas. Because Homoptera feed on plant juices, they are not of public health or veterinary importance, but in many cases they are very important agricultural pests. Both Homoptera and Heteroptera have sucking mouthparts, which are folded back beneath the head and thorax when at rest (Fig. 37.1). Members of the two suborders are easily distinguished if they have wings: Both pairs of homopterans' wings are wholly membranous, whereas heteropteran forewings (**hemielytra**) are divided into a heavier, leathery proximal portion and a membranous distal portion.

Hemipterans are exopterygotes with hemimetabolous development; consequently, nymphs have habitat and feeding preferences similar to those of adults. Like Homoptera, most heteropterans suck plant juices, but many are predatory, using their mouthparts to suck body fluids of smaller arthropods. Some are quite cannibalistic, at least in culture, and can acquire parasites (especially trypanosomatid flagellates) from their prey.[40] Although some normally plant-feeding forms suck blood in rare instances,[47] relatively few heteropterans obtain most or all of their nutrition from bird or mammal blood. Most of the ones who do are found in families Cimicidae and Reduviidae.

MOUTHPARTS AND FEEDING

Regardless of whether they feed on plant or animal juices, homopterans and heteropterans have similar basic mouthparts.[14] The labrum is short and inconspicuous, but the labium is elongated and forms a tube containing the mandibles and maxillae (Fig. 37.2). Maxillae enclose a food canal, through which fluid food is drawn, and a salivary canal, through which saliva is injected. Mandibles run alongside the maxillae in the labial tube and together with maxillae comprise the **fascicle.** Tips of the mandibles and maxillae may be barbed or spined.

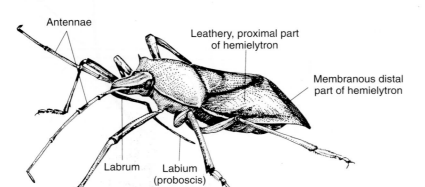

Antennae

Leathery, proximal part of hemielytron

Membranous distal part of hemielytron

Labrum

Labium (proboscis)

Figure 37.1   Diagram of typical bug (Hemiptera).

The labium forms a tube within which lies the fascicle (maxillae and mandibles).

Drawing by Ian Grant.

Operation of mouthparts in the bedbug *Cimex lectularius* and in the reduviid *Rhodnius prolixus* has been studied.[10, 23] Both are solenophages. *Rhodnius prolixus* applies the tip of its labial tube to a host's skin, anchors itself with the barbed tips of its mandibles, and slides each maxilla against the other, piercing the dermal tissue (Fig. 37.3). The route of maxillary penetration curves because the tip of one maxilla is hooked and the other is spiny. In *C. lectularius* the entire fascicle penetrates, and the labium folds at its joints (Fig. 37.4). In both cases feeding begins when a blood vessel is pierced. Some and perhaps all bloodsucking bugs have sensory receptors on the tips of the mandibles and maxillae that are undoubtedly essential to feeding success.[36] Not surprisingly, saliva of bloodsucking hemipterans contains a variety of anticoagulants.[34, 48]

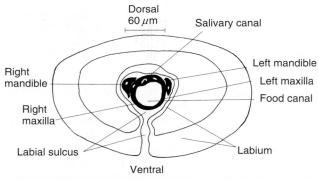

Figure 37.2 Cross section of the proboscis of the nonfeeding *Rhodnius prolixus* close to the base of the labium.

From M. M. Lavoipierre et al., "Studies on the methods of feeding of bloodsucking arthropods. I. The manner in which triatomine bugs obtain their bloodmeal, as observed in the tissue of the living rodent, with some remarks on the effects of the bite on human volunteers," in *Ann. Trop. Parasitol.* 53:235–250. Copyright © 1959.

Figure 37.4 Schematic diagram showing successive stages in the introduction of the fascicle of *Cimex lectularius* into a rodent's ear.

(*a*) The fascicle (mandibles and maxillae) is being thrust into the tissues. (*b*) Probing has commenced, and the flexible fascicle is shown bending in tissues. (*c*) The tip of the maxillary bundle has entered the lumen of a vessel, but the mandibles remain outside. Both the mandibles and maxillae enter and probe host tissues as a compact bundle (fascicle), and the labium is progressively bent to allow the fascicle to be projected deep into tissues.

From G. Dickerson and M. M. J. Lavoipierre, "Studies on the methods of feeding of blood-sucking arthropods. II. The method of feeding adopted by the bedbug (*Cimex lectularius*) when obtaining a blood-meal from a mammalian host," in *Ann. Trop. Med. Parasitol.* 53:347–357, 1959.

Like lice, both bloodsucking reduviids (Triatominae) and Cimicidae depend for part of their nutrition on endosymbiotic bacteria. These bacteria are contained in epithelial cells of the triatomine gut, and cimicids bear theirs in two disc-shaped **mycetomes** in their abdomen beside the gonads (see Fig. 37.7). *Triatoma infestans* has symbiotic bacteria

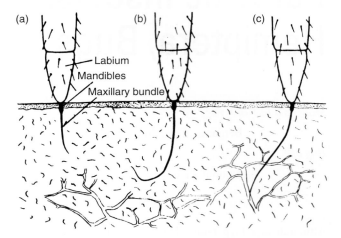

Figure 37.3 Schematic diagram showing successive stages in the introduction of the fascicle of *Rhodnius prolixus* into the skin of a rodent.

(*a*) The maxillary bundle is being thrust into tissues. (*b*) Probing has commenced, and the flexible maxillary bundle is shown bending sharply. (*c*) The tip of the maxillary bundle has entered the lumen of a vessel. Barbed mandibles act as anchors to the fascicle while maxillae are projected deep into tissues.

From M. M. Lavoipierre et al., "Studies on the methods of feeding of blood-sucking arthropods. I. The manner in which triatomine bugs obtain their bloodmeal, as observed in the tissue of the living rodent, with some remarks on the effects of the bite on human volunteers," in *Ann. Trop. Parasitol.* 53:235–250. Copyright © 1959.

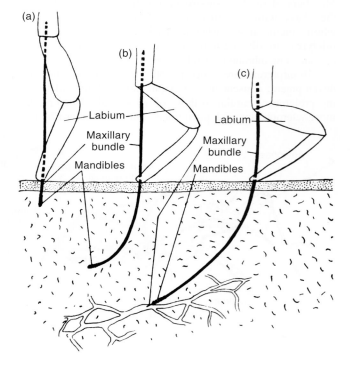

throughout many of its tissues, including muscles, gonads, and nervous system. There is no evidence that these bacteria do the bug any harm.[18] The mutualistic bacteria of the gut, at least, are essential for growth and maturation of the bugs; aposymbiotic triatomines, those that are artificially "cured" of their bacteria, can reach only the second, third, or fourth instars before death, depending on the species.[20]

FAMILY CIMICIDAE

Cimicids include a variety of small, wingless bugs that feed on the blood of warm-blooded animals, primarily birds and bats (Fig. 37.5). *Cimex lectularius, C. hemipterus,* and *Leptocimex boueti* are known as *bedbugs* and attack humans. *Cimex lectularius* is cosmopolitan but is found primarily in temperate zones, whereas *C. hemipterus* tends to be more tropical, and *L. boueti* is confined to West Africa. Of the 22 genera of cimicids, 12 are parasites of bats. Bird feeders of the group most often attack birds that commonly nest in caves, although swallows, especially the colonial cliff swallow that builds structurally complex mud nests, can support large populations of the swallow bug *Oeciacus vicarius.* Brown and Brown[7] reported up to 700 bugs per nest in large swallow colonies, and blood loss due to ectoparasitism significantly reduced the weight of nestlings.

It is believed that caves were probably the ancestral home of Cimicidae,[3] and it is not unreasonable to assume that humans acquired bedbugs during their cave-dwelling period. Bedbug specimens have been collected from Egyptian tombs dated older than 3000 years.[38] All three species that parasitize humans will also feed on bats. In addition, *Cimex* spp. will feed on chickens, and *C. lectularius* readily attacks rodents and some other domestic animals.

Although bedbugs are not known to transmit any human disease naturally, they can be extremely annoying. People who are chronically infested by bedbugs suffer loss of sleep, sores from infected bites, iron and hemoglobin deficiencies, and rarely mechanical transmission of hepatitis B virus.[29] *Cimex lectularius* and *Rhodnius prolixus* both have been experimentally infected with hepatitis B virus (HBV), which stayed in bedbugs for over a month, surviving molts, and was passed in feces.[5] HBV was detected in *R. prolixus* feces for two weeks. Hepatitis C virus does not survive in either insect.[43] *Cimex hemipterus* has been experimentally infected with human immunodeficiency virus (HIV), which remained viable in the bug for as long as eight days.[51] However, mechanical transmission could not be demonstrated and there was no virus in the insect feces. As is the case with numerous invertebrates, *Wolbachia* bacteria have been isolated from bedbugs.[39]

Bedbug infestations have increased dramatically over the past two decades, noticeably so in urban environments, for example in homeless shelters.[17] Insecticide resistance has been reported, and some entomologists believe this resistance is a major factor in the global resurgence of bedbug populations.[38] This resurgence has been blamed for serious economic losses in the tourism industry.[16] It would not be surprising for a student using this text to discover bedbugs in a college dormitory; in one study, nearly half the units in an Indianapolis high-rise building had infestations within 41 months of the first infestation and half the residents were not aware of the bugs in their apartments.[50]

For an excellent treatment of the taxonomy, ecology, morphology, reproduction, and control of cimicids, consult Usinger's monograph.[46]

Morphology

Cimicids are reddish brown bugs, up to about 8 mm long (see Fig. 37.5), and are flattened dorsoventrally. They do not stay on their hosts for any longer than the 5 to 10 minutes

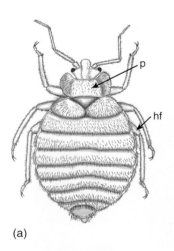

(a)

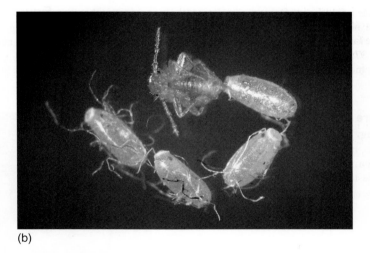

(b)

Figure 37.5   *Cimex lectularius.*
(*a*) Adult female; (*b*) eggs and an emerging first-instar nymph. *Cimex hemipterus*, the other bed bug species important o humans, has a narrower pronotum (p) and shorter hind femur (hf) than *C. lectularius*. The eggs of the two species can also be distinguished by their surface patterns.

(*a*) Drawn by John Janovy, Jr., from various sources; (*b*) courtesy of the University of Nebraska-Lincoln Department of Entomology.

of feeding. Their appendages are not particularly adapted for clinging to their hosts, but they enable the bugs to run rather rapidly. Adults have no wings, although rudimentary wing pads form in nymphs. Doubtless, this winglessness is an adaptation for inhabiting narrow crevices in which the bugs reside between feedings. Their antennae have four segments, and the distal two are much more slender than are the proximal ones. Cimicids have conspicuous compound eyes but no ocelli, and their first thoracic somite forms a rim around the posterior portion of their head. The tergum of the first thoracic somite **(pronotum)** of *C. lectularius* is about two and one half times broader than long, the pronotum of *C. hemipterus* is just more than twice as broad as long, and the pronotum of *L. boueti* is only a bit wider than the head. Scent glands opening on the ventral side of the third thoracic somite produce an oily secretion and give bedbug-infested dwellings a disagreeable odor. The secretion is probably a defense against predators.

Biology

Bedbugs are nocturnal, emerging from their daytime hiding places to feed on resting hosts during the night (Fig. 37.6). Peak activity of *C. lectularius* is just before dawn.[28] Bites cause little reaction in some people, whereas in others they cause considerable inflammation as a result of allergic reactions to the bug's saliva. The annoyance may disturb sleep, and persistent feeding may reduce a person's hemoglobin count significantly.[14]

Bedbugs can survive long periods of starvation. Adults commonly live without food for more than four months and can survive to at least 18 months.[3] Cannibalism is common.[11]

Cimicids practice a rather startling type of copulation known as *traumatic insemination,* employed among other insects only by related families Anthocoridae and Polyctenidae.[3] Males have a copulatory appendage, the **paramere,** which curves strongly to the left; the right paramere has been lost. The aedeagus is small and lies immediately above the paramere base. During copulation the paramere stabs into a notch **(paragenital sinus)** near the right side of the posterior border of the female's fifth abdominal sternite[46] (Fig. 37.8). Sperm enter a pocket **(spermalege)** from which they emerge into the hemocoel and make their way to organs

at the base of the oviducts, **seminal conceptacles** (Figs. 37.7 and 37.8). Seminal conceptacles are analogous but not homologous to spermathecae of other insects. From the conceptacles, sperm travel by minute ducts in the walls of the oviducts to the ovarioles and ova. Male cimicids mate with females repeatedly, and homosexual behavior is common. Evidently the last male to mate with a female contributes the most sperm.[44] In the laboratory, male *C. hemipterus* and female *C. lectularius* interbreed rather freely; the mating produces sterile eggs, however, and in mixed infestations *C. lectularius* females may lay mostly infertile eggs.[30]

Females lay from 200 to 500 eggs in batches of 10 to 50. Eggs hatch in about 10 days, and the five nymphal instars must each have at least one blood meal. A blood meal is also necessary before males will mate and females will oviposit. The time from egg to maturity is between 37 and 128 days, but this time is lengthened during periods of starvation.

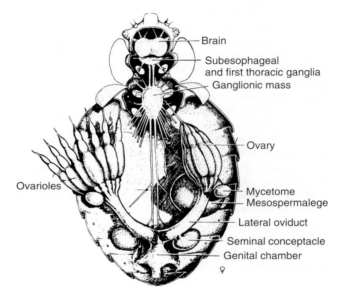

Figure 37.7 **Internal anatomy of *Cimex lectularius.***
Drawing shows female reproductive organs, mycetomes, and parts of nervous system.

From R. L. Usinger, *Monograph of Cimicidae (Hemiptera-Heteroptera)*. College Park, MD: Entomological Society of America, 1966. Drawing by G. P. Catts.

Figure 37.6 **Bedbugs in a domestic setting.**
Adults, nymphs, eggs, and dark fecal spots can be seen in this infestation of furniture from a Midwestern apartment complex.

Courtesy of Alvaro Romero, University of Kentucky.

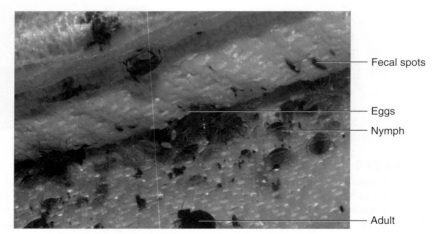

Epidemiology and Control

Their shape makes it possible for bedbugs to insinuate themselves into a variety of tight places. Their daytime sites can be seams of mattresses and box springs, wooden bedsteads, cracks in the wall, and spaces behind loose wallpaper (see Fig. 37.6).[19] These sites may be some distance from their host, which they find by following a temperature, and perhaps a carbon dioxide, gradient. They can be transported from one dwelling to another in secondhand furniture, suitcases, bedding, laundry, and other items. Only a single female is required to create the nucleus of a new infestation.

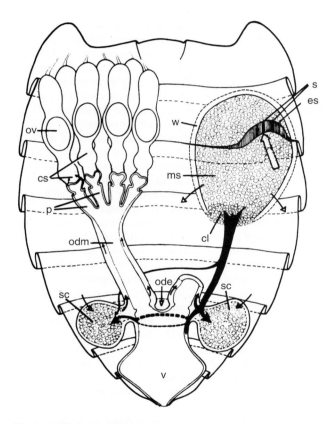

Figure 37.8   **Diagram of paragenital system and process of insemination in Cimicidae, based principally on *Cimex* type.**

Right ovary and nearly all of corresponding lateral oviduct omitted, and spermalege shown farther forward than is actually the case in *Cimex. Broad white arrow,* course followed by paramere of male in reaching ectospermalege (hatched and crossed by three black bands representing scars of copulation). *Black arrows,* normal routes of migration of spermatozoa from mesospermalege to bases of ovarioles. *Small arrows with white points,* migratory routes never or rarely used in *Cimex* but seen in other Cimicidae. *cl,* Conductor lobe; *cs,* syncitial body; *es,* ectospermalege; *ms,* mesospermalege; *ode,* paired ectodermal oviduct; *odm,* paired mesodermal oviduct; *ov,* oocyte; *p,* pedicel; *s,* scars or traces of copulation; *sc,* seminal conceptacle; *v,* vagina; *w,* wall of mesospermalege.

From R. L. Usinger, *Monograph of Cimicidae (Hemiptera-Heteroptera).* College Park, MD: Entomological Society of America, 1966. Drawings by G. P. Catts.

Transmission in public conveyances and gathering places occurs frequently.[14]

Control of bedbugs by application of residual insecticides to the areas of likely hiding places is usually effective, although resistance to some insecticides has been encountered.[38] A high level of domestic cleanliness certainly helps in control.

FAMILY REDUVIIDAE

Most reduviids are predators on other insects and, for this reason, are commonly called *assassin bugs.* They often are valuable for their predation on pest species; *Reduvius personatus* may even enter houses and feed on bedbugs! Most of these can but usually do not bite humans. Their bite may be quite painful. One of the authors has vivid memories of a college field trip on which he carelessly picked up an unidentified reduviid with his fingers—the reduviid remained unidentified because the author released it quite abruptly when it bit him.

One subfamily of Reduviidae, Triatominae, is of great public health significance because its members are vectors for *Trypanosoma cruzi,* the causative agent of Chagas' disease, for which an estimated 32 million people in Central and South America are seropositive.[41] The Triatominae characteristically feed on blood of various vertebrates. In contrast to an assassin bug's bite, the bite of Triatominae, as might be expected of forms that must suck blood for several minutes unnoticed by the host, is essentially painless. They are called *kissing bugs* because they often bite the lips of sleeping persons.

Morphology

Triatomines (Fig. 37.9) are relatively large bugs, up to 34 mm in length. They usually have wings, which are held in a concavity on top of the abdomen. Their head is narrow, and large eyes are located midway or far back on the sides of the head. Two ocelli may be present behind the eyes. Antennae are slender and in four segments. The apparently three-segmented labial tube folds backward at rest into a groove between the forelegs. The bugs can make a squeaking sound by rubbing their labium against ridges in this groove (stridulation).

Figure 37.9   **Specimen of *Triatoma dimidiata* discovered feeding on a parasitologist.**

Courtesy of Warren Buss.

Biology

Various reduviids characteristically frequent different sites; for example, some species are normally found on the ground, some in trees, and some in human dwellings. Eggs, numbering from a few dozen to a thousand depending on species, are deposited in the normal habitat of the adult. There are usually five nymphal instars. Triatomines do not seem to be very choosy about their food sources; whatever vertebrate is available in their habitat is apparently acceptable. Triatomines that inhabit human dwellings feed on humans, dogs, cats, and rats; other species depend more on wild animals.

Research on trypanosomiasis cruzi as well as on xenodiagnosis (chapter 5) demands a supply of lab-reared bugs, and much effort has gone into the development of rearing techniques, especially feeding stimuli. Temperature seems to be a stimulus for *Triatoma infestans;* in experimental feeders mammalian body temperatures evoked a feeding response, although crop filling did not depend on blood temperature.[24] Evidently, *T. infestans* can detect objects that are near mammalian body temperatures and orient toward them when seeking food.[25] *Triatoma infestans* is also somewhat particular about what it eats, being partial to citrated over heparinized blood (sodium citrate and heparin are both anticoagulants), but mouse odor added to the feeder does not make the blood more appetizing.[32]

Reduviids also release volatile chemicals, especially short chain acids, esters, and alcohols, from a variety of glands and these substances are generally thought to be involved in communication. Brindley's glands open on the metathorax dorsal surface and release compounds that evidently serve as alarm signals;[27] conversely, some triatomines are attracted to conspecifics' feces and in some cases to footprints.[49] The exact role of these communications relative to disease transmission, however, is not entirely clear, although insect chemical signaling systems are always potential targets for control. One study showed that triatomines are attracted to human skin odor, especially that emanating from the face, and that bacterial flora of the skin seemed to be a contributing factor.[33] For a review of the chemical ecology of triatomines, see Cruz-Lopez et al.[8]

Mating in triatomines can involve a fairly complex set of behaviors. In one laboratory study involving *Triatoma mazzottii,* nine steps were identified, including vigilance on the part of the male, female advancement, gyrations, copulation, and separation, all happening in about 10 minutes.[37] Only about 1 in 10 of the matings was completed, however, due to a combination of nonreceptiveness on the part of females and indifference on the part of males.[37]

Egg laying follows a circadian rhythm in *Rhodnius prolixus,* and lab-reared populations can be made to lay more or less synchronously using light-dark cycles.[1] The restricted timing of egg laying persists when the insects are transferred to total darkness, suggesting environmental control of population level ovulation and oviposition. Both egg laying and feeding in *Rhodnius prolixus* are likely under hormonal control; serotonin is secreted from tissues associated with the abdominal nerves and builds up in the hemolymph during feeding.[22] Blood meals are essential to egg production, but adults may lay eggs without feeding, provided nymphs have fed well.[31]

Epidemiology and Control

We discussed epidemiology of trypanosomiasis cruzi in chapter 5. Evidently, all species of triatomines are suitable hosts for *Trypanosoma cruzi,* but species differ in their susceptibility and presumably in their ability to serve as vectors. In experiments using infected mice and 11 species of triatomes, *Dipetalogaster maximus* and *Triatoma rubrovaria* passed the most trypomastigotes in their feces, and *T. vitticeps* passed the fewest, but all were infectable.[42] Importance of a particular species depends on its domesticity (**synanthropism**). The most common vectors are *Panstrongylus megistus, Triatoma infestans, T. dimidiata,* and *Rhodnius prolixus.* The relative importance of each varies with locality (see Fig. 5.14). The insects are nocturnal and hide by day in cracks, crevices, and roof thatching. Poorly constructed houses are thus a significant epidemiological factor. *Triatoma infestans* does not have to be alive to transmit *Trypanosoma cruzi* infections. Live trypomastigotes infective for mice were found in dead bugs for up to two weeks after one spraying campaign.[2]

Dogs, cats, and rats are important reservoir hosts around human habitations, and there is a wide variety of sylvatic reservoirs, most important of which are opossums, *Didelphis marsupialis.* Opossums are common and successful marsupials occurring from northern United States to Argentina. Other important reservoirs are armadillos, bats, squirrels, wild rats and mice, guinea pigs, and sloths. In one study, 3.6% of dogs in rural Oklahoma were seropositive for *T. cruzi,* leading the authors to conclude the disease was enzootic in that state.[6]

Transmission of *T. cruzi* to humans via triatomines does occur, although infrequently, within the United States. In one case, an 18-month-old child was diagnosed and treated after his mother found a bug in his crib. She "saved it because it resembled a bug shown on a television program about insects that prey on mammals."[15] The insect was positive, as were raccoons trapped near the residence. Triatomine bugs occur in the United States from New England to California. Not surprisingly, *Trypanosoma cruzi* has been found from coast to coast in wild mammals, including wood rats, raccoons, opossums, and skunks. Several cases of human infection have been diagnosed in Arizona.

The number of triatomines in a house increases with the number of people living there,[35] but triatomine populations can be reduced by reducing the number of hiding places for the bugs; that is, by improving construction and altering nearby environments. In one study, removal of stacked firewood from near houses and replacement of dirt floors with concrete nearly eliminated *Triatoma dimidiata* infestation.[52] Replacement of thatched roofs with sheet metal is of great aid, and even whitewashing of mud walls helps; but in one case in Brazil even repeated plastering of a house did not permanently reduce the population of *T. infestans* because the owner would not replace the roof tiles.[13] The bugs are surprisingly adept at finding and maintaining refuges; feces serve as a chemical signal for bugs to aggregate in particular shelters.[26] Reducing the number of other food sources for bugs around the dwelling, such as dogs, birds, and rats, also is of value in controlling triatomines.

As with bedbugs, residual insecticides around potential hiding places are effective in control, and paint with insecticide (chlorpyrifos) has been used with some success, especially on wood interiors.[21] However, nutritional state affects susceptibility of *T. infestans* to insecticide. Starved nymphs were nearly 200 times more resistant to DDT than were well-fed ones.[12] Precocene II, a natural product extracted from the plant *Ageratum* sp., shows promise as a fumigant against triatomines. It is cytotoxic to the corpora allata, preventing production of juvenile hormone. Precocene blocks oogenesis in adult females and causes immatures to molt precociously into sterile adults.[45] One intriguing approach, still in the experimental stage, involves genetic modification of symbiotic bacteria so that they express anti–*T. cruzi* antibodies, thus rendering the bugs refractive.[4] Major Chagas' disease control efforts, focused initially on *Triatoma infestans,* are underway in Central and South America. These efforts, known as the Southern Cone Initiative, involve spraying, follow-up community surveillance, and screening of blood donors and have been remarkably successful in some areas.[9, 41]

Learning Outcomes

By the time a student has finished studying this chapter, he or she should be able to:

1. Describe the breeding biology of *Cimex* species.
2. Describe some general methods for bedbug control.
3. Describe the ecological factors that promote triatomine infestation and explain why these factors also promote the spread of *Trypanosoma cruzi* infections.

References

References for superscripts in the text can be found at the following Internet site: www.mhhe.com/robertsjanovynadler9e

Additional Readings

Barrozo, R. B., P. E. Schilman, S. A. Minoli, and C. R. Lazzari. 2004. Daily rhythms in disease-vector insects. *Biol. Rhythm Res.* 35:79–92.

Dias, J. C. P., Z. C. Silveira, and C. J. Schofield. 2002. The impact of Chagas disease control in Latin America: a review. ***Mem. Inst. Oswaldo Cruz*** 97:603–612.

Ibarra-Cerdena, C. N., V. Sanchez-Cordero, A. Townsend Peterson, and J. M. Ramsey. 2009. Ecology of North American Triatominae. *Acta Tropica.* 110:178–186.

Pohlit, A. M., A. R. Rezende, E. L. Lopes Baldin, N. P. Lopes, and V. F. de Andrade Neto. 2011. Plant Extracts, Isolated Phytochemicals, and Plant-Derived Agents Which Are Lethal to Arthropod Vectors of Human Tropical Diseases—A Review. *Planta Medica* 77:618–630.

Ramsey, J. M., and C. J. Schofield. 2003. Control of Chagas disease vectors. *Salud Publica de Mexico* 45:123–128.

Rivero, A., J. Vezilier, M. Weill, A. F. Read, and S. Gandon. 2010. Insecticide Control of Vector-Borne Diseases: When Is Insecticide Resistance a Problem? *PLoS Pathogens* 6:e1001000

Chapter 38

Parasitic Insects: Fleas, Order Siphonaptera

The combined effects of Nero and Kubla Khan, of Napoleon and Hitler, all the Popes, all the Pharaohs, and all the incumbents of the Ottoman throne are as a puff of smoke against the typhoon blast of fleas' ravages through the ages.

—B. Lehane[35]

564   Foundations of Parasitology

Figure 38.1
Diagram of a flea.
Drawing by Larry S. R

How can this be? Ravages of *fleas?* But Lehane's claim does not seem an exaggeration when we consider that fleas transmit the dreaded plague, killer of millions of people from the dawn of civilization through the beginning of the 20th century. Of all of humanity's major diseases and important insect pests, the combination of flea and plague bacillus has had the greatest impact on human history.[37] We will return to the subject of plague later in this chapter.

The approximately 2500 species of fleas are small insects, from just less than a millimeter to a few millimeters long. Most are parasites of mammals, but approximately a hundred species regularly occur on birds. They are rather heavily sclerotized, bilaterally flattened, and secondarily wingless. Evolutionary loss of wings is a condition commonly found in parasitic insects. Some fleas are tan or yellow, but they are commonly reddish brown to black. Their mouthparts are of the piercing-sucking type; adults feed exclusively on blood. Larvae usually are not parasitic but feed on debris and materials associated with the nest or surroundings of the host, especially feces of adult fleas. Strong evidence suggests that fleas descended from a winged ancestor much like present-day scorpion flies (Mecoptera). In fact, several features of the jumping mechanism, which is well-developed in most fleas, seem to be homologous with flight structures of flying insects.[44]

Flea taxonomy still is a relatively unsettled discipline, and various authors disagree mainly on the numbers of families and the status of subspecies.[36]

MORPHOLOGY

A flea's head is broadly joined to its thorax and often bears a **genal ctenidium** (Fig. 38.1). Ctenidia are series of rather stout, peglike spines often found on the posterior margin of the first thoracic tergile (**pronotal ctenidium**) as well as on the head. Many species lack ctenidia, however. Ctenidia and

backwardly directed body setae are adaptations that help a flea retain itself among the fur or feathers of its host. Width of the space between adjacent spine tips in ctenidia of a particular flea species is correlated to diameter of the hairs of its usual host, being slightly wider than the maximal diameter of its host's hairs.[26] Thus, in backward movement, hairs tend to catch between ctenidial spines. Because of this resistance to being dragged backward, removal of a flea by host grooming or preening is relatively difficult. Obviously, these structures do not impede a flea's forward progress between hairs; neither do antennae, which fold back into grooves on the sides of the head. Antennae appear trisegmented, but the apparent terminal segment actually consists of 9 or 10 segments. Eyes, when present, are simple and have only a single, small lens. Ocelli are absent. Fleas bear a peculiar sensory organ near their posterior end called the **pygidium** that apparently functions to detect air currents.[2]

The legs are strong; hind legs are commonly much larger than are the other two pairs and are modified for jumping.

Jumping Mechanism

Many fleas are champion jumpers. Oriental rat fleas, *Xenopsylla cheopis,* can jump more than a hundred times their body length, and cat and human fleas are capable of a standing leap of 33 cm high.[47] In terms of proportionate body length, this would be the equivalent of a 6-foot human executing a standing high jump of almost 800 feet! It is not at all obvious how fleas accomplish this remarkable feat. *Xenopsylla cheopis* reaches an acceleration of $140 \times g$ in a little more than a millisecond, yet the fastest single muscle twitches (as in the locust, for example) require 15 milliseconds to reach peak force. The answer lies in use of flight structures that the flea inherited from its ancestors. Fleas have **resilin** in their **pleural arch,** an area between the internal ridges of the metapleuron

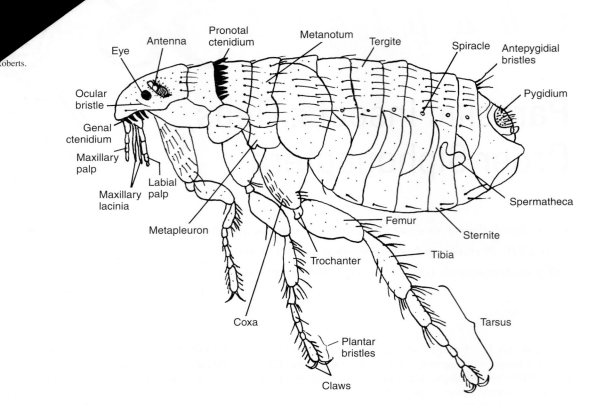

Roberts.

and metanotum (see Fig. 38.1). The pleural arch is homologous to the wing-hinge ligament in dragonflies, locusts, and scorpion flies.[47] Resilin is a stabilized protein with highly unusual elastic properties. It is a better "rubber" than rubber, releasing 97% of its stored energy on returning from a stretched position, compared with only 85% in most commercial rubber.

When a flea prepares to jump, it rotates its hind femurs up so that they lie almost parallel to the coxae, then rests on the tarsi of its front two pairs of legs and on the trochanters and tarsi of its hind legs. The resilin pad is compressed and is maintained in that condition by catch structures on certain sclerites (two on each side). In effect, the flea has cocked itself; to take off it must exert a relatively small muscular action to unhook the catches, allowing the resilin to expand. Then the flea rotates its femurs down toward the substrate and pushes off with its hind tarsi.

By using resilin, fleas have circumvented two major limitations of muscle: relatively slow rate of contraction and relaxation and poor performance at low temperature. Because resilin does not become deformed under prolonged strain, once the jumping mechanism is cocked, little energy is required to retain this state. A flea can lie in waiting, ready to hop aboard a host in a fraction of an instant. Interestingly, flea species whose preferred hosts are small or relatively accessible tend to have reduced pleural arches, whereas those that prefer larger hosts (such as deer, sheep, cats, and humans) have the largest pleural arches and are the best jumpers.[47]

Mouthparts and Mode of Feeding

Like Anoplura and Hemiptera, Siphonaptera have piercing-sucking mouthparts, but their structure is different from that of mouthparts of lice and bugs. The broad maxillae bear conspicuous, segmented palps (Fig. 38.2), as does the slender labium. The piercing fascicle comprises two elongated maxillary lobes (**laciniae**) and a median, unpaired **epipharynx.** Laciniae lie closely on each side of the epipharynx, and the fascicle is held in a channel formed by grooves in the inner side of the labial palps. A hypopharynx cannot be demonstrated (Fig. 38.3), and the labium is rudimentary. Lavoipierre and Hamachi[34] reported that several species of fleas are solenophages, but Rothschild and her coworkers believed that *Spilopsyllus cuniculi* is primarily a telmophage.[46] Laciniae are cutting organs, and piercing is achieved by back-and-forth cutting action of these structures. The epipharynx tip, but not the laciniae, enters a small blood vessel. Saliva is ejected into the area near the puncture, but it does not enter the vessel lumen. Very little damage is done by a penetrating fascicle, and, after withdrawal from the skin, hemorrhage is scant or absent.

DEVELOPMENT

Fleas undergo holometabolous development, and larvae thus are quite different in form and habits from adults. Although adult females usually oviposit while on a host, eggs are not sticky and so drop off the host's body, typically into the nest or lair, where there is a supply of detritus and flea feces on which larvae feed. High humidity tends to favor egg laying in adults and of course is apt to be the prevalent condition in nests and burrows. Eggs are relatively large (about 0.5 mm) (Fig. 38.4), providing many essential nutrients to larvae. Under favorable conditions, eggs hatch within 2 to 21 days; larval instars (usually three) require 9 to 15 days; and pupae complete development in as short a time as a week. Low

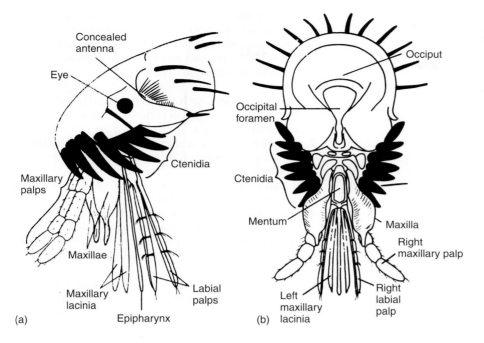

Figure 38.2   Head and mouthparts of a flea.
(*a*) Side view; (*b*) ventral view.

Reprinted with the permission of Simon & Schuster from R. F. Harwood and M. T. James, *Entomology in human and animal health,* 7th ed. Copyright © 1979 Macmillan Publishing Company.

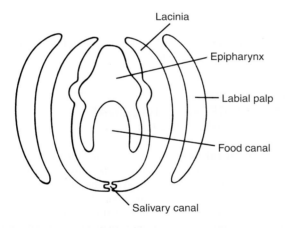

Figure 38.3   Diagrammatic representation of transverse section through flea mouthparts.

From R. R. Askew, *Parasitic insects.* New York: American Elsevier Publishing Company, 1971.

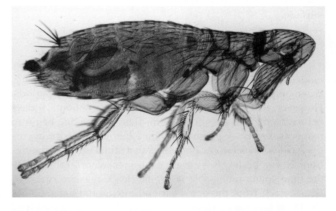

Figure 38.4   *Leptopsyllus segnis,* the European mouse flea, which is also common in parts of the United States.
Note the large size of eggs visible within this cleared specimen.
Courtesy of Jay Georgi.

temperatures and those as high as a host's body temperature retard development. Low temperatures can extend the larval period to more than 200 days and the pupal stage to nearly a year. Because larvae cannot close their spiracles, they are sensitive to low humidity. Larvae are white, legless, and eyeless, resembling maggots of some Diptera (Fig. 38.5). They have chewing mouthparts and stout body hairs. Pupae spin loose, silken cocoons from salivary secretions, often picking up debris from their surroundings and incorporating it into the cocoon.

In common with other lair parasites such as bedbugs and in contrast with lice, fleas can survive long periods as adults without food, particularly under conditions of high humidity. Unfed *Pulex irritans* have survived 125 days at 7°C to 10°C

and *Xenopsylla cheopis* for 38 days; periodically fed *P. irritans* may live up to 513 days and *X. cheopis* up to 100 days.[16] Periodically fed *Ctenophthalmus wladimiri* have survived for more than three years at 7°C to 10°C and 100% relative humidity. Such longevity has clear epidemiological importance because it allows flea-transmitted pathogens to survive long periods when vertebrate hosts are absent. Cases are known in which long survival occurs even in the face of highly adverse conditions: *Glaciopsyllus* spp. larvae, pupae, and some adults can withstand freezing in their host's nest and being covered with ice for nine months out of the year.[54]

Two genera of subfamily Spilopsyllinae, *Spilopsyllus* and *Cediopsylla,* are unusual in that their reproduction is closely controlled by their host's hormones.[46] *Spilopsyllus*

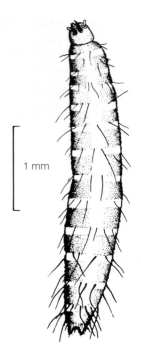

1 mm

Figure 38.5 **Third instar larva of *Spilopsyllus cuniculi*, the European rabbit flea, showing mostly the ventral surface.**

From R. R. Askew, *Parasitic insects.* New York: American Elsevier Publishing Company, 1971.

cuniculi, the **European rabbit flea,** is relatively sedentary, commonly attaching itself to its host's ears for long periods. However, it does not breed on adult rabbits. About 10 days before a pregnant doe gives birth, fleas on the doe begin to mature sexually. A pregnant rabbit experiences a rise in cortisol and corticosterone, hormones that stimulate flea maturation. Hormone levels also are high in newborn rabbits. By the time young rabbits are born, the flea's eggs are ripe, and fleas detach from the doe's ear and move onto her face. As she tends her young, fleas hop onto newborn rabbits and feed voraciously, mate, and lay eggs. After about 12 days, fleas leave the young and return to the doe.

Good evidence indicates that the growth hormone somatotropin, present in young rabbits, constitutes the stimulus for fleas to mate and lay eggs. Reproductive control of *C. simplex* by host hormones is essentially similar to that of *S. cuniculi*.[45] This remarkable coordination of flea reproduction with that of its host assures that flea eggs will be ripe at just the right moment to be deposited into a host's nest and assures larvae of a plentiful supply of food.[43]

HOST SPECIFICITY

In general, fleas are not very host specific, although they have preferred hosts. Most can transfer from one of their main hosts to another or to a host of a different species. Their common names (for example, *rat flea, chicken flea,* and *human flea*) refer only to their preferred host and do not imply that they attack that host exclusively. In the United States at least 19 different species have been recorded as biting humans,[20] but well over 50 genera are of medical importance globally as demonstrated or potential plague vectors.[36]

Fleas can be grouped into four categories according to the degree of attachment to a host:

1. Some rodent fleas, such as *Conorhinopsylla* and *Megarthroglossus* spp., are seldom on a host but occur abundantly in its nest.
2. Most fleas spend most of their time on a host as adults but can transfer easily from one host individual to another.
3. Female sticktight fleas, *Echidnophaga gallinacea,* attach permanently to the host skin by their mouthparts.
4. Female **chigoes,** *Tunga penetrans,* burrow beneath host skin and become stationary, intracutaneous, and subcutaneous parasites. Fleas that become permanently attached play little or no role in disease transmission.

There are some variations to these four categories. A species is known (*Uropsylla tasmanica,* on Tasmanian devils) in which larvae burrow beneath the skin and live as endoparasites. Larvae of *Hoplopsyllus* spp. on the arctic hare live as ectoparasites in the fur of the host. Bird fleas also exhibit some rather remarkable adaptations to their hosts. Southern fulmars, for example oceanic birds that spend vast amounts of time at sea, are the major hosts for *Glaciopsyllus antarcticus.* Live *G. antarcticus* do not occur in the nest material, suggesting that the fleas survive South Pole winters by going to sea with their hosts.[8, 54]

FAMILIES CERATOPHYLLIDAE AND LEPTOPSYLLIDAE

The **northern rat flea,** *Nosopsyllus fasciatus,* is a common parasite of domestic rats and mice (*Rattus* and *Mus* spp.) throughout Europe and North America and has been recorded on many other hosts, including humans. Although it may be of some importance in transmission of plague from rat to rat, it is not regarded as an important plague vector because it usually does not bite humans, and it is widespread in temperate climates where plague is not usually a problem.[41] **Ground squirrel fleas,** *Diamanus montanus,* are found in western North America from Nebraska and Texas to the Pacific coast and may be of some importance in transmission of plague in wild rodents.

Ceratophyllus niger and *C. gallinae* are bird fleas, although both will bite humans. *Ceratophyllus niger* is the **western chicken flea.** It can be distinguished easily from another common chicken flea, the **sticktight flea** *(Echidnophaga gallinacea),* in that *C. niger* is larger and does not attach permanently. **European chicken fleas,** *C. gallinae,* commonly parasitize a wide variety of other birds, especially passerines.

Leptopsyllus segnis is the **European mouse flea** (see Fig. 38.4), but it is common throughout the United States Gulf states and in parts of California. It is more common on species of *Rattus* than on *Mus.* Although it can be infected with plague, it is not an important vector because it does not readily bite humans.

FAMILY PULICIDAE

Pulex irritans, the **human flea** (Figs. 38.6 and 38.7), and other species of medical and veterinary importance belong to this family. *Pulex irritans* has been recorded from pigs, dogs, coyotes, prairie dogs, ground squirrels, and burrowing owls, but some records may refer to another species, *P. simulans,* which occurs in central and southwestern United States and in Central and South America.[50] Both species lack genal and pronotal ctenidia, and their metacoxae have a row or patch of short spinelets on the inner side of the podomere. However, maxillary laciniae of *P. irritans* extend only about half the length of the forecoxae, whereas those of *P. simulans* extend about three-fourths this distance. *Pulex irritans* can transmit plague, and it has been implicated in transmission from person to person in some epidemics. Kalkofen[29] found in a survey in Georgia that more than 80% of fleas on dogs were *P. irritans* and that dogs seemed to be the preferred hosts. Because dogs are susceptible to plague (see Kalkofen[29] for

references), this observation has important public health implications.

Echidnophaga gallinacea is an important poultry pest, but it also attacks cats, dogs, horses, rabbits, humans, and other animals. Its name is *sticktight flea* because it buries its fascicle in a host's skin and remains in place. Maxillary laciniae are broad and coarsely serrated (Fig. 38.8). This flea is widespread in tropical and subtropical regions and also occurs in the United States as far north as Kansas and Virginia.[15] Like those of *Tunga penetrans,* its thoracic segments are reduced, being shorter together than the head or the first abdominal segment; however, *E. gallinacea* has a patch of spinelets on the inner side of its metacoxa (absent from *T. penetrans*). It prefers to attach in areas with few feathers, such as the comb, wattles, around eyes, and around the anus of its host. An infestation causes ulcers, into which females deposit eggs. Larvae hatch in the ulcers but then drop to the ground to develop off the host, as in most other fleas. Heavy infestations may kill chickens.

Ctenocephalides canis (Fig. 38.9) and *C. felis* are **dog** and **cat fleas,** respectively. They can be distinguished from other common fleas by presence of a genal ctenidium with more than five teeth. In spite of their names, both species attack cats and dogs as well as humans and other mammals; *C. felis* is especially opportunistic, being reported also from horses, skunks, foxes, mongooses, koalas, and poultry.[20, 48] *Ctenocephalides felis* is more common than is *C. canis* on dogs in North America. Both species can be very annoying pests of humans, particularly when cats and dogs are kept on the premises. Oddly, they do not occur in the mid- to north-Rocky Mountain area. Cat fleas deposit most of their eggs at times their hosts are least active (midnight to very early morning) and most likely to be in their resting areas.[30] Flea feces (larval food) make up more than 40% of the debris that falls off infested cats and so are also concentrated in these areas.[31] The effect of this coordination is a localization of larval development sites.

Figure 38.6   Male *Pulex irritans,* **the human flea.**
The copulatory apparatus is visible in the abdomen. Rothschild[43] has called the genital organs of male fleas the most elaborate in the animal kingdom.

Courtesy of Jay Georgi.

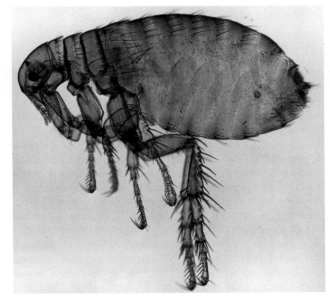

Figure 38.7   Female *Pulex irritans.*

Courtesy of Jay Georgi.

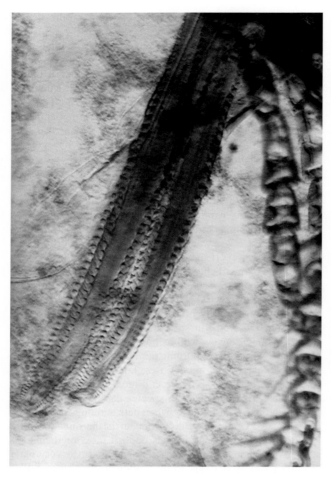

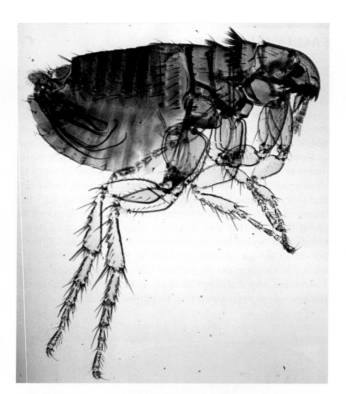

Figure 38.9 *Ctenocephalides canis,* **the dog flea.**

This species and *C. felis* bite humans frequently and are the source of much annoyance. They are the most common intermediate hosts of the tapeworm, *Dipylidium caninum,* of dogs and cats.

Courtesy of Jay Georgi.

Figure 38.8 **Maxillary laciniae of *Echidnophaga gallinacea,* the sticktight flea.**

The laciniae are broad and coarsely serrated, and the thoracic somites are much reduced compared with most other fleas. Maxillary palps are to the right in this photograph.

Photograph by Larry S. Roberts.

Like many insects, cat fleas and their feces are allergenic and may contribute to the allergenicity of house dust.[53] At least 15 different proteins from *C. felis* are allergenic.[19] Efforts to control *C. felis* result in expenditures of approximately $1 billion annually by frustrated pet owners in the United States alone.[48]

Xenopsylla cheopis (Fig. 38.10) is called the **oriental** or **tropical rat flea,** although it is almost cosmopolitan on *Rattus* spp., except in cold climates. In the United States it ranges as far north as New Hampshire, Minnesota, and Washington.[41] Like *Pulex irritans, X. cheopis* lacks both genal and pronotal ctenidia. It can be distinguished by location of its ocular bristle, which originates in front of the eye in *X. cheopis* and beneath the eye in *P. irritans.* Females can be recognized easily by their dark-colored spermatheca; *X. cheopis* is the only species in the United States with a pigmented spermatheca (compare Figs. 38.7 and 38.10).[41] *Xenopsylla cheopis* has enormous public health significance because it is the most important vector of plague and of murine typhus. *Xenopsylla brasiliensis,* an African

Figure 38.10 *Xenopsylla cheopis,* **the oriental, or tropical, rat flea.**

This flea is the most common vector of plague and murine typhus. The spermatheca is darkly pigmented and clearly visible.

Courtesy of Jay Georgi.

species that has become established in South America and India, appears to be an important plague vector in Kenya and Uganda. Some other species of *Xenopsylla* are implicated or suspected as vectors in various plague outbreaks.

FAMILY TUNGIDAE

Tunga penetrans is called **chigoe, jigger, chigger, chique,** and **sand flea** (Fig. 38.11). Some of these names are said to result from irritation the flea causes, prompting a host to "jig" about. This flea is apparently a native of Central and South America and the West Indies, from which it was introduced to Africa in the 17th century and again in the 19th century. It spread all over tropical Africa and then to India. Two recent cases in the United States involved children who were international adoptees.[14]

Female chigoes penetrate skin, most commonly around nail bases of hands and feet or between toes. Only a small aperture through the skin is left to communicate with the outside world. Males do not penetrate host skin, but copulate with a female after she has reached her final position. When she enters the skin, she is barely 1 mm long, but she gradually expands to about the size of a pea. Her body is enclosed in a sinus, into which she lays her eggs. After hatching, larvae exit through this aperture and develop on the ground. Presence of the female causes extreme itching, pain, inflammation, and often secondary infection (Fig. 38.12). Tetanus and gangrene occasionally are complications. Autoamputation has been attributed to results of infection with this flea and its secondary infection in Angola.[15] Surgical removal of fleas with careful sterilization and dressing of wounds is the recommended remedy.

Human tungiasis can be a significant health problem in poor communities, especially in warmer climates. Infestations may also exhibit seasonal fluctuations, with prevalence

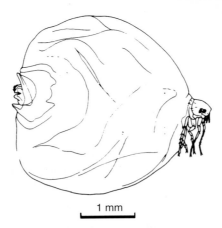

1 mm

Figure 38.11   **Engorged female *Tunga penetrans,* the chigoe, or jigger.**

This stage is found in a subcutaneous sinus that communicates with the outside by a small pore, through which larvae escape. Legs are degenerating by the time a female is engorged.

From R. R. Askew, *Parasitic insects.* New York: American Elsevier Publishing Company, 1971.

peaking at over 50% during the dry seasons; pets and rats evidently serve as reservoirs in endemic areas.[22, 23]

In addition to feeding on humans, *T. penetrans* attacks several other mammals, particularly swine. None of the 10 described *Tunga* species attacks birds; all occur on mammals, including domestic stock, rodents, anteaters, and armadillos (Fig. 38.13).[39]

FLEAS AS VECTORS

Plague

Plague, also known as *pest* and *black death,* is caused by a bacterium, *Yersinia pestis* (formerly known as *Pasturella pestis*). The bacterium releases two potent toxins that have identical serious effects.[28] Some animals, such as rats and mice, are more sensitive to these toxins than are others (rabbits, dogs, monkeys, and chimpanzees). Evidence indicates that the toxins act on mitochondrial membranes of susceptible animals, inhibiting ion uptake and interfering with normal functioning of the respiratory chain.[28] One toxin is a plasmid-encoded phospholipase that evidently is required by *Y. pestis* before it can survive in a flea's gut.[25]

The great pandemics of plague have probably had more profound effect on human history than has any other single infection.[37] The 14th century pandemic alone, for example, took 25 million lives, or a fourth of the population of Europe, and has been called the worst disaster that has ever befallen humanity (Fig. 38.14).[33] Most scientists assume that medieval black death was caused by *Yersinia pestis.* Strong evidence that is the case comes from DNA analysis of teeth from a 14th century grave in France.[42] Adequate treatment of this subject in relation to its importance is far beyond the scope of this book. Interested readers can avail themselves of numerous sources on plague, including highly readable accounts of plague and fleas in history by Lehane,[35] McNeill,[37] and Pollitzer.[40] A fascinating picture of human behavior during the terror of this universal catastrophe is given by Langer.[33] The last pandemic began in the interior of China toward the end of the 19th century, reached Hong Kong and Canton by 1895 and Bombay and Calcutta in 1896, and then spread throughout the world, including to numerous port cities in the United States. Between 1898 and 1908, more than 548,000 people per year died from plague in India.[40]

Between 1900 and 1972 there were 992 cases in the United States (416 of them in Hawaii), of which 720 were fatal.[41] The disease has decreased in incidence and severity in recent years: Between 1958 and 1972 there were 51 cases in the United States, of which only nine were fatal. A similar decline has occurred globally.[36] It is not clear if this improvement results entirely from better medical care.

The disease was formerly centered in seaports, from which it spread out in epidemics. Recent cases in the United States have been virtually all rural (campestral or sylvatic); that is, they were contracted after contacts with wild rodents in the countryside rather than with *Rattus* spp. in cities. *Yersinia pestis* is widely distributed in rodents and occurs across broad areas of virtually every continent.[36] It is unknown whether urban plague spread into wild rodents in the United States or whether it was already present in wild rodents

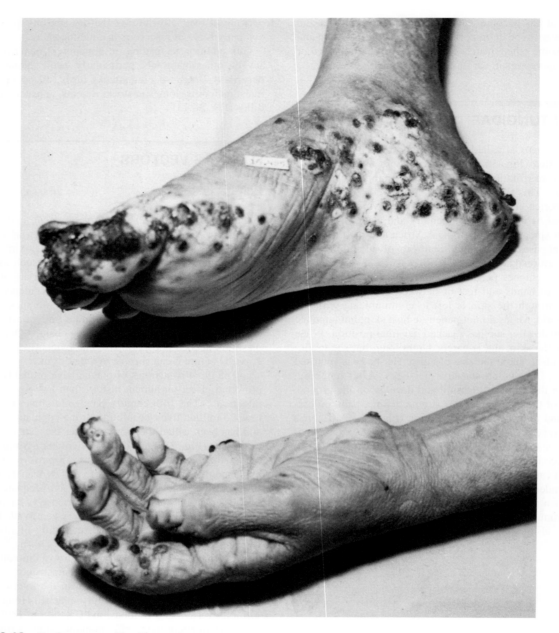

Figure 38.12   **Lesions caused by *Tunga penetrans*.**

From G. W. Hunter, J. C. Swartzwelder, and D. F. Clyde, *Tropical medicine,* 5th ed. W. B. Saunders, Co., 1976. Photo by Rodolfo Cespedes, Hospital San Juan de Dios, San Jose, Costa Rica.

before 1900. Although the black, or roof, rat, *Rattus rattus,* is common and abundant around human habitations, it appears not to be a reservoir of infection.[49]

Plague is essentially a disease of rodents, from which it is contracted by humans through flea bites, particularly those of *Xenopsylla cheopis.* Bacteria are consumed by a flea along with its blood meal, and the organisms multiply in the flea's gut, often to the extent that passage of food through the proventricular teeth is blocked. When the flea next feeds, a new blood meal cannot pass this obstruction and is contaminated by bacteria and then regurgitated back into the bite wound. Propensity of a particular flea species to have its gut blocked by growth of *Y. pestis* is an important, but not absolutely

necessary, determinant of its efficacy as a vector.[12] *Xenopsylla cheopis* is a good vector because it becomes blocked easily, feeds readily both on infected rodents and humans, and is abundant near human habitations.[37]

The three main clinical forms of plague are **bubonic, pneumonic,** and **septicemic;** meningeal infections are known but rare.[9] In bubonic plague, definite bubo formation occurs. **Buboes** are swollen lymph nodes in the groins or armpits; they are hard, tender, and filled with bacteria. Buboes sometimes reach the size of a hen's egg (Fig. 38.15) and may rupture to the outside. Over 80% of cases in the United States are first reported as bubonic. Pneumonic plague is a condition in which lungs are most heavily involved,

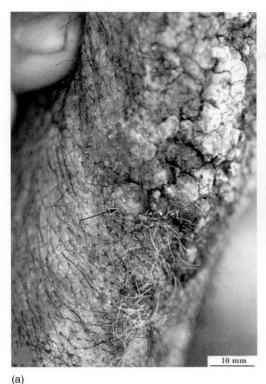

(a)

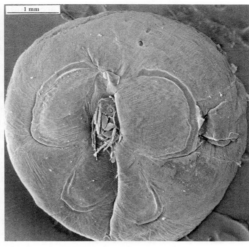

(b)

Figure 38.13   *Tunga trimamillata.*

(*a*) Lesion due to *Tunga trimamillata* in a pig's leg. *Arrow* points to the opening through which a female flea can be seen. (*b*) Scanning electron micrograph of a female *T. trimamillata* extracted from a lesion.

From S. Pampigliione, M. Trentini, M. L. Fioravanti, G. Onore, and F. Rivasi, "Additional description of a new species of *Tunga* (Siphonaptera) from Ecuador," in *Parasite* 10:9–15. Copyright © 2003. Reprinted by permission.

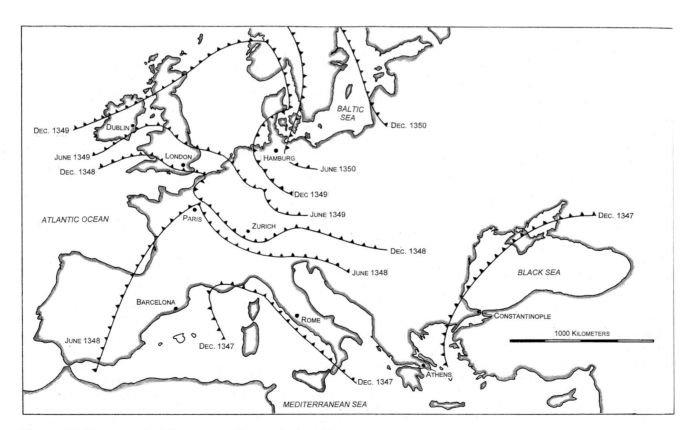

Figure 38.14   Spread of plague across Europe in the 14th century.

Progress of the epidemic is shown by lines indicating distribution of cases at six-month intervals from December 1347, through December 1350.

Redrawn by John Janovy Jr. from E. Carpentier, "Autour de la peste noire: Famines et epidémies dans l'histoire du XIVe siècle," *Annales, Economies, Sociétés, Civilisations* 17:1062–1092, 1962.

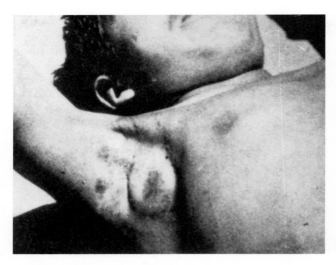

Figure 38.15 **Plague bubo in right axilla of human.**
AFIP neg. no. ACC 219900-7-B.

producing a pneumonialike disease that develops rapidly and is highly contagious to other people. Primary septicemic plague is a generalized blood infection with little or no prior lymph node swelling, apparently because the blood is invaded too rapidly for typical nodal inflammation to develop. However, secondary septicemic plague occurs when an infection breaks out of lymph nodes; *Y. pestis* can infect most organs, the result being massive tissue destruction.[9] Bubonic plague is fatal in about 25% to 50% of untreated cases; pneumonic and septicemic plague are usually fatal.

The incubation period after a flea's bite is usually two to six days, followed by a chill and rapidly rising temperature to 39.5°C and 40°C. Lymph nodes draining an infection site swell, becoming hemorrhagic and often necrotic. Damage to vascular and lymphatic endothelium in many parts of the body leads to petechial and diffuse hemorrhages. At first there is mental dullness, followed by anxiety or excitement and then delirium or lethargy and coma. If untreated, the disease may cause death within five days. A patient usually dies within three days after onset of primary septicemic plague. If a patient is to recover, the fever begins to drop in two to five days. Treatment with antibiotic and antitoxin is usually effective. A vaccine is available, although its effectiveness varies among individuals, regular boosters are required, and its use is recommended only for military and laboratory personnel in high risk situations.[9]

Conditions conducive to high rat and flea populations contribute to plague outbreaks. The disease may exist in rodent populations in acute, subacute, and chronic forms. Epidemics among humans usually closely follow epizootics, with high mortality among rats. When a rat dies, its fleas depart and seek "greener pastures." Meteorological conditions are important, doubtless because of their effects on flea populations. Plague is usually seen in temperate climates during summer and autumn and in the tropics during the cool months. Recent studies suggest that relative high late winter precipitation boosts the number of cases in a following summer.[13] Extreme heat and dryness inhibit spread of plague. Epidemics of pneumonic plague occur only during conditions of low temperatures and high humidity, which are conditions that favor survival of the bacilli in sputum droplets.

Campestral plague (so called because it is associated with animals of open areas rather than wooded or sylvatic country) is widespread among wild rodents and rabbits in the United States west of the 100th meridian. Cases among humans are reported sporadically, usually after a person has contacted wild rodents and their fleas. New Mexico has had the highest case rate. One Californian contracted bubonic plague with secondary pneumonic plague after hunting ground squirrels and was a source of 13 cases of primary pneumonic plague in other persons, with 12 deaths.[41] Nowadays, hospitals routinely put plague patients under respiratory isolation. In one recent instance, molecular markers were used to determine that patients had indeed contracted the disease from wood rats on their property instead of from an act of terrorism.[7]

A number of cases have been associated with skinning, cooking, and eating wild rabbits and hares; human victims may have been bitten by the rabbit's fleas.[1] There is one report of cases resulting from consumption of raw camel liver; *Y. pestis* was subsequently isolated from sick camels, as well as birds and fleas around the camel corral.[3]

Throughout recorded history, plague has been cyclical, smoldering in endemic foci and then giving rise to great outbreaks.[37] The world seems to be in a remission phase at present, but a vast and easily accessible supply of *Y. pestis* is still present in nature. There is also strong evidence that *Y. pestis* can pick up genes, including those for drug resistance, from bacterial species cooccurring in a flea's gut.[24] Given the widespread distribution of rodents and their fleas, plague bacillus seems to persist regardless of control measures. For example, a decade-long study accompanying control efforts in Tanzania implicated dogs as reservoirs, with over 6% seropositive, and reported insecticide resistence in *P. irritans*.[32]

Murine Typhus

Murine typhus, also called *endemic* or *flea-borne typhus,* is caused by *Rickettsia mooseri* (= *R. typhi*) which is morphologically indistinguishable from *R. prowazekii* (p. 552). It occurs in warmer climates throughout the world. It can infect a wide range of small mammals, including opossums, *Didelphis marsupialis,* but the most important reservoir is *Rattus norvegicus,* in which it causes slight disease symptoms. Murine typhus can be transmitted from rat to rat by *Xenopsylla cheopis; Nosopsyllus fasciatus; Leptopsyllus segnis;* the rat louse, *Polyplax spinulosa;* and the tropical rat mite, *Ornithonyssus bacoti.*

In humans the disease is a rather mild, febrile illness of about 14 days' duration, with chills, severe headaches, body pains, and rash. It tends to be more severe in elderly persons. *Xenopsylla cheopis* is considered the primary vector transmitting the disease to humans, either through a bite or through contamination of skin abrasions with flea feces by scratching. Ingestion of infected fleas and their feces also can produce infection in rats.[16] Rickettsias proliferate in midgut cells of fleas but do not kill them. Rupture of midgut cells releases rickettsias into a flea's gut.

Before 1945, incidence of murine typhus was high in the United States; it reached a peak of 5401 cases in 1945.

After institution of rat control programs, use of DDT, and increasing use of antibiotics, reported incidence dropped dramatically, ranging between 18 and 36 cases per year between 1969 and 1972. However, opossums are proliferating in many urban and suburban areas, creating a possible resurgence in the number of cases. In one Los Angeles study, city opossums were heavily infested (mean flea count = 104.7/animal) with the cat flea, *Ctenocephalides felis,* a species that readily bites humans. In that report the authors concluded that the biology of this focus differed substantially from the classical transmission cycle, and that cats, opossums, and *C. felis* may play an important role in the occurrence of human cases.[52]

Myxomatosis

The myxoma virus causes a disease in rabbits and is transmitted by several bloodsucking arthropods, including mosquitoes, fleas, and mites.[51] The principal vector in England is *Spilopsyllus cuniculi,* and the disease causes considerable losses in the domestic rabbit industry. The virus was apparently introduced from South America, where rabbits are relatively resistant to myxomatosis.[43] It was intentionally introduced into Australia to control the abundant rabbits there. However, the principal vectors in Australia are mosquitoes, which are not ideal vectors for rabbit control. Mosquitoes confer a selective advantage on an attenuated "field strain" of the virus and are most abundant during the warm months, when rabbits have the best chance of surviving the disease. Consequently, resistance to the virus in rabbit populations was unintentionally selected for. Introduction of *S. cuniculi* along with reintroduction of virulent virus strains subsequently offered more hope of better rabbit control.[51]

Other Parasites

Nosopsyllus fasciatus is a vector for the nonpathogenic *Trypanosoma lewisi* of rats (p. 76). *Ctenocephalides canis, C. felis,* and *Pulex irritans* serve as intermediate hosts of *Dipylidium caninum,* a common tapeworm of cats and dogs (p. 342). *Nosopsyllus fasciatus* and *Xenopsylla cheopis* can serve as vectors for the rat tapeworm, *Hymenolepis diminuta;* the mouse tapeworm, *H. nana,* can develop in *X. cheopis, C. felis, C. canis,* and *P. irritans* (p. 340). All of these fleas acquire tapeworms as larvae when they consume eggs passed in feces of a vertebrate host, retaining cysticercoids in their hemocoel through metamorphosis to adults. All three species can be transmitted to humans through inadvertent ingestion of infected fleas. Young children are especially at risk for such infections.

A filarial worm of dogs, *Dipetalonema reconditum,* which lives in subcutaneous, connective, and perirenal tissues, is transmitted by *C. canis* and *C. felis.* Microfilariae are picked up by fleas in their blood meal, develop to an infective stage in the flea's fat body in about six days, migrate to the head, and then pass to the wound when the flea next feeds. *Dipetalonema reconditum* is of slight or no pathogenicity, but its microfilariae may be easily confused with those of the serious pathogen *Dirofilaria immitis* (p. 453). For techniques for distinguishing the two, see Ivens, Mark, and Levine.[27]

CONTROL OF FLEAS

Many of us occasionally need to control fleas around our homes or on our pets. It is sometimes extremely important for public health reasons to control rat fleas and more importantly their hosts. For more complete instructions about and techniques for flea and rat control, consult Pratt and Stark[41] and Blagburn and Dryden.[4]

Within habitations one should keep debris that harbors larval fleas to a minimum, such as under carpets, in floor crevices, and in pet bedding. Some persistent insecticides may be used indoors. These tend to act especially on eggs and larvae. For example, diflubenzuron inhibits *C. felis* development in carpets for up to a year, but it breaks down when used outdoors.[21] Flea eggs are infertile when hosts are treated with certain pyripoxyfenor methoprene-based formulations.[11] More subtle control attempts include light traps with yellow-green filters to which fleas respond positively. One such device collected 86% of the live fleas released into a carpeted room, attracting them from as far away as 8 m.[10] Both oral and injectable treatment with the insect growth regulator lufenuron have been shown experimentally to control fleas on dogs and cats, respectively.[17, 18] In the case of cats protection lasted up to six months, depending on the dose.[17] Rodent bait containing imidacloprid has been used to control fleas in wild ground squirrel populations where these rodents were reservoirs of plague bacillus. Similar use of imidacloprid to control rat fleas in dwellings are successful, although flea populations rebound quickly once the bait is removed.[6]

It is important to keep areas where livestock are maintained as free from debris, manure, and other litter as possible. Various insecticidal flea powders for use on dogs and cats are available, but repeated application may be necessary because the animals easily pick up more fleas in outdoor areas not treated with insecticide. In recent years flea collars with slow-release vapors have proven effective.

Personal protection may be achieved by use of insect repellants. In areas in which *Tunga penetrans* is found, it is important to wear shoes.

▌ Learning Outcomes

By the time a student has finished studying this chapter, he or she should be able to:

1. Describe adaptations of fleas for jumping.
2. Describe adaptations of fleas for maintaining their position on hosts with feathers or hair.
3. Justify this statement: fleas have played a major role in human history.
4. Describe several ways in which fleas may affect human health.

▌ References

References for superscripts in the text can be found at the following Internet site: www.mhhe.com/robertsjanovynadler9e

▌Additional Readings

Ben Ari, T., S. Neerinckx, K. L. Gage, K. Kreppel, A. Laudisoit, H. Leirs, and N. C. Stenseth. 2011. Plague and Climate: Scales Matter. *PLoS Pathogens* 7:e1002160.

Heukelbach, J. A., and H. Feldmeier. 2004. Ectoparasites—the underestimated realm. *Lancet* 363:889–891.

Lehane, M. J. 2005. *The biology of blood-sucking insects*, 2nd ed. Cambridge, UK: Cambridge University Press.

Pampiglione, S., M. L. Fioravanti, A. Gustinelli, G. Onore, B. Mantovani, A. Luchetti, and M. Trentini. 2009. Sand flea (*Tunga spp.*) infections in humans and domestic animals: state of the art. *Med. Vet. Entomol.* 23:172–186.

Smith, C. R., J. R. Tucker, B. A. Wilson, and J. R. Clover. 2010. Plague studies in California: a review of long-term disease activity, flea-host relationships and plague ecology in the coniferous forests of the Southern Cascades and northern Sierra Nevada mountains. *J. Vector Ecol.* 35:1–12.

Chapter 39

Parasitic Insects: Diptera, Flies

Time is fun and we're having flies.

—Kermit the Frog

Among all orders of insects, Diptera stands out as by far the most medically important. Various flies contribute to disfiguring, debilitating diseases of many kinds, either as vectors of pathogenic organisms or as active parasites in their own right. The order is so vast, with approximately 120,000 species in up to 140 families (depending on the source), that we can do no more than introduce the subject in this text. Published information on mosquitoes alone would fill a small library.

The name *fly* applies to insects that have a pair of wings on their mesothorax and a reduced pair, known as halteres, on their metathorax. Halteres are knoblike appendages that function as gyroscopic balance organs. Of course, some parasitic flies have secondarily lost all wings. Flies are holometabolous, with obtect, coarctate, or puparious pupae. Dragonflies and mayflies and other insects whose common names are written as one word are not actually flies. Names of true flies such as house fly and bot fly are written as two words.

Because the order is large and its members vary considerably in feeding habits, both as larvae and as adults, the structure of mouthparts also is diverse. We can divide them into five general types—mosquito, horse fly, house fly, stable fly, and louse fly—differing in the structures and functions of palps and labium, mandibles, maxillae, and hypopharynx.[35] Mouthparts of some larvae are of medical interest.

Historically Diptera was divided into three suborders: Nematocera, Brachycera, and Cyclorrhapha. However, major taxonomic works combine the last two and make Cyclorrhapha an infraorder, Muscomorpha, a practice followed here.[15] Taxonomic characters most used in Diptera are differences in mouthparts, head sutures, ocelli, antennae, wing venations, tarsi, and placement of bristles (**chaetotaxy**). Male genitalia offer useful taxonomic characters at the genus and species levels.

SUBORDER NEMATOCERA

Antennae of species in this group have many segments and are filamentous. They may be plumose, especially in males,

but basically they are simple and longer than the head. The wings have many veins, which is a primitive character. Larvae are active, with a well-developed head capsule, and pupae often are free swimming. Most larval and pupal stages are aquatic, although some develop in bogs or wet soil.

Family Psychodidae

Two subfamilies are of medical importance: Psychodinae contain **moth flies,** which are of only slight parasitological significance; Phlebotominae consist of **sand flies,** of great importance in many parts of the world.

Subfamily Psychodinae
The body and ovoid wings are densely covered with hairs. The rooflike position of the wings when at rest suggests the appearance of a tiny moth; hence, their common name *moth fly.* These flies breed in substrates rich in organic decomposition. *Psychoda alternata,* the **trickling filter fly,** is often found in incredible numbers in sewage disposal plants; it and other species are common in cesspools and drains where larvae develop in gelatinous linings that commonly accumulate in pipes. Emerging adults may be so numerous as to constitute a genuine annoyance to householders. In addition, larvae of *P. alternata* have been reported in pseudomyiasis, and this species can be involved in mechanical transmission of nematodes parasitizing livestock.[82]

Subfamily Phlebotominae
In contrast to moth flies, sand flies are not so hairy and hold their wings at rest 60 degrees from the body, not rooflike. More importantly, their mouthparts are of the horse fly subtype, with cutting mandibles. Females feed on plant fluids and on blood by telmophagy, whereas males feed mainly on plant juices and never on blood. Fascicle structures in this group vary somewhat depending on the skin characteristics of their usual host.[52] Many species feed on reptiles or

amphibians, whereas others feed on birds and mammals, including humans.

Adult sand flies (see Fig. 5.17) usually feed at night or at twilight and early morning, although some are day feeders. They are weak flyers, capable of navigating only short distances, and are inactive when any wind blows. Because of their soft, delicate exoskeleton, their survival depends on avoiding hot, dry places. However, there are desert species that thrive by hiding in dry mud cracks, and burrows, emerging only during the few humid night hours. These hours often coincide with the timing of some human activities, such as fetching water.

There are two generally recognized Old World genera of sand flies, *Phlebotomus* and *Sergentomyia,* and one American, *Lutzomyia.* Although a few former members of *Phlebotomus* have been placed in new, recently established, genera, these species are not known to transmit diseases. *Lutzomyia* spp. occur as far north as Canada, but medically important sand flies are primarily south of Texas. *Lutzomyia diabolica* and *L. shannoni* are the only known anthropophilic sand flies in the United States. Both have been shown experimentally to transmit *Leishmania mexicana.* Other North American species feed mainly on reptiles and rodents.

To breed, phlebotomines require a combination of darkness, high humidity, organic debris on which larvae feed, and possibly a sleeping host. These requirements are met in animal burrows, crevices, hollow trees, under logs, and among dead leaves. Lek behavior has been observed in *Lutzomyia longipalpis,* with males setting up territories on anesthetized mice, maintaining these territories with wing fanning and aggressive behavior, and competing for females.[40] Females are chemically attracted to dead plant material and feces of both small mammals and larval sand flies.[23] They lay several eggs at a time. The tiny, white larvae feed on such matter as animal feces, decaying vegetation, and fungi. They have simple, chewing mandibles. The four larval instars require 2 to 10 weeks before pupation. Pupae develop in about 10 days.

Sand flies are good vectors of disease, which is surprising considering their weakness and fragility. They transmit leishmaniases (chapter 5), including those of nonhuman vertebrates, bartonellosis, and some viral diseases. Studies show that salivary secretions contribute far more than previously suspected to phlebotomine vector potential and leishmanial pathogenesis. For example, injection of 1/10 of a single *Lutzomyia longipalpis* salivary gland into mice, along with parasites, increased the severity of infections with *Leishmania major* and *L. mexicana.*[80] In the case of *L. braziliensis* infections in BALB/c mice, coinjection of salivary gland lysates produced a large cutaneous lesion filled with parasites, whereas skin lesions regressed in control mice.[53, 70] Saliva alters the course of *Leishmania* infections through injection of a vasodilator protein, **maxadilan**, modulation of cytokine production by antigenproducing cells, and inhibition of complement activation (see chapter 3). Saliva proteins also have been used experimentally as vaccines in mice, and immunization with saliva results in protection or reduced severity of infection. Rohoušová and Volf provide an excellent review of sand fly saliva components and their various roles in promoting infection and stimulating immunity.[69] Obviously, phlebotomine physical frailty is at least partially offset by biochemical sophistication.

The bacterium *Bartonella bacilliformis* causes a disease known as **Carrión's disease,** with two clinical forms, **Oroya fever** and **verruga peruana.** It is found in Ecuador, southern Colombia, and the Andean region of Peru, being transmitted by *Lutzomyia verrucarum* and probably *L. colombiana.* Oroya fever is a sometimes fatal, visceral form of the disease, accompanied by bone, joint, and muscle pains; anemia; and jaundice. Verruga peruana is a mild, nonfatal cutaneous form. The disease is named after Daniel Carrión, who inoculated himself with organisms obtained from a verruga patient and subsequently developed Oroya fever. Before he died of it, he recognized that the two entities were actually expressions of the same disease.

Sand fly fever is transmitted by *Phlebotomus papatasi, P. sergenti,* and other flies in much of the Old World. Also known as **papatasi fever** and **three-day fever,** it occurs in the Mediterranean region, eastward to central Asia, southern China, and India. Sand fly fever is a nonfatal, febrile, viral disease of short duration but with a long convalescence period. Sand flies acquire the virus when they feed, but, because males do not feed on blood and also are sometimes infected, it seems probable that transovarial transmission occurs. The fly, then, also is a reservoir. Sporadic epidemics occur, such as in Yugoslavia in 1948 when three-fourths of the population (1.2 million persons) acquired it.[35]

Family Culicidae

Mosquitoes are the most important insect vectors of human disease and the most common bloodsucking arthropods. They feed on amphibians, reptiles, birds, mammals, intrepid explorers, and homemakers. Some species exhibit considerable host specificity, while others have more catholic tastes. Mosquitoes have greatly affected the course of human events and continue to do so even today when we have an arsenal of insecticides at our disposal and a vast knowledge about these insects and diseases they carry. More than a million people die every year from malaria, and other mosquito-borne diseases cause incalculable misery, poverty, and debilitation. The annoyance of hordes of ravenous mosquitoes is in itself enough to affect real estate values, tourist industries, and outdoor activities. Humans are not alone in their mosquito-borne misery; significant agricultural losses occur as a result of attacks on domestic animals.[74] The world's mosquito fauna is rich and diverse, and populations often are enormous. Approximately 3500 species have been described, at least 175 of these in North America. As expected from ecological observations and theory, species diversity is greatest in the tropics, especially in the Neotropics and Southeast Asia. On tropical islands, mosquito diversity increases exponentially with size of the land mass and islands tend to have higher numbers of endemic species than do mainland countries of similar size.[26] Foley et al. give an extensive review of the global distribution of mosquito speices.[26]

Mosquitoes are readily differentiated from superficially similar Dixidae, Chaoboridae, and Chironomidae by a combination of slender wings with scales on the veins and margins and elongated mouthparts that form a proboscis (Fig. 39.1). The fascicle consists of six stylets in the mosquito subtype. Two mandibles, two maxillae, a hypopharynx, and a labrum-epipharynx are loosely ensheathed in the

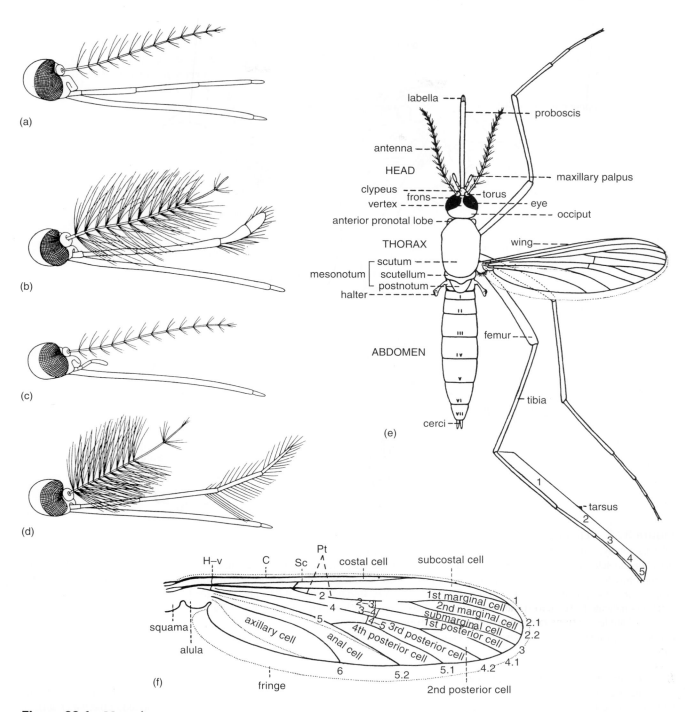

Figure 39.1   Mosquito anatomy.

(*a–d*) Heads and appendages: (*a*) female *Anopheles;* (*b*) male *Anopheles;* (*c*) female culicine; (*d*) male culicine. (*e*) Female culicine. (*f*) Wing with veins and cells labeled: *H-v,* humeral crossvein; *C,* costa; *Sc,* subcosta; *numbers* indicate longitudinal veins; *Pt,* petiole of vein 2.

From S. Carpenter and W. LaCasse, *Mosquitos of North America* (*North of Mexico*). Copyright © 1974 University of California Press, Berkeley, CA. Reprinted by permission.

elongated labium, which has a lobelike tip, or **labella.** Mosquitos insert their fascicle into a vessel or, more likely, into a pool of blood that accumulates when vessels are cut and pump blood into the food channel formed by their labrum-epipharynx and hypopharynx. Males lack mandibular stylets and do not feed on blood. Mosquito antennae are long and filamentous with 14 or 15 segments. Whorls of hairs on the antennae are quite plumose in males of most species. Male terminalia are taxonomic characters useful to experts in differentiating species.

Many species, particularly females, are fairly easily identified with the proper literature. Anyone studying mosquitoes will soon become familiar with thoracic sclerites and bristles (Fig. 39.2). Members of some species complexes,

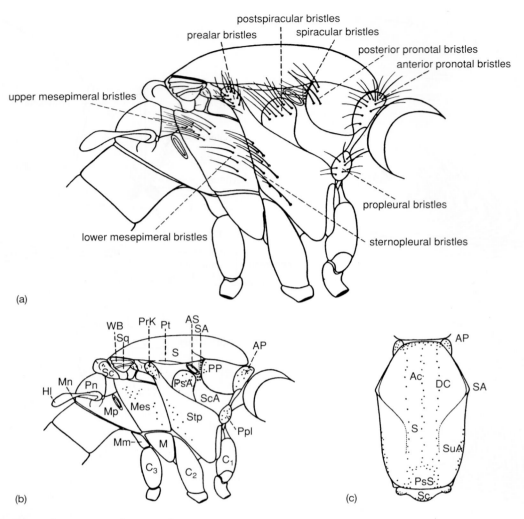

Figure 39.2 **Thoracic anatomy of a mosquito.**

(*a*) Chaetotaxy, with bristles named. Spiracular and postspiracular bristles are important identification characters at the generic level. (*b*) Thoracic sclerites: *AP*, anterior pronotum; *AS*, anterior spiracle; C_1, C_2, C_3, first, second, and third coxae, respectively; *Hl*, haltere; *M*, meron; *Mes*, mesepimeron; *Mm*, metameron; *Mn*, metanotum; *Mp*, metapleuron; *Pn*, postnotum; *PP*, posterior pronotum; *Ppl*, propleuron; *Prk*, prealar knob; *PsA*, postspiracular area; *Stp*, sternopleuron; *WB*, wing base. (*c*) Dorsal view of thorax with bristle locations indicated: *Ac*, acrostichal bristles; *AP*, anterior pronotal lobe; *DC*, dorsocentral bristles; *PsS*, prescutellar space; *S*, scutum; *SA*, scutal angle; *Sc*, scutellum; *Sq*, squama; *SuA*, supra-alar bristles.

From S. Carpenter and W. LaCasse, *Mosquitos of North America* (*North of Mexico*). Copyright © 1974 University of California Press, Berkeley, CA. Reprinted by permission.

however, can be differentiated only by techniques such as cross mating or use of genetic markers.[2] Subtle differences between mosquito strains can have epidemiological significance when various genotypes have different behaviors or physiological attributes.[2]

Mosquitoes undergo complete metamorphosis, with egg, larval, pupal, and adult stages (Fig. 39.3). Larval and pupal stages can develop only in water. Adults deposit eggs singly on water or soil or in rafts of eggs on water. They either hatch quickly or, in the case of those on soil, after a period of drought followed by flooding. Most floodwater mosquitoes hatch after the first flooding, but some remain for subsequent floodings, in some cases for up to four years.[92]

Most mosquito larvae hang suspended at the water's surface by a prominent breathing siphon, or **air tube** (Fig. 39.4). Most are filter feeders or browse on microorganisms inhabiting

solid substrates. Some larvae are predaceous on other insects, including other mosquitoes. Four larval instars precede the pupa. Pupae are remarkably active, breaking the water surface with a pair of trumpet-shaped breathing tubes on the thorax to respire. At the faintest disturbance they swim quickly to the bottom in a tumbling action. The pupation period is short, usually two to three days. When fully developed the skin on the thorax splits, and the adult quickly emerges to fly away. Adult females live for four to five months, especially if they undergo a period of hibernation. During hot summer months of greatest activity, females live only about two weeks. Males live about a week, but, under optimal conditions of food and humidity, their life span may extend to more than a month.

Behavior is an important determinant of the vector potential for all arthropods, including mosquitoes. Generally

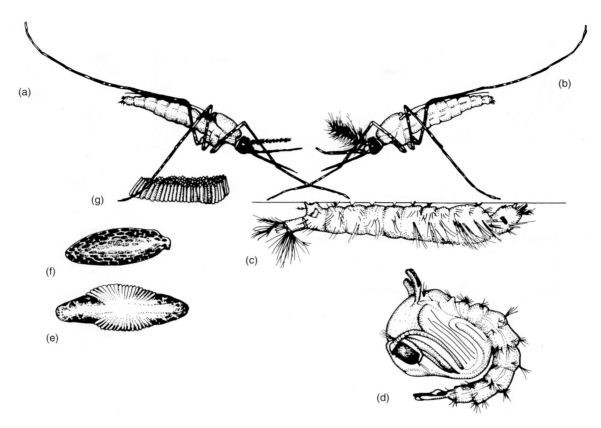

Figure 39.3   Life history stages of *An.*

(*a*) Adult female; (*b*) adult male; (*c*) larva; (*d*) pupa; (*e*) egg; (*f*) egg of *Aedes* sp.; (*g*) egg raft of *Culex* sp.; (*f*) and (*g*) are included for comparison.

Drawings by Ian Grant.

each vector species, because of its preferred breeding sites, feeding schedules, postfeeding behavior, and host choice, presents a unique problem in disease control. For example, *Anopheles gambiae* has been called "the most dangerous animal in the world" because it is such an excellent vector for *Plasmodium falciparum*.[17] This mosquito is strongly attracted to humans and is especially adept at breeding in places created by human activities. As a result, in parts of rural Africa, each villager may suffer 50 to 100 bites per night, with 1 to 5 of these mosquitoes carrying sporozoites.[17] However, *A. gambiae* rests on walls after feeding, and thus is vulnerable to residual insecticides. By contrast, *A. dirus* in Southeast Asia breeds in small pools away from human habitation and leaves a dwelling immediately after feeding; thus, it is quite difficult to control.

Culex quinquefasciatus, considered by some a subspecies of *C. pipiens,* also is synanthropic, often breeding in cesspools and latrines, and is vector for "most of the world's 90 million chronic filariasis cases."[17] One pit latrine in Zanzibar yielded 13,000 of these mosquitoes per night, and residents of one rural area near Calcutta suffered an average of one bite per minute per night.[17] *Culex quinquefasciatus* transmits avian malaria, too, and its introduction into Hawaii had a major negative impact on native birds that were parasitologically naïve.[28]

Physiological attributes are also important contributors to the medical significance of arthropods. It has long been known that mosquito species vary genetically in their susceptibility to infection by parasites such as *Plasmodium* spp., thus contributing to differences in vector potential. Recent studies aimed at explaining this variation suggest several stages at which specificity might be expressed. For example, ookinetes must penetrate the chitinous peritrophic membrane (PM) in order to transform into an oocyst (chapter 9), and mosquito species differ in the timing of PM formation.[5] At least some *Plasmodium* spp. ookinetes secrete chitinase, which is in turn activated by a mosquito trypsin.[6] Later sporozoite penetration of salivary glands may be controlled in part by lectins on gland cell membranes. And, once penetrated, some mosquito species' salivary glands permit sporozoite infectivity to develop, while others do not, although the reasons are still unclear.[60] Several mosquito genes may be responsible for susceptibility, and mechanisms controlling this susceptibility may be both specific and nonspecific.[94]

Like many pest insects, mosquitoes have developed insecticide resistance in several parts of the world. In *Culex pipiens,* so-called OP-resistance (to organophosphate insecticides) involves three genes, two of which are rare. Wide geographic distribution of rare OP-resistance alleles implies extensive mosquito migration.[11] Thus, control strategies must account for both microevolution, driven by insecticide use, and mosquito movement, a natural phenomenon that varies with species.[50]

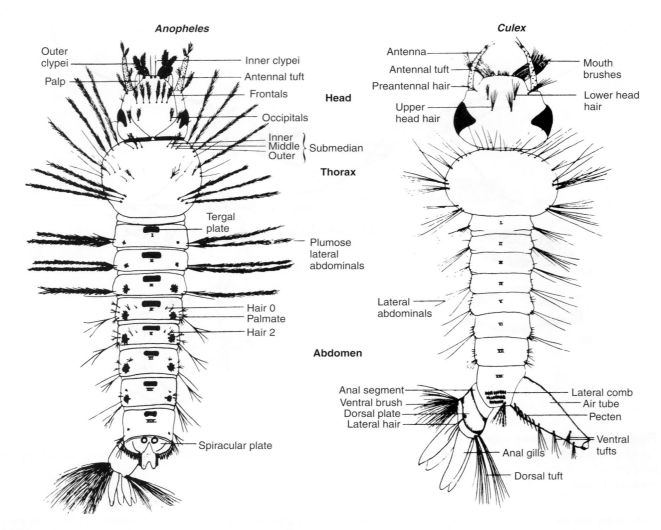

Anopheles

Outer clypei
Palp
Inner clypei
Antennal tuft
Frontals
Occipitals
Head

Inner
Middle } Submedian
Outer

Thorax

Tergal plate
Plumose lateral abdominals

Hair 0
Palmate
Hair 2

Abdomen

Spiracular plate

Culex

Antenna
Antennal tuft
Preantennal hair
Upper head hair
Mouth brushes
Lower head hair

Lateral abdominals

Anal segment
Ventral brush
Dorsal plate
Lateral hair
Lateral comb
Air tube
Pecten
Ventral tufts
Anal gills
Dorsal tuft

Figure 39.4 **Mosquito larvae, showing basic taxonomic characters.**

Courtesy of Communicable Disease Center, 1953, U.S. Public Health Service, Washington, DC.

Subfamily Culicinae

Adult members of Culicinae have a scutellum with a trilobed posterior margin, in dorsal view. The abdomen is densely covered with scales. Culicines lay eggs in rafts or singly on soil. Larvae have a prominent air tube and hang by it nearly perpendicular to the water surface. The subfamily has more than 30 genera and about 3000 species, most of which are in genera *Culex, Aedes,* and *Ochlerotatus.*

Genus *Culex.* *Culex* spp. females have rounded tips on their abdomens, and their palps are less than half as long as the proboscis. They have no thoracic spiracular or postspiracular bristles. Larvae have a long, slender, air tube bearing many hair tufts. Most *Culex* spp. are bird feeders but do not have a narrow host specificity. They overwinter as inseminated females. Several species are important vectors of bird malaria parasites and arboviruses, although their significance may vary geographically and according to whether the setting is urban or rural.

Culex tarsalis, a robust, handsome mosquito, is widespread and common in the semiarid western United States and in southern states as far northwest as Indiana. Its coloration is distinctive: nearly black with a white band on the lower half of each leg joint and a prominent white band in the middle of its proboscis. *Culex tarsalis* breeds in water in almost any sunny location. It is a bird feeder, most active at night, but is not reluctant to feed on humans and other mammals; hence, it is the main vector of **western equine encephalitis (WEE)** and also transmits **St. Louis encephalitis (SLE)** virus, mainly in rural settings. WEE is normally a bird infection, with no apparent symptoms, but it can be acquired by other hosts. Horses are particularly susceptible, with a high rate of mortality. Humans also can be infected; it is not as commonly fatal in humans as in horses but can be severe in children. In adults it results in fever and drowsiness; hence, it is sometimes called **sleeping sickness.** Rarely, following a coma, a person may have reduced physical capabilities.

Culex pipiens, the **house mosquito** (Fig. 39.5), is nearly worldwide in distribution. It is a plain, brown insect that breeds freely around human habitation, laying egg rafts in tin cans, tires, cisterns, clogged rain gutters, and any other receptacle of water. It enters houses readily and is

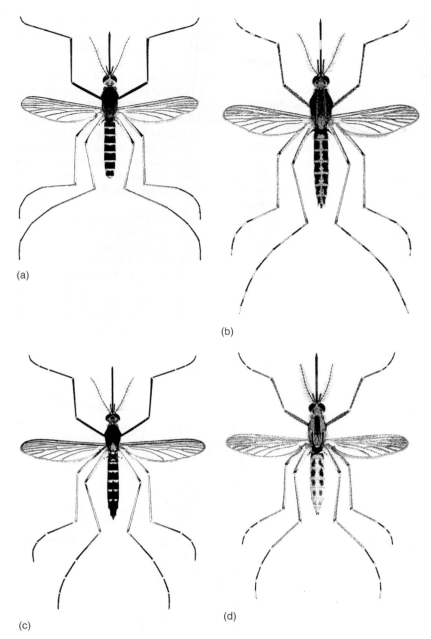

(a)

(b)

(c)

(d)

Figure 39.5   **Some important mosquitoes.**
(*a*) *Culex pipiens;* (*b*) *Ochlerotatus sollicitans;*
(*c*) *Aedes vexans;* (*d*) *Ochlerotatus dorsalis.*

From S. Carpenter and W. LaCasse, *Mosquitos of North
America* (*North of Mexico*). Copyright © 1974 University of
California Press, Berkeley, CA. Reprinted by permission.

a night feeder, causing consternation in many a bedroom. *Culex pipiens* actually is a complex of species with slight physiological differences, only some of which are understood. Members of this complex are important not only for their annoyance factor but also because they are major urban vectors of SLE virus and the filarial worms *Wuchereria bancrofti* and *Dirofilaria immitis* (chapter 29), although in the United States most dog heartworm is transmitted by *Aedes* species. *Culex pipiens* also transmits bird malaria and avian pox.

Genera *Aedes* and *Ochlerotatus*. *Aedes* has now been divided into two genera by elevation of subgenus *Ochlerotatus* to full generic status, a decision based mainly on structural details of genitalia.[67] Thus, some common and familiar species have been placed in this new genus. In both genera,

the posterior end of the female abdomen is rather pointed, and postspiracular bristles are present on the thorax. Larvae have siphons bearing only one pair of posteroventral hair tufts. Because nearly half of North American mosquitoes are in *Aedes* and *Ochlerotatus,* and many of the rest are *Culex,* the pointed abdomen of the female usually is all one needs for separating *Aedes* and *Ochlerotatus* species from those of *Culex* in the field. Keys to species of *Aedes* published prior to 2000 are usually still valid, but a list of those species now in *Ochlerotatus* allows one to assign these to their proper genus. See the Walter Reed Biosystematics Unit Web page for a list (http://wrbu.si.edu).

Species of both genera are notable for their ferocity. Most are diurnal or crepuscular in their activities, as contrasted with night-biting *Culex* spp. They lay their eggs singly on water, mud, soil in places likely to be flooded, or in small containers, depending on the species. Mosquitoes of these genera are not only among the most obnoxious of bloodsucking insects but also are extremely important medically because of the diseases they transmit.

Two species, *Oc. dorsalis* and *Ae. vexans* (see Fig. 39.5), are scourgemates in the western United States. Both are fierce daytime biters; at a single swat a person may kill half a dozen of each species. *Ochlerotatus dorsalis* is broadly distributed over most of the Holarctic region, North Africa, and Taiwan. It breeds in salt marshes as well as in fresh water. The range of *Ae. vexans,* aptly named, overlaps that of *Oc. dorsalis* and includes South Africa and the Pacific Islands. *Ochlerotatus dorsalis* is easily recognized as a straw-colored, medium-sized mosquito of great beauty but utmost persistence. *Aedes vexans* is brown to black with white bands encompassing *both halves* of each leg joint. *Ochlerotatus sollicitans* (see Fig. 39.5) is a flood-water mosquito found throughout most of the eastern two thirds of the United States and southern Canada, where it makes life miserable for biologists and others who frequent the marshes. Last but certainly not least, *Oc. taeniorhynchus,* the **black salt marsh mosquito,** gives up nothing in the biter reputation category. In the words of one experienced parasitologist, it "considers deet a trivial impediment not to be taken seriously." Deet is not only the major insect repellant used by the United States military, but also the active ingredient in best-selling repellants used by the American public.

Among the many other species in these genera are snow–water mosquitoes of the far north and western North American mountains. A difficult complex of species, they are all characterized by their immense numbers and ferocious appetites. Usually there is only one generation per year: Females lay eggs singly in low-lying areas destined to become flooded by melting snow water the following year. Although snow-water mosquitoes transmit no known diseases to humans and domestic animals, their presence in large numbers precludes carefree sport by those who venture into their domain.

Several species of *Aedes* and *Ochlerotatus* are tree-hole breeders. Species such as *Ae. aegypti* (Fig. 39.6) that have adapted to breeding in small containers or leaf axils appear to be derived from tree-hole breeders. *Ochlerotatus triseriatus* is a widespread tree-hole breeder east of the Rocky Mountains. It is similar in appearance to *Ae. vexans* but lacks white rings on the tarsi. It is an important vector of **California (La Crosse) encephalitis** virus. The most common tree-hole breeder in the United States is *Ae. hendersoni,* which is very similar to *Oc. triseriatus.*

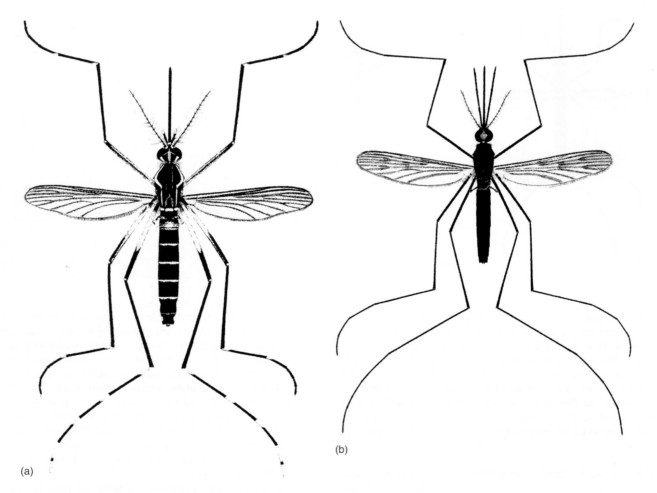

(a)

(b)

Figure 39.6   More important mosquitoes.

(*a*) *Aedes aegypti;* (*b*) *Anopheles quadrimaculatus.*

Aedes albopictus, the **Asian tiger mosquito,** another species that breeds in small containers, was discovered in Houston, Texas, in 1985. It evidently arrived in this country in a shipload of used tires and successfully colonized half of the 30 states that received used tires in 1985. Federal law now requires that all used tires from Asia be certified insect-free, but the used tire trade virtually ensures an eventual global distribution for *Ae. albopictus.* Since its introduction, *Ae. albopictus* has quickly spread to many areas east of the Rocky Mountains; it is also evidently replacing *Ae. aegypti* in parts of the southern United States.[66] It is a good vector for dengue, dengue hemorrhagic fevers, equine encephalitis, yellow fever, and La Crosse virus.[36]

The introduction has considerable public health importance, for three reasons: *Ae. albopictus* has been implicated in several dengue epidemics, some viruses can be transmitted (thus introduced) transovarially, and the species can breed in both urban and rural settings. *Ochlerotatus japonicus* is another Asian species that has invaded North America, especially the temperate zones from Canada to the southern United States. This species is an excellent vector of several viruses, including zoonotic ones such as West Nile, and because of its hardiness and breeding habits, again in small containers, it poses major potential health hazards for both humans and livestock.[81]

Many species of *Aedes* are vectors of a variety of virus diseases. The topic is too extensive to explore adequately here, but one species, *Ae. aegypti,* the **yellow fever mosquito,** is of extreme importance and wide distribution, being found within a belt from 40° N to 40° S latitude, except in hot, dry locations. *Aedes aegypti* is common in much of the southern United States; it is a beautiful mosquito, jet black or brown with silvery white or golden stripes on the abdomen and legs; the last tarsal segment is white. A lyre-shaped pattern covers the dorsal surface of the thorax. *Aedes aegypti* is a tree-breeding species in sylvatic situations, but, when associated with human habitation, it breeds freely in containers, cisterns, and other water storage units. As many as 140 eggs are laid singly at or near the waterline, and they can withstand desiccation for up to a year.

Aedes aegypti originated in Africa, from where it was widely distributed by the slave trade, being transported to much of the world via water barrels in ships. With the mosquito went a *Flavivirus,* which causes yellow fever, a devastating disease that has wrought havoc wherever it has emerged. After establishing in the New World, *Ae. aegypti* caused many epidemics. For example, the British army lost 20,000 of 27,000 men who attempted to conquer Mexico in 1741; the French lost 29,000 of 33,000 men trying to acquire Haiti and the Mississippi Valley.

France was willing to negotiate the Louisiana Purchase largely because of the presence of yellow fever in Louisiana and parts north. Many outbreaks hit coastal cities in the United States, such as Charleston, New Orleans, and Philadelphia, and gold-rush settlements in California. Yellow fever and malaria forced France to abandon completion of the Panama Canal and prevented the job from being attempted again until William Gorgas developed a program of mosquito control in Havana and then applied it to Panama (p. 144). Strangely, yellow fever has never established in Asia. Urban yellow fever is transmitted only by *Ae. aegypti,* but a sylvatic form existing in monkeys, both in Africa and

South America, is transmitted by other *Aedes* and *Haemogogus* mosquitoes.

Dengue is a *Flavivirus* disease also transmitted by *Ae. aegypti.* It is also called **breakbone fever** and **epidemic hemorrhagic fever.** The four distinct serotypes of dengue virus cannot be differentiated by symptoms. In uncomplicated cases a patient has fever, severe headaches, and pains in the muscles and joints; weakness and temporary prostration are common. Recovery is rapid. A hemorrhagic complication occurs occasionally, especially in indigenous Asian children of three to six years old. This condition ranges from a rash and mottled skin to severe hemorrhaging in the lungs, digestive tract, and skin. The mortality rate is up to 7% in those who are hospitalized; unhospitalized cases must have a higher rate.

The spread of *Ae. albopictus* is of major concern because of dengue, with current estimates of 30 million to 100 million cases a year and 2 billion people at risk.[17, 66] Dengue occurs from eastern Europe through most of Asia; North, Central, and South America; and the Caribbean. The current high volume of air travel contributes significantly to spread of dengue, especially through the tropics; for example, the number of reported cases of dengue hemorrhagic fever increased from 2,067 in 1967 to more than 600,000 in 1987.[66] Currently approximately 2.5 billion people are at risk, resulting in 50–100 million new cases a year, half a million of them severe and life-threatening.[90] The virus is endemic in more than a hundred countries.[8] Throughout much of Asia, dengue is mainly a childhood disease because vectors are domestic species such as *Aedes aegypti.*[63]

West Nile Virus (WNV) is transmitted by culicine mosquitoes, especially species of *Aedes, Ochlerotatus,* and *Culex. Aedes albopictus* is highly susceptible; *Cx. pipiens* is less so but is an excellent vector because of its behavior.[85] WNV causes West Nile fever, a disease that rapidly spread across the United States following its introduction in the late 1990s. WNF is most serious in elderly people, and numerous deaths have been reported, especially in the west and upper Midwest. Current statistics are readily available from the Centers for Disease Control and Prevention website, www.cdc.gov.

Other Culicine Genera. Anyone who studies mosquitoes seriously discovers a rich culicine fauna that presents many challenges for a biologist. *Toxorhynchites* spp., for example, are predatory tree-hole breeders and candidates for biological control agents focused on species of medical importance. In *Mansonia* and *Coquillettidia* spp., larval air tubes are sharply pointed, enabling them to pierce stems of aquatic plants to obtain air. Simply coating the water with oil will not prevent these insects from obtaining oxygen. Species in this complex are important vectors of **brugian filariasis** (chapter 29).

Genus *Culiseta* contains eight North American species and subspecies. They are large, brownish mosquitoes, some restricted to feeding on birds and mammals other than humans. However, *C. inornata* and *C. melanura* are involved in transmission of western and eastern encephalitis viruses. *Culiseta inornata* is the most widespread of the two, being found in southern Canada and conterminous United States, whereas *C. melanura* is restricted to the eastern and central United States.

Other genera of Culicinae are marginally important in the transmission of arboviruses.

Subfamily Anophelinae

Adult anophelines have a scutellum that is rounded or straight but never trilobed (except slightly in *Chagasia* spp.) in dorsal view. Abdominal sternites largely lack scales. Palpi of both sexes are almost as long as the proboscis (except in *Bironella* spp.). Larvae lack an air tube, and their dorsal surface bears branching hairs (see Fig. 39.4).

The subfamily contains three genera: *Bironella,* with seven species in New Guinea and Melanesia; *Chagasia,* with four species in tropical America; and *Anopheles,* with about 400 species, including 15 in North America. Resting and feeding postures are distinctive for *Anopheles* spp. When the mosquito is at rest, its head, proboscis, and abdomen are almost in a straight line; while feeding, its body is inclined at a sharp angle from the surface of its host. Because genera *Bironella* and *Chagasia* are of no medical importance, we will consider only *Anopheles* spp. in this chapter.

Genus *Anopheles*. Female *Anopheles* spp. (see Figs. 39.3 and 39.6) lay up to a thousand eggs, depositing them singly on water. The eggs have useful taxonomic characters, such as presence or absence of lateral floats, which are characteristically marked, and a lateral frill. Eggs must remain in contact with water to survive. Usually they hatch within two to six days and develop through four larval instars in about two weeks, followed by a three-day pupal stage. Development from egg to adult takes from three weeks to one month. Development of mosquitoes, like that of virtually all insects, however, is temperature dependent.

Preferred breeding sites vary tremendously among the species of *Anopheles,* a factor that must be understood before effective control of malaria vectors can be undertaken. Thus, some species breed most efficiently in stagnant mangrove swamps, others in sunny, partly shaded pools, and still others along the edges of trickling streams. A few are tree-hole breeders.

Taxonomy of genus *Anopheles* is complicated by the existence of several species complexes. For example, the species initially named *An. maculipennis* we now know to consist of at least seven subspecies (or perhaps sibling species), differing slightly in host preferences and egg characteristics. American representatives of this complex are *An. quadrimaculatus* (see Fig. 39.6), *An. freeborni, An. aztecus, An. earlei,* and *An. occidentalis.* What was formerly known as *An. gambiae* in Africa comprises six species, including some freshwater and some saltwater breeders. Reproductive isolation in and genetic barriers exist between all six species.[56, 84] Students seeking recent information on all aspects of *An. gambiae* biology—from molecular to ecological—should consult the October 4 2002, issue of *Science.*

Of all diseases transmitted to humans by insects, that caused by *Plasmodium falciparum* takes more lives and causes more suffering than the others put together (see chapter 9). This and the other malaria parasites of humans (*P. ovale, P. knowlesi, P. malariae,* and *P. vivax*) are all transmitted by species of *Anopheles.* Historically the primary vectors of this disease in North America were *An. quadrimaculatus* and *An. freeborni.* Both are still common on the continent, as is *Aedes aegypti,* but, like yellow fever, endemic malaria has been eradicated. *Anopheles* species also transmit bancroftian filariasis *(Wucheraria bancrofti),* primarily in rural settings.

The conquest of malaria and yellow fever in the United States stands among the greatest medical triumphs of this country. That victory, however, is due more to elevated standards of living than it is to elimination of mosquitoes. From experiences in many parts of the world, it is now obvious that insecticide resistance is a predictable result of insecticide use, a realization that has inspired a broad research effort to find alternate ways of controlling not only mosquitoes, but also black flies, tsetse flies, and bot flies. Some of these approaches being tested are ones that 50 years ago would have been considered "far out" or frivolous; others that have been tried for almost a century now are getting a modern reexamination.

Predatory fish are used in attempts to control mosquito larvae in many parts of the world, in settings ranging from rice fields to cisterns.[78] The infamous "mosquito fish" *Gambusia affinis* has been spread far and wide as a result, and such human-aided dispersal has the potential to reduce numbers of competing native fishes.[55] Australian and South Pacific copepods of genus *Mesocyclops* are also being tested as larval predators; these crustaceans have the advantage of occurring naturally in habitats ranging from lakes and streams to wells and tree holes.[7] Species of *Toxorhynchites* have been tested as predators on their fellow tree-hole breeder mosquitoes.[83] Biological control with nematodes was discussed by Platzer.[65] Other methods and devices that have been tried in recent years are planarians,[54] specially designed lids for water jars,[44] sustained release insecticide pellets,[47] digging of shallow channels to help flush tidal marshes,[18] use of copper linings in cemetery flower vases,[61] root and bark extracts of mangroves,[79] strains of the bacterium *Bacillus sphaericus,* and genetically engineered cyanobacteria containing *B. sphaericus* genes.[42, 93] Commercially available pellets containing *Bacillus thuringiensis* work reasonably well to control mosquito larvae in small ornamental ponds, bird baths, and similar containers.

Obviously we have a long way to go before we find completely effective ways to prevent insect transmission of malaria, filariasis, and certain viral diseases. For good general discussions of mosquito control and pesticide resistance problems, see papers by Catteruccia[9] and Nauen.[58]

Family Simuliidae

Simuliids (see Fig. 39.7) are commonly called **black flies,** although many species are gray or tan. They are small, 1 mm to 5 mm long. The prescutum of their mesonotum is reduced, giving them a humpbacked appearance, which explains their other common name, **buffalo gnat.** Their wings are broad and iridescent, with strongly developed anterior veins. Antennae are filiform, usually with 11 segments. Females' eyes are separated, whereas those of males are contiguous above the antennae. There are no ocelli. Mouthparts are of the horse fly subtype, although delicate (Fig. 39.7*b*). Serrated teeth on the edges of the mandibles are cutting structures, whereas recurved teeth on the maxillary lacinia serve to anchor mouthparts during feeding.[21]

Black flies are found worldwide but are most abundant in northern temperate and subarctic zones. Females of most species feed on blood as well as nectar, but males feed only on plant juices. Mating occurs in flight, when females fly

(a)

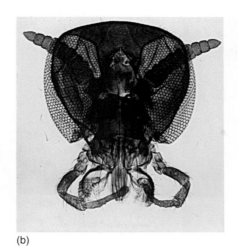

(b)

Figure 39.7 *Simulium* **sp.**

(*a*) Larva; (*b*) head. Note the short, cutting-type mandibles.

(*a*) Courtesy of Warren Buss; (*b*) courtesy of Jay Georgi.

into swarms of hovering males. A female simuliid produces 200 to 800 eggs, laying them on the water surface, where they rapidly sink. In some species females land at the water's edge, crawl down a rock or plant to deposit eggs underwater, and then crawl back out of the water and fly away.

Larval development can occur only in running, well-oxygenated water. Hence, black flies are most numerous near rivers and streams, although they are known to travel several miles when aided by winds. On hatching, larvae (see Fig. 39.7*a*), with their modified salivary glands, spin a silken mat on some underwater object. They attach to this mat with a hooked sucker at the posterior end of their abdomen. Thus, their head hangs downstream and, with fanlike projections around the mouth, filter protozoa, algae, and other small organisms and organic detritus from the passing water. Larvae are often so numerous they form a solid covering on a favorable location, such as a spillway or the downstream side of a rock or log. Larvae are capable of changing locations rapidly by stretching out, spinning a new mat and clinging to it with their mandibles, and then releasing the old mat and hooking onto the new one. The six or seven larval instars require 7 to 12 days under ideal temperature conditions and food availability, but this time may be greatly extended. Some species overwinter as larvae.

Before pupation, larvae spin flimsy cocoons around themselves. After molting, pupae remain nearly immobile, respiring through long filamentous gills on their anterior end. Number and arrangement of these filaments are of taxonomic importance. The pupal stage lasts from a few days to three or four weeks. To emerge, an imago first cuts a *T*-shaped slit in the pupal thorax and crawls through it. It quickly fills its air sacs with air extracted from the water, forming an internal balloon; it releases its hold on the substrate and shoots to the surface. One to six generations may mature per year, depending on the locality.

Classification and identification of simuliids are often difficult because of numerous complexes of sibling species. The most important genus is *Simulium,* with more than 1200 species. Also important medically are *Prosimulium* and *Cnephia* in North America and *Austrosimulium* in Australia and New Zealand. Black flies are fairly host specific. Few species will bite humans, but those that do are extremely vexatious.

In North America, *Prosimulium mixtum* can be very annoying, and *Cnephia pecuarum,* the **southern buffalo gnat,** has been known to ravage entire herds of livestock. *Simulium vittatum* is widespread in the United States and is particularly irritating to livestock. *Simulium meridionale,* the **turkey gnat,** torments poultry, biting them on the combs and wattles. All fishers and campers in the northern United States and Canada are familiar with the attacks of *S. venustum,* which often occur in such numbers as to ruin a vacation. Vast numbers of *S. arcticum* killed more than a thousand cattle in western Canada annually from 1944 to 1948. *Simulium colombaschense,* of central and southern Europe, killed 16,000 cattle, horses, and mules in 1923 and 13,900 in 1934.[35] Simuliids probably have been biting vertebrates for well over 100 million years; a Jurassic fossil pupa indistinguishable from that of a modern *Prosimulium* has been found (Fig. 39.8).[14]

Individuals react differently to black fly bites. Few people have little or no reaction; most develop local reactions in the form of reddened, itching wheals. **Black fly fever,** a combination of nausea, headache, fever, and swollen limbs, occurs in particularly sensitive persons. Deet (N, N-diethyl-3-methylbenzamide) and an extended-duration repellant formulation (EDRF) of deet are repellants of choice.[68]

Black flies are vectors of *Onchocerca volvulus,* the cause of human onchocerciasis, discussed in detail in chapter 29. The most common vector in Africa is *S. damnosum,* although *S. neavei* also is important. In the New World, *S. ochraceum, S. metallicum, S. callidum,* and *S. exiguum* are the most efficient vectors because of their preference for humans and the timing of their activity, coinciding with that of humans. *Onchocerca gutterosa* commonly is transmitted to cattle by *S. ornatum* in Europe. In Australia, *Simulium* and *Culicoides* spp. infect cattle with *Onchocerca gibsoni,* causing considerable loss to flesh and hides.

Figure 39.8 Fossil simuliid pupa, *Simulimima grandis,* from the middle Jurassic.

This pupa is virtually indistinguishable from that of a modern member of genus *Prosimulium.*

From R. W. Crosskey, "The fossil pupa *Simulimima* and the evidence it provides for the Jurassic origin of the Simuliidae (Diptera)," in *Syst. Entomol.* 16:401–406. Copyright © 1992. Reprinted by permission.

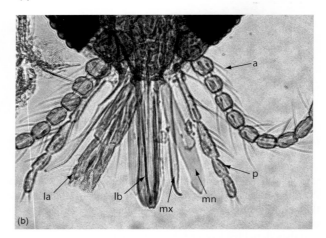

Figure 39.9 *Culicoides* species showing mottled wings and mouthparts.

(*a*) Adult female; (*b*) Mouthparts. Species of *Culicoides* and they differ in a variety of ways, most notably the patterns of mottling on the wings and the shapes of the palpal segments. a, antenna; la, labium; lb, labrum; mn, mandible; mx, maxilla; p, palp, Adult *Culicoides* are typically less than 1.5 mm in length.

Drawing and photograph by John Janovy, Jr.; drawing from a variety of sources; photograph of specimen collected by Harold Manter.

Mathematical models of onchocerciasis transmission suggest some fascinating interactions between humans and simuliid flies. Davies,[19] using data from a single village, predicted that a 99% effective vector control program would have to continue for 18 years to eradicate the worms and that infected immigrant flies posed a greater threat for reinfection than did infected resident humans. See Basáñez et al.[3] for an excellent review of onchocerciasis control strategies, with a focus on Africa.

The malarialike bird disease caused by *Leucocytozoon* spp. (chapter 9) is transmitted by various species of *Simulium.* Virtually any species of bird is likely to be infected with a species of *Leucocytozoon; L. simondi,* a severe pathogen of ducks and other anatids, is transmitted by *S. rugglesi, S. anatinum,* and other species.

Family Ceratopogonidae

This family comprises the **biting midges,** also called **punkies, no-see-ums,** and "sand flies." They are very small, usually less than 1 mm long, but what they lack in size they

make up for in ferocity. The majority are daytime feeders that cannot cope with blowing winds so they are most pesky on hot, still days. Their small size enables them to crawl through ordinary window screening; some species, particularly in the tropics, enter houses freely. Taxonomy of North American species of ceratopogonids has been treated by Wirth and Atchley.[91]

Most of the 60 or more genera feed on insects, a few feed on poikilothermic vertebrates, and members of only four genera feed on mammals, including humans: *Culicoides* (Fig. 39.9), *Forcipomyia, Austroconops,* and *Leptoconops.* Only females feed on blood. Biting midges are recognized by, in addition to their small size, their narrow wings, which have few veins, often are distinctly spotted, and are folded over the abdomen when at rest. Single species may be widespread; for example, *Culicoides furens* is common from Massachusetts to Brazil, throughout the West Indies, and on the Pacific coast from Mexico to Ecuador.

Ceratopogonids breed in a wide variety of situations. Larvae are aquatic or subaquatic or develop in moist soil, tree holes, decaying vegetation, and cattle dung. *Leptoconops* larvae have been found as deep as 3 feet in the soil. Some species breed readily in saltor brackish water, notably mangrove swamps and salt marshes. The life cycle is completed in between six months and three years.

Of the 1000 or so species of *Culicoides,* several that bite humans may be so annoying as to affect tourism in infested areas. Farm workers often are intensely annoyed, and domestic livestock are plagued by these tiny flies. An estimated 10,000 *C. nubeculosus* have been witnessed on a single cow.[59]

Three apparently nonpathogenic filarioid nematodes are transmitted to humans by ceratopogonids: *Mansonella perstans* and *M. streptocerca* in Africa and *Mansonella ozzardi* in South and Central America (see chapter 29). *Onchocerca cervicalis* of horses and *O. gibsoni* of cattle are transmitted by *Culicoides* spp., as are other filarioids of domestic and wild animals. Blood-dwelling protozoan parasites also use *Culicoides* spp. as vectors. *Hepatocystis* spp. in monkeys and other arboreal mammals, some species of *Haemoproteus* in birds, and various species of *Leucocytozoon* (see chapter 9) are transmitted by ceratopogonids.

Orbivirus, the viral etiological agent of **bluetongue,** is spread by *C. variipenis* in North America and by other species of *Culicoides* in Africa and Asia Minor. Bluetongue is a hemorrhagic disease of ruminants, including sheep. It causes some mortality, but infected animals suffer mainly from loss of weight, loss of wool, and reduced breeding. Ceratopogonids may also transmit encephalitis viruses, bovine ephemeral fever, and African horse sickness. Like simuliids, *Culicoides* spp. have been making life miserable for large animals ever since the Mesozoic.[76]

SUBORDER BRACHYCERA

In this group of mostly robust flies antennae are reduced to three apparent segments, the terminal one being drawn into a sharp point, or **style.** A flagellumlike **arista** may be present. Wing venation is reduced. Larvae are active, usually predaceous; their heads may be incomplete, retractable (able to be pulled into the thorax), or vestigial. Life cycles are aquatic or semiaquatic. Among Brachycera only Tabanidae and Rhagionidae have bloodsucking habits as adults; larvae of calliphorid genus *Protocalliphora* feed on blood.

Infraorder Tabanomorpha
Tabanomorph flies have a bulbous adult face and a retractable larval head.

Family Tabanidae
Horse flies and deer flies are widely distributed in the world, and their fierce questing for blood has earned them universal

animosity. They are large, powerful flies, from 6 mm to 25 mm long. Tabanids are mainly daytime feeders. Only females feed on blood; males lack mandibles and eat only plant juices. Eyes are widely separated in females but contiguous in males. About 4000 species are divided into 30 to 80 genera, depending on the authority.

Tabanid mouthparts (Fig. 39.10) are of the horse fly type. They are similar to those of Ceratopogonidae and Simuliidae but are stouter and stronger. The fascicle consists of six piercing organs: two flattened, bladelike mandibles with toothlike serrations; two more narrow maxillae, also serrated; a median hypopharynx; and a median labrum-epipharynx. In biting, mandibles cut in a scissorslike motion, whereas maxillae pierce and rend tissues, rupturing blood vessels. The fly feeds on a pool of blood that wells into the wound (telmophagy). The hypopharynx and labrum-epipharynx form a food canal. Some fierce-looking species with long mouthparts, such as those in Pangoniinae, actually are not blood feeders.

Tabanids usually breed in aquatic or near aquatic environments, although some complete larval development in soil. Females generally lay from a hundred to a thousand eggs at water's edge or on overhanging vegetation or rocks. At egg-laying time, such locations may be swarming with ovipositing flies. On hatching, larvae fall or crawl into water or burrow into mud. Many feed on organic debris, but others are voracious predators on insect larvae, worms, and other soft-bodied animals, including other horse fly larvae and even toads.

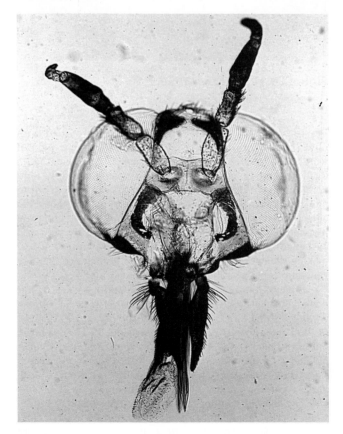

Figure 39.10   Head of *Chrysops* sp., showing the tips of the mandibles.

Courtesy of Warren Buss.

Figure 39.11 *Tabanus atratus.*
Courtesy of Warren Buss.

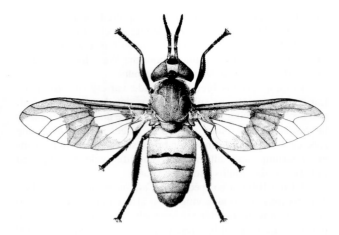

Figure 39.12 Fly of genus *Chrysops.*
From R. P. Lane and R. W. Crosskey (Eds.), *Medically important insects and arachnids.* Copyright © 1992 Chapman & Hall, London. Reprinted with permission.

Larvae have a small retractable head provided with powerful, sharp mandibles capable of inflicting a painful wound on the unwary. Their body has 12 segments and a tracheal siphon that retracts into their posterior end. In temperate zones, tabanids require about a year to develop to pupation. Larvae crawl into drier earth and pupate. Pupae are obtect and require from five days to two weeks to complete metamorphosis. Adults escape the pupal case by cutting a *T*-shaped opening in the dorsal thorax, crawling out, and making their way to the surface. In the tropics two or more generations per year may occur.

Several species of horse flies are serious pests of humans and livestock. *Tabanus quinquevittatus* and *T. nigrovittatus* are the large, gray, "greenhead" horse flies of most of the United States. *Tabanus atratus* (Fig. 39.11) is a huge, uniformly black horse fly of eastern North America; *T. lineola* and *T. similis* are smaller, striped flies also in the eastern states. In the western states, *T. punctifer* is a very large, black horse fly with a yellow thorax. Such species as *T. atratus* and *T. punctifer* seldom bite humans; they buzz so loudly when approaching that they seldom are allowed to land. *Haematopota americana, Hybomitra* spp., and *Silvius* spp. are also common western pests. *Diachlorus* spp. are common, aggressive pests in Central America, where they are called **doctor flies.** Unlike most tabanids they freely enter houses.

The name **deer fly** is applied to genus *Chrysops,* of which there are about 80 North American species. Deer flies (Fig. 39.12) usually are smaller than horse flies and have brown-spotted wings. Their flight is not as noisy as that of most horse flies, so they bite humans more commonly.

Medical importance of Tabanidae is two sided: the annoyance and blood loss occasioned by the bite and infections transmitted mechanically and biologically by the flies. Because of the large size of mouthparts, tabanid bites are quite painful. Most people have little or no allergic reaction to them, although such sequellae can occur. Their annoyance factor may seriously interfere with use of recreational areas, and field-workers and timber-workers may have lowered productivity as a result of harassment by these flies.

A serious problem for livestock is blood loss, interrupted grazing, and energy consumed in trying to escape the insects. Tethered or caged animals particularly suffer from these flies because they are unable to escape their tormentors, even briefly. One well-known parasitologist reported seeing a caged mule deer simultaneously being fed on by a dozen or more *Tabanus punctifer* and dozens of open, freely bleeding wounds covering the wretched animal's face. No tabanids are strictly host specific, but most have host preferences. Thus, *Haematopota,* with at least 300 species, feeds mainly on cattle and antelope, with which it may have evolved. Birds are attacked uncommonly.

Certain characteristics of tabanids enhance their capabilities to transmit pathogens:[48] (1) **anautogeny,** the necessity of a blood meal for development of eggs, stimulating hostseeking behavior; (2) **telmophagy,** through which blood-dwelling pathogens can enter the pools from which flies suck blood; (3) **relatively large blood meals,** enhancing the possibility that pathogens will be imbibed; (4) **long engorgement time,** enabling pathogens to infect a fly's tissues; and (5) **intermittent feeding behavior,** increasing the chances for mechanical transmission of pathogens.

Tabanids are involved in transmission of protozoan, helminthic, bacterial, and viral diseases of animals and humans. Among diseases caused by protozoa, two species of *Trypanosoma* are transmitted mechanically by tabanids.

Trypanosoma evansi, the causative agent of surra in many wild and domestic animals, is spread by species of *Tabanus* (chapter 5). Other vectors, such as stable flies, other genera of horse flies, and vampire bats, also can be involved; but *Tabanus* spp. appear to be the most effective vectors of this trypanosome. *Trypanosoma theileri* is a cosmopolitan parasite of cattle and antelopes. Cyclopropagative development occurs in the insect gut, so the tabanid species involved are actually true intermediate hosts. Examples of such species are *Haematopota pluvialis, Tabanus striatus,* and *T. glaucopis.*[48]

The African eye worm, *Loa loa* (chapter 29), is transmitted by species of *Chrysops.* There appear to be two strains of *L. loa,* one in monkeys in the forest canopy and one in humans. Night-feeding *Chrysops langi* and *C. centurionis* transmit the former, and diurnal *C. silaceus* and *C. dimidiatus* transmit the latter.

The arterial filarioid *Elaeophora schneideri* lives in head and neck vessels of American species of deer, elk, moose, and domestic sheep in the western states. It is symptomless in deer but in other hosts causes much distress, such as blindness, nervous dysfunction, necrosis, and head deformity. Species of *Tabanus* and *Hybomitra* are vectors of this worm.[37]

Bacterial infections known to be mechanically transmitted by tabanids are anaplasmosis and anthrax. Similarly, hog cholera and equine infectious anemia (swamp fever) viruses use horse flies as vectors.

Infraorder Muscomorpha

Except for mosquitoes, members of this group are the most important flies in veterinary and human medicine. Antennae are short and pendulous, usually with a conspicuous arista on the second segment. Three prominent ocelli are arranged in a triangle on the vertex and frons. The compound eyes are very large, separated in females and close together in males. Arrangement of bristles on the head and thorax is important in the taxonomy of flies. At the anal angle of the wing is a prominent lobe, the **squama,** or **calypter,** in medically important families.

Larval calyptrates are **maggots,** with an elongated, simple body, usually tapered toward their anterior end. A true head is absent; vertically biting mandibles are part of a conspicuous cephalopharyngeal skeleton. These sclerotized structures are important in taxonomy of larvae. Two spiracles are found on the posterior end. Each species has distinctive markings on the spiracular plates that are useful in identification. When fully developed, the third-stage larva undergoes **pupariation,** resulting in a pupalike surrounding case **(puparium)** made of the hardened third-stage larval tegument. After internal reorganization, adults emerge through a circular hole in the puparium. They push the operculum off the puparium by inflating a balloonlike organ, the **ptilinum,** in the head. The ptilinum extends from a frontal suture of the head, pushes off the operculum, and then withdraws back into the head; the frontal suture immediately heals.

Family Chloropidae

Eye gnats look very much like tiny house flies. The most important genus in this family is *Hippelates.* These flies are very small, 1.5 mm to 2.5 mm long, and are acalypterate.

They are called *eye gnats* because they are attracted to eye secretions as well as to other body secretions and free blood. They do not bite but feed like house flies, sponging up liquid food and vomiting liquid stomach contents onto their food, to be reeaten. In some species the labellum is provided with tiny spines that scarify a host's skin, also leading to infection by pathogens. *Hippelates* spp. are very persistent when hungry and hence may become intensely irritating.

Life cycles of Chloropidae all seem to be similar. About 50 eggs are laid on the surface of or slightly under soil, which must be loose and have an abundance of well-aerated organic matter. Larval development consumes 7 to 12 days under optimal conditions; the pupal period requires about six days, and adults age seven days before oviposition.[30]

Aside from the considerable annoyance caused by these flies, they are important vectors of disease. They congregate at wounds caused by others, such as tabanids and stable flies, thereby further contaminating the host. Although unproven, *Hippelates* spp. may aid mechanical transmission of the bacillum causing human **pinkeye,** or **bacterial conjunctivitis,** as well as of the spirochete *Treponema pertenue,* the etiological agent of yaws. Flies feeding at tips of teats spread a bacterial disease, **bovine mastitis,** from cow to cow. Control of eye gnats is difficult. The best system in use so far is a combination of attractant baits with a pesticide and efficient soil management.

Family Glossinidae

These are the infamous tsetse flies. The family contains a single genus, *Glossina,* and occurs only in Africa and two localities on the Arabian peninsula. It was once more widespread, however; four species have been found as fossils in Oligocene shales of Colorado.

***Glossina* Species.** Tsetse flies (see Fig. 5.5) are 7.5 mm to 14.0 mm long and brownish gray. When at rest, their wings cross like scissors. The palpi are almost as long as the proboscis, which protrudes from the front of the head. Mouthparts and thus feeding habits are much like those of stable flies. The base of the proboscis is swollen into a characteristic bulb. Tsetses are daytime feeders and are visually attracted to moving objects. Both sexes feed exclusively on blood of a wide variety of animals, including humans, and are particularly attracted to pigs.

Tsetse flies are larviparous and pupiparous, giving birth to a single, completely developed larva at intervals, producing from 8 to 20 in all. While in the oviduct, the larva feeds on secretions from specialized milk glands. Larvae are deposited on loose, dry soil, usually under shelter of some type. They have no locomotor structures but, by contraction and extension, bury themselves under a few centimeters of loose soil. Hardening of their integument to form a puparium occurs within an hour of larviposition. The integument darkens to brownish black; pupae are barrel shaped and have two prominent posterior lobes. Adults emerge within two to four weeks. The biology and influence of tsetses are beautifully illustrated by Gerster.[29]

Tsetse flies have chemoreceptors on their tarsi, among them sensillae that stimulate feeding when exposed to uric acid, leucine, valine, and lactic acid (stimulatory ingredients in human sweat).[87] *Glossina morsitans* females also respond to chemicals released by larvae before pupation by depositing larvae in moist sand where there are already larvae.[51] The result is aggregation in shady, relatively moist areas, especially during dry seasons.

Twenty-three species of *Glossina* usually are recognized and can be identified with the key in Lane and Crosskey.[41] All but three have been found capable of transmitting trypanosomes of mammals. Six of these are of outstanding medical importance: *G. palpalis, G. fuscipes,* and *G. tachinoides* are found along rivers and are primary vectors of Gambian sleeping sickness. *Glossina morsitans, G. swynnertoni,* and *G. pallidipes* are savanna species and principally transmit Rhodesian sleeping sickness. Probably all of these, as well as several other species, can transmit nagana to cattle (see chapter 5). Vale and coworkers discussed the use of traps to control tsetses (see also the epigraph quotes to chapter 5).[86]

Family Hippoboscidae

Louse flies look neither like lice nor flies but rather like six-legged ticks. In most species males are winged and females wingless, although in some, such as *Hippobosca* spp., both sexes are winged. Both sexes are bloodsuckers, with some species parasitizing mammals and others birds. Larvae are retained within the female, feeding on secretions from special glands, and when born are ready to pupariate.

The **sheep ked,** *Melophagus ovinus* (Fig. 39.13), is distributed worldwide except in the tropics. Its puparium is glued to a host's wool at any season of the year. Each female produces from 10 to 12 young. A ked's entire life is spent on its host; when removed, it dies in about four days. Heavy infestations cause emaciation, anemia, and general unthriftiness of sheep. The skin may be so scarred by bites as to lose its market value. Hippoboscids are not loath to bite humans, and sheep shearers particularly are vulnerable to their attacks. The bite is said to be as painful as a wasp sting.

Other genera and species are found on mammals and birds in various parts of the world. *Olfersia coriacea* has been observed to bite humans in Panama.[33] The **pigeon fly,** *Pseudolynchia canariensis,* is common on pigeons throughout most temperate regions and is the vector for *Haemoproteus columbae* (see chapter 9). Both sexes are winged. The pigeon fly is willing to bite people, with painful results.

Families Streblidae and Nycteribiidae

These two small, poorly known families are the **bat flies,** parasitic only on bats. Streblids may be winged or wingless, or they may have reduced wings. Compound eyes are small or absent. Six species are found in North America; most are associated with New World tropical bats. Nycteribiids are called **bat spider flies** (Fig. 39.14) because of their superficial resemblance to spiders. They are wingless, with their head folded back into a groove in the dorsum of the thorax. Five species are found in North America; most species feed on Old World bats. Both families are pupiparous.

Family Fanniidae

This family contains over 260 species, the large majority of which are in genus *Fannia. Fannia canicularis,* the **lesser house fly,** resembles *M. domestica* but is smaller, more slender and dark, and has only three brown longitudinal stripes on the thorax, rather than four as in houseflies. Biology of lesser house flies parallels that of *M. domestica. Fannia canicularis* will enter houses freely, but, unlike *M. domestica, F. canicularis* is not particularly attracted to food and so is not as efficient a vector as are house flies. **Latrine flies,** *F. scalaris,* breed in fresh dung, particularly that of swine, but seldom enter houses. Both *Fannia* species are known to cause accidental myiasis of the rectum of humans, presumably

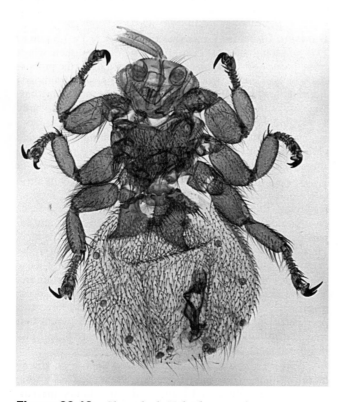

Figure 39.13　Sheep ked, *Melophagus ovinus.*
Courtesy of Warren Buss.

Figure 39.14　Bat spider fly, family Nycteribiidae.
Courtesy of Warren Buss.

when they lay eggs on the anus and larvae crawl into the rectum and begin developing. No medical problem is caused by this opportunistic occupation beyond consternation in a person who discovers maggots in his or her stool.

Family Muscidae

Members of this family are often **synanthropic;** that is, they live closely with humans. Many freely enter houses and readily avail themselves of whatever food and drink they may find there. They are smallto medium-sized flies, usually dull colored, with well-developed squamae and mouthparts.

Musca domestica. House flies are the most familiar of all flies as well as some of the most medically important. They are gray, 6 mm to 9 mm long, and have four conspicuous, dark, longitudinal stripes on the top of the thorax. Their distribution is nearly cosmopolitan but has changed markedly in societies with increased sanitation and decreased dependence on horses. House flies breed in all types of organic wastes except decaying flesh. Feces of any kind are preferred, although decaying milk around dairy barns, silage, slops around hog troughs, rotten fruit and vegetables, and so on are all stock in trade for house flies. Garbage cans are a favorite breeding place, and rotting garbage in the tropics can become transformed into an apparently equal mass of maggots, seemingly overnight.

Under ideal conditions, an egg can develop to adult in 10 days. One female deposits 120 to 150 eggs in each of at least six lots in its short lifetime. If all offspring of a pair of flies in April lived and reproduced, as did their succeeding generations, by August there would be 191,010,000,000,000,000,000 flies, which would cover the earth to a depth of 47 feet. With such reproductive potential a depleted population of flies can recover in a very short time.

House flies are efficient disease carriers for three primary reasons:

1. Their construction favors carrying bacteria. The multitude of tiny hairs covering most of the body readily collects bacteria, spores, and helminth eggs; and their mouthparts and six feet also have sticky pads that collect such matter.
2. They relish human food and excrement alike. While walking on food and utensils, they not only leave a trail of bacteria, but also while feeding they defecate and vomit the remains of their last meal. Helminth eggs, protozoan cysts, and bacteria survive the intestinal tract of flies and thus can be widely distributed from the site of their initial deposition.
3. Because of their synanthropy and powerful flight, house flies move about freely between indoor and outdoor attractions. Thus, it would appear that house flies are ideal vehicles for mechanical transmission of disease.

The list of diseases known to be transmitted by house flies is too long to be repeated here. Most are enteric diseases resulting from fecal contamination, such as typhoid fever, cholera, polio, hepatitis, shigellosis, salmonellosis, and other dysenteries, but the list also includes yaws, leprosy, anthrax, trachoma, tuberculosis, and infections with various worms, such as *Ascaris lumbricoides.* Several diseases of domestic animals also are transmitted by these pests; in one study 70% of the flies from a goat yard contained coccidian oocysts in their gut.[20]

Other Species of *Musca.* About 60 species have been placed in genus *Musca.*[16] In Australia the aggravating **bush fly,** *M. vetustissima,* occurs in incredible numbers. Its importance as a vector is much less than that of *M. domestica,* partly because it is not so willing to enter human habitation and partly because of the widely scattered human population in much of that continent. It is not unusual for a person walking through the central Australian desert to have at least a thousand of these flies on his or her back.

The **face fly,** *M. autumnalis,* is a native of Africa, Asia, and Europe but was introduced into the United States in 1950. It now occurs from coast to coast and well into Canada. It is a little larger than the house fly. Sides of the abdomen of females are black, and those of males are orangish. Larvae develop in cow dung. Adults feed on secretions around the eyes of cattle and other large ruminants. They serve as vectors of eye worms, *Thelazia* spp. (chapter 28). Annual losses in the United States are estimated at $60 million.

Stomoxys calcitrans. The **stable fly** is the only member of genus *Stomoxys* that occurs in North America; it is a cosmopolitan species, and there is molecular evidence that gene flow among *S. calcitrans* populations is high, suggesting equally high intercontinental mobility.[45] The stable fly is similar in appearance to *M. domestica* and is often mistaken for it. The gray abdomen is rather checked, and the long, slender proboscis protrudes in front of the head. A valuable reference to this fly is Zumpt.[96]

Stable flies are daytime biters. Both sexes feed on blood. Labella are equipped with rows of teeth that can readily pierce skin and underlying tissues. The flies then sponge up blood that wells into the wound. Stable flies avidly bite humans and other animals, especially cattle and horses. They prefer to breed in decaying vegetation rather than manure but are adaptable.

When this insect occurs in great numbers, its attacks on humans are intolerable, affecting tourist industries in some areas. *Stomoxys calcitrans* is one of the most important pests of livestock, causing weight loss and lowered milk production. Hides can be damaged by the bites, and adult cattle can be killed if bitten enough times. It has been calculated that 25 flies a day per cow is the economic threshold; more flies cause a recognizable loss.[74] A thousand flies have been observed on a cow at one time.

Several diseases are known or suspected to be transmitted by stable flies. Among these is *Trypanosoma evansi,* the agent of surra (chapter 5). Mechanical transmission of the *T. brucei* complex also occurs. Epidemic relapsing fever, anthrax, brucellosis, swine erysipelas, equine swamp fever, African horse sickness, and fowl pox are also transmitted by *S. calcitrans.* Stable flies are intermediate hosts of the horse stomach worm, *Habronema microstoma,* which infects horses when infected flies are swallowed.

In addition to insecticides, control measures include release of parasitoid wasps *(Muscidifurax raptor)* and solar-powered electrocuting traps designed to attract both house and stable flies.[64, 72] Vaccine development shows some long-term promise. In one study, *S. calcitrans* fed on rabbits that had been immunized with fly-derived antigens exhibited higher mortality and lower fecundity than control flies.[89]

***Haematobia irritans.* Horn flies** (also known as *Hydrotaea irritans*) are found in the Americas, Europe, Asia Minor, and Africa. They closely resemble stable flies but are more slender. They feed with their head toward the ground, whereas stable flies feed with their head up. Horn flies breed in fresh cow manure. They will bite humans but are not as active fliers as are stable flies. They may be vectors of bovine mastitis.[38]

Horn flies are mainly of veterinary importance, with loss to livestock approaching that caused by stable flies. We know of no diseases transmitted by this insect, with the exception of *Stephanofilaria stilesi,* a nematode of cattle. This skin parasite causes thickening and scabbing of epidermis, especially around the navel, thus attracting more horn flies.

Horn flies are among the ectoparasite targets of insecticides incorporated into cattle ear tags, but such use produces resistant horn fly populations within a few weeks.[12, 73]

Myiasis

Various species in the families to follow are agents of myiasis, which is a term given to infection by fly maggots. There are several categories of myiasis, as in other forms of parasitism. In **obligatory myiasis** the insect depends on a period of parasitism before it can complete its life cycle. In **facultative myiasis** a normally free-living maggot becomes parasitic when it accidentally gains entrance into a host. Other terms are useful in describing the location of larvae: **gastric, intestinal,** or **rectal** for invasion of the digestive system; **nasopharyngeal** for nose, sinuses, and pharynx; **cutaneous,** either **creeping,** when larvae burrow through the skin, or **furuncular,** if they remain in a boil-like lesion; **urinary** or **urogenital; auricular** for the ears; and **ophthalmic** for the eyes. Some larvae are intermittent bloodsuckers and are loosely included under the term *myiasis.* **Accidental myiasis** is usually enteric, occurring when eggs or larvae are eaten with contaminated food or drink. Such cases also are called **pseudomyiasis.**[95] Pseudomyiasis usually involves muscoid flies, such as *M. domestica* and *Fannia* spp.

See Holdsworth et al.[39] for a review of myiasis in ruminants, with special reference to treatment and prevention in agricultural settings.

Family Calliphoridae

This is the large family of blow flies, most of which help destroy carcasses. A few, however, are of great importance in causing myiasis. Blow flies are usually metallic green, blue, or copper color, although some are nonmetallic.

Common Species. *Calliphora vomitoria* is a large, metallic blue fly, probably the "blue-tailed fly" of song. It is conspicuous because of its large size and loud buzz as it flies. It can cause pseudomyiasis. Other species of *Calliphora* are facultative parasites.

Phaenicia and *Lucilia* spp. are metallic green with copper iridescence. *Phaenicia sericata* will breed in carrion as well as excrement and garbage. Along with *Lucilia cuprina,* it is important in **wool strike** in Australia. *Strike* is the term for the action of a fly laying its eggs or larvae on an animal. In wool strike the eggs are laid on the soiled wool on a sheep's rump. Maggots feed on feces and bacteria; their activities cause a great deal of irritation to the sheep, which leads to other complications. This is why tails are docked from lambs soon after they are born. A recent estimate of losses to sheep blow fly strike is $161 million annually in Australia alone, including $115 million in labor for control efforts, $14 million in wool loss, and $12 million in mortality.[57]

Sheep tend to develop some resistance to *L. cuprina,* but, without continuous exposure, that resistance tends to be short lived.[71] However, inflammatory responses in sheep bred for resistance to *L. cuprina* infection were more intense than those of susceptible strains.[62] Efforts to develop vaccines against myiasis and fly attacks on livestock have been partially successful. In one study, *L. cuprina* larvae grown on sheep immunized with fly peritrophic membrane antigens were half the size of those grown on control sheep.[22] On the other hand, immune suppression by *L. cuprina* excretory-secretory products also has been reported.[43]

Some species of calliphorid blow flies limit infections in wounds; laboratory-reared *P. sericata* were used in World War I to clean wounds in servicemen.[24] It has been reported as a facultative parasite in the ear canal and in open wounds. Other blow flies that are facultative parasites are species of *Phormia, Cochliomyia,* and *Chrysomyia.*

Cochliomyia hominivorax. This species is the **New World screwworm** (Fig. 39.15) that is one of the most important causes of myiasis in the world. It is an obligate parasite, occurring throughout the Neotropical region. It causes dermal myiasis in nearly any mammal, as well as nasopharyngeal myiasis in humans. Adult flies are a deep, greenish-blue metallic color with a yellow, orange, or reddish face and three dark stripes on the thorax. It is difficult to differentiate *C. hominivorax* from the **secondary screwworm,** *C. macellaria,* which is not an important myiasis-causing insect.

Screwworm larvae cannot penetrate intact skin, although mucous membranes of the face and genitalia are susceptible to their attack. Usually a preexisting wound, however small, attracts the fly. There is evidence that bacteria in wounds produce volatile compounds that attract gravid female screwworm flies.[10] Cuts from barbed wire or needle grass, castration and dehorning of calves, and insect and tick bites are all examples of sources of entry for screwworm maggots. Wounds from dog fights are commonly attacked. A noted parasitologist once removed more than a hundred screwworms from around the ear of an embattled dog in Trinidad. Screwworms in humans are not uncommon. Generally the more abundant they are in livestock, the greater the chances of human infection. Infection in the head can be fatal, and urogenital infection can be grossly deforming.

The New World screwworm cannot survive winter in cold climates, but summer migrations have brought it as far north as Montana and Minnesota. Its normal range is from Mexico to northern Chile and Argentina. Historically, severe epizootics have occurred in Texas cattle. More than 1.2 million cases were recorded in 1935 in that state alone.

The best control of this pest so far developed involves rearing flies in the laboratory, sterilizing the males, and freeing them to mate with wild females. Because females of this species usually mate only once, they thus cannot produce offspring after a sterile mating. *Cochliomyia*

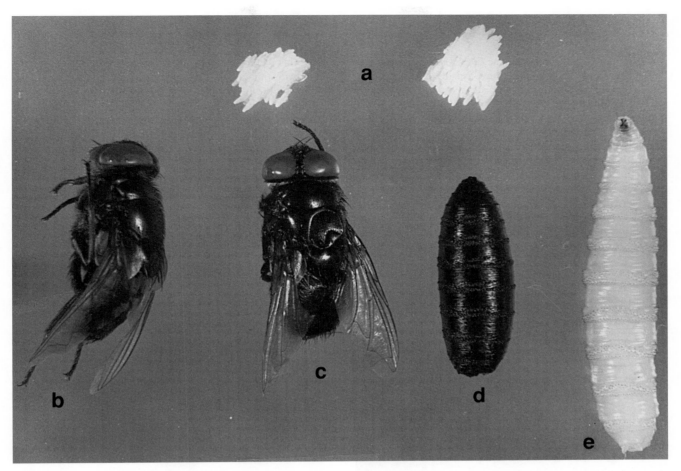

Figure 39.15   **Life history stages of the primary screwworm, *Cochliomyia hominivorax*.**
(*a*) Two egg clusters; (*b*) male; (*c*) female; (*d*) puparium; (*e*) larva.

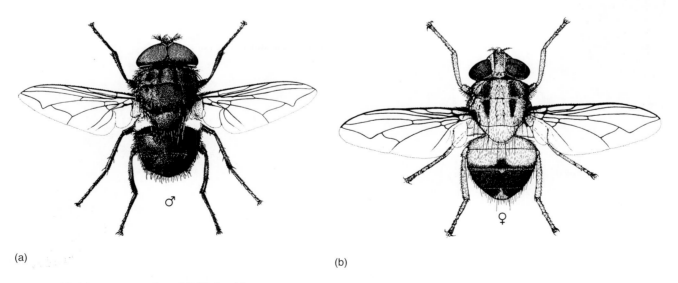

(a)   (b)

Figure 39.16   **Two species of Calliphoridae.**
(*a*) male *Chrysomya megacephala;* (*b*) female of the tumbu fly of Africa, *Cordylobia anthropophaga.*

hominivorax has been eradicated from the United States and in fact is present in North America only in the Mexican state of Yucatan.[46]

The **Old World screwworm,** *Chrysomya bezziana,* occurs in Africa, India, the Philippines, Australia, and the East Indies. Its biology, veterinary impact, and control strategies are almost identical to those of *Co. hominovorax.* Molecular data show there are two distinct races, one from sub-Saharan Africa and another from Asia.[32] Economic studies demonstrate that a major epidemic in Queensland alone would require 250 million sterile male flies a week and five years to eliminate.[1] Vaccines made from larval antigens evidently reduce larval growth in sheep.[75]

A related blow fly, *Chrysomya megacephala* (Fig. 39.16*a*), occurs naturally in Africa, India, the Philippines, and the East Indies but also has invaded the New World. It appears periodically in forensic entomology investigations.

Cordylobia anthropophaga. The **tumbu fly** (Fig. 39.16*b*) is a calliphorid restricted to Africa south of the Sahara. Adults are yellowish, as contrasted with metallic blue or green calliphorids we are accustomed to in the Northern Hemisphere. They are stimulated to lay eggs on soil that has been contaminated with urine. When first-stage larvae contact mammalian skin, they penetrate and begin to grow, causing furuncular myiasis. Reports of this parasite from other parts of the world probably reflect infection acquired in Africa and detected elsewhere. Many wild mammals are reservoirs.

Auchmeromyia luteola. The **Congo floor maggot** is a bloodsucking species found south of the Sahara. It is the only dipteran larva known to suck blood of humans. Eggs are laid on floor mats, dry soil, or crevices in huts. Larvae are quite resistant to desiccation. They feed like bed bugs: When a person is asleep on the floor or on a mat, maggots come out of hiding and pierce the skin with their powerful mouth hooks. Feeding

(a)

(b)

(c)

(d)

Figure 39.17 **A wood frog,** *Rana sylvatica,* **being consumed by** *Bufolucilia silvarum,* **a calliphorid fly.**
(*a*) Eggs on the frog's back; (*b*) second instar larvae in the lesion; (*c*) third instar larvae in the now dead frog; (*d*) frog bones remaining after fly pupation. Time elapsed between (*a*) and (*d*) is less than three days.
Photographs courtesy Matthew Bolek.

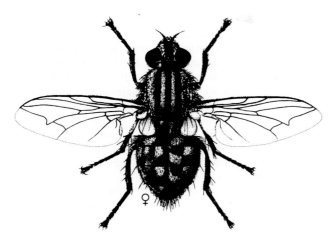

Figure 39.18   **A member of the genus *Sarcophaga*, illustrating the typical three-striped thorax and checkered abdomen.**

From Richard P. Lane and Roger W. Crosskey (Eds.), *Medically important insects and arachnids.* Copyright © 1992 Chapman & Hall, London. Reprinted by permission.

is completed in 15 to 20 minutes, after which the larvae return to hiding. They are not known to transmit any disease.

Bloodsucking maggots of birds are common. In the Northern Hemisphere, *Protocalliphora* spp. sometimes destroy entire broods of young birds. Other calliphorids attack amphibians. Figure 39.17 shows a wood frog being consumed by maggots within three days after eggs were laid on its back.

Family Sarcophagidae

These ubiquitous insects are known as **flesh flies** (Fig. 39.18). They are closely related to Calliphoridae, but, instead of being metallic, their abdomen is checkered gray and black. They often parasitize invertebrates, including insects and snails. Most sarcophagids are larviparous and normally breed in carrion, but females will deposit larvae in open wounds to become facultative parasites.

Sarcophaga hemorrhoidalis is widespread in the northern hemisphere and well into the tropics. It looks like a large house fly, but the tip of the male abdomen is red. Similarly, *Wohlfartia magnifica* is a facultative parasite of mammals in warmer zones of the Palearctic region. Fatal human cases have been recorded.

Cutaneous furuncular myiasis is caused by *Wohlfartia vigil* in Canada and the northern United States and by *W. opaca* in the western United States. Larvae are deposited on unbroken skin, which they quickly penetrate. Human infections usually occur in infants left unattended outdoors, although sleeping adults have been infected indoors. In the northern United States *W. vigil* is a serious pathogen of mink and fox kits in fur farms where newborns are often struck and soon die of the infection. Rodents and rabbits probably are reservoirs, as are carnivores.[25]

Family Oestridae

According to recent classifications, this family now contains several subfamilies that were formerly given family status.[15] Students seeking additional information about oestrids, especially in older literature, should expect to find it under family names, with *-idae* endings instead of subfamily level *-inae*.

Figure 39.19   Rice rat, *Oryzomys capito*, with larvae of *Cuterebra* sp. in the skin.

Courtesy of C. O. R. Everard.

Subfamily Cuterebrinae. Cuterebrinae are **skin bot flies.** The common genus *Cuterebra* is a large black or blue fly about the size of a bumble bee. Found from the north temperate to tropical zones of the New World, species of this genus parasitize rodents, lagomorphs, and marsupials. Often their host is disproportionately small, and it seems incredible that it can survive parasitism by such a large bot (Fig. 39.19). Eggs are laid on or near natural orifices. After hatching, larvae enter the body, tunnel under the skin, cut an air hole in it, and begin to feed. Larvae are densely covered with thick spines and in some cases grow as large as a human adult's thumb. When located in the scrotum, *Cuterebra* spp. often will castrate their rodent host. Human cases are rare, with entry made through the anus, nose, mouth, or eye.

Bot flies have been postulated as participants in evolutionary and ecological phenomena. For example, the most aggressive of three chipmunk species in the Colorado Rockies lives at the highest elevation but suffers severe pathological effects of *Cuterebra fontinella* infections at lower elevations, while a less aggressive, smaller species survives at lower elevations and is resistant to myiasis.[4] Previous explanations of the habitat separation were based on the higher species aggression, assumed to exclude the smaller species.

Parasite avoidance also has been suggested as a basis for postcalving reindeer migrations.[27]

Dermatobia hominis is the **human skin bot** (Fig. 39.20*a*). It is common from Mexico through most of South America. A forest-inhabiting fly, it develops in the skin of almost any warm-blooded animal, including birds. Adults resemble bluebottles. Unlike any other myiasis-causing fly, *D. hominis* does not lay its eggs directly on a host. Instead, it catches another insect, such as a mosquito, and glues its eggs to the side of it, with the operculated anterior end hanging down. At least 48 species of flies and one species of tick are carriers.[31]

When a carrier insect lands on warm skin, the eggs immediately hatch, and larvae drop onto the new host, penetrating unbroken skin. They bore into the dermis and remain there without further wandering. Development to the pupal stage requires about six weeks; pupation is in soil.

This fly commonly parasitizes humans, in whom it causes painful lesions. An infected traveler may return home from a distant place before noting the infection and having it diagnosed. A small incision in the skin allows the larva to be removed. It is readily recognized by two cauliflowerlike projections at the posterior end. One ocular case was treated with oral

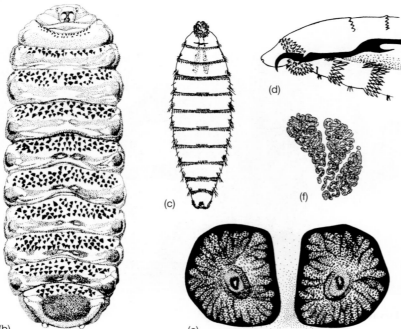

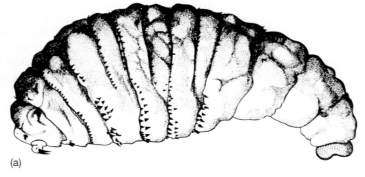

(a)

(b)

(c)

(d)

(e)

(f)

Figure 39.20 **Representative life cycle stages of bot flies.**

(*a*) Third-stage larva of *Dermatobia hominis;* (*b–e*) *Oestrus ovis:* (*b*) ventral view of third-stage larva; (*c*) first-stage larva; (*d*) mouthparts of first-stage larva, lateral view; (*e*) posterior spiracles of third-stage larva. (*f*) Posterior spiracle slits of third-stage *Cuterebra emasculator* larva.

From Richard P. Lane and Roger W. Crosskey (Eds.), *Medically important insects and arachnids.* Copyright © 1992 Chapman & Hall, London. Reprinted by permission.

ivermectin, and another, in the pre-auricular region, was suffocated with paraffin wax before being removed with forceps.[34, 88]

Subfamily Oestrinae.

Head maggots are about the same size and shape as honey bees and do not feed as adults. Larvae develop within sinuses and nasal passages of hoofed animals.

Sheep bots, *Oestrus ovis* (see Fig. 39.20*b–e*), are cosmopolitan parasites of domestic sheep and goats and related wild species. Females deposit active larvae in the nostrils of their host during summer or early autumn. Larvae rapidly crawl up into the sinuses, where they attach to the mucosa and feed. Often they are present in great numbers, causing considerable damage and pain to the host. By spring larvae are developed and crawl back down to the nostrils, where they fall or are sneezed out. Pupation in soil lasts from three to six weeks. Heavy infections can be fatal, but usually a host is only tormented, showing evidence of great distress by sneezing, shaking of its head, loss of appetite, and a purulent discharge from its nose.

Other head maggots are *Rhinoestrus purpureus* in horses of Europe, Asia, and Africa; *Gedoelstia* spp. in African antelopes; and *Cephenemyia* spp. in Old and New World deer.

Ophthalmomyiasis occasionally occurs in humans, usually because of strike by *Oestrus ovis* or *Rhinoestrus purpureus*. Larvae cannot develop beyond the first stage and usually do not last long. Inflammation and conjunctivitis may result. Head maggot strike in humans is zoonotic, being most common in shepherds and others who work closely with sheep or horses.

Subfamily Hypodermatinae.

Variously known as **cattle grubs, ox warbles,** and **heel flies,** these skin parasites are found in most of the northern hemisphere. They primarily infect cattle and Old World ungulates, including reindeer, but they have been known to parasitize horses and humans as well.

Hypoderma lineatum and *H. bovis* are two species that infect cattle. The former is common in Asia, Europe, and the United States, whereas the latter is slightly more northern in its distribution. Both look much like small bumble bees, with light and dark bands on the bodies.

Life cycles of the two species are similar. Both flies strike the hair of cattle, mainly on the hind legs. Although this is painless, cattle become agitated and even terrified and gallop back and forth to avoid the flies. This action is called *gadding* and gave rise to the term *gadfly,* sometimes applied to people. Larvae hatch within a week, penetrate the skin, and make a remarkable migration, first to the front end of their relatively huge host and then back to the lumbar region, where they develop until pupation. All aspects of the migration are not yet known, but *H. bovis* reach the spinal cord, usually in the neck, and burrow posteriorly between the periosteum and dura mater for a distance and then complete their journey through tissues to the back. *Hypoderma lineatum* rests for a time in the wall of the esophagus and appears not to invade the spinal cord. Both species, on arriving at the lumbar skin, cut a hole in it, reverse position, apply the spiracles to the hole, and begin to feed. When ready to pupate, grubs cut their way out, fall to the ground, and bury themselves. The entire life cycle requires about a year.

These flies cause considerable damage to their hosts, as well as to the cattle and dairy industry resulting primarily from the loss of weight, reduced milk production, and damage to hides. Warble fly has been nearly eradicated in Britain.[77]

Numerous cases of *Hypoderma* spp. in humans are recorded, mostly in people who have a close association with cattle. Such infections are thus zoonotic. Unlike those of *Gasterophilus* spp., larvae of these flies can successfully migrate and develop in humans. Usually they surface in the neck region, probably because of the upright position of humans. Results of migration can be dire, including partial or total paralysis of the legs. Ocular myiasis can occur, with loss of an eye.

Other species of warbles infect sheep and goats in Africa, Asia, and Europe. The reindeer warble, *Oedemagena tarandi,* is distributed over the range of caribou and domestic reindeer, causing considerable loss in young animals. Some Eskimos consider the fresh live grubs a delicacy to be eaten immediately upon slaughter of a caribou.

Subfamily Gasterophilinae.

This family comprises the **stomach bots** of equids, elephants, and rhinoceroses. Adult flies are similar to honey bees in size and appearance and are strong fliers. Their ovipositor is long and protuberant. Larvae of these species cause true enteric myiasis, attaching to the mucosa of the host's stomach.

Three species have been introduced into the United States from the Old World. They are parasites of horses, asses, and mules.

Gasterophilus intestinalis is called the **horse bot fly.** It is very common in North America and throughout most of the world. The female attaches approximately a thousand eggs to hairs of a horse, mainly on the knees. When the animal licks its hair, the warmth and moisture stimulate hatching. First-stage larvae immediately penetrate the tongue epithelium and tunnel their way down to the stomach, where they emerge and attach with powerful mouth hooks. Feeding on blood, they grows through two ecdyses to the third stage (Fig. 39.21). All instars have circles of strong spines on all but the last few segments. They remain attached until the following spring and early summer, when they detach and pass out with feces. Pupation takes place in loose earth, and after

Figure 39.21   **Third-stage larvae of the horse stomach bot,** *Gasterophilus intestinalis.*

Courtesy of Warren Buss.

three to five weeks adults emerge. Cogley and coauthors well illustrate migration in the oral cavity.[13]

Gastrophilus nasalis, the **throat bot fly,** has a similar life cycle except that eggs are attached to hairs under the jaw. The larvae hatch in four to five days without need of moisture, crawl along the jaw, and enter between the lips.

The **nose bot fly,** *G. haemorrhoidalis,* strikes a horse on the lips. The remainder of its life cycle is similar to that of *G. intestinalis* except that third instar larvae attach inside the anus for a short time before passing out.

Other genera are *Cobboldia, Platycobboldia,* and *Rodhainomyia* in elephants and *Gyrostigma* in rhinoceroses.

A few stomach bots cause little or no problems in horses, but a heavy infection may cause enough damage to the mucosa of the stomach and to the intestine during migration to kill the animal. Blockage of the pylorus also can occur.

Occasionally first instars will penetrate human skin and cause creeping myiasis, but they cannot mature or move deeper into the tissues.

Learning Outcomes

By the time a student has finished studying this chapter, he or she should be able to:

1. Diagram the basic anatomical features of a mosquito and indicate which features are likely to vary between taxa.

2. Draw the life cycle of a mosquito and describe how culicines and anophelines differ in their various life-cycle stages.

3. Explain why some mosquitoes make excellent vectors for various infectious agents and others might not serve this role so well.

4. Write the scientific names of some important Diptera and tell the role(s) that these species play in the transmission of infectious disease.

5. Tell the kinds of economic damage and host impact that muscoid flies produce in agricultural settings.

6. Define "myiasis" and tell how this condition might arise in both humans and domestic animals.

References

References for superscripts in the text can be found at the following Internet site: www.mhhe.com/robertsjanovynadler9e

Additional Readings

Farajollahi, A., D. M. Fonseca, L. D. Kramer, and A. M. Kilpatrick. 2011. "Bird biting" mosquitoes and human disease: A review of the role of *Culex pipiens* complex mosquitoes in epidemiology. *Inf. Gen. Evol.* 11:1577–1585.

Fresia, P., M. L. Lyra, A. Coronado, A. M. L. De Azeredo-Espin. 2011. Genetic structure and demographic history of New World Screwworm across its current geographic range. *J. Med. Entomol.* 48:280–290.

Mirzaian, E., M. J. Durham, K. Hess, and J. A. Goad. 2010. Mosquito-borne illnesses in travelers: A review of risk and prevention. *Pharmacotherapy* 30:1031–1043.

Otranto, D. 2001. The immunology of myiasis: Parasite survival and host defense strategies. *Trends Parasitol.* 17:176–182. An excellent review of cellular mechanisms that operate in immune reactions to myiasis, with special reference to potential vaccine development.

da Silva-Nunes, M., M. Moreno, J. E. Conn, D. Gamboa, S. Abeles, J. M. Vinetz, and M. U. Ferreira. 2012. Amazonian malaria: Asymptomatic human reservoirs, diagnostic challenges, environmentally driven changes in mosquito vector populations, and the mandate for sustainable control strategies. *Acta Tropica* 121:281–291.

Chapter 40

Parasitic Insects: Strepsiptera, Hymenoptera, and Others

"...Let wasps and hornets break through."

—Jonathan Swift (A Critical Essay Upon the Faculties of the Mind)

Thus far, we have considered orders of insects containing members of medical or veterinary significance. Remaining are several orders that are not covered often in parasitology texts but that are of biological interest and, particularly in the case of Hymenoptera, have considerable impact on human welfare. Half or more of hymenopterans are parasites of other insects, and many are extremely important natural controls of agricultural and forestry pests. Some of these species are prime candidates for use in integrated pest management schemes (see p. 608). Interest in biological control is not limited to agricultural pests. An encyrtid wasp was discovered parasitizing the tick vector of Lyme disease (see chapter 41).[22] Diversity of hymenopterans alone confirms the assertion that parasitism in its many forms is the most common way of life on earth. In a typical study, for example, Butler[9] found eight families and 74 species of parasitic arthropods in 46 species of caterpillars from a West Virginia deciduous forest.

Many insects discussed in this chapter are parasitic as larvae, growing inside or outside a host and eventually killing it. Such insects would seem to fit the conventional definition of *parasite* when they are small, but, because they invariably kill their host as they become larger during or at the completion of development, they seem to become predators at this point. Consequently, they are often referred to as **parasitoids,** rather than parasites. Other authors prefer to use the term **protelean** to describe all insects whose immature stages are parasitic and whose adults are free living. This designation would include some members of most orders covered in this chapter as well as certain Diptera (see chapter 39). We concur with Doutt's[17] interchangeable use of the terms *parasite* and *parasitoid.* Adults of protelean parasites are generally the dispersal mechanism and host-finders. Host-finding behavior of parasitoids is a major research focus among behavioral ecologists because of these insects' agricultural importance.

Although of less economic significance than Hymenoptera, all Strepsiptera are parasitic and demonstrate very

interesting adaptations to parasitism. We will consider Hymenoptera and Strepsiptera in some detail and briefly consider a few parasitic members of other orders.

ORDERS WITH FEW PARASITIC SPECIES

Order Dermaptera (Earwigs)

Dermaptera (1100 species) contains primarily the hemimetabolous, free-living earwigs. Members of only one genus, *Hemimerus,* can be considered parasitic, and some investigators place this genus in its own separate order.[35] The abdomen of adult free-living earwigs terminates in a pair of unsegmented, pincerlike cerci, whereas cerci of *Hemimerus* spp. are short and threadlike. *Hemimerus* spp. are parasites of pouched rats in tropical Africa. They are eyeless and wingless, and they feed on their host's epidermis and a fungus that grows thereon.[36]

Order Neuroptera (Lacewings)

Species of Neuroptera (5000 species) are holometabolous, as are those of the remaining orders to be considered. Most neuropterans are predators as larvae and sometimes as adults on other insects and mites; only about 190 species can be considered protelean parasites. They are beneficial to humans because they help control numerous pests. Most larvae are terrestrial, although members of Sisyridae have aquatic larvae. *Sisyra* and *Climacia* spp. larvae parasitize freshwater sponges, a food source apparently distasteful to most other animals. First instar larvae float until they encounter a sponge, enter ostia, and begin to feed. Larval Mantispidae parasitize egg cocoons of several families of spiders. When fully developed, mantispid larvae have a small head, large abdomen, and small legs, characteristics correlated with the fact that their food source is abundant and does not have to be caught and killed.

Order Lepidoptera (Butterflies and Moths)

Members of the large order Lepidoptera (120,000 species) are familiar to most people. Adults have comparatively large wings covered with tiny scales, and a long suctorial proboscis formed from parts of the maxillae, mandibles being absent or rudimentary. They are adapted for feeding on nectar or other extruded plant juices. Their polypodous larvae have strong mandibles adapted for chewing. Both adults and larvae are almost entirely phytophagous. A few species are ectoparasites of mammals as adults, and a few are protelean parasites of insects. Members of several families occasionally visit eyes of domestic ungulates (Fig. 40.1), where they feed on lacrimation; those of a different group visit eyes fairly frequently, feeding on lacrimation and fluid running down a host's cheeks.[10]

None of the 100 or so lacryphagous species have mouthparts that can sever tissue. Therefore these moths do not pierce the conjunctiva, but they do take leucocytes from infected eyes or in some cases feed on blood from wounds, with feeding behavior similar to that of sucking nectar.[4] However, a Noctuidae species, *Calyptra eustrigata,* has the ability to pierce vertebrate skin and feed on blood (Fig. 40.2).[3] Mouthparts of *C. eustrigata* actually are adapted to piercing fruit, but this enables the insects to attack skin.[4] At least seven species of *Calyptra* are known to pierce skin and suck blood; five of these are known to attack humans. These observations suggest that dependency on animal fluids may have evolved independently, once by way of nectar-feeding to lacryphagy, and separately by way of fruit- to skin-piercing.[4]

Cyclotorna monocentra in Australia has a very curious life history.[16] Its first instar larva finds, attaches to, and feeds on but does not kill nymphs or adults of a species of leafhopper (Hemiptera, Homoptera). When *C. monocentra* molts to the second instar, it somehow induces an ant (a species of *Irdomyrmex*) to pick it up and carry it back to the ant's colony. The caterpillar then ingratiates itself by providing the ants with a sweet secretion and by running its mouthparts over the ants' bodies. The caterpillar exacts payment for these services by feeding on ant larvae. Finally the lepidopteran larva emerges from the ant colony and pupates on a tree trunk.

Order Coleoptera (Beetles)

Judged by the number of described species, (300,000). Coleoptera is the most successful order of animals on earth. Less than 2% are parasites, but both mammalian ectoparasites and protelean parasites of insects are represented. As to be expected in such a large and speciose order, Coleoptera exhibits a wide range of morphological form, habit, and adaptation. In most species mandibles are well developed for biting and chewing, but in some they are adapted for piercing and sucking. Although most beetles can fly well, their forewings are hardened into sheathlike elytra. Most coleopterans that are protelean parasites are **hypermetamorphic,** leaving the host-finding task to first instar larvae.

Hypermetamorphosis is a condition in which different larval instars have dissimilar forms, and it is found among mantispid neuropterans, Strepsiptera, and several families of Diptera and Hymenoptera as well as in some Coleoptera. In all of these the first instar is rather heavily sclerotized and quite active. In Neuroptera, Strepsiptera, and Coleoptera, the larva is a campodeiform oligopod and is called a **triungulin** or **triungulinid** (Fig. 40.3c). The active dipteran and hymenopteran larvae are apodous, but they move about with the aid of thoracic and caudal setae; this larval type is called a **planidium** (see Fig. 40.3a, b). Subsequent instars of both types are typically much less sclerotized and have a smaller

Figure 40.1   **Eight *Lobocraspis griseifusa* (Lepidoptera, Noctuidae) suck tears from the eye of a banteng, *Bos banteng,* in northern Thailand.**

Note the proboscis of each moth extended to feed at the eye perimeter.

Courtesy of Hans Bänziger.

Figure 40.2   **A noctuid moth, *Calyptra eustrigata,* piercing the skin and sucking blood of a Malayan tapir.**

This is the only known genus of bloodsucking moth.

Courtesy of Hans Bänziger.

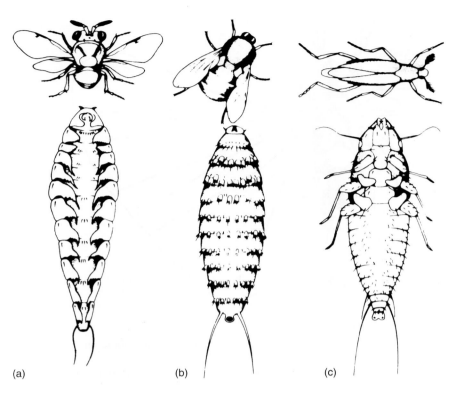

Figure 40.3   First instar larvae with corresponding adults (*not* drawn to same scale).

The larvae are active and must find a host. (*a*) Perilampidae (Hymenoptera) and (*b*) Acroceridae (Diptera) exemplify the planidium type of larva. (*c*) A triungulin larva (with well-developed legs) of Rhipiphoridae (Coleoptera).

From R. R. Askew, *Parasitic insects.* New York: American Elsevier Publishing Company, 1971.

(a)   (b)   (c)

head and much reduced means of locomotion. Planidia and triungulins are interesting instances of convergent evolution.

Examples of protelean parasitic beetles include aleocharine staphylinids, whose triungulins seek out puparia of Diptera, penetrate them, and feed on the pupa within. Larvae of some Meloidae feed on grasshopper eggs; others feed on eggs of solitary bees and then on honey and pollen stored in cells in which eggs are laid. In the case of *Meloe franciscanus,* which parasitizes the nests of the bee *Habropoda pallida,* triungulins cooperatively form round or oval aggregations on branch tips of plants that mimic female bees. In addition, these triungulin larvae produce a chemical mimic of the female bee sex pheromone to attract males.[37] When males bees contact the aggregation, the triungulins attach in less than 2 seconds. The beetles are eventually transported back to the bee's nest via phoresy.

A few beetles are symbiotic as adults. *Platypsyllus castoris* (Platypsyllidae) is an ectoparasite of beavers in both the Palearctic and Nearctic. It is a blind, obligate parasite and an ectoparasite in both adult and larval states, feeding on skin debris. Some species of Staphylinidae are apparently mutuals of marsupials and certain rodents, feeding on fleas and mites.[18] These beetles are rather large (5 mm to 16 mm) but are well tolerated by a host even as they cling to its hair in areas such as the base of the tail where a more noxious passenger would excite grooming attention.

ORDER STREPSIPTERA (STYLOPS)

Although Strepsiptera, commonly known as **stylops,** is a very small order (about 600 species), it is of great interest biologically and demonstrates some of the most extreme adaptations to protelean parasitism of any insects. Extant strepsipteran species have been recovered from Oligocene amber, strongly suggesting that their remarkable life cycles evolved during the age of dinosaurs.[24] The phylogenetic relationship of these insects has long been a mystery. Certain morphological characters led investigators to consider them related to beetles, but certain molecular studies indicate a close relationship to Diptera.[44] Other research suggests this apparent relationship to Diptera is due to a high nucleotide substitution rate and "long branch attraction" artifacts common to certain types of analysis methods for molecular sequence data, so that the so-called "Strepsiptera problem" remains unresolved.[23] A molecular phylogenetic analysis focused on relationships within Strepsiptera[31] indicates that the increase in the relative rate of molecular evolution for this insect order is correlated with an increased rate of morphological evolution, including changes associated with the transitions to parasitism. However, correlation does not establish causation, and other factors such as changes in generation time, lifespan, and other features related to reproduction may have been important factors influencing rates of molecular evolution.[31]

Morphology

Strepsipterans exhibit extreme sexual dimorphism. Males are small, robust, brown-to-black insects about 1.5 mm to 4.0 mm long (Fig. 40.4). Their forewings are small halteres bearing numerous sensory endings; hindwings are large, membranous, and fan shaped and are borne on a large, third thoracic somite. The compound eyes are large and protuberant. One or more antennal segments have lateral processes so that antennae appear branched (see Fig. 40.4). Mandibles are

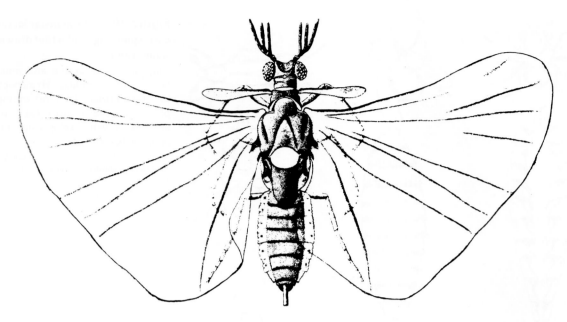

Figure 40.4 *Eoxenos laboulbenei,* **adult male (Strepsiptera).**
The mesothoracic wings have been reduced to halteres, and large, membranous metathoracic wings are borne on its very large metathorax.

From H. L. Parker and H. D. Smith, "Further notes on *Eoxenos laboulbenei* Peyerimoff with a description of the male," in *Ann. Entomol. Soc. Am.* 27:468–479, 1934.

present but are simple and sickle shaped; other mouthparts are reduced or absent. These characteristics reflect the fact that adult males spend their very short existence agitatedly seeking females to mate with; thus, they have more use for well-developed sense organs than for feeding appendages.

Typically adult females are parasitic, but exceptions occur in Megeidae and Myrmecolacidae, discussed later. They are vermiform, up to 20 mm to 30 mm long according to species, and have no wings, eyes, legs, or antennae (Fig. 40.5). The mouth and anus are tiny and nonfunctional, and the gut has no lumen. The head and thorax are fused to form a cephalothorax and are rather heavily sclerotized. Females breathe through a spiracle on each side of their cephalothorax, which protrudes from the host between two sclerites, usually between tergites of the host abdomen. The female abdomen is large and soft, lying within its host's body and ensheathed in cuticle of the last larval instar. Space between the abdomen and its larval cuticle forms a brood canal, which opens to the outside beneath the cephalothorax. Copulation is accomplished by insertion of a male's aedeagus into the brood canal. Sperm make their way through two to five genital openings in the female's abdomen and thence to her hemocoel; there they fertilize eggs, which are lying free in the hemocoel. This type of reproductive system is found in no other insects.

Development

Embryos develop and hatch within a female's hemocoel to produce triungulin larvae, which exit through the genital pores and brood canal. One female can produce 2000 or more triungulins, and polyembryony has been reported. Triungulins are only about 0.2 mm long, but they have a well-sclerotized cuticle, well-developed legs, and one or two pairs of long caudal filaments (Fig. 40.6). They die in a short time if they do not reach a new host.

Although the vast majority of triungulins perish, they possess some adaptations to increase their chances of survival. For example, triungulins of *Corioxenos antestiae* rest motionless, with the anterior part of their body raised, on their hind legs and central caudal bristles, which are bent forward beneath the abdomen. A movement in their vicinity, particularly if the moving object is black and orange (colors of their host, a pentatomid bug),[27] stimulates a triungulin to leap up to 10 mm vertically and 25 mm horizontally, and adhesive pads on its front two pairs of tarsi help maintain contact if it hits a host (see Fig. 40.6). Triungulins will attach to a wide variety of insects, but they soon detach unless the insect is a correct host. Other species are known, however, that will penetrate an unsuitable host and die therein.

Whereas parasites of Hemiptera and those of social Hymenoptera reach their hosts directly, stylops parasitic in solitary Hymenoptera have special problems because host larvae are hidden within nest chambers. A bee infected with *Halictoxenos jonesi,* for example, drags her abdomen among the stamens of flowers, depositing triungulins.[5] When another bee visits the flower, a triungulin attaches to it, is transported by the adult bee back to a cell containing its larva, and transfers to the larva while the adult bee is feeding its young. Once on its host, the triungulin penetrates the hemocoel through an intersegmental membrane and begins to feed. The second instar, however, differs vastly from the triungulin. It is a grublike organism without mouthparts and legs, and it feeds by absorbing nutrients through its cuticle. There are normally six larval instars, including the triungulin.[19] The sixth instar regains its mandibles and chews its way through the intersegmental membrane between sclerites of its host's abdomen.

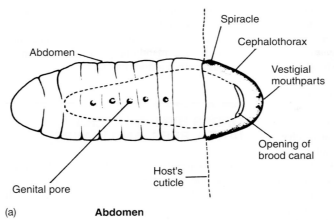

(a)

(b)

Figure 40.5 **Diagram of female strepsipteran.**

(*a*) Ventral view; (*b*) longitudinal section. The more heavily sclerotized cephalothorax protrudes from the host body between abdominal tergites, whereas the soft abdomen lies within the host's hemocoel and is covered by cuticle of the last larval instar. Initially closed, the brood canal is pierced by a male during copulation, and triungulins subsequently exit through that opening.

From R. R. Askew, *Parasitic insects.* New York: American Elsevier Publishing Company, 1971.

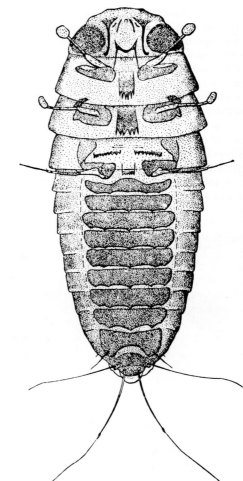

Figure 40.6 **First instar larva (triungulin) of strepsipteran *Corioxenos antestiae*.**

The triungulin characteristically lies motionless, resting on its hind legs and central caudal bristles. It is stimulated to leap in the air by movements of nearby black and orange objects, such as its host, a pentatomid bug. Adhesive pads on its forelegs help it stay on the host when it makes contact.

From T. W. Kirkpatrick, "Studies on the ecology of coffee plantations in east Africa. II. The autecology of *Antestia* spp. (Pentatomidae) with a particular account of a Strepsipterous parasite," in *Trans. R. Entomol. Soc. Lond.* 86:247–343, 1937.

If the parasite is a male, it emerges completely and pupates within the cuticle of the last larval instar. If it is a female, only the cephalothorax emerges with no obvious pupation.

Some strepsipteran familes have different patterns of parasitism or development. For example, in Mengeidae, females retain several characteristics that are believed to be ancestral: They emerge completely, pupate, and become free living. Although wingless, they have functional eyes, legs, and antennae. Mengeids are parasites of Thysanura (silverfish), an order of insects also with numerous primitive characters, but they probably colonized their hosts rather than coevolved with them because Strepsiptera is of more recent origin than Thysanura.[1] In Myrmecolacidae, males parasitize ants whereas females parasitize Orthoptera and Mantodea; this dimorphism in host parasitism by the sexes in unique among insect parasitoids. Males emerge from ants as free-living adults to mate with parasitic females.

Stylops differ from virtually all other protelean parasites of insects in that their host usually lives approximately a normal life span. Even so, strepsipterans may cause reproductive death of a host through parasitic castration.[29] Although there is some variation, depending on whether a host is infected in an early or late instar, secondary sexual characteristics tend to take on an intersex appearance because effects on the gonads may be profound. Williams[45] reported that the aedeagus and parameres were often reduced or absent in leafhoppers (*Dicranotropis muiri*) infected with *Elenchus templetoni* in Mauritius, and length of ovipositors and sheath was reduced in female hosts. Gonads of hosts with advanced larvae or extruded forms were much reduced or could not be found at all.

Strepsipterans parasitize at least 34 insect families in seven different orders, exhibiting far less host specificity than most other parasitoids.[26] At least one species infecting katydids encloses itself in a host epithelium bag, within which it undergoes subsequent molts. This process evidently shields the parasite from host defense responses, and may be the mechanism that allows strepsipterans to colonize such a wide diversity of hosts without becoming specialists themselves.[26]

Strepsipterans are particularly attractive as potential biological control agents for fire ants now spreading across much of southern United States (Fig. 40.7).[25] Strepsiptera may also be vectors for *Wolbachia,* bacteria that influence sex ratios in many invertebrate species (pages 441 and 608). They are usually transmitted through eggs within a species, but one study showed strepsipterans had the same bacterial strain as their planthopper hosts, suggesting transmission of bacteria between host and parasite.[33]

ORDER HYMENOPTERA (ANTS, BEES, AND WASPS)

Hymenoptera is the second largest order of insects, estimated to include more than 200,000 species; at least half of these are insect protelean parasites. Free-living hymenopterans such as ants, bees, and wasps are familiar to everyone. Although feared and avoided by most people because of their sometimes potent sting, they are nevertheless extremely important pollinators of flowering plants, including most of our vegetables and fruits. At least as important to humans, if not more so, is the role played by parasitic Hymenoptera in population control of other insects.

Morphology

Hymenopterans usually have a fairly heavily sclerotized cuticle and show a vast range of body form, size, and color. They include the smallest insect known (*Alaptus* spp., 0.21 mm long) and insects up to 115 mm (some ichneumons, including the ovipositor). They have two pairs of membranous wings, of which the forewings are the larger (Fig. 40.8) except in Symphyta. Hindwings are attached to forewings by

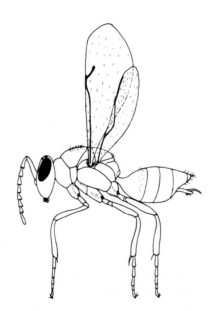

Figure 40.8   Chalcidoid hymenopteran (Encyrtidae). Members of this family commonly are parasites of scale insects (Hemiptera, Homoptera), and several species have been used successfully in biological control of important pests.[41]

Figure from Donald J. Borror and Dwight M. De Long, *An introduction to the study of insects,* 3d ed. Copyright © 1974 by Saunders College Publishing. Reproduced by permission of the publisher.

Figure 40.7   Male *Caenocholax fenyesi* (*arrow*) emerging from a fire ant (*Solenopsis invicta*) abdomen.

Photograph courtesy of Jerry Cook.

a row of small hooks on the leading edge of the hindwings. Wing venation tends to be reduced and may be absent in some minute species. Some forms are wingless. The first abdominal segment is fused to the thorax and is called the **propodeum.** The second abdominal segment in most Hymenoptera is constricted to a waistlike **pedicel,** or **petiole.** The head is remarkably free, with a small neck and large compound eyes. Antennae usually have 12 segments in males and 13 in females. Mandibles, maxillae, and labium are present and variously modified for different feeding habits. A **glossa** (considered a hypopharynx by some authors) is present, and it and some other mouthparts may be lengthened to form a tongue or proboscis for gathering nectar.

The **ovipositor** is modified to form a stinging organ in some species, but it is important to the biology of parasitic forms as well. In common with those of many other insects, the ovipositor of hymenopterans has been derived from pairs of abdominal segment appendages (Fig. 40.9). Its basal portions represent coxae of appendages of abdominal segments eight and nine and are called **valvifers.**[39] Coxal endites have been lengthened to become the ovipositor body and are called **valvulae.** Another process from the second valvifer (third valvula) has developed to become the ovipositor sheath.

In Hymenoptera second valvulae are fused and have longitudinal ridges that interlock with grooves on the first valvulae. This arrangement forms a tube through which eggs must pass, and the functional unit is called a **terebra.** The much broader third valvulae form a sheath on each side of the terebra; they are well endowed with sensory endings on their outer surfaces and are much less rigid than are terebra. Cutting ridges on first and second valvulae enable an insect to drill through host cuticle, through cells that contain a host, or even through hard vegetable matter before oviposition.

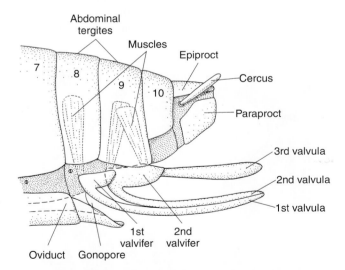

Figure 40.9   Diagram of generalized appendiculate ovipositor of a pterygote insect.
Some internal muscles and the position of the oviduct are shown. The valvulae and valvifers are all paired, although second valvulae are fused in Hymenoptera. They join with first valvulae to form an ovipositor tube (terebra), and the third valvulae form the ovipositor sheath.

Drawing by William Ober.

Because the terebral lumen is much smaller in diameter than eggs, eggshells must be elastic enough to travel down the ovipositor to gain entrance to a host's body.

Development

Parthenogenesis occurs throughout Hymenoptera. Three types are recognized: thelyotoky, deuterotoky, and arrhenotoky. In **thelyotoky** all individuals are uniparental (parthenogenetic), and virtually no males are produced. Some sawflies and parasitic Hymenoptera in several families are in this category. Some males are produced in **deuterotoky,** but all individuals are nevertheless uniparental. The most common condition is **arrhenotokous parthenogenesis** (haplodiploidy; see also chapter 27, p. 425), in which only males are uniparental and are haploid. Females come from fertilized eggs and are diploid. By some still obscure means, a mated, egg-laying female can influence whether a given egg will be fertilized. Needless to say, all of the foregoing types of parthenogenesis may lead to sex ratios that diverge strongly from the 50:50 that we normally expect. An interesting discussion of parthenogenesis in Hymenoptera is given by Doutt.[17]

Some parasitic Hymenoptera, especially members of Encyrtidae, exhibit polyembryony. Within an embryo, a series of cell divisions results in cell aggregations called "morulae," each of which is contained in a membrane.[31] The parasite may become a chain or branched sac filled with larvae. This chain later breaks up as larval development continues. All embryos from a single egg are of the same sex; female embryos develop from fertilized eggs and males from unfertilized eggs. In some species up to 3000 embryos may be formed from a single egg, and subsequent morphogenesis is controlled by host hormones.[2]

A few hymenopterans are hypermetamorphic and have a planidium larva. An example is *Perilampus hyalinus,* an American chalcid, which lays its eggs on foliage.[1] Planidia search for and penetrate caterpillars such as fall webworms (*Hyphantria* sp.). Once inside, they must find and penetrate larvae of another webworm parasite, a member of tachinid (Diptera) genus *Ernestia.* Thus, they are hyperparasites, a common condition among Hymenoptera. Hypermetamorphosis is found among Perilampidae, Echaritidae, and Ichneumonidae, but, of course, not all species are hyperparasites.

Classification and Examples

Hymenoptera is divided into two suborders, Symphyta and Apocrita, easily distinguished because Symphyta do not have a pedicel. The latter are mostly phytophagous, but the small family Orussoidea parasitizes larvae of cerambycid and buprestid beetles.

Two divisions of Apocrita, Aculeata and Parasitica (also called Terebrántes), are commonly recognized, but the latter is possibly polyphyletic. Some Aculeata are parasitic, and some Parasitica are not. In Aculeata the eighth and ninth tergites are reduced and are retracted into the seventh, so that the ovipositor (sting) seems to issue from the abdomenal apex. In Parasitica the eighth segment is not retracted into the seventh, and the ovipositor is exposed almost to its base. Aculeata superfamilies with parasitic forms are Bethyloidea and Vespoidea (some). In Parasitica, parasitic forms are Evanioidea, Trigonaloidea, Ichneumonoidea, Proctotrupoidea (Serphoidea), Chalcidoidea (many), and Cynipoidea (some).[1]

The arbitrary distinction between parasitism and predation in hymenopterans is well illustrated in Aculeata, which contains ants, solitary and social wasps, and bees. A number of wasps sting their prey (host) to paralyze it and then lay their eggs on the host, which provides food for young. Some entomologists distinguish *parasite* from *predator* on the basis of whether an adult wasp constructs a cell to house the paralyzed host. An intermediate position is occupied by members of Bethylidae, which drag the host to a sheltered position after paralyzing it and then oviposit on it. The female, who stands guard while larvae develop as ectoparasites of the host, sometimes bites the host and feeds on its body fluids. Additional females may lay eggs on the same host individual, and the insects guard their young cooperatively. Such behavior may demonstrate an early stage in evolution of maternal care practiced by social wasps. Various vespoid families exhibit a spectrum of maternal behavior, ranging from the typical parasitoid practice of laying an egg on prey and then abandoning it to building of a cell to house the prey and wasp young and finally to maternal care found in social wasps.

Ichneumonoidea comprises Ichneumonidae, Braconidae, and Aphidiidae (sometimes considered a subfamily of braconids). Ichneumonidae is a very large family with more than 3000 described species. They are all parasitic, sometimes on other ichneumonids, and most are endoparasitic. They tend to have a slender abdomen and often have a very long

ovipositor that is permanently extruded. The largest ichneumonids in the United States may exceed 40 mm, and their ovipositors may be twice that length. Such insects attack larvae of horntails, wood wasps, and wood-boring beetles, somehow detecting a host within its tunnel, which may be several centimeters below the wood surface. In Britain *Rhyssa persuasoria* is attracted by a substance produced by fungus that grows on frass (feces and wood pulp) in tunnels of its host, a wood wasp of genus *Sirex.* When the host's location has been determined, apparently by antennae, the ichneumonid raises its abdomen and places the tip of its ovipositor exactly in the location indicated by its antennae. Aided by sharp cutting ridges at the end of its ovipositor, the ichneumonid forces the ovipositor through the wood by pressure and twisting motions of its abdomen[1] (Fig. 40.10).

It is astonishing that such an apparently delicate ovipositor can penetrate wood to reach a host. *Pseudorhyssa alpestris,* a parasite of the alder wood wasp, *Xiphydria camelus,* has solved this problem of reaching a host in another way. *Xiphydria camelus* is also parasitized by *Rhysella curvipes,* which locates its host and oviposits much like *R. persuasoria. Pseudorhyssa alpestris,* however, simply locates *R. curvipes'* oviposition holes, inserts its ovipositor, and lays an egg in *R. curvipes'* host. When the *P. alpestris* hatches, it kills the larva of *R. curvipes* and takes over the host![41]

Members of Braconidae are similar to those of Ichneumonidae, although they are generally smaller (15 mm or less in length) and tend to be more heavy bodied. They also differ in that they pupate in silken cocoons on the outside surface of a host rather than within its body (Fig. 40.11). A number of species of ichneumonid and braconid adults are known to feed on a host, as well as to oviposit within it. *Polysphincta* spp. sting and paralyze their spider hosts and then lay an egg on the spider's opisthosoma and feed on its body fluids.[13] It is known that some species cannot mature their eggs or achieve full reproductive potential without first feeding on host body fluids,[8] and it is possible that many other species

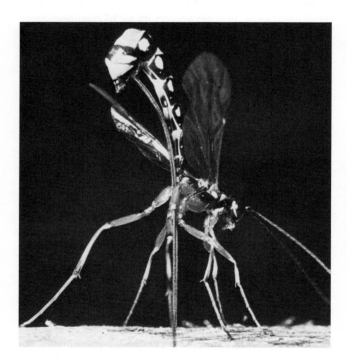

Figure 40.10 **Ichneumonid with the end of her abdomen raised to thrust her ovipositor through wood to a host, a wood-boring, larval beetle.**

Coxae of the hind legs help steady and support the ovipositor as it is thrust into the wood.

Photograph by L. L. Rue III.

Figure 40.11 **Larva of a tomato hornworm (Lepidoptera, Sphingidae) parasitized by the braconid *Apanteles* sp.**

The white objects on the dorsum of the caterpillar are the pupal cocoons of the wasp.

Photograph by O. W. Olsen.

have this requirement. Larvae of ichneumonids *Hyposoter* spp. can parasitize the tussock moth only if the host has been previously parasitized by a braconid, *Cotesia melanoscela.*[20] The first parasite apparently modifies the host defense system in such a way that the second can survive.

Developing larvae may modify their host environment to the advantage of the parasite. No more than one larva of the aphidiid parasites of pea aphids, *Aphidius smithi,* can develop in a single host; if more than one egg is deposited, the supernumerary parasites are somehow eliminated. However, superparasitized pea aphids have a greater food incorporation efficiency and growth rate than singly parasitized hosts; thus, early presence of more than one parasite larva modifies host physiology by some unknown mechanism.[12] Some investigators have proposed that the dominant egg releases substances that suppress development of supernumerary parasites.[38]

Chalcidoidea is a large superfamily that contains 18 families. Its members are very small (less than 5 mm), metallic green or black wasps with few wing veins. Some of them induce growth of plant galls. Parasitic chalcidoids attack a wide variety of hosts, the majority of which are in Lepidoptera, Coleoptera, Hymenoptera, Diptera, and Hemiptera. A few are parasites of ticks, mites, and egg cocoons of spiders. Trichogrammatidae and Mymaridae are parasites of insect eggs. Mymarids include the smallest insects known (0.2 mm); they are called *fairyflies* and are smaller than some protozoa (Fig. 40.12).

Several Encyrtidae have been very valuable in control of scale insects (Hemiptera, Homoptera). Encyrtids also illustrate the rather remarkable complexity of parasitoid evolution. *Ooencyrtus nezarae,* for example, prefers previously parasitized *Megacopta* (Hemiptera) over unparasitized ones, even though larval survival is lower in hosts that already contain *O. nezarae* larvae.[40] Evidently, the effort of boring a new hole in the host is a more expensive physiological burden to the parasite than is lowered survival of its offspring.

Ooencyrtus nezarae does not specialize in scale insects, however, and, in recent years, study of its host-finding behavior has helped clarify the role played by chemical cues. The term **kairomone** refers to pheromones that attract parasitoids to hosts, although such compounds obviously do not occur primarily for the benefit of parasites. For example, *Riptortus clavatus,* a heteropteran, develops only on certain host plants, but adult bugs scatter their eggs on nonhost plants, too.[30] Male bugs release pheromones when food is available, and these pheromones function not only to attract other adults, but also as aggregation signals for newly hatched nymphs. *Ooencyrtus nezarae* functions as an egg parasite and responds positively to these chemical signals; needless to say, *R. clavatus'* "Eat Here" sign is "read" by some undesirable guests.

Most Cynipoidea are phytophagous, and they induce formation of galls in plant tissues on which they feed. However, a number are parasitic on various other insects. *Charips victrix* is a hyperparasite of a braconid primary parasite of aphids (Fig. 40.13).

Evanioidea, Trigonaloidea, and Proctotrupoidea are less common parasites of a variety of insects. Trigonaloids parasitize larvae of social wasps (Vespoidea), and some are hyperparasites of ichneumonid and tachinid dipteran parasites of lepidopterans. Trigonalid females lay many eggs, not on prospective hosts but on vegetation. The eggs do not hatch until they are eaten by a caterpillar, but they can remain viable for several months.[1] Once eaten by a caterpillar, they hatch, penetrate the gut to the hemocoel, and then try to find a larval primary parasite in which to develop further. Evanioidea contains several families, one of which (Evaniidae) contains parasites of cockroach eggs. An evaniid egg is laid within one of the cockroach eggs. After hatching, the evaniid larva consumes the roach egg and then may eat other eggs in the cockroach egg case (ootheca). Proctotrupoidea is a somewhat larger group, including seven families.[1] They parasitize various insects, including homopterans, dipterans, neuropterans, and coleopterans. Pelecinidae (Fig. 40.14) have a long, attenuated abdomen, which may compensate for the short ovipositor when the female burrows through the ground in search of her host, larvae (grubs) of May beetles.

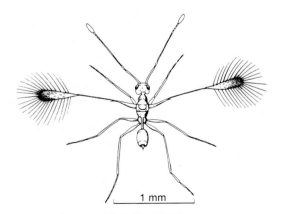

Figure 40.12   Fairyfly, *Mymar pulchellus,* female (Chalcidoidea, Mymaridae).

This is one of the larger species in the family, commonly parasitic in eggs of other insects.

From R. R. Askew, *Parasitic insects.* New York: American Elsevier Publishing Company, 1971.

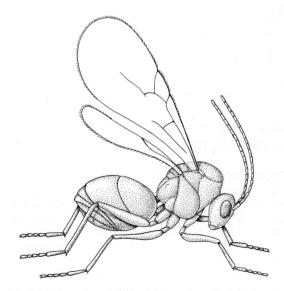

Figure 40.13   Female *Charips victrix* (Cynipoidea, Cynipidae), a hyperparasite on a braconid primary parasite of aphids.

Drawing by William Ober.

Figure 40.14 **Female *Pelecinus polyturator*** **(Proctotrupoidea, Pelecinidae).**

This striking insect is 2 in. or longer and shining black. The rare males are about an inch long and have a swollen abdomen. They are parasites of larvae (grubs) of May beetles (Scarabaeidae), which burrow in soil.

Drawing by William Ober.

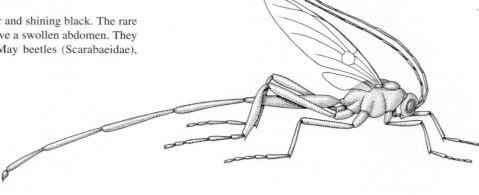

WOLBACHIA BACTERIA, VIRUSES, AND PARASITOID INSECTS

Strains of bacterial genus *Wolbachia,* an intracellular parasite of invertebrates, are now known to produce a wide variety of effects on hosts. Among insects, these effects have been studied extensively in parasitic hymenopterans, and they include inducement of cytoplasmic incompatibility between sperm and eggs, parthenogenesis, feminization, and male death.[15] In one braconid wasp species, *Asobara tabida,* infection with a particular strain of *Wolbachia,* known as wAtab3, is required for oogenesis in females, but other, coinfecting *Wolbachia* strains have no activity in this regard.[15]

Viruses also may influence host-parasite relationships in insects. For example, *Leptopilina boulardi* (Hymenoptera, Eucoilidae) parasitizes *Drosophila* larvae, typically injecting a single egg. Female wasps avoid fruit fly larvae with existing *L. boulardi* infections, but a virus known as LbFV alters this behavior so that wasps will inject eggs into parasitized hosts.[42] This behavioral change allows horizontal transmission of the virus among wasp larvae in multiple-infected flies, an effect interpreted as "consistent with the hypothesis that the virus manipulates the behavior of the parasitoid."[42]

An excellent review of the many fascinating interactions between parasitic insects, viruses, and *Wolbachia* bacteria is provided by Boulétreau and Fleury.[7]

BIOLOGICAL CONTROL

Examples presented in this chapter are but a few of the parasitoid/host combinations along with viral, bacterial, nematode, and other pathogens and pheromones now being studied as potential biological control agents. Major stimuli for this research effort are (1) pesticide resistance, (2) cost of agricultural chemicals, and (3) increasing regulations on pesticide use. We have learned, through what amounts to a massive experiment in evolution, that insecticides quickly select for resistant pest populations; thus, new chemicals must be developed continually. Cost of agricultural chemicals includes not only research, development, patenting, disposal, and associated legal expenses, but also fertilizer, hydrocarbon stocks from which new compounds are synthesized, and transportation. Finally chemical insecticides are toxic to some degree to other fauna, including humans; that is, they are nonspecific. In addition, they are often quite toxic to beneficial species, such as pollinators and natural predators of pest insects.

By contrast, at least in theory, once a successful biological control agent has been introduced, no additional cost accrues except occasionally for reintroduction of the parasite. Furthermore, a good biological control agent should keep the host population to a low but tolerable level in equilibrium. Evolution of greater resistance in a host population to the parasite is likely to be matched by adaptations in the parasite population to restore an equilibrium. However, initial costs of developing biological controls for practical use may be quite high because extensive research is required. Careful ecological investigations of both parasite and pest must be conducted before a foreign organism can be introduced into an area. Cryptic species or different biological races of a parasite may have different host preferences or biological characteristics that determine success or failure. Parasites that are successful in biological control usually require the following qualities (modified from Askew)[1]:

1. They must have a high host-searching capacity.
2. They must have a very limited range of hosts but be able to use a few other host species in addition to the target; that is, when a pest population is reduced, parasites should be able to maintain themselves on alternative hosts. Thus, a high enough parasite population must be available to counteract surges in the pest population.
3. Their life cycle must be substantially shorter than that of the pest if a pest population consists of overlapping generations, or their life cycle must be synchronized with the pest life cycle if the pest population is composed of a single developmental stage at any time.
4. They must be able to survive in all habitats occupied by the pest.
5. They must be easily cultured so that large enough numbers can be available for introductions.
6. They must control the pest population rapidly (some workers have suggested that control must occur within three years of the time of introduction).[11]

These conditions mean that, although biological control is theoretically rather straightforward, in practice it is often a quite difficult goal to achieve. The exploration and experimentation needed to find, domesticate, and develop wild insects with the listed properties can easily consume one's career. Hokkanen[21] estimates that only about one of seven biological control introductions results in positive economic results. Nevertheless, the approximately 300 instances of successful introductions have returned about 30 times their original investments.[21]

A massive body of literature on biological control agents of all kinds, many of them obscure insects a layperson would never recognize by name or by sight, is accumulating rapidly. For a wondrous eye-opening tour through this world, see the paper by Waage and Hassell,[43] the major three-volume set edited by Pimentel,[34] the theoretical work on integrated pest management edited by Kogan,[28] the volume on plants, parasitoids, and insect predators edited by Boethel and Eikenbarry,[6] and the second edition of DeBach and Rosen[14] (particularly useful for students).

Learning Outcomes

By the time a student has finished studying this chapter, he or she should be able to do the following:

1. A recurring theme of this book is the ubiquity of parasitism. Considering the parasites in this chapter, describe how arguments based on phylogenetic relationships support this theme.

2. It has been argued that parasitism is a more effective biological control strategy than employing predators. Critique this argument considering the reproductive strategies of parasites of insects described in this chapter.

3. Jonathan Swift wrote ("On Poetry," 1733)
   "So, naturalists observe, a flea
   Hath smaller fleas that on him prey;
   And these have smaller fleas to bite 'em,
   And so proceed, *ad infinitum.*
   Explain how this quotation is relevant to the parasitic hymenopterans.

References

References for superscripts in the text can be found at the following Internet site: www.mhhe.com/robertsjanovynadler9e

Additional Readings

Gauld, I. D., and J. Dubois. 2006. Phylogeny of the Polysphincta group of genera (Hymenoptera: Ichneumonidae; Pimplinae): a taxonomic revision of spider ectoparasitoids. *Syst. Entomol.* 31:529–564.

Traynor, R. E., and P. J. Mayhew. 2005. A comparative study of body size and clutch size across the parasitoid Hymenoptera. *Oikos* 109:305–316.

Chapter 41

Parasitic Arachnids: Subclass Acari, Ticks and Mites

There is a degree of originality about being an arachnologist, about being detectably superior to those who cannot distinguish a Pholcus *from a* Phalangium.

—Theodore Savory [68]

Ticks and mites are immensely important in human and veterinary medicine, many by causing diseases themselves and some by acting as vectors of serious pathogens. A large body of literature exists on all aspects of acarine biology, yet vast areas remain relatively unexplored. For a more extensive treatment of subjects covered in this chapter, see volumes written or edited by Dusbábek and Bukva,[14] Evans,[18] Houck,[29] Sauer and Hair,[67] Schuster and Murphy,[69] and Sonenshine.[78]

All ticks are epidermal parasites during their larval, nymphal, and adult instars; and many mites are parasites on or in skin, respiratory system, or other organs of their hosts. Some mites, although not actually parasites of vertebrates, stimulate allergic reactions when they or their remains come into contact with a susceptible individual. Ticks are found in nearly every country of the world; mites are even more ubiquitous, thriving on land, in freshwater, and in the oceans. Many millions of dollars are spent annually throughout the world in attempts to control these pests and diseases they transmit.

The general morphology of Acari, described in detail in chapter 33, can be summarized as follows. Segmentation is reduced externally, having been obscured by fusion. Tagmatization has resulted in two body regions: an anterior **gnathosoma,** or **capitulum,** bearing mouthparts, and a single **idiosoma,** containing most internal organs and bearing the legs. Most adult Acari have eight legs but some mites have only one to three pairs. The idiosoma is further divided into regions (see Fig. 33.14) as follows: The portion bearing legs is the **podosoma,** the first and second pairs of legs are on the **propodosoma,** and the third and fourth pairs are on the **metapodosoma.** That portion of the body posterior to the legs is the **opisthosoma.** Gnathosoma and propodosoma together comprise the **proterosoma,** and metapodosoma and opisthosoma together are the **hysterosoma.** These terms may seem somewhat confusing at first, but they are very useful in describing and therefore identifying acarines.

The capitulum mainly is made up of feeding appendages surrounding the mouth. On each side of the mouth is a **chelicera,** which functions in piercing, tearing, or gripping host tissues. Form of the chelicerae varies greatly in different families; thus, they are useful taxonomic features. Lateral to the chelicerae is a pair of segmented **pedipalps,** which also vary greatly in form and function related to feeding. Ventrally coxae of the pedipalps are fused to form a **hypostome;** a **rostrum,** or **tectum,** extends dorsally over the mouth. Some or all of these structures can be retracted in some acarines.

Ticks and mites, although basically similar, have distinct differences, as shown in Table 41.1. Mouthparts of Acari are modified for specialized feeding. In ticks the pedipalps grasp a fold of skin while chelicerae cut through it. As cutting progresses, the hypostome is thrust into the wound, and its teeth help anchor the tick to its host. Blood and lymph from lacerated tissues well into the wound and are sucked up. Soft ticks feed rapidly, leaving their host after engorging, whereas hard ticks remain attached for several days. Some hard ticks, particularly those with short mouthparts, secrete a cementing substance that hardens, further securing them to their host.

Table 41.1	Differences Between Ticks and Mites	
Ticks	**Mites**
Hypostome toothed, exposed	Hypostome unarmed, hidden
Large, easily macroscopic	Small, usually microscopic
Haller's organ (Fig. 41.1) present on first tarsi	Haller's organ absent
Peritreme absent	Peritreme present in Mesostigmata

611

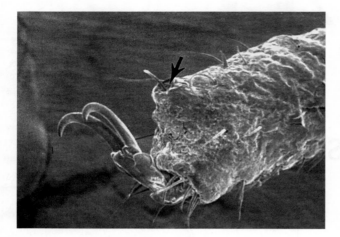

Figure 41.1   Haller's organ (*arrow*).

Courtesy of Tyler Woolley.

Mites feed on vertebrates in much the same way, but because their mouthparts are so small, most feed on lymph or other secretions rather than on blood. There is no toothed hypostome, and chelicerae vary from chelate cutters to hooks or stylets. Variations in mite feeding behavior are discussed in the sections on their respective groups.

Classification of Arachnida and Acari

Acarology is a highly active discipline in which biologists the world over are still describing many species annually while pursuing sophisticated work in chemical ecology, immunology, biological control, and epidemiology. Phylogenetic research is an indispensable part of this effort, but the taxonomic database on acarines is a massive one that grows monthly. Evans[18] elevates Acari to subclass level (class Arachnida) and recognizes two superorders and seven orders but admits that other schemes are equally acceptable, given the fluid state of arachnid systematics. Krantz[42] proposes two orders: Acariiformes, with medically important members in suborders Acaridida and Actinedida (all mites), and Parasitiformes, with medically important species in suborders Mesostigmata (mites) and Ixodida (ticks). Both orders also contain nonparasitic species. Varma[83] agrees with Krantz.[42] The major difference between classifications is the level (order, suborder) assigned to the taxa. This book uses a system based on Evans[18] mainly because this book gives a detailed morphological and historical presentation of the subclass, and thus it provides a reader with information needed to make decisions about the hierarchical levels presented by various authors.

ORDER IXODIDA: TICKS

Because of their large size and pesky habits, ticks have been recognized for centuries. Both Homer and Aristotle referred to them in their writings, but Linnaeus in 1746 was first to attempt to classify them among other animals. Evans[18] recognizes three superfamilies: Ixodoidea, hard ticks; Argasoidea, soft ticks; and Nuttallielloidea. The latter contains a single African species, *Nuttalliella namaqua,* of which only females are known, and will not be considered further here. Ticks' importance as agents and vectors of disease also has long been recognized. Pathogenesis attributable to these parasites appears in several ways:

1. *Anemia.* Blood loss in heavy infections can be considerable; as much as 200 pounds of blood can be lost from a single large host in one season.[30]
2. *Dermatosis.* Inflammation, swelling, ulcerations, and itching can result from a tick bite. These reactions often are caused by pieces of mouthparts remaining in a wound after a tick is forcibly removed, but constituents of tick saliva and secondary infection by bacteria probably are also involved.
3. *Paralysis.* A condition known as *tick paralysis* is common in humans, dogs, cattle, and other mammals when they are bitten near the base of their skull. This paralysis evidently results from toxic secretions and is quickly reversed when the parasite is removed.[83, 84] Mechanisms were reviewed by Gothe, Kunze, and Hoogstraal.[23]
4. *Otoacariasis.* Infestation of an ear canal by ticks causes a serious irritation to a host, sometimes accompanied by severe infection.
5. *Infections.* In addition to transmitting common pyogenic infections, ticks can transmit viruses, bacteria, rickettsias, spirochetes, protozoa, and filariae. These will be discussed further with their appropriate vectors.

Livestock losses from all arthropod infestations were estimated at $2.8 billion annually in the mid-1970s.[53] Losses to the cattle industry in Australia in the mid-1990s, due only to ticks, were estimated at $132 million annually, including $20 million in labor for control efforts, $63 million in meat loss, and $28 million in mortality.[51] But such economic impacts vary geographically for a number of reasons, including cattle breeds and socioeconomic factors. Control efforts are worth their expense, however; prior to 1906, *Babesia bigemina* (chapter 9) transmitted by *Boophilus microplus* caused annual losses of $100 million in the United States alone.[53] Since that time *B. bigemina* has been virtually eradicated in the United States by tick control efforts.

Biology

All ticks undergo four basic stages in their life cycles—egg, larva, nymph, and adult—all of which might require from six weeks to three years to complete. Ixodids have only one nymphal instar, but argasids have as many as five. Copulation nearly always occurs on the host. A male tick produces a spermatophore, which it places under a female's genital operculum. A blood meal is usually required for egg production, although exceptions are known. Also, some ticks are parthenogenetic. An engorged female drops to the ground, where she deposits eggs in soil or humus. A six-legged larva (Fig. 41.2) hatches from each of the 100 to 18,000 eggs and climbs onto low vegetation, where it quests for a host. On finding one, it feeds and then molts to an eight-legged nymph. If molting through all instars occurs on the same host animal, the tick is called a **one-host tick,** such as *Boophilus* species. If the nymph drops off, molts to adult, and attaches

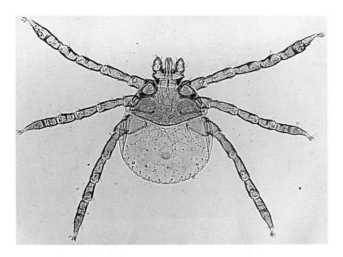

Figure 41.2   Six-legged larva of *Ixodes* sp.

Courtesy of Jay Georgi.

to another host, the tick is a **two-host tick.** Confinement of instars to one or two hosts is an adaptation to feeding on wide-ranging hosts.[28] Most ixodids are **three-host ticks,** whereas argasids, with their multiple nymphal stages, are **many-host ticks.** Clearly, use of a series of hosts increases opportunities for transmission of pathogens.

Some ticks are rather host specific, but most are opportunists that will feed on a variety of hosts. Ticks are hardy and can withstand periods of starvation as long as 16 years. Some may have a life span of up to 21 years.

Ticks exhibit a surprising amount of complex behavior, much of which is under chemical control. Tick pheromones that have been recognized influence aggregation, attachment, and reproduction, especially mate recognition and subsequent courtship. This research was reviewed by Sonenshine.[77]

Major pheromones, and behaviors they elicit, include guanine (aggregation), *o*-nitrophenol (searching and aggregation), methyl salicylate and pelargonic acid (clasping and attachment during mating), 2,6-dichlorophenol (attraction and potential mate recognition in males), cholesteryl oleate (mounting), and 20-hydroxyecdysone (copulation), although a particular behavior may depend on mixtures of pheromones as well as on a tick's developmental stage and nutritional state. Species also differ, as might be expected, in their aggregation behaviors, mating rituals, and responses to pheromones. The chemicals themselves emanate from the anus, coxal glands, and female genital aperture.

Mating behavior is typically specific, complex, and controlled at several stages by chemicals.[77] In *Dermacentor variabilis,* 2,6-dichlorophenol excites feeding males, which then detach, move toward females, and mount them. The male then begins probing for the female gonopore. Unless the male encounters a genital sex pheromone in the vulva, using sensillae on his chelicerae, he will stop the mating attempt. Species-specific aspects of this behavior occur at this last stage. Thus, a *Dermacentor andersoni* male cannot distinguish a female *D. variabilis* from one of his own species until he probes her genital opening.

In camel ticks, decision-making events occur at the beginning rather than end of the mating ritual. *Hyalomma dromedarii* and *H. anatolicum* males respond to different amounts of 2,6-dichlorophenol produced by females of their respective species. Complex behavior does not necessarily end with copulation. Male *Ixodes scapularis* (= *dammini*), for example, mate preferentially with feeding females, repel competing males, and may remain attached to (thus protective of) a female after she finishes feeding.[85]

At least some ticks also use olfaction to help find their hosts. In one study, ixodids *Amblyomma variegatum, Rhipicephalus sanguineus,* and *Ixodes ricinus,* as well as the argasid *Ornithodorus moubata,* were all attracted to somewhat diluted human breath and responded by walking upwind and exhibiting searching behavior.[52] The overall picture of tick chemical ecology is that of highly evolved systems with significant population level consequences.

Family Ixodidae

Hard ticks are divided into three subfamilies: Ixodinae, with a single genus *Ixodes;* Amblyominae, containing *Amblyomma, Haemaphysalis, Aponomma,* and *Dermacentor;* and Rhipicephalinae, with *Rhipicephalus, Anocentor, Hyalomma, Boophilus,* and *Margaropus.* We will discuss these genera individually.

Hard ticks are easily recognized as such, because their capitulum is terminal and can be seen in dorsal view. By contrast, in soft ticks the capitulum is subterminal and cannot be seen in dorsal view. Other Ixodidae characters include (1) presence of a large anterodorsal sclerite, the **scutum,** (2) eyes, when present, on the scutum; (3) pedipalps that are rigid, not leglike; (4) marked sexual dimorphism in size and often in coloration;(5) **porose areas** (regions with many small pits) on females' basis capituli; (6) coxae usually with spurs; (7) a pulvillus on the tarsus; (8) stigmatal plates behind the fourth pair of legs; (9) the posterior margin of the opisthosoma usually subdivided into sclerites called **festoons;** and (10) only one nymphal instar.

Ixodes Species

Ixodes is the largest genus of hard ticks, with over 200 species; nearly 40 species are known from North America. Most parasitize small mammals and are so small themselves that they are easily overlooked. Pronounced sexual dimorphism of their mouthparts, which are longer in females than in males, is a condition unknown in any other genus. Festoons are absent, and the anal groove is anterior to the anus. *Ixodes* spp. are three-host ticks.

Ixodes scapularis, the **blacklegged tick** (Fig. 41.3), is common in eastern and south-central United States. It feeds on a wide variety of hosts and can be a major pest on dogs. It bites humans freely, commonly resulting in a strong reaction with pain at the site and generalized malaise for a short time. *Ixodes pacificus* is found along the West Coast, in California, Oregon, and Washington, on deer, cattle, and other mammals. It also welcomes a human meal. Most active in spring, *I. pacificus* is the primary vector for both Lyme disease and granulocytic ehrlichiosis in western United States.[15] *Ixodes holocyclus* is the major cause of tick paralysis in Australia. Other *Ixodes* species are also common agents of paralysis in several parts of the world. Tick-borne encephalitis in Europe and Asia can be transmitted by *I. ricinus, I. persulcatus,* and *I. pavlovskyi.*

Species of *Ixodes* are vectors of Lyme disease, caused by the spirochaete *Borrelia burgdorferi.* Lyme disease takes its name from the town of Old Lyme, Connecticut, where an

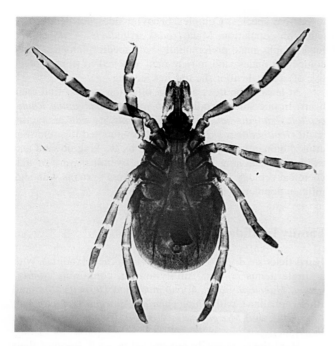

Figure 41.3 *Ixodes scapularis,* the black-legged tick.
Courtesy of Jay Georgi.

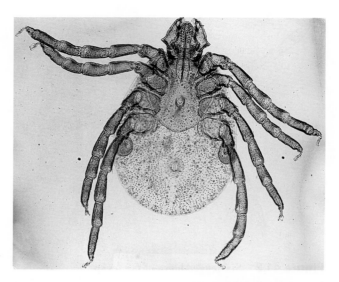

Figure 41.4 *Haemaphysalis leporispalustris,* the rabbit tick.

Courtesy of Jay Georgi.

epidemic of arthritis occurred during the mid-1970s.[33] Many of those afflicted had noticed a skin erythema (red swelling and rash) at the site of a tick bite prior to the onset of joint pains. This rash sometimes spread, and it was often accompanied by fever and headaches. We now know that neurological symptoms, facial paralysis, meningitis, and occasionally cardiac disease occur in some cases.

Half of all arthropod-transmitted infections in the United States are with *Borrelia burgdorferi,* making it the country's leading arthropod-borne disease. Recognition of this disease, its zoonotic nature, and its mode of transmission has stimulated taxonomic work on potential vectors. *Ixodes scapularis* is the primary vector in the United States, but seven other species of *Ixodes* are competent to transmit the disease.[33] Ticks formerly identified as *I. dammini,* occurring in northern and eastern United States, are now considered the same species as *I. scapularis,* which prior to the merger had a more southern distribution. Northern and southern tick strains can be distinguished by molecular techniques, however, and evidently their nucleotide sequences differ more than would be expected in intraspecific variation.[4] The northern strain is apparently expanding its range and may be primarily responsible for zoonotic outbreaks of Lyme disease.[62]

In northeastern United States, between 25% and 50% of *I. scapularis* (= *dammini*) are infected with the Lyme disease spirochaete.[33] This high prevalence is attributed to relatively unselective feeding habits of the tick, especially nymphal stages, and participation of rodent and deer reservoirs. Deer, cats, raccoons, opossums, chickens, and lizards all serve as hosts, but the ticks vary in feeding and egg production, depending on their hosts.[34, 50] Mice of genus *Peromyscus* are primary reservoirs; deer serve mainly as a large source of blood to maintain a high tick population. Transovarial and transstadial passage of *B. burgdorferi* have been reported.[43] Lyme disease is thus a classic zoonosis in which landscape

characteristics and human activities both play major roles in transmission.[56] *Borrelia burgdorferi* is widely distributed in the Northern Hemisphere,[33] and the spirochaete has also been isolated from other ticks, *Amblyomma americanum* and *A. maculatum,* and the flea *Ctenocephalides felis.*[81]

Ehrlichia spp. are obligate intracellular bacterial parasites that also are transmitted *by I. scapularis.*[9] Ruminants and horses can be infected with *E. phagocytophilia* and *E. equi.* In humans, similar or identical ehrlichias cause a disease known as **human granulocytic ehrlichiosis (HGE);** symptoms are acute fever, leukopenia, and elevated levels of liver transaminases in the serum.[12] Ehrlichiosis may be quite severe and even fatal when associated with opportunistic or secondary infections.[13]

Dogs are susceptible to numerous tick-borne infections. Several species of *Borrelia, Ehrlichia, Rickettsia,* and *Babesia,* including causative organisms of Lyme disease, human granulocytic ehrlichiosis, and Rocky Mountain spotted fever, are transmitted among dogs by ticks of genera *Ixodes, Dermacentor, Rhipicephalus,* and *Haemaphysalis.*[74] Transportation of pets and their contact with nonurban environments has potential for spreading these infections, not only among animals but also to their owners. Fipronil and permethrin remain tick control agents of choice for pets.[74]

Haemaphysalis Species

Haemaphysalis spp. are easily recognized by the second segments of their pedipalps, which are produced laterally into spurs. These small ticks often are overlooked unless they are engorged. About 150 species are known worldwide, with only two found in North America. Most are three-host ticks. There is little sexual dimorphism, and both sexes have festoons.

Haemaphysalis leporispalustris, the **rabbit tick** (Fig. 41.4), is common on rabbits from Alaska to Argentina. It occasionally feeds on domestic animals but rarely bites humans. Its main importance is as a vector of tularemia and Rocky Mountain spotted fever among wild mammals. *Haemaphysalis cordeilis,* the **bird tick,** is common on turkeys, quail, pheasants, and related game birds. It may transmit diseases among these hosts. It seldom is found on mammals and is not a pest on humans. Most species of *Haemaphysalis* occur in Asia, Africa, and the East Indies.

Dermacentor Species

Genus *Dermacentor* contains about 30 species of which at least seven are found in the United States. They are among the most medically important of all ticks, particularly in North America. These ticks are ornate, with punctations and colored markings, and they usually show sexual dimorphism of color. Both sexes have festoons. Eyes are well developed, and stigmatal plates have numerous shallow depressions called **goblets.** The sides of their basis capituli are parallel. Most species are three-host ticks, but a few are one-host.

Dermacentor andersoni, the **Rocky Mountain wood tick** (Fig. 41.5), is distributed throughout most of western United States west of the Great Plains and is most prevalent in mountainous, brushy terrain. Larvae feed on small mammals, especially chipmunks, ground squirrels, and rabbits; nymphs also feed on these hosts as well as on marmots, porcupines, and other medium-sized mammals. Adults may feed on any of these animals but seem to prefer larger hosts such as deer, sheep, cattle, coyotes, and humans. When hosts are plentiful, the entire life cycle may take a year, but if long waits occur between meals, the life cycle may extend up to three years. All stages can survive about a year without feeding. This species becomes most active in early spring, as soon as snow cover is off, and actively continues to seek hosts until about the beginning of July, when the ticks become dormant.

Dermacentor andersoni is a vector for several diseases that afflict humans: tick paralysis, Powassan encephalitis virus (mainly transmitted by *Ixodes spinipalpis*), Colorado tick

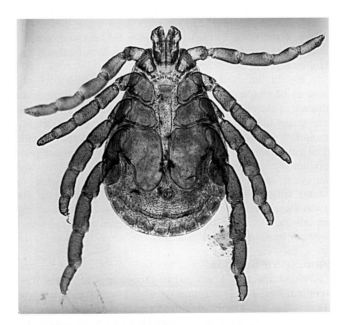

Figure 41.5 *Dermacentor andersoni*, the Rocky Mountain wood tick.

Courtesy of Jay Georgi.

fever virus, tularemia, Rocky Mountain spotted fever, and some nonpathogenic viruses. These infections and others, also such as anaplasmosis, also are transmitted to other mammals by *D. andersoni.*

Dermacentor variabilis, the **American dog tick,** is common throughout the eastern United States and is extending its range rapidly. Isolated foci are known in the Pacific Northwest, mainly along river valleys in Washington, Oregon, and Idaho. It prefers to feed on dogs but will attack horses and other mammals, including humans. This tick appears to occupy the same niche in eastern states that *D. andersoni* occupies in western states, although *D. variabilis* is more urban. Females lay 4000 to 6500 eggs on the ground; eggs hatch in about 35 days. The six-legged larvae feed on small rodents, especially voles and deer mice, and then drop off to molt. Nymphs feed again on these hosts for about a week, dropping off again to molt to adult. Adults prefer larger hosts.

Dermacentor variabilis is the principal vector of Rocky Mountain spotted fever in the central and eastern United States. Oddly enough, by far the majority of cases of this poorly named disease occur in the eastern half of the country. This tick also transmits tularemia and causes paralysis in dogs and humans.

Dermacentor occidentalis, the **Pacific Coast tick,** is very similar in morphology and biology to *D. andersoni.* Adults are found on many species of large mammals, including humans, in California and Oregon. *Dermacentor occidentalis* transmits Colorado tick fever virus, Rocky Mountain spotted fever, tularemia, anaplasmosis, and chlamydial infections resulting in abortion in cattle.

Dermacentor albipictus is called the **horse tick** or **winter tick** because it does not feed during the summer months. This one-host tick is widely distributed in the northern United States and Canada, where it feeds on elk, moose, horse, and deer. Infection can be so heavy as to kill a host. It rarely attacks cattle or humans, but moose may average over 30,000 ticks per individual, and moose die-offs have been associated with winter tick infestations.[66] Eggs are laid in spring and hatch in three to six weeks; larvae become dormant until cold weather stimulates them to seek a host. Once aboard, the tick remains through its instars and subsequent mating until it drops off in the spring. Because it is a one-host tick, *D. albipictus* is an inefficient vector for microorganisms.

Other species of *Dermacentor* throughout the world are vectors of several diseases caused by viruses and rickettsias.

Amblyomma Species

The hundred or so species in this genus mostly are restricted to the tropics, but *A. americanum* occurs well into temperate North America and migratory birds can carry several Neotropical species into Canada.[70] All appear to be one-host ticks and to have a one-year life cycle. *Amblyomma* spp. are fairly long, highly ornate ticks with long mouthparts. The second segment of the pedipalp is longer than the others, resulting in their mouthparts, including the hypostome, being longer than the basis capituli.

Larvae and nymphs are common on birds, although adults usually feed only on large animals. Immature stages will feed on almost any terrestrial vertebrate and also can be found alongside adults on the final host. Because all three

Figure 41.6 *Amblyomma americanum,* the lone star tick.
Courtesy of Warren Buss.

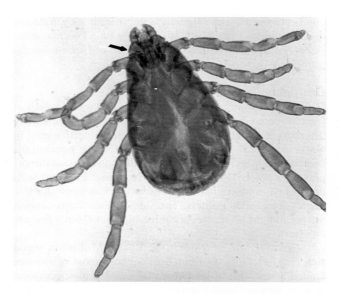

Figure 41.7 *Rhipicephalus sanguineus,* the brown tick.
Note the spurs on the basis capituli (*arrow*).
Courtesy of Warren Buss.

stages will readily bite humans, which is unusual among hard ticks, these ticks are exceptionally annoying.

Amblyomma americanum is the **lone star tick** (Fig. 41.6). Both sexes are dark brown with a bright silver spot at the posterior margin of their scutum. Males may have more than one spot. The tick ranges throughout much of southern United States and well into Mexico.[71] It has a wide variety of hosts, including livestock and humans, and is a vector for Rocky Mountain spotted fever and tularemia.

Aponomma Species

The few species of *Aponomma* parasitize reptiles, particularly in Asia and Africa. *Aponomma elaphensis* is known from rat snakes in Texas, and four species have been described from South America and Haiti. They are not known to be of medical or economic importance, although it has been speculated that they may transmit haemogregarines among reptiles.[27] They are eyeless.

Rhipicephalus Species

Continental Africa appears to be the place of origin and center of distribution of rhipicephaline ticks.[27] The approximately 60 species and subspecies of *Rhipicephalus* usually are restricted to forests, mountains, and semidesert regions or at least those with certain limits of rainfall. Most species show little host specificity, biting many mammals and even reptiles and ground-dwelling birds. Both two-host and three-host species are known. *Rhipicephalus* species are easily recognized by a combination of festoons and spurs on each side of their basis capituli. The only other genus with a similar basis capituli is *Boophilus,* which lacks festoons.

Rhipicephalus sanguineus (Fig. 41.7), the **brown dog tick** or **kennel tick,** is most widely distributed of all ticks, being found in practically all countries between latitudes 50° N and 35° S, including most of North America. It quickly becomes established whenever it is introduced.[28] *Rhipicephalus sanguineus* is a three-host tick because it leaves its host to

molt. All three stages feed mainly on dogs, mostly between their toes, in their ears, and behind the neck. The host range is very wide and occasionally includes human attacks, but this tick has a definite taste for dogs.

In some areas of Europe and Africa *R. sanguineus* is a major vector of boutenneuse fever *(Rickettsia connorii),* which usually is acquired by crushing a tick against the skin. In Mexico it transmits Rocky Mountain spotted fever. Other diseases of animals that are transmitted by this tick include *Borrelia theileri,* a spirochaete of sheep, goats, horses, and cattle; a highly fatal canine rickettsiosis caused by *Rickettsia canis;* and malignant jaundice, caused by *Babesia canis.* The protozoan *Hepatozoon canis* infects dogs when they swallow infected ticks. For a list of successful and unsuccessful experimental transmissions of diseases by *R. sanguineus* see Hoogstraal.[27]

By far the most important disease transmitted by any species of *Rhipicephalus* is **East Coast fever,** a protozoan disease of the red blood cells of cattle (chapter 9). This highly malignant infection mainly kills adult cattle, with mortality ranging from an average of 80% to 100% of infected animals. The causative agent, *Theileria parva,* is widespread geographically, and is transmitted by several ticks, but the most important vector seems to be the **brown ear tick** *R. appendiculatus.* Infected ticks transmit this disease only during the life cycle stage following an infected meal; thus, larvae acquiring an infection transmit it when they feed again as nymphs, and nymphs transmit *T. parva* as adults.

Anocentor Species

Only one species, *A. nitens,* is known in this genus. The **tropical horse tick** is found through South America up to Texas, Georgia, and Florida. It is very similar to *Dermacentor* but has 7 festoons rather than 11, its eyes are poorly developed, and it is inornate. It feeds primarily on horses and is not known to attack humans. It transmits *Babesia caballi,* a protozoan blood parasite in horses.

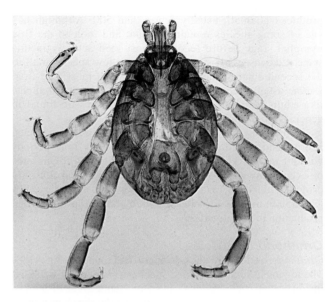

Figure 41.8 *Hyalomma* **sp.**

Courtesy of Jay Georgi.

Hyalomma Species

Few, if any, ticks are as difficult to identify as are species of *Hyalomma* (Fig. 41.8). This difficulty results partly from a natural genetic variation but also from a tendency toward hybridization. Furthermore, extrinsic factors, such as periods of starvation and climatic conditions, cause morphological variations. This group probably originated in Iran or the southern portion of the former Soviet Union and radiated into Asia, the Middle East, southern Europe, and Africa.[27] Frequently nymphs are carried from Africa to Europe by migrating birds.

These are fairly large ticks with no body ornamentation. The legs are banded. Eyes are present, and festoons are indistinct. These ticks very active, and it has been reported that in Africa and Arabian deserts they will come rushing from beneath every shrub when persons or other animals stop by.[49]

Hyalommas must be the hardiest of ticks, because they are found in desert conditions where there is little shelter away from hosts, where there are few small mammals available for larvae and nymphs to feed on, and where large mammals are far ranging. They usually are the only ticks existing in such places.

Adult *Hyalomma* usually feed on domestic animals. Occasionally they will bite humans, and, because they transmit serious pathogens, they are among the most dangerous of ticks. Immature forms often feed on birds, rodents, and hares that are reservoirs for viruses and rickettsias. For instance, **Crimean-Congo hemorrhagic fever** is carried between Africa and Europe by *H. marginatum* on migrating birds. Other viruses isolated from this species of tick are Dugbe virus and West Nile virus (also carried by mosquitoes), and *H. anotolicum* harbors Thogoto virus and a swine poxvirus. Rickettsial diseases that can be transmitted by *Hyalomma* spp. include Siberian tick typhus, boutonneuse fever, Q fever, and those that are caused by *Ehrlichia* spp. Malignant jaundice of dogs, caused by a protozoan *Babesia canis,* is transmitted by *H. marginatum* and *H. plumbeum* in Russia. Another protozoan, *Theileria annulata,* is transmitted to cattle by *H. anatolicum* in Eurasia.

Boophilus Species

Boophilus ticks resemble *Rhipicephalus* spp. in that their basis capituli is bilaterally produced into points. They differ, however, in lacking festoons and an anal groove. Because unengorged specimens are quite small and easily overlooked, they have spread to many parts of the world when cattle have been exported from endemic zones. Two hypotheses on their place of origin have been suggested: Either they came from the Indian subcontinent attached to Brahman cattle (zebu), or they originally were parasites of American bison or deer, then adapted to cattle, and thence were exported to other places.[27]

Taxonomy of this genus has been confused, but at least three species are clearly recognized. *Boophilus annulatus* is the most widely distributed of these. It is often called the **American cattle tick** because it was once widespread in the southern United States and is still common in Mexico, Central America, and some Caribbean islands. It is also known in Africa; the species known as *B. calcaratus,* from the Near East and Mediterranean region, is actually *B. annulatus.*[27] This tick has been eradicated from the United States but appears sporadically along the Mexican border, as cattle and deer carry it across.

Boophilus microplus is similar in biology to *B. annulatus* and also has been eradicated from the United States. It still is found in Mexico and Africa as well as Australia, Central and South America, Madagascar, and Taiwan. The common occurrence of parthenogenesis in this species aids its survival when harsh conditions restrict the size of a population.[80] Cattle are the primary hosts, but sheep, goats, horses, and other animals may be infested.

The **blue tick,** *B. decoloratus,* occurs widely in continental Africa. Mainly, it attacks cattle, but it bites many other animals, including humans.

Boophilus spp. are all one-host ticks. Larval, nymphal, and adult stages are all spent on the same host animal, a rarity among ticks. Engorged females drop off and lay 2000 to 4000 eggs during the next 12 to 14 days. Newly hatched larvae are quite active, crawling to the tips of grasses and other plants, where they often accumulate in great numbers. After reaching a host, they remain until after breeding and feeding. Obviating the finding of two or three hosts during its life has obvious survival value to these ticks.

Control, however, is greatly aided by the fact that all stages are to be found on the same animal. Dipping kills all parasitic stages at once. Unfed larvae die in about 65 days, so a pasture becomes tick free if cattle are dipped and kept off for this duration. Larvae, though, can be windblown for considerable distances.[47]

Species of *Boophilus* have been implicated as vectors of Crimean-Congo hemorrhagic fever and Ganjon viruses as well as Bhanja virus in Nigeria and Thogoto virus in Kenya. The rickettsia *Anaplasma marginale* is transmitted to cattle by all three species of *Boophilus* in Africa. Mortality ranges from 30% to 50% in infected animals. Experimentally, *B. decoloratus* can transmit *Trypanosoma theileri* among cattle.[3]

By far, the most important disease transmitted by a species of *Boophilus* is **Texas cattle fever,** also called **red-water fever.** The agent of this disease is a piroplasm, *Babesia bigemina* (see chapter 9). To transmit most of the aforementioned diseases, ticks must change hosts, usually under crowded conditions such as in pens or railroad cars.

However, because *B. bigemina* is transmitted transovarially, newly hatched ticks are already infected and capable of passing the disease on to cattle. Red-water fever, along with its tick vectors, was eradicated in the United States in 1939, but it persists in Central and South America, Africa, southern Europe, Mexico, and the Philippines.

The overall economic impact of tick parasitism is not easy to estimate accurately because of many contributing factors, including reduced weight gain, loss of milk production, costs of tick control, nutritional state and breed of cattle. In one study, the annual cost of all tick-borne diseases was estimated at $364 million in Tanzania alone.[39] Jonsson[37] provides an excellent review of studies attempting to discover the cost of *Boophilus microplus* infestation in Australia. This paper is a model for critical analysis of such work in general and is highly recommended reading for veterinary students especially.

Margaropus Species

The four rare species in this genus, including *Margaropus winthemi* and *M. reidi,* are found in East Africa and Sudan. They are parasites of giraffes and occasionally horses but are not known to be medically or otherwise economically important. For a review see Hoogstraal.[27]

Family Argasidae

The family of soft ticks has traditionally contained five genera: *Argas, Ornithodorus, Otobius, Nothoaspis,* and *Antricola,* with a total of about 180 species. After extensive phylogenetic analysis using structural, developmental, and behavioral characters, Klompen and Oliver[41] reduced the number of genera to four: *Argas, Ornithodorus, Otobius,* and *Carios,* the latter containing former genera *Nothoaspis,* and *Antricola,* as well as several others. Curiously, *Carios* species infect mainly bats and sea birds, including albatross (see below).

Soft ticks are easily distinguished from hard ticks by a number of characters, most obviously a subterminal capitulum in nymphs and adults (but not larvae) that cannot be seen in dorsal view. Their capitulum lies within a groove or depression called a **camerostome.** The dorsal wall of the camerostome extends over the capitulum and is called the **hood.** In addition, (1) there are no festoons or scutum; (2) sexual dimorphism is slight; (3) pedipalps are freely articulated and leglike;(4) there are no porose areas on the basis capituli; (5) eyes are on the supracoxal fold; (6) coxae lack spurs; (7) pulvilli are absent from the tarsi; (8) stigmatal plates are behind the third leg; and (9) there may be two to eight nymphal stages.

In general, argasid ticks inhabit localities of extremely low relative humidity. Those that occur in wet climates seek dry microhabitats in which to live. Unlike ixodid ticks, most argasids feed repeatedly, resting away from a host between meals. This behavior makes them difficult to collect because they hide in loose soil, crevices, birds' nests, and the like. Examination, including sifting, of soil or detritus of burrows, rodent nests, big game resting and rolling places, caves, and other animal lairs is usually required to find them.

Adult females lay eggs in their hiding places several times between feedings. Even so, the total number of eggs produced is small, usually fewer than 500. Although larvae of some species remain dormant and molt to the first nymphal stage before feeding, most feed actively in this stage. Likewise, first-stage nymphs of a few species molt to the second stage without feeding, although most feed first. Larvae usually remain on a host until molting, but nymphs, like adults, leave the host. Exceptions to this are found in genera *Otobius* and *Argas,* discussed later.

Hosts may be few and far between in desert habitats, and argasid ticks have adapted to potentially long periods without meals. They are capable of estivating for months or even many years without food. A blood meal not only provides nutrition for developing eggs but also triggers a ganglion in the brain to release a hormone that instigates egg production in their ovaries.[73]

Ornithodoros Species

Ornithodoros ticks are thick, leathery, and rounded (Fig. 41.9). The tegument in nonengorged specimens is densely wrinkled in fairly consistent patterns, allowing them great distention when feeding. Over 100 species of *Ornithodoros* parasitize mammals, including bats. It is unusual for them to feed on birds or reptiles, although *O. capensis* is found on marine birds in North America. Some species in this genus are very important in that they are vectors for relapsing fever spirochaetes, and the bite of several species is itself highly toxic and painful.

Ornithodoros hermsi is found throughout the Rocky Mountains and Pacific Coast states. It is an important vector of *Borrelia recurrentis,* the etiological agent of **relapsing fever,** which was first reported in North America from gold miners near Denver in 1915. The tick is primarily a rodent parasite. Its life cycle is typical for most species of *Ornithodoros.* Females lay up to 200 eggs in crannies and crevices like those in which adults hide. Larvae actively seek hosts and feed for about 12 to 15 minutes. After molting, the

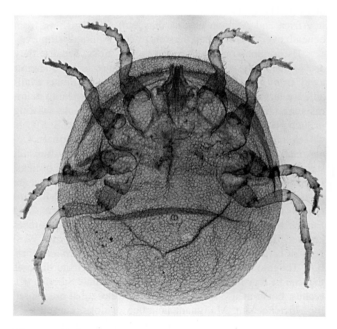

Figure 41.9 *Ornithodoros* sp.

Courtesy of Jay Georgi.

two nymphal instars each feed again and then molt to adult. The life cycle under laboratory conditions takes about four months but may be greatly prolonged in the absence of food.

Ornithodoros cariaeceus occurs from Mexico to southern Oregon, hiding in soil around bedding areas of large mammals, such as deer and cattle. It is greatly feared by many people because of its painful bite and the venomous aftereffects of its attack. Like many bloodsucking arthropods, it is attracted to carbon dioxide and thus can be trapped using dry ice for bait.

Ornithodoros moubata is an eyeless argasid found in widely dispersed arid regions of Africa. Closely related species are *O. compactus* in South Africa, *O. aperatus* in mideastern Africa, and *O. porcinus* from middle to southern Africa. *Ornithodoros porcinus* is mainly a parasite of burrowing warthogs but readily invades human habitation. It feeds on many mammals and birds and can survive starvation for at least five years. Larvae of these species do not feed but molt directly into the first nymphal instar. Some populations are parthenogenetic. An interesting physiological adaptation in this species complex is the absence of a passage between the midgut and hindgut, with the result that all waste matter must remain within the intestinal diverticuli during a tick's life.[16] A great deal is known about the biology, control, and other aspects of this group,[28] members of which are all important vectors of relapsing fever.

Ornithodoros savignyi is similar in appearance to *O. moubata*, except that it has eyes. It is found in arid regions of North, East, and southern Africa; the Near East; India; and Ceylon. It is mainly a parasite of camels but will bite practically any mammal and fowl. It does not invade human habitation, as does *O. porcinus*, but buries itself in shallow soil, awaiting its prey. It is quite bold, attacking across wide open areas if the ground is not too hot, and feeds quickly. Its bite is quite painful, but *O. savignyi* apparently does not transmit diseases in nature.

Otobius Species

These argasids are called **spinose ear ticks** because nymphs have a spiny tegument and usually feed within folds of the external auditory canal. *Otobius megnini* (Fig. 41.10) is widely distributed in warmer parts of the United States and in British Columbia. It also has been introduced into India, South Africa, and South America. It feeds mainly on cattle but attacks many domestic and wild mammals as well as humans.

Adult *O. megnini* do not feed. The adult capitulum is submarginal, but the hypostome is vestigial; its tegument is not spiny. The capitulum of larvae and nymphs is marginal; the hypostome is well developed; and the nymphal tegument, especially the second stage, is spiny. Eggs are laid on soil. On hatching, larvae contact a host, wander upward toward the head, and attach in the ear. There they remain through two molts, detaching and dropping to the ground for a final molt to the adult stage. Heavy infections can have serious, even fatal, effects on livestock.

Otobius lagophilus has a similar life cycle, except that its larvae and nymphs feed on the faces of rabbits in western North America.

Argas Species

Species of *Argas* are almost exclusively parasites of birds and bats. However, most of them have been known to bite

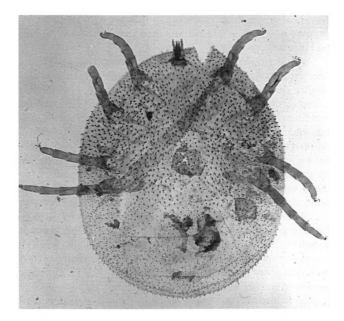

Figure 41.10   Nymph of *Otobius megnini,* the spinose ear tick.

Courtesy of Warren Buss.

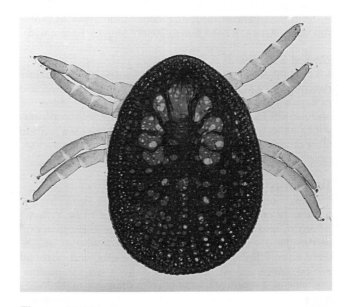

Figure 41.11   *Argas* sp.

Courtesy of Jay Georgi.

humans, although they apparently do not transmit diseases to them. *Argas* spp. (Fig. 41.11) are superficially similar to those of *Ornithodoros* but are flatter and retain a lateral ridge even when engorged. Furthermore, the peripheral tegument is typically sculptured in *Argas* spp., whereas that of *Ornithodoros*, although minutely wrinkled, does not show an obvious pattern without very high magnification. Eyes are absent.

Argas ticks have bed bug–like habits, feeding briefly by night and hiding by day in crevices and under litter. Eggs are laid in these hiding places, and hatched larvae eagerly seek a host. They usually remain attached, feeding for a few days before dropping off to molt into first-stage nymphs.

Nymphal stages feed in the same manner as do adults, engorging in less than an hour and then leaving the host to hide and digest their meal.

Argas persicus, the **fowl tick,** is primarily an Old World species, although it does exist in the New World, along with similar species *A. miniatus, A. sanchezi,* and *A. radiatus,* all parasites of domestic fowl and other birds. Under favorable conditions, these ticks may build up a huge population in a henhouse, and their nocturnal depredations can exhaust a flock or even kill individuals. Vagabonds or others who try to spend the night in a deserted chicken house are sometimes surprised by masses of ravenous fowl ticks that literally come out of the woodwork to attack them. Their bite is painful, often with toxic aftereffects, but such attacks on humans are rare.

The **pigeon tick,** *A. reflexus,* is a Near and Middle Eastern pest that has spread northward through Europe and Russia and eastward to India and other Asian localities. It has been reported in North and South America but probably was misidentified. *Argas reflexus* mainly attacks domestic pigeons, but, because these birds are closely associated with human habitation, this tick bites people more often than does *A. persicus.*

Argas cooleyi is commonly associated with cliff swallows and other birds in the United States. *Argas vespertillionis* is widely distributed among Old World bats and occasionally bites humans. The largest of all ticks is *A. brumpti,* a parasite inhabiting dens of the hyrax and some rodents in Africa. It is 15 mm to 20 mm long by 10 mm wide.

Carios Species

Like members of genus *Argas, Carios* species are mainly parasites of birds and bats. *Carios kelleyi* is sometimes found in large numbers in homes with bat colonies and residents of such houses may complain of bites.[22] Finally, in experiments with another species, *C. capensis,* collected from sea birds in Georgia, West Nile virus was transmitted to domestic ducks.[31]

Immunity to Ticks

Mammals can develop resistance to hard ticks.[44] Some strains of host species are naturally more resistant than others; in these species inflammatory lesions occur at attachment sites.[45] Rabbits, some domestic cattle, and goats all exhibit such reactions. These observations suggest vaccination may be a practical means of controlling economic losses due to acarine infestations (see also chapter 39 on immunity to myiasis).

Acquired resistance often is manifested as failure of the parasite, especially larvae and nymphs, to engorge fully.[20] Antigens responsible for these immune reactions are of both salivary gland and gut origin.[35, 46] Some salivary gland antigens are shared by *Ixodes scapularis, Dermacentor variabilis,* and *Amblyomma americanum.*[35] Adult *Rhipicephalus sanguineus, Dermacentor variabilis,* and *Amblyomma maculatum* elicit stronger immune reactions in rabbits than do either larvae or nymphs. Resistance is nonspecific and is passively transferrable with immune serum.[10] Unless rabbits are continually exposed, however, resistance is lost after about three months.

Because of the diversity of pathogens transmitted by ticks, and consequent diversity of antigens, vaccination against ticks themselves should be the best strategy for control of tick-borne diseases. Currently a single vaccine for cattle, developed from *Boophilus microplus* midgut protein Bm86 (a "concealed" antigen that the mammal never encounters directly), is commercially available.[55] When blood from an immunized animal is taken up by a tick, antibodies in the blood meal attack the gut lining, thus reducing vector populations. Vaccination programs based on Bm86 are most effective in combination with integrated pest management systems.[55]

ORDER MESOSTIGMATA

Mesostigmatid mites (Fig. 41.12) have a pair of respiratory spiracles, the **stigmata,** that are located just behind and lateral to the third coxae. Usually extending anteriorly from each stigma is a tracheal trunk, the **peritreme,** which makes it easy to recognize specimens belonging to this suborder. The gnathosoma forms a tube surrounding the mouthparts. A **tectum** is present above the mouth, and a ventral bristle-like organ, the **tritosternum,** usually is present immediately behind the gnathosoma. The palpal tarsus has a forked tine at its base. The dorsum of adults usually has one or two sclerites called **shields** or **dorsal plates.**

Family Laelapidae

The cosmopolitan family Laelapidae includes a large number of diverse genera. They are the most common ectoparasites of mammals, and some species parasitize invertebrates. Most species have pretarsi, caruncles, and claws on all legs. The dorsal shield is undivided. Their second coxa has a toothlike projection from its anterior border.

The **common rat mite,** *Echinolaelaps echidinus* (Fig. 41.13), transmits the protozoan *Hepatozoon muris* from rat to rat. Although laelapids are not known to transmit diseases to humans, they are suspected of causing dermatitis. The virus of epidemic hemorrhagic fever has been demonstrated in several species collected in rodent burrows in the Far East.[6]

Family Halarachnidae

Closely related to laelapid mites, halarachnids are parasites of the respiratory systems of mammals. They are easily recognized by a combination of morphological features: The dorsal shield is undivided and reduced; the sternal plate is reduced and has three pairs of setae; the female genital sclerite, or epigynial plate, is rudimentary; the male genital opening is in the anterior margin of the sternal plate; the tritosternum is absent; and the movable digit of the chelicera is more strongly developed than is the fixed digit.

Halarachne spp. are found only in the respiratory system of seals of family Phocidae, and *Orthohalarachne attenuata* parasitizes other families of Pinnipedia (Fig. 41.14). Several species of *Pneumonyssus* are found in primates, although infection in humans is unknown. Respiratory problems caused by these mites in captive monkeys and baboons are common[82] (Fig. 41.15). *Pneumonyssus caninum* inhabits nasal

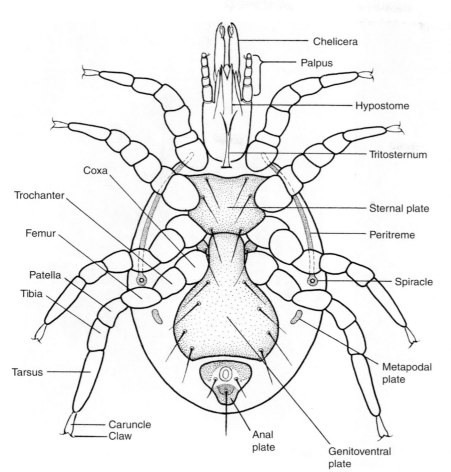

Figure 41.12   **Generalized mesostigmatid mite, ventral view.**

Courtesy of Communicable Disease Center, 1949, U.S. Public Health Service, Washington, DC.

Figure 41.13   *Echinolaelaps echidinus,* **the common rat mite.**

Ventral view of female.

From S. Hirst, "Mites injurious to domestic animals (with an appendix on the acarine disease of hive bees)," in *Brit. Mus. (Nat. Hist.) Econ. Ser.* 13:1–107. Copyright © 1922 The National History Museum, London.

, passages and sinuses of dogs and may cause central nervous system disorders.[5] *Raillietea auris* is the **cattle ear mite.** Evidently it feeds on secretions and dead cells of the external auditory meatus but is not known to be pathogenic.

Family Dermanyssidae

Dermanyssids are parasites on vertebrates and are of considerable economic and medical importance. The female's dorsal plate either is undivided or is divided with a very small posterior part. The sternal plate has three pairs of setae, and metasternal plates are reduced and lateral to the genital plates. A tritosternum is present. Chelicerae may be normal, with reduced chelae, or they may be quite elongated and needlelike. All legs have pretarsi, caruncles, and claws.

Dermanyssus gallinae, the **chicken mite** (Fig. 41.16), attacks domestic fowl, particularly chickens and pigeons, throughout the world. These mites hide by day in crevices near roosting places, emerging at night to feed. Their numbers may be so great as to kill the birds. Setting hens may abandon their nests, and young chicks may rapidly perish. This mite readily attacks humans, especially children, causing a severe dermatitis. Roosting pigeons may bring *D. gallinae* into proximity with human habitation, where wandering mites may discover a mammalian meal.[72] They are attracted to warm objects and so tend to accumulate in electric clocks and around fireplaces and water pipes.

The viruses of western and St. Louis equine encephalitis have been isolated from *D. gallinae.* It is unlikely that these mites play an important role in transmission of such diseases to mammals, but they may help keep up a reservoir of infection among birds.[76] Natural and experimental transmissions of fowl poxvirus have been demonstrated.[75] Experiments

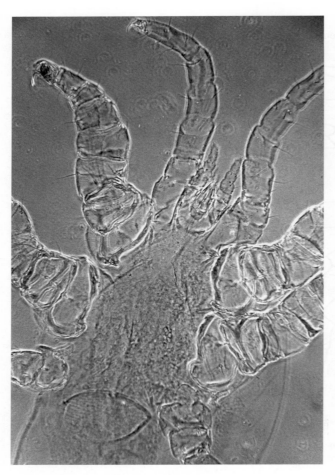

Figure 41.14 *Orthohalarachne attenuata* from the nasal passages of a northern fur seal.

Courtesy of Warren Buss.

Figure 41.15 Lung of a baboon, *Papio cynocephalus*, with nodules caused by the mite *Pneumonyssus* sp.

Courtesy of Robert Kuntz.

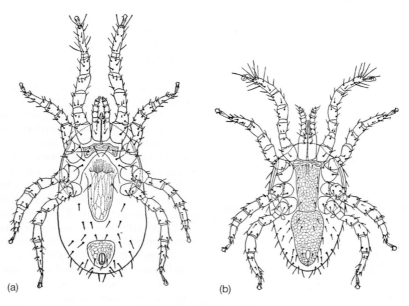

(a)   (b)

Figure 41.16 Ventral views of *Dermanyssus gallinae,* the chicken mite.
(*a*) Female; (*b*) male.

From S. Hirst, "Mites injurious to domestic animals (with an appendix on the acarine disease of hive bees)," in *Brit. Mus.* (*Nat. Hist.*) *Econ. Ser.* 13:1–107. Copyright © 1922 The National History Museum, London.

have shown that *D. gallinae* also can transmit Q fever and fowl spirochaetosis.

Liponyssoides sanguineus, the **house mouse mite,** prefers to feed on that host but will readily attack humans. This mite can transmit the rickettsialpox pathogen to humans. Rickettsialpox is a mild, febrile condition with a vesicular rash commencing three to four days after onset of fever. A scab develops at the site of the bite, and healing is slow. Besides fever, a patient has chills, sweating, backache, and muscle pains. Patients recover in one to two weeks; no fatalities are known. Q fever has been experimentally transmitted by this mite. The biology of the house mouse mite is summarized by Baker and coworkers.[1] Several mite species, along with *L. sanguineus,* have been collected from mice in pet stores.[61]

Family Macronyssidae

Macronyssids occur on numerous vertebrates, notably rodents, bats, and terrestrial reptiles. The female's dorsal plate (shield) is undivided; the sternal shield is wide and strongly sclerotized; and, the genital plate is much longer than wide. Chelicerae are of a piercing type, lacking barbs or hooks, and chelae are of about equal length. Radovsky[60] provides characters for distinguishing macronyssids as a family distinct from other groups, e.g., Dermanyssidae or Laelapidae, in which they were formerly included.

The **tropical rat mite,** *Ornithonyssus bacoti,* is found worldwide, in both temperate and tropical climates, where, as its name implies, it normally infests rats. It is a serious pathogen of laboratory mouse colonies, where it can retard growth and eventually kill young mice. When rats are killed or abandon their nests, these mites can migrate considerable distances to enter human habitation. They cause a sharp, itching pain at the time of their bite, and skin-sensitive persons may develop a severe dermatitis. Nonfeeding larvae hatch and rapidly molt to bloodsucking protonymphs. These nymphs molt to become nonfeeding deutonymphs, which in turn become feeding adults. Parthenogenesis is common, producing only males. The life cycle from egg to egg can be completed in 13 days. Adult females live about 60 days and produce approximately a hundred eggs.

These mites apparently do not transmit any pathogens to humans, although experimentally they can transmit plague, rickettsialpox, Q fever, and murine typhus. *Ornithonyssus bacoti* is intermediate host of the filarial nematode *Litomosoides carinii,* a parasite of cotton rats, *Sigmodon hispidus* and other *Litomosoides* species in related rodents.[54] Because rats, mites, and worms can easily be maintained in the laboratory, this host-parasite system has been used as a model for studies on filariasis, including drug testing.

The **northern fowl mite,** *Ornithonyssus sylviarum* (Fig. 41.17), is widespread in northern temperate climes and has been reported from Australia and New Zealand. It will bite humans and can be a nuisance to egg processors. It does not appear to be particularly pathogenic to fowl. *Ornithonyssus bursa* is the **tropical fowl mite.** It is ectoparasitic on chickens, turkeys, and some wild birds, including English sparrows.[25] It can be pathogenic to poultry, causing them to be listless and poorly developed. It bites humans but causes only a slight irritation.

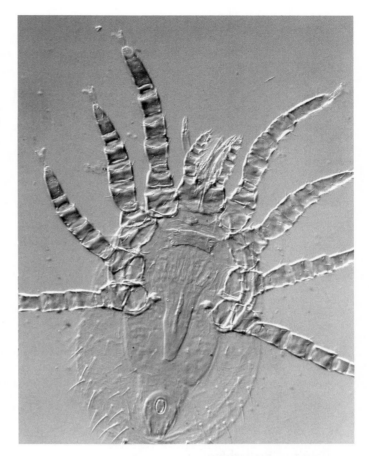

Figure 41.17   *Ornithonyssus sylviarum,* **the northern fowl mite.**

Courtesy of Barry OConnor, University of Michigan.

Family Rhinonyssidae

This family is considered by some authorities to be a subfamily of Dermanyssidae. All members are parasitic in respiratory tracts of birds. Rhinonyssids are oval in shape and have weakly sclerotized plates. All tarsi have pretarsi, caruncles, and claws. Stigmata are present with or without short dorsal peritremes. The tritosternum is absent.

These mites are viviparous, producing larvae in which a protonymph is already developed. Nearly every species of bird examined has nasal mites; many species have been described, and many more species undoubtedly are yet to be discovered.[57] Because of their blood- or tissue-feeding habits, these mites may be regarded as significant disease agents in wild bird populations. The **canary lung mite,** *Sternostoma tracheacolum* (Fig. 41.18), can sicken and kill captive canaries and finches.

ORDER PROSTIGMATA

In this order spiracles are located either between the chelicerae or on the dorsum of the hysterosoma. These mites usually are weakly sclerotized. Chelicerae vary from strongly chelate to reduced. Pedipalps are simple, fanglike, or clawed. There are phytophagous, terrestrial and aquatic free-living, and parasitic forms.

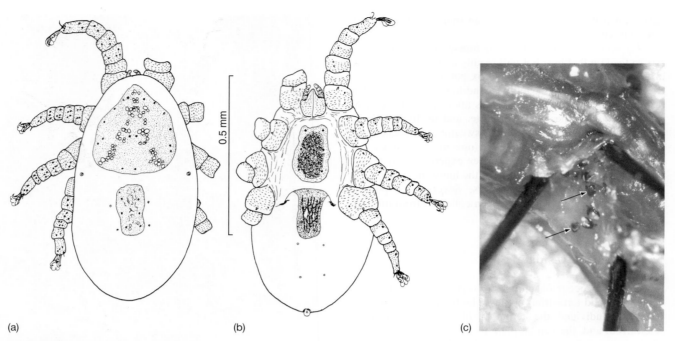

(a)   (b)   (c)

Figure 41.18   *Sternostoma tracheacolum,* **the canary lung mite.**

(a) Dorsal view; (b) ventral view; of female. (c) Several mites (arrows) in the respiratory tract of a small passerine bird.

(a) and *(b)* from D. B. Pence, "Keys, species and host list, and bibliography for nasal mites of North American birds (Acarina: Rhinonyssinae, Turbinoptinae, Speleognathinae, and Cytoditidae)," in *Special Publications of the Museum of Texas Tech University,* 8:1–148. Copyright © 1974 Texas Tech University Press. Reprinted by permission of Museum of Texas Tech University. *(c)* Courtesy of Barry OConnor, University of Michigan.

Family Cheyletidae

Cheyletid mites are small, measuring 0.2 mm to 0.8 mm long. Most are yellowish or reddish, oval, and plump, except for feather-inhabiting species, which are elongated. The propodosoma and hysterosoma are clearly delineated and usually have one or more dorsal shields. Eyes are present or absent. Strong peritremes, which usually surround the gnathostoma, are present. Chelicerae are short and styletlike; palpi are large and pincerlike.

In *Cheyletiella parasitivorax* the male genital opening is dorsal, a rare occurrence in arthropods. *Cheyletiella yasguri* of dogs and *C. blakei* of cats cause a mange dermatitis on their normal hosts. They also will feed on humans, although only temporarily but can cause dermatitis in people as well as their pets.

Family Pyemotidae

Pyemotid mites mainly parasitize insects that infest cereal crops. They are brought into contact with humans when people harvest grains or work with stored grains or sleep on straw mattresses. When these mites bite, they leave a small, itching vesicle that may become inflamed rapidly and cause considerable discomfort. Itching, headache, nausea, and internal pains may accompany severe attacks.

These are soft-bodied mites with tiny chelicerae and pedipalps. A wide space occurs between the third and fourth pairs of legs. Sexual dimorphism is marked. Females become enormously swollen when gravid.

Pyemotes tritici, the **straw itch mite,** normally is a parasite of various stored grain beetles, but it readily attacks humans. Males can be seen by the unaided eye only with difficulty, but gravid females reach nearly a millimeter in length. A female's body contains 200 to 300 large eggs, which hatch internally. Developing mites complete all larval instars before being born. The few males emerge first and cluster around their mother's genital pore, copulating with females as they emerge.

The **grain itch mite,** *P. ventricosus,* is similar in biology and pathogenicity. It normally infests boring beetle larvae, grain moths, and numerous other insects. Mites of genus *Pyemotes* have been recovered from amber of Eocene age, along with their bark beetle hosts.[38]

Family Psorergatidae

These are small- to medium-sized mites that are unarmored and have striated skin and peritremes. Because they are soft, they are susceptible to desiccation and are less numerous during dry periods. Chelicerae are minute and styletlike; pedipalps are simple and minute and are not used for grasping. The first pair of legs is modified for grasping hairs.

Psorobia ovis is the itch mite of sheep and is a serious pest in sheep-raising countries, including the United States, New Zealand, and Australia. It causes skin injury and fleece derangement. *Psorobia simplex* is found on laboratory mice, and *P. bos* is known from cattle in western United States.[36]

Family Demodicidae

These minute, cigar-shaped parasites are known as the **follicle mites.** They range in length from 100 μm to 400 μm and have short, stumpy, five-segmented legs on the anterior half of the body. The opisthosoma is transversely striated. Species of *Demodex* live in hair follicles and sebaceous glands of many species of mammals. Although numerous species have been described, it is probable that many more exist, especially in wild mammals, because they seem to be rigidly host specific.

Humans serve as hosts to two species. *Demodex folliculorum* (Fig. 41.19*a*) lives in hair follicles, whereas *D. brevis,* a stubbier species (Fig. 41.19*b*), inhabits sebaceous glands.[8] Both exist mainly on the face, particularly around the nose and eyes. All life stages may be found in a single follicle. These mites may penetrate the skin and lodge in various internal organs, where they elicit a granulomatous response.

Prevalence of these mites in humans is very high, from about 20% in persons 20 years of age or younger to nearly 100% in the aged. Infection usually is benign, although rarely there may be loss of eyelashes or granulomatous skin eruptions.[24] Follicle mites may be involved in introducing acnecausing bacteria into skin follicles of susceptible individuals. Both *Demodex* species evidently can also proliferate opportunistically in immunocompromised individuals.[32] An easy means of diagnosing both species in humans is to examine microscopically some oil expressed from the side of the nose.

Much more pathogenic is the **dog follicle mite,** *Demodex canis.* This species, together with some form of the bacterium *Staphylococcus pyogenes,* causes **red mange,** or **canine demodectic mange.** Infection in young dogs can be serious, even fatal. There is hair loss on the muzzle, around the eyes, and on the forefeet. The skin develops reddish pimples and pustules, becoming hot, thickened, and covered with a foul-smelling reddish-yellow exudate. Exact diagnosis depends on demonstrating a mite in skin scrapings. Treatment is difficult, and severely infected puppies may have to be killed. Symptoms may disappear gradually, and older, although perhaps still infected dogs show no further signs of disease, probably because of acquired immunity.

Other demodicids of importance are the **cattle follicle mite,** *D. bovis,* the **horse follicle mite,** *D. equi,* and the **hog follicle mite,** *D. phylloides.* All three cause a pustular dermatitis with nodules and loss of hair. Holes in the skin caused by these mites may reduce the value of hides.

Family Trombiculidae

This family contains the infamous **chigger mites,** which are all too common in most tropical and temperate countries of the world. They are unique among mites that attack humans in that only the larval stage is parasitic; nymphs and adults are predators. Over 1200 species have been described, many from larvae only; in numerous cases nymphs and adults are unknown or have not been associated with larvae.

In a generalized chigger life cycle the egg hatches in about a week, releasing a **prelarva,** or **deutovum,** which then molts into a larva (Fig. 41.20). Larvae have six legs and

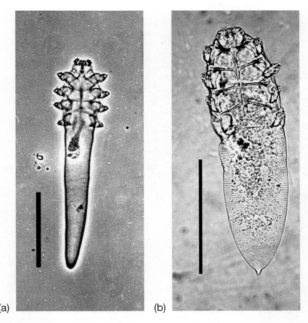

Figure 41.19   *Demodex* **species, the human follicle mites.**

(*a*) *Demodex folliculorum,* male, ventral view. (*b*) *Demodex brevis,* female, ventral view. This specimen was donated by a student taking a parasitology lab (she was thrilled that her mite would be in a textbook!). Bars = 100 μm.

(*a*) From C. Desch and W. B. Nutting, "*Demodex folliculorum* (Simon) and *D. brevis* Akbulatova of man: Redescription and reevaluation," in *J. Parasitol.* 58:169–177. Copyright © 1972. (*b*) Photograph by John Janovy, Jr.

are parasitic, sometimes crawling up on grass blades where they encounter a passing host, attach by chelicerae, and feed on partially digested skin. An engorged larva drops off its host and becomes a quiescent **protonymph,** also known as a **prenymph** or **nymphochrysalis.** Following some tissue reorganization, a **deutonymph** emerges from the protonymph at the next molt and actively feeds on insect eggs and soft-bodied invertebrates. The deutonymph is followed by another quiescent stage, the **tritonymph,** or **imagochrysalis,** which then molts into an adult. Males deposit a stalked spermatophore, which the female inserts into her genital pore. Most chiggers show little host specificity.

Taxonomy of trombiculids is based mainly on larvae. The larval body is rounded and usually is red, although it may be colorless. It bears a dorsal plate, or scutum, at the level of the anterior two pairs of legs; usually two pairs of eyes are near the lateral margins of the scutum. The scutum bears a pair of sensillae and three to seven setae. Chelicerae have two segments: The basal segment is stout and muscular, whereas the distal segment is a curved blade with or without teeth. Pedipalps consist of five segments. The fifth, or tarsus, bears several setae and opposes a tibial claw like a thumb. Numerous plumose setae are distributed on the body.

Adult chiggers are among the largest of mites, reaching a millimeter or more in length. There is a conspicuous constriction between the propodosoma and hysterosoma. Eyes may be present or absent. Both sexes are clothed with a dense covering of plumose setae, which gives them the appearance of velvet. Commonly they are bright red or yellow.

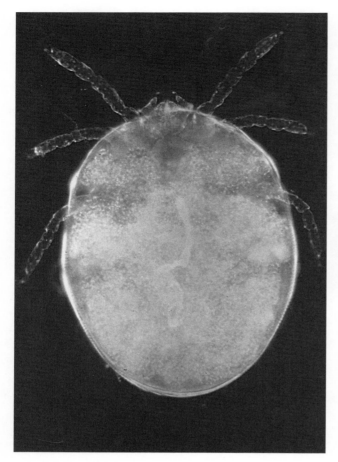

Figure 41.20   A chigger larva.

Courtesy of Mark Pope.

Figure 41.21   Larval *Arrenurus* sp. feeding on a damselfly.

Note the long stylostome penetrating the insect's body wall. The internal organs of the damselfly have been removed.

From B. L. Redmond and J. Hochberg, "The stylostome of *Arrenurus* spp. (Acari: Parasitengona) studied with the scanning electron microscope," in *J. Parasitol.* 67:308–313. Copyright © 1981.

There are two medical aspects of chigger bite: **chigger dermatitis** and transmission of pathogens. We will consider these separately.

Larval chiggers do not burrow into skin, as is commonly thought. After their mouthparts penetrate the epidermis, these mites inject salivary secretions that are proteolytic, killing and digesting host cells, which the parasite then sucks up along with interstitial fluids. Simultaneously host cells harden under the influence of other salivary secretions to become a tube, the **stylostome** (Fig. 41.21). The mite retains its mouthparts in the stylostome, using it like a drinking straw, until engorged, and then it drops off. Not all chiggers cause an itching reaction; those that do usually have detached before a host reaction begins. Some people are immune to their bites, whereas others may incur a violent reaction.

Most chiggers of medical importance in North America are of genus *Eutrombicula*. Of these, *E. alfreddugesi* is the most common species, ranging throughout the United States except in the western mountain states. It is most abundant in disturbed forest that has been overgrown with shrubs, vines, and similar second-growth vegetation. It feeds on most terrestrial vertebrates. *Eutrombicula splendens* is the most abundant chigger in the southeastern United States, especially in moist areas such as swamps and bogs. The two species overlap in many areas but are active in different seasons.

Other trombiculid species, some representing other genera, bite humans or are important pests of livestock. The **turkey chigger,** *Neoschoengastia americana,* causes discoloring of the skin and loss of feathers of turkeys and related birds, rendering them less fit for market.[19] *Euschoengastia latchmani* causes a mangelike dermatitis on horses in California.

Several species of *Leptotrombidium* are vectors of a rickettsial disease in humans called **scrub typhus** (tsutsugamushi disease). The microorganism *Rickettsia tsutsugamushi* is transovarially transmitted among mites; wild rodents, particularly species of *Rattus,* are reservoirs. This disease was first described from Japan and now is known from Southeast Asia; adjacent islands of the Indian Ocean and southwest Pacific; and coastal North Queensland, Australia. A primary lesion appears at the site of a chigger bite. It slowly enlarges to 8 mm to 12 mm and becomes necrotic in the center. By the fifth to eighth day, a red rash appears on the trunk and may spread to the extremities. Other symptoms are enlarged spleen, delirium and other nervous disturbances, prostration, and possibly deafness. Mortality rates range from 6% to 60%. Early treatment with broad-spectrum antibiotics usually is successful.

ORDER ORIBATIDA

In these mites the exoskeleton usually is strongly sclerotized or leathery and often deep brown (Fig. 41.22). Stigmata and tracheae are usually present, opening into a **porose area** (pitted region). Mouthparts are withdrawn into a tube, the **camerostome,** which may have a hoodlike sclerite over it.

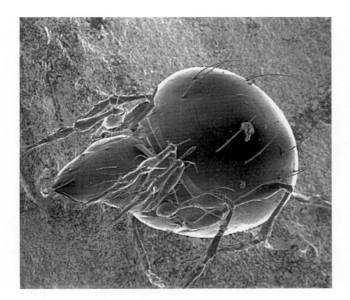

Figure 41.22   *Oppia coloradensis,* a beetle mite.

Courtesy of Tyler Woolley.

A complex of families within this group is called Oribatei. All feed on organic detritus and as such are among the dominant fauna of humus. They are of no direct medical importance, but many serve as intermediate hosts of *Moniezia expansa* (p. 343) and other anoplocephalid tapeworms. Members of this group are also intermediate hosts for *Bertiella studeri,* the primate cestode that sometimes infects humans (p. 343).

ORDER ASTIGMATA

Mites of order Astigmata totally lack tracheal systems; they respire through the tegument, which is soft and thin. They lack claws, which are replaced with suckerlike structures on their pretarsi (Fig. 41.23). Some of the most medically and economically important mites belong to this order.

Family Psoroptidae

Members of this family are very similar to those of family Sarcoptidae and are easily confused with them. However, unlike sarcoptids, they do not burrow into skin; instead, psoroptids pierce the skin at the bases of hairs, causing an inflammation that can become severe. Furthermore, Psoroptidae lack propodosoma vertical setae, which are present on Sarcoptidae.

Chorioptic mange is a condition of domestic animals caused by mites of genus *Chorioptes* (Fig. 41.24). Formerly each host species of *Chorioptes* was thought to harbor a distinct species of parasite, but recent molecular work indicates that infestations are mainly with two species, *C. bovis* and *C. texanus,* neither of which is particularly host specific.[17] *Chorioptes* spp. also infest wild ungulates, but it's not always clear which species is (are) involved.[48]

These mites usually inhabit the feet and lower hind legs of cattle and horses. In sheep, chorioptic mange of

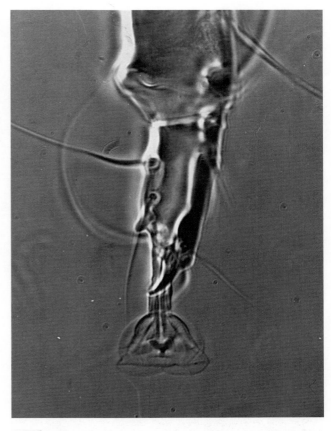

Figure 41.23   Pretarsus of *Chorioptes* sp., showing suckerlike modification.

Courtesy of Jay Georgi.

the scrotum is well known to cause seminal degeneration. In fact, surveys of sheep in the United States have shown *C. bovis* to be the most common arthropod parasite, although pathogenic results usually are rare. Pour-on formulations of doramectin are effective in control of *C. bovis* and a number of other ectoparasites of cattle.[65]

Psoroptic mange is caused by several species of mites on several species of hosts. *Psoroptes* spp. are distinguished by long legs that extend beyond their body and by the pedicel of the caruncle, which is segmented. They pierce skin and suck exudates. These fluids congeal to form scabs that provide a protective cover for the parasites; then the mites can reproduce under ideal conditions, increasing their numbers into millions in only a few days. Most domestic and a number of wild animals suffer from psoroptic mange. Wool production may be greatly inhibited by *P. ovis* in sheep. Unlike sarcoptic mange, *P. ovis* infests parts of the body most densely covered with wool. The species' biology was reviewed by Kirkwood.[40]

Mites very closely related to *Psoroptes* spp. are in genus *Otodectes.* Fairly common in cats, dogs, foxes, and ferrets, *O. cynotis* (Fig. 41.25) is usually found in the ears, although other parts of the head may be infested. Thousands of these mites swarming in the ears of a luckless host can cause desperate distress, with scabby, flowing ears and fitlike behavior. Current methods of mite control for companion animals are reviewed by Ghubash.[21]

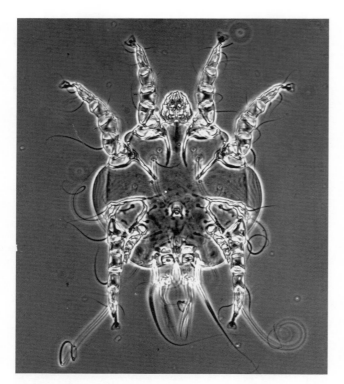

Figure 41.24 *Chorioptes* **sp., male.**

Courtesy of Jay Georgi.

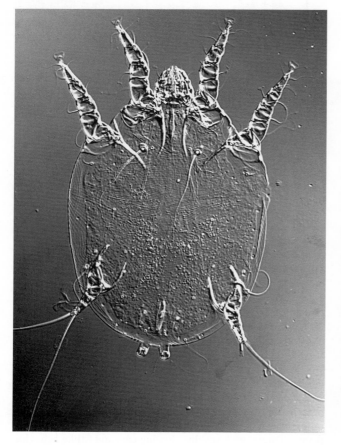

Figure 41.25 *Otodectes cynotis* **nymph.**

Courtesy of Jay Georgi.

Family Sarcoptidae

Sarcoptic mange, or **scabies,** may result from an infestation of the itch mite, *Sarcoptes scabiei* (Fig. 41.26). Although separate *Sarcoptes* spp. have been described from a wide variety of domestic animals as well as humans, they are morphologically indistinguishable and may represent physiological races or perhaps sibling species. Thus, *S. scabiei* var. *equi* has a predilection for horses but will readily bite riders as well.

Sarcoptids are skin parasites of mammals. Their body is rounded without a constriction separating the propodosoma from the hysterosoma. A propodosomal shield may be present or absent; either way, a pair of vertical setae projects from the dorsal propodosoma. The tegument has fine striae arranged in fields interrupted by scales, spines, or setae. The legs are very short and may or may not have claws or caruncles.

Scabies mites mate on their host's skin, males inseminating immature females. Immature females move rapidly over the skin and at this stage are probably transmissible between hosts. Males do not burrow into skin but remain on the surface, along with nymphs. A mature female uses long bristles on her posterior legs to lift her back end up until she is nearly vertical. She cuts rapidly with her mouthparts and claws, becoming completely embedded in two and one half minutes. She remains within the horny layer of the skin, forming tortuous tunnels for about two months. Scattered along the burrows are eggs, hatched larvae, ecdysed cuticles, and excrement. Eggs hatch in three to eight days, and larvae and nymphs emerge to wander on the skin surface.

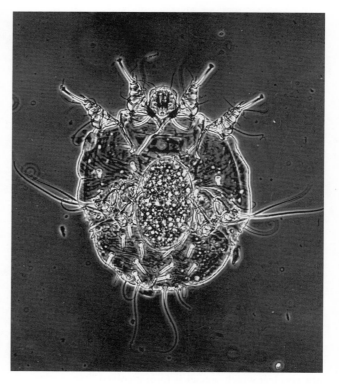

Figure 41.26 *Sarcoptes scabiei,* **the itch mite.**

Courtesy of Jay Georgi.

This tunneling and the secretory and excretory products produce an intense itching sensation in most infected persons. Usually a person does not notice any symptoms until the case is well advanced. A rash begins to show, and vesicles and crusts may begin to form in some cases. This disease has several names, such as **seven-year itch, Norwegian itch,** or simply *scabies*. Skin between the fingers, breasts, and shoulder blades and around the penis and in creases of the knees and elbows is most often infected. Scratching can cause bleeding and secondary infection. Transmission occurs primarily through physical contact between persons. A 17- to 20-year cycle of resurgence of scabies infection in the world apparently occurs for unknown reasons, possibly because of a changing immunity in the human population.[79] Scabies was well reviewed by Robinson.[64]

Sarcoptic mange in domestic and wild mammals is essentially the same as that in humans. Hairless or short-haired regions of the body are most affected. Secondary infection by bacteria is more common in animals other than humans, resulting in severe weight loss or failure to gain weight, loss of hair, and pruritic dermatitis.[58] Infection is readily passed on to humans who are in contact with mangy animals.

Cats may develop mange caused by *Notoedres cati*. *Notoedres* is very similar in appearance to *Sarcoptes* but is smaller and more circular. It affects rodents and dogs but apparently not humans. **Notoedric mange** usually begins at the tips of the ears and spreads down over the head, sometimes onto the body.

Family Knemidokoptidae

Knemidokoptid mites are very similar in morphology and biology to Sarcoptidae. They all are parasites of birds. *Knemidokoptes mutans* causes a condition known as **scaly leg** in chickens and small birds such as canaries, but has also been reported from owls and partridges. The mites burrow into skin and under scales of the feet and lower legs, which become distorted and covered with thick, nodular, spongy crusts. Infection may be so severe as to kill the birds, either directly or by secondary sensitization, which can involve internal organs. Scaly leg is highly contagious. Other species of *Knemidokoptes* occur on a wide variety of wild birds; *K. jamaicensis* has been reported from numerous small passerines, and *K. pilae* infests the face and legs of budgerigars.

Neocnemidocoptes gallinae, formerly placed in *Knemidokoptes*, is the **depluming mite** of chickens. This species embeds in skin at the bases of quills. Infected birds pluck out their feathers in an attempt to alleviate the itching, or feathers may fall out by themselves. Usually large patches of deplumation extend over the body.

For advice about control of mite pests in birds, see Rickards.[63]

Family Pyroglyphidae

Important mites in this family are species of *Dermatophagoides* (Fig. 41.27). They are not parasitic, but can cause a severe dermatitis on the scalp, face, and ears of humans. Mites in this family are responsible for **house dust allergy.** Several species, especially those in *Dermatophagoides*, are

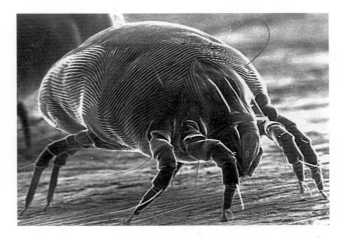

Figure 41.27 *Dermatophagoides* **sp., a house dust mite.**

From G. W. Wharton, "Mites and commercial extracts of house dust," in *Science* 167:1382–1383. Copyright © 1970 by the AAAS.

abundant denizens of house dust. When whole mites or their parts or excrement are inhaled, as happens to everyone every day, they can stimulate an allergic reaction in sensitive individuals.

Bee Mites

Members of two mite families, in two different orders, are of major economic importance because they parasitize honey bees. **Varroa mites,** *Varroa jacobsoni* (Mesostigmata: Varroidae), reproduce in brood cells and feed on pupae, although they may overwinter on adult workers by attaching to the abdomen, piercing the exoskeleton, and feeding on hemolymph.[2, 11] Resistance to Varroa mites has been reported.[26] Female **tracheal mites,** *Acarapis woodi* (Prostigmata: Tarsonemidae), lay eggs in a host's trachea. Eggs hatch in about five days; larvae molt in another four to five days, and females then disperse to other host individuals.[59] Both mite species can be devastating to bee colonies. Bee strains vary in both resistance and colony hygiene, and treatment ranges from acaricides to commercial vegetable oil products.[7] Bee mites affect not only honey production but also, indirectly although more importantly, pollination of several crops.

Learning Outcomes

By the time a student has finished studying this chapter, he or she should be able to:

1. Diagram the basic anatomical features of ticks and mites and indicate which features are likely to vary between taxa.

2. Draw the life cycle of a typical three-host tick.

3. Explain why some ticks make excellent vectors for various infectious agents and others might not serve this role so well.

4. Write the scientific names of some important ticks and mites and tell the role(s) that these species play in the transmission of infectious disease.

5. Tell the kinds of economic damage and host impact that ticks and mites produce in agricultural settings.

References

References for superscripts in the text can be found at the following Internet site: www.mhhe.com/robertsjanovynadler9e

Additional Readings

Baker, E. W., and G. W. Wharton. 1952. *An introduction to acarology*. New York: The Macmillan Co. An excellent reference for the beginner and professional alike, although a bit out-of-date.

McDaniel, B. 1979. *How to know the ticks and mites*. Dubuque, IA: William C. Brown Publishers. Useful, well-illustrated keys to genera and higher categories of ticks and mites in the United States.

Walker, J. B., J. E. Keirans, and I. G. Horak. 2000. *The genus Rhipicephalus (Acari, Ixodidae): A guide to the brown ticks of the world*. Cambridge, UK: Cambridge University Press.

glossary

A

ablastin (ā-BLAS-tin) Antibody that appears during infection with *Trypanosoma lewisi* in rats and inhibits parasite reproduction.

abundance Average number of parasites, of one species, per host (infected + noninfected) in a sample (= density).

abscess (AB-ses) Tissue necrosis in a localized area with increase in hydrostatic pressure from pus accumulation.

acanthella (ā-kan-THEL-ə) Developing acanthocephalan larva, between an acanthor and a cystacanth, in which definitive organ systems are developed.

acanthor (ā-KANTH-ər) Acanthocephalan larva that hatches from an egg.

accessory filament A darkly staining structure that runs longitudinally within the undulating membrane of certain trichomonads.

accessory piece (ak-SES-ə-rē pēs) One of the sclerotized parts of a monogenean copulatory apparatus.

accidental myiasis (mī-I-ə-səs) Presence within a host of a fly not normally parasitic. Also called **pseudomyiasis.**

accidental parasite Parasite found in other than its normal host. Also called an **incidental parasite.**

acephaline (ā-SEF-ə-lēn) Literally "lacking a head"; referring to gregarines without a septum in the anterior portion of the cell.

acetabulum (a-set-TAB-ū-ləm) Sucker, ventral sucker of a fluke; a sucker on the scolex of a tapeworm.

acraspedote (ā-KRAS-pə-dōt) Condition in cestodes in which the posterior edge of one proglottid does not overlap the anterior edge of the succeeding segment.

acquired immunity Immunity arising from a specific immune response, stimulated by antigen in the host's body (active) or in the body of another individual with the antibodies or lymphocytes transferred to the host (passive).

ACT (artemisinin-based combination therapy) An artemisinin derivative in combination with another malarial drug, such as piperaquine.

actinospore (ak-TIN-ə-spōr) Life-cycle stage of a myxozoan, possessing hook-like structures and released from annelid host.

acute infection (ə-KŪT) Rapidly developing infection that produces relatively severe symptoms.

adaptive immunity (Also acquired immunity) An immunity specific to the nonself material, requires time for development; present only in jawed vertebrates.

ADCC (antibody dependent cell-mediated cytotoxicity) Condition in which effector cells, such as neutrophils and eosinophils, are stimulated to attack an invader (antigen) in the presence of antibody to that antigen.

adenolymphangitis (AD-ən-ə-limf-an-JĪ-təs) Inflammation of lymph channels.

adhesive disc Suckerlike circular organ or organelle used for attachment.

adjuvant (AD-jə-vənt) Material added to an antigen to increase its immunogenicity, probably by enhancing expression of costimulators on antigen-presenting cells.

adoptive immunity Immune state conferred by inoculation of lymphocytes, not antibodies, from an immune animal rather than by exposure to the antigen itself.

adoral zone of membranelles (AD-ō-rəl) Fields or rows of cilia and their kinetosomes linked by electron dense fibrous networks into membranes and located to the "left" of or counterclockwise from the side of the oral area of the more complex ciliates.

aedeagus (ə-DĒ-ə-gəs) Copulatory organ or penis in insects and acarines.

aerotolerant (ERR-ō-TOL-ə-rənt) Able to live in an environment with oxygen.

agamete (a-GAM-ət) A life-cycle stage of Mesozoa, in which nuclei proliferate and form an aggregation of cells within a larva.

aggregated Description of the concentration of most parasites of a single species in a minority of hosts. Also called **overdispersed.**

ague (Ā-gū) Malarial fever.

ala (pl. alae; Ā-lə; Ā-lē) Term often applied to winglike structure on plants or animals: the lateral winglike expansions of the branchiuran carapace to form respiratory alae, cuticular winglike expansions of nematodes, and others.

algid malaria (AL-jəd) Form of malaria characterized by coldness of skin, extreme weakness, and severe diarrhea, caused by *Plasmodium falciparum.*

allograft (AL-lə-graft) Graft of a piece of tissue or organ from one individual to another of the same species.

allozyme (AL-ō-zīm) Enzymes that catalyze the same reaction and are coded for by genes at the same locus (alleles).

alternative pathway Pathway for activation of complement that does not depend on presence of fixed antibody. See **classical pathway.**

alveoli (al-VĒ-ə-lī) Pockets or spaces bounded by membrane or epithelium.

amastigote (ā-MAST-i-gōt) Form of Trypanosomatidae that lacks a long flagellum. Also called a **Leishman-Donovan (L-D) body,** as in *Leishmania.*

ameboma (a-mē-BŌ-mə) Granuloma containing active trophozoites, occasionally resulting from a chronic amebic ulcer; rare except in Central and South America.

amebula (a-MĒ-bū-lə) Daughter cell resulting from mitosis and cytokinesis of an encysted ameba.

amphid (AM-fid) Sensory organ on each side of the "head" of nematodes.

amphimictic (AM-fə-MIK-tik) Interbreeding and sexual reproduction through fusion of gametes.

amphistome (AM-fi-stōm) Fluke with the ventral sucker located at the posterior end.

anaphylaxis (AN-ə-fə-LAK-səs) Systemic form of immediate hypersensitivity, primarily mediated by IgE antibodies and release of pharmacologically active substances from mast cells and basophils.

anapolysis (AN-ə-POL-ə-sis) Detachment of a senile proglottid after it has shed its eggs.

anautogeny (AN-ä-TOJ-ən-ē) In some Diptera, necessity of a blood meal before eggs can develop within a female.

androgenic gland (AN-drə-JEN-ik) Gland located near the vas deferens in many Crustacea. Its secretions are responsible for development of male secondary sexual characteristics.

---

bat/āpe/ärmadillo/herring/fēmale/finch/līce/crocodile/crōw/
duck/ūnicorn/tüna/ə "uh" as in mammal, fishes, cardinal, heron,
vulture/stress as in bi-OL-o-gy, bi-o-LOG-i-cal

androgyny (an-DRO-jə-nē)   Condition in a hermaphroditic animal in which male organs mature before female organs. See **protandry.**

anecdysis (AN-ek-DĪ-səs)   Ecdysis in which successive molts are separated by quite long intermolt phases; referred to as *terminal anecdysis* when maximal size is reached and no more ecdyses occur.

anisogametes (AN-īs-ə-GAM-ēts)   Outwardly dissimilar male and female gametes.

Anisogamy (AN-Ī-SOG-ə-me)   Is the condition of having dissimilar male and female gametes.

annuli (AN-ū-lī)   Rings on the body of a parasite; not necessarily indicative of internal segmentation.

antennae (second antennae of crustaceans) (an-TEN-ē)   Second pair of appendages in Crustacea, with bases usually immediately posterior to antennules; primarily sensory but sometimes adapted for other functions; derived from appendages on primitive third preoral somite; no homologous appendage in insects.

antennules (first antennae) (an-TEN-ūlz)   Anteriormost pair of appendages of Crustacea; primarily sensory but often adapted for additional or other functions in particular species; derived from appendages on primitive second preoral somite; homologous to **antennae** of insects.

anterior attachment organs   Structures at the anterior end of Monogenoidea, consisting of glands and specialized tegument, sometimes opening on small lobes or in sacs.

anterior station   Development of a protozoan in the middle or anterior intestinal portions of its insect host, such as section *Salivaria* of Trypanosomatidae.

antibody (AN-tē-BOD-ē)   Immunoglobulin protein, produced by B cells (or plasma cells derived from B cells), that binds with a specific antigen.

antibody titer (TI-tər)   Measure of the amount of antibody present, usually given in units per milliliter of serum.

anticoagulant (an-tē-kō-AG-ū-lənt)   Substance that prevents blood clotting.

antigen (AN-tə-jən)   Any substance that will stimulate an immune response.

antigen challenge   Dose or inoculation with an antigen given to an animal some time after primary immunization with that antigen has been achieved.

antigenic determinant (AN-tə-JEN-ik dē-TERM-ə-nənt)   Area on an antigen molecule that binds with antibody or specific receptor sites on the sensitized lymphocyte; it "determines" the specificity of the antibody or lymphocyte. See **epitope.**

antigen-presenting cells   Cells such as dendritic cells and macrophages that "present" epitopes of antigens on their surface, in the cleft of MHC II proteins, resulting in activation of appropriate T cells.

antigenic variation (AN-tə-JEN-ik var-ē-Ā-shun)   See **variable antigen type.**

anting   In birds, purposeful application of ants to the feathers, or allowing ants to occupy feathers, presumably as a means of lice control.

apical complex (Ā-pē-kəl KOM-plex)   Dense ring and conelike structure, along with associated microtubules, micronemes, and rhoptries, at the anterior end of an apicomplexan sporozoite. (*apical* = at the apex.)

apical organ (Ā-pē-kal OR-gan)   Organ of unknown function at the apex of a cestode's scolex.

apicoplast (Ā-pē-kō-plast)   Membranous organelle in apicomplexans, considered a vestigal plastid.

apodeme (AP-ō-dēm)   Spinelike inward projection of the cuticle in arthropods on which a muscle inserts; a ridgelike projection is an **apophysis** (a-POF-ə-sis).

apodous larva (a-PŌD-əs LARV-ə)   Larva with no legs and with reduced head; usual in Hymenoptera, Diptera, some Coleoptera; requires maternal care or deposition in or on food source.

apolysis (a-POL-ə-sis)   Disintegration or detachment of a gravid tapeworm segment; also, the detachment of the hypodermis from the old procuticle in arthropods before molting.

apomorphic (ap-ō -MOR-fik)   Adjective that refers to the form of characters in particular characters whose form differs from that of the same characters in an outgroup.

areoles (AR-ē-ōls)   Raised and sometimes intricately sculpted areas on the surface of a nematomorph (hairworm).

arista (ə-RIS-tə)   Flagellumlike appendage on the antenna of a fly of the suborder Brachycera and some members of the Nematocera.

arrhenotoky (ə-REN-ō-TŌK-kē)   Parthenogenetic production of males. See **haplodiploidy.**

artemisinin (ÄR-tə-MIS-ən-ən)   Terpene extracted from *Artemisia annua,* which is active as an antimalarial drug.

arthropodization (är-thrō-PÄD-i-ZĀ-shən)   Evolutionary development of the combination of characteristics associated with Arthropoda, including a firm cuticular exoskeleton containing chitin.

ascaridine (əs-KAR-ə-dēn)   Protein of unknown function in the sperm of *Ascaris.*

ascaroside (əs-KAR-ə-sīd)   Glycoside found in *Ascaris,* made of the sugar *ascarylose* and a series of secondary monol and diol alcohols.

ascites (ə-SĪT-ēz)   Edema, or accumulation of tissue fluid, in the mesenteries and abdominal cavity.

autoantibody (AW-tō-ANT-ə-bod-ē)   An antibody made by an organism against one of its own proteins or other antigens.

autogamy (aw-TOG-ə-mē)   Form of selfing in ciliates, in which haploid pronuclei from a single individual fuse to restore the diploid condition.

autoimmunity (Ä-tō-im-MŪN-i-tē)   Immune response to one's own proteins or other antigens.

autoinfection (AW-tō-in-FEK-shən)   Reinfection by a parasite juvenile without its leaving the host.

autotrophic nutrition (AW-tō-TROF-ik)   Feeding that does not require preformed organic molecules as nutritive substances.

axial cells (AX-ē-əl)   Central cells of a dicyemid mesozoan.

axoneme (AX-ō-nēm)   Core of a cilium or flagellum, comprising microtubules.

axopodia (ax-ō-PŌD-ē-ə)   Unbranched pseudopodia that contain a slender axial filament composed of microtubules that extends into a cell's interior.

axostyle (AX-ō-stīl)   Tubelike organelle in some flagellate protozoa, extending from the area of the kinetosomes to the posterior end, where it often protrudes.

B

B cell   Type of lymphocyte that gives rise to plasma cells that liberate antibody to the antigen; so called because in birds they are processed through a lymphoid organ called the *bursa of Fabricius;* of primary importance in humoral immune response.

bacillary bands (BAS-ə-la-rē)   Lateral zones in the body wall of some nematodes, consisting of glandular and nonglandular cells of unknown function.

Baer's disc (BĀ-ərz)   Large ventral sucker of an aspidogastrean trematode.

ballonets (bal-ō-NETS)   For inflated areas within the "head" of nematodes of the family Gnathostomatidae; each is connected to an internal cervical sac of unknown function.

basal body (BĀ-səl)   Centriole from which an axoneme arises; also called a **kinetosome** or *blepharoplast.*

basis (basipodite) (BĀ-sis; ba-SIP-ə-dīt)   Joint of a crustacean appendage from which the exopod and endopod originate; that is, the joint between the coxa and the exopod and endopod.

basophil (BĀ-sə-fil)   Least numerous of polymorphonuclear leukocytes, so called because it stains with basic stains. Its function is similar to that of mast cells in tissues: release of pharmacologically active compounds mediated by IgE in immediate hypersensitivity.

benign tertian malaria   Malaria caused by *Plasmodium vivax.*

bicornuate pyriform apparatus (bi-KORN-u-ət)   See **pyriform apparatus.**

bilharziasis (bil-härz-Ī-ə-sis)   Disease caused by *Schistosoma* spp. Also called *schistosomiasis.*

binary fission   Mitotic division of one cell into two, usually applied to protozoa.

biodiversity (bī-ō-dī-VER-si-tē)   The variety of living organisms, usually within a particular region or ecosystem, often taken as a measure of the health of an ecosystem when compared to the number of taxa expected to be found.

biomass (BĪ-ō-mass)   The total mass, approximated by weight, of organisms in a sample or other unit of measure.

biota (bī-Ō-tə)   All of the organisms, including plants, animals, protists, fungi, and microorganisms in a region.

biological vector (VEK-tər)   Vector in which a disease organism lives or develops. Contrast with **mechanical vector.**

biramous appendage (bī-RĀM-əs ə-PEN-dij)   Appendage with two main branches from a common basal joint, characteristic of Crustacea, although not all appendages of a crustacean may be biramous.

black fly fever   Combination of symptoms resulting from sensitization to bites by black flies (Simuliidae).

blackhead   Disease of turkeys caused by a protozoan, *Histomonas meleagridis.* Also called *histomoniasis* or *infectious enterohepatitis.*

blackwater fever   Complication of falciparum malaria manifested by massive lysis of red blood cells with excretion of hemoglobin in the urine.

bladderworm (BLA-dər-wərm)   See **cysticercus.**

blastocyst (BLAST-ō-sist)   In cestodes, posterior portion of plerocercus metacestode into which the body can withdraw.

blastoderm (BLAST-ō-derm)   "Primary epithelium" formed in early embryonic development of many arthropods when the nuclei migrate to the periphery and undergo superficial cleavage; usually encloses the central yolk mass.

blue tongue   Virus disease of ruminants transmitted by biting midges (Ceratopogonidae).

book lungs   Respiratory structures in some arachnids, characterized by a highly folded set of membranes in an enclosed chamber.

bothridium (bäth-RID-ē-əm)   Muscular lappet on the dorsal or ventral side of the scolex of a tapeworm. Bothridia are often highly specialized, with many types of adaptations for adhesion.

bothrium (BÄTH-rē-əm)   Dorsal or ventral groove, which may be variously modified, on the scolex of a cestode.

bradyzoite (brā-dē-ZL-īt)   Small stage in various coccidia of the *Isospora* group that develops in a zoitocyst; similar to a **merozoite.**

breakbone fever   Another name for **dengue,** a virus disease transmitted by mosquitoes.

bubo (BU-bō)   Hard, swollen, bacteria-filled lymph nodes, usually in the arm pits or groin.

buccal cone (BUK-kəl)   Portion of the mouthparts of acarines composed of hypostome and labrum.

C

cadre (KAD-rē)   Sclerotized mouth lining of a pentastomid.

Calabar swelling (KAL-ə-bär)   Transient subcutaneous nodule, provoked by the filarial nematode *Loa loa.*

calcareous corpuscles   Small mineral bodies, primarily of calcium and magnesium carbonates, secreted in cells or excretory canals of many cestodes and some trematodes. Their function is unknown.

calotte (ka-LOT)   Light-colored area at the anterior end of a nematomorph (hairworm).

calypter (kə-LIP-tər)   Squama or lobe in the anal angle of a dipteran wing.

camerostome (kə-MER-ə-stōm)   Ventral groove in the propodosoma of soft ticks wherein lies the capitulum.

campestral (kəm-PES-trəl)   Characteristic of rural locations, especially open country and grasslands.

campodeiform (kam-pō-DĒ-ə-form)   Describes an insect larva with a well-defined head and thoracic appendages; typically predatory.

capitulum (kə-PIT-ū-ləm)   Anterior of two basic body regions of a mite or tick. Also called a **gnathosoma.**

capsule (KAP-sül)   In reference to the eggshell of flatworms, that portion composed of sclerotin, with precursors principally contributed by vitelline cells. Contrast with **coat.**

carapace (KAR-ə-pās)   Structure formed by posterior and lateral extension of dorsal sclerites of the head in many Crustacea, usually covering and/or fusing with one or more thoracic somites; considered as arising from a fold of head exoskeleton. Also a dorsal sclerotized plate often covering the idiosoma of acarines.

carcinoma (kär-si-NŌ-mə)   Malignant tumor originating in epithelial tissue.

Carrion's disease (KAR-ē-onz də-zēz)   Bacterial disease transmitted by sand flies. See also **Oroya fever** and **verruga peruana.**

cascade process (kas-KĀD PRAH-sess)   A series of linked events within a cell, in which one event triggers the next.

cathepsins (kə-THEP-sinz)   A family of cysteine peptidases, mostly intracellular, mostly in lysosomes.

caveolae (ka-VĒ-ə-lē)   Invaginated pits or vesicles in cell surfaces.

cecum (SĒ-kum)   Blind pouch or diverticulum of an intestine.

cell-mediated immunity (CMI)   Immunity in which antigen is bound to receptor sites on the surface of sensitized T lymphocytes that have been produced in response to prior immunizing experience with that antigen and in which manifestation is through macrophage response with no intervention of antibody.

cellular immune response (SEL-ū-lər)   Binding of antigen with receptor sites on sensitized T lymphocytes to cause release of lymphokines that affect macrophages, a direct response with no intervention of antibody. Also, entire process by which the body responds to an antigen, resulting in a condition of cell-mediated immunity.

cement glands (SĒ-ment)   Glands in a male acanthocephalan that produce secretions sealing the female reproductive tract after copulation.

centrolecithal egg (cen-trō-LES-ə-thəl)   Type of egg found in many arthropods, in which the nucleus is located centrally in a small amount of nonyolky cytoplasm, surrounded by a large mass of yolk. After fertilization and some nuclear divisions, the nuclei migrate to the periphery to proceed with superficial cleavage, the yolk remaining central.

cephaline (SEF-ə-lēn)   Referring to gregarines with a septum in the anterior part of the cell.

cephalogaster (sef-AL-ə-gas-tər)   Contractile organ in adult epicaridean isopods that functions in sucking blood and perhaps in respiration.

cercaria (ser-KAR-ē-ə)   Juvenile digenetic trematode, produced by asexual reproduction within a sporocyst or redia.

cerci (SER-sē)   Appendages on the 11th abdominal somite of some insects; usually sensory.

bat/āpe/ärmadillo/herring/fēmale/finch/līce/crocodile/crōw/ duck/ūnicorn/tŭna/ə "uh" as in mammal, fishes, cardinal, heron, vulture/stress as in bi-OL-o-gy, bi-o-LOG-i-cal

cercomer (SER-kō-mer) Posterior, knoblike attachment on a procercoid or cysticercoid. Usually bears the oncospheral hooks.

cerebral malaria Most common type of complicated falciparum malaria, characterized by severe headache, coma, and very high temperature, often ending in death.

chaetotaxy (KE-tō-tax-ē) Taxonomic study of the location and arrangement of bristles on an insect. Especially important in order Diptera.

Chagas' disease (SHÄ-gəs) Disease of humans and other mammals caused by *Trypanosoma cruzi.*

chagoma (shä-GŌ-mə) Reddish nodule that forms at the site of entrance of *Trypanosoma cruzi* into the skin.

chalimus (KAL-ə-məs) Specialized, parasitic copepodid, found in copepod order Siphonostomatoida; attached to its host by an anterior "frontal filament" that is secreted by the frontal gland.

challenge Dose of an antigen administered later after an initial immunizing dose.

chelate (KĒ-lāt) Condition of an arthropod appendage in which the subterminal podomere bears a distal process to form a pincer with the terminal podomere; sometimes (incorrectly) used to describe the subchelate condition.

chelicerae (kə-LIS-ər-ē) Anteriormost pair of appendages in chelicerate arthropods, which include spiders, ticks, and mites; generally the most important feeding appendages in these groups.

chigger (CHIG-ər) Mite of family Trombiculidae. Also sometimes applied to *Tunga penetrans,* the chigoe flea.

chigoe (chig-ō) A flea of genus *Tunga* that burrows into the skin of a host.

chitin (KĪ-tin) High molecular weight polymer of *N*-acetyl-glucosamine linked by 1,4-β-glycosidic bonds.

choanomastigote (kō-an-ō-MAST-ə-gōt) Like a promastigote but with the flagellum emerging from a collarlike process, as in *Crithidia* spp.

chorioptic mange (kōr-ē-OP-tik mānj) Disease caused by mites of genus *Chorioptes.*

chorioretinopathy (KO-rē-ō-ret-ən-OP-ə-thē) Disease process involving the choroid and retina of the eye.

chromatin (KRŌ-mə-tin) DNA combined with characteristic proteins in chromosomes.

chromatoid bar (KRŌ-ma-toyd) Masses of RNA, visible with light microscopy, in young cysts of *Entamoeba* spp.

chronic infection (KRON-ik) Long-lasting or recurring infection, often resulting from partial immunity, in which symptoms are relatively nonsevere although pathology may be progressive.

chyluria (kīl-ŪR-ē-ə) Lymph in the urine, characterized by a milky appearance.

ciliary organelles (SIL-ē-ar-ē or-gən-ELZ) Organelles of specialized function formed by fusion of cilia.

circomyarian muscles (SER-kō-mī-AR-ē-ən) Type of nematode muscle in which the sarcoplasm is completely surrounded by contractile (striated) fibrils.

circumsporozoite protein (SUR-cəm-spōr-ə-ZŌ-it) Protein in the surface of *Plasmodium* sporozoites against which antibodies are raised in some experimental vaccines.

cirri (SER-rī) Fused tufts of cilia in some protozoa that function like tiny legs. Also plural for **cirrus.**

cirrus (SER-əs) Penis or copulatory organ of a flatworm.

cladistics (kla-DIS-tiks) General method of recovering hypothesized evolutionary histories using shared apomorphic characters as criteria for grouping.

clamp Complex set of sclerotized bars, forming a "pinching" organ on the opisthaptor of a monogenetic trematode.

Claparedé organs (klap-er-ə-DĀ) See **urstigmata.**

classical pathway Pathway of complement activation that depends on presence of fixed antibody. See **alternative pathway.**

cleaning symbiosis Association between unlike species, commonly marine, in which individuals of one species clean parasites and diseased tissue from individuals of another.

cloaca (klō-Ā-kə) Common chamber into which digestive, reproductive, and sometimes excretory systems empty.

CO1 (SEE-OH-one) Subunit 1 of the enzyme cytochrome oxidase; the gene for this protein is used as a "bar code" to compare taxa, especially through use of phylogenetic algorithms.

coarctate pupa (kō-ARK-tāt PŪ-pə) Pupa in which the last larval cuticle is retained as a puparium.

coat In reference to the eggshell of many cestodes, portion contributed by the outer envelope, derived from embryonic blastomeres.

coccidiostat (kok-SID-ē-ō-stat) A chemical compound that will control coccidial infections when fed regularly to a host.

coelomocyte (sē-LOM-ə-sīt) Any of various cell types found in the pseudocoel of nematodes, coelom of annelids, or coelom and hemal system of echinoderms.

coelomyarian muscles (SĒL-ə-mī-AR-ē-ən) Nematode muscles in which the contractile portion of a cell extends in a *U* shape and partly surrounds the sarcoplasm.

coelozoic (sēl-ə-ZŌ-ik) Living in the lumen of a hollow organ, such as the intestine.

coenurus (sē-NŪR-əs) Tapeworm metacestode in family Taeniidae, in which several scolices bud from an internal germinative membrane; not enclosed in an internal secondary cyst.

colleterial glands (kō-lə-TER-ē-əl) Female accessory glands in insects that produce a substance to cement eggs together or material for an ootheca.

commensalism (kō-MEN-səl-izm) Kind of symbiosis in which one symbiont, the commensal, benefits, and the other symbiont, the host, is neither harmed nor helped by the association.

community All of the organisms (in our case parasites) of all species living in a particular habitat.

complement (KOMP-lə-mənt) Collective name for a series of proteins that bind in a complex series of reactions to antibody (either IgM or IgG) when the antibody is itself bound to an antigen; produces lysis of cells if the antibody is bound to antigens on the cell surface.

complement fixation test Immunological method used to detect presence of antibodies that bind (or fix) complement; standard diagnostic test for many infections.

concealed antigen (kon-SĒLD AN-tə-jen) A protein antigen that the host is never directly exposed to, but that be used to vaccinate a host against a vector because the antibodies act against the antigen once the host's blood is inside the vector.

concomitant immunity (kon-KOM-ə-tənt) Host is protected against reinfection, but the parasite eliciting the immunity remains alive and unaffected.

condyles (KON-dīlz) Bearing surfaces between arthropod joints that provide the fulcra on which the joints move.

congenital (con-JIN-ə-təl) Occurring concurrently with birth; applies to both infections and inherited conditions.

conjugation (kon-jə-GĀ-shən) Form of sexual reproduction in ciliates, in which partners exchange haploid nuclei, which then unite.

conoid (KŌ-noyd) Truncated cone of spiral fibrils located within the polar rings of some Apicomplexa.

constant region (Fc for crystallizable fragment) Part of an antibody molecule that is composed of a limited number of different amino-acid chains determining its class.

contagious (kon-TĀJ-əs) Capable of being transmitted through direct contact. Also used to describe population distributions that are **aggregated,** such as in an area.

contaminative antigen (kon-TAM-in-ə-tiv) Antigen borne by the parasite that is common to both host and parasite but that genetically is of host origin.

contractile vacuoles Organelles that function to eliminate excess water, especially from protozoans.

copepodid (ko-PEP-ə-did) Juvenile stage that succeeds naupliar stages in copepods, often quite similar in body form to an adult.

coracidium (kōr-ə-SID-ē-əm) Larva with a ciliated epithelium, hatching from the egg of certain cestodes; ciliated oncosphere.

cordons (KOR-dənz) Cuticular ridges at the anterior end of some nematodes.

costa (KOS-tə) Prominent striated rod in some flagellate protozoa that courses from one of the kinetosomes along the cell surface beneath the recurrent flagellum and undulating membrane.

cotylocidia (kot-ə-lō-SID-ē-ə) Larva of Aspidobothrea.

coxa (KOX-a) Most proximal podomere of an arthropod limb, sometimes called *coxopodite* in crustaceans.

coxal glands Excretory organs of arachnids, consisting of a sac, tubule, and opening on the coxa.

crabs Infestation with the crab louse, *Phthirus pubis*.

craspedote (KRAS-pə-dōt) Condition in cestodes in which the posterior edge of each segment overlaps the anterior edge of the succeeding segment. See **acraspedote.**

creeping eruption Skin condition caused by hookworm larvae not able to mature in a given host.

crura (KRÜ-rə) Branches of the intestine of a flatworm.

cryptogonochorism (KRIP-tō-gō-nō-KOR-izm) Separate sexes joined or associated to form the appearance of hermaphroditism.

cryptoniscus (krip-tō-NIS-kəs) Intermediate, free-swimming larval stage of isopod suborder Epicaridea, developing after microniscus; attaches to definitive host.

cryptozoite (krip-tō-ZŌ-īt) Preerythrocytic schizont of *Plasmodium* spp.

ctenidium (tē-NID-ē-əm) Series of stout, peglike spines on the head (genal ctenidium) and first thoracic tergite (pronotal ctenidium) of many fleas.

cutaneous (kū-TĀN-ē-əs) Pertaining to the skin.

cuticulin (kū-TIK-ū-lin) Protein component of arthropod exoskeletons.

cypris (SĪ-prəs) Postnaupliar larva of barnacles (crustacean subclass Cirripedia) in which the carapace largely envelops the body; so called because of its resemblance to ostracod genus *Cypris*.

cyrtocyte (SUR-tō-sīt) The terminal flagellated cell, or flame cell, in a protonephridial excretory system.

cyst (sist) Stage in a parasite's life cycle, occurring either outside a host or in tissues and sometimes offering resistance to unfavorable conditions. Also closed sac enveloped by a distinct membrane.

cystacanth (SIS-tə-kanth) Juvenile acanthocephalan that is infective to its definitive host.

cysticercoid (sis-tə-SER-koyd) Metacestode developing from the oncosphere in most Cyclophyllidea. It usually has a "tail" and a well-formed scolex.

cysticercosis (sis-tə-ser-KŌ-sis) Infection with one or more cysticerci.

cysticercus (sis-tə-SER-kəs) Metacestode with a fluid-filled bladder as the "tail."

cystogenic cells (sis-tō-JEN-ik) Secretory cells in a cercaria that produce a metacercarial cyst.

cytogamy (sī-TOG-ə-mē) Fusion of ciliates during conjugation but without exchange of nuclear material.

cytokine (SĪT-ə-kīn) Protein hormone produced by one cell type that modifies the physiological condition of the cells that produce it and/or other cells.

cytomeres (SĪT-ə-merz) In piroplasms, products of multiple fission in cells of the tick hosts, before differentiation and separation as secondary kinetes.

cyton (SĪ-ton) Cell body; contains nucleus and some other organelles but excludes processes extending from the cell. For example, the neurocyton is the nerve cell body, excluding the axon and dendrites.

cytophaneres (SĪ-tō-fan-ER-ēz) Fibers radiating out from a zoitocyst into surrounding muscle; found in some species of Sarcocystidae.

cytophore (SĪ-tə-fōr) A mass of cytoplasm surrounded by nuclei, in tissue giving rise to sperm in Platyhelminthes.

cytopyge (SĪT-ə-pīj) Permanent opening, found in many ciliates, through which indigestible material is voided.

cytotoxic T lymphocytes (CTLs) CD8$^+$ T cells that bind to and destroy target cells displaying certain antigens.

D

dauer juvenile (DOW-ər JÜ-vən-əl) Nematode juvenile in which development is arrested during unsuitable conditions and resumed when conditions improve.

decacanth (DEK-ə-kanth) Ten-hooked larva that hatches from the egg of a cestodarian tapeworm. Also called a **lycophora.**

defensins Antimicrobial peptides produced by certain cells in mammals and others with different structure in insects and plants that destroy a wide range of microbes with certain conserved molecules on their surface.

definitive host (də-FIN-ə-tiv) Host in which a parasite achieves sexual maturity. If there is no sexual reproduction in the life of the parasite, the host most important to humans is the definitive host.

dehiscence (dē-HIS-əns) Release of oocysts from a gregarine gametocyst.

deirid (DAR-id) Sensory papilla on each side near the anterior end of some nematodes.

delayed type hypersensitivity (DTH) Manifestation of cell-mediated immunity, distinguished from immediate hypersensitivity in that maximal response is reached about 24 hours or more after intradermal injection of the antigen. The lesion site is infiltrated primarily by monocytes and macrophages.

dendritic cells Important antigen-presenting cells (APCs), displaying epitopes of antigens on their surface to other cells of the immune response.

dengue (DEN-gē) Virus disease transmitted by mosquitoes.

density Average number of parasites, of one species, per host (infected + noninfected) in a sample (= abundance).

denticles (denticulate) (DENT-ə-klz; den-TIK-ū-lāt) Small, toothlike projections.

dermatitis (derm-ə-TĪ-tis) Infection or inflammation of the skin.

desmosome (DEZ-mə-sōm) Anchoring point of a cell to its neighbors, with thickening of apposing membranes and filamentous material between the cells.

determinant Epitope. The part of an antigen molecule displayed by APCs, to be recognized by specific antibodies or T-cell receptors.

deuterotoky (DÜT-ər-ō-tō-kē) Type of parthenogenesis in which all individuals are uniparental but in which both males and females occur.

deutomerite (dü-TOM-er-īt) Posterior half of a cephaline gregarine protozoan.

deutonymph (DÜT-ō-nimf) In the life cycle of some mesostigmatid mites, nonfeeding stage that molts into the adult.

bat/āpe/ärmadillo/herring/fēmale/finch/līce/crocodile/crōw/ duck/ūnicorn/tüna/ə "uh" as in mammal, fishes, cardinal, heron, vulture/stress as in bi-OL-o-gy, bi-o-LOG-i-cal

deutovum (dü-TŌV-əm)   Incompletely developed larva that hatches from an egg of a chigger mite.

developmental arrest   Cessation in development for a period of time in which the organism remains viable and capable of resuming development upon application of appropriate stimuli.

diapause (DĪ-ə-pawz)   Quiescent phase in arthropods in which most physiological processes are suspended.

diapolar cells (DĪ-ə-PŌL-ər)   Ciliated somatodermal cells located between the parapolar and uropolar cells of a mesozoan.

diecdysis (dī-ek-DĪ-sis)   Condition in which ecdysis processes are going on continuously and one ecdysis cycle grades rapidly into another.

dilator organ (DĪ-lā-tər)   Part of the male pentastome reproductive system, functioning in copulation.

dioecious (dī-Ē-shəs)   Having separate sexes; males and females are different individuals.

diplokarya (dip-lō-KER-ē-ə)   Process, in Microsporidia, in which nuclei of a multinucleate plasmodium may be associated in pairs, although apparently not as a part of sexual reproduction.

diplostomulum (dip-lō-STŌM-ū-ləm)   Strigeoid metacercaria in family Diplostomatidae.

diporpa (dī-PŌRP-ə)   Larval stage in the life cycle of the monogenean *Diplozoon*.

direct development   In arthropods refers to development in which a juvenile hatches from an egg, and the juvenile is not distinctly different from an adult except in size and maturity.

distal cytoplasm (DIS-təl SĪT-ō-plazm)   Distal cytoplasmic layer in tegument of Monogenea, Digenea, and Cestoidea.

distome (DĪ-stōm)   Fluke with two suckers, oral and ventral.

dorsal plate   Dorsal plate on the body of a mesostigmatid mite.

dourine (DOW-rēn)   Disease of horses and other equids caused by *Trypanosoma equiperdum*.

Duffy blood groups   Blood types categorized by certain surface antigens on erythrocytes that serve as coreceptors for invading *Plasmodium vivax*. Absence of the coreceptors (Duffy negative) precludes invasion by *P. vivax*.

dum-dum fever   Another name for visceral leishmaniasis or kala-azar.

duo-gland adhesive system   Platyhelminth tegument gland with two types of cells, one producing an adhesive substance and the other a releasing substance.

dynein (dī-NĒ-in)   A protein that serves as a "molecular motor" by interacting with cytoskeletal elements such as microtubules.

dyskinetoplasty (dis-kī-NĒT-ō-plas-tē)   Condition in which a trypanosomatid kinetoplast is nonfunctional, especially in terms of mitochondrial function.

dyspnea or dyspnoea (DISP-nē-ə)   Difficult or labored breathing.

E

East Coast fever   See **theileriosis**.

ecdysis (ek-DĪ-sis)   Molting or discarding of inexpansible portions of cuticle, after which there is an increase in physical dimensions of an animal's body before its newly secreted cuticle hardens.

echinostomiasis (ē-KĪN-ə-stōm-Ī-ə-sis)   Disease caused by infection with flukes of family Echinostomatidae.

eclipsed antigen (ē-KLIPST)   Antigen borne by the parasite that is common to both host and parasite but that genetically is of parasite origin.

ecological niche (ĒK-ə-loj-i-kal nitsch)   A set of environmental conditions which each species "occupies" and to which that species is uniquely adapted.

ectocommensal (EK-tō-kō-MEN-səl)   Commensal symbiont that lives on the outer surface of its host.

ectolecithal (EK-tə-LES-ə-thəl)   Condition in which yolk to nourish a developing embryo is contributed by cells separate from a female gamete; found in parasitic and some free-living flatworms.

ectoparasite (EK-tō-PAR-ə-sīt)   Parasite that lives on the outer surface of its host.

ectopic (ek-TOP-ik)   Infection in a location other than normal or expected.

edema (ē-DĒM-ə)   Accumulation of more than normal amounts of tissue fluid, or lymph, in the intercellular spaces, resulting in localized swelling of the area.

ehrlichiosis (err-lik-i-Ō-sis)   A disease caused by bacterial genus *Ehrlichia*, infecting white blood cells and transmitted by ticks.

elephantiasis (el-ə-fan-TĪ-ə-sis)   Permanently swollen body parts, usually limbs and scrotum, resulting from a lengthy filarial nematode infection.

embryophore (EM-brē-ə-for)   In reference to the eggshell of many cestodes, portion contributed by the inner envelope, derived from embryonic blastomeres.

encephalitis (en-cef-ə-LĪ-tis)   Infection of the brain, especially by viruses or amebas.

endemic (en-DEM-ik)   Normally present in a certain geographic area or part of an area.

endemnicity (en-dem-NIS-ə-tē)   Amount or severity of a disease in a particular geographic area.

endite (EN-dīt)   Medial process from the protopod.

endocommensal (EN-dō-kō-MEN-səl)   Commensal symbiont that lives inside its host.

endocytosis (EN-dō-si-TŌ-sis)   Ingestion of particulate matter or fluid by phagocytosis or pinocytosis; that is, bringing material into a cell by invagination of its surface membrane and then pinching off the invaginated portion of a vacuole.

endodyogeny (EN-dō-dī-OJ-ə-nē)   Same as endopolyogeny except that only two daughter cells are formed.

endolecithal (EN-də-LES-ə-thəl)   Condition in which yolk to nourish a developing embryo is deposited within a female gamete; found in animal groups other than some flatworms.

endoparasite   Parasite that lives inside its host.

endopod (endopodite) (END-ō-pod; en-DOP-ə-dit)   Medial branch of a biramous appendage.

endopolyogeny (EN-dō-pol-ē-OJ-ə-nē)   Formation of daughter cells, each surrounded by its own membrane, while still in the mother cell.

endopterygote (EN-dop-TER-ə-gōt)   Condition of internal wing bud development in an insect. Also insect in which the wing buds develop externally or any insect secondarily wingless but derived from such an ancestor; associated with holometabolous insects.

endosome (END-ə-sōm)   Nucleoluslike organelle that does not disappear during mitosis.

endosymbiosis   Symbiotic association in which a symbiont, such as a parasite, lives within the body of its host.

enteroepithelial (IN-ter-ō-ep-ə-THĒL-ē-əl)   Parasites occurring within cells of the intestinal tract, as in *Toxoplasma gondii* infections initiated when a cat ingests zoitocysts containing bradyzoites or oocysts containing sporozoites.

enzyme-linked immunosorbent assay (ELISA) (ē-LIS-ə)   Immunodiagnostic test designed to detect the presence of fixed antibody through linkage with an enzymatic reaction.

eosinophil (ē-ō-SIN-ō-fil)   Type of polymorphonuclear leukocyte very important in many parasitic infections; so called because it stains with acidic stains such as eosin.

eosinophilia (ē-ō-SIN-ō-FIL-ē-ə) Elevated eosinophil count in the circulating blood; commonly associated with chronic parasite infections.

epibiont (epē-BĪ-ont) An organism that lives on the surface of another, typically attached.

epicaridium (ep-ē-kar-ID-ē-əm) First larval stage of isopod suborder Epicaridea; attaches to a free-living copepod.

epicuticle (ep-ē-KŪT-i-kəl) Thin, outermost layer of arthropod cuticle; contains sclerotin but not chitin.

epidemic (ep-i-DEM-ik) Sharp rise in incidence of an infection or disease.

epidemic hemorrhagic fever (him-ō-RAJ-ik) Virus disease transmitted by mosquitoes. Also called **dengue.**

epidemiology (ep-i-DEM-ē-OL-ə-jē) Study concerned with all ecological aspects of a disease to explain its transmission, distribution, prevalence, and incidence.

epididymitis (ep-i-DID-ə-MĪT-əs) Inflammation of the epididymis.

epigenetic (ep-ə-jən-ET-ik) Describes factors other than DNA sequences that influence cellular differentiation.

epipharynx (ep-ē-FAIR-inks) Tube or tongue-like structure originating from the dorsal region of an arthropod's foregut.

epimastigote (ep-ē-MAST-ə-gōt) Trypanosomatid flagellate similar to a promastigote but with a short undulating membrane, such as in *Blastocrithidia.*

epimerite (ē-PIM-ər-īt) Attachment organelle of a gregarine.

epipod (epipodite) (EP-ē-pod; e-PIP-ə-dīt) Lateral process, from the protopod, usually with one or more joints. May be called an **exite.**

epitope (EP-ē-tōp) Antigenic determinant; portion of the antigen molecule displayed on the surface of an antigen-presenting cell (APC).

epizootic (ep-ē-zō-OT-ik) Massive infection rate among animals other than humans; identical to an epidemic in humans.

espundia (es-PÜN-dē-ə) Disease caused by *Leishmania braziliensis.* Also called *chiclero ulcer, uta, pian bois,* or *mucocutaneous leishmaniasis.*

estivoautumnal malaria (EST-ə-vō-ä-TUM-nəl) Malaria caused by *Plasmodium falciparum.*

eutely (Ū-te-lē) Cell or nuclear constancy; the adult has the same number of nuclei or cells as the first-stage juvenile. Eutely may exist in tissues, organs, or entire animals.

excretory/secretory (EX-krə-tō-rē-SĒ-krə-tō-rē) Adjective describing macromolecules such as antigenic proteins or glycoproteins released from a parasite.

exflagellation (EX-flaj-el-Ā-shən) Rapid formation of microgametes from a microgametocyte of *Plasmodium* and related genera.

exite (EX-īt) Lateral process or joint from the protopod, sometimes referred to as an **epipod.**

exoerythrocytic schizogony (EKS-ə-ē-ri-thrō-SIT-ək skiz-ÄG-ə- nē) Schizogony in *Plasmodium* infections that takes place in liver cells, before erythrocytic schizogony begins.

exopod (exopodite) (EX-ō-pod; ex-OP-ə-dīt) Lateral branch of a biramous appendage.

exopterygote (ex-op-TER-ə-gōt) Condition of external wing bud development in an insect. Also, any insect in which the wing buds develop externally; associated with hemimetabolous insects.

expression site Position on a chromosome where a gene, such as that for an antigen, is expressed.

extraintestinal (ex-trə-in-TEST-in-əl) Occurring in tissues outside the digestive tract.

extrusomes (EX-trü-sōmz) Protozoan organelles that, upon proper stimulus, fuse with the cell membrane and release their contents to the exterior.

F

facette (fa-SET) Funnel-shaped opening through the inner membrane complex of the egg of a pentastomid. It receives the product of the dorsal organ.

facultative symbiont Opportunistic symbiont, establishing a relationship with a host only if an opportunity presents itself but not physiologically dependent on doing so.

falx (fälx) Sickle-shaped field of kinetosomes at the anterior end of an opalinid flagellate.

fascicle (FAS-i-kəl) Stylet bundle or combination of mouthparts used to pierce the skin in a blood-feeding arthropod. Composition of a fascicle varies according to group.

favism (FĀV-əsm) Condition marked by hemolytic anemia, jaundice, and fever upon exposure to fava bean (broad bean) or its pollen, found especially in males of Mediterranean descent.

Fc, Fab Parts of antibody molecules. See **constant region** and **variable region.**

femur (FĒ-mər) Podomere of an insect or acarine leg fixed to the trochanter proximally and articulating with the tibia distally in insects and with the patella in acarines.

ferredoxin (fair-ə-DOX-ən) An iron-sulfur protein that acts as electron acceptor in metabolic reactions.

festoons (fes-TOONS) Sclerites on the posterior margin of the opisthosoma of certain hard ticks.

fever An increase in body temperature above normal.

fibrosis Process whereby fibrous connective tissue is deposited in an area; formation of scar tissue.

filopodia (fil-ə-PŌD-e-ə) Slender, sharp-pointed pseudopodia composed only of ectoplasm.

flabellum (fla-BEL-əm) Recurved process often found on the first two thoracic exopods of branchiuran crustaceans.

flagellar pocket (fla-JEL-ər) Depression, sometimes long and deep, from which a flagellum arises.

flame bulb (or cell) Specialized hollow excretory or osmoregulatory structure of one or several small cells containing a tuft of flagella (the "flame") and situated at the end of a minute tubule; connected tubules ultimately open to the outside. See **protonephridium.**

flare Redness caused by dilation of blood vessels.

food web (FŪD web) The more or less established feeding relationships between producers, predators, and prey in an ecosystem.

G

gametocyst (gə-MĒT-ō-sist) Cyst produced by some apicomplexan parasites. Sexual reproduction and spore formation occur within this cyst.

gametogony (ga-mə-TOG-ə-nē) Process by which gametes are produced in protozoa, especially during Apicomplexa life cycles.

gamont (GA-mont) Apicomplexan life-cycle stage that is committed to undergoing gametogenesis.

gena (JE-nə) Anterioventral portion of an insect head. For example, genal ctenidium is a row of heavy spines on the gena of a flea.

genital atrium (JEN-ə-təl ĀT-rē-əm) Cavity in the body wall of a flatworm into which male and female genital ducts open.

genital primordia Sex organs at their earliest stage of recognizable differentiation.

genitointestinal canal (JEN-ə-tō-in-TES-tin-əl) Duct connecting the oviduct and intestine of some polyopisthocotylean Monogenea.

bat/āpe/ärmadillo/herring/fēmale/finch/līce/crocodile/crōw/ duck/ūnicorn/tüna/ə "uh" as in mammal, fishes, cardinal, heron, vulture/stress as in bi-OL-o-gy, bi-o-LOG-i-cal

geophagia (geo-FĀ-j-ē-ə) Consumption of earth, clay, or chalk; a behavior that occurs with certain nutritional deficiencies.

germarium (jer-MAR-ē-əm) Fused mass of ova and vitelline cells, found in *Gyrodactylus* species. Also called **ovovitellarium.**

gid (GID) Disorientation caused by cysticerci in the brain; usually manifested by staggering or whirling.

glial cells (GLĒ-əl) Supporting, nonneuronal cells in the nervous system.

glossa (GLOS-ə) Tonguelike mouthpart in Hymenoptera (considered a hypopharynx by some authors).

glycocalyx (GLĪ-kō-KĀL-ix) Finely filamentous layer containing carbohydrate, found on the outer surface of many cells, from 7.5 nm to 200.0 nm thick.

glycosomes (GLĪ-kō-sōmz) Organelles found in *Trypanosoma* that contain enzymes of glycolysis and for oxidizing reduced NAD.

glyoxylate cycle (glī-OX-ə-lāt) Metabolic pathway that functions to convert fatty acids or acetate to carbohydrate.

gnathopod (NATH-ə-pod) Prehensile appendages of some Crustacea, such as the second and third thoracic legs of Amphipoda and the first thoracic legs of some Isopoda.

gnathosoma (nath-ə-SŌM-ə) Anterior of two basic regions of the body of a mite or tick. Also called **capitulum.**

gnathostome (NATH-ə-stōm) Vertebrates with jaws. Also, in copepods, gnathostomous mandibles are fairly short, broad, biting structures with teeth at their ends, and buccal cavities are large and widely open.

goblets (GOB-lets) Markings on stigmatal plates of certain hard ticks.

gonotyl (GŌN-ō-tīl) Muscular sucker or other perigenital specialization surrounding or associated with the genital atrium of a digenetic trematode.

granuloma, granulomatous tissue (gran-ū-LŌM-ə; gran-ū- LŌM-ə-təs) Repaired area of a body marked by fibrous connective tissue (fibrosis). Also fibrous connective tissue surrounding an antigen source.

gravid Condition of uterus with eggs or developing embryo.

ground itch Skin rash caused by bacteria introduced by invasive hookworm larvae.

gubernaculum (GÜ-bər-NAC-ū-ləm) Sclerotization in the cloacal wall of many nematodes that guides exsertion of the spicules.

Guinea worm *Dracunculus medinensis.*

gynandry (JEN-an-drē) In a hermaphroditic organism, maturation first of the female gonads and then of the male organs. Also called *protogyny.*

gynecophoral canal (gin-ə-KOF-ōr-əl) Longitudinal groove in the ventral surface of a male schistosome fluke.

H

Haller's organ (HAL-erz) Depression on the first tarsi of ticks; functions as an olfactory and humidity receptor.

haltere (HAL-tər) Vestigial wing on the metathorax of a fly of the order Diptera; necessary for balance during flight.

halzoun (hal-ZÜN) Disease resulting from blockage of the nasopharynx by a parasite. Also called **marrara.**

hamuli (HAM-ū-lī) Large hooks on the opisthaptor of a monogenetic trematode; referred to as *anchors* by some American authors.

haplodiploidy (HAP-lō-DIP-loi-dē) Reproduction in which males are haploid (parthenogenetically produced) and females are diploid (from a fertilized egg).

haptens (HAP-tenz) Molecules of small molecular weight (usually) that are immunogenic only when attached to carrier molecules, usually proteins.

haptocyst (HAP-tə-sist) One type of extrusome consisting of several separate parts in its nonextruded state and possibly containing a poisonous substance.

heat shock factor (HSF) Molecules that induce production of heat shock proteins, expressed in response to stress, such as increase in ambient temperature.

heat shock proteins (HSPs) Found in all living cells. Many are molecular chaperones, mediating protein folding, transport across membranes, assembly, and degradation. Upregulated to stimulate morphological and metabolic differentiation of some (perhaps all) parasites in response to changes in environmental conditions, such as from free-living to warm-blooded hosts.

hematuria Blood in the urine.

hemelytra (HIM-ē-LIT-rə) Front wing of an insect of order Hemiptera.

hemidesmosome (him-e-DEZ-mə-sōm) Desmosome anchoring a cell to a basement lamina rather than to another cell.

hemimetabolous metamorphosis (HIM-ē-mə-TAB-ō-ləs met-ə-MORF-ə-sis) In insects gradual metamorphosis in which nymphs are generally similar in body form to adults and become more like adults with each instar.

hemocoel (HĒM-ə-sēl) Main body cavity of arthropods, the embryonic development of which differs from that of a true coelom but that includes a vestige of a true coelom.

hemoflagellate (hem-ə-FLAJ-ə-lāt) A protistan parasite of family Trypanosomatidae that infects the blood and/or blood-forming organs.

hemoglobinuria (HĒM-ə-glōb-in-ŪR-ē-ə) Bloody urine.

hemolymph (HĒM-ə-limf) Fluid within the hemocoel of arthropods. Also pseudocoelomic fluid of nematodes.

hemozoin (HĒM-ə-ZŌ-ən) Insoluble product of hemoglobin degradation by malarial parasites.

hepatosplenomegaly (hē-PAT-ō-SPLEN-ō-meg-ə-lē) Swollen liver and spleen.

hermaphroditism (her-MAF-rM-di-tizm) Possession of gonads of both sexes by a single (monoecious) individual.

heterogonic life cycle (het-ər-ə-GŌN-ik) Life cycle involving alternation of parasitic and free-living generations.

heterokont (HET-er-ə-kont) Condition in which a flagellated protozoan has at least two flagella, with differing structures.

heterotrophic (het-er-ō-TRŌ-fik) Requiring both carbon and energy in the form of complex organic molecules.

heteroxenous (het-ər-ə-ZĒN-əs) Describes a parasite that lives within more than one host during its life cycle.

hexacanth (HEX-ə-kanth) Oncosphere; six-hooked larva hatching from the egg of a eucestode.

histomoniasis (HIS-tə-mōn-Ī-ə-səs) Poultry disease caused by infection with flagellates of genus *Histomonas*.

histozoic (HIS-tə-ZŌ-ik) Dwelling within the tissues of a host.

holoblastic cleavage (HŌ-lō-BLAS-tik) Each nuclear division in an early embryo that is accompanied or closely followed by complete cytokinesis, the nuclei being separated by cell membranes.

hologonic (HŌ-lō-gon-ək) Female gonads that produce gametes at any location in their structure.

holometabolous metamorphosis (HŌ-lō-mə-TAB-ə-ləs met-ə-MORF-ə-sis) Metamorphosis in an insect with a larva, pupa, and adult.

holophytic nutrition (HŌ-lō-FIT-ik) Formation of carbohydrates by chloroplasts.

holostome Trematodes with body form as in Strigeidae, with their body divided by a constriction into an anterior cup- or spoon-shaped portion and an ovoid posterior portion.

holozoic nutrition (HŌ-lō-ZŌ-ik)   Feeding by active ingestion of organisms or particles.

homeobox gene (HŌM-ē-ō-box jēn)   Gene containing a 180-base pair DNA segment (the homeobox) that specifies the binding sequences.

homeotic gene (HŌM-ē-ōt-ik)   Genes that code for transcription factors and specify major structural features such as segmentation and anterior-posterior axes.

homogonic life cycle (HŌ-mō-GŌN-ik)   Life cycle in which all generations are parasitic or all are free living. There is no (or little) alternation of the two.

homologous (hō-MOL-ə-gus)   Term used to describe structures occurring in different taxa but having a common genetic and evolutionary origin.

homothetogenic fission (HŌ-mō-thet-ə-JEN-ik FISH-shən)   Mitotic fission across the rows of cilia of a protozoan.

hood   Dorsal wall of the camerostome that extends over the capitulum.

host specificity   Degree to which a parasite is able to mature in more than one host species.

host switching   Colonization—in an evolutionary sense—of a new species of host by a parasite, in which the parasite is able to use that host to complete part of its life cycle (= host capture).

humoral immune response (HŪM-er-əl)   Binding of antigen with soluble antibody in blood serum. Also the entire process by which the body responds to an antigen by producing antibody to that antigen.

hydatid cyst (hī-DAT-id sist)   Metacestode of the cyclophyllidean cestode genus *Echinococcus,* with many protoscolices, some budding inside secondary brood cysts.

hydrocele (HĪ-drə-sēl)   Accumulation of fluid in any saclike cavity or duct, especially the tunica vaginalis of the testis or spermatic cord.

hydrogenosomes (hī-drə-JEN-ə-sōmz)   Small organelles in certain anaerobic protozoa that produce molecular hydrogen as an end product of energy metabolism.

hydrostatic skeleton   Skeleton in which a noncompressible volume of fluid serves as support.

hyperapolysis (HĪ-pər-ap-ō-LĪ-sis or HĪ-per-ə-POL-ə-sis)   Detachment of a tapeworm proglottid while still immature, before eggs are formed.

hyperendemic (HĪ-pər-en-DEM-ik)   Condition in which a disease or infection has high, usually seasonal transmission in a certain geographic area.

hyperinfection (HĪ-pər-in-FEK-shən)   Condition in *Strongyloides* infections in which filariform juveniles repenetrate mucosa of the small intestine and proceed with migration.

hypermetamorphosis (HĪ-pər-met-a-MORF-ə-sis)   Type of metamorphic development in which different larval instars have markedly dissimilar body forms.

hyperparasitism (HĪ-pər-PAR-ə-sit-izm)   Condition in which an organism is a parasite of another parasite.

hypnozoite (HIP-nō-ZŌ-īt)   Dormant exoerythrocytic form found in certain *Plasmodium* species.

hypobiosis (HĪ-pō-bī-Ō-səs)   See **developmental arrest.**

hypodermis (HĪ-pō-DER-mis)   Syncytial layer that secretes the cuticle in nematodes.

hypopharynx (HĪ-pō-FAR-inx)   Tonguelike lobe arising from the floor of the mouth in insects; variously modified for feeding in many groups.

hypostome (HĪ-pō-stōm)   Portion of the mouthparts of acarines; composed of fused coxae of pedipalps.

hysterosoma (HIST-er-ō-SŌM-ə)   Combination of metapodosoma and opisthosoma of the body of a tick or mite.

I

ick (ik)   Serious disease of freshwater fishes, caused by a ciliate protozoan *Ichthyophthirius multifiliis.*

icterus (jaundice) (IK-tər-əs; JĂN-dis)   Yellowing of the skin and other organs because of bile pigments in the blood.

idiosoma (ID-ē-ō-SŌM-ə)   Posterior of the two basic parts of the body of a mite or tick, bearing the legs and most internal organs.

imago (i-MAG-ō)   Adult or final instar in the development of an insect.

imagochrysalis (i-MAG-ō-KRIS-ə-lis)   Quiescent stage between nymph and adult in the life cycle of a chigger mite.

immediate hypersensitivity (hī-per-sen-sə-TIV-a-tē)   Biological manifestation of an antigen-antibody reaction in which the maximal response is reached in a few minutes or hours. Intradermal injection of the antigen produces local swelling and redness with heavy infiltration of polymorphonuclear leukocytes. Intravenous injection may produce anaphylactic shock and death.

immune cross reaction (im-MŪN)   Binding of an antibody or cell receptor site with an antigen other than the one that would provide an exact "fit"; that is, an antigen-antibody reaction in which the antigen is not the same one that stimulated the production of that antibody.

immunity (im-MŪN-ə-tē)   State in which a host is more or less resistant to an infective agent; preferably used in reference to resistance arising from tissues that are capable of recognizing and protecting the animal against "nonself."

immunogenic (IM-ū-no-JEN-ik)   Refers to any substance that is antigenic; that is, that stimulates production of antibody or cell-mediated immunity.

immunoglobulin (IM-ū-nō-GLOB-ū-lin)   Any one of five classes of proteins in blood serum that function as antibodies; abbreviated IgM, IgG, IgA, IgD, and IgE.

incidence (IN-sə-dens)   In epidemiology, number of new cases of a disease per unit time; that is, a rate measurement. Contrast with **prevalence.**

incidental parasite (in-se-DEN-tal)   Accidental parasite.

indirect development   In arthropods, refers to development in which larva or nymph hatches from an egg and is distinctly different in body form from the adult; that is, development with metamorphosis.

indirect fluorescent antibody test (IFA)   Technique to localize antigen in cells or tissues by binding antibody molecules with fluorescent substances and then combining them with a sample and viewing areas of fluorescence with a microscope.

indirect hemaglutination test (IHA)   Immunodiagnostic test in which red blood cells are coated with a specific antigen. In presence of antibody to that antigen, the red cells stick together, or agglutinate.

inflammation (in-flə-MĀ-shən)   Defense process of body including congestion of blood vessels, escape of plasma to interstitial tissue space, swelling, and warmth.

infraciliature (IN-frə-SIL-ē-ə-tūr)   All cilia, basal bodies, and their associated fibrils in a ciliate protozoan.

infracommunity   All parasites of all species living in a single host.

infrapopulation (IN-frə-POP-ū-lā-shən)   All individuals of a single parasite species in one host.

infusoriform larva (IN-fū-SŌR-ə-form)   Ciliated larva produced by an infusorigen within a dicyemid mesozoan.

infusorigen (IN-fū-SŌR-ə-jen)   Mass of reproductive cells within a rhombogen.

ingroup (IN-grüp)   Taxon being studied in a cladistic analysis of evolutionary history.

bat/āpe/ärmadillo/herring/fēmale/finch/līce/crocodile/crōw/ duck/ūnicorn/tüna/ə "uh" as in mammal, fishes, cardinal, heron, vulture/stress as in bi-OL-o-gy, bi-o-LOG-i-cal

innate immunity   An immune response that does not depend on prior exposure to the invader, present to some extent in all animals and at least some plants.

inner nuclear mass   Dense accumulation of nuclei in acanthocephalan embryos that gives rise to all internal organ systems of the worm.

instar (IN-star)   Molt stage in the life of an arthropod.

integrated pest management   A system of managing insect pests that relies on ecological knowledge, minimal pesticide use, and agricultural practices designed to reduce insect problems.

intensity (in-TEN-sə-tē)   Number of parasites of one species in an infected host (**infrapopulation**). Mean intensity is the average number of parasites per infected host.

interferons   Certain small proteins produced by vertebrate cells in response to viral infection or as cytokines in the immune response. Interferon-γ is a cytokine secreted by activated T cells. Important lymphocyte growth factors.

interleukin (IN-ter-LÜ-kin)   Cytokines produced by white blood cells and mediating their own activities or those of other white blood cells.

intermediate host (IN-ter-MĒD-ē-ət)   Host in which a parasite develops to some extent but not to sexual maturity.

intermittent parasite (IN-ter-MIT-tent)   Temporary parasite.

internal transcribed spacer   Nonfunctional RNA located between ribosomal RNAs on a single strand of transcribed RNA. Abbreviated as ITS, these spacers are removed during post-transcriptional editing, but they are also highly variable at low taxonomic levels thus useful to those studying phylogeny.

internuncial processes (in-ter-NUN-sē-əl)   Cytoplasmic channels that connect one part of a cell to another, such as those linking distal cytoplasm to tegumental cytons in many flatworms.

intralecithal cleavage (IN-tra-LES-ə-thal CLĒ-vaj)   Cleavage in which nuclei undergo several divisions within the yolk mass without concurrent cytokinesis; common in arthropods.

iodinophilous vacuole (ī-ō-din-OF-ə-lus VAK-ū-ōl)   Vacuole within a protozoan that stains readily with iodine.

isogametes (Ī-sō-GAM-ēts)   Outwardly similar male and female gametes.

isozymes (Ī-sə-zīmz)   Enzymes that catalyze the same reaction but are encoded by genes that are not at the same locus; that is, are not alleles.

ITS   See **internal transcribed spacer.**

J

jacket cells (JAK-et)   Ciliated somatoderm of an orthonectid mesozoan.

K

kairomone (KĪR-ə-mōn)   Chemical, produced by host, that attracts parasitoids.

kala-azar (ka-lə-Ā-zar)   Disease caused by *Leishmania donovani*. Also called *Dum-Dum fever* or *visceral leishmaniasis*. Term is Hindi for "black sickness."

karyomastigont (ker-ē-ō-MAS-ti-gont)   Combination of protozoan flagellum or flagella and cytoskeletal elements such as microtubules, all in association with the nucleus.

Katayama fever (kä-tä-YÄ-mä)   Acute schistosomiasis, especially schistosomiasis japonicum.

kentrogon (KEN-trō-gən)   Larva in crustacean order Kentrogonida that is attached to its host crab; formed after the cypris larva molts and its appendages and carapace are discarded.

keratitis (KER-ə-TĪ-təs)   Corneal inflammation of the eye.

kinete (KĪ-nēt or kə-NET)   In piroplasms, stage that develops from zygote and moves into tick cells (primary kinete) and stage that develops from cytomeres (secondary kinete) and moves into salivary gland cells to become a sporozoite.

kinetid (kī-NET-id)   Axoneme of a cilium of flagellum together with its basal fibrils and organelles. Also called **mastigont.**

kinetocyst (kī-NĒT-ə-sist)   One type of extrusome characterized by having a central core and an outer jacket in its nonextruded state.

kinetodesmose (kinetodesmata) (kī-net-ō-DEZ-mōs; kī-net-ō-des-MÄ-tə)   Compound fiber joining cilia into rows.

kinetoplast (kī-NĒT-ō-plast)   Conspicuous part of a mitochondrion in a trypanosome; usually found near the kinetosome.

kinetosome (kī-NĒT-ō-sōm)   Centriole from which an axoneme arises. Also called **basal body** or *blepharoplast*.

kinety (kī-NĒT-tē)   Row of cilia basal bodies and their kinetodesmose. All kineties and kinetodesmata in the organism are its infraciliature.

Koch's blue bodies (KŌKS)   Schizonts of *Theileria parva* in circulating lymphocytes.

K-strategist   Species of organism that uses a survival and reproductive "strategy" characterized by low fecundity, low mortality, and longer life and with populations approaching the carrying capacity of the environment, controlled by density-dependent factors.

Kupffer cells (KÜP-fer)   Phagocytic epithelial cells lining sinusoids of the liver.

L

labellum (la-BEL-əm)   Expanded tip of the labium in an insect.

labium (LĀB-ē-əm)   Mouthpart in insects composed of fused second maxillae; homologous to second maxillae of crustaceans.

labrum (LĀ-brəm)   Sclerite forming the anterior closure of the mouth in arthropods; specifically the free lobe overhanging the mouth.

lachryphagous (lak-rə-FĀG-əs)   Feeding on secretions from lachrymal glands.

lacunae (la-KÜ-nē)   Channels making up the lacunar system in Acanthocephala. Also, in developing wings of insects, canals that contain nerves, tracheae, and hemolymph.

lacunar system (la-KÜ-nər)   System of canals in the body wall of an acanthocephalan, functioning as a circulatory system.

landscape epidemiology   Approach to epidemiology that employs all ecological aspects of a nidus. By recognizing certain physical conditions, the epidemiologist can anticipate whether a disease can be expected to exist.

larva (LAR-və)   Progeny of any animal that is markedly different in body form from the adult.

late phase reaction   Phase of immediate hypersensitivity (IgE-mediated) in which eosinophils infiltrate an area of inflammation to kill parasites.

Laurer's canal (LÄ-rərz)   Usually blind canal extending from the base of the seminal receptacle of a digenetic trematode. It probably represents a vestigial vagina.

Leishman-Donovan (L-D) body (LĪSH-mən DON-ə-vən)   Amastigote in Trypanosomatidae.

leishmaniasis (LĪSH-mən-Ī-ə-sis)   Infection by a species of *Leishmania*.

lemniscus (lem-NIS-kəs)   Structure occurring in pairs attached to the inner, posterior margin of the neck of an acanthocephalan, extending into the trunk cavity. Its function is unknown.

leptotriches (LEP-tə-tri-chəz)   Tiny, slender cellular extensions, such as found in the interlocking cell processes, forming a weir in a flame bulb protonephridium.

lesion (LĒ-zhən) Abnormal condition in some tissue or organ, usually of a well-defined form and delimited in some way.

leukorrhea (lükō-RĒ-ə) White, puslike discharge resulting from infection.

lichenification (lī-KEN-i-fə-KĀ-shən) Pathological changes in which human skin becomes thickened and hard, caused by repeated scratching of an inflammatory lesion.

life cycle Ontogenetic history of an organism. Set of events, including growth and reproduction, that must occur before an organism can survive and reproduce. In the case of parasites, the life cycle also includes requisite hosts.

ligand (LI-gənd) A molecule that binds at a receptor site of another molecule, combining to form a biomolecule with a physiological function.

linguatulosis (lin-GWAT-u-lō-səs) Infection by a pentastome in the nasopharyngeal region. See **halzoun.**

liposome (LIP-ə-sōm) Artificial lipoid particle used to deliver antiparasitic drugs directly to macrophages (which eat the particles).

lobopodia (lō-bō-PŌD-ē-ə) Finger-shaped, round-tipped pseudopodia that usually contain both ectoplasm and endoplasm.

loculi (LOK-ū-lī) Shallow, suckerlike depressions in an adhesive organ of a flatworm.

lumen (LÜ-mən) Space within any hollow organ.

lunules (LÜN-ūlz) Small, suckerlike discs on the anterior margin of some copepods in family Caligidae, functioning as organs of adhesion.

lycophora (lī-KOF-ōr-ə or lī-kə-FOR-ə) Ten-hooked larva that hatches from the egg of a cestodarian tapeworm. Also called a **decacanth.**

lymph varices (limf VER-ə-sēz) Dilated lymph ducts.

lymphadenitis (limf-FAD-ən-ī-tis) Inflamed lymph node.

lymphedema (lĭm-fə-DĒ-mə) Swelling of tissues due to blockage of lymph vessels with resulting accumulation of lymphatic fluid; swelling often occurs in the arm or leg.

lymphocyte (LIMF-ō-sīt) Type of leukocyte vital in immune response. Several different types are known. See **B cell** and **T cell.**

lymphokine (LIM-fə-kīn) Cytokine released by a lymphocyte.

lymphokine-activated killer (LAK) cells Natural killer (NK) lymphocytes that are stimulated by lymphokines (especially IL-2 and IL-12) to lyse target cells.

lysosome (LI-sə-sMm) Intracellular vacuole or vesicle containing digestive enzymes (lysozymes).

lysozyme (LĪ-sə-zīm) Enzyme widespread in animal secretions that splits the glycosidic bond in mucopolysaccharides and mucopeptides in bacterial cell walls.

M

macroepidemiology (MACK-rō-ep-ə-dēm-ē-OL-ə-jē) Study of the effects of large scale factors, such as climate and culture, on distribution of disease in a population.

macrogamete (MA-krō-GA-mēt) Large, quiescent, "female" anisogamete.

macrogametocyte (MA-krō-gə-MĒT-ə-cīt) Cell giving rise to a macrogamete.

macrogamont (MA-krō-gə-MONT) Synonym of *macrogametocyte.*

macromolecule (MAK-rō-MOL-ə-kūl) Large molecules, typically polymers such as polypeptides and nucleic acids.

macroparasite (MA-krə-PAR-ə-sīt) Large parasite that does not multiply in the host of interest. Examples are cestodes, trematodes, and most nematodes in their definitive hosts.

macrophage (MA-krə-fāj) Important phagocytic cell and antigen-presenting cell, derived from monocyte.

macrophage migration inhibitory factor (MIF) Cytokine released by sensitized lymphocytes that tends to inhibit migration of macrophages in the immediate vicinity, thus contributing to accumulation of larger numbers of macrophages close to the site of MIF release.

magainins (mə-GĀ-nənz) Family of peptides from frog skin that kill bacteria, protozoa, and fungi.

maggots (MA-gəts) Larval diptera with vertically biting mandibles, with an elongated, simple body, usually tapered toward the head end, and without a true head.

major histocompatibility complex (MHC) (HIS-tə-com-PAT-ə-BIL-ə-tē) Cluster of genes encoding MHC proteins. Proteins they encode are highly variable between individuals, are carried in cell surfaces, and function in recognition of foreign cells by cells of the immune system.

major sperm protein (MSP) Sperm-specific protein that mediates locomotion of nematode sperm.

mal de caderas (mal-də-ka-DER-əs) South American disease in horses similar to surra and caused by *Trypanosoma equinum.*

malignant tertian malaria Malaria caused by *Plasmodium falciparum.*

Malpighian tubule (mal-PIG-ē-ən) Blind tubules opening into the hindgut of nearly all insects and some myriapods and arachnids and functioning primarily as excretory organs.

mamelon (MA-mə-lon) Ventral, serrated projection on the ventral surface of a male nematode of the genus *Syphacia.* Its function is unknown.

mandibles Third pair of appendages from the anterior in Crustacea; second pair in Insecta; they primarily function in feeding; derived from appendages on primitive fourth (first postoral) somite.

mange (mānj) Dermatitis caused by species of mites, often designated with the causative organism. For example, *Sarcoptes* causes sarcoptic mange.

marginal bodies Sensory pits or short tentacles between the marginal loculi of the opisthaptor of an aspidogastrean trematode.

marrara (mə-RA-rə) Nasopharyngeal blockage by a parasite. Also called **halzoun.**

mast cell Type of cell in various tissues that releases pharmacologically active substances with a role in inflammation.

mastigont (MAS-tə-gont) Axoneme of a cilium or flagellum together with its basal fibrils and organelles.

mastitis (mas-TĪ-təs) Infection of the udder of cattle.

mathematical models Sets of equations, or algorithms, intended to mimic natural processes.

Maurer's clefts (MÄR-rərz) Blotches on the surface of an erythrocyte infected with *Plasmodium falciparum.*

maxadilan (max-ə-DĪ-lən) A protein in sand fly saliva that functions to dilate blood vessels of a host.

maxillae (second maxillae) (MAX-ə-lē or max-IL-ē) Fifth pair of appendages in Crustacea, primarily feeding in function, derived from appendages on primitive sixth (third postoral) somite; homologous to **labium** in insects. The maxillae of insects are the third pair of head appendages, homologous to **maxillules** of Crustacea.

maxillipeds (max-IL-ə-pedz) One or more pairs of head appendages originating posterior to maxillae in Crustacea; derived from appendages on somites that were primitively posterior to gnathocephalon; usually function in feeding but sometimes adapted for other functions, such as prehension, in parasitic forms.

maxillopodan eye (max-ə-LOP-ə-dən) Naupliar eye of crustacean class Maxillopoda; has a *tapetum* (crystalline reflective layer).

---

bat/āpe/ärmadillo/herring/fēmale/finch/līce/crocodile/crōw/
duck/ūnicorn/tüna/ə "uh" as in mammal, fishes, cardinal, heron,
vulture/stress as in bi-OL-o-gy, bi-o-LOG-i-cal

maxillules (first maxillae) (MAX-ə-lulz) Fourth pair of appendages in Crustacea, primarily feeding in function; derived from appendages on primitive fifth (second postoral) somite; homologous to **maxillae** in insects.

mean intensity Average number of parasites per *infected* host in a sample.

mechanical transmission Transmission of a parasite, by a vector, but without necessary development of the parasite.

mechanical vector Vector that transmits disease organisms by mechanical means only. Contrast with **biological vector.**

median body (MĒ-dē-ən) Darkly staining structures found in *Giardia* species; of unknown function.

megacolon Flabby distended colon caused by chronic Chagas' disease.

megaesophagus Distended esophagus caused by chronic Chagas' disease.

Mehlis' glands (MĀ-ləs) Unicellular mucous and serous glands surrounding the ootype of a flatworm.

membranelle (mem-brən-EL) Short, transverse rows of cilia, fused at their bases, serving to move food particles toward the oral groove of a protozoan.

membranocalyx (məm-BRĀN-ə-KĀL-əks) Outer covering of schistosomules formed from fusion of laminae from vesicles in tegument, replaces original glycocalyx.

memory cells Long-lived lymphocytes with specific antibodies or receptors on their surface.

merogony (mer-OG-ə-nē) Multiple fission to produce merozoites; schizogony.

meront (ME-rənt) Asexual stage in the life cycle of some protozoa that undergoes merogony (schizogony) to form merozoites.

merozoite (mer-ə-ZŌ-īt) Daughter cell resulting from schizogony.

mesocercaria (mez-ə-ser-KAR-ē-ə) Juvenile stage of the digenetic trematode *Alaria*. It is an unencysted form between the cercaria and the metacercaria.

mesostomate Describes a trematode cercaria in which the common collecting ducts of the excretory system extend to the region of the midbody, there to fuse with the excretory bladder.

metacercaria (met-ə-ser-KAR-ē-ə) Stage between cercaria and adult in the life cycle of most digenetic trematodes; usually encysted and quiescent.

metacestode (met-ə-SES-tōd) Developmental stage of a cestode after metamorphosis of an oncosphere; juvenile cestode.

metacryptozoite (met-ə-krip-tə-ZŌ-īt) Merozoite developed from a cryptozoite.

metacyclic Stage in the life cycle of a parasite that is infective to its definitive host.

metacyst (MET-ə-sist) Cystic stage of a parasite that is infective to a host.

metacystic trophozoites (MET-ə-sis-tik trō-fō-ZŌ-itz) Small amebas that emerge from cysts.

metamere (MET-ə-mer) One of the segments in a metameric animal.

metamerism (met-AM-ər-izm) Division of the body along the anteroposterior axis into a serial succession of segments, each of which contains identical or similar representatives of all the organ systems of the body; primitively in arthropods, including externally a pair of appendages and internally a pair of nerve ganglia, a pair of nephridia, a pair of gonads, paired blood vessels and nerves, and a portion of the digestive and muscular systems.

metamorphosis (met-ə-MORF-ə-sis) Type of development in which one or more juvenile types differ markedly in body form from the adult; occurs in numerous animal phyla. Also applies to the actual process of changing from larval to adult form.

metanauplius (met-ə-NĀ-plē-əs) Later naupliar larvae of some crustaceans; that is, occurring after several naupliar stages but before another larval type or preadult in the developmental sequence.

metanemes (ME-tə-nēmz) Proprioceptors on epidermal cords of some nematodes.

metapodosoma (MET-ə-PŌD-ə-sō-mə) Portion of the podosoma that bears the third and fourth pairs of legs of a tick or mite.

metapolar cells Posterior tier of cells in the calotte of a dicyemid mesozoan.

metapopulation (MET-ə-pop-ū-LĀ-shən) All infrapopulations of a parasite species within a single host species in an ecosystem.

metasome (MET-ə-sōm) Portion of the body anterior to the major point of body flexion in many copepods; usually includes the cephalothorax and several free thoracic segments.

metraterm (MET-rə-tərm) Muscular, distended termination of the uterus of a digenetic trematode.

metrocytes (MET-ro-sīts) Cells accumulating inside *Sarcocystis* species' tissue cyst wall and eventually giving rise to infective bradyzoites.

microbivore (MĪ-krōbə-vôr) An organism that feeds on microorganisms; typically a small animal that specializes in feeding on bacteria, fungi, or protozoa.

microbodies (MĪ-krō-bod-ez) Spherical intracellular structures usually containing enzymes.

microenvironments (MĪ-krō-en-VĪ-ron-mənts) Discrete sets of ecological conditions occurring on a small or very small scale.

microepidemiology (MĪK-rō-ep-ə-dēm-ē-OL-ə-jē) Study of the effects of small scale factors, such as parasite strains, individual host responses, on distribution of disease in a population.

microfilaria (MĪK-rə-fi-LAR-ē-ə) First-stage juvenile of any filariid nematode that is ovoviviparous; usually found in the blood or tissue fluids of the definitive host.

microgamete (MĪK-rə-GAM-ēt) Slender, active "male" anisogamete.

microgametocyte (MĪK-rə-gam-ĒT-ə-sīt) Cell that gives rise to microgametes.

microgamont (MĪ-krō-gə-MONT) Synonym of *microgametocyte.*

microglial cells (mī-KRŌ-glē-əl) Macrophages in the central nervous system.

micronemes (MĪK-rə-nēmz) Slender, convoluted bodies that join a duct system with the rhoptries, opening at the tip of a sporozoite or merozoite.

microniscus (MĪK-rə-NIS-kəs) Intermediate larval stages of isopod suborder Epicaridea, parasitic on free-living copepods.

microparasite (MĪK-rə-par-ə-sīt) Small (or very small) parasite that multiplies within the host of interest. Examples are protistan and prokaryotic parasites.

micropore (MĪK-rō-pōr) Opening on the side of a sporozoite, functioning in food uptake.

micropredator Temporary parasite.

micropyle (MĪK-rə-pīl) Pore in the oocyst of some coccidia and in the egg of an insect.

microthrix (microtriches) (MĪK-rə-thrix; MĪK-rə- trich-ēz) Minute projections of the tegument of a cestode.

Miescher's tubules (MĒSH-ərz) Sarcocysts; tissue cysts of *Sarcocystis* spp.

mild tertian malaria Malaria caused by *Plasmodium vivax*.

miracidium (mir-ə-SID-ē-əm) First larval stage of a digenetic trematode; ciliated and often free swimming.

monoclonal antibody (MON-ə-KLŌN-əl AN-tē-BOD-ē) Antibodies made by a clone of cells and specific to a particular antigen.

monocyte (MON-ə-sīt) Phagocytic leukocyte with an oval or *U*-shaped nucleus. It differentiates into macrophages in various tissues, Kupffer cells in liver, and microglial cells in the central nervous system.

monoecious (mən-Ē-shəs) Hermaphroditic; individual that contains reproductive systems of both sexes.

mononuclear phagocyte system See **reticuloendothelial system.**

monophyletic (mon-ō-fī-LET-ik) Adjective describing a group of taxa that includes a hypothetical ancestral taxon and all its descendants.

monostome (MON-ə-stōm) Fluke that lacks a ventral sucker.

monoxenous (MON-ə-ZĒN-əs *or* MON-ox-ĒN-əs) Living within a single host during a parasite's life cycle.

monozoic (MON-ə-ZŌ-ik) Tapeworm whose "strobila" consists of a single proglottid.

morphogen (MŌRF-o-jen) A chemical substance that influences cellular differentiation and development of tissues.

mucocutaneous (MŪ-kō-kū-TĀN-ē-əs) Involving mucous membranes of the nasopharyngeal region, as in ulcerous lesions of certain leishmanial infections.

mucocyst (MŪ-kō-sist) Dark (electron-dense) elongated bodies perpendicular to the cell membrane in ciliates; form of extrusome.

mucron (MŪ-krən) Apical anchoring device on an acephaline gregarine protozoan.

murrina See **surra.**

muscularis mucosae (məs-kū-LAR-is mū-KŌ-sē) Smooth muscle fibers around the mucosa of the gut wall, surrounding the lamina propria and surrounded by the submucosa.

mutualism (MŪ-chu-əl-izm) Type of symbiosis in which both host and symbiont benefit from the association.

mycetome (MĪ-sē-tōm) Specialized organ in some insects that bears mutualistic bacteria.

myiasis (mī-Ī-ə-sis) Infection by fly maggots.

myocyton (MĪ-ə-sī-ton) Cell body of a muscle cell.

myxomatosis (MIX-ō-mə-tō-səs) A viral disease of rabbits, potentially fatal, depending on the host species.

myxospore (MIX-ə-spōr) Life-cycle stage of a myxozoan, possessing at least two valves and polar capsules and released from a vertebrate host.

myzorhynchus (MĪ-zo-RINK-əs) Apical stalked, suckerlike organ on the scolex of some tetraphyllidean cestodes.

N

nagana (nə-GA-nə) Disease of ruminants caused by *Trypanosoma brucei brucei* or *T. congolense.*

nasopharyngeal (nā-zō-far-IN-jē-əl) Occurring inside the mouth and/or nasal passages.

nauplius (NÄ-plē-əs) Typically earliest larval stage(s) of crustaceans; has only three pairs of appendages: antennules, antennae, and mandibles—all primarily of locomotive function.

neascus (nē-AS-kəs) Strigeoid metacercaria with a spoon-shaped forebody.

necrosis (nə-KRŌS-is) Cell or tissue death.

nematogen (nə-MAT-ə-jən) State in the life cycle of a dicyemid mesozoan.

neoplasia (nē-ə-PLĀ-shə) Tumor or process of tumor formation.

neosporosis (NĒ-ə-spōr-Ō-səs) Disease caused by infection with coccidian parasites of genus *Neospora.*

neoteny (nē-OT-ə-nē) Attainment of sexual maturity in the larval condition. Also retention of larval characters into adulthood.

neutrophil (NU-trə-fil) Most abundant of polymorphonuclear leukocytes; important phagocyte; so called because it stains with both acidic and basic stains.

niche (nitch) A set of environmental conditions required by and specific to a particular species. All the biotic and abiotic resources used by a population in a particular habitat.

nidus (NĪ-dəs) Specific locality of a given disease; result of a unique combination of ecological factors that favors the maintenance and transmission of the disease organism.

night soil Human excrement used as fertilizer for food crops.

nymphochrysalis (NIM-fə-KRIS-ə-lis) Nonfeeding, prenymph stage in the life cycle of a chigger mite.

nymphs (nimfs) Juvenile instars in insects with hemimetabolous metamorphosis. Also, juvenile instars of mites and ticks with a full complement of legs.

O

obligate symbiont Organism that is physiologically dependent on establishing a symbiotic relationship with another.

obtect pupa (OB-tekt PŪ-pə) Pupa with wings and legs tightly appressed to its body and covered by an external cuticle.

oligopod larva (ə-LIG-ə-pod) Usual larva in Coleoptera and Neuroptera, with a well-developed head and thoracic legs.

onchocercoma (ON-kō-sər-KŌ-mə) Subcutaneous nodule containing masses of the nematode *Onchocerca volvulus.*

oncomiracidium (ON-kō-mir-ə-SID-ē-əm) Ciliated larva of a monogenetic trematode.

oncosphere (ON-kə-sfer) Synonym of *hexacanth;* used interchangeably.

oocapt (Ō-ə-kapt) Sphincter muscle controlling release of oocytes from a flatworm ovary.

oocyst (Ō-ə-sist) Cystic form in Apicomplexa, resulting from sporogony; an oocyst may be covered by a hard, resistant membrane (as in *Eimeria*), or it may not (as in *Plasmodium*).

oocyst residuum (re-ZI-jü-əm) Cytoplasmic material not incorporated into the sporocyst within an oocyst; seen as an amorphous mass within an oocyst.

oogenotop (ō-ə-GEN-ə-tōp) Female genital complex of a flatworm, including oviduct, ootype, Mehlis' glands, common vitelline duct, and upper uterus.

ookinete (ō-ə-KĪN-ēt) Motile, elongated zygote of a *Plasmodium* or related organism.

oostegites (ō-OS-tə-gīts) Modified thoracic epipods in females of crustacean superorder Peracarida. They form a pouch for brooding embryos.

ootheca (ō-ə-THĒK-ə) Egg packet secreted by some insects; may be covered with sclerotin.

ootype (Ō-ə-tīp) Expansion of a flatworm female duct, surrounded by Mehlis' glands, where, in some flatworms, ducts from a seminal receptacle and vitelline reservoir join.

operculum (ō-PER-kū-ləm) Lidlike specialization of a parasite eggshell through which the larva escapes.

opisthaptor (Ō-pist-HAP-tər) Posterior attachment organ of a monogenetic trematode.

opisthomastigote (ō-PIS-thə-MAS-ti-gōt) Form of Trypanosomatidae with the kinetoplast at the posterior end. The flagellum runs through a long reservoir to emerge at the anterior. There is no undulating membrane. An example is *Herpetomonas.*

---

bat/āpe/ärmadillo/herring/fēmale/finch/līce/crocodile/crōw/
duck/ūnicorn/tüna/ə "uh" as in mammal, fishes, cardinal, heron,
vulture/stress as in bi-OL-o-gy, bi-o-LOG-i-cal

opisthosoma (ō-PIS-thə-Sō-mə) Portion of the body posterior to the legs in a tick or mite.

opsonization (OP-sən-i-ZĀ-shən) Modification of the surface characteristics of an invading particle or organism by binding with antibody or a nonspecific molecule in such a manner as to facilitate phagocytosis by host cells.

oral ciliature (OR-əl-SIL-ē-ə-tur) Cilia, typically including polykinetids and membranes, associated with a ciliated protozoan's mouth.

orchitis (or-KĪT-is) Inflammation of testes.

organelle (organ-ELL) A subcellular structure with defined function, for example mitochondrion, undulipodium, or Golgi apparatus.

oriental sore Disease caused by *Leishmania tropica.* Also called *Jericho boil, Delhi boil, Aleppo boil,* or *cutaneous leishmaniasis.*

Oroya fever Clinical form of Carrion's disease, caused by the bacterium *Bartonella bacilliformis* and transmitted by sand flies.

orthogon (ÄR-thə-gon) Describes the ladderlike arrangement of nervous systems in flatworms, in which two or more longitudinal trunks are cross-connected by a series of commissures.

otoacariasis (OT-ō-ak-ər-Ī-ə-sis) Infestation of the external ear canal by ticks or mites.

outgroup In cladistic analysis, taxon chosen that is related to the taxa in the ingroup and has ancestral (plesiomorphic) characters in common with the ingroup.

overdispersion Nonrandom dispersion of individuals in a habitat, such as when a minority of host individuals bears a majority of parasites. Also called **aggregation.**

ovicapt (Ō-vi-kapt) Sphincter on the oviduct of a flatworm.

ovijector (ŌV-ə-JEK-tər) Muscularized terminal uterus in some platyhelminths and nematodes for ejecting shelled embryos through the female genital pore.

ovipositor (Ō-vē-PAZ-əd-ər) Structure on a female animal modified for deposition of eggs. In many insects it is derived from segmental appendages of the abdomen.

ovisac External sac attached to the somite that bears openings of gonoducts in females of many Copepoda. Fertilized eggs pass into the ovisacs for embryonation.

ovovitellarium (Ō-vō-vit-ə-LAR-ē-əm) Mixed mass of ova and vitelline cells; found in monogenean genus *Gyrodactylus* and in a few tapeworms.

ovoviviparous (Ō-vō-vī-VI-par-əs) Describes reproduction in which embryos develop within the maternal body without additional nourishment from the parent and hatch within the parent or immediately after emerging.

P

paedogenesis (pē-dō-JEN-ə-sis) Reproduction by immature or larval animals caused by acceleration of maturation.

pandemic Very widely distributed epidemic.

pansporoblast (pan-SPŌR-ə-blast) Myxosporidean sporoblast that gives rise to more than one spore. Also called *sporoblast mother cell.*

Papatasi fever Virus disease transmitted by sand flies. Also called *sand fly fever.*

parabasal body Golgi body located near the basal body (kinetosome) of some flagellate protozoa, from which the parabasal filament runs to the basal body.

parabasal filament Fibril, with periodicity visible in electron micrographs, that courses between the parabasal body and a kinetosome.

paramastigote (PA-rə-MAS-ti-gōt) Form of trypanosomatid in which the kinetosome and kinetoplast are beside the nucleus.

paramere (PA-rə-mər) Copulatory appendage in male cimicid bugs.

paraphyletic (par-ə-fī-LET-ik) Adjective describing a group of taxa that includes a hypothetical ancestor but does not include all that ancestor's descendants.

parapolar cells Cells making up the ciliated somatoderm immediately behind the calotte of a mesozoan.

parasite *Raison d'être* for parasitologists.

parasite induced trophic transmission (PITT) Condition in which parasite transmission is enhanced by an effect that facilitates feeding on the present host by a succeeding one.

parasitic castration Condition in which a parasite causes retardation in development or atrophy of host gonads, often accompanied by failure of secondary sexual characteristics to develop.

parasitism Symbiosis in which a symbiont benefits from the association while the host is harmed in some way.

parasitoid Organism that is a typical parasite early in its development but that finally kills its host during or at the completion of development; often used in reference to many insect parasites of other insects.

parasitologist Quaint person who seeks truth in strange places; person who sits on one stool, staring at another.

parasitology (PA-rə-si-TOL-ə-jē) Study of the most common mode of life on earth.

parasitophorous vacuole (PAR-ə-sit-OF-ər-əs VAK-ū-ōl) Vacuole within a host cell that contains a parasite.

paratenic host (par-ə-TĒN-ik) Host in which a parasite survives without undergoing further development. Also known as **transport host.**

paraxial (crystalline) rod (par-AX-ē-əl) Rod that runs alongside the axoneme in the flagellum of a kinetoplastid flagellate.

parenchyma (pə-REN-kə-mə) Spongy mass of vacuolated mesenchymal cells filling spaces between viscera, muscles, or epithelia. In some flatworms the cells are cell bodies of muscle cells. Also the specialized tissue of an organ as distinguished from the supporting connective tissue.

parenteral (PÄR-ən-TE-rəl) Outside of intestine.

pars prostatica (parz prə-STAT-i-kə) Dilation of the ejaculatory duct of a flatworm, surrounded by unicellular prostate cells.

parthenogenesis (PAR-thə-nō-JEN-ə-sis) Development of an unfertilized egg into a new individual.

paruterine organ (par-ŪT-ər-in) Fibromuscular organ in some cestodes that replaces the uterus.

passive immunization Immune state in an animal created by inoculation with serum (containing antibodies) or lymphocytes from an immune animal, rather than by exposure to the antigen.

patent (PĀ-tent) Stage in an infection at which infectious agents produce evidence of their presence, such as eggs or cysts. Contrast with **prepatent.**

pathogenesis (PA-thə-JEN-ə-sis) Production and development of disease.

pathogenicity Capability of an agent to produce disease.

pattern recognition receptor Proteins expressed by host cells that recognize microbial proteins and subsequently mediate antimicrobial protein production by a host.

pedicel (petiole) (PED-ə-sel; PĒT-ē-ōl) Slender, second abdominal segment that forms a "waist" in most Hymenoptera.

pediculosis (pe-DIK-ū-lō-sis) Infestation with lice.

pedipalps (PĒD-ə-palps) Second pair of appendages in chelicerate arthropods, modified variously in different groups.

peduncle (PĒ-dun-kəl) Stalk. Tapering posterior part of the body of a monogenean just anterior to the opisthaptor.

pellicle (PEL-i-kəl)   Thin, translucent, secreted envelope covering many protozoa.

pellicular microtubules (pel-IK-ū-lur mīk-rə-TŪB-ūlz)   Part of the cytoskeleton of a protozoan, consisting of microtubules arranged, usually in a single layer, beneath the plasma membrane.

pelta (PEL-tə)   Curving sheet of microtubules surrounding the flagellar bases in trichomonads.

pentastomiasis (PENT-ə-stōm-Ī-ə-sis)   Human infection, either visceral or nasopharyngeal, with pentastomes.

pereiopod (pə-RĪ-ə-pod)   Thoracic appendage of a crustacean.

perikaryon (perikarya) (pe-ri-KAR-yən; pe-ri-KAR-yə)   Portion of a cell that contains the nucleus (karyon); sometimes called **cyton** or cell body; used in reference to cells that have processes extending some distance away from the area of the nucleus, such as nerve axons or tegumental cells of cestodes and trematodes.

peritreme (PE-rə-trēm)   Elongated sclerite extending forward from the stigma of certain mites, mainly in suborder Mesostigmata.

peritrophic membrane (pe-ri-TRō-fik)   Noncellular, delicate membrane lining an insect's midgut.

permanent parasite   Parasite that lives its entire adult life within or on a host.

peroxisomes (pe-ROKS-ə-sōmz)   Small organelles containing enzymes of the glyoxylate cycle, catalase, and peroxidases.

petechial (pə-TĒ-kē-əl)   Like a rash, with small red or hemorrhagic spots on skin or mucous membranes.

Peyer's patches (PĪ-ərz)   Lymphoid tissue in the wall of the intestine; not circumscribed by a tissue capsule.

phagocytosis (FA-gə-sī-TŌ-səs)   Endocytosis of a particle by a cell.

phagolysosome (FA-gə-LĪ-sə-sōm)   Vacuole in a cell in which a phagocytosed particle is digested.

phagosome (FA-gə-sōm)   Vacuole in a cell containing a phagocytosed particle.

phasmid (FAZ-məd)   Sensory pit on each side near the end of the tail of nematodes of subclass Rhabditia.

pheromone (FE-rə-mōn)   Substance produced by one animal that affects the physiological state or behavior of another individual.

phoresis (fō-RĒ-səs)   Form of symbiosis when the symbiont, the phoront, is mechanically carried about by its host. Neither is physiologically dependent on the other.

phyletics (fī-LET-iks)   Phylogenetic systematics, cladistics.

phylogenetic systematics (FĪ-lō-jən-ET-ik)   Cladistics, an analytical method to infer evolutionary histories.

phylogeny (fī-LOJ-ə-nē)   Evolutionary history of the origin and diversification of a taxon.

phylogeography (fī-lō-je-OG-ra-fē)   Interaction between evolution of taxa and the long-term geological processes affecting the places where these taxa live.

pian bois (PĒ-an BWA)   A form of cutaneous leishmaniasis, with flat, oozing lesions, found especially in Venezuela and Paraguay.

pica (PĪ-kə)   Craving and compulsive consumption of substances other than normal food; may occur during childhood, pregnancy, or sometimes with mental illness.

pinkeye   Bacterial conjunctivitis, sometimes transmitted by flies of genus *Hippolates*.

pinocytosis (pi-nō-si-TŌ-səs)   Form of "cell drinking" in which a cell takes in fluids by endocytosis.

pipestem fibrosis (fī-BRŌ səs)   Thickening of walls of a bile duct as the result of the irritating presence of a parasite.

piroplasm (PI-rə-plazm)   Any of class Piroplasmea, while in a circulating erythrocyte.

plague (plāg)   Zoonotic disease caused by infection with bacteria of species *Yersinia pestis*.

planidium (pla-NID-ē-əm)   First instar of hypermetamorphic, parasitic Diptera and Hymenoptera, which is apodous but moves actively by means of thoracic and caudal setae.

plasma cell   Progeny cell of a B lymphocyte; its function is to secrete antibody.

plasmodium (plaz-MŌ-dē-əm)   General term for a growing, typically multinucleate, sheet of cytoplasm forming the vegetative stage of some parasites.

plasmotomy (plaz-MOT-ə-mē)   Division of a multinucleate cell into multinucleate daughter cells, without accompanying mitosis.

platymyarian muscles (PLA-tē-mī-ÄR-ē-ən)   Nematode muscles in which the contractile portion is wide and shallow and lies close to the hypodermis.

pleopods (PLĒ-ə-podz)   Abdominal appendages of Crustacea.

plerocercoid (PLĒ-rə-SER-koyd)   Metacestode that develops from a procercoid. It usually shows little differentiation.

plerocercoid growth factor (PGH)   Substance produced by plerocercoids of *Diphyllobothrium mansonoides* that mimics effects of mammalian growth hormone.

plerocercus (PLE-rə-SER-kəs)   Tapeworm metacestode in order Trypanorhyncha in which the posterior forms a bladder, the blastocyst, into which the rest of the body withdraws.

plesiomorphic (PLEZ-ē-ə-MORF-ik)   Ancestral characters; characters possessed by members of both ingroup and outgroup.

plica polonica (PLĒ-kə pə-LŌN-i-kə)   Condition that develops in untreated head louse *(Pediculus humanus capitis)* infection consisting of hair matted together with exudate, fungal growth, and fetid odor.

pleurite (PLŪ-rīt)   Lateral sclerite of a somite in an arthropod.

podomere (PŌ-də-mer)   More or less cylindrical segment of a limb of an arthropod, generally articulated at both ends.

podosoma (pō-də-SŌ-mə)   Portion of the body of a tick or mite that bears the legs.

poecilostome (pē-SIL-ə-stōm)   Describes mouthparts borne by members of copepod order Poecilostomata (buccal cavity large, somewhat slitlike, with sickle-shaped mandibles). Also a member of order Poecilostomata.

polar capsule   Compartment bearing the polar filaments in myxozoans.

polar filament   Threadlike organelles in Myxozoa and Microspora.

polar granule   Refractile granule within a coccidian oocyst.

polar ring   Electron-dense organelles of unknown function, located under the cell membrane at the anterior tip of sporozoites and merozoites.

polaroplast (pō-LA-rə-plast)   Organelle, apparently a vacuole, near the polar filament of a microsporidean.

polyclonal activation (POL-ē-KLŌ-nəl)   Activation of several or many clones of lymphocytes; sometimes mechanism of immune evasion by parasites in which a host produces a large amount of antibodies, few or none of which are active against an invader.

polydelphy (PO-lē-DEL-fē)   Condition in which nematodes have more than two uteri.

polyembryony (PO-lē-EM-brē-ə-nē)   Development of a single zygote into more than one offspring.

polykinetid (PO-lē-ki-NE-təd)   Rows or fields of kinetids in ciliates linked by fibrous networks.

polymorphonuclear leukocytes (PMNs) (POL-e-mor-fō-NŪ-kle-ər LŪ-kə-sīts)   Leukocytes with variable, multilobular nuclei (neutrophils, basophils, eosinophils). Also called *granulocytes*.

polyphyletic (pol-ē-fī-LET-ik)   Adjective describing a group of taxa that do not share a common ancestor, hypothetical or otherwise.

bat/āpe/ärmadillo/**h**erring/f**ē**male/finch/l**ī**ce/crocodile/cr**ō**w/ duck/**ū**nicorn/t**ü**na/ə "uh" as in mammal, fishes, cardinal, heron, vult**u**re/stress as in bi-OL-o-gy, bi-o-LOG-i-cal

polypod larva (PO-lē-pod)   Caterpillar type of larva found in Lepidoptera and some Hymenoptera. It has thoracic appendages and abdominal locomotory processes (prolegs). Also called *cruciform.*

polyzoic (PO-lē-ZŌ-ik)   Strobila, when consisting of more than one proglottid.

population structure   Set of quantitative descriptors of a population, including prevalence, density (mean, abundance), variance of a frequency distribution, and curve of best fit.

porose area (PŌ-rōs *or* PŌ-rəs)   Sunken areas on the basis capituli of certain mites and ticks.

post-cyclic transmission   Transmission of an adult parasite, e.g., a helminth, typically by predation, to a new host in which it survives.

posterior station   Development of a protozoan in the hindgut or posterior midgut of its insect host, such as in the section Stercoraria of the Trypanosomatidae.

post-kala-azar dermal leishmanoid   Disfiguring dermal condition developing about one to two years after inadequate treatment of kala-azar.

praniza (prə-NĒ-zə)   Parasitic larva of isopod suborder Gnathiidea. It parasitizes fishes and feeds on blood.

precysts (PRĒ-sist)   Spherical stage in the life cycle of some amebas, containing much glycogen and occurring prior to cyst formation.

predation   Animal interaction in which a predator kills its prey outright. It does not subsist on the prey while the prey is alive.

preerythrocytic cycle   Exoerythrocytic schizogony of *Plasmodium* spp.

prelarva (PRĒ-lar-və)   Life-cycle stage that hatches from an arthropod egg then must molt to become an active larva.

prepatent (prē-PĀ-tənt)   Developmental stage in an infection before agents produce evidence of their presence.

premunition (prē-mū-NI-shən)   Resistance to reinfection or superinfection, conferred by a still-existing infection. The parasite remains alive, but its reproduction and other activities are restrained by the host response.

prenymph (PRĒ-nimf)   Nonfeeding, quiescent stage in the life cycle of a chigger mite.

presoma (PRĒ-sō-ma)   Proboscis, neck, and attached muscles and organs of an acanthocephalan.

prevalence (PRE-və-ləns)   In epidemiology, number of cases of a disease at a given time; that is, a static measurement. Contrast with **incidence.**

primary amebic meningoencephalitis (PAM) (PRĪ-mə-rē a-MĒ-bik mə-NIN-jō-en-sef-ə-LĪ-təs)   Acute, rapidly fatal illness resulting from brain infection with amebas such as *Naegleria fowleri.*

primite (PRĪ-mīt)   Anterior member of a pair of gregarines in syzygy.

procercoid (prō-SER-koid)   Cestode metacestode developing from a coracidium in some orders. It usually has a posterior cercomer.

procrusculus (prō-KRUS-kū-ləs)   Blunt outgrowths on the posterior half of a redia, perhaps locomotory in function.

procuticle (PRŌ-kū-tə-kəl)   Thicker layer beneath the epicuticle of arthropods that lends mass and strength to the cuticle. It contains chitin, sclerotin, and also inorganic salts in Crustacea. The layers within the procuticle vary in structure and composition.

proglottid (prō-GLO-təd)   One set of reproductive organs in a tapeworm strobila. It usually corresponds to a segment.

prohaptor (PRŌ-hap-tər)   Collective adhesive and feeding organs at the anterior end of a monogenetic trematode.

prokaryote (prō-KA-rē-ət)   Organism in which the chromosomes are not contained within membrane-bound nuclei.

prolegs (PRŌ-legz)   Unjointed abdominal appendages in the larva of Lepidoptera and some other insects.

promastigote (prō-MAS-tə-gōt)   Form of Trypanosomatidae with the free flagellum and the kinetoplast anterior to the nucleus, as in genus *Leptomonas.*

pronucleus (prō-NÜ-klē-əs)   Haploid gametic nucleus of a conjugating ciliate; termed *stationary* or *migratory,* depending on whether it remains in one of the conjugants or moves across the conjugants' fused membranes to fertilize another pronucleus. Also the haploid gametic nucleus of other organisms.

propodeum (prō-PŌ-dē-əm)   First abdominal segment of hymenopterans, fused to the thorax.

propodosoma (PRŌ-pō-də-SŌ-mə)   Portion of the podosoma that bears the first and second pairs of legs of a tick or mite.

propolar cells (PRŌ-pō-lər)   Anterior tier of cells in the calotte of a dicyemid mesozoan.

prosoma (PRŌ-sō-mə)   Anterior tagma of arachnids, consisting of cephalothorax; fused imperceptibly to opisthosoma in Acari.

prostate gland cells   Unicellular gland cells surrounding the ejaculatory duct of many flatworms.

protandry (prō-TAN-drē)   Maturation first of male gonads and then of female organs within a hermaphroditic individual. Also called *androgyny.*

protease (PRŌ-tē-ās)   An enzyme that breaks down proteins.

protelean parasite (prō-TEL-ē-ən)   Organism parasitic during its larval or juvenile stages and free living as an adult, usually changing form with each stage.

proterosoma (PRŌ-te-rə-SŌ-mə)   Combination of the gnathosoma and propodosoma of the body of a tick or mite.

protogyny (prō-TOJ-ə-nē)   Synonym of *gynandry.*

protomerite (prō-TOM-ə-rīt)   Anterior half of a cephaline gregarine protozoan.

protonephridium (PRŌ-tō-nə-FRID-ē-əm)   Excretory system that is closed at the inner end by a flame cell or solenocyte and opens by a pore at the distal end.

protonymph (PRŌ-tə-nimf)   Early, bloodsucking stage in the life cycle of some mesostigmatid mites.

protopod (protopodite) (PRŌ-tə-pod; prō-TOP-ə-dīt)   Coxa and basis together.

protopod larva   Larva found in some parasitic Hymenoptera and Diptera; limbs are rudimentary or absent; internal organs are incompletely differentiated; requires highly nutritive and sheltered environment for further development.

protoscolex (PRŌ-tə-SKŌ-leks)   Juvenile scolex budded within a coenurus or a hydatid metacestode of a taeniid cestode.

pseudapolysis (SÜD-ə-POL-ə-sis)   Synonym of *anapolysis.*

pseudobursa (SÜD-ō-BUR-sə)   Lobelike structures on each side of the anus of *Trichinella* sp., presumably to facilitate copulation.

pseudocoel (SÜD-ō-SĒL)   Body cavity found in several animal phyla, such as Nematoda, derived from persistent blastocoel.

pseudocyst (SÜ-də-sist)   Pocket of protozoa within a host cell but not surrounded by a cyst wall of parasite origin.

pseudointestine (sü-dō-in-TES-tin)   Granular mass of cells in a nematomorph (hairworm) larva, thought to contribute to later cyst formation.

pseudolabia (SÜ-də-LĀ-bē-ə)   Bilateral lips around the mouth of many nematodes of order Spirurata; they are not homologous to the lips of most other nematodes but develop from the inner wall of the buccal cavity.

pseudomyiasis (SÜ-də-mī-Ī-a-sis)   Presence within a host of a fly not normally parasitic.

pseudosuckers (SÜD-ō-SUK-ərz)   Accessory suckers on each side of the oral sucker in some strigeiform trematodes.

pseudotubercles (SÜD-ō-TÜ-bər-kəlz)   Localized granulomatous reactions around the eggs of schistosomes in host tissue; resemble reactions around tuberculosis bacteria (tubercles).

ptilinum (ti-LĪ-nəm)   Balloonlike organ in the head of teneral dipterans that pushes off the operculum of the puparium.

pulmonary (PUL-mo-ner-ē)   Relating to or located in the lung.

pupariation (pū-pa-rē-Ā-shən)   Formation of a puparium by the third-stage larvae of certain families of Diptera.

puparium (pū-PA-rē-əm)   Pupal stage of certain families of Diptera.

pygidium (pī-JID-ē-əm)   Sensory organ on a posterior tergite of fleas; apparently detects air currents.

pyriform apparatus (PIR-ə-form)   Embryophore formed in some anoplocephalid cestode eggs; may have two horn-like extensions (bicornuate) at one end.

pyogenic (PĪ-ə-JEN-ik)   Pus producing.

pyrogenic (PĪ-rə-JEN-ik)   Substance that causes a rise in body temperature; causes fever.

Q

quartan malaria (KUAR-tən)   Malaria with fevers recurring every 72 hours. Caused by *Plasmodium malariae.*

quinone tanning (kwi-NŌN)   Chemical reactions leading to sclerotization of arthropod exoskeletons and at least some helminth egg shells.

quotidian malaria (kuo-TID-ē-ən)   Malaria with fevers recurring every 24 hours. Found in cases of overlapping infections.

R

rachis (RĀ-kis)   Central, longitudinal, supporting structure in the ovary of some nematodes.

ray bodies   Gametocytes of *Babesia* spp.

reactive oxygen intermediates (ROIs)   Cytotoxic oxygen molecules or compounds, such as superoxide radical and hydrogen peroxide, released by certain defense cells, that can kill invasive microorganisms or parasites.

reactive nitrogen intermediates (RNIs)   Cytotoxic nitrogen compounds, such as NO and NO_2^-, released by certain defense cells.

red-water fever   Disease in cattle caused by *Babesia bigemina.*

redia (RĒ-dē-ə)   Larval, digenetic trematode, produced by asexual reproduction within a miracidium, sporocyst, or mother redia.

residual body (re-ZID-ū-əl)   Cytoplasmic fragments remaining after formation of gametes or sporocysts in apicomplexans.

reservoir (REZ-ər-vuar)   Living or (rarely) nonliving means of maintaining an infectious agent in nature that can serve as a source of infection for humans or domestic animals.

residuum (rē-ZI-jü-əm)   Cytoplasmic material not incorporated into either sporocysts or sporozoites during maturation of coccidian oocysts.

resilin (rə-ZIL-in)   Elastic protein that releases up to 97% of stored energy upon release from stretched condition.

rete system (RĒ-tē)   Highly branched system of tubules in acanthocephalans, lying on longitudinal muscles or between longitudinal and circular muscles; thought to assist in contraction stimuli.

reticuloendothelial (RE) system (rə-TIK-ū-lō-EN-də-THĒL-ē-əl)   Phagocytic cells of the mononuclear phagocyte system, such as macrophages, Kupffer cells, and microglial cells.

retrofection (RE-trə-FEK-shən)   Process of reinfection, whereby juvenile nematodes hatch on the skin and reenter the body before molting to third stage.

RFLP (RIF-lip)   Restriction fragment length polymorphism; DNA fragments, of various lengths, resulting from digestion by endonuclease enzymes.

rhabdites (RAB-dīts)   Rodlike bodies embedded in the tegument in most free-living flatworms; their function may be adhesion or predator repulsion.

rhabditiform (rab-DIT-ə-form)   Word used to describe first-stage juveniles of some nematodes whose esophagus has a terminal bulb separated by an isthmus from the anterior portion (corpus).

rhombogen (ROM-bə-jən)   Stage in the life cycle of a dicyemid mesozoan.

rhoptries (RŌP-trēs)   Elongated, electron-dense bodies extending within the polar rings of an apicomplexan.

Romaña's sign (rō-MÄN-yəz sīn)   Symptoms of recent infection by *Trypanosoma cruzi,* consisting of edema of the orbit and swelling of the preauricular lymph node.

Romanovsky stain (RŌ-mən-OV-skē)   Complex stain, based on methylene blue and eosin, used to stain blood cells and hemoparasites. Wright's and Giemsa's stains are two common examples.

rostellum (räs-TEL-əm)   Projecting structure on scolex of tapeworm, often with hooks.

rostrum (tectum) (RÄS-trəm; TEK-təm)   Dorsal part of capitulum projecting over chelicerae in acarines.

r-strategist   Species of organism that uses a survival and reproductive "strategy" characterized by high fecundity, high mortality, and short longevity. Populations are controlled by density-independent factors.

ruffles   Slender projections of the exterior surface of a dicyemid mesozoan.

rugae (RŪ-gə)   Transverse ridges across the large ventral holdfast of an aspidobothrian of family Rugogastridae.

S

Saefftigen's pouch (SĀF-ti-gənz)   Internal, muscular sac near the posterior end of a male acanthocephalan. It contains fluid that aids in manipulating the copulatory bursa.

salivarium (sa-li-VA-rē-əm)   Chamber in buccal cone of acarines into which salivary ducts open.

sand fly   Member of dipteran subfamily Phlebotominae, family Psychodidae; sometimes also applied to Simuliidae (New Zealand) and Ceratopogonidae (Caribbean).

saprophytic (SAP-rə-FIT-ik)   Plant living on dead organic matter.

saprozoic nutrition (SAP-rə-ZŌ-ik)   Nutrition of an animal by absorption of dissolved salts and simple organic nutrients from surrounding medium. Also refers to feeding on decaying organic matter.

sarcocystin (SÄR-kə-SIS-tin)   Powerful toxin produced by zoitocysts of *Sarcocystis.*

sarcoma (sär-KŌ-mə)   Malignant tumor arising from a mesodermal tissue.

sarcoptic mange (sär-KOP-tik mānj)   Disease caused by mites of genus *Sarcoptes.* Also called **scabies.**

satellite   Posterior member of a pair of gregarines in syzygy.

scabies (SKĀ-bēz)   Disease caused by mites of genus *Sarcoptes.* Also called **sarcoptic mange.**

scarabaeiform (SKA-rə-BĒ-ə-form)   Describes grublike larvae with lightly sclerotized cuticle; found in some coleopteran families.

schistosomule (shis-tə-SOM-ūl)   Juvenile stage of a blood fluke, between a cercaria and an adult; migrating form taking the place of a metacercaria in the life cycle.

schizeckenosy (shiz-ə-KEN-ə-sē)   System of waste elimination found in some mites with a blindly ending midgut; the lobe breaks free from the ventriculus and is expelled through a split in the posterodorsal cuticle.

bat/āpe/ärmadillo/herring/fēmale/finch/līce/crocodile/crōw/ duck/ūnicorn/tüna/ə "uh" as in mammal, fishes, cardinal, heron, vulture/stress as in bi-OL-o-gy, bi-o-LOG-i-cal

schizogony (shiz-ÄG-ə-nē *or* **skiz-ÄG-ə-nē)**   Form of asexual reproduction in which multiple mitoses take place, followed by simultaneous cytokineses, resulting in many daughter cells at once.

schizont (SHIZ-änt *or* **SKIZ-änt)**   Cell undergoing schizogony, in which nuclear divisions have occurred but cytokinesis is not completed; in its late phase sometimes called *segmenter.*

Schüffner's dots (SHÜF-nerz)   Small surface invaginations that appear as stippling on the membrane of an erythrocyte infected with *Plasmodium vivax* after Romanovsky staining.

sclerite (SKLER-it)   Any well-defined, sclerotized area of arthropod cuticle limited by suture lines or flexible, membranous portions of cuticle.

sclerotin (SKLER-ə-tən)   Highly resistant and insoluble protein occurring in the cuticle of arthropods; also thought to occur in structures secreted by various other animals, such as in the eggshells of some trematodes, in which stabilization of the protein is achieved by orthoquinone crosslinks between free imino or amino groups of the protein molecules.

scolex (SKŌ-leks)   "Head" or holdfast organ of a tapeworm.

scoliosis (skM-lē-LŌ-əs)   Lateral curvature of the spine.

scopula (SKO-pū-lə)   Organelle composed of kinetosomes and producing a stalk in ciliates.

screwworm   Larvae of flies, such as *Cochliomyia hominivorax,* that enter wounds and develop as maggots in the subcutaneous tissues.

scrub typhus (TĪ-fəs)   Rickettsial disease transmitted by certain chigger mites.

scutum (SKÜ-təm)   Large, anteriodorsal sclerite on a tick or mite.

sea lice   Fish parasites of crustacean family Caligidae, not lice (which are insects).

secondary endosymbiosis (or secondary symbiogenesis)   Result when a eukaryote that originated from primary symbiogenesis becomes permanently resident as an organelle within another. An example is the apicoplast in Apicomplexa.

secondary response   Stronger immune response stimulated by a challenge antigen exposure; due to presence of memory cells.

segmenter   Mature schizont (meront) of *Plasmodium* spp. before cytokinesis.

SEM (ESS-Ē-EM)   A photograph taken with a scanning electron microscope; may also refer to the technique of scanning electron microscopy.

sensilla (pl. sensillae; sin-SIL-ə; sin-SIL-ē)   Ciliary sense organs in the tegument of some flatworms and other organisms, consisting of cilia attached to nerve endings.

septicemia (SEP-ti-SĒM-ē-ə)   Systemic infection in which a pathogen is present in the circulating blood.

sequestration (se-kwes-TRĀ-shən)   Act of setting apart or isolating. Erythrocytes containing trophozoites of *Plasmodium falciparum* have proteins in their surface that cause them to bind to venular endothelium and to each other, sequestering them from circulating blood.

serial homolog   Series of segments in which each repeats the genetic expression of the genes in the segment (somite) before it.

sexual selection   Evolutionary process in which reproductive success is based on traits possessed by one of the sexes, typically the male.

sickle cell anemia   Condition that causes the red blood cells to collapse (sickle) under oxygen stress. The condition becomes manifest when an individual is homozygous for the gene for hemoglobin-S (HbS).

siphonostome (si-FON-ə-stom)   Mouth type in some copepods in which the labium and labrum form a conical, siphonlike extension surrounding the mandibles.

sister group   One of only two taxa that share a most recent common ancestor.

skin test   Immunodiagnostic test that depends on an inflammatory reaction when an antigen is injected subcutaneously.

sleeping sickness   African trypanosomiasis and mosquito-borne, virus-induced encephalitis.

slime ball   Mass of mucus-covered cercariae of dicrocoeliid flukes, released from land snails. Also a term of derogation applied to really disgusting persons.

solenophage (sō-LEN-ə-fāj)   Blood-feeding arthropod that introduces its mouthparts directly into a blood vessel to feed.

somite (SŌ-mīt)   Body segment or metamere; usually used in reference to arthropods.

sowda   Name used in some regions for severe dermatitis caused by *Onchocerca volvulus.*

sparganum (spär-GA-nəm)   Cestode plerocercoid of unknown identity.

spermalege (SPER-mə-lēj)   Organ that receives sperm in the female cimicid bug during copulation.

spermatheca (sperm-ə-THĒ-kə)   Sclerotized structure in a female insect that receives and holds sperm.

spermatodactyl (spər-MAT-ə-DAK-təl)   Modification in some Acari of chelicera, which functions in transfer of sperm from male's gonopore to copulatory receptacles between third and fourth coxae of female.

spermatophore (spər-MAT-ə-for)   Formed "container" or packet of sperm that is placed in or on the body of a female, in contrast to the sperm in copulation which are conducted directly from male reproductive structures into a female's body.

spicule (SPIK-ūl)   Tiny needlelike structure, such as copulatory spicules in nematodes.

spiracle (SPI-rə-kəl)   Opening into the respiratory system in various arthropods.

spironucleosis (spī-rō-NŪK-lē-Ō-sis)   A potentially fatal intestinal disease of poultry caused by *Spironucleus meleagridis;* other *Spironucleus* species infect salmon.

spondylitis (SPON-də-LI-tis)   Inflammation of one or more spinal vertebrae.

sporadin (SPŌR-ə-dən)   Mature trophozoite of a gregarine protozoan.

sporoblast (SPŌR-ə-blast)   Cell mass that will differentiate into a sporocyst within an oocyst.

sporocyst (SPŌR-ə-sist)   Stage of development of a sporozoan protozoan, usually with an enclosing membrane, the oocyst. Also an asexual stage of development in some trematodes.

sporocyst residuum (SPŌR-ə-sist rē-ZI-jü-əm)   Cytoplasmic material "left over" within a sporocyst after sporozoite formation; seen as an amorphous mass.

sporogony (spōr-ÄG-ə-nē)   Multiple fission of a zygote; such a cell also is called a **sporont.**

sporont (SPŌR-ənt)   Undifferentiated cell mass within an unsporulated oocyst.

sporophorous vesicle (spō-ROF-ər-əs VES-i-kəl)   Membranous envelope containing spores in some Microsporidia.

sporoplasm (SPŌR-ə-pla-zəm)   Amebalike portion of a microsporan or myxosporan cyst that is infective to the next host.

sporoplasmosome (spōr-ə-PLAZ-mə-SŌM)   Electron-dense bodies, of unknown function, in myxozoan sporoplasms.

sporozoite (SPŌR-ə-ZŌ-īt)   Daughter cell resulting from sporogony.

squalamine (SKWĀL-ə-mēn)   Antimicrobial steroid from sharks.

squama (SKUA-mə)   Prominent lobe in the anal angle of a dipteran wing.

stable endemic malaria   Describes malaria epidemiology in which transmission occurs throughout the year, mosquito reproduction remains constant, adults become generally resistant, and children are at greatest risk.

stenostomate (ste-NOS-təm-āt)   Condition in trematode cercariae in which the collecting tubules of the excretory system pass to the anterior and then to the posterior to join the excretory bladder.

sternite (STER-nīt)   Main ventral sclerite of a somite of an arthropod.

stichosome (STIK-ə-sōm)   Column of large, rectangular cells called *stichocytes,* supporting and secreting into the esophagus of most nematodes of family Trichuridae.

Stieda body (STĒ-də)   Plug in the inner wall of one end of a coccidian oocyst.

stigma (pl. stigmata; STIG-mə; stig-MÄ-tə)   Operculumlike area of an eggshell through which the miracidium of a schistosome fluke hatches. Also an arthropod spiracle.

strike   Deposition of fly eggs or larvae on a living host.

strobila (STRŌ-bi-lə)   Region of a tapeworm behind the scolex; chain of proglottids of a eucestode.

strobilation (STRO-bə-LĀ-shən)   Formation of a chain of zoids by budding, as in the strobila of a tapeworm.

strobilocercoid (STRŌ-bə-lō-SER-koid)   Cysticercoid that undergoes some strobilation; found only in *Schistotaenia.*

strobilocercus (STRŌ-bə-lō-SER-kəs)   Simple cysticercus with some evident strobilation.

strongyliform (stron-JIL-ə-form)   Esophagus type in nematode J$_3$s in which the bulb is not separated from the corpus by an isthmus.

style   Terminal segment of the antenna of a brachyceran dipteran. It is drawn into a sharp point.

stylops (STĪ-lops)   Member of insect order Strepsiptera.

stylostome (STĪ-lə-stōm)   Hardened, tubelike structure secreted by a feeding chigger mite.

subacute infection (SUB-ə-kūt)   Infection in which pathogenic conditions are extended over time; not as clinically severe as acute infections.

subchelate (səb-KĒL-āt)   Condition of an arthropod appendage in which the terminal podomere can fold back like a pincer against the subterminal podomere.

subpellicular microtubules (SUB-pel-IK-ū-lər MĪ-krō-TÜB-ūls)   Tubular cytoskeletal structures that lie beneath the plasma membrane of trypanosomatid flagellates and apicomplexan sporozoites.

subperiodic   Term used to describe a strain of *Wuchereria bancrofti* that shows no periodicity or shows a diurnal periodicity; some authorities consider it a separate species, *W. pacifica.*

substiedal body (səb-STĒ-dəl)   Additional plug material underlying a Stieda body.

superphylum   A taxon that includes a group of related phyla.

suprapopulation (SÜ-pra-POP-ū-lā-shən)   All of the parasites of a single species, regardless of developmental stages (eggs, larvae, juveniles, adults), that occur in an ecosystem.

suramin (SÜR-a-min)   Antitrypanosomal drug.

surra (SU-rə)   Disease of large mammals caused by *Trypanosoma evansi.*

sutural plane (SÜ-chur-əl)   Seam between two halves of a myxozoan spore.

swarmer   Daughter trophozoites resulting from multiple fissions of *Ichythophthirius multifiliis* and a few other protozoa.

sylvatic (sil-VA-tik)   Existing normally in the wild, not in the human environment.

symbiology (SIM-bi-OL-ə-gē)   Study of symbioses.

symbionts (SIM-bī-änt *or* SIM-bē-änt)   Organisms involved in symbiotic relationships with other organisms, the hosts.

symbiosis (SIM-bī-OS-əs *or* SIM-bē-OS-əs)   Interaction among organisms in which one organism lives with, in, or on the body of another.

symmetrogenic fission (si-ME-trə-JEN-ik)   Mitotic fission between the rows of flagella of protozoa.

synanthropism (si-NAN-thrə-pizm)   Habit of an organism of living in or around human dwellings.

synapomorphy (si-NAP-ə-mor-fē)   Shared derived characters that set a taxon apart from related taxa.

syncytium (sin-SI-shəm)   Multinucleate mass of protoplasm whose nuclei are not separated from each other by cell membranes. Adj., **syncytial** (sin-SI-shəl).

syngamy (SIN-gam-ē)   Fusion of gametes that are whole cells.

synlophe (SIN-lōf)   Pattern of ridges on the cuticle of a nematode.

systematics   Classification of organisms according to their phylogenetic relationships; study of variation and evolution of organisms.

syzygy (SIZ-ə-jē)   Stage during sexual reproduction of some gregarines in which two or more gamonts connect.

T

T cell   Type of lymphocyte with a vital regulatory role in immune response; so called because they are processed through the thymus. Subsets of T cells may be stimulatory or inhibitory. They communicate with other cells by protein hormones called *cytokines.*

tachyzoite (TAK-ē-ZŌ-īt)   Small, merozoitelike stages of *Toxoplasma.* They develop in the host cells' parasitophorous vacuole by endodyogeny.

tagmatization (TAG-mə-tə-ZĀ-shən)   Specialization of metameres in animals, particularly arthropods, into distinct body regions, each known as a *tagma* (pl. **tagmata**).

tangoreceptor (TAN-gō-rē-SEP-tər)   A surface neural receptor sensitive to touch or pressure.

tantalus (TAN-tə-ləs)   Larva of crustaceans in subclass Tantulocarida.

tarsus (TÄR-səs)   Most distal podomere of an insect or acarine limb; articulates proximally with the tibia and usually is subdivided into two to five subsegments in insects.

taxonomy (tak-SÄN-ə-mē)   Study of the principles of scientific classification; ordering and naming of organisms.

tectum (TEK-təm)   Dorsal extension over the mouth of a crustacean or acarine. Also called a **rostrum.**

TEM (TĒ-Ē-EM)   A photograph taken with a transmission electron microscope; may also refer to the technique of transmission electron microscopy.

tegument (TEG-ū-mənt)   Surficial covering of a multicellular organism, an integument.

telamon (TEL-ə-mon)   Ventral sclerotization of the cloaca in some nematodes; helps to guide exsertion of the spicules.

telmophage (TEL-mə-fāj)   Blood-feeding arthropod that cuts through skin and blood vessels to cause a small hemorrhage of blood from which it feeds.

telogonic (TEL-ə-GON-ik)   Condition of a nematode gonad in which the germ cells proliferate only at the inner end and then must traverse the remaining length of the gonad before expulsion.

temporary parasite   Parasite that contacts its host only to feed and then leaves. Also called an **intermittent parasite** or **micropredator.**

teneral (TEN-ə-rəl)   Newly emerged adult arthropod that is soft and weak.

tenesmus (TE-nez-məs)   Straining to empty the bowels or bladder without emptying feces or urine.

---

bat/āpe/ärmadillo/herring/fēmale/finch/līce/crocodile/crōw/
duck/ūnicorn/tüna/ə "uh" as in mammal, fishes, cardinal, heron,
vulture/stress as in bi-OL-o-gy, bi-o-LOG-i-cal

terebra (tə-RĒ-brə)   Functional unit of a hymenopteran ovipositor, formed from first and second valvulae.

tergite (TER-gīt)   Main dorsal sclerite of a somite of an arthropod.

tertian malaria (TER-shən) (tertian ague)   Malaria in which fevers recur every 48 hours. Caused by *Plasmodium vivax, P. ovale,* and *P. falciparum.*

tetracotyle (TET-rə-CÄ-təl)   Strigeoid metacercaria in family Strigeidae.

tetrathyridium (TET-rə-thī-RI-dē-əm)   Only metacestode form known in tapeworm cyclophyllidean genus *Mesocestoides.* Large, solid-bodied cysticercoid.

thalassemia (thal-ə-SĒM-ē-ə)   Group of heritable anemias caused by reduction or failure in synthesis of one of the globin chains in hemoglobin; most common in individuals of Mediterranean, Southeast Asian, or African ancestry.

theileriosis (TĪ-lər-ē-ŌS-əs)   Disease of cattle and other ruminants, caused by *Theileria parva.* Also called *East Coast fever.*

thelyotoky, thelytoky (THĒ-lē-ō-TŌ-kē, THĒ-lē-TL-kē)   Type of parthenogenesis in which all individuals are uniparental and essentially no males are produced.

thrombus   Blood clot in a blood vessel or in one of the cavities of the heart.

tibia (TIB-ē-ə)   Podomere of an insect or acarine leg that articulates proximally with the femur in insects and patella in acarines and distally with the tarsus in insects or with the metatarsus or tarsus in acarines.

titer (TĪ-tər)   Concentration of a substance in a solution as determined by titration.

Toll-like receptors (TLRs)   Receptors on a cell surface that recognize certain molecular patterns on microbes, the binding of which results in release of antimicrobial substances, such as defensins.

tomite (TŌ-mīt)   Daughter cell produced by some ciliates, through multiple fissions, typically within a cyst.

toxosome (TOX-ō-sōm)   Extrusome (ciliate pellicular organelle) that may release toxic substances evidently as a defensive mechanism.

trabecula (trə-BEK-ū-lə)   In general anatomical usage, septum extending from an envelope through an enclosed substance, which, together with other trabeculae, forms part of the framework of various organs; here referring specifically to cell processes connecting the perikarya of cestode and trematode tegumental cells with the distal cytoplasm. Also called **internuncial process.**

tracheal system (TRĀ-kē-əl)   System of cuticle-lined tubes in many insects and acarines that functions in respiration; opens to outside through spiracles.

transport host   Paratenic host.

triactinomyxon (TRĪ-ak-TIN-ə-MIX-ən)   Stage in the life cycle of a myxozoan, formerly assigned to a separate class.

tribocytic organ (TRI-bə-SI-tik)   Glandular, padlike organ behind the acetabulum of a strigeoid trematode.

trichocyst (TRIK-ō-sist)   Extrusome that produces a fiber, functioning in the mechanical resistance to predators.

trichogon (TRĪK-ə-gän)   Spiny male larva of a rhizocephalan cirripede that comes to lie within a special receptacle in a female.

tritonymph (TRĪ-tə-nimf)   Third nymphal stage in most acarines.

tritosternum (TRĪ-tə-STER-nəm)   Ventral, bristlelike sensory organ just behind the gnathosoma of a mesostigmatid mite.

triungulin (triungulinid) (trī-UN-gū-lən; trī-UN-gū-LI-nəd)   First instar larva of some parasitic, hypermetamorphic Neuroptera and Coleoptera and of Strepsiptera, which is an active, campodeiform oligopod.

trochanter (TRŌ-kan-tər)   Podomere of an insect or acarine leg that articulates basally with the coxa and distally with the femur; usually fixed to the femur in insects.

trophont (TRŌ-fənt)   Stage in the life cycle of gregarines.

trophosome (TRŌ-fə-sōm)   Food storage organ in mermithid nematodes; also site of mutualistic bacteria in certain marine worms.

trophozoite (TRŌ-fə-ZŌ-īt)   Active, feeding stage of a protozoan, in contrast to a cyst. Also called *vegetative stage.*

trypomastigote (TRI-pə-MAS-tə-gōt)   Form of Trypanosomatidae with an undulating membrane and the kinetoplast located posterior to the nucleus. An example is *Trypanosoma.*

tsetse fly (TSET-sē *or* SET-sē)   Bloodsucking fly of genus *Glossina.*

tubovesicular membrane network (TÜB-ō-ves-IK-ū-lər)   Network of membranes from the parasitophorous vesicle of *Plasmodium falciparum* to the host cell membrane.

tungiasis (tung-Ī-ə-səs)   Disease resulting from infestation with fleas of genus *Tunga.*

tumor necrosis factor (TNF)   Cytokine, major mediator of inflammation, produced mainly by macrophages, that activates polymorphonuclear cells and stimulates macrophages to produce other cytokines. In large concentrations it causes fever.

typhus (TĪ-fəs)   Disease caused by infection with bacteria of genus *Rickettsia* and transmitted by fleas or lice.

U

ulcer (UL-sər)   Area of inflammation that opens out to skin or a mucous surface.

ultrastructure   Structure of an organism or cell at the electron microscopic level.

undulating membrane (UN-dū-LĀT-ing)   Name applied to two quite different structures in protozoa. In some flagellates it is a finlike ridge across the surface of a cell, with the axoneme of a flagellum near its surface. In some ciliates it is a line of cilia that are fused at their bases, usually beating to force food particles toward the gullet.

undulating ridges   Undulatory waves in the surface of some protozoa, probably aided by subpellicular microtubules; means of locomotion in some species.

undulipodium (UN-dū-lə-PŌD-ē-əm)   Flagellum or cilium, typically possessing the 9 + 2 microtubule arrangement.

uniramous appendage (Ū-nē-RĀ-məs)   Arthropod appendage that is unbranched, characteristic of living arthropods other than Crustacea, although some crustacean appendages are uniramous.

unstable malaria   Describes malaria epidemiology where transmission is interrupted by a cool or dry season, epidemics occur, resistance is not generated, and symptoms often are serious.

urban   Peculiar to human environments, as contrasted with that found normally around wild animals.

urn   Region near the center of an infusoriform larva of a dicyemid mesozoan.

uropolar cells (Ū-rə-PŌ-lər)   Somatoderm cells at the posterior end of the trunk of a dicyemid mesozoan.

urosome (Ū-rə-sōm)   Portion of the body posterior to the major point of body flexion in many copepods; usually includes one or more free thoracic segments and abdomen.

urstigmata (UR-stig-MA-tə)   Sense organs between the coxae of the first and second pairs of legs on some mites. Apparently they are humidity receptors. Also called **Claparedé organs.**

uterine bell (ŪT-ər-ən)   Structure in female acanthocephalans that allows fully developed, shelled embryos to pass out of the body and that retains undeveloped ones.

V

vagabond's disease   Darkened, thickened skin caused by years of infestation with body lice, *Pediculus humanus humanus.*

valvifers (VALV-i-fərz)   Basal portions of the ovipositor in Hymenoptera, derived from coxae of segmental appendages.

valvulae (VALV-ū-lē)   Processes from the valvifers to form the body of the ovipositor (terebra) and the ovipositor sheath (third valvulae) in Hymenoptera.

variable region (Fab for antibody-binding fragment)   Portion of an antibody molecule that binds to an antigen. Constrast with **constant region.**

variable antigen type (VAT)   Applied to certain trypanosomes, any one of numerous antigenic types expressed on the surface of the organisms and "seen" by the immune system of the host. See **variant-specific surface glycoprotein.**

variant-specific surface glycoprotein (VSG)   Glycoprotein on the surface of certain trypanosomes recognized by the host's immune system. Each VSG is responsible for one VAT.

vector (VEK-tər)   Any agent, such as water, wind, or insect, that transmits a disease organism.

veins (vānz)   Blood vessels conducting blood toward the heart in any animal. Also more heavily sclerotized portions of wings of insects, which are remains of lacunae.

ventriculus (vin-TRIK-ū-ləs)   The stomach of an arthropod.

vermicle (VER-mə-kəl)   Infective stage of *Babesia* in a tick.

vermiform (VER-mə-form)   Wormlike.

verruga peruana (vər-UG-ə PE-rü-AN-ə)   Clinical form of Carrion's disease, caused by the bacterium *Bartonella bacilliformis* and transmitted by sand flies.

vesicular disease (ve-SIK-ū-lər)   Any disease of the urinary bladder, such as vesicular schistosomiasis.

vestibulum (ves-TI-bū-ləm)   Cavity leading into another cavity or passage, such as in ciliate order Vestibulifera.

villipodia (VIL-ə-PŌD-ē-ə)   Processes on the pseudopodium of nematode sperm.

virulence (VIR-ə-ləns)   Degree of pathogenicity of an agent; how much damage the agent can cause.

viviparity (vī-vi-PAR-ə-tē)   Bearing of living young (instead of laying eggs) with nutritional aid of maternal parent during development.

vulva (VUL-və)   Opening of uterus to exterior in nematodes; external female genitalia in mammals.

W

weir (WĒ-ər)   Filtration apparatus in flame-cell protonephridia of flatworms.

wheal (WĒL)   Swelling caused by release of serum into tissues by immediate hypersensitivity at a dermal site, usually accompanied by **flare,** a redness caused by dilation of the blood vessels.

whirling disease   Disease of fishes, caused by the protozoan *Myxobolus cerebralis.*

Winterbottom's sign   Swollen lymph nodes at the base of the skull, symptomatic of African sleeping sickness.

X

xenodiagnosis (ZĒN-ə-dī-əg-NŌ-səs)   Diagnosis of a disease by infecting a test animal.

xenograft (ZĒN-ə-graft)   Graft of a piece of tissue or organ from one individual to another of a different species.

xenoma (zē-NŌM-ə)   Combination of an intracellular parasite and its hypertrophied host cell.

xenosomes (ZĒN-ə-sōmz)   Body or organelle living within a cell that contains its own DNA and is capable of reproducing itself, once having functioned as a free-living organism. Examples are zooxanthellae and zoochlorellae.

xiphidiocercaria (zī-FID-ē-ə-ser-KA-rē-ə)   Cercaria with a stylet in the anterior rim of its oral sucker.

Y

yaws   Bacterial disease caused by the spirochete *Treponema pertenue,* often transmitted by flies.

yellow fever   Virus disease transmitted by the mosquito *Aedes aegypti.*

Z

zoid (ZŌ-id)   Member of a colonial organism.

zoitocyst (zō-ĪT-ə-sist)   Tissue phase in some of the coccidia of the *Isospora* group. They usually have internal septae and contain thousands of bradyzoites. Also called *sarcocyst* or **Miescher's tubule.**

zoonosis (ZŌ-ə-NŌ-sis)   Disease of animals that is transmissible to humans. Some authors subdivide the concept into zooanthroponosis, an infection humans can acquire from animals, and anthropozoonosis, a disease of humans transmissible to other animals.

zooxanthellae (ZŌ-ə-zan-THEL-ē)   Dinoflagellate protozoa living mutualistically in cells of certain marine animals. Among other benefits, their hosts derive nutrients from photosynthetic reactions of zooxanthellae.

zygotic meiosis (zī-GOT-ik mī-Ō-səs)   Meiosis that occurs in a cell, typically a protozoan, immediately following zygote formation; thus, all stages in the life cycle other than the zygote are haploid.

bat/āpe/ärmadillo/herring/fēmale/finch/līce/crocodile/crōw/ duck/ūnicorn/tüna/ə "uh" as in mammal, fishes, cardinal, heron, vulture/stress as in bi-OL-o-gy, bi-o-LOG-i-cal

Index

A